Biology and Evolution of the Mollusca

Biology and Evolution of the Mollusca

Volume Two

Winston F. Ponder

David R. Lindberg

Juliet M. Ponder

CRC Press
Taylor & Francis Group
Boca Raton London New York

CRC Press is an imprint of the
Taylor & Francis Group, an **informa** business

CRC Press
Taylor & Francis Group
6000 Broken Sound Parkway NW, Suite 300
Boca Raton, FL 33487-2742

First issued in paperback 2021

ISBN-13: 978-0-8153-6184-8 (hbk)
ISBN-13: 978-1-03-217354-2 (pbk)
DOI: 10.1201/9781351115254

Contents

Summary of contents of Volume 1

About the Authors

Winston F. Ponder MSc, PhD, DSc is a Senior Fellow at the Australian Museum (Sydney), and an Honorary Associate at the University of Sydney. Before retirement in 2005 he was a Principal Research Scientist at the Australian Museum, where he worked in the malacology section for 37 years and was instrumental in organising and building up the extensive collection. While at the museum he also held several honorary positions at a number of Australian universities. In addition to pursuing research programs, he has also been active in transferring expertise to new generations of young zoologists by way of training research students, hosting overseas collaborators, running workshops and organising a number of conferences. It was when running an intensive mollusc course at the University of Wollongong over several years, that the idea for this book dawned. Winston's research extends broadly across marine and freshwater molluscs and includes observations of living animals, microanatomy, biometrics, cladistics and molecular analysis. He, like his collaborator David Lindberg, has been involved in the revolution in molluscan phylogenetics that has taken place over the last 20 years. His publications include 250 peer-reviewed papers and book chapters and two edited books on molluscs and one on invertebrate conservation. Recently he produced (with two others) an online interactive key and information system of Australian freshwater molluscs, has been involved with the molluscan part of the online Australian faunal directory, and was editor of the journal *Molluscan Research* for 14 years. He organised or co-organised 12 national or international meetings and presented papers at many more. He has undertaken study trips and field work in various parts of the world over more than 40 years but field studies have mainly focused on Australian marine and freshwater habitats. His awards include the Hamilton Prize (Royal Society of N.Z., 1968), Fellow of the Royal Zoological Society, NSW, Silver Jubilee award, Australian Marine Sciences Association (2008), and the Clarke Medal, Royal Society of NSW (2010).

David R. Lindberg is Professor Emeritus of Integrative Biology, Curator Emeritus in the UC Museum of Paleontology, and a former member of the Center for Computational Biology at UC Berkeley. He has authored over 125 peer-reviewed papers and edited or authored three books on the evolutionary history of nearshore marine organisms and their habitats. At Berkeley, he served as major advisor to 21 PhD graduate students and six post-doctoral researchers. He also served as Chair of the Department of Integrative Biology, Director of the UC Museum of Paleontology, and Chair of the UC Berkeley Natural History Museums. In addition to providing graduate seminars in evolution and organismal biology, he regularly taught a marine mammal course, an invertebrate zoology course with laboratories, and two semester principles of phylogenetics course. Prof. Lindberg has conducted research and field work along the rocky shores of the Pacific Rim for over 45 years. In addition to his research and teaching, Prof. Lindberg was actively involved in K–16 outreach projects at the UC Museum of Paleontology, and focused on the use of the internet to increase access to scientific resources, and the training of teachers in principles of evolutionary biology, science, and global change.

Juliet M. Ponder MA Grad Dip Ed has spent 55 years married to the senior author of this book and developing a variety of interests. She excelled at school and began studying science at Auckland University where she met Winston. She was blown away by his passion for molluscs and decided that, if such dedication was required to be a scientist, she didn't have it. However, she was happy to support his work and to this end began developing competence in scientific writing and illustrating and even scuba diving. She and Winston have two children and, while rearing them, Juliet acquired qualifications in fine art, community welfare and education. She worked part-time for 10 years painting lifelike models in the Australian Museum but, as her children grew older, she moved into full time community work and then teaching for 15 years. Retirement from paid work in 2004 gave her the opportunity to learn computer graphics and develop a method of colour coding illustrations to clarify similarities and differences between groups of animals. The drawings in this book combine her love of art, science and education.

Introduction to Volume Two

This two-volume work is devoted to the Mollusca, one of the most diverse and important groups of animals. It provides an overview of the diversity, utility, physiology, and functional morphology of molluscs, as well as their evolutionary history and relationships. Also, we highlight some current areas of research and flag some areas urgently in need of work.

The first volume provides general introductory material on molluscs and overviews of the history of malacology, molluscan physiology and genomics, and the structure and function of each of the body systems and processes. It also includes a chapter on natural history that gives a general account of various aspects of molluscan biology not otherwise covered elsewhere in the book. There is also a chapter on the significance of molluscs to, and their interactions with, humans, and the last chapter covers aspects of current and future research relevant to the topics covered in the volume.

In this second volume we discuss the relationships of molluscs to other invertebrates and to each other, review their early fossil history, and provide comprehensive overviews of each of the major molluscan groups. Where necessary, the chapters dealing with the major groups include a brief summary of the information covered in the system chapters, and additional detail is often provided about some systems that are particularly significant. Lastly, we discuss aspects of current and future research relevant to the areas covered in this volume and some more general aspects relevant to the future of malacology.

The Appendix provides a summary of the classification of all major extant classes (i.e., those that contain at least some living representatives) to the family level, with both living and fossil taxa included. Extinct major groups (classes) are reviewed in Chapter 13.

The bulk of this volume comprises the chapters dealing with the extant classes of molluscs. We have tried to present the most up-to-date information on these groups, including their phylogenetic relationships. This information is, however, far from static, and readers should be aware that phylogenies and classifications can quickly change. We recommend that names and classifications provided in the chapters and in the Appendix be checked for changes using online databases such as the World Register of Marine Species (WoRMS).

We do not provide a glossary, but most terms used are explained in the text, and many of these can be located using the comprehensive index. Most information provided is extensively referenced, and all cited references are listed in the bibliography associated with each volume.

The extraordinary range of body form and function exhibited by molluscs has been addressed in part in the first volume by way of the chapters dealing with the shells and external body and those reviewing the individual organ systems. In this volume we examine molluscan fossil history, focusing on their early evolution in the Paleozoic. From humble beginnings, they have evolved into the remarkable array of modern molluscs which range from minute spicule worms, to chitons, tusk shells, clams, limpets, snails, slugs, octopuses, and squid. In the chapters dealing with these groups, we outline their classification and review their external morphology, organ systems, habits, biology, ecology, fossil history, and human uses.

From the beginning, ordering the incredible molluscan diversity into a classification has been a preoccupation of those who have studied molluscs. Until recently shells and anatomy were the main tools used to build those classifications; then molecular studies came to the fore. The latter are often seen as a panacea, but they are not necessarily, especially when it comes to understanding divergences in deep time. One has only to look at the molecular phylogenies produced in recent years to see the often-considerable disagreements in branching patterns in different studies. We hope that the data provided in this book will encourage those engaged in phylogenetic studies to look for congruence between their branching patterns and the plausibility of gains and losses in morphological features and physiological attributes.

None of this work would have been possible without the huge body of research on which it draws heavily. Our role has been to attempt to distil this work, but in so doing it has been impossible to be comprehensive as even though thousands of references are listed, these are but a fraction of the available works. So, dear colleagues, we apologise in advance if we have not cited a particular paper or book you favour.

Once again we acknowledge the huge help from volunteers Doris Shearman and Rosemary Coucouvinis (see introduction to Volume One). Both have been invaluable in checking references, organising the bibliographic database, and proof reading. Doris also greatly assisted with editing and checking the manuscript.

Figures incorporating colour photographs of living animals are provided in the taxon chapters and were largely from three sources. A large number were kindly provided via Philippe Bouchet of the Muséum national d'Histoire naturelle (MNHN) and mostly selected and sent by Philippe Maestrati. These photographs were taken by several photographers who are identified by the museum acronym following their name in the captions. Other photographers who generously provided multiple photographs for this volume were Dr Terry Gosliner, Denis Riek (www.roboastra.com/), and Ria Tan (www.wildsingapore.com/). Photographs were also provided by Julia Sigwart, Ron Schimek, John Buckland-Nicks, Nina T. Mikkelsen, Emanuel Redl, Carmen Cobo, Kevin Kocot, John Walker, Steven Smith, Peter Middelfart, Cristine Huffard, Jenna Judge, Gerald and Buff Corsi, Jann Vendetti, Julian Finn, Monterey Bay Aquarium Research Institute (MBARI), Roy Caldwell, D. Chowdery, Alison Miller, Edie Widder (NOAA), Gary MacDonald, Peter Batson and Janet Voight, Chong Chen, Scott Johnson, Jeanette Johnson, Yuki Tatara, Hiroshi Fukuda, S. Groves, Stijn Ghesquiere, Tomoki

Kase, Prof. Brian Eversham, Clay Bryce, Ian F. Smith, Hugh Jones, Barbara Buge, Katja Peijnenburg, Keith Hiscock, Rüdiger Bieler, Masanori Taru, Gustav Paulay, Bastian Brenzinger, Mat Nimbs, Rosemary Golding, Lars Peters, N. Yotarou, Barry Roth, David Lochlin, David G. Robinson, and Vince Kessner.

REVIEWERS OF CHAPTERS

Chapter 12. G. Giribet; Chapter 13. J. Frýda, P. Wagner; Chapter 14. J. Sigwart; Chapter 15. P. Mikkelsen and B. Morton (an early draft), R. Bieler, J. Taylor, G. Oliver; Chapter 16. J. Sigwart; Chapter 17. M. Reid (an early draft), S. O'Shea, D. Fuchs, C. Klug, R. Hoffmann, A. King, R. Caldwell; Chapter 18. B. Marshall, C. Hickman, C. Chen; Chapter 19. E. Strong, Y. Kantor, A. Hallan, F. Criscione; Chapter 20. W. Rudman, F. Kohler, I. Burghardt, T. Gosliner, I. Hyman; Chapter 21. B. Mishler.

Appendix—the following sections were reviewed: Scaphopoda, G. Steiner; Bivalvia, J. Taylor, J. Cope; Gastropoda, P. Bouchet; Neogastropoda, A. Fesodov.

12 Molluscan Relationships

Detailed treatments of the early history of molluscan classification and other studies are given by Simroth (1892–1894), Lameere (1936), and Hyman (1967) and so are not repeated in detail here. Although molluscs were used and eaten by humans in prehistoric and historic times, an understanding of their place among other animals and their classification did not come to anything approaching a modern concept of the group until Cuvier (1795).

Aristotle (384–322 BCE[1]) recognised two groups of molluscs: Ostrachodermata for those with shells and Malachia for the cephalopods. Although the Polish scholar J. Jonston (or Jonstonus) (1603–1675) first coined the name Mollusca in 1650 (to include cephalopods and barnacles), the name did not come into general usage until it was used and redefined by the great Swedish naturalist Carl von Linné (e.g., Linnaeus 1758). Linnaeus included all 'invertebrates', other than insects, as Vermes, which he divided into several groups, one of which was Mollusca, and in which he included soft-bodied animals such as slugs, coleoid cephalopods, and pteropods that we still recognise as molluscs, but also some non-molluscs including certain cnidarians, tunicates, polychaetes, and echinoderms. Most shelled molluscs were included in another group, Testacea (including chitons, snails, limpets, bivalves, and *Nautilus* but also barnacles and the serpulid polychaetes). In contrast, Cuvier (1795) devised a concept of molluscs (mollusques) which included the subgroups Céphalopodes (including Foraminifera), Gastéropodes (including slugs and snails and also parasitic copepods) and Acéphales (bivalves, tunicates, brachiopods, and barnacles). This scheme was modified a little by Duméril (1806) and later again by Cuvier (1817); both included the pteropods (as Ptéropodes) as a separate group, although non-molluscan groups were still included in the Mollusca[2]. For example, barnacles were thought to be related to chitons and grouped as the Nematopoda, but by the 1830s barnacles were recognised as crustaceans and foraminiferans were excluded from cephalopods. Tunicates were removed in the mid-1860s, but brachiopods remained with the molluscs until close to the end of the 19th century.

12.1 THE HYPOTHETICAL ANCESTRAL MOLLUSC

'In the post-Hennig world, cladograms have replaced HAMs and exposed our ignorance' (Runnegar 1996, p. 78).

Many treatments of molluscan phylogeny have used archetypes, *Baupläne*, or other images of reconstructed common ancestors to postulate what the first mollusc looked like. These figures are known as *hypothetical ancestral molluscs* (HAMs), with over 40 different examples in the literature. The first was proposed by the great evolutionary biologist Thomas H. Huxley (Huxley 1853), and they continue to appear in many invertebrate textbooks today. The HAM is also known as the Archimollusc (Salvini-Plawen 1972, 1981) or 'archaeomollusk' (Yochelson 1978), and there have been several concepts over the years (Figure 12.1). Because the HAM is a device to convey features that the molluscan ancestor might have possessed, HAM concepts have evolved according to prevailing ideas about molluscan evolution (Lindberg & Ghiselin 2003). One trend in the evolving HAM concept was the expansion and enlargement of the posterior mantle cavity and its structures, making these supposed common ancestors of all molluscs much more gastropod like. In other renditions, there was a substantial loss of organs and other features, particularly the radula and gonads, or the organs appear in the juvenile condition, although the animal is represented as an adult. Overall, the classical textbook version of HAM might serve better as a hypothetical conchiferan ancestor rather than a common ancestor of all molluscs. Lindberg and Ghiselin (2003) noted that major discoveries were slow to impact on these images in standard textbooks, and they recommended the extinction of the HAM concept.

Haszprunar (1992a) proposed a new HAM concept based on the idea that aplacophorans are 'basal' molluscs (e.g., Salvini-Plawen 1972, 1985a, 2003). In this concept, the ancestral mollusc was small, rather worm-like, and covered with spicules instead of a shell.

Attempts to define the molluscan ancestor also depended on ideas about what the sister taxon of the Mollusca is (see Section 12.2.2). Concepts of ancestral molluscs thus included designs based on shared ancestry with turbellarian flatworms, reduced annelids, or even entoprocts (see Ghiselin 1988; Haszprunar 1996 for reviews).

None of the HAM ideas discussed above are likely to be an accurate approximation of the first mollusc. Indeed, the molluscan ancestor may have been larger than the small, worm-like ancestor favoured by some and perhaps more like a chiton as suggested by some ancient fossils (see Chapter 8).

12.2 HYPOTHESES OF PHYLOGENETIC RELATIONSHIPS OF THE MAJOR MOLLUSCAN GROUPS

The grouping of molluscan classes into higher taxon groups has been a popular pastime among molluscan workers, with several names proposed for those groupings (see Box 12.1 and Figure 12.2). For example, polyplacophorans and aplacophorans are often grouped as the Aculifera, with the remaining classes contained within the Conchifera in recent phylogenies (e.g., Haszprunar et al. 2008; Kocot et al. 2011; Smith et al. 2011; Vinther et al. 2011).

---

[1] Before Common Era, an alternative to BC.
[2] Blainville (1825) changed Mollusca to Malacozoa, from which the term malacology is derived.

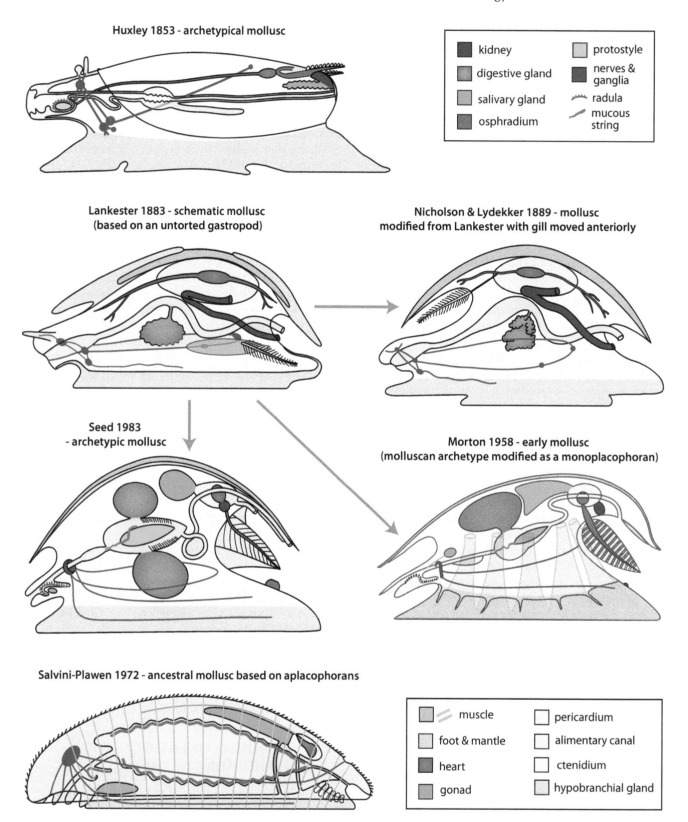

Huxley 1853 - archetypical mollusc

■ kidney		□	protostyle
■ digestive gland		■	nerves & ganglia
■ salivary gland			radula
■ osphradium			mucous string

Lankester 1883 - schematic mollusc (based on an untorted gastropod)

Nicholson & Lydekker 1889 - mollusc modified from Lankester with gill moved anteriorly

Seed 1983 - archetypic mollusc

Morton 1958 - early mollusc (molluscan archetype modified as a monoplacophoran)

Salvini-Plawen 1972 - ancestral mollusc based on aplacophorans

■ muscle		□	pericardium
□ foot & mantle		□	alimentary canal
■ heart		□	ctenidium
■ gonad		□	hypobranchial gland

FIGURE 12.1 Examples of the changing concepts of the hypothetical ancestral mollusc (HAM). The arrows indicate changes in HAM thinking based on new findings and theories when they were proposed. The illustrations are redrawn and modified from the authors mentioned in the figure. See text for explanation and Lindberg and Ghiselin (2003) for more details.

BOX 12.1   SOME NAMES PROPOSED FOR GROUPINGS OF MOLLUSCS ABOVE THE CLASS LEVEL

Below we list many of the names used to group the molluscan classes. The list includes the reference in which the term was first proposed and the taxa included in the original concept. For some names, the concepts have been changed since their original introduction with more or fewer taxa included, but these modified usages are mainly not detailed below (see Runnegar & Pojeta 1974; Salvini-Plawen 1980, 1981; Ax 2000; Kocot et al. 2011; Kocot 2013; Giribet 2014). This list is not comprehensive but includes most of the names used in recent literature.

Adenopoda – all living groups of molluscs other than Caudofoveata (Salvini-Plawen 1972).

Aculifera – Polyplacophora + Aplacophora (Hatscheck in Blumrich 1891), although Stasek (1972) included only Aplacophora. This proposed grouping was based on shared characters of the nervous system, the presence of sclerites and epidermal papillae (see Scheltema 1993), and also supported by ciliary ultrastructure (Lundin et al. 2009). Several extinct taxa are also assignable to this grade or group (see Chapter 9). Equivalent to Amphineura.

Amphineura – Aplacophora + Polyplacophora (Ihering 1876a). Equivalent to, and predates, Aculifera.

Ancyropoda – Scaphopoda + Bivalvia (Hennig 1979). Equivalent to, and predates, Loboconcha.

Aplacophora – Solenogastres + Caudofoveata (Ihering 1876a).

Cephalomalacia – Cephalopoda + Gastropoda + Scaphopoda (Keferstein 1862).

Conchifera – Mollusca excluding Aculifera (i.e., Monoplacophora, Cephalopoda, Bivalvia, Scaphopoda, and Gastropoda) (Gegenbauer 1878).

Cyrtosoma – Monoplacophora + Gastropoda + Cephalopoda (Runnegar & Pojeta 1974).

Diasoma – Rostroconchia + Bivalvia + Scaphopoda (Runnegar & Pojeta 1974). Not equivalent to Loboconcha which does not include the rostroconchs.

Dorsoconcha – Polyplacophora, Monoplacophora, Gastropoda, and Bivalvia (Stöger et al. 2013).

Eumollusca – all molluscs other than Aplacophora. Originally introduced by Roule (1891) and subsequently by Ax (1999). Equivalent to, and predates, Testaria.

Galeroconcha – Tryblidiida (i.e., Monoplacophora) + Bellerophontida (Salvini-Plawen 1980).

Ganglioneura – all Conchifera other than Monoplacophora (Hennig 1979).

Hepagastralia – all molluscs other than Solenogastres (Salvini-Plawen & Steiner 1996; Haszprunar 2000).

Heterotecta – Solenogastres + Polyplacophora (Salvini-Plawen 1980).

Loboconcha – Scaphopoda + Bivalvia (Salvini-Plawen 1980). Equivalent to Ancyropoda.

Mollusca – originally included several non-molluscan groups including brachiopods, ascidians, and barnacles (Cuvier 1798).

Placophora – proposed as a class for the chitons. Equivalent to the earlier name Polyplacophora (Ihering 1876a).

Pleistomollusca – Gastropoda + Bivalvia (Kocot et al. 2011).

Rhacopoda – Gastropoda + Cephalopoda (Hennig 1979).

Scutopoda – introduced as a 'subphylum' for Caudofoveata only (Salvini-Plawen 1978).

Serialia – Polyplacophora + Monoplacophora (Giribet et al. 2006).

Testaria – Conchifera + Polyplacophora (Salvini-Plawen 1972). Equivalent to Eumollusca.

Variopoda – Scaphopoda, Cephalopoda, and the two aplacophoran classes (Stöger et al. 2013).

Visceroconcha – Gastropoda + Bellerophontida + Cephalopoda (Salvini-Plawen 1985a) – originally as Vesceroconcha but emended by Haszprunar (1988f).

There have been several alternative hypotheses of the relationships of the major molluscan groups (classes) to one another that have been derived from recent analyses involving morphological, molecular, developmental, and fossil data. The main competing ideas involve the relationships of (1) the aplacophoran groups, (2) the Monoplacophora, (3) the Scaphopoda, and (4) the Cephalopoda.

The aplacophoran taxa (Solenogastres and Caudofoveata) and Polyplacophora are of particular interest because they are often considered to represent the sister taxon (Aculifera) of the Conchifera. There are several hypotheses regarding the relationships of the aplacophoran groups. Until recently the main ones were: (1) they are a paraphyletic grade with the Caudofoveata the sister taxon to other molluscs (Testaria) (Salvini-Plawen 1972, 1980, 1981; Haszprunar 2000; Salvini-Plawen 2003; Haszprunar et al. 2008; Todt et al. 2008b) and (2) they form a monophyletic group (Aplacophora) that is sister to the Testaria (e.g., Hyman 1967; Scheltema 1993). More recent molecular phylogenies (Kocot et al. 2011; Smith et al. 2011; Vinther et al. 2011; Smith et al. 2012) have placed the aplacophoran taxa as a monophyletic group with the polyplacophorans in the Aculifera and sister to the Conchifera. Other molecular data (Wilson et al. 2010) suggested that the Caudofoveata were derived, possibly paedomorphic, and sister to the cephalopods, while the solenogasters were grouped outside the molluscs with the Sipuncula (Annelida).

The Monoplacophora has traditionally been included in the Conchifera and the chitons in the Aculifera (see Section 12.2.1 and Figure 12.2), although two molecular studies using Sanger-based approaches have shown monoplacophorans and chitons form a clade named Serialia (Giribet et al. 2006;

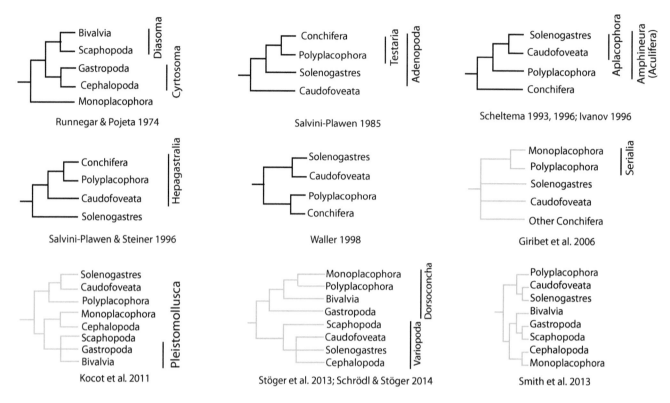

FIGURE 12.2   Some of the alternative hypothesised relationships among the Mollusca, with molecular analyses in blue.

Wilson et al. 2010) (see Section 12.2.1 and Figure 12.2). In contrast to these results, the only other phylogenomic analyses incorporating monoplacophorans (Smith et al. 2011, 2012) favoured a sister taxon relationship with cephalopods.

Traditionally, Polyplacophora and Aplacophora were combined into Amphineura based on similarities in their nervous systems (Ihering 1876a, b; Spengel 1881; Hoffmann 1929–1930), and the same group was named Aculifera, characterised by having a mantle with cuticle-bearing spicules (see Scheltema 1988a; Ivanov 1996; Scheltema 1996). In schemes with paraphyletic aplacophorans at the base of the molluscan tree, the remaining molluscs form a clade Testaria, with Polyplacophora the sister group to Conchifera (e.g., Salvini-Plawen 1980, 1985a, 1900, 2003; Wingstrand 1985; Salvini-Plawen & Steiner 1996). More recently Haszprunar (2000) proposed Hepagastralia for all molluscs other than Solenogastres, based mainly on gut morphology.

The idea that bivalves and scaphopods were related was previously popular and is indicated by the grouping Diasoma (as a subphylum), along with the extinct Rostroconchia (Runnegar & Pojeta 1974). These classes all have a shell primitively open at both ends with flow-through ventilation and possess a relatively straight gut. Runnegar and Pojeta (1974) also proposed another subphylum, Cyrtosoma, to include the Monoplacophora, Gastropoda, and Cephalopoda (Figure 12.2). A morphological analysis of molluscan relationships by Waller (1998) showed that scaphopods were not related to bivalves (and rostroconchs) and should be included in the Cyrtosoma and that Monoplacophora should be excluded from that group (Figure 12.2). These latter

conclusions relating to scaphopods have been supported by other more recent analyses using molecular data.

12.2.1   ACULIFERA – CONCHIFERA, TESTARIA, AND SERIALIA HYPOTHESES

As seen from the previous section and Figure 12.2, the phylogenetic relationships of the molluscs have been expressed as three competing hypotheses:

- The Aculifera (chitons + aplacophorans) + Conchifera (all other classes) hypothesis
- The Testaria (or Eumollusca) (chitons + all other classes) hypothesis (with aplacophorans basal)
- The Serialia (chitons + monoplacophorans) hypothesis, with all other molluscs forming a separate clade

Today, most workers consider living molluscs to be divisible into two groups, the Aculifera and the Conchifera. The Aculifera includes the Caudofoveata and Solenogastres (considered by some to be sister taxa comprising the clade Aplacophora) and the Polyplacophora (chitons). The conchiferan taxa are the Monoplacophora, Bivalvia, Scaphopoda, Cephalopoda, and Gastropoda. The primary distinction between aculiferan and conchiferan molluscs is the formation of single or paired calcified structure(s) (i.e., the shell) that cover the dorsal surface of their bodies. Both aplacophorans and chitons produce calcium carbonate spicules, and chitons produce eight shell plates. In the aplacophorans, these spicules cover the body surface while in chitons they are restricted to an epidermal band (the girdle), which surrounds the eight dorsal shell plates. Other characters that unite the Aculifera are epidermal papillae (extensions of the mantle into the body cuticle, thought to be secretory in function) and a suprarectal commissure in the nervous system (Scheltema 1993; Scheltema et al. 1994; Ivanov 1996).

Other controversies regarding the Aculifera are (1) whether the Caudofoveata and Solenogastres represent a grade or clade and (2) whether the Caudofoveata and Solenogastres are closer to the base of the molluscan tree than the polyplacophorans (Figure 12.2). Those that argued for Caudofoveata and Solenogastres being a basal grade (e.g., Salvini-Plawen 1980; Salvini-Plawen & Steiner 1996) placed the Caudofoveata as the most basal extant group (i.e., they are the sister to all remaining molluscs, the Hepagastralia, Salvini-Plawen 1996; Haszprunar 2000), followed by the Solenogastres, Polyplacophora, and lastly, the Conchifera. This hypothesis renders the Aculifera paraphyletic, and for this reason, Salvini-Plawen (1980) argued against that name. In contrast, Scheltema (1993) and Ivanov (1996) thought that aplacophorans were monophyletic and were the sister taxon to the Polyplacophora and together (as the Aculifera) were the sister taxon to the Conchifera. Lastly, Waller (1998), like Scheltema and Ivanov, argued for a monophyletic Aplacophora but considered the Polyplacophora to be the sister taxon of the Conchifera, thereby making the Aculifera biphyletic.

Regardless as to whether the Caudofoveata and Solenogastres represent a grade or clade, many molluscan morphologists have erroneously considered these two groups to represent the earliest molluscan 'bauplan' and to lie near the base of the molluscan tree simply because they are worm-like (Todt et al. 2008b). Yochelson (1978) suggested that the worm-like morphology of the aplacophorans was secondarily simplified rather than primitive, although he continued to regard the Polyplacophora as a primitive group. Lindberg and Ponder (1996) made a similar argument and suggested that the Caudofoveata and Solenogastres represented separate lineages, derived through paedomorphosis, from extinct or living molluscan lineages. These matters are discussed further in Chapter 13.

A relationship between the Monoplacophora and Polyplacophora was first noted by Knight and Yochelson (1958). The shell muscles have a similar configuration in both taxa with eight-fold shell-attachment muscle groups, so some have suggested that this configuration was a feature of their common ancestor (Wingstrand 1985), although just what that ancestor looked like is still open to speculation (e.g., Eernisse & Reynolds 1994). Assuming a common ancestor, various hypotheses have been proposed to explain the origin of the very different shells of monoplacophorans and chitons. Suggestions include independent derivations (e.g., Scheltema 1988a), that the single shell of monoplacophorans resulted from the fusion of multiple valves (Haas 1981), or that chiton valves were derived from the segmenting of a single shell (Runnegar & Pojeta 1974).

In a molecular analysis based on two mitochondrial genes and three nuclear genes, with a single (18S rRNA) monoplacophoran sequence included, the Monoplacophora were placed within the Polyplacophora, and the aplacophorans were located among the other 'Conchifera' (Giribet et al. 2006). Solenogastres was the sister taxon to Patellogastropoda, and the Caudofoveata to the Cephalopoda, although these groupings were poorly supported. Based on this analysis, Giribet et al. (2006) proposed a new taxon, the Serialia, to include the Monoplacophora and Polyplacophora – taxa with serially repeated gills and eight pairs of pedal retractor muscles. In a subsequent paper, Wilson et al. (2010) included additional sequence data for several gene fragments for a second monoplacophoran species. In this second analysis, the Serialia were again recovered, but the monoplacophorans were positioned as the sister taxon of the Polyplacophora rather than being nested among them, and statistical support for the Serialia was strong. The Caudofoveata were once again identified as the sister taxon of the Cephalopoda, still with weak support, while the Solenogastres were placed outside the Mollusca as part of a clade that included Sipuncula and Annelida. So far only two studies have sequenced multiple monoplacophoran taxa. Kano et al. (2012) used sequences from three monoplacophoran taxa and recovered a well-supported Serialia clade in their 18S, 18S+28S, and five-gene (18S, 28S, 16S, H3, COI) analyses. Stöger et al. (2013) also recovered a Serialia clade in their Maximum Likelihood and Bayesian analyses of the same molecular markers for six monoplacophoran taxa. In a follow-up analysis including gene order and sequence data from three complete monoplacophoran mitochondrial genomes, Stöger et al. (2016) recovered Serialia in some analyses, but their results were inconclusive because of sensitivity to taxon sampling.

Kocot et al. (2011) used transcriptome[3] data and obtained a very different result with the two aplacophoran taxa grouping together (Aplacophora) as sister to the Polyplacophora (i.e., they recovered Aculifera and Conchifera); however they did not have a monoplacophoran in their analysis. A more recent phylogenomic study (Kocot et al. 2017) did include a single monoplacophoran. A strongly supported Aculifera was recovered in every analysis, and in one of these (Kocot 2017, Fig. 7), the monoplacophoran taxon was recovered as the sister taxon of the Aculifera. In the remaining analyses, the monoplacophoran was sister to the remaining conchiferan taxa in three of them, sister to the gastropods, or sister to the cephalopods (in one analysis each).

Not only has there been debate about the relationship of molluscan classes, but also ideas regarding relationships within the classes have changed considerably over the last few decades, as discussed in the appropriate chapters.

---

[3] The transcriptome is the RNA molecules in a cell or tissue and may include just mitochondrial RNA (mRNA) or all RNA.

12.2.2 THE SEARCH FOR THE MOLLUSCAN SISTER GROUP

Phylogenetic studies based on either morphological or molecular data vary in their conclusions as to the sister taxon of molluscs. While it has been suggested that this may result from the burst of rapid evolution in the Cambrian, our current analytical methods and the datasets employed also need to be improved to uncover these ancient relationships.

The molluscs, along with several other phyla, are included in the Lophotrochozoa (Figure 12.3). Although not supported by the most recent molecular analyses, lophotrochozoans have previously been divided into subgroups with the molluscs placed in the Trochozoa (molluscs, annelids, brachiopods, phoronids, and nemerteans) and then in the Eutrochozoa (molluscs, annelids, and nemerteans) (see Box 12.2). Potential candidates for the molluscan sister taxon include several of these phyla, as detailed below.

BOX 12.2 SOME HIGHER GROUPINGS OF ANIMALS

Bilateria can be divided into three main groups, Lophotrochozoa, Ecdysozoa, and Deuterostomia (Figure 12.3):

Xenacoelomorpha – Acoel flatworms (**Acoela**) and xenoturbellidans.

Deuterostomia – the blastopore[4] gives rise to the anus – includes echinoderms, hemichordates, cephalochordates, urochordates, and vertebrates (or craniates).

Protostomia – the blastopore gives rise to the mouth – includes molluscs and the remaining animals. The protostomes can be divided into three unresolved taxa: the **Spiralia, Ecdysozoa**, and **Chaetognatha** (or arrow worms). The Ecdysozoa includes arthropods, tardigrades, nematodes, priapulidans, etc. They have a modified spiral cleavage and moult. Arrow worms are predatory, dart-shaped animals with tripartite bodies.

Spiralia – are recognised as having spiral cleavage, and the mesoderm is derived from the 4d cell. Relationships are unresolved among the spiralian clades Cycliophora, Dicyemida, Orthonectida, Gnathifera, and Platytrochozoa. The Platytrochozoa consists of two clades – Rouphozoa (Platyhelminthes + Gastrotricha) and Lophotrochozoa.

Lophotrochozoa includes brachiopods, phoronids, bryozoans, entoprocts, nemerteans, molluscs, and annelids. Previously proposed subgroups within Lophotrochozoa are *Platyzoa* (Rouphozoa, Gnathifera), *Polyzoa* (Entoprocta, Cycliophora, Bryozoa), *Lophophorata* (Brachiopoda, Bryozoa, Phoronida), *Brachiozoa* (Brachiopoda and Phoronida), *Eutrochozoa* (Mollusca, Annelida, Nemertea, Orthonectida, and Rhombozoa, although it has been used in a wider sense), and *Trochozoa* (Polyzoa, Brachiozoa, Eutrochozoa), unified by the possession of a trochophore larva and, in many adults, the possession of lophophore feeding tentacles.

(For further details see Halanych et al. 1995; Giribet 2002; Giribet 2008b; Helmkampf et al. 2008; Edgecombe et al. 2011; Dunn et al. 2014; Giribet 2016b).

Proposed sister taxa include turbellarian flatworms (e.g., Salvini-Plawen 1972), reduced annelids (see Ghiselin 1988; Haszprunar 1996 for reviews) and, based on supposed similarities in their larvae, entoprocts (Wanninger et al. 2007; Haszprunar & Wanninger 2008) and sipunculans (Scheltema 1993). Brachiopods have also been suggested, based on both mitochondrial, nuclear, and genomic data (Stechmann & Schlegel 1999; Paps et al. 2009; Luo et al. 2015) and nemerteans based on mitochondrial genomic data (Podsiadlowski et al. 2009). One outgroup suggestion even included cnidarians, this idea being born through the observation of apparent strobilation in *Clio*, a thecosome pteropod (van der Spoel 1973) (see Chapter 8). This led to speculation that molluscs might be related to cnidarians through their vague similarity to Conulata, an extinct group often included in the Cnidaria (Pafort-van Iersel & van der Spoel 1979).

A distinct anterior head and ventral foot are features that set many molluscs apart from other animals. Superficially similar features in sipunculan and entoproct larvae have been identified and have fuelled debates about the relationships of those groups with molluscs (e.g., Haszprunar 1996; Scheltema 1996; Haszprunar & Wanninger 2008), but a closer examination of these supposedly homologous characters discounts these putative relationships. In the ectoprocts, the larval 'foot' is derived from the central region of the neurotroch, and the anus is located at the terminal end (Nielsen 1979). Molluscan trochophores lack a neurotroch (Rouse 1999), and the molluscan foot is derived from an ectodermal thickening that forms behind the mouth (Raven 1964). The location of the anus on the terminal end of the 'foot' is even more problematic as in the Mollusca the anus is always situated above and independent of the foot. Even in the footless, worm-like Caudofoveata, the anus lies above the pedal nerves which innervate the foot of all other molluscs. The molluscan foot is also distinctive in having regular dorsoventral pedal muscles.

It is now generally agreed that molluscs are lophotrochozoans (Figure 12.3), but which lophotrochozoan ancestor is most closely related to molluscs remains uncertain, although a growing body of evidence (morphological, molecular, fossil) is focusing on the Brachiozoa (see also Chapter 13).

A mineralised shell is not a unique molluscan character as shells are produced by brachiopods (Figure 12.4), some crustaceans (barnacles), and some tube-living annelids.

---

[4] The fate of the blastopore is more variable than the protostome/deuterostome dichotomy suggests. In some protostome taxa the blastopore gives rise to the anus or ultimately closes and the openings are *de novo* structures (Martín-Durán et al. 2016).

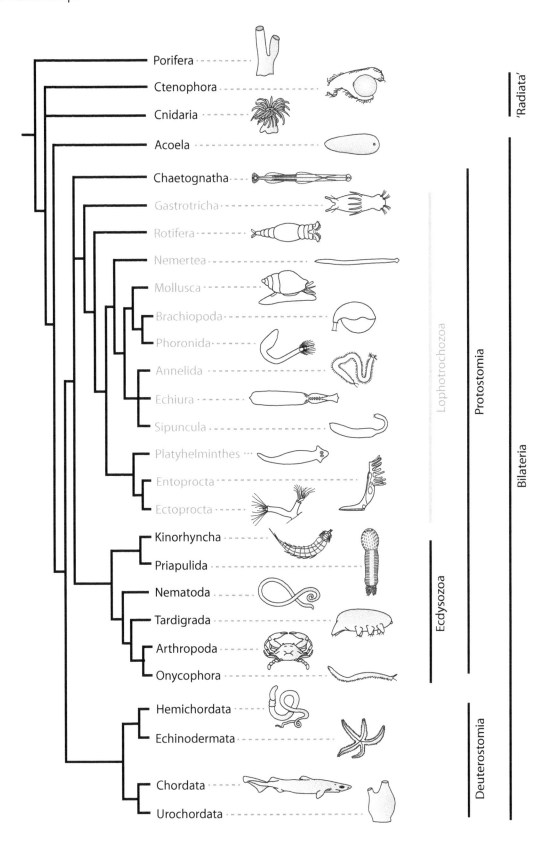

FIGURE 12.3 A hypothesis of relationships of animal phyla. Lophotrochozoans are shaded in blue. Relationships based largely on analyses of nuclear ribosomal genes by Paps et al. (2009).

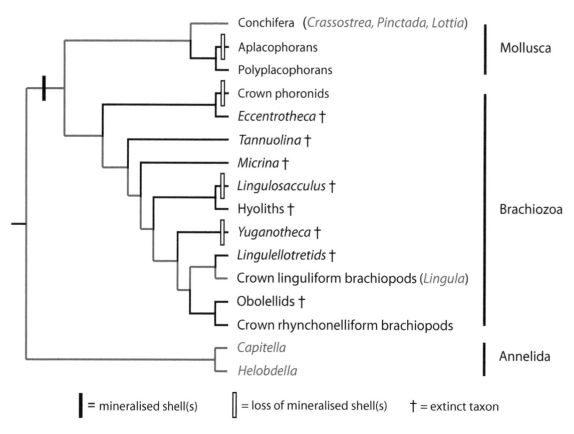

FIGURE 12.4 Mineralised shell(s) in major molluscan and brachiozoan clades. Mineralised shell(s) would have occurred in the common ancestor of molluscs and brachiozoans and were later lost at least four times. In molluscs, mineralised shell loss would have occurred at least once (or possibly twice) in the aplacophorans and on three occasions in the brachiozoans, once in the phoronids and twice in the putative stem brachiopods *Lingulosacculus* and *Yuganotheca*. Lophotrochozoan relationships (red lines and taxa) from phylogenomic analyses in Luo, Y.J. et al., *Nat. Commun.*, 6, 9301, 2015, brachiozoan phylogeny from Moysiuk, J. et al., *Nature*, 541, 394–397, 2017. See Chapter 13 for information on some of the extinct groups.

The homology of the molluscan coelom is also important when considering molluscan relationships. The plesiomorphic molluscan coelom is a *gononephrocoel* – that is, a coelomic space within both the gonad and kidney – but its homology with the coeloms of other phyla is not clear, as it is uncertain if it independently evolved or is a plesiomorphic condition (see Haszprunar 1992b for discussion; Bartolomaeus 1997) (see Chapters 7 and 8 for more information on the molluscan coelom).

The radula is often cited as a uniquely molluscan character, but bilaterally symmetrical, mineralised buccal structures also occur in other spiralian taxa including Annelida and Gnathifera and several extinct putative taxa (e.g., some halkieriids) (see Chapter 13). Each of these structures is characteristic of their respective lineages with some shared similarity (Chapter 13, Figure 13.10) suggesting both convergence as well as the possibility of plesiomorphic buccal structures in some lineages.

Summaries of earlier studies relating to the search for the molluscan sister group can be found in Haszprunar (1996), Lindberg et al. (2004), and Haszprunar et al. (2008). The search is, of course, ongoing with recent molecular analyses reconstructing the annelids, phoronids and brachiopods, or nemerteans (or various combinations of these) as potential molluscan sister taxa (Table 12.1).

Three main trends have occurred independently in two or more major groups of molluscs with fundamental effects on more than one organ system of the body. These are briefly outlined below.

12.3 REPETITION OF ORGANS

Multiple pairs of organs are found in some molluscs, a phenomenon known as *serial repetition*. It includes shell muscles in chitons, monoplacophorans, protobranch bivalves, and *Nautilus*; ctenidia and auricles in *Nautilus* and monoplacophorans; ctenidia alone in chitons; kidneys in monoplacophorans and *Nautilus*; and gonads in monoplacophorans. *Nautilus* has two pairs of pedal retractors, kidneys, gills, and atria, but the numbers of repeated organs in monoplacophorans do not correspond, with zero to two pairs of auricles, three to six pairs of ctenidia, three to seven pairs of kidneys, and one to three pairs of gonads. In chitons, only the shell muscles and ctenidia are serially repeated, but these two systems do not correspond to each other.

There have been two main theories to account for organ repetition in molluscs. It has been thought to be a primitive feature indicating metamery (e.g., Naef 1926; Lemche & Wingstrand 1959; Wingstrand 1985) or that the repetition is

TABLE 12.1

Sister Taxon Relationships Suggested by Selected Molecular Analyses of Molluscan Relationships over the Last Ten Years

Authority	Sister Taxa
Dunn et al. (2008, figure 1)	Annelida (Phoronida (Brachiopoda, Nemertea)
Dunn et al. (2008, figure 2)	Annelida (Brachiozoa, Nemertea)
Giribet et al. (2009, figure 6.1)	Entoprocta
Paps et al. (2009, figure 1)	Brachiozoa
Edgecombe et al. (2011, figure 1)	Annelida, Brachiozoa, Nemertea
Smith et al. (2011, figures S2, S3, S4, S5, S6)	Brachiopoda, Nemertea
Smith et al. (2011, figures S7, S9)	Nemertea (Annelida, Brachiopoda)
Kocot et al. (2011, figure 2)	Annelida
Laumer et al. (2015, figure 1)	Nemertea (Brachiozoa, Annelida)
Kocot et al. (2017, figures 1, 3, 5, 6)	Brachiozoa (Annelida, Nemertea)
Kocot et al. (2017, figures 4, 7)	Brachiozoa

Note: Brachiozoa includes Brachiopoda and Phoronida

due to secondary duplication (e.g., Morton & Yonge 1964; Haszprunar & Schaefer 1996), as there is no support for segmentation due to the lack of correspondence of the repeated organs and no evidence of segmentation in the nervous system.

In the 1950s, the discovery of a living monoplacophoran (*Neopilina*) (Lemche 1957; Lemche & Wingstrand 1959) that possessed several gills, shell muscles, and kidneys gave rise to the notion that molluscs were primitively segmented and rekindled the idea that annelids might be the molluscan sister taxon.

Serial repetition of shell muscles may be a primitive feature of molluscs (Haszprunar & Wanninger 2000), with a reduction in the number of these muscles having occurred independently in different molluscan groups, as shown in the fossil record. This is notably the case for cephalopods (Kröger & Mutvei 2005) (see Chapter 17) and protobranch bivalves (e.g., Driscoll 1964).

Based on outgroups, the lack of segmentation is plesiomorphic in the Lophotrochozoa as only Annelida are segmented. Serial replication of organs also occurs in the gonads of Nemertea, the proglottids of the Cestoda, and in the muscle systems of some bryozoans and brachiopods (Bulman 1939; Jebram 1986; Giribet 2016a).

12.4 ANO-PEDAL FLEXURE AND A CHANGE IN BODY ORIENTATION

This topic has been covered in Chapters 1, 3, 5, and 8, but the main points of this important aspect of molluscan evolution are repeated here for convenience.

Chitons, living monoplacophorans, and aplacophorans have the anteroposterior axes (as determined by the position of the mouth and anus) coinciding with their body orientation (i.e., the main body axis), and they have a linear alimentary system. Because of this body configuration, their dorsoventral axes are shorter than their anteroposterior axes. The other major groups (classes) exhibit ano-pedal (or anal-pedal) flexure, with the gut curved into a U-shape, although it is less marked in bivalves than in other groups. This resulted in the anteroposterior axis becoming shorter and the dorsoventral axis longer, so the latter became the longest body axis in gastropods, scaphopods, and cephalopods. The consequences of this change were profound, resulting in greater lateral compaction of the viscera and dorsoventral elongation. In bivalves, this enabled shortening of the body and its enclosure in a two-valved shell, while in gastropods and cephalopods a single conical or coiled shell was developed and in scaphopods a tube.

12.5 REDUCTION AND PAEDOMORPHOSIS

Paedomorphosis occurs when juvenile or larval traits are retained in adults, and such events are thought to have played an important role in molluscan evolution. There are two ways paedomorphosis can occur. The first is by *progenesis* where sexual maturation occurs earlier relative to the rest of development, and the other is by *neoteny* (or juvenilisation) where somatic development is retarded relative to reproductive maturity. These processes often result in simplification or reduction in organs and are two of the developmental changes termed *heterochrony* that involve changes in the timing of developmental events (see Chapter 8).

Today, most workers consider living molluscs to be divisible into two groups, the Aculifera and the Conchifera. Relationships within these two clades continue to be investigated, but consensus on a single hypothesis of relationships within these groups remains elusive (Sigwart & Lindberg 2015). In addition, our current understanding of the Aculifera and Conchifera does little to elucidate the morphology of the earliest mollusc. Assuming they were worms because there are worm-like aplacophorans in the Aculifera is without merit. Only a well-supported outgroup or an extraordinary fossil occurrence is likely to assist in resolving this issue. In the former case, molecular data appears to be homing in on three taxa, Brachiozoa, Nemertea, and Annelida. Of these three, the fossil record is likely to be useful only with the Brachiozoa, but care must also be exercised here to guard against overly broad diagnoses of character state homology and the 'shoe-horning' (Gould 1989) of inimitable fossil taxa into living taxa.

13 Early History and Extinct Groups

13.1 INTRODUCTION

Molluscan fossils have been involved in every aspect of the development of the fields of palaeontology, palaeobiology, evolution, and phylogeny. Fossils are the remains and traces of once-living organisms, but in the past naturalistic and supernaturalistic explanations were given for their existence. Over 300 years ago, the term 'fossil' referred to any object found buried in the earth and included geological objects such as crystals and concretions, archaeological items, and the remains of dead organisms. The latter category was especially problematic because many of the molluscan fossils strongly resembled living taxa, suggesting an organic origin for these remains. Others were more fragmentary or were the remains of extinct taxa such as ammonites (which at one time were thought to be representations of decapitated snakes). Such strange objects suggested a non-organic origin of fossils, one such explanation being that they had grown spontaneously within the rock. Extensive beds of fossil molluscs found at high elevations in mountains, far from the ocean, also prompted the notion that they had spontaneously appeared there.

The idea of the spontaneous generation of molluscs and other fossils within solid rock traces its origins to Aristotle (384–322 BCE[1]) and his writings on animal reproduction and generation. Aristotle thought that molluscs reproduced exclusively by spontaneous generation and that they would suddenly appear where and whenever conditions were suitable for their lifestyle. For example, when mountainous areas with briny soils and abundant limestone were inundated by water, conditions became appropriate for molluscs, and they would spontaneously generate, live out their lives, and then die, remaining embedded in the rocks in which they had first appeared. This view was commonly held until Nicolaus Steno (1638–1686) established the foundations of palaeontology. Steno made observations of conditions surrounding living organisms and used these observations to test the idea that fossils had actually grown within the rocks in which they were found. Many of his arguments involved molluscan fossils. For example, he noted that tree roots in softer soils were relatively straight and regular in their growth pattern while the roots of trees growing in harder soils were contorted and irregular. He pointed out this was not the case for fossil molluscs which were often similar to one another regardless of the hardness of the rock in which they had supposedly grown. He also noted that their growth within the rock should have, but did not, crack the rocks. The observations and arguments by Steno established fossils as the remains of once-living organisms. It was not until about 200 years later that molluscs again played a major role in the next advances in the study of fossils. From the time of Steno, there remained the question as to why extensive fossil beds of marine snails were found at high elevations in mountains far from marine habitats. Catastrophic flood stories, prominent in many religious traditions, were often invoked to account for their distribution. During the Renaissance, Leonardo da Vinci (1452–1519) argued that the fossil molluscs in the mountains did not represent a death assemblage from a single event but rather the different beds comprised different taxa, which in turn represented events at different times.

Charles Lyell (1830) observed that younger strata had a higher percentage of fossils of living mollusc species than those in the older strata, which had few, if any, living taxa present (Stanley et al. 1980). Lyell later inferred from this pattern that over time, extinction removed older taxa, while the 'origination of fresh species' gave rise to a greater abundance of living taxa in younger strata (Lyell 1881 p. 5). The percentage of extinct to living taxa was the first basis for the recognition of the Pliocene, Miocene, and Eocene time periods.

Charles Darwin, a friend of Lyell, made extensive geological observations in South America during his global circumnavigation aboard the *Beagle* between 1831 and 1836. Darwin (1838–1843) noted elevated shell beds in Argentina, along the coast of Tierra del Fuego, and along the coast and in the mountains of Chile. His experience of strong earthquakes and observations of the extent to which they caused parts of the coast to be uplifted led him to propose an earthquake-driven uplift mechanism for the origin of these elevated beds.

Today molluscs provide one of the most important datasets in palaeobiological research and are crucial for understanding and analysing factors and patterns in the evolutionary history of life on Earth. The molluscan fossil record also plays a critical and unique role in advancing our understanding and reconstruction of the evolutionary relationships within molluscs and between the molluscs and other lophotrochozoan taxa. This is driven primarily by the evolution of the molluscan exoskeleton (shell and/or spicules), which has ensured that the Mollusca are well represented in the metazoan fossil record. Only the aplacophoran taxon Caudofoveata remains unrecognised in the fossil record, although the supposed occurrence of putative Solenogastres in the Silurian remains debatable (see Section 13.3.4.1.3). Techniques for reconstructing soft tissue morphology (Sutton 2008) and recovering microfossils and other microscopic skeletal elements are rapidly improving. Within the last thirteen years, even putative developing lophotrochozoan embryos have been discovered from the Cambrian of China, and putative molluscan-like radulae from the early Cambrian of Canada have been reported (Butterfield 2006, 2008). Thus, the potential to discover aplacophoran spicules and other molluscan elements and microfossils

---

[1] Before Common Era, an alternative to BC.

remains high, and it is likely that our knowledge in these areas will continue to increase.

13.1.1 THE FIRST MOLLUSCS?

The first appearance of molluscs in the fossil record is controversial. According to Parkhaev and Demidenko (2010), the first known molluscs (*Purella*) appeared in the uppermost zone of the Nemakit-Daldynian (542–534 Ma).[2] Beginning with the Tommotian (534–530 Ma) putative molluscs, thought to have been prominent in Cambrian 'small shelly fossil' assemblages, have been identified as Polyplacophora, Monoplacophora, Bivalvia, and Gastropoda. In contrast to the aplacophoran groups, these four groups and the remaining molluscan taxa can be followed through the fossil record. For some taxa, the affinities and relationships between the fossils and living taxa appear straightforward. For example, bivalves, most gastropods, cephalopods, and monoplacophorans are recognised and allocated to extinct and living groups. As one goes deeper in time, the interpretation and affinities become more difficult – especially when extinct stem groups are involved.

Determining taxon affinities in the earliest portions of the fossil record is fraught with potential difficulties. For Cambrian 'molluscs', typically only a shell is available for comparison with the six extant molluscan morphologies – cephalopods (shell with septa), gastropods (limpet or coiled shell), bivalves (bivalved shell), scaphopods (tusk-like shell), polyplacophorans (multiple shell plates), and monoplacophorans (limpet shell); living aplacophoran molluscs have spicules but lack shells. Besides gross shell morphology, several other characters may be resolvable, including shell microstructure, muscle attachment areas, and external and internal sculpture.

The presence of other shelled taxa in the Cambrian such as brachiopods, hyoliths, tentaculitans, and some arthropods further complicates accurate allocation. In these situations, extra caution must be exercised to avoid 'shoe-horning' specimens into more familiar taxa. Gould (1989) highlighted this phenomenon with another Cambrian group, the Burgess Shale arthropods. In this example, diverse morphologies of arthropods were originally identified as trilobites or as members of extant taxa, rather than understood for the unique lineages they represented. Numerous potential 'shoe-horning' opportunities exist in Cambrian shells, especially with groups such as the siphonoconchs, helcionellidans, pelagiellidans, and sachitidans.

Recent palaeontological and molecular work in one of the putative molluscan sister taxa, the Brachiozoa, is illustrative here. The Brachiozoa (brachiopods + phoronids) were once thought to be a relatively low disparity group. Brachiopods were well known and delimited, bivalved animals with a lophophore and, typically, a peduncle; the group first appeared at the Nemakit-Daldynian boundary (542.0 Ma). In contrast,

phoronids are shell-less, tubular animals[3] with U-shaped guts and a lophophore. The U-shaped gut necessitates that both groups undergo a folding similar to ano-pedal flexure in some molluscs (see Chapters 8 and 12). Over the last 20 years, brachiozoan morphological disparity has been substantially increased by proposals to include additional groups in the phylum. Tommotiidans, hyoliths, and tentaculitans are all tubular animals living in shells (blind tubes) with hypothesised U-shaped guts, and were formerly placed in the Mollusca as well as in other lophotrochozoan groups. Other recently discovered 'brachiozoan' taxa, such as *Lingulosacculus* and *Yuganotheca*, are more brachiopod-like but lack shells or have agglutinated shells (Balthasar & Butterfield 2009; Moysiuk et al. 2017).

Based on a morphological analysis, Vinn and Zatoń (2012) concluded that tentaculitans clustered with the Brachiozoa rather than Mollusca or Bryozoa, while the re-examination of hyoliths from the Burgess Shale by Moysiuk et al. (2017) reported a putative lophophore in this taxon, again suggesting brachiozoan rather than molluscan affinities. Tommotiidans have long been problematic, and reconstructions before the discovery of partial, tubular scleritomes (Skovsted et al. 2008) often featured the individual sclerites arranged in a similar way to the plates of polyplacophorans (e.g., Evans & Rowell 1990). In addition, the discovery of a bivalved larval shell in some tommotiids suggests further brachiozoan affinities. While not all workers accept these new interpretations and alternative relationships, these data and the hypotheses they support require a re-examination of other putative early molluscs, especially given molecular analyses of lophotrochozoan relationships which place the brachiozoans as the sister taxon of the Mollusca rather than the Annelida (Paps et al. 2009; Luo et al. 2015).

Allocation of Cambrian taxa to the Mollusca was initially straightforward with few controversies (Runnegar & Pojeta 1974a), although both new specimens and the re-examination of existing data have sometimes been in conflict with traditional concepts. For example, Dzik (2010) made a cautionary call regarding putative early Cambrian monoplacophoran-like limpets which, based on the muscle attachment patterns, may actually represent brachiopod valves and not the ancestors of cephalopods, and Butterfield (2006) cogently argued that the Burgess Shale taxa *Odontogriphus omalus* and *Wiwaxia corrugata* are not stem group molluscs but rather jawed, segmented worms which could represent two phyla. Last, the startling discovery by Roger Thomas and colleagues (Thomas et al. 2010; Thomas & Vinther 2012) of pelagiellidans with paired clusters of chaetae calls into question the recognition of this spirally coiled, septate shell as a mollusc. Dzik and Mazurek (2013) reported similar structures in *Aldanella* and suggested the transfer of the pelagiellidans to the Hyolitha rather than them representing one of the earliest occurrences of the Gastropoda (see Section 13.3.2.2.7). If correct, this assignment would also substantially increase the disparity

---

[2] Nemakit Daldynian 'molluscs' were initially thought to be latest Ediacaran, although they are now considered to be Terreneuvian (Cambrian) by most workers.

[3] We regard animals with U-shaped guts as 'tubular' rather than worm-like or vermiform because of their lack of a posterior anus.

of the brachiozoan taxa by including spirally coiled shells in the group. An alternative view was presented by Vendrasco and Checa (2015) who noted that many helcionelloids also had shell pores, which are more typical of brachiopods than of molluscs, and suggested that chaetae and shell pores might be plesiomorphic character states shared by Mollusca, Brachiopoda, Bryozoa, and Annelida, which were subsequently lost in crown molluscs. Many of the aforementioned debates centre on the homology (and envisioned importance) of specific structures, often unconstrained by outgroups or using imaginary cartoon creatures to reconstruct the correspondence of the molluscan body plan in ancient conchs (Lindberg & Ghiselin 2003b). This is especially inappropriate when the cartoon represents, among other features, a gastropod with a detached head (Figure 13.1).

The potential position of a morphologically diverse brachiozoan clade as sister to the Mollusca necessitates a careful reconsideration of character states – what is a mollusc and what is something else? For example, if sister taxa, several calcareous shell microstructures become potentially plesiomorphic within the ancestor of the two groups and not diagnostic of either (Carter 1985, Figure 26; Malakhovskaya 2008; Li et al. 2017b). Moreover, a diverse array of extinct body plans also provides additional considerations for the last common ancestor of molluscs and brachiozoans. Although focused on the Burgess Shale Cambrian arthropod disparity, the conclusions of Briggs et al. (1992) are illustrative of this problem (Figure 13.2). If only the extant morphologies are considered, taxa (body plans and their associated character states) are reduced by more than half (20 versus 46 taxa). Extinction rates in the three surviving clades ranged between 0% and 87.5% (Figure 13.2), and the disparity is reduced (Briggs et al. 1992). Extinction applies similar sampling constraints on molecular data which exists only for a small subset of living taxa.

The above discussion highlights the difficulties of determining what constitutes a mollusc, much less a definitive origin(s) of the phylum. Therefore, we have taken a conservative approach and consider alternative narratives for the Cambrian appearance of the Mollusca.

13.1.2 SHELL MORPHOLOGY

Palaeontologists have searched for diagnostic shell morphologies by which to recognise molluscs in the fossil record. Molluscan shells are predominately the product of accretionary growth processes that add calcium carbonate in a protein matrix to the growing edge of the shell. Some molluscs such as polyplacophorans and aplacophorans also secrete intracellular spicules and scales. Definitions of hard part structures (e.g., spicule, scale, plate, conch) are given in Table 13.4. A multitude of morphologies are produced by variation in the rates of shell secretion around the edge of the shell. Through this relatively simple process, molluscs have generated an amazingly diverse range of morphologies both between and within different groups and lineages. These morphologies included coiled shells, limpets, tubes, and bivalves. The shell has also been lost numerous times in gastropods and more than once in coleoid cephalopods. Unfortunately, the early forms are not unique to individual molluscan clades or even to molluscs themselves, and this has resulted in many difficulties in interpreting their early history. For example, Smith and Caron (2010) proposed that the Burgess Shale animal *Nectocaris* was an early, shell-less, stem cephalopod although previous workers had identified it as either an arthropod or chordate, and obvious molluscan synapomorphies are absent (see Mazurek & Zaton 2011). Scaphopods and the multivalved Polyplacophora are unique in having a limited morphological diversity compared to the other major molluscan groups. Simple morphologies, such as the often unsculptured tubes of scaphopods, also present problems for palaeontologists because of their lack of diagnostic characters and potential confusion with calcified tubes of other living and extinct lophotrochozoan taxa (e.g., some polychaetes).

The most recognisable putative molluscan form in the fossil record is the single bilaterally symmetrical shell or conch. This ancestral shell is thought to have been subsequently decalcified along the dorsal midline during development to produce the pair of valves of the Bivalvia, each with a prodissoconch. While rostroconchs might appear to be intermediate in this sequence, they have a single protoconch associated

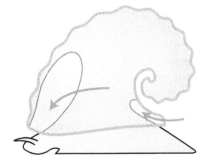

Hypothetical Cambrian mollusc
(apocephalic and endogastric)

Stem hypothetical mollusc
(ligocephalic and endogastric)

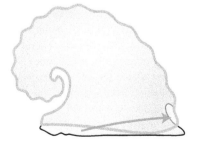

Stem hypothetical mollusc
(ligocephalic and exogastric)

FIGURE 13.1 Left: Hypothetical ancestral mollusc cartoon commonly used to reconstruct anatomical features and functions; note free head (apocephalic), proposed water currents, and large anterior and small posterior mantle cavities. Middle and right figures reconstructed with the plesiomorphic attached head (ligocephalic), estimated extent of posterior and lateral mantle cavities and hypothesised water currents.

Extinct & Extant   Extant Only

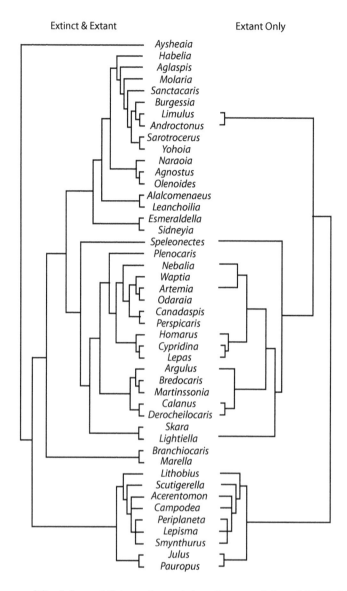

FIGURE 13.2 Unrooted phylogeny of Cambrian and living arthropods based on morphology. Modified from Briggs, D.E. et al., *Science*, 256, 1670–1673, 1992.

with one valve and a calcified groove joining the two valves. Polyplacophorans are unique among the molluscs in having exoskeletons that consist of multiple valves, although in the fossil record, the disarticulated valves of what are assumed to be early chitons can be difficult to recognise as molluscan or envisioned in life position. Therefore, the rare discovery of whole specimens (the scleritome) is critical in recognising these disarticulated bits as being part of a single animal. Once the overall pattern is recognised, the ratio of head and tail-like valves to mid-valves in bulk samples can be very informative for estimating both the diversity and abundance of fossil chitons.

Because molluscan palaeontologists have to rely to a large degree on shell morphology, this limited character set has produced a proliferation of higher taxon names because supposed differences in shell morphology are the sole determinant of many extinct higher taxa. The limitations of this approach in understanding the diversity and relationships of

fossil molluscs are readily apparent when one considers the degree of morphological disparity that can occur in both small and large clades, e.g., Neritimorpha (~450 species) and Caenogastropoda (~75,000 species). At the other end of the spectrum is rampant morphological convergence such as seen in the limpet-like morphologies that have independently evolved in living Mollusca over 54 times (Vermeij 2016). As previously noted, this convergence is particularly problematic when trying to interpret molluscan relationships in the Cambrian where limpet morphologies dominate (also see Section 13.3.2).

13.1.2.1 Beyond Shell Morphology

Aware of the problems that convergent shell morphologies present, many palaeontologists have sought and described non-traditional morphological characters from fossils, including shell microstructure, anatomical impressions from muscles, and even some soft anatomy features and structures.

By the mid-20th century, palaeontologists had recognised that phosphatic casts of the internal surface of bivalve shells contained impressions that allowed the identification of the type of shell microstructure (Runnegar 1983, 1985) (see Chapter 3). Both phosphatic and fine-grained sediments such as silt, mudstones, and clays can preserve these structures, including nacre, prismatic, and crossed lamellar, as far back as the Cambrian (Kouchinsky 2000; Feng et al. 2003; Feng & Sun 2003). While exceptional preservation of the original shell structure components is more common in Mesozoic and Cenozoic fossils, original shell structure has been found in some early Paleozoic specimens (i.e., Ordovician bellerophontians and bivalves). Sometimes, combinations of certain shell microstructures, such as crossed lamellar and nacre, have been documented in extinct taxa although this combination is unknown in living bivalves. Phosphatic and fine-grained sediments can also preserve anatomical impressions such as muscle attachment surfaces (muscle scars – see below and Chapter 3), pallial lines, radula, veins, digestive systems, and even gills. Typically, the most illuminating fossil characters come from the exceptional preservation of anatomy, such as found in the *Lagerstätten* deposits. These are thought to occur where anoxic conditions suppress bacterial decomposition, and impressions or traces of the body parts are incorporated into the fossil via the presence of rapidly accumulating fine sediment. Unfortunately, Lagerstätten that provide insights into molluscan ancestry and relationships are rare.

The Precambrian Ediacara Hills (700 Ma) in South Australia, Doushantuo Formation (600–555 Ma) in Guizhou Province, China, and other localities worldwide are well known for the preservation of numerous strange, soft-bodied organisms. When these fossils have been referred to extant phyla, the vast majorities are assigned to the Cnidaria, but putative annelids have also been identified (Bottjer 2002) as well as a putative mollusc – *Kimberella* (see Section 13.3.1). Alternatively, Seilacher (1984a) and others have argued that the Ediacaran faunas represented an early experiment in metazoan diversification that largely underwent extinction at the Cambrian explosion (Hoyal Cuthill & Han 2018).

Cambrian Lagerstätten have been problematic. These localities include the Chengjiang (525 Ma) in Yunnan Province, China, the Emu Bay shale (525 Ma) in South Australia, the Sirius Passet Formation (518 Ma) in Greenland, and the Burgess Shale (505 Ma) in British Columbia, Canada. Skeletonised body fossils such as brachiopods and trilobites are present in the Chengjiang formation, but the only supposed molluscs found in abundance are hyoliths, which are now considered putative brachiozoans. The well-known Burgess Shale fauna, popularised by Stephen J. Gould in *Wonderful Life* (1989), is also rich in skeletonised body fossils including brachiopods, arthropods, and echinoderms (Hagadorn 2002a). Again, the only putative molluscs occurring here are hyoliths (Briggs et al. 1994). A similar absence of unequivocal molluscs also occurs in the Emu Bay shale (525 Ma) in southern Australia and in the Sirius Passet Formation in Greenland, although the latter yielded the controversial *Halkieria* (see below).[4]

The absence of unequivocal molluscs from Cambrian Lagerstätten is puzzling. Helcionelloidans, gastropods, and bivalves are known from numerous non-Lagerstätten localities around the world from the early Cambrian on (Parkhaev 2008); however, they are virtually non-existent in Lagerstätten faunas. Whether their absence is due to ecological or taphonomic factors, or a combination of these and other factors, is not known. However, P. Wagner (pers. comm., Feb. 2019) has pointed out that the fact that aragonitic hyolithid shells are present argues against mineralogy-based taphonomy, but there could have been a size-based taphonomic filter. Lastly, although hyolithids often occur with molluscs, they also occur without them, suggesting that they had a broader tolerance for Cambrian environments.

Molluscs are known from subsequent Lagerstätten, of which the Silurian Herefordshire Lagerstätten of England (426.2–422 Ma) is perhaps the most important to date. This locality has provided fossilised gastropod soft tissue and organ morphology (Sutton et al. 2006) as well as *Acaenoplax hayae* – a possible early shelled solenogaster (Sutton et al. 2004) (see Chapter 14). The Herefordshire specimens are preserved as three-dimensional fossils within calcareous nodules, and the fossils are computer-reconstructed from serial images recorded as the specimens are literally ground out of the matrix (Sutton et al. 2001a).

Soft-body structures of cephalopods are abundant at several Carboniferous localities, including the Mississippian Bear Gulch beds of central Montana, USA (339.4–318.1 Ma) and the Pennsylvanian Mazon Creek material of northern Illinois, USA (309.2–302.0 Ma). Cephalopods from Mazon Creek are so well preserved that using a scanning electron microscope (SEM) it has been possible to compare the ultrastructure of fossil ink from these specimens with that from living specimens (Doguzhaeva et al. 2007a).

The Mesozoic Posidonia Oil Shale of southwestern Germany (Lower Jurassic 183–175 Ma) is the earliest Lagerstätten to preserve the belemnite animal; aspects of ammonite and coleoid anatomy and bivalves with colour patterns are also present (Etter & Tang 2002). The Middle Jurassic La Voulte-sur-Rhône (164.7–161.2 Ma) formation in southern France contains some of the best-known fossil cephalopod anatomy (Etter 2002a) (Figure 13.3).

Another Jurassic locality is the Oxford Clay of central England (164.7–161.2 Ma), which has revealed over 50 species of bivalves (Tang 2002) and gastropods; scaphopods are also abundant but not diverse. Cephalopods are again especially well preserved and include ammonites, nautiliforms, belemnites, squid and sepiids. Lastly, there is Solnhofen in southern Germany (150.8–145.5 Ma). Best known for the fossil feathered bird *Archaeopteryx*, molluscs from Solnhofen include bivalves, gastropods, ammonites, belemnites, nautiliforms, and sepidans – many still in association with their substrata and habitats (Etter 2002b).

---

[4] If hyoliths are in fact brachiozoans, the pattern of molluscan absence in these communities is strengthened.

FIGURE 13.3   *Vampyronassa rhodanica*. A pyritised coleoid from the Middle Jurassic (Callovian) (164.7–161.2 Ma). Voulte-sur-Rhône, France. Public Domain. Photograph by William Stoddar (https://commons.wikimedia.org/wiki/File:Vampylarge.JPG).

These, and other exceptionally preserved molluscan fossil localities, provide important morphological and palaeoecological insights into diverse groups of extinct putative molluscs. In addition, co-occurring taxa at these localities provide insights into ecological interactions with molluscs and other organisms, including bivalve/algal associations in the Mississippian Bear Gulch beds (Hagadorn 2002b) and the distinctive hypothesised feeding techniques of ichthyosaurs on belemnites in the Jurassic seas of Germany (Wiesenauer 1976). With the possible exception of the Silurian Herefordshire Lagerstätten, these amazingly resolved glimpses into the past have failed to produce uncontested stem taxa that would facilitate the connection of the disparate morphologies that define the living molluscan groups (Lindberg & Ponder 1996). This markedly contrasts with some other groups where Lagerstätten have been crucial in contributing to our understanding of the relationships of the crown taxa, as for example with the aforementioned Burgess Shale Arthropoda (Briggs & Fortey 1989).

Some physiological traits can be inferred both from hard parts and exceptionally well-preserved fossils. For example, hypothesised water flow patterns into and out of the mantle cavity have been used to test alternative shell orientations for Cambrian and other extinct taxa (Knight 1952; Lindberg & Ghiselin 2003b) (Figure 13.1). If the shell is altered in such a way that it has the potential to direct water, the remaining and more difficult question is – in which direction did it flow? In crown molluscs outflow control is plesiomorphic (Lindberg & Ponder 2001). Altered surfaces include grooves, holes, trains, folds, and notches. The attempts to link living taxa to Paleozoic fossils carry with them inferences about anatomy. Examples include the hot vent limpet-like *Neomphalus* being linked with euomphaloideans (McLean 1981a, b), fissurellids linked with euphemitid bellerophontians based on shell microstructural similarities (MacClintock 1967; McLean 1984b), or patellogastropods linked to platyceratoideans (Ponder & Lindberg 1997). Protoconch characters have also been used extensively in gastropods (Bandel & Frýda 1998; Frýda et al. 2008a; Frýda et al. 2009; Nützel 2014) and provide both characters and inferences of reproductive mode (Shuto 1974; Jablonski & Lutz 1983). In very rare cases soft

tissue can be preserved, giving insights into the anatomy of long extinct taxa. Anatomical characters from a supposed Silurian platyceratoidean led Sutton et al. (2006) to link them with patellogastropods, and fossilised intestinal tracts in the bellerophontiform 'monoplacophoran' *Cyrtodiscus nitidus* demonstrate the potential presence of torsion in that taxon (Horný 1998), which suggests bellerophontian rather than monoplacophoran affinities.

13.1.2.2   Muscle Scars

Muscles in living molluscs are discussed in Chapter 3. Two main groups of muscles concern us here because they leave distinctive scars on the shell – the dorsoventral and oblique foot retractor muscles (or shell muscles) and the buccal muscles. The latter are the main retractor muscles associated with the buccal mass and are typically involved in the retraction of the odontophore, and hence the radula, where present.

Horný (1965) recognised two putative apex/muscle scar relationships in molluscs. In the tergomyan condition the shell apex is located outside the 'muscle ring', and within it in the cyclomyan state. This dichotomy was questioned by Peel (1991a) and examined by Schaefer and Haszprunar (1996), who concluded there is no major difference in the position of the apex relative to the muscle ring between the two states. These terms do usefully demarcate two distinct muscle organisation character states (see below), which are defined in Table 13.1.

Not all shells with dorsal serial muscle scars are molluscs, as similar muscle scar patterns are also found in Cambrian inarticulate brachiopods (Dzik 2010) with some, including *Lenaella, Moyerokania, Scenella*, and *Kirengella* being previously treated as monoplacophorans. Some brachiopod taxa may be distinguished from monoplacophorans by a medial pair, or pairs, of muscle scars (e.g., as in *Craniops*), although in some fossil shells, a single pair of anterior medial scars could be mistaken for buccal musculature.

There are three major patterns in molluscan musculature: (1) multiple right and left retractor muscles (e.g., polyplacophorans, monoplacophorans, protobranch bivalves, and solenogasters, albeit greatly reduced). This pattern results in the bilateral presence of muscle units and is non-homologous

TABLE 13.1

Some Terms Used to Describe the Relationship of the Shell, Musculature, and Body Relative to the Shell in Molluscs

Tergomyan	Foot musculature of paired dorsoventral retractors with paired oblique muscles. Buccal muscle scars often present.	Polyplacophora, Monoplacophora, Protobranchia, Solenogastres
Cyclomyan	Foot musculature of paired dorsoventral retractors often fused into horseshoe-shaped muscle band; paired oblique muscles absent. Putative buccal muscle scars often present.	Helcionelloida, Cyrtolitones, Rostroconchia

Both these terms have been used as ordinal names (Tergomya, Cyclomya). Here they are used as descriptors of muscle scar patterns found in both stem and crown molluscs.

We introduce two new terms here:

Ligocephalic	Head region dorsally attached to the shell. Buccal muscle scars often present.	Monoplacophora, Polyplacophora, Bivalvia, Rostroconchia?
Apocephalic	Head region not attached to the shell. Buccal muscle scars absent except those integrated with shell muscle(s).	Cephalopoda, Scaphopoda, Gastropoda

The two terms below describe coiling direction relative to the head of the mollusc. They were first used in fossil cephalopod descriptions before being extended to gastropods and to early Cambrian molluscs.

Endogastric	Earliest shell whorls positioned over the posterior region of the body. This coiling direction supposedly enables the head to withdraw before the foot.
Exogastric	Earliest shell whorls positioned over the anterior region of the body. This coiling direction supposedly requires the foot to withdraw before the head.

with 'muscle bundles' in gastropods (see below). (2) In addition to the plesiomorphic dorsoventral retractor muscles, oblique retractor muscles are present in polyplacophorans, monoplacophorans, solenogasters, and protobranch bivalves. These muscles are combined into distinct bundles in polyplacophorans and monoplacophorans and are both bundled and separate in the protobranchs and solenogasters. The dorsoventral pedal muscles are lost in heterodont bivalves, leaving the oblique pedal retractors and protractors and the autapomorphic adductor muscles. Muscle proliferation in the ancestor(s) of these groups probably increased functionality and control of foot movement giving great mobility over diverse surfaces and, in ancestral protobranchs, some burrowing capability. In crown molluscs, increasing motility is often associated with shell reduction or sometimes either loss or replacement with spicules or multiple shell plates, an apparent parallel theme within Sachitida as well (see Section 13.3.2.2.2.1). (3) In stem gastropod and cephalopods, a paired retractor muscle configuration is typically present. In both fossil and living taxa, these paired muscles can expand posteriorly on each side of the shell, forming either a partial or complete horseshoe-shaped muscle band (e.g., Archinacelloidea, Hipponicidae). The muscle band may be traversed by blood sinuses suggesting separate muscles (e.g., Patellogastropoda), but these are not to be confused with duplications of retractor muscles as in polyplacophorans, monoplacophorans, protobranch bivalves, and solenogasters. Instead, the divisions between bundles are superficial and only deep enough to allow venous blood to move from the central visceral mass to the mantle edge for oxygenation. Such muscle scar morphology is: (1) probably homoplastic in numerous Cambrian lineages, especially

among limpet-shaped groups, (2) unlikely to reflect either a torted or non-torted state of the former occupant, and (3) while it can assist in the difficult task of identifying early 'Monoplacophora', it is much less useful in assisting with the identification of possible ancestors of the various conchiferan groups. In several early coiled lineages there was a subsequent reduction of the right retractor muscle and its eventual loss so that, as in gastropods, only the left retractor muscle remains.

Besides the pedal retractor muscles, multiple paired buccal muscles that manipulate the mouth and radula often terminate on the shells of the shelled groups. These muscles and their size appear to be correlated with the robustness of the radular apparatus (e.g., Polyplacophora and 'Aplacophora'), and they may be closely associated with the dorsoventral retractor muscle scars. The similarity of the placement of the buccal muscle attachment scars in various groups of molluscs is surprising given the vast differences in morphology. In polyplacophorans, monoplacophorans, both aplacophoran groups, and bivalves the head region is not detached from the shell, but rather it is continuous with it, providing attachment surfaces directly above the head region. In scaphopods, gastropods, and cephalopods the head is detached from the calcareous shell and is connected to the body by a short peduncle (neck), and attachment of buccal muscles to the shell must be done through this 'neck'. In these groups, the buccal musculature extends posteriorly to attach to the shell.

In the 'aplacophoran' groups, the radular retractor muscles terminate near the third and fourth pair of dorsal-ventral retractors, while in chitons and monoplacophorans the posterior components of the mouth muscles terminate with the first pair of dorsal-ventral retractors (see Chapter 14). The radula is absent in all bivalves, but buccal muscles are present in at least some protobranchs and still attach to the dorsal surface of the shell between the oblique anterior retractors and dorso-ventral retractors (Heath 1937). In scaphopods, cephalopods, and gastropods the posterior retractor muscles extend through the 'neck' to the foot region where they integrate with the pedal retractor muscles and then attach to the shell.

While detached heads do not inform us as to whether the mollusc was torted or not, their distribution in crown taxa and consideration of other lophotrochozoans suggests that the earliest molluscs had head regions attached to the shell, which prohibits the presence of a primarily anterior mantle cavity. A detached head and anterior mantle space are sometimes featured in hypothetical ancestral mollusc reconstructions, as shown in Chapter 12 and Lindberg and Ghiselin (2003b). Detached heads are correlated with the presence of ano-pedal flexure in the molluscan body plan.

Life in a narrow tube is difficult without a detached head and foot complex capable of extending and retracting beyond the confines of the aperture. And unlike most tube-dwelling lophotrochozoans, the secretory relationship between the body surface (mantle) and the shell prevents the organism from moving independently within the tube.

13.1.3 Phylogenetics and Fossils

Numerous studies have shown the importance of including fossils in phylogenetic analyses (Wagner 1999, 2001; Giribet 2002; Waller 2006), but such analyses are uncommon (Neige et al. 2007), and full molluscan group analyses including fossil taxa are rare (Runnegar 1996; Sigwart & Lindberg 2015). Instead, most have focused on individual taxa and include: cephalopods (Landman 1989; Monks 1999; Moyne & Neige 2004; Sutton et al. 2015), placophorans[5] (Vendrasco et al. 2004; Sigwart & Sutton 2007a), rostroconchs (Wagner 1997), bivalves (Waller 1998; Carter et al. 2000; Harper et al. 2000), and gastropods (Wagner 1999, 2002). As discussed above, the limits of shell characters and the rarity of exceptionally well-preserved fossils which could expand the character matrix have undoubtedly placed limits on analyses for some groups. As demonstrated by Wagner (2000), morphological character space may become exhausted through geological time. Wagner's result implies that biases in character selection are not at fault for the lack of characters but suggests biological factors, such as ecological restrictions, internal constraints, or long-term selective pressures, reduce the number of character states.

Working with ancient faunas affects how monophyletic groups are identified, as what is a monophyletic group in the past is not necessarily a monophyletic group in living taxa. For example, in the hypothetical cladogram in Figure 13.4, the 'Archaeogastropoda' was once a clade composed of the Eogastropoda, Vetigastropoda, and stem Neritimorpha. This

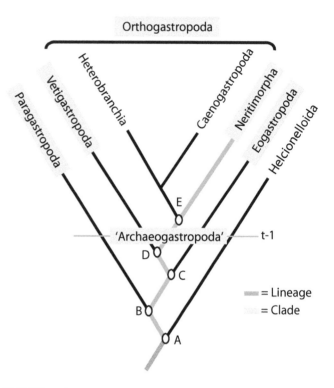

FIGURE 13.4 Hypothetical cladogram illustrating changing clade definitions through time and showing one example of a lineage. A-E = ancestral nodes, t-1 = an arbitrary time in the past.

remained true until the divergence of the lineage that gave rise to the heterobranchs and caenogastropods, which resulted in a paraphyletic 'Archaeogastropoda'. This origination did not change the character states that diagnosed the former clade, and while this name is now no longer used in gastropod classification, this former 'primitive order' remains important in discussions of the 'archaeogastropod' level of organisation or grade within the Orthogastropoda (Graham 1985; Haszprunar 1993). Monophyletic status can also depend on how monophyly is defined. Clades are defined as containing all and only the descendants of a common ancestor. For example, in Figure 13.4 the Orthogastropoda contains C and all of its descendants, and in the past, the clade Archaeogastropoda contained B and all of its descendants. Monophyletic clades are a synchronic concept, a snapshot of a lineage at a single time slice (e.g., t-1 in Figure 13.4). Lineages, which are also monophyletic, connect an ancestor with its descendants and extend through time rather than being defined only at a particular slice of time, for example, the neritimorph lineage (A → Neritimorpha) in Figure 13.4. Thus, a lineage is a diachronic concept; a series of connected replicators through time. See Mishler (2010) for further discussion of the synchronic and diachronic concepts.

13.2 ORIGINATIONS, EXTINCTIONS, AND RECOVERIES

Molluscan diversity is typically estimated using taxonomic ranks, with the family and genus-level being the most commonly used indicators when looking across large expanses of time. Changes in molluscan diversity patterns occur because

---

[5] We use the term 'placophoran' informally to refer to the Solenogastres, Polyplacophora, Caudofoveata, and Monoplacophora, which likely share plesiomorphic character states. The formal taxon name Placophora is a synonym of Polyplacophora.

of the interaction between origination and extinction rates although these rates vary in different molluscan groups (Sepkoski et al. 2002). For example, ammonites had high turnover rates, speciating rapidly and going extinct quickly. This volatility may have contributed to this highly diverse clade almost going extinct at the Permian/Triassic boundary and its final appearance at the end of the Cretaceous. In contrast, the Neogastropoda began a spectacular diversification in the Lower Cretaceous and continued unfettered into the Cenozoic. Whatever the factor(s) were that contributed to the extinction of the ammonites, they apparently had relatively little effect on neogastropod extinction rates.

Rather high rates of molluscan origination occurred during the Upper Cambrian–Lower Ordovician, the Middle Triassic, Paleocene, and Pliocene (Sepkoski et al. 2002), but if periods of high origination rates were accompanied by high rates of extinction, such as possibly occurred during the Lower Cambrian, the overall increase in diversity was minimal. Although such events are less common, there have also been periods (e.g., Oligocene) when both the relative origination and extinction rates dropped, and here again, a corresponding change in overall diversity may not be evident. Thus, in understanding diversification and extinction patterns, it is important to examine the relative relationship between origination and extinction rates by epoch.

Origination rates have exceeded extinction rates in 21 of the 30 Phanerozoic epochs, making the Mollusca one of the most diverse and abundant groups on Earth today. During the Paleozoic period, overall molluscan origination and extinction rates were about equal, however during seven of the 16 Paleozoic epochs, extinction rates exceeded origination rates, albeit only slightly. During the Mesozoic, origination rates were about 1.5 times that of extinction rates, and in only two of eight epochs did extinction rates exceed origination rates. During the Cenozoic, origination rates increased to more than twice the extinction rate, and overall, extinction rates never exceeded origination rates (see Table 13.2). The high diversity of the Cenozoic may be an artefact of sampling methods and an effect called the 'pull of the Recent' because of the more

comprehensive sampling of living taxa (Raup 1977; Jablonski et al. 2003). The stratigraphic ranges of living taxa are typically extended from their first occurrences to the modern day, including intervals in which they are not known in the fossil record. This increases taxon richness for living taxa in these intervals but not for extinct ones, resulting in artificially low diversity in intervals without living taxa. Correcting for the 'pull of the Recent' with standardised sampling and more robust counting methods reveals only a modest rise in diversity after the Mid-Cretaceous. During the Neogene, taxon diversity was twice as high as it was during the Mid-Paleozoic (see Table 13.2). This pattern exists at both global and local scales as well as at high and low latitudes, suggesting that the ratio of global to local taxon richness has changed little, and a latitudinal diversity gradient was present in the early Paleozoic (Alroy et al. 2008).

The greatest origination rate was during the Lower Silurian when it was over four times the extinction rate and during the Lower Triassic and Paleocene when origination rates were over three times the extinction rates. Most of the Cenozoic also saw origination rates in excess of twice the extinction rates (see Table 13.2).

Extinction events are typically divided into two categories – background extinction and mass extinctions. Background extinctions are thought to represent the 'normal' winnowing of taxa that occurs over geological time. Mass extinctions are significant increases in background extinction rates and typically occur on much shorter timescales. Some mass extinctions appear to have been caused by extra-terrestrial impact events such as the Cretaceous–Paleogene (K–Pg) mass extinction, while others are thought to be driven by large-scale tectonic events, volcanism, climate change, and oceanographic changes. While over 15 mass extinction events have been proposed, the most spectacular of these are often called the 'Big Five' mass extinctions – Ordovician–Silurian (444 Ma), Upper Devonian (385–359 Ma), Permian (251 Ma), Upper Triassic (228–199 Ma), and the K–Pg (66 Ma). The molluscs show mixed responses to the Big Five events, and only the Upper Devonian, Permian, and Upper Triassic events show marked decreases in molluscan diversity. The Permian event has long been recognised as a major molluscan extinction horizon marking the final occurrence of the Rostroconchia, most Bellerophontia and ammonites, and numerous other molluscan groups.

In addition to three of the Big Five events, relatively high rates of molluscan extinctions occurred during the Lower Cambrian, Upper Silurian, Eocene, and Pliocene. With the Lower Cambrian, Upper Silurian, and Pliocene, these periods also show relatively high origination rates as well, and therefore there is no marked decrease in overall molluscan diversity. In contrast to the eight epochs where molluscan originations exceeded extinction rates by more than a factor of two, extinction rates have exceeded origination rates by more than a factor of two on only three occasions, and these are all associated with mass extinctions (see Table 13.2).

Increased rates of origination immediately followed mass extinctions. These recovery events are often called diversity

---

TABLE 13.2

Epochs Where Molluscan Origination and Extinction Rates Exceed One Another by at Least a Factor of Two

Origination Rate > Extinction Rate (n = 21 epochs)	Extinction Rate > Origination Rate (n = 9 epochs)
Silurian (l) 4.5x	Permian (u) 3.7x
Paleocene 3.8x	Devonian (u) 2.8x
Triassic (l) 3.6x	Triassic (u) 2.2x
Triassic (m) 2.2x	
Miocene 2.2x	
Oligocene 2.0x	
Eocene 2.0x	
Carboniferous (m) 2.0x	

Data are from Sepkoski (1998). l = Lower, m = Middle, u = Upper.

rebounds and are thought to represent increased diversification of surviving lineages as they move into habitats and niches recently vacated by unsuccessful lineages. Two of the three highest rates of origination relative to extinction in the Mollusca occur following mass extinctions – in the Lower Triassic following the Permian mass extinction and in the Paleocene following the K–Pg mass extinction.

13.3   THE MOLLUSCAN FOSSIL RECORD

The following overview of the early history of the Mollusca covers their putative origin and early history up to the Permian–Triassic extinction event and the establishment of the 'Modern Evolutionary Fauna' (Sepkoski 1981). Brief reviews of fossil (and crown) taxa in each class through the Mesozoic and Cenozoic are presented in Chapters 14–20. Estimates of first occurrences and stratigraphic ranges of the taxa discussed here have been gleaned from the Paleobiology Database (www.paleobiodb.org/) and the compendium of fossil marine animal genera by Sepkoski et al. (2002). We have also updated the dates used here from the primary literature when ranges were extended or reduced, although our search was not comprehensive for every group and should be viewed accordingly. Stratigraphic stages follow the ICS International Chronostratigraphic Chart 2018-08 (Cohen et al. 2013; updated) unless we could not confidently resolve regional faunal stages, and then we report the original text of the author.

The literature on early molluscan history is extensive and diverse, and our overview is, of necessity, superficial. As discussed in Section 13.1.1 we have attempted to review the early history of molluscs from traditional, current, and alternative perspectives and to present and synthesise our own views on early molluscan morphology, origins, and relationships. While also being informed by molecular studies of surviving taxa, we have relied primarily on stratigraphy to verify morphological character distributions and to propose alternative scenarios of relationships and affinities. Our assignment of various taxa to the Mollusca, and to one of the surviving classes, has been cautious. Lastly, we relied on outgroups, mostly living and often outside the Mollusca, from which to determine putative shared characters in early taxa. While we have included some new interpretations and ideas for discussion in this review, we echo Erwin and Valentine (2013) who concluded their treatment of lophotrochozoan origins with the promise that we have yet to hear the final word on the relationships of these intriguing animals.

Because of the origination and diversification of the 'modern evolutionary fauna' following the Permian extinction (Sepkoski 1981), we only treat Paleozoic molluscan palaeontology in this chapter (i.e., the Cambrian and Paleozoic faunas). Discussions of taxa comprising the modern fauna may be found in Chapters 14–20.

13.3.1   Ediacaran

Many Russian workers consider that the earliest molluscs are from the latest Ediacaran (635–542 Ma) of Siberia in Russia. These small (<5 mm) bilaterally symmetrical shells are initially partially coiled but rapidly open into an expanded, limpet-like aperture. The number of whorls rarely exceeds 1.5, and these minute specimens, along with a variety of sclerites and plates, are commonly called 'small shelly fossils' (SSF). Their occurrence is associated with the sudden appearance of composite mineralised skeletons, including those of many of the major groups extant today such as echinoderms, brachiopods, arthropods, etc., as well as putative molluscs. Several hypotheses for this event include increasing levels of calcium carbonate in seawater, increasing body sizes, and the evolution of skeletons as responses to increased predation pressure. Regardless of the factor or factors responsible for the appearance of mineralised skeletons, the ancestors of the skeletonised organisms were already present in the Ediacaran, although identifying them has been problematic as most of the multicellular organisms from that period bear little resemblance to the taxa of the Cambrian explosion. Even so, the origins of the Mollusca are somewhere there, and one suggested possible stem mollusc is *Kimberella*.

Kimberella (Figures 13.5 and 13.6) was first collected from the Ediacaran of South Australia and originally identified as a 'problematic fossil possibly belonging to the Siphonophora' (Glaessner & Daily 1959). It was formally described by Glaessner and Wade (1966) and compared to the cnidarian taxon Cubozoa (box jellyfish) and the Hydrozoa (medusae) (Figure 13.6). Additional (>800), better-preserved specimens from the Ediacaran of the White Sea region of Russia (555.3 ± 0.3 Ma) became available in the early 1990s. These new specimens revealed a bilaterally symmetrical animal with rigid parts, and the fine sediments in which they were preserved revealed details of the external morphology and internal anatomy as well as the mode of locomotion and feeding traces. The presumptive shell of *Kimberella* was not mineralised, and although rigid it remained flexible, as demonstrated by its deformation in numerous fossils. The ventral portion of the body bore a fringed foot-like structure that extended beyond the putative shell. *Kimberella* also appeared to have had dorsoventral musculature arranged in a segmented pattern and weaker transverse ventral musculature. Examination of this additional material led to a reinterpretation of the taxon by Fedonkin and Waggoner (1997) who considered it a benthic bilaterian, possibly related to molluscs. Follow-up studies by Ivantsov and Fedonkin (2001) described and illustrated specimens with internal anatomy, the putative dorsal shell, and additional trace fossils that purported to show both its path on the substratum and feeding marks (Figure 13.5, right). A second species of *Kimberella*, *K. persii*, has been described from the Ediacaran of Iran (Vaziri et al. 2018).

These characteristics and the Precambrian occurrence of the fossils led some palaeontologists and malacologists to accept *Kimberella* as the first mollusc (Fedonkin & Waggoner 1997). It has been argued that *Kimberella* shares many features with monoplacophoran molluscs such as the segmented musculature, a ventral foot surrounded by respiratory structures – ctenidia in monoplacophorans and the foot fringe in *Kimberella* – while the fan-shaped rasp marks near

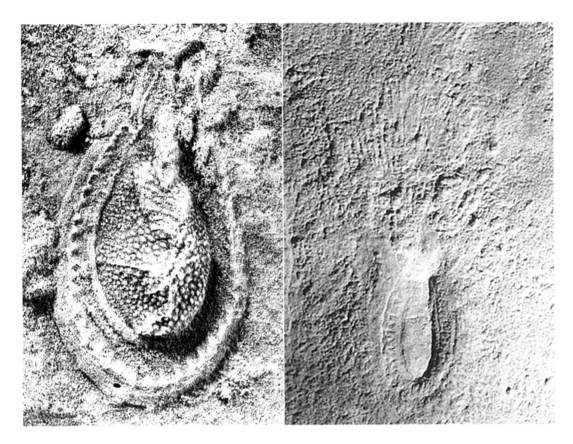

FIGURE 13.5 *Kimberella quadrata* from the White Sea Region of Russia. The figure on the right shows possible feeding traces. (Courtesy of M. Fedonkin.)

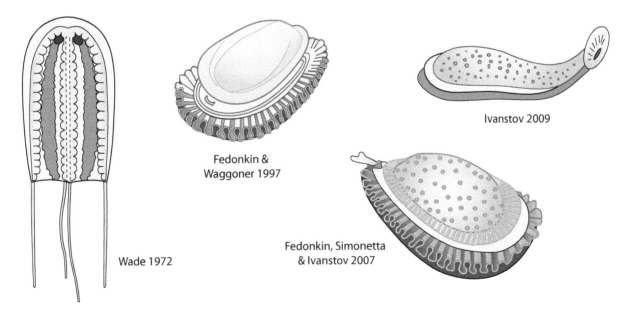

Wade 1972

Fedonkin & Waggoner 1997

Ivanstov 2009

Fedonkin, Simonetta & Ivanstov 2007

FIGURE 13.6 Some reconstructions of the Ediacaran taxon *Kimberella*. Originally it was reconstructed as a cnidarian (Wade 1972), before being reinterpreted as a mollusc 25 years later (Fedonkin & Waggoner 1997). Additional specimens initially refined the molluscan reconstruction (Fedonkin et al. 2007), but this was called into question when specimens with an elongated, worm-like body and a compressible head (Ivantsov 2009) were found. Today it is regarded as a possibly distantly related lophotrochozoan lineage.

Kimberella trails were considered as evidence it possessed a radular-like structure, although none has been yet reported in the hundreds of specimens examined, and the rasping marks extended in front of the supposed feeding swath of the animal, not behind it as in feeding molluscs. In the early 2000s, additional specimens became available that showed an elongated, worm-like body with a compressible head region (Ivantsov 2009). These new specimens necessitated the re-examination of the putative molluscan relationships of *Kimberella,* leading Ivantsov (2010, 2012) to conclude that *Kimberella* was not a mollusc but may have been a related lophotrochozoan lineage appearing between 558 and 555 Ma (see Erwin et al. 2011).

Whether putative molluscs such as helcionelloidans first appeared in the Ediacaran or Cambrian is determined by the preferred age assignment of the Nemakit-Daldynian (542–534 Ma), which is based on trace fossil occurrences and the beginning of a negative carbon isotope excursion. Russian workers typically regard this period as the latest Vendian (Rozanov et al. 2008), whereas other workers consider it to represent the first stage of the Cambrian (Budd 2003; Gradstein et al. 2004).

TABLE 13.3
Cambrian Chronostratigraphy

System/Period	Series/Epoch*	Stage/Age*	Beginning
Cambrian	Furongian	Stage 10	490 Ma
		Jiangshanian	494 Ma
		Paibian	497 Ma
	Miaolingian	Guzhangian	501 Ma
		Drumian	505 Ma
		Wuliuan	509 Ma
	Series 2	Stage 4	514 Ma
		Stage 3	521 Ma
	Terreneuvian	Stage 2	529 Ma
		Fortunian	541 Ma

* Some Epochs and Stages have yet to receive formal names.
Based on International Chronostratigraphy Chart (2018-08) [www.stratigraphy.org/ICSchart/ChronostratChart2018-08.pdf]

13.3.2 CAMBRIAN

The first supposed occurrences of the earliest crown Mollusca are found during the Cambrian, but recognition of these taxa is complicated by a plethora of other skeletonised SSF (Matthews & Missarzhevsky 1975; Dzik 1994; Maloof et al. 2010; Erwin & Valentine 2013). Besides putative molluscs, these assemblages include poriferan (sponge) spicules, pieces of archaeocyathan[6] walls, putative annelid tubes, stem brachiozoans, hyoliths (see Section 13.3.2.2.1), and a vast morphological array of sclerites including shelly plates. Complete or partial scleritomes (see Table 13.4) are rarely preserved but provide insights into both the arrangement and complexity of the armour of these early animals which include chancelloriids, halkieriids, wiwaxiids, and tommotiids (Conway Morris & Caron 2007; Skovsted et al. 2009). The vast majority of the SSF are calcareous, although many are secondarily phosphatised and some phosphatic tubes, valves, and spicules also occur (Bengtson & Runnegar 1992; Kouchinsky et al. 2012). As noted above, the first SSFs appear in the latest Ediacaran or earliest Cambrian, depending on the chosen time scale, and disappear during the Middle Cambrian (Erwin & Valentine 2013) (see Table 13.3).

Regardless of whether the SSFs first appear in the latest Ediacaran or earliest Cambrian, they are the first component of the Nemakit-Daldynian skeletonisation event (Fortunian), a relatively short period of time (approximately 10 million years) in which most stem and crown taxa of most living skeletonised phyla appeared (Cloud 1948; Stanley 1976; Budd & Jensen 2000; Budd 2003). This period saw major events in the history of metazoan biomineralisation, including the

---

[6] The Archaeocyatha was an extinct group of sessile, reef-building organisms that lived during the early Cambrian.

TABLE 13.4
Exoskeleton Terminology

Term	Definition	Composition	Morphology	Examples
Scleritome	The complete set of hard parts (sclerites) which make up the exoskeleton of an organism	Phosphatic, calcium carbonate, chitinous	Varied	Mollusca, Brachiopoda, Arthropoda, Annelida, Hyolitha, Bryozoa, Kinorhyncha
Sclerites	A component of an exoskeleton. Sclerites may occur individually or aggregated; aggregated sclerites may be articulated or unarticulated	Phosphatic, calcium carbonate, chitinous	Often with internal shell pores	Mollusca, Brachiopoda, Arthropoda, Annelida, Hyolitha, Bryozoa
Molluscan-specific terminology				
Spicules	Typically small needle-like structures	Calcium carbonate; typically aragonitic	Solid or partially hollow; intracellular growth	Aplacophorans, Polyplacophora
Scales	Small plate-like structures	Calcium carbonate; typically aragonitic	Solid; intracellular growth	Aplacophorans, Polyplacophora
Conch	A single (univalve) shell	Calcium carbonate; aragonitic and calcitic	Accretionary growth; often with internal shell pores	Monoplacophora, Scaphopoda, Cephalopoda, and Gastropoda
Plates and valves	Multiple sclerites, often articulated	Calcium carbonate; aragonitic and calcitic	Accretionary growth, often with internal shell pores	Polyplacophora (plates or valves), Bivalvia (valves)

appearance of the primary skeletal minerals (phosphate, carbonate, silica), a diverse array of microstructures (fibrous, granular, nacreous, prismatic and crossed lamellar structures), and morphological forms including spicules, tubes, conchs, and sclerites (Runnegar 1989; Bengtson & Conway Morris 1992; Feng et al. 2003) (see Table 13.4).

The exoskeleton structures are typically derived from ectodermal tissues but may not be homologous between groups. For example, the absence of chitin in the polyplacophoran spicule matrix suggests that spicule biomineralisation differs substantially from the biomineralisation of chiton shell plates where, as in most molluscs, chitin provides the scaffolding for shell formation (Levi-Kalisman et al. 2001; Treves et al. 2003).

The 'sudden' appearance of skeletonised bodies in the fossil record has been attributed to multiple biological, geophysical, and geochemical mechanisms (Marshall 2006; Maloof et al. 2010; Peters & Gaines 2012). These include the crossing of an oxygen threshold which supported larger animals and their more complex physiologies, changes and innovation in developmental pathways, environmental perturbations associated with global changes in climate (e.g., Gaskiers glaciation), and changes to the geophysical setting of the Earth (Marshall 2006; Erwin & Valentine 2013) (Figure 13.7). The treatment by Marshall of the so-called 'Cambrian explosion' examined whether potential factors and drivers were both necessary and sufficient to account for different aspects (disparity, diversity, timing, duration, etc.) of the event and found only the origin of predation to be both

necessary and sufficient to explain an increase in diversity. Some have argued that, correlated with this selection pressure, the evolution of body armour or skeletonisation marked the beginning of an arms race between predator and prey (Stanley 1973; Vermeij 1987; 1989; Bengtson 2002). The appearance of all these metazoan taxa undoubtedly produced significant changes in the marine trophic food webs of the Cambrian which may have been partially sustained by parallel increases in diversity of planktonic microfossils (e.g., acritarchs) during Stage 3 of the Cambrian (see Table 13.3) (Nowak et al. 2015). Acritarchs are also thought to have had a significant role in the Great Ordovician Biodiversification Event (see Section 13.3.3).

As with most major events in the history of life, it is probable that multiple drivers (both biotic and abiotic), and the complex interactions which they engender, were responsible for the 'Cambrian explosion'. For example, along with the formation of enormous evaporite basins during the late Neoproterozoic (~1000–~541 Ma), glaciations cooled the oceans and salinity fell. The coincidence of falling temperatures and salinity may have led to significant increases in dissolved oxygen, thus allowing for metazoan respiration in calcite- and silica-saturated oceans (Knauth 2005). This and other events triggered scenarios suggesting that before the Cambrian oceans were low in calcium, but as concentrations rose calcium was initially secreted as a waste product resulting in the availability and potential use of calcium carbonate as a skeletal material (Vermeij 1989). A parallel pattern in calcification intensity in the early Cambrian is also seen in Cyanobacteria (Riding

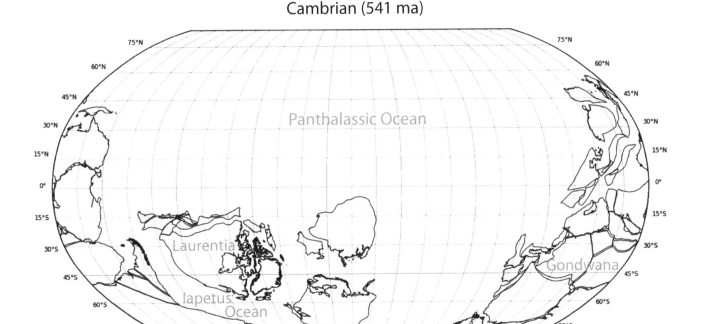

FIGURE 13.7 Palaeogeographic reconstruction of tectonic plate positions during the Cambrian (Terreneuvian) period.[1]

---

[1] Palaeogeographical reconstructions in this and subsequent maps were made with GPlates 2.0 (http://www.gplates.org/). Outlines represent plate boundaries and not necessarily coastlines; epeiric sea (or epicontinental seas) covered areas of some plates as well.

2006a, 2006b). The importance of geochemical and physical environmental perturbations during the Neoproterozoic has also been argued by Peters and Gaines (2012) who suggested that the Great Unconformity, which stratigraphically precedes the 'Cambrian explosion', is causally linked to the evolution of biomineralisation. They hypothesise that chemical weathering and reworking of continental crust during the Ediacaran and early Cambrian increased oceanic alkalinity during a period of intense expansion of shallow marine habitats.

13.3.2.1 Biomineralisation

Some of the Cambrian groups (e.g., Sachitida) have taxa with sclerites and shell plates. These are easily envisioned to fuse, giving rise to solid shells and fused sclerites in *Maikhanella*, and their apparent transition into solid shells in *Purella* has been argued to support this transformation (Bengtson 1992, 1993). Moreover, the presence of both plates and spicules in the Polyplacophora suggests the possibility that an intermediate step of this transformation sequence is preserved today in a group that originated in the Cambrian (Pojeta 1980; Salvini-Plawen 1985a; Eernisse & Reynolds 1994; Vinther 2009).

These scenarios for the formation of the molluscan shell date from the late 1800s, but it is the seminal paper on the evolution of calcareous hard parts in primitive molluscs by Winfried Haas (1981) that serves as the benchmark for this discussion (see Chapter 3 for details of shell formation). After a detailed study and review of spicule and plate formation in Polyplacophora and spicule and scale formation in the Solenogastres and Caudofoveata, Haas concluded that spicule formation was plesiomorphic in molluscs (with no outgroups) and that chiton shell plates and conchiferan shells were derived from the same cellular organisation that originally produced spicules. The evolutionary scenario required that individual invaginated epithelial cells that had secreted spines deep within the epithelium remained at the surface of the epithelium in plate formation and that a cuticle layer sealed these cells off from the environment. This allowed for the precipitation of calcium carbonate under the cuticle – the general mechanism by which both plates and conchs are formed in chitons and conchiferans, respectively. This conclusion is not unanticipated, as Haas found no independent assessment of the evolutionary direction or the polarity of the calcification characters he studied. With their spicule-covered bodies, solenogastres and Caudofoveata were assumed to represent the most primitive living molluscs, and the transformation into chiton plates and ultimately into the shells of conchiferans was predetermined to a large extent by the phylogeny he followed. Thus, the scenario was more an explanation of what was assumed to have happened, rather than an independent test of calcification patterns in molluscs.

Haas's scenario for the transformation from spicule to plate secretion did not address another important step required in this conversion. Solenogastres and Caudofoveata are covered by spicules, which correspond to the individual secretory cells densely distributed over the dorsal epithelium. Chiton plates are also dorsal, but like conchs, their growth depends on the ability to add material to the shell edge – accretionary growth. Because calcification is an intracellular process, it must be isolated from the environment and requires a shielding layer (cuticle or periostracal) and shell-secreting cells that line the growing edge of the mantle. Thickness is added by shell-secreting cells in the dorsal mantle surface. Accretionary growth from calcification centres isolated by cuticle also occurs in brachiopods (a potential sister taxon), bryozoans, serpulid and sabellid polycheate worms, and barnacles (Bourget & Crisp 1975; Mukai et al. 1997; Williams 1997). While the Haas scenario adequately explains what occurs early in the ontogeny of chiton and conchiferan shell formation, it does not address the required reorganisation of the individual spicule calcification centres and the evolution of these specialised tissues along the mantle edge.

If spicule formation was secondary, not primary, the evolution of molluscan hard parts is more parsimonious with the fossil record and outgroup comparison. If the common ancestor of molluscs secreted a dorsal cuticle, the addition of the calcified layer or shell required only the proliferation of cells capable of calcium secretion between the epithelium and cuticle at the edge of the dorsum. Away from the shell margins, the role of the cuticle in sealing the crystallisation chamber for shell formation was no longer necessary, and additional shell material could be ventrally added as required in the controlled environment between the existing shell and epithelium. Loss of these marginal calcification centres may have been associated with becoming worm-like, a morphology that enables much greater mobility and range of movement than possible in a body covered with fixed plates or valves. A spiculate body covering might result if elongation was accompanied by strong selection to increase the number of dorsal calcification centres, thereby reducing their size.

This alternative hypothesis is supported by the Paleozoic fossil record with the first occurrence of unequivocal molluscs being shell-bearing taxa – rostroconchs, bivalves, and then gastropods – rather than the supposedly more primitive sclerite-bearing taxa (Figure 13.8).

Furthermore, the putative stem molluscan taxa such as *Odontogriphus* (see Section 13.3.2.2.2) lack spicules, and the sclerites of *Wiwaxia* and halkieriids differ markedly from molluscan spicules both in their formation and composition (see Section 13.3.2.2.2). Therefore, calcium carbonate spicules must have evolved later in an aplacophoran or polyplacophoran lineage. Indeed, the earliest aplacophoran-like animal does not appear until the Silurian, 120 Ma after the appearance of conch and plate-bearing molluscs.

While a later appearance of unequivocal stem molluscs is better supported by the fossil record, are there additional data that can be considered? In the Haas scenario, only the calcification processes within molluscs are considered and examined, but for the hypothesis to be tested, insights are required into calcification mechanisms in lophotrochozoan outgroups.

In the Lophotrochozoa three non-molluscan groups produce calcified shells – Brachiopoda, Bryozoa, and Annelida. At various times, based on different datasets and analyses, all three have been considered as possible molluscan sister

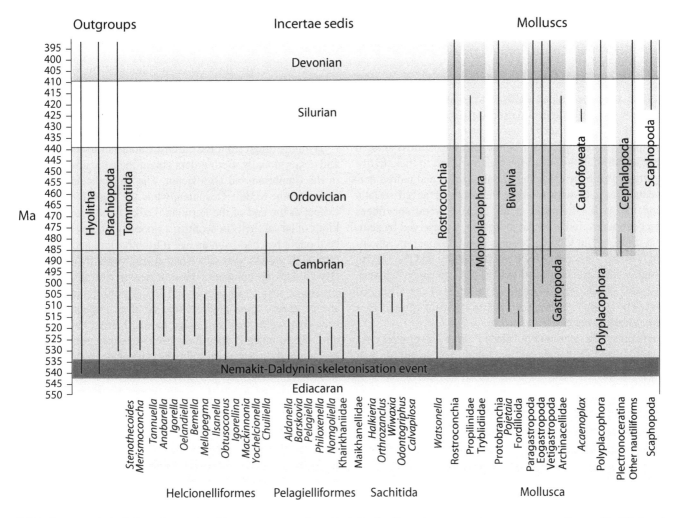

FIGURE 13.8 Early Paleozoic stratigraphic occurrences of small shelly fossils, outgroups, and crown group molluscs. Modified and updated from Maloof, A.C. et al., *Geol. Soc. Am. Bull.*, 122, 1731–1774, 2010.

taxa. All produce plate-like structures, not spicules, and enlarge these structures through accretionary growth, like the majority of molluscs. Studies of the calcification process that produces annelid worm tubes are few, but the calcification processes that produce the brachiopod shell and bryozoan test are well documented. In the Brachiopoda and Bryozoa, the calcium carbonate is laid down between a cuticle and the epithelium by specialised secretory cells found at the growing margins (Mukai et al. 1997; Williams 1997). This is the same basic organisational pattern seen in the production of chiton plates and conchiferan shells and in most non-molluscan groups examined, except for some annelid tubes which are produced from compacted and moulded granular secretions (Simkiss & Wilbur 1989; Taylor et al. 2010). For example, in the polychaete genus *Pomatoceros* (Serpulidae), calcareous granules are formed in intracellular vesicles or calcium-secreting glands. They are then expelled into an acid mucopolysaccharide matrix, which is moulded around the body of the worm and solidifies on contact with sea-water (Simkiss & Wilbur 1989). This type of calcification has not been reported in other lophotrochozoan phyla. A second secretion process, also found in serpulids, more

closely resembles that found in molluscs; the calcareous tube is formed by extracellular mineralisation mediated by an organic matrix secreted by a secretory epithelium. This calcification mechanism is also inferred for some Mesozoic and living Sabellidae (Vinn et al. 2008).

Thus, based on outgroup comparisons, spicules in molluscs would appear to be secondarily derived and not primitive. In order to continue to advocate the Haas scenario, it must be argued that the plesiomorphic state of calcification in the Lophotrochozoa was lost in the lineage leading to molluscs and replaced by spicule production. It would then have to be re-acquired in the common ancestor of the chitons and conchiferans, and spicule production subsequently lost in the conchiferan ancestor (see also Chapter 12).

13.3.2.2 Taxa

Many of the earliest putative crown molluscs in the Cambrian record are morphologically similar to extant gastropods and bivalves (Runnegar & Pojeta 1974b; Dzik 1994; Parkhaev 2007, 2008; Maloof et al. 2010; Erwin & Valentine 2013). As it is highly improbable that gastropods and bivalves arose *de novo* from a lophotrochozoan ancestor in the earliest

Cambrian, the search for and identification of molluscan stem groups have been intensely debated for over 20 years. The following is a review of these groups of putative molluscs.

13.3.2.2.1 Hyolitha

Hyoliths are cone-shaped shells with a flattened side and a circular to triangular aperture in cross-section; septa are present in some taxa, and accretionary growth lines are present on both the shell and 'operculum'. Shell microstructure is crossed lamellar. Musculature has been reconstructed to include five bilaterally symmetrical longitudinal muscles and paired dorsoventral muscles with multiple insertions arranged serially along the shell (Runnegar et al. 1975). Several specimens have been recovered with gut morphology preserved by sediment infilling; gut looping consists of two distinct morphologies. The first is a relatively straight section, and the second is highly folded and accordion-like. The transition between these two morphologies occurs at the apex of the U-shaped bend. The straight dorsal section has been interpreted as a rectum leading back to the aperture.

Hyoliths are commonly placed in two groups, the Hyolithida and Orthothecida (see Table 13.5), based on shell morphology. Hyolithida have irregular apertures and a pair of apertural projections called *helens* that curve over the dorsal surface of the shell. The helens articulate in sockets on the operculum, and their function is not known. The interior surface of the Hyolithida operculum has raised dorsal (and sometimes ventral) opercular processes. In the Orthothecida the aperture is planar, sometimes with indentations, and the interior surface of the operculum lacks processes.

Hyoliths range from the earliest Cambrian to the Permian and occur globally. Both orders obtain their greatest diversity in the Cambrian and Ordovician. The Orthothecida became extinct in the Middle Devonian, while the Hyolithida became extinct at the end of the Permian. Dzik (1978) identified two kinds of larval shells in hyoliths, a smooth globose form which he thought hatched from an egg (Orthothecida) and a pointed form with growth lines which he felt developed in the plankton (Hyolithida). He compared these structures with similar larval structures seen in Paleozoic and living gastropods. Hyoliths are generally considered sessile, benthic, epifaunal organisms (but see below), and range in length from about 10–40 mm. Taxonomically they have been variously considered to be pteropods, gastropods, cephalopods, or operculate worms

---

TABLE 13.5

Classification of the Hyolitha

Brachiozoa

Brachiozoa includes the brachiopods and phoronids, and the hyoliths were placed there based on well-preserved material in the Burgess Shale (Moysiuk et al. 2017).

(Class) Hyolitha

(= Hyolithomorpha, Orthothecimorpha)

Cambrian (Terreneuvian)–Permian (Lopingian) (530–252 Ma)

Cone-shaped conchs with distinctive larval shell and an operculum. Septa and apertural spines (helens) present in some taxa.

(Order) Hyolithida	
Cambrian (Terreneuvian)–Permian (Lopingian) (530–252 Ma) Conch oval to subtriangular in cross-section; dorsal re-entrant present in some taxa; with external 'operculum' resting on a ventral extension of the aperture (ligula); apertural spines (helens) present. Families include Hyolithidae, Angusticornidae, Aimitidae, Australothecidae, Carinolithidae, Crestjahitidae, Doliutidae, Nelegerocornidae, Pauxillitidae, Parakorilithidae, Sulcavitidae, and Similothecidae.	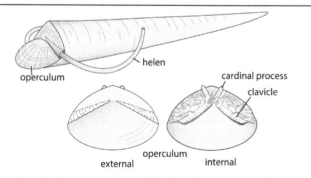 *Joachimilites*, redrawn and modified from Devaere et al. (2014).
(Order) Orthothecida (?= Tetrathecida)	
Cambrian (Terreneuvian)–Devonian (Middle) (530–388 Ma) Conch cross-section highly variable ranging from circular to triangular, with or without a longitudinal furrow on the ventral surface producing a heart- or kidney-shaped cross-section; lacks the ventral ligula and helens; has a retractable 'operculum'. Families include Orthothecidae, Allathecidae, Circothecidae, Gracilithecidae, Spinulithecidae, Turcuthecidae, and Tetrathecidae.	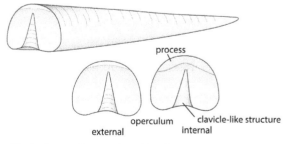 *Nephrotheca*, redrawn and modified from Devaere et al. (2014).

(Fisher 1962). Hyoliths have often been treated as an extinct class of molluscs (Downie et al. 1967; Marek & Yochelson 1976), while Runnegar et al. (1975) placed them in a separate phylum. A recent analysis of Burgess and Spence Shale hyoliths by Moysiuk et al. (2017) revealed soft tissue structures suggestive of a lophophore and a deep ventral visceral cavity. These and other features led the authors to conclude taxonomic affinities with brachiopods, phoronids, tommotiids, and other lophophorate taxa. The gut morphology in material from Noire, France, has also been described (Devaere et al. 2014).

Landing and Kröger (2012) suggested the possibility that small juveniles of the Cambrian taxon 'Allatheca' degeeri may have been nektic/planktic while both hatchlings and larger individuals (>17 mm) were negatively buoyant and benthic. More recently, Martí Mus et al. (2014) have reconstructed hyoliths as relatively mobile organisms. Their study of the muscle insertion patterns on conchs, helens, and 'opercula' (Martí Mus & Bergström 2005) suggests an articulated skeleton capable of functioning as oars to move the organism over the substratum, as well as operating as stabilisers and struts to lift the anterior end off the substratum. The location of epibionts on the hyolith shell further supports this interpretation (Galle & Parsley 2005).

13.3.2.2.2 *Coeloscleritophora and Sachitida*

Sclerite-bearing Cambrian lophotrochozoans have been known for over 100 years (Walcott 1911), but a name for these taxa was first proposed by Bengtson and Missarzhevsky (1981). They grouped the Wiwaxiidae, Halkieriidae, Sachitidae, Siphogonuchitidae, and Chancelloriidae in the globally distributed taxon Coeloscleritophora (see Table 13.6). This was based on these taxa having tissue-filled, hollow sclerites which lacked accretionary growth lines. Grouping the bilateral halkieriids with the sponge-like chancelloriids was controversial but has been supported by the more recent detailed microstructure studies of Porter (2008). At that time *Wiwaxia* was the only supposedly benthic, bilateral member of the group with a complete scleritome, albeit, unlike the other coeloscleritophorans, it was unmineralised. Bengtson (1992) later described cap-shaped shells formed from merged sclerites in the Cambrian fossil *Maikhanella*, which was also allocated to the Coeloscleritophora.

The discovery of the complete scleritome of *Halkieria evangelista* (Conway Morris & Peel 1990, 1995) provided additional insights into the complexity of coeloscleritophoran animals and revealed a diverse pattern of sclerite morphology as well as the presence of anterior and posterior plates with accretionary growth leading to consideration of a possible relationship with the Mollusca (Conway Morris & Peel 1995). More recently, Vinther and Nielsen (2005) compared sclerite morphology and scleritome patterns in *Halkieria* with girdle scales and spicules in chitons, concluding there was sufficient similarity to declare *Halkieria* a mollusc, and Vinther (2009) compared the aesthete pores of polyplacophorans with the pores in the sclerites of the halkieriid *Sinosachites*.

Conway Morris and Caron (2007) united the halkieriids and wiwaxiids based on the discovery of *Orthrozanclus*, a Burgess Shale fossil which possessed an anterior plate, as in the halkieriids, combined with *Wiwaxia*-like sclerites. They combined these two groups in the stem group 'Halwaxiida'. Conway Morris and Caron (2007) also proposed two alternative hypotheses for placing *Orthrozanclus* in early lophotrochozoan phylogeny. In the first hypothesis, *Odontogriphus* and the Ediacaran *Kimberella* were considered stem group molluscs and the sachitidans their sister group. In the second hypothesis, sachitidans were monophyletic, and *Odontogriphus* and *Kimberella* were treated as stem group lophotrochozoans; stem group molluscs were not identified in the second hypothesis.

The Burgess Shale *Odontogriphus* has figured prominently in discussions and scenarios of molluscan evolution. Originally described only as a lophophorate (Conway Morris 1976), the discovery of additional, better-preserved specimens led to the reinterpretation of this taxon as a stem group mollusc (Caron et al. 2006). Despite the lack of sclerites and a shell, structures suggesting a radula, ctenidia, muscular foot, and possible salivary glands were identified. Butterfield (2006) criticised this placement, in particular critiquing the identification of two primary molluscan characters – the putative radula and ctenidia (but see the response by Caron et al. 2007). Butterfield concluded that *Odontogriphus* represented a stem group lophotrochozoan, while *Wiwaxia*, which shared a similar, possibly plesiomorphic, feeding apparatus was a stem group polychaete. This view was, however, disputed by Eibye-Jacobsen (2004), who concluded that *Wiwaxia* was not closely related to Annelida.

Vinther et al. (2017) extended the Sachitida into the Ordovician with the description of *Calvapilosa kroegeri* from the Fezouata biota of Morocco.

13.3.2.2.2.1 Scleritome Reduction

Conway Morris and Caron (2007) produced a partial outline of lophotrochozoan phylogeny that featured relationships between Halwaxiida (i.e., Sachitida), Mollusca, Annelida, and Brachiopoda. In their first hypothesis (Figure 13.9A), the 'halwaxiids' were placed as the sister taxon of the Mollusca and formed a lophotrochozoan grade with the Annelida and Brachiopoda, respectively. In the second hypothesis (Figure 13.9B), the 'halwaxiids' were placed between the molluscs and annelids. In both hypotheses, the unarmoured *Kimberella* and *Odontogriphus* are outgroups to both the clade (hypothesis 1) and grade (hypothesis 2) of 'halwaxiids' and molluscs. A comparison of the genomes of molluscs, annelids, and brachiopods by Paps et al. (2009) and Luo et al. (2015) (Figure 13.9C) necessitates a re-examination of outgroup relationships in the hypotheses of Conway Morris and Caron. Placing brachiozoans as the sister taxa of molluscs (Figure 13.9D) suggests that mineralisation of body armour would be present in their last common ancestor[7] and calls into question the placement of *Kimberella* and *Odontogriphus* as non-mineralised molluscan ancestors.

---

[7] Luo et al. (2015) reported that although *Lingula* builds its shell from calcium phosphate and molluscs from calcium carbonate, they share shell formation-related genes and mechanisms.

TABLE 13.6

Classification of the Coeloscleritophora

Lophotrochozoa

(Class) **Coeloscleritophora**

Taxa with hollow sclerites that have a microstructure consisting of a thin, possibly organic, outer layer overlying a single layer of aragonite fibres
oriented parallel to the long axis of the sclerites; fibre bundles inclined, producing a scaly upper surface of the sclerite; scale-forming projections absent
from the lower surface (Porter 2008). There are two very different body forms of Coeloscleritophora – the sac-like Chancelloriida and the slug-like
Sachitida. If this relationship is correct, the diversity of body form is similar to living Tunicata, which includes the sac-like, sessile Ascidiacea and the
motile, bilaterally symmetrical Larvacea. Complete scleritomes are rare; many of the taxa are known only from disarticulated sclerites and are a
substantial part of the Cambrian small shelly fossil fauna.

(Order) **Chancelloriida** Cambrian (Terreneuvian)–Cambrian (Furongian) (541–487 Ma) These sac-like animals were somewhat similar to sponges and tunicates. The chancelloriids are not considered further here. See Bengtson and Collins (2015) and Porter (2008) for further treatment.	 Disarticulated sclerite, redrawn and modified from Walcott (1924). 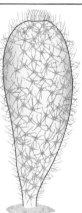 *Chancelloria*, redrawn and modified from Bengtson and Collins (2015).
(Order) **Sachitida** (= Thambetolepida; Diplacophora, 'Halwaxiida') Cambrian (Terreneuvian)–Cambrian (Miaolingian) (530–505 Ma) The Sachitida are bilaterally symmetrical, seemingly motile, slug-like animals typically covered by sclerites and one or more plates; when present, two or three forms of sclerites make up the scleritome; plates, when present, typically enlarged by marginal accretion; early accretionary growth replaced by fused sclerites in some taxa (*Maikhanella*). Sclerites appear to have been shed and replaced by larger ones during ontogeny. Mineralised mouthparts occurred in some taxa. These slug-like creatures have been considered molluscs by some workers. The classification below is largely inspired by the phylogeny of Vinther et al. (2017). (Superfamily) **Halkierioidea** Cambrian (Terreneuvian)–Cambrian (Series 2) (530–513 Ma) Typically three forms of mineralised sclerites; with two plates. Family Halkieriida	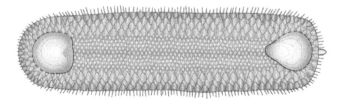 *Halkieria*, redrawn and modified from Vinther and Nielsen (2005).
(Superfamily) **Siphogonuchitoidea** Cambrian (Terreneuvian)–Ordovician (Lower) (530–478 Ma) Simple mineralised scleritome generally with two sclerite forms. Plates sometimes composed of fused sclerites after initial accretionary shell. Mineralised(?) mouthparts present in some taxa. Disarticulated sclerite plates have been referred to the Cambrian limpet taxon Maikhanellidae, which was treated as a monoplacophoran by Bouchet et al. (2017). Families Siphogonuchitiidae and Maikhanellidae (= Purellidae).	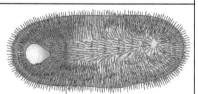 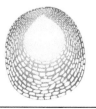 *Maikhanella*, redrawn and modified from Bengtson (1992). *Calvapilosa* (Siphogonuchitidae), redrawn and modified from Vinther et al. (2017).

(Continued)

TABLE 13.6 (CONTINUED)

Classification of the Coeloscleritophora

(Superfamily) **Wiwaxioidea** new name Cambrian (Series 2)–Cambrian (Miaolingian) (513–505 Ma) Sclerites demineralised; without plate; mineralised mouthparts present. Family Wiwaxiidae	*Wiwaxia*, redrawn and modified from Briggs et al. (1994).
(Superfamily) **Orthrozancloidea** new name Cambrian (513–488.3 Ma) With a single anterior plate and long lateral sclerites. Family Orthrozanclidae	*Orthrozanclus*, redrawn and modified from Conway Morris and Caron (2007).
(Superfamily) **Odontogriphoidea** new name Cambrian (Series 2)–Cambrian (Miaolingian) (513–505 Ma) Lacks plates and sclerites; mineralised mouth parts present. Molluscan affinities have been suggested (see text). Family Odontogriphidae	*Odontogriphus*, redrawn and modified from Briggs et al. (1994).

While a hypothesis with the sachitidans and molluscs as sister taxa (Figure 13.9D) is not falsified by placing brachiozoans as the living sister taxa of the molluscs, we suggest that the polarity of biomineralisation within the sachitidans is not one of increasing mineralisation, as required by the placement of *Kimberella* and *Odontogriphus* as putative outgroups, but rather one of scleritome reduction. Scleritome reduction and loss is an omnipresent trend in molluscs and other groups (see Chapter 3) and should be considered as a potential trend in sachitidans as well, rather than assuming the traditional mineralisation scenario for the group as discussed in Section 13.3.2.1.

Thus, within the sachitidans, there would have been a demineralisation of sclerites along with plate reduction and/ or loss (*Orthrozanclus, Wiwaxia*) and lastly, complete sclerite loss in *Odontogriphus* (Figure 13.9D). These possible losses suggest to us increasing motility of the sachitidans. In addition, this topology better fits the stratigraphic appearances of sachitidan taxa. Stem molluscs may also have undergone sclerite loss, while mineralised plates and valves, which are plesiomorphic and shared with brachiozoans, diversified within the phylum. In contrast to the sachitidans, the siphogonuchitids do not appear to have been part of the trend for increasing motility and are instead cap-shaped with a single solid valve present in the early apical region while the later shell is composed of sclerites (Bengtson 1992). For this reason, we treat any relationship of the siphogonuchitids with the

sachitidans as uncertain. Additional molluscan apomorphies would include the radula (see Chapter 5), a chambered heart (see Chapter 6), and the mantle cavity (see Chapter 4).

The work of Bengtson, Conway Morris, Caron, Peel, Vinther, and their colleagues has brought new views and data to our consideration of possible molluscan origins. Some reviews imply that the identity and relationships of the molluscan stem groups are now known and stable (Telford & Budd 2011; Vinther 2015), but there remain numerous issues in the interpretation, stratigraphy, and comparisons of these taxa and their traits. Placing these taxa as stem molluscs primarily rests on two traits – the presence of a radular-like structure in a 'pharynx' and a shell and/or scleritome. Significant conjecture remains as to the method of formation of the scleritome of these groups, whether or not they were mineralised, and the homology of molluscan shells and spicules with those of the sachitidans (Conway Morris & Caron 2007); the absence of spicules or plates in *Odontogriphus* is also problematic. As discussed by Todt et al. (2008b), the molluscan radula has particular significance in scenarios of molluscan evolution, but it is critical to distinguish special similarity (apomorphies) from overall or functional similarity when comparing the feeding apparatus of various lophotrochozoans (Sober 1991).

Chitinous and mineralised feeding structures are common in the Lophotrochozoa, including the Mollusca, Annelida, Rotifera, and Gnathostomulida (Brusca et al. 2016), and although separated by over half a billion years of evolution,

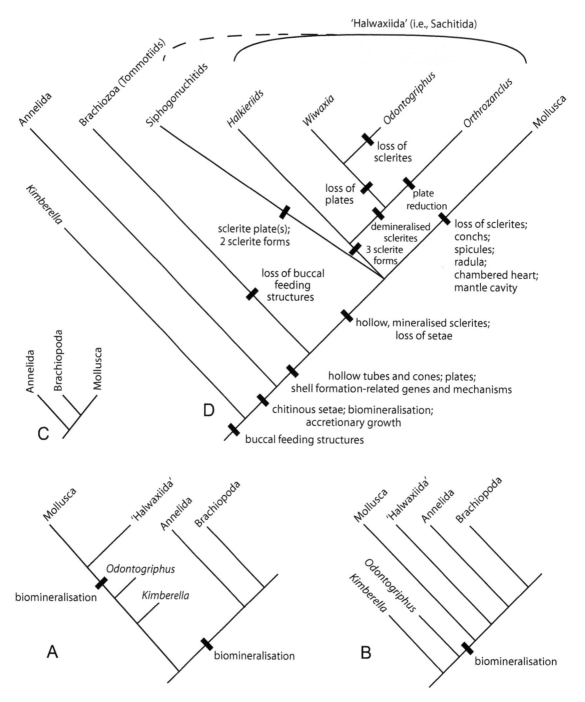

FIGURE 13.9 Outline of a partial lophotrochozoan phylogeny based on Conway Morris, S. and Caron, J.-B., *Science*, 315, 1255–1258, 2007. (a) Hypothesis 1. (b) Hypothesis 2. (c) The relationship among molluscs, annelids, and brachiopods used in the construction of hypothesis 3. (d) Alternative hypothesis 3; modified from hypothesis 1 with brachiozoans (e.g., tommotiids) as the sister taxon of the Mollusca. (a) and (b) redrawn from Conway Morris, S. and Caron, J.-B., *Science*, 315, 1255–1258, 2007 (c) redrawn from Luo, Y.J. et al., *Nat. Commun.*, 6, 9301, 2015.

they can often be surprisingly similar in these different phyla (Figure 13.8). As argued by Butterfield (2006) and Todt et al. (2008b), the putative radula of *Odontogriphus* and *Wiwaxia* is considered well outside the bauplan of the molluscan radula, which is considerably broad to begin with (see Chapter 5). As pointed out by Conway Morris (2006) and others, many comparisons of the shell, sclerites, and radular structures of the Cambrian taxa are often made with apparently unrelated

taxa that are substantially younger. Smith (2012) has recently argued for potential homologies between the molluscan radula and the feeding structures found in *Odontogriphus* and *Wiwaxia*, but there remain substantial differences such as the lack of wear on the anterior-most 'teeth' of *Odontogriphus* and *Wiwaxia* and questionable comparisons with living taxa with highly derived tooth morphologies (Figure 13.10).

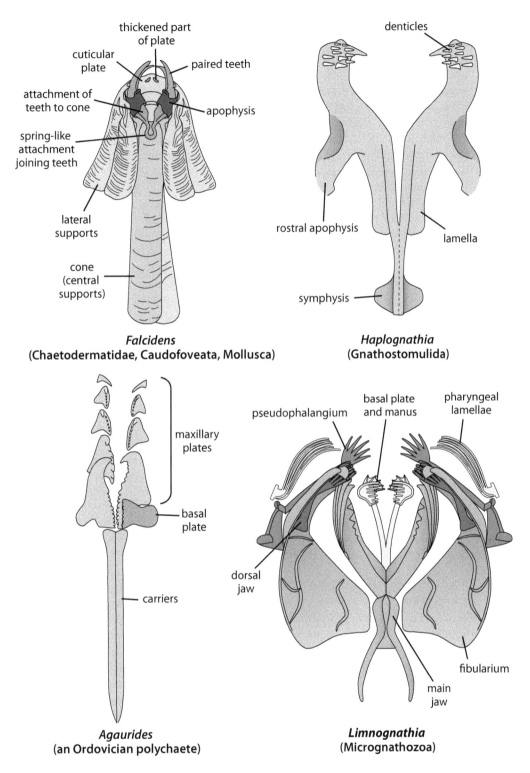

FIGURE 13.10 Examples of chitinous and mineralised feeding structures in lophotrochozoan taxa. Redrawn and modified from the following sources: *Falcidens* (Schander, C. et al., *Mar. Biol. Res.*, 1, 79–83, 2006), *Haplognathia* (Sørensen, M.V. and Sterrer, W., *J. Morphol.*, 253, 310–334, 2002), *Agaurides* (Kielan-Jaworowska, Z., *Acta Palaeontol. Pol.*, 7, 291–332, 1962), *Limnognathia* (Sørensen, M.V., *J Morphol.*, 255, 131–145, 2003).

While these putative 'stem group' taxa appear to fulfil the early molluscan evolutionary scenario which envisions an ancestral dorsoventrally flattened animal with a broad, ciliated ventral foot for locomotion, a ventral mouth with a radula for feeding and dorsally covered by a cuticle, with the shell and/or sclerites making up a scleritome (e.g., *Halkieria*, *Wiwaxia*, *Odontogriphus*) (Vinther & Nielsen 2005; Caron et al. 2006; Telford & Budd 2011; Vinther 2015), their stratigraphic occurrences are typically after the appearance of more likely molluscan morphologies (Figure 13.8). For example, while the coeloscleritophoran sclerites are present in the Cambrian record approximately 5 Ma before the helcionelloidans, the first occurrences of the supposed stem group molluscs (*Wiwaxia*, *Odontogriphus*) are almost 20 Ma later. If these scleritic animals are members of ancestral lineages leading to molluscs, one would expect them to occur earlier in the fossil record than the supposed descendant lineages. There are several possible explanations for this gap. Perhaps conditions for preservation in different habitats or at different times favoured conchs while excluding sclerites, or vice versa. The lack of univalve and bivalve stem group molluscs from the Middle Cambrian Burgess Shale Formation in Canada suggests that taphonomic biases may have occurred, although it could equally well be due to molluscs avoiding the area because of their ecological preferences. Also, the relatively short period of phosphatic preservation (521–542 Ma) may have contributed to the differential occurrence of groups during the early Cambrian. Thus, perhaps these and other stem group taxa were present but were not preserved. Alternatively, if the fossil record, while not complete, does accurately reflect the sequence of the origin of molluscan stem group taxa, the univalve and bivalve taxa are the earliest molluscan stem groups and the sclerite-bearing animals a more distant lophotrochozoan lineage (Lindberg & Ponder 1996; Benton et al. 2000).

13.3.2.2.3 Univalves and Bivalves

Many of the earliest 'molluscs' in the Cambrian record have shells morphologically similar to crown gastropods and bivalves, being either limpet-like (e.g., *Anabarella*, *Barskovia*, *Bemella*, *Igorella*, *Oelandiella*, *Purella*), coiled (e.g., *Aldanella*, *Latouchella*, *Pelagiella*), or consisting of two valves (e.g., *Fordilla*, *Pojetaia*) (Runnegar & Pojeta 1974b; Dzik 1994; Parkhaev 2007, 2008; Maloof et al. 2010; Erwin & Valentine 2013). In contrast to the sachitidans, most of these shells are small (less than 5 mm in length) (Runnegar 1983), although a few exceed 1 cm in length (Martí Mus et al. 2008). While some of these early conchs and valves may be components of a large scleritome, others may represent lophotrochozoan or other taxa. The appearance of bivalves and rostroconchs also marked diversification events from the plesiomorphic epifaunal state into infaunal habitats.

13.3.2.2.3.1 Helcionelloida

The earliest limpet-like 'molluscs' include the Helcionelloida which date from the Nemakit-Daldynian age/stage of the Cambrian (534 Ma) (Peel 1991b; Gubanov 1998; Gubanov et

al. 1999; Gubanov & Peel 2000; Parkhaev 2008), and while most of these taxa were extinct by 501 Ma, *Chuiliella* survived until the Lower Ordovician (Tremadocian) (477 Ma) (Gubanov & Peel 2001). Helcionelloidans had a worldwide distribution and are bilaterally symmetrical cap- and horn-shaped shells. They include the Helcionellidae, Coreospiridae, Securiconidae, Stenothecidae, and Yochelcionellidae. Most workers assume them to be the ancestors of living conchiferans, although Parkhaev (2006b, 2007, 2008) considered them to be stem gastropods (Figure 13.11).

Helcionelloidan morphology is variable and includes limpets and loosely symmetrically coiled conchs, typically with less than three-quarters of a whorl. One exception is the Coreospiridae where coiling typically exceeds a single whorl. Shell sculpture consists primarily of raised axial ridges (costae), which in some taxa can be quite well developed (e.g., *Igorella*, *Obtusoconus*, and some Securiconidae). While many groups have broad oval apertures, members of the Securiconidae, Stenothecidae, and Yochelcionellidae show lateral compression and narrowing of the aperture (Pojeta & Runnegar 1985; Gubanov 1998). For example, in most helcionellids, apertural length is less than twice apertural width, while in stenothecids apertural length is three to five times apertural width (Waller 1998). Parietal trains[8] and emarginations are also common in these taxa (Peel 1991c; Parkhaev 2001). In Yochelcionellidae the train may become closed ventrally with growth producing a snorkel-like structure (see Table 13.7).

It has also been suggested that some helcionelloidans may actually be 'protoconchs' of larger (>5 mm) taxa. Martí Mus et al. (2008) reported 2–3 cm specimens from the Lower Cambrian Rio Huso group in the Montes de Toledo region of central Spain, which resembled *Scenella* but had a coarsely sculptured helcionelloidan-like conch affixed at the apex. The larger and extremely thin shells continued the earlier concentric costate sculpture over the presumptive dorsal surface. More recently, similar specimens have been reported from the Lower Cambrian Hawker Group of the Flinders Ranges in South Australia by Jacquet and Brock (2015), further suggesting the possibility that not all helcionelloidans were small. While the size and position of these possible 'protoconchs' are suggestive, further work is needed. Questions remain, including how a protoconch, which typically reflects egg shape and size (especially in lecithotrophic taxa), would become laterally compressed – a feature of many helcionelloidan taxa – and the discrepancy between *Scenella* and helcionelloidan muscle scar patterns.

Pores are found in the shells of many Cambrian helcionelloidan taxa, including *Auriculaspira*, *Auricullina*, *Postacanthella*, *Tuberoconus*, *Igorella*, and *Daedalia* (Feng & Sun 2006; Parkhaev 2006a). In some taxa such as *Auricullina*, the exterior surface of the pore is marked by a raised ridge encircling the pore, forming small tubercles (Kouchinsky 2000). Shell pores in the helcionelloidans differ markedly from the canal systems found in halkieriid sclerites (Vinther 2009) and polyplacophoran plates (see Chapter 14).

---

[8] A vaulted posterior projection of the aperture.

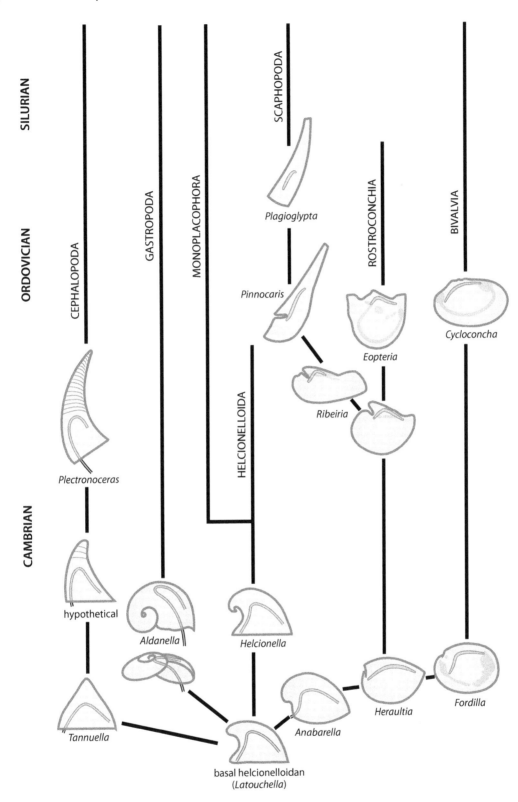

FIGURE 13.11 One view of the phylogeny of the early Paleozoic molluscs, showing their hypothetical derivation from a helcionelloidan ancestor and showing muscle scars and pallial lines. Based in part on Runnegar, B.N. and Pojeta, J., *Science*, 186, 311–317, 1974b. Note: It is now thought that the early Cambrian genus *Aldanella,* shown in this figure as an early gastropod, is not a mollusc (see text).

In most helcionelloidan shells where shell microstructure is preserved, it is in the form of phosphatic replicated and replaced microstructures visible in exfoliated and broken sections of the shell (Runnegar 1985; Kouchinsky 2000). Even if the shell in the fossil is not preserved, the shell microstructures that make up the shell can be determined by their distinctive morphologies on the internal moulds or steinkerns (MacClintock 1967; Runnegar 1985). Of the six major shell structure groups (I–VI) recognised by Carter (1985) (see Chapter 3), only two, prismatic and laminar, are

TABLE 13.7

Classification of the Helcionelloida, a Group of Putative Molluscs Often Considered to Be an Extinct Early Class

(Class) **Helcionelloida**

(= Helcionellida)

Cambrian (Terreneuvian)–Ordovician (Lower) (530–478 Ma)

Small cap-shaped to planispiral conchs, some with strong growth(?) increments or costae. Some limpet-like forms may be plate-like sclerites from sachitids (see above) or polyplacophoran valves (e.g., Carinopeltidae).

This group is treated as an order of the 'subclass Archaeobranchia' (in Gastropoda) by Bouchet et al. (2017).

    (Order) **Helcionellida**

      (= Helcionelliformes)

(Superfamily) **Helcionelloidea** Cambrian (Terreneuvian)–Ordovician (Lower) (530–478 Ma) Aperture ovoid, simple, and complete. Families Helcionellidae, Carinopeltidae (= Igarkiellidae), and Coreospiridae (= Archaeospiridae, Yangtzespirinae, and Latouchellidae)	*Latouchella* (left) and *Helcionella* (right) *Mackinnonia* (left) and *Helcionella* (right)
(Superfamily) **Yochelcionelloidea** Cambrian (Terreneuvian)–Ordovician (Lower) (530–478 Ma) Substantial lateral apertural narrowing accompanied by a posterior groove or train and its modification into a snorkel in some taxa. Families Yochelcionellidae, Stenothecidae, and Securiconidae (= Rugaeconidae, Trenellidae)	'*Latouchella*' (left) and *Eotebenna* (right) *Yochelcionella*

Images redrawn and modified from Peel (1991c)

known from putative Cambrian molluscs. A prismatic outer layer has been reported in stenothecids, trenellids, coreospirids, and helcionellids while in yochelcionellids there is an internal prismatic layer (Vendrasco et al. 2010). Inner lamellar layers include semi-nacreous (stenothecids), lamellofibrillar (onychochilids), and foliated (rostroconchs); in yochelcionellids the laminar layer is on the outside of the shell. Unspecified laminar structures are also present in the trenellids, coreospirids, and helcionellids. Crossed structures, the dominant form of shell structures in living molluscs, have not been reported in these putative stem molluscs, but lamellofibrillar is very similar to crossed structures (e.g., crossed lamellar). Lamellofibrillar structures are the first to appear in the earliest Cambrian of Yunnan, China (Feng et al. 2003; Feng & Sun 2003; Vendrasco et al. 2010) while nacreous structures appear later in the record, being reported in *Anabarella* and *Watsonella* (Runnegar 1983, 1985; Carter 2001). Most of these microstructures are inferred to have been originally aragonitic, but calcitic shell structures are known from *Eotebenna, Mellopegma, Ribeiria,* and *Anabarella* and may have been common during the calcitic seas of the Middle Cambrian (Vendrasco et al. 2011b).

Putative muscle scars have been reported in a variety of helcionelloidans (Wen 1979; Geyer 1994; Parkhaev 2002b, 2004; Parkhaev 2014). While muscle scars are important characters in molluscan palaeontology, they are often weakly distinguished and difficult to locate in many fossil taxa and especially on steinkerns. They are typically recognised by changes in shell texture or slightly depressed areas on the internal surface of the conch or elevated areas on internal moulds. A single pair of symmetrical muscles has been reported in several limpet-like genera of 'Yangtzeconiacea' by Wen (1979), although their strap-like morphology and apical position make a functional interpretation difficult. Parkhaev (2014) described single pairs of symmetrical muscles in the helcionelloidans *Bemella, Oelandiella, Anabarella,* and *Anhuiconus;* Vendrasco et al. (2010) reported and illustrated a single pair of muscle attachment scars near the apex of *Yochelcionella,* although three pairs of muscles were reported for *Bemella communis* (Parkhaev 2014, 2017). This limited sample suggests that helcionelloidans did not have multiple dorsoventral retractor muscles as seen in monoplacophorans but rather only a single pair. The attachment points of these muscles on the shell vary widely (apex, 'columella', overhead region), and this variability far exceeds the variation seen across conchiferans. Moreover, there is evidence of surprising convergence within molluscs and even between phyla. For example, Dzik (2010) illustrated remarkably similar muscle scar patterns in Cambrian 'monoplacophorans' and brachiopod valves.

Helcionelloidans have been considered to be gastropods (Knight & Yochelson 1958; Golikov & Starobogatov 1975; Parkhaev 2000, 2001; Parkhaev 2008), monoplacophorans (Knight 1952; Runnegar & Pojeta 1974b; Pojeta & Runnegar 1976; Wen 1981; Runnegar & Pojeta 1985), or to represent a separate molluscan class (Yochelson 1978; Wen 1984; Peel 1991b; Geyer 1994; Gubanov & Peel 2000). We consider helcionelloidans to represent an extinct paraphyletic class of either molluscs or of unknown affinities. Within its ranks are lineages which range from limpet-like to coiled and tubular taxa (see Section 13.3.2.2.8 and Figure 13.11). One strong trend in the group is the lateral compression of the conchs. This compression was typically accompanied by the formation of a train or tube on the narrow side of the conch, which began after a period of uniform aperture expansion. While train formation requires the partial folding of the accretionary tissue, formation of a tube would appear to have required both extension and folding and subsequent fusion as the conch again returned to uniform apertural expansion. The formation of the tube gave these taxa two apertures, one dorsal and one ventral.

13.3.2.2.3.1.1 Endo- and Exo-Gastric Coiling

Categorising conchs as either endo- or exogastric has featured in both helcionelloidan and cephalopod systematics and palaeobiology (see Chapter 17). In helcionelloidans endo- and exogastric morphologies are used to infer putative water flow patterns around and through shells, especially those with holes, slits, and sinuses (Pojeta & Runnegar 1976; Peel 1991b; Parkhaev 2008). All crown molluscan groups with coiled shells are endogastric except Cephalopoda, in which both coiling directions are present (Figure 13.12), although endogastric shells are often hypothesised for the cephalopod ancestor (see Chapter 17).

Both exo- and endogastric shells occur in the Nautilida (e.g., exogastric *Nautilus,* while some Discosorida are endogastric [see Chapter 17]). Among coleoids, *Spirula* is endogastric, while most ammonites have been reconstructed as exogastric. Coiling preference in nautiliforms and ammonites is thought to be partly related to shell hydrodynamics and performance in the water column. For example, Stridsberg (1985) suggested that endogastric shells were mechanically better for swimming than the exogastric shells in oncoceratid cephalopods. In curved nautiliforms and ammonites which lack coiling, shells are considered exogastric if the anterior side is convex and endogastric if the posterior side is convex (see Chapter 17, Figure 17.5). For example, the curved tubes of baculitid ammonites have been reconstructed as exogastric according to Klug and Lehmann (2015).

13.3.2.2.3.1.2 Scaphopodisation

One of the key features of the scaphopodisation scenario by Peel (2006) was the formation of anterior and posterior apertures on conchs (see Chapter 16). He used the shell ontogeny of living scaphopods as a model for the formation of these tubular shells. In scaphopods, the developing shell envelops the larvae laterally and ultimately fuses along the ventral margin, forming a cylinder around the organism and leaving a small posterior opening (Wanninger & Haszprunar 2001)

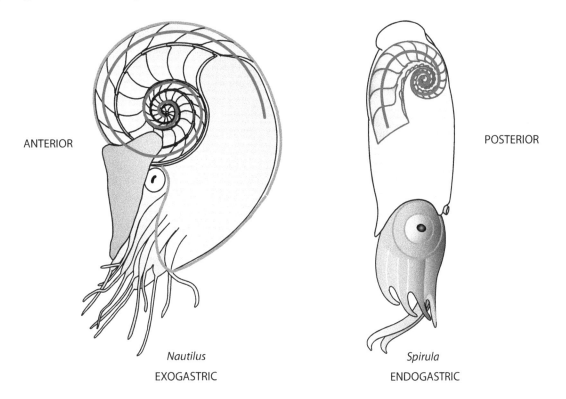

ANTERIOR

Nautilus
EXOGASTRIC

POSTERIOR

Spirula
ENDOGASTRIC

FIGURE 13.12 Examples of exogastric and endogastric coiling in cephalopods. Figures reconstructed from various sources.

(see also Chapter 8). Subsequent accretion of the adult shell or teleoconch then occurs along the anterior margin of the fused cylinder. The lateral expansion of the mantle epidermis and its subsequent ventral fusion as it envelops the larva coincides with ano-pedal flexure. This changes the orientation of the anus by about 90°, and the foot extends anteriorly rather than ventrally relative to the original embryo axes. The anus is now in the anterior third of the shell even though the posterior portion of the mantle never fuses (see also Chapter 8). This developmental pattern is identical to that seen in gastropods except for the complete enveloping of the posterior portion of the gastropod embryo (see Chapter 8, Figure 8.33). Even the fusion of the developing lateral and posterior shell margins and the transition to apertural accretionary growth are alike.

While the helcionelloidans *Yochelcionella* and *Eotebenna* appear to share similar shell ontogeny with the Rostroconchia, there is no evidence for ventral fusion of the lateral shell edges, although both taxa express strong lateral compression of the teleoconch. Instead, the helcionelloidans appear to undergo scaphopodisation by shell elongation along the dorsoventral axis combined with a temporary enfolding of a portion of the posterior secretory epidermis near the apex during shell accretion, thereby forming a tube. The foot is assumed to remain in its ventral position relative to both the larval and adult axes.

Determining possible shell ontogeny associations of the taxa *Janospira*, *Jinonicella*, and *Rhytiodentalium* is more difficult because of the lack of living taxa with similar morphologies. Based on gross morphology they probably underwent ventral fusion similar to living scaphopods, although later in their ontogeny, thereby generating a longer (or more coiled) early shell. Regardless of the affinities of *Janospira*, *Jinonicella*, and *Rhytiodentalium*, if these are burrowing morphologies, the convergent shell forms of the taxa identified by Peel suggest multiple incursions into the sediment by several molluscan lineages during the early Paleozoic. The elongate tubular shells of these taxa appear to be derived through different growth trajectories, and the ventral mantle fusion of scaphopods may have had its earliest inception in one of these possible stem groups. Scaphopodisation must also have produced significant changes in internal anatomy in these and other lineages, as it did in the scaphopods (see Chapters 4, 5, and 16).

13.3.2.2.3.2   Merismoconchia
The Merismoconchia was erected by Wen (1979) for small cap-shaped shells with apparent tripartite morphology. A transverse ridge divides the shell into two (not three parts as originally thought), a small anterior rostrate segment and a much larger body segment (Wen 2008; Devaere et al. 2013) (see Table 13.8). Wen also mapped internal organs, including muscle scars, gut pattern, and nephridia on the interior surface of the second and body segments, but these are thought more likely to be preservation artefacts (Devaere et al. 2013). Merismoconchs are known from the Meishucunian (= Nemakit-Daldynian, in part) (542.0–516.0 Ma) of China, and Tommotian (530.0–516.0 Ma) of southern France, and

TABLE 13.8
Classification of Merismoconchia

Merismoconchia	
Cap-shaped shells with two (initially reported as three) distinct body segments. Cambrian (Terreneuvian) – Cambrian (Series 2) (530–516 Ma) Family Merismoconchidae	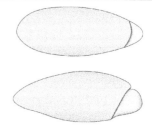

Merismoconcha, redrawn and modified from Devaere et al. (2013).

possibly the Lower Cambrian of Siberia and Mongolia (Wen 2008; Devaere et al. 2013).

Wen (2008) compared the merismoconchs to polyplacophorans, noting that whereas each of the eight polyplacophoran valves has two pairs of muscles, in merismoconchs dorsoventral muscles were reportedly attached to the two 'valves'. The multiple 'organs' and circular gut Wen identified (but see above) also drew comparisons with the Monoplacophora and led Wen to propose plate reductions in the chitons that gave rise to the merismoconchs, with further plate fusions producing the monoplacophoran lineage.

Material from the Cambrian of France examined by Kerber (1988) and Devaere et al. (2013) showed no trace of muscle scars and imprints of organs although they were better preserved than the original Chinese material. As noted above, only two shell segments were present in the French specimens (and in several Chinese specimens), and Devaere et al. (2013) concluded that the third putative segment was an artefact. They suggested that the anterior chamber might be homologous with the molluscan protoconch. There is a resemblance of the anterior segment of the Merismoconchia to the distinct, thickened, anterior portion of the conch of the Silurian *Tryblidium reticulatum* illustrated by Lindström (1884), but this anterior enlargement is absent in the apparently closely related *Tryblidium unguis*, also described by Lindström from the Silurian of Götland.

While the Merismoconchia may represent a molluscan lineage, we are uncertain of their affinities within the phylum.

13.3.2.2.3.3   Stenothecoida
Stenothecoidans[9] are bivalve organisms that range from the Terreneuvian to the Miaolingian of the Cambrian (Kouchinsky 2001). The shells are asymmetric with inequivalve valves, and a simple tooth and socket have been reported in two taxa – *Cambridium* and *Serioides* (Pelman 1985) (see Table 13.9). Putative muscle scars have also been reported (Yochelson 1969), but these are loop-like and more closely resemble intestinal loops of orthothecid hyoliths (Rozanov & Zhuravlev 1992).

---

[9] Not to be confused with the helcionelloidan family Stenothecidae.

TABLE 13.9

Classification of Stenothecoida

Stenothecoida (= Probivalvia)
(Superfamily) Cambridioidea
Bivalved, asymmetric shells with
  inequivalve valves and a simple tooth and
  socket; loop-like muscle scars have also
  been reported in some taxa.
Cambrian (Series 2) – Cambrian
  (Miaolingian)
(516–501 Ma)
Families Stenothecidae and Cambridiidae

	Stenothecoides, redrawn from
	Yochelson (1969).

	Cambridium, redrawn from Knight and
	Yochelson (1960).

Rasetti (1954), Aksarina (1968), and Yochelson (1969) considered stenothecoidans to be an independent, extinct molluscan class, although Rozov (1984) considered them to be closer to brachiopods than to molluscs and proposed a separate phylum Stenothecata. Runnegar and Pojeta (1974b) suggested that stenothecoidans might be bivalved monoplacophorans, while Waller (1998) considered Stenothecoida to represent the sister taxon of the Rostroconchia + Bivalvia.

Dzik (1981a) argued that stenothecoidans were inarticulate brachiopods with calcareous shells and characters that unequivocally indicated molluscan affinities could not be identified. Thus, the stenothecoidans are problematic lophotrochozoans, located outside the current concept of Mollusca (Waller 1998).

13.3.2.2.4 Monoplacophora

Early Paleozoic monoplacophorans are the subject of much conjecture. Historically, the group has been an assortment of various supposedly untorted limpets and some planispirally coiled conchs. Pojeta and Runnegar (1976) and Peel (1991c) consider most Cambrian cap-shaped taxa, the coiled Helcionelloida, and some, if not all, of the bellerophontian taxa, to be monoplacophorans. Other workers, including Knight and Yochelson (1960), Starobogatov and Moskalev (1987), and Parkhaev (2002a), limited the diagnosis of Monoplacophora to cap-shaped taxa and considered the remaining Helcionelloida and bellerophontian taxa to be torted gastropods. Because these positions are based on the interpretations of a small suite of muscle insertion characters, it is difficult to test either view.

Limpet-like shells are common in Cambrian and Ordovician deposits. Based on muscle scar morphology, many of these have been assigned to the Monoplacophora, but in a revision of putative monoplacophoran ancestors of cephalopods, Dzik (2010) has argued that many supposed tergomyan taxon are probably stem brachiopods rather than molluscs. These include genera placed in four monoplacophoran families, including *Kirengella* (Kirengellidae), *Hypseloconus* (Hypseloconidae) and *Pygmaeoconus* (Pygmaeoconidae), and *Lenaella*, *Nyuella*, *Romaniella*, and *Moyerokania* (Scenellidae). These taxa are characterised by a putative posterior to central apex, multiple pairs of muscle scars (sometimes showing the fusion of several attachment points), and a pair of medial muscle scars. Here we agree with Dzik and regard this distinctive muscle scar pattern as brachiopod rather than molluscan.

Another problematic group of limpets, the Archinacelloidea, have also been included in the monoplacophorans by different workers (Peel & Horný 1999). As in helcionelloidans, the muscle scar morphologies of assigned taxa are highly diverse and include a single pair of small apical muscles (*Barrandicella*), horseshoe-shaped muscle scars with or without medial muscle scar (*Archinacellopsis*), complete muscle circles (*Archinacella*, *Archinacellina*), and strap-like lateral muscles (*Archaeopraga*). Unlike monoplacophorans, distinct muscle attachment points associated with muscle segments are absent. Also absent are blood sinuses through the muscle bands as found in patellogastropods. The number of types of muscle patterns far exceeds that seen in fossil and living patellogastropods. We agree with Peel and Horný (1999) and consider the Archinacellida to represent one or more independent secondary flattenings within eogastropod lineage(s). The Archinacellida (families Archinacellidae and Archaeopragidae) range from the Furongian (Cambrian) to Pridoli (Silurian) (492.5–418 Ma).

Because we are uncertain of the phyletic relationships of the cap-shaped helcionelloidans with molluscs, we restrict the monoplacophorans to those molluscan taxa with distinct bilateral, serial muscle segments attached to the dorsal surface of the conch. The number of serial pairs of muscles in the Monoplacophora varies between four and eight. This interpretation is similar to that of Geyer (1994) and Vendrasco (2012).

In Monoplacophora, two superfamilies are recognised, namely Tryblidioidea and Neopilinoidea (see Appendix for details of classification). The Cyrtonelloidea are sometimes included in the Monoplacophora (e.g., Bouchet et al. 2017) but are here treated as bellerophontians within the Gastropoda (see Appendix). In the tryblidioidians, the anterior-most pairs of muscles are often fused on either side of the shell, and there are prominent buccal muscle scars and a diaphragm scar. The latter originates from the anterior-most muscle bundles and extends towards the centre of the shell; it is absent in Neopilinoidea. While known only from fossil monoplacophorans, similar scars are formed from the attachment of a coelomic partition on the anterior plates of the Polyplacophora (Plate 1897; Lemche & Wingstrand 1959). In chitons, this partition isolates a blood sinus around the radula into which the aorta empties (Plate 1897). A similar partition is found in the aplacophoran *Limifossor* (Heath 1905a), and according to Heath, they occur in aplacophoran taxa with a well-developed radula. We infer from these observations of living 'placophoran' taxa that the tryblidiid monoplacophorans (*Pilina, Tryblidium, Archaeophiala*) had more massive radulae than seen in Neopilinoidea. These putative diaphragm scars are also correlated with the large radular attachment muscle scars seen in tryblidiid taxa. Based on the change in habitats documented over geological time, Stuber and Lindberg (1989) suggested that monoplacophorans had more robust radulae in the Paleozoic, concluding that the morphology of the radulae of living monoplacophorans would probably be inefficient in Paleozoic near-shore habitats. Living neopilinids have substantially reduced radular muscle attachment areas. The earliest monoplacophoran, *Pilina*, first appeared in the Upper Cambrian (483 Ma), and the group reached its greatest diversity between the Ordovician and Silurian. Associated fauna include cephalopods, rostroconchs, gastropods, trilobites, and brachiopods (Wen & Yochelson 1999).

13.3.2.2.5 *Polyplacophora*

The polyplacophorans first appear in the late Cambrian (*Preacanthochiton, Chelodes, Matthevia*) (Bergenhayn 1960; Runnegar et al. 1979; Pojeta et al. 2010b) but the polyplacophoran affinities of some of these fossils are disputed by Sirenko (1997). In contrast, Wen (2001) proposed that the Terreneuvian (Cambrian) appearance of *Paracarinachites* represents the first polyplacophorans. Conway Morris and Peel (2009) have allocated *Paracarinachites* to the Cambroclaves – a group

of enigmatic, phosphatised, hollow spine-shaped sclerites thought to represent palaeoscolecidan worms.

Early undisputed chitons belong to the Paleoloricata, their shell valves lacking an articulamentum layer and therefore also lacking both insertion plates and sutural laminae which articulate adjoining valves in modern chitons. The articulamentum shell layer and its associated valve components first appeared in the Devonian Multiplacophorida (Vendrasco et al. 2004). It is also present in the eight-valved Neoloricata (crown group chitons) which first appeared in the Lower Carboniferous (Dinantian 359–326 Ma) (Sirenko 2006).

The late Cambrian–Ordovician fossil record is rich with putative polyplacophoran plates (Vendrasco & Runnegar 2004; Pojeta et al. 2010b), but demonstrating clear polyplacophoran affinities of these individual plates is difficult. Such assignments have been bolstered by the discovery of articulated specimens, ratios of different valve morphotypes, and valve microstructure (absence of the articulamentum layer) (e.g., Runnegar et al. 1979; Pojeta et al. 2003; Sutton & Sigwart 2012). In addition, Pojeta et al. (2010) demonstrated that aesthete pores are present in the shell plates of the earliest paleoloricates. Even if restricted to these better-characterised taxa, the overall morphological variation of the Paleoloricata is surprisingly diverse given the relatively conservative morphology of living Neoloricata (see Chapter 14). For example, the Paleoloricata include taxa with triangular, upright, sharply pointed plates (Mattheviidae) (Figure 13.13), rectangular plates (Septemchitonidae), and more typical chevron-shaped plates as seen in living neoloricates (*Echinochiton*) (Pojeta & Dufoe 2008). Most paleoloricates are reconstructed with a narrow girdle surrounding the plates, but in *Echinochiton* the valves are also surrounded by large hollow valve spines, similar to those found in some Multiplacophorida (Vendrasco et al. 2004b). In transverse section, individual intermediate plates vary from inverted V-shapes to inverted U-shape and approach semi-circular in some Septemchitonidae, if reconstructions by Dzik (1994) are accurate (Figure 13.22). These encircling valves suggest a greatly reduced or absent foot.

Vendrasco et al. (2009) skilfully combined three Chinese Cambrian small shelly taxa (*Ocruranus, Eohalobia,* and another probably incorrectly assigned to the Ordovician *Gotlandochiton*) as a possible stem polyplacophoran. The shell microstructure layers of *Ocruranus* are similar to those found in *Pelagiella* (see Section 13.3.2.2.7), but if the plate reconstruction is correct, the order of the layers is reversed. In

Matthevia, stem aculiferan

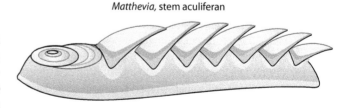

FIGURE 13.13 A reconstruction of the Paleozoic paleoloricate *Matthevia*. Redrawn and modified from Vinther, J., *J. Nat. Hist.*, 48, 2805–2823, 2014.

addition, the morphologies of these Cambrian valves (especially the putative head [*Eohalobia*] and tail [*Ocruranus*] valves) differ from the valves of any known putative fossil or living chiton. *Ocruranus* and *Eohalobia* have been considered possible brachiopods or halkieriids (Liu 1979; Conway Morris & Caron 2007) (see below), and *Ocruranus/Eohalobia* may represent sclerites of another plated brachiozoan.

Unlike most Cambrian molluscan taxa, the Polyplacophora have been subject to several phylogenetic analyses (Cherns 2004; Sigwart & Sutton 2007a; Pojeta et al. 2010b; Sutton et al. 2012). Most recover monophyletic Polyplacophora, Neoloricata, and Multiplacophorida. The Paleoloricata are often paraphyletic or polyphyletic and placed among aplacophoran taxa (e.g., Sigwart & Sutton 2007b; Sutton et al. 2012). Stratigraphically the groups are distinct, with the Paleoloricata in the early Paleozoic, followed by the multiplacophoridans (Neoloricata) with crown Neoloricata appearing in the Upper Paleozoic (Cherns 2004).

It has been suggested that the most likely Cambrian ancestor of Polyplacophora is *Wiwaxia* or *Halkieria*. Sirenko (1997 p. 16) argued that the shell plates were probably derived from spicules like those of *Wiwaxia*, as shown by the similarity of the shape of the plates in the 'most primitive Ordovician Chelodidae … but also by several characters of anatomy of chitons and some features of their ontogenesis'. The shell microstructure of the Paleozoic paleolicate, *Matthevia variabilis*, was reported as differing greatly from that of living chitons, showing a vesicular structure (Runnegar et al. 1979; Carter & Hall 1990), but this was later reinterpreted as oblique sections of aesthete pores (Pojeta et al. 2010a). There are no other reports of shell microstructure in Paleozoic or Mesozoic chitons. Reported calcium phosphatic shell structure in the Paleozoic chiton *Cobcrephora* (Bischoff 1981) – the so-called Phosphatoloricata – may be the result of secondary replacement (Carter & Hall 1990), or they are doubtful polyplacophorans (Hoare 2000; Pojeta et al. 2010a). No other mollusc has this type of shell structure.

The confusion surrounding the early Paleozoic 'placophorans' has resulted in different classifications being proposed. Smith and Hoare (1987) recognised three subclasses – Paleoloricata, Phosphatoloricata, and Neoloricata – but the former two were combined into a single subclass (Paleoloricata) by Sirenko (1997), as had been proposed earlier (Bergenhayn 1955; Van Belle 1983). All living chitons are classified in Neoloricata, and fossils range back into the Lower Devonian, but all Paleoloricata are from the Paleozoic. The latter group is quite diverse, with Sirenko (1997) listing 14 families arranged in four orders and five suborders, but these were substantially reduced by Pojeta et al. (2010a).

The classification of polyplacophorans is provided in the Appendix.

13.3.2.2.6 Cephalopoda

As also discussed in Chapter 17, the earliest known cephalopod taxon is *Plectronoceras*, from the Furongian (Cambrian) of China (493–491 Ma) (Chen & Teichert 1983b). The conchs are relatively large compared to earlier Cambrian shelled taxa, approximately 1.5 cm in length, slightly curved endogastrically, and with a siphuncle along the posterior edge of the shell. The aperture is oval in section due to lateral compression of the shell. The septa are thin and closely spaced and the body chamber about a quarter to a third of the length of the shell. In addition to China, they are known from West Antarctica and North America (Landing & Kröger 2009). The first cephalopods are associated with tropical, shallow-water carbonate platforms (stromatolitic and oolitic limestones), as well as with deeper water/lower energy, massive limestones (Landing & Kröger 2009). Chen and Teichert (1983a p. 650) characterised the shallow water setting as both well-oxygenated and turbulent. Other early cephalopods were thought to be nektonic, living above an anoxic benthos.

According to Landing and Kröger (2009), the entire history of stem Cephalopoda in the Cambrian is confined to the final two million years. Chen and Teichert (1983a) thought they arose only shortly before their first occurrence as fossils, while others have hypothesised hyolith, helcionelloidan, or monoplacophoran ancestors from the early or mid-Cambrian (e.g., Flower 1954; Yochelson et al. 1973; Dzik 1981b; Kobayashi 1987; Kröger 2007; Kröger et al. 2011) (also see Chapter 17 for further discussion). Most current evolutionary scenarios derive cephalopods from a benthic monoplacophoran-like mollusc with a high conical shell and septa (e.g., Kröger et al. 2011). Putative ancestors were found in taxa often attributed to Hypseloconidae, including *Tannuella* (530–518 Ma), *Knightoconus* (501–497 Ma), and *Shelbyoceras* (493–488 Ma). Stinchcomb and Echols (1966) questioned the affinities of some of these hypseloconids (*Tannuella* and *Shelbyoceras* spp.) in cephalopod phylogeny, and we have argued above that some at least are better treated as brachiopods (see Section 13.3.2.2.4). Another Cambrian septate taxon was the coiled *Aldanella* (Aldanellidae) (Parkhaev 2008), which is also probably not a mollusc (see Section 13.3.2.2.7).

The development of septa is typically discussed in terms of the evolution of buoyancy control in cephalopods (Boyle & Rodhouse 2005), however the septa in the taxa suggested as cephalopod ancestors (*Tannuella*, *Knightoconus*, *Shelbyoceras*) are not pierced by a siphuncle with which to control the movement of liquid between chambers to affect buoyancy (see Chapter 17). Thus, when septa first appeared in the cephalopod ancestor, they may not have been associated with buoyancy. At least two alternatives exist. In most gastropods, a single septum forms to seal off the adult teleoconch from the protoconch after metamorphosis (e.g., Smith 1935). Some living and fossil taxa (see Cook et al. 2015) create additional septa, often in response to wear or loss of the early whorls. These include a few fissurelloideans, some of the tube-like caenogastropod vermetids (e.g., Savazzi 1996), the coralliophiline muricid *Magilus*, a Miocene melongenid, a species of *Melongena* (Vermeij & Raven 2009), and a few turritellids (Andrews 1974; Waite & Allmon 2013). Several freshwater (Simone et al. 2012) and terrestrial (Gude 1905; Pilsbry 1909) heterobranchs also produce septa (see Chapter 3). Septate gastropods are also widely distributed in the fossil

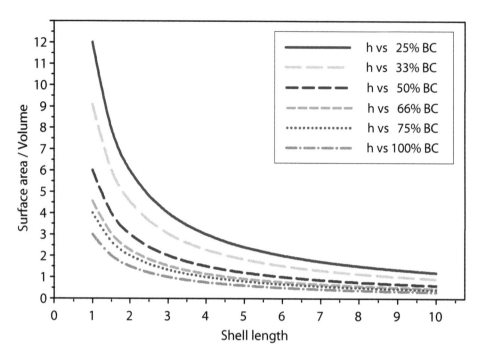

FIGURE 13.14   Relationships between apertural surface area and body volume in conical shells, with variation in body cavity sizes due to the presence of septa. h = height of conical shell, BC = body cavity volume.

record (Yochelson 1971; Gubanov et al. 1995; Cook et al. 2015); Wagner (2002) found septation to be a plesiomorphic character in gastropods.

Another related feature of septa is their ability to increase the surface-area-to-volume ratio in tubular shells. Because only the final body chamber is occupied, the body mass can be substantially reduced relative to the potential respiratory surfaces exposed at the aperture (Figure 13.14). This advantage decreases with increasing shell size but provides a four- to six-fold increase in apertural-area-to-body-volume ratios at smaller sizes, especially for shells where 50% or less of the total volume makes up the body chamber. For example, a conch 4.4 units in height with a body chamber occupying 33% of the total shell volume has the same potential respiratory area to body mass volume as a conch 1.5 units high with the entire shell occupied by the animal. How much these and other factors contributed to the Cambrian appearance of solid septa in the cephalopod ancestor cannot be known, but the functionality of intermediate states needs to be considered and explored.

While the scenario of a monoplacophoran with a tall conic shell developing septa and then a siphuncle is appealing, other anatomical issues accompanying this choice of taxon must be considered, some of which have been obscured by the continued use of hypothetical ancestral cephalopod cartoons (see Chapter 17, Figure 17.45 for examples). Because HAM[10] cartoons typically represent an ancestral gastropod rather than an ancestral mollusc (Lindberg & Ghiselin 2003a), the hypothesised anatomical configurations can become morphological chimeras when crammed into fossil shells. For example,

reconstructions of the cephalopod ancestor generally feature multiple segmented muscles, an anterior mantle cavity, a distinct, free head, and sometimes even gastropod-like cephalic tentacles (Yochelson et al. 1973; Dzik 1981b; Teichert 1988; Kröger 2007) or an operculum (Dzik 1981b), none of which would be found in a monoplacophoran-like ancestor.

While muscle scars are unknown for the putative ancestral taxa *Tannuella*, *Knightoconus*, and *Shelbyoceras*, other 'hypseloconids' show a continuous ring-like muscle scar not broken into separate muscle bundles as in monoplacophorans. The head region of monoplacophorans is attached to the anterior region of the shell. Without a detached head region, extension and retraction beyond the anterior region of the shell are impossible, and an anterior dorsal mantle cavity cannot exist. Thus, the proposed simple transition from a tall, septate monoplacophoran to a buoyant, tentaculate cephalopod requires substantial anatomical modifications either preceding, or concordant with, the evolution of the first cephalopod. The detachment of the head alone requires the reorganisation of some of the buccal and mouth musculature and modification of anterior blood sinuses. In addition, jet propulsion is only possible in shelled cephalopods (such as *Nautilus*), with a detached head, with water flowing into the mantle cavity and being forced out through the funnel with piston-like movements of the free head (Chamberlain 1987; Bizikov 2002) (see also Chapter 17).

These constraints and other anatomical ambiguities suggest a reconsideration of ancestral scenarios rather than imaginary monoplacophorans. Tightly coiled taxa may be especially interesting because their shell morphology places potential limits on their internal anatomical arrangements. Tightly coiled conchs suggest motile molluscs capable of extension and retraction of the head-foot complex, requiring detached

---

[10] Hypothetical Ancestral Mollusc – see Chapter 12.

heads, which are found in scaphopods, gastropods, and cephalopods. This character was the basis for a taxon Cephalomacia or headed molluscs (Kopf-Weichthiere) (Keferstein 1862). Here we refer to the condition in which the head region is dorsally attached to the shell as *ligocephalic* and the state where the head region is not attached as *apocephalic* (see Table 13.1). Detached heads are also correlated with ano-pedal flexure which sees the anus displaced ventrally during development (see Chapter 8). Ano-pedal flexure also serves as an exaptation for many features seen in cephalopods.

The detaching of the head was significant in molluscan evolution and presumably occurred independently in different lineages. Both ligocephalic and apocephalic crown groups appear geologically simultaneously in the fossil record, thus providing no suggestion of an ancestral state. Developmentally, the head region forms relatively late in ontogeny; in the veliger phase in gastropods (Buckland-Nicks et al. 2002), during metamorphosis in scaphopods (Wanninger & Haszprunar 2002a), and in middle and late embryos (four–six months) in *Nautilus* (Shigeno et al. 2008), probably indicating that undifferentiated heads are plesiomorphic.

13.3.2.2.7 Gastropoda

Coiled gastropod-like conchs are among the earliest small shelly fossils to appear in the Terreneuvian (535 Ma) of the Cambrian (Parkhaev 2008). Most are known as steinkerns and include Pelagiellidae, Aldanellidae, and Khairkhaniidae. All appear near the end of the 'Nemakit-Daldynian skeletonisation' event (Maloof et al. 2010) and are contemporary with the helcionelloidans but predate the appearance of the Sachitida. All are multi-coiled, and their overall morphology appears similar to extant gastropod shells. The khairkhaniids are planispiral, dextral, or sinistral, with circular apertures. Aldanellids are typically dextrally coiled with more elliptical apertures and an axial fold sculpture; septa, protoconchs, and muscle scars are known for *Aldanella* (Parkhaev & Karlova 2011). The spiral pelagiellids are low spired and dextral like aldanellids and have elliptical apertures. They also have the greatest variability in whorl expansion rates. These taxa are globally distributed, including Antarctica, Australia, New Zealand, North America, China, Europe, Middle East, Asia, and Siberia, and their distribution and abundance caused Brasier (1989) to suggest a pelagic or epiplanktonic habit for the group. Many taxa were lost during the End-Botomian extinction event (517 Ma), and none survive after the Dresbachian extinction event (502 Ma).

Pelagiellidae and Aldanellidae were placed in the Orthostrophina in the Paragastropoda by Linsley and Kier (1984) who remarked that the noted Paleozoic gastropod specialist J. Brookes Knight (1952) doubted that they were gastropods. Yochelson (1975) also rejected the molluscan affinities of Aldanellidae, and Linsley and Kier (1984 p. 241) further commented that 'the long axis of the aperture (*Pelagiella* and *Onychochilus*) is oriented at approximately right angles to elongated apertures of modern gastropods'. The discovery of clusters of chaetae in *Pelagiella* (Thomas et al. 2010; Thomas & Vinther 2012) and *Aldanella* (Dzik & Mazurek

2013) provides further evidence of the non-molluscan affinities of these taxa and led Dzik and Mazurek (2013) to place the Pelagiellida in the Hyolitha (see Section 13.1.1).

The exclusion of both helcionellidans and Pelagiellida from the gastropod fossil record leaves several mimospiridan taxa as the earliest possible gastropods. Wen (1979, 1990) reported representatives of the Onychochilidae and Euomphaliformii from the earliest Cambrian of China. Both were considered hyperstrophic and included in the Paragastropoda by Linsley and Kier (1984). A 'para-gastropod' (i.e., 'resembling') remains appropriate. Whether these and other early coiled molluscs are gastropods (torted) or not has been the subject of discussion, often accompanied by cartoons of imagined animals (Knight 1952; Horný 1965; Peel 1991c; Parkhaev 2008; Parkhaev & Karlova 2011). Like the cephalopod example above, many of these hypothetical animals are constructed from gastropod parts, including an anterior mantle cavity, a free head, and cephalic tentacles. Typically, the model is rotated in the candidate shell to see if it functions better as a gastropod or monoplacophoran. The assumption of gastropod character states for the ancestral anatomy limits outcomes as character states of monoplacophoran morphology have not been tested. One notable exception is the work of Morris (1991) who used water flow experiments to conclude that some paragastropods were untorted. Consequently, we treat paragastropods as a separate class (see Table 13.10). See the Appendix for gastropod classification.

Paired muscle scars have been argued to provide evidence of torsion (Runnegar 1981; Peel 1991b; Horný 1992). Vendrasco (2012) and others pointed out that torsion, the defining synapomorphy of gastropods (see Chapters 8 and 18), is difficult to convincingly infer without detailed anatomical information, although some general patterns may be deduced from muscle attachment scars and have been used to infer muscle tracts for head and foot retractors (Peel 1991b; Harper & Rollins 2000; Parkhaev 2008). In a few very rare cases, anatomy (such as gut tracts) can be preserved (Horný 1998; Sutton et al. 2006) (see Section 13.3.2.2.7.1).

While torsion remains a difficult character to identify in putative early gastropods, living in closed tubes requires that ano-pedal flexure takes place during ontogeny[11] bringing the larval telotroch (anus) from the apex of the blind tube to the apertural opening (see Chapter 8, Figure 8.33). After this event, the anus would remain in a posterior mantle cavity in juxtaposition with the back of the foot, proximity which may have affected water circulation in and out of the cavity. Torsion brings the anus and mantle cavity to lie dorsally over the detached head in gastropods and would probably enhance water circulation. In cephalopods and scaphopods, the morphological changes related to their unique habitats (water column and burrowing respectively) also provide alternative solutions to a constricted posterior mantle cavity. It is not known if ano-pedal flexure evolved separately in each group or in an ancestor shared by two

---

[11] As noted in Section 13.3.2.2.6, ano-pedal flexure occurs only in apocephalic molluscs.

TABLE 13.10

Classification of Paragastropoda

(Class) **Paragastropoda**

Cambrian (Series 2)–Mississippian (Middle) (513–345 Ma)

Spirally coiled conchs with bulbous protoconchs; coiling geometry includes dextral and sinistral forms, many with moderately high whorl expansion rates. Withdrawing into the paragastropod shell requires a free head and ano-pedal flexure of the body but not necessarily torsion. Thus, paragastropods as presently constituted are probably paraphyletic and may include both non-torted and torted (i.e., gastropod taxa). This grouping is treated as one of several 'Paleozoic Basal Taxa that are certainly Gastropoda' by Bouchet et al. (2017).

(Order) **Mimospirida**

(= Hyperstrophina)

Geological range as in Paragastropoda

Shell hyperstrophic to depressed-orthostrophic, commonly with angulation on the outer part of upper whorl surface marking the possible inhalant or exhalant channel; long axis of aperture converging towards the apex of the depressed spire; shell walls thick, outer layers calcitic, inner layers thick, aragonitic.

(Superfamily) **Clisospiroidea**	
(= Mimospiroidea)	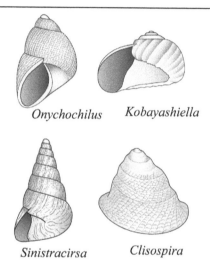
Geological range as in Paragastropoda	
Shells hyperstrophic, with highly prosocline tangential apertures. Shell form varying from high spired to moderately low spired. The depressed spire often has a sharp ridge on the upper whorl face.	
Families Clisospiridae and Onychochilidae	
	Examples of Onychochilidae and Clisospiridae (*Clisospira* only). Redrawn and modified from Knight et al. (1960).

or all three. Waller (1998) considered ano-pedal flexure to be a synapomorphy of gastropods + scaphopods + cephalopods and listed additional possible synapomorphies for the group. He also noted that the pedal retractor muscles of these three taxa all insert in the shell in a single close cluster rather than being dispersed over the shell, although most molecular trees fail to recover such a group (Sigwart & Lindberg 2015).

Frýda et al. (2008a) suggested that the earliest torted gastropods were late Cambrian bellerophontians and euomphaliforms, including *Strepsodiscus* and *Schizopea*. They reached this conclusion primarily based on teleoconch and protoconch morphology of Paleozoic gastropods (Figure 13.15). While teleoconch morphology varies broadly both within and between Paleozoic gastropod taxa just as in living groups, protoconch morphology is much more conservative. While non-gastropod groups have cap-shaped protoconchs, in gastropods the protoconch is more cup-shaped to tubular, with the aperture typically narrower than the length (Ponder & Lindberg 1997). In living taxa, Sasaki (1998) recognised four types: (1) symmetrically uncoiled, (2) paucispiral, (3) multispiral, and (4) globular. Globular was restricted to living terrestrial Neritimorpha and will not be considered further here.

A symmetrical, relatively open coiled protoconch with a bulbous initial chamber that slightly constricts with growth is found in Bellerophontia, Macluritida, Perunelomorpha, and Euomphaloidea (Frýda 1999a; Frýda et al. 2006). In patellogastropods the uncoiled protoconch also has a bulbous initial chamber that slightly constricts with growth, but there is no hint of coiling until the teleoconch is formed. A straight to loosely coiled protoconch with a bulbous initial chamber that constricts with growth is a probable synapomorphy of the Eogastropoda (Ponder & Lindberg 1996) (see also Chapter 18). The paucispiral protoconch found in vetigastropods is more tightly coiled (usually less than two whorls), and the chamber continues to enlarge with apertural growth (i.e., there is no constriction). Multispiral protoconchs are found in many caenogastropod groups and are distinguished by having a second growth phase (protoconch II) which results from larval planktotrophy and signals an extended larval life (Jablonski & Lutz 1983). In the 'cyrtoneritimorphs', protoconch I is orthoconic and similar to symmetrical open-coiled protoconchs. In addition, protoconch II is distinctive among multispiral groups in having open coiling and in being fish-hook-shaped. This group has been thought to be either ancestral to the Neritimorpha (Frýda et al. 2009) or a separate early

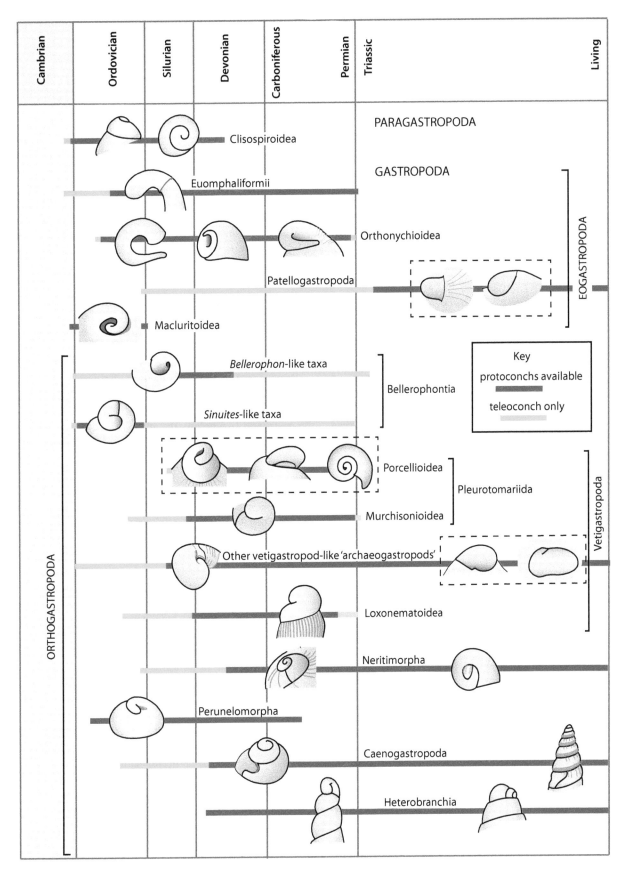

FIGURE 13.15 Paragastropod and gastropod protoconch morphologies through time. Redrawn and modified in part from Frýda J. et al., Paleozoic Gastropoda, pp. 239–270, in Ponder, W.F. and Lindberg, D.R. (eds.), *Phylogeny and Evolution of the Mollusca*, University of California Press, Berkeley, CA, 2008a.

gastropod lineage (Frýda & Heidelberger 2003) (see Section 13.3.5.1.1).

Based on the fossil record, the earliest unambiguous gastropods (Eogastropoda) do not appear until the Miaolingian (Cambrian) (501 Ma) (e.g., *Sinuopea, Schizopea, Dirhachopea*). Their origins may lie within the paragastropod groups with apertural notches or constrictions suggesting exhalant control of the water flow through the mantle cavity. Other characters probably included a basic docoglossate radular morphology and associated musculature, an openly coiled protoconch, a subradular organ, lateral statocyst position, lack of skeletal support in the ctenidial filaments, and paired foot/head retractors. While there has been little disagreement regarding the gastropod status of the Euomphaliformii, the same cannot be said for the Bellerophontia, and whether all, some, or none of the bellerophontians are torted has been controversial (Wahlman 1985; Peel 1993; Harper & Rollins 2000).

13.3.2.2.7.1 Bellerophontia (= Amphigastropoda)

Bellerophontians first appear in the latest Cambrian (Furongian, 488.3 Ma). They are bilaterally symmetrical (planispiral) shells, sometimes with high whorl expansion rates, and a prominent median sinus, channel, or slit in the outer lip that gives rise to a selenizone. Limited shell structure studies have identified aragonitic shells with complex crossed lamellar microstructure. Nacre, found in many vetigastropod lineages, is not known (MacClintock 1967), although the nature and placement of the apertural flutes and embayments and their placement on the shell are highly suggestive of a bilaterally symmetrical mantle cavity similar to that seen in the vetigastropod Fissurellidae, with exhalant flow control (Lindberg & Ponder 2001) (see also Chapters 4 and 18). In a few taxa this idea is reinforced by the finding of shell structures within the last whorl that divide the internal space into two symmetrical parts (Rohr et al. 2003). Apertures with flutes and channels are also prevalent in both the euomphaliforms and paragastropod onychochilids and continue to crown taxa such as Fissurelloidea and Pleurotomarioidea. In asymmetrically coiled taxa such as Pleurotomarioidea, they are not restricted to the median anterior edge of the aperture as in the bellerophontians but are located more laterally on the aperture. However, a bilaterally symmetrical teleoconch with a median selenizone is present in post-Silurian Porcelliidae and is thought to be an apomorphic shell character for that group (Frýda 1997; Frýda et al. 2019). Wagner (2002) provided a phylogenetic analysis of many of these anisostrophically coiled taxa and concluded that the vetigastropod lineage traces its origin to a bellerophontian taxon and that slit-less trochoidean and apogastropod-like morphologies evolve just as frequently as pleurotomarioid-like ones in these stem gastropods.

According to Frýda (1999a), *Bellerophon* has a small, bilaterally symmetrical early shell, indicating a planktotrophic larval stage. There are few other reports of bellerophontian protoconch morphology. Some indicate a protoconch of about half a whorl with an inflated apex region, while others

had two to three complete whorls, and based on these differences Dzik (1981a) suggested that bellerophontians consisted of two groups, a position also held by Frýda (1999c). Wagner (2002) also recognised two groups of bellerophontians based on adult shell characters that coincided with data from the muscle scars, one group having multiple, monoplacophoran-like muscle scars and the other gastropod-like muscle scars. Given the non-correspondence of both muscle scar and protoconch characters, it is not surprising that multiple scenarios for bellerophontian evolution have been proposed (Figure 13.16).

Fossilised intestinal contents have been reported from the Lower Ordovician *Cyrtodiscus nitidus* (Horný 1998). They appear in one or more assumed intestinal loops, some with a terminal 'rectum'. The fossilised gut lies immediately behind the sediment infilling of the last whorl and could indicate that torsion had occurred. The variability of the depth of infilling and its composition caused Horný to caution that the gut was obviously displaced by infilling and did not represent the original position, and therefore any conclusions would be premature. It remains an important find and demonstrates some of the extraordinary detail the fossil record can sometimes provide.

Runnegar (1981) inferred that bellerophontians were untorted but considered them to be a monophyletic group distant from both gastropods and monoplacophorans, and that the gastropod-like columellar muscle scars of some reflected parallelism with gastropods that allowed untorted bellerophontians to retract deep into the shell.

Besides the bilaterally symmetrical muscle scars on the columellar region of the shell, some presumed bellerophontian taxa also have paired central scars over the assumed head region (e.g., *Cyrtolites, Sinuilopsis*) (Horný 1965, 1996). These circular scars are far enough back in the body cavity that they would not conflict with the assumption of a detached head and may represent the buccal (i.e., radular) retractor muscle attachment areas seen in other molluscan taxa (Graham 1959, 1964; Lemche & Wingstrand 1959; Wingstrand 1985). The muscle scars of *Cyrtolites nitidus* have been interpreted as monoplacophoran-like because of the three pairs of muscle scars; two pairs are typical of bellerophontian columellar region scars while the third one is centrally located (Horný 1996). While the lateral pairs of muscle scars are typical, the central scars are bifid and unlike either gastropod or monoplacophoran muscle attachment patterns. Alternatively, these channel-like scars in the central area could represent anastomosing vessels and sinuses in the roof of the mantle cavity as seen in patellogastropods and some fissurelloideans (Hickman & Lindberg 1985; Lindberg & Squires 1990). Such an interpretation would also support a torted morphology with anterior auricles.

As noted above, much of the argument as to whether bellerophontians are gastropods or monoplacophorans is based on the muscle scar patterns. There are three hypotheses: (1) bellerophontians are untorted, exogastric monoplacophorans, or (2) are torted, endogastric gastropods, or (3) a mixture of both. See Peel (1991b); Wahlman (1992); Frýda (1999a, 2012); Frýda et al. (2008a) for reviews. The idea that bellerophontians are untorted monoplacophorans is largely based

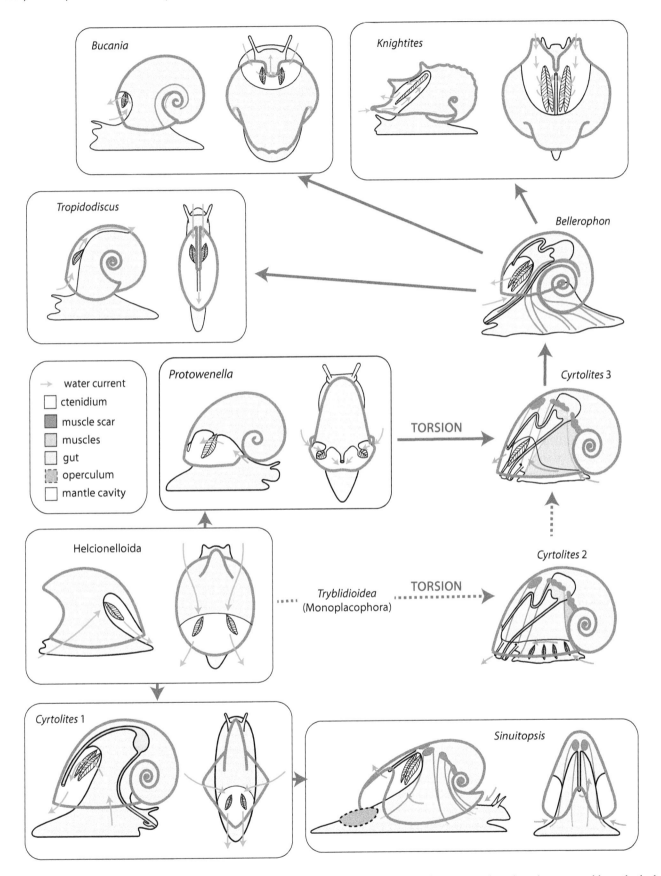

FIGURE 13.16 Some of the alternative reconstructions in the literature of bellerophontian anatomy based on the supposed hypothetical ancestral helcionelloidan. Both torted and untorted models are imagined; for example, see the three different interpretations for *Cyrtolites*. Redrawn and modified from various sources.

on comparisons of the muscle scars of cyrtonellids with cyrtolitid bellerophontians. The muscle scar patterns are similar (Horný 1965), but the central dorsal muscle scars of the cyrtonellids are much more anteriorly placed, suggesting an attached head, while the central scars of cyrtolitids are located further back into the last whorl, a position suggestive of a detached head.

In response to the assertion by Wenz (1940) that cyrtonellid 'monoplacophorans' and cyrtolitid bellerophontians were both untorted, Knight (1947) countered that although both shared similar musculature, the similarity was due to convergence and not common ancestry. Knight described bellerophontian muscle scars which fitted his expectations for an anatomically primitive gastropod (such as a vetigastropod). This placement in the Gastropoda was not new but dated from Koninck (1883). Before this, they had been variously thought to be brachiopods, cephalopods, bivalves, or either 'heteropod' or 'pteropod' gastropods.

Knight (1952) proposed that secondary shell deposits on the ventral parietal surface of the shell indicated their gastropod affinities, although not all workers have agreed with this assessment, and some (e.g., Pojeta & Runnegar 1976; Harper & Rollins 1982) argued there were strong functional arguments to support the non-gastropod hypothesis of Knight (1947). The secondary shell deposits (parietal inductura or callus) are often thick, and similar structures are known only on a few gastropods. It is difficult to imagine how they would function on an untorted coiled snail where the coil was carried over the head. In a torted snail, the parietal inductura would rest on the muscular foot. Frýda and Gutiérrez-Marco (1996) also noted that secondary parietal shell deposits were similar in position, form, and ornamentation in both sinuitid and euphemitid bellerophontians and proposed that their formation indicated that the shell must have been partially enveloped by mantle tissue. More recently, Harper and Rollins (2000) also concluded that bellerophontians are a clade of gastropods.

Within gastropods, Ponder and Lindberg (1997) placed the bellerophontians in the vetigastropods. MacClintock (1967), McLean (1984b), and Golikov and Starobogatov (1975) were even more specific, linking the euphemitid bellerophontians to Fissurellidae based on shell microstructure, external bilateral symmetry, and other shell and anatomical characters. Based in part on the presence of secondary shell deposits on the exterior of the shell, Linsley (1978) and McLean (1984b) reconstructed the bellerophontian genera *Euphemites* and *Retispira*, respectively, with internal shells.

The muscle scar evidence cited for the above competing hypotheses of relationships led to the third hypothesis: that bellerophontians include both monoplacophorans and gastropods, because of parallel evolution of shell form (Yochelson 1967; Wahlman 1992) or because gastropods evolved amid untorted bellerophontians (Knight 1952). The recognition of the differences in head region muscle scars between taxa with attached and detached heads suggests this compromise is no longer necessary (see discussion of cyrtonellid versus cyrtolitid muscle scars above), and based on this criterion we regard the bellerophontians as stem vetigastropods. The addition of

bellerophontians to the vetigastropods further expands an already impressive range of morphology in the group.

According to Knight et al. (1960), the main groups of what they considered bellerophontians arose in the Upper Cambrian and include: (1) Cyrtonellidae[12] which continued to the Devonian and had whorls with little overlap and a V-shaped apertural sinus; (2) the Sinuitidae which included tight and looser coiling taxa with a broad, U-shaped sinus, which persisted to the Guadalupian (Permian); and (3) the Bellerophontidae which had a narrow, median slit and tightly coiled whorls and were extant until the Middle Jurassic (176 Ma). Some bellerophontians had a flared aperture (Bucaniinae); in others (such as *Tremanotus*) the slit is divided into exhalant holes rather like those in abalone shells. Other taxa such as *Chalarostrepsis* and *Temnodiscus* had rapidly expanding whorls, sometimes becoming limpet-like with an internal shelf (as in, for example, *Pterotheca* and *Cycotheca*) similar to that formed in some Neritidae (Neritimorpha) and in the caenogastropod slipper limpets (Calyptraeidae). Sometimes the open coiling is so extreme it is probable these taxa were not very mobile (Rohr et al. 2003).

13.3.2.2.8 Bivalvia

Tiny bivalved molluscs are among the earliest molluscs appearing in the Cambrian record. Both *Fordilla* and *Pojetaia* first appear in the Terreneuvian (529–521 Ma). *Fordilla* is present up to the Botomian extinction event (517 Ma), while *Pojetaia* persists through most of the Miaolingian (~501 Ma). Both taxa are widely distributed with fossil occurrences in North America, China, Greenland, Europe, Siberia, and Turkey; *Pojetaia* also occurs in Australia (Elicki & Gürsu 2009). Other putative Cambrian Miaolingian bivalve taxa include *Tuarangia* of Europe and New Zealand (504–501 Ma) and *Camya* in Europe (509–497 Ma). *Arhouriella* occurs in Epoch 2 of Morocco (521–514 Ma) (Geyer & Streng 1998). While some authors have expressed doubts regarding the bivalve affinities of these latter three taxa (see Runnegar & Pojeta 1992), they are more often accepted as such (e.g., Cope & Kříž 2013). After these first occurrences in the Cambrian Epoch 2 and the Miaolingian, the bivalve fossil record is remarkably depauperate (Cope 2000).

These early bivalves are small, less than 5 mm in length (Cope & Kříž 2013). *Pojetaia* has up to three hinge teeth per valve, similar to some living protobranch taxa, while *Fordilla* has only a single tooth in each valve. Articulated specimens of both *Fordilla* and *Pojetaia* are known, as are their muscle scars, opisthodetic ligament morphology, and shell microstructure, which are generally similar to these features in Devonian nuculoideans (Runnegar & Bentley 1983; Carter 1990; Runnegar & Pojeta 1992; Pojeta 2000). These and other studies have led to general agreement that these taxa represent the earliest stem members of the Bivalvia. Elicki and Gürsu (2009) discussed various Cambrian taxa initially thought to be bivalves and indicated their taxonomic placement.

---

[12] Treated as Mollusca *incertae sedis* herein.

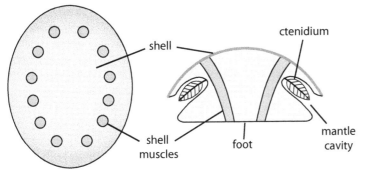

Hypothetical ancestral monoplacophoran
with several paired dorsoventral shell
muscles, head and radula.
Multiple ctenidia in mantle grooves.

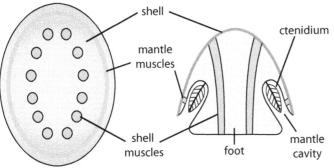

Lateral compression and dorsal elongation.
Movement of shell muscles away from shell
margin and narrowing of foot.
Development of mantle muscles.

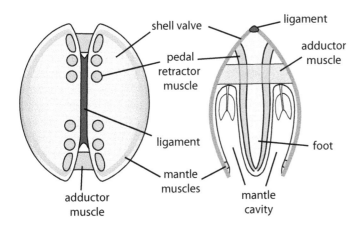

Further lateral compression, longitudinal
split and ligament development.
Head and buccal structures lost.
Adductor muscles developed and animal
completely enclosed by shell.
Ctenidia reduced to a single pair.

FIGURE 13.17   One scenario of the origin of a bivalve from a monoplacophoran ancestor. Redrawn and modified from various sources.

Geyer and Streng (1998) described two prong-like centrally located hinge teeth in *Arhouriella* in addition to an amphidetic ligament, making it distinct from *Fordilla* and *Pojetaia*. *Tuarangia* is well characterised (e.g., shell and hinge structure, muscle scars, ligament) and based on over 100 specimens, but its placement in the Bivalvia remains controversial. In the original description of *Tuarangia*, MacKinnon (1982) described numerous taxodont-like hinge teeth, an amphidetic ligament, and foliated calcite shell structure, suggesting placement in the Pteriomorphia. Based on similar shell structure, Runnegar and Pojeta (1992) considered *Tuarangia* to be a laterally compressed 'monoplacophoran' closely related to *Pseudomyona*; Wagner (1997) found the same relationship. The latter taxon is often treated as a helcionelloidan, although both Hinz-Schallreuter (1995) and Carter et al. (2000) suggested that *Tuarangia* had a closer affinity with Bivalvia than Monoplacophora; Carter et al. (2011) also treated it as a bivalve.

Owen et al. (1953) proposed that the bivalved shell evolved from a simple, domed univalved shell (such as a helcionelloidan) with a single centre of calcification (Figure 13.17). As also discussed in Chapter 15, the shell became laterally compressed, and the mantle expanded laterally into two lobes, each with its own centre of calcification but narrowly connected dorsally. The shell had a periostracum, and the ostracum (shell) consisted of two layers, with the periostracum extending between the two mantle lobes but with the shell in the middle area 'largely uncalcified'. In this model, the adductor muscles were derived from the cross fusion of mantle muscles with the ligament extending between them on both sides of the umbo of the shell (i.e., an amphidetic ligament). A modification of this theory has bivalves descended from unspecialised ribeiriidan rostroconchs that were posteriorly elongate (Pojeta & Runnegar 1985; Runnegar & Pojeta 1985), with a flexible dorsal margin which evolved into the ligament,

TABLE 13.11

Siphonoconcha, an Extinct Group of Bivalved Brachiopods that Were Thought to Be a Separate Class of Molluscs

(Phylum) **Brachiopoda** **Siphonoconcha** Cambrian (Series 2) (516–513 Ma) Proposed by Parkhaev (1998) for early Cambrian bivalve-like fossils including *Apistoconcha,* these are now considered stem brachiopods (Li et al. 2014).	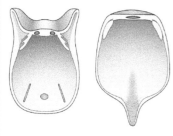 *Apistoconcha,* redrawn and modified from Parkhaev (1998).

but both of these models fail to account for the bivalved larval shell that characterises bivalves.

Living molluscan bivalve taxa are often related back to these putative Cambrian ancestors, but it is possible that some Cambrian 'bivalves' are separate derivations of the bivalved shell, which is demonstrably homoplastic within the Mollusca, Lophotrochozoa, and Protostomia. 'Bivalved' morphologies are present in Brachiopoda, such as the 'Siphonoconcha' (see Table 13.11), proposed by Parkhaev (1998) as a class of molluscs for early Cambrian bivalve-like fossils including *Apistoconcha,* but these are now considered stem brachiopods (Li et al. 2014). There are also numerous extinct and living arthropods that are bivalved, including the Cambrian Bradoriida, Phosphatocopida, and Ordovician Leperditicopida (Vannier et al. 2001; Legg et al. 2013) and living Ostracoda, Diplostraca (clam shrimps), and Ascothoracida (parasitic barnacles) (Brusca et al. 2016). Some even have adductor muscles, including the Diplostraca and Ascothoracida. The presence of so many bivalved organisms in the Cambrian fossil record (Briggs 1977; Popov 1992) has contributed to the difficulty in differentiating bivalved molluscs from non-molluscs (Elicki & Gürsu 2009). Bivalve-like forms are also common among the earliest putative molluscs, including the Stenothecoida (see Section 13.3.2.2.10) and the almost bivalve-like *Pseudomyona* and *Eotebenna* (Runnegar & Pojeta 1985; Gubanov et al. 2004) as well as the more recent origin of the bivalve shell in the sacoglossan gastropod group Juliidae (see Chapter 20). Despite these reservations, most workers, and the weight of evidence, suggest that some or perhaps most of the Cambrian taxa usually considered bivalves are indeed members of that class.

Cambrian bivalves are often associated with low-energy, carbonate platforms with reduced sedimentation rates (Cope & Kříž 2013). Both Tevesz and McCall (1979) and Morton (1996) argued these earliest bivalves were epifaunal, while Cope (1996b) considered them only to be able to survive as infauna in fine sediments. Morton (1996) suggested that the earliest bivalves were small, lived on the surface of sediments,

and the inhalant water entered the shell along the anteroventral margins and exited posteriorly. Food collection was done by the foot, and the bivalve was usually oriented on its side. While adult modern nuculoideans are too specialised to be considered models of ancestral bivalves, Morton (1996), following Reid et al. (1992), considered that their surface crawling juveniles would be similar, as these would feed using the foot rather than the derived palp proboscides (see also Chapters 5 and 15).

Over their long history, marine bivalves have experienced periods of elevated extinction and origination, as well as periods of relative evolutionary quiescence. Bivalves exhibit moderate rates of extinction and origination through the Phanerozoic, but these rates decline over time. Prominent peaks in extinction occurred during the late Cambrian, End-Ordovician, late Devonian, End-Permian, End-Triassic, and End-Cretaceous. A similar decline in rates is also observed at broader taxonomic scales (Raup & Sepkoski 1982; Valen 1984; Foote 2003) and may result from losing extinction-prone lineages over time (Roy et al. 2009a). Regions differ little in the severity of extinction experienced by marine bivalves, but they differ markedly in the timing and the processes of recovery (Raup & Jablonski 1993; Jablonski 1998).

The effects of extinction on diversity dynamics have been intensively studied in marine bivalves because of their relatively complete fossil record, the considerable biological variation among taxa, and their diversity and abundance in shallow marine environments today and in the past. These studies have shown that diversity-dependent processes[13] were most pronounced following mass extinctions but also operated consistently throughout the history of the clade. Geographic range size is the most consistent predictor of bivalve survival, although traits like feeding mode and life habit may also be important but are probably more dependent on the particular context of environmental change. Perhaps surprisingly, bivalve body size is largely decoupled from extinction risk (Harnik & Lockwood 2011).

See Chapter 15 for further information on bivalves and the Appendix for their classification.

13.3.2.2.9 *Rostroconchia*

Rostroconchs look superficially like bivalves but differ in having a single, cap-shaped early shell (Figure 13.18) which subsequently grows laterally, forming extensions or valves on each side of the body. Valve contact along the anterior, ventral, and posterior margins may include a gap or be tightly appressed. In some taxa, the posterior margins may be elongated into a rostrum. Unlike bivalves, there is no dorsal hinge. Instead, some or all of the calcified shell layers are continuous across the dorsal margin. The musculature of bivalves and rostroconchs also differs (Figure 13.20). Rostroconchs are divisible into two groups – the Ribeiriida and Conocardiida. A posteroventrally

---

[13] Diversity-dependent processes are outcomes of interspecific competition and are thought to influence the dynamics of both speciation and extinction (Rabosky 2013).

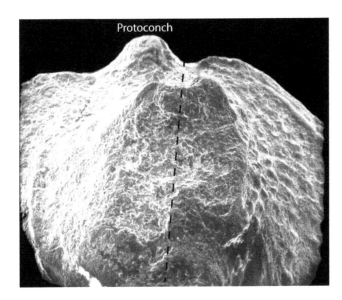

FIGURE 13.18 Posterior view of the protoconch on the left plate of the rostroconch *Pseudoconocardium lanterna* (USNM 209292). Dashed line indicates midline between plates. (Courtesy of B. Runnegar.)

directed transverse shell partition (the *pegma*) extends across the anterior region of most ribeiriidan and conocardioidean rostroconchs (Pojeta & Runnegar 1985) which connects both valves and provides additional muscle attachment points. The pegma leaves a characteristic notch in the anterior dorsal margin of internal moulds (Runnegar et al. 1978).

The earliest rostroconchs, if represented by the genus *Watsonella,* are found in Terreneuvian (Cambrian) (~530 Ma) strata in Siberia. The genus *Ribeiria* first appears during the Miaolingian (Cambrian) (498.5 Ma) in Australia. The group diversified during the Lower Ordovician, becoming about equal in abundance to bivalves (Wagner 1997). Members of the Conocardioidea (Figure 13.19) mostly survived the End-Ordovician extinction event although a few riberioids continued into the Llandovery (Silurian). All rostroconchs became extinct by the end of the Permian. The classification of rostroconchs is summarised in Table 13.12.

Rostroconch taxa were first described in the early 1800s (Pojeta & Runnegar 1976). While most early workers considered them to be molluscs (typically Bivalvia), they have also been thought to be notostracan crustaceans (Kobayashi 1933). Pojeta et al. (1972) noted the molluscan characters (protoconch, calcified shells with accretionary growth, and prominent muscle scars) of rostroconchs and proposed a new class of bivalved molluscs – the Rostroconchia. Rostroconch shells also reflect the full diversity of molluscan shell structures, with a prismatic outer layer and nacre, foliate, and crossed lamellar layers reported in different taxa (Kouchinsky 1999; Vendrasco et al. 2010). Muscle scar morphology is cyclomyan-like in early rostroconchs such as *Ribeiria* (see Section 13.3.2.2.4). The horseshoe-shaped scar is central on the dorsolateral inner surface of the shell and opens anteriorly; the dorsal portion of the muscle band is the largest and tapers towards the anterior. A single pair

FIGURE 13.19 Four views of the conocardioidean rostroconch *Apotocardium.* (Courtesy of B. Runnegar.)

of muscle attachment areas lies just beyond the anterior ends of the horseshoe-shaped scar on the lateral surface of each plate, and an additional pair of muscle attachment areas occurs on the anterior pegma (Pojeta & Runnegar 1976). Muscle scar morphology became more complex in the Conocardioidea, with both band fusion and fragmentation; anterior shell gaps also became more pronounced and ornate. In most Conocardioidea the dorsal margin is also modified by a reduction in the number of shell layers continuous between the shell plates, suggesting greater shell plate flexibility. A small cap-shaped protoconch is found on the left shell plate.

Rostroconchs are thought to have originated within the helcionelloidans (Vendrasco et al. 2011b). Key exaptations within that helcionelloidan lineage were strong lateral compression as seen in the Stenothecidae and Yochelcionellidae, and posterior shell extensions ('trains') which ultimately formed snorkel-like structures, the latter requiring mantle extension, slit-like indentation, and fusion of the mantle tissue. The development of 'snorkels' probably facilitated the formation of the separate, albeit dorsally fused, shell plates in rostroconchs. The supposed earliest rostroconchs (*Watsonella*) lacked anterior pegma (Runnegar 1996), and Peel (2004) suggested that ribeirioideans and conocardioideans arose from separate helcionelloidan ancestors, but as pointed out by Vendrasco (2012), this scenario would require a stunning amount of morphological convergence. Wagner

TABLE 13.12

Classification of Rostroconchia, an Extinct Class of Molluscs

(Class) **Rostroconchia**

Cambrian (Terreneuvian)–Permian (Lopingian) (530–252 Ma)

Pseudo-bivalves with a single univalved protoconch from which extended two rigidly fixed lateral valves with no functional hinge. The rostrum is prominent.

 (Order) **Ribeiriida** (= Ischyrinoida)

Cambrian (Miaolingian)–Silurian (Wenlock) (530–428 Ma)

Shell layers continuous across the dorsal margin, with an anterior pegma and dominant posterior growth; anterior and posterior median muscles connected.

(Superfamily) **Ribeirioidea** Range as for order. Ribeiriidans with anterior and posterior shell gapes; lacking radial ornament. Families Technophoridae and Ribeiriidae	*Technophorus*, redrawn and modified from Wagner (1997). 
(Superfamily) **Ischyrinioidea** Cambrian (Stage 2)–Ordovician (Upper) (530–444 Ma) Rostroconchs with a dominant anterior growth component resulting in protoconch at centre or posterior on the shell; there are two pegmas and radial ornament. Family Ischyriniidae	*Ischyrinia*, redrawn and modified from Wagner (1997). 

(Order) **Conocardiida**

Cambrian (Miaolingian)–Permian (Lopingian) (501–252 Ma)

Rostroconchs with external and internal ribs, the latter expressed as marginal denticles on the inside edge of the commissure, and with an anterior gape and dorsal clefts.

(Superfamily) **Conocardioidea** Cambrian (Furongian)–Permian (Lopingian) (488–252 Ma) Anteriorly elongate with shell divided into posterior rostrum, median body, and anterior snout; hood absent. Families Bransoniidae and Conocardiidae	*Pseudoconocardium*, redrawn and modified from Wagner (1997). 
	Redstonia, redrawn and modified from Wagner (1997). 
	Oxyprora, redrawn and modified from Mazaev (2015).  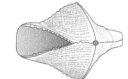
(Superfamily) **Eopterioidea** Cambrian (Miaolingian)–Upper Ordovician (501–446 Ma) Shell posteriorly elongate with anterior or anterior and posterior dorsal clefts; anterior, ventral, and posterior shell gapes continuous; rostrum rudimentary or lacking. Family Eopteriidae	*Eopteria*, redrawn and modified from Billings (1865) and Pojeta and Runnegar (1976). 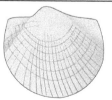

(Continued)

TABLE 13.12 (CONTINUED)

Classification of Rostroconchia, an Extinct Class of Molluscs

(Superfamily) **Hippocardioidea**	
Ordovician (Upper)–Permian (Cisuralian) (456–280 Ma) With one or more hoods around the rostral area; hoods consisting of right and left halves. Families Hippocardiidae and Pseudobigaleaidae	*Hippocardia*, redrawn and modified from Wagner (1997). 

(Order) **Anetshellida**
Permian (Cisuralian)–Permian (Guadalupian) (273–268 Ma)
Shell cap-shaped, with rostrum positioned posteriorly between the apex and the posterior margin of the aperture; rostral structure separated externally from the apex by a transverse septum. There are multiple paired muscles.

(Superfamily) **Anetshelloidea** As for order. Family Anetshellidae	*Anetshella*, redrawn and modified from Mazaev (2012).

(1997) produced a phylogenetic analysis of rostroconchs which suggested that *Watsonella* and *Heraultipegma* were nested within the *Anabarella*-like helcionelloids and were sister to the earliest bivalves, while the earliest rostroconchs were sister to *Pseudomyona* in a second *Anabarella*-like clade. Perhaps even more surprising is the apparent evolution of a limpet form among the rostroconchs – the Permian Anetshellida (Mazaev 2012).

Pojeta and Runnegar (1976) suggested the possibility of deposit feeding in rostroconchs, and Pojeta (1979) suggested that the loss of rostroconch diversity after the Lower Ordovician was due to competition with bivalves.

Runnegar and Pojeta (1974a) and Pojeta and Runnegar (1976) suggested that rostroconchs gave rise to both bivalves and scaphopods, but more recent studies have discounted these relationships. Based on morphological and molecular data respectively, Waller (1998) and Steiner and Dreyer (2003) argued that scaphopods were more closely related to cephalopods and only distantly related to bivalves. Differences in putative pedal musculature between protobranchs (oblique and dorsoventral retractors) and rostroconchs (cyclomyan-like muscle bands) (Figure 13.20) suggest independent lineages.

13.3.3 Ordovician

Following the End-Cambrian extinction, the Ordovician Period (Table 13.13 and Figure 13.21) is known as the Great Ordovician Biodiversification Event (GOBE) (Droser & Finnegan 2003; Webby et al. 2004) and has been considered one of the most significant events in the evolutionary history of marine metazoans. During this period, marine familial diversity tripled; the resulting fauna came to dominate and remained relatively unchanged through the rest of the Paleozoic (Droser et al. 1997). The GOBE followed a series of Cambrian–Ordovician extinction events (Sepkoski 1986; Fortey 1989), which saw the extinction of the majority of the Cambrian fauna and, from the

survivors, the rapid diversification of new animals that were primarily suspension feeding and pelagic (Servais et al. 2008, 2010). The GOBE also saw a marked increase in the stacking or tiering of invertebrate taxa above the sediment/water interface and an increase in bioturbation (Ausich & Bottjer 1982; Bottjer & Ausich 1986; Droser & Bottjer 1989) suggesting increased burrowing activity. It is also surmised that there was an increase in water column food resources and increased competition for suspension-feeding space (Signor & Vermeij 1994). Other palaeoecological changes included the appearance of carbonate hardgrounds (Wilson et al. 1992; Taylor & Wilson 2003) and the transition from trilobite- to brachiopod-dominated communities (Droser & Sheehan 1995). Potential drivers of this faunal turnover include tectonic activity which increased provinciality, as well as a warming climate and elevated CO_2 levels which are thought to have increased nutrient levels (Botting & Muir 2008). The Ordovician had some of the highest sea levels seen in the Paleozoic, and there were multiple transgression events. Prior to the End-Ordovician extinctions, both ecological breadth and morphological disparity were similar to the recent fauna (Bambach et al. 2007; Bush & Bambach 2011), although these increases had been neither global nor instantaneous but happened at different times in different regions. Therefore, it is unlikely there was a single cause but rather multiple geological and ecological factors, and their interactions were probably responsible for the diversification (Droser & Finnegan 2003). Just as the GOBE followed the Cambrian–Ordovician extinction event, it ended in another one. The End-Ordovician extinction event was the second largest extinction in the history of metazoans, involving as much as 60% of all marine species (Sepkoski 1981; Sheehan 2001). Global cooling, leading to the glaciation of Gondwana followed by dropping sea level, were probably driving agents for the event.

The Ordovician Period includes the first appearance of scaphopods and impressive diversifications in other molluscan classes, as outlined below.

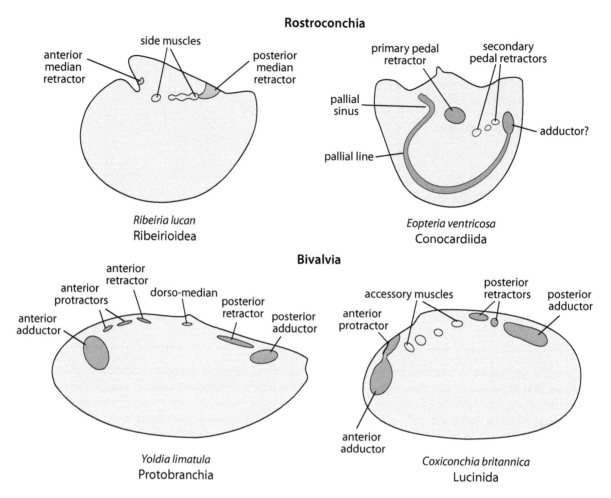

FIGURE 13.20 Comparison of the muscle scar patterns in rostroconchs with two bivalve taxa. Modified from Polechová, M., *Est. J. Earth Sci.*, 64, 84–90, 2015.

TABLE 13.13

Ordovician Chronostratigraphy

System/Period	Series/Epoch	Stage/Age	Beginning
Ordovician	Upper	Hirnantian	445 Ma
		Katian	453 Ma
		Sandbian	458 Ma
	Middle	Darriwilian	467 Ma
		Dapingian	470 Ma
	Lower	Floian	478 Ma
		Tremadocian	485 Ma

Based on International Chronostratigraphy Chart (2018-08) [www.stratigraphy.org/ICSchart/ChronostratChart2018-08.pdf]

13.3.3.1 Taxa

13.3.3.1.1 Gastropoda

Gastropod evolution in the Ordovician shows phases of relative stability separated by periods with high levels of turnover, resulting in diversification (Wagner 1995; Frýda & Rohr 2004). The gastropod groups of the Ordovician are divisible into three major taxa – the Paragastropoda, Eogastropoda, and Vetigastropoda. Members of the paragastropods include the Onychochilidae and Clisospiridae. They have been considered either hyperstrophic or sinisterly coiled, with a smooth, conical protoconch. The eogastropods include the Euomphaliformii and Macluritoidea. They are characterised by an openly coiled protoconch with a bulbous, initial portion. The open whorl protoconch is shared with the Perunelomorpha which also first appears in the Lower Ordovician (Frýda 2012) and has been considered an early caenogastropod by Frýda and Bandel (1997). The Perunelomorpha differ from the planispiral or low trochispiral shells of the euomphaloideans and macluritoideans[14] in having trochispirally rather than planispirally coiled shells. A third group known as the Cyrtoneritimorpha[15] also had a distinctive open coiled protoconch similar to those of the Euomphaloidea, Macluritoidea, Orthonychioidea, and Perunelomorpha. In the cyrtoneritimorph taxa, the initial teleoconch is relatively straight giving the protoconch and initial teleoconch a fishhook-like

---

[14] Adult *Maclurites* apparently lived more like an epifaunal suspension-feeding bivalve. The apertural morphology underwent significant morphological change probably associated with a switch from grazing when young to suspension feeding (Novack-Gottshall & Burton 2014).

[15] We combine this group with the Orthonychioidea in the Appendix.

Ordovician (467 Ma)

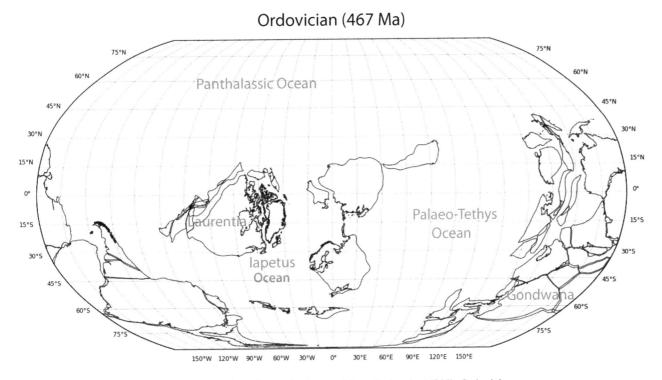

FIGURE 13.21   Palaeogeographic reconstruction of tectonic plate positions during the Middle Ordovician.

appearance (Frýda et al. 2008a). See Appendix and Chapter 18 for further details of stem and crown groups.

Groups such as the euomphaloideans, macluritoideans, and vetigastropods showed the greatest diversity during the first part of the Ordovician. After the drop in both extinction and origination rates in the Darriwilian (467–458 Ma) the euomphaloideans and macluritoideans waned, while the bellerophontians, and especially the vetigastropods (both with and without a selenizone), increased in richness, while the Mimospirina changed little over the period (Frýda & Rohr 2004).

The Ordovician also includes the diversification of the Archinacelloidea – a group of limpet-like shells with a diverse array of muscle patterns, including horseshoe-shaped muscle bands with and without apical muscles, possible buccal muscles, and asymmetric apical attachment areas (Peel & Horný 1999). The Archinacelloidea have been suggested to be the earliest patellogastropods[16] (Yochelson 1988), but there is little evidence for this view except for their limpet morphology. Instead, the diversity of muscle scar patterns within the group suggests that they probably comprise at least three distinct groups – one containing *Archinacellina* and *Archinacellopsis* with cyclomyan-like musculature (horseshoe-shaped muscles with putative buccal muscles), another including *Archinacella* and *Barrandicella* with apical muscles and with or without horseshoe-shaped muscles, and the third containing *Archaeopraga* with its parallel strap-like lateral muscles. Thus, this group appears to include at least three lineages which have independently undergone secondary shell flattening.

After their appearance in the Cambrian, bellerophontian generic diversity remained low until the Middle Ordovician and then reached its zenith in the Upper Ordovician. A move into a wider range of substrata accompanied this diversification, and increased morphological diversification, including secondary flattening (e.g., *Pterotheca*) (Peel 1977; Wahlman 1992).

13.3.3.1.2 Bivalvia

It is not known which of the Cambrian bivalve taxa survived the End-Cambrian extinction as there is no evidence of the known lineages having done so. Ordovician originations of new lineages were initially low with those in the Tremadocian having included a limited number of protobranch and autobranch taxa. The protobranchs appeared first, represented by two indeterminate praenuculids, while autobranchs included two cyrtodontoidean pteriomorphians (Cyrtodonta, Pharcidoconcha) and the heteroconch *Babinka* (Cope 2004). The Floian saw diversification in the nuculoideans and heteroconchs and the early appearances of the solemyidans, trigoniidans, and afghanodesmatinans. While heteroconch diversity fell in the Upper Ordovician, the remaining taxa showed high origination rates coupled with low extinction rates until the Himantian, when the heterodont anomalodesmatans first appeared (Cope 2004).

13.3.3.1.3 Rostroconchia

In contrast to the bivalves, the rostroconchs showed their greatest diversification during the earliest Ordovician (Tremadocian). Most ribeiriidan lineages were extinct by the beginning of the Floian and only a handful of genera (e.g., *Pinnocaris, Technophorus,* and possibly *Jinonicella*) continued into the Silurian. The conocardioideans remained at low diversity throughout the Ordovician (Cope 2004).

---

[16] See Chapter 19 for discussion of patellogastropod origins.

Silurian *Carnicoleus gazdzickii*

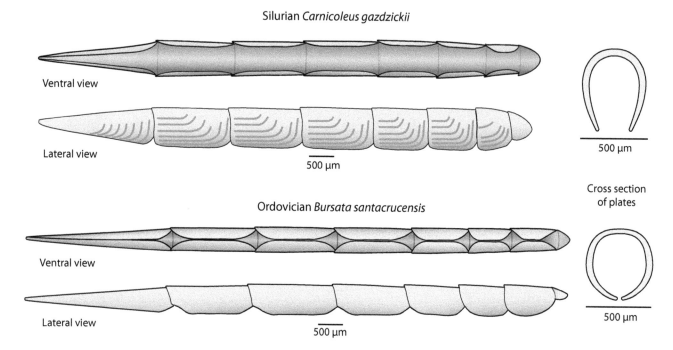

FIGURE 13.22   Reconstruction of shell plates of early Paleozoic solenocaridid Septemchitonina. Redrawn and modified from Dzik, *Acta Palaeont. Pol.*, 39, 247–313, 1994.

13.3.3.1.4   Cephalopoda

The Ordovician Period included the largest radiation of nau-tiliforms in the fossil record, especially during the Floian (Frey et al. 2004). This radiation included the greatest diversifica-tion in morphology ever seen in cephalopods (Teichert 1988), much of which has been linked to buoyancy control (Crick 1988) (see Chapter 17). While globally distributed in a diver-sity of habitats, nautiliforms were most abundant and reached their maximum diversity and size on shallow marine carbonate platforms deposited at low latitudes in warm climatic condi-tions (Flower 1976b). The most important Ordovician group in diversity and abundance was the Orthoceratida. They are considered the ancestors of the Bactritida, from which the ammonites and coleoids probably arose. Another Ordovician group, the Endoceratida, included large nautiliforms such as *Endoceras* and *Cameroceras* (>10 m long), although the majority of Ordovician nautiliforms were much smaller (<1 m). The first major extinction of nautiliforms occurred at the end of the Ordovician Period (Himantian), resulting in the extinction of numerous taxa including the Ellesmeroceratida which origi-nated in the Cambrian (see Chapter 17 for further discussion).

13.3.3.1.5   Polyplacophora

Following the first occurrence of *Matthevia* and *Listrochiton* in the Cambrian, the Ordovician saw further diversification of the paleoloricates (Cherns 2004; Cherns et al. 2004). In the Lower Ordovician, this diversification was primarily in the Mattheviidae and Preacanthochitonidae, with additional orig-inations in the Gotlandochitonidae and Helminthochitonidae (Hoare 2000). Originations and extinction rates were low for the remaining Ordovician with additions to the septemchi-tonids and alastegiids. Only three polyplacophoran lineages

survived the End-Ordovician extinction – the mattheviid gen-era *Chelodes*, *Gotlandochiton*, and *Helminthochiton*.

The Ordovician also saw the first occurrence of possible footless chitons (Figure 13.22). Based on laterally compressed chiton valves from the Ordovician of Poland and closed valves in the Silurian, Dzik (1994) hypothesised that there was a trend in foot reduction in the septemchitonid paleoloricates. Sigwart and Sutton (2007a) provided more evidence for the ori-gin of this morphology after examining multiple body fossils of '*Helminthochiton*' *thraivensis* from the Upper Ordovician of Scotland. They concluded that the ventral spiculate girdle of this paleoloricate was so expansive there was not enough space for a typically polyplacophoran foot; Sutton and Sigwart (2012) reached a similar conclusion. A paleoloricate genus *Echinochiton*, with large, hollow spines radiating outward from the eight shell plates, also first appeared in the Ordovician.

Lower Ordovician polyplacophoran radiations were cen-tred on the low-latitude Laurentian margin (Cherns et al. 2004), while *Chelodes* was associated with the Gondwanan margin (Runnegar et al. 1979). Palaeoenvironments include stromatolitic, shallow marine carbonates and dolomites (Runnegar et al. 1979; Stinchcomb & Darrough 1995).

13.3.3.1.6   Scaphopoda

The first occurrence of scaphopods in the fossil record is controversial, due largely to their simple and conservative morphology (Yochelson 2004). Proposed first occurrences include the Ordovician (Pojeta & Runnegar 1979), Silurian (Rohr et al. 2006), Devonian (Haas 1972), and even as late as the Mississippian (Yochelson 1978). Peel (2006) argued that scaphopods were probably derived from a helcionelloidan or conocardioidean rostroconch lineage in the late Paleozoic.

Given their occurrence in soft sediments that permit burrowing, their rarity in the fossil record is surprising. The scaphopod fossil record improves after the Mississippian, but it is only in Cenozoic deposits that occasional concentrations are found. The early Paleozoic records represent the Dentaliida, the first occurrence of the Gadilida being in the Permian (Artinskian). Here we regard the report of Pojeta and Runnegar (1979) of *Rhytiodentalium* from the Middle Ordovician as representing the first record of a scaphopod. We base this on the overall morphological similarity and range of variation of the Ordovician, Silurian, and Devonian specimens with the undisputed Carboniferous taxa and the origination times of putative sister taxa (gastropods and cephalopods) (Sigwart & Lindberg 2015). This scenario substantially reduces the need for 'ghost' taxa spanning 137 Ma (Norell 1992; Cavin & Forey 2007). Scaphopod shell microstructure has been observed in Devonian specimens (Haas 1972). See Chapter 16 and Section 13.3.2.2.3.1.2 for discussion of scaphopodisation.

13.3.3.1.7   *Tentaculita (= 'Tentaculites')*

Tentaculitans are small (1–30 mm in length), cone-shaped, ringed fossils which first appear in the Lower Ordovician. They reached their greatest diversity and distribution during the Devonian but did not survive the Upper Devonian mass extinction. They were first noted in the late 1700s and formally described in 1820 as crinoid fragments (Bouček 1964). Subsequently, they have been thought to be related to Annelida (Bouček 1964); Bryozoa (Dzik 1993); Phoronida (Vinn & Mutvei 2005); Brachiopoda (Towe 1978); Mollusca – pteropods (Shrock & Twenhofel 1953) or cephalopods (Blind 1969); Cnidaria (Herringshaw et al. 2007); and Hyolitha (Lyashenko 1955). Ljashenko (1957), Lardeux (1969), and Drapatz (2010) suggested that they are an extinct class of Mollusca, but molluscan affinities are unlikely, given their shell morphology and microstructure.

The conical shells are divisible into two regions. The proximal or juvenile region includes the earliest part of the shell and typically differs in sculpture from the distal or adult region of the shell. The proximal section begins with the initial chamber or bulb, which has two distinctive morphologies. Two main groups are recognised (Table 13.14). In the Dacryoconarida an initial bulb-shaped chamber transforms into the proximal section with a slight constriction, while in the Chonioconarida an open hollow tube extends posteriorly from the bulb; in some reconstructions, an apical spine closes and extends from this tube. Shell sculpture includes concentric growth lines, longitudinal ribs, and rings, separately or in combination. Septa are present in the juvenile region of some taxa.

The shell structure of tentaculitans comprises two main layers – an outer secreted by accretion at the aperture and a secondary, inner layer which lines the chambers and forms the septa. Both layers contain multiple sub-layers, and the outer shell layer is penetrated by numerous pores. The inner layer is more compact and non-porous. TEM studies of fractured shells have revealed a calcitic microstructure of ridges and grooves and a cross-bladed fabric otherwise known only from articulate brachiopods (Towe 1978). Vinn and Mutvei

(2009) suggested that tentaculitans, along with problematic tubeworm groups such as the cornulitids, microconchids, trypanoporids, *Anticalyptraea*, and *Tymbochoos*, form a monophyletic group based on their homologous shell microstructure. More recently, Vinn and Zatoń (2012), based on a phenetic analysis, concluded that the tentaculitans were more closely related to the Brachiozoa.

Tentaculite habitats have been reconstructed as both benthic and pelagic. Generally, the heavier shelled taxa (Chonioconarida) are thought to have been benthic but gave rise to the possibly pelagic Dacryoconarida by becoming miniaturised, reducing the shell layers and losing septa. The geological occurrence of these two groups also supports a pelagic habit for this latter group. Organic remains of tentaculitans have been recovered from samples (Filipiak & Jarzynka 2009; Devaere et al. 2014) and may prove important in resolving their phylogenetic relationships. We concur with Yochelson (2000) and others who have argued against their inclusion in the Mollusca and suspect that their affinities are with other shell-forming lophotrochozoans such as the brachiozoans.

13.3.4   SILURIAN

During the Silurian (Table 13.15 and Figure 13.23) the climate stabilised and warmed after the chaotic time of the Himantian glaciation, and the melting glaciers produced a substantial rise in sea level. The transgressive oceans of the Silurian combined with the relatively flat palaeocontinents produced a rich diversity of environmental settings. Coral reefs appeared, as did several major evolutionary events in fish, including the radiation of jawless fish and the diversification of freshwater and jawed fish. It is also during the Silurian that the first evidence of a major diversification of terrestrial life is preserved, including vascular plants and arthropods. These features suggest increased productivity associated with the invasion of terrestrial habitats by plants, thereby making possible the first terrestrially derived eutrophication of the near-shore marine realm. The appearance of armoured fish and the existence of large arthropod predators such as the eurypterids during this period are an indication of increased predation pressure (Signor & Brett 1984; Vermeij 1987).

Molluscs responded to these changing factors in the Silurian in multiple ways but overall were much more stable than in earlier periods. In the cephalopods, the Endoceratida, which contained the giant *Endoceras* and *Cameroceras*, failed to survive the Silurian while other nautiliform orders such as the Tarphyceratida, Discosorida, Oncoceratida, and Orthoceratida slowly decreased in size and diversity although they remained one of the dominant predatory animals in the oceans despite increased competition from jawed fish (Placodermi and Acanthodii) which appeared in the Llandovery (Silurian).

13.3.4.1   Taxa

13.3.4.1.1   *Bivalvia*

Silurian bivalves exhibit little provincialism, with their distributions often cosmopolitan (Cope & Kříž 2013). This pattern was thought to be due to their relatively long pelagic larval life

TABLE 13.14

Classification of Tentaculita – a Group Sometimes Included in Mollusca

Brachiozoa

Brachiozoa includes the brachiopods and phoronids, and the tentaculitans were placed there based on shell structure and morphology, especially the shared micro-lamellar layers, cross-bladed fabric, and pseudopunctae (Vinn & Zatoń 2012).

(Class) **Tentaculita**

(= Cricoconarida, 'Tentaculites')

Cambrian (Series 2)–Jurassic (Upper) (513–235 Ma)

Conical fossils with simple, circular apertures; the apex is pointed to bulbous. External sculpture typically composed of strong annulations; conch calcitic with three layers and brachiopod-like microstructure. The 'tubeworms' Microconchida and Cornulitida are sometimes included in Tentaculita.

(Order) **Chonioconarida** Silurian (Ludlow)–Devonian (Upper) (426–371 Ma) Thin conchs with several septa closing off large, cone-shaped embryonic chamber. Includes the genus *Tentaculites*.	*Homoctenus*, redrawn and modified from Wei et al. (2012) and Bouček (1964). 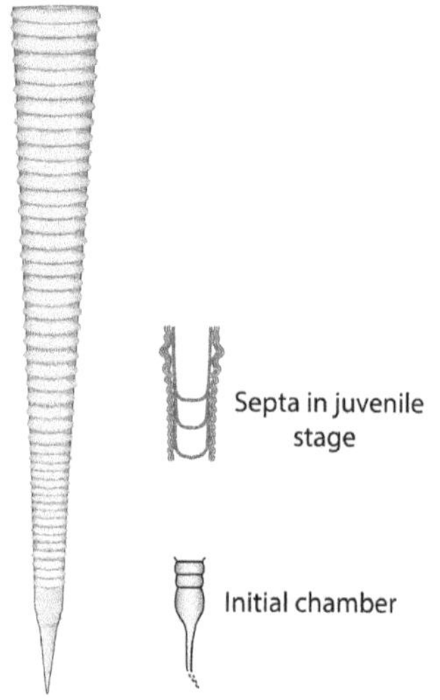 Septa in juvenile stage Initial chamber
(Order) **Dacryoconarida** Silurian (Llandovery)–Mississippian (Lower) (444–354 Ma) Thin conchs without septa; embryonic chamber bulbous; some taxa lack strong external sculpture.	*Nowakia*, redrawn and modified from Bouček (1964). 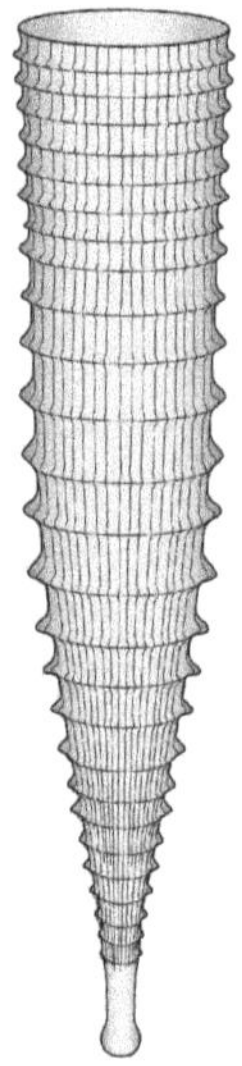

and the relatively small distances between the basins, islands, and continents (Kříž 2011). Silurian bivalves are also characterised by the evolution of numerous new free-burrowing and epibyssate forms from Ordovician infaunal byssate ones. The percentage of non-burrowing attached genera increased rapidly during the Silurian from 6.7% in the Llandovery to 10.6% in the Wenlock, 19.4% in the Ludlow and 24.5% in the Pridoli (Kříž 1984). The Silurian also includes the origin of *Archanodon*, the oldest known genus of freshwater bivalves (Chamberlain et al. 2002), and the discovery of the bivalves

at the oldest known methane seep which hosted a metazoan fauna (Jakubowicz et al. 2017) (see also Chapter 15).

13.3.4.1.2 Gastropoda

Silurian gastropod evolution was not as subdued as in cephalopods and bivalves, and the period was one of increasing diversity in most gastropod clades (Frýda et al. 2008a). While the majority of Silurian gastropods represent bellerophontians, euomphaliforms, murchisonioideans, and porcellioideans, other groups such as the loxonematoideans, perunelomorphs, and cyrtoneritimorphs also diversified (Frýda et al. 2008a), some possibly representing early crown lineages. For example, some Silurian gastropod genera (*Bucanospira*, *Codonocheilus*, *Craspedostoma*, *Spirina*, *Temnospira*, *Auriptygma*, *Kjerulfonema*, *Morania* and *Stylonema*) include some possible early caenogastropods (Sepkoski et al. 2002). However, the characteristic protoconch II is only present in *Auriptygma* (J. Frýda, pers. comm., 2018). While close-coiled vetigastropod-like protoconchs first occur in the Silurian, definitive crown caenogastropod protoconch morphology is not found until the Devonian (see Section 13.3.5.1.1) (Frýda et al. 2008a).

By the Pridoli (Silurian) two stem pleurotomarioid clades (Eotomarioidea and Trochonematoidea) were present, containing the bulk of vetigastropod morphologies and

representing the first major diversification of slit-bearing gastropods (Wagner 1999). Overall spire height also increased, especially in the loxonematoidean and subulitoidean groups (Frýda et al. 2008a). The earliest unequivocal fossil record for so-called cyrtoneritimorphs is latest Silurian–Devonian (Frýda et al. 2008a), but the putative relationship of this group to later stem and crown neritimorphs (Nützel et al. 2007; Frýda et al. 2009) is not recognised here (see Section 13.3.5.1.1, Chapter 18, and Appendix).

While the actual origin of crown caenogastropod, heterobranch, and neritimorph groups probably occurred later, there is little doubt that the Silurian set the stage for their appearance. The work of Bandel, Frýda, and Nützel has emphasised the evolution of egg size and larval characteristics as determined from protoconch morphology, and the origin of these three taxa also corresponds to the apparent origin of planktotrophic larvae (Jablonski & Lutz 1983; Ponder & Lindberg 1997; Hickman 1999; Frýda 2012; Nützel 2014). Based on the developmental timing of cell fates in gastropod embryos, Lindberg and Guralnick (2003) proposed that nutrient increases in marine systems by a diversifying terrestrial flora, along with changing predation pressure of the Silurian, were possible drivers of the reorganisation of the gastropod developmental pathway that resulted in the evolution of feeding larvae. These are some of the same factors suggested by Vermeij (1995) to correlate with other Phanerozoic macroevolutionary changes.

13.3.4.1.3 Monoplacophora

A new monoplacophoran morphology appeared in the Silurian. Since their first appearance in the latest Cambrian the monoplacophoran tryblidioidians (*Pilina*) had been

---

TABLE 13.15
Silurian Chronostratigraphy

System/Period	Series/Epoch	Stage/Age	Beginning
Silurian	Pridoli		423 Ma
	Ludlow	Ludfordian	426 Ma
		Gorstian	427 Ma
	Wenlock	Homerian	430 Ma
		Sheinwoodian	433 Ma
	Llandovery	Telychian	439 Ma
		Aeronian	441 Ma
		Rhuddanian	444 Ma

Based on International Chronostratigraphy Chart (2018-08) [www.stratigraphy.org/ICSchart/ChronostratChart2018-08.pdf]

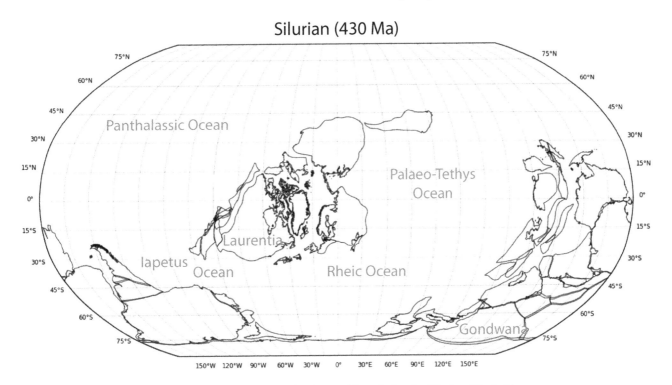

Silurian (430 Ma)

FIGURE 13.23 Palaeogeographic reconstruction of tectonic plate positions during the Silurian (Wenlock).

relatively thin and small, but in the Silurian of Götland, Sweden, they achieved what became the iconic monoplacophoran morphology. Described by Lindström (1884), these large shells (around 50 mm long) had strong growth phases present in the shell and on the outer surface. These relatively massive shells contrasted markedly with early monoplacophorans, but their muscle scars were identical to Cambrian *Pilina,* having paired oblique and dorsoventral retractors and a complex set of buccal muscles.

The Silurian also produced the Drahomiridae with seven sets of dorsal paired muscle scars and several large and small buccal muscle scars. Both adults and juveniles have been reported *in situ* on fragments of orthoconic nautiliform shells (Horný 2005).

13.3.4.1.4 Aplacophora and Polyplacophora

The Silurian represents the recovery of the 'placophorans' after the End-Ordovician extinction event as well as the first appearance of possible transitional fossils that bridge class-level morphology within molluscs. The first putative aplacophoran fossil, *Acaenoplax hayae* (Figure 13.24), was described from the Herefordshire Lagerstätte of England (425 Ma) (Sutton et al. 2001a). This vermiform fossil was interpreted as an aplacophoran with 18 rows of spicules interspersed among seven dorsal plates and a single posterior ventral plate. Thus, this taxon exhibits characters of both polyplacophoran (dorsal shell plates and spicules) and aplacophoran (spicules and a vermiform body without a foot). Steiner and Salvini-Plawen (2001) pointed out morphological inconsistencies with extant molluscan morphology and suggested *Acaenoplax hayae* was more likely a semi-sessile annelid and noted an overall similarity with some living tube-dwelling annelids. Sutton et al. (2001b) responded, noting that while the ventral surface of *A. hayae* might appear somewhat annelid-like, the aragonitic shell plates, spicules, serial rather than segmented organisation, and the posterior cavity indicated molluscan and not annelid affinities. In a subsequent paper, Sutton et al. (2004) documented the molluscan affinities of *A. hayae.* An additional putative molluscan character is the semi-circular curvature of many of the illustrated fossil remains. This curvature in Solenogastres, Polyplacophora, and Caudofoveata is caused because the longitudinal muscles contract to bend the animal into a semi-circle at death (Scheltema 1992). In chitons, these same muscles enable them to roll up into a ball when dislodged. While annelid worms also have longitudinal muscles,

they do not take on this characteristic form in death (Briggs & Kear 1993).

While the mosaic of 'placophoran' characters in *Acaenoplax* is confusing from a neontological perspective, they are fascinating from an evolutionary one. Further study of the posterior 'mantle cavity' has increased the uniqueness of this animal and provides additional characters for consideration. Dean et al. (2015) reconstructed the posterior mantle cavity of two new specimens of *A. hayae.* Using tomographic analysis of sequential thin sections through the fossils, they reconstructed a mantle cavity unlike that found in any known molluscan taxon (see also Steiner & Salvini-Plawen 2001). Based on Sutton et al. (2001a), the original description of the *Acaenoplax* mantle cavity was unusual because it is underlain by a ventral plate which, with the dorsal plate, encloses the posterior portion of the animal (Dean et al. 2015). These new reconstructions show the putative mantle cavity surrounded by three bilaterally symmetrical structures, which include three pairs of papillate lobes, three pairs of subspheroidal projections above each lobe, and a medial pair of lobes without papillate surfaces. Lastly, there is a single dorsomedial lobe above the central lobes. These lobes extend beyond the edges of the dorsal and ventral plates, and none of these structures can be satisfactorily homologised with those in the mantle cavities of any 'placophoran' or other molluscan group. Thus the molluscan affinities of this strange armoured Silurian 'worm' remain uncertain (e.g., Vinther & Nielsen 2005; Todt et al. 2008a).

Sutton et al. (2012) described a second 'placophoran', *Kulindroplax perissokomos* (Figure 13.25), from the Wenlock Series Lagerstätte fauna of England (also about 425 Ma). *K. perissokomos* has seven valves which resemble those of paleoloricates, and, as in that group, the valves do not articulate on each other, and the head valve is the smallest. Densely packed spicules are present along the broad lateral surface of the body below the plates. Neither a foot nor radula appears to be present; however, there appears to be a posterior mantle cavity extending under the final two valves with four 'gill elements', which are neither paired nor resemble ctenidia or caudofoveate respiratory mantle cavity folds. Sutton et al. (2012) consider *K. perissokomos* to be a stem aplacophoran; however, there is little to support this allocation other than the worm-like body plan and lack of a foot, although both Ordovician and Silurian stem polyplacophorans (paleoloricates) are also rather worm-like but not footless (Dzik 1994; Sutton & Sigwart 2012). Whether *K.*

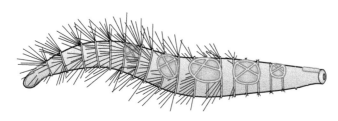

FIGURE 13.24 *Acaenoplax hayae,* which has been suggested to be a possible early aplacophoran. Redrawn and modified from Sutton, M.D. et al., *Nature,* 414, 602, 2001a.

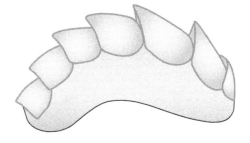

FIGURE 13.25 *Kulindroplax perissokomos,* an apparently footless 'placophoran' from the Silurian of England. Drawn from reconstruction in Sutton, M.D. et al., *Nature,* 490, 94–97, 2012.

perissokomos is an aplacophoran or a highly divergent polyplacophoran is uncertain as the boundaries between these two groups appear to be blurred. More of these remarkable finds are needed to resolve the distinction between aplacophorans and possible 'footless' chitons.

Silurian 'placophorans' were also documented by Cherns et al. (2004), who described a diverse Silurian paleoloricate assemblage from Götland, Sweden, which probably co-occurred with *Acaenoplax*[17] in carbonate shelf environments in shallow, near-shore seas. Cherns (2004) also provided the first cladistic analyses of paleoloricates. Sigwart and Sutton (2007a) also included numerous paleoloricates in their cladistic analysis of the relationships of the Aculifera and related

---

[17] Cherns et al. (2004) introduced the family name Heloplacidae to include four Silurian plated genera including *Acaenoplax.*

TABLE 13.16
Devonian Chronostratigraphy

System/Period	Series/Epoch	Stage/Age	Beginning
Devonian	Upper	Famennian	372 Ma
		Frasnian	383 Ma
	Middle	Givetian	388 Ma
		Eifelian	393 Ma
	Lower	Emsian	408 Ma
		Pragian	411 Ma
		Lochkovian	419 Ma

Based on International Chronostratigraphy Chart (2018-08) [www.stratigraphy.org/ICSchart/ChronostratChart2018-08.pdf]

taxa, but despite these analyses having multiple shared taxa, their results are quite disparate.

13.3.5 DEVONIAN

In the Devonian (Table 13.16 and Figure 13.26) increases in potential molluscan predators continued with the appearance of both ray-finned (Actinopterygii) and lobe-finned bony fish (Sarcopterygii) which joined the already ecologically diverse placoderms (see Chapter 9). In the Devonian we see additional freshwater bivalves, the first stem neritimorphs, unequivocal caenogastropods, and the first heterobranchs, thus completing the suite of major crown gastropod groups (Frýda et al. 2008a). The Upper Devonian was marked by the Frasnian–Famennian biodiversity crisis which perhaps resulted in the extinction of 31% of Devonian bivalve genera (Bretsky 1973).

It also includes the first freshwater molluscan faunas represented by viviparid-like gastropods and *Modiomorpha*-like clams in the northern hemisphere (Solem & Yochelson 1979). The earliest ammonites also appeared, which, like all coleoids, are thought to have shared a common ancestor with a Devonian Bactritida lineage (see Chapter 17).

13.3.5.1 Taxa
13.3.5.1.1 Gastropoda

Based on protoconch morphology, the earliest undoubted caenogastropods are subulitids which appeared in the Lower Devonian (Frýda et al. 2008a). Earlier reports based on teleoconch morphology (high spired and fusiform, with and without siphonal canals) are suspected to relate to convergent morphologies in non-caenogastropod taxa (e.g., Loxonematoidea and Subulitoidea) (Frýda 1999b; Nützel et al. 2000; Wagner

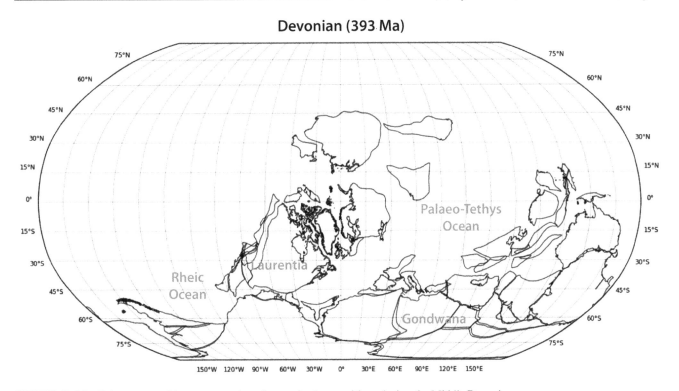

Devonian (393 Ma)

FIGURE 13.26 Palaeogeographic reconstruction of tectonic plate positions during the Middle Devonian.

2002). The origin of heterobranchs was also marked by a heterostrophic protoconch (see Chapters 3 and 20). The earliest stem heterobranchs include the Kuskokwimiidae (Frýda & Blodgett 2001) and Soleniscidae (Bandel & Heidelberger 2002).

The origin of Neritimorpha is more problematic as the earliest distinctive neritimorph protoconchs date only from the Triassic (Bandel & Frýda 1999; Bandel 2000), although with both caenogastropod and heterobranch lineages in Devonian strata, stem neritimorphs must have also been present by this time (given the phylogenetic relationships of these three clades), but their identity remains obscure. Two candidate groups have been identified: the Cycloneritimorpha with a caenogastropod-like protoconch and the Cyrtoneritimorpha with an open fishhook-shaped protoconch (Bandel & Frýda 1999; Frýda 1999a; Frýda et al. 2008a). The Cycloneritimorpha is now considered a synonym of Neritimorpha (see Appendix) and the open whorled protoconch of the older (Ordovician) Cyrtoneritimorpha we treat as a synonym of Orthonychioidea, considered to be an extinct, independent lineage of early gastropods (Frýda et al. 2008b) and which we include in the Eogastropoda (see also Chapter 18).

The platyceratids were allied with cycloneritidans by Bandel (1992). *Platyceras* and its allies first appear in the Silurian of North America, Europe, and China. These limpet-like gastropods have been found attached to the aboral cup or calyx of crinoids where they probably fed on the faecal material of the host. If this reconstruction is correct it is one of the earliest known occurrences of coprophagy in molluscs (Frankenberg & Smith 1967). They appear not to have been obligate coprophages as some species are over 10 cm in length, too large to be epizoic on the co-occurring crinoids and therefore must have had alternative feeding habitats (Bowsher 1955; Morris & Felton 1993; Baumiller & Gahn 2002). Based on protoconch morphology Frýda et al. (2009) suggested that Paleozoic platyceratids were a diphyletic group and the 'platyceratids' with tightly coiled protoconchs (that he included in the 'Cycloneritimorpha') may have given rise to the neritimorphs. We have (somewhat tentatively) included the platyceratoideans in the Eogastropoda in the Appendix.

13.3.5.1.2 Polyplacophora

The polyplacophorans continued their diversification in the Devonian with the appearance of the Multiplacophorida, and this coincided with the first appearance of the Neoloricata. Whereas most chitons have only eight shell plates, the multiplacophoridans had as many as 17, which appear to be formed by sagittal divisions of the original intermediate plates. This extinct stem group of Neoloricata had the shell plates surrounded by a border of spine-like plates. They occurred during the Devonian–Permian and have been reported from North America and Europe. They were first described in the late 1800s, and the partially articulated and disarticulate valves were thought to represent barnacle plates (Hall & Clarke 1888) and as a result were named accordingly (e.g., *Strobilepis*, *Protobalanus*, and *Hercolepas)*, until Name (1926) acquired articulated material of *Protobalanus* and

concluded they were not barnacles. He did, however, reject the Polyplacophora as a placement because the specimen had only seven plates and because of the presence of plate spines. If the intermediate median plates of *Protobalanus* were independent and separated the two lateral sets of intermediate plates, the plate count for *Protobalanus* is 17. Hoare and Mapes (1995) suggested that they were related to chitons, but their partially articulated specimens did not allow for accurate reconstruction of the animal, and their illustrations of the shell plates demonstrated affinities to Neoloricata. Moreover, the large spines surrounding the body appeared to be derived from plate precursors and not girdle spines as they had the same shell morphology as the plates and were hollow and riddled with pores similar to the pores that innervate the aesthetes on the surface of the dorsal plates. Vendrasco et al. (2004) reported the first articulated multiplacophoridan and established the arrangement of the plates as well as providing a cladistic analysis and systematic treatment of the group. It appears that this extinct group of chitons is a branch of the Neoloricata that experimented with plate fission and the production of large marginal spines. The plate fission seems analogous to the division of intermediate valves seen in the living *Schizoplax brandtii* (Kaas & Van Belle 1985c). Also, accompanying the fragmentation of the plates was the appearance of the articulamentum shell layer. This layer provides the articulating surfaces between the valves in neoloricates. Similar modified shell plate spines are also found in the eight-plate Ordovician paleoloricate *Echinochiton* (Pojeta et al. 2003). Vendrasco et al. (2004) provide an excellent review of this Devonian to Permian group of chitons.

13.3.5.1.3 Bivalvia

The diversification begun in the Silurian continued into the Lower Devonian (Babin 2000). Kříž (1979) estimated a 59% increase over Silurian generic diversity; some of this origination occurred in early crown taxa giving the bivalve fauna a more modern aspect. Palaeotaxodonts, pteriomorphians, and anomalodesmatans were especially abundant and diverse during the Devonian. The increase in diversity experienced a downturn in the Middle Devonian when extinction rates exceeded origination rates (Sepkoski et al. 2002). The Devonian also saw global dispersal of the earlier, more endemic bivalve faunas, resulting in increased cosmopolitanism (Babin 2000). For example, Rode (2004), in an analysis of *Leptodesma* (Pterineidae), concluded that dispersal in the Middle and Upper Devonian was more likely to be responsible for speciation in this group than vicariance – a pattern also seen in Devonian trilobites and phyllocarid crustaceans. This increase in taxon ranges may have also had a role in ameliorating the impacts of the Upper Devonian biodiversity crisis (Rode 2004).

Bivalve aggregations at methane seeps, first reported in the latest Silurian (Jakubowicz et al. 2017) continued into the Devonian with the Modiomorphidae (Cardiata) being the most abundant taxon. In the Devonian, modiomorphid taxa (e.g., *Ataviaconcha*) formed large aggregations similar to those formed by living cold-seep and hydrothermal vent bivalves (Hryniewicz et al. 2017); also of note was the observation that the shell morphologies (relatively large elongated shells with allometric growth) of these earliest seep taxa appear convergent with those of other unrelated chemosynthetic bivalve taxa that subsequently came to inhabit these unique environments. Modiomorphid taxa are also unique in combining nacre and crossed lamellar shell microstructures in the same shell, a combination not found in living bivalve taxa (Carter & Michael 1978).

13.3.5.1.4 *Cephalopoda*

The Devonian saw the origination of the predominately straight-shelled Bactritidae and the coiled ammonites and the migration of these latter cephalopods into the water column during the Devonian Nekton Revolution (Klug et al. 2010, 2015b). This event followed the divergence of stem coleoids from nautiliforms and also mirrored the increase in fish diversity during the same period (Young et al. 1998). The early ammonites were loosely coiled (e.g., *Metabactrites*), but the Devonian history of the group was marked by an increase in coiling, which included both the juvenile and adult shells (Klug et al. 2015b). Increased coiling parameters were also accompanied by an increase in sutural complexity (Ubukata et al. 2014), as well as more variation in internal characters. For example, the Devonian clymeniidan ammonites had a dorsal siphuncle, similar to nautiliforms, rather than the characteristic ventral siphuncle as in other ammonites. De Baets et al. (2012) showed that the size of the ammonite embryonic shells decreased during the Devonian, suggesting a smaller egg and hatching size. Combined with the concurrent increase in adult shell size, De Baets et al. suggested this represented a change in life history strategy in Devonian ammonites; a switch from a K reproductive strategy (a few large eggs) to an R strategy (numerous smaller eggs), the latter requiring less maternal investment and the possibility of earlier feeding by juveniles or paralarvae (see Chapter 8).

Nautiliforms began a precipitous decline during the Middle Devonian, perhaps due to competition with the recently evolved ammonites and predation by durophagous[18] fish. Also, like the ammonites, nautiliform shells became increasingly more tightly coiled. Signor and Brett (1984) documented a 15% reduction in smooth and finely sculptured nautiliform taxa whereas moderate to strong sculpture increased from 8% to 20%, and very strong sculpture went from nonexistent to 11%. Similar changes were also documented in the ammonites (Ward 1981). Lastly, some Devonian nautiliform taxa, which at the beginning of the period appeared to have had relatively low energy buoyancy regulation, became extinct, while subsequent cephalopod morphotypes displayed more energy intensive buoyancy regulation (Kröger 2008a). The Devonian cephalopod record marks an important transition for the group. The appearance of the ammonites, followed by the radiation of durophagous predators, dramatically changed the selective environment for the nautiliforms. Their generic diversity was reduced by about 70% during the End-Devonian extinction event (Sepkoski et al. 2002), but a substantial recovery occurred in the Carboniferous.

See Chapter 17 for further details on the fossil history of cephalopods.

13.3.6 CARBONIFEROUS

The Carboniferous (Table 13.17 and Figure 13.27) was a period of global coal formation derived from the extensive lowland swamps and forests that covered the landscape. Among the molluscs, terrestrial gastropods first appeared among the rich vegetation (and litter) in a warm and humid climate with an atmosphere rich in oxygen (>30%) (Graham, 1995). Freshwater bivalves and gastropods were also present, and the marine Paleozoic molluscan fauna continued its diversification despite the Upper Devonian mass extinction. Shallow, warm seas covered the equatorially located continents, forming numerous shallow basins. Reef-building organisms such as bryozoans and both rugose and tabulate corals were abundant and diverse, while the sea floor was dominated by brachiopods. During the Pennsylvanian, a southern ice sheet formed over Gondwana as the continent moved south and the average global temperature dropped about 12°C (Feulner 2017). Ice sheet formation would also have been accompanied by a drop in sea level, resulting in the loss of many shallow seas.

---

[18] Shell crushing.

TABLE 13.17

Carboniferous Chronostratigraphy

System/Period		Series/Epoch	Stage/Age	Beginning
Carboniferous	Pennsylvanian	Upper Pennsylvanian	Gzhelian	304 Ma
			Kasimovian	307 Ma
		Middle Pennsylvanian	Moscovian	315 Ma
		Lower Pennsylvanian	Bashkirian	323 Ma
	Mississippian	Upper Mississippian	Serpukhovian	331 Ma
		Middle Mississippian	Viséan	347 Ma
		Lower Mississippian	Tournaisian	359 Ma

Based on International Chronostratigraphy Chart (2018-08) [www.stratigraphy.org/ICSchart/ChronostratChart2018-08.pdf]

Carboniferous (323 Ma)

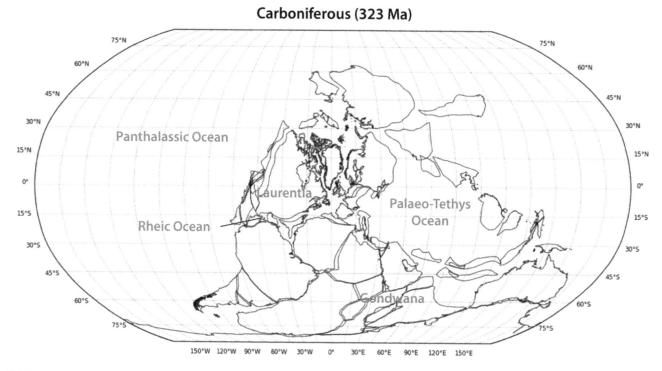

FIGURE 13.27   Palaeogeographic reconstruction of tectonic plate positions during the Lower Pennsylvanian.

13.3.6.1   Taxa

13.3.6.1.1   Bivalvia

Diverse bivalves occurred in swampy freshwater habitats which later became coal measures. These included the pteriomorphian group Myalinidae and the heteroconch Amnigeniidae (see Chapter 15). In the marine realm, brachiopods dominated habitats shared with bivalves, which remained subordinate in overall taxonomic diversity. In seep communities, the protobranch solemyids were relatively rare among abundant brachiopods, which were represented by a single rhynchonellid species (Peckmann et al. 2001). After an initial decrease in generic diversity in the early Carboniferous, bivalves increased in numbers and importance while brachiopods decreased over the same period (Babin et al. 1992), but the bivalve dominance was short-lived as the Permian extinction approached. Perhaps more importantly, this diversification in taxa and habitats within major stem marine bivalve clades in the Carboniferous (e.g., the protobranch Nuculida and Solemyida, the pteriomorphian Aviculopectinoidea, Pterioidea, and Mytiloidea, and the heterodont Anthracosioidea) may have facilitated lineage survival across the Paleozoic–Mesozoic boundary. During the Upper Carboniferous diversification geometric mean bivalve size increased by over 20 mm, although mean bivalve size remained stable (Kosnik et al. 2011). In contrast, the geometric mean of gastropod size decreased by about eight mm during the same period.

13.3.6.1.2   Rostroconchia

Conocardiid rostroconchs underwent their largest generic diversification event in the Carboniferous. Previous

diversification events had been primarily period-restricted. For example, only three of the ten genera that first appeared in the Silurian remained extant beyond that period, and none of the six genera first recorded in the Devonian survived the End-Devonian mass extinction. Similar to the Silurian diversification, the Carboniferous included first occurrences of 13 genera, four of which remained extant into the Permian. The Carboniferous diversification also included the appearance of minute species in the Silurian genus *Hippocardia* and the Carboniferous *Pseudoconocardium* (Wagner 1997).

13.3.6.1.3   Cephalopoda

Carboniferous cephalopods were predominately tightly coiled as the earlier Paleozoic orthoconic and cyrtoconic nautiliform taxa became increasingly rare following the Lower Pennsylvanian extinctions of the Actinoceratia and Oncoceratida. Goniatitidan ammonites, which had first appeared in the Devonian, remained dominant during the Carboniferous and were later joined by early representatives of the Ceratitida. Fossils of a supposed octopod body were found in the Pennsylvanian deposits at Mazon Creek, Illinois, USA (Kluessendorf & Doyle 2000), where other soft-bodied cephalopods occur, including *Jeletzkya*, thought to represent the earliest known crown group squid (Johnson & Richardson 1968; Allison 1987; Doguzhaeva et al. 2007a). Not surprisingly the Carboniferous also records the first ink sacs in cephalopod body fossils (Doguzhaeva et al. 2003, 2004) and the first occurrence of Spirulida (Doguzhaeva et al. 1999). Overall, the Carboniferous is a period of transition for the cephalopods, with the loss of the earlier nautiliform shelled taxa, the increasing diversity of ammonites, and the first appearance of shell-less cephalopods (Nishiguchi & Mapes 2008) (see Chapter 17 for further details).

13.3.6.1.4 Polyplacophora

The Mississippian saw the second largest origination of polyplacophoran genera of the Phanerozoic, being exceeded only in the Miocene; in both cases, origination rates were more than twice that of extinction rates (Sepkoski et al. 2002). This event also marked the first major radiation of the eight-plated Neoloricata. Prior to this, the only known neoloricates were multiplacophoridans (see Section 13.3.5.1.2). Numerous complete articulated chitons have been found in the Carboniferous, including over 100 specimens of *Glaphurochiton concinnus*, some with preserved radulae (Yochelson & Richardson 1979). Hoare and Mapes (1995) described a new Carboniferous (Pennsylvanian) multiplacophoridan genus *Diadeloplax*, and Vendrasco et al. (2004) described an exceptionally well-preserved specimen of another multiplacophoridan, *Polysacos vickersianum*, from the Carboniferous of Indiana, USA, which provided a more accurate reconstruction of the body plan of that group.

13.3.6.1.5 Scaphopoda

While earlier reports of scaphopods in the Paleozoic have been controversial (see Section 13.3.3.1.5), the earliest unequivocal appearance of the crown scaphopod taxon Dentaliida dates from the Mississippian (Yochelson 1999).

13.3.6.1.6 Gastropoda

Marine gastropods showed few changes in diversity at the family rank during the Carboniferous (Erwin 1990b). One significant extinction was the Perunelomorpha, an early putative caenogastropod group with an open coiling protoconch (Frýda 1999a). For most gastropods generic and familial diversity was relatively stable, including bellerophontians, Pleurotomarioidea, Murchisonioidea, and Trochoidea, and several groups of uncertain affinities including Lophospiridae (Trochonematoidea), Loxonematoidea, Subulitoidea, and the probable eogastropod Euomphaloidea. Species diversity was also high by Paleozoic standards. For example, the pleurotomarioidian genus *Worthenia* was especially diverse with over 100 species and was one of 30 co-occurring Carboniferous genera (Frýda et al. 2008a). Characteristic vetigastropod characters of crown taxa were prevalent by the Carboniferous, including nacreous shell structure (Batten 1972) and trochoidean protoconch morphology (Bandel et al. 2002). Neritimorphian diversity also appeared stable through the Carboniferous. Because the first occurrence of crown neritimorphs is Mesozoic, the affinities of the Carboniferous *Naticopsis* were questioned, but the analyses of shells and opercula by Kaim and Sztajner (2005) showed its inclusion in the Neritimorpha. A Carboniferous or earlier origination of the group is also suggested by the first occurrences of the putative terrestrial neritimorph taxon *Dawsonella* (Solem & Yochelson 1979) (see below). These patterns appear to be global, and the Carboniferous gastropod faunas were cosmopolitan in their distributions. For example, early Carboniferous gastropods reported by Yoo

(1994) from New South Wales, Australia, include abundant Bellerophontoidea, Euomphaloidea, Vetigastropoda (Pleurotomarioidea, Trochoidea), Neritimorpha (*Naticopsis*), Caenogastropoda (Loxonematoidea, Subulitoidea), and numerous Heterobranchia – a taxonomic diversity similar to early Carboniferous faunas in Europe and North America.

The Carboniferous also marks the first appearance of terrestrial gastropods, with at least two excursions into the lush vegetation of the period – the neritimorph *Dawsonella* (Solem & Yochelson 1979; Kano et al. 2002) and possible Eupulmonata stem lineages (the ellobioidean Carychiinae and early stylommatophorans) (Tracey et al. 1993; Bandel 1997). Mordan and Wade (2008) provided a systematic listing of the earliest fossil occurrence of the various terrestrial heterobranch families and pointed out that the identifications of Carboniferous stylommatophorans by Solem and Yochelson (1979) were in error. Unambiguous stylommatophoran taxa do not appear until the Jurassic–Cretaceous boundary, suggesting morphological convergence by the later-appearing stylommatophorans with the earlier terrestrial groups. Bandel (1997) recognised the Carboniferous terrestrial genera *Anthracopupa, Maturipupa,* and *Dendropupa* as 'carychiids', although, as with the stylommatophorans, subsequent convergence by the crown taxa is a reasonable alternative hypothesis. This latter scenario is further supported by heterobranch phylogeny (Wägele et al. 2008; Jörger et al. 2010), which requires numerous clades, appearing substantially later in the fossil record, to be present but unpreserved. This includes all 'lower heterobranchs', Nudipleura, Euopisthobranchia, Hygrophila, Siphonarioidea, etc., and while many of the taxa in this highly diverse group lack shells, they are present in some, including the stratigraphically useful holoplanktonic pteropods that do not appear in the fossil record until the Cenozoic. These absences from the record suggest that Carboniferous terrestrial taxa bearing similar shells to carychiines (Ellobiidae) are convergent. Whether they belong to a distantly related stem heterobranch group or to another group of gastropods has yet to be determined. These three reported terrestrial taxa are first known from the Pennsylvanian, approximately 50 million years after the establishment of terrestrial plants. The late Paleozoic invasion of land by gastropod molluscs, and perhaps the even later evolution of herbivory, follows a general trend seen in the evolution of herbivory in both terrestrial and marine systems (Labandeira 1998; Vermeij & Lindberg 2000; Labandeira 2002), with a substantial delay between the colonisation of land by plants and the colonisation and evolution of herbivorous organisms. Even after their first appearance in the Pennsylvanian, gastropod grazing patterns on fossil leaves are not known from the late Paleozoic, although insect feeding tracks and other traces on fossil plant material are well documented in the Carboniferous (Labandeira 1998, 2002). This absence suggests that early terrestrial diets were probably based on bacterial and fungal resources, food sources shared with freshwater and marine habitats and still utilised today by many terrestrial gastropods (see Chapters 5 and 20).

The co-occurrence of moist, terrestrial environments (e.g., swampy fern forests) and the high oxygen concentration of the atmosphere in the Carboniferous may have facilitated the terrestrial invasion by gastropods, the only molluscan class to achieve this habitat transition. Such evolutionary transitions between ecosystems are rare (Vermeij & Dudley 2000) and are thought to occur when low-intensity competition and predation exists in the new ecosystem. The gastropod experiment was not without challenges. An extinction event, the Carboniferous Rainforest Collapse (CRC), occurred during the Upper Pennsylvanian (~305 Ma). This event involved the aridification of the continents and the collapse of the vast tropical rainforests, and where they survived, they were restricted to small relictual patches among new floras and communities (Sahney et al. 2010). Effects on amphibians were particularly

devastating as the hot and humid climate changed to cool and arid (Sahney et al. 2010). Amphibians, slugs, and snails have convergent physiology and habitat requirements, and the CRC was probably also damaging to these early terrestrial gastropod experiments with expatriation, and perhaps extinction, common and widespread.

13.3.7 PERMIAN

The Permian (Table 13.18 and Figure 13.28) marine fauna was similar to that of the Carboniferous. Corals, stromatolites, sponges, bryozoans, brachiopods, and foraminiferans formed reef ecosystems in the warm shallow waters. Cephalopods were common predators, along with a great diversity of fish, including agnathans (jawless fish), chondrichthyans (such as sharks), and many types of bony fish. This collection of marine species represents the last of the Paleozoic evolutionary fauna, which first rose to dominance in the Ordovician, some 200 million years previously.

13.3.7.1 Extinction

The End-Permian mass extinction, approximately 252 Mya, was the largest in the history of multicellular life, with up to 90% of all marine species becoming extinct (Benton 2003; Erwin 2006; Sahney & Benton 2008). This extinction had far-reaching effects on molluscan evolution and marked the transition from the brachiopod-dominated Paleozoic fauna to the mollusc-dominated modern fauna (Gould & Calloway 1980; Sepkoski 1981). Accompanying this change in taxonomic structure was also a change in the ecology of marine ecosystems (Bambach et al. 2002; Bottjer et al. 2008). Prior to the extinction, about two-thirds of marine animals were sessile

TABLE 13.18

Permian Chronostratigraphy

System/Period	Series/Epoch	Stage/Age	Beginning
Permian	Lopingian	Changhsingian	254 Ma
		Wuchiapingian	259 Ma
	Guadalupian	Capitanian	265 Ma
		Wordian	269 Ma
		Roadian	273 Ma
	Cisuralian	Kungurian	284 Ma
		Artinskian	290 Ma
		Sakmarian	295 Ma
		Asselian	299 Ma

Based on International Chronostratigraphy Chart (2018-08) [www.stratigraphy.org/ICSchart/ChronostratChart2018-08.pdf]

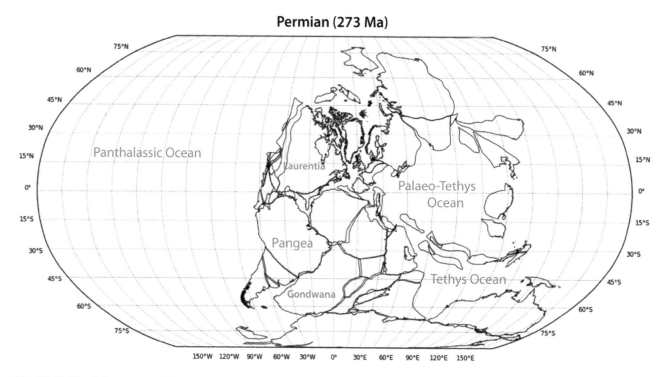

Permian (273 Ma)

FIGURE 13.28 Palaeogeographic reconstruction of tectonic plate positions during the Permian (Guadalupian).

and attached to the sea floor, but after the extinction, this dropped to 50% as motile invertebrates became just as common, many of which were gastropods (Wagner et al. 2006). Before the End-Permian mass extinction, both complex and simple marine ecosystems were equally common, and following recovery from the extinction, complex communities outnumbered simple communities by three to one (Wagner et al. 2006). This change and the increased levels of predation pressure in these more complex communities may have set the stage for the Mesozoic marine revolution (Vermeij 1977).

Between one and three pulses of extinction have been suggested for the End-Permian (Jin et al. 2000; Sahney & Benton 2008), and an array of causal mechanisms has been proposed (Erwin 1990a; Erwin 2006) (see below). Song et al. (2013) documented two pulses separated by a 180,000-year recovery period. The first pulse occurred in the latest Permian and marked the loss of numerous taxa, but primarily many calcareous algae, all rugose corals, some sponges, all trilobites, most radiolarians, and all fusulinid foraminiferans. The second pulse actually occurred in the earliest Triassic, and smaller foraminiferan, ostracod, brachiopod, bivalve, gastropod, ammonite, and conodont taxa suffered the greatest losses. Gastropods, bivalves, and ammonites also dominated the recovery originations between the two extinction pulses. Stratigraphic resolution of rostroconchs is not sufficient to determine if they also had two staged extinctions. As well, it appears that the second pulse was the one responsible for the collapse of the Paleozoic fauna.

Because the two pulses affected taxa differently, Song et al. (2013) concluded that different environmental causes were probably responsible but did not propose any candidate drivers. Others have linked earlier extinctions in the Guadalupian epoch with gradual environmental change, while the final pulse at the Permian–Triassic boundary is thought to have resulted from a catastrophic event (Ward et al. 2005; Algeo et al. 2012). Several lines of evidence suggest that shallow-water bottom communities were metabolically poisoned, but it remains unclear whether this was due to high carbon dioxide levels (hypercapnia) (Knoll et al. 2007), a severe deficiency of oxygen (anoxia) (Wignall & Twitchett 2002), hydrogen sulphide (euxinia) (Cao et al. 2009), or some combination of these (Clapham & Payne 2011). There is less agreement on what might have caused the toxicity. Possible causes include an impact event, the onset of massive volcanism (e.g., the Siberian Traps), a greenhouse effect triggered by methane hydrate gasification from the deep sea and/or the evolution of a new metabolic pathway in methanogenic microbes (Rothman et al. 2014). These catastrophic events and their short- and long-term effects would have been overlain with ongoing global changes (Kring 2000; Clapham & James 2008), which included glaciations and accompanying sea level changes, ocean acidification, increasing terrestrial aridity, the formation of the Pangaea supercontinent, shifts in ocean circulation, etc.

The End-Permian extinctions were so severe that many of the surviving lineages took 5 Ma, and some as long as 10 Ma, to recover from the decimation and return to pre-extinction diversity levels (Benton 2003). This slow recovery rate, compared to other mass extinctions, is thought to be due to residual environmental stresses that continued into the Lower Triassic (Woods et al. 1999; Payne et al. 2004), although one surviving ammonite lineage made an extraordinary recovery in only about one million years (see Section 13.3.7.2.3).

13.3.7.2 Taxa

13.3.7.2.1 Bivalvia

Bivalves were one of the most widespread Permian groups and experienced relatively low extinction rates overall (60%) compared to other molluscs. Biakov (2015) recognised three distinctive bivalve faunas in each of the major Permian basins. In the Boreal Superrealm, bivalves were a dominant benthic group although they had relatively low familial diversity. Taxa included the Inoceramus-like Kolymiidae (Pterioida) and numerous protobranch Nuculida, while pterinopectinids (Pectinida) and the heterodont Carditida and Lucinida were less common. In the more equatorial Tethyian Superrealm, the bivalve fauna was more diverse, but bivalves were much less dominant in these communities which were dominated by brachiopods. Among the bivalves, pterioideans were again dominant and included Posidoniidae, Alatoconchidae, Myalinidae, Pterineidae, and Isognomoniidae. Other groups included Parallelodontidae (Arcida), Ostreida, numerous Pectinida, including Entoliidae, Annuliconchidae, Pterinopectinidae, and Bakewellidae, and, in the Trigoniida, Schizodidae. Some Alatoconchidae were extremely large and are suspected of being photosymbiotic (Isozaki & Aljinović 2009). In the southern-most Gondwanan Superrealm bivalves again were not very diverse, and there was high endemism, particularly in Pectinida (Eurydesmatidae) and in the Pterioida, the Inoceramus-like ambonychiids. Overall, this fauna was most similar to the Boreal Superrealm, and bipolar distributions were not uncommon. The larger epifaunal bivalves also provided hard substrata for other 'invertebrate' taxa and communities (Biakov 2015).

Several bivalve taxa have been identified as 'disaster taxa' (in the sense of Rodland & Bottjer 2001) following the Permian extinction; these having been characterised as highly abundant, widespread, and ecologically dominant and included the pectinid genera Claraia and Eumorphotis, the myalinid Promyalina, and the trigoniid Unionites (Petsios & Bottjer 2013). While the overall extinction rate was low for the class, in some lineages loss was more severe and recovery substantially slower. For example, pectinoideans gradually declined from 23 genera in the Guadalupian to about five genera (a 78% reduction) in the earliest Triassic. After this decline the pectinoideans would not return to their earlier diversity until the Upper Triassic, 35 Ma later (Newell & Boyd 1995).

Bivalves were neither as diverse nor as abundant as brachiopods in Paleozoic faunas but became a major component of the modern fauna following the End-Permian extinction. Rudist bivalves (see Chapter 15) would replace the reef builders (including corals, bryozoans, brachiopods, and echinoderms) lost at the Permian extinction in the Mesozoic (Flügel 1994).

13.3.7.2.2 Gastropoda

Like bivalves, gastropods had relatively low extinction rates (66%). Most gastropod clades experienced two extinction pulses. The first, at the end of the Guadalupian, affected all major Permian gastropod clades, including the Bellerophontia, Euomphalina, Pleurotomariina, Trochina, Neritopsina, Loxonematoidea, Subulitoidea, Murchisonioidea, and Heterobranchia (Erwin 1990a). The second pulse, at the End-Lopingian, was not as strong as the first, although all but the Loxonematoidea saw further declines in diversity. In the first pulse, about half of the Permian gastropod genera were lost, and origination rates were low. Extinction rates were lower in the final pulse, but another third of generic diversity was lost, and originations fell to one of the lowest rates seen in the Paleozoic (Erwin 1990a). Bellerophontians survived the End-Permian extinction, but they continued to decline in diversity and were extinct by the Upper Triassic.

Payne (2005) examined gastropod size across the Permian–Triassic boundary. His data suggested that maximum size was relatively stable in the Permian, but minimum adult shell size decreased in both the Cisuralian and Lopingian after a slight recovery in the Guadalupian. Maximum adult size underwent an abrupt, precipitous decline across the Permian–Triassic boundary while minimum adult size had already begun to increase, but Lower Triassic gastropods generally remained small. Both maximum and minimum size gradually increased in the Lower Triassic, and within 10 Ma, gastropod size was once again equivalent with Guadalupian and Lopingian values. Payne (2005) proposed that two scenarios were consistent with the patterns of size decrease across the Permian–Triassic boundary: (1) size-selective extinction at the species level and (2) within-lineage size decrease.

Besides their small size, Lower Triassic gastropod origination rates only slightly exceeded extinction rate, producing a relatively modest rise in diversity (Erwin 1990a). When examined separately, the caenogastropods showed an early and pronounced radiation (Nützel 2005), and within four million years, new originations outnumbered Paleozoic survivors (Pan & Erwin 2002) with most Lower Triassic caenogastropod genera originating after the End-Permian extinction (Nützel & Erwin 2002).

13.3.7.2.3 Cephalopoda

Ammonites were reduced by two extinction events in the Permian. The first, in the Guadalupian, reduced the level of morphological differences in the group. This decline was relatively gradual and was apparently environmentally driven, but the second, the terminal Permian extinction, was non-selective and catastrophic following water column toxicity. Morphological differences in ammonites across the Permian–Triassic boundary declined by almost 60% (McGowan & Smith 2007), and within one million years after the final pulse, the pre-extinction morphospace was reoccupied, but with the divergent shell parameters distributed differently among the new and surviving lineages. This extraordinary recovery occurred due to the diversification of a small surviving lineage of Ceratitida represented by only three genera, one of which would give rise to most Triassic ammonites (Brayard et al. 2007).

At the beginning of the Permian, there were four major clades of ammonites (Saunders et al. 2008; Brayard et al. 2009). After the final extinction pulse, only two remained, the Prolecanitida and the Ceratitida, and unlike other molluscs, this lineage reached its greatest diversity within the Lower Triassic (Brayard et al. 2009). This difference has been explained as the presence of residual toxicity in benthic habitats (gastropods and bivalves) compared to its absence in the water column (ammonites). But as Marshall and Jacobs (2009) pointed out, both *Nautilus* and *Vampyroteuthis*, members of two ancient cephalopod lineages, are low-oxygen specialists. The water column might also have had residual toxicity, and the surviving Ceratitida could have been plesiomorphically adapted to low-oxygen conditions, leading to their more rapid recovery.

13.3.7.2.4 Other Classes

Scaphopod species diversity was reduced by 85% across the Permian–Triassic boundary (data from Reynolds 2002). Rostroconchs had already been on the decline, and only the Conocardiida had survived beyond the Ordovician, with the final three genera (*Bransonia*, *Conocardium*, and *Pseudoconocardium*) disappearing at the end of the Permian. The loss of polyplacophoran taxa was continuous through the Permian (Cherns 2004). Almost all paleoloricates were extinct by the end of the Silurian, and the Permian extinction marked the final occurrence of that group. For neoloricates, J. Sigwart (pers. comm., 2016) estimated that generic diversity was reduced by 89% (from 28 to three taxa) by the event, including the loss of the multiplacophoridans. Monoplacophorans are unrecorded in the Permian but obviously survived.

14 Polyplacophora, Monoplacophora, and Aplacophorans

14.1 INTRODUCTION

In this chapter, we introduce the aplacophorans, polyplacophorans, and monoplacophorans, loosely called the 'placophoran'[1] groups, which have been regarded as the most 'primitive' living members of the molluscan tree, although not without controversy (see Chapter 13 for discussion).

Members of this informal level of morphological organisation (or grade) are bilaterally symmetrical and typically have elongate bodies. Monoplacophorans have a single shell and polyplacophorans have eight valves (in some fossils up to 17). Two living groups ('aplacophorans') lack shell plates altogether and instead have calcium carbonate spicules, these being the Solenogastres (= Neomeniomorpha) and Caudofoveata (= Chaetodermomorpha).

With the increasing application of molecular systematics, the collection of 'placophoran'-like fossil taxa, and their supposedly basal position, their relationships with the other molluscan groups has generated considerable controversy, as outlined in Chapters 12 and 13.

Well-preserved Cambrian fossils of *Wiwaxia* and *Odontogriphus* both show a few oral structures that may have been used for grazing on algal mats. These mouthparts have been likened to radulae (e.g., Caron et al. 2006) (see Chapter 13) and used to formulate hypotheses regarding the evolution of early radulae. According to this idea, the early radula comprised a few rows of hardened, wide, denticulate teeth held together with a flexible cuticle, and was used to scrape detritus or larger food items. This is a similar arrangement to the radulae found in some caudofoveates and solenogasters (e.g., Salvini-Plawen 2003; Scheltema et al. 2003; Todt et al. 2008b) and, according to this hypothesis (which we do not support), the specialised radulae seen in modern aplacophorans and chitons evolved from this primitive condition. Specialised offshoots from this early configuration included the pincer-like structures seen in caudofoveates that have evolved to pick up individual foraminiferans and diatoms, and in solenogasters, the development of rows of hooks and other structures suited to carnivorous feeding. In marked contrast, chitons evolved an extremely long radula specialised for scraping hard substrata, with some teeth in each tooth row hardened by metals (Todt et al. 2008b).

---

[1] The term 'placophoran' is here used informally, as in Chapter 13, and by Lindberg and Ponder (1996), and Parkhaev (2008), as a general term to conveniently encompass all the groups dealt with in this chapter as well as similar fossil taxa. The formal name Placophora has been used, particularly by some European malacologists, for Polyplacophora alone.

14.2 POLYPLACOPHORA (CHITONS, PLACOPHORA, LORICATA, AMPHINEURA IN PART)

14.2.1 INTRODUCTION

Polyplacophorans, or chitons, are small to rather large, entirely marine, dorsoventrally flattened animals that are elongate-oval in shape. Dorsally they have eight articulating shell valves (plates) surrounded by a thick girdle that may be covered with spines, scales, or hairs (Figure 14.1). Ventrally there is a broad, oval, creeping foot. Most chitons are between 5 and 50 mm in length, but some are larger, and one, the 'gumboot' chiton, *Cryptochiton stelleri*, which lives on the north west Pacific coast of North America, grows to about 36 cm in length.

Polyplacophorans are ecologically important as they are often abundant in the littoral zone and, from the evolutionary viewpoint, are sometimes considered the most primitive living molluscs. They have a long fossil history from at least the Upper Cambrian (see Section 14.2.5 and Chapter 13).

There have been several accounts of chiton anatomy, the most comprehensive being those of Plate (1897, 1899, 1901), Hyman (1967) and Eernisse and Reynolds (1994) while reviews of aspects of their physiology and biology include those of Boyle (1977) and Pearse (1979).

14.2.2 PHYLOGENY AND CLASSIFICATION

Polyplacophoran monophyly is well established (e.g., Okusu et al. 2003), except for the analysis by Giribet et al. (2006) where a monoplacophoran was nested within the group. Subsequent analyses providing better data (Wilson et al. 2010) have again demonstrated monophyly, with the monoplacophoran being the sister taxon, and together they formed the higher taxon Serialia (see Chapter 12 for discussion).

While extensively studied at the species level, the higher classification of chitons was in flux until recently (e.g., Okusu et al. 2003; Sirenko 2006; Todt et al. 2008b). Today, the phylogenetic relationships of the major chiton clades are still argued, but the membership of species and families within those clades is not in doubt (Sigwart et al. 2013). While their classification incorporates shell, girdle, radular, egg hull, sperm, and neural (aesthete) characters and is in good agreement with molecular phylogeny, earlier classifications were based mainly on features of the shell valves, with other characters such as girdle and radula being mostly utilised at generic or species levels.

The higher-level groups generally recognised in the 20th-century literature on living chitons (Neoloricata or Loricata)

FIGURE 14.1 Photos of living chitons. (a) *Leptochiton cascadienses* (Leptochitonidae) San Juan Island, Washington. Courtesy of J. Sigwart. (b) *Ferreiraella* sp. (Abyssochitonidae), Guadeloupe, French West Indies. Courtesy of L. Charles - MNHN. (c) *Tripoplax regularis* (Ischnochitonidae), Monterey, California (© Gerald and Buff Corsi/Focus on Nature, Inc.). (d) *Stenoplax* sp. (Ischnochitonidae), Panglao, Philippines. Courtesy of P. Maestrati - MNHN. (e) *Tonicia lebruni* (Chitonidae), Falkland Islands. Courtesy of J. Sigwart. (f) *Acanthochitona fascicularis* (Acanthochitonidae), Azores. Courtesy of J. Sigwart. (g) *Cryptoplax lavaeformis* (Cryptoplacidae), Papua New Guinea. Courtesy of P. Maestrati - MNHN. (h) *Cryptochiton stelleri* (Mopaliidae), Vancouver, BC, Canada. Courtesy of J. Sigwart.

were the 'suborders' Lepidopleurina, Acanthochitonina, and Ischnochitonina, the latter being the most diverse. An additional monotypic suborder, Choriplacina, is recognised for *Choriplax* (Gowlett-Holmes 1987; Kaas & Van Belle 1994; Kaas et al. 1998). The almost exclusive use of valve morphology for higher classification (e.g., Kaas & Van Belle 1994; Kaas et al. 1998) was because it was the only character-set available for fossil chitons. The higher taxon Chitonida was introduced for chitons with elaborate extracellular hull processes surrounding their eggs (Sirenko 1993), and this grouping includes most of the living chitons. Egg hull morphology, gill placement and morphology (e.g., Russell-Hunter 1988; Sirenko 1993), and sperm ultrastructure (Buckland-Nicks 1995) have been found to correlate rather well with classifications based on shell morphology (Todt et al. 2008b). Based on all these characters living chitons were found to form two main lineages (Sirenko 1993; Buckland-Nicks 1995; Sirenko 2006). These two lineages were recognised as orders by Sirenko (1997, 2006), namely the Lepidopleurida and Chitonida, the latter having two suborders, Chitonina and Acanthochitonina. Within Acanthochitonina, two groups were recognised – Mopalioidea and Cryptoplacoidea – and two groups also within Chitonina – Chitonoidea and Schizochitonoidea. This classification was largely corroborated by the first molecular analysis of polyplacophorans (Okusu et al. 2003). This analysis recovered the two main groups, Lepidopleurida and Chitonida, and generally supported Chitonina, but some classical higher taxonomic groups, such as Ischnochitonina and Acanthochitonina and several families, were not monophyletic. The analysis supported the monophyly of Chitonida. Lepidopleurida were basal, and Chitonida formed three lineages:

1. Those with egg hulls having rounded to weakly hexagonal cupules, abanal gills, and type I sperm (clade A of Okusu et al. 2003)
2. Those with egg hulls having strongly hexagonal cupules with flaps, abanal gills, and type I sperm (clade B of Okusu et al. 2003)
3. Those with spiny egg hulls of various shapes, adanal gills, and type II sperm (clade C of Okusu et al. 2003), equivalent to Chitonoidea of Sirenko (1997)

Sirenko (2006) revised his earlier classifications (Sirenko 1993, 1997) and recognised four orders – Chelodida, Septemchitonida, Lepidopleurida, and Chitonida.

BOX 14.1   HIGHER CLASSIFICATION OF POLYPLACOPHORA

(Class) **Polyplacophora** (= Placophora)
(Subclass) **Paleoloricata**†
(Order) **Chelodida**†
(Suborder) **Septemchitonina**†
(Subclass) **Neoloricata**
(Order) **Multiplacophorida**†

(Order) **Lepidopleurida**
(Suborder) **Cymatochitonina**†
(Suborder) **Lepidopleurina**
(Order) **Chitonida**
(Suborder) **Chitonina**
(Suborder) **Acanthochitonina**

(† = extinct taxa)
See Appendix for detailed classification.

14.2.2.1   Sister Group Relationships

Despite attempts to resolve molluscan phylogeny using morphological and molecular data, there is no consensus view regarding the position of Polyplacophora within the molluscan tree (see Chapters 12 and 13), in part due to poor sampling for basal taxa (particularly aplacophorans and monoplacophorans) and the considerable extinction that has occurred. Three main ideas about sister group relationships of polyplacophorans have emerged: (1) they are the sister to aplacophorans, forming the Aculifera (= Amphineura), synapomorphies being the elongate body form and girdle spicules (Scheltema 1996); (2) aplacophorans are paraphyletic, and the remaining molluscs (Testaria) contain polyplacophorans as the basal taxon (e.g., Salvini-Plawen 1969; Wingstrand 1985; Haszprunar 2000); (3) Polyplacophora is the sister to all the extant molluscs, with aplacophorans derived and either monophyletic or diphyletic. This latter hypothesis is rarely suggested (e.g., Lindberg & Ponder 1996) but is supported by the Serialia concept (see Chapter 12).

The possibility that stem aplacophorans may have had plate-like structures similar to those of polyplacophoran valves was supported by the discovery of foot-less chitons (*Kulindroplax* and *Phthipodochiton*) (Sigwart & Sutton 2007; Sutton & Sigwart 2012). In addition there is also a putative fossil aplacophoran, the Silurian *Acaenoplax hayae* (Sutton et al. 2001a, 2004), which has both spicules and plates (see Chapter 13), and a putative aplacophoran larva with seven transverse dorsal spaces lacking spicules has been described (Scheltema & Ivanov 2002). The molluscan affinities of both the larva (G. Rouse in Nielsen et al. 2007) and Silurian fossil have been disputed (Steiner & Salvini-Plawen 2001). If stem aplacophorans did have plates and spicules, as some evidence seems to suggest, then these structures are plesiomorphic and shared with extinct 'placophoran' lineages as proposed by Lindberg and Ponder (1996).

14.2.3   Morphology

Works on the general morphology of Polyplacophora, as well as those detailing anatomy and histology, include the historical accounts of Plate (1897, 1899, 1901), Wingstrand (1985), and the synopses of Hyman (1967) and Eernisse and Reynolds (1994). For details of many features involved in chiton taxonomy see Kaas and Van Belle (1985a, 1985b, 1985c) and Sirenko (1993, 1997, 2006).

14.2.3.1 Shell and Girdle

External chiton anatomy has been reviewed and standardised by Schwabe (2010). All living chitons have eight dorsal aragonitic shell valves (plates) (other than very rare anomalous specimens) (Dell'Angelo & Tursi 1990; Dell'Angelo & Schwabe 2010; Torres et al. 2018), but valve number and configuration were more variable in Paleozoic taxa (see Chapter 13). The shell valves are greatly reduced or even internal in a few living species (e.g., *Cryptochiton*, *Cryptoconchus*), and some of these have an elongate, somewhat worm-like body with non-overlapping valves (e.g., *Cryptoplax*).

The anterior and posterior shell valves differ from the intermediate valves in size and shape. Typically, they vary from convex to acutely arched and are composed of two primary layers – the dorsal *tegmentum* and ventral *articulamentum*. The outermost layer of the tegmentum is called the *properiostracum* and is equivalent to the periostracum in other shelled molluscs. The innermost layer of the articulamentum is called the *myostracum* (shell areas where the muscles attach), and together the shell layers are called the *hypostracum*. The tegmentum and articulamentum are of taxonomic importance: the tegmentum, which is often brightly coloured, may be sculptured and is often divided into central and lateral areas. The articulamentum extends anteriorly beyond the tegmentum to form a pair of *apophyses* on valves 2–7 (the intermediate and posterior valves) that slot beneath the preceding valve; they occur in all neoloricate chitons. *Insertion plates* (also called sutural lamellae) extend laterally under the girdle and bear slits in some taxa (Figure 14.2). Insertion plates are not present in all chitons, and Sirenko (2006) has argued that they evolved independently in different taxa and have also been lost multiple times.

These articulating valve features integrate the shell valves and form a flexible set of armour that allows the chiton to fit irregular surfaces and to roll up when dislodged (Connors et al. 2012). The early fossil chitons (Paleoloricata) lack an articulamentum, having a single layered shell, and they lack insertion plates and apophyses. Although their valves formed a series, they would not have articulated (Sigwart & Sutton 2007). The shell valves are penetrated by sensory aesthetes and, in some chitons, ocelli (see Section 14.2.3.7).

The lack of insertion plates, as in *Leptochiton*, is considered a plesiomorphic feature (e.g., Sirenko 1997; Todt et al. 2008b), and, when present, insertion plates without slits are thought to be more primitive than those with slits. Slits in the insertion plates correlate with the derived lateral innervation of the aesthetes in the tegmentum (Eernisse & Reynolds 1994).

The entire shell is aragonite, as are the girdle spicules and scales (reviewed by Haas 1981; Scheltema 1988a; Carter & Hall 1990; Eernisse & Reynolds 1994). The inner layers of the shell valve microstructure are cross-lamellar but of a type different to that in other conchiferan molluscs in that 'the third-order lamellae are rod-like and do not comprise well-defined, laminar second-order lamellae' (Carter & Hall 1990 p. 30). Connors et al. (2012) reported this layer as 'composite prismatic'. Also, the tegmentum varies from composite prismatic to irregular spherulitic prismatic while the ventral shell layers are comprised of crossed lamellar, irregular spherulitic prismatic and irregular simple prismatic structures (Haas 1972, 1976). In a subsequent assessment of the shell structure of *Acanthopleura granulata*, Carter and Hall (1990) found that the articulamentum has a mainly rod-type crossed lamellar to crossed acicular microstructure while the myostracum ranges from irregular simple prismatic to homogeneous.

The tegmentum also has a high protein and pigment content (Peebles et al. 2017) and on its dorsal surface is the organic properiostracum secreted by epidermal pockets at the mantle edge (Checa et al. 2017). It is usually very thin and difficult to distinguish. In *Tonicella* the properiostracum is composed of three layers. It was suggested that the properiostracum was secreted by the aesthetes (Baxter et al. 1990), but this supposition has not been verified.

The girdle (perinotum) surrounds the dorsal shell valves or, sometimes, overlaps or covers them. The girdle is covered with a chitinous cuticle, and it often contains shelly scale, or spicule-like sclerites, or corneous processes that may be hair-like; there is also a fringe of marginal spines that are especially obvious in juveniles (Eernisse & Reynolds 1994). The dorsal surface of the girdle also has small sensory structures similar to the aesthetes (see Section 14.2.3.7), and ventrally the girdle is usually covered with minute calcareous scales. The calcareous girdle structures are all aragonitic. The epidermis forming the surface of the girdle is covered by a thick glycoproteinaceous cuticle in which the spines, scales, or corneous protrusions are embedded. These structures are secreted by clusters of secretory cells, similar to those in aplacophorans (Hoffmann 1949). Each cluster comprises tall, spaced cells that surround one or two low epidermal cells (Eernisse & Reynolds 1994).

In most species the girdle is limited to a narrow border around the shell valves; in others it envelops the valves leaving only a small dorsal part visible, or, rarely, it completely covers them. In a few species, the anterior portion of the girdle may be broadened and elongated as a hood to trap prey such as small arthropods (see Section 14.2.4.3).

14.2.3.2 Head-Foot, Muscles, and Mantle

The ventral oval foot (Figure 14.4) has a large, creeping, ciliated sole capable of powerful adhesion. At its anterior end is the simple head which does not possess a neck and is thus not free from the body. The head consists only of the ventral mouth and is separated from the foot by a transverse groove. The mouth is surrounded by a simple fold (mouth lappets). The head lacks eyes and tentacles, except a few genera (e.g., *Placiphorella*) have tentacle-like elaborations of the mouth lappets, which provide support during feeding (McLean 1962; Todt et al. 2008b). The epidermal cells of the foot are ciliated with interspersed secretory and adhesive gland cells (Eernisse & Reynolds 1994).

Chiton buccal muscles have been described by Plate (1897), Hyman (1967), Graham (1973), and Wingstrand (1985), and the reader is referred to those sources and to Chapter 5 for details. These muscles are red as they are rich in myoglobin. Plate (1897) recognised 38 pairs of muscles and six unpaired

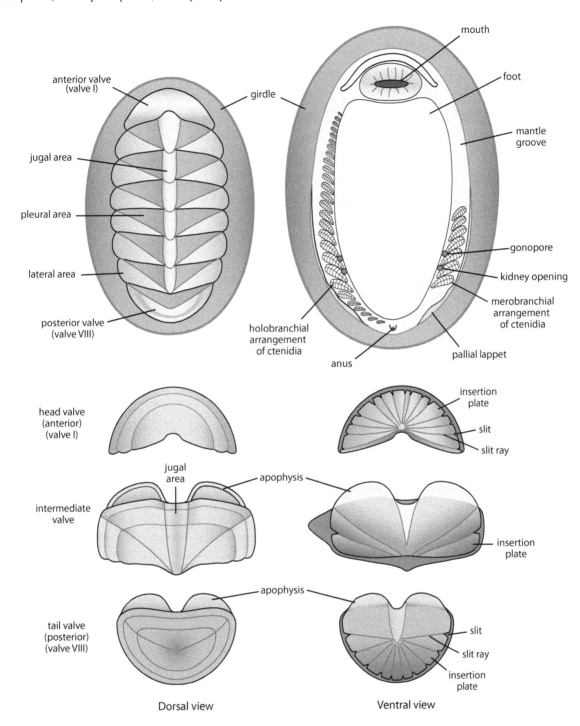

FIGURE 14.2 External body and shell morphology of a typical chiton. The ventral view (top right) shows a different configuration of the gills on each side, those on the left side of the figure represent an example of the adanal condition, and those on the right the abanal condition. Redrawn and modified from Beesley, P.L. et al. (eds.), *Mollusca: The Southern Synthesis, Part A*, CSIRO Publishing, Melbourne, VIC, 1998a.

muscles in *Acanthopleura echinata*. Wingstrand (1985, p. 61) noted their similarity to those in monoplacophorans, stating that the 'entire radula [sic] apparatus of the Polyplacophora shows so many features identical to those of the tryblidians [i.e., monoplacophorans] that a derivation from a similarly shaped apparatus in a common ancestor appears unavoidable'.

Besides the buccal muscles, the main muscles in chitons are those involved with the head and the shell valves.

The shell muscles are a complex system involving several groups of muscles that repeat under all the intermediate valves but are somewhat different under the anterior and posterior valves (Wingstrand 1985). More details of the shell muscles are given in Chapter 3. Their articulated shells combined with their complex musculature enable chitons to roll into a ball when dislodged (e.g., Connors et al. 2012).

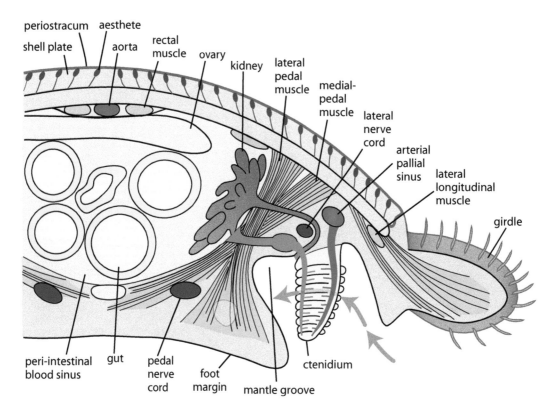

FIGURE 14.3 Diagrammatic transverse section through a chiton showing aspects of the anatomy, including the musculature. Redrawn and modified from Lemche, H. and Wingstrand, K.G., *Galathea Report*, 3, 9–71, 1959.

14.2.3.3 Mantle Groove (Mantle Cavity)

The mantle cavity in chitons is a deep groove (the mantle or pallial groove) running around the foot and, within it, ctenidia (gills) are arranged serially in the mantle groove (Figures 14.2 to 14.4). The mantle groove also contains the posterior anal opening, the paired genital openings, and epithelial patches with distinct sensory functions (see Section 14.2.3.7). The epithelium of the mantle groove consists of ciliated cells interspersed with mucous goblet cells which become more numerous toward the foot (Eernisse & Reynolds 1994).

Each gill is attached to the side of the animal near the lateral nerve cords and consists of a median axis to which the lamellae are affixed on each side (Figure 14.3); these decrease in length distally. Depending on the species, the lamellae in each row are either paired or alternate with each other along the axis. The structure of the gills shows that their function is for respiration, and they are innervated and have a rich blood supply. The lamellae are covered with a columnar epithelium that has dense microvilli, and in the middle of each lamella is a band of strongly ciliated cells (Fischer et al. 1990).

Water enters the mantle groove in variable locations via one or two small raised areas of the girdle. It is driven through the cavity by the action of the lateral cilia on the ctenidia and exits posteriorly where the anus is located. The gills are connected by interlocking cilia to form a continuous curtain on each side that effectively separates the mantle groove into a ventral inhalant cavity and an upper exhalant cavity (Yonge 1939b; Russell-Hunter 1988) (Figure 14.4). A few taxa such as

Mopalia have a posterior cleft in the mantle at the posterior exit point (Eernisse & Reynolds 1994).

The gill rows may meet posteriorly as in Lepidopleurina or may be separated posteriorly as in other chitons (Figures 14.2 and 14.4). The number of gills ranges from 6 to 88 pairs (Okusu et al. 2003), with different taxa having characteristic numbers and arrangements. The gill number is not constant within a species as it increases with body size during growth and in some individuals may even differ a little on different sides of the body (Hyman 1967), although the size of the gill lamellae is similar regardless of age or gill size, with smaller gills possessing fewer lamellae (Eernisse & Reynolds 1994). In the post-settlement juvenile, the first pair of gills to appear is just behind the pair of renal openings (nephridiopores). During growth, the addition of gills may be *adanal* – with the gills added both anteriorly and posteriorly to this pair, as in most chitons, but in approximately a third of living chitons it is *abanal*, with extra gills added only anteriorly (Sirenko 1993; Eernisse & Reynolds 1994) (Figure 14.2). About four-fifths of those with the adanal condition are similar to those with abanal gills in having a substantial gap between the left and right posterior-most gills; thus it is only in the Lepidopleurina and relatively few Chitonida that the gills extend to the anal papilla (Eernisse & Reynolds 1994). The direction of growth in the gill row and gill size can vary substantially in some species (Sigwart 2008).

14.2.3.4 Digestive System

Studies on the anatomy and function of the gut of chitons, include Haller (1882); Plate (1897, 1899, 1901); and Fretter

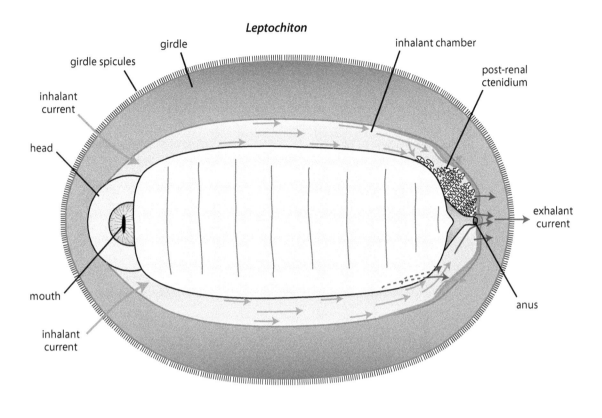

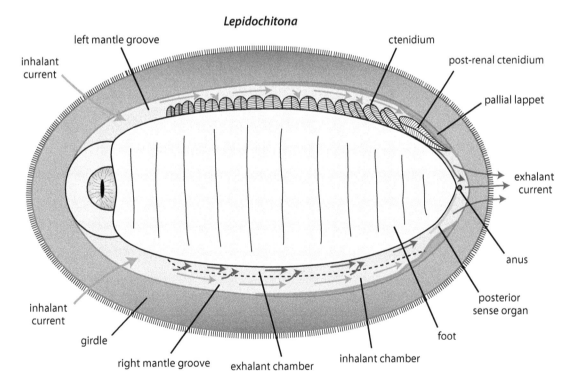

FIGURE 14.4 Ventral views of two chitons showing ciliary currents; upper *Leptochiton asellus* (Lepidopleurida) and lower *Lepidochitona cinerea* (Chitonida). The ctenidia are removed from the right side of the animal in both. Redrawn and modified from Yonge, C.M., *Q. J. Micros. Sci.*, 81, 367–390, 1939b.

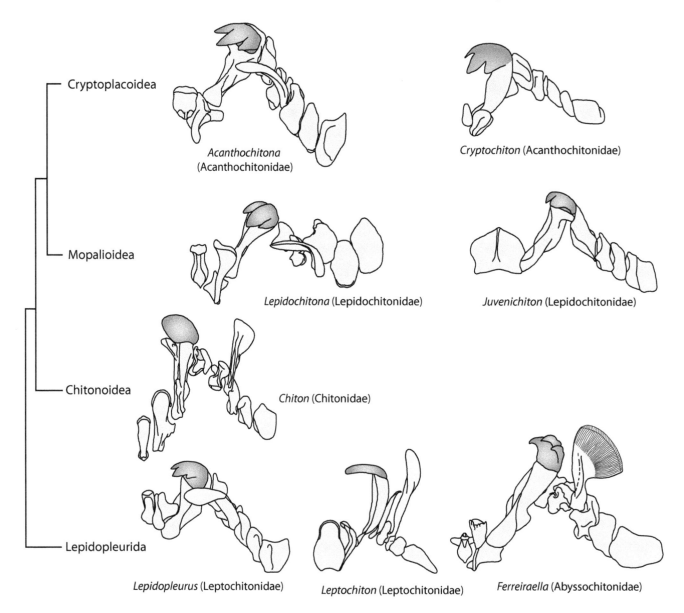

Cryptoplacoidea

Acanthochitona
(Acanthochitonidae)

Cryptochiton (Acanthochitonidae)

Mopalioidea

Lepidochitona (Lepidochitonidae)

Juvenichiton (Lepidochitonidae)

Chitonoidea

Chiton (Chitonidae)

Lepidopleurida

Lepidopleurus (Leptochitonidae)

Leptochiton (Leptochitonidae)

Ferreiraella (Abyssochitonidae)

FIGURE 14.5 Some examples of half rows (showing central tooth and right side) of polyplacophoran radulae. Redrawn and modified from various sources. The radulae of the species illustrated do not necessarily reflect the range of variation in the genera they belong to or of the higher taxon they belong to.

(1937), and on gut physiology, Greenfield (1972). No comprehensive ultrastructural study of the chiton gut has been undertaken, and Eernisse and Reynolds (1994) summarised the state of knowledge at that time.

The ventral mouth is surrounded by a complex musculature and a veil comprising mantle tissue. The mouth opens to a short buccal tube into which open the subradular and radular sacs. The muscular, sensory subradular organ lies in the subradular sac and, when protruded during feeding, is assumed to 'taste' potential food (see Section 14.2.3.7). The radular sac lies above the subradular sac and is typically long and coiled; it extends posteriorly over the top of the stomach. The radula (see below) has a subradular membrane and is supported by the odontophore.

The odontophoral apparatus is very similar to that in monoplacophorans (Wingstrand 1985). Both groups are the only molluscs to possess a pair of fluid-filled radular vesicles,

although Wingstrand (1985 p. 63) noted that sections of a lepidopleurid species showed that the vesicles were largely filled by an 'invasion of small-celled tissue' and suggested that the vesicles were the homologues of the anteromedial cartilages in patellogastropods. In both groups, a pair of lateral and medial cartilages partially enfold each vesicle, with the medial cartilages bound by a ventral approximator muscle (= unpaired median radular muscle). Graham (1973) and Wingstrand (1985) compared the complex buccal musculature of chitons, monoplacophorans, and patellogastropods, and these comparisons are summarised in Chapter 5.

The chiton radula (Figure 14.5) is not as variable as it is in some other classes. Most have 17 teeth or plates in each row, rarely 13 or 15, and the long radular ribbon has from 25 to 150 tooth rows (Eernisse & Reynolds 1994). Thus, while radular characters have been used in chiton taxonomy (Saito 2004), they are apparently of limited value at higher taxonomic levels (e.g., Okusu et al. 2003).

The second lateral radular teeth are disproportionately large, hardened, and capped with magnetite ($FeO·Fe_2O_3$) (Kisailus & Nemoto 2018). The utilisation of magnetite is rare in animals, otherwise being known in some insects and birds. These teeth have one to four cusps or are spade or shovel-like. The magnetite is incorporated into an organic matrix (see Chapter 5 for more details). Patellogastropod radulae also use iron oxides to impregnate their teeth but not magnetite (see Chapter 5).

The magnetite-tipped second lateral teeth, along with the fifth and eighth pairs of lateral teeth, are the first to appear in juveniles (Eernisse & Kerth 1988). The fifth lateral teeth are elongate and may at least function partly to protect the epithelium of the mouth region from contact with the sharp second laterals (Eernisse & Kerth 1988).

Mineralisation occurs progressively along the radular sac, with most iron mineralisation occurring posteriorly, followed by calcification (Kim et al. 1986). The process of mineralisation in chiton radulae has been intensively studied (e.g., Macey et al. 1994; Macey & Brooker 1996; Lee et al. 1998; Brooker et al. 2003; Wealthall et al. 2005) and is described in Chapter 5.

The radula is moved over the odontophore at its most anterior end, and the teeth, which articulate like a zipper, are alternately splayed open or rolled back into a tube. The feeding stroke usually combines longitudinal scraping with 'rolling the opposed teeth together to scrape or tear off chunks of food, which then are moved up internally into the overlying esophagus' (Eernisse & Reynolds 1994 p. 70).

The bulk of the digestive system lies below the gonad and makes up a large part of the body mass. The gut is essentially a 'straight through' system, having an anterior mouth and a posterior anus, and between them there are an asymmetric stomach and an extensively coiled intestine (see Chapter 5, Figure 5.52). Intestinal coiling can vary among species and appears to be correlated with diet (Sigwart & Schwabe 2017).

Behind the radula lies the buccal cavity (sometimes called a pharynx) with a mucous groove and ciliated infoldings, the latter of unknown function. The buccal cavity receives secretions from a pair of laterally situated buccal glands (called salivary glands by Fretter 1937) which may be simple or, in larger chitons, compound.

Posteriorly the buccal cavity opens to the oesophagus into which discharge, through narrow openings, a pair of diverticula, the buccal (or pharyngeal) glands (also variously called sugar glands, oesophageal glands, or salivary glands). The buccal glands are filled with glandular villi that produce carbohydrate-digesting enzymes (Meeuse & Fluegel 1958, 1959) and are probable homologues of the oesophageal diverticula in monoplacophorans (Wingstrand 1985). The oesophagus is rather short and separated from the stomach by a sphincter. The large, muscular stomach is surrounded by the two lobes of the digestive gland. A longitudinal dorsal channel runs from the oesophageal opening to the middle part of the stomach with a pair of ciliated bands on either side. The stomach is constricted in the mid-left side, and, just posterior to

this, the ciliated bands and dorsal channel continue into the anterior intestine. The left lobe of the digestive gland opens to the dorsal channel in the constricted middle region, and the more posteriorly located right lobe of the digestive gland opens to the channel immediately before the intestinal aperture. Ventrally the stomach is a large sac which is distended when full but smaller and wrinkled when empty.

As in other molluscs, the digestive gland comprises basophilic and digestive cells, the latter with many food vacuoles and covered with microvilli, and these cells absorb and store nutrients. The basophilic cells occur singly or in small groups among the digestive cells. They have endoplasmic reticulum and Golgi apparatus typical of protein-secreting cells as well as secretion granules and vacuoles containing spherites. These cells presumably produce extracellular digestive enzymes. Both the basophilic cells and digestive cells usually contain near-spherical peroxisomes which seem to be involved in the production of catalase (Lobo-da-Cunha 1997). These organelles are delimited by a single membrane and are also common in the digestive gland of bivalves (Owen 1973) and gastropods (Lobo-da-Cunha et al. 1994).

The intestine can be divided into anterior and posterior portions separated by a complex valve, with the anterior intestine containing the dorsal channel, and histologically it generally resembles the stomach (Fretter 1937). The valve consists of anterior and posterior sections. Muscular contraction of the anterior valve nips off a part of the food string which is rotated by the posterior valve into a faecal pellet (Fretter 1937). The posterior intestine has several to many loops and is lined with ciliated epithelium in which there are many mucous cells. The rectum is a short ciliated tube that passes through the muscles of the body wall and opens at the anus.

14.2.3.5 Renopericardial System

The circulatory system of chitons consists of a dorsal heart, few main blood vessels, and mainly haemocoelic sinuses that extend through the body. The heart consists of two lateral auricles and a long ventricle and lies within a large pericardium attached to the body wall below the two posterior shell valves. As in all molluscs, other than cephalopods, the blood vessels are not lined with endothelium but are surrounded by connective tissue (Eernisse & Reynolds 1994). Within the gill lamellae, blood passes through the venous sinuses into 'capillary sinuses' to the efferent branchial channels (Fischer et al. 1990) and from there to the auricles. Blood enters the auricles via the auricular pores which vary from two to several pairs, and there may also be a single pore at the posterior end where the two auricles join behind the ventricle (Eernisse & Reynolds 1994). Between each auricle and the ventricle, there are one to four pairs of muscular valves (auriculoventricular ostia). Structurally, the heart is simple with an epicardium, a myocardium that is thinner in the auricles than the ventricle, and a pericardium which has a thin muscular wall (Økland 1980, 1981). Podocytes are located in the auricular epicardium.

The dorsal aorta gives off vessels as it extends anteriorly from the ventricle to just behind the first shell valve where it passes through a vertical diaphragm and opens into the cephalic sinus surrounding the buccal mass. This diaphragm is also present in some solenogasters (Heath 1905a) and at least some monoplacophorans (Lemche & Wingstrand 1959). Several blood channels connect with the cephalic sinus and run along the foot, including the large, so-called visceral artery that runs along the radular sac and then ramifies in the viscera (Hyman 1967). Other sinuses that connect with the cephalic sinus include the large median pedal sinus and smaller paired pedal, neuropedal, and neurolateral sinuses, the latter surrounding the lateral nerve cords and running along the inner side of the mantle groove. Venous blood moves from the visceral sinus to the pedal sinuses, and posteriorly the blood in these sinuses collects in the afferent branchial sinuses via a posterior transverse sinus (Hyman 1967; Eernisse & Reynolds 1994). Blood from the gonad, and from the pedal sinus and pedal nerve sinus, passes through the kidney before reaching the heart via the branchial vein, but blood from the visceral cavity, mantle, and pallial sinus does not (Heath 1905b).

The blood contains haemocyanin, and myoglobin occurs in the muscles associated with the buccal mass (Terwilliger & Read 1969), and sometimes other tissues contain haemoglobin-like compounds (Eernisse et al. 1988). Haemocytes and phagocytes occur in chiton blood, with the latter associated with immune responses, and fixed phagocytic cells in connective tissue also remove foreign material (Crichton et al. 1973).

The excretory system of chitons has been reviewed by Hyman (1967), Andrews (1988), and Eernisse and Reynolds (1994) and is compared with that of other molluscs in Chapter 6. Podocytes in the auricular wall are assumed to be the site of ultrafiltration as in other molluscs, although the podocyte pedicels of chitons do not possess slit diaphragms (Økland 1980). The contractile pericardium probably aids in excretion by circulating the pericardial fluid and assisting the movement of the ultrafiltrate to the kidneys by way of the pair of renopericardial ducts lined with very long cilia (Økland 1981).

The kidneys lie ventrolaterally along the length of the floor of the visceral cavity, varying from a simple straight tube to doubled back tubes (Hyman 1967). The tubes give off simple to branching diverticula associated with the blood sinuses. The kidney epithelial cells are cuboidal and possess microvilli and secretory vacuoles with granular contents (Andrews 1988). In *Nuttallina* (Mopalioidea), these cells have a highly infolded basal cell membrane with that and other ultrastructural features suggesting active transport (Eernisse & Reynolds 1994). Excretory granules found in these cells are released into the lumen of the kidney. The kidney opens to the exterior in the posterior part of the mantle groove by a narrow duct that leads from a swollen sac-like part of the kidney.

The protonephridia of larval *Lepidochitona* are generally similar to those of other molluscs (Bartolomaeus 1989; Baeumler et al. 2011), and the ontogeny of the kidney development was described by Baeumler et al. (2012).

14.2.3.6 Reproductive System

Almost all chitons have separate sexes, although two brooding species, in which self-fertilisation is common, are hermaphroditic (Eernisse 1988; Buckland-Nicks & Eernisse 1993). There is also one reported case of 'occasional hermaphroditism' in which 1.5% of sampled individuals of the mopaliid *Plaxiphora aurata* were hermaphrodite (Scarano & Ituarte 2009). Most chitons are free spawners, with eggs and sperm being shed into the water column where the embryos develop as free-swimming lecithotrophic trochophore larvae. Mantle groove brooding has been reported in over 40 species (Sirenko 2015), with the brooded young released as either well-developed trochophores or crawl-away juveniles (see Section 14.2.4.4).

The mature gonad occupies much of the dorsal part of the body, lying just anterior to the pericardium. During development it is initially paired, but fusion into a single structure occurs in all chitons except *Nuttallochiton* and *Notochiton* (Hyman 1967). Internally the lateral and ventral walls of the gonad are folded. The pair of ciliated gonoducts arise dorsally from the posterior part of the gonad and run along the anterior portion of the pericardium before bending laterally to open in the mantle groove a little anterior to the kidney openings (Hyman 1967) and corresponding to the junction of the sixth and seventh valves (Plate 1901). In some species, part of the sperm ducts is swollen; this swelling has been called a seminal vesicle, but it is not known to store sperm (Hyman 1967). In many chitons, a glandular section of the oviduct may become sac-like, and gland cells in that part of the duct secrete 'gelatinous or mucous material' to form egg strings or stick egg masses to stones (Hyman 1967). In turbulent flows these gelatinous accumulations of eggs are probably broken up and the eggs dispersed.

As in other molluscs, there are differences in egg size between species with different developmental modes (Sirenko 2015). Mature egg diameters (without the hull) range between 150 and 240 μm in free spawning species, while in species that spawn eggs in mucous masses or strings, egg diameter ranges from 300 to 620 μm. In brooding species that release trochophores, egg diameter varies between 210 and 400 μm; in species which release juveniles, egg diameters range between 270 and 800 μm.

Chiton eggs are surrounded by a thick egg hull[2] that is produced primarily by microapocrine secretions from the ovum; in Chitonida additional secretions from the follicle cells are also involved (Buckland-Nicks 2014). The variation in hull morphology is of value taxonomically at not only the species level but also at higher levels as there are considerable differences between the egg hulls of the Lepidopleurida and Chitonida, with eggs of the former having a thick jelly-like hull and the latter having more robust and elaborately sculptured hulls (Selwood 1970; Richter 1986; Buckland-Nicks & Reunov 2009) (Figure 14.6).

The hulls of Lepidopleurida, the most plesiomorphic, may be simple and smooth, as in *Leptochiton* (Buckland-Nicks 2008) or perforated by large pores as in another lepidopleuran, *Deshayesiella* (Pashchenko & Drozdov 1998) and, in a primitive member of the Chitonida, the callochitonid *Callochiton* (Buckland-Nicks & Hodgson 2000) (see below). Other

---

[2] The term *chorion* is often used for chiton egg hulls, but we do not follow this practise because the hull material is also derived from the oocyte in some chiton species (see below). True chorions, as found in vertebrates, are formed only from follicle cells.

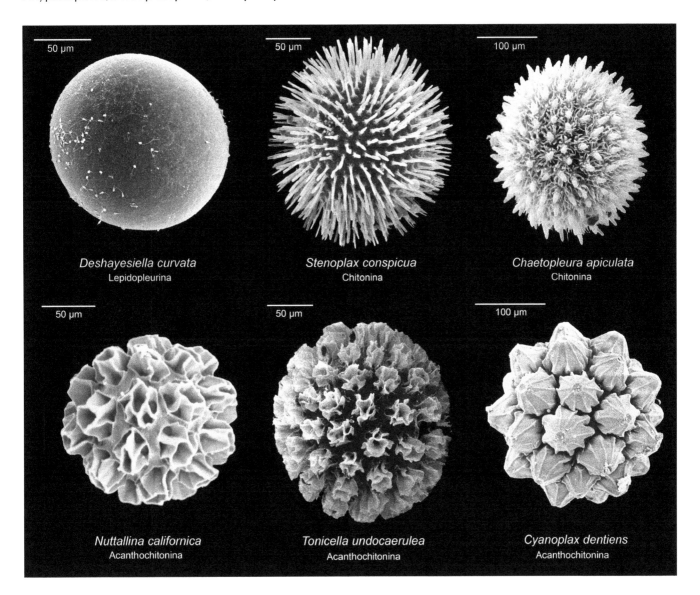

FIGURE 14.6   Examples of chiton egg hulls. (Courtesy of J. Buckland-Nicks.)

Chitonida have more derived egg hulls, some with spine-like projections with narrow bases (Chitonina), cup-like processes (cupules), or projections with broad bases (Acanthochitonina). There are considerable species-specific differences (Pearse 1979; Sirenko 1993; Buckland-Nicks 2006; Sirenko 2006; Buckland-Nicks 2008). The function of these complex egg hulls may include slowing sinking in the water column, predator deterrence, adhesion in jelly strings, or spacing to allow oxygenation (Buckland-Nicks 1993; Eernisse & Reynolds 1994). The ornament on the egg hulls of brooding species is less pronounced, although they show the general pattern typical of the group to which they belong (Eernisse & Reynolds 1994). It seems, however, that the most important function of the egg hull ornament appears to be to direct sperm to localised areas to aid fertilisation (Buckland-Nicks 1993).

While it has been known for some time that the chiton egg hull processes are primarily secreted by the egg (Selwood 1968; Richter 1986), it was only recently shown that the egg hull formation occurs by microapocrine secretion

(Buckland-Nicks & Reunov 2009). This involves secretions from microvilli of the oocyte or follicle cells and not by the more usual merocrine secretion as previously suggested (Selwood 1970; Richter 1986). Apocrine secretions in which apical blebs (containing the secretions) are liberated from the cell are known in the reproductive tracts of insects and mammals but have only rarely been reported in molluscs (in gastropods) (Buckland-Nicks & Reunov 2010; Buckland-Nicks 2014).

The study of Buckland-Nicks and Reunov (2010) on the mechanism of egg hull formation in *Callochiton* showed it differed from that of previously studied chitons. At the start of vitellogenesis, large vacuoles are released from the oocyte into the intercellular space between the follicle cells and the oocyte. Long microvillar projections from the oocyte are then extended into the intercellular space and reach the follicle cells. The long microvilli break down and release material that mixes with other secretions from short oocyte microvilli which together make up the jelly-like part of the hull

(Buckland-Nicks & Reunov 2010). The follicle cells collect the oocyte secretions and draw them out into radial stripes.

The eggs of Chitonida (other than *Callochiton*) differ from those of Lepidopleurida in having a rather stable fibrous layer covering a thin jelly-like inner layer, and it is this fibrous layer that forms the spines or cupules. In contrast to some earlier observations, the inner jelly-like layer is formed mainly by the oocyte from microapocrine secretions released from the tips of oocyte microvilli, while the fibrous layer is primarily created by the follicle cells (Buckland-Nicks & Reunov 2009). As in *Callochiton*, long microvilli from the oocyte radiate from the oocyte. These, and short oocyte microvilli, both disintegrate releasing their secretions, with the two types of microapocrine secretions mixing in the intercellular space. At this stage, long projections extend from the follicle cells[3] into the intercellular space, and these become coated with a cone-like mix of secretions. The projections then contract, drawing the secretions across the intercellular space and forming radial stripes, thus creating the moulds for the spines. In *Callochiton* the hull does not develop further, but in those chitons that produce spine-like projections on the hull, these projections create the moulds for forming the individual spines (Buckland-Nicks & Reunov 2009) or, presumably, the cupules in other chitons. The next stage is the formation of a thin vitelline layer formed by way of exocytosis of vesicles from the oocyte (Buckland-Nicks & Reunov 2010). The hull then develops a network of fibres and pores that extend through the hull, and membrane cups are formed that are aligned with the base of a pore structure in the membrane of the mature egg. The follicle cells adhere to the hull until ovulation when they are lost.

The follicle cells are therefore important in hull formation, helping to shape the mucopolysaccharides and proteins that make up the outer hull into cups, cones, or spines. The number of these structures corresponds to the number of follicle cells surrounding each egg during its formation (Eernisse & Reynolds 1994).

Thus in all studied chitons, the secretions that form the jelly-like layer are derived from the oocyte, and these form the entire hull in Lepidopleurida and *Callochiton*. In other Chitonina these secretions only contribute to the inner layer of the spines while the 'intermediate layer' of the hull is formed by proteinaceous secretions produced by the follicle cells and released by microapocrine secretion (Buckland-Nicks & Reunov 2009; Buckland-Nicks 2014). In contrast, the follicle cells of *Callochiton* and probably all Lepidopleurida just manipulate the oocyte secretions.

The sperm of the lepidopleurid *Leptochiton asellus* is similar to that of other free spawning molluscs in not having an elongated nucleus and in possessing a large acrosome (Buckland-Nicks et al. 1988a, b; Buckland-Nicks et al. 1990).

In marked contrast, spermiogenesis in Chitonida involves the elongation of the nucleus into a long anterior filament with a small acrosome at its tip. Intermediate sperm morphologies appear present in other lepidopleurids, such as *Deshayesiella curvata* and *Hanleya hanleyi*, which have eggs with smooth hulls but sperm with smaller acrosomes on short nuclear filaments (Buckland-Nicks 2006).

The formation of an acrosome independent of the Golgi body in the Chitonida is the only known example where small proacrosomal vesicles migrate to the anterior part of the spermatid (Buckland-Nicks et al. 1990). The sperm of members of the Lepidopleurida (Figure 8.5) and Callochitonidae (Hodgson et al. 1988) have a more normal midpiece with five to six spherical mitochondria arranged in a ring behind the nucleus. All other chitons have asymmetrically arranged midpiece mitochondria, and some also have one or more placed laterally, although their function is unknown (Eernisse & Reynolds 1994).

The Chitonida are unique among metazoans in their unusual interaction with the egg hull structure which has been the subject of some investigations. Sperm fertilises the egg at specific sites – either in the centre of cupules or at junctions of the bases of the cone-like processes. The breakdown of two granules within the terminal acrosome creates enzymatic activity facilitating the penetration of the egg hull by the anterior filament which then pierces the vitelline membrane. The inner acrosomal membrane is then exposed enabling the fusion of the anterior filament tip with a single egg microvillus. Simultaneously the plasma membrane is raised around the base of the microvilli to form a tube, which is surrounded by a small fertilisation cone, through which chromatin from the sperm nucleus passes into the egg (Buckland-Nicks et al. 1988a, b, 1990; Hodgson et al. 1988; Buckland-Nicks 1995; Buckland-Nicks & Hodgson 2000). This unique fertilisation method ensures that only chromatin is injected into the egg, thereby excluding the mitochondria, centrioles, nuclear membrane, sperm plasma membrane, and the flagellum (Buckland-Nicks 2014).

14.2.3.7 Nervous System and Sense Organs

The chiton nervous system consists of a circumoesophageal (or cerebrobuccal) nerve ring that lacks distinct cerebral ganglia and gives off paired, medullary lateral and ventral (pedal) nerve cords (tetraneurous condition) (Hyman 1967; Eernisse & Reynolds 1994; Moroz et al. 1994; Faller et al. 2012; Sigwart & Sumner-Rooney 2015) (see Chapter 7, Figure 7.1). Several connectives run between the ventral nerve cords, and in many species also between the lateral and ventral nerve cords, to create a ladder-like nervous system. The lateral nerve cords join posteriorly and supply each gill with two nerves, one associated with the venous and the other with the arterial blood sinuses (Fischer et al. 1990). A pair of nerves given off from the lateral part of the circumoesophageal nerve ring run anteriorly to the paired buccal ganglia which are connected by a commissure. Two other nerves from the same part of the nerve ring give rise to the subradular ganglia, which are also connected and lie between the subradular and radular sacs. The second pair of nerves from the buccal ganglia

---

[3] Similar follicle cell projections are seen during oogenesis of patellogastropods (Buckland-Nicks & Howley 1997; Hodgson & Eckelbarger 2000) and may be found in other molluscs with jelly-like egg envelopes. The egg hull of limpets differs from that of these chitons in remaining homogeneous, similar to the situation in *Leptochiton asellus* (Hodgson et al. 1988), and does not develop substructure like that in lepidopleurids (e.g., Pashchenko & Drozdov 1998).

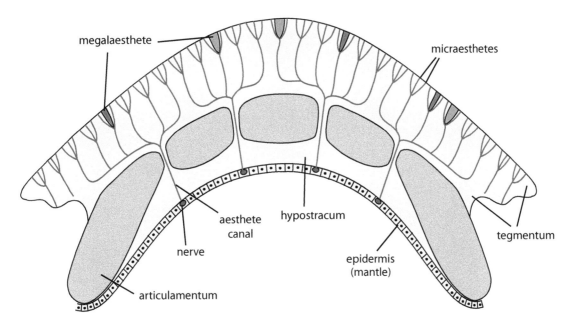

FIGURE 14.7 Diagrammatic longitudinal section of a chiton valve. Redrawn and modified from Nowikoff, M., Z. wiss. *Zool.*, 88, 154–186, 1907.

forms the ventral oesophageal commissure (which may have a small supraradular ganglion at its posterior limit) that lies between the oesophagus and the radular sac. There is also a pair of subradular ganglia connected to the posterior part of the cerebrobuccal ring.

The most important sense organs are the numerous and various types of aesthetes (= esthetes) and, in a few species, ocelli that penetrate the tegmentum through pores (e.g., Fischer & Renner 1979; Currie 1992) (Figures 14.7 and 14.8). The pores vary in size within a single valve and are termed *micropores* and *macropores*, which reflect the presence of *micraesthetes* and *megalaesthetes* (Schwabe 2010) respectively. The function of the aesthetes has been disputed, with early workers suggesting mechanoreception. However, some have a minute lens and retina-like structure and are clearly ocelli. Ocelli or *intrapigmented aesthetes* (ocelli with pigment in the aesthete tissue and covered by a lens in the shell) can respond to shapes. Species with *extrapigmented aesthetes* (the pigment is in the tegmentum, and the aesthete is covered by a modified apical cap) are also photosensitive and may have a crude kind of spatial vision as well (Li et al. 2015; Kingston et al. 2018). Pigmented aesthetes without a lens also occur in some species. Besides being photosensitive, aesthetes may also function as mechanoreceptors, chemoreceptors, and as secretory cells. Baxter et al. (1987, 1990) suggested that the larger megalaesthetes were primarily chemosensory and perhaps also (as noted above) secreted the properiostracum, while smaller micraesthetes were secretory. No evidence of micraesthete properiostracal secretion has subsequently been found and was not observed by Fischer (1988), but he suggested that the secretory cells of megalaesthetes might assist in preventing desiccation, predation, and epibiont fouling.

Each aesthete (Figure 14.8) consists of a narrow bundle of microvillus cells, nerves (from the mantle nerve cord),

secretory cells, photoreceptive cells, and peripheral cells that pass through the shell valves to terminate at the outer surface. The main part of the aesthete (the megalaesthete) gives off several branches (micraesthetes) and these, and the megalaesthete, terminate at the shell surface. The micraesthetes are simple epidermal papillae while the larger megalaesthete contain a cluster of several sensory cells below a cuticular thickening. There can be several hundred or even thousands on a single valve, and densities of up to 4,200 micraesthetes in a square millimetre have been recorded (Eernisse & Reynolds 1994). The reader is referred to the detailed accounts of aesthetes microstructure given by Eernisse and Reynolds (1994) and Schwabe (2010) for further information.

There is structural (and possibly functional?) similarity between polyplacophoran aesthetes and the caecae in the shells of articulate brachiopods (Pérez-Huerta et al. 2009), but they are thought not to be homologous by Reindl and Haszprunar (1994).

The dorsal surface of the girdle also has some sensory structures. Minute projections, often referred to as stalked nodules, contain dendritic nerve endings and arise from epidermal papillae (Fischer et al. 1980, 1988; Leise 1988) and occur around the bases of calcareous spicules (Leise 1988; Eernisse & Reynolds 1994). While most are mechanoreceptors (e.g., Fischer et al. 1988), photoreceptor cells are also found in epidermal papillae (Fischer et al. 1980).

In Lepidopleurina there is a pair of pigmented anterior 'Schwabe organs' that lie lateral to the head region (Sigwart et al. 2014) (Figure 14.1). Schwabe organs are light sensitive and appear to be derived from the larval ocelli (see below) (Sumner-Rooney & Sigwart 2015; Sirenko 2018). Schwabe organs are also important in terms of taxonomy, systematics, and development.

Other supposed sense organs in Lepidopleurina include the 'branchial' and 'lateral' sense organs found in the mantle

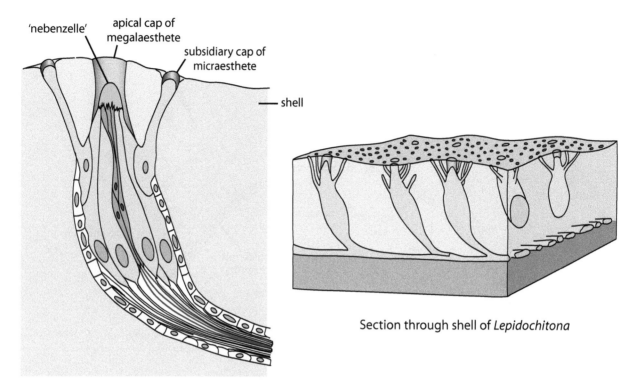

Lepidochitona

Section through shell of *Lepidochitona*

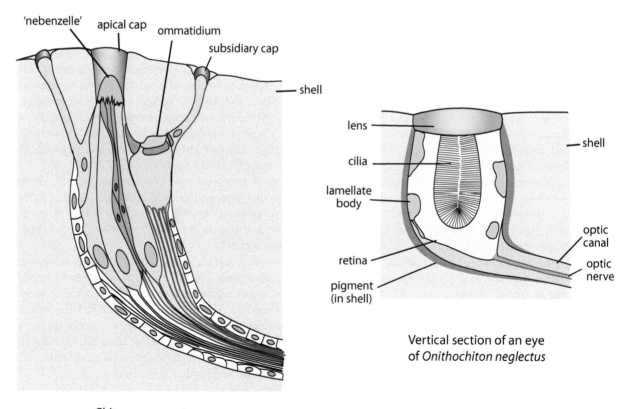

Chiton marmoratus

Vertical section of an eye
of *Onithochiton neglectus*

FIGURE 14.8 Examples of chiton aesthetes. Redrawn and modified from the following sources: *Lepidochitona* (Boyle, P.R., *Cell Tissue Res.*, 153, 383–398, 1974), *Onithochiton* (Boyle, P.R., Z. Zellforsch. *Mikrosk. Anat.*, 102, 313–332, 1969), and *Chiton* (Baxter, J.M. et al., *J. Zool.*, 220, 447–468, 1990).

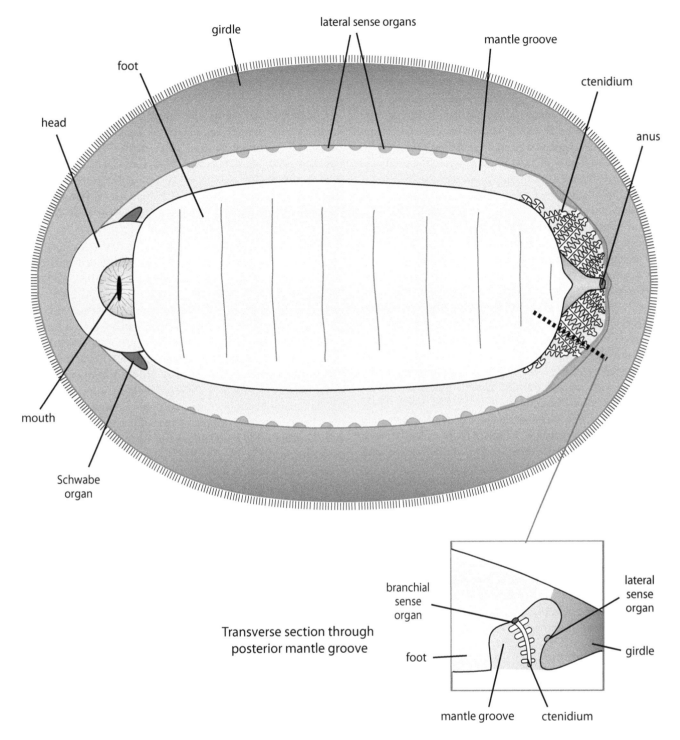

FIGURE 14.9 The mantle groove sensory organs of a generalised Lepidopleurida. Redrawn and modified from Sigwart, J.D. et al., *Front. Zool.*, 11, 7, 2014 and Yonge, C.M., Q. J. *Micros. Sci.*, 81, 367–390, 1939b.

groove. The branchial sense organs (Figure 14.9) are on the efferent (outer upper side) of the ctenidial axis in the inhalant chamber of the mantle groove, unlike the osphradia in Chitonida which are in the exhalant chamber. Within the epithelium of these organs, supporting cells surround free nerve processes that arise from the underlying efferent ctenidial nerve. The lateral sense organs in Lepidopleurina (Figure 14.9) are found anteriorly in the inhalant chamber and consist

of a series of small patches on the outer wall of the mantle groove lying separate, or anterior to, the ctenidia. They contain sensory cells innervated by the lateral nerve cord in addition to supporting and secretory cells. Haszprunar (1987a) argued that osphradia might have been secondarily lost in lepidopleurids because of the posterior gills. He did not consider that the branchial and lateral 'sense organs' were true sensory organs because they were poorly developed.

Previously, only members of Chitonida were thought to have a pair of true osphradia, but Lindberg and Sigwart (2015) noted that the putative chiton osphradia were innervated by the lateral nerve cords rather than ctenidial nerves and suggested that these structures were actually posterior sense organs (PSO) as in the 'aplacophorans'. In chitons, each PSO is a stripe of pigmented, raised epithelia lying posterior to the gills on the roof of the mantle groove. The PSO is yellowish-brown due to pigment granules in the supporting cells. These cells have microvilli, and the sensory cells are ciliated (Haszprunar 1987a). The posterior location of the PSO in the exhalant current suggests a role in reproduction, possibly in the coordination of spawning (Haszprunar 1987a).

Larval sensory organs include the ocelli and a unique ampullary system. The ampullary cells are four pairs of FMRFamide-positive neurons located beneath the apical tuft in the trochophore and with connections to the cerebral ganglia (Haszprunar et al. 2002; Voronezhskaya et al. 2002) (see Chapter 7). The pair of ocelli is found posterior to the prototroch in the trochophore larvae and has the molecular characteristics of true eyes (Vöcking et al. 2015). Each ocelli has a cup-shaped cluster consisting of eight to nine granular pigment cells with short microvilli and a few sensory cells with long microvilli. These microvilli fill the cup-like space forming a rhabdomeric photoreceptor (Rosen et al. 1979; Fischer 1980). These ocelli persist through metamorphosis and can be seen until the valves become too opaque (Eernisse & Reynolds 1994); in Lepidopleurina they are elaborated to form the Schwabe organs (Sumner-Rooney & Sigwart 2015; Sirenko 2018).

The muscular subradular organ lies within the subradular sac but is protruded from it before feeding with observations suggesting that it 'tastes' the substratum. Ultrastructurally there are ciliated cells interspersed with microvillus epithelial cells with secretory granules. The epithelium has a nerve supply from the subradular ganglia (Boyle 1975).

The cerebral commissure and ventral nerve cords have neurosecretory tissue lying between these structures and lacunae of the blood system that may be involved in neuroendocrine regulation of gametogenesis (Vicente & Gasquet 1970).

14.2.3.8 Chromosomes

Chromosome data for chitons have mainly been provided by Nakamura (1985) and Yum (in Todt et al. 2008b), jointly covering 22 taxa. The diploid chromosome number ranges from 12 to 26. Some families show more variation than others: Ischnochitonidae 24, Chitonidae 24–26, and Acanthochitonidae 16–24. The chromosome arm morphology also varies (Todt et al. 2008b).

14.2.4 BIOLOGY, ECOLOGY, AND BEHAVIOUR

14.2.4.1 Habits and Habitats

Chitons are exclusively marine and are distributed worldwide. While most live in the rocky intertidal or shallow sublittoral, some live in the deep sea to over 7,000 m. A few species live exclusively on algae, some only on coralline algae or marine angiosperms such as seagrass. Others are endemic to sunken wood, in multiple radiations in the deep sea.

Many chitons are well adapted to intertidal life on hard shores. They cling to hard surfaces using their oval, muscular foot, aided by the ventral surface of the muscular girdle and mucous secretion from the sole. When dislodged they can curl up into a ball, providing protection for the foot and mantle grooves.

Compared with many other intertidal molluscs, there are relatively few studies on the physiological adaptations of chitons to the intertidal zone (e.g., Boyle 1977; Focardi & Chelazzi 1990; McMahon et al. 1991; Carey et al. 2012, 2013) even though numerous species occupy different parts of the shore (e.g., Otaïza & Santelices 1985) and have different physiological profiles and feeding habits (e.g., Carey et al. 2013).

14.2.4.2 Locomotion

Movement in chitons is expedited by coordinated muscular waves of opposed foot contractions (Sampson 1985; Voltzow 1988; Kuroda et al. 2014) that enable them to move forward or backward (see also Chapter 3), with the speed varying from slow to relatively fast (~10 cm/min) in a few taxa such as *Ischnochiton* (Boyle 1977). Monotaxic retrograde waves are used by chitons for forward locomotion and direct waves for backward movement (Kuroda et al. 2014). A few genera with extensive girdles, such as *Cryptoplax*, have worm-like bodies that enable them to crawl into tight crevices, and at least one 'fleshy' species (*Notoplax cuneata*) can move through soft substrata using peristaltic waves of contraction along the body (Boyle 1977).

Not all chitons behave in the same way, with different movement patterns helping to minimise zonal overlapping during feeding excursions and thereby reducing interspecific competition for food. (e.g., Chelazzi et al. 1983a). Trail following occurs in some chitons (e.g., Chelazzi et al. 1990, 1993); some intertidal species return to home resting points after feeding excursions, and some exhibit aggressive behaviour when the 'home' is invaded (Chelazzi et al. 1983b; Chelazzi & Parpagnoli 1987). Some taxa use magnetoreception for orientation (Sumner-Rooney et al. 2014).

14.2.4.3 Feeding and Diet

Sigwart and Schwabe (2017) identified seven major ecological feeding strategies in the polyplacophorans – carnivorous, detritivorous, epizoophagous, herbivorous, omnivorous, spongiovorous, and xylophagous. These diets include sponges, bryozoans, barnacles, diatoms, foraminiferans, newly settled larvae, and algal crusts scraped from the substratum with the radula (Eernisse & Reynolds 1994). A few taxa feed directly on foliose algae and marine angiosperms (e.g., *Stenochiton*), while some deep-water species feed on detritus and others (such as several species of *Leptochiton* and *Ferreiraella*) directly on decaying wood or overlying bacterial biofilm (e.g., Sirenko 1998; Duperron et al. 2012; Sigwart 2016). A few species are associated with hydrothermal vents (e.g., Saito & Okutani 1990).

Several taxa from three superfamilies – *Placiphorella* (Mopaliidae), *Loricella* (Loricidae), and *Craspedochiton* (Acanthochitonidae) – are predators, capturing small crustaceans, worms, or brittle stars by trapping them under the

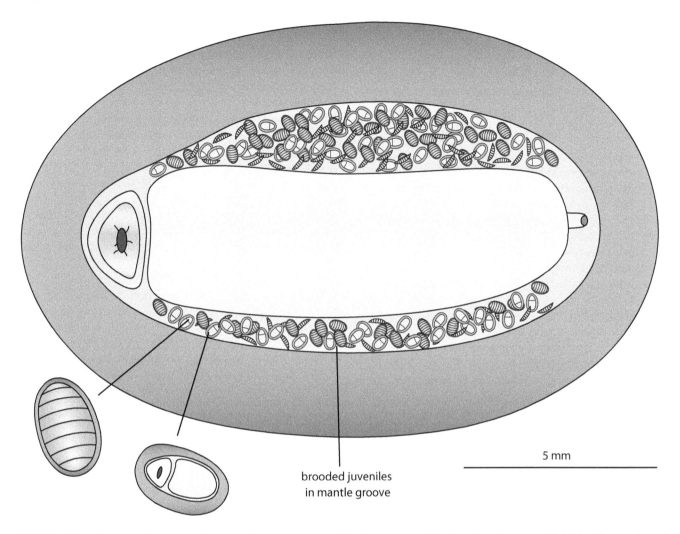

FIGURE 14.10 Brooding in *Ischnochiton mayi*. Redrawn and modified from Pearse, J.S., Polyplacophora, pp. 27–86, in Giese, A.C. and Pearse, J.S., *Reproduction of Marine Invertebrates. Molluscs: Pelecypods and Lesser Classes.* Vol. 5, Academic Press, New York, 1979.

anterior portion of the girdle, a habit they have all acquired independently (Saito & Okutani 1992; Eernisse & Reynolds 1994). These taxa can also supplement their diet by grazing (Saito & Okutani 1992). Another carnivorous chiton, *Lepidozona*, feeds on tiny benthic animals and foraminiferans (Latyshev 2004).

The grazing habit of most chitons appears to have been adopted early in their evolution. Furongian (Cambrian) to Lower Ordovician chitons are commonly found with stromatolites and oncolites, suggesting that they probably fed on the cyanobacterial and algal mats that formed these structures. By the Middle Ordovician their habitats had diversified (Pojeta et al. 2010), including the evolution of the foot-less chitons, the biology of which is a mystery (see Chapter 13).

14.2.4.4 Reproduction, Development, and Growth

Most chitons practise broadcast spawning with eggs and sperm released into the water. Unfertilised eggs are typically shed into water singly or in gelatinous strings and fertilised in the water column, but a few species brood eggs in the mantle grooves. In these brooding species, the eggs are maintained in the mantle groove of the female and fertilised by sperm moving along it, aided by the respiratory currents. The brooded

young may be released as developed trochophores (Creese 1986) or crawling juveniles (Pearse 1979). A few others, such as *Stenoplax heathiana*, attach a gelatinous mass containing their (fertilised?) eggs to rocks, and their larvae emerge at a late stage of development (Heath 1899) (Figure 14.10).

Chitons exhibit no known mating behaviour, but female spawning generally occurs at night (Boyle 1977) and is usually induced by the release of sperm. Some chitons release very large numbers of eggs, in some species over 100,000 (Hyman 1967). Spawning is determined by environmental cues (moon, tide, day length) in some studied taxa (e.g., Boyle 1977; Yoshioka 1989; Barbosa et al. 2009).

During embryonic development, chitons have a typical spiralian pattern of equal cleavage (Heath 1899; Grave 1932; Biggelaar 1996). Watanabe and Cox (1975), Buckland-Nicks et al. (2002), Lord (2011) and Sirenko (2018) provide detailed accounts of chiton larval development. The trochophore larvae are lecithotrophic and, in most species, free-swimming for periods ranging from minutes to a few days. Chiton trochophores possess unique features which include details of the prototroch and the larval eye position. The prototroch is unique in Mollusca in consisting of two or three irregular rows of trochoblasts, although their cell lineage origin (1q and

2q) is the same as that of *Patella*[4] (Henry et al. 2004). Chiton trochophores uniquely have their larval eyes (ocelli) in a posttrochal (below the prototroch) position (rather than pretrochal – above the prototroch), and they are derived differently in development (from second-quartet micromeres 2a, 2c) from those in other spiralians which arise from first-quartet micromeres (1a, 1c) (Henry et al. 2004) (see also Chapter 8).

The trochophore develops directly into a juvenile chiton, there being no veliger stage. Besides a change of body shape, other changes during metamorphosis include calcification of the girdle spicules and then the shell valves (Kniprath 1980; Leise 1984) and radular development (Eernisse & Kerth 1988).

The cells that give rise to the epidermal sclerites (spicules) in the larva are derived from different cells than the embryonic shell valves (Henry et al. 2004). The sclerites extend both pre- and posttrochally and originate from cells 1a, 1d, 2a, 2c, 3c, and 3d while the initially six (Kniprath 1980), then seven shell valve precursors are all posttrochal (see Chapter 8, Figure 8.33) and derived from 2c, 2d, and 3d (Henry et al. 2004). The eighth shell valve, the most posterior, appears later in development during metamorphosis at settlement, when the body elongates and is flattened dorsoventrally (Sirenko 2018). Each rudimentary shell valve is covered by long microvilli from bordering epithelial cells, and this covering was termed the 'stragulum' (Latin for cover) by Kniprath (1980). The stragulum completely covers the crystallisation chamber of the shell valve field, shielding it from the external seawater (see Chapter 3, Figure 3.15). The shell valves are originally rod-shaped and grow anteriorly and posteriorly with the tegmentum being formed first, then the hypostracum (Kniprath 1980).

The idea that chiton shell valves were originally formed by spicule fusion (Blumrich 1891; Pojeta 1980; Salvini-Plawen 1985a) is not supported by their development. Carter and Hall (1990) suggested that the valves may have originated from a modification of spicule calcification to form subcuticular shell plates, rather than by spicule fusion.

The muscles and dorsal shell valves are not simultaneously developed, in contrast to the situation in segmented animals such as annelids (Friedrich et al. 2002; Wanninger & Haszprunar 2002b).

14.2.4.5 Associations
Diseases, parasites, and other symbionts of chitons were reviewed by Lauckner (1983) and are outlined in Chapter 9.

14.2.4.6 Predators, Defence, and Behaviour
Chitons are prey for other molluscs such as carnivorous whelks and octopuses, as well as other predatory animals including fish, birds, starfish, and crabs. The larger species are sometimes eaten by humans, including early humans (Marean et al. 2007). Besides the shell valves, and spines in a few species, defensive strategies include clamping, crypsis, rather rapid movement away from light (for example when a rock they are sheltering under is overturned), or, when dislodged, rolling into a ball.

---

[4] No comparisons with other taxa were made.

14.2.5 Diversity and Fossil History

This relatively small group has been estimated to have about 920 living species (Schwabe 2005), with about 120 of these being members of Lepidopleurina and the remainder members of Chitonida.

Polyplacophorans evolved from unknown 'placophoran' ancestors in the Cambrian and have a reasonably good fossil record (see Chapter 13), with the oldest fossil recognised as a polyplacophoran being from the Furongian (Cambrian) (Yates et al. 1992; Pojeta et al. 2010). They were fairly morphologically stable during their history, as shown by some intact fossils. The articulamentum is lacking in all chitons before the Devonian. The status of some of the Cambrian taxa as polyplacophorans is not universally accepted (see Chapter 13 for details and discussion).

Because the lateral teeth of the chiton radula contain magnetite, the strength of the field measured in an exemplar has become the 'chiton tooth standard' for calibrating 'anhysteretic remnant magnetisation' (ARM) strength of magnetite in nature, including that acquired by iron fixing bacteria, other molluscs, geological sediments, and even magnetite biomineralisation in the human brain (Kirschvink et al. 1992).

14.3 THE MONOPLACOPHORA (= TRYBLIDIA, NEOPILINIDA, TERGOMYA IN PART)

14.3.1 Introduction

Monoplacophorans were known as Paleozoic fossils since the late 18th century and until the mid-1950s were thought to have become extinct during the Devonian Period in the Mid-Paleozoic (about 375 Mya). Thus, in 1952, when the Danish research ship *Galathea* found living monoplacophorans in water 3,570 m deep off the coast of Costa Rica (Lemche 1957), it was considered one of the great biological discoveries of the 20th century, ranking alongside the few similar discoveries such as a living coelacanth in 1938 and the Wollemi Pine in Australia in 1995. Their discovery caused a flurry of excitement with several publications appearing in prestigious journals and culminating in a monograph detailing their anatomy (Lemche & Wingstrand 1959). Later Wingstrand (1985) re-examined the original material and described in detail another species, *Vema ewingi*, from even deeper water (5,607–6,489 m) off Peru. Since then several new taxa have been described, including some minute species, with over 35 known living taxa distributed worldwide between 175 and 6,489 m (Sigwart et al. 2018). Only two of these additional species have been described anatomically in detail (Schaefer & Haszprunar 1996; Ruthensteiner et al. 2010).

14.3.2 Phylogeny and Classification

The name Monoplacophora has been variously used by palaeontologists for limpet-like and coiled univalve fossils and either restricted to a few taxa or encompassing a much more diverse assemblage (see Chapter 13 for details). Because of

FIGURE 14.11   Living monoplacophorans. Left – *Laevipilina hyalina.* Santa Rosa-Cortes Ridge, California, USA (32° 59.0' N, 119° 32.8' W), 367 to 389 m. (Courtesy of G. Rouse.) Right – *Neopilina* sp. at 'Utu' seamount, American Samoa (12.2740° S, 168.3670° W, approximately 3,837 m) showing apparent radular scratches within trackways. Image from Sigwart, J.D. et al., *Mar. Biodivers.*, 48, 1–8, 2018. Original video data from the NOAA Office of Ocean Exploration and Research. Reproduced here under Creative Commons Attribution 4.0 International Licence (http://creativecommons.org/licenses/by/4.0/).

the different concepts surrounding the name, various alternatives have been proposed and calls made for the rejection of the 'wastebasket' taxon Monoplacophora. For example, Peel (1991b) used the higher taxon (Class) Tergomya (see Chapter 13) based on univalves interpreted as being exogastrically curved or coiled, including the tryblidiids, the extinct Cambrian group Helcionelloidea, and living monoplacophorans. In an attempt to further refine the group, the name Tryblidia (as an order or a class) has been used for living monoplacophorans. This name is based on the Silurian fossil *Tryblidium* (Tryblidiidae) which has a flat, limpet-like shell with the apex at the anterior-most point. Internally there is a set of muscle scars suggesting a similar arrangement to the shell muscles found in modern monoplacophorans. The exterior is covered in thick lamellae, and the shell is punctured by many pores reminiscent of chitons (Knight 1952; Knight & Yochelson 1958, 1960), although such structures are unknown in the very thin shells of living monoplacophorans. Besides the muscle attachments, there is little that is similar in *Tryblidium* to modern monoplacophorans, which differ in having a more arched, thin shell, with the interior lacking obvious scars and the apex set near, but behind, the anterior edge (Figures 14.11 and 14.12).

BOX 14.2   HIGHER CLASSIFICATION OF MONOPLACOPHORA

(Class) **Monoplacophora**
(Order) **Tryblidiida** †
(Order) **Neopilinida**

(† = extinct taxa)
See Appendix for detailed classification.

Tryblidium was originally proposed for fossil limpets with eight pairs of symmetrical muscle scars, and the group was

thought by most palaeontologists to be related to patellogastropod limpets and thus gastropods. Dall (1893 p. 287) cautioned that the resemblances could result from convergence – 'it is almost inconceivable that the Silurian form should have any closely allied recent representative' and 'a peculiar disposition of the organs which might, indeed, have paralleled in some particulars the organisation of some of the Chitons of that ancient time'. It took another 45 years before a similar observation was made (Wenz 1938–1944) and a further 18 years before the finding of living monoplacophorans confirmed Dall's original insights.

Lindberg (2009) argued that there was evidence in the asymmetry of mesoderm-derived structures to suggest living monoplacophorans appeared to be primitive because of secondary simplification as a result of paedomorphosis. Certainly, reduction and simplification can be seen in the tiny (<1 mm) brooding monoplacophoran *Micropilina arntzi* (Haszprunar & Schaefer 1997), including no postoral tentacles, only three pairs of gills and kidneys and one pair of gonads. Also, there is just a single pair of radular cartilages, but the number of shell muscles remains the same, and the nervous system is very like that of larger monoplacophorans.

14.3.2.1   Sister Group Relationships

The recognition of more fossil species and genera extended the group back into the Ordovician and Cambrian (see Chapter 13). Initially, monoplacophorans were recognised by multiple, symmetrical muscle scars in limpet-shaped shells, but similar configurations of muscle scars have been found in high conic and also coiled shells. Whether these fossils are monoplacophorans, gastropods, or brachiopods remains controversial (see Chapter 13 for discussion).

Pores in the shells of *Tryblidium* caused Knight and Yochelson (1958 p. 40) to comment they were 'similar to those carrying nerves to the aesthetes of some Amphineura. For this reason, and because of other similarities, we feel that these two classes of molluscs and the Gastropoda are clearly

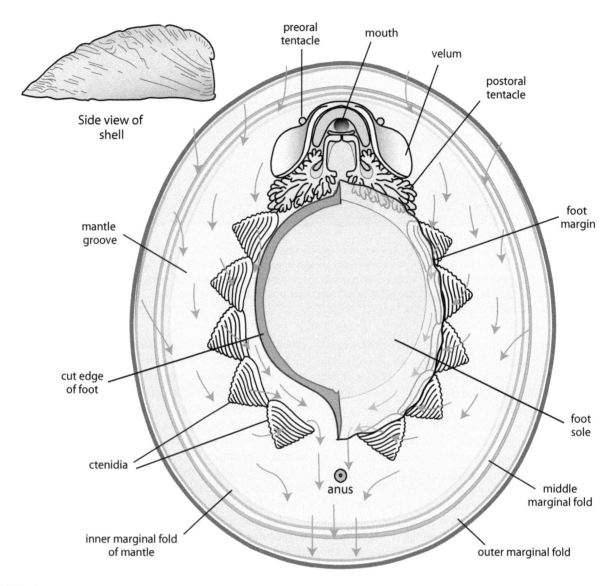

FIGURE 14.12 Ventral view of *Neopilina* showing the head-foot morphology, the assumed direction of water currents, and a lateral view of the shell. Redrawn and modified from Lemche, H. and Wingstrand, K.G., *Galathea Report*, 3, 9–71, 1959.

related'. Forty-eight years later, Giribet et al. (2006) reported a monoplacophoran nested inside Polyplacophora based on a short sequence of 28S rRNA and proposed the name Serialia for Monoplacophora + Polyplacophora. This result was questioned by some (e.g., Steiner in Haszprunar 2008; Wägele et al. 2009), but additional data suggested a relationship between chitons and monoplacophorans in some analyses (Wilson et al. 2010; Stöger et al. 2013) but not all (Smith et al. 2011) (see Chapter 12 for discussion of the Serialia concept).

14.3.3 MORPHOLOGY

14.3.3.1 Shell and Protoconch

Shells in living and fossil monoplacophorans typically have the apex at the anterior end of the shell, and, notably in some fossil species, it may overhang the anterior edge. The shape of the aperture varies from almost circular to pear-shaped in some fossil taxa. The height of the shell varies from rather flat

to a low conical shape. Living species are typically less than 1 cm in length, but most fossil taxa we include in this class ranged between 2 and 5 cm.

In living species, the outer surface of the shell is covered by a thin periostracum that is more similar to that of other conchiferans than to chitons (Poulicek & Jeuniaux 1981). An outer prismatic shell layer and an inner foliated layer lie beneath the periostracum (see summary in Haszprunar and Schaefer 1997), which is the same configuration of shell layers present in the few fossil monoplacophorans reported on, despite their thicker shells. Examination of *Tryblidium reticulatum* from the Silurian of Götland, Sweden, by Hedegaard (1996) indicated that the inner shell layer is foliated, thus disputing an earlier interpretation by Erben et al. (1968) who reported it as being recrystallised stacked nacre. The shell structure of living monoplacophorans was most recently examined by Checa et al. (2009). They found that only one (*Veleropilina zografi*) of the three species they examined had part of the internal surface covered with what they considered a layer of true nacre. The rest of the internal surface of the shell of that species, and the other two species they examined, was covered by an aragonitic material consisting of lath-like, instead of brick-like, crystals, arranged into lamellae that resemble the foliated calcite of bivalves. They proposed the name foliated aragonite for this structure. They argued that these crystals apparently lacked preformed interlamellar membranes and were not a form of nacre, although they could have been derived from it. They considered all previous reports of nacre in living monoplacophorans to actually represent an aragonitic foliated layer. The outer prismatic layer consists of large hexagonal crystals that are unusually prominent in some species.

Lemche and Wingstrand (1959) illustrated a spirally coiled 'protoconch' they found on a single specimen of *Neopilina galatheae*, although all subsequent shells of living monoplacophorans that have retained the earliest ontogenetic stage, and all known Paleozoic examples, have cap-shaped protoconchs. The spirally coiled structure of Lemche and Wingstrand was mistakenly based on an area of shell repair (Wingstrand 1985). Wingstrand (1985) illustrated a protoconch of a supposed juvenile monoplacophoran, but this was also in error, being a bulbous protoconch (previously figured by Menzies 1968) belonging to a lepetelloidean limpet. Illustrated cap-shaped monoplacophoran protoconchs are highly symmetrical, being 123 to 150 μm in diameter, this size comparing favourably with the known mature egg diameters (200–350 μm) (Haszprunar & Schaefer 1997). In larger species, the protoconch is lost as the shell grows.

Shell pores are present in the shells of at least some fossil triblidiid monoplacophorans as 'microscopic perforations that branch on approaching the outer shell surface in [a] manner suggesting minute openings that carry nerves for aesthetes (or shell eyes) in many Polyplacophora' (Knight & Yochelson 1960 p. 179). Shell pores have not been found in living monoplacophorans, but some have shallow pits on the outer surface (Warén & Gofas 1996).

14.3.3.2 Head-Foot and Mantle

The animal has a simple head which does not possess a neck and is thus not free from the body. The head consists only of the ventral mouth and is separated from the foot by a transverse groove. The anterior part of the head consists of the hood-like oral lappets (or 'velum')[5] that are usually expanded laterally, with these lateral portions especially thickened and giving rise to a 'median velar ridge' (Figure 14.2). The entire ventral surface of the oral lappets is ciliated, with especially strong ciliation along its borders.

In some taxa, preoral tentacles arise in the groove on the anterior side of the oral lappets and are absent in others.

There is a single pair in *Vema ewingi* (Wingstrand 1985) and *Neopilina galatheae* (Lemche & Wingstrand 1959) and ten pairs in *Neopilina rebainsi* (Moskalev et al. 1983).

The mouth opening is a horizontal slit surrounded anteriorly by a thick, V-shaped anterior lip and posteriorly by a substantially smaller posterior lip with postoral tentacles, the latter showing a variety of morphologies and configurations (Moskalev et al. 1983). The mouth is heavily muscularised with five pairs of muscles (one transverse) on either side of the mouth. Two pairs terminate on the anterior shell while the posterior mouth muscles terminate adjacent to the first pair of dorsal-ventral retractor muscles.

Lemche and Wingstrand (1959) identified a pair of small tentacles on either side of the mouth of *Neopilina galatheae*. These ciliated tentacles comprised several cell types, including secretory cells. At the base of each tentacle was a small concentration of 'dark, granulate cells' (Lemche & Wingstrand 1959, figure 70) (see Chapter 7, Figure 7.41).

The anterior lip is cuticularised, the cuticle extending into the mouth where it is continuous with the jaw and preoral cuticular plate on the floor of the buccal cavity. The posterior lip is also covered by a thin cuticle which extends into the subradular sac.

The postoral tentacles arise from a ridge that extends posteriorly and then laterally behind the posterior lip. The part immediately behind the posterior lip is highly ciliated and separated from the lip by a distinct cleft. Distally these ridges curve laterally and ramify into numerous poorly ciliated tentacles. While this arrangement is typical of most monoplacophorans, in *Neopilina veleronis* the ridges are poorly developed and the tentacles represented only by a single pair of tubercles, although a transverse row of tentacles lies just behind each ridge (Moskalev et al. 1983).

A groove between the inner bases of the oral lappets and the outer edges of the tentacular ridges has been called the 'feeding furrow'. The parts of the lappets and tentacular ridges adjacent to the furrow are strongly ciliated, unlike the furrow itself.

While the oral lappets are presumably homologous to the similar structure in chitons, other structures associated with the monoplacophoran mouth such as the preoral tentacles, postoral tentacular tufts, and 'feeding groove' are absent in polyplacophorans, although homologies have been suggested with structures in other conchiferans (Lemche & Wingstrand 1959; Salvini-Plawen 1980). In particular, Lemche and Wingstrand (1959) suggested that the anterior preoral tentacles were homologues of the cephalic tentacles of gastropods as those tentacles are innervated from the cerebral ganglia. They also proposed that the postoral tentacle complex was homologous with the inner labial palps of bivalves and possibly cephalopod arms and scaphopod captacula.

The foot (Figure 14.12), located in the middle of the ventral surface, is approximately circular with the sole musculature developed peripherally but undeveloped in the centre (Figure 14.13). The foot is particularly small in species with depressed shells (e.g., *Monoplacophorus zenkevitchi*, *Veleropilina seisuimaruae*) and corresponds to only about a third or slightly more of the shell length in the latter species (Kano et

---

[5] Lemche and Wingstrand (1959) referred to this structure as the 'velum', because they believed it to be homologous with the velum of larval gastropods and bivalves and argued that if that was the case, the lateral portions of the monoplacophoran 'velum' would be homologous to the outer labial palps in adult bivalves, although there is no evidence to support this homology (Haszprunar & Schaefer 1997).

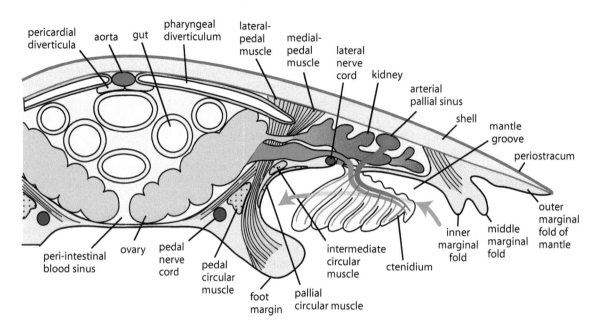

FIGURE 14.13 Diagrammatic transverse section through *Neopilina*. Redrawn and modified from Lemche, H. and Wingstrand, K.G., *Galathea Report*, 3, 9–71, 1959.

al. 2012). A report of an anterior pedal gland by Lemche and Wingstrand (1959) in *Neopilina* is incorrect (Haszprunar & Schaefer 1997), there being no pedal gland in any monoplacophoran other than some epithelial gland cells in the otherwise ciliated sole epithelium.

All living monoplacophorans have eight pairs of shell retractor muscles (see Chapter 3 for details), although the anterior-most two pairs are fused in *Micropilina arntzi* (Haszprunar & Schaefer 1996). These paired shell muscle bundles run from the foot to the shell and surround the visceral mass. Each muscle originates from a single point on the shell and consists of two parts, the medial pedal muscle which passes dorsally over the ventral nerve cord and ramifies in the centre of the foot, and the latero-pedal muscle which passes on the outside of the ventral nerve cord and ramifies in the marginal part of the foot. The Paleozoic tryblidiid monoplacophorans, such as *Tryblidium* and *Pilina*, also had eight pairs of shell muscles, the scars of which comprise one major oval to circular scar surrounded by the scars of several smaller muscles. Two of these latter small scars probably correspond to the oblique muscles in modern monoplacophorans, but their attachment points are further removed from those of the shell muscles in *Neopilina* and *Vema* (Lemche & Wingstrand 1959; Wingstrand 1985). Corresponding to each shell muscle in *Neopilina*, there are anterior and posterior pairs of oblique muscles, and pairs of mantle muscles and branchial muscles are also present (Lemche & Wingstrand 1959).

Muscles that encircle the foot and the roof of the mantle groove, which do not directly attach to the shell, connect with a small anterior pair of muscles (muscle y of Lemche & Wingstrand 1959; Wingstrand 1985). Retractor muscles of the anterior body include the preoral muscle and the anterior and posterior oral muscles. All three attach to the anterior shell and pass into the head region. There is also an oral

dilator muscle that attaches to the anterior part of the shell (Wingstrand 1985). At least some of these anterior muscles are strongly developed (judging from their large scars) in Paleozoic monoplacophorans (Figure 14.12).

14.3.3.3 Mantle Groove (Mantle Cavity)

A wide mantle groove encircles the animal between the lateral sides of the foot and the ventral mantle edge (Figures 14.12 and 14.13). It ranges from rather wide (compared with that in most chitons) to very wide, as in *Monoplacophorus zenkevitchi* (Moskalev et al. 1983) and *Veleropilina seisuimaruae* (Kano et al. 2012). In most taxa it contains five or six pairs of gills but as few as three in the minute *Micropilina arntzi* (Warén & Hain 1992) (see Table 14.1). These gills are somewhat muscular, and examination of living specimens shows that their movements appear to generate the respiratory currents, at least in part (Lowenstam 1978). The gills are also ciliated but lack the distinctive lateral cilia seen in other ctenidia (see Chapter 4). There are, however, rather long cilia on the outer bulb-like tips of the finger-like filaments, suggesting that cilia also play an important (but unknown) role. Beating or vibrating ctenidia are also found in nuculanoidean protobranch bivalves.

The gill histology (thick epithelium, small blood spaces, dense ciliation) suggests that they are not effective respiratory surfaces (Schaefer & Haszprunar 1997). Instead, this function is apparently served by the extensive surface of the mantle groove, immediately under which lies the venous pallial sinus and the outer wall of the kidneys, with both areas probable respiratory surfaces.

The kidneys are each associated with a gill and lie in the roof of the mantle groove in juxtaposition to both the venous and arterial mantle sinuses (Figure 14.13). Thus, the gills are each associated with a respiratory area on the roof of the mantle groove, which they ventilate. The importance of the

TABLE 14.1

Numbers of Paired Organs in Some Living Monoplacophorans

	Neopilina galatheae	*Vema ewingi*	*Laevipilina antarctica*	*Micropilina minuta*	*Micropilina arntzi*
Shell muscle bundles	8	8	8	8	8 (anterior two pairs fused in *M. arntzi*)
Gills	5	6	5	4	3
Kidneys	6	7	5	4	3
Gonads	2	2	2	2	1
Gonoducts	2	3	2 (female), 3 (male)	2	1
Atria	2	2	2	heart absent	heart absent
Lateropedal connectives	6	8	8	4	4

Modified from Haszprunar and Schaefer (1996)

kidneys in respiration relative to the gills is indicated by the larger number and size of the vessels connecting the kidneys and the auricles, compared with those of the gills. The location of the kidneys in the roof of the mantle groove differs considerably from their arrangement in Polyplacophora where they are in the visceral mass but is similar to the situation in some bivalves and gastropods, where the kidneys sometimes also form respiratory surfaces.

14.3.3.4 Digestive System

The ventral mouth opens to the buccal cavity which is laterally and ventrally lined with cuticle. Just behind the anterior lip lies the single thick jaw that is extended laterally into a projection on either side. This structure lies anteroventrally to the anterior end of the radula. Immediately behind it, the posterior lip opens to the subradular pouch which contains a large subradular organ and is glandular laterally. The small, ductless salivary glands open to the anterior-most part of the buccal cavity a little dorsal to the upper edge of the jaw.

The odontophoral muscles, cartilages, and bolsters are very like those in chitons (see Section 14.2.3.4 and Chapter 5), although the radular cartilages are substantially smaller in size and confined to the anterior third of the vesicles. Thus there is a pair of hollow, fluid-filled vesicles to which are attached the two pairs (lateral and medial) of cartilages, although a pair of small cartilages, detached from the medial cartilage, was present in *Neopilina galatheae* (Wingstrand 1985).

The monoplacophoran radula (Figure 14.14) is docoglossate and primarily stereoglossate and thus similar to that of chitons and patellogastropods, although it lacks the metallic hardening of some teeth seen in those taxa. Each row of monoplacophoran radular teeth has a central tooth, three pairs of lateral and two outer pairs of teeth usually interpreted as marginals (e.g., Warén & Gofas 1996) – a total of 11 teeth compared with 17 teeth in a row in chitons. The lateral teeth are stepped in configuration, the first and second pair of lateral teeth being lined up with the central tooth and the third lateral pair slightly posterior and lateral to the first two pairs. The cusps of each of the inner-most marginal teeth in most species are finely and deeply divided making them fan-shaped or rake-like, an exception being species of *Veleropilina* which

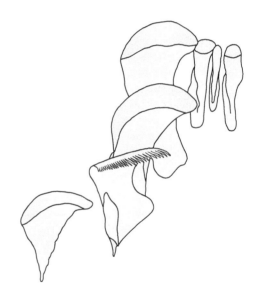

FIGURE 14.14 Half row (left) of the radula of *Neopilina galatheae* as an example of a monoplacophoran radula. Redrawn and modified from Lemche, H. and Wingstrand, K.G., *Galathea Report*, 3, 9–71, 1959.

have several short cusps (Warén & Gofas 1996). The central teeth are unicuspid, except in *Micropilina arntzi* which are tricuspid. The three pairs of lateral teeth and the outer marginal teeth have a thickened cutting edge that stains more intensely than the rest of the tooth. In the larger species (of *Neopilina* and *Vema*) the lateral teeth cutting edges are smooth or almost smooth with only indistinct denticles, at least in adults, and the outer marginals have a smooth or almost smooth cutting edge in all described species (McLean 1979; Warén & Gofas 1996).

The oesophagus consists of anterior, mid, and posterior parts. The anterior part is simple except for a broad ciliated dorsal food channel bordered by a pair of longitudinal folds. The midoesophagus has two pairs of large, lateral, thin-walled oesophageal pouches that were incorrectly termed pharyngeal pouches in earlier descriptions (Haszprunar & Schaefer 1997), and the posterior pair was incorrectly called the 'dorsal coelom' by Lemche and Wingstrand (1959). These pouches are lined with a thin epithelium with microvillus borders and sparse cilia. In the tiny *Micropilina arntzi* there are two pairs of moderate-sized pouches and in *Micropilina minuta* only a single large pair (Ruthensteiner et al. 2010). The posterior oesophagus is a simple tube that opens to the stomach.

The rather small stomach is triangular in shape and located in the middle of the body. The oesophagus opens anteroventrally, and the two lobes of the digestive gland communicate with the anterior part of the stomach via a single large opening. The digestive glands are lined with two cell types as in other molluscs, and, unusually, the lumen contains large food particles (Haszprunar & Schaefer 1997). The stomach lacks a gastric shield, sorting area, and typhlosoles. A caecum is located posteriorly and is short except in *Laevipilina antarctica*, where it is long. Reports of a protostyle in the larger taxa (*Neopilina* and *Vema*) are inaccurate; instead, the lumen of the caecum is filled with long microvilli from the epithelial cells (Haszprunar & Schaefer 1997).

The intestine departs the stomach posteriorly, a little to the left. It is long and composed of several (usually six) anticlockwise loops. Internally it is circular, with the ciliated epithelium lacking folds, and contains a faecal string. The rectum, which is not differentiated from the intestine, runs through the pericardium and opens at the anus, borne on a short papilla in the middle of the posterior mantle groove. Non-feeding specimens observed in the laboratory empty their gut within 24 hours (Wilson et al. 2009).

14.3.3.5 Renopericardial and Reproductive Systems

The main body cavity is a haemocoel while the coelomic spaces (lined with mesodermal endothelium) are the pericardium, kidneys, and gonads.

The heart lies posterodorsally in the body in a large pericardium, except in the tiny *Micropilina* which lacks a heart and probably also the pericardium (Haszprunar & Schaefer 1997). The rectum passes through the pericardium and ventricle, giving the false appearance of a divided heart. The anterior aorta runs from the ventricle to the buccal region. Two pairs of auricles that collect the blood from the three posterior kidneys and their walls contain podocytes. There are multiple (three to seven) pairs of kidneys, with larger-sized taxa having the largest number (Table 14.1). The kidneys vary in shape, are not interconnected, and open next to the ctenidial bases (Figure 14.15). The kidney walls are weakly muscular and are lined with large cells with microvillar borders, with vacuoles and infolded bases. There are no renopericardial ducts, although these were incorrectly reported by Lemche and Wingstrand (1959) and (tentatively) by Wingstrand (1985).

The main body cavity contains several types of haemocytes with pore cells (rhogocytes) the most common, while blood cells, amoebocytes, and fibroblasts are also found. The latter produce the collagen fibres that make up much of the body volume (Haszprunar & Schaefer 1997).

Monoplacophorans differ from the aplacophorans and polyplacophorans in having dorsal rather than ventral gonads. All examined monoplacophorans, except *Micropilina arntzi*, have two pairs of gonads and are gonochoristic. *M. arntzi* has a single gonad and is a simultaneous hermaphrodite. This minute brooding species is also the only monoplacophoran known to have a glandular female urogenital opening.

There are two pairs of gonoducts, one from each gonad, except in males of *Laevipilina antarctica* which have three pairs, two from the anterior testis. Each gonoduct empties to the lumen of the corresponding kidney in the middle part of the body.

Eggs are typically large (200–350 μm), and spermatozoa of *Laevipilina antarctica* are the primitive type (see Chapter 8), with a short, rounded head and a single flagellum (Healy et al. 1995).

14.3.3.6 Nervous System and Sense Organs

The monoplacophoran central nervous system is ganglionic, but the ganglia are indistinct and the main nerves are cord-like (Haszprunar & Schaefer 1997; Sigwart & Sumner-Rooney 2015) (Figure 14.15). Whereas chitons have a lobed brain (Sumner-Rooney & Sigwart 2018), in Monoplacophora a well-developed cerebral commissure connects the well-spaced cerebral ganglia, from where a paired nerve runs to the anterior mantle margin surrounding the head. The oral region is innervated by a pair of nerves which originate at the median side of the cerebral ganglia.

The pedal ganglia lie inside and adjacent to the cerebral ganglia, connected by the labial (or postoral) commissure, and the buccal connectives run from them dorsally. Two nerves from near the junction of the pedal connective and the pedal ganglia run to a small subradular nerve or ganglion that innervates the subradular organ in the larger species; no ganglion is present in the smaller species. The small buccal ganglia lie anterior and dorsal to the cerebral commissure in most species (Sigwart & Sumner-Rooney 2015).

Two prominent pairs of nerve cords extend posteriorly, the lateral nerve cords on the outer side of the shell muscles and the ventral (pedal) nerve cords on the inner side. Connections between these nerve cords correspond to the gaps between the shell muscles, at least in *Vema* and *Neopilina*. The ventral nerve cords form a loop, being connected posterior to the circumoesophageal ring by the interpedal commissure and fusing posteriorly at the back of the foot. In *Micropilina* the nerve cords are particularly thick laterally. A Y-shaped connective at the top of the ventral nerve cords gives rise to two anterior branches – the cerebropedal connectives and a thin connective between the pedal ganglion and the main ventral nerve cord. This connective has been found in all examined monoplacophorans. The lateral nerve cords are also connected posteriorly and, anteriorly, join directly with the cerebral ganglia.

The ring-like condition of the ventral nerve cord is unlike the parallel-sided ladder arrangement in chitons and more like that in patellogastropods. Unlike chitons, monoplacophorans have a pair of statocysts that lie slightly anterior to the pedal commissure; they contain multiple statoconia and are open to the external environment via a narrow duct (Lemche & Wingstrand 1959).

All living monoplacophorans lack eyes and osphradia, but the subradular organ (licker) is large and located in the

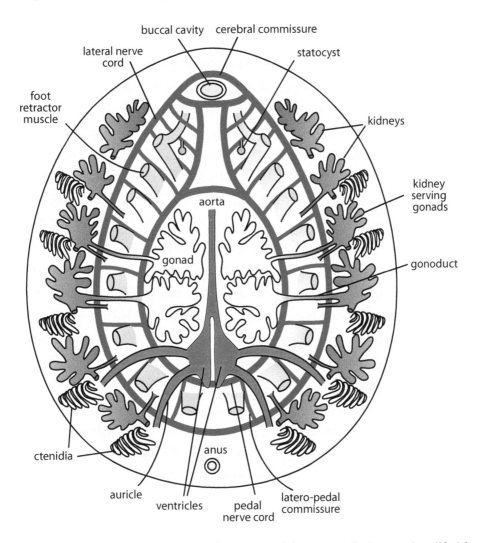

FIGURE 14.15 Diagrammatic dorsal view of *Neopilina* showing features of the anatomy. Redrawn and modified from Lemche, H. and Wingstrand, K.G., *Galathea Report*, 3, 9–71, 1959.

posterior part of the sublingual cavity. Its epithelium is ciliated and has a microvillus border.

The preoral and postoral tentacles are possibly sensory. Based on the presence of a high epithelium and rich innervation at the tip of each preoral tentacle, Lemche and Wingstrand (1959) suggested that the preoral tentacles might serve as a sensory organ involving chemoreception or even photosensitive structures (see Chapter 7, Figure 7.41). Postoral tentacles were not discussed. In *Laevipilina* the postoral tentacles and the epithelium of the mantle groove and the head have a high population of symbiotic bacteria in bacteriocytes (Haszprunar et al. 1995; Schaefer & Haszprunar 1996).

14.3.4 BIOLOGY, ECOLOGY, AND BEHAVIOUR

14.3.4.1 Habits and Habitats

Living monoplacophorans are found worldwide; some species live on hard surfaces in the deep sea, including manganese or phosphate nodules, and others are apparently found on soft sediments (Menzies et al. 1959), including sand (Schrödl et al.

2006). Ivanov and Moskalev (2007) have summarised habitat information for 19 monoplacophoran species, which have also been found on the continental shelf and seamounts at depths between 174 and 6,500 m (Figure 14.16). Early Paleozoic taxa are associated with relatively shallow-water faunas (<100 m) and vent communities, but moved into deeper water beginning in the Silurian (Warén & Bouchet 2001; Stuber & Lindberg 1989; Lindberg 2009). The shallowest recorded living monoplacophoran is *Laevipilina hyalina* which occurs in water 174 to 388 m deep on phosphate nodules on the Santa Rosa-Cortes Ridge, off southern California (Lowenstam 1978; McLean 1979; Wilson et al. 2009).

Seven living monoplacophorans occur in abyssal depths (4,000–6,000 m), but more than half occur in less than 1,000 m of water (Schwabe 2008; Lindberg 2009). Body size increases as depth increases, with the larger taxa (>5 mm) living at depths of over 3,000 m, and their size is comparable to the early Paleozoic tryblidiids which lived in shallow water (Figure 14.16). Smaller taxa remain present beyond 3,000 m, substantially increasing the variance in body size with depth.

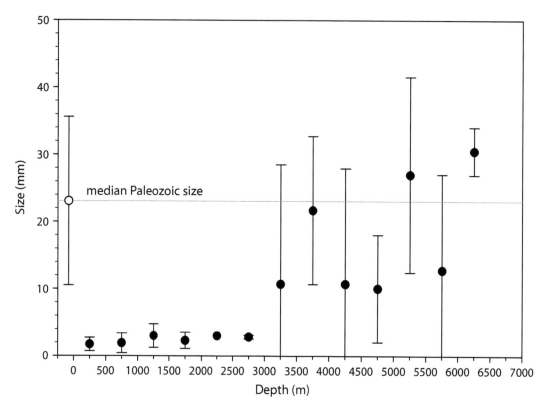

FIGURE 14.16 Bathymetric size distribution of living monoplacophoran species. Size data are median size grouped by 500 m bin ± standard deviation. Solid circles are living species; open circle = median size and standard deviation for Paleozoic (Cambrian to Devonian) taxa. Modified and updated from Lindberg, D.R., *Evolution: Education and Outreach*, 2, 191–203, 2009 and Schwabe, E., *Zootaxa*, 1866, 205–222, 2009.

14.3.4.2 Feeding and Diet

The stomach, digestive gland, and intestine contain the remains of intact foraminiferans, diatoms, xenophyophores, and dinoflagellates (Haszprunar & Schaefer 1997) or, in the case of *Vema ewingi*, diatoms, radiolarians, pelagic foraminiferans, sponge spicules, and 'innumerable bacteria-size particles' (Menzies et al. 1959). Schrödl et al. (2006) reported mineral particles, detritus, foraminiferans, diatoms, nematodes, and polychaete bristles in the stomach contents of *Laevipilina antarctica*. These reports appear to indicate deposit-feeding (e.g., Warén & Hain 1992), although *Neopilina* has been suggested to be a specialist feeder on xenophyophore protozoans (Tendal 1985). Ivanov and Moskalev (2007) recovered *Neopilina starobogatovi* from living gorgonarians (*Parastenella* sp.).

The habitat changes of monoplacophorans through time probably involved shifts in diet and consequently modifications to radular morphology (Stuber & Lindberg 1989). Thus, the radular teeth found in the deep-water, living monoplacophorans could not have been effectively utilised by their Paleozoic predecessors, which probably grazed on bacterial and algal films in shallow-water habitats.

14.3.4.3 Reproduction, Development, and Growth

Micropilina arntzi broods developing embryos (Haszprunar & Schaefer 1996), but based on egg size and protoconch morphology, other taxa are thought to be broadcast spawners with lecithotrophic development.

Sperm morphology (Figure 8.5) is typical of external fertilising species – i.e., ectaquatic (in the water column) or entaquatic (in the mantle groove). The high number of sperm cells in *Laevipilina antarctica* suggests the sperm are ectaquatic (Schaefer & Haszprunar 1996).

Reproductive and developmental studies are limited to observations on the brooded young in *Micropilina arntzi* (Haszprunar & Schaefer 1996) and the recovery of viable gametes from *Laevipilina hyalina* (Wilson et al. 2009), who reported that egg diameters ranged between 80 and 120 μm. They also observed a single egg being expelled which cleaved after 9 h at 12°C but failed to continue further development. In general, their development is thought to be typical of molluscs. In the brooding species the sole and the mantle edge appear early in development. Anteriorly there is a yolky mass that decreases with age, and the gut is originally a thick, straight tube lacking a mouth opening. The intestine commences coiling in late embryos. The shell muscles develop before the intestinal coiling and gonopericardial structures (heart, kidneys, gonads), with the gonads and statocysts appearing late in development. The posterior gills appear first, and they are added anteriorly.

14.3.4.4 Associations

Unidentified bacterial symbionts are maintained in bacteriocytes, the lateral roof of the mantle groove, the head, the proximal areas of the ctenidia, the lateral wall of the foot, and on the postoral tentacles (Haszprunar et al. 1995). Nothing is known of parasites or internal symbiotic bacteria, notably the gut bacteria, which are undoubtedly present.

14.3.4.5 Predation, Defence, and Other Behavioural Attributes

Little is known about these aspects of the biology of these animals, with only limited observations of species observed alive after recovery to the surface (Lowenstam 1978; Urgorri et al. 2005; Schrödl et al. 2006; Ivanov & Moskalev 2007; Wilson et al. 2009). Lowenstam (1978) observed specimens of *Laevipilina hyalina* (as *Vema* sp.) found on phosphorite nodules in water 348–402 m deep that stayed alive for 25 days. He noted (p. 232) that

'…strong incandescent light induced defensive behaviour similar to that of patellacean gastropods: an animal raised itself high above the foot, then tilted the shell either sideways or more commonly anteriorly, and thereafter rotated the shell in alternate directions sideways, as far as 90° from the body axis at rest. Movements of the shell were accompanied by acceleration in the beating of the gills. The animals then moved slowly to more shaded areas of the cobbles'.

The gill beating was reduced in intensity when the experiment was repeated using infrared or ultraviolet-filtered light, which may be associated with a reduced temperature increase with these light sources. Wilson et al. (2009) kept animals alive in the laboratory for nine days, although they did not feed. Recently, Sigwart et al. (2018) published a high-definition video of monoplacophorans on a seamount near American Samoa, taken using a remotely operated vehicle. These images show both occupied and unoccupied trackways on the substratum, and when occupied, apparent radular feeding marks are visible within the trackways (Figure 14.11).

14.3.5 Diversity and Fossil History

While some Paleozoic taxa (notably Tryblidiidae) are undoubtedly related to the living taxa, the characters that distinguish some Paleozoic 'monoplacophorans' from 'gastropods' (and vice versa) are open to alternative interpretations and have been the subject of much debate (see Chapter 13).

The living monoplacophorans distributed around the oceans of the world have been placed in seven genera and two families (Haszprunar 2008; Kano et al. 2012), and some undescribed species are known (Schwabe 2008; Sigwart et al. 2018). Besides an undescribed *Micropilina* from the Middle Pleistocene of Sicily (Taviani et al. 1990), there are no recorded Mesozoic or Cenozoic monoplacophoran fossils. They are common as early Paleozoic fossils but gradually decline during the late Ordovician and Silurian. This absence from the fossil record is puzzling but, given their rarity as living organisms, their mostly minute size and delicate shells, and because they live in habitats not very conducive to fossilisation, it is perhaps not really surprising. The very considerable gap in the fossil record and the general similarity of the shell, including the microstructure and muscle scars of Paleozoic tryblidiids to living monoplacophorans, have

enhanced the reputation of these animals as living fossils that have changed little in almost half a billion years.

14.4 CAUDOFOVEATA AND SOLENOGASTRES: THE APLACOPHORAN GROUPS

14.4.1 Introduction

Two small groups of entirely marine, mainly small-sized, 'spicule worms' are collectively known as aplacophorans; these are the mostly minute, largely deep-water Solenogastres (= Neomeniomorpha) and the cnidarian-feeding Caudofoveata (= Chaetodermomorpha). Both groups have cuticle-covered bodies armed with aragonitic sclerites, giving them a characteristic shiny appearance. Both groups are characterised by their worm-like bodies that are circular (or nearly circular) in section and covered with calcareous spicules. They have a poorly defined head, a very reduced to absent foot, and a small posterior mantle cavity. In size, they range from the caudofoveate *Chaetoderma felderi*, which reaches 40 cm in length, to minute interstitial solenogasters of about half a millimetre long (Todt 2013).

Aplacophorans are more diverse than generally thought with around 280 Solenogastres and 130 Caudofoveata named (Todt 2013), but many additional species in collections await description.

In this first section, we briefly review the features that Caudofoveata and Solenogastres have in common (see summary in Box 14.3).

BOX 14.3   SUMMARY OF SOME OF THE COMMON FEATURES THAT LINK THE TWO GROUPS OF APLACOPHORANS AND SEPARATE THEM FROM OTHER MOLLUSCS

Based on Scheltema (1993) and Ivanov (1996).

Cylindrical body shape
Isolation of circumoral sensory region
Monocrystalline spicules
Absence of shell field
Body wall a skin/muscle sac that incorporates the mantle
Simple, straight gut lacking a long intestine
Reduction of first pair of coelomoducts
Gonads open to pericardium and U-shaped gametoducts which (usually) open to mantle cavity
Posterior sense organs (PSO) dorsal to mantle cavity
Ganglionated tetraneural nervous system
Reduced posterior mantle cavity
No kidneys
Covered with chemically simple cuticle
Serial latero-ventral musculature
Distichous radula
Dorsal posterior sense organ(s)
Reduction or loss of foot

14.4.2 PHYLOGENY AND CLASSIFICATION

A caudofoveate was the first aplacophoran to be described, albeit incorrectly as an echinoderm, by the Swedish zoologist, Lovén, in 1844. Aplacophorans were recognised as molluscs three decades later (in 1875), although not by all malacologists (e.g., Thiele 1902; Hoffmann 1929–1930). Aplacophoran taxa have become well known morphologically, primarily due to the detailed studies of Lutfried Salvini-Plawen (1939–2014) and Amélie Scheltema (1928–2015) in the late 1900s that built on some earlier anatomical studies. They do not resemble other molluscs externally, but their internal anatomy and embryology show their molluscan affinities.

BOX 14.4   HIGHER CLASSIFICATION OF THE APLACOPHORAN GROUPS

(Class) **Caudofoveata** (= Chaetodermatomorpha)
(Order) **Chaetodermatida**
(Class) **Solenogastres** (= Neomeniomorpha)
(Superorder) **Aplotegmentaria**
(Order) **Neomeniida**
(Order) **Pholidoskepia**
(Superorder) **Pachytegmentaria**

See Appendix for a more detailed classification.

As outlined in Table 14.2, there are substantial anatomical differences between the two groups known as Caudofoveata and Solenogastres. Distinctive features of Solenogastres include having a rudimentary foot and lateroventral muscle bands, both of which are lacking in Caudofoveata. Solenogastres are hermaphroditic while Caudofoveata are gonochoristic. Caudofoveata have a pair of ctenidia in the mantle cavity and have a digestive gland, with both of these features lacking in Solenogastres. Both groups have elaborated oral structures which differ considerably.

Overall, Solenogastres are anatomically more diverse than Caudofoveata. As an illustration of this, Scheltema (1992) listed 43 characters used to describe solenogaster taxa, 31 of which were internal anatomical features, but gave only nine external or hard part characters to describe caudofoveates, indicating a much more conservative internal anatomy in that group.

The Caudofoveata and Solenogastres are either thought to be sister taxa and grouped in a monophyletic Aplacophora (e.g., Hyman 1967; Scheltema 1988a, 1993, 1996; Ivanov 1996), or Aplacophora is considered paraphyletic (e.g., Salvini-Plawen 1972, 1980, 1981, 1985a, 2003; Salvini-Plawen & Steiner 1996; Haszprunar 2000; Haszprunar et al. 2008; Todt et al. 2008b). In both these scenarios, they are thought to occupy a basal position in the molluscan phylogenetic tree, but a few workers consider them unrelated groups located away from the base (e.g., Yochelson 1978; Lindberg & Ponder 1996; Giribet et al. 2006; Wilson et al. 2010). We treat them here as two separate, apparently not closely related higher taxa (classes), and we briefly outline issues concerning their relationships below.

Aplacophorans have been more difficult to work with than most other molluscan groups because of their small size, no shell, and frequently poor sclerite preservation due to acidic preservatives. Also, histology is often required for identification and classification, so the level of detail available is generally superior to other groups of molluscs (Todt et al. 2008b). Species identification often also requires information on the radula and copulatory stylets (Scheltema & Schander 2000).

As noted above (and in Chapter 13), there is no clear consensus on the position of the aplacophoran taxa on the molluscan tree. The two main questions are:

TABLE 14.2

Some of the Main Characters Distinguishing Caudofoveata and Solenogastres

Caudofoveata	Solenogastres
Head shield (= oral shield)	Preoral sensory organ
Differentiation of body into three sections	No marked differentiation into sections
No foot – cuticle fused mid-ventrally	Foot groove present and narrow, keel-like foot
No pedal gland	Large anterior pedal gland
Peristaltic locomotion with retraction of anterior part of body	
Two ctenidia	Ctenidia absent
Unpaired digestive gland	Serial digestive gland
Stomach present (i.e., separate from digestive gland)	No distinct stomach
Paired PSOs	Single PSO
No elaboration of gonochoristic reproductive system	Elaborated hermaphroditic reproductive system, including copulatory spicules
Paired genital openings	Unpaired genital opening
Nervous system with lateral and ventral cords fused posteriorly	Nervous system with lateral and ventral cords not fused posteriorly
Few cross connections between nerve cords	Many cross connections between nerve cords
Larva a trochophore	Pericalymma larva

Modified from Scheltema (1993)

1. What are their relationships to other molluscs – are they basal or secondarily simplified (derived) from a (shelled?) molluscan ancestor?
2. What are the relationships of the two aplacophoran groups – are they monophyletic, paraphyletic, or separately derived?

The aplacophoran body with its covering of cuticle and calcium carbonate spicules superficially resembles several lophotrochozoan worm groups suggested as molluscan sister taxa (Annelida, Platyhelminthes, and Sipuncula), and therefore it is sometimes argued that the worm-like aplacophoran morphology is plesiomorphic and aplacophorans are the most primitive living molluscs (e.g., Salvini-Plawen 1985a, 1990, 2003; Haszprunar 2000). The worm-like morphology was considered a symplesiomorphy (shared primitive character) by some (e.g., Haszprunar 2000) and that Caudofoveata and Solenogastres represent an aplacophoran grade of evolution, a view shared by some others (e.g., Haszprunar 2000; Haszprunar et al. 2008). In this view, the sclerites or spicules presumably arose as a protective covering that transformed into plates or the shell in other molluscan lineages (see Chapters 3, 12, and 13). Spicules are otherwise known in chitons, while morphologically different structures termed spicules are known from various heterobranch gastropods (some nudibranchs and acochlidians) and a few bivalves (Carter & Aller 1975).

An alternative view is that the worm-like shape of aplacophorans is derived (e.g., Scheltema 1996), body elongation being an adaptation to an epizoic life (Solenogastres) or burrowing in sediments (Caudofoveata), a view more in agreement with brachiozoans as the molluscan sister taxon (but see Scheltema 1996 for a contrasting interpretation). In this scenario the worm-like shape is combined with the marked reduction of the foot in solenogasters and its loss in caudofoveates, giving the body a near circular cross section. The development of a mouth shield in caudofoveates is also probably related to their burrowing habits. The worm-like form may have evolved from a chiton-like ancestor (see Chapter 13). Such modification and reduction (Scheltema 1993; Ivanov 1996; Vinther et al. 2011) may have occurred in one or two lineages. If derived once, they can be grouped in a monophyletic Aplacophora, usually as subclasses (Pelseneer 1906; Hyman 1967; Scheltema 1978, 1988, 1993, 1998). Alternatively, two class-level taxa are grouped in a paraphyletic Aplacophora (Salvini-Plawen 1985a; Salvini-Plawen & Steiner 1996; Haszprunar 2000). Another idea is that one or both groups were derived by way of major modification and simplification from one or more conchiferan lineages in a similar way that secondary worm-like morphologies have evolved in a few groups of gastropods (some internal parasitic Eulimidae and several groups of 'opisthobranchs') and the bivalve Teredinidae. Such modifications would be expected to leave traces of their ancestry, but there are few in aplacophorans.

To date, molecular analyses have been unable to satisfactorily resolve which of the above scenarios is the most

likely (e.g., Scheltema 1998; Todt et al. 2008b; Kocot 2013; Todt 2013) (see Chapter 12). Some recent phylogenomic and molecular studies (Kocot et al. 2011; Smith et al. 2011; Vinther et al. 2011) tend to favour a sister group relationship of the aplacophoran taxa (Kocot 2013; Todt 2013), and thus a monophyletic Aplacophora (e.g., Scheltema 1988a), rather than supporting the Solenogastres and Caudofoveata as two independent basal groups.

Whether the evolution of the worm-like morphology in aplacophorans resulted from a single evolutionary event (Scheltema 1993; Ivanov 1996; Scheltema 1996) or from convergence remains uncertain. While most 'worms' are dorsoventrally compressed, the cylindrical body form is typically a derived character state, being widespread in burrowing and parasitic metazoans (Ivanov 1996). Thus while this body form may have been independently derived in the two aplacophoran taxa, it has been treated as an apomorphy for the Aplacophora (Ivanov 1996). As pointed out by Salvini-Plawen (1981, 1985a, 1990), the acquisition of a cylindrical body shape necessitated some modifications including the restriction of the mantle cavity to the posterior end of the body. There are also several other synapomorphies of Aplacophora listed in Box 14.3, and, as discussed below, several of these characters are plesiomorphic and others paedomorphic so the apparent similarities are not as convincing as they might initially appear.

Some authors (e.g., Salvini-Plawen 1972, 1980, 1990; Scheltema 1988a; Ivanov 1996) argued that the aplacophoran groups were basal molluscs, although Salvini-Plawen considered them paraphyletic and Scheltema and Ivanov treated them as a monophyletic Aplacophora. Thus the morphology of aplacophorans, especially that of solenogasters, has been used as a model for the ancestral condition of molluscs (Haszprunar 2000; Salvini-Plawen 2003), although some considered that the basal status of the aplacophoran groups was poorly justified.

The idea that the Caudofoveata and Solenogastres are secondarily simplified derived taxa is not new but is rarely considered in discussions regarding the relationships of these taxa. Yochelson (1978 p. 184) thought that the 'Aplacophora' were 'secondarily simplified rather than primitive'. This idea was revived by Lindberg (1988a), Lindberg and Ponder (1996), Buckland-Nicks and Hadfield (2005), and Scheltema (1993). Lindberg (1988b) argued that heterochronic processes, mainly operating on structures derived from the 4d cell lineage, confused the distinctions between primitive and paedomorphic characters. Scheltema (1993) suggested that aplacophorans had evolved from a polyplacophoran-like ancestor by progenesis.[6] In post-metamorphic juvenile Solenogastres the mesoderm has not differentiated into gonad, pericardium, and coelomoducts (Thompson 1960a), and a paedomorphic effect in adults is seen in the tubular reproductive tracts and the lack of differentiation into distinct kidneys, pericardium, and gonads. The bipartite radula may also result from paedomorphosis as in development the odontoblasts are first paired

---

[6] The acceleration of sexual maturation relative to the rest of development.

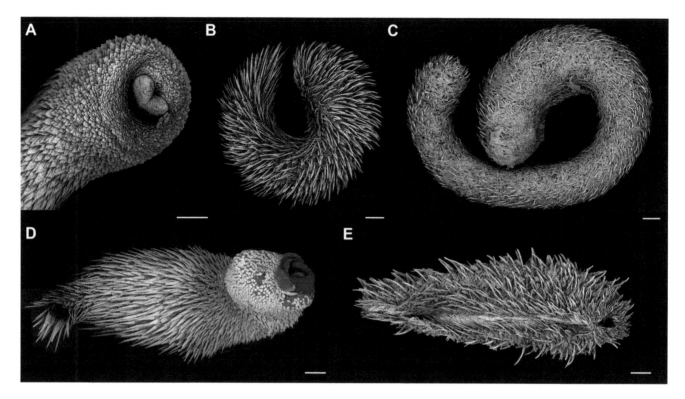

FIGURE 14.17 Scanning electron micrographs of representative specimens of Caudofoveata and Solenogastres. (A) *Falcidens* sp. (Caudofoveata), New Zealand. (B) Pruvotinidae sp. (Solenogastres), Iceland. (C) Simrothiellidae sp. (Solenogastres), Iceland. (D) *Spathoderma alleni* (Caudofoveata), Iceland. (E) Pruvotinidae sp. (Solenogastres), Weddell Sea, Antarctica. Scale bars = 100 μm. (Courtesy of K. Kocot.)

and then fuse during development in the radular sac where the radula is secreted (e.g., Raven 1966). If development ceases before fusion is complete, a bipartite radula would result. If aplacophorans are indeed paedomorphic offshoots, the sister taxa still need to be determined, although it is probable they reside in ancient chiton-like ancestors (Scheltema 1993; Lindberg & Ponder 1996; Vinther et al. 2011; Sutton & Sigwart 2012). Regrettably, there are no undoubted aplacophoran fossils, although some fossil organisms have been attributed to them (e.g., Sutton et al. 2001a; Sutton et al. 2012) but not without controversy (e.g., Steiner & Salvini-Plawen 2001) (see Chapter 13).

As also pointed out by Todt (2013), developmental and morphological studies on various systems will assist our understanding of the evolution of the aplacophoran taxa. Some examples include the ultrastructure of locomotory cilia (Lundin & Schander 1999, 2001a, b), which support the basal position of both chitons and the aplacophoran groups. Their cilia are like the basic metazoan type with paired ciliary rootlets orientated at nearly 90° to each other, and there is no accessory centriole. Paired ciliary rootlets are not found in monoplacophorans, bivalves or gastropods (Lundin & Schander 2001b), or scaphopods (Lundin et al. 2009). Other TEM studies on sperm morphology provide support for separate origins of the aplacophoran groups, with Caudofoveata most closely similar to the Polyplacophora while the internally fertilising Solenogastres have sperm that convergently share characters with neritimorph gastropods (Buckland-Nicks & Scheltema 1995).

14.4.3 Morphology

The cuticle is covered with aragonitic sclerites (spicules) that are secreted extracellularly by individual epidermal cells with a crystallisation chamber formed by surrounding cells. The spicule morphology differs on different parts of the body and is species-specific, showing considerable differences in shape, size, and thickness (Figure 14.17). For taxonomic purposes, spicules are sampled from standard regions of the body as they differ in different locations.

The spicules of both groups of aplacophorans differ from those of chitons in having a monocrystalline structure with almost no organic matrix. This feature can also be considered a synapomorphy of Aplacophora (Ivanov 1996).

14.4.3.1 Head-Foot and Mantle

In both groups, the surface of the body is covered with a structurally simple cuticle (Scheltema 1993). This cuticle is secreted by the epidermis making up the mantle, which also covers the whole body, or almost the entire body, except the pedal groove in Solenogastres. Successive series showing closure of the mantle along the mid-ventral line is seen in aplacophorans (Ivanov 1996). The body wall musculature usually consists of three layers in both groups – an outer circular, middle oblique, and inner longitudinal.

Mantle musculature is a separate layer included in the 'skin-muscle body wall', a condition not seen in other molluscs, and while Solenogastres have several sets of lateroventral muscle bands these are absent in Caudofoveata (Figure 14.18).

Small epidermal papillae which appear to discharge their contents to the exterior occur and thus may be excretory (Scheltema et al. 1994). Similar epidermal papillae are known in chitons.

The head region of caudofoveates is dominated by the sensory, cuticle-covered head shield, while that of solenogasters has a papillate, sensory supraoral vestibule (Scheltema et al. 1994). The mouth is at the anterior extremity in caudofoveates and is anteroventral in solenogasters. Caudofoveates lack a foot or pedal groove, but the foot is represented by a narrow ciliated strip in a pedal groove in solenogasters. In the latter case, the strip transforms into a keel when it contracts on preservation (Nielsen et al. 2007).

A shallow, posterior mantle cavity is a feature common to both groups but may be convergent due to this being a consequence of their worm-like shape.

14.4.3.2 Digestive System

Both aplacophoran groups share several gut characters (Scheltema 1981). The foregut lacks jaws and contains a radular apparatus that is similar in both groups, consisting, as in other molluscs, of a radula, radular sac, odontoblasts, a radular membrane, and a pair of odontophore bolsters to which muscles that move the radula are attached. The radula is distichous and lacks articulation, the radular membrane is divided or fused, and there is no subradular membrane. Radular formation is like that in other molluscs, except that the teeth have no separate tooth base, being continuous with the underlying radular membrane (Wolter 1992). The radular membrane(s) has a few to >40 rows of teeth that may be monostichous, distichous, or polystichous. As in other molluscs, these teeth and the membrane are secreted in a radular sac. The distichous type (two teeth per row), with a bipartite membrane, as seen in some Solenogastres and Caudofoveata, has sometimes been considered the most primitive (Scheltema 1988a, 1989, 1994; Eernisse 1988, Scheltema et al. 2003); with the unipartite membranes seen in other molluscs being derived. Others (Sirenko & Minichev 1975; Salvini-Plawen 1988, 2003; Ivanov 1990) considered the unipartite state ancestral and the bipartite condition derived. The radula comprises an organic matrix rich in chitin (Peters 1972; Salvini-Plawen & Nopp 1974; Wolter 1992) and, in some taxa at least, hardened by deposition of minerals (Cruz et al. 1998; Todt et al. 2008b). In caudofoveates, the distal radular teeth are lost as the radula grows but are retained in a ventral sac in Solenogastres.

Other shared features include the cuticularised foregut, paired tubular foregut glands (often present), and a ciliated dorsal structure (band, groove, or typhlosole) that runs through the midgut to the ciliated intestine (Figure 14.18). Despite these similarities, there are also considerable differences in gut morphology in the two groups. The undivided midgut of the cnidarian-feeding Solenogastres lacks a digestive gland, protostyle, or gastric shield, all of which are present in many of the microphagous Caudofoveata (Salvini-Plawen 1980; Scheltema 1981). The midgut of the Solenogastres is a simple tube with regular shallow lateral pockets caused by the latero-ventral muscles (Salvini-Plawen 1969). This gut morphology in Solenogastres may not be primitive but instead may be secondarily simplified as a result of their feeding habits.

The foregut goblet cells are found in both groups but may not be homologous. Foregut glands are important taxonomically in Solenogastres (see Chapter 5).

14.4.3.3 Renopericardial and Reproductive Systems

The tube-like pseudocoelom is differentiated into gonads, pericardium, and gonoducts, but in both groups it lacks kidneys. The rather large pericardium is located posteriorly, with the gonads opening into it. It contains a heart comprising a ventricle and (usually) a pair of auricles. Except for a dorsal aorta or sinus emerging from the anterior end of the ventricle, the blood circulates in an open haemocoelic system, the ventral part of which is divided off by a horizontal septum. Heath (1905a) reported that the aorta opened into the head sinus through a vertical membrane which separates the stomach and radula; a second opening in this membrane drains blood from the head region to the pedal sinus. Haemocyanin is present in the blood of at least several caudofoveates but is absent in solenogasters (Lieb & Todt 2008).

The reproductive system is generally similar in both groups. Unlike other molluscs, the gonads open into the anterior end of the pericardial lumen via a pair of gonopericardial ducts, and the gametes leave the pericardium by way of U-shaped gametoducts which open to the mantle cavity (Figure 14.18).

Caudofoveates have separate sexes and shed eggs and sperm (ectaquasperm) into the water column while solenogasters are internally fertilising hermaphrodites, with derived spermatozoa (introsperm) (Scheltema et al. 1994). Given their different reproductive strategies, it is unsurprising that the Solenogastres have developed accessory reproductive structures and that these are lacking in caudofoveates.

In the absence of kidneys, waste is probably excreted mainly via the epidermis in which the epidermal papillae may play a significant role.

14.4.3.4 Nervous System

The nervous system of both groups is of a tetraneural plan similar to those of polyplacophorans and monoplacophorans. In both groups, paired (usually fused) cerebral 'ganglia' give off a pair of connectives to form a circumoesophageal nerve ring. Arising from the cerebral ganglia are four main nerve tracts, paired lateral and ventral longitudinal nerve cords that run along the body and are connected with many cross commissures, resulting in a ladder-like system (see Chapter 7). These cords usually originate independently from the cerebral ganglion in Solenogastres, but they arise from a single pair of connectives in caudofoveates. The Solenogastres have a simpler nervous system in the head region than Caudofoveata; this is arbitrarily assumed to be either more primitive (Scheltema 1993, 1996) or reduced (Salvini-Plawen 1985a; Salvini-Plawen & Steiner 1996). There are two pairs of anterior ventral ganglia, with the larger anterior pair in Solenogastres giving off nerves to the pedal pit and pedal glands.

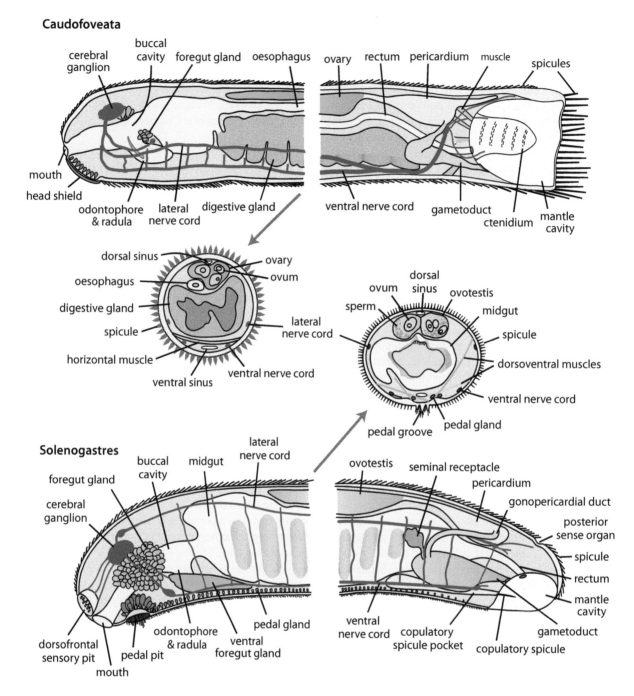

FIGURE 14.18 Comparative anatomy of aplacophorans. Redrawn and modified from von Salvini-Plawen, L., *Malacologia*, 9, 191–216, 1969.

A commissure passes over the rectum at the posterior end of the lateral cords, and in Caudofoveata the ventral and lateral cords fuse anteriorly to this commissure. The suprarectal commissure forms a suprarectal ganglion which innervates the mantle cavity, receives the nerves from the posterior sense organ[7] (Lindberg & Sigwart 2015) (paired in Caudofoveata and single in Solenogastres), which was thought to be an osphradium (Haszprunar 1986, 1987b) and, in caudofoveates, innervates the ctenidia.

14.4.3.5 Development

The few studies on the embryology and development of aplacophorans (Baba 1938; Thompson 1960a; Okusu 2002; Nielsen et al. 2007) show that their development includes spiral cleavage and a trochophore larva. The trochophore in both groups has a prototroch with one ring of compound cilia, a telotroch. Solenogasters have a pericalymma larva, but caudofoveates do not.

14.4.3.6 Habits and Habitats

All aplacophorans are marine but are seldom abundant. They are found at depths as shallow as 3 m, but are commoner in

---

[7] In most Solenogastres the PSO occurs just above the mantle cavity opening while in others it is within the mantle cavity (Ivanov 1996).

deeper water and are even found in the deepest parts of the ocean (>7,000 m). They are found in low energy habitats, including soft sediments in deep water and hard substrata in the sublittoral. Of those that live in shallow water, most Solenogastres are associated with coral reefs or live in subtidal meiobenthos communities (Todt et al. 2008b).

The habits and zoogeographical distribution of Caudofoveata and Solenogastres differ. Most caudofoveates are found in the northern hemisphere; they are selective microphages and burrow into soft sediments. Most species of Solenogastres are known from the southern hemisphere and mostly live on soft corals on which they feed, but a few are interstitial, gliding over sediments but not burrowing, and are detritophages or microphages (Ivanov 1996). Only Solenogastres have been found at hydrothermal vents.

14.5 CAUDOFOVEATA (OR CHAETODERMOMORPHA, SCUTOPODA)

14.5.1 Introduction

The members of this burrowing, worm-like group (Figure 14.19) range in length from less than a millimetre to 14 cm (e.g., *Chaetoderma productum*).

14.5.2 Phylogeny and Classification

Only three or four families of Caudofoveata are recognised (see Appendix), so they are much less taxonomically diverse than Solenogastres. Recent molecular work based on mitochondrial genes by Mikkelsen et al. (2018) identified two major clades, the Prochaetodermatidae and the Chaetodermatidae + Limifossoridae.

Before recent molecular work, taxa were mainly based on radular, mouth shield, and body shape characters (e.g., Salvini-Plawen 1975; Todt et al. 2008b). *Scutopus* and *Psilodens* (family Limifossoridae) may be basal genera because some species retain traces of a ventral suture, but the phylogenetic relationships within the group are not well resolved.

The radular configuration is thought by some to be the most primitive (distichous pairs of teeth with median denticles) as is the midgut morphology (Todt et al. 2008b).

14.5.2.1 Sister Group Relationships

Salvini-Plawen and others (e.g., Salvini-Plawen 1972, 1990) thought Caudofoveata to be the sister to the rest of the molluscs. With this topology, the name Scutopoda has been used for the Caudofoveata alone, with the equivalent supraclass level including all the other molluscs being called Adenopoda (Salvini-Plawen 1981, 1985a) In the Salvini-Plawen and Steiner (1996) tree the two aplacophoran taxa were unresolved at the base of the Mollusca, or Solenogastres were basal.

Molecular phylogenetic analyses have produced seemingly strange results, with the Caudofoveata identified as the sister taxon of the Cephalopoda in some (Giribet et al. 2006; Dunn

et al. 2008; Wilson et al. 2010). In one analysis, caudofoveates are the basal taxon in a clade with Solenogastres and cephalopods as sister taxa (Stöger et al. 2013).

14.5.3 Morphology

The body is more differentiated than that of Solenogastres. It consists of three regions: an anterior *anterium*, a middle *trunk*, and a posterior *posterium*. The anterium includes the mouth and oral shield and is demarcated from the trunk by a slight constriction and a change in spicules. The trunk makes up the majority of the animal and includes the internal organs except for the intestine, pericardium, gonoducts, and mantle cavity, all of which are contained in the posterium. The transition between the trunk and posterium is marked by a narrowing of the body. In the family Prochaetodermatidae, the posterium can be narrow with a terminal knob that has elongate sclerites. Most Caudofoveata are beige to brownish in colour.

Most of our knowledge of the internal anatomy comes from a few early studies (e.g., Wirèn 1892a, b; Heath 1905a, 1911; Thiele 1913; Hoffmann 1929–1930; Lummel 1930; Hyman 1967; Salvini-Plawen 1969, 1971, 1972, 1975; Scheltema et al. 1994), with many of the taxonomic works focusing on the body, spicules, and radula.

14.5.3.1 Spicules

The body is covered with calcareous (aragonitic) sclerites (spicules) attached to the fibrous cuticle. These are flattened, blade-like, sometimes with secondary ornamentation, and their morphology changes along the body.

14.5.3.2 Head-Foot and Mantle

The mantle epidermis surrounds the body, fusing ventrally, so a ventral groove and foot are absent, except for a trace of a ventral suture, as noted above, in some species of *Scutopus* and *Psilodens*, which is innervated from the ventral (pedal) nerve cords (Todt et al. 2008b). The mouth is partly or entirely surrounded by a sensory, paired oral shield (= foot shield), which lacks spicules and is covered by thick cuticle. The oral shield is formed from anterior gut epithelium that becomes external during development (Scheltema 1981).

In some caudofoveates the longitudinal muscles of the body wall are arranged in four thick bands along the narrower part of the body (the trunk), while anteriorly the circular muscles are thickened and form a constriction of the body, separating off the anterium. The various head muscles are attached to this ring of muscle. Caudofoveates lack the serial sets of lateroventral muscle bands seen in solenogasters.

14.5.3.3 Mantle Cavity

The small posterior mantle cavity of caudofoveates contains a pair of bipectinate ctenidia (see Chapter 4, Figure 4.7) which are probably homologous with the ctenidia of other molluscs. They are attached just behind the openings of the pericardioducts. The anus also opens to this cavity.

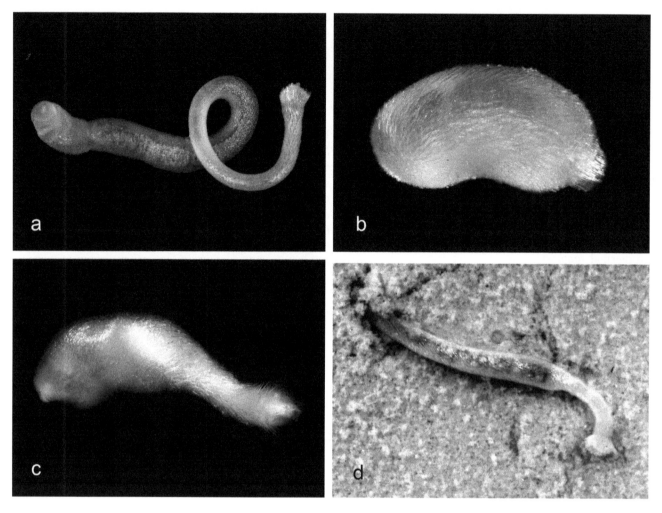

FIGURE 14.19 Photos of living Caudofoveata. (a) *Falcidens crossotus* (Chaetodermatidae), North Sea. Courtesy of N. Mikkelsen. (b) *Limifossor holopeltatus* (Limifossoridae), Southern Atlantic Ocean. Courtesy of N. Mikkelsen. (c) *Claviderma amplum* (Prochaetodermatidae), Southern Atlantic. Courtesy of N. Mikkelsen. (d) *Falcidens ryokuyomaruae* (Chaetodermatidae) burrowing. Courtesy of E. Redl.

14.5.3.4 Digestive System

The mouth is closed by a sphincter and is partially or entirely surrounded by the oral shield (see above). It opens to an oral cavity lined with secretory and sensory epithelium and is separated from the buccal cavity by a sphincter. Jaws are lacking except in Prochaetodermatidae where they are large, lateral, cuticular, and paired. The cuticle-lined buccal cavity is muscular and contains a ventral radula (see below) supported by the radular membrane, beneath which lie a pair of bolsters and their associated musculature. A pair of ventral glands or clusters of glands may open to the buccal cavity (see Chapter 5).

The radula and odontophore are highly diverse and more developed than those seen in the Solenogastres. As in most other molluscs the distal radular teeth are utilised in feeding, in contrast to the situation in Solenogastres. Arguably, the most primitive configuration is seen in *Scutopus* (Limifossoridae) where the few large pincer-like radular teeth are fixed to the radular membrane, and the odontophore is simple and not capable of much movement other

than sliding back and forth, although it does possess the ability to close the teeth. The large odontophore of *Limifossor* (Limifossoridae) is capable of more complex movements, and the split radular membrane has more numerous, curved, pointed teeth.

Prochaetoderma (Prochaetodermatidae) has a bipartite rasping radula with several transverse rows and is somewhat reminiscent of a gastropod radula (Scheltema 1981). It also has the most modified midgut (Figure 14.20).

Falcidens (Falcidentidae or Chaetodermatidae) and *Chaetoderma* (Chaetodermatidae) also have a highly modified midgut, and both have a reduced radula consisting of a single pair of teeth which act like pincers in the capture of prey (Ivanov 1990). Large ventral and lateral supports for the teeth are present.

An oesophagus is usually present, and the midgut is divided into a dorsal tubular duct that opens to the digestive gland (digestive diverticulum) and a ventral sac, the stomach. The single digestive gland appears to have developed as a lobe from the stomach as its epithelium is continuous.

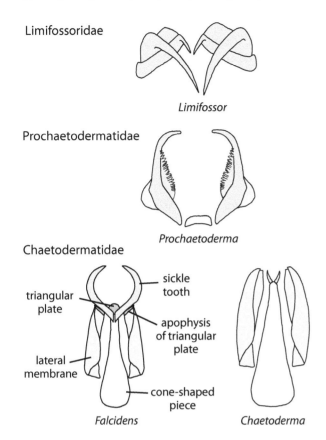

FIGURE 14.20 Examples of caudofoveate radulae. Redrawn and modified from the following sources: *Limifossa* (Scheltema, A.H., *Malacologia*, 20, 361–383, 1981); *Prochaetoderma* (Scheltema, A.H., *J. Nat. Hist.*, 48, 2855–2869, 2014); *Falcidens* and *Chaetoderma* (Ivanov, D.L., Evolutionary morphology of Mollusca. The radula in the class Aplacophora, pp. 159–198, in Shileyko, A.A. (ed.), *Archives of the Zoological Museum of Moscow State University*, vol. 28, Moscow, 1990). The radulae of the species illustrated do not necessarily reflect the range of variation in the genera they belong to or of the higher taxon they belong to.

A simple gut is seen in *Scutopus*, while in the more modified forms (Chaetodermatidae) the stomach base is cuticularised and may form a gastric shield. A differentiated style sac contains a mucoid rod (protostyle). Some, notably *Prochaetoderma*, show simplification and reduction with a single type of digestive cell, shorter digestive gland, no protostyle, and, unlike most other genera, no dorsal ciliated typhlosole.

In two genera (*Prochaetoderma* and *Falcidens*) a convoluted intestine with a long, thin posterium has developed independently.

14.5.3.5 Renopericardial and Reproductive Systems

The heart lies within a large pericardium and consists of a ventricle and (usually) paired auricles formed as invaginations of the pericardial wall. The ventricle wall is strongly folded, and its cells are rich in mitochondria, suggesting a transporting function (Scheltema et al. 1994). Oxygenated blood is collected in a sinus at the base of the efferent channels and is passed to the auricles. The only blood vessels are the dorsal

aorta and short vessels that supply blood to the odontophore. The haemocoel is separated ventrally by a ventral horizontal septum, and anterior and posterior vertical septa are also sometimes present.

Haemocyanin is present in the blood of at least some caudofoveates (Lieb & Todt 2008).

14.5.3.6 Nervous System and Sense Organs

The caudofoveate brain is complex, with considerable variability within the group. The paired cerebral hemiganglia are fused in Chaetodermatidae, resulting in one large cerebral ganglion, but are not fused in Prochaetodermatidae (Sigwart & Sumner-Rooney 2015). The anterior and posterior sections are separated by a band of neuronal somata (Faller et al. 2012). Anterior to the cerebral ganglion are three or more pairs of frontal swellings that innervate the mouth and the oral shield (Heath 1911) (see Chapter 7, Figure 7.1). These have been called 'precerebral ganglia'; however they do not have a central neuropil and are composed almost entirely of neural somata (Sigwart & Sumner-Rooney 2015). They also contain neurites (Faller et al. 2012; Sigwart & Sumner-Rooney 2015). At least one species (*Falcidens crossotus*) also has a cluster of somata on the anterior part of each brain hemisphere. These somata are exclusively serotonin-like immunoreactive, but this feature was not found in other caudofoveates examined by Faller et al. (2012).

Another unusual feature of the caudofoveate nervous system is the swellings found along the ventral and lateral nerve cords. They are not ganglionic and do not correspond with any other serial body structure, so are therefore not a feature of segmentation (Shigeno et al. 2007; Faller et al. 2012; Sigwart & Sumner-Rooney 2015).

The development of the anterior chemosensory network, which includes the oral shield and the frontal swellings of the cerebral ganglion, may be related, at least in part, to locating prey in the sediment.

14.5.4 Biology, Ecology, and Behaviour

14.5.4.1 Habits and Habitats

Caudofoveates use their oral shield to burrow shallowly in the upper 2–3 cm of the sediment. They are found from around 30 to 7,300 m in depth and sometimes occur in soft-bottom habitats in high densities. Some members of the family Prochaetodermatidae are considered opportunistic, being abundant and even dominating some deep-sea communities (e.g., Scheltema 1985), have broad geographic and depth ranges, and are omnivorous (Scheltema & Ivanov 2009).

Species of *Scutopus* and *Psilodens* (Scutopodidae or Limifossoridae) can exist in physically stressed environments which show a range of temperature, salinity, and oxygen levels (Scheltema 1992).

14.5.4.2 Feeding and Diet

Some caudofoveates feed on organic detritus (e.g., *Scutopus* and *Chaetoderma*), while the prochaetodermatids feed selectively on foraminiferans or other small organic particles, or organisms such as minute polychaetes, and thus can be considered micro-omnivores (Scheltema & Ivanov 2009; Todt 2013). The buccal structures in prochaetodermatids, which include the jaws and radula, are protruded from the mouth. The jaws grasp food which is then rasped by the radula. The worn anterior teeth fall off and are replaced by new teeth. No other aplacophorans, as far as known, are omnivorous (Scheltema & Ivanov 2009).

14.5.4.3 Reproduction, Development, and Growth

Caudofoveates are dioecious and are broadcast spawners with large yolky eggs and lecithotrophic development. Their sperm are highly modified ectaquasperm with two flagella and, despite their modification, are generally similar to the sperm of other external fertilising molluscs.

Developmental studies on *Chaetoderma* (Nielsen 1995, 2004; Nielsen et al. 2007) reveal a lecithotrophic trochophore with a prototroch and a telotroch. A pair of protonephridia are developed (Nielsen et al. 2007), despite the lack of kidneys in adults. In the late developmental stages, a ventral suture and seven dorsal transverse rows of spicules are present (Todt et al. 2008b). The ventral suture disappears in the adult.

Nothing is known regarding longevity, but *Prochaetoderma yongei* was inferred to grow to adult size in two months and to reach sexual maturity in a year (Scheltema 1987).

14.5.4.4 Associations

We are not aware of any associations having been recorded.

14.5.5 DIVERSITY AND FOSSIL HISTORY

There are about 130 named species of caudofoveates, but many more are undescribed (Todt 2013). The two most primitive caudofoveate genera, *Scutopus* and *Limifossor*, are the least diverse while the micro-omnivorous or microcarnivorous *Falcidens* and *Chaetoderma* are much more speciose, as are members of the genus *Prochaetoderma*.

Typical caudofoveates have no known fossil history (see Chapter 13 for a discussion on fossil taxa attributed to aplacophorans).

14.6 SOLENOGASTRES (OR NEOMENIOMORPHA)

14.6.1 INTRODUCTION

The Solenogastres (Figure 14.21) are found from shallow subtidal to abyssal depths where they are often associated with Cnidaria (hydroids, anemones, and soft corals) on which they mostly feed. They are more speciose than caudofoveates. Most species average less than 5 mm in length, but some reach over 300 mm, with the smallest less than a millimetre long. Shallow-water taxa can be brightly coloured, but most reflect a silvery sheen from the numerous spicules that cover their bodies.

A narrow, ciliated, non-muscular foot with an anterior pedal gland is present in a groove along the ventral body surface. They can creep on the colonial cnidarians on which they feed and on other hard or soft substrata using a mucous strand secreted by an anterior pedal pit. The tiny posterior mantle cavity lacks gills. The digestive system is simple with no separate digestive gland and stomach. A radula is usually present in the tube-like muscular foregut but is absent in some taxa. Solenogastres are simultaneous hermaphrodites and have seminal receptacles for sperm storage.

There are several overviews of solenogaster morphology (e.g., see Hyman 1967; Salvini-Plawen 1971, 1978; Scheltema et al. 1994), and a general review was provided by Todt et al. (2008b).

14.6.2 PHYLOGENY AND CLASSIFICATION

Four orders were recognised by Salvini-Plawen (1978) (see Box 14.4). These are distinguished mainly by external characters, including cuticle thickness, sclerite morphology, and types of foregut glands. A morphological phylogenetic analysis of solenogaster genera (Salvini-Plawen 2003) obtained poor resolution but resulted in the order Cavibelonia being monophyletic and derived, while Pholidoskepia was part of a basal polytomy.

14.6.2.1 Sister Group Relationships

As discussed in Section 14.4.2, sister taxon relationships of the Solenogastres are controversial. Morphologists generally regard the Caudofoveata as the sister taxon, irrespective of whether or not the Aplacophora is paraphyletic or monophyletic. In a paraphyletic Aplacophora, the Solenogastres have either been treated as the second branch on the molluscan tree, next to the Caudofoveata (e.g., Salvini-Plawen 1972), or the most basal clade of living molluscs and sister taxa of the Hepagastralia, a clade containing all other living molluscs (Salvini-Plawen & Steiner 1996; Haszprunar 2000). Molecular data are even more ambiguous with several apparently spurious placements, such as sister to the gastropods (Giribet et al. 2006), cephalopods (Stöger et al. 2013), or even outside the Mollusca (Wilson et al. 2010). Some of these results were confounded by exogenous contamination (Okusu & Giribet 2003).

14.6.3 MORPHOLOGY

14.6.3.1 Spicules

As in caudofoveates, the spicules are secreted from a single epidermal cell and are aragonitic, but they show a greater range of variation in their morphology. They can be solid or hollow, flat and scale-like or leaf-like, rimmed and trough-like, hook-shaped or needle-like. The different spicule morphologies may be scattered over the body of the animal or concentrated in specific regions. Some species show an ontogenetic change from solid sclerites to hollow needles; the latter condition is thus considered derived.

14.6.3.2 Body

In Solenogastres, the head is recognisable only by the mouth, and the rest of the body is undifferentiated with a narrow furrow, the pedal groove, along the ventral surface. A narrow,

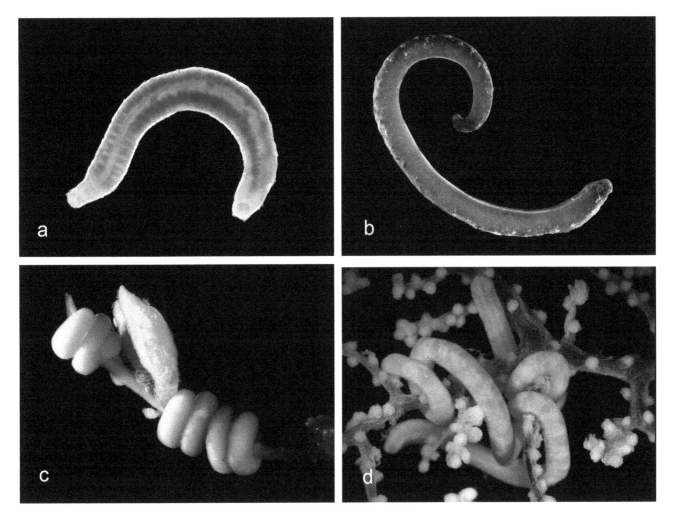

FIGURE 14.21 Photos of living Solenogastres. (a) *Wirenia argentea* (Gymnomeniidae), Norway. Courtesy of E. Redl. (b) *Epimenia sp.* (Epimeniidae), Vanuatu. Courtesy of D. Brabant - MNHN. (c) *Anamenia* sp. (Strophomeniidae) on a gorgonian octocoral, Spain. Courtesy of C. Cobo. (d) unidentified solenogaster on a gorgonian octocoral, Panglao, Philippines. Courtesy of P. Lozouet - MNHN.

non-muscular foot lies within this groove and is represented only by one to several folds of ciliated epidermis. The sides of the groove are lined with numerous pedal gland cells. At its anterior end, the pedal groove expands into the pedal pit, with dense concentrations of pedal glands opening into that cavity.

The body musculature usually consists of three main layers – an outer circular layer, a middle layer composed of two orthogonal bands (at right angles to each other), and an inner longitudinal band. In some taxa these three layers are not present. Lateroventral longitudinal muscles occur on each side of the foot, the contraction of which results in the body assuming a characteristic crescent shape. There are also two serial sets of lateroventral muscles that extend from the body wall to the vicinity of the pedal groove. Where the surface of the gut lies against these muscles, it becomes indented and produces a lobate surface.

14.6.3.3 Mantle Cavity

The mantle cavity contains the anus, gonopore(s), and the openings of the paired copulatory dart sacs. Ctenidia or secondary gills are absent, but the respiratory surface within the mantle cavity consists of ciliated folds or papillae along the roof. In a few taxa, brood pouches are formed from invaginations of the mantle cavity (Figure 14.23). A dorso-terminal posterior sense organ is usually present, immediately adjacent to the mantle cavity (see Section 14.6.3.6).

14.6.3.4 Digestive System

Foregut structures include the mouth, buccal cavity, foregut glands, radula, and sometimes an oesophagus. The anteroventrally located mouth opens into an expansive muscular oral cavity. The oral cavity opens through a sphincter into the buccal cavity, which may be protrusible, and typically contains the radular apparatus (see below). Multiple kinds of foregut ('salivary') glands open into the buccal cavity. These have been used taxonomically, especially the multicellular lateroventral and dorsal glands (Salvini-Plawen 1972, 1978; Handl & Todt 2005) (see Chapter 5 for details). Handl and Todt (2005) concluded that the most primitive type was the so-called *Wirenia*-type of lateroventral foregut glands, seen in the pholidoskepian Gymnomeniidae, which lack a duct or lumen.

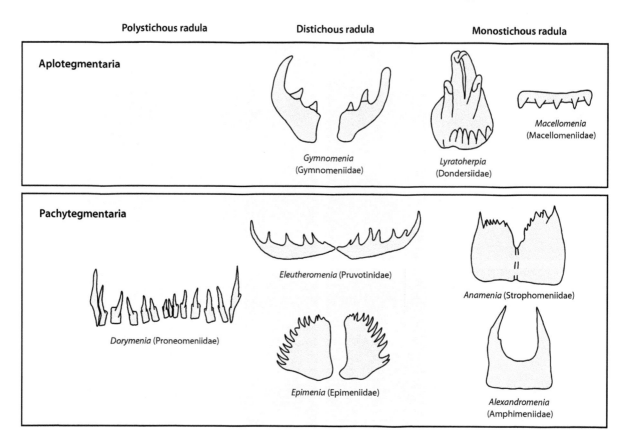

FIGURE 14.22 Examples of solenogaster radulae. Redrawn and modified from the following sources: *Gymnomenia* (Scheltema, A.H., *Malacologia*, 20, 361–383, 1981); *Dorymenia* (Scheltema, A.H., *J. Morphol.*, 257, 219–245, 2003); *Lyratoherpia, Eleutheromenia, Macellomenia*, and *Alexandromenia* (Scheltema, A.H., *The Aplacophora: History, taxonomy, phylogeny, biogeography, and ecology*. Ph.D. dissertation, University of Oslo, 1992); *Anamenia* (Ivanov, D.L., Evolutionary morphology of Mollusca. The radula in the class Aplacophora, pp. 159–198, in Shileyko, A.A. (ed.), *Archives of the Zoological Museum of Moscow State University*, vol. 28, Moscow, 1990); *Epimenia* (Baba, K., *Venus*, 10, 91–96, 1940b). The radulae of the species illustrated do not necessarily reflect the range of variation in the genera they belong to or of the higher taxon they belong to.

About 20% of Solenogastres lack a radula, and in those that do, the buccal cavity is lined with strong circular muscle forming a buccal pump, presumably for suctorial feeding (Scheltema et al. 1994). The radula may be distichous (two teeth per row), monostichous (one tooth per row), or polystichous (four or more teeth per row). In all three configurations, the individual teeth tend to be multicuspid with the cusps fine to prominent. As the radula grows, it bends ventrally and ends in a blind ventral pocket where even the oldest teeth are retained. Thus the teeth used in feeding in adults are not the most distal as in most other molluscs including caudofoveates. *Gymnomenia* has a relatively tiny radula that is thought to be one of the most primitive radulae seen in aplacophorans (Scheltema 1981) (Figure 14.22).

The buccal cavity opens through a short oesophagus into the midgut, which is a combination of the stomach and digestive gland. In some species, an extension of the midgut, the dorsal caecum, extends anteriorly into the head region.

Unique among molluscs, the solenogaster midgut is a single structure, not divided into a stomach or oesophageal gland and intestine. The rectum is short, narrow, and simple and opens into the posterior mantle cavity.

14.6.3.5 Renopericardial and Reproductive Systems

The heart is situated posteriorly in the pericardium and consists of an anterior ventricle and posterior auricle(s), both formed as paired invaginations of the pericardial wall (Hyman 1967). A dorsal aorta extends anteriorly from the ventricle, but otherwise the circulatory system is open, with ventral and dorsal haemocoels defined by horizontal muscle layers. Haemocyanin has not been found in solenogaster blood (Lieb & Todt 2008).

The paired gonads are dorsal of the digestive system, as in chitons. The gonads are hermaphroditic, and fertilisation is assumed to be internal. Eggs are formed along the medial walls of the gonads, while the sperm originates along the lateral walls. The eggs are large (250 to 400 μm) and yolky, and the filiform spermatozoa are of the introsperm type, indicative of internal fertilisation. They have a unique extracellular striated cone on top of the acrosome (Buckland-Nicks & Scheltema 1995).

The reproductive tract is more complex than in caudofoveates. Paired gonoducts exit the pericardium posteriorly, then turn anteriorly to lie below the pericardium. There are often seminal vesicles for the storage of endogenous sperm.

Near the anterior end of the pericardium, the gonoducts turn again posteriorly, and at this point, single or multiple seminal receptacles for exogenous sperm storage are usually found. The remaining portions of the gonoducts are lined with glandular epithelium (the shell glands) which open separately or through a common opening into the mantle cavity. Spermatophores have been found in the lumen of the lower gonoducts in *Gymnomenia* (Gymnomeniidae) (Scheltema et al. 1994). Many neomenioids have copulatory spicules in deep pockets in the mantle cavity, and accessory copulatory spicules may lie at the mantle cavity edge. Other pockets in the mantle cavity may act as extensions to gonoduct openings or brood pouches (Scheltema et al. 1994). Pairing and copulation have rarely been observed. Fertilised eggs are released singly, embedded in mucous egg ribbons, or brooded within the mantle cavity, thus providing a range of dispersal abilities from pelagic to crawl-away.

There is no kidney, with excretion thought to be via the epidermis. Because there is no system to modify the primary urine, solenogasters are not effective at regulating their internal osmotic pressure (Todt 2013).

14.6.3.6 Nervous System and Sense Organs

The nervous system of Solenogastres consists of a fused cerebral ganglion complex that gives rise posteriorly to two lateral nerve cords and two ventral nerve cords. The kidney-shaped, cerebral complex ('brain') is less complex than that of caudofoveates (Scheltema et al. 1994; Todt et al. 2008a; Faller et al. 2012) (Chapter 7, Figure 7.3). Details of the immunoreactive cells in the complex are given by Faller et al. (2012). The nerve cords are medullary, and commissures at these nodes connect the lateral and ventral cords and cross-connect the right and left ventral nerve cords. There are slight swellings of the cords where the commissures originate (Salvini-Plawen 1981). The layer of cell bodies around the central neuropil is thickened at these swellings, but because somata surround the medullary cords on either side, these swellings are not classified as ganglia (Sigwart & Sumner-Rooney 2015). The lateral nerve cords are connected posteriorly at the suprarectal commissure or ganglion.

The two ventral nerve cords are joined at regular intervals by pedal commissures throughout the length of the body.

Pholidoskepia has a previously undescribed dorsal unpaired neurite bundle that runs from the buccal nervous system to the dorsal ciliary tract of the midgut while in other solenogasters an unpaired neurite bundle runs along the pedal fold (Faller et al. 2012).

Typically, a single or double dorso-terminally located posterior sense organ (previously thought to be an osphradium) is located externally above the opening into the mantle cavity (Lindberg & Sigwart 2015; Sigwart & Sumner-Rooney 2015). Innervation of this sense organ is from paired nerves from the suprarectal ganglion suggesting that it was originally paired. In a number of taxa, the dorso-terminal posterior sense organ is a retractable knob, while in others it appears as a sunken pit (Haszprunar 1987b), and in some it is lacking.

14.6.4 Biology and Ecology

14.6.4.1 Habits and Habitats

Solenogastres have been reported at depths ranging from shallow subtidal to 4,500 m. While a few solenogasters live in shallow subtidal sandy sediment along the Atlantic coasts of Spain, France, Bermuda, Florida, and Panama, most taxa are found on continental shelves, and their diversity decreases with depth, although numbers can be high (Todt 2013). Members of the Epimeniidae are unusual in typically occurring at depths less than 100 m. Some, such as the 30 cm long *Epimenia* and the slightly smaller *Anamenia*, are found in tropical Indo-West Pacific coral reefs. At least six species of *Helicoradomenia* and a single representative of the Simrothiellidae occur at hydrothermal vents.

14.6.4.2 Feeding and Diet

Solenogastres are carnivorous with many feeding on cnidarian polyps, including hydroids, alcyonaceans, and sea anemones. They are not harmed by the cnidocysts of their prey, but it is not known if they produce protective compounds. Others feed on polychaetes (Todt & Salvini-Plawen 2005), and some appear to scavenge gelatinous zooplankton (Todt 2013). Generally, diets have been determined from gut contents or assumed by their association with prey species of Cnidaria (such as hydroids and alcyonaceans). Sasaki and Saito (2005) observed *Neomenia yamamotoi* feeding on sea anemones. Both hydroids and Actiniaria (sea anemones) occur at hydrothermal vents with the solenogaster *Helicoradomenia*, although nematocysts have not been found in its gut, and its diet may be mainly sulphur-fixing bacteria (Scheltema & Kuzirian 1991).

14.6.4.3 Reproduction, Development, and Growth

Solenogasters are hermaphrodites, and evidence suggests that Solenogastres primarily copulate and outcross rather than self-fertilise. The function of the complex copulatory apparatus of solenogasters and their reproductive behaviour is unknown, including any mechanisms utilised for mate attraction.

Brooding of developing larvae in the mantle cavity has been reported in several Solenogastres, including species of *Epimenia* (Figure 14.23) and *Halomenia*. Okusu (2002) noted that *Epimenia babai* coiled their bodies around alcyonarian corals, placing the mantle cavity opening against the surface of the coral. She further observed that alcyonarian corals, the food source of *Epimenia*, live in areas of strong currents, and this behaviour may assist in maintaining their eggs within the mantle cavity.

Most shallow-water species appear to spawn during the summer and autumn months, and, in *Epimenia babai*, Okusu (2002) induced spawning by simulating darkness. During her study, about 20,000 eggs were laid by three individuals over three months in the laboratory. The fertilised, uncleaved eggs were deposited in groups of 20–50 in paired mucous sheets. Copulation was not observed, and one individual, which had been isolated for over 90 days, continued to lay viable eggs, suggesting either sperm storage or self-fertilisation.

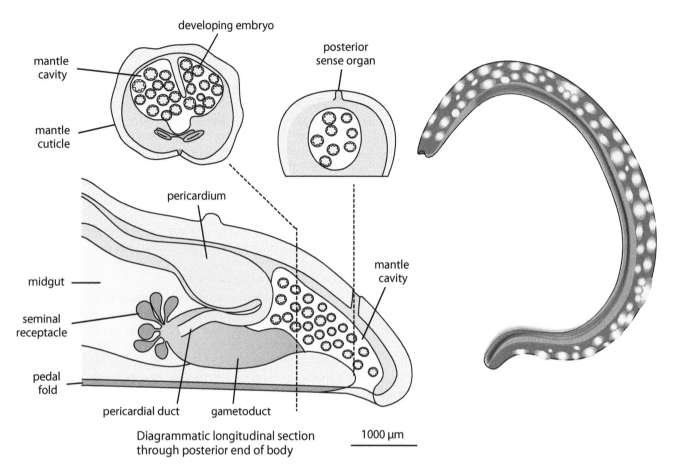

FIGURE 14.23 Brooding in *Epimenia australis*. Modified from Scheltema, A.H. and Jebb, M., *J. Nat. Hist.*, 28, 1297–1318, 1994 and notes supplied by A. Scheltema. Drawing of the living animal of *E. australis* is based on a photograph by R. Willan published in Beesley, P.L. et al. (eds.), *Mollusca: The Southern Synthesis, Part A*, CSIRO Publishing, Melbourne, VIC, 1998a.

Development has been studied in a few species, but early development is known only for *Epimenia babai* (Baba 1940a; Okusu 2002), with other observations made on embryos found in the mantle cavity (Pruvot 1890; Heath 1918; Baba 1938; Thompson 1960a). In *Epimenia babai* early cleavage is spiral, unequal, and holoblastic. Two polar lobes are formed during early cell divisions, the second being incorporated into the D macromere (Okusu 2002). The lecithotrophic larvae have differing numbers of rows of ciliated prototrochs. In *E. babai* the free-swimming larvae hatch from their egg capsules after three days, and the trochophore larvae possess an enlarged swimming test (pericalymma). This pericalymma[8] (or test cell) larva consists of three distinct regions: the apical cap, trunk, and caudal regions. The apical cap is crowned by long compound cilia of the apical tuft, and at its midline, it is encircled by a row of compound prototroch cilia. The remaining surfaces of the apical cap are finely ciliated. The test cells can be large and regularly arranged in rows as in *Neomenia carinata* (three rows above the prototroch and two below)

(Thompson 1960a), or the cells are small and irregularly scattered over the surface of the apical cap as in *Epimenia babai* (Okusu 2002). The larval trunk emerges from the pseudo-blastopore opposite the apical tuft. The caudal region is also finely ciliated with longer compound cilia around the edge (the telotroch). The apical–caudal axis of the larvae becomes the anterior–posterior axis of the adult.

In *Epimenia babai* metamorphosis began after seven days and took about 48 hours to complete (Okusu 2002). The unciliated trunk region of the larvae gives rise to ectodermal structures, such as the cuticle, epidermis, and spicules.

The development of *Wirenia argentea* was recently studied by Todt and Wanninger (2010). The fertilised eggs were deposited in capsules and hatched after about 45 hours as ciliated larvae which swam in the water column and developed into pericalymma larvae. Development and metamorphosis in this Norwegian species were slow, the whole process taking about 40 days to develop a foregut and two months for the hindgut and radula. At that stage yolk reserves were depleted. Development proceeded with the elongation of the trunk region and formation of the digestive and nervous systems. The nervous system formed from depressions on the apical cap. The anterior-most depressions formed the cerebral ganglia, while in other species

---

[8] Pericalymma larvae are also found in protobranch bivalves and scaphopods, but the homology of these larval forms is doubtful (e.g., Todt & Wanninger 2010).

additional posterior depressions formed the pedal ganglia and ventral nerve cords (Thompson 1960a). Mesodermal structures, including the pericardium, heart, and gonoducts, were not fully differentiated until after metamorphosis and settlement (Thompson 1960a). As the trunk elongated, a ciliated region appeared on the ventral surface of the larvae and subsequently formed the foot groove. The remaining trunk ectoderm became papillate in appearance and secreted the cuticle and spicules. The trunk ectoderm ultimately enveloped the apical cap and caudal region, and the yolk-rich apical cap cells are thought to be a source of nutrition in some species. Except for the nervous system, the apical cap and caudal region contributed nothing to the adult body.

Reports of protonephridia in the larvae are contradictory; they were reported in *Wirenia* (Todt & Wanninger 2010), but they were not noted in the earlier accounts of solenogaster development.

In most solenogaster larvae there is no evidence of external metamerism (Thompson 1960a; Okusu 2002; Todt & Wanninger 2010).

Growth studies of Solenogastres are rare. Scheltema (1987) reported that at least one taxon living at 2,000 m, and at temperatures near 0°C, was sexually mature within a year of settlement.

14.6.4.4 Associations

The deep-water occurrence of many taxa and their recovery by dredges, grabs, cores, and other deep-sea sampling apparatus make the documentation of associations difficult for these taxa. Thus, associations are best known for shallow-water species within the range of direct observation. The only well-documented associations of Solenogastres are those with their cnidarian food sources (see Section 14.6.4.2). There is a report of a rotifer living within the mantle cavity of *Dorymenia* (Scheltema 1985).

Two vent species of *Helicoradomenia* have been shown to harbour both epibiotic and endocuticular bacteria (both γ-proteobacteria and α-proteobacteria) (Katz et al. 2006). In *Neomenia carinata*, bacteria were found associated with the brush border of the mantle cavity epithelium and were engulfed by these cells (Scheltema 1992).

14.6.5 Diversity and Fossil History

There are approximately 280 named species of Solenogastres (Todt 2013), and taxonomists who work on this group have pointed out there are probably many undescribed species in deep-sea samples and museum collections.

Given the lack of a solid shell, the absence of a spicule matrix, the difficulty of fossilising soft body parts, and the deep-water habitat of living species, it should not be surprising that the fossil record of aplacophorans is both sparse and contentious. However, if one allows for greater morphological diversity in the stem taxon, some interesting possibilities exist. For example, if stem aplacophorans had plate-like structures similar to those of polyplacophoran valves, the discovery of foot-less chitons (*Kulindroplax* and *Phthipodochiton*) (Sigwart & Sutton 2007; Sutton & Sigwart 2012) suggests the possibility of foot loss in paleoloricate chitons. In addition, the Silurian *Acaenoplax hayae* (Sutton et al. 2001a, 2004) also lacks a foot but has both plates and spicules. Besides the original report of *Acaenoplax* from England, Cherns (2004) reported *Acaenoplax* valves from the Silurian of Götland, Sweden, where they apparently co-occurred with paleoloricates in shallow, nearshore carbonate shelf habitat. Additional study of the English material by Dean et al. (2015, p. 839) revealed a novel respiratory system not 'closely comparable to that of either extant aplacophoran group'. See Chapter 13 for further discussion.

15 The Bivalvia

15.1 INTRODUCTION

Bivalves, comprising the oysters, mussels, and clams, are the second largest group of molluscs. They are headless, their shell consists of a pair of laterally compressed, hinged valves connected by a ligament, and the mantle cavity typically surrounds the entire body. The mantle cavity contains the gills (ctenidia) and labial palps, both of which are highly significant in the evolution of the group.

Bivalves (Figures 15.1 and 15.2) are an extremely diverse group with about 50,000 living species (Gosling 2003) that range in adult size from 0.5 mm to giant clams that reach 1.4 m in length with valves which can each weigh up to about 115 kg.

Several detailed treatments of bivalve evolution, morphology, and systematics are available, with a number of books being devoted to this class (Franc 1960; Cox et al. 1969a; Morton 1990c; Morse & Zardus 1997; Johnston & Haggart 1998; Harper et al. 2000b; Gosling 2003; Mikkelsen & Bieler 2007; Huber 2010; Dame 2012; Huber et al. 2015).

The main characters that separate bivalves from other molluscs are the possession of a bivalved shell throughout their ontogeny, the lack of a head and associated buccal apparatus, the possession of adductor muscles (typically a pair, sometimes one) that close the shell valves, a horny dorsal ligament, U-shaped kidneys, and, typically, a digging foot that either has a narrow sole or lacks one.

15.2 CLASSIFICATION

Bivalves are divided into two main groups, the Protobranchia and the more typical bivalves with sheet-like gills, the Autobranchia (see Box 15.1). In the past, various attempts to interpret bivalve evolution have involved not only the shell, especially the hinge and the ligament, but also the form of calcium carbonate (aragonite or calcite) and its crystalline structure, the gills and their ciliation, and the internal stomach morphology have also been used. The latest classifications reflect recent phylogenetic analyses based either on morphology, molecular data, or both (Box 15.1).

BOX 15.1 HIGHER CLASSIFICATION OF BIVALVES

(Class) **Bivalvia** (= Acephala, Pelecypoda, Lamellibranchia, Lamellibranchiata)
　(Subclass) **PROTOBRANCHIA** (= Palaeotaxodonta)
　(Orders Nuculida, Solemyida, Manzanellida, and Nuculanida)
　(Subclass) **AUTOBRANCHIA** (= Autolamellibranchiata; called lamellibranchs by some authors [e.g.,

Yonge 1959a; Purchon 1987a, 1990], but that term has usually been used to encompass all bivalves)
　(Infraclass) **Pteriomorphia**
　(Living orders Mytilida, Arcida, Ostreida, Pteriida, and Pectinida, Limida, and some extinct groups – see Appendix for details)
　(Infraclass) **Heteroconchia** (= Eulamellibranchia of Thiele)
　(Cohort) **Palaeoheterodonta**
　(Orders Unionida and Trigoniida)
　(Cohort) **Heterodonta**
　(Subcohort) **Archiheterodonta** (Order Carditida and the extinct Actinodontida)
　(Subcohort) **Euheterodonta**
　(Megaorder) **Anomalodesmata** (= Poromyata)
　(Orders Pholadomyida, Poromyida, Pandorida, and Thraciida)
　(Megaorder) **Imparidentia** (= Cardioni, Euheterodonta [in part], Neoheterodontei)
　(Living orders Lucinida, Cardiida, Solenida, Hiatellida, Pholadida, and two extinct orders including the 'rudists')
　A more detailed classification is provided in the Appendix.

The higher-level groups in earlier bivalve classifications differed considerably from the example given here. There has been disagreement over how many and at what rank major subdivisions should be recognised. Thiele (1929–1935), Franc (1960), and Cox (1960a) all recognised three main groups as subclasses (Cox, Franc) or orders (Thiele). Both Franc and Cox recognised Protobranchia for forms that had protobranch ctenidia, and both authors also accepted a taxon for forms with usually filibranch ctenidia (see Section 15.3.6.1) and with a tendency towards reduction or elimination of one of the adductor muscles (Filibranchia of Franc, Pteriomorphia of Cox). The remaining taxa were placed by Cox (1960a) in the subclass Heteroconchia, a taxon broadly equivalent to Thiele's Eulamellibranchia. Yonge (1959a) advocated treating the septibranchs as a third subclass and Franc (1960) formally excluded the septibranchs from his Eulamellibranchia. Earlier, Thiele (1929–1935) included within his (order) Eulamellibranchia the (suborders) Schizodonta (= Palaeoheterodonta), Heterodonta, and Anomalodesmata.

A major modification of bivalve classification was that of Newell (1969) in the *Treatise on Invertebrate Paleontology*. His concept of the Pteriomorphia was very like that of Franc (1960) and Cox (1960a), but he split the Protobranchia into two taxa of equal rank (the Palaeotaxodonta with taxodont hinge teeth, and the mostly edentulous Cyrtodonta). Newell also doubted the monophyly of Thiele's concept of

FIGURE 15.1 Examples of a living protobranch and some pteriomorphian bivalves. (a) *Solemya* sp. (Solemyidae), Guadeloupe, French West Indies. Courtesy of P. Maestrati - MNHN. (b) *Pinna bicolor*, New South Wales, Australia. Courtesy of J. Walker. (c) Limidae, Panglao, Philippines (Pinnidae). Courtesy of S. Tagaro - MNHN. (d) *Bathymodiolus* sp. (Mytilidae), northwestern Atlantic. Courtesy of *NOAA Office of Ocean Exploration and Research, Northeast U.S. Canyons Expedition 2013*. (e) *Bractechlamys antillarum* (Pectinidae), Guadeloupe, French West Indies. Courtesy of L. Charles - MNHN. (f) *Pedum spondyloideum* (Pectinidae), New South Wales, Australia. Courtesy of S. Smith. (g) *Lopha cristagalli* (Ostreidae), Philippines. Courtesy of T. Gosliner. (h) *Saccostrea glomerata* (Osteidae), New South Wales, Australia. Courtesy of J.M. Ponder.

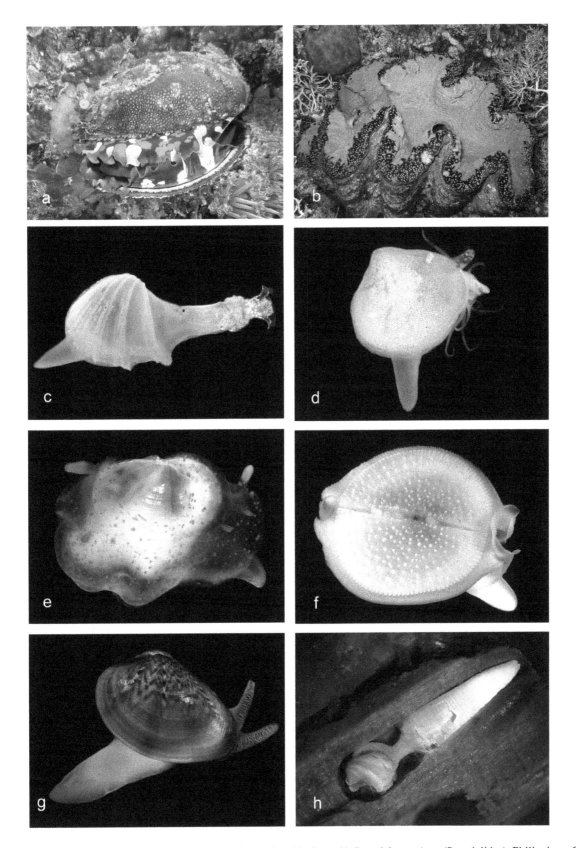

FIGURE 15.2   Examples of a living pteriomorphian and heterodont bivalves. (a) *Spondylus varians* (Spondylidae), Philippines. Courtesy of T. Gosliner. (b) *Tridacna gigas* (Cardiidae), Indonesia. © Cristine Huffard (c) *Cardiomya ornatissima* (Cuspidariidae), Guadeloupe, French West Indies. Courtesy of P. Maestrati - MNHN. (d) *Poromya* sp. (Poromyidae), Guadeloupe, French West Indies. Courtesy of P. Maestrati - MNHN. (e) *Scintillona* sp. (Galeommatidae), New South Wales, Australia. Courtesy of D. Riek. (f) *Ephippodontoana mcdougalli* (Galeommatidae), South Australia. Courtesy of P. Middelfart. (g) *Eumarcia fumigata* (Veneridae), New South Wales, Australia. Courtesy of D. Riek. (h) *Xyloredo nooi* (Xylophagaidae), Monterey, California. Courtesy of J. Judge.

Eulamellibranchia (Thiele 1929–1935), which had been accepted by most previous authors, and he split the eulamellibranchs into three subclasses – Palaeoheterodonta, Heterodonta, and Anomalodesmata. Thus, until recently, most classifications have recognised three groups of heteroconch bivalves, one of which was Anomalodesmata, although Franc (1960) modified this arrangement by removing the septibranchs from his 'Anomalodesmacea', regarding them as a separate higher taxon of equal rank to Eulamellibranchia. Another noteworthy deviation from the accepted ideas of bivalve evolution was the entirely unsupported notion (Nevesskaja et al. 1971; Starobogatov 1992) that the septibranchs, together with the Cambrian fordilloideans, were rostroconchs, an extinct group with bivalved shells (see Chapter 13).

Purchon (1963) classified bivalves based on morphological criteria, particularly stomach anatomy. He divided them into two subclasses, Oligosyringia, including two orders, Protobranchia and Septibranchia, and Polysyringia, with the latter 'group' including all the 'lamellibranch' bivalves (other than the septibranch pholadomyidans). The polysyringians were subdivided into three orders – Gastrotriteia, Gastrotetartika, and Gastropempta – based on stomach morphology (see Chapter 5 and Figure 15.26 for further details).

There have been significant changes in several new classifications published since the review of bivalves in *Treatise on Invertebrate Paleontology* in 1969. Schneider (2001) summarised changes up to the year 2000, and a further summary was provided by Giribet (2008a). While the higher groupings remained largely intact, the subdivision of these has varied considerably, largely as a result of molecular analyses. Some of the more significant changes are summarised below in the accounts of the major groups.

The classification we have adopted (see Box 15.1 for an outline and the Appendix for details) mainly follows Carter et al. (2011) and Bieler et al. (2014). Other classifications have been recently provided by Bieler and Mikkelsen (2006) and by R. Bieler, J. G. Carter, and E. V. Coan in Bouchet et al. (2010). Bivalve phylogeny, and hence classification, has until very recently been unstable, mainly because of results from discordant molecular analyses and varying interpretations in placing some extinct groups. Molecular analyses (e.g., Adamkewicz et al. 1997; Campbell 2000; Sharma et al. 2012; González et al. 2015; Combosch et al. 2017) and morphological analyses (e.g., Purchon 1978, 1987a; Morton 1996; Waller 1998) are generally largely concordant in the recognition of major groups. Giribet and Wheeler (2002) analysed both morphological and anatomical data, as did Bieler et al. (2014). In some phylogenies using mitochondrial DNA, anomalous results have been produced. For example, the Palaeoheterodonta are at the base of the autobranch branch, with the Pteriomorphia + Heterodonta forming a clade, named by Plazzi et al. (2011) as Amarsipobranchia. No morphological data, including the fossil record, support the idea that palaeoheterodonts are basal autobranchs.

Protobranch bivalves (see Section 15.7) have a simple gill and, except for the solemyoideans, deposit feed using their elongate labial palp proboscides. While the basal position of the protobranch bivalves is rarely disputed, the details of their relationships to the mostly suspension-feeding autobranch bivalves with their highly modified gills (see Section 15.3.6.1) remain unresolved (see Section 15.7.1 for details).

In the most recent classifications of bivalves, all the non-protobranch bivalves belong to the Autobranchia, which in turn are divided into two large groups, the Pteriomorphia and Heteroconchia. These two groups are monophyletic in most molecular analyses.

Besides the greatly enlarged W-shaped ctenidial filaments, autobranch bivalves have a byssal gland in the posterior part of the foot, at least in the larva, and special laterofrontal cilia on the tips of the gill filaments which are involved in capturing particles in suspension feeding. The Pteriomorphia (pearl oysters, true oysters, scallops, and mussels; see Section 15.8.1) include infaunal, or partially so, taxa, but many are epifaunal, being byssally attached or cemented. The Heteroconchia comprises the Palaeoheterodonta and Heterodonta. The former group includes the 'brooch shells' (Trigoniidae) (see Section 15.8.3.1) and the freshwater mussels (see Section 15.8.3.2) while the Heterodonta contains the remaining bivalves (including clams, cockles, venus shells, razor shells, and shipworms). These are divided into two groups: the Archiheterodonta, which includes only the Carditida (see Section 15.8.4.1), and the Euheterodonta (see Section 15.8.5) which comprises the remaining heterodonts.

The Anomalodesmata are a diverse, ancient group having members as divergent as the watering pot shells and septibranchs; the latter have either very reduced gills or have lost them altogether on becoming carnivores (see Section 15.8.6.1.2). The Anomalodesmata (see Section 15.8.6) was considered a separate subclass until recently but is now included in the euheterodonts as a group separate (see Section 15.8.6) from the remaining euheterodonts (the Imparidentia).

15.3 MORPHOLOGY

Bivalves are typically bilaterally symmetrical, although some aspects of their internal anatomy are not. There are, however, significant departures from bilateral symmetry in some, mainly attached epibenthic taxa such as scallops, oysters, and anomiids (see Section 15.4.1.2.1). In a few bivalves, the shell can be internal, reduced (or very rarely almost absent), and in 'shipworms' (Teredinidae) the animal is worm-like.

The bivalve body is usually encased within the two valves of the shell. The mantle cavity contains the gills and labial palps and surrounds the foot and much of the visceral mass, which is primarily situated dorsally and continues ventrally into the foot. It contains the heart, gut (including the digestive gland), major muscles, central nervous system, kidneys, and gonads.

Several structures have dominated bivalve studies – the shell, along with its hinge and ligament and the periostracum, the gills and palps, the mantle and its elaborations including the siphons, the foot, and the muscles that drive the foot and close the shell. We deal with each system in more detail below.

15.3.1   THE SHELL

The bivalve shell consists of two dorsally hinged valves, usually with interlocking teeth (the hinge) and always with a horny ligament that connects the two valves dorsally and is under compression when the valves are closed. The shell grows by way of accretion to the margins via the mantle edge. The interior of each valve has scars at the points where the various muscles were attached to it in life. These include the adductor muscles (usually two, sometimes one) that close the valves, and the pedal (and byssal if present) muscles. The line of attachment of the mantle to the shell is represented by the pallial line, and in most bivalves that possess siphons there is a posterior embayment of the pallial line, the pallial sinus. The muscle attachment to the pallial line 'migrates' as the shell grows.

Some bivalves have unequal-sized or differentially shaped valves (*inequivalve*). One may be slightly smaller than the other and fit within the larger valve as, for example, in the arcoidean *Cucullaea* or the myoidean Corbulidae, or they may differ in the degree of inflation, such as in *Pecten*, most oysters, pearl oysters, Myochamidae, and Thraciidae. Many cemented taxa such as oysters, Chamidae, and Cleidothaeridae have a deeper attached valve and a flatter unattached one. The left valve of the free-living *Pecten* is flat or concave, and the right is convex. In less obvious examples of inequivalve shells, many tellinids have the posterior end of both valves twisted a little to the right, and the auricles of some pteriomorphians such as pectinids and pteriids are often differently shaped on each valve.

15.3.1.1   Orientation

Some bivalves are so modified that the anterior–posterior and dorsal–ventral axes are not readily identified. Anatomical markers such as the mouth, anus, adductor muscles, and foot can be used reliably. The mouth is anterior and the foot anteroventral, and the byssus (if present) is associated with the foot. Special shell structures associated with the byssus, such as the pectinid *ctenolium*, are therefore also anteroventrally located, while some pteriomorphians (e.g., scallops, pearl oysters) have *auricles*, wing-like extensions from the dorsal shell that often have a byssal notch.

Regarding shell positional markers, the hinge is dorsal and the ligament usually posterior to the beaks; the pallial sinus, if present, is always posterior. In many bivalves, the posterior end is longer than the anterior, but this is not always the case as, for example, in most protobranchs, many tellinoideans, and galeommatoideans, where the anterior end is longer.

Other markers on the shell of some heterodonts such as venerids and lucinids include the *lunule*, which is located anterior to the umbones, and the *escutcheon* posterior to the umbones (Figure 15.3).

In the great majority of bivalves, the umbones are inclined anteriorly (*prosogyrus*), but in some the umbones are not inclined (*orthogyrus*) or are posteriorly inclined (*opisthogyrous*). In some, notably Mytilidae and Pinnidae, the umbones are situated close to or at the anterior extremity of the shell. The few that have opisthogyrous umbones include some anomalodesmatans such as *Thracia*, *Myadora*, and *Cuspidaria*. If the umbones are in the middle of the shell, this is a state called *equilateral*. If they are anterior or posterior, the condition is *inequilateral*.

The constraint imposed by the ligament sometimes induces spiral growth with a high expansion rate as seen, for example, in the fossil oyster *Exogyra* and the heterodont Glossidae.

In some pteriomorphians (a few arcids and some Mesozoic Bakevelliidae, and in a mytilid, *Modiolus americanus*) the shell is twisted, a phenomenon called bivalve torsion (McGhee 1978; Salvazzi 1984). These are all byssate shallow infaunal or epifaunal, with the arcid *Trisidos* being the best-known example (Tevesz & Carter 1979; Morton 1983b) (Chapter 3, Figure 3.12).

15.3.1.2   Sculpture

Bivalve sculpture on the outer surface of the valves is of three main types – commarginal (concentric), radial, and, less commonly, various forms of oblique sculpture (e.g., Ubukata 2005). Commarginal sculpture is secreted periodically by the entire mantle margin while radial sculpture is secreted continuously at specific points along the margin. The secretion points of oblique sculpture migrate along the mantle edge as the bivalve grows (Checa & Jiménez-Jiménez 2003). The sculpture of many bivalve shells can be highly elaborate and sometimes complex. It ranges from simple growth lines to elaborate frills or heavy radial ribs and even to spines. The more unusual oblique sculptural types include divaricate and antimarginal. Divaricate ribbing occurs in a few members of several unrelated groups (e.g., some Nuculanidae, Lucinidae, Cardiidae, Galeommatoidea, Tellinoidea, Veneroidea, Solecurtidae, and Pholadomyoidea). This consists of oblique ribbing that runs in opposite directions on different parts of the valve and may functionally assist in burrowing as ratchets (Stanley 1969; Seilacher 1972).

The shell margins are smooth or variously crenulated or serrated (Vermeij 2013).

15.3.1.3   Shell Structure

The diverse microscopic crystalline structure of bivalve shells has been the subject of several major investigations and is described below in some detail. The summary below is largely based on the comprehensive accounts of Taylor (1973), Carter and Lutz (1990), and the more recent account in Bieler et al. (2014).

Bivalve shell structures are divisible into four major groups: (1) nacre, (2) prismatic, (3) foliated, and (4) crossed lamellar structures. These structural types are illustrated and described in more detail in Chapter 3. With finer resolution, these four structural types can be further divided into subgroups such as simple or composite prismatic, complex crossed lamella, or irregular foliated. Homogeneous shell structures also occur; these have no identifiable elements other than minute granules and are therefore recognised by their lack of structure.

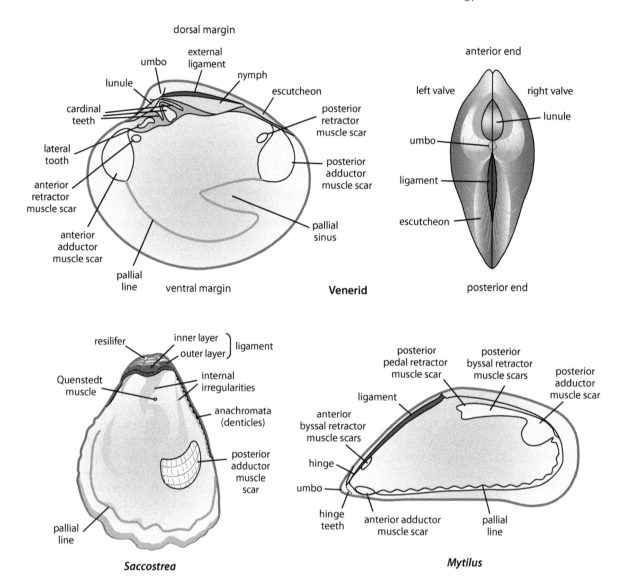

FIGURE 15.3 Bivalve shell features and terminology. Redrawn and modified from various sources.

Homogeneous structures are likely derived from more organised structures such as prismatic and crossed lamellar. For example, the amino acid composition of the organic matrices that generate the major shell structure groups also differs between shell structures. However, the amino acid composition of the homogeneous structures in venerid bivalves is most similar to those of the crossed lamellar structures, suggesting that the 'homogeneous' form, in that group at least, is derived from the cross lamellar type (Shimamoto 1993).

Certain bivalve shell structures predominate in, or are restricted to, a particular calcium carbonate mineralogy (i.e., calcite or aragonite). Thus, nacre and crossed lamellar structures are found in aragonitic layers, while foliated structures are mostly calcitic. Prismatic structures may be aragonitic or calcitic, and while both calcitic and aragonitic homogeneous structures occur in bivalves, the latter form is more common.

There are two forms of bivalve nacreous structures – sheet nacre and lenticular or columnar nacre. Sheet nacre is present in the Nuculoidea, Mytiloidea, Pinnoidea,

Pterioidea, Unionoidea, Trigonioidea, Pandoroidea, and Pholadomyoidea. However, several of these groups also have taxa with columnar nacre which is found in the Nuculidae, Pteriidae, Mytilidae, Unionidae, Trigoniidae, Pandoridae, and Lyonsiidae (Taylor et al. 1969, 1973). Nacre and crossed lamellar structures are aragonitic, but they seldom co-occur in the same shell. When they occur together, most examples are from extinct species within Mytiloidea, Trigoniidae, Pholadomyidae, and the extinct families Cyrtodontidae, Aviculopectinidae, and Edmondiidae (Carter 1990d; Newell & Boyd 1990; Carter 2001).

Some internal nacre has been found in an early crassatelloidean, the Middle Devonian *Eodon* (Eodonidae) (Carter & Lutz 1990; Carter et al. 2000), but, apart from some anomalodesmatans, living heterodonts lack nacre.

The reason that various lineages lost the prismato-nacreous microstructure is unclear. This combination is mechanically superior to other shell microstructures (Taylor & Layman 1972), although homogeneous microstructure is strong under compression. Homogeneous shell structure has a lower organic content (Taylor & Layman 1972; Harper 2000), which may mean it requires less energy to produce than other microstructures, especially nacre (Palmer 1992).

Foliated structures are found only in the Pteriomorphia (Ostreoidea, Pectinoidea, Anomioidea, and Limoidea) and appear to be derived from calcitic prismatic structure (Taylor 1973). The co-occurrence of nacre and foliated structures in the same shell is also rare and occurs only in some Paleozoic aviculopectinids (Carter 1990b). Foliated and crossed lamellar structures commonly co-occur, especially in the Pectinoidea. Inner crossed lamellar and outer prismatic shell structures predominate in most Heterodonta, with nacre shell structure restricted to three clades in the anomalodesmatan Pholadomyoida (Pandoridae, Lyonsiidae, and Pholadomyidae) (Taylor 1973). In the Veneridae, homogeneous layer structures, apparently derived from cross lamellar structure (see above), are also common.

The earliest bivalves – *Fordilla* and *Pojetaia* – have internal foliated aragonite (Vendrasco et al. 2011a) overlain by a prismatic layer in the latter (Carter 1990a). This suggests that nacre originated independently from that seen in other classes of molluscs (Vendrasco et al. 2011a). The excellent bivalve fossil record enables documentation of the way shell structures have changed in different lineages through time. Nacre in the Protobranchia appears in the Middle Ordovician (470 Mya), although since the early Paleozoic there had been an increase in the diversity of shell structures within that group (Figure 15.4). For example, in the Nuculoida, Carter (1990a) has shown that most early shells were composed of nacreous shell structure, while a few taxa had shells with crossed lamellar and homogeneous structure. Taxa with crossed lamellar shells were extinct after the Triassic, and nuculoidean taxa with homogeneous shell structure became more dominant than nacreous taxa after the Cretaceous. Both of these transitions took place during time periods associated with mass extinction events – the End Triassic extinction (219–201 Mya) and the Cretaceous–Paleogene extinction (66 Mya).

In the Pectinoidea, the nacre shell structure became reduced through time, and since the Cretaceous–Paleogene boundary had been replaced by lineages with foliated and crossed lamellar shell structure (Carter 1990b). Within the Heterodonta, the anomalodesmatan Thraciidae was predominately nacreous in the Cretaceous but today lacks nacre, although nacre is retained in some pholadomyoideans (see above). Prismatic and crossed lamellar structures have also changed through time. From the Miocene, the outer shell layers in Veneridae, which consist of prismatic, crossed lamellar and/or homogeneous structures, have diversified into at least 17 combinations. One of the structures (spherulitic prismatic) is known only in a Miocene taxon. In contrast to the diversity in the outer shell layer, inner shell layers in these same taxa appear to have been in stasis and in 12 venerid subfamilies have been unchanged since the Miocene (Hikida 1996).

Changes in the shell structure composition in different groups make it difficult to identify the plesiomorphic shell structures in bivalves. Mapping shell structures on a phylogeny of living bivalves (Figure 15.4) suggests that nacre shell structure could be homoplastic with at least four unique originations (Nuculoidea, Pteriomorphia [Mytiloidea, Pinnoidea, Pterioidea], Palaeoheterodonta, and Anomalodesmata), although a case can also be made for retention from one or two derivations.

A glance at some of the phylogenetic trees based on living bivalves might suggest that crossed lamellar structure, rather than nacre, was plesiomorphic. However, when the fossil distribution of shell structures in bivalve clades is considered, it becomes apparent that losses and reductions have been common, especially in nacre, and appear to have occurred in parallel in several groups (e.g., Nuculoidea and Pectinoidea).

Both adaptive and non-adaptive scenarios have been proposed to explain why bivalve shell structures have changed through time. Different shell structures have different functional and biomechanical properties (Taylor & Layman 1972), and these are subject to selection. Because calcite is more stable at low temperatures than aragonite, certain shell structures are more likely to be physiologically favoured at different temperature regimes, both geographically and bathymetrically. Thus, global temperature changes, and changes in seawater chemistry (e.g., ocean acidification), may also have played a role in shell structure change. Calcite is more resistant to dissolution than aragonite, and thus there was a selective advantage in having calcitic shells in the 'corrosive' seas of the Jurassic to Cretaceous (Harper et al. 1997). Hautmann (2006) observed that several groups of epifaunal bivalves (Ostreidae, Gryphaeidae, Pectinidae, Plicatulidae, and the extinct Buchiidae) replaced their aragonitic shells with calcitic shells as conditions changed from the 'aragonite sea' to a 'calcite sea' at the beginning of the Jurassic.

15.3.1.4 Periostracum

The periostracum, the organic layer covering the exterior of the shell (see Chapter 3), may be thin and almost invisible to thick and heavy and, while usually smooth, there can be hairs or frills over the surface. Hairs and branching structures on the periostracum are produced by the byssal gland and attached to the outer shell surface by the foot (Ockelmann 1983; Choo et al. 2014).

Periostracal processes may extend well beyond the edges of the shell valves as in *Solemya* and, in the lucinid *Rastafaria*, the periostracum is extended into long tubes through which the foot penetrates as it probes anoxic sediments (Taylor & Glover 1997). In another lucinid, *Lucina pensylvanica*, the periostracum is elaborated into calcified lamellae (Taylor et al. 2004). These structures can be useful in taxonomy; for example, the branching patterns of some processes can be species-specific as in some mussels (Mytilidae). Harper (1997) surveyed the bivalve periostracum, particularly its thickness, and noted that it has been important in enabling certain specialised life habits and in facilitating different kinds of shell

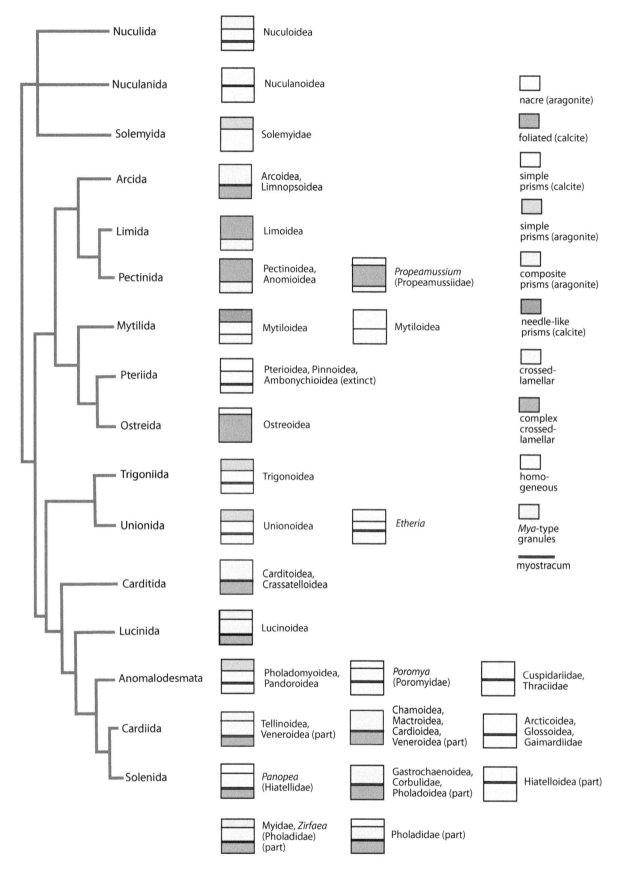

FIGURE 15.4 Bivalve shell structure and its phylogenetic distribution. The diagram represents shell layers with the outermost upper. Data largely from Taylor, J.D. et al., *Zoology*, 22, 255–294, 1973, with updates from Bieler, R. et al., *Invertebr. Syst.*, 28, 32–115, 2014.

ornamentation, with finer shell ornament patterns only possible with a thin periostracum.

The bivalve periostracum typically has two layers, an outer proteinaceous layer and an inner (probably mucoid) one (Morton & Peharda 2008). The outer layer is secreted by the periostracal groove on the inner surface of the outer mantle fold, and the inner layer is secreted more distally on the same fold (Yonge 1982).

Calcified periostracal structures that are often spicule-like are present in the periostracum of some veneroidean bivalves (Glover & Taylor 2010), and it has been suggested (unconvincingly) that they are homologous with cuticular spicules found in aplacophorans and polyplacophorans (Carter & Aller 1975). Calcified periostracal structures are rare in living taxa but may have been an ancestral feature in some bivalve groups (Schneider & Carter 2001), including the Mytiloidea, Myoida, Cardiida (in the extinct Kalenteridae), and Anomalodesmata (Carter & Aller 1975). The functional significance of these structures is unknown, although G. Oliver (pers. comm., 2018) noted that spicules in some venerids such as Pitarinae appear to facilitate the adhesion of sediment to the siphons, a possible antipredator adaptation. The spicules are often arranged in radial rows and are a notable feature of many anomalodesmatans (Harper 1997).

15.3.1.5 Ligament

The two shell valves are connected by a non-mineralised ligament which lies on the hinge line, is largely organic, and usually brown or black in colour. Its elastic properties enable the shell to open, as it is under compression when the valves are closed and serves to separate them when the adductor muscles relax.

The nature of the ligament is important in bivalve classification, although different ligament configurations can be convergent and related to life habits. The type of ligament is not only important in the taxonomy of living bivalves, but it can also be deduced from shell morphology in fossils.

The evolution of the ligament and its supporting structures was discussed by Waller (1990). His view that the ligament in adult bivalves was primitively located behind the beaks was in contrast to some other phylogenetic scenarios which proposed that the primitive ligament lay beneath the beaks. According to Waller, the primitive bivalve ligament had a three-layered structure – periostracum, lamellar, and fibrous aragonite layers (see below) – with only the periostracal part continuous with the shell layers.

A complex nomenclature has been developed to describe the different kinds and positions of ligaments. The main terms are listed below and illustrated in Figures 15.5 and 15.6.

The ligament consists of two basic parts:

- A calcified inner ligament (the fibrous ligament), which lies ventral to the axis of rotation of the valves and responds to compression. It is composed of fine fibres of aragonite with the fibres orientated perpendicular to the growth surface of the layer.

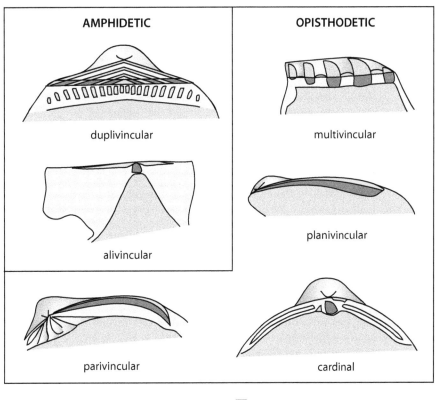

□ lamellar layer ■ fibrous layer

FIGURE 15.5 The main types of ligaments in bivalves. Redrawn and modified from Ubukata, T., *Paleobiology*, 26, 606–624.

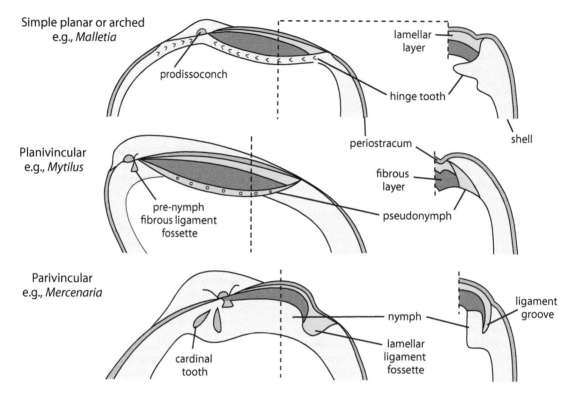

FIGURE 15.6 Ligament structure and terminology. Redrawn and modified from Waller, T.R., The evolution of ligament systems in the Bivalvia, pp. 49–71 in B. Morton (ed.), *The Bivalvia. Proceedings of a memorial symposium in honour of Sir Charles Maurice Yonge (1899–1986), Edinburgh, 1986*. Edinburgh, Scotland, UK, Hong Kong University Press, 1990.

- A non-calcified outer ligament (the lamellar ligament) which lies above the axis of rotation of the valves and responds to tension. This is often divided into two differently coloured sublayers.

Unlike earlier scenarios of bivalve evolution, Waller (1990) argued that the lamellar ligament did not originate from part of the primary shell but was derived from repair material secreted in response to periostracal splitting along the dorsum. Some evidence for such fractures is seen in conocardioidean rostroconchs (Pojeta 1985). Waller further argued that while the fibrous ligament might have been a modified primary shell layer, it may have been absent in the earliest Cambrian bivalves. In living nuculoidean families, the equivalent part of the ligament has a granular structure rather than fibrous, suggesting that it is either a novel apomorphic structure or represents the plesiomorphic condition of the fibrous ligament (Waller 1990).

Ligaments may be visible along the dorsal shell margin (i.e., they are *external ligaments*) where they are attached directly to the shell margins, or they may be supported by ridges (*nymphae* or *pseudonymphae*). Nymphae are sometimes extended below the hinge line inside the shell. An *internal ligament* (or *resilium*) is not visible externally, and the structure for its attachment can be called the *resilifer*, which may be a spoon-like *chondrophore*. In some anomalodesmatans an accessory calcareous plate (the *lithodesma* or *ossiculum*) may reinforce the internal ligament.

BOX 15.2   LIGAMENT TERMINOLOGY

Based on these ligament support systems and ligament structure, Waller (1990) recognised three basic types of ligaments:

Simple ligaments. These are seen in most protobranchs and are attached to unmodified shell margins. They can be straight or arched and are referred to by Waller (1990) as *simple planar* or *arched*. The other ligament types evolved separately from this condition.

Ligaments supported by a nymph. Nymphae evolved twice, once in protobranchs (Ctenodontidae and Solemyidae) and separately in heteroconchs.

Fibrous ligament broken up into several elements and lacking true nymphae. This type is characteristic of many Pteriomorphia.

There is a complex nomenclature used to describe ligaments. The main terms used are briefly outlined below and illustrated in Figure 15.6.

Location – an *amphidetic* ligament is evenly distributed on either side of the umbones while an *opisthodetic* ligament is mainly posterior.

Form – an *alivincular* ligament is flattened, lies between the cardinal areas of the valves, and the lamellar layer is both anterior and posterior to the fibrous layer (e.g., *Ostrea*, *Mytilus*). A *parivincular* ligament is usually cylindrical, clearly visible externally, and located posteriorly (e.g., *Tellina*).

Growth – alivincular ligaments grow from their ventral surfaces, and their dorsal surfaces are worn away. *Multivincular* ligaments grow by the mantle adding successive fibrous layers (*resilia*) posteriorly from the umbones (e.g., as in *Isognomon*), each of which is separated by the lamellar layers. *Duplivincular* ligaments form from amphidetic ligaments and also produce multiple layers of ligament in a series of dorsoventrally arranged layers (e.g., *Arca*).

15.3.1.6 Hinge

The hinge line is the dorsal border and point of articulation of the valves and includes the ligament and hinge teeth. The different kinds and arrangement of hinge teeth are important in bivalve classification. The various terms, including names like taxodont and heterodont, are described in Box 15.3 and illustrated in Figure 15.8.

BOX 15.3   THE MAIN TYPES OF BIVALVE HINGE DENTITION

Taxodont hinge teeth – short straight or curved teeth that extend across most or all of the dorsal margins. Where this is evolutionary primary it is called *ctenodont* and where it has been derived secondarily from another type of dentition it is called *pseudoctenodont* or *pseudotaxodont*.

Isodont hinge teeth – two equal-sized teeth (together with a resilium pit). Found in some Pectinoidea. Narrow ridges, few in number, may emerge near the resilium in some pectinoideans – these are termed *crura*. A somewhat similar condition to isodont occurs in some unionoideans.

Dysodont hinge teeth – a few small, weak teeth or denticles rather close to the beak. Found in some mytilidans.

Actinodont hinge teeth – these radiate from the beak and are known only in some early fossil bivalves (see Steinova 2012). Evolved into heterodont and (via lyrodesmatoid) into schizodont (Figure 15.8). *Parallelodont* and *cyrtodont* hinge types are modifications of the actinodont condition (Figure 15.8).

Schizodont hinge teeth – small number of strong teeth radiating from beaks, one of which is bifid (Trigoniida only). Some Unionida lack the bifid teeth, which is a somewhat similar condition that has been called schizodont. In many unionoideans, *pseudolateral* teeth arise near the beaks and extend near the dorsal margin, and *pseudocardinal* teeth are rather short teeth arising near the beaks.

Heterodont hinge teeth – comprised of cardinal teeth that lie just below the beaks and lateral teeth that extend close to the dorsal shell margins but not arising near the beaks.

Pachydont – a modified heterodont hinge with strong, blunt teeth mostly derived from the cardinals.

Microdont – a series of short vertical teeth resembling a taxodont hinge but derived from a heterodont condition.

Cyrtodont, edentulous – teeth absent.

Hinge-plate – the hinge teeth project from the hinge-plate, a widened part of the dorsal shell margin.

The most complex hinges, the heterodont type, comprise both lateral and cardinal teeth with sockets for the teeth in the opposite valve. Thus, the details of the hinges in each valve differ, and while these differences are typically fixed, there are many examples of hinge transposition in individuals or, sometimes, species of heterodont bivalves. Transposed hinges have been recorded in members of the living families Carditidae, Condylocardiidae, Crassatellidae, Astartidae, Trapezidae, Cardiidae, Tellinidae, Veneridae, Sphaeriidae (Matsukuma 1996), and Chamidae (Hamada & Matsukuma 1995; Matsukuma et al. 1997) (Figure 15.7) as well as in some extinct families and in the palaeoheterodont Unionidae.

15.3.1.6.1   The Evolution of the Bivalve Hinge

Scenarios of how the bivalve hinge evolved depend on interpretations of the relationships of Paleozoic bivalves. The following account is based on the scenario offered by Waller (1998) (outlined in Figure 15.8) which was built on a wealth of earlier studies. The major changes occur in the Ordovician, commencing with taxa that have a taxodont hinge with chevron-shaped teeth such as in several groups of Middle Ordovician (and probably Lower Ordovician) presumed protobranchs (e.g., Praenuculidae and Malletiidae) but are unusual in having the anterior teeth enlarged. Cope (1996a) used the generally similar Middle Ordovician *Cardiolaria* to argue that greater valve opening without valve dislocation was made possible by the larger anterior teeth along with a posterior shift of the hinge axis. He further suggested these changes were probably associated with the evolution of a suspension-feeding gill (i.e., the origin of the Autobranchia) and the associated need to void pseudofaeces.

The Middle Ordovician genus *Inaequidens* (?Malletiidae) is similar to the above genera except that the anterior hinge teeth are more irregular and somewhat nested. The opisthodetic ligament is like that assumed to be present in other protobranch-like fossils. It is the simple planar or arched type found in living protobranchs and is housed in a groove, but there is no nymph (Waller 1990). Members of this genus have an impressed anterior adductor muscle scar which lies at the base of the broad anterior hinge-plate, a feature also seen in bivalves that shows the next stage in hinge development, notably Lower Ordovician genera like *Tironucula* and *Ekaterodonta* (Tironuculidae), as well as others. These taxa have the plesiomorphic ligament as in the previous taxon but

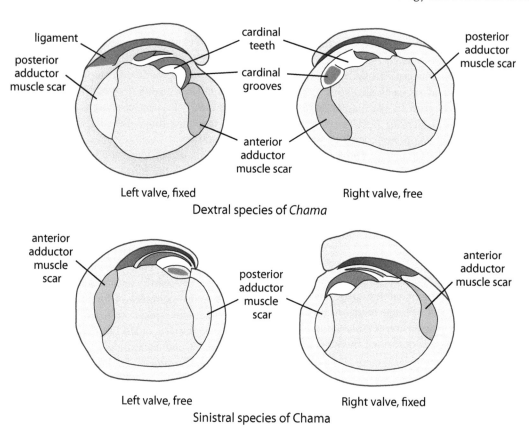

FIGURE 15.7 Hinge transposition in two species of *Chama*. Redrawn and modified from Fischer, F.P., *Manuel de conchyliologie et de paleontologie conchyliologique ou histoire naturelle des mollusques vivants et fossiles*, Savy, Paris, 1880–1887.

show what appears to be a transition between the taxodont, with many similar vertical hinge teeth, and actinodont hinge types. The latter have both chevron-shaped and lamellar teeth, with the lamellar teeth formed from the elongated dorsal parts of the posterior chevron teeth. Whether such bivalves were protobranchs or autobranchs is debatable, although Waller (1998) argued that the tironuculid ligament system and the position of the anterior adductor linked them with autobranchs, and if this is so, the taxodont hinge type is presumably plesiomorphic for autobranch bivalves.

The next stage in the series is represented by another Ordovician bivalve, *Copidens* (Pojeta & Gilbert-Tomlinson 1977). The muscle scar and ligament show a similar configuration to those in tironuculids, with the anterior teeth also distinctly chevron-shaped. However, the posterior teeth are simple lamellae with no chevron-shaped teeth remaining. This configuration represents the grade of hinge found in primitive actinodontidans. The more advanced anodontopsoideans retained this type of hinge but developed true parivincular ligaments with nymphae (Waller 1990) and gave rise to the Heteroconchia and a heterodont dentition. Such scenarios, however, need to be further tested with more detailed phylogenetic analyses that incorporate fossils.

The Pteriomorphia may have originated from an ancestor representing the condition seen in the tironuculids, rather than that represented by *Copidens*, because another Lower Ordovician bivalve, *Catamarcaia*, has a duplivincular

ligament system, and its posterior hinge retains remnants of chevron teeth below lamellar teeth (Cope 1997a).

15.3.1.7 Pallial Line and Sinus

The pallial line is formed where the mantle retractor muscles are attached to the shell. This line may terminate at the adductor muscles or may extend dorsal to those muscles, indicating cross-fusion of the mantle in that area (Owen 1958). When present, the pallial sinus is formed in a similar way to the pallial line, indicating the line of attachment of the siphonal muscles to the shell. The depth of the sinus is usually an indication of the length of the siphon.

A pallial line is not obviously present in most other mollusc shells, exceptions being some gastropod limpets and many rostroconchs (Waller 1990).

15.3.1.8 Shell Reduction, Accessory Shell Plates, and Secondary Shells

Shell reduction is rare in bivalves, and no known species has completely lost its shell. Many galeommatids have their shells partially to fully enclosed in a sac formed by lobes of the mantle, and in some of these the shells are reduced or even rudimentary (e.g., *Phlyctaenachlamys* and *Chlamydoconcha*), resulting in a bivalve slug.

Shipworms (Teredinidae) have a small anterior globular shell (Figure 15.9) used for burrowing into wood, and most of the

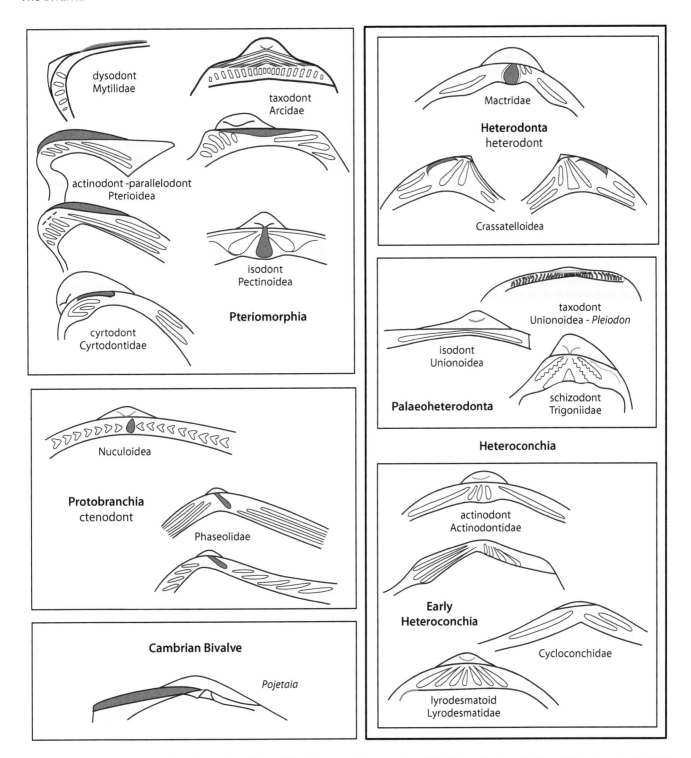

FIGURE 15.8 Bivalve hinge diversity and evolution. Most figures redrawn and modified from Scarlato, O.A. and Starobogatov, I., *Phil. Trans. R. Soc.* B, 284, 217–224, 1978; *Pojetaia* redrawn and modified from Runnegar, B.N. and Bentley, C., *J. Paleont.*, 57, 73–92.

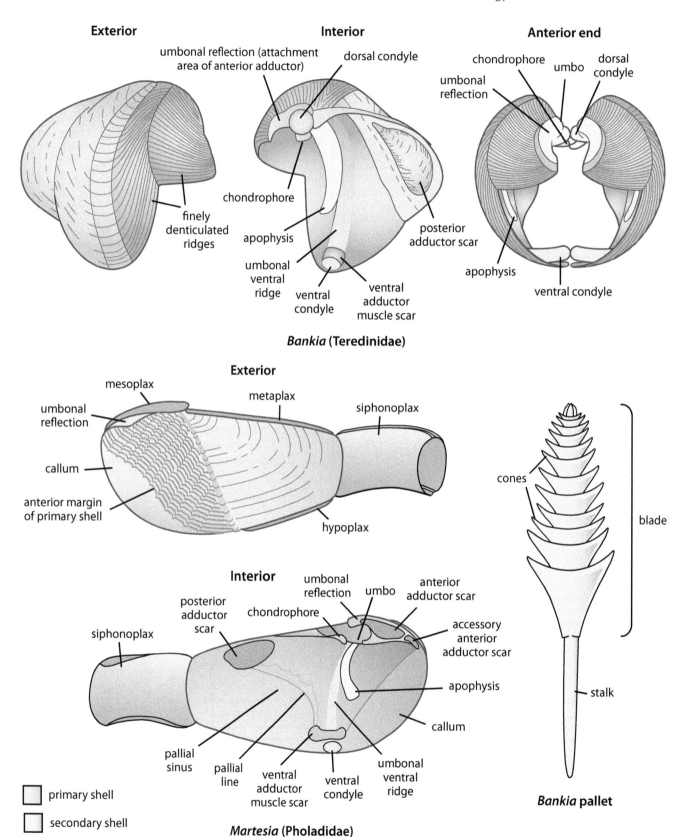

FIGURE 15.9 Shell morphology of *Bankia* as an example of Teredinidae and *Martesia* as an example of Pholadidae and a pallet from a species of *Bankia*. *Bankia* redrawn and modified from Turner, R.D., *A Survey and Illustrated Catalogue of the Teredinidae (Mollusca: Bivalvia)*, Museum of Comparative Zoology, Cambridge, UK, 1966, *Martesia* redrawn and modified from Turner, R.D., *Johnsonia*, 3, 33–162, 1954.

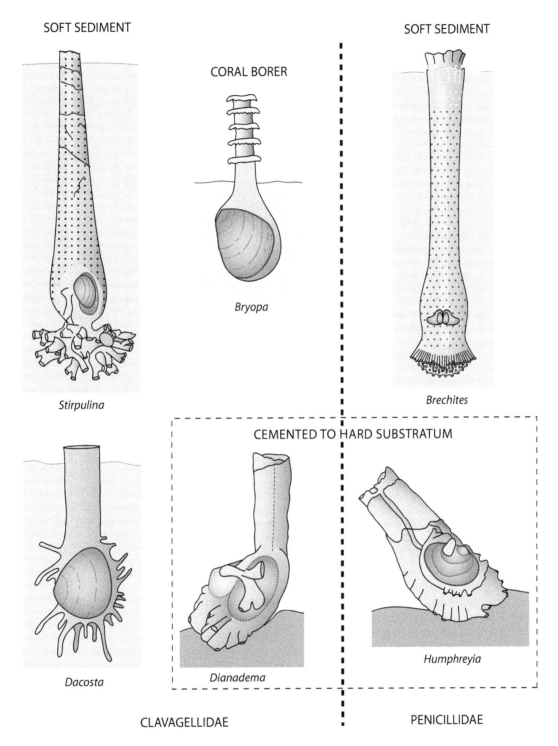

SOFT SEDIMENT

Stirpulina

CORAL BORER

Bryopa

SOFT SEDIMENT

Brechites

CEMENTED TO HARD SUBSTRATUM

Dacosta

Dianadema

Humphreyia

CLAVAGELLIDAE

PENICILLIDAE

FIGURE 15.10 Examples of clavagelloideans showing their secondary shells. Redrawn and modified from Morton, B., *Rec. West. Aust. Mus.*, 24, 19–64, 2007.

animal is encased in the long, worm-like body that lies within a burrow lined with a calcium carbonate tube (see Section 15.4.1.4).

A related family, Pholadidae, the rock borers, have accessory shell plates developed ventrally and dorsally which enclose their siphons and the pedal gape (Figure 15.9). Both teredinids and pholadids have a peg-like *apophysis* (Figure 15.9) to which the pedal retractor muscles are attached. An apophysis is lacking in the related Xylophagaidae.

The 'watering pot shells' (Clavagellidae) have a very reduced true shell while the animal, again largely contained within the siphons, is covered by a tubular secondary shell or 'crypt' (Figure 15.10), analogous to the shelly lining of the teredinid burrow. How these strange tubes have evolved from an ordinary bivalve morphology has been the subject of several papers by Brian Morton (see Morton 2007 for review) and most recently by Harper and Morton (2004). The question arises as to how these 'shells' form and, more significantly, how they grow. Morton suggested that a secondary shell was secreted in a single event after very rapid growth of the juvenile, resulting in all shells being about the same size. No other growth occurs except for some minor posterior (i.e., at the siphonal end) additions forming 'ruffle'-like structures (e.g., Morton 2002a). Harper and Morton (2004) showed that the shell was in fact secreted by the mantle in the normal way after extensive, rapid growth of the fused mantle lobes forms a long siphon-like structure that is quickly calcified.

15.3.2 Muscles and Their Scars

Other than the pallial line scars formed by the mantle muscles, the adductor muscles and other muscle attachments, notably the pedal and byssal (if present) muscles, leave impressions in the inner surface of each valve. The adductor muscles have nothing to do with the shell-foot muscles seen in other groups but are derived from mantle muscles (see below). They are not unique to bivalves as similar adductors are found in bivalved crustaceans and also in bivalved gastropods. The possession of two (*dimyarian*) adductor muscles is the plesiomorphic condition in bivalves. Where these muscles are equal in size the state is known as *isomyarian* (or homomyarian); when there is a relatively larger posterior adductor muscle it is called *heteromyarian*. When one adductor muscle (always the anterior in autobranchs but the posterior in *Nucinella*, a protobranch) is lost, this condition is called *monomyarian* (Yonge 1953a) (see Section 15.4.1.2.1 and Figure 15.12).

Such changes in the adductor muscles are seen in many pteriomorphians, for example, the anisomyarian Mytiloidea, some of which, such as the boring genera *Botula* and *Lithophaga*, are secondarily isomyarian. The Limopsidae has examples of anisomyarian taxa and at least one (fossil) taxon that is monomyarian (Heinberg 1979). Some entirely monomyarian taxa are the Pectinoidea, Limoidea, Entolioidea, and Anomioidea. The palaeoheterodont Mullerioidea contains examples of heteromyarian and monomyarian taxa and the heterodont Dreissenoidea, *Sphenia* (Myoidea), and *Entodesma* (Lyonsiidae) contain some heteromyarian taxa, while the cardiid *Tridacna* is monomyarian. All of these groups have an exclusively or predominantly posterior inhalant water flow and have achieved the heteromyarian condition independently. The only monomyarian protobranch, *Nucinella* (Manzanellidae), differs entirely from other monomyarian taxa in having lost the posterior adductor (Allen & Sanders 1969).

The highly modified 'watering pot shells' (the anomalodesmatan Clavagelloidea) have reduced adductor muscles or have lost them altogether, i.e., they are *amyarian* (e.g., Morton 2002a) (see also Section 15.8.6.1.2).

The generally accepted theory for the origin of the adductor muscles is that they originated from the cross-fusion of radial muscles in the mantle in the embayments near the hinge (Yonge 1953a; Owen 1958) (see also Chapter 13, Figure 13.17). This theory is consistent with their innervation (see below) but is questioned by Waller (1998), who pointed out that the two adductor muscles

develop asynchronously in bivalve larvae, the anterior one first, and neither is associated with the development of the mantle muscles during ontogeny (Raven 1966). Instead, the adductors develop from mesodermal cells inside the tissue that will be attached to the pallial line. Thus, the posterior adductor may have formed as a result of cross-fusion of pallial muscles, not in the dorsal embayment but rather those located more ventrally, below the rectum. The cross-fusion of dorsal radial pallial muscles forms a novel secondary posterior adductor muscle in the razor clam *Ensis*, but that muscle lies above the rectum, not below it like the posterior adductor (Waller 1998).

The paired pedal muscles are the homologues of the dorsoventral and oblique shell muscles of other classes. There are four pairs of pedal retractor muscles (dorsoventral and oblique) and two pairs of protractors (oblique) in some protobranchs (Figure 15.11), but most bivalves have only two pairs of retractors (dorsoventral) inserted next to the anterior and posterior adductor muscles and a single, anteriorly attached pair of protractors (oblique) (see Chapter 3, Figure 3.52). In a few bivalves (notably oysters) these are lost along with the foot. In byssate bivalves such as mussels (Mytilidae), the byssus has large muscles attached to it which insert on the shell, and these are derived from much-modified pedal retractor muscles.

In some protobranchs (e.g., *Yoldia*, *Nucula*, *Saccella*) five pairs of muscles attach on the ventral body wall adjacent to the mouth and terminate on the dorsal body wall; in *Saccella* the mouth muscles terminate on the shell adjacent to the anterior pedal protractors. The contraction of these muscles may assist the passage of food from the buccal cavity into the oesophagus (Heath 1937). The palps are associated with a separate well-developed musculature.

15.3.3 The Foot

In many groups the bivalve foot is a digging structure while in those that live a permanently attached life (e.g., oysters), it is reduced or lost.

In protobranchs the foot has a sole fringed with papillae. The foot of autobranchs may or may not have a narrow sole and does not possess papillae.[1] A narrow, distinct sole is present in pteriomorphians, but in heterodonts the sole is absent or, if present, is usually only visible when the foot is pressed against the substratum (Waller 1998).

In many autobranch bivalves, the foot is modified for burrowing, being hatchet-shaped and sole-less. A few bivalves, notably juveniles and small-sized taxa, especially galeommatoideans, can effectively crawl on hard surfaces using their foot.

A posterior pedal gland is found in most bivalves, and this forms the byssal gland in autobranchs (see next section). An attachment thread is produced by the glochidium larvae of freshwater mussels (unionidans), although this is probably not

---

[1] Small papilla-like structures occur on the edges of the sole of *Neotrigonia*, but these are probably not homologous with the papillae of protobranchs.

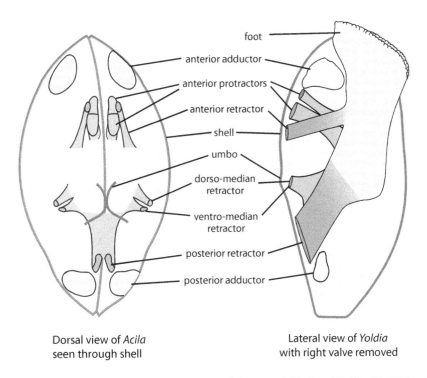

foot
anterior adductor
anterior protractors
anterior retractor
shell
umbo
dorso-median retractor
ventro-median retractor
posterior retractor
posterior adductor

Dorsal view of *Acila*
seen through shell

Lateral view of *Yoldia*
with right valve removed

FIGURE 15.11   Pedal muscles of the protobranch bivalves *Acila* (Nuculidae, Nuculoidea) and *Yoldia* (Yoldiidae, Nuculanoidea). Redrawn and modified from Heath, H., *Mém. Mus. r. his. nat. Belg.*, 10, 1–26, 1937.

a byssus but a new structure as it is produced from a gland anterior to the foot anlage (Waller 1998).

Anterior pedal glands are usually absent in bivalves; they do occur in a few taxa but are probably independently derived in each group that possesses them (Waller 1998). In *Neotrigonia* and some protobranchs they are small and lateral and in *Gastrochaena* are well-developed and specialised.

15.3.4   The Byssus – Structure and Function

In autobranch bivalves the byssus is secreted by the byssal gland. This gland is near the posterior end of the foot; it is the only pedal gland found in most larval and juvenile autobranchs but is often absent in adults (Stanley 1972). In many autobranch larvae, postlarvae, or juveniles, the byssal gland produces one or a few threads used for temporary attachment. In some groups, adults of various families retain and utilise a byssus (see Section 15.4.1.2), although it is lost in many. The larval byssus is used by some larvae as a flotation device, assisting in dispersal, and for attachment on settlement. A posterior pedal gland is also present in protobranch bivalves but does not secrete a byssus, although it is a possible homologue of the autobranch byssal gland (Waller 1998).

Byssal threads are thin but very strong and composed of complex proteins (see below). They attach the bivalve to various hard substrata or act as an anchor in soft substrata (e.g., Bieler et al. 2005), and with some mytilids they form 'nests' in which the adults lie (see also Section 15.4.1.2).

The byssal threads making up the byssus are initially secreted by the byssal gland as a liquid. The secretion passes into a groove on the ventral surface of the foot and hardens on contact with water. Two main types of byssal apparatus are found: a long strip of byssal material secreted from a long channel in the foot (arcids only) or one or more threads fixed individually to the substratum.

The process of byssal formation and the structure of the byssal threads have been well studied in the 'common blue mussel' *Mytilus edulis*. It begins with the mussel using its foot to probe for a suitable surface, and then, on that surface, it creates a watertight vacuum. A gel secreted by the byssal gland makes the contact point (the 'attachment plaque'), and then the adhesive thread is formed, again with the aid of the foot. This initially liquid thread coagulates and quickly hardens as soon as it comes into contact with water. In *Mytilus* nine types of glandular cells secrete different parts of the byssal thread, making it a very complex structure (e.g., Harrington et al. 2018). A byssus thread can be separated into three main regions, the point of attachment (the plaque), the thread, and a thicker stem nearest the foot (Figure 15.13). The stem typically gives rise to several threads (as in *Mytilus*) and is the point of origin of the byssus in the byssal gland. Some retractor muscles cluster around the base of the byssal gland in the foot, providing a powerful anchor and allowing for some rotation and tension control.

In *Mytilus* (*M. edulis* and *M. galloprovincialis*), the byssus comprises many (sometimes several hundred) threads, each around 2–4 cm long and about 50 μm in width, with the attachment plaque 2–3 mm wide. Each thread is stiffer at the distal end (where it is attached to rocks) than at the proximal end where it emerges from the byssal gland. The byssal thread is composed of a protective polyphenolic protein cuticle that surrounds a collagen core. The core consists of three fibrous collagen-like proteins (preCol-D, preCol-NG, and preCol-P), two of which (preCol-D, preCol-P) form gradients

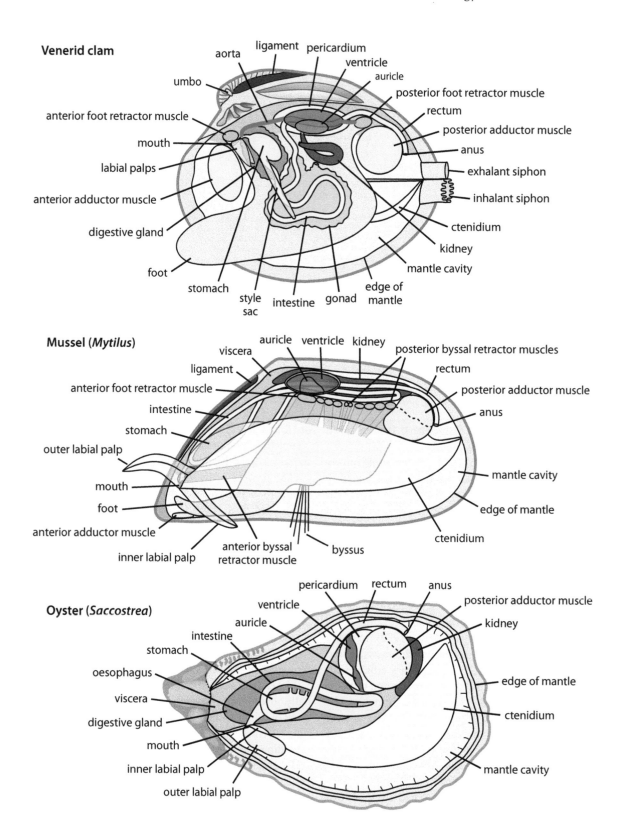

FIGURE 15.12 Comparative anatomical diagrams of a venerid (isomyarian), mytilid (anisomyarian), and an ostreid (monomyarian), all with the left valve removed. The left ctenidium of the venerid has been removed. Redrawn and modified from several sources.

along the thread which presumably account for the dissimilar mechanical properties of different portions of the thread (Coyne & Waite 2000; Lucas et al. 2002; Waite et al. 2004). Certain metal ions appear to be an essential component of the

elasticity of the byssal thread (Vaccaro & Waite 2001; Lucas et al. 2002).

The byssal thread itself consists of two parts – a proximal smooth, elastic section and a corrugated and stiff distal

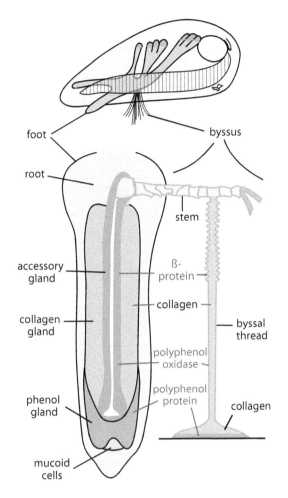

foot

byssus

root

stem

accessory
gland

ß-
protein

collagen

collagen
gland

byssal
thread

polyphenol
oxidase

polyphenol
protein

phenol
gland

collagen

mucoid
cells

FIGURE 15.13 Formation, structure, and composition of a byssal thread of *Mytilus*. Chemical components are labelled in purple. Redrawn and modified from Waite, J.H., *Biol. Rev.*, 58, 209–231, 1983.

part (Vaccaro & Waite 2001; Lucas et al. 2002). The distal portion of the mussel byssus absorbs the shock from wave action, and the attachment plaque glue forms a very firm attachment. There is a ten-fold decrease in stiffness between the distal and proximal parts of the thread in mussels (Waite et al. 2004). The adhesive pad of the byssal thread (the plaque) has attracted attention because it is a more effective underwater adhesive than any other discovered so far. The protein involved was described by Waite and Qin (2001), and this and other byssal proteins are of great interest because of their potential practical applications.

A comparison of the byssus of two species of *Mytilus* showed that, although mostly very similar, there are a few important differences (Lucas et al. 2002). Thus, when even closely related species can differ, it suggests that significant differences probably occur in proteins and functional properties in other byssate bivalves, given the gross morphological differences that exist (Figure 15.14).

Byssal attachment has led to profound changes in the morphology of some groups of bivalves as outlined in Section 15.4.1.2.1.

15.3.5 MANTLE AND SIPHONS

In bivalve evolution, modifications to the mantle edge have accompanied major changes. It usually consists of three lobes – inner, middle, and outer – the latter being responsible for the secretion of the shell. Most bivalves make contact with the external environment at the mantle edge, and this is where most sense organs are located. These sensory structures range from simple sensory cells to sensory tentacles and even eyes (see Chapter 7 and Section 15.3.10).

While it is generally stated that the bivalve mantle margin is comprised of three folds, Morton and Peharda (2008) argued that primitively it had only two folds – a condition seen in some living taxa such as *Arca*, with the outer fold secreting the periostracum on its inner surface and the shell on its outer surface (Waller 1980; Morton & Peharda 2008). Two folds have also been noted in some bivalve veligers, such as those of *Ostrea* and *Nucula*. It is thus probable that the pleisomorphic bivalve condition of the mantle edge may have been two folds with the middle fold arising from the outer part of the inner fold.

The evolution of mantle fusion in bivalves was reviewed by Yonge (1957). The mantle is plesiomorphically entirely open, but in many bivalves it is fused to varying extents. There is no mantle fusion in nuculid protobranchs, Arcoidea, and some monomyarian pteriomorphians (Pectinoidea, Pterioidea, and Anomioidea). In such taxa, the inhalant and exhalant parts of the mantle edge are separated simply by the apposition of the inner lobes of the mantle which are greatly enlarged in some pteriomorphians to form a so-called 'pallial curtain' or 'velum'.

Despite the simple mantle in most protobranchs, some do show some mantle fusion. For example, *Solemya* has the ventral mantle edge fused (although the inhalant and exhalant apertures are not separated by a fusion), and in some nuculanoideans there is mantle fusion in the formation of the siphons (see below). In palaeoheterodonts the inner lobes of the mantle edge are fused to form a separate exhalant aperture, but the ventral edge of the inhalant opening is not fused. In lucinids, the inhalant aperture is sometimes separated from the pedal gape while in thyasirids there is no fusion. The fusion between the inhalant and exhalant openings, and between the inhalant and pedal openings, occurs in other heterodont bivalves. In some euheterodonts, there is fusion involving both the middle and outer mantle folds over all or part of the ventral margin to form a restricted anterior pedal opening. In some bivalves, such as pharids, solenids, several anomalodesmatans, pholadids, tellinids, and mactrids, there is a small fourth mantle opening (the 'fourth pallial aperture') concerned with cleansing between the inhalant and pedal openings (e.g., Atkins 1937c; Yonge 1948a; Harper et al. 2006).

The mantle edge has extended to form short to long, separate or fused siphons in many heterodonts and some other bivalves including burrowing nuculanoidean protobranchs and a few pteriomorphians. In nuculanoideans the siphons are formed by an extension of the posterior inner mantle folds and range from being connected by cilial junctions to complete fusion (Yonge 1939a, 1957). In some siphonate burrowing Mytilidae such as *Lithophaga*, *Botula*, and *Arenifodiens* the

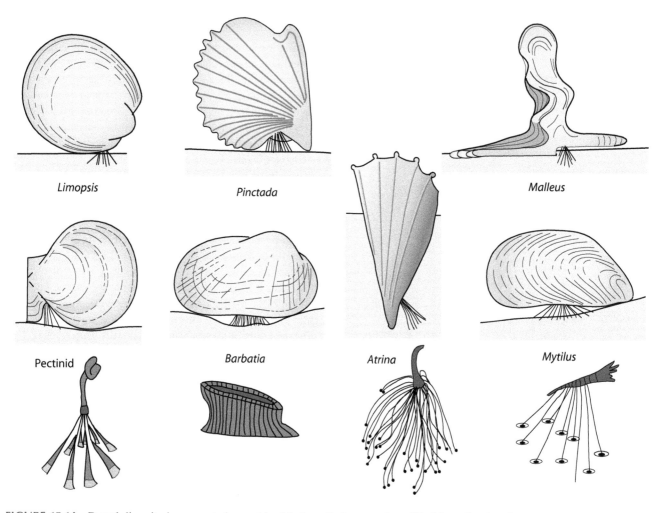

FIGURE 15.14 Byssal diversity in some pteriomorphian bivalves. Redrawn and modified from Stanley, S. M., *J. Paleont.*, 46, 165–212, 1968, Waite, J.H., *Biol. Rev.*, 58, 209–231, 1983, and Bromley, R.G. and Heinberg, C., *Palaeogeogr. Palaeoclimatol. Palaeoecol.*, 232, 429–453.

inhalant and exhalant siphons are separated by fusion but the ventral margin of the inhalant siphon is not fused. Siphons are also formed from the inner mantle folds in some galeommatoideans and tellinoideans, but, in other eulamellibranchs and anomalodesmatans, other mantle folds may be incorporated in the siphons (Figure 15.15). Incorporating the middle mantle fold brings with it the development of sensory tentacles (e.g., Veneroidea, Solenoidea, Pholadoidea, Myoidea, Pandoroidea), and sometimes eyes (e.g., Cardioidea, *Laternula*), surrounding the siphonal openings. Incorporation of the inner surface of the outer fold results in the siphon being covered in periostracum as in Mactroidea, Hiatelloidea, and Myoidea (Yonge 1957). With all folds incorporated including the outer surface of the outer fold, the siphon is encased in shell, as in the Cuspidariidae.

The siphons (Figures 15.16 to 15.18) in some bivalves are longer than the shell length and in others (e.g., *Mya*, *Panopea*) are so long and bulky that they cannot be retracted inside the shell. The muscles making up the siphons of some heterodont bivalves were described by Duval (1963). Longitudinal muscles usually make up the main muscular component of the siphon and comprise the bulk of the siphonal tissue. Circular

muscles lie on the inner and outer sides of the longitudinal muscle mass, and a third band may divide the longitudinal muscles into two blocks. Radial muscle fibres criss-cross through these muscles from the outer to inner epithelial layers. The longitudinal muscles retract the siphons while the circular muscles constrict, controlling the diameter of the siphon, and the radial muscles regulate the thickness of the siphonal wall. These muscle arrangements are similar, no matter which mantle lobes develop the siphons.

However, at least in the heterodonts not all siphonal extension is solely through antagonistic muscles working in concert with haemocoelic spaces among the muscles. In *Mya*, for example, the extension of the long, fused siphons is aided by water forced into them from the mantle cavity through the contraction of the adductor muscles (Chapman & Newell 1956).

Some burrowing bivalves that lack siphons use mucus-lined tubes to draw in water from their position deep in the sediment. Examples are Solemyidae, Lucinidae (Figure 15.18), and Thyasiridae, which have bacterial symbionts in their gills (see Section 15.4.4.1 and Chapter 9). Other asiphonate bivalves are epifaunal or shallow infaunal burrowers.

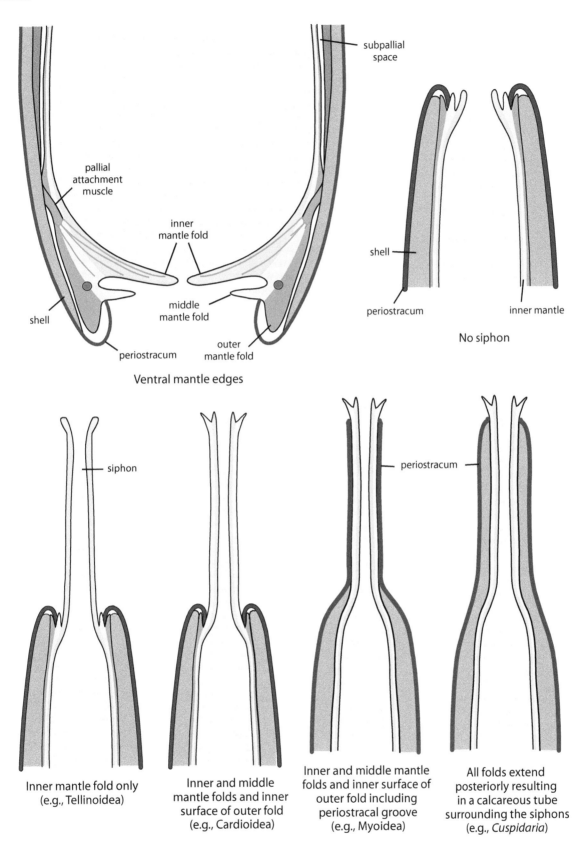

FIGURE 15.15 Mantle fusion and siphon formation in autobranch bivalves. Redrawn and modified from Yonge, C.M., *J. Mar. Biol. Assoc. U. K.*, 27, 585–596, 1948b, Yonge, C.M., *Pubb. Staz. Zool. Napoli*, 29, 151–171, 1957, and Morton, J.E. and Yong, C.M., Classification and structure of the Mollusca, pp. 1–58, in Wilbur, K.M. and Yonge, C. M. (eds.), *Physiology of Mollusca*, Vol. 1, Academic Press, New York, 1964.

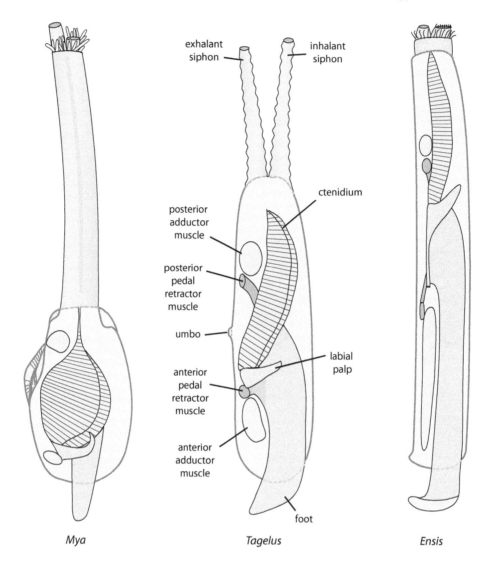

FIGURE 15.16 The siphons of the deep-burrowing bivalves, *Mya* (Myidae; Myoidea), *Tagelus* (Solecurtidae; Tellinoidea), and *Ensis* (Pharidae; Solenoidea). Redrawn and modified from Yonge, C.M., *Univ. Calif. Publ. Zool.*, 55, 421–438, 1952b (*Mya, Tagelus*) and Yonge, C.M. and Thompson, T.E., *Living Marine Molluscs*, Collins, London, 1976 (*Ensis*).

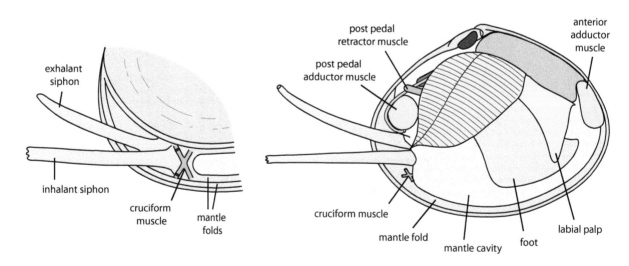

FIGURE 15.17 The siphons and cruciform muscle of the tellinoidean *Macoma*. Redrawn and modified from Yonge, C.M., *Phil. Trans. R. Soc. B*, 234, 29–76, 1949.

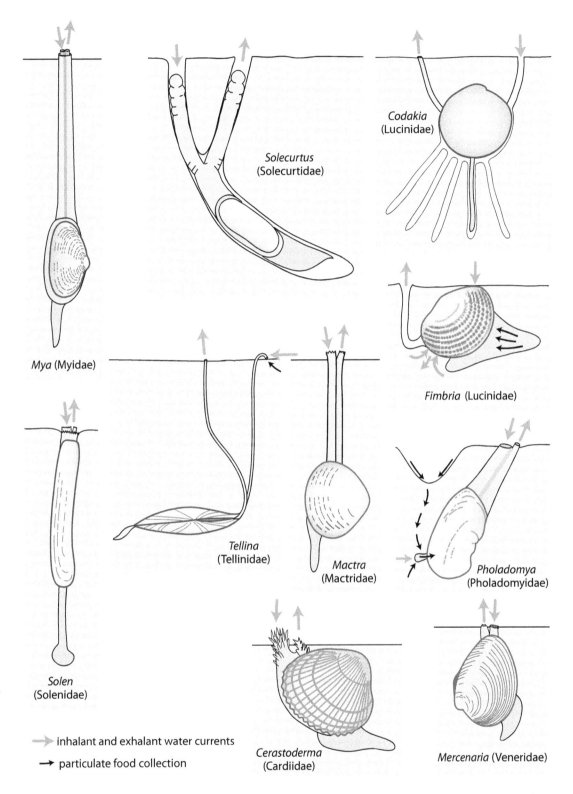

→ inhalant and exhalant water currents
→ particulate food collection

FIGURE 15.18 Examples of differences in siphons, life orientation and habit in infaunal heterodont bivalves and the main water currents generated by the gills and, sometimes, by ciliated pedal epithelium. Redrawn and modified from the following sources: *Mya* and *Solen* (Yonge, C.M., *Phil. Trans. R. Soc. B*, 234, 29–76, 1949), *Tellina* and *Mercenaria* (Stanley, S. M., *J. Paleont.*, 42, 214–229, 1968), *Solecurtus* (Bromley, R.G. and Asgaard, U. *Solecurtus strigilatus*: A jet-propelled burrowing bivalve, pp. 313–320 in B. Morton (ed.), *The Bivalvia. Proceedings of a Memorial Symposium in Honour of Sir Charles Maurice Yonge (1899–1986), Edinburgh, 1986.* Hong Kong University Press, Hong Kong., 1990), *Pholadomya* (Morton, B., *Videnskabelige Meddelelser fra Dansk naturhistorisk Forening*, 142, 7–101, 1980), *Fimbria* (Morton, B., *Rec. Aust. Mus.*, 32, 389–420, 1979a), *Mactra* (Morton, J.E. and Miller, M.C., *The New Zealand Sea Shore*, Collins, London, UK, 1968), *Codakia* (Taylor, J.D. and Glover, E.A., Functional anatomy, chemosymbiosis and evolution of the Lucinidae, pp. 207–225 *in* E.M. Harper, Taylor, J. D., and Crame, J. A. (eds.), *Evolutionary Biology of the Bivalvia*, the Geological Society, London, UK, 2000), and *Cerastoderma* (Johnstone, J., *L.M.B.C. Memoirs*, 2, 1–84, 1899).

A small pair of crossed muscles, the cruciform muscles (Graham 1934), is found only in tellinoideans and is located between the base of the inhalant siphon and the pedal opening (Figure 15.17). A small, probably mechanoreceptor, sense organ is associated with the posterior part of each muscle (Odiete 1978; Frenkiel & Mouëza 1980; Pichon et al. 1980) (see also Chapter 7).

15.3.6   Mantle Cavity

In bivalves the mantle cavity surrounds almost the entire body. Besides the visceral mass and foot, the mantle cavity contains the ctenidia, used primarily for respiration in protobranch bivalves, but in the great majority of autobranch bivalves greatly enlarged and modified for food collection as well, as described in some detail in the next section and in Chapter 5. The mantle cavity also contains two pairs of labial palps (see Section 15.3.6.2) and the chemosensory osphradium (see Section 15.3.10). Well-developed cilial tracts on the inner surface of the mantle and on the surface of the body, including the foot, are involved in the collection and ultimate rejection of waste particles (see Section 15.3.6.3.1) and, in some taxa, can play a role in food collecting and generating feeding currents (e.g., Figure 15.18).

15.3.6.1   The Ctenidia – Their Structure and Ciliation

Bivalves have a single pair of bipectinate ctenidia – one on each side of the body. They are respiratory organs and, in autobranch bivalves, collect food by suspension feeding (as described in Chapter 5).

In protobranch bivalves the triangular ctenidial filaments are short and broad; the ctenidia are narrow and do not extend most of the length of the mantle cavity as they do in autobranch bivalves – this is the *protobranchiate* condition. The protobranch gills effectively divide the posterior mantle cavity into two chambers, but all the connections (other than the gill axis) between the filaments and the sides of the body are by way of ciliary, not tissue, junctions. The inhalant area (or chamber) is ventral and lateral to the gills, and the exhalant part is dorsal to the filaments (Figure 15.19). The gills of nuculanoidean protobranchs are muscular, being modified for pumping water through the mantle cavity, in which they are aided by being orientated parallel to the anterior–posterior axis. Well-developed muscles run through the gill axis and extend into the gill leaflets (Figure 15.20). The nuculanoidean gills in effect form 'septal membranes perforated by four rows of ciliated pores' (Yonge 1939a, p. 143) which pump water through the mantle cavity. The filaments are firmly connected by ciliary junctions. In contrast, the gills of Nuculoidea and Solemyoidea, as well as chitons and primitive gastropods, are orientated obliquely to the longitudinal body axis, and this is presumably the plesiomorphic bivalve condition (Figure 15.19).

In all classes the ctenidial filaments alternate along the gill axis, the one exception being the nuculoidean and solemyoidean protobranchs where the filaments lie opposite each other along the gill axis. These two protobranch groups also have a unique configuration of the frontal cilia (Type A of Atkins 1937b) which may be related to the different arrangement of the filaments in that the cilial stream is continuous across the gill axis from the outer to inner opposite filaments (Figure 15.20).

The change from protobranchiate to filibranch gills was a major modification involving filament elongation and reflection resulting in W-shaped filaments. This change correlated with a switch from food particles being primarily collected by the palps and/or the foot to food collection by the gills, a feeding change assisted by plesiomorphic ctenidial cleansing mechanisms being enhanced and modified for suspension feeding. The modification of the ctenidial ciliation to collect particles also required associated changes in ciliation to enable them to be moved to and along food-collecting gutters. These changes occurred in parallel with different modifications in different groups. The result was a massive improvement in food-collecting efficiency and was probably the most important driver of bivalve diversification and increase in body size in the Lower Ordovician (Cope 1996b).

The axes of the pair of bivalve gills are attached to the viscera which form the roof of the mantle cavity, and the lamellae hang down into the mantle cavity. The whole ctenidium can be called a *holobranch*, and each of the two halves is called a *demibranch*. In autobranchs, the part of each filament attached to the axis is called the *descending lamella*, and the part bent upwards is the *ascending lamella* (Figure 15.21). The surfaces of the filaments are complexly ciliated with long lateral cilia generating the inhalant water flow bearing oxygen and food particles, while others capture particles or pass the potential food to the labial palps and the mouth via food-collecting tracks or grooves. The novel structure of the ctenidia of autobranch bivalves physically separates the inhalant and exhalant chambers of the mantle cavity – their external surfaces are surrounded by the large inhalant (= branchial) chamber while the lamellae enclose a dorsal exhalant (= suprabranchial) chamber in the interior of the gill. The upper ends of the ascending lamellae may or may not be attached dorsally, depending on the type of gill (see below).

Each demibranch comprises many (often hundreds) of filaments. These filaments are interconnected and arranged in various, sometimes complex, ways. Bivalve gill morphology has long been used as a major character in classifying bivalves (see Section 15.3.6.1.4). The main characters used are the shape of the filaments (the difference between the protobranch and autobranch types of gills), the connections between the filaments, and details of the ciliation. With autobranch gills, the relative size of the demibranchs (outer demibranchs are lost in some) and how the gill is attached to the mantle and visceral mass are additional characters used at various levels in bivalve phylogeny (and hence classification). We outline in the next section aspects of these characters in more detail as the gills are one of the most important structures in autobranch bivalve evolution.

15.3.6.1.1   Junctions Between the Filaments

The main types of gills recognised, largely according to the nature of the junctions between the filaments (Figures 15.21 and 15.22), are:

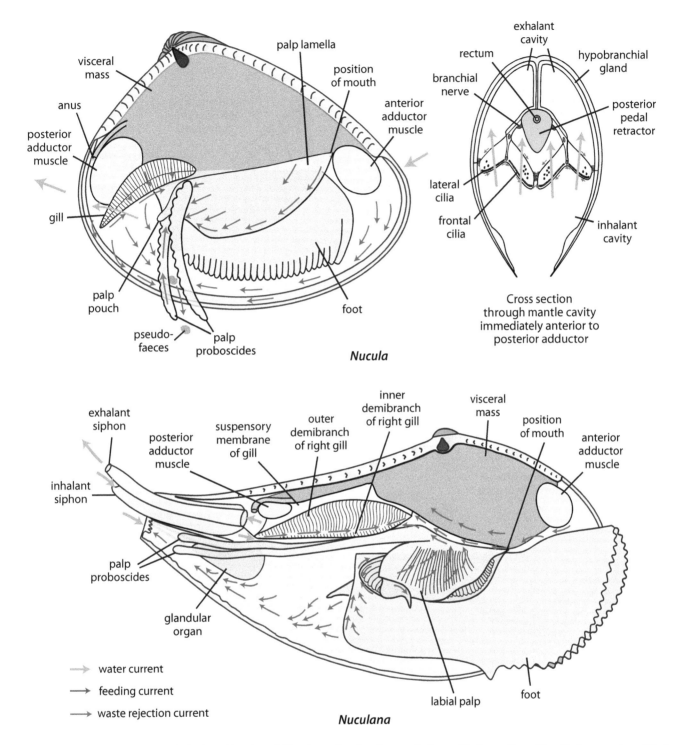

FIGURE 15.19 Two protobranch bivalves, *Nucula* and *Nuculana*, showing the morphology of their mantle cavities. Redrawn and modified from Yonge, C.M., *Phil. Trans. R. Soc. B*, 230, 79–147, 1939a (*Nucula*) and Atkins, D.G., *Q. J. Microsc. Sci.*, 79, 181–308, 1936 (*Nuculana*).

- *Filibranch* ctenidia are seen in many Pteriomorphia (for example many scallops, mussels, arcs, glycymerids, and some pearl oysters [Pteriidae]) and trigoniids (Palaeoheterodonta). The filaments are simple, with adjacent filaments held together by specialised interfilamentary cilial junctions ('cilial discs') (Figure 15.22). These are on small, raised,

muscular connective spurs on the edges of the filaments. Differences in the spacing of the spurs result in minute changes in the spacing between the filaments and thus the pore size of the filter. In some filibranch bivalves there may also be a few simple tissue connections. Gills with cilial disc connections are called *eleutherorhabdic*, whether or not they also

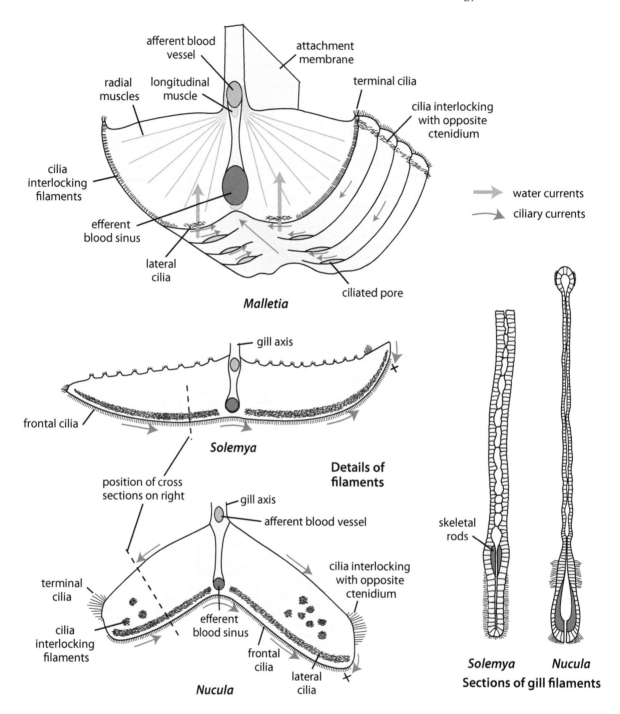

FIGURE 15.20 Examples of the gill filaments of three protobranch bivalve taxa. Redrawn and modified from the following sources: gill filaments (Yonge, C.M., *Phil. Trans. R. Soc. B*, 230, 79–147, 1939a); sections of gill filaments (Ridewood, W.G., *Phil. Trans. R. Soc. B*, 195, 147–284).

have some tissue connections. Gills that lack any cilial disc connections and only have tissue connections are called *synaptorhabdic* (Ridewood 1903).

- *Pseudoeulamellibranch* ctenidia are found in Ostreidae and Pinnidae, as well as in some Pteriidae, Pectinidae, and certain other related families. These gills have the adjacent filaments held together by ciliary junctions and some tissue junctions. This

condition has also been found in some Thyasiridae (Heterodonta) where interlamellar junctions may be present (Dufour & Felbeck 2006).

- *Eulamellibranch* gills are found in most other bivalves, including the freshwater mussels (Unionida: Palaeoheterodonta) and Limidae (Pteriomorphia). These have extensive interfilamentary tissue which occupies most of the area between the filaments

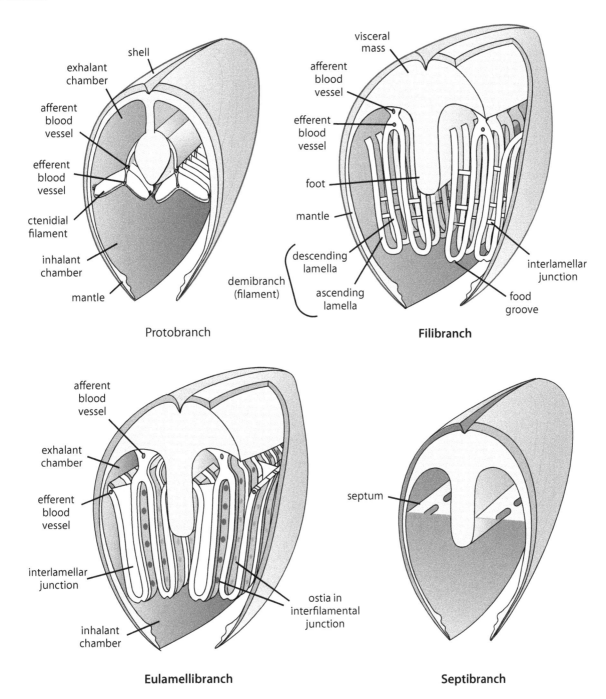

FIGURE 15.21 The main types of bivalve gills. Redrawn and modified from Westheide, W. and Rieger, R., *Spezielle Zoologie – Erster Teil: Einzeller und Wirbellose*, Gustav Fischer, Stuttgart, 1996.

leaving only small pores (ostia) penetrating an otherwise continuous barrier (Figure 15.22). There may be interlamellar junctions.

The evolution from the filibranch to eulamellibranch condition involves a change from ciliary junctions to tissue junctions between filaments, and this change has occurred several times independently in bivalve evolution. While most pteriomorphian bivalves have filibranch gills, some have achieved the pseudoeulamellibranch condition which is simpler than the typical eulamellibranch type seen in heterodont bivalves.

The pseudoeulamellibranch gills of some oysters are rather complex, with the interlamellar tissue arranged in parallel rows in the inner and outer demibranchs, a unique arrangement which forms a series of water tubes with quadrate openings (Waller 1998).

The ascending lamella may be attached to the body. In filibranchs such as scallops and mussels, only the descending lamellae are attached to the body, and the ascending lamellae are free dorsally. In contrast, in bivalves with eulamellibranch gills the ascending limb of the inner (= medial) lamella is attached along the upper part of the foot and the

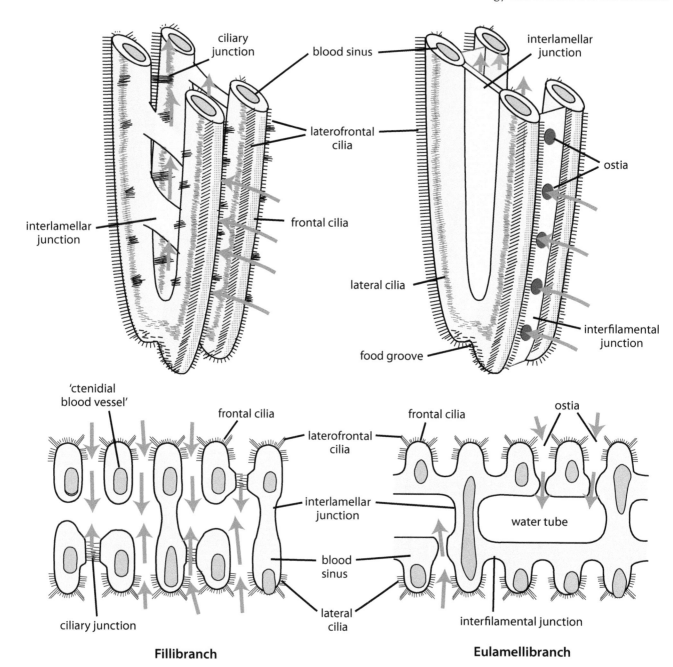

FIGURE 15.22 Comparison of the pattern of ciliation and tissue connections in filaments of gills with the filibranch and eulamellibranch conditions. Upper figures show details of the ventral part of two gill filaments, and the lower figures show sections through them. Upper figures redrawn and modified from Pechenik, J.A., *Biology of the Invertebrates*, 2nd edition, WCB, Dubuque, IA, 1991, lower figures redrawn and modified from Barnes, R.D., *Invertebrate Zoology*, 2nd edition, W.B. Saunders, Philadelphia, PA, 1968.

ascending limb of the outer (= lateral) lamella is attached to the mantle.

15.3.6.1.2 Arrangement of Filaments

How the filaments are aligned along the gill axis in autobranch bivalves differs in two main ways (Figure 15.23).

- *Simple ctenidia* – the filaments are arranged in a straight line with the gill surface appearing to be flat.

- *Plicate ctenidia* – the filaments are clustered into groups causing the surface of the lamellae to become corrugated – i.e., it consists of a series of parallel ridges (plications) and grooves with each ridge comprised of a cluster of filaments.

15.3.6.1.3 Uniformity of Filament Size

Plicate ctenidia may have uniform filaments or two kinds of filaments (Figure 15.23).

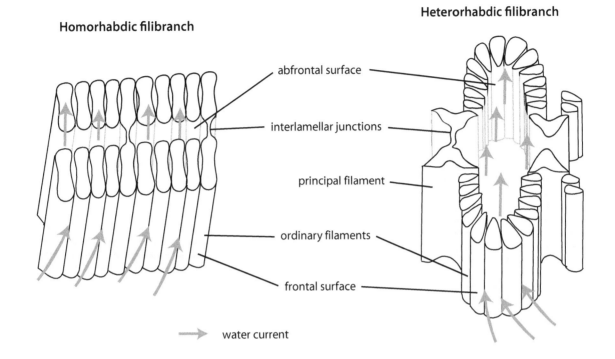

Homorhabdic filibranch

Heterorhabdic filibranch

abfrontal surface

interlamellar junctions

principal filament

ordinary filaments

frontal surface

→ water current

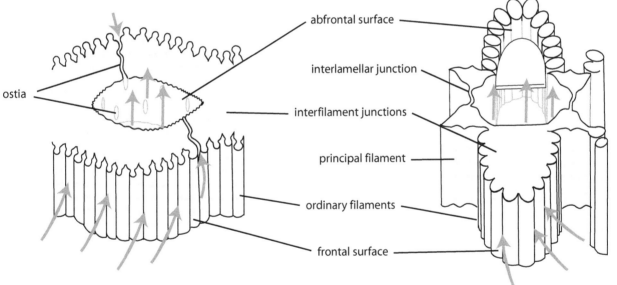

Homorhabdic eulamellibranch

Heterorhabdic pseudoeulamellibranch

abfrontal surface

interlamellar junction

interfilament junctions

principal filament

ordinary filaments

frontal surface

ostia

FIGURE 15.23 Sections through the demibranchs of the principal gill types in suspension-feeding bivalves. Redrawn and modified from Dufour, S.C. and Beninger P.G., *Mar. Biol.*, 138, 295–309, 2001.

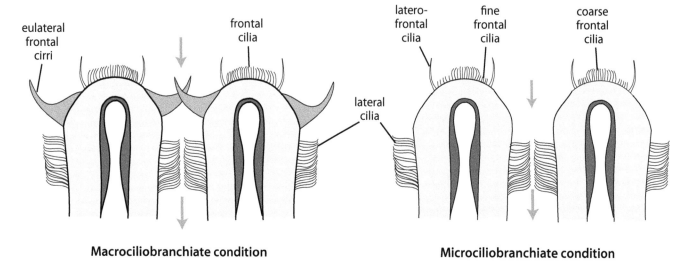

FIGURE 15.24 Micro- and macrociliobranchiate patterns of ciliation on the distal tips of autobranch bivalve ctenidial filaments (demibranchs). Redrawn and modified from Atkins, D.G., *Q. J. Miscrosc. Sci.*, 80, 345–436, 1938b.

- *Homorhabdic* ctenidia – all the filaments are the same size. This gill type occurs in many autobranchs.
- *Heterorhabdic* ctenidia – there are two types of filaments. Many, but not all, plicate ctenidia are heterorhabdic. Most filaments are very narrow and together make up the plications of the lamellae while those at the bases of the grooves between the plications are larger and called the *principal filaments*. These provide most of the structural support, having a heavier skeletal rod than the ordinary filaments, and maintain the shape of the gill. There are tissue connections (interlamellar junctions) between the adjacent principal filaments on the other lamella. The heterorhabdic condition appears to have been convergently derived from homorhabdic taxa in many bivalve lineages including many pteriomorphians, Etheriidae in the Unionida, and many heterodonts, including some Anomalodesmata.

15.3.6.1.4 Ciliation and Water Currents

Ciliation on the filaments varies in detail between taxa and has been studied extensively, notably by Daphne Atkins in the 1930s (Atkins 1937b, a, 1938a, b, c) who investigated 117 species in 51 families. Details on the process of food capture and the recent literature dealing with that are reviewed in Chapter 5. There are three main kinds of cilia (Figure 15.24) – frontal cilia on the outer surface of the filaments, lateral cilia on the surfaces adjacent to other filaments that generate the water current, and laterofrontal cilia, which (except in many pteriomorphians – see below) are long and involved in food capture.

Frontal cilia are also found on the ctenidia of non-bivalve groups where they have a cleansing function. In protobranch bivalves, particles brought into the mantle cavity in the inhalant water current are carried to the tips of the gill filaments by the frontal cilia where they are rejected, while in autobranchs the captured particles are transported to food tracts or grooves.

Laterofrontal cilia are found only on bivalve gill filaments. These cilia beat towards the frontal surface of the filament and trap particles which are then moved by the smaller frontal cilia. The presence of laterofrontal cilia in protobranchs, as well as basal pteriomorphians (Mytiloidea) and heteroconchs, could indicate that ancestral bivalves used their gills for feeding (Waller 1998), and if so, this ability was largely lost in the ancestors of living protobranch bivalves.

Laterofrontal cilia are long in protobranchs and most other bivalves, but they have become secondarily short in many pteriomorphians. This observation led Atkins (1938b) to propose that bivalves could be divided into two distinct groups. In her *macrociliobranchiate* condition, the laterofrontal tracts are composed of compound *eulaterofrontal cirri*[2] together with one or more rows (usually two rows except for *Nucula sulcata*, which has four rows of subsidiary *prolaterofrontal* cilia [Owen 1978]). This group includes the protobranch bivalves, all the eulamellibranch families, and the Mytilidae. *Neotrigonia* is also a macrociliobranch (Morton 1987b). In the second group (the *microciliobranchiate* condition), the laterofrontal tracts consist of a single row of cilia, the *microlaterofrontal* cilia. Included in this group were all the filibranch bivalves except Mytilidae and Trigoniidae. The Ostreoidea and Pinnoidea (both with pseudoeulamellibranch gills) were originally considered macrociliobranchiate, but close examination of the compound cilia (cirri) in ostreids showed that they were not typical of macrociliobranchs, having an arrangement that probably evolved independently, and thus should be considered modified microciliobranchs (Owen 1978). The remaining microciliobranch families are characterised by the possession of laterofrontal tracts consisting of simple cilia only, although the Pinnidae have minor modifications (Atkins 1938b).

---

[2] Cirri are structures composed of fused or partially fused cilia.

The different frontal cilial arrangements are important functionally. In the microciliobranchiate groups, the laterofrontal ciliated tracts are simple, and food collection and transport by the gill mainly depend on water currents rather than direct involvement of cilia, while in those bivalves with compound eulaterofrontal cirri (the macrociliobranchiate groups and, convergently, the Ostreidae), the compound laterofrontal cilia play a major role in collecting food particles (Owen 1978).

Gill structure and the associated ciliation within the microciliobranchiate pteriomorphians are much more diverse than in other bivalves. In the pteriomorphians it varies from simple, filibranch, and homorhabdic to plicate, pseudoeulamellibranch, and heterorhabdic. The Arcidae and Glycymerididae have simple, short laterofrontal cilia. Water currents are apparently not important in the collection and transport of particles in arcoideans, but they are in anomiids, pectinids, and limids. In the Anomiidae the gills are homorhabdic, filibranch, and simple; the laterofrontal tracts are a single row of cilia on either side of the frontal cilia. There are two tracts of frontal cilia, one beating ventrally and the other dorsally. However, despite this, particles are rarely carried dorsally, and Atkins (1937b) concluded that the gills of anomiids seemed mainly concerned with rejection as nearly all particles were transported ventrally to the free margins of the demibranchs where the currents are directed posteriorly. Atkins concluded anomiids feed largely on particles brought by the main water current directly to the dorsal grooves of the gills where they are moved anteriorly to the mouth.

On the backs of the gill filaments in protobranchs (the abfrontal surface), abfrontal cilia are particularly well-developed. The role of the cilia (and mucocytes) on this side of the filaments was investigated by Beninger and Dufour (2000) who pointed out that the original cleansing role of these cilia in protobranchs is not required in autobranch bivalves, with their resultant reduction or loss in nearly all groups other than in homorhabdic filibranchs such as mytilids.

The gills of pectinids are filibranch but are plicate and heterorhabdic, with the principal filaments forming gutters in the grooves between adjacent plicae. In the principal filaments, frontal cilia beat dorsally, in contrast to the ventrally beating frontal cilia of the narrow filaments that make up the plicae. The cilia on principal filaments, which are U-shaped in section, transport suspended particles dorsally to the anteriorly directed currents between the bases of the two demibranchs and along the edges of all the ascending lamellae (Owen & McCrae 1976). However, if the gill is contracted (e.g., when irritated), the principal filaments become T-shaped in section, plication of the gill is much reduced, and, as a consequence, food collection by water currents cannot operate, and material on the gill surface is carried ventrally by the frontal cilia of the ordinary (narrow) filaments and rejected.

In limids, the gills are eulamellibranch, markedly plicate and heterorhabdic. The U-shaped groove of the principal filaments has the ridges with the frontal cilia more developed than in pectinids so that it forms an almost tubular gutter in which the cilia beat dorsally. The filaments that form the

crests of the plicae are also enlarged and have cilia that beat ventrally. Contraction of the gill results in the plicae becoming closely apposed resulting in the enlarged apical filaments making up most of the exposed surface of the gill while the principal filaments become flattened. In the relaxed state, particles are presumably drawn into the channels in the principal filaments to be carried dorsally in suspension (as in pectinids), but in the contracted state these water currents cannot operate, and material will instead be carried ventrally on the crests of the plicae and rejected (Owen 1978).

The gills of oysters (e.g., *Ostrea*) are pseudoeulamellibranch, plicate, and heterorhabdic; the frontal currents are dorsal on the principal filaments, ventral on the apical filaments while intermediate filaments have dorsally and ventrally beating cilial tracts (Yonge 1926; Atkins 1937a). However, the principal filaments differ from those described above with U-shaped grooves, in having an inverted V or U shape in section in both the contracted and relaxed state. They are thus convergently like the principal filaments of some heterodont eulamellibranchs (see below). Because of the shape of the principal filaments and the presence of ostia on each side of these filaments, the mode of food gathering seen in other pteriomorphians such as pectinids and limids (described above) would not be possible. Thus, oysters feed using compound laterofrontal cirri to collect particles filtered from the water which are then transported by cilial action to the mouth via the ventral edges of the demibranchs.

The gills of all heterodonts have the eulamellibranch condition, and they may be heterorhabdic or homorhabdic. Frontal cilia are present and abfrontal cilia are weak or absent. The relatively small spaces between the ascending and descending lamellae open to the main exhalant (or suprabranchial) chamber in the dorsal part of the gill. In a heterorhabdic plicate gill, this chamber is subdivided into vertical channels by the largely fused principal filaments of the two adjacent lamellae of each demibranch.

The arrangement of the demibranchs, and consequently, the feeding currents, is also important. The outer demibranch is often reduced in length or consists of only the descending lamellae, or it is reflected dorsally or lost. Such rearrangements result in different cilial patterns on the gills as shown in Figure 15.25.

The gill filaments of both protobranchs and autobranchs are richly supplied with mucocytes, particularly on their frontal, but also the abfrontal, surfaces. There remains some doubt as to the exact role that mucus plays in feeding (see Chapter 5).

15.3.6.2 The Labial Palps

Two pairs of flattened labial palps lie on either side of the mouth in almost all bivalves and overlap the anterior end of the gill lamellae. The facing surfaces of each palp are covered with ciliated ridges where sorting occurs (see Chapter 5 for details).

The palps of most protobranchs are much larger than those of autobranchs and differ in possessing elongate palp proboscides on the outer pair. In the Solemyidae there is only a single pair of palps which is much reduced and lacks proboscides. In the possibly related *Nucinella*, there are two pairs of reduced

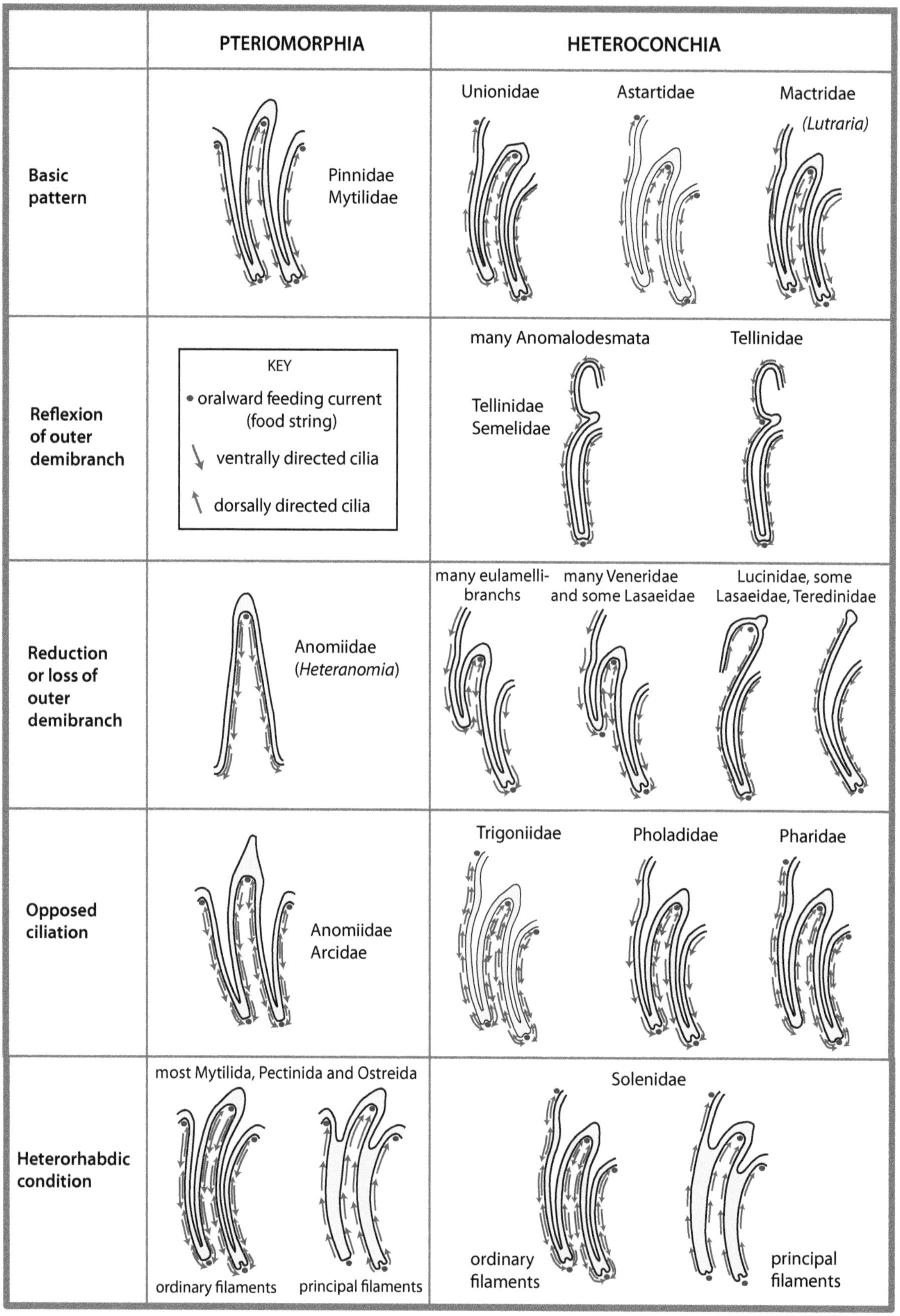

FIGURE 15.25 Ciliary sorting mechanisms on the frontal surfaces of the ctenidial filaments of eulamellibranch bivalves. The diagrammatic sections through the ctenidia are redrawn and modified from the following sources: Astartidae (Saleuddin, A.S.M., *Proc. Malacol. Soc. Lond.*, 36, 229–257); Trigoniidae (Morton, B., *Rec. Aust. Mus.*, 39, 339–354, 1987b); remainder (Atkins, D.G., *Q. J. Microsc. Sci.*, 79, 375–421, 1937b).

palps that also lack proboscides (Allen & Sanders 1969; Taylor et al. 2008; Oliver & Taylor 2012). The palps have been lost in two tiny 'suctorial' bivalves (see Section 15.4.2).

In other molluscs the homology of the palps with labial structures is not clear because they differ considerably from the various elaborations around the mouth seen in the other classes. In all bivalves, including protobranchs, the palps and the anterior ends of the gills are in intimate contact. Waller (1998) argued that originally there was no contact between palps and gills, suggesting that they may have evolved as food-gathering organs independent of the gills, a scenario supported by the study of juvenile bivalves. In this scenario, food particles collected by cilia on the foot, or from water driven into the mantle cavity by those cilia, are passed to the labial palps. In a few groups (e.g., Nucinellidae [Allen & Sanders 1969] and Lucinoidea [Allen 1958]), cilia of the epithelium associated with the anterior adductor muscle may also assist in passing food brought in by an anterior inhalant current to the palps. Important additional steps in the evolution of bivalves included lateral body compression and the dorsoventral elongation of the foot, resulting in the mouth being raised well off the substratum and the modification of the palps to sort particles before ingestion (Waller 1998).

15.3.6.3 Other Features of the Mantle Cavity

15.3.6.3.1 Mantle Rejection Tracts and Pseudofaeces

In most autobranch bivalves, special ciliary rejection tracts in a shallow groove inside the mantle edge carry waste particles bound in high-viscosity mucus which accumulate as pseudofaeces in the posterior inhalant part of the mantle cavity. These mantle rejection tracts have very long composite cilia (Beninger & Veniot 1999; Beninger et al. 1999). The pseudofaeces are ejected by occasional 'sneezes' caused by sudden valve closure which results in water being ejected from the mantle cavity via both inhalant and exhalant apertures. Exceptions are the heterorhabdic filibranch bivalves, such as scallops, limids, and anomiids, which only collect food in the dorsal grooves between the gills, and material carried to the ventral edges of the gills is rejected. In these bivalves, there is no groove along the mantle edge, and pseudofaeces can be ejected at any point along the open mantle edge by valve clapping (Beninger et al. 1999).

15.3.6.3.2 Hypobranchial Glands

Morton (1977) reviewed the distribution of hypobranchial glands in bivalves. These mucus-secreting structures are absent in most bivalves, including the nuculanid protobranchs, but present in the nuculid and solemyid protobranchs and a few autobranch bivalves. When present they are typically found on the posterior part of the mantle above the ctenidia and below the rectum, where they assist in binding particles prior to their rejection. With the effective division of the bivalve mantle cavity into separate exhalant and inhalant chambers this function was no longer required. However, a few autobranchs do possess a hypobranchial gland (or analogous structure); these include a few filibranchs such as *Placuna*, some anomiids, and the heterodont *Fimbria* (Lucinoidea).

In the freshwater *Corbicula* (Cyrenoidea) and *Sphaerium* (Sphaerioidea), the interlamellar junctions of the gill that form a marsupium in these brooding bivalves are lined with mucous cells, which Morton (1977) hypothesised might be derived from a hypobranchial gland.

15.3.7 Digestive System

Bivalves lack a buccal apparatus (odontophore, radula, and jaws) and salivary glands and have a simple oesophagus without an oesophageal gland. The simplicity of the anterior gut is compensated for by the complexity of the stomach where the food is processed and mixed with enzymes. The bivalve stomach has been intensively studied, and an overview of the structure and function is given in Chapter 5. In general, bivalve stomachs are large and complex, with sophisticated ciliary sorting mechanisms and a style sac. In protobranchs, the style sac contains a 'protostyle', a rotating mass of mucus-bound waste, but in autobranchs, this sac houses a crystalline style, a rotating hyaline rod which liberates enzymes into the stomach. A review of aspects of the structure of the digestive system of bivalves is given in Bieler et al. (2014).

Purchon (1959) recognised five main types of bivalve stomachs (Figure 15.26) and published several subsequent studies detailing bivalve stomach morphology. The summary below is slightly modified from the detailed overview of Purchon (1990), although this varies in some details from the review in Bieler et al. (2014).

1. *Stomach Type I (nuculoid type)*: This type of stomach is found only in protobranchs. It has only a few openings to the digestive gland, by way of which enzymes enter the stomach where extracellular digestion occurs. Also, the intestinal groove and major typhlosole do not extend into the stomach, being confined to the style sac. While Stomach Type I is probably similar to the stomach of the deposit-feeding ancestral bivalves, the stomach morphology seen in living protobranchs is not thought to be directly ancestral to that seen in autobranch bivalves.

2. *Stomach Type IV (Gastrotetartika)*: This is the most basic stomach morphology seen in autobranch bivalves and is found in many. In response to the adoption of suspension feeding, the main changes included an increase in the number of digestive gland ducts opening to the stomach. The intestinal groove and major typhlosoles extend into the stomach, possibly protecting the digestive gland duct apertures. They form an uninterrupted arc from the opening of the style sac across the stomach floor to the opening of the left caecum or left pouch. Unlike the situation in protobranchs, fine particles in the stomach pass into the digestive gland ducts and are intracellularly digested in the digestive gland diverticula. Coarse particles in the stomach pass continuously into the intestine. These changes allow continuous food processing and efficient digestion.

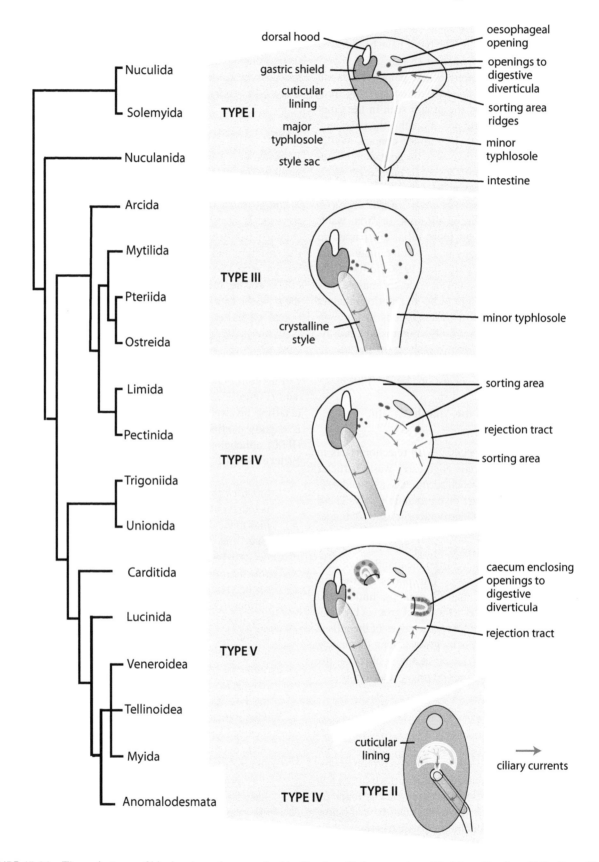

FIGURE 15.26 The main types of bivalve stomachs recognised by Purchon. Redrawn and modified from Purchon, R.D., *Proc. Malacol. Soc. Lond.*, 33, 224–230, 1959.

Purchon hypothesised that his Stomach Types II, III, and V arose by modification from Stomach Type IV.

3. *Stomach Type III* (Gastrotriteia): This stomach type is found in most (six superfamilies) pteriomorphians. The main feature is that the major typhlosole has a long narrow tongue which extends into the stomach, and the intestinal groove runs around it on both sides while the sorting area curves around it.

4. *Stomach Type V* (Gastropempta): In this stomach type the intestinal groove lies next to the major typhlosole as it forms a circular flange which passes into the right caecum and another typhlosole passes into the left caecum – thus protecting the openings of the digestive gland ducts lying inside these caeca. This stomach type is found in the Myoida and many euheterodonts. Other euheterodonts (Lucinidae and Thyasiridae) (Allen 1958) and *Donax* spp. and *Chama* spp. (Purchon 1960) have a Type IV stomach.

5. *Stomach Type II* (septibranch type): The stomach is modified as a muscular gizzard. This very derived stomach arose in the anomalodesmatan septibranchs. The stomach of *Parilimya* (Parilimyidae) is intermediate between this type and the Type IV of Purchon (Morton 1982b), and Allen and Turner (1974) have shown that stomachs in verticordiids range from relatively complex (with sorting areas) in *Lyonsia* to very simplified in *Lyonsiella*.

Although, as pointed out by Bieler et al. (2014), Purchon's recognition of 'stomach types' does not recognise the reality that bivalve stomachs are a complex of largely independent characters (with 51 separate characters being coded for the stomach and style sac in the analysis they conducted), they are nevertheless a useful concept for outlining the structural diversity in these complex organs.

Digestion occurs in large paired digestive diverticula made of many narrow tubules (see Chapter 5 for more details). The intestine is looped within the visceral mass (which extends into the dorsal foot) and crosses over the dorsal side of the posterior adductor muscle, opening dorsally into the exhalant area in the posterior part of the mantle cavity. The extent of intestinal looping varies considerably in different bivalve taxa from one or a few loops to numerous loops.

15.3.8 RENOPERICARDIAL SYSTEM

Aspects of the renopericardial system of bivalves are described in more detail in Chapter 6. The vascular system is bilaterally symmetrical and consists of a mid-dorsal heart with paired auricles and a ventricle usually perforated by the rectum. In nuculids the ventricle is bilobed, apparently indicating that it was originally penetrated by the rectum, and the rectum lies above it. A similar situation is found in the nuculanoidean Malletiidae but the rectum lies below the ventricle (Villarroel & Stuardo 1998). In other protobranchs and in many autobranch bivalves, the rectum penetrates the ventricle, but in

some autobranchs it does not. The heart is not developed in the tiny suctorial bivalve *Draculamya* (Oliver & Lützen 2011).

The blood of protobranchs contains haemocyanin, but this respiratory pigment is lacking in autobranch bivalves. Two groups, the arcids and some archiheterodonts, have independently acquired haemoglobin in their blood (see Chapter 6), and haemoglobin has also been reported (e.g., Read 1962; Alyakrinskaya 2002) from a few other bivalves but not investigated. These include *Amygdalum* (Mytilidae) (Oliver 2001), Poromyidae, some tellinids, the galeommatoidean *Barrimysia* (Goto et al. 2018), *Pharus legumen* (Pharidae – as *Solen* in the early literature [Yonge 1959b]), and *Xylophaga* (Xylophagaidae). In the latter example, it occurs in erythrocytes. In other taxa, it may either be in erythrocytes or dissolved in the blood.

The kidneys are U-shaped, paired, and, in autobranchs, are often separate from the reproductive system. In protobranchs, however, the short gonoduct opens to the proximal portion of the kidney and shares a common opening into the posterior exhalant chamber of the mantle cavity (see Chapter 6, Figure 6.5).

15.3.9 REPRODUCTIVE SYSTEM

The gonads are paired (although they are often difficult to distinguish) and, in autobranchs, open to the exhalant chamber of the mantle cavity by way of a short, simple duct or, in some, open to a urinogenital duct (Figure 15.27). In protobranchs, as noted above, the gametes exit via the renal pore.

Some bivalves are gonochoristic whereas others are hermaphroditic. In the latter case various types of hermaphroditism occur – simultaneous, consecutive (mostly protandric, very rarely protogynous), rhythmical consecutive, or alternative (see Chapter 8 for further explanation).

Sperm morphology has provided important characters for the phylogenetic studies and is reviewed in Bieler et al. (2014).

Some populations of normally gonochoristic species of the freshwater bivalve *Corbicula* may also be hermaphrodites (e.g., Morton 1983a; Houki et al. 2011). *Corbicula fluminea* is very unusual among molluscs in producing only eggs when it first matures (protogynous), followed later by sperm; this markedly contrasts with nearly all other sequential hermaphrodites (which are protandrous). *Corbicula fluminea* later becomes a simultaneous hermaphrodite and remains so throughout its life (McMahon 1991). The protogynous species of *Corbicula* practise androgenesis, a rare phenomenon in animals, where the oocytes are fertilised by sperm that have not undergone a reduction (meiosis) in their chromosomes. Following fertilisation, the entire maternal nuclear genome is extruded from the egg as two polar bodies, but the mitochondria and other organelles are retained. Thus the maternal nuclear genome does not participate at all in the development of the zygote (Pigneur et al. 2012) (see also Chapter 8), and the offspring are identical paternal clones (Komaru et al. 1998; Pigneur et al. 2012).

In general, members of the genus *Corbicula* show a considerable range of sexuality and life history. The probable

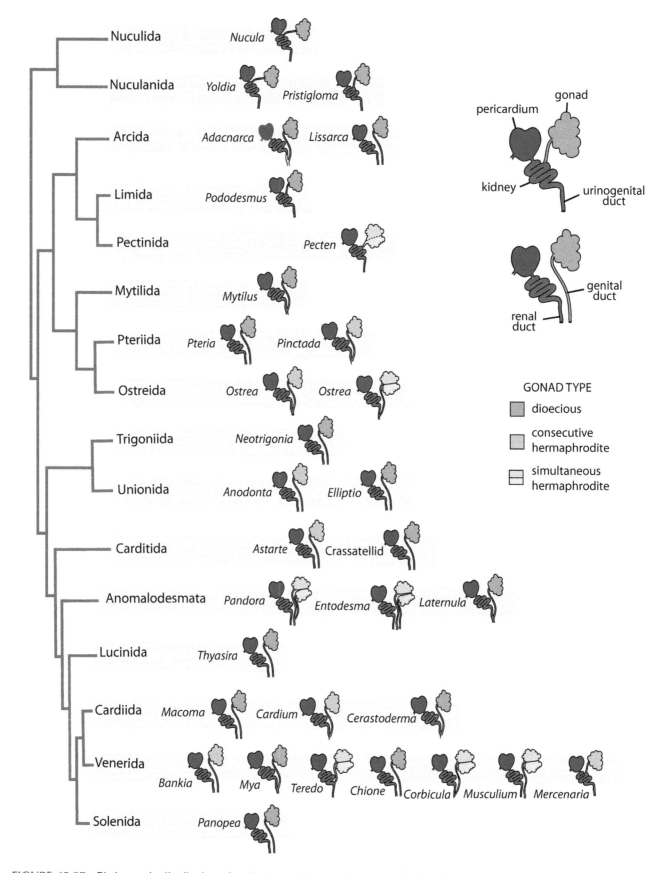

FIGURE 15.27 Phylogenetic distribution of genital morphology and strategies in the Bivalvia. Cartoon scheme from Mackie, G.L., Bivalves, pp. 351–418 *in* Tompa, A.S. Verdonk, N.H., and van den Biggelaar, J.A.M. (eds.), *Reproduction. The Mollusca*, Vol. 7, Academic Press, New York, 1984; data from numerous sources.

primitive condition is a dioecious, free-spawning, and dispersive life history as seen in the lacustrine *C. sandai* and the brackish-water species *C. japonica*. While lacustrine species from Sulawesi and Sumatra are apparently dioecious and brood their larvae, all other freshwater species also brood but are hermaphroditic (see reviews in Glaubrecht et al. 2006; Pigneur et al. 2012) and androgenetic, as described above.

In some species of *Corbicula* self-fertilisation is indicated by embryos within the follicles of the gonad (Ikematsu & Yamane 1977; Kraemer et al. 1986), although in most species the embryos are incubated in the gills, including in hermaphrodite taxa (e.g., Glaubrecht et al. 2006).

Polyploidy is uncommon in bivalves but androgenetic *Corbicula* lineages range from diploidy to tetraploidy and retain ploidy throughout their life cycle with unreduced sperm (Pigneur et al. 2012). Other polyploid bivalves include some individuals of the marine mussel *Mytilus trossulus* (González-Tizón et al. 2000) which reproduces sexually, as do the freshwater sphaeriids, some of which are polyploids (Lee & Ó Foighil 2002). The gynogenetic[3] marine genus *Lasaea* also has polyploid species (Ó Foighil & Thiriot-Quiévreux 1991).

Further aspects of reproduction in bivalves are explored in Section 15.4.3 and Chapter 8.

15.3.10 Nervous System and Sense Organs

Bivalves lost their head and any associated sensory structures that may have been present in their ancestors. Instead sensory structures (e.g., tentacles and sometimes eyes) are located on the middle fold of the mantle margin and siphons (if present), as these are the parts of the animal in contact with the external environment. Additional sensory structures are the statocysts, osphradia, and proprioceptors (stretch receptors in muscles). Adult 'cephalic' or 'branchial' eyes are found in some pteriomorphian bivalves (including some Arcoidea, Limopsoidea, Mytiloidea, Anomioidea, Ostreoidea, and Limoidea) but have little functional significance and are probably retained larval eyes (Morton 2001). These and the mantle and siphonal eyes found in bivalves are described in Chapter 7.

Because the osphradia retain their plesiomorphic position on the axis of the ctenidium, they are located in the exhalant suprabranchial chamber of the mantle cavity. Thus the osphradia are not in a suitable position to detect incoming chemical stimuli although their cytology indicates that they retain a chemosensory function, probably mainly in relation to reproductive activities (Haszprunar 1987c).

Bivalve nervous systems are reviewed by Bullock and Horridge (1965). There are typically three pairs of bilaterally symmetrical ganglia, in autobranchs the largest being the visceral ganglia. The cerebropleural ganglia innervate the anterior adductor muscle and the anterior mantle, branchial eyes (if present), the mouth region, including the labial palps, and the anterior gut and stomach. The pedal and byssal muscles are innervated by the pedal ganglia which lie in the foot muscles in protobranchs but are in the visceral mass in autobranchs.

The visceral ganglia, which are commonly pigmented, generally comprise the fused visceral and parietal ganglia, although they are sometimes separate. They innervate the mantle, siphons (if present), posterior adductor and posterior foot retractors, gills, osphradia, the renopericardial organs, gonads, and intestine.

A pair of statocysts lies near the pedal ganglia, with connections to the cerebral ganglia. Both single statoliths and multiple statoconia have been recorded from bivalve statocysts.

Protobranchs have an adoral sense organ that lies on the outer side of the labial palps. It is not involved in feeding and is most likely chemosensory (Schaefer 2000). An anterior ciliated sense organ in the enlarged middle fold of the mantle edge is found in nuculanids (Yonge 1939a). Nuculids have a tubular sense organ (Stempell's organ) that lies above the anterior adductor. It is innervated by the cerebral ganglion and is probably a mechanoreceptor (Haszprunar 1985).

Abdominal sense organs are associated with the posterior adductor muscles and are found in many pteriomorphians and palaeoheterodonts. These and other bivalve sense organs are described in Chapter 7.

15.4 BIOLOGY AND ECOLOGY

15.4.1 Habits and Habitats

Bivalves live in a wide variety of habitats, with the majority benthic, marine, and infaunal. While most infaunal taxa burrow into soft sediments, some bore into wood, coral, or rock. Others attach to hard surfaces by using byssal threads or by cementing, and a few can swim intermittently. While many groups are found in estuarine environments, only five living groups have successfully and independently transitioned to fresh water (see Section 15.6.2.1) although some additional extinct groups of freshwater bivalves are also known (see Section 15.6.2.1 and the Appendix).

Many changes in bivalve morphology were induced by adaptive responses to different habits – for example burrowing in soft sediments or boring in hard substrata. Associated changes in anatomy include a foot modified for burrowing, fused mantle edges, and siphon development. Epifaunal bivalves exploit the surface of hard substrata using byssal attachment and later cementation (see Section 15.4.1.2). Anatomical changes in these lineages typically led to a reduction of the foot and often the development of the heteromyarian or monomyarian condition (see Section 15.4.1.2.1).

Burrowing bivalves are important bioturbators in many soft-sediment assemblages, and their feeding and movement is thought to affect sediment structure and community development (e.g., Droser & Bottjer 1989).

15.4.1.1 Locomotion

Most bivalves move using their foot. In recently settled juveniles, the foot is used to crawl to suitable areas for attachment

---

[3] The embryo contains only maternal genetic material as the sperm only activates the egg and does not fuse with the egg nucleus.

or burrowing. Surface crawling is also seen in adults of some small-sized bivalves such as galeommatoideans and some other small heterodonts. A few shallow burrowers can move about in the surface sediments, and some, such as cardiids, have a large, very active foot capable of twisting and turning and flipping the animal over when it is out of the sediment. In adults that burrow into deeper sediments, the foot is usually used just for digging. While cemented taxa (e.g., oysters – see Section 15.4.1.2) are obviously incapable of movement, many byssally attached pteriomorphians cannot reattach if dislodged. However, a number of byssate forms retain the ability for the animal to move using their often-reduced foot, and some, such as mussels (e.g., *Mytilus*) have retained a sufficiently large foot that makes at least smaller individuals capable of crawling to a new location and reattaching if detached.

Pectinids and limids are mostly byssally attached as adults, but some have abandoned that lifestyle and are free-living, and a few can swim. Swimming has been achieved by modifying the valve flapping which was ancestrally used to dislodge waste in the mantle cavity and, in the case of limids, helped by a rowing action of the long mantle edge tentacles (see Section 15.8.1.2 and Chapter 3).

Planktomya henseni is a bivalve that was thought to be a planktonic species, but it proved to be the planktonic phase of a teleplanic larva of a galeommatoidean (Allen & Scheltema 1972; Gofas 2000).

15.4.1.2 Attachment

Bivalves can attach to hard surfaces using a byssus or by cementing. A byssus is secreted in most autobranch larvae, but this ability has persisted in the adults in many pteriomorphians and only rarely in heterodonts (see Section 15.3.4 and below for more detail). The retention of an adult byssus has profoundly altered the biological attributes and the morphology of many taxa during their evolution (Yonge 1962b), as discussed in the next section.

A functional byssus is found in adults of the following groups (Yonge 1962b) – Pteriomorphia: Arcoidea (absent in some burrowing forms such as Glycymerididae and *Anadara* and its relatives), Mytiloidea (used in connection with boring in some), Pterioidea (absent in a few such as *Vulsella*), Pinnoidea, Pectinoidea (absent in some, including swimming taxa such as *Pecten* and *Amusium*, and in cemented taxa such as Spondylidae and Plicatulidae); Limoidea (absent in some free-living taxa); Heterodonta: well-developed in Dreissenoidea and Tridacninae (Cardiidae, Cardioidea). Small byssal glands, which form a byssus with a few threads, are found in some heterodonts including Carditoidea, Sphaerioidea, Glossoidea, Gaimardioidea, Galeommatoidea, some Cardioidea, a few Veneroidea (notably *Venerupis*), and some Trapeziidae, Hiatelloidea (in the crevice dwelling *Hiatella*), Myoidea (some Corbulidae and a few Myidae but absent in adult burrowing myids), Gastrochaenoidea (lacking in most), and Pandoroidea (some Lyonsiidae, absent in most). The Anomiidae (other than *Enigmonia*) uniquely attach to the substratum by way of a calcified byssus protruded through an opening in the lower valve (Yonge 1977; Eltzholtz &

Birkedal 2009). Byssal attachment (see also Section 15.3.4) has a wide range of variation in adult bivalves, especially in pteriomorphians (Figure 15.14). Use of the byssus by epifaunal bivalves to attach to hard substrata is the most obvious and is referred to as the *epibyssate* condition. Some partially or fully infaunal species, such as members of the Pinnidae, some Malleidae (*Malleus*), some mussels (Mytilidae), and myoideans, are *endobyssate*, using the byssus to attach to objects in the sediment. Some mytilids and limids use the byssus to build 'nests', sometimes forming colonies, as in the mytilids *Musculus* and *Musculista*.

Bromley and Heinberg (2006) recognised five kinds of byssal attachment that could be defined in relation to their position and the morphology of the shell. Among dimyarian bivalves these range from those with the hinge line parallel or nearly so to the substratum (e.g., arcids such as *Arca* and *Barbatia*), to those where the hinge line is at a considerable angle and where the byssus may emerge some distance from the anterior end (e.g., *Mytilus*) or very near to it (e.g., *Limopsis*). This latter configuration is also seen in some monomyarian autobranch bivalves (e.g., *Lima*). In members of these three groups, the foot has a long, flat sole which can make firm contact with the substratum, and in mytilids and limopsids the anterior adductor muscle is reduced. Their fourth and fifth categories are found only in some monomyarian bivalves. In one the hinge axis is parallel with the substratum, and the byssus emerges through a hole in the hinge (e.g., *Malleus*). In the other, seen for example in many pectinids and pteriids, asymmetrical shell projections (auricles) extend from the umbo of the right valve, and the byssus emerges at the base of the anterior one. The anterior auricle is usually longer, and, with the byssus and the anterior part of the right valve, it makes contact with the substratum in a tripod-like fashion.

Cementation of one of the valves to the substratum (see also Chapter 3) has occurred independently in many families during bivalve evolution (Harper 2012). It developed, for example, in many of the Mesozoic rudists (see Section 15.8.5.4), some of which formed reefs, and, among living bivalves, it is seen in pteriomorphians (e.g., Dimyidae, Ostreidae, Pectinidae, Placunidae, Plicatulidae, Spondylidae), heterodonts (Chamidae, Clavagellidae, Cleidothaeridae, Myochamidae), a few freshwater unionidans, mainly in Africa (Etheriidae) (Yonge 1979; Bogan & Hoeh 2000), and a cyrenid (*Pososostrea*) (Bogan & Bouchet 1998). In this latter case the cemented species, which lives in the ancient Lake Poso in Sulawesi, Indonesia, appears to have evolved recently from *Corbicula* (Rintelen & Glaubrecht 2006).

While the mechanisms involved in cementation are not well understood in some groups, in oysters the larval foot secretes a substance that glues the left valve to the substratum with subsequent cementation by way of the mantle (Harper 1992). In spondylids and cemented scallops the settled post-larva is first byssally attached, and cementation via the mantle follows (Yonge 1979).

Cementation mostly occurred in bivalves after the Paleozoic (Harper 1991, 2012), peaking in the Upper Triassic and Jurassic. This means of attachment may have been a response to predation – the so-called 'Mesozoic marine revolution' (Vermeij 1977) – notably driven by the concurrent appearance of crushing predators such as crabs and fish. This idea was tested by Harper (1991) who found that byssally attached bivalves were subjected to significantly higher levels of predation than cemented ones. The same response to predation may be at least a partial explanation for the more recent cementation in freshwater bivalves in Africa and parts of southern Asia that live with crabs, while their co-occurrence with only freshwater crayfish in the Americas, Europe, and northern Asia is not correlated with a similar response (Bogan & Hoeh 2000).

15.4.1.2.1 The Morphological Consequences of Attachment

While many arcids have a well-developed adult byssus, they have retained their original isomyarian condition, probably because they have maintained an anterior inhalant water flow (Yonge 1962b). This retention of two equal-sized adductor muscles is in marked contrast to many other bivalves with a well-developed adult byssus that have undergone major transformations in their morphology. The most notable modifications have resulted in changes of symmetry and the reduction (anisomyarian) or loss (monomyarian) of the anterior adductor muscle in the pteriomorphians (Figure 15.28). Cementation also caused major alterations to the orientation of the animal and resulted in a monomyarian condition in Ostreoidea and the palaeoheterodont Etheriidae, although some other cemented taxa such as Chamidae and Cleidothaeridae retain an isomyarian state.

Byssal attachment can maintain a vertical orientation (as in most mytilids), or the adoption of an inequivalve shell can enable the animal to lie flatter against the substratum, as in many pterioideans and pectinoideans. The anomiids, with a calcified byssus, lie horizontal to the substratum, as do most cemented bivalves. Heteromyarian bivalves are often triangular in shape while many monomyarians have acquired a more or less circular outline, typically with an inflated attached valve and flatter upper valve.

Attachment, by whatever means, often brings with it risks of being dislodged by wave action. Streamlined shell morphologies and the orientation of the shell relative to the waves or currents can reduce the likelihood of this occurring (e.g., Kauffman 1969; Stanley 1970, 1975).

Some attached bivalves have rugose or spinose upper valves that in part may deter predators directly and/or aid in facilitating camouflage by providing extra surface area for the attachment of algae and encrusting animals.

15.4.1.3 Secondary Soft-Bottom Dwellers

Some bivalves belong to groups which have become adapted to living epifaunally on hard surfaces and have lost the ability to burrow. Some have secondarily taken up life on, or in, soft sediments, a move requiring new modifications. These evolutionary changes were reviewed by Chinzei et al. (1982) and Seilacher (1984b) who found that bivalves had adapted to this change in habitat in two main ways. The first was by

becoming mobile, either by enlargement of the reduced foot (as in some Arcidae), or, in the groups where the foot was lost, it involved swimming, as in some pectinids and limids. The second method affected mechanical stabilisation of the body by changes in the shape, weight, or size of the shell and is the commonest method employed. Occupation of soft-bottom habitats is usually achieved by the juveniles attaching to a small object such as a small rock or shell fragment and then growing into shapes and/or orientations that provide stability (Seilacher 1984b). Such groups include a number of living and fossil oysters and the fossil rudists (Gili et al. 1995; Seilacher 1998; see Section 15.8.5.4), some of which produce elongate shells partly buried in mud. In pinnids and some fossil oysters, both valves are elongate and embedded in the sediment, except for the youngest part of the shell which protrudes into the water column. In others, the shell was expanded laterally to help prevent it sinking into the sediment. This was achieved in some oysters by extending the entire shell margin and in *Malleus* by extending the shell auricles. However, when soft sediments were invaded by some now extinct groups of oysters and rudists, they developed differently. For example, in oysters the originally attached valve became cup-shaped or very elongate cone-shaped, enabling the smaller lid-like upper valve to remain free of the sediment. The true rudists lost the ligament and thus resemble brachiopods in this respect, a shell form rare among non-rudist bivalves. Some living and fossil oysters approach a brachiopod morphology, but these all retain the ligament (Seilacher 1998).

The mode of growth of a functional ligament induced shell coiling in early rudists (Skelton 1985), as it does in *Chama* and some oysters, enabling the lower valve to elongate while the upper valve could become flattened by overgrowing the earlier parts of the ligament. In early rudists the ligament, which is much narrower than in oysters, moved from the shell margin during their evolution so that it no longer functioned to open the valves (Seilacher 1998). This loss of ligament function enabled the development of the lower valve into an uncoiled column-like structure of unlimited length and the upper valve into a lid that accreted around its entire margin during growth. This was a major step in rudist evolution but created the problem as to how the upper valve was opened to enable feeding and respiration. Some rudists may not have opened their shells but instead drawn in water through canal systems in the shell (Skelton 1976). Alternative suggestions as to how they may have opened their shells include a hydraulic mechanism of some sort or (as yet unproven) the involvement of modifications to the adductor muscles (Seilacher 1998). In an entirely different group, the Pholadoidea (pholads, teredos), the ligament has been lost, and the shell opening function has been taken over by the large anterior adductor muscle, which is attached above the hinge line.

The large hinge teeth in rudists grew into long, spine-like processes on the upper valve that slotted against the wall of the lower shell or fitted into corresponding sockets in the lower valve. In parallel with this, the muscle scars also became tooth-like (myophores) in the upper valve and socket-like in the lower one and are often difficult to distinguish from the

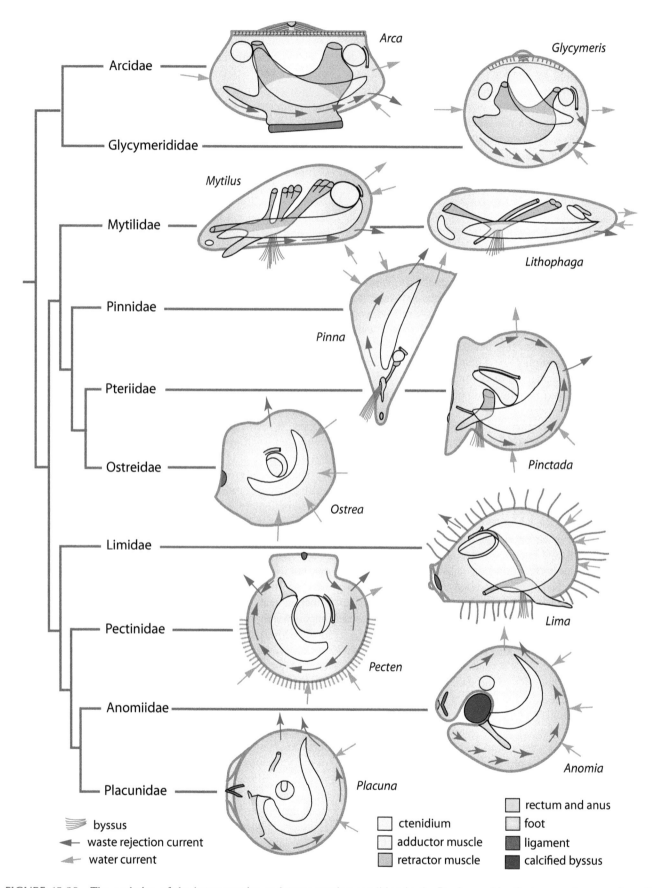

FIGURE 15.28 The evolution of the heteromyarian and monomyarian condition in the Pteriomorphia. Cartoons redrawn and modified from Yonge, C.M., *Trans. R. Soc. Edinburgh*, 62, 443–478, 1953a; Yonge, C.M., *J. Mar. Biol. Assoc. U. K.*, 42, 113–125, 1962; and Yonge, C.M. and Thompson, T.E., *Phil. Trans. R. Soc. B*, 276, 453–527, 1976. Tree topology based on Lemer, S. et al., *Proc. R. Soc. B*, 283, 20160857, 2016.

peg-like hinge teeth. Some have suggested that many of the evolutionary novelties of rudists were driven by possession of symbiotic zooxanthellae (e.g., Seilacher 1998), although this is by no means universally accepted. The rudists are discussed further in Section 15.8.5.4.

15.4.1.4 Burrowers and Borers

Most burrowing bivalves utilise siphons to reach the surface of the sediment so they can take in water for respiration and feeding (see Section 15.3.5). Their foot is modified for digging. In protobranchs, they have a rather broad sole, the edges of which can be extended to use as an anchor when digging. In autobranch bivalves, the foot is laterally compressed and 'hatchet shaped'. The mechanism in burrowing is basically similar in all autobranch bivalves and involves the following steps (see Chapter 3, Figures 3.39 and 3.40). The adductor muscles relax, and the ligament forces the valves to gape, anchoring the shell in the burrow while the foot is forced down into the substratum. The distal part of the foot then dilates through an increase in haemocoelic pressure forming an anchor, against which the shell is drawn down by contraction of the pedal retractor muscles. The adductor muscles contract and force the blood into the foot and, in many bivalves, the contraction forces a jet of water down into the sand which liquefies the sediment and assists burrowing. The efficiency of this mechanism is greatly enhanced by mantle fusion along the ventral part of the bivalve as in the deep-burrowing lineages. Non-siphonate burrowers such as the arcoideans *Anadara* and *Glycymeris*, the primitive heterodont *Astarte*, and the palaeoheterodont *Neotrigonia* are relatively inactive and live largely in the subtidal. Mantle fusion and the development of siphons better equipped heterodont burrowing bivalves for intertidal and shallow subtidal environments (Stanley 1968).

In some burrowers the ligament is too weak to open the valves sufficiently, this being largely achieved instead by pressure in the mantle cavity due to siphon and/or foot withdrawal. Thus, in some deep burrowers such as *Mya*, the retraction of the siphons forces the valves to gape, without which the siphons could not retract. Here the siphonal muscles act antagonistically with the adductors, an arrangement utilised in some boring taxa, as described below.

Silurian members of the Lucinoidea were the first deep-burrowing autobranch bivalves to appear in the fossil record (Liljedahl 1992). Typical heterodont deep burrowers did not appear until the Permian or later, with, for example, solecurtids first appearing in the Carboniferous and solenids and myids in the Jurassic.

Some borers may have evolved from sediment burrowers or bivalves that nestled in crevices in hard substrata, just as some do today. The division between nesters and burrowers is sometimes blurred, as for example in Petricolidae, with species which nestle in crevices, some which burrow into sand or mud, others which are soft rock borers. Species of *Petricola* bore into calcareous substrata, including dead coral, with the aid of chemicals (Morton & Scott 1988). A few species such as the small-sized *Hiatella* nestle in crevices in rock

or other softer substrata, including algal holdfasts, and some can bore into calcareous rocks. In marked contrast, another hiatellid is the large, deep-burrowing *Panopea*. The petricolids *Rupellaria* and *Petricolaria* and the myid *Platyodon* are unspecialised borers and can burrow into firm substrata such as peat, clay, and soft rock. More specialised borers such as the pholadids, teredinids, and gastrochaenids have obvious modifications for boring. For example, most have a large pedal gape through which the sucker-like foot can clamp on the burrow wall enabling the shell to be moved rasp-like against it.

While some boring bivalves (e.g., pholads and teredinids) exclusively use the movement of their shells to grind their way through the substratum, other taxa use this method in conjunction with chemical dissolution (Kleemann 1996; Dorgan 2015) (see below).

Ansell and Nair (1969) investigated mechanical boring and observed that the processes involved were similar to movements in related non-boring taxa. Mechanical borers all use the shell as the boring tool, but the method varies according to the ancestral habits of different groups. Yonge (1963) recognised two main groups of mechanical borers. The first contains taxa derived from byssate ancestors (mytiloideans, veneroideans, hiatelloideans, gastrochaenoideans, and cardioideans), in which the main force driving the shells in their abrading action is provided by contractions of the pedal or byssal retractor muscles. The ligament provides the outward pressure which forces the shell against the burrow wall. The second group contains those taxa derived from deep burrowers such as myoideans and pholadoideans, in which the adductor muscles provide the major force in the abrading movements of the shell. In marked contrast to the highly specialised pholadoideans, the myoidean burrowers are little modified. In both, the basic digging movements utilised by these mechanical burrowers have been modified by adding a rocking action (about the dorsoventral axis) so the coarsely sculptured valves act as rasps that widen and deepen the burrow. This rocking is achieved in part by a reduction or loss of the ligament but also a reciprocal action of the adductor muscles which causes the valves to gape anteriorly as the large posterior pedal retractor muscle contracts (Ansell & Nair 1969) (see Chapter 3, Figure 3.41).

Additional modifications of the pholadoideans (Figure 15.9) include a reduced hinge, a peg-like apophysis (see Section 15.3.1.8), a large pedal gape with a disc-shaped, sucker-like foot which enables it to clamp onto the wall of the burrow, the complete loss of the ligament, the formation of extra shelly plates (see Section 15.3.1.8), and a unique musculature. The pholads are typically regarded as mechanical borers, but this is not always the case as some taxa also employ chemical dissolution (e.g., *Penitella*, *Parapholas*, and *Jouannetia*) (Kleemann 1996), some having large mantle glands which facilitate their boring into carbonate substrata (Smith 1969; Morton 1985a).

Some pholadoideans are specialised wood borers, including five of the 27 living and fossil genera of Pholadidae. Members of a closely related group, the Xylophagaidae, in particular the genus *Xylophaga*, are mainly found in wood

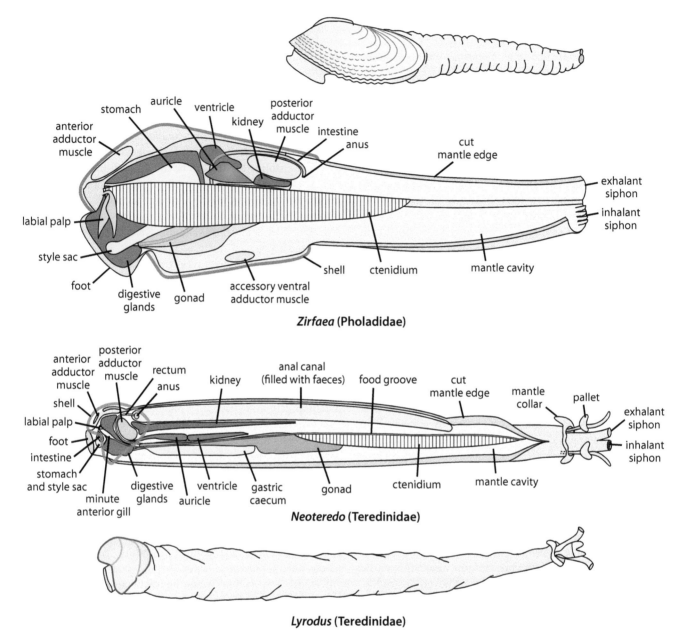

FIGURE 15.29 Comparative morphology of a pholadid and a teredinid. Redrawn and modified from Turner, *A Survey and Illustrated Catalogue of the Teredinidae (Mollusca: Bivalvia)*, Museum of Comparative Zoology, Cambridge, UK, 1966.

transported into the deep sea, although at least one species (*X. dorsalis*) is also found in wood on the continental shelf (e.g., Norman 1976).

All but one of the 15 genera of 'shipworms' (Teredinidae) are wood borers (Hoagland & Turner 1981). Teredinids have elongated bodies necessitating significant reorganisation of their internal organs (Figure 15.29); they have only a small rasping shell at the anterior end and a pair of short siphons posteriorly. They also secrete calcium carbonate to line their burrow, and the siphonal retractor muscles are attached to this calcareous tube. They have also evolved a unique pair of pallets (Figure 15.9), calcareous structures of varying morphology with which they close their tube. While some species of the teredinid *Bankia* reach a metre in length, the

largest teredinid, *Kuphus polythalamia*, commonly reaches a metre in length but can be up to one and half metres long and harbours symbiotic bacteria in its gills (Distel et al. 2017). Initially *K. polythalamia* is a wood borer, but the animal inside its calcareous tube soon becomes embedded in mud in mangrove swamps (Shipway et al. 2018).

The oldest record of a wood-boring bivalve is an unidentified taxon found in fossil driftwood from the Lower Jurassic (Pliensbachian) (190.8–182.7 Ma) of north western Germany (Vahldiek & Schweigert 2007).

The hydrodynamic forces of the body fluids are more important in the least specialised borers, such as the veneroidean *Petricola*, than they are in burrowing taxa. However, in all borers, the muscles involved in the boring actions are enlarged and act against a fulcrum, usually the foot (Ansell & Nair 1969).

In the byssate forms borers mostly evolved from endobyssate burrowers, and in some lineages they merely translated burrowing actions into boring. However, some mechanical borers also employed chemical erosion. Pholadids have been thought to be mechanical borers (e.g., Ansell & Nair 1969), but as noted above, at least a few taxa appear to use some chemical means (Morton 1990b). However, utilising chemical erosion means these taxa are restricted to carbonate substrata, notably coral.[4] Chemical boring is facilitated by the secretion of Ca^{2+}-binding micropolymers and acid-secreting proton pumps (Dorgan 2015). In all, some members of seven families of bivalves bore into living or dead coral. Chemicals are secreted by the mantle in the date mussels (members of the Lithophaginae and the Gastrochaenidae), which comprise the majority of bivalve borers. One or a few members of Pectinidae, Tridacninae (Cardiidae), Petricolidae, Pholadidae, and Clavagellidae are also coral borers (see reviews by Morton [1990b] and Kleemann [1990b]). While most bore into dead coral, a few such as some Lithophaginae (Mytilidae), Pholadidae, and the one boring species of *Tridacna* (*T. crocea*), bore into living coral with some having very intimate relationships (see also Section 15.4.4). One such species is the pectinid *Pedum spondyloideum*, which is embedded in massive living coral colonies and uses chemical erosion (see Section 15.4.5). Many taxa line their burrows with a layer of calcium carbonate, and the ability to chemically dissolve the calcareous layer allows the burrow to be relined as the occupant grows.

The oldest suspected boring bivalve is the mytiloidean *Corallidomus* from the Ordovician of North America (Pojeta & Palmer 1976). This taxon was found in association with a calcitic stromatoporoid sponge and is thought similar to some extant date mussels (e.g., *Lithophaga*, *Botula*). Extant date mussels live attached to the burrow using their byssus, and as the animal grows the flask-shaped chamber they create is enlarged by a chelating agent secreted by the mantle (Kleemann 1973). *Lithophaga* and other boring mytilids use chemical means to bore into coral, limestone, or shells, as do species of the clavagellid genus *Bryopa* which bore chemically into dead coral (Savazzi 2000; Morton 2005).

The boring 'giant clam', *Tridacna crocea*, uses a combination of mechanical and chemical dissolution to penetrate the coral substratum in which it lies, as do some gastrochaenids (Morton 1990b).

The large majority of boring bivalves have bilaterally symmetrical shells, but, in some the two shell valves are asymmetrical. These include *Claudiconcha* (Petricolidae), *Jouannetia* (Pholadidae), *Clavagella*, and *Bryopa* (Clavagellidae) that bore in either soft rock or coral. Such asymmetry appears convergently only in adults at terminal growth and is associated

with either the cementation of one valve to the substratum in the petricolid and clavagellids or, as in *Jouannetia*, otherwise functions to prevent one valve from moving within the borehole. It has been suggested that the cemented valve may help prevent the dislodgement of the bivalve from the shallow borehole following exposure due to the erosion of the substratum (Savazzi 2005).

15.4.2 Feeding

As previously noted (Chapter 4 and Figure 15.30; also see Section 15.4.3 and Figure 15.34), the original mode of feeding in bivalves was probably deposit feeding on material gathered by cilia on the foot. Long processes involved in deposit feeding on the labial palps, as seen in living protobranch bivalves, are assumed to have evolved later. These long palp 'proboscides' probe the sediment and collect potential food which is sorted on the ciliated ridges on the inner surface of the palps and the rejected material is expelled from the mantle cavity. Potential food particles in the inhalant current may get caught up on the palps, but the gills are mainly used only for respiration, and particles entangled in them are mostly rejected by the frontal cilia and eventually discarded from the mantle cavity. However, some of these particles have been shown to be used for food (see below).

The extraordinary modifications of the autobranch ctenidium (see 15.3.6.1) resulted from bivalves taking up suspension feeding. The autobranch mantle cavity contains a pair of very large gills used to capture food particles suspended in the inhalant water current (suspension or filter feeding – see Chapter 5). We can only speculate as to how the transformation from the protobranch to the autobranch condition came about as no intermediate gill conditions occur in living bivalves. It probably involved a progressive elongation of the filaments with the rejection currents on the gill filaments being co-opted for food gathering as happens to some degree in at least a few living protobranchs (e.g., Reid et al. 1992). A large surface area was required for efficient food collecting, and this was effectively doubled by each of the elongated filaments being folded upwards to form a long, upside down V-shape. The autobranch gill further evolved to maximise the efficiency of suspension feeding through major modifications in structure and ciliation (see Section 15.3.6.1). The food is bound in mucus in strings that are carried by cilia along food grooves on the edges of the gills, to the mouth region. Here some particles from the gill are sorted on the ciliated labial palps before they enter the mouth.

While most autobranchs employ suspension feeding, there have been some departures from this. One large and successful group of heterodont bivalves, the Tellinoidea, with the exception of the suspension-feeding Donacidae (e.g., Ansell 1981), are mainly deposit feeders (Yonge 1949), with some feeding on both deposits and suspended particles (e.g., Pohlo 1969). They vacuum surface deposits using the inhalant siphon, with this change in feeding strategy necessitating modifications to sorting mechanisms, including an increase in the size and sorting efficiency of the labial palps.

---

[4] The trace fossil, *Gastrochaenolites*, which is thought to represent the work of boring bivalves, has been reported from Miocene gneiss boulders from Spain (Rodríguez-Tovar et al. 2015), suggesting that those metamorphic substrata may have been vulnerable to bivalve boring.

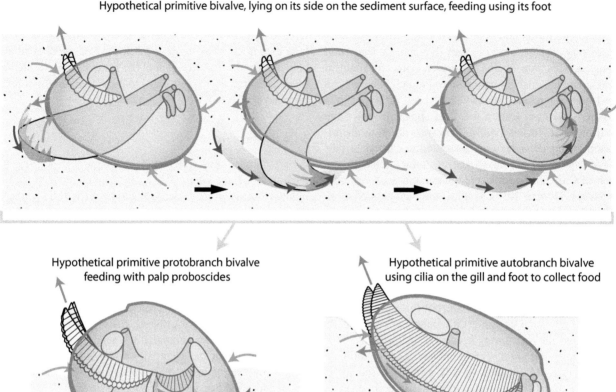

FIGURE 15.30 A scenario for the evolution of feeding in hypothetical primitive bivalves. Redrawn and modified from Morton, B., The evolutionary history of the Bivalvia, pp. 337–359 in J. D. Taylor (ed.), *Origin and Evolutionary Radiation of the Mollusca*, Oxford University Press, Oxford, UK, 1996.

In several groups, notably the Lucinidae, the gills are modified to house symbiotic chemosynthetic bacteria which provide some to all of the nutritive requirements of the host (see Section 15.4.4.1 and Chapter 5). A marked departure from the normal autobranch pattern is seen in septibranchs, with gill reduction and loss (see Section 15.8.6.1.2) associated with a carnivorous habit. This group feeds on small animals, mainly crustaceans, that are sucked into the mantle cavity or actively captured (see Section 15.8.6.1.2 for more details). Pectinoidean propeamussiids also capture and consume small crustaceans (Morton & Thurston 1989).

The galeommatoidean bivalve *Entovalva* lives inside holothurians and is attached to them by a large sucker-like foot. Its mode of feeding is not yet known, but diatoms in the gut suggests at least some suspension feeding, and the large surface area of the body might indicate absorption (Bristow et al. 2010). A somewhat similar species, *Austrodevonia sharnae*,

lives externally on a holothurian and is a suspension feeder (Middelfart & Craig 2004). The feeding habits of the deepsea montacutid bivalve '*Mysella*' (= *Kurtiella*) *verrilli* are unknown, but the mouth and oesophagus have become suctorial with hypertrophied musculature, and the palps are lost (Allen 2000). It may live suctorially in a soft-bodied host as does another montacutid, *Draculamya* (Oliver & Lützen 2011), that has also lost the palps and has a 'puncturing organ' possibly derived from the byssal apparatus.

15.4.3 LIFE HISTORY AND GROWTH

The sexuality of bivalves is discussed in Section 15.3.9. Sexual dimorphism is usually not externally apparent and when it does occur, is generally minor. In a few taxa, females are more swollen due to brooding chambers in the gills. A so-called *marsupium*, a special compartment for the embryos

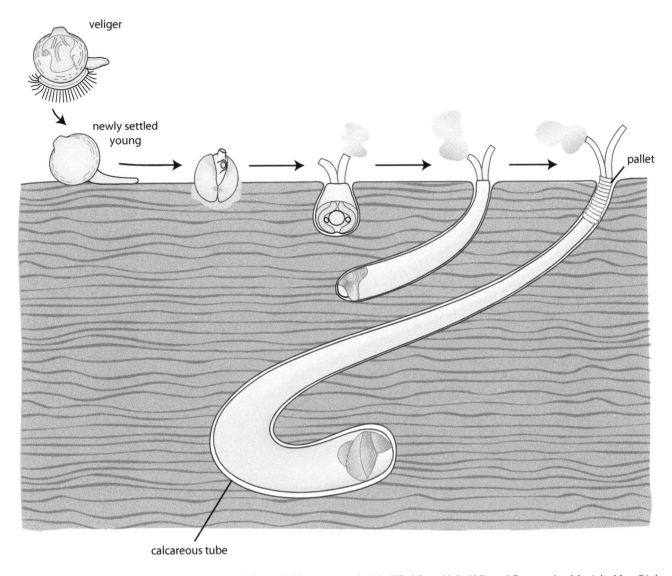

FIGURE 15.31   Settlement and growth of a typical teredinid – not to scale. Modified from Nair, N.B. and Saraswathy, M., *Adv. Mar. Biol.*, 9, 335–509, 1971.

in the shell, separated by a layer of shell material, is found in the carditids *Thecalia* and *Milneria* (see Chapter 8, Figure 8.18) (Yonge 1969; Coan et al. 2000) and in a species of *Acar* (Arcidae) (Oliver & Holmes 2004). Extreme sexual dimorphism in bivalves is very rare. Dwarf males occur in the teredinid genus *Zachsia* – shelled larvae are released from the female into the plankton where they eventually settle on living rhizomes of the seagrass *Phyllospadix*, but those that enter the mantle cavity of females become males (Turner & Yakovlev 1983). All other known cases of extreme sexual dimorphism in bivalves are a few instances within the Galeommatoidea. Some have dwarf males which, in a few cases, are permanently attached to the female and are little more than a testis (Morton 1976, 1981b; Jespersen & Lützen 2006) (see Chapter 8, Figure 8.20).

Protobranch eggs are large and yolky, while those of most autobranchs are small and not very yolk-rich. Fertilisation is usually external, but in brooding species it occurs in the mantle cavity. Cleavage patterns are spiral, and polar lobes and unequal cleavage patterns are present throughout the group. Autobranch embryos developing in the water column go through both trochophore and veliger ('spat') larval stages. The veliger stage is probably convergent rather than homologous with the somewhat similar gastropod veliger stage. The initial uncalcified shell grows laterally in two distinct lobes to envelop the body. Larval autobranch bivalves have a byssal gland that secretes one or a few threads that may assist with flotation while planktonic but later attach the juvenile to the substratum. This occurs in shallow burrowers (e.g., Cardiidae, where juveniles attach to rocks or algae) and even in many highly modified deep-burrowing forms like myids, solenids, pholads, and teredinids (Yonge 1962b) (Figure 15.31); *Mya* has relatively large (up to about 7 mm) juveniles living on the surface prior to burrowing (Kellogg 1899).

Spawning is often broadcast, with eggs and sperm being released freely into the water column. Typically, males release sperm which triggers egg release by the females with fertilisation occurring externally in the water column. In

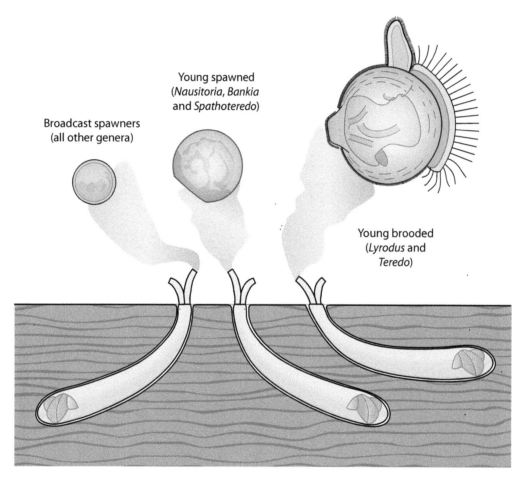

Broadcast spawners
(all other genera)

Young spawned
(*Nausitoria, Bankia*
and *Spathoteredo*)

Young brooded
(*Lyrodus* and
Teredo)

FIGURE 15.32   Life history strategies of teredinids. Modified from Turner, R.D. and Johnson, A.C., Biology of marine woodboring molluscs, pp. 259–301 *in* E. B. G. Jones & Eltringham, S. K. (eds.), *Marine Borers, Fungi and Fouling Organisms of Wood*, O.E.C.D., Paris,
1971.

many bivalves, the eggs are released into the mantle cavity
where they are fertilised by sperm drawn in with the inhalant water current. The embryos are held within the mantle
cavity, typically in brood chambers formed from pouches in
the gills (Figure 15.33 and Chapter 8, Figures 8.18 and 8.19),
the suprabranchial cavity, sometimes in special brood chambers, or, rarely, free in the mantle cavity as seen in many filibranchs, *Entovalva*, and *Teredo* (Mackie 1984). Reproductive
modes can vary, even in closely related taxa, as for example in
Teredinidae (Figure 15.32).

In *Ostrea* the eggs are first fertilised in the ctenidia which
later rupture, releasing the embryos into the mantle cavity
where they develop further. Many brooding bivalves release
their young as swimming veliger larvae while others retain
them longer and release them as juveniles. Ctenidial brood
pouches in eulamellibranchs vary in position – they can
occasionally be in the suprabranchial chamber only but are
mainly in the interlamellar spaces of either both pairs or one
pair of demibranchs (see Chapter 8, Figure 8.18). Specialised
brood pouches are found in some bivalves, notably in all but
the most basal freshwater Sphaeriidae where the pouches are
modifications of gill filaments (e.g., Heard 1965; Mackie et al.
1974; Heard 1977; Lee & Ó Foighil 2003) (Figure 8.19). In a
few bivalves, brood pouches are formed from the mantle, as

in the cardiid *Cerastoderma elegantulum* (Ockelmann 1958)
and the galeommatoid '*Montacuta*' *percompressa* (Chanley
& Chanley 1970).

Only a few bivalves are ovoviviparous – these include a few
species scattered through the major groups. The protobranch
Nucula delphinodonta has eggs within a sand-encrusted
egg capsule which is often attached to the shell (Drew 1901;
Scheltema & Williams 2009). At least two species of *Musculus*
(Mytilidae) produce an egg ribbon (Ockelmann 1958), while
the heterodonts *Phacoides pectinatus* (Lucinidae) (Collin
& Giribet 2010) and *Abra tenuis* (Tellinidae) produce benthic jelly-encased spawn (Gibbs 1984; Holmes et al. 2004).
Turtonia minuta (Turtoniinae, Veneridae) produces a firm
egg capsule containing several eggs (Oldfield 1964) (Figure
15.33).

All autobranch bivalves have one or more larval stages, usually a trochophore that becomes a veliger (see Chapter 8, Figure
8.35) in which the larval shells are formed. In the freshwater
Unionida, the veliger larva is modified as a specialised parasite
(a glochidium, lasidium, or haustorium – see Section 15.8.3.2.2,
and Figures 15.40 and 15.41) that attaches to fish. In protobranchs, the non-feeding larva is a specialised trochophore, the
pericalymma (see Chapter 8, Figure 8.31) and the larval shells
are developed at that stage (Zardus 2002). In autobranchs, the

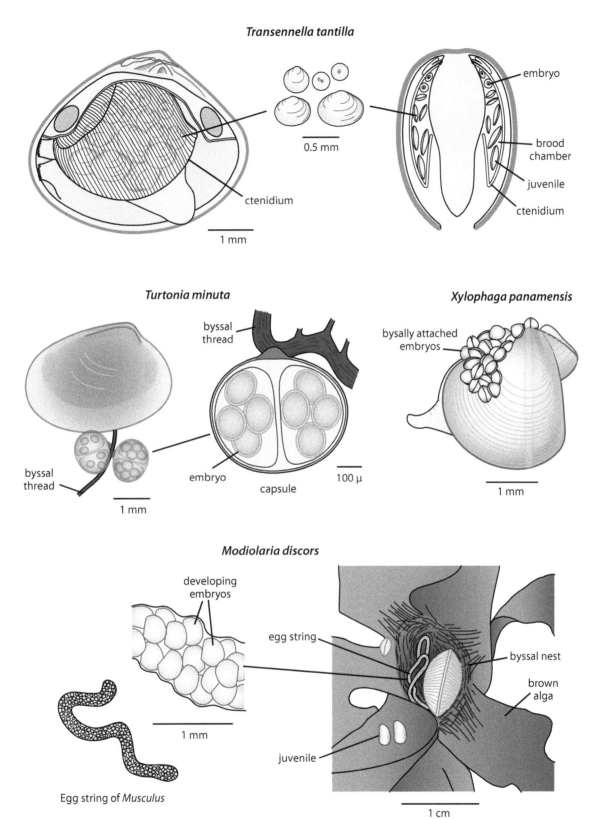

FIGURE 15.33 Examples of brood protection and spawn in bivalves. Redrawn and modified from the following sources: *Transennella* (Hansen, B., *Videnskabelige Meddelelser fra Dansk naturhistorisk Forening*, 115, 313–324, 1953); *Turtonia* (Oldfield, E., *Proc. Malacol. Soc. Lond.*, 31, 226–249; Oldfield, E., *Proc. Malacol. Soc. Lond.*, 36, 79–120, 1964); *Xylophaga* (Knudsen, J., *Galathea Report*, 5, 163–209, 1961); *Modiolaria* (Thorson, G., *Meddelelser om Grønland*, 100, 1–71, 1935).

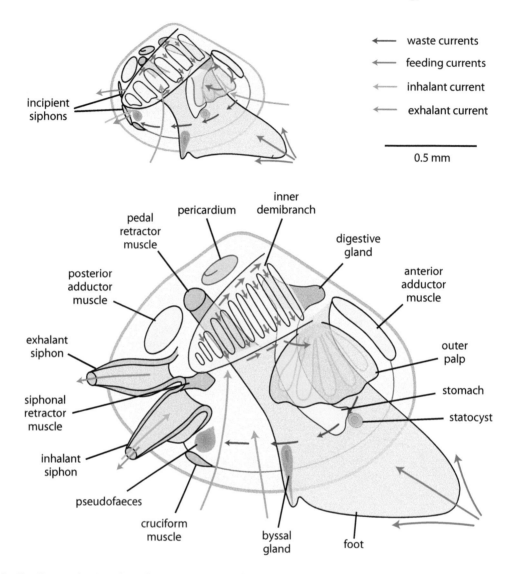

FIGURE 15.34 Feeding mechanisms in early (above) and later (below) post-larval stages of the tellinid *Macoma balthica*. Redrawn and modified from Caddy, J.F., *Can. J. Zool.*, 47, 609–617, 1969.

larvae may be planktotrophic with a feeding veliger stage, lecithotrophic with a (usually short) non-feeding veliger stage, or may be brooded. In some sphaeriids, the brooded larvae may be represented by several developmental stages while in others they are synchronous (e.g., Guralnick 2004).

Growth is rapid in post-larval protobranchs but decreases with age, though rates may not necessarily be slow, especially in continental shelf species. Life spans of protobranchs are commonly one to two decades, but deep-sea representatives may grow more slowly and live longer (Zardus 2002).

In autobranch bivalves other than direct developers, metamorphosis occurs when the pediveliger settles (Figure 15.31). However, a second metamorphosis occurs in some watering pot shells (*Brechites* and *Humphreyia*) at the end of a mobile juvenile phase where the individual becomes immobile, cements to the substratum, and transforms into its highly unusual adult form (Morton 2002b, a) (see Section 15.8.6.1.2).

Post-larval bivalves often engage in pedal feeding (Figure 15.34), assumed to be the ancestral mode (see Section 15.4.2), before switching to ctenidial feeding.

15.4.4   ASSOCIATIONS

15.4.4.1   Bacterial and Algal Symbionts

Solemyoideans, lucinids, and the vent- and seep-inhabiting vesicomyids all inhabit reducing environments and derive their nutrition from endosymbiotic chemosynthetic bacteria which live in their modified gills (Roeselers & Newton 2012) (see also Chapters 5 and 9). The acquisition of bacterial endosymbionts by the rather aberrant solemyoideans has attracted much interest as the bacteria provide most or all of the necessary nutrition to the bivalve, and this has resulted in their much-reduced to absent gut (Reid 1990). More recently, bacterial symbionts have also been described from the gills of the possibly related manzanellids (Oliver & Taylor 2012). Bathymodiolinae (Mytilidae) also have bacterial symbionts (Duperron 2010) while among heterodont bivalves, gill-inhabiting bacteria are found in Lucinidae (Taylor & Glover 2010), the vent- and seep-inhabiting *Calyptogena* (Vesicomyidae) (Glossoidea) (Krylova & Sahling 2006), some Thyasiridae (Taylor et al. 2007a), in the galeommatoidean *Syssitomya*

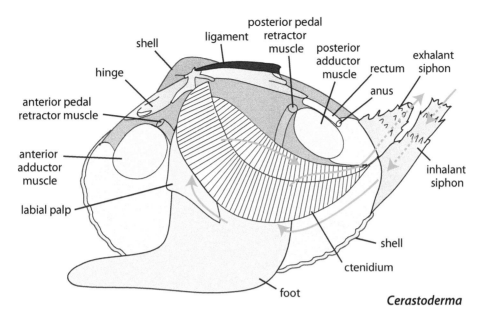

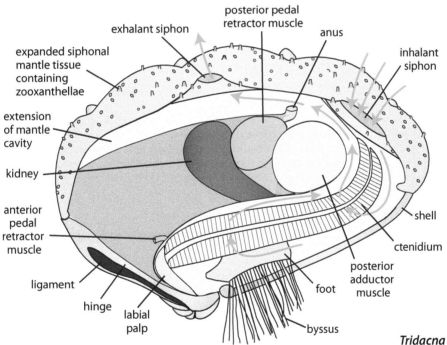

FIGURE 15.35 Comparison of *Tridacna* with an unmodified cardiid, *Cerastoderma*. The large area of mantle exposed dorsally that houses the zooxanthellae in *Tridacna* is derived from tissue around the posterior siphons in *Cerastoderma*. The green arrows show water currents. Redrawn and modified from Yonge, C.M., *Proc. Zool. Soc. Lond.*, 123, 551–561.

(Oliver et al. 2012), and in the cyamioidean *Atopomya* (Oliver 2013).

Symbiotic relationships with zooxanthellae ('photo-symbiosis') in bivalves were first discovered in the giant clam *Tridacna* (Yonge 1936) and this association has been reviewed by Savazzi (2001) and Kirkendale and Paulay (2017). The body of *Tridacna* has been rotated within the shell to maximise the area of zooxanthellae-bearing mantle tissue exposed to sunlight (Figure 15.35). The animal is byssally attached near the umbo, while the zooxanthellae-bearing fused mantle and its siphonal openings, which were

originally posterior, are exposed on top between the widely gaping valves. Zooxanthellae symbionts were subsequently discovered in some small tropical cardiids (Kawaguti 1983; Jones & Jacobs 1992; Kirkendale 2009), and this relationship is suspected in living and fossil members of a few other groups of cardiids based on their specialised shell features (Seilacher 1990; Kirkendale & Paulay 2017). Notable among those are the shell 'windows' in the unusually flattened cardiid, *Corculum* (Watson & Signor 1986; Carter & Schneider 1997). In these relationships, the 'algae' can be farmed in the mantle directly exposed to sunlight or indirectly in tissues

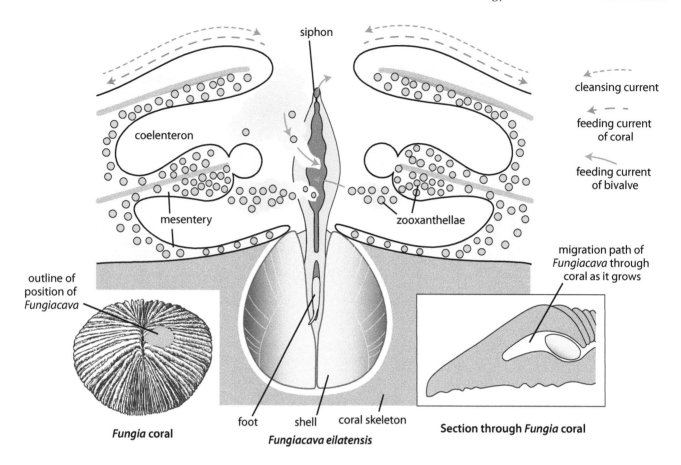

FIGURE 15.36 Diagram showing the mytilid *Fungiacava eilatensis* embedded in the coral *Fungia*. The left lower insert shows the location of the bivalve in the coral, and the right insert shows it in its burrow. The solid blue arrows show feeding currents, the dotted blue arrows incoming water currents and the dotted red arrows are cleansing currents. Main diagram redrawn and modified from Goreau, T.F. et al., *J. Zool.*, 160, 159–172, 1970; *Fungia* from William and Robert Chambers Encyclopedia (1881); and lower right diagram is redrawn and modified from Savazzi, E., *Palaont. Z.*, 56, 165–175, 1982.

below a translucent shell. A report of zooxanthellae in the trapeziid *Fluviolanatus subtortus* (Morton 1982c) is erroneous (Kirkendale & Paulay 2017). The isotopic signatures of shells of symbiont-bearing species differ when compared with 'normal' bivalves, but there are limits on how these can be used in fossils, due to limitations on the preservation of the organic matrix (Dreier et al. 2014).

Some of the extinct reef-forming rudists (see Section 15.8.5.4) may have harboured symbiotic 'algae' (Vogel 1975; Seilacher 1998; Kirkendale, 2017), perhaps explaining aspects of their large size and strange morphologies, although this view is not universally accepted (e.g., Lewy 1995). It is possible that diverticula within the shell contained folds or tubules of tissue with 'algal' symbionts and that light was able to penetrate thin shell structures in at least a few rudist taxa (e.g., Gili et al. 1995).

The Permian Alatoconchidae were the largest Paleozoic bivalves, reaching up to one metre in size, and their large spreading valves are thought to have harboured symbiotic photosynthetic organisms (Isozaki & Aljinović 2009).

Further details regarding photosymbiosis in bivalves and other molluscs are given in Chapter 5.

15.4.5 RELATIONSHIPS WITH OTHER ANIMALS

Various commensal and parasitic relationships of other animals with bivalves are outlined in Chapter 9. Some bivalves are themselves in symbiotic relationships with other animals. Notably, the very diverse galeommatoideans are associated in various ways with other animals such as burrowing anemones, crustaceans, polychaetes, sipunculans, or holothurians (e.g., Boss 1965; Morton 1988; Goto et al. 2012; Li et al. 2012, 2016).

Some bivalves have an intimate association with living coral (see also Section 15.4.1.4), a relationship requiring larval and adult adaptations. Experiments have shown that, unlike other veliger larvae, the larvae of *Lithophaga bisulcata*, a species mainly associated with the coral *Stephanocoenia*, are not harmed by the coral nematocysts of that species but are stung by other coral species (Scott 1988). One of the most derived coral-boring associates is a globular mytilid, *Fungiacava eilatensis* (Figure 15.36). This species lives in a cavity which opens into the coelenteron of a fungiid coral where the long siphon collects not only phytoplankton from the seawater but zooxanthellae extruded from the host coral (Goreau et al. 1970).

Other coral associates include the pectinid *Pedum spondyloideum* which lives embedded in massive coral colonies, particularly *Porites* but also other genera. The relationship may be a mutualistic one, as their presence has been shown to reduce the effects of predation by the coral-eating crown of thorns starfish *Acanthaster planci*, possibly by the expulsion of jets of water (DeVantier & Endean 1988). The *Pedum* larva and adults are both immune to the corals' defences (Yonge 1967a). This pectinid often occurs in high densities and uses chemical means to interact with other individuals and with their host corals, as well as in bioerosion (Kleemann 1990a).

15.4.5.1 Predators of Bivalves

One reason bivalves are often considered ecologically important is that they are a very significant source of food for a wide range of predatory animals, notably many fish, some asteroids, many gastropod whelks, various shorebirds, some marine mammals (e.g., walruses, sea otters, seals) (see Chapter 9), and, of course, humans (see Chapter 10).

Bivalves have evolved defensive strategies such as thickened shells, strong attachment (byssus or cement), infaunal habits, and, particularly, deep-burrowing. Some burrowers such as solenids can burrow rapidly to escape predators or, as with the shallow-burrowing cardiids, can, when disturbed, leap away using their foot. A few, such as some scallops and limids, can swim away from a predator. With burrowing bivalves, the siphons are the part of the animal most exposed to predation. These structures are sensory and sometimes even equipped with eyes and usually can be rapidly withdrawn. If they are damaged or partly removed, they can be regenerated. A few such as the limid *Limaria fragilis* (Gilmour 1967; Morton 1979b) and the galeommatid *Galeomma* (Morton 1973) have long mantle tentacles that produce a noxious secretion when damaged or autotomised.

15.5 ECONOMIC VALUES

Bivalves often play a very important ecological role in many marine and some freshwater habitats because of the body size, abundance, and biomass of some taxa (see also Chapter 9). Despite being a much smaller group, they have an economic value considerably greater than gastropods (see Chapter 10). Several pteriomorphians, in particular oysters, scallops, mussels, and pearl oysters, are very significant in aquaculture. Many pteriomorphians (mainly Mytilidae, Ostreidae, and Pectinidae) are very important as human food items and comprise the great bulk of bivalves consumed, although some arcoideans, solenids, myids, cyrenids, venerids, mesodesmatids, mactrids, cardiids, and hiatellids (*Panopea*) are also fished and/or farmed in reasonably large numbers and are significant food items in various parts of the world, notably in Asia and Europe (see Chapter 10). Commercial applications inspired by the byssus are also likely (e.g., Harrington et al. 2018).

Other issues about bivalves such as over-fishing and other conservation concerns, invasive and pest species, and human health issues are discussed in Chapter 10.

15.6 DIVERSITY AND FOSSIL HISTORY

15.6.1 THE FIRST BIVALVES

The origin of bivalves is discussed in Chapter 13. Their sister group is usually thought to be the extinct rostroconchs (e.g., Pojeta & Runnegar 1976; Pojeta 1978; Waller 1998; Carter et al. 2000). The first undoubted bivalves appeared in the Cambrian (Series 2) where they are represented by the genera *Fordilla* and *Pojetaia* (Pojeta 2000), the former co-occurring with the earliest rostroconch, *Ischyrinia* (Landing 1991). *Pojetaia* extends into the Miaolingian when a few other bivalve genera also appear, with one (*Tuarangia*) having taxodont teeth. How these Cambrian bivalves are related to other Paleozoic bivalves is uncertain. Various scenarios have been suggested – for example, that *Pojetaia* gave rise to 'palaeotaxodont' (protobranch) bivalves (Praenuculidae) while *Fordilla* may have given rise to mytilidans (isofilibranchs) (Runnegar & Bentley 1983). Carter et al. (2000) undertook a cladistic analysis of early bivalves and found that both these genera were basal to all others but represented a paraphyletic grade. They differ from all later bivalves in having larger nacre tablets, different details of their pedal and adductor muscles, and a discontinuous pallial line. However, largely because there is a gap in the bivalve fossil record between the Miaolingian and the Lower Ordovician, the relationships of these Cambrian bivalves remain obscure (see Chapter 13). There have also been a variety of other Cambrian taxa mistakenly identified as bivalves (see Pojeta 2000 and Schneider 2001 for details). For example, one fossil thought to be a Miaolingian bivalve, *Lamellodonta* (Lamellodontidae) (e.g., Newell 1969), has been shown to be a brachiopod (Havlicek & Hriz 1978).

All the major lineages of bivalves appeared in the Lower Ordovician (see Section 15.6.2) and must have been derived from Cambrian bivalves of the protobranch grade. Given the major differences between protobranchs and autobranchs (see Section 15.7.1), some workers (e.g., Waller 1990) have argued that extant protobranch bivalves cannot be considered similar to the ancestor of autobranch bivalves.

15.6.2 THE EVOLUTIONARY RADIATION OF BIVALVES

In marked contrast to their slow start in the Cambrian, their diversification from one or more Cambrian protobranch grade lineages in the Ordovician was the most important in the history of bivalves. The most significant radiation occurred in the Lower Ordovician and may well have coincided with the evolution of the autobranch gill which enabled bivalves to tap into the rich suspended food supplies developing in the water column, significantly expanding their potential to colonise many habitats previously inaccessible to them.

The Lower Ordovician bivalves quickly adopted infaunal to epifaunal habits and increased in size. These first autobranch bivalves were previously largely grouped in a palaeoheterodont Modiomorphacea (Newell 1969), but that grouping has since been shown to be para- or polyphyletic with representatives of basal members of several major lineages (e.g., Carter et al. 2000; Schneider 2001). Now they are regarded as having evolved into about 17 families which included the stem members of all the major bivalve lineages recognised today (Pojeta 1978; Morton 1996; Cope & Babin 1999; Carter et al. 2000; Schneider 2001). These early bivalves mostly lived in shallow-water sediments and were restricted to Gondwana (Cope & Babin 1999). However, by the Middle Ordovician they had become more widespread, and by the Upper Ordovician many more epifaunal groups had evolved, notably the heterodont modiomorphoids and the pteriomorphians. This burst of Paleozoic bivalve evolution saw the diversification of pteriomorphians and archiheterodonts which quickly occupied most shallow marine environments. They may have displaced the protobranch bivalves which moved into deeper water. The infaunal lineages underwent major radiations during the Mesozoic and Cenozoic, and siphonal development resulted in the independent evolution of several groups of deep burrowers and borers (Stanley 1968), a habit that, as already noted, afforded enhanced protection from predators.

The early Paleozoic autobranch bivalves were all suspension-feeding, epifaunal, or shallow infaunal forms. True infaunal siphonate taxa appeared later in the Paleozoic and became more abundant following the extinction event at the end of the Permian. Their diversity, along with that of the epifaunal taxa, greatly increased through the Mesozoic, but the major extinction event at the end of the Cretaceous markedly reduced the variety of epifaunal bivalves, although the siphonate infaunal taxa continued to diversify during the Cenozoic. The diversity of protobranch bivalves remained fairly stable throughout bivalve history, and they were relatively unaffected by extinction events.

Bivalves are ideal subjects for evolutionary studies because changes in their shell reflect changes in the mantle cavity, mantle, and musculature and can be readily traced in fossils using the muscle and pallial line scars.

Besides the changes to the gills, another important adaptation in the early autobranchs was the development of a byssus in the larvae. This larval character was retained in some adult lineages, allowing the external surfaces of hard substrata such as rock to be colonised, and resulted in new adaptive radiations (Morton 1996). This enabled several lineages to move from an originally shallow, infaunal existence to an epibenthic habit and, as described above, had significant consequences for bivalves. Switching between these two habitats has occurred in the evolution of various bivalve lineages, and several living families have examples of closely related infaunal and epifaunal species (e.g., Figure 15.38). Another form of attachment, cementation (see Section 15.4.1.2), first evolved in Paleozoic oysters and scallops which, as a consequence, lost their byssal apparatus.

Attachment drove radical changes in the body orientation of many pteriomorphian bivalves, resulting in anisomyarian and monomyarian taxa (see Section 15.4.1.2.1). These trends commenced early in bivalve evolution, with the first byssally attached adult forms being present in the Lower Ordovician

(Pojeta 1978). In byssally attached forms, there is a tendency for a reduction in the size of the anterior end and an enlargement of the posterior end, along with associated changes in the relative size of the adductor muscles. This anterior reduction culminated in the loss of the anterior adductor in some lineages (see Section 15.4.1.2.1). Byssal attachment or cementation allowed bivalves to live successfully in high energy coastal intertidal and shallow subtidal situations.

In the heterodonts, *Tridacna*, which may have first appeared in the Upper Cretaceous, has undergone a major reorientation of its body to maximise the exposure to sunlight of the expanded mantle containing symbiotic zooxanthellae (see Section 15.4.4.1, Figure 15.35, and Chapter 5). Some other bivalves, notably the protobranch Solemyidae and Manzanellidae, some Mytilidae and Thyasiridae, and all members of the Lucinoidea and Vesicomyidae, have symbiotic autotrophic bacteria housed in their gills which provide most, or sometimes all, of their nutritional requirements (see Section 15.4.4.1 and Chapter 5). While most other autobranch bivalves retained their suspension-feeding habits, one large group, the Tellinoidea, successfully adapted this feeding mode to take in surface deposits using the inhalant siphon, an innovation that resulted in a major radiation of the group. In addition, a few taxa became microcarnivores (see Section 15.4.2).

Cemented true oysters first appeared in the Upper Triassic while two other unrelated cementing families, Chamidae and Cleidothaeridae, appeared in the Upper Cretaceous and Miocene respectively.

Anomalodesmatans contain the bizarre 'watering pot shells' of the Clavagelloidea (see Section 15.8.6.1.2), that possess adventitious tubes and first appeared in the Upper Cretaceous. Some are attached to hard substrata, some even resemble vermetid gastropods, while others are embedded in soft sediment.

The Paleozoic radiation of bivalves was depleted by the Permo-Triassic mass extinction, but several lineages survived and radiated in the Mesozoic. Some groups such as lucinoideans and trigonioideans differentiated extensively, their habits expanding to include attached forms, deep burrowers, and borers. The Anomalodesmata were never very speciose but diversified markedly in habits and morphology, producing not only the bizarre clavagelloideans (Figure 15.10) mentioned above, but also the highly modified septibranchs (see Section 15.8.6.1.2) (Morton 1996).

Many of the modern bivalve groups arose and radiated during the latter part of the Mesozoic and Cenozoic, mainly in shallow marine environments. These included diverse shallow burrowers that often strengthened their shells with ribbing, deeper burrowers in soft sediments with streamlined, smooth shells, and burrowers of harder substrata (clay, rock, coral, and wood) (see Section 15.4.1.4). Attached forms (see Section 15.4.1.2) also diversified during this period, with some pectinids and limids becoming free-living and capable of short bursts of swimming.

While much bivalve evolution and radiation was occurring in shallow water, two other environments were also significant. In the deep sea, protobranchs, two pteriomorphian lineages, the arcoideans and propeamussiids, and the anomalodesmatan septibranchs, underwent substantial marine radiations.

Bivalves also invaded freshwater habitats several times through their history, as discussed in the next section.

15.6.2.1 Freshwater Bivalves

Although there were earlier incursions into freshwater habitats, the major non-marine bivalve radiation was undertaken by the unionidans (see Section 15.8.3.2) and commenced in the Triassic, their success possibly related to the evolution of their parasitic larvae. The exclusively freshwater sphaerioids first appeared in the Jurassic when the group was represented by the Pseudocardiniidae, a freshwater group, while the Sphaeriidae first appeared in freshwater deposits in the Cretaceous.

A few mytilids and arcids live in fresh water, and cyrenids and dreissenids also have marine and freshwater taxa. The Dreissenidae, which includes the invasive freshwater 'zebra mussel' *Dreissena*, first appeared in marine rocks in the Cretaceous of North America. The Jurassic Ferganoconchidae is a possible member of the Dreissenoidea and inhabited freshwater habitats. The mainly brackish-water cyrenoideans first originated in the Jurassic and invaded freshwater habitats, probably in the Paleogene, with the chiefly Asian genus *Corbicula* being particularly successful since it appeared, presumably in the Upper Cretaceous.

Another brackish-freshwater incursion has been made by a clade in the Corbulidae comprising the genera *Lentidium*, *Erodona*, and *Potamocorbula* (Hallan et al. 2013).

Other incursions into fresh water occurred in extinct groups of uncertain affinities although most have been included at some time, even if tentatively, in the Unionida (e.g., Cox et al. 1969b). However, these taxa existed much earlier than the members of what is currently thought to comprise the unionoidean freshwater radiation. These earlier freshwater bivalves included the pteriidan Myalinidae and the medium-sized euheterodont Anthracosiidae, both of which inhabited Carboniferous forest swamps and are found in coal deposits, particularly in Britain. The large-sized members of the heteroconch Amnigeniidae lived during the Upper Devonian and Carboniferous and resembled unionoideans such as *Anodonta*. The related Palaeomutelidae are medium-sized non-marine bivalves previously associated with both protobranchs and autobranch bivalves, although more recently classified as heterodonts in the order Actinodontida, along with the Amnigeniidae (see Appendix). They are abundant in non-marine beds in the Lopingian (Permian) of Eastern Europe and had a shell similar to that of *Unio* and a unique type of taxodont hinge (Cox et al. 1969b; Silantiev 1998). The euheterodont Prilukiellidae (= Microdontidae) is comprised of a few, small-sized, Permian freshwater taxa with 'microdont' dentition.

Several groups of mainly marine bivalves also have one or a few freshwater species (see Table 15.1).

TABLE 15.1

Extant Bivalve Families Containing Freshwater Species

Order	Family	Genera	Species
Arcida			
Arcoidea	*Arcidae	1	4
Mytilida	*Mytilidae	4	6
Mytiloidea			
Unionida	#Etheriidae	1	1
Etherioidea			
	#Iridinidae	6	41
	#Mycetopodidae	12	39
Unionoidea	#Unionidae	142	620
	#Margaritiferidae	3	12
Hyrioidea	#Hyriidae	17	83
Anomalodesmata			
Pandoroidea	*Lyonsiidae	1	1
Cardiida	*Cardiidae	2	5
Cardioidea			
Cyrenoidea	Cyrenidae	3	6
Sphaerioidea	#Sphaeriidae	8	196
Dreissenoidea	Dreissenidae	2	4
Tellinoidea	*Donacidae	2	2
	*Solecurtidae	1	1
Solenida			
Solenoidea	*Pharidae	2	2
Pholadida			
Myoidea	*Corbulidae	1	1?
Pholadoidea	*Teredinidae	1	1

Modified from Bogan (2008)

Those with mainly marine species are marked with *; those exclusively freshwater are marked #

15.6.3 PATTERNS OF BIVALVE DIVERSITY

Bivalve diversity gradients have been better studied than those of other groups of molluscs (see reviews in Crame 2000a, b, c; Roy et al. 2009b). They are influenced by various factors such as sea temperature, primary productivity, and history, resulting in both latitudinal and longitudinal gradients (see Chapter 9, Figure 9.18). There is a general trend of higher diversity in the tropics, although there are local anomalies. Crame (2000c) concluded that various factors, including fluctuating sea level changes over the last few tens of millions of years, led to the evolution of new taxa in marginal tropical regions which were then locally concentrated in 'hotspots'. A notable example is Australia which accumulated tropical faunas, in addition to the older temperate ones, during its movement north during the Cenozoic.

Not all bivalve groups show the same diversity patterns. In a study on eastern Pacific bivalves, Jablonski et al. (2000) found that suspension feeders and autobranch deposit feeders conformed to the general bivalve pattern with an increase in diversity towards the tropics, but epifaunal and infaunal taxa do so at different rates, and protobranchs do not exhibit a diversity gradient, despite species turnover.

15.7 THE PROTOBRANCH BIVALVES

Protobranchs are a diverse group of mostly small-sized, shallow-burrowing marine bivalves often found in the deep sea, where they are sometimes abundant. Some taxa are also found in shallow coastal and estuarine environments, including the so-called nut shells (Nuculidae) and the 'date shells' (Solemyidae).

With few exceptions, protobranchs are deposit feeders and digest food extracellularly. These features, together with their lecithotrophic larval development, make them well-adapted to cold and oligotrophic habitats (Zardus 2002).

Several significant characters separate protobranchs from other bivalves, and these were presumably shared by their common ancestor. Their large labial palps have long extensions employed in deposit feeding, and their simple gills have triangular filaments used primarily for respiration. This so-called protobranchiate gill is morphologically similar to the plesiomorphic conchiferan ctenidium (see Chapter 4), and the foot sole is flattened and fringed with short processes. A more complete list of distinguishing characters is given in Table 15.2. Many have a hinge composed of numerous similar small teeth (taxodont condition). The interior of the shell is nacreous in some Solemyoidea and Nuculoidea, groups that lack siphons, and water currents pass from anterior to posterior. In contrast, the nuculanoideans lack nacre, have posterior mantle fusion, and form posterior siphons through which inhalant and exhalant water currents pass (Figure 15.37).

Bacterial symbionts that reside in the gill have been acquired by solemyoideans and the (possibly related) manzanellids (see Chapter 5 and Section 15.7.2).

15.7.1 PHYLOGENY AND CLASSIFICATION

The earliest protobranch bivalves are considered ancestral to all other groups. The name Protobranchia was replaced by an extended concept of Palaeotaxodonta in schemes put forward by some palaeontologists (Cox et al. 1969b; Pojeta 1987; Cope 1996b, 1997b), but most authors have used Protobranchia as the name for this group (e.g., Nevesskaja et al. 1971; Boss 1982; Purchon 1990; Waller 1990, 1998; Prezant 1998; Coan et al. 2000; Zardus 2002; Bieler & Mikkelsen 2006; Bouchet et al. 2010; Carter et al. 2011). Carter et al. (2011) recently split bivalves into two 'grades', Euprotobranchia for the Cambrian taxa and Eubivalvia for all other bivalves, including the protobranchs, and this classification has been adopted in the Appendix.

Yonge (1959a, p. 214) argued that 'there is clearly greater affinity between the Nuculidae and Nuculanidae than between these two and the Solemyidae'. The two groupings were treated as orders or even subclasses, for example by Newell (1969), in the *Treatise of Paleontology*, where they were called Palaeotaxodonta and Cyrtodonta, the former including the nuculoideans and the nuculanoideans, the latter the Solemyoidea and the extinct Praecardioida. A slight modification of that classification was followed by Zardus (2002), who

TABLE 15.2

The Main Features Common to Protobranch Bivalves Contrasted with Those in Autobranch Bivalves

Character	Protobranchia	Autobranchia
Foot	With well-developed sole surrounded by papillae	Sole narrow or absent and papillae absent, except in *Neotrigonia*
Pedal muscles	Often five to six pairs	Usually two to three pairs
Ctenidia	Small (larger in solemyids), posteriorly located, and with triangular filaments used primarily for respiration	Large, with W-shaped filaments and extending along most of the body; secondarily reduced or absent in septibranchs. Suspension feeding the norm
Labial palps	Large, with palp proboscides (secondarily reduced or absent in solemyids). Involved in deposit feeding	Small, lacking proboscides in autobranchs; secondarily modified and muscular in septibranchs. Involved in food processing
Stomach	Muscular, triturating, with small sorting area and chitinous girdle, secondarily reduced to absent in solemyids; digestion extracellular	Large with extensive sorting area and gastric shield; convergent with protobranchs in septibranch taxa; digestion mainly intracellular
Style sac	With protostyle	With crystalline style
Nervous system	With visceral ganglia smaller than, or about equal to, cerebral ganglia and located anterior to posterior adductor muscle	Visceral ganglia larger than cerebral ganglia and ventral to posterior adductor muscle
Adoral sense organ	Present	Absent
Larva	Pericalymma; non-feeding	Veliger; non-feeding or feeding
Haemocyanin	Present	Absent

(Data mainly from Yonge 1939a; Villarroel & Stuardo 1998; Simone 2009).

recognised two orders, Solemyoida and Nuculoida, the latter including the nuculanoideans.

Purchon (1956, 1959) recognised the 'Gastroproteia' based on a unique stomach morphology in protobranch bivalves. This concept was refined by Villarroel and Stuardo (1998) who recognised three subtypes within Protobranchia: one in nuculids with three or four ciliary sorting areas and a wide extension of the typhlosole; another in nuculanids and malletiids with three ciliary sorting areas, a small extension of the minor typhlosole, and a larger gastric shield than in nuculids; and the third in solemyids and manzanellids that lacked distinct sorting areas and typhlosoles.

Phylogenetic studies in the last decades have produced conflicting results about the monophyly of protobranch bivalves and the main groupings within them. There is currently no

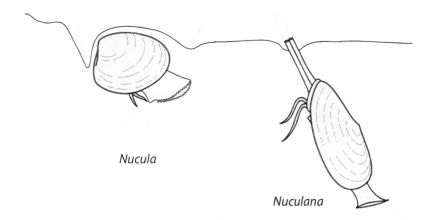

Nucula

Nuculana

FIGURE 15.37   Two protobranch bivalves in life position, showing the main water currents and the palp proboscides. Redrawn and modified from Stanley, S. M., *J. Paleont.*, 42, 214–229, 1968.

clear resolution, as indicated below in our summary of the main findings in recent years.

A basal monophyletic Protobranchia was recovered by Sharma et al. (2012) and Smith et al. (2011), the latter using EST[5] data. In contrast, some recent morphological (Waller 1998; Carter et al. 2000; Simone 2009) and several molecular studies (Giribet & Wheeler 2002; Giribet 2008a; Plazzi et al. 2011) recovered the protobranch bivalves as a paraphyletic group, and the groupings differed from those of earlier classifications. In the most recent comprehensive molecular analysis of protobranchs by Combosch et al. (2017) they were not monophyletic, with solemyids clustering with a scaphopod and nucinellids being at the base of the autobranch bivalves. This result, however, was not consistent and requires further analysis although we reflect it in our classification by removing these two groups from their traditional grouping with the Nucleiformii.

The Nuculiformii uniquely have the gill filaments in each demibranch arranged opposite each other, resulting in a unique pattern of ciliary currents (see Section 15.3.6.1). The Nuculanoidea was considered the probable sister group to the rest of the autobranch bivalves (Giribet 2008a), and a similar result was obtained in the morphological analysis of Simone (2009), although an earlier morphological analysis recovered a monophyletic Protobranchia (Waller 1998).

In some molecular analyses, nuculanoideans appear within the autobranchs with either archiheterodonts (Giribet & Distel 2003) or Pteriomorphia (Plazzi & Passamonti 2010; Plazzi et al. 2011). This result led Plazzi et al. (2011, p. e27147) to make the highly unlikely suggestion that, in their tree, 'the eulamellibranch condition seems to have reverted to the ancestral protobranchiate state in the superfamily Nuculanoidea'. Similar results were obtained by Wägele et al. (2009) and Wilson et al. (2010). As in the most recent classifications (Bieler & Mikkelsen 2006; Bouchet et al. 2010; Carter

et al. 2011; Bieler et al. 2014), we recognise the protobranchs as a subclass and the monophyly that it assumes.

While it is likely that the ancestors of autobranch bivalves are included among the known extinct taxa (either included within protobranchs or not), the substantive number of characters that differentiate the two groups suggest it is unlikely that any living protobranch lineage gave rise to autobranch bivalves (see Section 15.7 and Table 15.2).

There are several important characters (see Table 15.2) that appear to group living protobranchs into a probable monophyletic assemblage, comprising the two groups recognised by Giribet (2008a) and others (see above). However, those groups are not supported in both the most recent molecular analyses (Smith et al. 2011; Sharma et al. 2012), where solemyids are sister to both nuculanids and nuculids. Consequently, we do not recognise the nuculoidean + solemyoidean grouping despite significant shared characters, notably the arrangement of the ctenidial filaments (a character not used by Simone [2009]), that define that group. The Silurian *Janeia* has shell features in common with both solemyoideans and nuculoideans (Liljedahl 1984).

15.7.2   BIOLOGY AND ECOLOGY

Protobranchs are mostly deposit feeders in soft sediments, using the elongate processes on their large labial palps to gather food. However, the degree to which they are selective in their diet is difficult to determine. The Solemyidae inhabit reducing environments, including hydrocarbon seeps, and farm endosymbiotic chemosynthetic bacteria in their gills from which they derive most of their nutrition (e.g., Stewart & Cavanaugh 2006). As a consequence, they have a much reduced or absent gut and labial palps.

All protobranchs have a lecithotrophic larval stage (see Chapter 8), the pericalymma, which is unique among bivalves. This type of larva is found in a few other groups, notably sipunculans and aplacophorans, and may well have arisen independently in each (Nielsen 2001). It is thus unclear

---

[5] Expressed Sequence Tag.

if it is plesiomorphic for bivalves or a synapomorphy of protobranchs.

The pericalymma remains in the plankton for only a short time and presumably has low dispersal ability. It settles and directly develops into the juvenile, there being no veliger stage.

Solemyids live in a U-shaped burrow in anoxic mud but can swim by forcibly ejecting water from the mantle cavity with piston-like movements of the foot. The calcareous material in their shell is much reduced, the shell being formed mainly from periostracum.

15.7.3 DIVERSITY AND FOSSIL HISTORY

Bivalves, that are assumed to be anatomically similar to protobranchs, originated in the early Cambrian (Series 2), but modern groups arose later with the nuculoideans and nuculanoideans known from the Ordovician and solemyoideans appearing later, in the Devonian (Cox et al. 1969b). Today protobranchs are found throughout the oceans of the world but are most diverse in the deep sea. There are about a dozen mostly small-sized families usually recognised among living protobranchs, which are arranged in five superfamilies (see Appendix).

The predominance of protobranchs in the deep-sea benthos may result from a deep-sea origin or, as already noted, their displacement from shallow waters by autobranch bivalves.

15.8 THE AUTOBRANCH BIVALVES

The divergence of autobranch and protobranch bivalves occurred early in bivalve evolution and was marked by several fundamental changes as noted above and summarised in Table 15.2, mainly involving, or the result of, radical changes in the morphology of the gills in concert with a shift from deposit feeding to suspension feeding (see Section 15.3.6.1).

It appears likely that the autobranch ancestor divided into two main lineages, the pteriomorphians and heteroconchs, which had separate evolutionary histories for most of their existence. However, Cope (1996b) saw links between the hinges of some early 'palaeotaxodonts' (i.e., protobranchs) and early palaeoheterodonts such as *Glyptarca* and Actinodontida which he considered ancestral to pteriomorphians and heterodonts. He argued that another palaeoheterodont group, the Modiomorphoida, gave rise to the Anomalodesmata. This idea that Pteriomorphia is derived is also supported by a few molecular phylogenies but gains little or no support from the fossil record, comparative anatomy, or most molecular phylogenies.

We accept that the autobranch bivalves can be divided into two major groups, Pteriomorphia and Heteroconchia, and we provide an overview below.

15.8.1 PTERIOMORPHIA

The pteriomorphian bivalves are morphologically and taxonomically very diverse and include some of the most important commercial and familiar bivalves such as scallops (Pectinidae), oysters (Ostreidae), pearl oysters (Pteriidae),

mussels (Mytilidae), and arcs (Arcidae). Recent classifications of this group differ from each other in details of higher classification, but the diversity reflected in their classifications, at least for living pteriomorphian bivalves, is similar. For example, Bouchet et al. (2010) recognised five orders of living pteriomorphian bivalves, comprised of ten superfamilies and 23 families while Carter et al. (2011) recognised four orders, with 13 superfamilies containing 24 families.

Most pteriomorphians are byssate, many are epifaunal, and several groups, including the oysters (Ostreidae), cement onto the substratum. They range from isomyarian to monomyarian, and there is a great diversity of shell morphology. Shells may be calcite, aragonite, or both, some have a nacreous shell interior, they may be equivalved or inequivalved, and extensions of the dorsal hinge margin ('auricles') are common. The hinge is variable, being taxodont (notably in Arcoidea), with a few reduced teeth (dysodont) or lacking teeth. Several families have a reduced anterior adductor muscle (anisomyarian condition) or have lost that muscle (the monomyarian condition).

The mantle edges are mostly unfused and, in the great majority of pteriomorphians, are not developed into siphons. The foot is often reduced and is absent in some cemented taxa. The gills are mostly filibranch but in some taxa are pseudoeulamellibranch, and eulamellibranch in one group (Limoidea) (see Section 15.3.6.1 and next section below).

15.8.1.1 Phylogeny and Classification

Thiele (1929–1935) included the taxodont pteriomorphians (Arcoidea) in the 'Taxodonta' along with the protobranch bivalves, and the remainder of the pteriomorphians were grouped in Anisomyaria. The name Pteriomorphia has been almost universally used after Cox (1960a) and Franc (1960) used the name, and the latter author divided the group into Taxodonta and Anisomyaria. Newell (1969) in the *Treatise* classification recognised Arcoida, Pterioida, and Mytiloida as orders within the 'subclass' Pteriomorphia. Since the *Treatise*, these groupings were further refined in some classifications as reflected in the increased number of ordinal-level taxa now recognised, particularly for extinct groups (e.g., Carter et al. 2011) (see Appendix). Pojeta (1971) advocated raising mytilidans to subclass level, mainly because they lack a duplivincular ligament and because they have a fossil record from the Lower Ordovician. Waller (1978) divided the pteriomorphians into three 'superorders' within the Autobranchia – the Isofilibranchia comprising the mytilidans, the Prionodonta comprising the arcoideans, and the Pteriomorphia comprising the remaining groups which were grouped into three orders. The most recent classifications include all these taxa in Pteriomorphia.

There are six main groups of living pteriomorphians recognised in nearly all classifications, although, as indicated by the discussion above, their ranking and arrangement differ. These groups are:

Order Mytilida ('mussels') – are anisomyarian, sometimes isomyarian, or rarely monomyarian, with equivalve but very inequilateral shells, and the ligament is alivincular or elongate and opisthodetic. The shell is aragonitic, the structure nacreous, cross lamellar, or both, and sometimes an outer calcitic layer is present. Usually byssally attached. The mantle edge never bears eyes and is with or without well-developed siphons, but the middle fold of the mantle is fused to form a separate exhalant aperture or siphon. The gills are filibranch, and the foot is present in adults.

Order Arcida ('arcs') – are isomyarian to anisomyarian or, rarely, monomyarian, with circular to trapezoidal, usually equivalve shells, and the ligament is duplivincular. The shell is aragonitic with cross lamellar structure. Often byssally attached. The mantle edge is open all round, there being no siphons, and eyes are usually developed on the outer mantle fold. The gills are filibranch, and the foot is present in adults. Unlike other pteriomorphians, Arcidae have intracellular haemoglobin in their blood.

Order Pterioida or suborder Malleidina ('pearl oysters' and relatives) – are anisomyarian or monomyarian, with shells variable in shape, usually inequivalve, and the ligament is duplivincular, alivincular, multivincular, or elongate and opisthodetic. The shell has an outer calcitic layer and an inner aragonitic nacreous layer. Adults usually fixed by a byssus. The mantle edge is open, and the middle fold is extended to form a curtain-like 'velum'; there is no mantle fusion or siphon extensions, and eyes are absent. The gills are typically filibranch, although pseudoeulamellibranch in Pinnoidea, and the foot is present in adults.

Order Ostreida ('oysters') – are monomyarian or anisomyarian (the latter only in Dimyoidea), the shell is variable in shape, inequivalve, without a byssal notch, and the ligament is duplivincular, alivincular, or multivincular. The shell is composed of foliated calcite which is sometimes overlain with a simple calcitic layer. Adults are cemented or lie free. The mantle edge is open, the middle fold extended to form a curtain-like 'velum', there is no fusion or siphonal extensions, and eyes are lacking. The gills are filibranch or pseudoeulamellibranch (ostreids), and the foot is absent in adults. This group was included in Pterioida by Bieler and Mikkelsen (2006).

Order Pectinida ('scallops') – are monomyarian, and the shell is subtriangular to subcircular, often inequivalve, many having a byssal notch, ligament is duplivincular or alivincular. Byssally attached, cemented or free. Shell with or without an outer calcitic layer, foliated calcite is often present, and crossed lamellar aragonite is variably developed. The mantle edge is open, with the middle fold extended to form a curtain-like 'velum', there is no fusion or siphonal extensions, and eyes are present on the middle fold. The gill is filibranch, and the foot is present in adults.

Order Limida ('file shells') – are monomyarian, the shell is usually ovate to subtrigonal in shape, usually equivalve, and the ligament is alivincular. The shell has an outer calcitic layer and an inner cross lamellar aragonitic layer. Adults are usually fixed by a byssus or are sometimes free. The mantle edge is open, the middle fold is extended to form a curtain-like 'velum', and there are no fusion or siphonal extensions, but long tentacles are present, and there are often eyes on the middle fold. The gills are eulamellibranch, and the foot is present in adults.

In earlier classifications, the last three groups were included in Pterioida.

15.8.1.2 Biology and Ecology

All pteriomorphians are suspension feeders (see Chapter 5). Most are epifaunal and either byssally attached or cemented (e.g., oysters, spondylids – see Section 15.4.1.2). As noted above, some limids and mytilids build protective nests from byssal threads, sometimes in sediment.

A few pectinids and limids are free-living and capable of swimming (see below), but many are byssally attached. Some mytilids and arcoideans have become secondarily infaunal and shallow burrowers, but a few highly specialised mytilids burrow deeper and have rather long siphons. Some other mytilids bore into coral and limestone.

Morton and Scott (1980) argued that extant mytilids became infaunal (Figure 15.38) in three separate lineages – one *Botula*, a coral borer, is a member of the Modiolininae while another burrowing genus, *Arcuatula*, is a member of the Crenellinae along with *Musculus* and relatives, some of which live embedded in ascidians or in sediment. Morton (1982a) derived the lithophagines from this latter group, commencing with *Gregariella* which occupies vacated *Lithophaga* bore holes. Anteriorly and ventrally the fused inner mantle folds of *Gregariella* form glandular swellings which produce mucoid secretions with chemicals which may be involved in enlarging the burrow. Secretions from similar glands in lithophagines dissolve the calcium carbonate of the coral in which they burrow, and others inhibit the regrowth of coral around the siphonal aperture (Morton & Scott 1980). A siphonate modioline mytilid, *Arenifodiens vagina*, is the only mytilid known to burrow deeply into soft sediment (Wilson 2006).

Most pteriomorphians are marine, but a few have invaded fresh water. Among living taxa these include several Asian mytilids and members of the arcid genus *Scaphula*, from South East Asia. Living pterioideans are marine except for a few that live in brackish water, but some members of the extinct Myalinidae (Ambonychioidea) lived in both brackish and fresh water.

Arcoideans range from being epifaunal to shallow burrowing. The epifaunal taxa are byssally attached, their shells are usually elongate, and the byssus is large and solid and associated with strong musculature. Many have mantle eyes, including some in Arcidae that are compound (see Chapter 7). Some are nestlers, others such as *Anadara* are cockle-like shallow

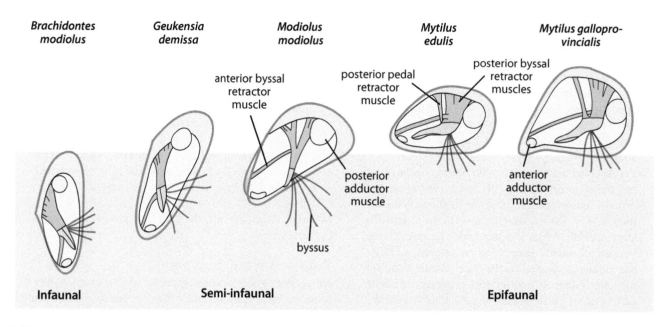

FIGURE 15.38 A sequence of mytilids showing a transition from infaunal to epifaunal habits. Modified from Stanley, S. M., *J. Paleont.*, 46, 165–212, 1972 and Seed, R., Shell growth and form in the Bivalvia, pp. 23–67 in Rhoads, D.C. and Lutz, R. A. (eds.), *Skeletal Growth of Aquatic Organisms*, Plenum, New York, 1980.

burrowers, and *Litharca* is possibly a true borer (Thomas 1978). *Trisidos* has the shell twisted 90° through its middle (Tevesz & Carter 1979; Morton 1983b).

The trend towards infaunal life in arcoideans[6] is associated with an increase in rib strength, inequivalve condition, and inflation. Thus infaunal arcoideans have quadrate, inflated shells which resemble cardiids (cockles) as they are often heavily radially ribbed and have interlocking crenulations on the shell margins (Oliver & Holmes 2006). Most retain a byssus of only a few threads, so byssus retractor muscles are not developed. Arcids with this morphology range from semi-infaunal to infaunal and include genera such as *Anadara*, *Scapharca*, and *Cunearca*. It is also seen in some Noetiidae (*Noetia* and *Eontia*) (Oliver & Holmes 2006) and in Cucullaeidae, where the byssus may be functional only in juveniles (Morton 1981c).

Other burrowing arcoideans included the limopsids and glycymeridids. Limopsids have ovoid, variously sculptured shells and are semi-infaunal to shallow burrowers, usually with varying degrees of byssal attachment, but some lack a byssus (Oliver 1981). All glycymeridids have lost the byssus and have thick, near-circular shells which are smooth to weakly radially ribbed, and they are rather poor burrowers.

Limids show a trend towards shell reduction and hypertrophy of the tentacles. Their tentacles are unusual in being very long (sometimes much longer than the shell), conspicuous, often brightly coloured, and arising from the middle mantle fold. Some limids cannot fully contract into their shells, and the tentacles remain exposed. Entire tentacles or parts of tentacles can be autotomised, and epidermal gland cells on

them can secrete distasteful viscous mucus to deter predators (Gilmour 1967). Autotomy occurs at one of the numerous transverse septa along each tentacle. These septa also regulate the hydrostatic pressure of blood in the tentacles enabling them, in unison with muscles and interseptal blood spaces, to perform complex movements (Gilmour 1963, 1967). In swimming limids, the tentacles can perform a 'rowing motion' which provides thrust (Gilmour 1967; Morton 1979b) and accounts for at least 13% of the distance obtained during each valve clap in *Limaria fragilis* (Donovan et al. 2004). The complex sensory structures associated with these tentacles are described by Owen and McCrae (1979). Another unusual feature of limids is the rotation of the foot during development in a process somewhat analogous to gastropod torsion, with the loss of the original pedal retractor muscles and the formation of new ones following a rotation of 180° (Gilmour 1990).

Thinning of the shell assists with swimming, as does another trend towards reliance on muscular rather than ciliary action to discharge pseudofaeces. This latter trend required the anterior-ward direction of the exhalant current, which has, in turn, necessitated increasing folding and fusion of the oral lips to prevent the dislodgement of food from the oral grooves (Morton 1979b).

Swimming occurs in several groups of pectinoideans, facilitated by a streamlined shell which has a well-developed internal ligament (resilium) and a substantial central adductor muscle of which a relatively large part is striated 'quick' muscle. These features were present in the Late Paleozoic *Pernopecten* (Waller 2006), indicating that the swimming habit is ancient in scallops. Interestingly, the byssally attached forms, typified by *Chlamys* with a distinct byssal notch, evolved later from several pectinoidean lineages, and the group as a whole was well diversified by the Late Mesozoic

---

[6] A recent molecular analysis (Feng et al. 2015) has cast doubt on most of the extant familial groupings in Arcoidea.

(Waller 2006). One pectinid genus, *Pedum*, lives embedded in massive coral (see Section 15.4.5).

15.8.1.3 Diversity and Fossil History

The oldest pteriomorphians are mytilidans and arcoid-like forms from the Lower Ordovician (Cox et al. 1969b; Pojeta 1971; Cope 1997b, 2000). The oldest mytilidans are contained in the Modiolopsoidea, a paraphyletic group (Carter et al. 2000).

The arcoid Parallelodontidae were significant in the late Paleozoic and Mesozoic (Amler 1989) and were similar to modern arcoideans in having trapezoidal, quadrate, and modioliform shells (Oliver & Holmes 2006). Parallelodonts, as their name suggests, have subparallel hinge teeth, rather than sub-vertical as in most modern arcoideans. They are mostly extinct except for a Japanese species, *Porterius dalli* (Newell 1969), although genera such as *Bathyarca*, *Bentharca*, *Samacar*, and *Deltaodon* may possibly also be extant parallelodonts or, due to their narrow hinge-plate, are secondarily similar (Oliver & Holmes 2006).

The earliest arcoideans were strongly asymmetric, and relatively streamlined burrowing forms like *Glycymeris* were derived from such forms (Oliver & Holmes 2006), in contrast to some scenarios proposed (e.g., Morton 1996, figure 29.8).

The Arcidae and Cucullaeidae first appeared in the Jurassic from parallelodont ancestors, and the Noetiidae and Glycymerididae first appeared in the Cretaceous (Oliver & Holmes 2006). The Limopsidae originated in the Lower Cretaceous and underwent a rapid radiation to the Upper Cretaceous but showed little subsequent diversification (Oliver 1981).

Members of the paraphyletic Pterineidae, which first appeared in the Middle Ordovician, probably gave rise to various pteriomorphian clades including the oysters, pterioideans, and pinnids (e.g., Waller 1998; Carter 2004). The pinnids probably evolved from Devonian pterineids while the isognomonids and pteriids evolved later from the same stem group (Carter 2004), the former in the Triassic from the Bakevelliidae (Tëmkin 2006), a group that arose from the Pterineidae in the Carboniferous and became extinct in the Eocene. The pterineids became extinct at the end of the Permian.

Some molecular phylogenies suggest that the Pinnoidea are also basal to the Ostreoidea (e.g., Giribet & Wheeler 2002), but others do not (e.g., Giribet & Distel 2003) with the former scenario more compatible with the fossil record (Carter 2004). The oysters (Ostreoidea) (i.e., Gryphaeidae and Ostreidae) have been shown to be monophyletic with molecular (Campbell 2000; Steiner & Hammer 2000) and morphological (Malchus 2008) data, a situation previously doubted (Carter 1990c). The first true oysters evolved after the Carboniferous, and oyster-like shells (including left valve cementation) referable to the Ostreoidea but with retention of an anterior adductor and aragonitic nacre, are known from the Upper Triassic (Carter 1990c). The true oysters have lost the

FIGURE 15.39 *Inoceramus* – a specimen from Greenland on display in the Greenland Institute of Natural Resources. It is 178 cm long. Photo by Mike Beauregard. Reproduced here under Creative Commons Attribution 2.0 Licence (https://creativecommons.org/licenses/by/2.0/legalcode).

anterior adductor and calcitic foliated shell structure (Carter 2004; Malchus 2008).

Members of the family Inoceramidae are an important, large-sized group of fossil pterioideans typified by the genus *Inoceramus* (Figure 15.39), members of which somewhat resemble some pteriids, albeit much larger. They were distributed world-wide in the Cretaceous; some reached more than a metre and a half in size and are the largest bivalves known.

The earliest pectinoidean (*sensu* Carter 1990b; see also Waller in Waller & Stanley 2005) is the Paleozoic *Pernopecten* which had a 'pectinoidean resilifer'. The mainly Triassic Entolioidea provide a link between the Pernopectinidae and other pectinoideans. Molecular analyses indicate that the Spondylidae and Pectinidae are more closely related to each other than to the Propeamussiidae, and this is reflected in the Carter et al. (2011) classification where they are placed in separate superfamilies (see Appendix). The ability to swim was an early adaptation in some lineages in the Pectinoidea (see Section 15.8.1.2).

15.8.2 HETEROCONCHIA

This higher grouping contains the Palaeoheterodonta and Heterodonta. These bivalves typically have schizodont or heterodont hinges, although hinge reduction or loss has occurred multiple times. The first heteroconchs appeared in the Lower Ordovician and diversified during that period. These ancient heteroconchs were lumped together as the Modiomorphoida, but that grouping is now considered at least paraphyletic (e.g., Carter et al. 2000; Schneider 2001). Carter et al. (2011) recognised a restricted concept of the modiomorphoids as an order (Modiomorphida) of the euheterodonts in their 'Megaorder' 'Cardiata' (i.e., Neoheterodontei).

15.8.3 PALAEOHETERODONTA

This group includes two very distinct clades, the marine 'brooch shells' (Trigoniida, Trigoniidae) and the freshwater mussels (Unionida).

There has been some dispute about the monophyly of palaeoheterodonts based on morphological differences. For example, the gills are filibranch in Trigoniidae but eulamellibranch in Unionida, there are short siphons in the Unionida but not trigoniids, and the morphology of the foot differs (see below). Differences such as these led Morton (1987b) and Salvini-Plawen and Steiner (1996) to conclude that trigoniids were more closely related to pteriomorphians than to unionidans. This conclusion, however, is not supported by recent molecular or morphological (e.g., Carter et al. 2000) analyses which show trigoniids to be sister to the heterodonts. This relationship is reflected in them being grouped in the Heteroconchia, along with the heterodonts, in the most recent classifications.

There are differences between trigoniids and Unionida, and there are also some significant similarities. The shells of both unionidans and trigoniids are equivalved and composed of aragonite, with internal nacre and usually just a few hinge teeth. Common features in the sperm morphology of unionidans and trigoniidans also suggest that the two groups are related (Healy 1989a), and this contention is also strongly supported by molecular data (see Giribet 2008a for references).

Because living members of the palaeoheterodont groups are so different, we treat them separately below.

15.8.3.1 Trigoniida

Typified by the Trigoniidae, popularly known as 'brooch shells', members of Trigoniida have nacreous shells strongly ornamented with radial or concentric ribs which often bear tubercules and they have a schizodont hinge. The animals lack any mantle fusion but have some minor siphonal elaboration, and the gills are filibranch (Tevesz 1975; Morton 1987b). This group was abundant in the Mesozoic and today is represented by a single living genus, *Neotrigonia*, which is found only in Australian waters.

The trigoniid hinge is very characteristic. It is large and can occupy nearly a third of the shell volume. The large teeth are characteristically transversely ridged, a character lacking in the ancestral Myophoriidae.

15.8.3.1.1 *Phylogeny and Classification*

The phylogeny of this group is rather poorly resolved, and it is very likely that the classification presented in the Appendix (largely following Carter et al. 2011) is inflated, with five superfamilies, all but one extinct. An alternative classification from the year before (Bouchet et al. 2010) recognised three superfamilies.

15.8.3.1.2 *Biology and Ecology*

Trigoniids are shallow burrowers. They lack siphons and have a very muscular foot fringed with short papillae that is capable of leaping, employing similar movements to those seen in some cardiids (Tevesz 1975; Morton 1987b) (see Chapter 4).

Neotrigonia is dioecious, and at least one species is suspected to have non-planktotrophic larvae (Ó Foighil & Graf 2000).

15.8.3.1.3 *Diversity and Fossil History*

Members of the Trigoniida are found from the Silurian, but the group is most diverse during the Mesozoic. Typical trigoniids appear to have been derived from the Myophoriidae in the Permian and greatly diversified in the Jurassic. They reached their greatest diversity in the Cretaceous, with most genera not surviving the extinction event at the end of the Mesozoic. The trigoniids were widespread in the Mesozoic when they were the cockles of that time and were more advanced than the shallow-burrowing, suspension-feeding bivalves of the Paleozoic (Stanley & Waller 1978). The family is now confined to five or six living species of *Neotrigonia* in Australia. *Neotrigonia* probably evolved from *Eotrigonia*, which ranges from the Eocene to the Miocene in Australia (Darragh 1986).

15.8.3.2 Unionida

The Order Unionida ('freshwater mussels' or 'naiads') are a large group of freshwater bivalves found on all continents except Antarctica and comprise the most significant freshwater radiation of bivalves. Its members brood their young, and the larvae are obligate parasites, nearly always of fish (e.g., Wächtler et al. 2001).

The shell interior is often nacreous, and the schizodont hinge is composed of a few, often large, oblique teeth. There are short posterior siphons, and the gills are eulamellibranch. All are shallow burrowers except for three genera which cement their shells to rocks and have been called 'river oysters'. One of these is a species in the African family Etheriidae, and two other cementing genera were previously included in that family but are now known to be members of other families (Bogan & Hoeh 2000) (see below).

Like the trigoniids, the Margaritiferidae and Unionidae have no mantle fusion, but Hyriidae, Etheriidae, Iridinidae, and Mycetopodidae have some fusion so that the exhalant aperture is separated and, in some, also the inhalant aperture. These openings are elaborated to some extent as short to very short 'siphons' (Bogan & Roe 2008).

15.8.3.2.1 *Phylogeny and Classification*

Based on their larval characteristics, Unionoidea has until recently been thought to contain two superfamilies: Unionoidea with the families Unionidae, Hyriidae, and Margaritiferidae, and Muteloidea containing Mutelidae, Mycetopodidae, and Etheriidae. However, a recent classification using molecular methods (Graf & Cummings 2006b) rearranged these as Unionoidea (Unionidae, Margaritiferidae) and Etherioidea (Hyriidae, Iridinidae (= Mutelidae), Mycetopodidae, and Etheriidae). Unlike previous classifications, Carter et al. (2011) split the living members of the Order Unionida into two suborders, Unionidina (with families Unionidae, Margaritiferidae, Etheriidae, Iridinidae, and Mulleriidae [with Mycetopodinae as a subfamily]) and Hyriidina (family Hyriidae only).

Several recent papers have addressed the phylogeny of Unionida, the most comprehensive being those of Graf and Cummings (2006a), Bogan and Roe (2008) (who provided a useful review of the classification of the group), and Hoeh et al. (2009). While recent analyses generally agree that there are two major groups – usually treated as Unionoidea and Etherioidea (i.e., Mullerioidea) – the position of Hyriidae varies. It is sister to both groups at the base of the tree (Hoeh et al. 2001) or at the base of the 'Etherioidea' (Graf & Cummings 2006b). The analyses presented by Hoeh et al. (2009) also generally supported the former hypothesis. The positioning of the Hyriidae as the sister to other Unionida presumably led Carter et al. (2011) to include Hyriidae in a separate suborder, although whether this taxonomic inflation was necessary is arguable. Thus, the six recognised families are grouped in either two superfamilies (e.g., Bogan & Roe 2008) (Unionoidea and Etherioidea) or three (Carter et al. 2011) (Unionoidea, Mullerioidea [= Etherioidea], and Hyrioidea).

15.8.3.2.2   Biology and Ecology

All members of this group live in fresh water. They are the most endangered group of freshwater animals (Graf & Cummings 2006b; Bogan 2008) (see Chapter 10).

Most unionidans are shallow burrowers in sediment, but some nestle beneath rocks or lodge in burrows in thick mud. Some form shoals in rivers and can comprise several species and genera living together. In parts of the Tennessee River, Alabama, USA, 69 species were recorded from just one area in the early 20th century, but alterations to the river system have resulted in 32 of these disappearing (Lydeard et al. 2004).

Members of the Unionida are dioecious or hermaphroditic, the latter condition occurring sporadically through the group. Embryos are brooded until the larvae are ready for release. A short-term brooding strategy is plesiomorphic for the group, with long-term brooding having arisen twice within North American unionoideans (Graf & Ó Foighil 2000b). The proportion of the gill used for brooding (the marsupium) also varies from the probable plesiomorphic condition, where some of the outer demibranch is used, to the whole outer demibranch or both inner and outer demibranchs (Graf & Ó Foighil 2000b).

The radiation of unionoideans was probably aided by the evolution of parasitic larvae (e.g., the glochidium) capable of attaching to fish (Figure 15.40) when released from the brood chambers in the gills of the female to aid in movement upstream. The Iridinidae and Etheriidae have another type of modified larva, the lasidium (Graf & Ó Foighil 2000b) (Figure 15.41), which also attaches to fish and is thought to be derived from the glochidium. The differences between these larvae are summarised as follows:

Glochidia valves range in size from 75 to 350 μm and are simple or hooked. There is a single adductor muscle, and they attach to their host by forming a cyst. They are produced by all Unionida except Mullerioidea.

Lasidia range in total size from 85 to 150 μm, and the shell is univalved and uncalcified. The body is trilobed and usually has a long thread. It attaches by

forming a cyst but in a modified form (haustoria); the attachment is by way of tubular appendages (Wächtler et al. 2001; Graf & Cummings 2006b). The haustorium has only been described in one species (*Mutela bourguignati*: Iridinidae), with data for many other members of Iridinidae lacking (Graf & Cummings 2006b). The lasidia larva is known only in members of the superfamily Mullerioidea.

An attachment thread found in many glochidia is probably not a byssal thread, as sometimes stated, because it is produced by a gland anterior to the anlage of the foot (Waller 1998). The early postlarvae do, however, have a byssus.

In some cases among unionids, specific host fish are attracted to the vicinity of the mussel by lures. A variety of siphonal elaborations in some species of North American unionids attract the specific host fish to which the larvae will attach, including mimicking the fish itself (Bogan & Roe 2008). In some hyriids, the larvae are released bound in worm-like mucous strings (Klunzinger et al. 2013), with the hosts not being specific.

Specificity of the fish host varies according to the mussel taxon (e.g., Bogan & Roe 2008). In some species it is highly specific; in others many species may be suitable hosts. Glochidia can often attach to unsuitable hosts and form a cyst but will subsequently die (e.g., Meyers et al. 1980; Bauer & Vogel 1987). Indeed, some fish can develop an immune response to glochidia (Jansen et al. 2001; Dodd et al. 2006; Rogers-Lowery & Dimock 2006).

The glochidia or lasidia encyst in the epidermis of the host fish (or rarely amphibians), usually on the gills or fins but also the body, and are surrounded by a thin membrane that originates from the host (Wood 1974; Silva-Souza & Eiras 2002). They remain attached in this way until their complete metamorphosis (Lefevre & Curtis 1910). The time taken to metamorphose, and at what time they are released, depend on temperature and the unionoidean taxon (e.g., Bauer 1994).

There have been suggestions that different glochidial morphology determined the preferred location on the fish; those with small hooks or lacking them attaching to the gill filaments, while those with large hooks can attach to scales, skin, and fins (e.g., Berrie & Boize 1985; Pekkarinen & Englund 1995). For example, in a European study, the glochidia of *Anodonta* were found on ten fish species, mainly on the fins, and *Unio* glochidia were on 17 fish species, mainly on the gills (Blažek & Gelnar 2006). Exotic fish are also often parasitised, as are some amphibians (Watters & O'Dee 1998). Glochidium attachment is very rarely observed on decapod crustaceans (e.g., Walker 1981).

While some unionids are relatively short-lived, some live for decades, and others, such as *Margaritifera margaritifera*, have a lifespan exceeding a hundred years (Bauer 1992; Haag & Rypel 2011).

15.8.3.2.3   Diversity and Fossil History

Unionids have a long fossil history, from the Upper (or possibly Middle) Devonian. There are about 800 living species arranged in about 181 genera (Bogan & Roe 2008).

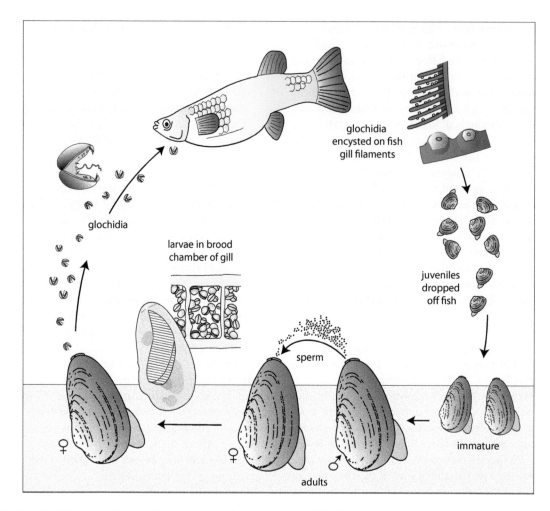

FIGURE 15.40 The life cycle of a unioniid (not to scale). Redrawn and modified from various sources.

Of the unionoidean mussels, the Margaritiferidae contains around 12 species in Europe, North America, some parts of Asia, Morocco, and North Africa. The Unionidae also occurs in Europe, North America, all of Asia, Sub-Saharan Africa and the Nile River, northern Madagascar, Central America, and New Guinea. The unionids are by far the largest family of freshwater mussels, with ~620 species (Bogan 2008), with the greatest diversity in southeastern USA and significant diversity in South East Asia.

The Hyriidae, with about 83 species (Bogan 2008), has a disjunct distribution, being found in South America, Australia, New Guinea, Solomon Islands, and New Zealand, and are thought to have a Gondwanan origin (Graf & Ó Foighil 2000a).

Members of Mullerioidea are more restricted in distribution than unionoideans or Hyriidae. The 41 species (Bogan 2008) known from the Iridinidae occur in Sub-Saharan Africa and the Nile River. The approximately 39 species (Bogan 2008) of Mycetopodidae are found only in Central and South America. The Etheriidae comprises a single species, *Etheria elliptica*, found in Africa and Madagascar, with two genera previously included in the family now excluded. These are *Pseudomolleria* from southern India, transferred to the

Unionidae, and *Acrostaea* from Columbia, now included in the Mycetopodidae (Bogan & Hoeh 2000; Hoeh et al. 2009). These three genera cement their shells to the substratum (hence their common name 'river oysters'), but where *Etheria* has two adductor muscles, *Acostaea* and *Pseudomulleria* have a single adductor muscle (monomyarian condition), the posterior, a condition otherwise only seen in some pteriomorphian bivalves.

15.8.4 HETERODONTA

Heterodonts are the largest group of living bivalves. Most are infaunal, but many exhibit marked adaptations to diverse habits (see below). The 'heterodont' type of hinge characterises the group. These hinges are somewhat variable and typically composed of at least two kinds of teeth – a set of one or two elongate lateral teeth, on one or both sides of the umbo, that run approximately parallel to the shell margin, and usually one to three short cardinal teeth that radiate from the umbo. However, hinge teeth can be reduced or even absent. The shells of heterodonts are always aragonitic, and, besides some anomalodesmatans, nacre is absent. The gills are always eulamellibranch except in septibranchs, where they are

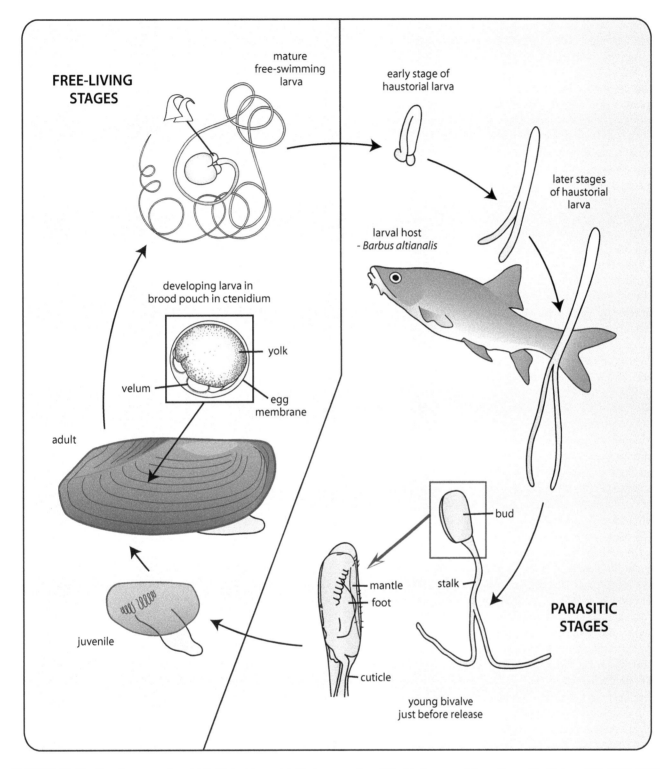

FIGURE 15.41 Development of an iridinid (not to scale). Redrawn and modified in large part from Fryer, G., *Nature*, 183, 1342–1343, 1959, and Fryer, G., *Phil. Trans.* R. Soc. B, 244, 259–298, 1961.

secondarily very reduced or lost in concert with a carnivorous diet (see Section 15.8.6.1.2 and Chapter 5).

Most heterodonts are infaunal, and the majority are suspension feeders (see Section 15.4.2 and Chapter 5), with occasional deviations from that lifestyle. As noted above, many members of the Tellinoidea have become specialised in deposit feeding, using their siphons to suck up surface detritus which they sort on their enlarged labial palps.

Various groups rely on symbiotic organisms – for example, symbiotic zooxanthellae (Cardiidae: Fraginae and Tridacninae) (see Section 15.4.4.1) - for some or all of their nutrition. Symbiotic sulphur-oxidising bacteria (see Chapter 5), which provide much of the nutrition to the host, are housed in the gills of Lucinidae, the vent- and seep-inhabiting *Calyptogena* (Vesicomyidae) (Glossoidea) and some Thyasiridae (see Section 15.4.4.1).

Gut bacteria also play an important role in digestion in many heterodont bivalves, notably in the wood-boring Teredinidae and Xylophagaidae. The stomachs of members of both families possess a special wood-storing caecum where the bacteria break down wood fibres. The degree of specialisation of wood-feeding differs in different genera, with suspension feeding also playing a role, and the gills and palps are reduced in the more derived teredinids (Hoagland & Turner 1981).

The classification of this group is summarised in the Appendix. In the relatively recent classification of Bieler and Mikkelsen (2006), the Heterodonta was divided into five orders – the Carditoida, Veneroida (which included the Lucinoidea), Myoida, and Anomalodesmata, and their arrangement was essentially that adopted by Bouchet et al. (2010). Giribet (2008a) recognised that Heterodonta comprised two main lineages, one of which he named Archiheterodonta, and the other, named Euheterodonta by Distel (2003), has also been recovered in other analyses (e.g., Taylor et al. 2007b). Included within the euheterodont bivalves in these phylogenies was the Anomalodesmata, a group long treated as a separate subclass. In a slightly later classification Carter et al. (2011) recognised two 'subcohorts', Carditioni (equivalent to Archiheterodonta) and Cardioni (equivalent to Euheterodonta), the latter being divided into two 'infrasubcohorts', Lucinida and Cardiida. Of these, the latter was subdivided into three 'Megaorders', Cardiata, Poromyata (= Anomalodesmata), and Solenata.

The classification we present in the Appendix is a compromise between the most recent versions. Archiheterodonta contains only the order Carditoida with two living superfamilies (Crassatelloidea and Carditoidea) and several extinct taxa. The euheterodonts include the oldest known heterodont lineages and first appear in the Lower Ordovician. *Babinka* is characterised by multiple pedal muscle scars on the shell, and it has been suggested that it is related to lucinoideans. However, the earliest fossil that is possibly a lucinid is the Silurian *Ilionia*, which apparently occupied a comparable habitat and had similar burrowing habits to living lucinids, making it the first known deep-burrowing autobranch bivalve (Liljedahl 1992).

The Anomalodesmata, although previously treated as a separate subclass, is usually nested within Heterodonta in recent molecular analyses but remains monophyletic. This group is distinguished by highly divergent characters, such as the possession of nacre in some taxa and the development of the septibranch condition in others (see Section 15.8.6).

The various groups of heterodonts are outlined below, and the ages of the taxa down to superfamily are given in the Appendix.

15.8.4.1 Archiheterodonta

All archiheterodonts are marine, mostly infaunal with a few byssate forms; they do not have distinct siphons and thus have an entire pallial line (i.e., they lack a pallial sinus), and the living taxa do not have nacre.

15.8.4.1.1 Phylogeny and Classification

Carditids were recognised as the most primitive of living heterodonts based on their shell and anatomy (Yonge 1969). With other crassatelloideans, they have a simple periostracum (Bottjer & Carter 1980) and unique sperm morphology, with a reduction of one of the two centrioles during spermiogenesis reported in members of the families Carditidae and Crassatellidae (Healy 1995), a character that may be a synapomorphy of Archiheterodonta. Molecular data (Taylor et al. 2007b; Giribet 2008a; González et al. 2015; González & Giribet 2015; Combosch et al. 2017) support the idea that this group is the sister group to the euheterodonts.

The group comprises the single order Carditoida which contains one superfamily, Crassatelloidea, comprising four living families, Crassatellidae, Carditidae, Astartidae, and Condylocardiidae (see Appendix), although Condylocardiidae and Carditidae were not well resolved in the only molecular analysis containing both taxa (González & Giribet 2015) suggesting that condylocardiids may represent one or more paedomorphic carditid lineages.

15.8.4.1.2 Biology and Ecology

Most archiheterodonts are marine, shallow infaunal burrowers, and they are all suspension feeders.

Members of this clade have been called 'bloody clams' because they have extracellular, high molecular weight haemoglobin in their blood (Terwilliger & Terwilliger 1985; Taylor et al. 2005) (see Chapter 6). All euheterodonts lack extracellular respiratory blood pigments, although a few have intracellular haemoglobin in haemocytes (see Chapter 6).

15.8.4.1.3 Diversity and Fossil History

Crassatelloideans have a fossil history reaching back to the Devonian or earlier (Taylor et al. 2007b). Devonian crassatelloideans include *Eodon* (Eodonidae), with some likely members of this clade from the Ordovician, including *Copidens* and *Ananterodonta* (Carter et al. 2000). The ancestor of this group was probably the cycloconchids, which include the Silurian *Actinodonta* (Carter et al. 2000).

15.8.5 EUHETERODONTA

The euheterodonts are by far the largest group of heteroconch bivalves (see the Appendix for an overview of their classification). They are found throughout the world, mostly in shallow marine environments. A few taxa, such as the wood-boring Xylophagaidae, and the seep- and vent-associated Vesicomyidae, are found mainly in deep water.

The euheterodonts include most familiar burrowing bivalve taxa, the so-called 'clams', with more than 40 families including the very large family Veneridae (the 'venus clams'), the true 'cockles' (Cardiidae), a family which includes the epifaunal to nestling 'giant clams' (until recently known as a separate family Tridacnidae), mactrids or 'trough shells'

(Mactridae), the 'tellins' (Tellinidae), and many other groups (see Appendix).

The most basal euheterodonts comprise two groups previously treated as Lucinoidea, the Thyasiroidea and Lucinoidea *s.s.*, members of which farm symbiotic bacteria in their gills, a habit they apparently acquired independently (Williams et al. 2004; Taylor et al. 2007a, b). The previous concept of the Lucinoidea is polyphyletic with the other groups included being the Ungulinidae (now Ungulinoidea near the Veneroidea) and the Cyrenidae (Cyrenoidea) (Williams et al. 2004; Taylor et al. 2009).

While most of the above groups are shallow burrowers, the heterodonts also include deep-burrowing clades such as the 'soft-shelled clams' (Myidae), 'razor shells' (Solenidae), and Solecurtidae. Members of the Pholadidae (the 'piddocks' or 'angel wings') burrow into soft rock or wood, but a few pholadids, the xylophagaids, and the 'ship worms' (Teredinidae) are highly specialised wood borers. The Chamidae (the 'jewel box clams') are the only exclusively cemented family of euheterodonts, and only a few members of the group are byssally attached as adults. Some members of the very diverse, mostly small-sized Galeommatoidea are commensals with a wide range of invertebrates. Some galeommatoideans cover their shells with flaps of the mantle, and a few have become almost slug-like with their reduced shells encased in the reflected, bag-like mantle. Two groups of euheterodonts have radiated in fresh water, the Cyrenidae and Sphaeriidae (see Section 15.6.2.1).

The extinct Hippuritoida, which includes the reef-building rudists, were probably derived from another extinct order, Megalodontida, and, despite some suggestions to the contrary, are not closely related to the Chamidae. We discuss the rudists separately in Section 15.8.5.4.

15.8.5.1 Phylogeny and Classification

Euheterodonta cannot be readily defined using morphological characters, and its present composition is mainly based on molecular analyses. It includes the 'veneroidan' families traditionally included in Heterodonta, except for those now included in Archiheterodonta. Molecular analyses (see Taylor et al. 2007b; Giribet 2008a; Sharma et al. 2012) have shown that some long-held ideas about bivalve classification required changing. These include, as noted above, the recognition of paraphyly in the long-held concepts of 'Lucinoidea' (now Thyasiroidea, Ungulinoidea, and Lucinoidea) and also 'Corbiculoidea' (now Cyrenoidea and Sphaerioidea). Cardiidae now includes (as a subfamily) what was previously a separately recognised family (Tridacnidae) and superfamily (Tridacnoidea). In addition, the 'order' Myida, as previously recognised, is polyphyletic (see below).

In molecular phylogenies of euheterodont bivalves, there is disagreement as to the most basal members of the clade and the detail of the phylogenetic arrangements. The morphological analysis of Carter et al. (2000) placed the lucinoideans as the sister to the palaeoheterodonts outside the Heterodonta. However, molecular studies (e.g., Giribet & Wheeler 2002;

Taylor et al. 2007b, 2009; Giribet 2008a) have them as euheterodonts and not related to crassatellids as sometimes suggested (e.g., Morton 1996). The most basal living euheterodonts vary between molecular analyses. In the combined analysis of Giribet and Wheeler (2002), the Hiatellidae is basal while in the molecular investigation of Taylor et al. (2007b), it is the Thyasiridae. The Cardiidae (Cardioidea) form a distinct clade in molecular analyses and, in the Taylor et al. (2007b) analysis and several others, they are sister to the Tellinoidea. In the Sharma et al. (2012) study they form a strongly supported clade at the base of the euheterodonts and below a lucinid + thyasirid clade that makes up the next branch on their tree.

Taylor et al. (2007a) named a large monophyletic group within Euheterodonta as the Neoheterodontei, which included the euheterodonts other than the Thyasiridae and Lucinidae (and Anomalodesmata). Apart from the exclusion of the Cardiidae, this grouping was also recovered in the Sharma et al. (2012) analysis and is more or less equivalent to Cardiata of Carter et al. (2011).

The concept of the 'order' Myida,[7] which included Pholadoidea, Myoidea, Gastrochaenoidea, and Hiatelloidea and which was used until relatively recently (e.g., Cox et al. 1969b; Morton 1996; Amler 1999), is polyphyletic in molecular analyses, although the details are not concordant. For example, in the Taylor et al. (2007b) analysis, the representatives of the myoid groups that form a 'well-supported' clade (together with a dreissenid) are Pholadoidea (Teredinidae and Pholadidae) and Myoidea (Myidae and Corbulidae), while in the Sharma et al. (2012) analysis, containing a much smaller sampling of euheterodonts, *Teredo* and *Mya* form a subclade in a clade also containing a corbulid and a galeommatid, while *Dreissena* groups with *Glossus*. The representatives of Gastrochaenidae and Hiatellidae occupy different positions on the tree in these analyses. These results are reflected in the Carter et al. (2011) classification with gastrochaenids included in the order Cardiida, but Hiatelloidea is placed in their 'megaorder' Solenata, albeit in a separate order (Hiatellida). Thus the dreissenids have a variable position within the heterodonts in recent molecular trees. Taylor et al. (2007b) noted that the byssate living members of that group might have had burrowing ancestors with rather long siphons.

Two members included in the old concept of Myida are unrelated to true myoideans; these are Hiatelloidea (including the small, nestling, and boring *Hiatella* and the large deep-burrowing *Panopea*), a group actually related to another deep-burrowing lineage, Solenoidea, long separated from the myids in classifications. The Gastrochaenidae was also shown not to belong to Myoidea, but its placement within euheterodonts is not well resolved (Taylor et al. 2007b, 2009).

The relationship of the small-sized, often commensal, galeommatoideans to other heterodonts is unclear. They are sometimes placed near crassatelloideans or lucinoideans largely because of their anterior inhalant siphon. To date, their relationships have not been well resolved in molecular analyses. The Gaimardiidae, a small, southern group of

---

[7] Now Pholadida (see Appendix).

byssate bivalves that live attached to algae, were included in a separate superfamily (Gaimardioidea) by Cox et al. (1969b) or treated as a subfamily of Cyamiidae (Ponder 1971; Ponder & Keyzer 1998; Bieler & Mikkelsen 2006) while Britton and Morton (1979) treated Cyamiidae and Gaimardiidae as separate families in Cyamioidea. No typical cyamiid has been included in any molecular phylogeny to date, but the supposed cyamioidean *Basterotia* (Basterotiidae) was grouped with the galeommatoideans in the Taylor et al. (2007b) analysis in which *Gaimardia* was well separated.

Tellinoidea formed a well-supported clade in the Taylor et al. (2007b) analysis, and those results suggested that some families may not be monophyletic.

The freshwater Sphaeriidae and the brackish and freshwater Cyrenidae (previously Corbiculidae) were grouped in a single superfamily, but they are now treated as two – Cyrenoidea (previously Corbiculoidea) and Sphaerioidea. Indeed, molecular analyses have shown that they form two entirely separate groups (Park & Ó Foighil 2000; Giribet & Distel 2003; Taylor et al. 2007b). *Glauconome* (Glauconomidae), that was included in Veneroidea or Solenoidea, grouped with Cyrenidae in recent molecular analyses (Mikkelsen et al. 2006; Taylor et al. 2007b; Sharma et al. 2012). While sperm morphology does not support a close relationship between cyrenids and glauconomids (Healy et al. 2006), their stomach morphology is similar (Purchon 1987a). Thus, there is good evidence that the sphaeriids, cyrenids, and dreissenids are all independently derived freshwater radiations (Park & Ó Foighil 2000; Taylor et al. 2007b) (see also Section 15.6.2.1).

Although Ungulinidae were included in Lucinoidea in the past, molecular analyses have shown that they are not related to lucinids, and there are also substantial morphological differences (Williams et al. 2004; Taylor et al. 2007b).

In the Taylor et al. (2007b) analysis, the representatives of Mactroidea (Mactridae, Mesodesmatidae, Lutrarinae) comprised a well-supported clade, but in the molecular analyses to date the relationships of this group within the euheterodonts are variable. The Taylor et al. (2007b) analysis supported the Allen (1985) hypothesis that mesodesmatids are related to tellinoideans and that they did not merit separate superfamily status as proposed by Yonge and Allen (1985).

The Trapezidae is a small warm-water family of byssate, nestling species usually included in the Arcticoidea (e.g., Cox et al. 1969b). Taylor et al. (2007b) showed that *Trapezium* was related to a group of families including Veneridae, Arcticidae, and Cyrenidae, but a close relationship with *Arctica* was not supported, a finding also indicated by differences in sperm morphology (Healy et al. 2006).

The Hemidonacidae contains a single genus *Hemidonax* with only five living species (Ponder et al. 1981). Placing this group has been controversial, with suggested relationships with tellinoideans, crassatelloideans, and cardioideans. Molecular data, while not resolving any clear affinity, has shown that they are not crassatelloideans, tellinoideans, or cardioideans but instead group with petricolids, venerids, and chamids (Taylor et al. 2007b) or with solenids (Sharma et al. 2012) but have long branch lengths separating them.

Sperm morphology similarly failed to align the family with other euheterodonts (Healy et al. 2008). Carter et al. (2011) provided a separate superfamily for the group.

The relationships of the cementing Chamidae have also been somewhat controversial with various hypotheses involving several groups of heterodonts as sister taxa (e.g., Kennedy et al. 1970), including the extinct rudists (Odhner 1919; Newell 1965; Yonge 1967b). Molecular results generally show support for a relationship with Cardiidae, but sperm morphology (Hylander & Summers 1977) and shell structure (Kennedy et al. 1970; Harper 1998) suggest links with Veneroidea and possibly the rudists (Taylor et al. 2007b).

The Veneridae is the largest extant family of bivalves with over 800 species. Its monophyly has been confirmed in molecular analyses (Mikkelsen et al. 2006; Taylor et al. 2007b). Several subfamilies are recognised.

The mainly deep-sea, chemosymbiotic vesicomyids are usually placed in the Glossoidea (e.g., Cox et al. 1969b), but molecular analyses suggest that they are sister to Veneridae (Giribet & Distel 2003; Mikkelsen et al. 2006; Taylor et al. 2007b) or, in a tree based on only 18S sequences, related to *Arctica* (Arcticidae) (Taylor et al. 2007b).

15.8.5.2 Biology and Ecology

The Euheterodonta is a large and diverse group, and this diversity is reflected in their biology and ecology. Nearly all euheterodonts have siphons (and hence a pallial sinus), although some have a siphon but no pallial sinus (e.g., Lucinoidea, Thyasiroidea, and some galeommatoideans), but the homology of these structures with the siphons of other euheterodonts, particularly in galeommatoideans, is questionable. Shell morphology is highly derived in some lineages, in particular some burrowing taxa such as the pholads, teredinids, and solenids.

Most euheterodont bivalves are found in soft sediments where they are shallow to deep burrowers and often dominate these habitats. The shallow-burrowing cardiids and venerids are highly speciose, whereas deep-burrowing taxa such as the myids and solenoideans are less so. Other diverse groups, including the lucinoideans and tellinoideans, are burrowers that live in intermediate depths in the sediment. A few families, such as the Hiatellidae, contain taxa with a range of habits, in this case including deep burrowers (*Panopea*) and nestlers (*Hiatella*). Euheterodonts are poorly represented in the deep sea, although vesicomyids, which often reach a large size, can be abundant in seep and vent habitats.

Some euheterodonts, including the pholadids, burrow into wood, coral, or soft rock, and in shallow marine and estuarine waters teredinids (shipworms) have an important role in breaking down wood and by so doing can become pests (see Chapter 10). Members of the related Xylophagaidae are also specialised wood borers.

A few euheterodonts have secondarily acquired epifaunal habits and become byssally attached or cemented to the substratum. Among these are included the cemented oyster-like chamids and the byssally attached mussel-like Dreissenidae.

While the great majority of euheterodonts are suspension feeders, many tellinoideans feed on deposits, while the

Lucinidae, Thyasiridae, and Vesicomyidae have chemosymbiotic bacteria housed in their gills which supply nutrition to their host (see Chapter 5). These feeding habits have resulted in modifications to the bivalve host with, for example, the lucinids having mantle folds (Figure 15.42) that act as secondary gills due to the decreased respiratory capacity of the bacteria-laden gills (Taylor & Glover 2006). Lucinids are the most diverse group of chemosymbiotic bivalves and one of the most ancient groups of euheterodonts. The thyasirids, previously thought to be closely related to lucinids, are now included in their own superfamily, and the two groups are paraphyletic at the base of the euheterodont tree. While many thyasirids have symbiotic bacteria like lucinids, some do not (Dufour 2005; Taylor et al. 2007a) (Figure 15.42).

A few euheterodonts, notably the 'giant clams' (*Tridacna* and *Hippopus*), incorporate in their tissues symbiotic zooxanthellae which supplement their food supply (see Chapter 5).

The diverse Galeommatoidea, some of which are commensals with various invertebrates, include normal-looking bivalves as well as highly modified forms. The more bizarre include taxa that have spread their valves to become limpet-like while others have enclosed their reduced valves in mantle extensions resulting in a slug-like form.

15.8.5.3 Diversity and Fossil History

The heterodont radiation commenced in the Paleozoic with some Ordovician bivalves assigned to Heteroconchia or Heterodonta (Pojeta 1971; Morris & Eagar 1978; Cope 1997b, 2002). Phylogenies for Paleozoic and early Mesozoic bivalves (Carter & Lutz 1990; Carter et al. 2000) indicate relationships substantially different from current molecular phylogenies. Thus, the relationships of these taxa are by no means certain (Taylor et al. 2007b).

With regard to the basal members of the euheterodont clade (see Section 15.8.5.1), thyasirids are known from the Lower Cretaceous (Taylor et al. 2007a), although there are probable Jurassic members (Taylor et al. 2007b), and there is a possible Triassic member (Cox et al. 1969b). Anomalodesmatans and lucinids were respectively the second and third most basal taxa in the Taylor et al. (2007b) analysis. The lucinids have a long and rather well-known fossil record, back to at least the Pridoli or Ludlow (Silurian) (Taylor & Glover 2006), with possible records from the Ordovician, while Cardiidae date from the Upper Triassic (Taylor et al. 2007b).

There was a significant diversification of heterodonts in the late Mesozoic, including many of the neoheterodont families, although some of those originated in the Paleogene (Crame 2002; Taylor et al. 2007b).

The origins of various euheterodont groups were briefly reviewed by Taylor et al. (2007b), and the ages of the fossils attributed to the superfamilies are provided in the Appendix. Cardiids first appeared in the Upper Triassic and although a derivation from carditids has been suggested (e.g., Schneider 1995; Schneider & Carter 2001), molecular analyses do not support such a relationship.

Cyrenids first appeared in the Middle Jurassic, probably derived from arcticid-like bivalves (Matsukawa & Nakada 2003; Gardner 2005), and soon moved into the brackish and freshwater habitats they still occupy. Venerids probably evolved from an 'arcticoid' ancestor in the Middle to Upper Jurassic (Casey 1952; Gardner 2005), with this scenario supported by molecular results as the Veneridae, Arcticidae, and Vesicomyidae are all closely related. A close relationship between arcticoideans, veneroideans, and cyrenids was suggested by Gardner (2005) who treated them as a single superfamily.

Families attributed to the Tellinoidea diversified in the Cretaceous but appeared earlier in the Upper Triassic (Cox et al. 1969b; Pohlo 1982) (see Appendix). Pohlo (1982) reviewed the relationships of the group and suggested that they were probably derived from the archiheterodont Astartidae, a result not supported by molecular data (e.g., Taylor et al. 2007b). Many tellinids are deposit feeders with long siphons, but the earliest members of the group (Tancrediidae) were probably suspension feeders with short siphons, rather like the present day donacids (Pohlo 1982).

Pholadids have a rather long fossil history from the Middle Jurassic, with the wood-boring teredinids first appearing in the Upper Cretaceous (Kelly 1988). A possible derivation from the anomalodesmatan Pholadomyoida was suggested by Kelly (1988), but molecular data do not support that idea.

Mactrid fossils first appear in the Lower Cretaceous (Aptian). Derivations from Arcticidae (Saul 1973) and Cyrenidae (Bernard 1895; Vokes 1946) have been suggested, and both are compatible with molecular phylogenies (Taylor et al. 2007b).

The solenids were placed in a separate 'megaorder' (Solenata) by Carter et al. (2011), possibly partly on the assumption that they were derived from the Paleozoic Orthonotidae, some of which had very similar shells (Runnegar 1974). A derivation from the Mesozoic Tancrediidae (usually placed in Tellinoidea) (Skelton et al. 1990; Morris et al. 1991) has also been proposed.

It has been suggested (e.g., Carter et al. 2006) that gastrochaenids evolved from the late Paleozoic to Mesozoic Kalenteridae as both have calcified spines in the periostracum. This feature is common in anomalodesmatans, and Morris et al. (1991) referred permophorids to that group. However, such spines have a wide occurrence within autobranchs (Glover & Taylor 2010), and the relationships of permophorids are controversial (Schneider & Carter 2001). The sister taxon of gastrochaenids remains uncertain.

15.8.5.4 The Rudists

One of the strangest groups of bivalves were the rudists (Hippuritoidea) – an extinct group that mostly had asymmetric valves, with one attached to the substratum (see Section 15.4.1.3). They are the only major group of bivalves to have become extinct and have the most extreme range of morphologies seen in any group (Donovan 1992) (Figure 15.43).

Rudists are now thought to be euheterodonts (e.g., Carter et al. 2011), although there was much early speculation about the relationships of these fossils, and it was not until the mid-1800s they were eventually recognised as molluscs and even

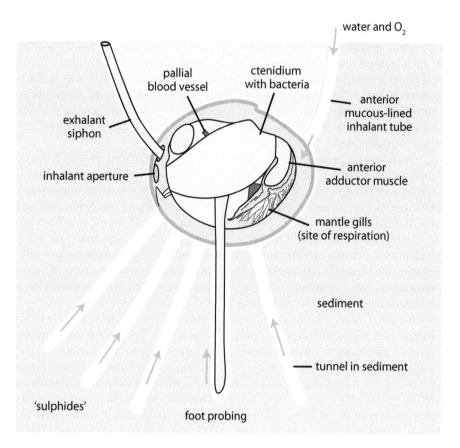

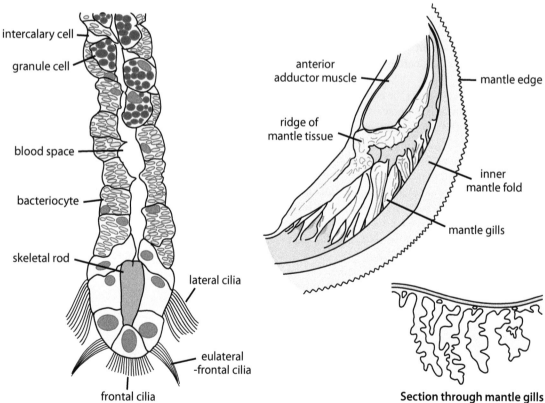

FIGURE 15.42 A typical lucinid (*Codakia*) showing life habit, mantle (secondary) gills, and part of a bacteria-laden gill filament. Redrawn and modified from Taylor, J.D. and Glover, E.A., Functional anatomy, chemosymbiosis and evolution of the Lucinidae, pp. 207–225 in Harper, E.M., Taylor, J. D., and Crame, J. A. (eds.), *Evolutionary Biology of the Bivalvia*, the Geological Society, London, UK, 2000.

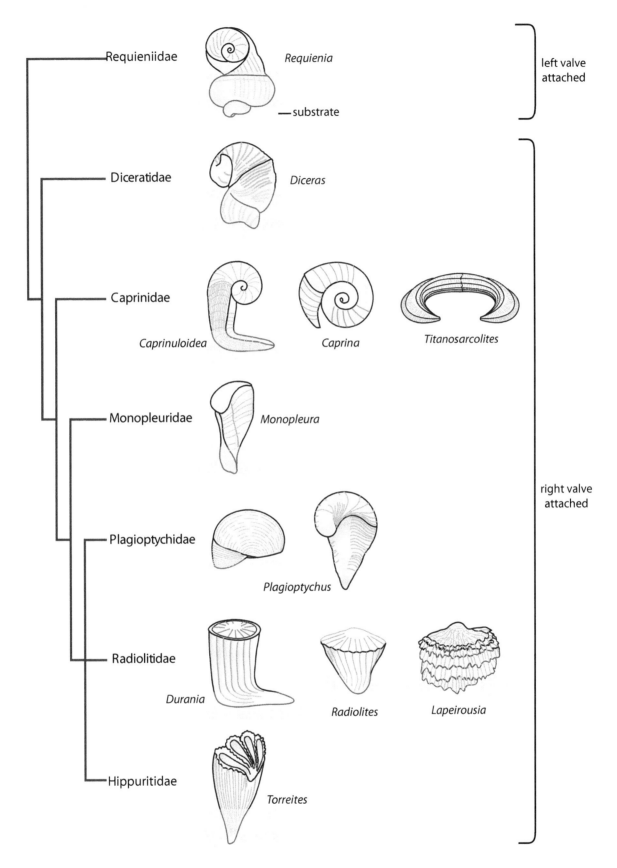

FIGURE 15.43   A simplified phylogeny of rudists (Hippuritoidea) showing some of their diverse shell morphologies. Cartoons redrawn and modified from various sources. Phylogeny based on Skelton, P.W. and Smith, A.B., A preliminary phylogeny for rudist bivalves: Sifting clades from grades, pp. 97–127 in Harper, E.M. Taylor, J.D., and Crame, J.A. (eds.), *Evolutionary Biology of the Bivalvia, the Geological Society*, London, UK, 2000.

longer until some taxa were transferred from the cephalopods to the bivalves (Johnson 2002). They flourished from the Upper Jurassic through the Mesozoic, and although two extinction events depleted their diversity, these events were followed by increased diversification (Jablonski 1996). They were particularly abundant in the Cretaceous where they formed massive and extensive shallow-water reefs in the tropical Tethys Ocean until their complete extinction at the end of the Cretaceous. At their height, in the areas where they occurred, they were the most important tropical reef-building organisms. It seems probable that at least some of their success was due to toleration of the rather extreme conditions of the Cretaceous when tropical seas were more saline and 6–14°C warmer than today. Because most corals would not have flourished in these conditions, suggestions that they out-competed corals are probably exaggerated (e.g., Gili et al. 1995; Johnson 2002). The massive accumulations of porous rudist shells in areas such as the Middle East and the Gulf of Mexico became important oil reservoirs (e.g., Johnson 2002). The rudists had begun to decline about 2.5 million years before the end of the Cretaceous but survived until the end of that period (Steuber et al. 2002).

The rudists probably evolved from megalodontids (Skelton & Smith 2000), a group that originated in the Silurian and coexisted with the true rudists through much of the Mesozoic, but the derivation of the megalodontids themselves is unclear.

Jurassic rudists typically had wide, coiled bases, with both valves of similar shape, and they were often elongate (Figure 15.43). Later rudists were more erect, with cone or pillar-like forms with narrow bases and more ornamentation (Kauffman & Johnson 1988). The non-attached valve was coiled in early forms but became smaller and less coiled. The Cretaceous reef-building rudists had a lid-like upper valve, and the lower (attached) valve was long and conical or even tubular, some more than a metre long.

The hippuritids were among the most modified of the rudists and are arguably the most highly modified bivalves. Their shell consisted of a small operculum-like left valve which capped a much larger, more or less cylindrical or conical, right valve. The lid-like left valve had two long tusk-like pegs (modified hinge teeth) on its inner surface and a socket that received the single peg-like tooth of the attached right valve. The right valve had deep sockets that received the teeth of the left valve. This unique dentition is called *pachydont*. Two other peg-like or plate-like projections, the myophores, had the two adductor muscles attached to them. The arrangement of the adductor muscles suggests that the valves were almost immovable and probably only separated by a very narrow gape (Skelton 1976), although this was disputed by Seilacher (1998). The small left valve possessed a complex canal and pore system, and the right valve contained a series of longitudinal canals and cavities, making it much lighter than it would have been if composed of solid calcium carbonate. The body was contained within a narrow, short cavity in the right valve in hippuritids, being sealed off behind by septae (tabulae) as the animal grew.

How these highly modified bivalves functioned is questionable. Two grooves (oscules) in the left valve probably provided an exit for faeces and pseudofaeces. Water was probably drawn through the pores in the left valve into the mantle cavity where it passed through the gill and out over the mantle edge. Skelton (1976) suggested that the mantle itself may have been involved in food collection, given the likely small size of the gill. Another suggestion is that the complex system of canals on the upper valve may have contained tissue with 'algal' symbionts (e.g., Seilacher 1998), but while this is a possibility in a few rudists, in many it is highly unlikely because thick shell overlaid the canals (e.g., Gili et al. 1995) which were more likely to have been used as water canals.

The ligament in rudists is invaginated into a ligamental groove, which first appeared in the Caprotinidae (Perkins 1969) and was unmodified in the basal Requiniidae. This ligamental groove facilitated more uncoiled shell designs (Yonge 1967b; Steuber 1999). In the most modified families (Radiolitidae and Hippuritidae), the absence of a functional ligament allowed upright growth to occur (Seilacher 1998; Steuber 1999) (see also Section 15.4.1.3).

The modern bivalve most resembling some of the early Hippuritoida is *Chama* (Chamidae), as noted by Yonge (1967b), but *Chama* differs from them in details of shell structure and was thought by Kennedy et al. (1970) to have evolved from carditids. However, recent molecular analyses place the chamids closer to cardiids and other veneroideans, where the rudists are also located (e.g., Carter et al. 2011).

15.8.6 ANOMALODESMATA (= POROMYATA)

This entirely marine group has been treated as a separate higher taxon, often at subclass rank, in most bivalve classifications until recently, when they were included within the euheterodonts. Anomalodesmatans are morphologically and biologically diverse and include such divergent taxa as 'watering pot shells' (Clavagellidae), *Chama*-like Cleidothaeridae, and septibranchs.

Most anomalodesmatan shells lack hinge teeth (i.e., are edentulous), but some have hinge teeth that are assumed to be secondarily derived (Yonge & Morton 1980). Cuspidariids, laternulids, thraciids, and cleidothaerids have spoon-shaped chondrophores housing the internal ligament with this arrangement perhaps evolving separately in the three major clades (Harper et al. 2006). Lying within the internal ligament is usually a small shelly plate, the *lithodesma*. It has been suggested that this structure may either aid in the alignment of the valves or increase thrust when they open (Yonge 1976; Yonge & Morton 1980). Somewhat similar structures have been reported from a few other bivalves, but they are commonly seen only in the anomalodesmatans.

The shells of all anomalodesmatans are entirely aragonitic, and many are prismatonacreous (Taylor et al. 1973), a condition often considered plesiomorphic in bivalves and shared by various other bivalve taxa such as some protobranchs and a range of pteriomorphian taxa (Taylor 1973) (see Section 15.3.1.3) but is not found in any other living heterodonts. A secondary homogeneous shell structure is seen in several taxa (Cuspidariidae, Thraciidae, some poromyids, and some lyonsiids) and has independently evolved at least three times (Harper et al. 2006).

As noted above, calcareous spicules are a feature of the periostracum of many anomalodesmatans, and this feature is thought to be plesiomorphic, with subsequent loss in several lineages (Harper et al. 2006).

The mantle margins are fused and, as in most heterodonts, some have three apertures, but a fourth aperture[8] of uncertain function, near the base of the siphons, is found in many anomalodesmatans but absent in septibranchs and clavagellids. It may facilitate removal of pseudofaeces (Atkins 1937c) or the release of pressure when the valves are clamped shut (Morton 1980).

The outer surface of the mantle of some anomalodesmatans has *arenophilic glands* (Prezant 1981, 1985; Morton 1987a). These occur in radial rows on the outer surface of the periostracum and discharge tufts of sticky mucoid threads which glue sand grains and other debris to the outside of the shell (Harper et al. 2006).

15.8.6.1.1 Phylogeny and Classification

In most recent molecular analyses of bivalves, the 'Anomalodesmata' is nested within the Euheterodonta, although the position has varied (Giribet & Distel 2003; Taylor et al. 2005, 2007b; Harper et al. 2006) A different result was found by Sharma et al. (2012) which had this group as sister to the euheterodonts. This position between the archiheterodonts and euheterodonts is similar to that reported by other authors (Campbell 2000; Dreyer et al. 2003; Harper et al. 2006), including in the combined morphological and molecular analysis of Giribet and Wheeler (2002). Thus, there appears to be support for the anomalodesmatans being the sister group to the euheterodonts rather than being nested within them, and this scenario appears to be better supported by the morphological evidence. The highly anomalous results that were obtained in two analyses, where the group formed a clade that was sister to the Heterodonta + Pteriomorphia (Adamkewicz et al. 1997; Plazzi et al. 2011), are not considered further.

Relationships within the Anomalodesmata have been investigated by Dreyer et al. (2003) using molecular data, Harper et al. (2006) using both molecular and morphological data sets, with the latter analysis indicating that the main carnivorous lineages were probably monophyletic, and Williams et al. (2017) using mitochondrial genome data.

The position of the 'septibranch' anomalodesmatans has been controversial. Cuspidariids were considered to be close to protobranchs by Runnegar (1974), using shell morphology, and Purchon (1956), based on supposed stomach similarities. Such findings led to the notion that the septibranchs should be treated as a separate order or suborder (e.g., Knudsen

1970; Barnard 1974; Bernard 1979; Allen & Morgan 1981) or even a subclass (Franc 1960). Sperm morphology and some other characters suggested they had affinities with myoideans (Salvini-Plawen & Haszprunar 1982; Healy 1996). However, the idea that the septibranchs were a separate group from the other anomalodesmatans has been shown not to be the case (Morton 1982b; Harper et al. 2000a, 2006). Various features of their anatomy, such as the pallial muscles, arenophilic radial mantle glands (Sartori et al. 2006) (see Chapter 3), and the gill structure, suggested to Morton (2003b) that the septibranchs are linked with the Pholadomyidae.

15.8.6.1.2 Biology and Ecology

Anomalodesmatans are probably the most biologically and morphologically diverse of all groups of bivalves. While most are shallow-burrowing suspension feeders, some are carnivores, feeding on small crustaceans and polychaetes (e.g., Morton 1981a) sucked into their inhalant siphon (see Chapter 5, Figure 5.5). This carnivorous feeding mode is practised by a few, mostly deep-water, families (collectively known as septibranchs) with a reduced gill or entirely lost gill filaments, as discussed in more detail below. One of these, a poromyid, is byssally attached (Leal 2008). Food is sucked in via the extendible inhalant siphon, the gills are reduced or lost, and a muscular pumping septum (hence the term 'septibranch') divides the mantle cavity into inhalant and exhalant chambers (Figure 15.45). The stomach has developed a gizzard and secretes proteases (Harper et al. 2006).

The ctenidia of the non-septibranch anomalodesmatans are *heterorhabdic* and deeply plicate with the inner demibranch complete and the outer with only the descending lamellae (Harper et al. 2006). In the carnivorous taxa, the gill bases are modified as muscular septa (Allen & Morgan 1981) which are perforated by simple openings (ostia), through which water is pumped by the muscular activity of the septum, from the inhalant to the exhalant chambers of the mantle cavity. A transitional series is shown in Figure 15.44 that gives an idea of the possible course of evolution in gill reduction and loss. Thus the euciroids, lyonsiellids, and verticordiids may possess reduced gills in addition to having a thin septum with many pores (Allen & Turner 1974). However, the gills are almost absent in cuspidariids and poromyids, and the septa are thick and muscular with relatively few pores (Allen & Morgan 1981). As a result, in a series from Verticordiidae through Poromyidae to Cuspidariidae there is reduction in the gill, an increase in the thickness and muscles of the septum, and a reduction in the number of pores (Allen & Morgan 1981; Harper et al. 2006).

The carnivorous habits of the septibranchs are facilitated by highly sensory siphonal tentacles on the extremely expandable inhalant siphon. Their reduced, very modified ctenidia cannot engage in suspension feeding, and, in most species, the reduced labial palps lack any sorting capability. These bivalves have a much more muscular gut than other bivalves, including the much-simplified stomach with a muscular gizzard (the Type II stomach of Purchon 1987b). Intermediate

---

[8] A fourth pallial aperture is also seen in some other heterodonts, including Tellinoidea, Mactridae, Solenidae, and some hiatellids (Harper et al. 2006).

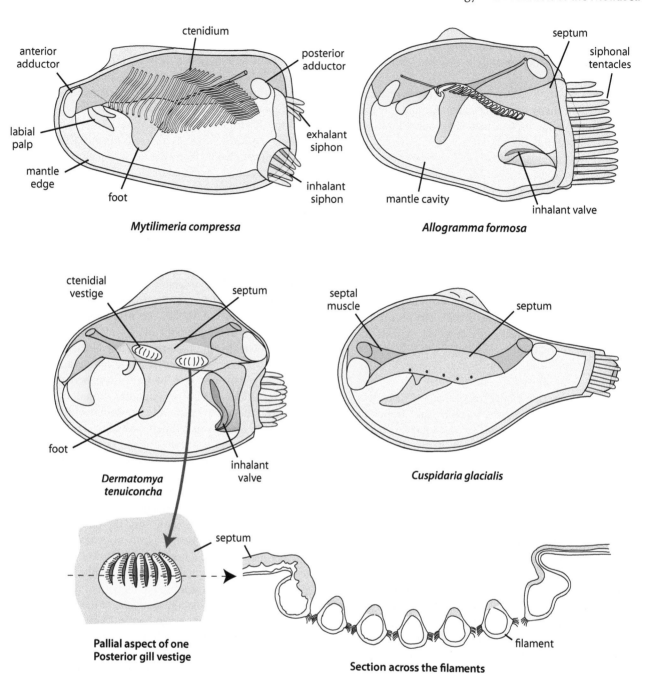

FIGURE 15.44 A possible transitional series between Lyonsiellidae (top row), one of which (*Mytilimeria*) has a functional ctenidium, to (middle row) vestigial ctenidia in a poromyid, and the septibranch condition in a cuspidariid. All have the left valve removed. Illustrations redrawn and modified from the following sources: *Mytilimeria* and *Allogramma* (Allen, J.A. and Turner, R.D., *Phil. Trans. R. Soc. B*, 268, 401–536, 1974) (as *Lyonsiella* spp.); *Dermatomya* (as '*Poromya oregonensis*'); and *Cuspidaria* (Ridewood, W.G., *Phil. Trans. R. Soc. B*, 195, 147–284, 1903).

stomach morphologies in the suspension-feeding anom-alodesmatans (euciroids, lyonsiellids, and some verticordiids) are transitional between the normal heterodont type (the Type IV stomach of Purchon 1987b).

The Parilimyidae was suggested by Morton (1982b) to be carnivorous because *Parilimya* has long siphons, small palps, and a modified stomach, but there is no direct evidence from stomach contents. If parilimyids are carnivorous, Harper et al. (2006) suggested that carnivory must have evolved twice within the anomalodesmatans, because they are phylogenetically distinct from septibranchs.

Another group of anomalodesmatans, the Clavagelloidea (the 'watering pot shells'), contains two families, the Clavagellidae and Penicillidae (Morton 2007), both having

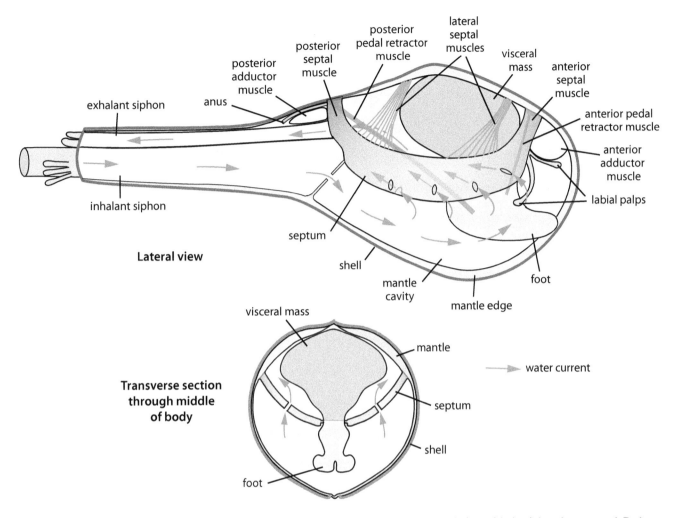

FIGURE 15.45 *Cuspidaria cuspidata*, an advanced septibranch. The upper figure is a lateral view with the right valve removed. Redrawn and modified from Reid, R.G.B. and Reid, A.M., *Sarsia*, 56, 47–56, 1974.

the animal enclosed in an adventitious tube or 'crypt' (Figure 15.10). Clavagellids have an internal ligament and are dimyarian, while penicillids have an external ligament and are essentially amyarian. Clavagellids have the left valve of the true shell included in the adventitious crypt, and the right valve is free inside it, whereas penicillids have both valves fused into the crypt. In both families, the base of the crypt has tubes or even a perforated plate through which water is sucked into the mantle cavity by the pumping action of the modified foot. This mode of feeding may have been derived from ancestral anomalodesmatans that feed on material collected via the pedal gape, such as seen in living Pholadomyidae (Morton 1980). At least two groups of clavagelloideans have independently become cemented (Morton 2003a), and the clavagellid *Bryopa* is a borer (Savazzi 2000).

15.8.6.1.3 Diversity and Fossil History

Anomalodesmatans have existed since the Lower Ordovician (Cope 1997b; Sanchez & Vaccari 2003), with significant radiations in the late Paleozoic and Mesozoic when they comprised a major part of the burrowing bivalve fauna (Runnegar 1974; Morris et al. 1991; Harper et al. 2006). They declined in relative abundance since the Mesozoic, possibly due to competition from heterodont bivalves.

Morton (1985b) argued that anomalodesmatans, unlike many other bivalve lineages, diversified anatomically but never showed much diversity at the species level. He suggested this might be due to their occupation of narrow marginal niches.

16 The Scaphopoda

16.1 INTRODUCTION

This small class of molluscs is commonly known as 'tusk shells' or 'elephant tusk shells' because of their characteristic tusk-shaped, curved, tubular shell. They are marine, benthic, infaunal micro-predators. These bilaterally symmetrical animals are surrounded by the shell which is open at both ends. The larger opening is ventral and the smaller opening dorsal, while the concave part of the shell is anterior and the convex side posterior. Sizes range from about 2.5 to 150 mm in length, with most around 40 to 60 mm.

Largely because of their unique shell morphology, they have long been recognised as a different group. The name Scaphopoda (from the Greek *skaphos* which means scoop, trough, or vessel) is commonly interpreted as 'shovel-footed', a reference to the morphology of the foot of these animals.

16.2 PHYLOGENY AND CLASSIFICATION

Two orders are recognised (see Box 16.1), Dentaliida and Gadilida, the former generally of substantially larger size than the latter. The two orders also differ in shell shape, radular features, and anatomy (Palmer 1974; Steiner 1992a; Shimek & Steiner 1997).

> ### BOX 16.1 THE MAJOR GROUPS OF SCAPHOPODS
>
> Order **Dentaliida** – the shell is usually longitudinally ribbed or smooth, being widest at the anterior aperture. The foot has a pair of lateral lobes and retracts by contracting or bending into the shell.
> Order **Gadilida** – the shell is typically smooth, polished, and with the widest portion some distance behind the aperture. The foot terminates in a fringed disc and retracts by inverting within itself.
>
> There are two suborders of Gadilida, Entalimorpha, and Gadilimorpha.
> See the Appendix for detailed classification.

Most Dentaliida were placed in Dentaliidae in the genus *Dentalium*, and gadilids were mostly included in the Gadilidae in the genus *Cadulus* until relatively recently. Particularly since the 1980s, more families and genera have been used to reflect minor anatomical and other morphological differences. Lists of the supraspecific taxa (Steiner & Kabat 2001) and species names (Steiner & Kabat 2004) are available.

Several morphological and a few molecular treatments of scaphopod phylogeny have been undertaken and are reviewed by Reynolds and Steiner (2008).

16.2.1 SISTER GROUP RELATIONSHIPS

Some of the disagreement regarding conchiferan phylogeny revolves around the placement of the Scaphopoda on the molluscan tree (see Chapter 12 for discussion).

Previously the extant sister group of scaphopods was thought to be the bivalves, with the two classes forming a clade Diasoma (e.g., Runnegar & Pojeta 1974; Steiner 1992a). Common features were a burrowing retractile foot, a reduced head, similar attachment to their shell, and shell structure, although unlike bivalves, scaphopods have a buccal apparatus and radula and have lost the ctenidia. A comparison of the anatomical attributes of both protobranch bivalves and scaphopods was carried out by Simone (2009) who found support for Diasoma, although few outgroups were included in his analysis. While early observations on scaphopod development also suggested bivalve affinities, this too has been disputed.

The extinct rostroconchs (notably the ribeiriid rostroconchs) have been proposed as a likely sister taxon (e.g., Pojeta & Runnegar 1976) (see Chapter 13), but this was disputed by Steiner (1992a) on morphological grounds, notably because of an unlikely major change in body axes. The reinterpretation of scaphopod body axes by Sigwart et al. (2017), which clarified the ventral position of the foot, did not falsify Steiner's criticism, and the putative relationship between scaphopods and rostroconchs remains suspect.

Recent studies of shell formation, and the architecture of the central nervous system, suggest affinities with gastropods and cephalopods (Wanninger & Haszprunar 2001; Sumner-Rooney et al. 2015). In addition, the scaphopod gut is U-shaped like that of gastropods and cephalopods, but that of bivalves is more linear, as probably was the rostroconch gut. Some molecular analyses suggest that scaphopods are a sister group to cephalopods (e.g., Giribet et al. 2006), the so-called helcionellid concept (Steiner & Dreyer 2003; Passamaneck et al. 2004), while phylogenomic studies have them either as a sister to Gastropoda (Smith et al. 2011) or Bivalvia + Gastropoda – a grouping named Pleistomollusca (Kocot et al. 2011). This lack of resolution is perhaps a result of the actual sister taxon being extinct.

An interesting shared character possessed by scaphopods and the aculiferan taxa is that cilia possess two roots whereas the rest of the conchiferans, including monoplacophorans (Lundin and Schander 2003), have one. Some bivalves also have paired rootlets, but Lundin et al. (2009) have argued these are not homologous with the aculiferan condition.

Given the loss of the gill and marked reduction and simplification of the pericardial complex, the ancestral scaphopod may have been minute. The significant differences in the structure and mechanics of the foot (see Section 16.3.2) in the two extant groups of scaphopods can be best explained by this

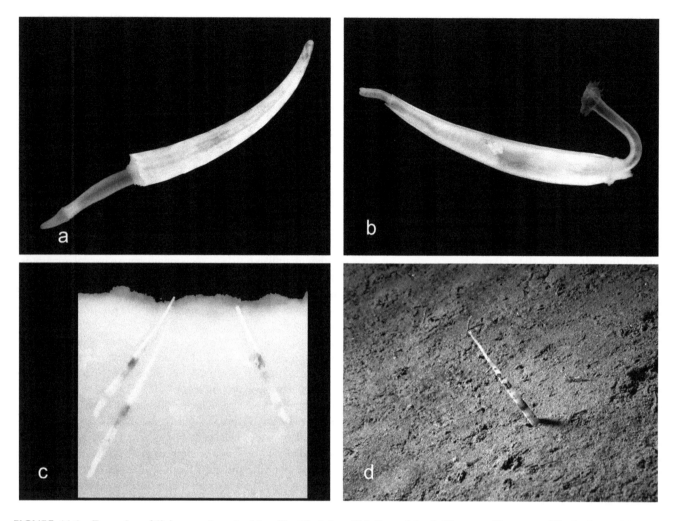

FIGURE 16.1 Examples of living scaphopods. (a) unidentified dentaliid, Santo Island, Vanuatu. (Courtesy of D. Barbant - MNHN.) (b) *Cadulus* sp. (Gadilidae), Madang, Papua New Guinea. (Courtesy of L. Charles - MNHN.) (c) *Rhabdus rectius* (Dentaliidae) that have burrowed into translucent agar, Bamfield, BC, Canada. (Courtesy of J. Sigwart.) (d) *Rhabdus rectius* (Dentaliidae) *in situ*, Cowlitz Bay, Washington. (Courtesy of R. Schimek.)

organ being relatively small and simple in the ancestor and modified separately as the two lineages evolved and became more specialised (Figure 16.2).

16.3 MORPHOLOGY

16.3.1 SHELL

The tusk-shaped, tubular shell (Figure 16.1) has a small dorsal (apical) aperture and a larger ventral aperture, with the concave side of the shell being anterior. These axes were misinterpreted until recently (Sigwart et al. 2017), with the larger end of the shell formerly considered anterior instead of ventral.

Generally, the shell of members of the Dentaliida is longitudinally ribbed or smooth, and the ventral aperture is at the widest point while the shells of Gadilida typically lack sculpture, are polished, and the widest portion is some distance behind the aperture. Scaphopod shell colour is usually white or yellowish to orange, but a few species are green or brown.

Many species have transparent or semi-transparent shells and can be identified to sex based on visible gonads (McFadien-Carter 1979; Lamprell & Healy 1998).

Scaphopod shell structure varies little and has been described by Smith & Spencer (2016). The shell is aragonitic, and in most taxa, the shell comprises three layers, an outer prismatic or irregular crossed lamellar layer, a middle crossed lamellar layer, and an inner prismatic or irregular crossed lamellar layer. The innermost layer has been lost in one gadilid family (Gadilidae) and in two dentaliid families (Gadilinidae, Fustiariidae). In the Dentaliida this loss has been accompanied by the appearance of an irregular crossed lamellar outer layer, instead of prismatic. The only other variation in scaphopods is a prismatic inner layer in the Rhabdidae (Dentaliida) and Pulsellidae (Gadilida). Except for the irregular crossed lamellar outer layer in the Dentaliida, all shell structure variation in the scaphopod is homoplastic between the Dentaliida and Gadilida (Smith & Spencer 2016 and references therein). The plesiomorphic shell structure of scaphopods most likely consisted of an outer prismatic layer, a

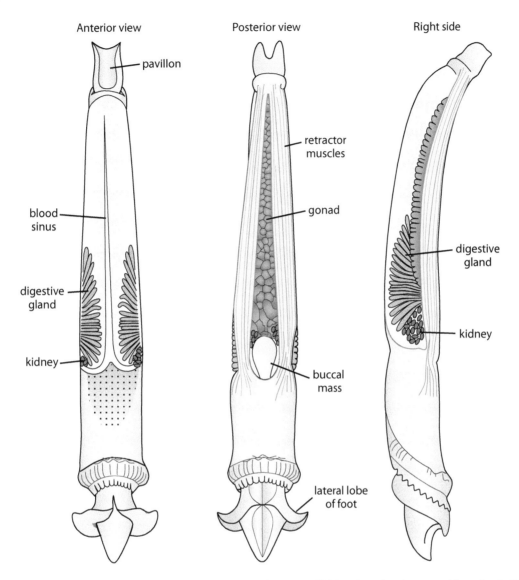

Anterior view   Posterior view   Right side

FIGURE 16.2 Anterior, posterior, and lateral views of the animal of the dentaliid *Antalis vulgaris* removed from its shell. Redrawn and modified from de Lacaze-Duthiers, H., *Ann. sci. nat.* Zool., 6, 225–228, 319–385, 1856–1857.

middle crossed lamellar layer, and an inner irregular crossed lamellar layer.

The protoconch (see Section 16.4.4) is lost early in the development of the adult shell. As the shell grows the apical (dorsal) end, which is often notched, is periodically decollated, thus increasing the width of that aperture but with the resulting loss of ontogenetic information (Reynolds 2002, 2006).

16.3.2 HEAD-FOOT, CAPTACULA, AND MANTLE

Scaphopods live with their ventral aperture pointed down in the substratum and the apical (dorsal) end of the shell projecting above the substratum. The simple head consists of a large snout (also called a proboscis or oral cone) and lacks eyes. The body (Figure 16.2) is suspended from the concave anterior part of the shell and is surrounded by the mantle cavity which extends posteriorly around the viscera on the lateral sides, and dorsally to the apex. The mantle is entirely within the shell,

and its ventral margin is mainly responsible for secreting the shell; the dorsal margin dissolves the juvenile shell as the animal grows. Dorsally, the mantle is a sleeve of tissue (the pavillon) which is open along the posterior side, and its epithelium has many sensory receptors (Steiner 1991). In some species, it secretes a secondary shell pipe by extending beyond the apical aperture (Palmer & Steiner 1998). A valve, vertical in Gadilida and horizontal in Dentaliida, separates the pavillon from the mantle cavity and regulates water flow (Steiner 1991; Shimek & Steiner 1997).

The ventral mantle forms a thick ring of tissue at the aperture and is supported by a ring of cartilage-like connective tissue which is an insertion site for mantle muscles. A sphincter muscle controls the size of the ventral aperture (Steiner 1991). The outer face of the thick fold forming the ventral mantle edges is lined with glandular epithelium in Dentaliida, and in most Gadilida this region has short sensory papillae. In both taxa, the mantle cuff is delimited by the periostracal groove. On the inner side of the central fold, the epithelium is simple, while beyond the fold it is glandular (Steiner 1991; Shimek & Steiner 1997).

The foot of Dentaliida is convergently similar to that of protobranch bivalves (Shimek & Steiner 1997; Simone 2009) and, for burrowing, extends from the ventral aperture to a maximum of about half the shell length (usually about one-third). The dorsal part of the foot forms a trough which deepens distally. In Dentaliida, epipodial lobes lie slightly proximal to the end of the foot. The lobes are continuous ventrally, but there is a dorsal gap. They are simple or slightly crenulate, not papillate. In some taxa, when the foot retracts it bends into the shell and can even become S-shaped (Shimek & Steiner 1997). In Gadilida, the foot terminates in a weakly muscular papillate disc covered with thin ciliated epithelium and surrounded by papillae on the edges. There is also one to several terminal papillae in the middle part of the disc. The foot retracts by introverting within itself (Steiner 1992b; Shimek & Steiner 1997).

The foot epithelium of Dentaliida is ciliated laterally and posteriorly but not anteriorly, and the epipodial lobes are also unciliated. The terminal disc of Gadilida is ciliated, but the rest of the foot is not.

The body musculature is mainly related to the foot. In Dentaliida the foot contraction is effected by longitudinal, circular, radial, and oblique muscles that surround a small haemocoelic space and thus form a muscular hydrostat, while in Gadilida the pedal muscles pass through the large pedal haemocoel, and the foot is extended by hydraulic pressure and is thus a haemal hydrostat (Shimek & Steiner 1997). The haemocoel expands behind the foot in Dentaliida. In both groups, the pair of main pedal retractor muscles continues posteriorly to attach near the apical part of the shell, while the outer muscles that make up the wall of the foot merge with the body wall (Figure 16.3).

Many (often several hundred) thread-like tentacles (*captacula*) (Figures 16.4 and 16.5) emerge from two bands of tissue, left and right, between the foot and the mantle. These captacula are extended by cilial action and retracted by muscles, with both circular and longitudinal elements involved (Shimek & Steiner 1997). These highly specialised feeding tentacles probe through the sediment, and their club-shaped terminal distal ends attach to food items which are then conveyed to the mouth (Morton 1959; Dinamani 1964; Gainey 1972; Poon 1987; Shimek 1988; Shimek & Steiner 1997; Reynolds 2002).

The attachment of particles to the captacula is facilitated by sticky mucus (Shimek & Steiner 1997). In the captacula of the dentaliid *Graptacme calamus*, Byrum and Ruppert (1994) showed that a longitudinal band of cilia transports small food particles along the filament, but in some taxa, the filaments (not the club-like end) are unciliated (see Section 16.4.3). The terminal club-like swelling of each captaculum has a ganglion and three kinds of gland cells (Figure 16.5), at least one of which secretes mucus involved in food transport and another produces the adhesive mucus (Byrum & Ruppert 1994).

16.3.3 Mantle Cavity

The mantle cavity is of the *peripedal* type, as in chitons, bivalves, and monoplacophorans (Salvini-Plawen 1980), but lacks ctenidia or other kinds of gill (Figure 16.6). The cavity is large and surrounds the lateral and ventral parts of the body, and the anus and kidney apertures open to it. Water is drawn into the mantle cavity through the apical and ventral apertures by cilial action, especially from cilia on a series of transverse ridges on the middle part of the mantle cavity. These ciliated bands run transversely along both the body and mantle walls

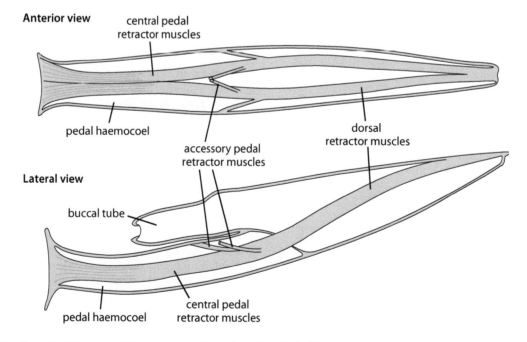

FIGURE 16.3 Simplified diagram of the major central pedal muscles of *Gadila*. Redrawn and modified from Shimek, R.L. and Steiner, G., Scaphopoda, pp. 719–781 in Harrison, F.W. and Kohn, A.J. (eds.), *Microscopic Anatomy of Invertebrates. Mollusca 2*, Vol. 6B, Wiley-Liss, New York, 1997.

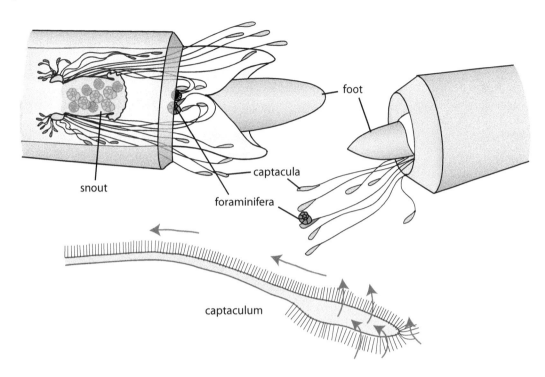

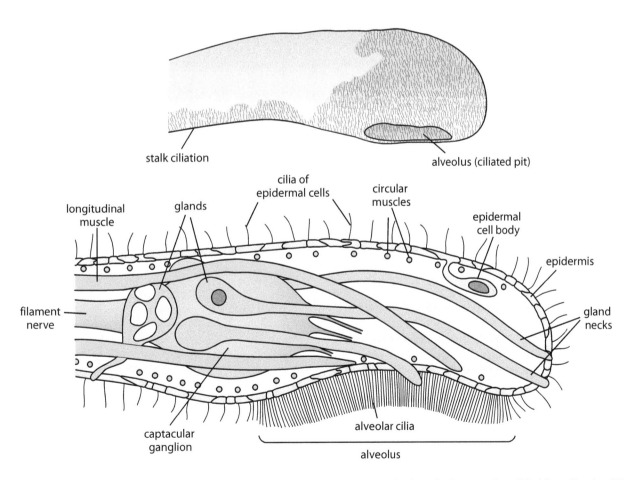

FIGURE 16.4 Feeding mechanism in *Tesseracme quadrapicalis* (= *Dentalium conspicuum*) showing a dorsal (upper left) and dorso-lateral view (right) of the head and foot and a single captaculum. Redrawn and modified from Dinamani, P., *Proc. Malacol. Soc. Lond.*, 36, 1–5, 1963.

FIGURE 16.5 Upper figure a lateral view of the distal end of a captaculum of *Antalis dentalis* drawn and modified from Beesley, P.L. et al., (eds.), Mollusca: The southern synthesis. Part A, CSIRO Publishing, Melbourne, VIC, 1998a. Lower figure is a diagrammatic lateral view showing the internal structure of the captacula of *Graptacme calamus* redrawn and modified from Byrum, C.A. and Ruppert, E.E., *Acta Zool.*, 75, 37–46, 1994.

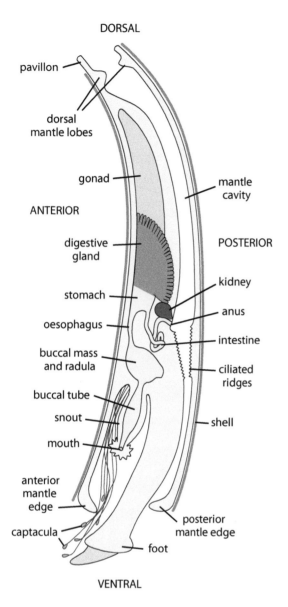

FIGURE 16.6 Diagrammatic longitudinal section of a dentaliid scaphopod. The anus is on a pulsatile anal bulb, on either side of which lie the two kidney openings and a pair of slit-like openings to the haemocoel. Redrawn and modified from Pelseneer, P., Mollusca, pp. 1–355 in Lankester, E.R. (ed.), *A Treatise on Zoology*, V. Adam and Charles Black, London, 1906.

and range from one in tiny species to around 30 in larger species (Reynolds 2002). Respiration occurs via the surface of the mantle cavity, which is especially highly vascularised below the ciliated ridges. There is no continuous flow of water, but instead muscular action, at intervals, expels exhalant deoxygenated water (which also carries waste) through the apical aperture. The captacula may also carry out a cleansing function within the mantle cavity (Reynolds 2002).

16.3.4 DIGESTIVE SYSTEM

A general account of the scaphopod digestive system is given in Chapter 5. The U-shaped gut (see Chapter 5, Figure 5.44) commences with the mouth at the distal end

of the snout. In Dentaliida the mouth has elaborated labial appendages, but in Gadilida these are smaller and simple (Steiner 1992a). The mouth opens to a short prebuccal area, which, particularly in Dentaliida, is expanded laterally into glandular pouches that secrete enzymes and can temporarily store food (Figures 16.4 and 16.9) before it is passed to the radula. In the prebuccal cavity, at the proximal end of the snout, there is a single dorsal jaw present in some taxa, and, immediately behind that, a very large odontophore and radula are located on the floor of the buccal cavity. An odontophoral cartilage (or bolster) supports the radula and is comprised of two lateral parts connected by a narrower transverse section. This single, but flexible, unit has several large odontophoral muscles attached to it (Shimek & Steiner 1997).

The cup-shaped subradular organ, which is probably involved in taste, lies on the floor of the buccal cavity in front of the radula. Lateral to the subradular organ are subepithelial gland cells that comprise the foregut glands (Reynolds 2002), but there are no salivary glands.

Relative to body size the radula is the largest found in molluscs. There are consistently five teeth per row (Figure 16.7), comprising a central plate (i.e., tooth), a single pair of lateral teeth, and a pair of lateral plates (teeth). The shape of the teeth differs in function and shape in the two orders. In Dentaliida the rigid and rather inflexible radula acts as a ratchet or occasionally as a grinding organ. The large central teeth are wider than they are high and hold the lateral teeth erect, and those pull material into the gut. In Gadilida the central teeth are small and narrower than they are high. The radula functions to cut and crush hard prey. It is flexible and asymmetric and can close

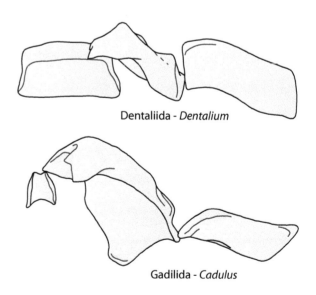

FIGURE 16.7 Half rows of the radula of a dentaliid and a gadilid scaphopod. Upper figure redrawn and modified from Thiele, J., *Handbuch der Systematischen Weichtierkunde. Volume 2, Teil 3, Scaphopoda, Bivalvia, Cephalopoda*, Gustav Fischer, Jena, Germany, 1931, lower figure redrawn and modified from Scarabino, V., *Mém. Mus. nat. hist. nat.*, 167, 189–379, 1995.

rather like a zip. The radular teeth are highly mineralised with mainly calcium and iron, the latter being both ferric and ferrous, which form complexes with phosphate (Reynolds 2002).

Food is broken up by the radula before swallowing and digestion. It is then passed through the oesophagus to the stomach. The oesophagus is a straight tube that enters the right side of the stomach after passing through the muscular diaphragm that separates the peri-intestinal blood sinus from those located more dorsally (Shimek & Steiner 1997). The stomach is muscular, approximately pyriform, and, internally, there is a cuticular gastric shield, a posterior sorting area, and a vestigial gastric caecum. The lobed digestive gland is paired in Dentaliida, and each opens laterally to the digestive diverticula in the dorsal end of the stomach (Sigwart et al. 2017). The single (left) digestive gland in Gadilida has a single opening to the stomach. Enzymes are secreted by the digestive gland and enter the stomach where extracellular digestion occurs. The dorsal portion of the digestive gland overlies the gonad. The digestive gland tubules are lined with both triangular basophilic and secretory cells. The intestine exits the stomach on the upper right side and receives mucus-bound waste from the stomach and digestive gland. It runs ventrally to pass through the muscular septum in front of the stomach where it usually loops three times (the number of loops varies among species, and the range is two to five) (Steiner 1994), the loops lying between the septum and the buccal mass. The rectum differs histologically from the intestine; it passes between the kidneys and opens at the anus at about the middle of the mantle cavity. Faecal material is a thread-like unconsolidated mass. An approximately spherical rectal gland of unknown function opens into the middle part of the rectum, consisting of a mass of ciliated tubules that may be involved in lipoid excretion (Shimek & Steiner 1997).

Digestion commences in the oesophagus, with the enzymes from the prebuccal pouches. Many enzymes are secreted by the digestive gland into the stomach where much of the digestion occurs, being aided by mixing enabled by movements of the muscular stomach walls. The fine food particles enter the digestive gland by way of muscular movements of the stomach, where further digestion and absorption occurs. Absorption of nutrients probably also occurs in the intestine (Reynolds 2002).

16.3.5 Renopericardial System

The pericardial complex is very reduced, being a small, simple vesicle lying posterior to the stomach and attached to it anteriorly. A shallow anterior invagination is thought to be a rudimentary, weakly muscular ventricle (Reynolds 2002). Blood enters the putative ventricle by way of pores between the pericardium and the stomach wall (Shimek & Steiner 1997). Podocytes are present in the pericardial wall (Reynolds 1990b), but the mechanics of ultrafiltration and excretion are uncertain (Shimek & Steiner 1997; Reynolds 2002).

Blood passes around the body in a series of sinuses and lacunae (see Shimek & Steiner 1997 for a detailed account). There are five major sinuses, the pedal sinus, buccal sinus, intestinal sinus, perianal sinus, and abdominal sinus (see

Chapter 6, Figure 6.6). Blood is circulated through haemocoelic sinuses by the movements of the foot and its associated muscles. The blood is transparent and contains leucoblasts, leucocytes, and amoebocytes (Shimek & Steiner 1997; Reynolds 2002). Myoglobin is apparently present in the buccal muscles, but there are no reports of haemocyanin being present in the blood (Reynolds 2002).

A pair of small pores open to the mantle cavity into the haemocoel, and these probably allow some release of blood on a sudden retraction of the animal (Shimek & Steiner 1997; Reynolds 2002).

The paired kidneys are in the perianal sinus and extend dorsally beneath the stomach to meet in the midline, although their lumina do not connect. The kidneys each open to the mantle cavity by way of a large pore lateral to the anus. Renopericardial ducts have been reported but not confirmed, and the renal epithelium consists of two types of vacuolated cells (Reynolds 1990a; Shimek & Steiner 1997).

16.3.6 Reproductive System

The sexes are normally separate, but hermaphrodite individuals are sometimes found (Reynolds 2006). The single large, lobed gonad occupies much of the dorsal part of the body and in both males and females connects with the right kidney by way of a short gonoduct (Reynolds 2002). Sperm emerge into the mantle cavity from the right kidney opening and are then expelled via the apical opening into the water column. Sperm are unmodified and of the aquasperm type with a unique type of acrosomal pit (Lamprell & Healy 1998). Eggs are shed either through the apical or ventral openings. In some dentaliids at least the spawn are liberated in a gelatinous strand and come to rest on the surface of the sediment after release (Shimek & Steiner 1997). Because the rather large, yolky eggs do not float, it has been suggested that they may be fertilised before release (Shimek & Steiner 1997).

16.3.7 Nervous System and Sense Organs

The tetraneural nervous system comprises the oesophageal nerve linking the pedal and cerebral ganglia, a pair of pedal and a pair of visceral nerves (Shimek & Steiner 1997; Reynolds & Steiner 2008; Sigwart & Sumner-Rooney 2015; Sumner-Rooney et al. 2015). The central nervous system ganglia consist of paired cerebral and pleural ganglia that lie on the anterior side beneath the buccal region and a pair of pedal ganglia in the foot. A pair of visceral ganglia lies further dorsally, near the anus, and are connected by the visceral nerves from the pleural ganglia. The cerebral ganglia are connected by a short, wide commissure that fuses the pair (Faller et al. 2012; Sigwart & Sumner-Rooney 2015).

Scaphopods have no eyes or osphradia, although they do have an extraordinary and novel sensory system in the captacula (see above). The only other noteworthy sensory organs are a pair of statocysts that lie posterior to the pedal ganglia and contain numerous statocones, and a probably chemoreceptive subradular organ.

Ciliary sensory receptors are known from the pavillon, the anterior mantle edge, and captacula (Shimek & Steiner 1997; Reynolds 2002). Pigmented epithelial sensory structures called the 'Steiner organ' have also been identified in the anterior mantle edge of *Rhabdus rectius* (Steiner 1991) although the innervation of these putative sensory structures has yet to be characterised (Sumner-Rooney et al. 2015).

The ability of the captacula to locate prey in the sediment is undoubted, but little is known of the sensory mechanisms involved. Bilyard (1974) experimentally demonstrated that *Antalis entalis* could distinguish living foraminiferans from empty tests and hypothesised that the captacula could detect the extended cytoplasm of living foraminiferans. Anatomically, a nerve runs along each captaculum and terminates in a small ganglion in the distal captacular bulb, and a few minor nerves run to the bulb epithelium from this ganglion. The alveolus, or groove, in the captacular bulb is apparently the main sensory part of the captaculum.

16.4 BIOLOGY AND ECOLOGY

16.4.1 HABITS AND HABITATS

Scaphopods are found in all the major oceans where they are infaunal, living in offshore sediments and rarely intertidally. They are also known from the deep sea and are recorded to around 6,000 metres (Reynolds 2002). Most scaphopods spend most of their lives buried in sediment and are often even below the surface of sediments, thus avoiding predators. Certain gadilidans have been shown to burrow up to 40 cm at a rate of 1 cm/sec (Shimek 1989, 1990). Many Dentaliida burrow with the concave side just below the substratum (Reynolds 2006). Thus, the way they are often depicted with the apical end extending from the sediment, while sometimes occurring, especially during spawning, is not typical.

While most scaphopods tend not to be abundant organisms, most occur in patchy distributions and can be locally highly abundant. One well-studied species, the north east Pacific *Rhabdus rectius*, can be abundant in silty sediments with sometimes over 50 per m² (Shimek 1990).

16.4.2 LOCOMOTION

Scaphopods move and burrow by extending their muscular foot. The epipodial lobes or disc that surround the distal end of the foot are held against the sides of the foot as it is thrust forward and then the disc or flaps expand (Figure 16.8). The expanded disc or flaps act as an anchor for the foot which the scaphopod can pull against by contraction of the powerful pedal retractor muscles, thus drawing the shell deeper into the sediment (Trueman 1968; Shimek & Steiner 1997).

In Dentaliida, the pedal lobes are usually depicted as having concave upper surfaces (as in Morton 1959 [Figure 16.8]), but observations on living material of two taxa by J. Sigwart (pers. comm., March 2019) suggests that they are much more flexible and inflatable.

Foot extension in Dentaliida is enabled largely by hydrostatic muscle action as haemocoelic spaces are small (Figure 16.8). The pedal muscles are arranged antagonistically. In Gadilida the haemocoelic space in the foot is large (Figures 16.3 and 16.8.) and continuous with the main body haemocoel, and the pedal retractors lie free within it. Thus, in this group the foot is extended by using hydraulic pressure, and its protraction can be extremely rapid, much more so than in the Dentaliida. In some Dentaliida, the foot can extend to about the length of the shell, but in a number of Gadilida the foot can extend to twice the shell length (Reynolds 2002).

16.4.3 FEEDING AND DIET

As outlined in Chapter 5, scaphopods feed on foraminiferans and other interstitial organisms, such as ostracods, mites, newly settled bivalves and gastropods, diatoms, detritus, etc. Most species are selective microcarnivores that feed on foraminiferans (Poon 1987; Shimek 1990). Langer et al. (1995) reported that the deep-sea scaphopod *Fissidentalium megathyris* ingested more than 17 foraminiferan species, including agglutinated, porcelaneous, and perforate taxa, while non-foraminiferans never exceed 1% of the total food items ingested. In contrast, the north east Pacific dentaliid *Rhabdus rectius* is more of a generalist, consuming almost anything in the right size range (Shimek 1990) including a substantial amount of sediment (Shimek 1988). Some, but not all, foraminiferan feeders also have sediment in their gut (Steiner 1994).

Feeding commences by the foot creating a cavity around the ventral aperture. The numerous captacula penetrate the sediments and capture prey by way of adhesive secretions from the club-shaped tip (Shimek & Steiner 1997). The captaculae carry food into the mantle cavity, and it is then picked up by the mouth. The food is then stored in the buccal pouch (Figure 16.9) before mastication with the radula prior to it being swallowed.

As reviewed by Byrum and Ruppert (1994), ciliation patterns on the captaculae may account for the differences in feeding reported in the literature, even within families. Those with cilia along the length of the captaculae had sediment as a major component of their diet, including the rhabdid *Rhabdus rectius* and the dentaliids *Tesseracme quadrapicalis* and *Graptacme calamus*, while those species with cilial tufts along the filament, such as species of *Gadila* (Gadilidae), mainly feed on foraminiferans but also take up some sediment. Taxa that lack cilia on the filaments are foraminiferan specialists and include the pusellid *Pulsellum salishorum* and the dentaliid *Antalis entalis*.

Small 'sediment' particles are carried to the mouth or mantle cavity by cilia on the captacula (if they are present), but larger items are moved by retraction and movement of the captaculae. Food accumulates within the mantle cavity and is selected and moved into the mouth or rejected as waste from the anterior shell aperture. Selection before ingestion may be carried out by the subradular organ or the captacula or both. When eating, gadilids hold the prey in the buccal sphincter and use the radula (and jaw if present) to knock pieces off, rotating it between radular strokes (Shimek 2008). Fracturing

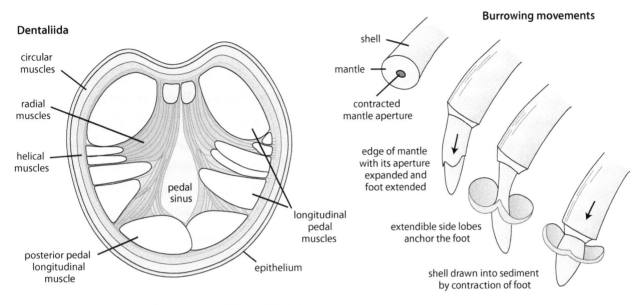

Dentaliida

circular muscles

radial muscles

helical muscles

pedal sinus

longitudinal pedal muscles

posterior pedal longitudinal muscle

epithelium

Transverse section of middle part of foot of *Rhabdus*

Burrowing movements

shell

mantle

contracted mantle aperture

edge of mantle with its aperture expanded and foot extended

extendible side lobes anchor the foot

shell drawn into sediment by contraction of foot

Successive stages in burrowing of *Antalis,* **showing action of foot and extensible lobes**

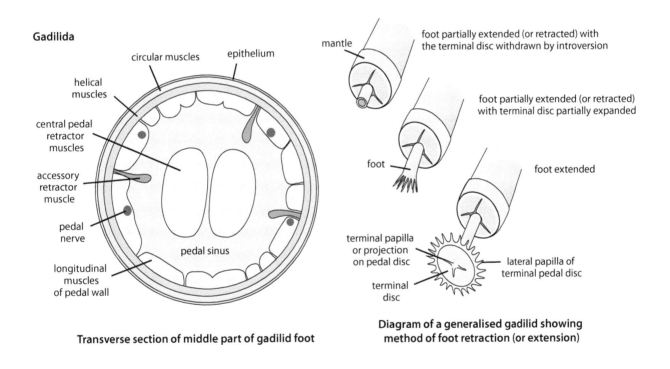

Gadilida

circular muscles

epithelium

helical muscles

central pedal retractor muscles

accessory retractor muscle

pedal nerve

pedal sinus

longitudinal muscles of pedal wall

Transverse section of middle part of gadilid foot

mantle

foot partially extended (or retracted) with the terminal disc withdrawn by introversion

foot partially extended (or retracted) with terminal disc partially expanded

foot

foot extended

terminal papilla or projection on pedal disc

lateral papilla of terminal pedal disc

terminal disc

Diagram of a generalised gadilid showing method of foot retraction (or extension)

FIGURE 16.8 Comparison of the musculature, pedal sinus, and foot morphology in Dentaliida and Gadilida. Figures on upper left and bottom right are redrawn and modified from Shimek, R.L. and Steiner, G., Scaphopoda, pp. 719–781 Harrison, F.W. and Kohn, A.J. (eds.), *Microscopic Anatomy of Invertebrates. Mollusca 2*, Vol. 6B, Wiley-Liss, New York, 1997; figure at top right redrawn and modified from Morton, J.E., *J. Mar. Biol. Assoc. U. K.*, 37, 287–297, 1959; and bottom left from Steiner, G., *J. Molluscan Stud.*, 58, 385–400, 1992b.

of the foraminiferan tests by the radula before digestion in the stomach appears to be typical, and the condition of chemically altered tests in the stomach seems positively correlated with the degree of mechanical damage to the test, suggesting that damaged specimens are more susceptible to dissolution and etching than undamaged ones (Langer et al. 1995).

The foot aids in feeding by constructing a cavity in the sediment. Also, especially in dentaliids, the foot may collect sediment and transport it to the mantle cavity where some may be selected for ingestion. In gadilids, the partially inverted foot forms a tube, and can grasp foraminiferans (Shimek 1988, 1990).

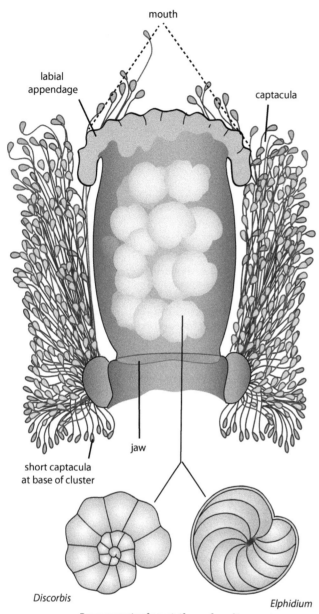

Representative foraminiferans found in snout

FIGURE 16.9 Snout and retracted captacula of the dentaliid *Antalis entalis* viewed from above, after removal of overlying mantle wall. Ingested foraminiferans are temporarily stored in the snout. Redrawn and modified from Morton, J.E., *J. Mar. Biol. Assoc. U. K.*, 37, 287–297, 1959.

16.4.4 Reproduction, Development, and Growth

The female releases her eggs in batches through the apical opening or sometimes one at a time through the ventral opening. The yolky, fertilised egg undergoes spiralian cleavage and develops a polar lobe extrusion. The development of scaphopod eggs has been a model system in experimental embryology, with the first study (Délage 1899) just before the beginning of the 20th century, but there have only been a few studies of the subsequent larval development. Initially, there is a free-swimming trochophore with several ciliated bands (see Chapter 8, Figure 8.31), but in the later trochophore the prototroch cells

are arranged in three rows. The apical organ bears numerous long cilia but is reduced and then lost in the veliger stage, and the velum develops from the prototroch before the larva settles as a post-metamorphosed young adult on the sea floor.

The foot rudiment appears early, then differentiation of the shell gland and mantle. The larval shell (protoconch) is initially bulbous and saddle-shaped and then changes by growth into a tubular form. This latter part has been called protoconch B, the initial part protoconch A (for discussion see Reynolds 2002). The mantle fuses along the ventral surface of the larvae producing a characteristic median fusion line on the embryonic shell.[1] The prototroch is shed during metamorphosis, and the protoconch stops growing. The adult shell then begins to form as well as the foot, captacula, and the buccal apparatus (see Reynolds 2002 for a review of development). Wanninger and Haszprunar (2001) described engrailed expression in the larval development in *Antalis* where it occurs at the margin of the protoconch close to the mantle edge in the larva. A pair of protonephridia are present in the larva and juveniles but become non-functional in the juvenile of the dentaliid *Antalis* after 13 days (Ruthensteiner et al. 2001). By that stage, paired anlagen of the adult kidneys have developed.

The young are able to feed a few days after metamorphosis (Moor 1983; Wanninger & Haszprunar 2001).

16.4.5 Associations

Symbiotic bacteria are found at the vegetal pole of fertilised oocytes of at least two Dentaliida species, and these are incorporated into the developing embryo (Geilenkirchen et al. 1971). Their role in the adult is unknown.

Trichodine ciliates occur in the mantle cavity as do some bacteria, presumably as commensals (Reynolds 2002; Reynolds 2006).

Possible mutualistic relationships between anemones or barnacles and scaphopods are known. *Fissidentalium actiniophorum*, a deep-water species living on the abyssal plain of the north eastern Pacific, has a symbiotic (mutualistic) sea anemone living on the posterior (concave) side of the shell (Shimek 1997; White et al. 1999), and similar relationships are known in a few other scaphopods (Reynolds 2006). Barnacles and a solitary coral have also been recorded attached to living scaphopods (Reynolds 2002; Reynolds 2006).

A few scaphopod parasites have been recorded and include ciliates, trematodes (Ptychogonimidae), a haplosporidium (*Minchinia dentali*), and parasitic associations with copepods, nematodes, and a 'turbellarian' are known (Reynolds 2002; Reynolds 2006).

Several specially adapted 'straight' or 'symmetrical' hermit crabs (Pylochelidae) occupy scaphopod shells as do various sipunculans (Reynolds 2006).

Narrow canals in the outer shell layers are thought to be produced by a boring cyanobacterium, *Hyella* (Reynolds 2002).

---

[1] This surface will become the posterior of the shell after the body reorients during ano-pedal flexure.

16.4.6 PREDATION, DEFENCE, AND OTHER BEHAVIOURAL ATTRIBUTES

Scaphopods are fed on by fish and crabs and predatory gastropods. Their burrowing behaviour presumably reduces predation (Shimek 1990; Reynolds 2006).

16.5 DIVERSITY AND FOSSIL HISTORY

Reynolds (2002) examined latitudinal diversity which showed marked gradients in all four major ocean basins. Diversity reached a peak of around 50 species in each basin, this being around the equator in the Pacific, but at about 20°N in the Atlantic. Diversity peaks for Dentaliida at around 500–800 m and Gadilida at 1,200–2,000 m (Reynolds 2002; Reynolds 2006).

There are approximately 565 living species (Caetano & Santos 2010), and about 500 additional species are known from the fossil record (Reynolds 2006). Members of the class first appear in the Ordovician (see Chapter 13), and the taxon has maintained a slow but steady rate of increase in morphological diversification since then. Records of Ordovician scaphopods have been controversial (Yochelson 2004).

Fossils that resemble scaphopod shells include polychaete tubes (with some recent polychaete tubes described as scaphopods in the past), and even some small fossil cephalopods. In the latter instance, there are a few nautiliform taxa in the Orthoceratidae (subfamily Kionoceratinae) that have been mistaken for scaphopods as they also have longitudinal ribbing and very narrow elongate shells that somewhat resemble scaphopods. They differ in usually being larger, with thinner shell walls, and in having internal septa with a siphuncle (e.g., Yochelson 2004; Yochelson & Holland 2004; Kues et al. 2006; Yochelson et al. 2007) and with only one opening.

Yochelson (2004) speculated that scaphopods developed their specialised habits and feeding mode after calcareous foraminiferans became abundant. He also suggested that their limited morphological diversity could be a consequence of the stability of their restricted habitat and their specialised food source.

16.5.1 THE MAJOR GROUPS OF SCAPHOPODS

The two major groups of scaphopods, Dentaliida and Gadilida, are outlined in Section 16.2 and Box 16.1. A more detailed classification is given in the Appendix.

16.6 HUMAN USES

Shells of *Antalis pretiosum* and *Dentalium neohexagonum* were used as shell money by the natives of the North American Pacific Northwest and California with some dating to over 6,000 years ago (Erlandson et al. 2005) (see Chapter 10). They were also used to make belts, headdresses, vests, necklaces, and other ornaments and are still sometimes used today (Clark 1960; Dubin & Jones 1999).

17 The Cephalopoda

17.1 INTRODUCTION

Cephalopods include the living coleoids (octopuses, squid, cuttlefish, and 'vampire squid'), the 'chambered nautilus', and a diverse array of extinct taxa including 'nautiliforms'[1] ammonites[2] (= 'ammonoids'), and some coleoids including the belemnites. Nearly all of the living examples of these remarkable animals are predatory; although some are pelagic drifters, many swim permanently, and most benthic species are also capable of active swimming. If only the approximately 800 living species are considered, it is a relatively small class, but it has a rich and diverse fossil record. Living cephalopods represent a considerable amount of biomass in the oceans and are an important constituent of the marine fishery catch (Clarke 1996a).

This exclusively marine group shows many modifications from the general molluscan body plan. They are characterised by dorso-ventrally elongated bodies that are usually oriented horizontally, so the originally anterior surface is dorsal and the posterior surface ventral. They are well equipped for active predation, with a mouth armed with beak-like jaws and surrounded by arms and/or tentacles derived from the foot (hence the name cephalopod – head-foot). They have sophisticated sense organs, including highly developed large eyes and brains that make these animals the most intelligent of all invertebrates. The sensory and neural centres are concentrated in the head (cephalisation). Many squid can swim rapidly, and these are the most motile of all non-vertebrate animals. In coleoids, sophisticated nervous and muscular control over chromatophores in the skin enable virtually instant colour changes, while texture and body shape changes also occur in many, especially octopods. The inhalant water enters the mantle cavity through lateral openings, and a mobile siphon-like structure, the *funnel*, serves as the exhalant aperture through which the water can be forcibly ejected during jet propulsion swimming. In coleoid cephalopods, both movement and respiration are facilitated by a complex set of muscles, connective tissue, and collagen fibres in the contractile mantle wall. Many can swim rapidly using jet propulsion, and decabrachians (= decapods)[3] can also hover and swim more slowly and precisely by undulations of the fins along the body.

Most living cephalopods are between 6 and 70 cm in body length (including the tentacles), but the giant squid (*Architeuthis*) and colossal squid (*Mesonychoteuthis*) (see Section 17.13.1.5) are the largest living invertebrates with a body size (mantle length) of over two metres (up to 13 m in length including the tentacles) while the smallest cephalopods are around 2 cm long (mantle length around 1 cm) (von Boletzky 2003a). The largest extinct molluscs are also cephalopods, and these include the largest known fossil invertebrates. The coiled shell of one ammonite (the Cretaceous *Parapuzosia seppenradensis*) reached about 2.5 m across while the straight shell of the Ordovician *Endoceras giganteum* reached up to 5.7 m in length (Klug et al. 2015a).

Living cephalopods include both active swimmers (squid and cuttlefish) and bottom dwellers (various octopods). Many pelagic species of squid have very streamlined bodies for efficient locomotion. These abundant animals are important food sources for larger fish, marine mammals, seabirds (see Chapter 9), and humans (see Chapter 10). Most octopods are benthic predators, but some are pelagic, many living in the deep sea. As relatively large predators, cephalopods play a significant role in marine ecosystems and feed mainly on fish, crustaceans, worms, and other molluscs (e.g., Rodhouse & Nigmatullin 1996) (Figures 17.1 to 17.4).

Cephalopods are well represented in the fossil record, with around 11,000 extinct species described. Not all are shelled forms, as preserved bodies with reduced or remnant shells are occasionally found.

Except for the 'chambered nautilus' (*Nautilus*), all other living cephalopods (the Coleoida) have either internalised or lost the external shell, but two large groups of extinct cephalopods, the nautiliforms and ammonites, which were common in Paleozoic and Mesozoic seas, had external shells. The interior of all calcareous cephalopod shells is chambered. In externally shelled cephalopods, the body of the animal is contained within the last, open section of the shell while the chambers are utilised as a buoyancy device (see Section 17.4.2.3). In coleoids, the shell became internal early in their history and is rudimentary in modern squid, highly modified in cuttlefish, and much reduced or lost in octopods (see Sections 17.3.1 and 17.13.1).

In *Nautilus*, the opening of the shell (the aperture) is covered by a large, tough, fleshy hood which acts as a protective shield when the animal is threatened. Many coleoids, lacking an external shell, use speed, camouflage, and/or inking (see Section 17.4.3), the latter strategy involving the release of an opaque cloud of ink as part of their alarm response.

Squid and fish share many similarities and thus are often cited as examples of convergent evolution (Packard 1972), although there are also considerable differences as discussed by O'Dor and Webber (1986). Some of these differences and similarities are briefly outlined in Box 17.1.

---

[1] We use the term nautiliform instead of 'nautiloid' to include the taxa traditionally treated as 'Nautiloidea' (i.e., cephalopods other than coleoids and ammonites) and which include the first cephalopods. This term, as we use it, includes both straight and coiled shelled forms.

[2] We use this term to informally refer to any member of Ammonitia (i.e., those previously referred to 'Ammonoidea').

[3] This name is often used for members of Decabrachia (= Decapodiformes) (see Hoffmann 2015). The name Decapoda was commonly used in the past, but this is currently used for a group of crustaceans.

FIGURE 17.1 A large ammonite and examples of living *Nautilus* and some Octobrachia other than incirrate octopods. (a) *Nautilus belauensis* (Nautilidae), Palau. © Gerald and Buff Corsi / Focus on Nature, Inc. (b) *Parapuzosia seppenradensis* (Desmoceratidae), Upper Cretaceous, Germany. Courtesy of J. Vendetti. (c) *Argonauta* (Argonautidae), Australia. Courtesy of J. Finn. (d) *Vampyroteuthis infernalis* (Vampyroteuthidae), California. © 2007 MBARI. (e) *Opisthoteuthis* sp. (Opisthoteuthidae), Monterey Bay, California. © 2013 MBARI. (f) *Cirrothauma murrayi* (Cirroteuthidae), Monterey Bay, California. © 2015 MBARI.

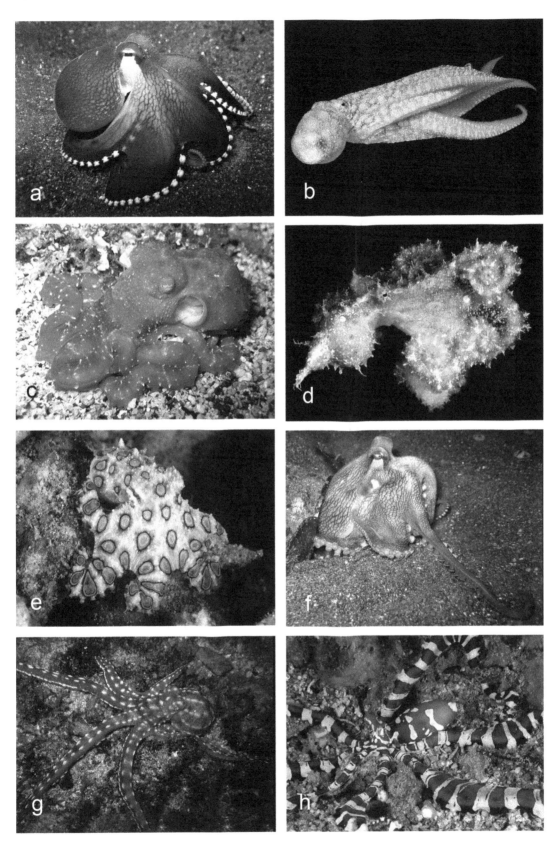

FIGURE 17.2   Examples of living incirrate octopods (all Octopodidae). (a) *Amphioctopus marginatus*, Indonesia. © Christine Huffard. (b) *Abdopus abaculas*, Moorea, French Polynesia. © Roy Caldwell. (c) *Octopus luteus*, Indo-West Pacific. Courtesy of T. Gosliner. (d) *Abdopus abaculas*, Moorea, French Polynesia. © Roy Caldwell. (e) *Hapalochlaena* sp., Indo-Pacific. Courtesy of D. Cowdery. (f) *Amphioctopus marginatus*, Philippines. Courtesy of T. Gosliner. (g) *Callistoctopus ornatus*, Indo-Pacific. © Christine Huffard. (h) *Wunderpus photogenicus*, Indo-Malayan Archipelago. Courtesy of D. Cowdery.

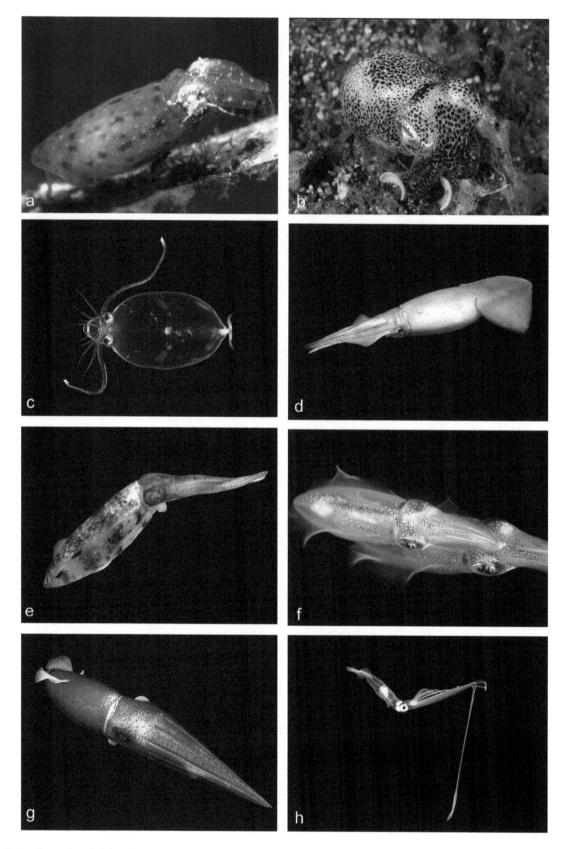

FIGURE 17.3 Examples of living decabrach coleoids other than Sepiidae. (a) *Xipholeptos notoides* (Idiosepiidae), southeastern Australia. Courtesy of A. Miller - Australian Museum. (b) *Euprymna* sp. (Sepiolidae), Indo-Pacific. Courtesy of D. Cowdery. (c) unidentified cranchiid, Northern Gulf of Mexico. Courtesy of E. Widder and NOAA Deep Scope Expedition 2004 (public domain). (d) *Dosidicus gigas* (Ommastrephidae), California. © 2009 MBARI. (e) *Sepioteuthus sepioidea* (Loliginidae), Bermuda. Courtesy of T. Gosliner. (f) *Sepioteuthis lessoniana* (Loliginidae), Indo-Pacific. Courtesy of D. Cowdery. (g) *Histioteuthis heteropsis* (Histioteuthidae), California. 2013 MBARI. (h) *Chiroteuthis calyx* (Chiroteuthidae), California. © 2013 MBARI.

FIGURE 17.4   Examples of living cuttlefish (Sepiidae). (a) *Sepia pharaonis*, western Indian Ocean. © Gerald and Buff Corsi / Focus on Nature, Inc. (b) *Sepia latimanus*, Indo-West Pacific. Courtesy of D. Cowdery. (c) *Sepia latimanus*, Indonesia. © Christine Huffard. (d) *Metasepia pfefferi*, tropical western Pacific. © Christine Huffard.

BOX 17.1   COMPARISONS OF SQUID AND FISH

Squid and fish anatomy and body plans are very different.

Squid are exclusively marine, but fish are common in both marine and freshwater habitats.

Squid and fish have a similar size range (approx. 0.01 to 10 m in body length).

The main means of propulsion differs; fish use muscular movements of the body and fins while squid use jet propulsion for rapid movements and fins for slower swimming. Fast-moving species in both groups have streamlined bodies and fins to provide lift. Both squid (and cuttlefish) and fish can move by using their fins.

Even the most efficient squid use more energy and move more slowly than many fish (O'Dor & Webber 1986).

In both squid and fish, species evolved that can 'fly' for tens of metres over the water.

Some squid and many fish exhibit schooling behaviours.

All cephalopods are carnivorous (including some feeding on minute animals), while fish have a wide range of diets, including some being predators.

Both have adaptations to control buoyancy, although the mechanisms are different. Fish mainly use a gas-filled swim bladder, while shelled coleoids (*Spirula* and *Sepia*) use gas-filled chambers in their shell. Some squid use ammonium (NH_4) ions in the blood to decrease density, while various fish use lipids for the same purpose. A 'swim bladder'-like structure is known in the octopod *Ocythoe tuberculata* (Packard & Wurtz 1994).

Both extract oxygen from the water efficiently, mainly using their gills. While squid (and other cephalopods) use haemocyanin as the oxygen carrier, fish have the more efficient haemoglobin.

Most shallow-water coleoids are semelparous, grow rapidly, reproduce (one to several spawning periods), and then die.[4] In contrast, most fish are iteroparous, growing and reproducing over several years (as do *Nautilus* and *Vampyroteuthis*).

Most fish and squid have a very active lifestyle, but their metabolism differs. Squid convert most of the protein obtained from their food into muscle, which they degrade when extra protein is needed for energy. Unlike fish, little is stored as lipid or glycogen, the vertebrate energy sources.

Squid and fish have superficially similar, highly developed, efficient, lens eyes and muscular control of eye movements. There are, however, several differences in the structure and physiology of their eyes (see Chapter 7).

Both fish and squid have complex statocysts for orientation, although they are structurally different.

Both squid and fish have chromatophores to control their colouration, but the structure and function of the chromatophores are very different. In squid, the chromatophores are operated by muscles that rapidly expand or contract the pigment sacs and are mainly regulated by the nervous system. In fish, the chromatophores are under hormonal control and change colour more slowly through the movement of pigment granules along microtubules within the cell.

Some squid and cuttlefish possess a lateral line system analogous to that in fish (Budelmann & Bleckmann 1988).

17.2 PHYLOGENY AND CLASSIFICATION

The class Cephalopoda has been traditionally divided into two or three subclasses. Older classifications used Dibranchiata and Tetrabranchiata, based on the number (or assumed number) of ctenidia, but most classifications recognise three major groups, 'Nautiloidea', 'Ammonoidea', and 'Coleoidea'. The earliest group of cephalopods is contained within the nautiliforms, and its ancestors gave rise to the ammonites and coleoids. Thus, the traditional concept of 'Nautiloidea' is paraphyletic, and fossil taxa contained in that grouping have been reclassified. Although there is some agreement regarding the phylogenetic relationships and classification of many of these fossil taxa, classifications can differ. For example, some workers (Zhuravleva 1972; Salvini-Plawen 1980) recognised the 'nautiloids' (i.e., the nautiliforms) as two subclasses – Orthoceroidea and Nautiloidea. In other classifications, several subclasses or superorders are recognised. In some classifications, six higher taxa ('subclasses' or 'superorders') are recognised (e.g., Teichert 1988; Wade & Stait 1998) and in others seven or eight (Shevyrev 2005, 2006). There are also alternative classifications that involve some major rearrangements.

A summary of the higher classification we are using is given in Box 17.2. A more detailed outline of the classification we have adopted is provided in the Appendix, and an overview account of each major group is given in Section 17.11.

BOX 17.2 OUTLINE OF THE HIGHER CLASSIFICATION OF CEPHALOPODS

The groups listed below marked with * are variously treated as subclasses or superorders.

Note that names including 'cer' are variously spelt 'cer' or 'cerat' in the literature. We follow the recommendation of the ICZN Article 29.3 (see Bouchet & Rocroi 2005 pp. 8–9) and use 'cerat' to form the names. Thus, for example, Ellesmero**cer**ida should be spelt Ellesmero**cerat**ida. The commonly used endings are also problematic. The 'oidea' ending is universally used for superfamilies but has been widely used by cephalopod workers for higher groups. Consequently, we have adopted the use of endings that conform with general zoological usage.

(Class) **Cephalopoda**

(Subclass) **'Palcephalopoda'** (paraphyletic) (= 'Nautiloidea' in part). We refer to this group informally as nautiliforms.

(Cohort) **Plectronoceratia**

(Orders Plectronoceratida, Protactinoceratida, Yanheceratida), late Cambrian

(Cohort) **Ellesmeroceratia*** (paraphyletic) (= Ellesmerocer(at)oidea, Multiceratoidea)

(Orders Ellesmeroceratida, Ascoceratida, Oncoceratida, Discosorida [sometimes misspelt Discocerida]), Bisonoceratida, late Cambrian to Lower Carboniferous

(Cohort) **Orthoceratia*** (= Orthocer(at)oidea)

(Orders Orthoceratida, Dissidoceratida, Pseudorthoceratida, Lituitida), Lower Ordovician to Triassic[5]

(Cohort) **Endoceratia*** (= Endocer(at)oidea)

(Orders Endoceratida, Intejocer(at)ida), Lower Ordovician to Middle Silurian

(Cohort) **Actinoceratia*** (= Actinocer(at)oidea)

(Order Actinoceratida), Lower Ordovician to Upper Carboniferous

(Cohort) **Nautilia*** (includes *Nautilus*)

(Orders Bassleroceratida, Tarphyceratida [= Barrandeocerida], Nautilida), Lower Ordovician to present

(Subclass) **Neocephalopoda**

(Cohort) **Bactrita***

(Order Bactritida), Lower Devonian to Upper Triassic

(Cohort) **Ammonitia*** (= Ammonoidea) (the ammonites)

(Orders Agoniatitida, Goniatitida, Clymeniida, Ceratitida, Ammonitida), Lower Devonian to Upper Cretaceous

(Cohort) **Coleoida*** (= Coleoidea) (cuttlefish, squid, octopuses)

(Superorder) **Belemnitia**

---

[4] This is not true for a number of deep-water taxa – see Section 17.7.3.

[5] There is one possible record from the Lower Cretaceous (Doguzhaeva 1995).

(Orders Hematitida, Donovaniconida, Aulacoceratida, Phragmoteuthida, Diplobelida, Belemnitida), Upper Triassic to Upper Cretaceous

(Superorder) **Decabrachia** (= Decapodiformes) (cuttlefish and squid)

(Orders Spirulida, Sepiida, Myopsida, Oegopsida, Sepiolida), Lower Cretaceous to present

(Superorder) **Octobrachia** (= Octopodiformes) (octopuses, vampire squid)

(Order) Vampyromorpha, Middle Jurassic to present

(Suborders Vampyromorphina, Loligosepiina)

(Order) Octopoda, Lower Jurassic to present (Suborders Teudopseina, Cirrata, Incirrata)

The vast majority of nautiliforms are extinct, which is why the few living species of 'pearly' or 'chambered' *Nautilus* are often called living fossils. The living species all have a spiral, nacreous shell with interconnected internal chambers, but some of the early fossil forms had narrow curved or straight cone-like shells. Nautiliform cephalopods first appeared in the Furongian (Cambrian) and underwent a rapid diversification in the Ordovician.

The head of modern *Nautilus* (Figure 17.5) is covered with a fleshy hood and surrounded by numerous short, sucker-less tentacles, and there are supposedly primitive features such as a pair of large laterally placed pinhole eyes. The mantle cavity contains two pairs of ctenidial gills, and there is no ink sac. Whether these features were shared with all extinct nautiliforms is not known but is unlikely. To date, no impressions of the tentacles, siphon, or head of fossil nautiliforms have been found, but it is probable they had sucker-less arms like those in modern *Nautilus*, as suckers seem to be a coleoid innovation. Some may have possessed fewer arms (possibly about ten) judging from a few trace fossils that probably represent activity by nautiliforms (e.g., Flower 1955a). They had a beak similar to that of *Nautilus* as fossil beaks are not uncommon and are known as *rhyncholites* (Teichert et al. 1964b) or *rhyncholiths* (upper jaw) and *conchorhynchs* (lower jaw) (Klug 2001).

The groups lumped together as nautiliforms dominated the ancient Paleozoic seas, along with the trilobites, with over 2,500 species described from fossils. They were overtaken by the ammonites in the Mesozoic but, unlike the latter group, one small group of nautiliforms survived the End Cretaceous extinction event and are represented in the living fauna by a few species in two genera of Nautilidae.

The nautiliforms and ammonites were included in the Tetrabranchia because it was supposed that they had two pairs of ctenidia like *Nautilus*. In fact, there is no good evidence indicating how many pairs of ctenidia the extinct groups had (e.g., Klug et al. 2012), other than a few reports from coleoids

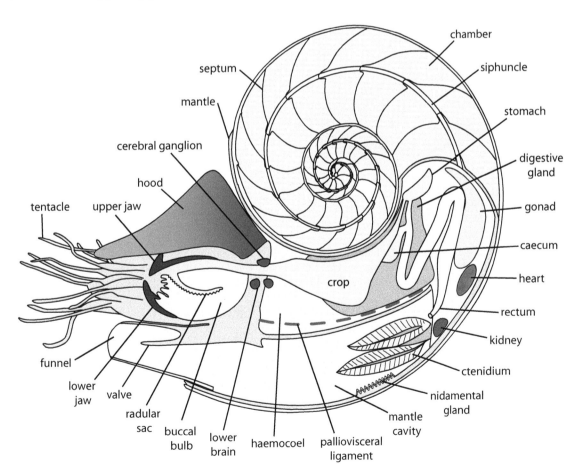

FIGURE 17.5 Diagrammatic body plan of *Nautilus*. Redrawn and modified from Griffin, L.E., *Mem. Nat. Acad. Sci.*, 8, 101–230, 1900, and Naef, A., *Ergeb. Fortschr. Zool.*, 3, 329–462, 1913.

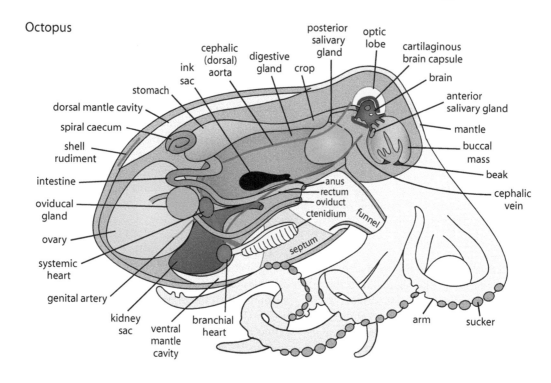

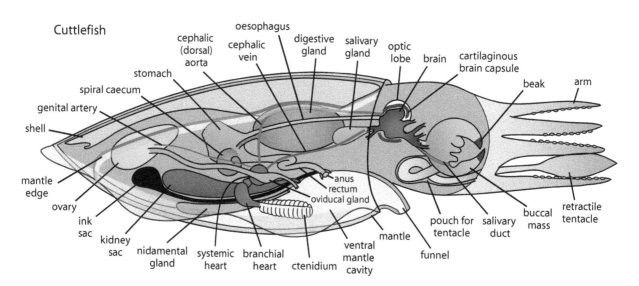

FIGURE 17.6 Comparison of the body plans of a female octopus and cuttlefish. Redrawn and modified from Naef, A., *Ergeb. Fortschr. Zool.*, 3, 329–462, 1913, and Boyle, P.R., Cephalopods, pp. 115–139 in Poole, T. (ed.), *UFAW Handbook on the Care and Management of Laboratory Animals: Amphibious and Aquatic Vertebrates and Advanced Invertebrates*, Vol. 2, Wiley Blackwell, Oxford, UK, 1999.

where they appear to have generally been similar to those in the living taxa (Reitner 2009).

The two pairs of ctenidia in *Nautilus* correspond to two pairs of retractor muscles, kidneys and atria, and these paired organs have been considered a remnant of metamery (e.g., Naef 1926; Lemche 1959; Wingstrand 1985). Others considered them to be secondarily duplicated (e.g., Morton & Yonge 1964). It is now known that such serial repetition of several organs occurred in monoplacophorans and also in the ctenidial gills and shell muscles of chitons (see Chapters 3, 12, and 14). In nautiliforms, there has been a tendency for reduction of the multiple pairs of muscles seen in early taxa (Kröger & Mutvei 2005), and this reduction in muscles and in gills is evident in coleoids (Figure 17.6).

17.3 MORPHOLOGY

17.3.1 Shell

The shells of cephalopods are either calcareous (aragonitic) chambered shells or horny remnants derived from such shells. Many extinct forms possessed an external shell, but species of *Nautilus* are the only living cephalopods to retain an external shell. A few living coleoids have internal calcareous shells (the cuttlebone of cuttlefish and the spiral shell of *Spirula*). A fragile shell-shaped egg case is produced by the 'paper nautilus' *Argonauta* to house its eggs and secondarily as a buoyancy device (see Section 17.4.2.4). It is not homologous with the original shell, as it is secreted by modified arms instead

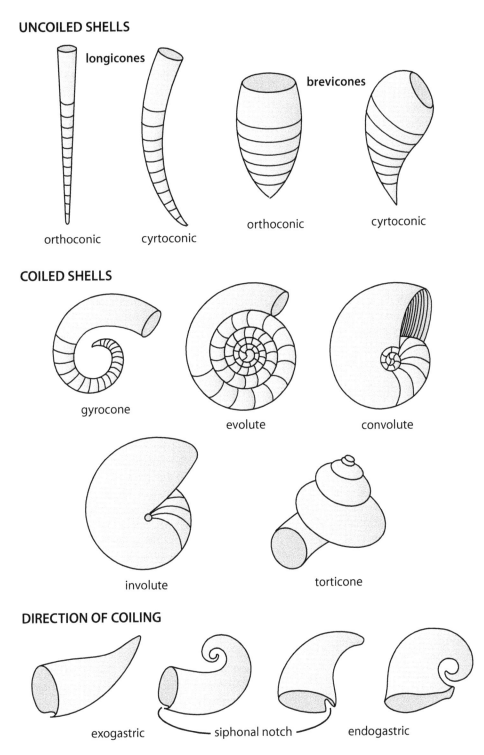

FIGURE 17.7 Terminology employed to distinguish the shapes of fossil cephalopod shells. Redrawn and modified from Teichert, C., Morphology of Hard Parts, pp. K13–K53 in Moore, R.C. (ed.), *Treatise on Invertebrate Paleontology Part K*, Vol. Mollusca 3, Geological Society of America and University of Kansas Press, Lawrence, KS, 1964b. See also Box 17.3.

of the mantle. A chitinous elongate shell remnant (the pen or *gladius*) is found in squid and *Vampyroteuthis*. Cirrate octopods have a partially calcified internal shell that is saddle- to horseshoe-shaped, and many benthic incirrate octopods have two thin, chitinous rods (the stylets) impregnated with calcium phosphate (e.g., Marquez & Re 2009) representing an incrementally growing shell remnant that can be used

to estimate age (e.g., Barratt & Allcock 2010; Doubleday et al. 2011). Some sepiidans and incirrate octopods lack shells altogether.

The chambered shells of living and fossil cephalopods exhibit great diversity in size and shape, and a rather complex terminology has been used to describe them (see Box 17.3 and Figure 17.7).

BOX 17.3   EXAMPLES OF SOME TERMINOLOGY USED TO DESCRIBE CEPHALOPOD SHELLS[6]

Septum/septa – the internal mineralised wall(s) dividing the shell into chambers. In many nautiliforms and primitive coleoids, the septa are saucer-shaped discs. In nautiliforms (except the ascoceratidans) the septa are curved inwards (i.e., towards the protoconch) (*procoelous*), but in ammonites only the first formed septa are curved inwards, the remainder being curved towards the aperture (*opisthocoelous*).

Camera – sealed chamber (space) between two septa. Although this term is commonly used in the palaeontological literature, we refer to the camerae as 'chambers'.

Suture – the line where the septum is connected to the inner surface of the shell; these are folded in ammonites, often forming complex patterns (see Section 17.12.2.2).

Phragmocone – the chambered part of the shell; its primary role is buoyancy.

Siphuncle – a shelly tube containing living tissue that runs through the chambers, connecting them (see Section 17.4.2.3.1).

Body (or living) chamber – houses the body in externally shelled species and is the most recently formed part of the shell.

Hyponomic sinus ('siphonal notch') – the assumed location of the funnel and thus the morphologically posterior side of the animal (functionally the ventral side in longitudinally oriented coleoids).

Ectocochleate – shell external.

Endocochleate – shell internal.

SHAPE AND COILING TERMINOLOGY

Figure 17.7. Each type of shell morphology has arisen more than once in the evolution of this group.

Uncoiled Shells

Orthocone – elongate long, straight, tapering shells. This shell form is found among the plectronoceratians, ellesmeroceratians, endoceratians, actinoceratians, orthoceratians, bactritans, Cretaceous heteromorph ammonites, and belemnites.

Cyrtocone – curved horn-shaped shells that may be narrow or wide. Found in some nautiliforms, Devonian and Cretaceous heteromorph ammonites, and Cenozoic spirulids.

Two main kinds of orthocones and cyrtocones are recognised: *longicones* with very long, tapering shells and *brevicones* with shorter, wider shells.

Coiled Shells

Gyrocone – open coiled.

Evolute – tightly coiled with whorls not overlapping.

Convolute – tightly coiled with whorls partially overlapping.

Involute – tightly coiled with whorls completely overlapping.

Torticones – a conical spiral.

Coiling Direction

Endogastric – the posterior side is on the direction of coiling. Examples are Discosorida, some Ellesmeroceratia and Endoceratia, and the coleoid *Spirula*.

Exogastric – the posterior side is on the opposite side of the direction of coiling. This is the assumed coiling direction in ammonites and most nautiliforms (Orthoceratia, Nautilia, and most Ellesmeroceratia).

The shell coiling and shape patterns sometimes occurred in combination – for example some (e.g., the Middle Ordovician *Lituites*) were initially coiled before becoming long and straight. Both orthoconic and coiled shells coexisted, with some of the straight non-coleoid taxa extending into the Triassic and orthoconic coleoids to the end of the Cretaceous. Tightly coiled nautiliform shells appeared in the Silurian, but the group that gave rise to *Nautilus* appeared later, in the Lower Devonian (ca. 400 Mya).

Some orthoconic nautiliforms filled the older chambers of the shell with calcite as a counterweight to balance the shell. Mutvei (2018) questioned whether these deposits were secreted by the living tissues of the animal in Carboniferous colorthoceratid[7] orthoconic shells and produced evidence that, in these taxa at least, calcifying bacteria may have produced the cameral deposits. While this may account for some deposited material, cameral deposits in many taxa occurred as a result of deposition by the living cephalopod. Other counterbalancing strategies were employed later in cephalopod evolution (see Section 17.4.2.3 and Figure 17.21). Accidental breakage of these elongate shells during life may have been common (Flower 1955b).

Nautilus has reddish radial markings on its shell, and some fossil nautiliforms had similar striped patterns, as well as other patterns such as longitudinal bands, wavy bands, and chevrons (Mapes & Davis 1996). Ammonite shells also had colour patterns in addition to their sculpture, including spots, zig-zags, transverse or longitudinal stripes, and sometimes combinations of these, or a single colour. These colourations were possibly of behavioural significance or for camouflage (Mapes & Davis 1996). Interestingly, no Paleozoic ammonites have been reported with a colour pattern, suggesting that they may have all been monochromatic, despite numerous

---

[6] Not including the *Argonauta* egg case.

[7] A group of uncertain relationships (see Appendix).

instances of colour patterns being preserved in co-occurring nautiliform fossils (Mapes & Davis 1996; Mapes & Larson 2015).

17.3.2 HEAD, ARMS, AND FUNNEL

The homologies of the head-foot of cephalopods have long been contentious. Shigeno et al. (2008 p. 15) state that 'With regard to orientation, shape and axis formation of body plans, the cephalopod head is one of the most complicated examples in bilaterian evolution'. What we know about the homology of the funnel and arms of cephalopods comes from the study of embryos and adults of *Nautilus* and coleoids and from a relatively small number of fossils with preserved impressions of the animal (see Section 17.3.5). From developmental data, we know that the head is completely merged with the foot, from which the arms, tentacles, and funnel are derived. The arms or tentacles surround the mouth. There are approximately (depending on species and sex) 90 tentacles in *Nautilus* (Fukuda 1987). They are divided into three groups: a pair of ocular tentacles, a variable number of labial tentacles arising from lobes surrounding the buccal mass, and 19 pairs of digital tentacles that surround and extend beyond the labial tentacles. In both males and females some labial tentacles are modified as secondary sexual structures. The tentacles are muscular but do not have suckers, and each has a slender terminal filament (*cirrus*) that is retractable into a basal sheath. The muscular cirri of the labial and digital tentacles have adhesive ridges on each side, but the cirri of the ocular tentacles are not adhesive and are probably sensory (Kier 1987). These cirri in *Nautilus* are not homologous with the structures called cirri seen on the arms of some coleoids (see below).

Decabrachian coleoids have eight suckered or hooked arms and (usually) a pair of partly to completely retractile tentacles. *Vampyroteuthis* has eight arms and two retractile filaments. Octopods have only eight arms (Figure 17.8). In all living coleoids, a web of tissue lies between the arms. The extent of the web is variable, ranging from a narrow ridge between the base of each arm to extending like an umbrella along most of the arm length. The arms have different functions, ranging from catching and holding food to walking (as

in benthic octopods), clinging to objects, or, in the case of males, transferring spermatophores to the female.

Based on fossil impressions and arm hooks of extinct early coleoids (e.g., Johnson & Richardson 1968; Jattiot et al. 2015; Fuchs & Hoffmann 2017), including belemnites, and on embryological evidence based on the number of arm buds of octobrachians, decabrachians, and *Nautilus*, the first coleoids are thought to have had ten equal-sized arms (Figure 17.8). The eight arms with two retractile tentacles seen in all modern decabrachians is thus a derived condition, and octopods lost a different pair to those that became the tentacles in decabrachians (Figure 17.8 and see below). An intermediate condition is seen in vampire squid which have ten arms, two of which are modified (the 'filaments') and probably represent the two lost arms in octopods, but there is no universal agreement about this. For example, based on their innervation, Young (1967) thought that the filaments were homologous with the preocular tentacles of *Nautilus* and were thus not arms, but this has since been discounted (Young & Vecchione 1996). Moreover, homologies of the tentacles and arms of coleoids with those of *Nautilus* remain uncertain.

The arms of *Nautilus* and coleoids function as muscular hydrostats. They have complex muscular systems involving transverse, longitudinal, and oblique muscles with the transverse muscles responsible for elongation, while contraction of the longitudinal muscles shortens the arms. These latter muscles are also involved in bending (Fukuda 1987; Kier 1988; Kier & Stella 2007).

In most coleoids the arms bear suckers, although some decabrachians have hooks, and others (cirrate octopods and *Vampyroteuthis*) have filaments called cirri (Figure 17.9). The suckers of decabrachians and octobrachian cephalopods differ in structure – inside the sucker rim, decabrachians have a horny (not chitin; Rudall 1955) ring which often develops teeth, and in several squid families these develop into hooks (Figure 17.10). Octopod suckers have an outer cup, the *infundibulum*, and an inner chamber, the *acetabulum*. They lack the horny inner ring seen in decabrachians but instead have a chitinous inner ring around the acetabulum (Hunt & Nixon 1981). The decabrachian suckers are attached to the arms by narrow stalks while those of octopods are attached by wide bases (Figure 17.10).

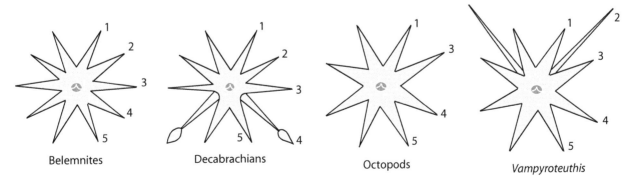

FIGURE 17.8 Configuration of the arms and tentacles in coleoid cephalopods. Redrawn and modified from Vecchione, M. et al., *Lethaia* 32, 113–118, 1999.

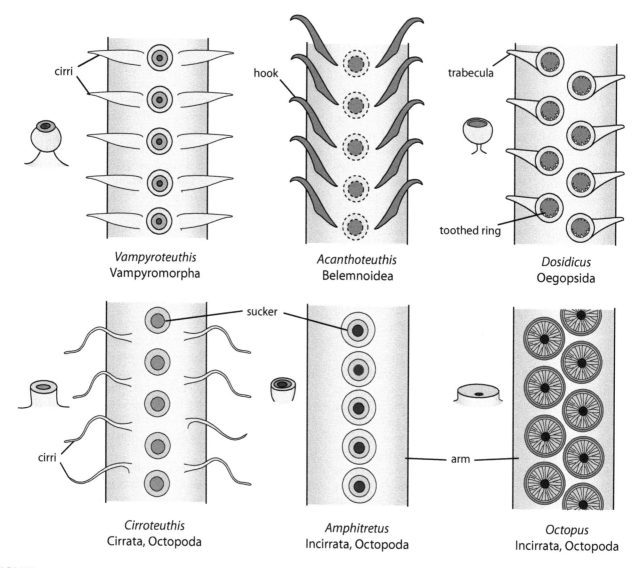

FIGURE 17.9 Examples of coleoid arm structures. All in ventral (i.e., oral) view and each with side views of a typical sucker. Redrawn and modified mainly from Fuchs, D. et al., *Neues Jahrb. Geol. Paläontol.*, 270, 245–255, 2013a.

Cirri[8] are tactile filamentous structures that form two rows on the arms of cirrate octopods and *Vampyroteuthis*. They are thought to be homologues of the two rows of hooks (or, more likely, their bases) seen in belemnites (Hoffmann et al. 2017) and also the stubby protuberances (*trabeculae*), associated with the suckers in some squid (Young et al. 1998; Fuchs et al. 2010; Fuchs et al. 2013a). In belemnites and early octobrachians the suckers were arranged in a single row (Fuchs et al. 2010), as they are in *Vampyroteuthis*[9] and also in cirrate and some incirrate octopods. The double row of suckers seen in most incirrate octopods and decabrachians came about by the single sucker row becoming zig-zag as the double suckers are not strictly opposite one another (Figure 17.9).

The funnel of *Nautilus* is quite mobile and used for both respiration and jetting. It is not a tube as in coleoids but instead a triangular fold rolled into a tubular shape. It has two functions: generating respiratory flows and controlling and directing jetting. The *Nautilus* funnel retractor muscles are weak and attach to the head retractors. In coleoids the funnel is used mostly for jetting control, and the retractor muscles are well developed and attach directly to the shell or shell rudiment (gladius) enabling the funnel to remain fixed in position during jetting. It can also bend in any direction and change shape (Wells 1987). When jetting, the funnel of *Nautilus* is rolled into a narrow tube and is held in position by the funnel cartilage and the funnel retractor muscles. Large, thin folds extend in from the funnel and continuously undulate to provide the respiratory current, a function performed by the mantle in coleoids. These folds contain both aerobic and anaerobic muscle fibres in about equal numbers (Baldwin 1987; Bizikov 2002). The mantle wall in *Nautilus* does not play a role in moving water in either respiration or jetting. In octopods the homologues of the head retractor muscles form a sheath around the viscera and do not serve to retract the head (Figure 17.11).

---

[8] The tentacles of *Nautilus* are also called cirri but are an entirely different structure.

[9] The Middle Jurassic *Proteroctopus* (Kruta et al. 2016) has biserial suckers, but its affinities are uncertain.

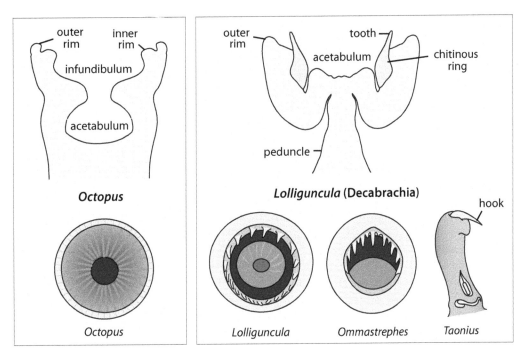

FIGURE 17.10 Structure of the suckers on the arms of coleoid cephalopods. Redrawn and modified from Kier, W.M. and Smith, A.M., *Biol. Bull.*, 178, 126–136, 1990 (*Octopus*), and Santi, P.A. and Graziadei, P.P.C., *Tissue Cell*, 7, 689–702, 1975 (*Lolliguncula*).

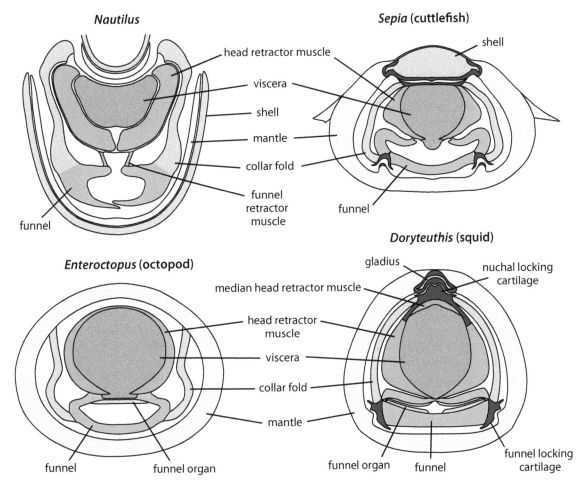

FIGURE 17.11 Transverse sections of the mantle collar and funnel in *Nautilus* and representatives of the main groups of coleoids. Redrawn and modified from Bizikov, V.A., *The Shell in Vampyropoda (Cephalopoda): Morphology, Functional Role and Evolution*, VNIRO Publishing, Moscow, 2008.

17.3.2.1 Embryological Evidence for the Derivation of Head Structures

Theories regarding the derivation of the cephalopod arms and funnel have been controversial, with both cephalic and pedal origins suggested. Shigeno et al. (2008) demonstrated that the foot, central nervous system, mantle, and shell in *Nautilus* are somewhat similar to the body plans of monoplacophorans and basal gastropods.

The tentacles/arms are derived from the foot region and surround the mouth during development – a condition unique in molluscs. This derivation is achieved during embryogenesis by the head being wrapped in epidermal tissue of pedal origin and the pedal region moving anteriorly so it is eventually in front of the head (von Boletzky 2003b; Shigeno et al. 2008). The tentacles develop from simple buds along the anteroposterior axis. In middle- to late-stage embryos, the body changes to the form seen in juveniles (and adults) forming a head complex that is, as described below, a mixture of various cephalic and pedal components.

Embryos of the coleoid ancestor may have had five pairs of arm buds on each side (Naef 1928). There are similarities with this pattern to the embryonic tentacle buds of *Nautilus* which also have five 'compartments' of bud-like 'anlagen', although the posterior compartments (4 and 5) are continuous (Shigeno et al. 2008). The five-compartment arrangement may well be the arrangement shared by the common ancestor of *Nautilus* and coleoids, but in *Nautilus* the first and second compartments develop into part of the hood, and the posterior compartments (3–5) develop into the multiple digital tentacles. Based on these observations, Shigeno et al. (2008) suggested that the multiplication of the tentacles in *Nautilus* was a secondary phenomenon, but this is by no means settled.

The ocular and buccal (labial) tentacles are, in living cephalopods, unique to *Nautilus* and originate separately from the other tentacles. The ocular tentacles differentiate during development from a part of the 'cephalic compartment', taking six months to appear during the long embryonic development (Shigeno et al. 2008). The anlagen of the buccal tentacles appear from the posterior ventral tissue at the buccal mass and are also slow to develop. Consequently, developmental origins of both ocular and buccal tentacles differ from those of the other (digital) tentacles.

In coleoid embryos the arm bases are formed from the base of each arm anlage and grow to cover the head as a secondary head cover, thus integrating the arms and head, and they also form the eyelids (Naef 1928; von Boletzky 2003b; Shigeno et al. 2008). *Nautilus* embryos do not have tissue that covers the whole head during embryogenesis although in three-month-old embryos similar arm base-like structures are found from the second, fourth, and fifth arm compartments. These do not, however, play a significant role in the development of adult structures (Shigeno et al. 2008).

The fleshy hood of *Nautilus* is formed from three different areas during embryonic development: 'the main frontal part of the collar-funnel compartment, the anterior region of the eyes, and the two anterior compartments of the tentacle buds' (Shigeno et al. 2008 p. 13). While these same areas are present in coleoid embryos, they do not develop a hood or any similar structure. Shigeno et al. (2008) suggested that the hood, which is analogous to the operculum of gastropods in its function, is secondarily derived.

Other differences between *Nautilus* and coleoids in the development of the head-foot structures include the collar, which is much smaller in coleoids than in *Nautilus*, and the funnel 'anlage' which is located more anteriorly in *Nautilus* than in coleoids.

There has been less dispute over the embryological origin of the funnel and collar than the arms, with most workers indicating an origin from the foot, although von Boletzky (2003b) suggested it was derived from the posterior arm rudiments. Shigeno et al. (2008) rejected the previously preferred pedal origin of the funnel in cephalopods as the anlagen in *Nautilus* is located laterally along the mantle and distinct from the pedal area. The collar-funnel in *Nautilus* arises from the posterior part of the hood-collar compartment, which, according to Shigeno et al. (2008 p. 15) may have been derived from an 'intermediate zone between the head-foot and visceral mass in the monoplacophoran ancestor'. Naef (1928) suggested that the collar-funnel in cephalopods may have been derived from a gastropod-like epipodium, a suggestion also entertained by Shigeno et al. (2008, Figure 11).

17.3.3 Epidermis (Skin)

Coleoid skin is very complex, and it is involved in several vital functions including ionic and water regulation and respiration as well as crypsis and communication displays. These displays involve the skin in producing colour and texture changes that are both elaborate and rapid, especially in octopods. Structures embedded in the skin that achieve these are the chromatophores that produce colour changes by expanding and contracting, the iridophores that reflect light, and white leucophores. These complex structures are described in Section 17.4.4.

The rather thin, transparent epidermis consists of columnar cells among which are scattered mucous, sensory, and ciliated cells. Connective tissue, muscle fibres, and nerves lie beneath the epidermis

Some deep-water taxa have light organs (see Section 17.4.5), and a few oceanic squid have a covering of scale-like 'dermal cushions' or tubercles on the mantle (Clarke 1960; Budelmann et al. 1997; Young & Vecchione 2009), as do the octopods *Ocythoe* and *Graneledone* (S. O'Shea, pers. comm., 2019).

Two skin structures, Hoyle's and Kölliker's organs, are found in pre-hatching embryos and early juveniles and assist hatching (Budelmann et al. 1997). These structures are absent in later juveniles.

The skin in *Nautilus* is simpler than in coleoids because it is covered by the external shell and lacks the special structures noted above.

Coleoid paralarvae may absorb nutrients through the skin as part of their energy requirements (Boucaud-Camou & Roper 1995).

17.3.4 HEAD AND FUNNEL RETRACTOR MUSCLES

The head, funnel, and associated retractor muscles of various cephalopods are shown in Figure 17.11 and Figure 17.12.

Nautilus has two pairs of retractor muscles, a pair of very large cephalic retractors and a pair of much smaller funnel retractors that merge with the ventral surface of the cephalic retractors. Decabrachian coleoids have two pairs of (presumably homologous) retractor muscles (cephalic and funnel retractors), while octopods have three pairs, the additional ones being a second pair of cephalic retractors (Wells 1988).

Nautilus utilises the two powerful head retractors to create the water pressure necessary for jetting by making plunger- (or piston-) like movements of its head. The rapid backwards movement of the head into the mantle cavity forces water

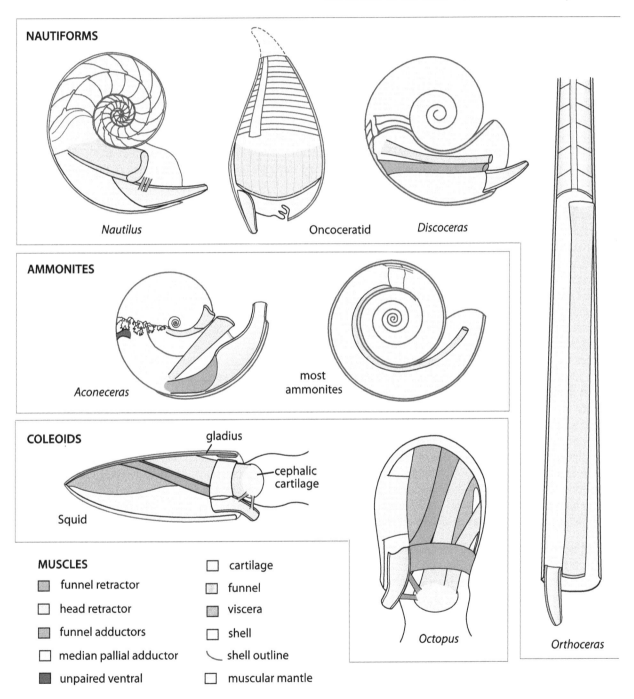

FIGURE 17.12 The major muscles associated with the shell, head, and funnel in cephalopods. Redrawn and modified from the following sources: figures in top row, including *Orthoceras*, and most ammonites (Mutvei, H., *Stockholm Contrib. Geol.*, 11, 79–102, 1964); *Aconeceras* (Doguzhaeva, L.A. and Mutvei, H., Attachment of the body to the shell in ammonoids, pp. 43–63 in Landman, N.H., Tanabe, K., and Davis, R. A. (eds.), *Ammonoid Paleobiology: From Anatomy to Ecology. Topics in Geobiology*, Plenum Press, New York, 1996); *Octopus* and squid (Mangold, K.M. et al., Organisation générale des céphalopodes, pp. 7–69 in Grasse, P.P. (ed.), *Céphalopodes. Traité de Zoologie*. Vol. 5, Fascicule 4, Masson, Paris, France, 1989).

through the funnel, but this plunger movement can only be sustained for short periods of time. The structure of the muscle shows two fibre types; about 90% of the fibres have few mitochondria and utilise anaerobic glycolysis for short bursts of jetting while the other ~10% of fibres contain numerous mitochondria and utilise aerobic metabolism for sustained activity (Baldwin 1987; Bizikov 2002). Because coleoids use their muscular mantle for both jetting and ventilation, the head retractor muscles are reduced, serving mainly to provide support for the viscera (Wells 1988; Bizikov 2002). In some juvenile and adult squid, they enable the head to be pulled into the mantle cavity. The ability to retract the head is limited by its width relative to the mantle cavity opening. Those in which it can retract not only have a narrow head but also a simple ridge-and-groove locking apparatus (e.g., onychoteuthids). Taxa with the head and mantle linked by complex locking cartilages (e.g., ommastrephids) cannot retract the head. Cranchiids, which have a fused mantle and nuchal region, can withdraw most of the head inside, leaving only their 'telescopic' eyes emerging (S. O'Shea pers. comm., 2019).

The muscular systems of extinct groups like the ammonites can only be reconstructed from muscle attachment areas on the interior of the body (last) chamber of the shell. Some examples of such reconstructions are shown in Figure 17.12 and are further discussed below and in the sections dealing with those taxa.

There are four main types of muscle scars in nautiliforms; the first of these is the *oncomyarian* condition – a 'buttress ring' of muscle scars considered to be the ancestral condition and recorded in Furongian (Cambrian) and some younger taxa. In other nautiliforms the main head retractor muscles are reduced to one pair. These may exhibit the *dorsomyarian* condition where the pair of retractor muscles is attached close together on the dorsal side of the body chamber, seen in various taxa including orthoceratids and endoceratids. The *ventromyarian* condition has the pair of retractor muscles attached close together on the ventral side of the body chamber and seen in more derived taxa including tarphyceratids and ascoceratids. The fourth condition is described as *pleuromyarian* in which the retractor muscles are attached to opposite sides of the body chamber; this is seen in all Nautilida.

17.3.4.1 Muscular Attachments, Shell Growth and the Addition of Septa

Ectocochleate cephalopods must anchor their bodies to the inner surface of the body chamber wall. As noted above, *Nautilus* has a pair of large head (cephalic) retractor muscles, and a small pair of funnel retractors are attached ventrally to these muscles (Wells 1988; Bizikov 2002), not to the shell. The mantle is attached to the shell by way of a myoadhesive epithelium arranged in three attachment bands ('aponeurotic bands') (Griffin 1897). The muscles and mantle attachment bands are attached to a 'thin, annular band of calcium carbonate' (Ward et al. 1981 p. 483), called the 'mural ridge' or 'annular elevation', that lies in front of the last septum in the body chamber (e.g., Mutvei & Doguzhaeva 1997). It can often be seen in moulds of fossil nautiliforms (Kröger &

Mutvei 2005), and the head retractor muscles are attached on the lateral sides of this ridge. These and other attachments form imprints on the inner shell surface that can be detected in fossils and thus provide information about the animal (see Section 17.3.5). The scar from the aponeuroses in *Nautilus* forms a posterior band in the body chamber (e.g., Blind 1976; Mutvei et al. 1993; Klug et al. 2008).

To summarise, the retractor muscles, longitudinal mantle muscles, and the subepithelial muscles (that attach the posterior part of the body to the septum) all originate from an area just below the septum which forms the slightly raised band on the shell (the annular elevation). These attachments must be detached and reattached as septa are added.

The process of septal formation in *Nautilus* is complex. As septa are a major feature of cephalopods, we describe their formation here in some detail. This description is based on the account in Klug et al. (2008), which in turn is based largely on accounts by Blind (1976) and Ward et al. (1981). The following steps are involved and illustrated in Figure 17.13:

(1) The commencement of new chamber formation begins with the secretion of the mural ridge in front of the septum and new shell is added at the apertural edge, with the apertural mantle attachment being shifted anteriorly.

(2) The posterior part of the body detaches from the septum and moves forward, leaving a space that fills with liquid. This movement is enabled by the contraction of small radial muscles attached to the septum which then cause the mantle attached to the septum (the 'septal mantle') to separate from it (Figure 17.13A). Transverse muscle fibres then lift the margin of the septal part of the mantle off the mural ridge where the septal mantle muscles are attached (Figure 17.13B).

(3) Muscle fibres attached to the shell at the mantle (the 'myoadhesive band' or 'myoadhesive ligament') contract and pull the septal mantle forward (Figure 17.13C).

(4) The cephalic retractor muscle attachments then creep forward along with the mantle myoadhesive band and the palliovisceral band (which form a unit in *Nautilus*), assisted by contraction of longitudinal muscles in the mantle (Figure 17.13D).

(5) The posterior part of the mantle is then attached to a newly formed mural ridge at the position of the new septum and is followed by the formation of a conchiolin membrane at this location. This membrane is then covered with a shell layer to form the new septum (Figure 17.13E).

There have been various models proposed to explain differences in the shape of the septa, particularly the major differences between nautiliforms and ammonites, including the lack of the mural ridge in ammonites. These models are briefly discussed below in Section 17.12.2.2 and in detail by Klug et al. (2008) and Klug and Hoffmann (2015).

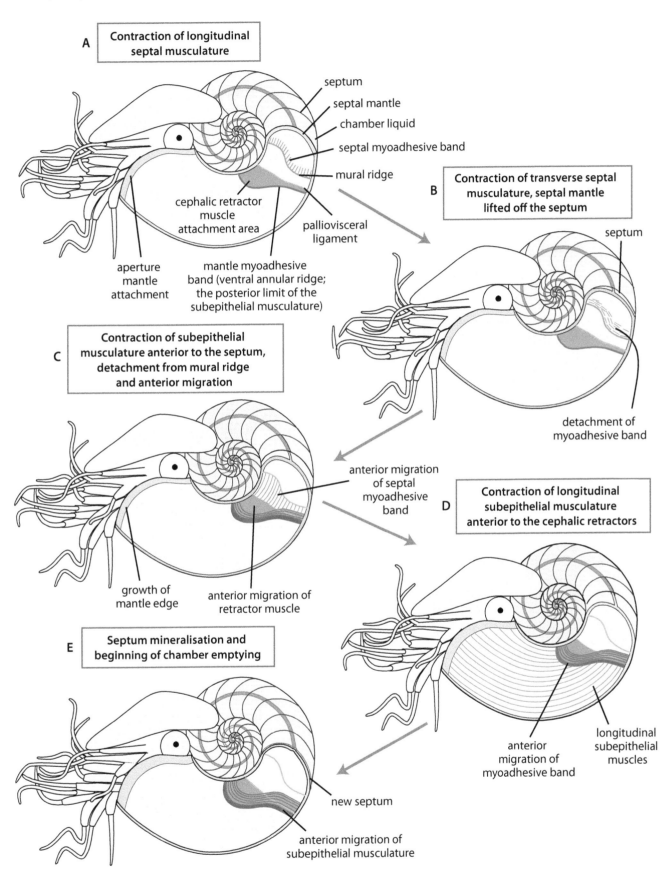

A Contraction of longitudinal septal musculature

septum
septal mantle
chamber liquid
septal myoadhesive band
mural ridge

cephalic retractor muscle attachment area

palliovisceral ligament

aperture mantle attachment

mantle myoadhesive band (ventral annular ridge; the posterior limit of the subepithelial musculature)

B Contraction of transverse septal musculature, septal mantle lifted off the septum

septum

detachment of myoadhesive band

C Contraction of subepithelial musculature anterior to the septum, detachment from mural ridge and anterior migration

anterior migration of septal myoadhesive band

growth of mantle edge

anterior migration of retractor muscle

D Contraction of longitudinal subepithelial musculature anterior to the cephalic retractors

anterior migration of myoadhesive band

longitudinal subepithelial muscles

E Septum mineralisation and beginning of chamber emptying

new septum

anterior migration of subepithelial musculature

FIGURE 17.13   Process of septal formation in *Nautilus* schematically depicting the mantle movements and the involvement of its musculature. Redrawn and modified from Klug, C.A. et al., *Lethaia*, 43, 465–477, 2008.

BOX 17.4   MUSCLE AND BODY IMPRESSIONS IN AMMONITES

Paired dorsal scars – lie in front of the last suture and dorsally or dorso-laterally on the body chamber. They are usually considered homologues of the cephalic retractor muscles in *Nautilus*, but they are much smaller. Also, because of their position, they could not have formed a muscular sheath for the viscera above the mantle cavity as do the cephalic retractors in *Nautilus*.

Unpaired mid-dorsal scar – initially paired small scars in the first chamber but unpaired in subsequent chambers. It is thought to be the attachment site of the palliovisceral ligament (see Section 17.5.3), as in *Nautilus*.

Unpaired ventral scar – lies in front of the ventral suture lobe. This muscle or ligament may have assisted in maintaining the shape and position of the 'circumsiphonal invagination' in the posterior part of the body during its forward migration in the body chamber during septal formation.

Paired lateral scars – lobe-like scars that lie along either side of the posterior half of the body chamber from the last suture. Each is subdivided into a dorsal and ventral part which may correspond to cephalic and funnel retractor muscles respectively. These have been found in a few ammonites, only those with a relatively short body chamber.

Lateral sinuses – deep indentations of the paired lateral sinus-like scars that extend from the shell aperture to about halfway along the body chamber. They have been likened to the pallial sinus of bivalves and may represent the attachment point of the funnel lobes.

Annular elevation – a narrow zone in front of the last suture in some genera. It is probably equivalent to the septal and mantle muscle attachment bands in *Nautilus*.

The imprints of muscle attachments are preserved in numerous fossil nautiliforms (e.g., Kröger & Mutvei 2005 and references therein) and ammonites (e.g., Doguzhaeva & Mutvei 1996; Richter 2002). The size, shape, and number of scars differ between taxa and are used in defining higher groupings. Early Paleozoic nautiliform cephalopods show multiple muscle scars, with about nine–ten pairs in ellesmeroceratidans, and four pairs (of cephalic retractors) in Tarphyceratida (e.g., Kröger & Mutvei 2005). The Paleozoic Oncoceratida and Discosorida have 7–25 pairs of retractor muscle scars, the largest being the ventral pair. Most nautiliforms (including *Nautilus*) and the extinct Tarphyceratida and Barrandeoceratida have either a single lateral pair, a single ventral pair, or ventral and lateral pairs of muscle scars. An

exception is *Estonioceras* (Tarphyceratida) which has four pairs. Orthoceratians have a paired or a single dorsal muscle scar (Kröger & Mutvei 2005).

In ammonites, there are several muscle scars, and their positions (and possibly their functions) differ in different groups. Doguzhaeva and Mutvei (1996) identified some muscle and body impressions in ammonites, although not all of them are seen in all ammonites. These are listed in Box 17.4.

While the muscles and other attachments in ammonites and *Nautilus* can be equated with reasonable certainty, questions remain as to whether they were functionally similar. Not only do the scars vary in size, but the body chamber in ammonites was narrower and usually longer than in *Nautilus*, occupying from a little under half a whorl to about two whorls of the shell. Thus, in ammonites, which apparently lacked a hood or operculum-like structure, the animal may have been able to withdraw into the body chamber well beyond the aperture (e.g., Monks & Young 1998).

The multiple muscle scars (the oncomyarian condition) of ellesmeroceratidans have usually been interpreted as marking the points of attachment of the head and/or funnel retractor muscles. These scars extend down towards the shell aperture and range from six to ten pairs in ellesmeroceratidans but are more variable in oncoceratidans and discosoridans (Mutvei 2013). These scars are interpreted as mantle muscle scars by Mutvei (2013). In *Nautilus* embryos about 40 pairs of well-developed longitudinal mantle muscles are attached to the shell near the aperture. It has also been suggested that the multiple muscle scars in ellesmeroceratidans, oncoceratidans, discosoridans, and ascoceratidans may have represented mantle tentacles involved in microphagous feeding (Mutvei 2013).

17.3.5   THE FOSSIL RECORD OF SOFT TISSUES

Although the most commonly encountered impressions of the soft body are muscle scars on the shell (see above), other parts of the anatomy are also sometimes fossilised. Many fossil beaks and 11 radulae of cephalopods have been found (see Section 17.5.2). The oldest radula is from the Ordovician (Gabbott 1999) and the oldest beaks from the Devonian (Mapes 1987). Most reports of beaks are from ammonites (Kennedy et al. 2002), with only a few from nautiliforms (Müller 1974; Mapes 1987). As noted earlier, these are known as rhyncholites. All known beaks from the Upper Paleozoic are either entirely chitinous or have some calcareous material around the tips as in *Nautilus*, but in the Mesozoic, some ammonites modified the chitinous lower jaw with one (*aptychus*) or two (*aptychi*) massive calcareous plates (see Section 17.12.2.4), which are much more abundant as fossils than the organic upper jaw (Klug & Lehmann 2015).

Nishiguchi and Mapes (2008) and Donovan and Fuchs (2016) reviewed the records of coleoid body fossils. The oldest are *Gordoniconus* with a substantive shell from the Early Carboniferous; *Jeletzkya*, with ten hook-bearing arms, and *Pohlsepia*, from the Upper Carboniferous; none of these taxa had a shell. There are many reports of Mesozoic coleoids with preserved arms (with hooks), mantle tissue, beaks, radulae, and even ctenidia (e.g., Reitner 2009). Fossil examples of delicate structures such as the excretory, reproductive, and digestive systems in coleoids are reviewed by Donovan and Fuchs (2016); the preserved remains of these animals are more abundant than they are for other cephalopods.

Ten arms are usually presumed to be plesiomorphic for neocephalopods but there is little direct evidence for this. Ten arms have also been recorded in sphaerorthoceratids (Stürmer 1985; Engeser 1990a, b), an orthoceratidan group possibly related to Bactrita (Kröger & Mapes 2007). Definitive arm impressions in ammonites are non-existent, so the number and nature of the arms they possessed (if any) is unknown (Klug & Lehmann 2015). Keupp (2007) suggested that some ammonites may have had a reduced number of arms (possibly six) as a result of their large, calcified aptychus being utilised as an operculum.

The remains of ink sacs have been recorded in some Mesozoic fossil coleoids, the oldest from the Upper Carboniferous of the USA (Doguzhaeva et al. 2004, 2007a; Nishiguchi & Mapes 2008). Ink sacs have not been found in any fossil nautiliforms (they are absent in *Nautilus*), and they are usually thought to be absent from ammonites (e.g., Lehmann 1985; Lehmann 1988) although questionable evidence for possible ink has been found in an Upper Triassic ceratitid ammonite (Doguzhaeva et al. 2004; Doguzhaeva et al. 2007b).

Lehmann (1985) reported the remains of gills in the body chambers of three ammonites but their preservation was too poor to determine the number of gills; their shape was apparently similar to those in *Nautilus*. Impressions of gills are also known from Mesozoic coleoids (Bandel & Leich 1986; Mehl 1990) and a few ammonites (Klug & Lehmann 2015).

Lehmann (1985) supposed that the eyes of ammonites were of the pinhole type like those of *Nautilus*, but there is no direct evidence to support this. Klug et al. (2012) described the remains of eye capsules in baculitid ammonites, suggested that they possessed lens eyes, and speculated that the pinhole eye was a derived feature of nautilids – a view supported by studies on *Nautilus* genes (Ogura et al. 2013).

Information about the animal comes not only from fossilised remains but from impressions in the sediment made by the animal in life. The most famous of these trace fossils are the trails and impressions of arms thought to be those of the orthoconic nautiliform *Orthonybyoceras* from the Upper Ordovician (Corryville beds, USA) (Flower 1955a). Some were interpreted as skid marks where the animal landed on the sediment, and their abundance suggested to Flower that the animals had congregated to feed. There were also indications that the animals swam forwards rather than backwards. Horseshoe-shaped groups of impressions may have been made by the tentacles clinging to the substratum to avoid being swept away – these impressions indicated relatively few arms, probably ten. This remains one of the very few observations of arm numbers in a fossil nautiliform and, if it is correct,

suggests that the arms of modern *Nautilus* are secondarily duplicated, a conclusion with some support from embryological studies (see Section 17.3.2.1).

Information about the digestive system (other than the buccal apparatus) comes from the fossilised remains of the undigested hard parts of the food in the gut (soft-bodied food would not be preserved) and their position in the body. This information also provides clues about the diet. Ammonites have been found to contain remnants of crinoids, smaller ammonites, and orthoconic nautiliforms or, in the case of the small (2 cm diameter) *Arnioceras*, minute prey such as foraminiferans and ostracods (Nixon 1988b). The fossilised food remains of the ammonite *Baculites* suggest that it fed on plankton (Kruta et al. 2011). Some large actinoceratian nautiliforms apparently fed on ammonites (Mapes & Dalton 2002), and the gut of a coleoid from the Lower Carboniferous contained fish scales (Landman & Davis 1988; Mapes et al. 2007).

Impressions of blood vessels on the interior of parts of the shell in fossils can give some clues about the circulatory system (e.g., Klug et al. 2007) and preserved siphuncular tissue of Permian ammonites, showing the associated blood vessels, has been described (Tanabe 2000) (see Section 17.4.2.3.1). Fossil coleoids also preserve such information (Donovan & Fuchs 2016).

Specifically distinct statoliths made of aragonite found in living coleoids are also encountered as fossils, with records back to the Jurassic (Clarke & Maddock 1988; Clarke & Hart 2018). Belemnite hooks are also common as fossils and were specialised for different functions on the arms (Hoffmann et al. 2017). The larger ones are called megahooks, the small ones microhooks.

17.3.6 MANTLE WALL

In *Nautilus*, the mantle wall is thin, but in coleoids, which all lack an external shell, it is thick and muscular. This thick, contractile wall is capable of the powerful contractions needed for jetting and ventilation. In coleoids mechanical support of the mantle wall is provided by a complex mesh of muscle fibres and connective tissue (see Chapter 3) and, in many, either a substantive (as in cuttlefish) or a reduced (most oegopsid squids) internal shell.

In all coleoids, circular and radial muscles in the mantle wall interact antagonistically. In octopods, these muscles lie between inner and outer layers of longitudinal muscles, but in decabrachians (squid and cuttlefish) longitudinal mantle muscles are weak or lacking, and instead there is a network of connective tissue fibres among the muscle fibres (see Chapter 3, Figure 3.54). The mantle wall of most squid differs from that of octopods in having a collagenous outer tunic. Thompson (2001) showed that the connective tissue network of collagen fibres among the mantle muscles in squid might have evolved more than once in different lineages. In species that attain neutral buoyancy by storing low-density fluids in the mantle, this network was modified into a reticulated meshwork of collagen fibres and, again,

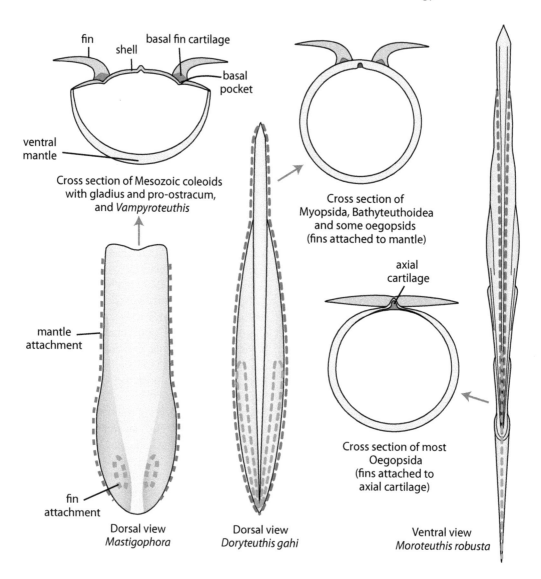

FIGURE 17.14 Types of mantle and fin attachment to the gladius in various coleoids. Redrawn and modified from Fuchs, D. et al., *Lethaia*, 49, 433–454, 2016.

this may have evolved independently in different lineages. In some neutrally buoyant taxa, the collagen is replaced by a thin layer of longitudinal muscle fibres.

Fins, an autapomorphy of coleoids, protrude from the mantle either laterally or dorsally, this positioning being related to their attachment to the underlying gladius (Figure 17.14) (Fuchs et al. 2016). The fins are important in locomotion, stabilisation, and manoeuvring. Their means of attachment to the body and the gladius is summarised in Figure 17.14.

17.3.7 MANTLE CAVITY

The mantle cavity of living cephalopods functions differently to that of other molluscs because the water currents for ventilation are driven by muscular contractions of the funnel (*Nautilus*) or mantle (coleoids), rather than by ctenidial ciliation. Also, the water is pumped through the cavity and over the ctenidia from dorsal to ventral – the reverse of other

molluscs (see Chapter 4). How this change may have come about is discussed in Section 17.9.

During normal respiratory movements in coleoids, mantle contraction and dilation occur only in the anterior part of the mantle, while jetting involves the whole mantle cavity (Packard & Trueman 1974). Octopods contract the circular mantle muscles to force water from the mantle cavity during respiratory expiration, but this is not the case in decabrachians. Instead, during slow ventilation, expiration is caused by the inward movement of the collar flaps in a way that resembles *Nautilus* (Wells & Wells 1985a), together with an elastic contraction of the mantle connective tissue (e.g., Bone et al. 1994).

The water flow pattern during ventilation is similar in both *Nautilus* and coleoids (Figures 17.15, 17.16) with water entering the mantle cavity through the narrow slits between the collar folds and the mantle. It then passes over the ctenidia to the ventral part of the mantle cavity from where it is expelled

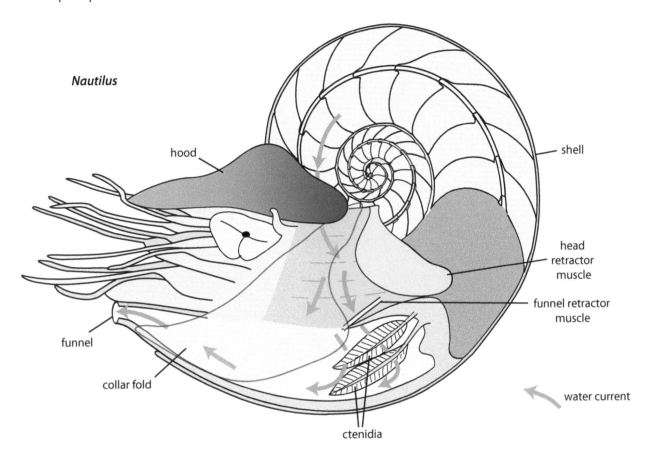

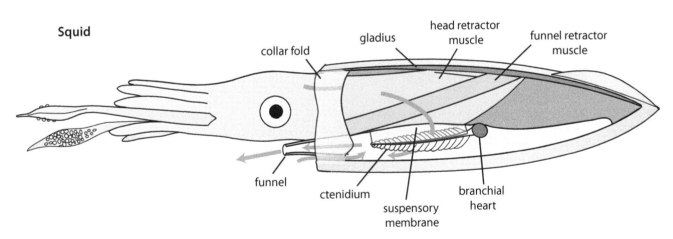

FIGURE 17.15   Diagrams showing the mantle cavity, the main water currents, and the major muscles of *Nautilus* and a squid. Redrawn and modified from Bizikov, V.A., *Am. Malacol. Bull.*, 17, 17–30, 2002.

through the funnel. The collar in coleoids acts mainly as a valve, opening and closing the slits on either side of the head. The much larger collar folds in *Nautilus* not only act as valves, but their movements also generate the flow of water through the mantle cavity, a function performed in coleoids by the mantle wall.

Nautilus has two pairs of ctenidia, while all other cephalopods have one pair. Unlike all other molluscan ctenidia (see Chapter 4), those of cephalopods are unciliated, with water movement over the surfaces of the ctenidial filaments

maintained by muscular movements of the mantle in coleoids and by broad inner flap-like extensions of the funnel in *Nautilus*. Contractile vessels and muscular movements of the ctenidium itself assist in moving blood through the ctenidial filaments, increasing their respiratory efficiency. Coleoids differ from *Nautilus* and other molluscs because the blood flow through their ctenidia is enhanced by the pumping of accessory hearts (see Chapter 6). Coleoid ctenidia can withstand high-flow velocities and pressures because they are attached to the mantle along the entire length of the afferent side of

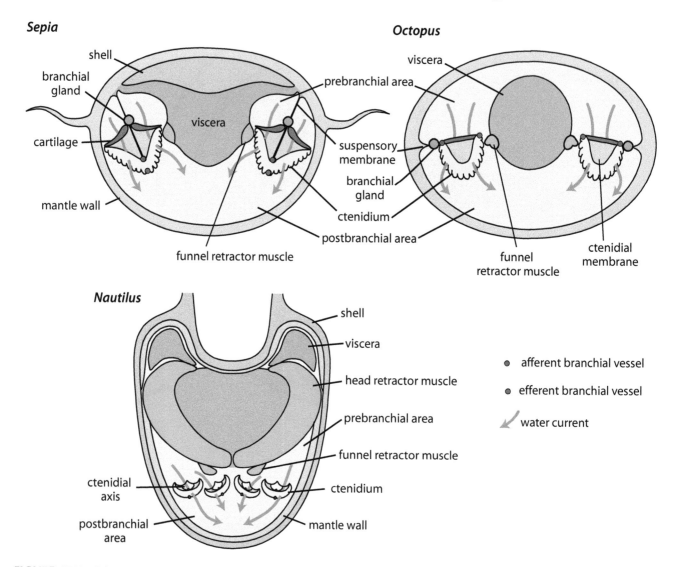

FIGURE 17.16 Diagrammatic transverse section through the mantle cavity of *Sepia*, *Octopus*, and *Nautilus* showing water flow through the ctenidia. Redrawn and modified from Wells, M.J., The mantle muscle and mantle cavity of cephalopods, pp. 287–300 in Trueman, E.R. and Clarke, M. R. (eds.), *Form and Function. The Mollusca*. Vol. 11, Academic Press, New York, 1988, (upper figures) and Bizikov, V.A., *Am. Malacol. Bull.*, 17, 17–30, 2002 (*Nautilus*).

the axis, with the other surface being free. In *Nautilus* the ctenidia are attached only at their posterior ends. The detailed structure of the cuttlefish (*Sepia*) ctenidium is illustrated in Chapter 4, Figure 4.11, and the shape of filaments of a range of cephalopods is shown in Figure 17.17.

The ctenidium of *Sepia* is complex, with a system of primary, secondary, and tertiary folds arranged at right angles to each other, giving a branching appearance, with each containing a network of blood vessels (Schipp et al. 1979; Young & Vecchione 2002). Both *Nautilus* and *Sepia* ctenidia have the same basic organisation (Young & Vecchione 2002), but octopods differ markedly both in structure and blood flow (Wells & Wells 1982; Young & Vecchione 2002) (Figure 17.17). The ctenidia of *Vampyroteuthis* and decabrachians have cartilaginous support (see below), but this is absent in most octopods and in *Nautilus*. Deep-sea squid have ctenidia with large

surface areas enhancing respiratory efficiency and a thin covering of skin to facilitate oxygen extraction, but, as a result, they are fragile (Seibel et al. 1997).

In a study of coleoid, particularly octobrachian, ctenidia, Young and Vecchione (2002) provided a detailed account of the ctenidia of cirrate octopods and Vampyromorpha (*Vampyroteuthis*). They showed that octobrachian ctenidia (including *Vampyroteuthis*) differ from what they consider the 'primitive cephalopod gill', such as that seen in *Nautilus* and decabrachians, by having 'septa' along the axes of the primary and secondary lamellae. These septa determine the type of folding of the filament surfaces, and when present the lamellae form 'tree-like folds' rather than the 'fan-like folds' seen in other cephalopods.

The circulation pattern in the ctenidia in *Vampyroteuthis* (the sister group to Octopoda) is unique and may be an

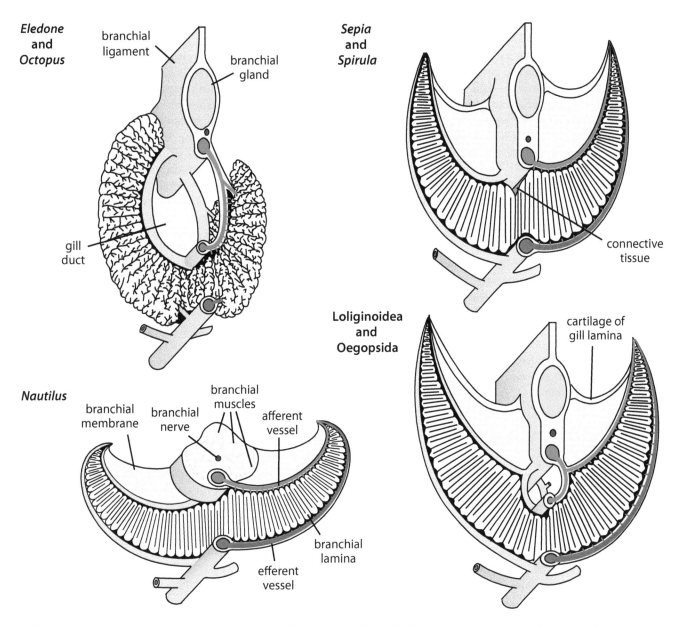

FIGURE 17.17 Cephalopod ctenidial filament morphology; a comparison of *Nautilus*, a squid (*Loligo* and an oegopsid); an octopod (*Eledone*) and the cuttlefish (*Sepia*) and *Spirula*. Redrawn and modified from Haas, W., *Abhandlungen der Geologischen Bundesanstalt*, 57, 341–351, 2002.

adaptation to life in the deep-sea low-oxygen environments it inhabits. Furthermore, Young and Vecchione (2002) suggested that octobrachian ctenidia, in general, were adapted to a low-oxygen environment. These adaptations occurred despite decabrachians generally having greater metabolic demands for oxygen. Fast-moving squid need about seven times more oxygen than a rapidly crawling octopus (Wells & Smith 1987), but the squid compensates for this by moving large quantities of water through the ctenidia during locomotion. *Octopus* moves water more slowly through the ctenidia and is more efficient at extracting oxygen (around 50% at rest, compared to 4–12% in *Nautilus* [Wells 1988] and 5–10% in the squid *Lolliguncula* [Wells 1990]).

Ventilation rates in *Octopus* depend on the level of oxygen in the water, increasing with decreasing oxygen levels, with oxygen uptake remaining fairly constant (average of 46%) (Wells & Wells 1985b).

Nautilus ctenidia are adapted to function in low-flow conditions (Packard et al. 1980; Wells & Wells 1985a; Bizikov 2002). They effectively divide the mantle cavity into branchial (inhalant) and postbranchial (exhalant) chambers, with drag exerted on water passing through the cavity (Bizikov 2002). For this reason, the mantle cavity functions very differently during respiration and jetting. During respiration, water is slowly moved through the cavity by the undulating actions of the funnel folds, and the volume of the cavity does not change significantly. In marked contrast, while jetting the pressure in the mantle cavity increases by one or two orders of magnitude, and the ctenidia are forced against the ventral mantle wall to allow the water to pass (Bizikov 2002). Also, with each piston-like thrust of the head, the volume of water in the mantle cavity changes dramatically. Oxygen extraction by the ctenidia during jetting falls to 4–7% compared with the normal rate of 9–12% (Wells 1988) and may explain why *Nautilus* can only maintain jet swimming for a few minutes at most (Bizikov 2002). During vertical migrations, mechanisms are employed to avoid anaerobiosis in hypoxic conditions encountered (Neil & Askew 2018). Immediately following activity, oxygen extraction reaches 20% or more, and, during this period (which may last for a few minutes), ventilation is increased by head and hood movements (Wells 1987). The ctenidia are probably not displaced during this phase as the funnel remains wide open, allowing a considerable flow through the mantle cavity to occur under low pressure (Bizikov 2002). The inability of the ctenidia of *Nautilus* to work in the high pressures and flows created during jetting is a constraint overcome in coleoids (see above), while *Nautilus* respires while undertaking slow swimming but not while engaged in short periods of jetting activity, primarily when required for escape.

17.4 BIOLOGY AND ECOLOGY

17.4.1 Habits and Habitats

Cephalopods are found in all oceans of the world in benthic and pelagic habitats, from the lower littoral to abyssal depths, but only a few species in a few genera (notably *Idiosepius* [Idiosepiidae] and some sepiolids) can tolerate brackish water. In living cephalopods, about 35% are planktonic, 16% benthic, 19% nektobenthic, 5% benthopelagic, and 25% nektonic (Barskov et al. 2008) (see Table 17.1). Some species are very abundant, with some estimates of coleoid biomass being as high as that of marine fish.

Roper and Young (1975) reviewed the vertical distributions of pelagic cephalopods. Many species undertake diurnal migration in the water column, moving into deeper water in the daytime and shallow water at night (Figure 17.18).

Many squid, most cuttlefish, and many octopus live in relatively shallow water in the photic zone. To avoid predators that hunt visually, they rely on colour changes and have other adaptations such as crypsis or speed, as well as their own visual abilities. While most octopods are benthic in shallow water, some burrow into sediments, and others live in extreme habitats such as deep water, hydrothermal vents, or cold seeps. Many other coleoids, including *Vampyroteuthis*, some squid and cirrate, and a few incirrate octopods, have adapted to life in the cold, deep ocean. Deep-sea squid are frequently translucent, their eyes may be very large, stalked or even unequal in size, and they often have photophores (see Section 17.4.5). To achieve neutral buoyancy, they typically reduce their body density through high concentrations of ammonium ions in gelatinous tissue (see Section 17.4.2.4). Some are substantially larger than their shallow-water relatives, with extreme examples being the 'giant' and 'colossal' squid (see Section 17.13.1.5).

A few living cephalopods can tolerate hypoxic or even dysoxic conditions for periods of time, notably *Vampyroteuthis* and *Nautilus*. This capability may have helped their ancient predecessors to survive some previous extinction events involving anoxia (see Chapter 13). *Vampyroteuthis* drifts in the oxygen-minimum zone of the deep sea and feeds mainly on detritus that comes into contact with its arms (Hoving & Robison 2012). *Nautilus* lives in the tropical central Indo-Pacific, ranging from southern Japan to tropical Australia. They live in deep water associated with 'drop-offs' at the outer edges of coral reefs. In the daytime they reside in dark, cold, deep water (about 300–650 m), probably to avoid predation, and migrate to about 300–100 m (or sometimes to as shallow as five metres) below the surface at night when they feed (e.g., Saunders & Ward 1987; Dunstan et al. 2011).

17.4.1.1 The Habits and Habitats of Extinct Ectocochleate Cephalopods

Reconstruction of the habits and habitats of fossil taxa can be done with reasonable accuracy by using data such as the sedimentary environment, isotopic signatures, and associated fauna, as well as information from the fossil shell, preserved anatomy, and by experiments with models. Shell morphology

---

TABLE 17.1

Diversity and Habitats of Living Cephalopods

Taxa	Genera	Benthic	Benthopelagic	Nectobenthic	Nectonic	Planktonic
			Habit			
Octopoda	42	22 (50%)	6 (15%)	-	-	14 (35%)
Sepiida	21	-	1 (5%)	18 (85%)	1 (5%)	1 (5%)
Myopsida	84	-	-	-	42 (50%)	42 (50%)
Oegopsida	10	1 (10%)	-	9 (90%)	-	-
Total	157	23 (16%)	7 (5%)	27 (19%)	43 (25%)	57 (35%)

Data from Nesis (1975) and reproduced by Barskov et al. (2008)

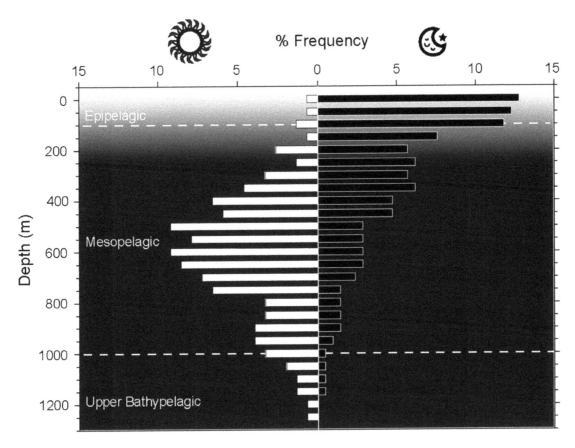

FIGURE 17.18 Diurnal distribution of pelagic squid. Data from Roper, C.F.E. and Young, R.E., *Vertical Distribution of Pelagic Cephalopods*, Vol. 209, Smithsonian Institution Press, Washington, DC, 1975. Original.

used with computer modelling can provide information on probable life orientation and buoyancy, for example in nautiliforms and ammonites.

Cephalopods underwent their first major diversification in the Lower Ordovician (see Section 17.10.1), where the shell morphotypes have been associated with particular ecological habits. Small, broad shells are mostly found in shallow water, including reef environments, and are thought to have belonged to slow-moving swimmers. Large orthocones (straight, tapering shells) with well-developed siphuncles are mainly found in sediments deposited in shallow seas and are thought to have been taxa that lived at or near the bottom. Small, fragile orthocones are known from black shales and belonged to animals that were probably pelagic in the open ocean (Westermann 1999; Kröger 2008b). There has also been considerable speculation on the habits of ammonites (e.g., Westermann 1996) which, based on their morphology and the nature of the rocks in which they are found as fossils, appear to have lived in many different marine environments, from shallow benthic to open-ocean waters, living as pelagic drifters or swimmers.

In a major review of the habits of Paleozoic cephalopods, Barskov et al. (2008) used differences in the shell morphology to ascertain their hydrostatic and hydrodynamic features and how they may have lived in various marine habitats (benthic, benthopelagic, nektobenthic, and planktonic). The first ellesmeroceratidans, discosoridans, and oncoceratidans all had a cyrtoconic shell and, according to Barskov et al. (2008),

were ecologically flexible and apparently occupied a wide range of 'adaptive zones', although initially they were benthopelagic. In contrast, those groups with an orthoconic shell developed ways of modifying their orientation and stability using deposits in the interseptal spaces and/or in the siphuncle (endoceratians, actinoceratians, and orthoceratians) and were mainly nektobenthic. Bactritans, which lacked such modifications, are thought to have been planktonic. Those groups that initially had a planispiral shell (nautilians, ammonites) were the best adapted for a pelagic life and underwent major diversification.

Barskov et al. (2008) argued that Ordovician cephalopods occupied a wide variety of habitats but the number of pelagic forms decreased in the Silurian, with a corresponding increase in bottom-dwelling (benthic and benthopelagic) forms. The range of habitats occupied in the Ordovician was regained in the Devonian, although with a higher proportion of pelagic taxa. Interestingly, in both the Ordovician and most of the Devonian the proportion of benthic and pelagic taxa did not markedly change with the appearance of new taxa – for example, ammonites in the Devonian. By the Upper Devonian (Frasnian), a major change in cephalopods occurred, with a halving of the benthic and benthopelagic forms (probably due to anoxic conditions) and a tripling of the planktonic taxa. These dramatic changes were largely due to the appearance of clymeniid ammonites and the disappearance of the nektobenthic orthoceratians. Another major reduction in cephalopod diversity occurred in the extinction event at the end of the Devonian followed by a radiation from the nektobenthic ammonites that survived. According to Barskov et al. (2008), during the Carboniferous and for the rest of the Paleozoic there were no fully benthic shelled cephalopods.

The Carboniferous–Guadalupian (Permian) ammonites and nautilians had planispiral shells and included pelagic, benthopelagic, and nektobenthic forms. By the end of the Guadalupian a small increase in the number of pelagic taxa had occurred, largely because of the evolution of the ceratitid ammonites. This latter group gave rise to the Mesozoic ammonite radiation which followed the major extinction at the end of the Permian.

Barskov et al. (2008) concluded that the ecological structure of cephalopod communities did not markedly change with the appearance of new groups, but, progressively through the Paleozoic, there was a reduction in the benthic taxa with an increase in abundance and morphological diversity of the pelagic forms. Associated with these trends was a reduction in the number of higher groups.

The shells of most ammonites were unlikely to have withstood depths greater than 200–400 m without imploding (Westermann 1973) but a few thick-shelled taxa (such as *Lytoceras*) may have been able to live in considerably deeper water, perhaps as deep as 800 m (Westermann 1996). *Nautilus* implodes below about 800–900 m and *Spirula* below 1,500 m (Westermann 1973). With their tightly packed septa and their complex suture patterns, ammonites have been thought to have had stronger shells than nautiliforms, although their shells are typically thinner. Most ammonites probably lived in shallow water, so shell strengthening might have been in response to other factors, such as predation.

Shell morphology and biomechanics can also provide clues to habits; shells that are hydrodynamically streamlined for cutting through the water presumably belonged to animals that swam more efficiently than those animals with shells with prominent external sculpture or which were globose or not otherwise streamlined. Bizarre coiling patterns appeared in some ammonite lineages, particularly the heteromorph ammonites, leading to much speculation as to their life habits (see Section 17.12.2.3).

Shell coiling has appeared multiple times in cephalopod evolution and has several adaptive advantages over the standard straight or curved nautiliform shell (Kröger 2005). A coiled shell is mechanically stronger (e.g., Kröger 2002) and, for swimming taxa, enhances manoeuvrability (e.g., Saunders & Shapiro 1986; Crick 1988; Jacobs & Chamberlain 1996; Westermann 1996), is reasonably hydrodynamically efficient (Chamberlain 1980; Jacobs & Chamberlain 1996), and is more efficient as a buoyancy device (Crick 1988).

The external shell of *Nautilus* provides a substratum for epizoites as was also the case with extinct cephalopods where brachiopods, bryozoans, and other molluscs (such as oysters) were sometimes attached during life (e.g., Hautmann et al. 2017). These attached animals were thus moved about in the water column and have been called pseudoplankton (e.g., Gabbott 1999). However floating empty cephalopod shells

are also used as a substratum, as is often observed in modern *Spirula* shells and cuttlebones.

17.4.2 Locomotion and Buoyancy

17.4.2.1 Locomotion

Locomotion in cephalopods can be accomplished in several ways including by 'jet propulsion', and in the case of many coleoids, by the movement of lateral fins or, in the case of octopods and a few sepiids, by using their arms. As outlined above, jetting is enabled by a rapid discharge of water from the mantle cavity via a funnel (*hyponome* or 'siphon'), a muscular tube-like structure derived from the foot.

The daily vertical migration undertaken by *Nautilus* is due to a combination of jetting and buoyancy changes in the shell (Saunders & Ward 1987; Dunstan et al. 2011). *Nautilus* swims forwards by propelling water through the backwardly pointing funnel. In this position the funnel lies beneath the shell and extends through an embayment in the middle of the apertural lip, the *hyponomic sinus*. This forward jetting is much less efficient than the fast, backwards jetting used to escape predators. A similar notch is seen in most extinct nautiliforms and Paleozoic ammonites, suggesting this mode of propulsion evolved early in cephalopod history, probably in the Ordovician. This sinus was lost in Mesozoic ammonites, leading to speculation regarding the fate of the funnel in ammonites (see Section 17.12.2).

As noted above, ammonites and nautiliforms evolved various shell shapes to minimise resistance while moving through the water. For example, many early nautiliforms had conical shells which presumably provided some streamlining. Ammonites probably had more efficient control of direction during escape responses, such as sharp edges to their coiled shells, but just how fast these responses were, and how efficiently they moved during normal locomotion, is not known. For further discussion on locomotion in the extinct nautiliforms and ammonites see Section 17.4.2.3, and for more details on ammonite swimming see Section 17.12.2.3.

Squid are particularly highly adapted for rapid jet propulsion. Their muscular mantle wall comprises much of the body mass, and its rapid, synchronous contraction is facilitated by giant nerve fibres and stellate ganglia (see Chapter 7). Squid have a cartilaginous ridge on each side of the anterior mantle wall (the mantle-locking cartilage) which slots into a corresponding cartilaginous groove on the wall of the funnel (the funnel-locking cartilage). These interlocking ridges and grooves close and lock the lateral mantle openings while the contracting circular muscles in the mantle cavity wall force water out through the funnel. In some particularly active groups, such as some Ommastrephidae, the locking cartilages are permanently fused. Such fusion is also seen in the slow-moving cranchiids so is not necessarily an adaptation just for rapid movement.

In squid the funnel is very manoeuvrable and can be turned in any direction, allowing backward or forward motion, but the fastest movement is backwards, typically as an escape response, with the funnel facing straight forward. Some fast-moving squid can even become airborne, with observations confirmed on several species (Maciá et al. 2004; Jabr 2010).

Effective, fast jet propulsion is metabolically very costly to maintain. Migratory oceanic squid mainly use jetting when seeking their relatively sparse prey and sustain high speeds through possessing a range of anatomical, physiological, and biochemical adaptations. For example, they have ten times the density of mitochondria in their muscles than *Nautilus*, and both aerobic and anaerobic metabolism are optimised (O'Dor & Webber 1991).

Sepiidans and squid also use their lateral fins to move backwards or forwards with much greater precision than by jetting. However, although undulating fins can increase efficiency, they can also limit speed. Octopods use jetting but can also use their arms to 'walk', and 'walking' also occurs in a few cuttlefish taxa (Roper & Hochberg 1988). Some squid have adopted alternative strategies such as neutral buoyancy (see Section 17.4.2.4), a strategy requiring extra volume and thus extra drag.

Various deep-sea cirrate octopods have been observed resting on the seafloor, crawling using their arms, or drifting or swimming using their fins or mantle pumping (Villanueva et al. 1997). Some cirrate families such as Grimpoteuthidae are largely pelagic, and others such as Opisthoteuthidae and Cirroctopodidae are benthic. Adult vampire squid mainly swim by using their fins, but juveniles mainly jet, despite possessing a set of fins that are later resorbed and replaced by the adult fins (Seibel et al. 1998). Fin swimming may have been the main means of locomotion in early coleoids as they probably had inefficient jet propulsion because the weak contractions of their thin mantle wall may have been used primarily for ventilating the ctenidia (Young & Vecchione 2002).

17.4.2.1.1 Ontogenetic Changes in Swimming Behaviour in Squid

The eggs hatch as swimming 'paralarvae', and the different ontogenetic stages of squid (paralarvae to adults) encounter very different conditions reflected in changes to their fins and jet propulsion system as they grow (Bartol et al. 2008). When they hatch as paralarvae, they typically have small fins, rounded bodies, and a relatively large funnel opening while juveniles and adults usually have larger fins, streamlined bodies, and relatively small funnel apertures. Experimental studies by Bartol et al. (2008) showed that juvenile and adult squid had more diverse swimming strategies than the paralarvae which mainly used jetting. In the juveniles and adults, there was more reliance on the fins, and the combination of fin and jet was shown to be more efficient than jetting alone. Interestingly, the efficiency of jet propulsion in the paralarvae was significantly higher than in juveniles or adults.

17.4.2.1.2 Metabolic Costs of Locomotion

Squid are the fastest-swimming invertebrates, but the metabolic cost of attaining these high speeds is much higher than for fish swimming at similar speeds. Experiments on *Loligo* by O'Dor et al. (1994) have shown that hovering consumes twice as much energy as resting on the bottom. These costs may be reduced in oceanic squid such as loliginids by 'soaring' on ocean currents to reduce energy expenditure. Maximum jetting is used relatively rarely and is typically associated with upward movement in the water column or to escape from predators. Temperature and various biotic factors, such as age, the presence of food, and activity, all affect metabolic rates. For example, in *Octopus* temperature differences as small as 1°C greatly influence growth through differences in metabolic rate (André et al. 2009).

In coleoids, metabolic rate decreases with depth (Seibel et al. 1997). These trends are more extreme in coleoids than in fish or crustaceans because of their very active lifestyles in the photic zone. There are, however, modifications that help achieve metabolic efficiency in the deep sea. Thus, deep-sea coleoids have a less active lifestyle and more fatty tissue than shallow-water species. The fat helps to achieve neutral buoyancy and thus less energy expenditure and a reduced need for food intake. The lowest metabolic rate reported for any coleoid is in *Vampyroteuthis infernalis* which functions normally at around 3% oxygen saturation compared with about 20% in species inhabiting the surface layers. The physiology of *Vampyroteuthis* is thus in marked contrast to that of most cephalopods (and most other animals) which cannot tolerate oxygen levels below about 10%, although a few, such as *Nautilus*, can tolerate oxygen saturations as low as 5% (Wells & Wells 1995; Seibel et al. 1999).

Most deep-sea coleoids, including *Vampyroteuthis*, have extensive respiratory surfaces on their ctenidia and a high haemocyanin concentration in their blood (Childress & Seibel 1998). Compared to other cephalopods, *Vampyroteuthis* haemocyanin is also more efficient at binding and transporting oxygen (Seibel et al. 1999; Seibel 2007).

Metabolic costs are lowest in *Nautilus* and highest in squid, the latter having a greater efficiency in ATP[10] conversion and faster rates of substrate breakdown, both anaerobically and aerobically. In *Nautilus*, the level of octopine dehydrogenase activity (which reversibly links pyruvate to arginine as an alternative to lactate formation – see Chapter 2) is many times less than in squid, as is alpha-lycerophosphate dehydrogenase activity (O'Dor & Webber 1991). In the regulation of these pathways, there is also a shift from octopuses (which are primarily AMP[11]-based) to squid (which have an additional NADH[12]-based regulation) (Storey & Storey 1983). Energy production and consumption in squid is primarily aerobic (Hochachka et al. 1983). Cephalopods mainly rely on carbohydrate (i.e., glycogen) metabolism (Hochachka et al. 1983; Storey & Storey 1983), but this alone is not adequate, and the catabolism of amino acids from protein is also important, especially in very active squid (O'Dor & Webber 1991). This latter process produces ammonia as a waste product, which in some deep-water squid is sequestered for buoyancy (see Section 17.4.2.4).

The anaerobic utilisation of glycogen reserves is limited by the amount of muscle phosphagen (arginine phosphate), with

---

[10] See Chapter 2.
[11] See Chapter 2.
[12] Nicotinamide adenine dinucleotide

octopine the main anaerobic end product. For a squid muscle to recover from 'oxygen debt', arginine must be rephosphorylated and pyruvate recycled to glycogen *in situ* (O'Dor 1988). In normal to high activity, octopine is recycled within muscle although some may be lost to the blood when the animal is stressed (O'Dor & Webber 1991). Thus, the energy used by muscles involved in pumping during jetting is supplied by both aerobic and anaerobic pathways, involving pools of high energy phosphate (Hochachka 1985).

17.4.2.2 Buoyancy

The ability to achieve buoyancy is a key innovation in cephalopods, and evolution in the group has resulted in many different mechanisms to regulate it and to achieve equilibrium (Teichert 1988; Hoffmann et al. 2015), as outlined below.

17.4.2.3 The Shell as a Buoyancy Device

The ability of the shell to act as a buoyancy device is a hallmark of the cephalopods and is seen in all groups, including some coleoids (*Sepia* and *Spirula*) with internal calcareous shells. The linked internal chambers are wide and obvious in *Nautilus* and *Spirula* but very narrow and tightly packed in cuttlefish (*Sepia*) shells ('cuttlebones' or *sepion*). In the chambered part of the shell (the phragmocone) of most nautiliforms, ammonites, and *Spirula*, the chambers are connected by a siphuncle, a narrow, tissue-lined tube, but in cuttlefish (*Sepia*) the siphuncular tissue is spread out thinly over the ventral surface of the sepion (Figure 17.19).

The origin of the gas in the shell chambers may arise by simple diffusion from tissues (e.g., Denton & Gilpin-Brown 1966). Because they use mainly nitrogen gas in their shell for buoyancy (Boucher-Rodoni & Mangold 1994), ammonia release via the ctenidia is the lowest in *Sepia* and *Nautilus*, but the mechanism involved in the production of gaseous nitrogen is not yet understood. Symbiotic bacteria have been suggested (Boucher-Rodoni & Mangold 1994), but as the only pathways known to be used by bacteria that release nitrogen are anaerobic (Pernice et al. 2007), how this might be achieved remains

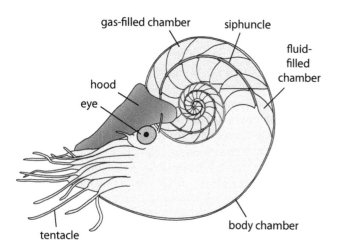

FIGURE 17.19 *Nautilus* showing the fluid and gas-filled chambers used to control buoyancy. Redrawn and modified from Denton, E.J. and Gilpin-Brown, J.B., *Adv. Mar. Biol.*, 11, 197–268, 1973.

a mystery. Little is known about the levels and composition of dissolved gases in the blood of cephalopods although studies on the gas in the chambers of *Nautilus* have been carried out (Greenwald & Ward 1987; Ward & Greenwald 2009).

It has been demonstrated that the internal pressure in the chambers of a *Nautilus* shell is insufficient to prevent the shell imploding below about 800–900 m, and the main attribute that keeps the shell intact down to this depth is the strength of the shell itself (e.g., Westermann 1973). There has to be a trade-off, as shell thickening means greater weight, and this interplay between the need for both buoyancy and shell strength has resulted in many modifications to cephalopod shells. The shell structure is also an important aspect of shell strength, with the development in ammonites of a three-layered construction – inner and outer prismatic layers with a nacreous layer sandwiched between.

According to Tendler et al. (2015), the main aspects of ammonite shell evolution are shell economy (i.e., the amount of shell material used), rapid growth, compactness, and hydrodynamic efficiency. Following each major extinction event, every new wave of ammonites responded in a similar way with these 'tasks' remaining more or less constant because of biomechanical constraints, resulting in much convergent evolution.

How cephalopod buoyancy first evolved remains a mystery (see Section 17.9). The earliest cephalopods may have been benthic or floated or bounced off the bottom with their head pointing downwards, perhaps as they fed on minute animals on the bottom or in the water column.

As noted above, the tissue in the siphuncle (see Section 17.4.2.3.1) connecting the chambers in the phragmocone achieves buoyancy by maintaining the appropriate volume of fluid within each chamber. As a new chamber is completed it is initially filled with water, but the siphuncle tissue draws water out of the chambers into the bloodstream. Pumping liquid from the chambers may have developed early as a buoyancy device in the elongate shells of early nautiliforms, but these shells would then have been vertically aligned in the water and rather unstable. These early nautiliforms may well have been able to swim but probably only for short bursts. The endoceratian nautiliforms (see Section 17.11.1.4) achieved stability by secreting calcareous deposits in the posterior chambers of the shell that acted as counterweights (Figure 7.20), allowing them to become horizontally oriented while swimming, and some may well have been semi-permanently or permanently pelagic. The change in body orientation resulted in a reorientation of the animal with the anterior side becoming functionally the dorsal side, whereas the posterior side, where the siphuncle was located in early cephalopods, became the ventral side. In other cephalopod lineages, different methods were evolved independently to alter the centre of gravity and thus achieve the same end. These included shortening the posterior part of the shell, coiling of the shell, calcareous deposits in the siphuncle, and moving the shell chambers to lie over the body chamber by overlapping them (Figure 17.20). Ammonites did not adopt a counterbalancing strategy but instead used variations in shell coiling and elongation of the body chamber to achieve the same

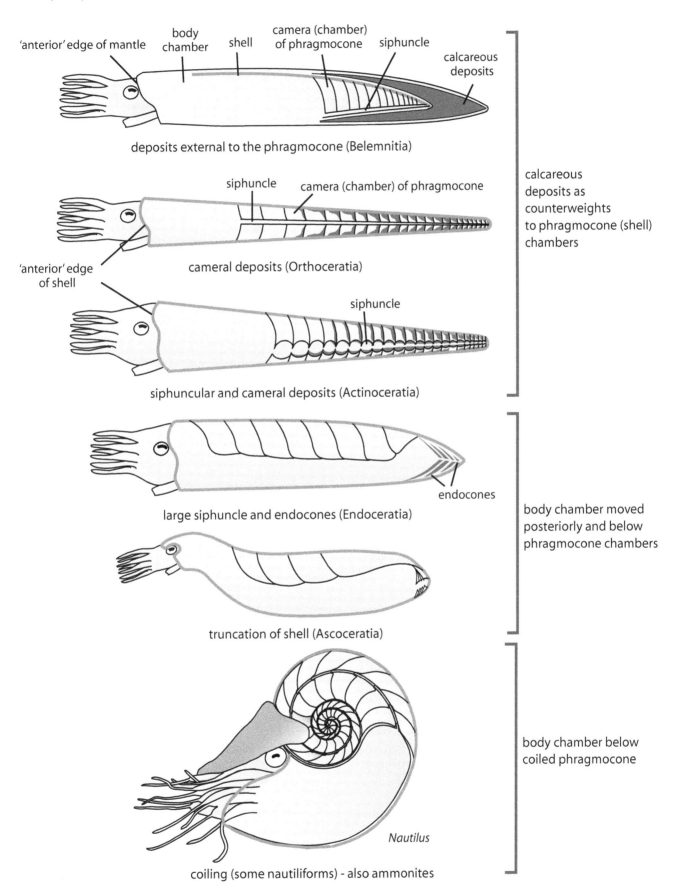

FIGURE 17.20 Various morphologies employed by cephalopods to enable the shell to float horizontally. Note that the reconstructions of head, arms, and funnel in all but *Nautilus* are imaginary. Redrawn and modified from Moore, R.C. et al., *Invertebrate Fossils*, McGraw-Hill, New York, 1952.

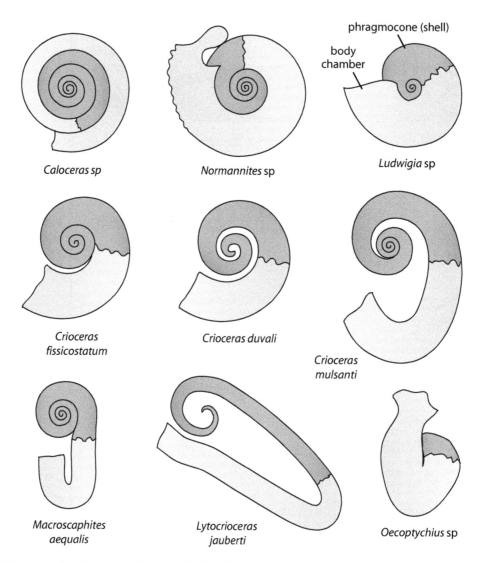

FIGURE 17.21 Orientation of various ammonites when floating. The light grey area indicates the body chamber. Modified from Trueman, A.E., *Q. J. Geol. Soc. Lond.*, 96, 339–383, 1941.

end (e.g., Lehmann 1981; Jacobs & Chamberlain 1996) (Figure 17.21). The orientation of ammonite shells was reconstructed by Naglik et al. (2016) using 3D models of various taxa, and a broad range of orientations was confirmed. These same models were used to investigate buoyancy (Tajika et al. 2015; Naglik et al. 2016). More bizarre solutions to orientation are seen in the weirdly coiled nautiliforms and heteromorph ammonites (e.g., Westermann 1996). Lemanis et al. (2015) found that ammonite hatchlings could potentially swim with at least partial buoyancy provided by just one or two emptied chambers, thus demonstrating the potential for hatchling dispersal.

Coleoids internalised the shell and, over time, reduced its size and weight by reducing the amount of calcification. Early coleoid groups, including the Belemnitida (the belemnites), had a two-layered posterior calcified extension, the *rostrum*, which was secreted on the outside of the phragmocone, possibly as a counterbalance. These additions to the original shell made the thickened parts of belemnite shells more prone to successfully fossilising than the thinner shells of their

cousins. The belemnite rostrum is composed of low-magnesium calcite which is more resistant to dissolution/diagenetic alteration, compared to the aragonitic phragmocone (Saelen 1989). Pelagic coleoids that lost the chambered shell (squid, octopuses) must continually swim to maintain their level in the water column or employ different buoyancy methods by using low-density chemicals (e.g., ammonium, see Section 17.4.2.4) in their blood and/or tissues.

17.4.2.3.1 The Siphuncle and its Role

The chambered shell with its connecting siphuncle is unique to the Cephalopoda and is the key innovation that led to buoyancy (e.g., Kröger 2003). *Nautilus* and *Sepia* have very different shell morphologies and have different approaches to buoyancy regulation. The shell chambers of *Nautilus* are simple and wide whereas those of *Sepia* are very narrow spaces separated by thin lamellae (Figure 17.22), enabling strong capillary forces to operate. The siphuncle contains living tissue connected to the main body of the animal in the

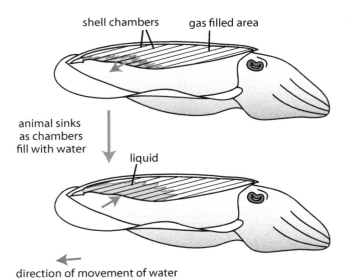

shell chambers   gas filled area

animal sinks
as chambers
fill with water

liquid

direction of movement of water

FIGURE 17.22 Buoyancy mechanism in *Sepia*. Redrawn and modified from Denton, E.J. and Gilpin-Brown, J.B., *J. Mar. Biol. Assoc.* U.K., 41, 319–342, 1961a.

body chamber. It may be a tubular extension of the body (as in *Nautilus* (Figure 17.23), *Spirula*, and most of the extinct shelled species) or can be thinly spread over the open ventral side of the chambers (e.g., *Sepia* and related genera such as *Sepiella* and the ascoceratidans). *Nautilus* maintains near-neutral buoyancy equilibrium and changes only to compensate for abnormal situations, such as injury or growth of epizoans on the shell (Ward et al. 1981; Ward 1986). Buoyancy adjustments in *Sepia* are frequent, and these are associated with behavioural changes with greater energy costs (Denton & Gilpin-Brown 1961a, 1973). These differences in *Nautilus* and *Sepia* are reflected in major differences in their metabolic rates as well as in the morphology of their shells.

The siphuncle runs through the centre of the shell chambers in many nautiliforms, but in both ammonites (except in clymeniids) and coleoids it is located 'ventrally' (i.e., on the posterior side). The siphuncle in nautiliforms is typically wider than in ammonites. In many Ellesmeroceratia, Endoceratia, and Actinoceratia it is over a third to a little over half the diameter of the interior of the body chamber. In those nautiliforms with wide siphuncles its structure is typically more complex (Kröger 2003), and, surprisingly, the Cambrian plectronoceratian siphuncle is 'the most advanced among nautiloids' (Mutvei et al. 2007, p. 1331), with a completely calcified connecting ring traversed by many pore canals which presumably contained extensions from the siphuncular epithelium.

In *Nautilus*, the initial part of the siphuncle in each chamber is encased in a porous calcareous tube (the septal neck), and the remainder is surrounded by a porous horny tube. The porosity of the calcareous part of the tube in *Nautilus* and other shelled cephalopods is due to the shell crystals not being oriented.

Ward (1982) found that, in *Nautilus*, a larger siphuncular surface area was correlated with a higher rate of fluid

regulation in the chambers. Consequently, Ward supposed that the relative size of the siphuncle might indicate the efficiency of buoyancy regulation in extinct cephalopods. There are, however, constraints on how wide the siphuncle can become in a chambered shell like that of *Nautilus* or ammonites, because if it is too large it would both weaken the shell and occupy too much of the volume of the chambers to make buoyancy effective (e.g., Westermann 1971; Hewitt & Westermann 1997; Kröger 2003; Hoffmann et al. 2015). The evidence suggests that efficient buoyancy changes were undertaken in cephalopods with both wide or narrow siphuncles. Such changes may not be related directly to siphuncle size but instead to the amount of energy required for buoyancy changes. There is probably a trade-off between the surface area available to pump fluid and the leakage of liquid and ions back into the chambers, with the latter being greater from larger siphuncular surfaces than from smaller surfaces (Kröger 2003). The constraints resulting from changes in siphuncular and other shell morphologies were addressed differently in different early shelled cephalopod lineages.

The siphuncle can perform several functions, which include removing the liquid in freshly formed chambers, preventing the passive back-flow of liquid, regulating the amount of liquid in the chambers, and, in some fossil taxa, providing space for calcium carbonate deposits (Kröger 2003).

Regulation of the amount of fluid present in the chambers enables the animal to approximate neutral buoyancy, although this does not happen quickly. Buoyancy regulation in living cephalopods is due to an osmotic process by the siphuncular epithelium (Denton & Gilpin-Brown 1961b, 1966; Denton et al. 1967). The animal can increase the salt content and acidity of the blood flowing through the siphuncle so the resulting osmotic gradient causes water in the chamber to move to the more concentrated blood, thus increasing the space in the chamber and increasing buoyancy. The reverse process decreases buoyancy. Extracting liquid from a chamber first requires active transport of the Na$^+$ and Cl$^-$ ions from the chamber liquid through the siphuncular epithelium (e.g., Denton & Gilpin-Brown 1966; Denton 1971; Guex & Rakus 1971). This process results in a reduction of the salinity of the liquid in the chamber which then passes into the blood vessels in the siphuncle by osmosis. The active pumping driving this is powered by ATPase in the epithelia of the siphuncle (Greenwald et al. 1982, 1984; Mangum & Towle 1982). A new chamber, initially fluid-filled, is emptied until equilibrium is reached between osmotic extraction and passive back-flow of fluid through the epithelium. In adult *Nautilus* this equilibrium is maintained when the liquid level of the chambers is between 4 and 11% (Denton & Gilpin-Brown 1966). Fluid exchange through the siphuncular epithelium achieves changes of buoyancy by adjustments in the fluid content of the chambers. The fluid level in the shell chambers of ammonites and ellesmeroceratian and plectronoceratian nautiliforms may have been somewhat higher than in *Nautilus* (Kröger 2003 and references therein).

The common idea that the spaces in the shell chambers are filled with gas under pressure is incorrect – it is actually

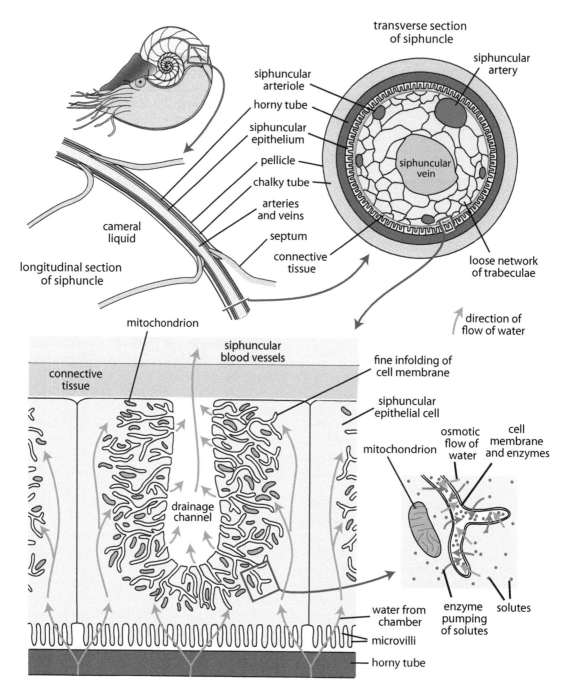

FIGURE 17.23 Structure and function of the siphuncle in *Nautilus*. Upper figures redrawn and modified from Denton, E.J., *Proc. R. Soc. B*, 185, 273–299, 1974 and Ward, P. et al., *Sci. Am.*, 243, 190–203, 1980, lower figures from Ward, P. et al., *Sci. Am.*, 243, 190–203, 1980.

a partial vacuum (Jacobs 1992, 1996). The gas in the shell chambers in *Sepia* was analysed by Denton and Taylor (1964) who demonstrated that partial pressures are never higher than those expected in the animal tissue. In the older chambers, it was 97% nitrogen and close to 0.8 atmospheric pressure. The newer chambers have more oxygen (to about 17%) and low levels of CO_2. Somewhat similar results were found in the chambers of *Nautilus*, which contained mostly nitrogen and smaller amounts of oxygen and argon (Denton & Gilpin-Brown 1966).

The siphuncular epithelium and the processes just described are generally similar in *Nautilus* and *Sepia* (Denton

& Gilpin-Brown 1973; Greenwald et al. 1982), but there are some structural differences, as there are between *Nautilus* and *Spirula* (Tanabe 2000). Some extraordinarily well-preserved fossils show the siphuncular tissue of ammonites and extinct coleoids was also very similar (Barskov 1996; Tanabe 2000; Donovan & Fuchs 2016). Because there appear to be no great differences between the major groups of cephalopods, the siphuncle is assumed to have performed a similar function throughout the group (Kröger 2003).

With increasing water depth there are increased energy costs to pumping water from the chambers, and early workers assumed that the osmotic pumping mechanism would not function below 250 m (Brunn 1943; Denton & Gilpin-Brown 1966). To explain how this apparent constraint might be overcome Denton and Gilpin-Brown (1966) proposed a concept, based on their observations on *Sepia*, they called *decoupling*. They noticed that ion exchange in the very closely spaced shell chambers was delayed because diffusion was markedly slowed due to high capillary pressure. They maintained that the decoupling spaces acted as buffers against liquid back-flow while the cuttlefish changed its position in the water column. Thus, decoupling involves a separation of fluid in the chamber from the siphuncular tissue, and this could be achieved by internal structural changes in the chambers or even by changing the shell orientation. As pointed out by Jacobs (1996) and Jacobs and Chamberlain (1996), there is no need to propose such a mechanism regarding the osmotic pump, as the supposed constraint was a fallacy based on a simple view of siphuncular structure and function. The ultrastructure of the siphuncular tissue (Greenwald et al. 1982, 1984) showed that the osmotic work is performed in minute channels in the cells, not on the outer surface (Figure 17.23), enabling the siphuncle to function as a hyperosmotic pump. These observations (on *Nautilus*) showed that the chamber liquid is pumped via minute channels in the epithelial cells which themselves act like microscopic decoupling spaces, allowing the transport of fluids through the cells even in a hyperosmotic environment. According to Jacobs (1996), these findings rendered the concept of decoupling invalid, but Kröger (2003) argued that the observations on *Sepia* suggest that decoupling spaces create a system with different salinity in different spaces which enhance the standing-gradient model[13] involved in fluid transport via the siphuncle and thus reduce energy costs. In a decoupling space much smaller than the entire chamber, a suitable ion concentration can be achieved much more rapidly than in the entire chamber.

Given the higher energy requirements of pumping at considerable depths, decoupling mechanisms were probably important in reducing these energy costs, at least in some extinct taxa (Kröger 2003). Jacobs and Chamberlain (1996) argued that with their narrow siphuncle, ammonites would, like *Nautilus*, not have used buoyancy changes as the main means of vertical migration. Kröger (2003) suggested that shelled cephalopods with both narrow and wide siphuncles could be capable of fast buoyancy changes using adaptations that created decoupling spaces and reduced the energy costs of buoyancy regulation. As an example, Kröger (2003) suggested that, in some extinct nautiliforms, the pores in the thick, porous connecting rings of the siphuncles were possible decoupling spaces.

In subadults or adults the most posterior shell chambers are of little use in buoyancy regulation because of their relatively small size. In *Spirula* these posterior chambers are filled with fluid (Denton et al. 1967), but in some extinct nautiliforms they were filled with shell material and were even shed in

some (e.g., the nautiliform ascoceratidans and the ammonite *Ptychoceras*). In other groups the siphuncle ceased to function in the posterior chambers – for example, many early ellesmeroceratidans developed chambers in the siphuncle, or it was filled with aragonite (Kröger 2003). In contrast, the siphuncle in some nautiliform groups produced deposits that were often very complex, suggesting that they were probably involved in buoyancy and stability regulation (e.g., Mutvei 1964; Wade & Stait 1998; Kröger 2003).

Since the end of the Paleozoic, the trend was towards a more energy-efficient buoyancy regulation system that involved a narrower, simpler siphuncle than that seen in many earlier cephalopods. The position of the siphuncle may also be important; a ventral siphuncle (as in the Bactrita, ammonites, and coleoids) enabled the dewatering of chambers in any orientation while a centrally located siphuncle would make dewatering in a horizontal position more difficult. However, this simplistic difference is rendered more complex in taxa with the posterior parts of the siphuncle filled with endocone deposits or those that possess long septal necks.

17.4.2.4 Other Buoyancy Strategies

Some 16 families of deep-water squid achieve neutral buoyancy by storing ammonium in their body tissues, and this strategy apparently has arisen independently several times in squid evolution (Voight et al. 1994). This method of buoyancy is usually achieved by storing ammonium ions in vacuoles in the body tissues, including the mantle muscles, a gelatinous outer layer, or (as in Cranchiidae) in a unique coelomic cavity (Voight et al. 1994) (Figure 17.24). The adoption of this buoyancy strategy may not be difficult because, as noted above, squid produce large quantities of ammonia, as their major energy source involves amino acid catabolism (Boucher-Rodoni & Mangold 1994). It has been argued that squid were exapted for ammonium storage because they have low blood pH which enables the efficient removal of the ammonia from the cells where it is produced, and this is coupled with minimal H^+ ion transfer between muscle cells and the blood (Voight et al. 1994). Normally, in other coleoids and most other molluscs, this waste product is released by diffusion through the ctenidial epithelia.

An unusual buoyancy strategy was recently described in the octopod *Argonauta*, the females of which produce a calcareous shell-like case (the 'paper nautilus') that serves as a receptacle for the eggs. Finn and Norman (2010) observed that, besides serving as an egg case, females use the shell to control buoyancy at various depths by capturing air at the sea surface. This pocket of air is then contained by the broad arms, and the animal then dives to a depth where the buoyancy provided by the air is counteracted by the weight of the body, including the eggs if they are present. This strategy has enabled the argonautids to become secondarily pelagic by using an entirely different device and behaviour compared to other cephalopods.

Yet another buoyancy method is a 'swim bladder'-like structure attached to the viscera in the octopod *Ocythoe tuberculata* (Packard & Wurtz 1994).

---

[13] The epithelia performing solute-linked standing-gradient osmotic water transport have long, narrow channels open at one end and closed at the other. They are found in various transporting epithelia, including those in the vertebrate gut.

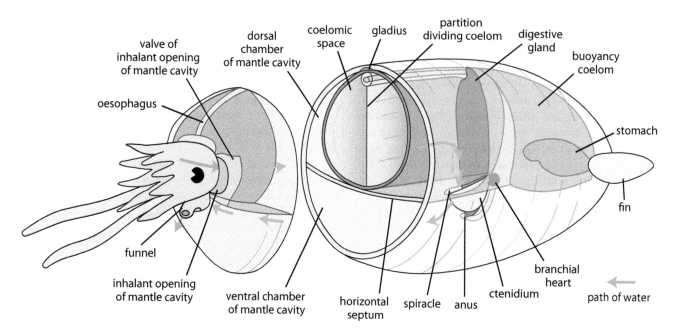

FIGURE 17.24 The cranchid *Megalocranchia fisheri* cut transversely to show the buoyancy coelom and the water circulation used in ventilation. Redrawn and modified from Clarke, M.R. *Nature*, 196, 351–352, 1962.

17.4.3 THE INK SAC AND INKING

The ink sac (Figure 17.25) is formed from a diverticulum from the posterior portion of the gastric caecum. It consists of an ink gland and a reservoir and is controlled by sphincters. The ink gland produces melanin that can be expelled through the funnel of the mantle cavity in response to a threat. The 'ink' is in suspension in the discharged exhalant water. This response (inking) is typically initiated to create a 'smokescreen' or 'pseudomorph' decoy when the animal is attempting to escape. The screen may be black, brown, red, or even bioluminescent, depending on the species. If the ink (which is composed mainly of melanin) is mixed with mucus, it can hang as a small cloud (pseudomorph) in the water to distract a potential predator. If the animal simultaneously changes colour, this will further distract the predator. Alternatively, a large cloud may be produced that acts as a 'smokescreen'. Chemicals in such screens may also interfere with the olfaction and taste senses of the predator. The ink of some squid, octopuses, and cuttlefish contains small amounts of amino acids, including some that have the potential to cause sensory disruption in crustaceans and fish. It also contains chemicals (such as dopamine) that act as alarm signals to conspecifics, besides having some toxic properties (see Derby 2007; Derby et al. 2007 for reviews). Interestingly, some species of the highly poisonous genus *Hapalochlaena* (blue-ringed octopuses) have a much-reduced ink sac in adults (Huffard & Caldwell 2002), although it is functional in young juveniles, and the ink sac has been lost independently in some octobrachian lineages (Strugnell et al. 2014).

The presence or absence of an ink sac is also related to water depth (Robson 1931; Voss 1988; Allcock & Piertney 2002), but observations on deep-sea squid by Bush and Robison (2007) have shown that they often release ink in areas well beyond the reach of light. They do this in a variety of inking types categorised as pseudomorphs, pseudomorph series, ink ropes, clouds/smokescreens, diffuse puffs, and mantle fills. Species typically produced one kind, but all the species observed could produce multiple kinds. Escape behaviour was usually associated with pseudomorphs and pseudomorph series while the animal normally remained adjacent to, or among, ink ropes, clouds, and puffs, which have also been suggested as a means of intraspecific communication (Bush & Robison 2007) perhaps associated with chemical signals (Gilly & Lucero 1992). Species of *Heteroteuthis* (Sepiolidae) release bioluminescent fluid with their ink (Dilly & Herring 1978).

Ink and ink sacs have been recorded from several fossil coleoids (e.g., Nishiguchi & Mapes 2008). The considerable pharmacological potential of coleoid ink has been discussed by Jose et al. (2018).

Clarkeiteuthis, a Lower Jurassic diplobelid, had an ink sac and a pair of ctenidia, the latter generally similar to those of octopods and vampyromorphans in outline and the number of filaments, but the structural details are not known (Reitner 2009).

17.4.4 COLOUR CHANGES – CHROMATOPHORES AND OTHER SPECIAL SKIN STRUCTURES

In cephalopods, as in many other animals, colour is produced by both structural and pigmented elements. Structural colouration is created by the scattering of light by colourless structures while pigmented material selectively absorbs light with the remaining part of the spectrum reflected. Despite the apparent colour blindness of most coleoids (see Chapter 7), there is an impressive range of body colour patterns for camouflage and signalling.

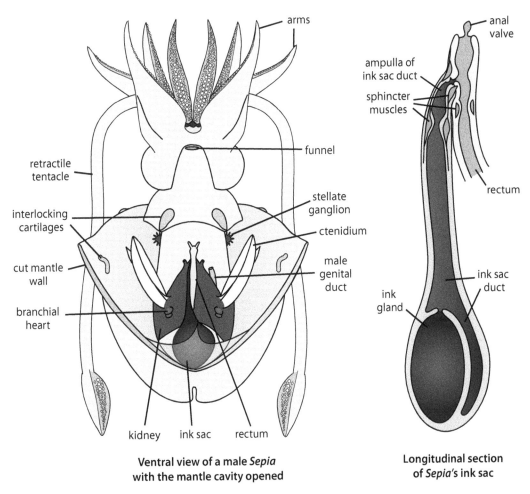

FIGURE 17.25 A ventral view of the cuttlefish *Sepia* showing the location of the ink sac, and the ink sac is shown in detail on the right. Left figure redrawn and modified from Tompsett, D.H., Sepia, pp. 1–191, in Daniel, R.J. (ed.), *L. M.B. C. Memoirs on Typical British Marine Plants and Animals.* University Press of Liverpool, Liverpool, 1939 and right from Mangold, K.M. et al., Organisation générale des céphalopodes, pp. 7–69 in Grasse, P.P. (ed.), *Céphalopodes. Traité de Zoologie.* Vol. 5, Fascicule 4, Masson, Paris, France, 1989.

Colour pattern changes in coleoids are largely controlled by the chromatophore system which is under neuromuscular control. Typically, there are also structural reflector cells (iridophores and leucophores) lying deeper in the skin beneath the chromatophores. The colour observed thus depends on which skin elements reflect the light falling on the skin, whether chromatophores, structural reflectors, or a combination of both. It is the interplay between these structures, and their ability to change, that enable most coleoids to produce their amazing colour and pattern repertoire.

Nautilus lacks chromatophores, iridophores, and the other specialised cells described below.

17.4.4.1 Chromatophores

These numerous small (up to about 1.5 mm in diameter when expanded) pigment-filled structures lie in the skin and are in part composed of cells containing contractile fibres innervated from the brain. The chromatophores give coleoids the ability to almost instantaneously change colour and patterns in response to danger or other stimuli. They are entirely under neuromuscular control and are not controlled

by hormones, as in analogous structures in other animals (Messenger 2001).

Depending on the species, there are typically two or three colour classes of chromatophores: red, yellow-orange, and brown-black. Each consists of a pigment sac with many radial muscles attached to its periphery (Figure 17.26). Contraction of these muscles causes the pigment sac to increase in size rapidly, and when they relax the cell returns to its original tiny dimensions. Selective expansion and contraction of different groups of chromatophores produce patterns such as stripes, bands, or spots. The chromatophores can also rapidly expose or hide colours produced by underlying coloured elements, notably iridophores.

17.4.4.2 Structural Reflectors (Iridophores, Silvery Reflectors, and Leucophores)

The reflective tissues in coleoids are just some of the wide variety that evolved in aquatic animals (Herring 1994). In coleoids there are two main kinds of structural reflectors: those that produce iridescence and those that produce 'diffuse reflectance' (e.g., those that produce white markings)

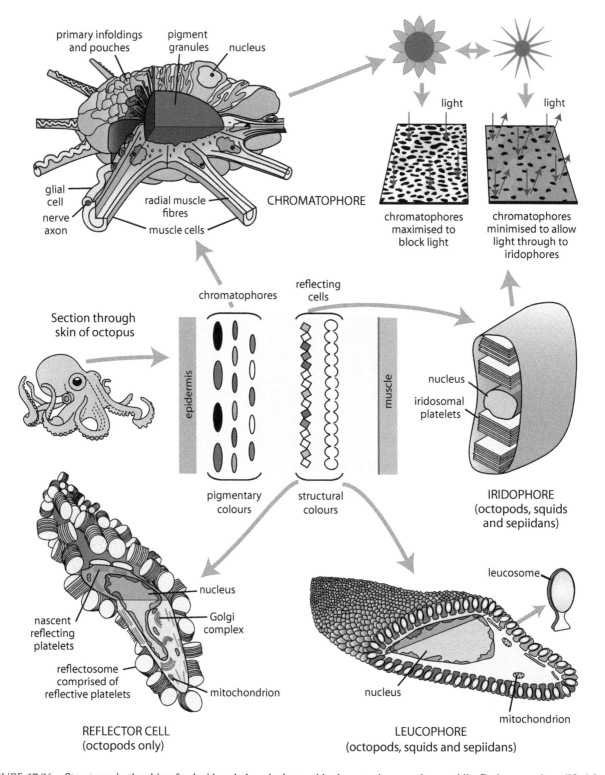

FIGURE 17.26 Structures in the skin of coleoid cephalopods that enable them to change colour rapidly. Redrawn and modified from the following sources: upper left (Cloney, R.A. and Florey, E., Z. *Zellforsch. Mikrosk. Anat.*, 89, 250–280, 1968), upper right and centre of middle row (Fleming, K., *Aust. Geog.*, 96, 94–93, 2009), middle right (Mäthger, L.M. et al., *J. Exp. Biol.*, 204, 2103–2118, 2004), lower left (Brocco, S.L. and Cloney, R.A. *Cell Tissue Res.*, 205, 167–186, 1980), lower right (Cloney, R.A. and Brocco, L., *Am. Zool.*, 23, 581–592, 1983).

(Mäthger et al. 2009). The former type includes iridophores and silvery reflectors, the latter leucophores.

Iridophores (= iridocytes) are found in the skin of octopuses, squid, and cuttlefish and are capable of red, orange, yellow, green, and blue iridescence. These tiny (<1 mm) colourless cells range in shape from oval to very elongate and consist of many parallel lamellae that contain platelets (Figure 17.26), reflecting light by thin-film interference (e.g., Mäthger et al. 2004).

They may be active and capable of changing colour or passive with a constant colour (Budelmann et al. 1997). Active iridophores change their optical properties (i.e., from non-iridescent to iridescent or by changing the iridescent colour) by changing their ultrastructural properties (Cooper et al. 1990; Hanlon et al. 1990). Shorter wavelengths (e.g., blue light) are reflected if the plates are very thin, but light with longer wavelengths (e.g., red) is reflected if they are thicker. The angle at which the light hits the plates is also important (i.e., the direction of the light and the orientation of the stacked plates relative to the surface of the skin); more oblique angles will reflect light with shorter wavelengths than do reflectors at less oblique angles. Squid can alter their iridescence depending on their behaviour, with maximum iridescence seen when they are exhibiting antagonistic behaviour (Hanlon 1982), but these changes are much slower than those of chromatophores, taking from several seconds to minutes (Mäthger et al. 2009). It is not known for certain how these alterations in the active iridophores are achieved, but it is possible that the proteins (reflectins) making up the plates undergo modifications, or there is an adjustment in the thickness of the plates (Mäthger et al. 2009). These changes can be induced experimentally by specific chemical stimuli (Mäthger et al. 2004). A much faster way iridescence can be altered is by expanding or contracting the overlying chromatophores to hide or expose the iridophores.

Iridophores produce the blue and green colours in coleoids – one of the most spectacular examples being the rings and lines of the blue-ringed octopuses (*Hapalochlaena* spp.). Some squid with transparent bodies have lateral stripes of iridophores that are thought to be involved in both signalling and crypsis (Mäthger & Denton 2001).

Silvery reflectors are responsible for the silver used to conceal the eyes and ink sac by reflection (Hanlon & Messenger 1996), and they consist of irregularly spaced layers of wavy plates, a structure which facilitates the reflection of a broad spectrum of colours. The orientation of the plates is such that light is reflected from almost any angle and is the same intensity as the background light, resulting in almost perfect camouflage (Mäthger et al. 2009). Some squid have a red-reflecting iridophore stripe along each side of the mantle and head that may be involved in coordinating movement within a school (Mäthger et al. 2009).

Leucophores, found in some coleoids, reflect mainly white light and form white spots. They are comprised of assemblages of leucosomes (Figure 17.26) and reflect the ambient wavelengths of light. Most squid lack these structures, which are mainly found in octopods and cuttlefish.

The interplay and optical interactions between the superficial chromatophores and the more deeply buried iridophores can produce almost instantly a range of colours that covers the visible spectrum (Mäthger & Hanlon 2007; Mäthger et al. 2009). This interplay may result in a change in the reflected spectrum of the iridescence from the iridophores due to their light being filtered as it passes through the thinly expanded chromatophores (Mäthger & Hanlon 2007), or it may enhance the colour of the chromatophores or block the underlying iridophores completely. The chromatophores can also produce

contrast to highlight the iridescence (e.g., as in the blue rings of the blue-ringed octopus) (Mäthger et al. 2009).

Coleoids can see polarised light (see Chapter 7). Light from iridophores viewed at an oblique angle is polarised and may play a role in communication, as it can be regulated and is not visible to cephalopod predators (teleosts, elasmobranchs, and marine mammals) as they are not polarisation sensitive (e.g., Mäthger & Hanlon 2007; Mäthger et al. 2009).

17.4.5 Luminescent Organs

A general account of bioluminescence in cephalopods is given in Chapter 9. It is common in deep-sea coleoids where it is used in countershading as well as in prey capture and defence, rather than having a role in reproductive activities or communication (Rees et al. 1998). Possible communication using bioluminescence may occur in some species, as for example reported in the squid *Taningia danae* (Kubodera et al. 2007).

Luminescence is generated by light organs (*photophores*), present in all orders of living coleoids and in nearly two-thirds of the genera of decabrachians, especially oegopsid squid, but they are absent in most inshore taxa (Hanlon & Messenger 1996). Photophores are mostly located on the ventral surface, especially below the eyes and on the arms, and they direct the light downwards for countershading. There are two main kinds of photophores (Figure 17.27), and both have multiple independent origins (Young & Bennett 1988). The first type is a pocket containing luminescent bacteria that can be covered or uncovered to release the light (Figure 17.27). Bacterial photophores are found in sepiolids and loliginids and are located internally (see Chapter 9 for more details). In the second type, luminescence is produced by cellular secretions from special tissues arranged as light organs, typically on the outer surface of the animal (Figure 17.27). The light organs are arranged differently in different species.

How photophores are controlled is poorly known, with some apparently under nervous control, but others may be controlled indirectly by changes to overlying chromatophores or by other means. The light organs can produce either a steady light (as used in ventral counter-illumination) or intermittent flashes for signalling (Hanlon & Messenger 1996). Short flashes probably reduce the threat of predation, and, in some instances, the light may act as a lure.

The deep-sea coleoid *Vampyroteuthis infernalis* has two ways of producing bioluminescence: one from photophores that cover much of its body and the other by releasing bioluminescent fluid from the arm tips that forms a luminescent cloud surrounding the animal (Robison et al. 2003). If threatened, 'vampire squid' also use a combination of writhing body movements, an expanded body form, and bright flashes from its arm tips to disorientate a potential predator.

A luminescent cloud is also produced by some squid (e.g., *Heteroteuthis dispar*) that secrete bioluminescent mucus from glands adjacent to the ink sac. The glowing cloud probably serves an analogous purpose to the ink cloud produced by shallow-water squid (Young 1977). Bioluminescence is

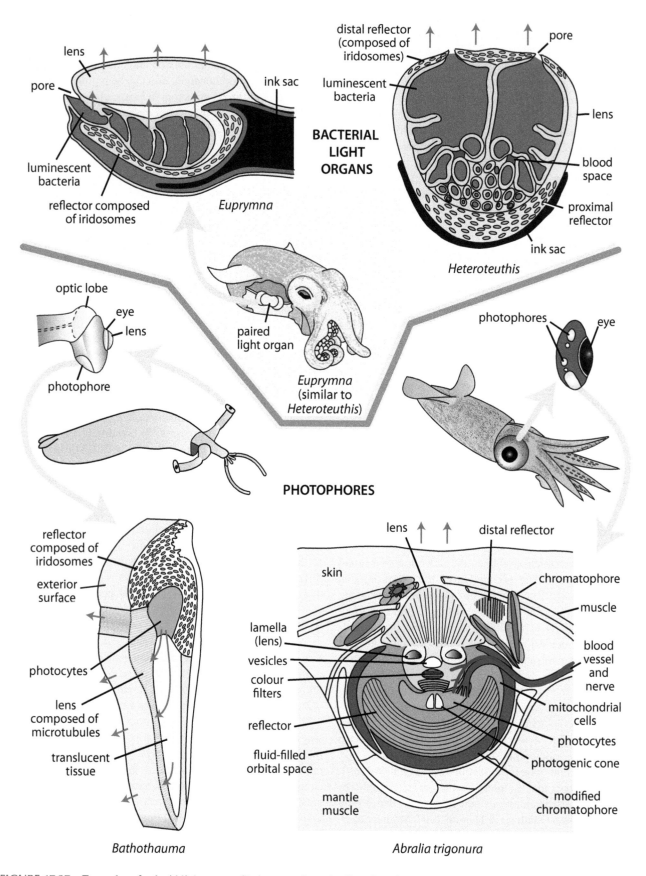

FIGURE 17.27 Examples of coleoid light organs. Red arrows show the direction of generated light. Redrawn and modified from the following sources: top row (Dilly, P.N. and Herring, P.L. *J. Zool.*, 186, 47–59, 1978), *Euprymna* (Jones, B.W. and Nishiguchi, M.K., *Mar. Biol.*, 144, 1151–1155, 2004), *Bathothauma* (Dilly, P.N. and Herring, P.L. *J. Zool.*, 172, 81–100, 1974), *Abralia* (Young, R.E. and Arnold, J.M., *Malacologia*, 23, 135–163, 1982).

produced by modified suckers along the length of the arms of the only cirrate octopod (*Stauroteuthis syrtensis*) known to be bioluminescent, and this may function as a lure to attract their crustacean prey (Johnsen et al. 1999). These suckers are modified for this purpose and have lost the ability to grasp. The pelagic incirrate amphitritid octopods have a photophore that encircles the mouth. Only two genera of incirrate octopods (*Japetella* and *Eledonella*) include bioluminescent species (Johnsen et al. 1999).

The largest photophores known in any animal belong to the squid *Taningia danae*. These are at the tips of the second (dorso-lateral) pair of arms and are up to 5 cm in length. They have been seen to flash rapidly if the squid is threatened (Wood 2003; Kubodera et al. 2007).

There is considerable anatomical variation in cephalopod photophores, especially in the presence and position of lenses, filters, and reflectors even within a family (e.g., Cranchiidae) (Herring et al. 2002); they usually have a thick layer of reflective plates at the base of the photophore. Some squid have several types of photophore. In the simplest kinds (e.g., in *Stauroteuthis*, *Vampyroteuthis*, and *Spirula*), the reflectors are formed from multiple layers of collagen rods, as they are in the simplest photophores of the midwater enoploteuthid squid *Abralia* and *Watasenia* (see Mäthger et al. 2009 for references). *Abralia* and a number of other squid also have more complex photophores, such as the one shown in Figure 17.27. These have, in addition to lenses, additional reflective areas, such as light guides and filters (Herring 1988; Young & Bennett 1988). Light is reflected through the photophore by the basal reflectors increasing the amount of emitted light, and the wavelength of this light can also be fine-tuned by selective filters and reflectors. Some squid can even regulate the spectral composition of their bioluminescence to match the down-welling light (Young & Mencher 1980), possibly by muscular distortion of the photophore iridophores (Arnold et al. 1974; Young & Arnold 1982). Some complex photophores, such as those below the eyes of the squid *Bathothauma* (Figure 17.27) and *Pterygioteuthis*, are covered with a transparent epidermis and are composed of a deep-seated cellular photogenic region, as well as a reflector region of iridophores and a 'light guide' region. The light produced by the photogenic tissue is reflected by the reflector layer which prevents light from reaching the surface other than via the light guide region (Arnold & Young 1974; Dilly & Herring 1974).

17.5 DIGESTIVE SYSTEM

The cephalopod digestive system is described in Chapter 5, and some details of the jaws and radula are briefly described below. The remainder of the gut consists of the oesophagus (which may be expanded into a crop), a muscular stomach which triturates the food and mixes it with enzymes, and a large caecum, where the food is sorted on ciliated leaflets and where most of the digestion takes place. Absorption occurs in the large digestive gland. The gut is characterised by muscular action and enzymes and extracellular digestion, thus

relying less on cilia than most other molluscs (see Chapter 5 for details).

17.5.1 THE JAWS

The most prominent parts of the mouth are the large parrot beak-like jaws (Figure 17.28) which, in some extinct cephalopods, are greatly modified (see Section 17.12.2.4).

17.5.2 THE RADULA

Several fossil radulae have been found *in situ* in ammonites, and a few in fossils of coleoids and nautiliforms. These are generally similar to those of living coleoids (Figure 17.29). Ammonite radulae (Figure 17.30) (Keupp et al. 2016) are very similar to those of modern coleoids (Figure 17.31), especially octopods (Lehmann 1967, 1981; Nixon 1988a, 1996; Kruta et al. 2011, 2014a), but one fossil coleoid (a possible early belemnite) has 11 teeth in each row (Doguzhaeva et al. 2007a). In contrast, *Nautilus* has 13 teeth in each row, four of which are marginal plates (Nixon 1988a). The few known radulae of nautiliforms vary considerably (Figure 17.29). In two species of *Paleocadmus* from the Middle Pennsylvanian (Carboniferous) of the USA (Solem & Richardson 1975; Saunders & Richardson 1979), the radula is similar to that of *Nautilus*, but that of the orthoceratian *Michelinoceras* from the Upper Silurian of Bolivia has only seven teeth in each row. An older unidentified orthoceratian from the Upper Ordovician of South Africa (Gabbott 1999) has five or possibly seven teeth in each row (Figure 17.29).

Radulae are known from at least 13 genera of ammonites (Kruta et al. 2015; Keupp et al. 2016). Most have nine teeth in each row including a pair of marginal plates, but others lack the plates. The morphology of the teeth varies considerably, as shown in Figure 17.30. Keupp et al. (2016) noted that there is a correlation between multicuspid teeth and the evolution of modified jaws, both appearing during the Toarcian (Lower Jurassic), a period marked by waves of extinctions. The radula of the Late Cretaceous uncoiled heteromorph ammonite *Baculites* is thought to be highly modified for catching small planktonic organisms (Kruta et al. 2011).

The radula is relatively small; in living coleoids, there are seven to nine teeth per row (Figure 17.31), and, in *Spirula* and in some cirrate octobrachs, the radula is lost.

17.5.3 FEEDING

Cephalopod feeding is reviewed in Chapter 5.

Coleoid cephalopods are mostly active carnivores and use their arms (and/or tentacles) to capture prey, although a few are microphagous, including *Spirula* which feeds on small crustaceans (Ohkouchi et al. 2013; Hoffmann & Warnke 2014), and *Vampyroteuthis* (Robson 1930; Hoving & Robison 2012). *Spirula* has no radula, and their horny beaks (modified jaws) are presumably used to break up their crustacean prey.

The saliva produced by the salivary glands can contain toxins and can be injected into the prey through punctures made

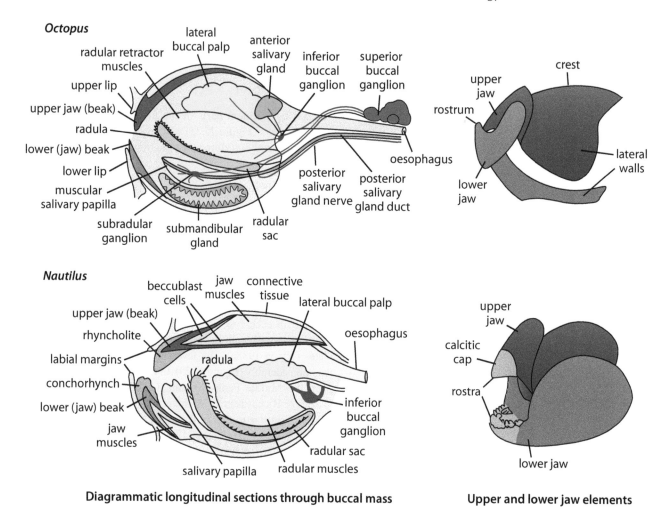

Diagrammatic longitudinal sections through buccal mass   **Upper and lower jaw elements**

FIGURE 17.28   The jaws and buccal mass of *Nautilus* and *Octopus*. Redrawn and modified from the following sources: upper left (Boyle, P.R. et al., J. Zool., 188, 53–67, 1979), lower left (Tanabe, K. and Fukuda, Y., *Lethaia*, 16, 249–256, 1983; Tanabe, K., and Fukuda, Y., Mouth part histology and morphology, pp. 313–322 in Saunders, W.B. and Landman, N. H. (eds.), *Nautilus: The Biology and Paleobiology of a Living Fossil. Topics in Geobiology,* Springer, New York, 1987), upper right (Nixon, M., Morphology of the jaws and radula in ammonoids, pp. 23–42 in Landman, N.H. Tanabe, K., and Davis, R.A. (eds.), *Ammonoid Paleobiology: From Anatomy to Ecology. Topics in Geobiology,* Plenum Press, New York, 1996), lower right (Seilacher, A., *Am. J. Sci.,* 293-A, 20–32, 1993).

by the beak or radula. Some active components have been identified in the saliva, including various enzymes and some toxic compounds. These include the neurotoxin tetrodotoxin which is probably obtained via food or by way of symbiotic bacteria (see Section 17.7.8) and others that are synthesised, including tachykinins and cephalotoxins, as well as biogenic amines such as serotonin and octopamine (see Cooke et al. 2015 for a recent review).

Many octopods drill holes in shelled molluscs, not with their radula but by using rasp-like teeth on the salivary papilla (Nixon 1979, 1980).

The lens-less *Nautilus* eyes are capable of only minimal vision (Muntz 1986, 1987a). Consequently, *Nautilus* probably relies mainly on touch and its olfactory organs to locate food (e.g., Basil et al. 2005), which mainly consists of small fish, crustaceans, and some scavenged items (Saunders 1985).

Shallow-water coleoids hunt their prey and avoid predators using mainly visual cues. Many pelagic coleoids actively hunt larger swimming prey, while benthic cephalopods, such as a number of octopods, eat crabs, worms, fish, and molluscs. In the deep sea, where visual hunting is only possible with bioluminescent prey, they often employ an energy-efficient 'sit-and-wait' strategy.

Extinct cephalopods (nautiliforms, belemnites, and at least some ammonites) were probably also mostly carnivores, although it has been recently suggested that the large orthoconic endoceratid nautiliforms were plankton feeders (Mironenko 2018) (Figure 17.32). Ammonite gut contents show that at least some ate small to very small invertebrates (Klug & Lehmann 2015), and examination of the radulae of two heteromorph Ancyloceratina suggests that members of that group (which possessed a massive lower jaw – see

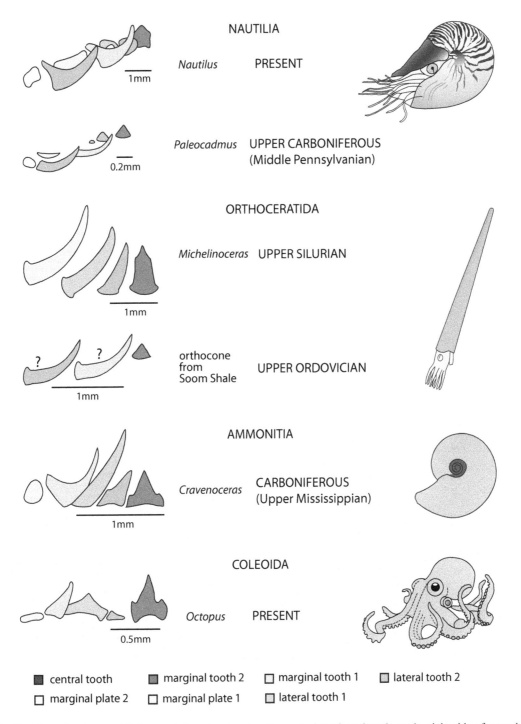

FIGURE 17.29 Radulae of extinct and living cephalopods, showing the central tooth and teeth on the right side of a tooth row and the probable homologies of the teeth. Radulae redrawn and modified from Gabbott, S.E. *Palaeont.*, 42, 123–148, 1993 with figures derived from three sources (Nixon, M., The buccal mass of fossil and Recent Cephalopoda, pp. 103–122, in Clarke, M.R. and Trueman, E.R. (eds.), *Paleontology and Neontology of Cephalopods. The Mollusca*, Vol. 12, Academic Press, New York, 1988a; Nixon, M., *J. Zool.*, 236, 73–81, 1995; Tanabe, K. and Mapes, R.H., *J. Paleont.*, 69, 703–707, 1995). Figure 17.30 for more detail on ammonite radulae and Figure 17.31 for coleoid radulae.

Section 17.12.2.4) may have been plankton feeders (Kruta et al. 2011, 2015; Tanabe et al. 2015). The early nautiliforms probably also fed on small invertebrates, perhaps including trilobites and other nautiliforms, while the hooked arms of belemnites (see Section 17.13.1.1) suggest that they were predators on more active prey such as fish and other cephalopods (Fuchs & Hoffmann 2017; Hoffmann et al. 2017) (Figure 17.32).

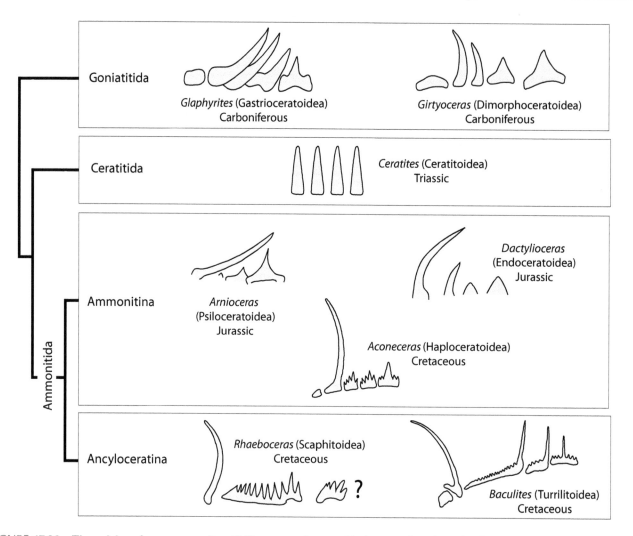

FIGURE 17.30 The radulae of some ammonites. Half rows are shown, with the central teeth on the right. Radulae redrawn and modified from Kruta, I., Ammonoid radula, pp. 485–505, in Klug, C., Korn, D., De Baets, K., Kruta, I., and Mapes, R.H. (eds.), *Ammonoid Paleobiology: From Anatomy to Ecology. Topics in Geobiology.* Springer, Dordrecht, the Netherlands, 2015 with figures derived from the following sources: Tanabe, K. and Mapes, R.H., *J. Paleont.*, 69, 703–707, 1995; Doguzhaeva, L.A. et al., *Lethaia*, 30, 305–313, 1997; Kruta, I. et al., *Lethaia*, 47, 38–48, 2014b.

17.6 RENOPERICARDIAL SYSTEM

In contrast to other molluscs, and as described in Chapter 6, in most cephalopods the coelomic spaces (pericardium, kidney, and gonad) are greatly expanded. In *Nautilus* and many decabrachian coleoids the coelom is divided into two connected chambers by the *genital septum*: the large dorsal gonadal part of the coelom and the ventral pericardial coelom, the latter enclosing the branchial heart appendages (pericardial glands) and the heart (Martin 1983). The kidneys, another part of the coelomic complex, are also connected (Figure 17.33). The pericardial coelom opens to the mantle cavity by way of two pores at the posterior ctenidial bases. In octopods, however, the pericardial part of the coelom is very much reduced, with only the branchial appendages lying within it (Figure 17.33). *Vampyroteuthis* has enlarged pericardial and genital coelomic spaces, but there is no separation between them.

The highly developed, efficient circulatory system of decabrachian coleoids differs from that of other molluscs in being largely a closed system (i.e., with both venous and arterial vessels interconnected with capillaries). In decabrachians the capillary system is well developed in contrast to the numerous blood sinuses present in not only *Nautilus* but also in *Vampyroteuthis* and octopods. The virtually closed circulatory system of decabrachians probably evolved as they increasingly relied on locomotion by jet propulsion (Young & Vecchione 2002).

The efficiency of the circulatory system in coleoids is increased by the addition of a pair of accessory hearts at the base of the ctenidia (see Chapter 6). These ensure rapid movement of blood through the ctenidia to supply the oxygen needed in these very active animals. The branchial hearts, derived from the vena cava, receive deoxygenated blood and pump it into the ctenidia. These structures are also involved in ultrafiltration, producing urine as well as playing a role in metabolising heavy metals and the breakdown of excess blood proteins. They are also thought to have an immunological function (Budelmann et al. 1997). Contractile 'branchial

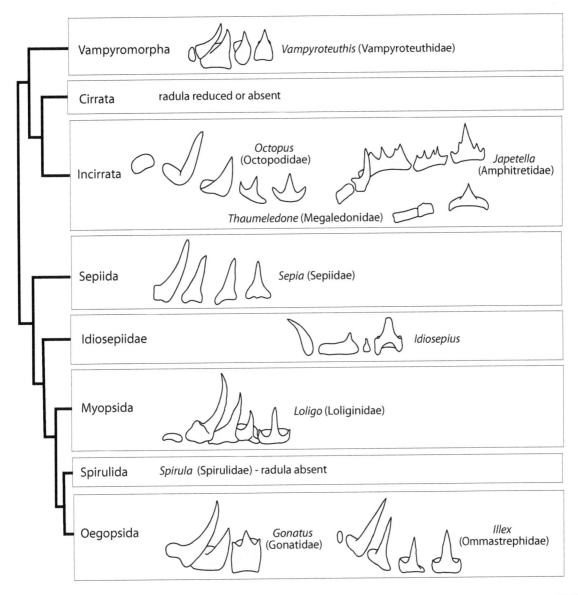

FIGURE 17.31   Examples of radulae of some living coleoids. Radulae redrawn and modified from the following sources: Hylleberg, J. and Nateewathana, A., *P.M.B.C. Res. Bull.*, 56, 1–9, 1991; Nixon, M., *J. Zool.*, 236, 73–81, 1995; Kruta, I., Ammonoid radula, pp. 485–505, in Klug, C., Korn, D., De Baets, K., Kruta, I., and Mapes, R.H. (eds.), *Ammonoid Paleobiology: From Anatomy to Ecology. Topics in Geobiology.* Springer, Dordrecht, the Netherlands, 2015.

hearts' are seen in embryonic *Nautilus* (Arnold 1987) but are not differentiated in adults.

Like most molluscs, cephalopods use haemocyanin as the respiratory pigment in the blood. Its efficiency is not as great as that of haemoglobin, but highly active squid use various physiological 'tweaks' to overcome its limitations (O'Dor & Webber 1991). One of these is a very efficient cardiovascular system. The power output needed for a squid heart at 12°C to deliver the required volume of oxygen-carrying blood exceeds the power output of mammalian hearts at 37°C, a feat that may be achieved in part by the mantle muscular hydrostat (Shadwick et al. 1990). It has also been suggested (Wells et al. 1988) that as much as a fifth of the oxygen used by some squid may be sourced through their thin skin, which perhaps explains why the most aerobic

muscle fibres are located in the inner and outer layers of the mantle.

Cephalopod excretory systems are highly modified compared with other molluscs (see Chapter 6). Coleoids have a pair of kidneys, and *Nautilus* has two pairs. Although cephalopod kidneys are homologous with those of other molluscs, they are greatly modified. They lie partly behind the posterior wall of the mantle cavity in *Nautilus* but are more exposed within the mantle cavity of coleoids. The pericardium envelops venous vessels involved in filtration; thus, unlike most other molluscs, there is no filtration associated with the heart (see Chapter 6 for additional details).

As described in Chapter 6, in *Nautilus* the pericardial and renal appendages are located on the four contractile afferent branchial veins (the venae cavae). In coleoids the contractile

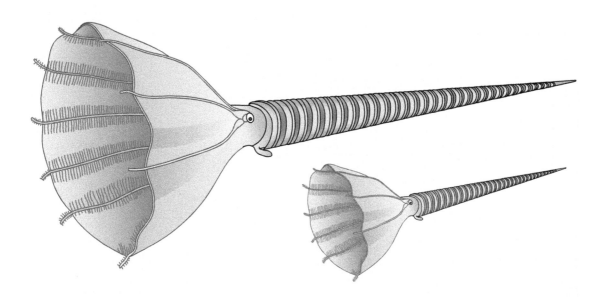

FIGURE 17.32 Endoceratids feeding on planktonic organisms as imagined by A. A. Mironenko. Figure redrawn and modified from Mironenko, A.A., *Hist. Biol.*, 2018, 1–9, 2018.

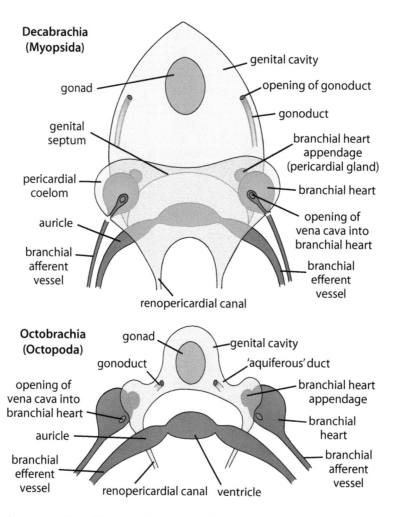

FIGURE 17.33 Diagrammatic representation of the coelomic system (without the renal system) in a decabrachian and an octopod. Redrawn and modified from Naef, A., *Ergeb. Fortschr. Zool.*, 3, 329–462, 1913.

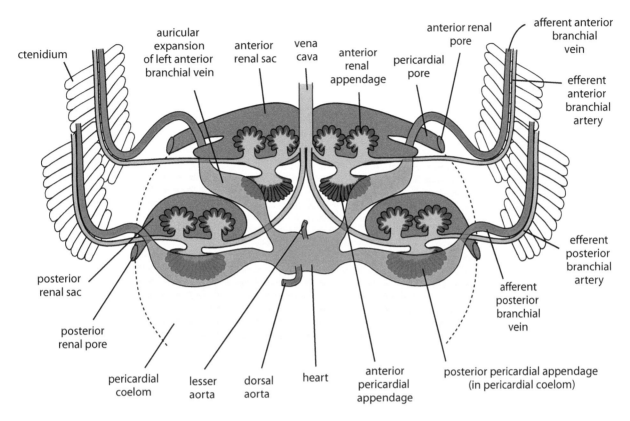

FIGURE 17.34 Renal sacs (kidneys) and neighbouring organs of *Nautilus*, viewed from the dorsal side. Redrawn and modified from Griffin, L.E., *Mem. Nat. Acad. Sci.*, 8, 101–230, 1900.

pericardial appendages are the homologues of the branchial heart appendages and, like them, contain podocytes involved in filtration. A renal appendage lies within each of the four renal sacs (kidneys) in *Nautilus* (Figure 17.34 and Chapter 6).

17.7 COURTSHIP, REPRODUCTION, DEVELOPMENT, AND GROWTH

Nearly all cephalopods have separate sexes. They have a single gonad, and males transfer spermatophores to females, often following elaborate courtship. Egg capsules are laid, and the yolk-rich eggs range in size from small to large. There is no true larval stage, and embryos develop directly into juveniles, although there may be a pelagic phase. The eggs and young are sometimes brooded, or they can be benthic or pelagic. The 'shell' of the 'paper argonaut'[14] (*Argonauta*) is a receptacle to contain the egg capsules. It is secreted by modified arms and is thus not a true shell. Two rare examples of hermaphroditic individuals in octopuses are known, *Octopus vulgaris* (Pickford 1947) and *Enteroctopus megalocyathus* (Ortiz & Ré 2006).[15]

Although fertilisation is internal in some coleoids, in most decabrachians it is essentially external – i.e., the eggs are fertilised outside the body cavities, in the arms or mantle cavity

as the eggs are deposited (Hanlon & Messenger 1996). When fertilisation occurs within the mantle cavity, it can be referred to as 'confined' external fertilisation (Naud & Havenhand 2006). True internal fertilisation involves the sperm entering into the body (usually the oviduct) and generally occurs as the eggs pass along the oviduct. Such placement of the spermatophore in the oviduct occurs, for example, in *Octopus vulgaris* (e.g., De Lisa et al. 2013), but in another octopod, *Eledone*, spermatophores actually enter the ovary (Fort 1937), while in *Heteroteuthis* (Sepiolidae), but not in other decabrachians, sperm pass from the seminal receptacle into the visceropericardial coelom in which the ovary lies, and the eggs are fertilised (Hoving et al. 2008).

17.7.1 REPRODUCTIVE SYSTEM

Although there is a single gonad in adults, it is paired during early development in octopods but not in decabrachians. The gonad opens to a gonoduct with associated accessory glands. Copulatory organs are present in the male, these being a modified arm (*hectocotylus*) or one or two penis-like organs (see below). The mesodermally derived gonad and the upper part of the duct are reasonably uniform throughout the group, but the ectodermally derived accessory structures show considerable modification in different groups. The ectodermal part of the gonoduct is developed into glandular structures producing the material that envelops the gametes and forms the egg capsule or spermatophore. In *Nautilus* males, the gonoduct is

---

[14] Also known as a 'paper nautilus'.

[15] We thank Dr S. O'Shea for drawing our attention to these cases.

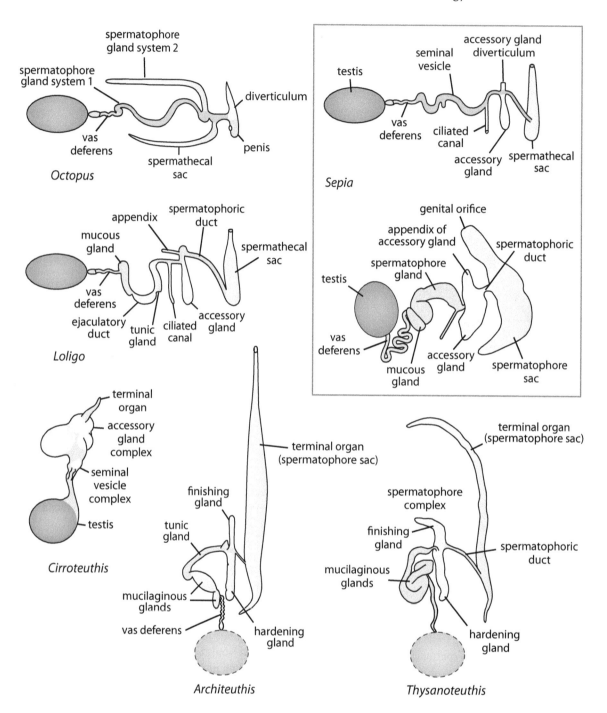

FIGURE 17.35 Examples of male coleoid cephalopod reproductive systems. Modified from the following sources: *Octopus* (Mann, T.R.R. et al., *Proc. R. Soc. B*, 175, 31–61, 1970), *Loligo*, *Cirroteuthis*, and *Architeuthis* (Hoving, H.J.T. et al., *J. Zool.*, 264, 153–169, 2004), and *Thysanoteuthis* (Nigmatullin, C.M. et al., *J. Zool.*, 224, 271–283, 1991). Note that two interpretations of the male system of *Sepia* are shown. The upper one is the highly stylised form used for *Octopus* and *Loligo* by Runham, N.W., Mollusca, pp. 311–383 in Adiyodi, K.G. and Adiyodi, R.G. (eds.), *Reproductive Biology of Invertebrates. Asexual Propagation and Reproductive Strategies. Reproductive Biology of Invertebrates*, Vol. 6, John Wiley & Sons, Chichester, 1993, and the other is *Sepia grahami* from Reid, A., *Proc. Linn. Soc. N.S.W.*, 123, 159–172, 2001.

on the right side while in coleoids it is on the left. In females, it is either paired (as in most oegopsid squid, *Idiosepius*, and incirrate octopods), or there is a single duct on the right side (*Nautilus*) or left side (sepiids, myopsidan squid, and cirrate octopods) (Budelmann et al. 1997). The male gonoduct (Figure 17.35) has seminal vesicles as well as accessory glands which make up the spermatophore organ. The spermatophore is formed in a mucous gland and then transferred to a storage area (*Needham's sac*) before release through the genital opening.

The spermatophores of most coleoids have a complex ejaculatory apparatus (Figure 17.36), but this is absent in *Nautilus* and cirrate octopods. The apparatus is triggered following the transfer of the spermatophore to the female. Fossil

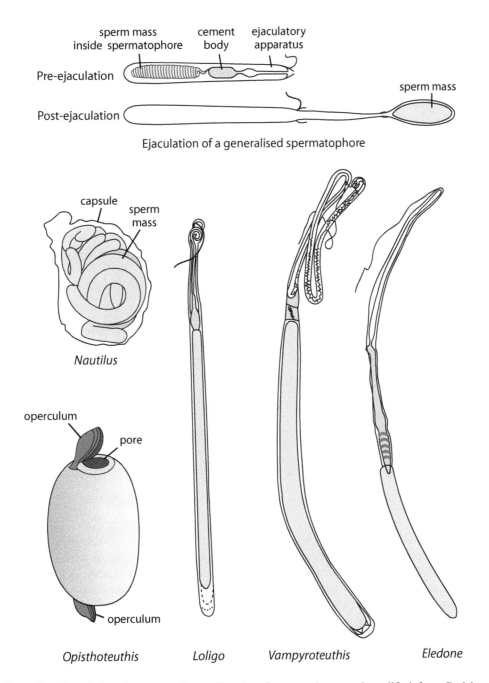

sperm mass
inside spermatophore
cement body
ejaculatory apparatus

Pre-ejaculation

sperm mass

Post-ejaculation

Ejaculation of a generalised spermatophore

capsule
sperm mass

Nautilus

operculum

pore

operculum

Opisthoteuthis   *Loligo*   *Vampyroteuthis*   *Eledone*

FIGURE 17.36 Examples of cephalopod spermatophores. Top two figures redrawn and modified from Budelmann, B.-U. et al., Cephalopoda, pp. 119–414, in Harrison, F.W. and Kohn, A.J. (eds.), *Microscopic Anatomy of Invertebrates: Mollusca 2. Mollusca*, Vol. 6A, Wiley-Liss, New York, 1997, based on Marchand, W., *Z. wiss. Zool.*, 86, 311–415, 1907. Lower figures from Young, R.E. and Vecchione, M., *Am. Malacol. Bull.*, 12, 91–112, 1996 based on the following sources: *Nautilus* (Mikami, S. and Okutani, T., *Venus*, 40, 57–62, 1981), *Opisthoteuthis* (Villanueva, R., *Mar. Biol.*, 275, 265–275, 1992), *Vampyroteuthis* (Hess, S.C., *Comparative morphology, variability and systematic applications of cephalopod spermatophores (Teuthoidea and Vampyromorpha)*, Ph.D. dissertation, University of Miami,), *Loligo* and *Eledone* (Marchand, W., *Zoologica*, 67, 171–200, 1912).

spermatophores have been reported from an Upper Jurassic coleoid by Keupp et al. (2010).

The female distal oviduct (Figure 17.37) forms the oviducal gland which produces a gelatinous material that envelops the eggs. In those coleoids with nidamental glands, the gelatinous material is transported to the oviducal opening via the accessory nidamental glands, and their secretion is incorporated in the egg capsules. Separate nidamental glands

are found in squid and cuttlefish but not in *Vampyroteuthis* or octopods, where they are incorporated in the oviducal gland (Budelmann et al. 1997). Bacteria, apparently symbiotic, are found in the accessory nidamental glands, and some are transferred to the egg capsules (see Section 17.7.6).

Copulation in many coleoids occurs by way of a modified arm – the hectocotylus (see below). The hectocotylus occurs in only 58% of coleoid families and subfamilies (Nesis

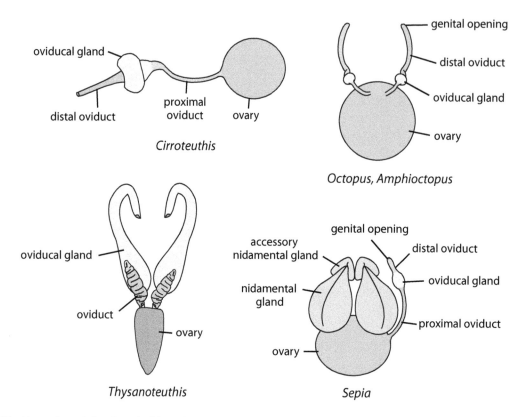

FIGURE 17.37 Examples of female coleoid cephalopod reproductive systems. Redrawn and modified from the following sources: *Cirroteuthis* (Voss, G.L. and Pearcy, W.G., *Proc. Calif. Acad. Sci.*, 47, 47–94, 1990), *Amphioctopus* (Huffard, C.L. and Hochberg, F.G., *Mollouscan Res.*, 25, 113–128, 2005), *Thysanoteuthis* (Nigmatullin, C.M. et al., *J. Zool.*, 224, 271–283, 1991), *Sepia* (Budelmann, B.-U. et al., Cephalopoda, pp. 119–414, in Harrison, F.W. and Kohn, A.J. (eds.), *Microscopic Anatomy of Invertebrates: Mollusca 2. Mollusca*, Vol. 6A, Wiley-Liss, New York, 1997).

1995), and, in the remainder, the spermatophores are transferred by a 'penis' or 'terminal organ' – better referred to as a *pseudopenis* (or 'pseudophallus') and found in some squid, *Vampyroteuthis*, and cirrate octopuses. The hectocotyli are formed from different arms in different groups – in incirrate octopods it is the right or left third (ventro-lateral) arm, in sepiids one or both dorsal arms, while in most other decabrachians it is one of the ventral arms (Budelmann et al. 1997). They vary considerably in structure and are often useful for distinguishing species. The hectocotylus can be autotomised in some pelagic incirrate octopods.

In *Nautilus*, modified cirri transfer the spermatophore, the main one being the *spadix* located on the right side in males. This large, solid, intromittent organ becomes erect during spermatophore transfer (i.e., copulation) (Arnold 2010). The *antispadix* is a group of slightly modified cirri on the opposite side of the head to the spadix (Saunders & Spinosa 1978). A glandular area (the '*organ of Hoeven*'), located below the buccal area in males, is of unknown function. In females, the *organ of Valenciennes* lies in the same area and is the equivalent of the seminal receptacle, as this is where the large spermatophores are deposited.

In coleoids, spermatophores are passed to a seminal receptacle by way of the hectocotylus or a pseudopenis. In octopods, the seminal receptacle is part of the oviducal gland, but in most decabrachians it is a pouch beneath the mouth or, in sepiolids, in the mantle cavity. In cuttlefish and the squid *Loligo*, fertilisation is external, occurring in the arms by sperm released from seminal receptacles located in the ventral buccal membrane. In some squid, a seminal receptacle is lacking, and, even when one is present, the eggs are fertilised externally, either as they are laid in the mantle cavity or in the arms. Unlike squid and cuttlefish, in octopods fertilisation is internal in the oviducal gland and egg care is practised in some species (see Section 17.7.3.1), while in others tiny pelagic eggs develop into the planktonic paralarvae.

17.7.1.1 Sexual Characters and Sexual Dimorphism

A few taxa are highly sexually dimorphic with body size differences between males and females the most obvious feature. In many sepiids and loliginids, the males are larger than the females while in oegopsids and octopods males are usually much smaller than females, although exceptions occur. In some coleoids, sexual dimorphism is very pronounced, as for example in *Argonauta*, where the male is much smaller than the female. In contrast, in *Nautilus* only small sexual differences in the shape and size of the shell can be seen with the female slightly smaller and narrower than the male (e.g., Willey 1902; Saunders & Spinosa 1978). In many ammonite

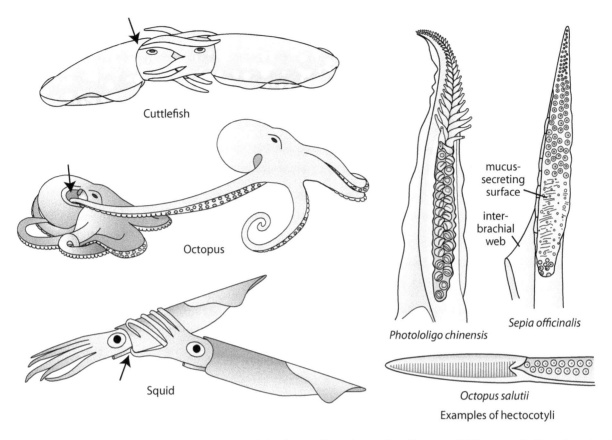

Cuttlefish

Octopus

Squid

mucus-
secreting
surface

inter-
brachial
web

Sepia officinalis

Photololigo chinensis

Octopus salutii

Examples of hectocotyli

FIGURE 17.38 Examples of body orientation during copulation in coleoids and examples of hectocotyli. The arrows indicate the point where the male is inserting the spermatophore. Cuttlefish and octopus from various sources; squid based on *Doryteuthis opalescens* (McGowan, J.A., *California Fish and Game*, 40, 47–54, 1954). Hectocotyli redrawn and modified from the following sources: *Photololigo* (Beesley, P.L. Ross, G.J.B., and Wells, A. (eds.), Mollusca: The southern synthesis, part A, CSIRO Publishing, Melbourne, VIC, 1998a), *Sepia* (Tompsett, D.H., *Sepia*, pp. 1–191, in Daniel, R.J. (ed.), L. M.B. C. *Memoirs on Typical British Marine Plants and Animals.* University Press of Liverpool, Liverpool, 1939), *Octopus* (Naef, A., Die Cephalopoden. *Fauna und Flora des Golfes von Neapel, Monographie* 35, Systematik, I Teil, 1 Band, Fascicle I: 1–148, translated in A. Mercado, 1972, Cephalopoda, Jerusalem, Israel Program for Scientific Translations, 1921; Naef, A., Die Cephalopoden. *Fauna und Flora des Golfes von Neapel, Monographie* 35. Systematik, Part I, vol. 1, Fascicle II: 149–863, translated in A. Mercado, 1972, Cephalopoda, Jerusalem, Israel Program for Scientific Translations, 1923).

taxa, particularly in the middle and late Mesozoic, there appears to have been pronounced sexual dimorphism with the assumed females up to four times larger than males (see Section 17.12.2). In another extinct group, some belemnites were also sexually dimorphic. This interpretation is based on the presence of large hooks (megahooks) which are thought to have been possessed by males to grasp the female during mating (Fuchs & Hoffmann, 2017). The development of an epirostrum has also been suggested to be a sexually selected character in belemnites (Stevens et al. 2017).

Besides the development of the hectocotylus or pseudopenis and the common occurrence of a difference in body size between sexes, there are several other characteristics of male maturation in coleoids. These include modifications of non-hectocotylised arms for grasping the female, the elongation and enlargement of some arms, the enlargement of certain suckers on some arms, and displays such as the development of special photophores and male colouration or, possibly, in some squid, tail elongation. Similarly, females may develop 'sexual' photophores on the arms or around the mouth, while other taxa show arm modifications, including the loss of

suckers at the tips of some arms or, rarely, the enlargement of suckers (Nesis 1995).

17.7.2 COURTSHIP AND MATING

Most of what we know about coleoid reproductive behaviour is based on direct observation of shallow-water taxa, mainly sepiids, incirrate octopods, and loliginids, or otherwise on inference. A summary of these observations is offered below, and a more detailed review is provided by Hanlon and Messenger (1996).

Pheromones that attract males are released by the freshly laid eggs of at least some coleoids (e.g., Zatylny et al. 2002; Cummins et al. 2011; De Lisa et al. 2013). Copulation in coleoids may occur either head to head or head to tail, depending on where the spermatophore is lodged (Figure 17.38). There are several sites for spermatophore attachment in coleoids (Nesis 1995), although for many taxa these are not known. They include a site on the buccal membrane or in a seminal receptacle under the mouth (as in spirulids, sepiids, and many of their relatives and some squid such as a number of loliginids and thysanoteuthids). Similar sites occur in other coleoids including the head in front of the eyes in *Vampyroteuthis*, at the anterior end of the gladius inside the mantle cavity in enoploteuthids, inside the mantle cavity near the oviduct opening(s) in various squid, including many Loliginidae, in a pouch in the mantle cavity near the oviduct opening (Sepiolidae), in the oviduct(s), oviducal glands, or ovary in some Chiroteuthidae and all Octopoda, or in cuts made for the purpose on the outer mantle in some Pholidoteuthidae, Onychoteuthidae, Octopoteuthidae, Histioteuthidae, and Cycloteuthidae. Head-to-head copulation occurs when there is transfer to the buccal membrane or buccal seminal receptacle and when the male uses a pseudopenis to lodge the spermatophore in the mantle cavity of the female. Placement in this latter location by a male using a hectocotylus occurs with the male holding the female from below. In those species where the spermatophores are transferred into cuts[16] on the female mantle, it is possible that they may do this in either a parallel position or perpendicular to one another, but no direct observations have been made. In *Idiosepius* the spermatophores are placed externally on the head, in between the ventral-most pair of arms (M. Reid, pers. comm.).

In the deep water that the vampire squid inhabits, encounters with the opposite sex are rare, so the female may retain spermatophores for a long time before fertilising the few large eggs which it produces in multiple batches (Hoving et al. 2015).

Courtship may be very brief to long and complex. In most octopods and some squid (e.g., *Illex*) it is very brief, while in some shallow-water coleoids, mating includes courtship rituals that may consist of colour changes, body movements, shape changes, or combinations of these.

In loliginid squid, antagonistic male competition is marked, and both sexes typically mate multiple times. So-called 'sneaker' behaviour (e.g., Hanlon & Messenger 1996; Hanlon et al. 2002; Zeidberg 2009) may occur, in which a smaller male, perhaps mimicking a female, will sneak in and deposit spermatophores while a larger male is paired with a female or is discouraging rivals. In these squid, spermatophores can be deposited in the seminal receptacle near the mouth or near the opening of the oviduct in the mantle cavity. It has been shown that the sperm in these smaller squid can be larger than those in the larger squid (Iwata et al. 2011), resulting in successful fertilisation.

Consumption of unused sperm by the female has been demonstrated in the sepiolid *Sepiadarium austrinum*. In this species, the contents of the spermatophores are used to provide nutrients to assist with egg production (Wegener et al. 2013).

17.7.3 Life History

Nautilus can live up to 20 years during which it can spawn several times (i.e., it is iteroparous). It (and possibly all nautiliforms) produces a few well-developed large offspring (K-selected) with each breeding event. Intracapsular development time may be more than a year (Tanabe et al. 1991), reaching maturity in

three to four years, although there is still little known about *Nautilus* growth rates other than in captive specimens.

The size of Mesozoic and Cenozoic nautiliforms at hatching was generally similar to that of *Nautilus*, and there is an assumption that other aspects of the reproduction and life history of extinct nautiliforms were similar to that of their living representatives. If so, they had a long lifespan, numerous spawning events, and little evidence of sexual dimorphism. The embryonic shells of nautiliforms were six to ten times larger than ammonite larval shells (ammonitella) (Figure 17.49). Thus, ammonites were different, as they hatched at a size similar to that of some coleoids.

Small individuals of *Nautilus* secrete a new chamber every two to three weeks compared to every 13 to 15 weeks in larger individuals (Ward 1985). Despite their often-large size, ammonites may have had high fecundity and a planktonic phase, as shown by their minute embryonic shell (Tajika et al. 2018). Thus, it is assumed that ammonites began life as tiny planktonic hatchlings less than 1 mm in diameter (e.g., Laptikhovsky et al. 2018) and then probably grew rapidly. They mainly lived in warm shallow waters (e.g., Westermann 1996) and may have had a higher metabolism than *Nautilus* which lives in rather deep cool water, so growth rates could have been faster. Three post-embryonic growth stages can be recognised in ammonites (Bucher et al. 1996); a neanic stage immediately after hatching, a juvenile, and then a mature stage. Growth accelerated during the neanic stage, was more or less constant during the juvenile stage, and slowed at the beginning of maturity.

Ammonite lifespans have been estimated to be one to two years for small, shallow-water species, five to ten years for most taxa, and possibly 50 to 100 years for some 'mesopelagic giants such as *Lytoceras*' (Westermann 1996 p. 669). A relatively short lifespan and possible semelparous reproductive strategy (mass spawning followed by the death of parents) may explain why many ammonites evolved and diversified rapidly. There remains, however, debate as to whether ammonites were semelparous, iteroparous, or both. Thus, some workers have suggested that ammonites grew quickly and died after spawning, as in most coleoids, but this is disputed by others.

In contrast to ammonites, long stratigraphic ranges and low diversity may be the hallmark of an iteroparous reproductive strategy in nautiliforms. Stephen and Stanton (2002) tested this idea with Carboniferous fossil taxa in the USA and found that the stratigraphic ranges of nautiliform taxa averaged about 4.3 times longer than those of co-occurring ammonites.

Belemnites had small larval shells and may have had a high fecundity and a planktonic phase (Tajika et al. 2018). Similarly, many modern coleoids also have a juvenile planktonic phase. As noted above, many modern coleoids are semelparous, dying after reproduction, although they may engage in several spawning events over a relatively short period (weeks or months) before death. These observations are, however, based mainly on shallow-water or commercially important species, and there is evidence that some oceanic and many deep-sea species do not necessarily conform to this pattern (e.g., Collins & Rodhouse 2006; Hoving et al. 2014). A notable exception is *Vampyroteuthis*, which can have several reproductive events

---

[16] Presumably made by the sucker hooks or the beak.

before death (Hoving et al. 2015). The young *Vampyroteuthis* uniquely go through three stages of development: the youngest stage resembles miniature adults but has a single pair of fins near the eyes; a second pair then develops, so the intermediate stage has two pairs of fins, while in the final (adult) stage the first pair of fins is lost, and only the second pair remain (Pickford 1949). Also, the young mainly use jet propulsion for movement, while adults mainly use their fins (Seibel et al. 1998).

Cuttlefish and some squid attach their large eggs to objects on the seafloor, sometimes with some attempt to hide them. The eggs are not guarded after deposition is completed (see Section 17.7.3.1), and the females die soon after spawning.

Some coleoids have a two-phase growth pattern with initial rapid growth followed by a slower growth rate which continues for the rest of their life, while in others the growth is constant through most of their life, although often slowing just before senescence (e.g., Moltschaniwiskyj 2004). Whether larger-sized coleoids live longer than smaller ones remains unclear, with different studies producing contradictory results; Wood and O'Dor (2000) analysed data from a wide range of coleoid taxa and found that larger species do take longer to reach maturity and that temperature is a major factor in growth rates.

17.7.3.1 Spawn and Parental Care

A detailed review of spawning in cephalopods is provided by Nesis (1995). Egg capsules are partially formed in the oviduct, but final moulding of the capsules occurs with the arms in coleoids and probably also in *Nautilus* (Arnold 1987).

Nautilus has large egg capsules which are sac-shaped, about 4 cm long, and fixed singly or in groups of two or three to the substratum. Each contains a single egg. The tough outer proteinaceous layer has fine slits enabling seawater to enter and bathe the embryo. The young embryo is about 2 cm long and contains much yolk. When the young nautilus hatches it is about 3 cm long, and its shell has about seven septa. *Nautilus* eggs are among the largest eggs of any invertebrate, attaining about 2 cm maximum diameter.

There is evidence that ammonites hatched at cooler temperatures, i.e., in deeper water, suggesting that they laid their egg masses on the sea floor (Lukeneder et al. 2010; Linzmeier et al. 2018). They often had small eggs and high fecundities (Laptikhovsky et al. 2018), although environmental perturbation influenced egg size (Laptikhovsky et al. 2013). There is also evidence for ovoviviparity in at least one ammonite from the Early Cretaceous (Mironenko & Rogov 2016) while De Baets et al. (2012) calculated the potential number of eggs that could be stored in the body chamber of early female ammonites during their transition from straight to coiled shells.

Modern coleoids have yolky eggs ranging from about 1 mm and usually up to about 10 mm, although octopods have a greater range (1 to 16 mm). The number of eggs spawned ranges from less than 100 to a few hundred thousand. Cuttlefish and other sepiidans also have relatively large eggs, from about 1 to 5 mm, although the eggs of at least one species of *Sepia* reach about 35 mm in diameter. The number of capsules cuttlefish spawn ranges from 35 to 500. In contrast, some squid spawn from several thousand to a few hundred thousand eggs, with egg sizes ranging from about 1 to 2.7 mm (Boyle & Rodhouse 2005). Some large squid such as *Architeuthis* and *Dosidicus* spawn several times and have the highest fecundities known in cephalopods, producing millions of small eggs (1 to 2 mm) (Hoving et al. 2004; Nigmatullin & Markaida 2009).

Coleoid egg masses are variable (Figure 17.39) and include small globular structures, each containing one egg and laid in connected clusters as in the sepiolids *Rossia* and *Sepietta*, oval 'capsules' containing large eggs and laid individually or in clusters as in *Sepia*, and clusters of elongate capsules each containing numerous eggs such as those of loliginid squid. The egg capsules of loliginid squid and cuttlefish are laid as benthic egg masses and, like most decabrachians, they do not brood or otherwise care for their young. Ommastrephid squid and at least one thysanoteuthid lay pelagic egg masses containing many small eggs, while gonatid squid brood the eggs in their arms, and some enoploteuthids apparently release single eggs.

Benthic octopods lay clusters of oval, stalked capsules containing single eggs (Figure 17.39) and exhibit two kinds of breeding strategy. Most have large, yolky eggs, bundles of which are attached to the substratum, while some lay many tiny eggs with little yolk. In argonautid octopods, the eggs are laid in a shell-like egg container occupied by the female (see Section 17.3.1).

Incirrate octopod females guard their eggs after they are laid, sometimes for months, until they hatch. Records of particularly long brooding times include that of *Bathypolypus*, which guards its eggs for 400 days (about 13 months) (Wood et al. 1998), while the deep-sea octopod *Graneledone boreopacifica* has been recorded brooding for 53 months (Robison et al. 2014). Female *Vampyroteuthis* brood for over a year before the eggs hatch.

During brooding, incirrate octopods aerate and clean the eggs. Many benthic species do this in a 'den', but others, including pelagic species, carry the eggs in their arms or arm webs and, in one case, within the mantle cavity. Recently, large 'nurseries' with hundreds to thousands of brooding octopods have been discovered off the Pacific coasts of California and Costa Rica (Daley 2018; Hartwell et al. 2018). Although the pelagic *Argonauta* carries its eggs in the 'shell' (see above), the related *Ocythoe* is ovoviviparous, retaining the eggs within the oviduct.

Brooding is thought to be typical of pelagic incirrate octopods. For example, females of the small pelagic, sexually dimorphic, short-armed incirrate octopod family Amphitriteidae hold their egg mass near the mouth with their suckers (Young 1972). Cirrate octopods do not brood but instead produce a few large eggs which are released singly and fall to the sea floor (Collins & Villanueva 2006) or are attached to gorgonian or antipatharian corals (S. O'Shea pers. comm., 2019), where they develop and the young eventually hatch.

17.7.3.2 Development

The zygote undergoes partial cleavage, the cleavage occurring at the animal pole and not in the mass of yolk below

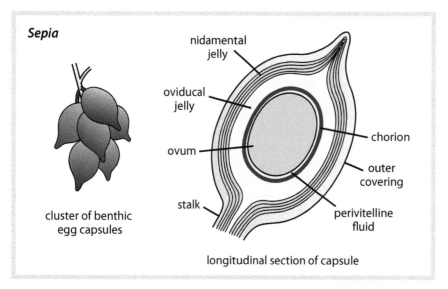

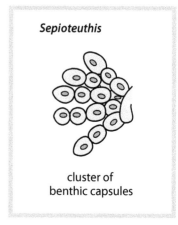

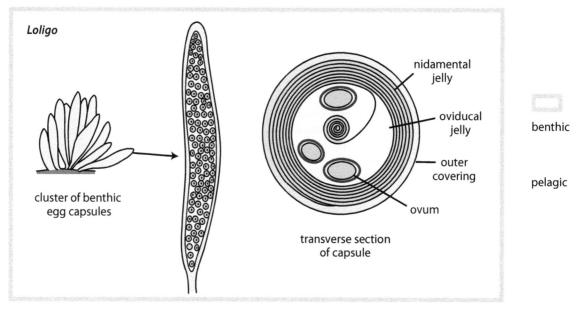

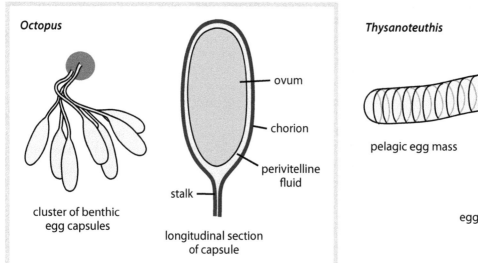

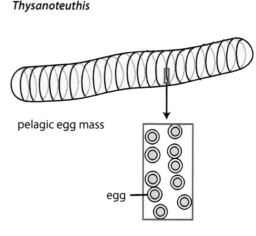

FIGURE 17.39 Egg capsule diversity in coleoid cephalopods. Redrawn and modified from Nesis, K.N., *Ruthenica*, 6, 23–64, 1995 with the sections of *Sepia*, *Octopus*, and *Loligo* capsules from Fioroni, P., Cephalopoda, Tintenfische, pp. 1–181, in Seidel, F. (ed.), *Morphogenese der Tiere, Erste Reihe: Deskriptive Morphologenese, Lieferung 2: G5–1*, Gustav Fisher Verlag, New York, 1978.

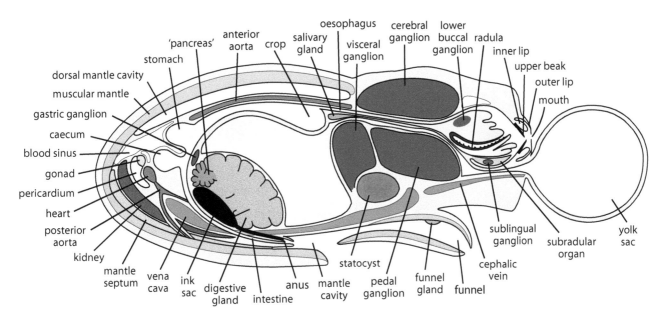

FIGURE 17.40 Longitudinal section of a nearly mature embryo of *Octopus vulgaris*. Redrawn and modified from Naef, A., *Die Cephalopoden. Embryologie*, Vol. 35, R. Friedlander & Sohn, 1928.

(see Chapter 8). Small blastomeres then form, with their main segments maintaining continuity with the yolk (von Boletzky 2003b). Gastrulation occurs next, with the marginal blastopore lip forming an envelope over the yolk sac followed by the development of blood lacunae and muscles that circulate the blood (von Boletzky 2003b; von Boletzky & Villanueva 2014). The yolk sac (Figure 17.40) size is related to egg size, being larger in species with large eggs. The remaining yolk is enveloped in the body of the embryo before hatching. Hatching is aided with a 'hatching gland' on the mantle, which produces an enzyme that forms an opening in the egg capsule (von Boletzky & Villanueva 2014).

Coleoid development is direct, there being no separate larval morphology (Figure 17.41), but even in benthic species the young are often pelagic (paralarvae).

Development in *Nautilus* from the blastoderm stage is similar to that of coleoids, although simpler (Arnold 1987). The early cleavage pattern in *Nautilus* is unknown (Arnold 2010), although it probably resembles that of coleoids. The egg sac in *Nautilus* is very large relative to that in coleoids and, unlike even the large-egged coleoids, has a vascular system associated with it. This large egg sac enables a long developmental period with the young hatching from the egg capsules as miniature 'adults'.

17.7.4 THE NERVOUS SYSTEM, SENSE ORGANS, AND BEHAVIOUR

As detailed in Chapter 7, cephalopods have a very highly developed nervous system, coupled with highly modified versions of the usual array of molluscan sensory equipment (eyes, statocysts, olfactory organs, touch receptors) and locomotory dexterity. The usual view of the cephalopod brain is that the ganglia encircling the oesophagus (as seen in some other molluscan groups, notably higher gastropods) are enlarged, concentrated, and fused to form a brain, but Shigeno et al. (2008) argued that embryological evidence indicates that in the early stages of development the central nervous system of cephalopods was not arranged in ganglia; rather it was a cord-like system similar to that seen in some vetigastropods.

The differences in the brain of *Nautilus* and the larger and more complex brain of coleoids are detailed in Chapter 7. As in vertebrates, the cephalopod brain is partitioned into different areas that control particular functions. Some of the upper brain lobes are involved in memory and learning (see Chapter 7) and the consequent complex behaviour and capacity to distinguish a variety of similar objects and remember these details. *Vampyroteuthis* has the most primitive nervous system of any coleoid. Among octopods, the cirrate octopods are the most primitive, particularly regarding the poorly developed vertical lobes, suggesting this group diverged early in octopod evolution, perhaps in the Triassic (see Section 17.13.1). In contrast to the cirrate octopods, decabrachian coleoids (including squid, cuttlefish, and *Spirula*) all have a similar brain structure with well-developed vertical lobes, as well as other brain centres, suggesting that the decabrachian brain had developed early in the evolution of the group, presumably at least by the Lower Jurassic (Young 1988).

In decabrachians, giant neurons run from the brain along the mantle wall on either side to large stellate ganglia, a configuration which enables the extremely rapid reflexes needed to control the muscular contraction of the mantle during respiration and jetting (Llinás 1999) (see Chapter 7).

In coleoids, the arms, suckers, and cirri are sophisticated sensory structures, as are the complex statocysts and olfactory organs, but their large eyes are the most obvious sensory equipment.

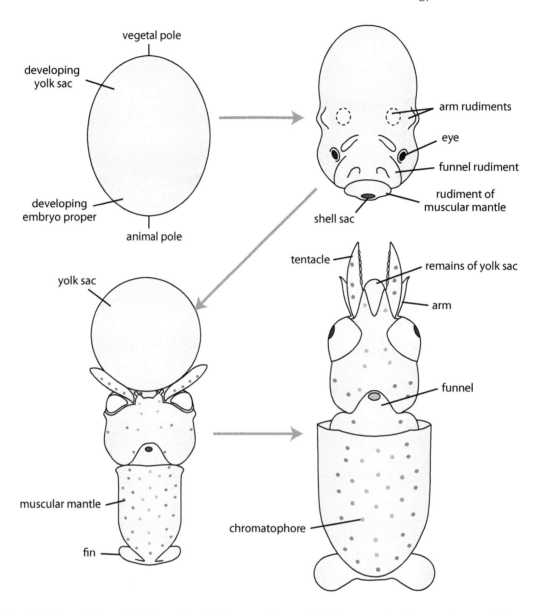

FIGURE 17.41 Life stages of the squid *Loligo vulgaris*. The top and bottom left figures are developmental stages in the egg capsule; the bottom right figure is the newly hatched paralarva. Redrawn and modified from Naef, A., *Die Cephalopoden. Embryologie*, Vol. 35, R. Friedlander & Sohn, 1928.

In *Nautilus* the eyes are large but, being of the pinhole camera type as noted in Section 17.3.5, their relatively primitive construction is probably secondary. They lack a lens and are open to the water outside by way of a small opening, and a vertical slit, the iris groove, runs ventrally from this opening. Cilia carry mucus across the opening and into the groove, preventing particles entering the eye opening (Muntz 1987b).

In coleoids, the eyes have an iris, lens, often a cornea, and muscles that control eye movements (see Chapter 7 for more details). They thus resemble vertebrate eyes in structure and function, although those that lack a cornea have an opening to the exterior (Figure 17.42). Coleoid eyes form images and can distinguish brightness, shape, size, orientation and are sensitive to polarised light. The eyes of deep-sea giant squid (*Architeuthis* and *Mesonychoteuthis*) are the largest known in any animal, being up to at least 27 cm in diameter (Nilsson

et al. 2012), although, relative to their body size, they are proportional to those of many other squid (Schmitz et al. 2013). Some deep-water squid have eyes of different size and orientation on the right and left sides of the body that appear to perform different functions (Thomas et al. 2017).

In contrast to *Nautilus*, coleoids with their more elaborate and efficient eyes are highly visual, not only as predators but socially; for example, they change colour for sexual displays and in response to aggression and other changes in mood.

The olfactory tentacles of *Nautilus* (often incorrectly called rhinophores)[17] and the olfactory organs of coleoids both develop from the posterior part of the cephalic compartment, indicating their homology (Shigeno et al. 2008). *Nautilus* has

---

[17] Rhinophores are found in some heterobranch gastropods, including Nudipleura, Aplysiida, and Sacoglossa, but these are not homologous.

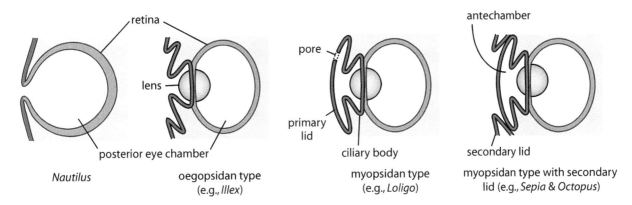

FIGURE 17.42 The types of eyes in living cephalopods. Redrawn and modified from Mangold-Wirz, K.M. and Fioroni, P., *Zool. Jahrb. Anat. Bd Syst.*, 97, 522–631, 1970.

a pair of olfactory tentacles, one below each eye, opening to the exterior by a pore. They contain a folded epithelium with a single type of sensory cell. These organs have been shown to detect odours from up to ten metres away. Lesser chemosensory abilities are possessed by the lateral digital tentacles and possibly the preocular tentacles (Basil et al. 2005). The olfactory sense in *Nautilus* is enhanced by the development of olfactory lobes in the brain that are larger than in coleoids (Young 1965). In coleoids, the olfactory organs are on either side of the head, posteroventral to the eyes, and vary from being pits to knobs and contain from three to five types of sensory cells (Budelmann et al. 1997).

Many coleoids have ciliated cells arranged in lines on their head and arms. Experiments with *Sepia* and the squid *Lolliguncula* have shown that they are equivalent to the lateral lines of fish, in having a mechanoreceptive function (Budelmann & Bleckmann 1988).

17.7.5 THE ENDOCRINE SYSTEM

A general description of the endocrine system in molluscs is given in Chapter 2. There are several endocrine-secreting areas in cephalopods (Budelmann et al. 1997) including:

- Juxtaganglionic tissue between the inferior buccal ganglion and the buccal sinus
- Neurosecretory cells in the superior buccal ganglion, subpedunculate, and olfactory lobes of the brain and in the posterior salivary glands
- Paraneural structures in the subpedunculate and paravertical tissues of the brain
- The neurosecretory system associated with the vena cava and pharyngo-ophthalmic vein
- The optic glands.

In cephalopods, the *optic glands* (also called optic body or spherical body) control gonad maturation and are important endocrine glands associated with the central nervous system. They are small, nearly spherical bodies on the upper posterior edges of each optic tract that are innervated by immunoreactive neurons from the olfactory and basal-dorsal lobes.

They are present only in coleoids but may be represented in *Nautilus* by a small area of tissue near the junction of the olfactory and optic lobes (Budelmann et al. 1997).

The secretory cells of the optic glands are modified neurons called the *stellate cells* and are thus unlike the dorsal body and juxtaganglionar cells of gastropods. They release a gonadotropic hormone and are involved in the control of feeding and defence against foreign proteins as well (Froesch 1979), and their long cell processes run along capillaries. There is also evidence that FMRFamide-like peptides are involved in the nervous control of the optic glands.

17.7.6 LEARNING AND INTELLIGENCE

Cephalopods, by virtue of their large brains, are by far the most intelligent molluscs, surpassing all other invertebrates and even some vertebrates in this regard (see Chapter 7). The abilities of some coleoids are well documented, with experiments, involving octopuses and cuttlefish in particular, showing they are capable of significant learning feats (Hanlon & Messenger 1996), cognition (e.g., Darmaillacq et al. 2014), and conditional discrimination, and arguably possess a form of intelligence and consciousness (e.g., Mather 2011, 2012) (see Chapter 7).

17.7.7 BEHAVIOUR

Cephalopods are by far the most behaviourally sophisticated molluscs. The behaviour of a number of octopuses, squid, and cuttlefish, and *Nautilus* has been extensively studied allowing some general conclusions to be drawn, at least regarding shallow-water cephalopods (see Chapter 7).

Cuttlefish can produce many body patterns mainly used for crypsis, but they regularly use only some of the possible combinations. Pattern changes are also used for signalling between individuals, including during courtship and hunting, with some patterns possibly distracting or attracting prey (Hanlon & Messenger 1988, 1996; Adamo et al. 2006).

Coleoids have some sophisticated defence strategies, including escape by jetting or inking, using cryptic, startling, or disruptive colouration (Figure 17.43), shape and surface

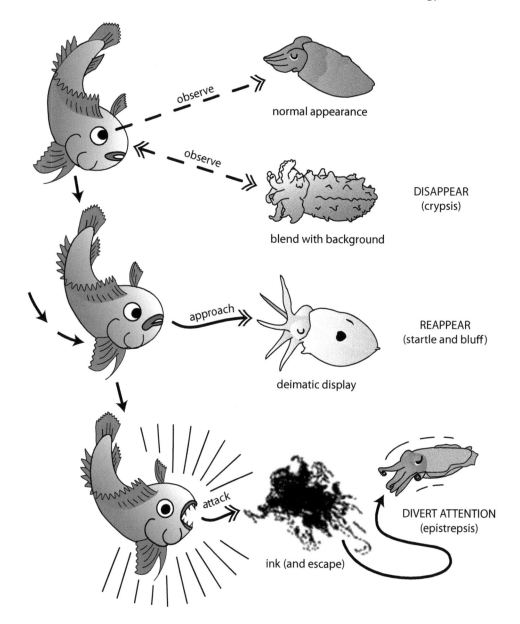

FIGURE 17.43 Responses to threat (from predatory fish) by cuttlefish. Redrawn and modified from Packard, A., Visual tactics and evolutionary strategies, pp. 89–104 in Wiedmann, J. and Kullmann, J. (eds.), *Cephalopods Present and Past. 2nd International Cephalopod Symposium: O.H. Schindewolf Symposium*, Tubingen 1985. Stuttgart, E. Schweizerbart'sche Verlagsbuchhandlung, 1988.

changes, injection of toxic secretions, and, in some octopods, arm autotomy.

Octopods usually use crypsis or escape to avoid predators, but some tropical species such as the mimic octopus (*Thaumoctopus mimicus*) can create body forms and colours that resemble flatfish or sea snakes or when motionless can mimic sponges, colonial ascidians, or tube worms (Norman & Hochberg 2005; Hanlon et al. 2008). A similar species, *Wunderpus photogenicus* (Hochberg et al. 2006), is not a mimic.

Inking (see Section 17.4.3) is a common response to threat in many shallow-water cephalopods. Deep-water octopods, including all cirrate octopods and *Vampyroteuthis*, lack ink sacs, but when threatened some display a 'ballooning response', increasing their overall size and even becoming near spherical (Villanueva et al. 1997).

17.7.8 Symbiotic Relationships

A variety of symbiotic bacteria is known from cephalopods (e.g., Boucher-Rodoni & Mangold 1994; Grigioni et al. 1999), including the luminescent *Vibro* bacteria in the photophores of certain sepiolid and loliginid squid (see Section 17.4.5 and Chapter 9), those that produce tetrodotoxin in blue-ringed octopus (*Hapalochlaena*) salivary glands (Hwang et al. 1989), and pigmented bacteria in accessory nidamental glands (see below). Betaproteobacteria and spirochaetes are associated with the excretory organs of *Nautilus*, although their role is unclear (Pernice et al. 2007; Pernice & Boucher-Rodoni 2012). There is also a possibility of bacteria being involved in nitrogen production in shelled cephalopods such as *Nautilus* and *Sepia* (Boucher-Rodoni & Mangold 1994), but this has not been rigorously demonstrated.

Bacteria in the accessory nidamental glands of loliginid and idiosepiid squid are also present in the egg capsules, embryo, and yolk of two loliginid species and in the eggs of one idiosepiid species (Pichon et al. 2005). Of the potentially symbiotic bacteria present in the accessory nidamental glands, Pichon et al. (2005) identified members of the phyla Proteobacteria (Alphaproteobacteria and Gammaproteobacteria) and Bacteroidetes. This was further investigated by Collins et al. (2012) who also identified Proteobacteria (Rhodobacterales and Rhizobiales) and Verrucomicrobia, with the Rhodobacterales genus *Phaeobacter* dominant in the 'bobtailed squid' (*Euprymna*). Suggested roles of these bacteria include the production of toxic compounds and antibiotics involved in egg protection, as well as carotenoid pigments (Pichon et al. 2005).

Dicyemid mesozoans are found in the excretory organs (renal appendages and pancreatic appendages) of mainly benthic decabrachians and octobrachians from temperate seas (e.g., Furuya 2006) (see Chapter 9). Chromidinid ciliates are found in these same excretory structures in mainly oceanic decabrachian coleoids (squid and cuttlefish); occasionally the ciliates and dicyemids occur together (Furuya et al. 2004). A dicyemid has also been found in the branchial heart appendages of the sepiolid *Rossia pacifica* (Furuya et al. 2004).

17.7.9 Predators

Adult coleoids have many predators (see Chapter 9), including humans (see Chapter 10), but a much wider range of predators eat the small paralarvae and juveniles. Coleoids are significant components of the diets of many fish (Smale 1996), seabirds (e.g., albatross, petrels) (Croxall & Prince 1996), seals (Klages 1996), and whales; an estimated 80% of all odontocete whale species regularly feed on cephalopods (Clarke 1996b). This situation was markedly different in the Mesozoic where cephalopods were among the larger predators (e.g., Tajika et al. 2018).

'Vampire squid' have been found in the stomach contents of large, deep-water fish as well as some whales and sea lions, attesting to the depths these air-breathing mammals can dive or perhaps indicating that some of the very deep-water taxa may also migrate at times into shallower water.

Fossilised stomach contents of plesiosaurs, ichthyosaurs, mosasaurs, and sharks have been found to contain remains of belemnites and ammonites. Ammonites often show shell repair, indicating unsuccessful attacks from predators such as large fish or crustaceans, marine reptiles, or other cephalopods (e.g., Massare 1987; Hengsbach 1996; Hoffmann & Keupp 2015).

17.8 ECONOMIC VALUES

Cephalopods are economically important, with squid fisheries being a substantial component of landed marine 'fish' catches (see Chapter 10). Many species of squid and octopods are eaten, but the high ammonium content of many deep-sea species makes them unsuitable for human consumption. There is also evidence that some cephalopod populations may have increased due to reductions in finfish stocks

(Caddy & Rodhouse 1998) while some non-commercial taxa have declined due to, at least in part, damage from fishing (S. O'Shea pers. comm., 2019).

17.9 THE ORIGIN OF CEPHALOPODS

A great deal of attention has been paid by palaeontologists to the origin of cephalopods. The acquisition of a chambered shell and a means of controlling the fluid content with the siphuncle (see Section 17.4.2.3.1) to achieve neutral buoyancy were the key innovations that led to the success of early cephalopods (see also Chapter 13). Using the shell as a flotation device was either concurrent with, or followed by, development of a funnel (hyponome, siphon) to enable effective movement by jetting.

Cephalopods are usually thought to have evolved from cone-shaped aseptate monoplacophoran-like ancestors (e.g., Yochelson et al. 1973; Pojeta & Runnegar 1976; Teichert 1988; Kröger 2007) similar to *Hypseloconus* and *Tannuella*. Other taxa, such as *Knightoconus* and some others that lived just before the first cephalopods appeared (Yochelson et al. 1973; Stinchcomb 1980; Kobayashi 1987), are unlikely to be cephalopod ancestors, as taxa closely resembling *Knightoconus* are now thought to be brachiopods (Dzik 2010) (see Chapter 13). The suggestion that the Cambrian (Series 2) *Nectocaris* is an early squid (Smith & Caron 2010) is rejected (see Kroger et al. 2011 and Chapter 13).

Several authors have argued that it is unlikely that cephalopods arose from an ancestor with continuous septa[18] because a diagnostic feature of cephalopods is that the shell chambers were connected by a tissue-containing tube, the siphuncle (see Section 17.4.2.3.1). Thus the ancestral cephalopod formed septa perforated by a strand of tissue (the siphuncle) (e.g., Chen & Teichert 1983a; Kobayashi 1987; Peel 1991). A modification of this latter idea by Jell (1978) and Pojeta (1980, 1987) has a snorkel-bearing 'monoplacophoran' like *Yochelcionella* converting the snorkel into a siphuncle. Unfortunately, nothing is known about the terminal part of the shell of Cambrian cephalopods, although as suggested by Starobogatov (1974) an initial function of the septa may have been to seal off the top of the shell after decollation or damage, which is the norm in the Cambrian material. In the very early cephalopods there is no evidence of a calcareous tube separating the assumed fleshy siphuncle from the chambers which would presumably have been fluid-filled. The osmotic strength of the fluid in the chambers may well have been regulated (if at all) simply via the epithelial–water interface, as osmotic adjustments are common in animal epithelial tissues (see Chapter 6). They may have also served to release some of the ammonia that probably built up in the tissues (see below). Once the water in the chambers was sealed from the external seawater, any reduction in its volume as a result of uptake due to osmotic adjustments would result in a partial vacuum being created as a by-product. An increase in the salt concentration in the

---

[18] For a discussion on the formation of septa with regard to cephalopod origins and their formation in outgroups, see Chapter 13.

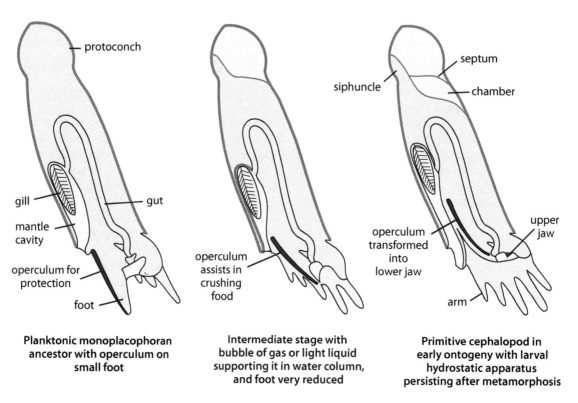

FIGURE 17.44 A speculative idea for the evolution of ancestral cephalopod from elongated 'planktic monoplacophorans possibly related to the circothecid hyoliths' according to Dzik, J., *Acta Paleont. Pol.*, 26, 1981b, p. 162. Redrawn and modified from Dzik, J., *Acta Paleont. Pol.*, 26, 161–191, 1981b.

water within the chamber(s) would have increased its density, while a reduction would have decreased it (salt water is about 3.5% heavier than fresh water), but these changes would have been small compared with the lift generated by an empty space created by water uptake. Thus, buoyancy may well have been a by-product of unrelated events, but, once realised, the cephalopods gained an adaptive advantage.

An intriguing question concerns the initial advantage to the ancestor in producing perforated septa. Various ideas have been suggested. Dzik (1981) proposed that septal development was initiated in the newly hatched young to assist with buoyancy during a pelagic phase. Dzik (1981, 1994, 2010) argued that cephalopods evolved from operculate, tall-shelled septate 'planktic monoplacophorans' or septate circothecid hyoliths (Figure 17.44). He also argued that the siphuncle is not homologous to any adult structures in other molluscs and that it may have originated from a larval shell-attachment muscle. Jacobs and Chamberlain (1996) preferred the possibility of the chambers forming around an adult muscle, analogous to that observed connecting shell chambers in a specimen of the bivalve *Spondylus* (Owen 1878), and that this muscle evolved into the siphuncle. Barskov et al. (2008) did not accept Dzik's (1981) derivation of the siphuncle from a larval muscle because he saw it as incompatible with its considerable width in the earliest cephalopods and because it contains well-developed blood vessels. Dzik's (1981) proposed derivation of cephalopods is refuted by Landing and Kröger (2012), and the ideas of Dzik, and Jacobs and Chamberlain (1996), regarding the derivation of the siphuncle from a muscle are entirely speculative and, to us, not convincing.

Starobogatov (1974) suggested that septa were introduced as a necessary first step to enable decollation of the apical parts of the shell, a fairly common process seen in several groups of conchiferan molluscs, notably some gastropods, a few bivalves, and in early cephalopods. He saw the development of a siphuncular connection and the septa essentially as a result of an incomplete series of decollations. Favouring this interpretation, Barskov et al. (2008) argued that early shelled cephalopods differed from monoplacophorans in having three independent zones of the shell-secreting inner mantle (shell walls, septa, and posterior siphuncular section) rather than just one – the shell walls. Some Furongian (Upper Cambrian) cephalopods have siphuncular septa blocking the siphuncular tube, indicating that the siphuncular tissue must have secreted these structures as it withdrew back along the siphuncle. Such structures do not occur in later cephalopods, but their presence is an indication that in these early cephalopods the siphuncle may have functioned somewhat differently.

The origin of the septa and siphuncle has inspired other speculation, including that of Donovan (1964) who argued that a buoyancy mechanism might have first evolved by liquid being secreted at the apex of the shell followed by the partitioning of this liquid-filled space, as seen in some bivalves and gastropods. Reabsorption of the liquid would result in a space being left that was probably filled with water vapour and gas diffused from the remaining liquid. The addition of multiple chambers would follow. If a part of the mantle was mainly involved in liquid absorption and this remained uncalcified, it would form a primitive siphuncle. Martin (1983) suggested a similar sequence of events but where gas, rather than liquid, was involved as a result of the Bohr effect. This is where oxygen bubbles from the blood in conditions with a high rate of aerobic glycolysis and low oxygen consumption, which is known in cephalopods (Redfield & Goodkind 1929). This could result in the displacement of fluid or tissue from the upper part of the shell and would provide some buoyancy. The addition of a septum could follow but with an aperture for the gas-forming tissue. This proto-siphuncle would allow development and slight modification of the osmotic mechanism (sodium pump) already present in the mantle tissue (see Chapter 2). Thus, the siphuncle may have evolved as a result of a small area of ventral mantle, along with some blood vessels, remaining attached to the initial septum separating the proto-conch from the rest of the shell and thus retaining a narrow connection with the distal body as new septa were formed.

There would have been considerable functional advantages of the siphuncle + septa (the phragmocone) for the early cephalopods. As in later cephalopods, albeit much more inefficiently, the spaces in the chambers reduced overall density while at the same time the septal walls provided extra strength to the otherwise thin and fragile shells. Whether these developments were precipitated as a response to increased predation in the Cambrian or provided an advantage in feeding or some other factor is unknown.

The development of the phragmocone has been likened to other major evolutionary innovations such as, for example, insect wings, that enabled a move into a new adaptive zone. When combined with the lift generated by forcefully expelling water from the mantle cavity (as described below), cephalopods were able to escape benthic predators by moving into what was, at that time, the relatively safe water column (Signor & Vermeij 1994). The ability to carry out rapid movements by jetting would have occurred later.

The ancestral cephalopods probably had a short mantle cavity or groove containing more than one pair (perhaps several pairs) of ctenidia attached only by their proximal ends (see below).

Cephalopods differ from nearly all other molluscs in having a respiratory current in the mantle cavity directed from above the ctenidia to below, with the afferent side of the gills remaining dorsal.[19] This switch in the direction of flow of the respiratory current is difficult to explain and has attracted little comment. Given that this flow is opposite to that seen in the ancestors of cephalopods (by outgroup comparison) it is of interest to speculate on how it came about. Yonge (1947) noted this switch had occurred and hypothesised that it was due to a change from cilial to muscular ventilation. In the first Cambrian cephalopods the ctenidia and circulation pattern may well have been similar to that in other molluscs. We suggest that the reversal resulted from the (possibly multiple pairs of) shell muscles being used to contract the head-foot when the shell was starting to elongate. This plunger-like mechanism caused the water in the mantle cavity to be forced out (as it

is in *Nautilus*), and presumably it provided some lift, pushing the animal off the sea floor. This response, possibly initially to threats, may have been sufficient, as a secondary consequence, to move the animal, with its thin, light shell, off the substratum into the water column as a primitive precursor to jetting.

As the shell continued to elongate, the aperture became smaller and the mantle cavity became deeper, presumably resulting in more efficient jetting. At this stage a reduction in ctenidial ciliation with the eventual loss of the current-forming lateral cilia may have co-occurred, assuming that these cilia were present in the ancestral cephalopods, as they are absent in living monoplacophorans. The task of generating a respiratory current in the deepening mantle cavity was progressively taken over by a precursor of the funnel flaps seen in *Nautilus*. These may have been formed from the reduced foot (as a result of the narrow shell opening) which became a pair of flattened flaps that could aid in directing the exhalant stream of water from the mantle cavity. These undulating flaps, presumably initially developed as a mechanism for supplementary mantle cavity respiratory ventilation and to improve exhalant control, must have evolved before the first Ordovician nautiliforms. With the implementation of the plunger and development of the ventilation flaps, the circulation pattern reversed. An analogous adaptation is seen in heterodont bivalves with water being rapidly expelled from the mantle cavity in the reverse direction to normal to enable the removal of pseudofaeces.

The funnel itself probably came a little later as an anterior extension of the ventilation flaps, as a means of further improving exhalant control, especially when jetting. This could have had additional early uses, such as blowing away sediment to uncover prey. Given their independent evolution, it is possible that the configuration of the funnel (or its equivalent), and associated structures, differed in the different groups of Ordovician cephalopods.

The efficiency of jet propulsion was enhanced in early nautiliforms, not only by acquiring a muscular funnel but also by the narrowing and elongation of the mantle cavity, coupled with the freeing of the head from its direct attachment to the shell and the development of head retractor muscles. This freeing of the head enabled its rapid withdrawal into the body chamber, so the head provided a piston-like action to forcefully expel water in sufficient quantity to create a jet stream. Body chamber elongation occurred independently in several lineages and even in some Cambrian nautiliforms as in, for example, *Tanycameroceras* and *Huaiheceras* (Chen & Teichert 1983a). Also, the efficiency of the plunger may have been greater in the straight body chamber of an orthoconic shell than in the curved chamber of a coiled or curved shell, perhaps explaining why several orthoconic lineages developed.

As noted below, the arms may have evolved from pedal papillae, with possibly the preoral velum also being involved. It is widely assumed that early and all later cephalopods had arms because they are present in *Nautilus* and coleoids. While there are numerous examples of fossil coleoids with impressions of their preserved muscular arms, strangely there are almost no other fossil records of arms. Also, the morphology of the arms in *Nautilus* and coleoids show many differences so

---

[19] A convergent dorsal to ventral flow is also known in at least one caudofoveate aplacophoran – see Chapter 3.

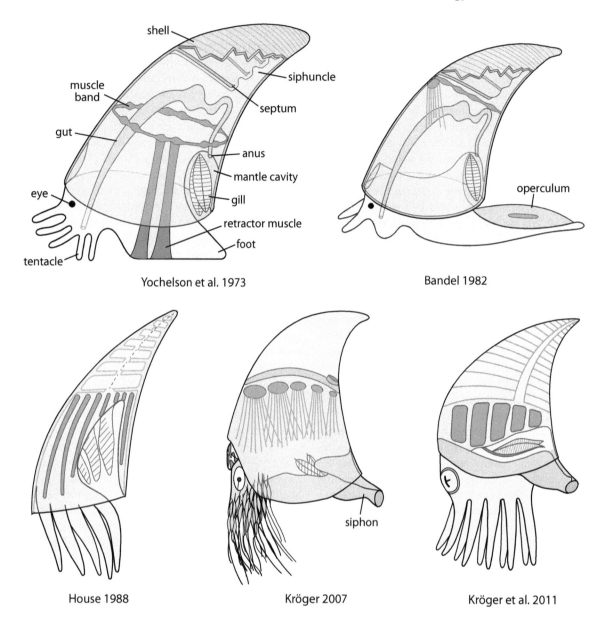

FIGURE 17.45 Some examples of hypothetical ancestral cephalopods that have been proposed. Redrawn and modified from the literature sources indicated in the figure.

an independent origin cannot be rejected. Although we do not reject the idea that cephalopods evolved arms early in their history, we suggest there is another possibility – that arms evolved, perhaps independently, in at least two lineages but either coleoid-like or nautilus-like arms may not have been present in all cephalopods. Other structures may have independently evolved for grasping food or perhaps funnelling plankton or other small animals into the mouth. Parallels are seen in gymnosome pteropods, with arms, including some with suckers, being evolved in that group, which in some ways are convergent with coleoids.

What the animal of the ancestral cephalopod was like and how it lived is open to speculation (Figure 17.45 for some alternative reconstructions). It has been suggested by some that these very early cephalopods drifted off the bottom and by others (e.g., Ebel 1999) that they were benthic. Assuming they arose from a high-spired 'monoplacophoran' like a

cone-shaped endogastric limpet, and based on comparison with outgroups, they presumably had the following features:

- Judging from the earliest fossil cephalopods, the shell of the first cephalopod was small, thin, elongate, conical, and curved endogastrically with one or a few chambers, about 0.5 mm apart and penetrated by a largely unconfined, narrow siphuncle formed from an extension of the body wall (epithelial tissue surrounding connective tissue and blood spaces).
- In the ancestral cephalopod the head was probably directly attached to the shell (the apocephalic condition) and became free later (the ligocephalic condition) (see discussion in Chapter 13), a necessary step to enable plunger-like movements to force water out of the mantle cavity. The head was simple, lacking arms

or a distinct neck and eyes. The lack of eyes contrasts with most reconstructions and the belief that cephalic eyes are a shared apomorphy with gastropods. The retinal structure of gastropods and cephalopods is different, suggesting that they have independently evolved, perhaps from the paedomorphic retention of simple shared ocelli in their larval stages. Thus, cephalic eyes may not have appeared early in cephalopod history. The hypothetical ancestor proposed by Bandel (1982) (Figure 17.45) not only has eyes, it has a pair of cephalic tentacles, but these are a gastropod synapomorphy (Ponder & Lindberg 1997).

- The small oval foot had a reduced sole and lacked an operculum. In marked contrast, the hypothetical ancestor envisaged by Bandel (1982) had a large, elongate foot, a gastropod feature, while that of Dzik (1981) (Figure 17.44) had an operculum that he suggested became the lower jaw element. The foot may have had papillae or short tentacles, somewhat similar to the tentaculate fringe on the foot of protobranch bivalves or gadilidan scaphopods. The mouth possibly had a preoral velum which may eventually have been incorporated into the head-arms complex, perhaps along with the pedal papillae. Developmental data indicate a pedal origin of the arms, but these were not present around the mouth in the cephalopod ancestor (cf. some hypothetical ancestor reconstructions – Figure 17.45). A reconstruction by Kröger (2007) of an early ellesmeroceratidan shows a well-developed funnel and funnel flaps. These structures were not present in the first cephalopods but must have developed later as part of the significant modifications imposed on the head-foot and the necessity for respiratory ventilation and exhalant control (see above).

- Lecithotrophy or direct development. Nothing is known about the size of the larval shell (protoconch) in Cambrian cephalopods, but members of the same group in the Lower Ordovician had large protoconchs, suggesting lecithotrophic development. As most data on the early helcionelloidean and 'monoplacophoran' limpets are from internal moulds, nothing is known about their protoconchs.

- It is generally considered that the narrowing of the shell necessitated the formation of a posterior mantle cavity, a feature assumed to be already present in the elongate 'monoplacophoran' (i.e., helcionellid) ancestor, but this was probably not the case. A deep posterior mantle cavity, sufficiently enclosed for water to be forced through the funnel for jetting, must have arisen later (e.g., Mutvei et al. 2007). The mantle cavity of the first cephalopods was probably located mostly laterally in a deep mantle groove, and the deep posterior mantle cavity developed later in concert with a long body chamber (Kröger 2007).

- The number of pairs of ctenidia is unknown, but there would have been at least one pair, or possibly, as suggested by *Nautilus*, two. Potential outgroups (chitons and monoplacophorans) have multiple pairs of ctenidia so this is also a possibility. Associated with each pair of ctenidia were paired kidneys and auricles. The ctenidia may have been ciliated, with the lateral cilia driving the pallial water current as in most other molluscs, but it is also possible that they were not ciliated. The unciliated ctenidial gills in modern monoplacophorans function more or less as ventilators, and the mantle wall and kidneys have largely taken over the respiratory function (see Chapter 4). Modern cephalopods have unciliated ctenidia. The water current was probably anterior to posterior through the shallow mantle cavity and flowed over the ctenidia in the normal way (efferent to afferent side). Along with the changes to the orientation of the mantle cavity in early cephalopods (see above), this water flow over the gills changed, necessitating major modifications to their respiratory surfaces.

- The mantle was thin, unmodified, and probably a significant respiratory surface.

- A U-shaped gut had the anus opening to the midline of the probably groove-like posterior mantle 'cavity'. Enzymes capable of dealing with at least a partly animal diet were probably present.

- A radula and a small dorsal jaw were present. It is highly unlikely that the first cephalopods were active carnivores with a highly modified gut and buccal apparatus – these attributes would come later. They may well have fed on small crustaceans and other tiny benthic invertebrates, for which the radula required no special modifications. The upper jaw is well developed in monoplacophorans, but there is no lower jaw, although that part of the mouth is covered with cuticle. It is probable that a little later in cephalopod evolution this cuticle lining thickened to provide a biting surface, with the eventual development of a lower jaw.

- Juvenile octopods have a paired gonad rudiment, but all adult cephalopods have a single (right) gonad, a feature shared with scaphopods but not gastropods. Living monoplacophorans differ in having two pairs of gonads with separate gonoducts.

- The nervous system had a circumoesophageal ring that was not particularly concentrated or enlarged. Development of a well-developed central nervous system came later as mobility and active carnivory developed.

- A pair of osphradia lay in the inhalant stream on the roof of the mantle groove. These are present in *Nautilus* as the interbranchial papillae (see Chapters 4 and 7). Osphradia are also present in gastropods and some bivalves but are assumed to be lost in modern monoplacophorans and scaphopods (Sigwart & Lindberg 2015).

- Several pairs of retractor muscles may have been present. Multiple muscle scars have been shown to be present in some Upper Cambrian high-conical monoplacophorans (e.g., Stinchcomb 1980), and the first cephalopods are also assumed to have had multiple muscle scars as they are known in late Cambrian (Mutvei et al. 2007) and Early and Middle Ordovician ellesmeroceratidans (eight–ten pairs) (Kröger & Mutvei 2005). It is not known exactly how many muscle scars were present in Cambrian cephalopods.

- The paired kidneys associated with each pair of ctenidia were simple renal sacs, possibly with small nephridial glands, and associated with the afferent system as in other molluscs. They lay at least partly in the mantle roof, as they do in *Nautilus* and modern monoplacophorans. Podocytes were present in the pericardium and/or auricles. The kidney later became highly modified in response to the adoption of a carnivorous diet, as seen in *Nautilus*, and more so in coleoids.
- The vascular system had two or more pairs of auricles, a single ventricle, and, apart from some arterial vessels, was an open vascular system with haemolymph spaces. The blood pigment was haemocyanin.

17.10 DIVERSITY AND FOSSIL HISTORY

Cephalopods first appeared in the Furongian (late Cambrian) of China (Kröger et al. 2011) (see Chapter 13). Their shells are mostly known from fragmentary specimens and were subsequently described from several parts of the world (see above). Not all were conical, some being very elongate and narrow, and others grew to several centimetres in length, although their maximum diameter was generally less than 10 mm. Towards the end of the Cambrian they were diverse, with 130 species in 36 genera, with eight families and four orders recognised (Chen & Teichert 1983a), and they had apparently radiated into most available ecological niches. About 90% of all described Cambrian cephalopods were found in China and they are now placed in two major groups, Ellesmeroceratia and Plectronoceratida (see the Appendix).

The first recorded Cambrian cephalopods lived in a 'well-oxygenated, somewhat turbulent, shallow-water, environment' (Chen & Teichert 1983a p. 650) in warm seas. Cephalopods also soon occupied different habitats such as the outer shelf, in shallow waters, and stromatolite reefs, where they became a dominant component of the fauna. Some were apparently nektonic, occupying parts of the ocean where the bottom was anoxic.

At the end of the Cambrian, there was extensive cooling and anoxia in the oceans, resulting in widespread extinctions. Only two Cambrian cephalopod lineages (both Ellesmeroceratida) are known to have survived the extinction (see Section 17.10.1 below), but the overall global marine biodiversity tripled during the Ordovician (Harper 2006; Kröger & Zhang 2009). With the advantage of their buoyant shells that enabled them to float in the water column, cephalopods rapidly radiated into numerous habitats and taxa. Many arthropods, including most

of the predatory anomalocaridids that dominated Cambrian seas, became extinct during the Cambrian–Ordovician extinction event, which may have also allowed cephalopods to diversify. Thus, in the Ordovician they became the only large predators that moved about in the water column, as that ability had not yet evolved in either cartilaginous or bony fish.

17.10.1 THE ORDOVICIAN RADIATION AND BEYOND

Cephalopods underwent a major radiation in the Ordovician when they were by far the largest swimming organisms, with some reaching several metres in length. They persisted to the end of the Mesozoic and were the dominant predatory animals in the oceans through the Ordovician and much of the Silurian. They were overshadowed by large predatory actinopterygians (ray-finned fish) that appeared in the Silurian but coexisted with them and Chondrichthyes (sharks and rays). During the Mesozoic, marine reptiles and a major group of actinopterygians, the teleosts, were additional competitors. We trace some of the major events in the post-Cambrian history of the cephalopods in more detail below.

In the extinction event at the end of the Cambrian, most of these early cephalopods disappeared, with only two lineages (*Ectenolites* and *Eoclarkoceras*) surviving to radiate in the Lower Ordovician. In the major nautiliform lineages in the Ordovician, the various shell morphologies evolved independently to overcome the limitations imposed by the ancestral shell morphology. The relatively rapid evolution of the major groups that appeared in the Ordovician involved each lineage adopting different approaches to buoyancy and control of orientation when swimming (Barskov et al. 2008). Some of the Lower Ordovician taxa looked very similar to their Cambrian ancestors, with curved, conical shells, a marginal siphuncle, and narrowly separated septa. There were changes to shell shape, notably coiling, external ornamentation, and, most importantly, the arrangement of the septa and modifications to the siphuncle. Thus, some shells ranged from short to long straight cones while others were curved or openly coiled, or variously tightly coiled. In others the septa were wider apart, and the siphuncle became more central and modified. Overall body size increased, and some reached considerable size. The internal features of the shell also showed substantial modifications, particularly in the structure of the siphuncle which, in some lineages, acquired the ability to precipitate shell material within the siphuncle (*endosiphuncular* deposits). These distinct shell morphologies possessed different hydrostatic and hydrodynamic properties and hence the ability to manoeuvre and alter buoyancy, as well as cope with the physiological costs of maintaining that buoyancy (Kröger 2013). Shell strengthening and the associated metabolic costs had to be balanced with the ability to withstand attacks from predators. Modifications also included the development of internal ballast in some nautiliform groups (e.g., some actinoceratian and endoceratians), where the shell was so heavy that they must have been bottom dwellers.

By the Middle Ordovician, cephalopods had achieved the greatest morphological diversification in their entire

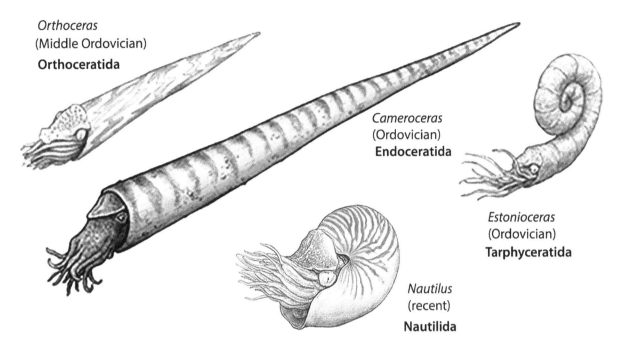

Orthoceras
(Middle Ordovician)
Orthoceratida

Cameroceras
(Ordovician)
Endoceratida

Estonioceras
(Ordovician)
Tarphyceratida

Nautilus
(recent)
Nautilida

FIGURE 17.46   Three Ordovician nautiliforms with animals as imagined by Mr Satoshi Kawasaki, with the interpretations influenced by living *Nautilus* (shown for comparison). Reproduced with permission from the artist.

history. While different body plans may have developed in the Cambrian progenitors, most of this diversity was lost in their near extinction at the end of the Cambrian.

The shell morphologies in the Ordovician evolved to such an extent that these cephalopods are grouped into several higher taxa (orders, superorders, or subclasses depending on the classification) (see Section 17.11.1 and the Appendix). Thus, nearly all the major groups (see Section 17.12.2.5 and the Appendix) evolved from the Lower Ordovician radiation of the Cambrian ellesmeroceratidan ancestors. One of these groups, the Nautilida, is represented today by two closely related genera (*Nautilus* and *Allonautilus*), and another, the Bactrita, gave rise to the ammonites and the coleoids (Chen & Teichert 1983a; Nishiguchi & Mapes 2008) (see Section 17.12.1), with the remaining groups and the ammonites becoming extinct.

In the transition from the Cambrian to the Ordovician, reduced oxygen levels may have triggered or enhanced changes in ventilation (pumping of water by muscular pedal flaps) and an increase in the respiratory surface of the ctenidium. Any cilia on the ctenidia (if they were present) would have become unnecessary for ventilation, enabling the surface of the ctenidia to increase in complexity and to provide the larger surface area for respiration required to support the increase in size and mobility in Lower Ordovician nautiliforms. This, in turn, would have increased the necessity for even more efficient respiration and ventilation and the parallel development of a more sophisticated vascular system. Further development of the 'protofunnel' flaps and associated modifications and enhancement of the muscular system for pumping water into the mantle cavity would also have occurred. Nevertheless, with many early nautiliforms, the volume of water expelled from the mantle cavity must have been small

compared to the size of the shell (and animal), so their swimming speed was presumably slow (Young et al. 1998).

The features of their head-foot are not known, although often imagined in the literature. Some examples of such speculation are given in Figure 17.46.

Among the groups that first appeared in the Lower Ordovician were the Endoceratida characterised by heavy endosiphuncular deposits in their large, slender, straight, or curved shells. The nautilian Tarphyceratida were the first cephalopods with coiled shells while the Orthoceratia had straight, slender shells with narrow siphuncles. A few early members of other groups also appeared in the Lower Ordovician, including two with short, curved shells, the exogastrically curved Oncoceratida and the endogastrically curved Discosorida. During the Middle Ordovician the Orthoceratia may have given rise to the nautilian Lituitida, a group characterised by its highly modified siphuncle morphology, and to the Bactrita, the ancestors of the ammonites and coleoids. At the same time the Actinoceratia, with large, slender shells with very wide siphuncles and heavy endosiphuncular deposits, appeared. The Upper Ordovician saw the first appearance of the Ascoceratida.

By the time the Bactrita evolved in the Lower Devonian, most features that characterise cephalopods were developed and refined (Engeser 1996). By the Upper Devonian the ammonites, which were rare in the Lower Devonian, became common and diverse. This trend increased through the Carboniferous, and there was a corresponding decrease in nautiliform diversity with only two nautiliform groups surviving through the Carboniferous. Only a single ammonite family survived the extinction at the end of the Permian, but this gave rise to the huge radiation of Mesozoic ammonites (see Section 17.12.2.5).

The first coleoids appeared in the middle Carboniferous but remained rare (as fossils at least), but in the Lower Jurassic one major group, the belemnites, that originated in the Triassic (Iba et al. 2012), began a major radiation that continued through the Mesozoic. Modern coleoid lineages first appeared in the fossil record in the Triassic, with octobrachian and decabrachian crown groups appearing before the end of the Cretaceous.

The ammonites, belemnites, and other coleoids were important components of the pelagic fauna in Jurassic and Cretaceous seas, but ammonites and belemnites disappeared during the major extinction event at the end of the Cretaceous. The modern coleoids underwent significant radiations in the Paleogene, and fossils are rather common. After the Cretaceous, nautilidans were common and widespread. Some Cenozoic nautilidans were somewhat ammonite-like, with complex, rather than simple, sutures, various coiling patterns, and some even had external ornamentation. Similarly, some Cenozoic coleoids had robust internally chambered shells and resembled belemnites. Coexisting with these early Cenozoic cephalopods were nautilidans similar to the living *Nautilus*, and coleoids with reduced or absent shells, both of which persisted through the Cenozoic to the present day. The only living coleoids with a well-developed phragmocone are *Spirula* and cuttlefish (Sepiidae). Other shelled coleoids and nautilidans, other than *Nautilus*, declined in diversity and disappeared during the Miocene, possibly due to the evolution of echo-locating cetaceans which first appear in the fossil record in the late Oligocene (Geisler et al. 2014). The coleoid shells with their partially empty chambers produce loud echoes, whereas soft tissues do not, enabling shell-bearing taxa to be more easily located by cetacean predators (e.g., Lindberg & Pyenson 2007).

There are around 800 living species of cephalopods, but there are over 20,000 known fossil species (Lindberg et al. 2004), mostly comprised of nautiliforms and ammonites. Thus, living cephalopods are much less diverse than bony fish (ca. 30,000 species). Nevertheless, there is still evidence that an evolutionary 'arms race' between fish and coleoids greatly influenced the morphology and biology of coleoids (Tanner et al. 2017).

More information about each group of extinct cephalopods is provided in the next section.

17.11 THE MAJOR GROUPS OF CEPHALOPODA

17.11.1 THE NAUTILIFORMS

As outlined above, nautiliforms arose in the Furongian (Cambrian) and suffered a significant extinction in the ice age at the end of the Cambrian. The survivors underwent a spectacular radiation in the Ordovician followed by a major extinction in another major glaciation at the end of the Ordovician. This latter event caused the loss of about a quarter of all animal species, including most of the earliest nautiliforms comprising the Ellesmeroceratida. Another major group, the Endoceratida, which contained the largest nautiliforms, died

out before the end of the Silurian. Four other major groups of nautiliforms (Ellesmeroceratia, Actinoceratia, Orthoceratia, and Nautilia) survived into the Devonian although they were slowly declining up to that time. The Ellesmeroceratia (represented by the Discosorida) were reduced in diversity at the end of the Devonian, although the group persisted at least until the Lower Carboniferous,[20] and the Actinoceratia died out in the Carboniferous, but members of the two remaining groups (Orthoceratia and Nautilia) continued through to the Triassic. Only the Nautilia survived into the Jurassic and beyond. Thus, the history of nautiliforms is largely Paleozoic, with their place taken in the Mesozoic by the ammonites.

The main groups lumped together as nautiliforms have been rather well established for some time (e.g., Teichert & Moore 1964; Teichert 1967), but the ranks used for them vary considerably from orders through superorders to subclasses, and how the names have been formed also differs. Over 30 ordinal-group or higher taxa have been proposed for nautiliforms, but less than half of these are generally accepted (Shevyrev 2006).

While uncertainties remain in their phylogenetic relationships, the evolutionary trends within each clade are rather well established. The extinct nautiliforms have, however, received less attention than the ammonites, probably because they evolved more slowly and have fewer readily diagnostic shell morphologies at the species level.

The classification outlined in Box 17.2 and given in more detail in the Appendix is slightly modified and updated from some earlier classifications (e.g., Teichert & Moore 1964; Dzik 1984; Wade & Stait 1998). Each major group is briefly discussed below.

Some earlier classifications divided the cephalopods into those with an external shell (the Ectocochlia) and those with an internal shell or lacking a shell (Endocochlia). Another two-group classification was proposed by Lehmann and Hillmer (1980) who divided the cephalopods into the Palcephalopoda (the majority of the nautiliforms) and Neocephalopoda. This division has not been generally accepted by taxonomists working on living cephalopods. The concept of Neocephalopoda was extended by Engeser (1996) to include some orthoceratian taxa with similar early ontogeny to ammonites and coleoids. In this extended concept, the Neocephalopoda contains the Coleoida, Ammonitia, Bactrita, and some members of the Orthoceratida. They have similar small (1.5 to 3.0 mm long), spherical embryonic shells that show no trace of growth lines, an assumed R-selected reproductive strategy (many small, planktonic offspring), the first formed septum differs from the following ones, and there are also differences in the formation of the *caecum* (the initial closed part of the siphuncle) and the position and morphology of the siphuncle (Engeser 1996). The grouping is also supported by some fossil radulae (e.g., Nixon 1988a) (see also Section 17.5.2), and, on that basis, a split into two very similar groups was proposed (Lehmann 1967) and

---

[20] There is one unconfirmed record from the Lower Cretaceous (Doguzhaeva 1995).

later modified (e.g., Berthold & Engeser 1987; Engeser 1996). In this scheme, the Angusteradulata (Ammonitia + Coleoida) and Lateradulata (most nautiliforms) are divided on the basis of their radular morphology (Mehl 1984; Engeser 1996; Gabbott 1999). Despite some questions regarding the early evolution of Neocephalopoda, there does seem to be a case for its recognition as a monophyletic group, although the radula may not be completely diagnostic.

The Palcephalopoda (which includes the remaining nautiliforms) have large embryonic shells, presumably a reflection of their large yolky eggs as seen in living *Nautilus*. The palcephalopods are a paraphyletic group as they contain the stem taxon that gave rise to all cephalopods and the ancestor of the neocephalopods.

The ranks applied to the major groups of cephalopods in different classifications vary. For example, the Neocephalopoda could be treated as an 'infraclass' or subclass and the next level of groups within it as subclasses or superorders respectively.

17.11.1.1 Ellesmeroceratia and Plectronoceratia

The first undoubted cephalopods had appeared in the Furongian of NE China, about 515 Mya (see Chapter 13). They were small (mostly 2–6 cm long), with curved chambered shells with simple, close-set septa, and the siphuncle was on the concave side. Shells of some later ellesmeroceratidans were straight or nearly straight. It is not known whether the first cephalopods (*Plectronoceras* and its allies *Paleoceras* and *Ectenolites*) were able to drift off the substratum or were entirely benthic, as suggested by Ebel (1999).

Mutvei et al. (2007) demonstrated these Upper Cambrian ellesmeroceratian-like nautiliforms fall into at least two distinct lineages, the plectronoceratians and the ellesmeroceratians, with both assumed to have multiple muscle scars (Kröger & Mutvei 2005; Kröger 2007). These two groups were characterised mainly by different connecting siphuncular rings. The ellesmeroceratians had connecting rings similar to those of early nautilians (the tarphyceratidans) in having a calcified outer porous layer and an inner organic (conchiolin) layer. In contrast, in the first group to appear, the plectronoceratians, the siphuncle initially had no connecting rings preserved so it may have been organic (Webers et al. 1991); calcified connecting rings perforated by numerous pore canals developed later. This siphuncular connecting ring structure is similar to that seen in orthoceratian and actinoceratian nautiliforms (e.g., Mutvei 2002a,). Mutvei et al. (2007) suggested that the plectronoceratians were probably more capable of vertical migration than the ellesmeroceratidans because of their mechanically stronger connecting rings, which were probably also more permeable to fluid. These two groups have sometimes been treated as separate major groups of equal rank, as we have done (see Box 17.2 and Appendix).

Two other groups erected for late Cambrian cephalopods, the Yanheceratida and Protactinoceratida, are often treated separately, or they are contained within a group along with the plectronoceratians, as we have done in the classification in Box 17.2 and the Appendix. The typical

Ellesmeroceratia[21] arose a little later in the Cambrian and, either directly or indirectly, gave rise to all the other groups of cephalopods in the Ordovician.

Ellesmeroceratians and plectronoceratians both have small orthoconic or cyrtoconic shells not adapted for jet-powered swimming. Since the beginning of their fossil record, nautiliforms had the necessary structures that would have provided at least the potential for buoyancy regulation. Despite this, their ability for jet-powered swimming probably appeared later in nautiliform evolution, as discussed above.

The shells of Ordovician Ellesmeroceratia were larger than those of Cambrian taxa but still rather small (10–15 cm long). Their thin shells probably could not have withstood the pressure in deep water; fossils of this group are found in deposits from shallow warm seas. In several later taxa (notably oncoceratidans), adults have a constricted aperture, leading to suggestions that the body chamber may have been used for brooding (Mutvei 2013).

While muscle attachment scars indicate that the Ordovician ellesmeroceratians had multiple muscles attached to their shell (Kröger & Mutvei 2005; Kröger 2007) (see also Section 17.3.4.1), nothing else is known about their soft-body anatomy, and the only information about the shell muscles of the Cambrian cephalopods is that *Balkoceras*, *Palaeoceras*, and possibly *Plectronoceras* have traces of oncomyarian muscle scars (Flower 1964).

The body chamber in ellesmeroceratidans had two sinuses, and Kröger (2007) suggested that the anterior one on the convex side of the curved cone was for the head and the posterior on the concave side was for the funnel and that the animal moved in an anterior–posterior direction. This type of movement in ellesmeroceratidans (and in oncoceratidans) is at variance to earlier interpretations that often had these animals moving dorso-ventrally. Orthoceratians however, do appear to have moved dorso-ventrally (Kröger 2007).

Ellesmeroceratidans may well have been benthic, slowly crawling or bouncing over the sediment in search of food. Limited buoyancy control may have been accomplished by using the narrow spaces between the crowded septa (rarely over 1 mm). Two Cambrian subgroups (Yanheceratida and Protactinoceratida) and some typical ellesmeroceratians secreted calcareous material in the siphuncles but not the chambers (Chen & Teichert 1983a).

The Ascoceratida (Ordovician–Silurian) is a small but distinctive and unusual group of small-sized nautiliforms (Figure 17.47). They were included in the orthoceratians until recently when they were moved to the Ellesmeroceratia (Mutvei 2013). Uniquely, the fragile shell consists of two parts, a juvenile (longiconic) and an adult (breviconic) portion, with the juvenile part shed. The siphuncle also differs markedly in the two sections (Figure 17.47). The mature part of the shell is more inflated, and the septa are on the dorsal side of the shell above

---

[21] The Ellesmeroceratida includes Plectronoceratoidea (or Plectronocteratida) in some commonly used classifications, such as that adopted in the *Treatise on Invertebrate Paleontology* (Teichert et al. 1964a) and by Wade and Stait (1998).

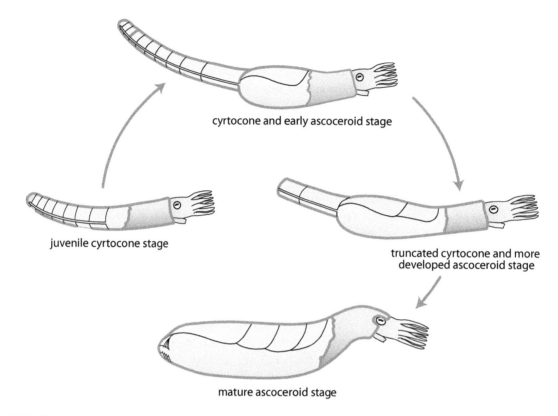

cyrtocone and early ascoceroid stage

juvenile cyrtocone stage

truncated cyrtocone and more
developed ascoceroid stage

mature ascoceroid stage

FIGURE 17.47   Ontogenetic changes in *Ascoceras* (Ascoceratida). Note that the head features are entirely speculative. The relationship between the juvenile and adult forms was recognised by Lindström (1890). Redrawn and modified from Furnish, W.M. and Glenister, B.F., Nautiloidea - Ascocerida, pp. K261–K277, in Moore, R.C. (ed.), *Treatise on Invertebrate Paleontology Part K. Mollusca*, Vol. 3, Geological Society of America and University of Kansas Press, Lawrence, KS, 1964.

the body chamber, presumably providing stable buoyancy to the assumed pelagic adult.

17.11.1.2   Orthoceratia

The Orthoceratia (= Orthocerida, Michelinocerida) were most common and diverse through the Ordovician to the Devonian but were represented by only a few taxa in the Triassic. There is also a record of an orthoceratian from the Lower Cretaceous (Doguzhaeva 1995). They probably arose from the Ellesmeroceratida, but their origin is not well understood, and the group is probably polyphyletic (Engeser 1990a,b).

The first orthoceratians are from the earliest Ordovician (Kröger 2008b). Like the Endoceratia (see below) most had a long, straight or slightly curved shell, but the internal structure of the shell differed from endoceratians. The chambers are widely spaced and connected by a thin, tubular, siphuncle with a small, spherical initial chamber lacking a scar (*cicatrix*) left by the original organic plate during embryonic development (Kröger 2006, 2008b). Many have deposits in the chambers (i.e., cameral deposits) to balance the elongate shell. The exterior of the shell was smooth or had longitudinal or annular ribs.

The internal ballast probably enabled orthoceratians to swim horizontally in the water column using jet propulsion for short bursts and perhaps enabled them to feed on prey such as arthropods. Their swimming was probably not very

efficient; they may have spent much of their time lying on the sea floor, and some may have been permanently benthic.

The shells of orthoceratians have few characters, making their classification difficult. The group may not be monophyletic, although their embryonic shells may help to clarify their relationships (e.g., Kröger & Mapes 2004), as discussed below.

The order Orthoceratida, as currently recognised, share a spherical protoconch with the probable ancestors of the ammonites and coleoids, the Bactrita (Engeser 1996). A relationship is also supported by radular data from fossils (Mehl 1984; Gabbott 1999), which indicate that at least some orthoceratians may have had a radula more like coleoids and ammonites than *Nautilus*. Also, soft tissue associated with *Michelinoceras* from Bolivia suggests there were ten arms present, with two differentiated as tentacles (Mehl 1984). Thus, while some taxa treated as orthoceratians should probably be included in Neocephalopoda (see above), a suggestion that the Lower Devonian Lamellorthoceratidae and some other Lower Devonian orthoconic taxa were early coleoids with fins and internal shells (Bandel & von Boletzky 1988; Bandel & Stanley 1989) is not generally accepted.

The major taxa within the orthoceratians fall into two main groups (Kröger 2008b): orthocones with a conical apex that has a cicatrix (Pseudorthoceratida) and orthocones with a

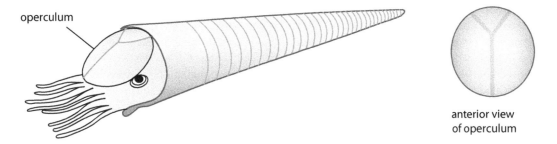

operculum

anterior view
of operculum

FIGURE 17.48   Reconstruction of opercular position of the aptychopsis plates in an orthoconic nautiliform. Note that the head and tentacles are hypothetical. Redrawn and modified from Turek, V., *Lethaia*, 11, 127–138, 1978.

small spherical apex that lacks a cicatrix (Orthoceratida).[22] The Orthoceratida has been a lumping ground for a varied assortment of taxa but was restricted by Kröger and Isakar (2006) to those orthoceratidans with a spherical apex and a simple siphuncle and which ranged from the Lower Ordovician to the Upper Triassic. The Dissidoceratida (Silurian and Lower Ordovician) had orthoconic shells, and their siphuncles were tubular and slightly expanded. Some had rod-like endosiphuncular and cameral deposits (Kröger 2008a).

A number of nautiliforms have been found with operculum-like plates called an *aptychopsis* in their apertures that somewhat resemble the aptychi of ammonites (see Section 17.12.2.4), but, unlike aptychi, they are not formed from the lower jaw. Instead, they are flat, circular structures composed of three parts that fit the aperture (Stridsberg 1984), possibly homologous with the hood of *Nautilus* and, like that structure, functioned as an operculum (Figure 17.48).

17.11.1.3   Discosorida

The Discosorida (Middle Ordovician to Upper Devonian) superficially resembled the contemporary nautiliform Oncoceratida, but they had different origins and differed in the internal structure of the shell. Discosorid shells were endogastrically curved (lower [siphuncle] side concave) or, in more derived families, exogastrically curved (lower [siphuncle] side convex). The aperture was usually simple but was constricted in some derived members, opening as a series of slits. The siphuncle was centrally to marginally located, the septal openings were reinforced with internal thickening, and some had additional endosiphuncular deposits. Such deposits were pronounced in the Ordovician taxa but were reduced in later members of the group.

The Discosorida probably evolved from ellesmeroceratidans, first appeared in the early Middle Ordovician, reached their peak during the Middle Ordovician, and declined in the Upper Ordovician. Another diversification occurred in the Silurian (Wenlock, Ludlow), after which the group declined, although it persisted until the Upper Devonian. These animals were probably benthic or moved in the water column close to the bottom. If they swam, they were probably oriented with their head down. There are no records of body preservation.

17.11.1.4   Endoceratia

The Endoceratia are distinguished from other nautiliforms by their wide siphuncle which contains layered, calcareous, hollow conical deposits (endocones) in its posterior part that were presumably used to help counterbalance the long, straight shell, a strategy contrasting with the cameral deposits seen in some other groups. Some moved the body chamber below the last enclosed chamber (Figure 17.21) which would have also assisted with maintaining a horizontal orientation in the water.

This group contains not only the largest nautiliforms but the largest known Paleozoic fossil invertebrate. Some Ordovician endoceratians were well over three metres long; a few, such as the Middle Ordovician *Cameroceras*, reached five metres in length (O'Donoghue 2008). While it is sometimes assumed that they were fast and streamlined predators, the relatively small body chamber (about a quarter or less of the shell length) suggests that the volume of water in the mantle cavity would have been inadequate to propel these animals through the water at high speed, assuming that they used jet propulsion. We consider it more likely they were slow drifters, floating by way of neutral buoyancy and kept level in the water by counterweights in the narrow end of the shell. While they may have been capable of a short burst of speed to catch large prey, the musculature associated with any jetting apparatus (such as funnel flaps or similar structures) was probably inadequate to sustain prolonged jetting. Instead, we suggest that these and other nautiliforms, especially those with relatively small body chambers relative to shell size, may have fed on the rapidly diversifying and increasing plankton biomass of the Ordovician (Ausich & Bottjer 1982; Signor & Vermeij 1994) (Figure 17.32). Feeding on the drifting and slow-moving members of the water column community has been proposed for some ammonite taxa (see Section 17.12.2.3) and remains common in living coleoid paralarvae as well as many adult deep-sea octopods (Villanueva et al. 2017) (see also Chapter 5).

17.11.1.5   Actinoceratia

This moderately large group (ca. nine families) of mainly straight, medium to large nautiliforms mostly reached about 60 to 90 cm in length, but some, like the Silurian Huroniidae, were much larger. The group is characterised by their wide siphuncle that extends into the chambers with a complex

---

[22] Also in lituitids (Nautilia).

system of radial canals. Both endosiphuncular and cameral deposits were present (Teichert 1964c).

The sediments in which their fossils are found indicate these animals occurred in shallow to rather deep water. They may have led a partially benthic lifestyle interspersed by swimming. They were probably predators, and given their complex, wide siphuncles they may have been able to regulate their buoyancy more effectively than most other contemporary nautiliforms.

This group may have been derived from taxa similar to the Furongian *Protactinoceras* (Protactinoceratida) (e.g., Barskov et al. 2008), although actinoceratians do not appear in the fossil record until late in the Lower Ordovician. They do not diversify much until the Middle Ordovician, where they reached their greatest diversity, before declining in the Lower Devonian.

17.11.1.6 Nautilia

Variations of this name have been used for all the nautiliform taxa, but it is now used in a more restricted sense for the group which includes *Nautilus*. This group usually contains four higher taxa (normally given the rank of order) (e.g., Wade & Stait 1998), the Nautilida (Silurian to present), Oncoceratida (Ordovician to Carboniferous), Tarphyceratida (Ordovician to Silurian), and Barrandeocerida (Ordovician to Devonian). The crook-like lituitids with coiled early whorls are sometimes included in the tarphyceratids, or they are treated as a separate order (Lituitida). We include Oncoceratida in the Ellesmeroceratia following Mutvei (2013) who also acknowledged the questionable placement of Tarphyceratida. The siphuncle of tarphyceratidans has a similar connecting ring to that of ellesmeroceratidans (including oncoceratidans) and, like members of that group, also has multiple muscle scars,[23] but otherwise they differ in having a long body chamber and an aperture that is not constricted at maturity (Mutvei 2013). We have retained Tarphyceratida in Nautilia (see Appendix).

The order Nautilida ranges from the late Paleozoic and includes the six species of living *Nautilus* (Ward & Saunders 1997). Another genus, *Allonautilus*, was created by Ward and Saunders (1997) for two species and, although made a synonym of *Nautilus* by Harvey et al. (1999), is currently recognised by most workers.

Devonian nautilidans ranged in shape from curved (cyrtoconic) to loosely (gyroconic) and tightly coiled (convolute or involute) shells that resembled *Nautilus*. Following a decline in the Upper Devonian, there was another diversification in the Carboniferous followed by subsequent declines and diversifications in the Permian and Triassic. By the Carboniferous there was a considerable range of shell morphology, with a few curved or loosely coiled shells and most taxa tightly coiled, often with highly sculptured exteriors and even spikes in some Permian taxa. Though fewer in number, the Permian nautilidans survived the End Permian extinction event more successfully than the ammonites. Some of the Upper Triassic nautilidans developed wavy sutures similar to those of the

ammonite *Goniatites*. The nautilidans underwent another smaller diversification in the remainder of the Mesozoic, and five genera in three families survived the extinction event at the end of the Cretaceous that eliminated the ammonites. There was a small diversification in the Paleogene, and again, two families (Hercoglossidae and Aturiidae) developed wavy sutures (e.g., Teichert 1988). Cenozoic nautilidans declined sharply from the middle of the Miocene with oceanic cooling; the living nautilids are restricted largely to tropical seas.

17.12 BACTRITA AND AMMONITIA (= AMMONOIDEA)

17.12.1 Bactrita (= Bactritoidea)

Bactritans had straight to slightly curved shells like those of orthoceratians. This small paraphyletic group of cephalopods (two families) is important because it is thought to contain the ancestors of both ammonites and coleoids, which, with the Bactrita, make up the Neocephalopoda. They had a small-sized, spherical or egg-shaped protoconch (e.g., Kröger & Mapes 2007) (Figure 17.49) and are thought to have produced many eggs in contrast to the few large eggs produced by nautiliforms. The Bactrita may have been derived from near basal Orthoceratia in the Ordovician (Holland 2003) (see Section 17.11.1.2).

In the early neocephalopods and later shelled taxa, several septa were formed before hatching, and the liquid was emptied from the first chamber. The point when the young hatched from the egg capsule was marked by a *nepionic constriction* in the embryonic shell, and the young hatchlings presumably lived in the plankton for a time. The post-embryonic shells had a typical four-layered structure of periostracum, outer prismatic layer, nacreous layer, and inner prismatic layer (Engeser 1996).

The shells of bactritans are orthoconic, narrow, long to short, and usually smooth. The siphuncle was narrow and, unlike most nautiliforms, lay near or against the ventral wall, as in nearly all ammonites. The positioning of the siphuncle caused a small indentation in the suture line called the siphonal (or ventral) lobe. The septa were concave and widely spaced, and the septal sutures were simple, except for the often small, V-shaped siphonal lobe. Engeser (1996) listed several apomorphies that probably separated the Bactrita from their ancestor. These included the narrow siphuncle that extended through the first septum into the first chamber as a closed off caecum, the likelihood of a reduced radula, and possibly the possession of ten arms. In the ancestor of the Bactrita, the narrow siphuncle was probably at or near the centre of the shell, but it moved ventrally. Because some key characters are rarely preserved in fossils, the placement of some taxa is uncertain.

The absence of any modifications for balancing in bactritans indicates that they swam or floated in a head-down position and were probably planktonic (Holland 2003). Nothing is known of their body, other than muscle scars and some trace fossils associated with *Bactrites* shells suggesting they may have had about ten arms (Teichert 1964a).

---

[23] Although fewer in number; it is not known if these are the attachments of mantle muscles or head-funnel retractors (Mutvei 2013).

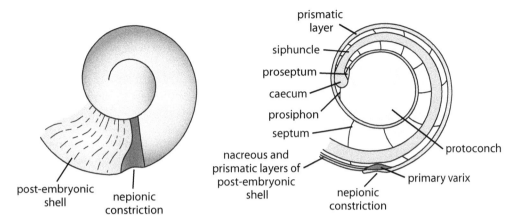

ammonitella (embryonic shell) of an ammonite

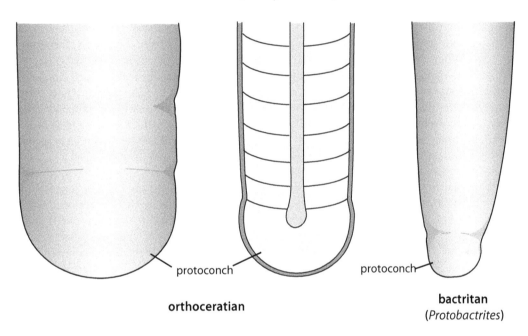

orthoceratian

bactritan
(*Protobactrites*)

FIGURE 17.49 Comparison of the embryonic shells of an orthoceratian, bactritan, and ammonite drawn approximately to scale. Upper figures redrawn and modified from Landman, N.H. Early ontogeny of Mesozoic ammonites and nautilids, pp. 215–228, in Wiedmann, J. and Kullmann, J. *Cephalopods Present and Past. 2nd International Cephalopod Symposium: O. H. Schindewolf Symposium*, Tübingen 1985. Stuttgart, Germany, E. Schweizerbart'sche Verlagsbuchhandlung, 1988; right lower figure redrawn and modified from Kröger, B. and Mapes, R.H., *Paläont. Z.*, 81, 316–327, 2007; left and middle lower figure redrawn and modified from Kröger, B., *Lethaia*, 39, 129–139, 2006.

Two families are recognised. They arose in the Lower Devonian, diversified during the Devonian, and persisted until the end of the Permian. The first ammonite evolved from one lineage (Bactritidae) in the Lower Devonian with ancestral forms becoming increasingly tightly curved and then coiled (Figure 17.54). Thus, the main difference between Bactritidae and ammonites is the orthoconic shell of the former, with taxa regarded as primitive ammonites possessing cyrtoconic shells.

The second family, the Parabactritidae, had shorter shells with more closely spaced septa. They arose at the beginning of the Carboniferous, shortly after gave rise to the coleoids, and disappeared at the end of the Cisuralian (Permian). Deposits on the shell were initially thin but soon became thicker, culminating in the rostral development seen in Hematitida and later in the calcitic deposits in belemnites. The transition from an external to an internal shell must have been a gradual one, and with some fossils, it is difficult to tell which they were.

There has been controversy regarding the best way to treat bactritans. For example, in the classifications proposed by Engeser (1996) and Shevyrev (2006), 'Bactritoidea' and 'Ammonoidea' are treated as separate higher groups (subclasses). Kröger and Mapes (2007 p. 326) stated:

'Late Paleozoic bactritoids are not simply straight ammonites but developed some unique characters that fully justify the erection of a high-level taxon equal in rank to the

Orthoceratida and Ammonoidea. Thus, we suggest that the initial beginning of the Ammonoidea can be defined and separated from the Bactritoidea by the beginning of curvature of the post-embryonic conch.'

Alternatively, the two bactritan families (Bactritidae and Parabactritidae) could be treated as basal ammonites and coleoids respectively, which would no longer necessitate the recognition of this paraphyletic group. We have adopted a conservative approach in maintaining this paraphyletic group at an equivalent rank to ammonites and coleoids (see Appendix).

17.12.2 AMMONITIA (AMMONITES)

Members of this extraordinarily diverse, extinct group had external coiled shells that were highly varied in morphology and size, and the septa separating the shell chambers had elaborately complex sutures. The largest reached 2.5 m in diameter (see below), with several being over a metre in diameter.

According to Kröger and Mapes (2007 p. 326), ammonites 'unquestionably evolved along the lineage from *Bactrites* to *Lobobactrites* to *Anetoceras*, which is a primitive ammonoid'. As outlined above, this evolutionary transition involved a morphological change from an orthoconic shell (as in *Bactrites*) to an openly coiled ammonite (*Lobobactrites*) and eventually tightly coiled ammonites, and it occurred quite rapidly in the Lower Devonian (over a period of about 26 Ma). *Lobobactrites* is considered the probable ancestral ammonite because it has a rather large, egg-shaped protoconch like other early ammonites (Erben 1964; Teichert 1988). Thus, the most primitive ammonites initially had curved shells, but these quickly became coiled, although some secondarily uncoiled taxa occurred in several lineages. The first true ammonites are placed in the 'order' Agoniatitida (= Anarcestina) (Erben 1964; Klug & Korn 2004). The relatively large protoconchs of early ammonites had reduced in size by the end of the Lower Devonian, and the adult shell became more tightly coiled. By the end of the Devonian, they all had a small, more or less spherical protoconch and a tightly coiled shell.

Ammonite shells can be separated from those of Bactrita and nautiliforms by their generally thinner shells with a marginal siphuncle (it runs just inside the edge of the outer part of the shell spiral), which is typically narrower than in nautiliforms, and the septal necks are oriented backwards (not forwards as in *Nautilus*).

Ammonites underwent an amazing radiation and are very well represented in the fossil record. Species evolved quickly, and because of their complex suture patterns and other aspects of their shell morphology, they can be more readily identified than most nautiliforms. Ammonites had a worldwide distribution as many were oceanic. This, coupled with their pelagic embryonic development, resulted in many having wide distributions making them excellent index fossils. This utility is further enhanced by the rapid turnover of species and genera, with many persisting for only short periods of time (a few million years). Thousands of generic and specific names have been proposed for members of this large and very

diverse group. An outline of their classification is provided in the Appendix. A comprehensive, although dated, treatment is available in the *Treatise on Invertebrate Paleontology*.

A coiled shell, a ventral (= siphonal or external) lobe, and the suture line making a zig-zag pattern are features characteristic of, but not unique to, ammonites, as quite a few nautiliforms have a coiled shell, some have a marginal siphuncle, a few have a ventral lobe, and others have simple sutural zig-zag patterns, similar to the basal goniatite ammonites. In contrast to most nautiliforms, the sutural patterns in most ammonites are complex but especially so in the Ammonitida, where they became progressively more intricately folded (see Section 17.12.2.2). Some ammonites are sculptured externally, whereas nautiliform shells are usually smooth, although there are exceptions, as noted in Section 17.11.1.6. The external surface of ammonite shells varies considerably from smooth (other than growth lines) to various patterns of ridges, ribs, spines, or knobs. Strongly sculptured forms were particularly prevalent in the Cretaceous.

Ammonites had nine teeth in each radular row, as do coleoids (see Section 17.5.2). Judging from their protoconchs, ammonite eggs were small and probably produced in large numbers. There is evidence to suggest that shallow-water taxa may have been faster-growing than nautiliforms, reaching maturity in about five years or less, while deep-water species may have grown more slowly (Bucher et al. 1996). As noted earlier (Section 17.7.1.1), Jurassic ammonite shells were commonly dimorphic (Davis et al. 1996; De Baets et al. 2015; Klug et al. 2015b); the larger shells (*macroconchs*) are assumed to be females and were often several times larger than the small male shells (*microconchs*). These differences can be so pronounced that the different forms were often first described as different taxa. The assumed female macroconchs presumably matured later than the smaller male microconchs, thus paralleling the later maturation of females seen in many coleoids, although in the great majority of coleoids males and females have similar body sizes. Further evidence that the macroconchs were females is from presumed egg sacs in the body chambers of a few specimens (Lehmann 1981). The microconchs often had lappets while some macroconchs had apertural constrictions or collars.

Ammonites are classified based on several sets of characters, with the ornamentation and structure of the septa and their sutures being particularly important (see below). Ammonite shells differ from those of nautiliforms in more ways than just their sutural complexity. The nautiliform hyponomic sinus (the notch that would have held the funnel [the hyponome]) is absent in many ammonites, and no fossil remains show any trace of such a structure (Klug & Lehmann 2015). In some there is a projection, which may be long, in this position (i.e., the middle of the outer lip of the aperture), which is termed a rostrum, although it is a completely different structure from the rostrum in coleoids. Some ammonites have lateral projections from the shell aperture (*lappets*) that are never seen in nautiliforms. These flanges protrude from the edge of the aperture in what are thought to be adult males (microconchs) in some sexually dimorphic taxa (Davis et al. 1996). They are never present on the larger female shells (macroconchs). The purpose of these flanges is unknown, but one (unlikely) suggestion is that they may be used in sexual display (e.g., Davis et al. 1996).

Shell form was typically planispiral, but the so-called 'heteromorph' ammonites, which appeared both in the Triassic and during the Lower Cretaceous, had a bizarre variety of shapes and coiling patterns, including some with the initial part of the shell coiled and the last part straight (e.g., the Late Cretaceous *Baculites* and a few species of *Hamites*), some others were helically coiled (e.g., *Turrilites*, *Bostrychoceras*, *Proturrilitoides*) or irregularly coiled (e.g., *Nipponites*). Some developed a hook-like end bent in the opposite direction to the initial coil (e.g., *Worthoceras*, *Scaphites*, *Acrioceras*, *Ancyloceras*), while others were bent to form several parallel lines (e.g., *Polyptychoceras*), a tight U-shape (e.g., *Hamulina*), or straight or slightly curved tubes, a few up to two metres long (e.g., *Baculites*). The uncoiled ammonites are easily distinguishable from some superficially similar nautiliforms in having complexly folded sutures. The heteromorph ammonites are not a monophyletic group, as different forms arose independently from different lineages during ammonite evolution, both in the Triassic and Cretaceous.

Ammonites in the Lower and Middle Jurassic were relatively small (mostly less than 20 cm in diameter, and rarely over 23 cm) until the Upper Jurassic and Lower Cretaceous. *Lytoceras taharoaense* from the Jurassic of New Zealand is 1.5 m in diameter, while the Late Cretaceous North American *Parapuzosia bradyi* grew up to 1.6 m in diameter (Stevens 1988), and the giant *Parapuzosia seppenradensis* (Figure 17.1b) from the Cretaceous of Germany is the largest ammonite, reaching 2.5 m in diameter.

The form of the final body chamber of some ammonites is quite different from the other chambers, with the change brought about by inflation and constriction of the aperture. Some heteromorphs formed a U-shaped final chamber. These modifications are often such that it is difficult to see how the animal functioned, especially as models indicate that the chamber would have faced upwards when floating (Monks & Young 1998). There have been some observations of fossil egg masses in these chambers, and the suggestion has been made that they are modified as egg chambers, perhaps analogous to the egg case of *Argonauta* (Lewy 1996, 2003). Because relatively few ammonites have these modified terminal chambers, Lewy (1996) suggested that they must have had two breeding strategies, one involving larger, yolky eggs attached in bundles to the substratum and the other involving many tiny eggs laid by the female in the terminal (last or body) chamber which also accommodated the body of the female. Lewy drew an analogy with the incirrate octopods, with the former strategy being practised by most of them, but the argonautids use a secondary shell (secreted by modified arms) to contain the eggs. Ammonites with a modified final chamber are often very widely distributed, while those that do not form such a chamber typically differentiate by forming local lineages (Lewy 1996, 2002b). Lewy (1996) further suggested that the adult did not feed while the young developed in the chamber

and that the hatchlings may have fed on her body, citing as evidence observations of juveniles in the body chamber of some fossils. These ideas have been discounted by the majority of cephalopod workers (e.g., Donovan 2012).

Some ammonite fossils are remarkably well preserved, retaining their original nacreous shell material, and a few even retain colour patterns (Mapes & Davis 1996). Strangely, useful impressions of the head and arms of ammonites are unknown despite their shells sometimes being fossilised in sediments that preserve such tissues in other animals. Beaks (jaws) and radulae are sometimes preserved (see below), with records of a possible ink sac (see Section 17.4.3) and traces of the gut, including food (Klug & Lehmann 2015). It has been assumed that ammonites had either numerous arms (as in *Nautilus*) or about ten (as in coleoids), but the absence of fossil traces suggests these may have lacked substantial musculature as impressions of the well-developed arms of coleoids fossilise occasionally. Speculative drawings of ammonite animals showing various numbers and lengths of arms are common in the literature (some examples are given in Figure 17.50), but there is an almost complete lack of fossil evidence to indicate what the head-foot features were actually like. A recent critical assessment of the available data suggested the possibility there were perhaps ten short arms capable of being retracted into the shell (Klug & Lehmann 2015).

There are several other possible forms of the external body of ammonites. Two highly speculative examples are a weblike structure for gathering planktonic food supported largely by a hydrostatic 'skeleton' or structures derived from funnel flaps that may have been expanded externally to be used as paddles, analogous to the wings of pteropods. While such possibilities may be fanciful, perhaps traces of the external body are present in some fossils but are not being recognised as such because of the focus on arms.

Muscle scars are sometimes preserved on the interior of shells, near the back of the body chamber close to the septal wall (see Section 17.3.4.1). In ammonites, the muscle scars are much smaller than in nautiliforms, being very small and paired or a narrow ventral band (Doguzhaeva & Mutvei 1993, 1996; Landman et al. 1999). These weak muscles suggest that ammonites did not use the nautiliform plunger technique to create a jet of water for movement. If this was the case, ammonites probably created a flow of water in their mantle cavity using alternative means. One suggestion is that the mantle in ammonites may have been more like that of the muscular coleoid mantle, with contractions producing respiratory and locomotory currents, and that the animal may have even been capable of extending beyond the shell (Jacobs & Landman 1993; Jacobs & Chamberlain 1996). Evidence for this idea is said to be the small muscle scars and the lateral embayments seen on the shells of some taxa (see Section 17.3.4.1). The latter are interpreted as attachments for muscles that facilitated the eversion of the funnel and part of the mantle and are thus analogous to the pallial sinus of bivalves. This idea was extended further to suggest that some ammonites may even have had fins. Given any lack of direct evidence for these views and the functional difficulties of the mantle pumping

Bostrychoceras
(Upper Cretaceous)

Schloenbachia
(Upper Cretaceous)

Nipponites
(Upper Cretaceous)

Crioceratites
(Lower Cretaceous)

Scaphites
(Cretaceous)

Pavlovia
(Jurassic)

Oxynoticeras
(Lower Jurassic)

Echioceras
(Lower Jurassic)

Oxynoticeras
(Lower Jurassic)

Soliclymenia
(Upper Devonian)

Ceratites
(Triassic)

Goniatites
(Devonian and Carboniferous)

FIGURE 17.50   A selection of ammonites with animals as imagined by Mr Satoshi Kawasaki. Reproduced with permission of the artist.

inside an external shell, it seems unlikely to us that the mantle was muscular. Alternatively, respiratory and locomotory currents may have been entirely created by flaps similar to those associated with the funnel in *Nautilus*. There remains some doubt as to whether ammonites possessed a funnel or, if they did, whether it was used for jet propulsion. The hyponomic notch seen in most nautiliforms is found in some of the older ammonites suggesting that they may have been able to direct a jet stream backwards, but other ammonites lack this notch and may not have been able to do so. In species that develop a rostrum, this would have been an obstruction to the funnel. Also, the long body chambers (which in some heteromorphs were twisted or even convoluted) and weak retractor muscles strongly suggest that it was unlikely that ammonites had an efficient method of jet propulsion. An alternative and

even more speculative means of propulsion may have been modified arms or funnel flaps, or even mantle flaps, acting as paddles. Useful data are lacking, and we concede that we do not know.

The apparently weak muscle attachment to the body chamber has been suggested by some authors to indicate that the ammonite animal could not have been firmly fixed to its body chamber, and thus the animal could not have been very active. This observation led to suggestions that some ammonite shells may have been semi-internal or internal (e.g., Doguzhaeva & Mutvei 1993), an idea that led to the proposition that octopods evolved from ammonites by losing the internal or semi-internal shell (e.g., Lewy 1996, 2003). Evidence in support of this hypothesis includes the similarity of the radula. This idea has not gained general acceptance (e.g., Hewitt & Westermann

2003), although it has been entertained as a possibility by some (e.g., Monks & Palmer 2002).

While some ammonites had relatively streamlined shells suggesting they may have been effective swimmers (although perhaps only as efficient as *Nautilus*), many others must have moved through the water slowly and inefficiently due to their prominent sculpture or cumbersome coiling patterns, both of which would have created significant drag. Calculations based on experiments suggest that ammonites with shells having intermediate degrees of compression may have been able to swim efficiently at relatively slow speeds, while more compressed forms could swim more rapidly. In general, ammonites would have been poor high-speed swimmers and unlikely to be capable of chasing swimming prey, although it is possible that some narrow, involute keeled ammonites may have achieved this (Jacobs & Chamberlain 1996).

No matter what their swimming abilities, the initial change from a straight bactritan shell to a coiled shell must have improved swimming capability and manoeuvrability (Klug & Korn 2004) as well as initiating a shift in the downward-facing aperture to an upward orientation.

Beaks and radulae have been found associated with a few examples of ammonites from each major radiation (e.g., Nixon 1996). As indicated above, the number of teeth per row in the radula (Figures 17.29 and 17.30) is fewer in ammonites than in the nautiliforms (nine instead of 13) and thus more like that of coleoids, particularly octopods (Lehmann 1981; Nixon 1996; Doguzhaeva et al. 1997). Unlike coleoids, some ammonites appear to have mineralised radular teeth (Doguzhaeva et al. 1997).

The beak-like jaws from Carboniferous and Permian goniatite ammonites have small upper and large lower elements. In contrast, Mesozoic ammonites had a massive lower jaw (the aptychus – see Section 17.12.2.4). Coleoid and at least some ammonite jaws are similar in that both the upper and lower beaks are composed of an organic matrix, although in some ammonites the pointed biting tips (the rostra) are weakly calcified, and this calcification differs from that seen in fossil nautiliform beaks (Doguzhaeva et al. 1997). In ammonites, the inner and outer walls of the beaks are about the same length and posteriorly separated only narrowly, with no room for muscle attachment. In contrast, in nearly all coleoids except *Vampyroteuthis* and *Tremoctopus* (Clarke 1986), these structures are widely separated posteriorly, allowing considerable space for muscle attachment. This observation may indicate that ammonites did not use their jaws for strong biting, and the jaw may simply be an evolutionary relict from their nautiliform ancestor, some of which were probably active predators.

Ammonites lived for about two to five or six years (Bucher et al. 1996), compared with *Nautilus* which becomes sexually mature at 10–15 years and lives for about another five years (Landman & Cochran 1987). Their relatively short lifespan is thus another feature where ammonites show more similarity with coleoids than with nautiliforms.

It has been calculated that in ammonites new septa formed at intervals of from one to three weeks to as little as two days. As a new chamber is formed, the adductor muscles must be detached and then reattached to the new septal wall, a process perhaps facilitated by their small size. In contrast, *Nautilus* forms chambers more slowly (e.g., 50–80 days, Landman et al. 1989). Checa and Garcia-Ruiz (1996) discussed various theories accounting for the intricately folded sutural lines in ammonites and proposed that a 'viscous-fingering model' provided a useful way of explaining most features of septal formation.

17.12.2.1 The Embryonic Shell (Ammonitella)

In many ammonites, the small fertilised eggs were presumably liberated into the water column with no parental care, to develop, hatch, and live there for a time. Irrespective of how the eggs were liberated or brooded, the small size of the 'larval' shell suggests they were produced in large numbers. Spawning may well have been followed by the death of the adult (i.e., they may have been semelparous).

These ammonitella 'larval' shells are unique to ammonites. They consisted of the protoconch and a few shell chambers connected by the siphuncle (Figure 17.49). The ammonite protoconch is typically only about 0.5 to 0.8 mm in maximum diameter (Lehmann 1981), with a range of about 0.3 to 1.6 mm, while the diameter of the ammonitella ranges from 0.5 to 2.6 mm (Landman et al. 1996). The protoconch is separated by a simple suture (the *proseptum*) followed by the first main septum (the *primary septum*) and then additional septa. The siphuncle penetrates all the septa and bulges a little into the protoconch interior as the so-called *caecum* (Figure 17.49). These small-shelled 'larvae' almost certainly lived for a time in the plankton, and it has been suggested that they may even have been planktotrophic (e.g., Weitschat & Bandel 1991).

As noted above, at least one ammonite may have adopted brood protection (Mironenko & Rogov 2016), although this has been suggested for those taxa with modified body chambers and constricted apertures, with some fossils containing brooded juveniles.

17.12.2.2 Ammonite Sutural Patterns

The actual septa in ammonites are relatively flat, and the intricately folded patterning occurs where the septum meets the outer shell at the suture. The suture lines are hidden by the outer shell layers so are only visible internally, usually when ammonites are preserved as internal moulds. The first few sutures that develop in the ammonitella larval shell are simple, with the sutural lines quickly becoming increasingly folded as the shell grows.

The function of the complexly folded sutures has been the subject of debate with several, not necessarily mutually exclusive, hypotheses suggested. These were recently discussed in detail by Klug and Hoffmann (2015), and some ideas are listed below.

- They provide buttress-like strength to the outer shell wall enabling it to resist hydrostatic pressure (e.g., Batt 1989; Hewitt 1996) with the complexity of the sutures indicating at what depth particular species may have been able to live. This idea is not well supported by more recent studies (e.g., Daniel et al. 1997; Hewitt & Westermann 1997).

- They provide a stabilising effect on fluid slopping about in the chambers of the moving ammonite (e.g., Kulicki 1979), especially in large species. Also, the complex septal configurations may have enhanced buoyancy regulation through increased surface tension effects (Daniel et al. 1997).
- They provide extra surface area for the temporary anchorage of the posterior mantle (Lewy 2002a, 2003), although this idea is not widely accepted (Hewitt & Westermann 2003). It has also been suggested that the septal folding provides increased attachment area for the retractor muscles. This latter idea is incorrect because the main muscles are attached to the shell anterior to the septa (Klug & Hoffmann 2015).
- As septal folding reflects the shape of the posterior mantle, it has been suggested that the larger surface area may be related to more efficient metabolism (e.g., respiration), but this idea has little support (see Klug & Hoffmann 2015 for discussion).

The sutural folds can be complex, with the lines forming saddles (or peaks) and lobes (or valleys), and their detailed configuration is critical for taxonomy, with a complex nomenclature developed to describe the many configurations. We will not elaborate on this further except to note that, other than the simple agoniatitic (nautilid-like) sutures in the Middle Devonian agoniatitidans (Klug & Hoffmann 2015), there are three basic suture patterns (Figure 17.51). These are:

1. Goniatitic (goniatite) – these are irregular zig-zag patterns of several undivided lobes and saddles, this being the characteristic pattern seen in most Paleozoic ammonites.
2. Ceratitic (ceratite) – these are regular wavy patterns with the saddles rounded and simple, but the lobes have subdivided saw-toothed tips giving them a serrated appearance. This pattern is typical of most Triassic ammonites but also reappears in the Cretaceous 'pseudoceratites'.
3. Ammonitic (ammonite) – these have the lobes and saddles greatly subdivided (fluted) to form complex feather-like or fern-like patterns, but the endpoints are usually rounded (i.e., not saw-toothed). This sutural pattern is typical of Jurassic and Cretaceous ammonites but is found in some ammonites from the Permian onwards.

The number of lobes that comprise these patterns varied through ammonite history. In the Devonian and Permian ammonites, there were only three lobes on either side of the median line. These lobes increased to four in Triassic taxa and five in those from the Jurassic. Cretaceous taxa had either four lobes (the heteromorphs) or five and six (tetragonitids) (Lehmann 1981).

17.12.2.3 Life Habits

Their amazing diversity has led to considerable debate regarding the supposed life habits of ammonites. While palaeontologists have made numerous, often highly speculative, attempts at construction, these based on shell morphology and the type of strata in which the fossils were deposited (e.g., Westermann 1996), much of the biology of these animals remains largely in the realms of imagination, in part because of the lack of firm data on head-foot features. Nevertheless, some general features of their biology are reasonably clear, as outlined below.

Many ammonites were probably pelagic and oceanic, and, at the other end of the spectrum, a few may have lived as part of the benthos. Because ammonite shells tend to be thinner than those of nautiliforms, they may have been unable to live as deep as *Nautilus* (and presumably many extinct nautiliforms). Numerous fossils of assumed oceanic species are found in rocks laid down in oceanic sediments ('black shales') where the anoxic conditions were not suitable for benthic life. Many of these have streamlined shells and were probably efficient swimmers. Like *Nautilus*, most ammonites are usually assumed to have moved by jet propulsion, albeit possibly weak (see discussion in Section 17.12.2), although some may have been benthic (following a planktonic juvenile stage), with the animals perhaps moving about on the bottom, possibly aided by arms (e.g., Ebel 1990, 1992). Colour patterns are retained on the shells of some ammonites (see above), giving further clues as to their lifestyles (Mapes & Davis 1996).

There is some direct evidence regarding the food of a few species. Preserved stomach contents indicate that they lived on small benthic crustaceans, molluscs, and echinoderms; soft-bodied invertebrates and small fish were probably included in their diet as well, and some may have been scavengers (e.g., Klug & Lehmann 2015). It is clear that some were macrocarnivores and others fed on minute animals such as ostracods. Ammonites were themselves preyed on by marine reptiles and elasmobranchs, and some fossils show tooth marks on their shells.

Considerable attention has gone into working out how ammonites were oriented in life, as this would give clues to their habits. Their orientation can be inferred from examination of fossil shells and using experimental models. The heteromorph ammonites are of particular interest in this regard. *Nautilus* (and other nautiliforms) have shorter body chambers than most ammonites, suggesting the possibility that the ammonite animal may have been able to retract some distance into the shell (Monks & Young 1998). Monks and Young (1998) went further, suggesting that in the bizarrely coiled heteromorphic ammonites the animal was relatively small compared with the body chamber and, by shifting within the chamber, they could affect the orientation of the floating shell. This idea has similarities to the concept of a mobile 'octopus-like' animal within the body chamber (Lewy 1996, 2002b, 2003) (see also Section 17.12.2), although how the shell could have been secreted in this (unlikely) scenario is problematic. It has been suggested that heteromorph shell orientation could be achieved by moving water in the chambers (Kakabadzé & Sharikadzé 1993), although this method is likely to have been

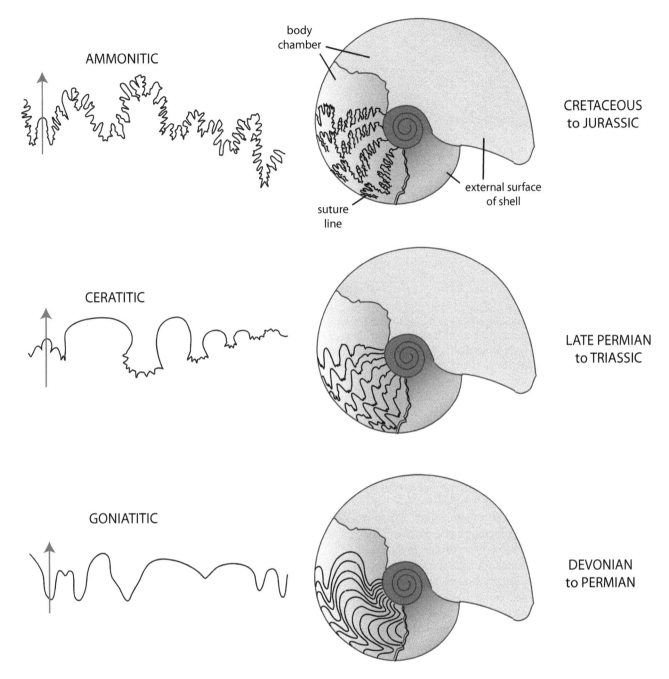

FIGURE 17.51 The three main kinds of ammonite sutural patterns. The blue arrows point in the direction of the edge of the aperture. Redrawn and modified from Lehmann, U., *The Ammonites: Their Life and Their World*, Cambridge University Press, Cambridge, UK, 1981 and various other sources.

too slow to be effective. The heteromorph funnel, if present in adults, would not have been able to produce an effective jet of water (although this may have been possible for juveniles), and their shell morphology would allow only slow movement at best. They may have been planktonic drifters catching zooplankton (e.g., Ward 1986; Westermann 1996) assuming that their shell openings were oriented away from the seafloor. If they could tilt their shell, they might have been able to feed on the seafloor (e.g., Monks & Young 1998). Others may have lived on the seafloor permanently as adults while Arkhipkin (2014) suggested that the hook-shaped last whorls

of Ancyloceratina were used to anchor these animals to large algal fronds. Other speculation concerning heteromorph ammonites includes ideas like the extension of the mantle beyond the aperture (Jacobs & Landman 1993), the mantle enveloping the shell (Doguzhaeva & Mutvei 1993) (Figure 17.52), or, as noted above, the use of the shell as a brood chamber (Lewy 1996).

17.12.2.4 Aptychus and Jaws

An aptychus (plural aptychi) is a symmetrical, plate-like, calcified (aragonite) structure found associated with Mesozoic

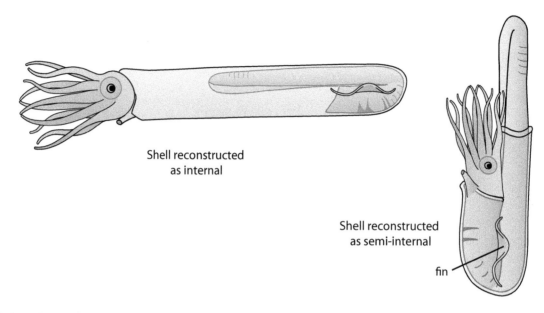

Shell reconstructed
as internal

Shell reconstructed
as semi-internal

fin

FIGURE 17.52 Hypothetical reconstructions of the body of the heteromorphic lytoceratid ammonite *Ptychoceras* and its relationship to the shell, with the suggestion that the shell may be either internal or semi-internal instead of the usual assumption that it was an external shell. Redrawn and modified from Doguzhaeva, L.A. and Mutvei, H., Structural features in Cretaceous ammonoids indicative of semi-internal or internal shells, pp. 99–114, in House, M.R. (ed.), *The Ammonoidea: Environment, Ecology, and Evolutionary Change*, Clarendon Press, Oxford, UK, 1993.

ammonites. Uncalcified structures thought to be aptychi have also been described from Upper Devonian ammonites (Frye & Feldmann 1991), but most reports of aptychus-like structures from the Paleozoic have not been substantiated. The name aptychus is used for various types of calcareous (and a few non-calcareous) structures presumed to have served as opercula or as jaws (mandibles) of ammonites. There is a detailed nomenclature and terminology associated with these structures (e.g., Frye & Feldmann 1991; Moore & Sylvester-Bradley 1996). The main terms are: *diaptychus* – aptychi composed of two equal parts (diaptychi were originally classified as a bivalve). In some texts, the term aptychus is restricted to this kind. Some aptychi are a single plate-like structure called an *anaptychus*. Anaptychi are rarer than diaptychi and are found in ammonites from the Devonian to the Cretaceous while diaptychi are only found in Mesozoic ammonites. Some authors have argued that anaptychi may have functioned as an operculum, although this interpretation is now disputed (Lehmann et al. 2015).

Nixon (1996) reviewed the data available on jaws associated with ammonite shells. Unmodified jaws have been found in Anarcestina and Goniatitina, but some of the latter have an anaptychus. Members of the Phylloceratina and Lytoceratina have lower jaws with calcified tips but are otherwise unmodified. Such jaws are often fossilised separately from the ammonite shells and are called *conchorhynchs*. This type of fossil jaw has also been found associated with some nautiliforms. The Ancyloceratina had horny jaws with some thin calcification and with a thicker layer of calcification on the larger lower jaw. Ammonitina had a large lower jaw with an outer calcified layer (which may be thick), and this jaw was very large and formed aptychi in at least some genera, while the upper jaw was

small and uncalcified. At least in some of the earlier ammonitines (e.g., the Lower Jurassic *Psiloceras* and *Eoderoceras* as well as the later *Pleuroceras*), and some of the later (e.g., the Cretaceous *Lytoceras*), the calcified lower jaw was an anaptychus while in others it formed a diaptychus. The transformation from one to the other was achieved by the formation of 'ligamental flexibility' between the two halves (Seilacher 1993), as depicted in Figure 17.53.

Being only rarely preserved *in situ* the aptychi are usually found as separate objects, so it is unclear as to what species of ammonite many forms of aptychi belong. The ammonite aptychi were once thought to be a calcified version of the fleshy hood of *Nautilus* that covers the animal when it retracts and thus similar to the aptychopsis seen in some orthoceratian nautiliforms. This latter structure is, however, a three-part calcareous 'operculum' (Figure 17.48), and the aptychi are quite different, as described below.

There has been considerable controversy over what the ammonite aptychus was formed from and what its function was. The function of closing the shell like the *Nautilus* hood or a gastropod operculum seems obvious, but this interpretation has been disputed. It has now been shown that the aptychus formed the lower part of the jaw apparatus (e.g., Lehmann 1981; Frye & Feldmann 1991) (Figure 17.53). Diaptychi have been found *in situ* in the body chamber of some Mesozoic ammonites, providing some evidence of their function. They are curved and are sometimes associated with the much smaller, uncalcified upper jaw. Despite the apparent anomaly of the large size of these structures relative to the cephalopod body (i.e., the body chamber) and the much smaller upper jaw, they are undoubtedly a highly modified lower jaw. Two main ideas as to their function have been suggested. One is that it

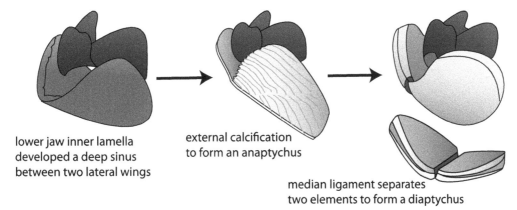

lower jaw inner lamella developed a deep sinus between two lateral wings

external calcification to form an anaptychus

median ligament separates two elements to form a diaptychus

Evolutionary series showing how the lower jaw changed into a broad aptychus, probably no longer used for biting

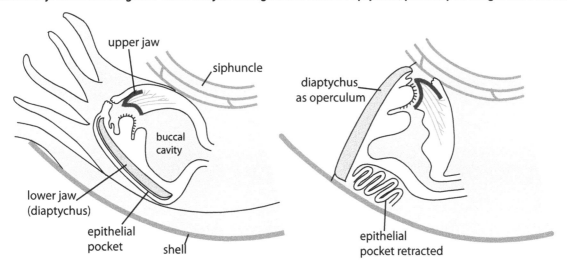

The aptychus could have acted as an operculum - sliding out from an epithelial pocket

FIGURE 17.53   Derivation and function of the ammonite jaw apparatus. Redrawn and modified from Seilacher, A., *Am. J. Sci.*, 293-A, 20–32, 1993.

no longer functioned as a jaw but instead provided the shield-like lid to the aperture. They are often as wide as the body chamber and, for this reason, were assumed to function as an operculum (Figure 17.53), but the fit with the aperture was not exact. Another idea was that these huge jaws, rather than being used for biting as in most other cephalopods, including some other Mesozoic ammonites (Nixon 1988a), were used to pump water or sediment into the mouth where small animals were extracted from the sludge (e.g., Lehmann 1972; Morton & Nixon 1987; Tanabe et al. 2015). The heavy calcification of the jaws would have also changed the balance of the animal, possibly enabling it to keep the buccal mass closer to the sub-stratum while feeding (Morton & Nixon 1987). The shovel-like jaw may have functioned like a dust-pan, disturbing and scooping up small animals on the seafloor which were then ingested (Lehmann 1981), or it may have served in extract-ing planktonic organisms from the water column (Kruta et al. 2011). Thus the large, modified jaws could have been used for microphagous feeding and in some taxa may also have functioned as an operculum (e.g., Lehmann 1981; Seilacher 1993; Keupp 2007).

17.12.2.5   Phylogeny, Classification, Diversity, and Fossil History of Ammonites

The ammonites underwent significant radiations, but we lack a good understanding of the evolutionary forces driving this group, largely because the main features of the non-calcified parts of these animals remain unknown. While the develop-ment of increasingly complex suture patterns and different shell shapes and coiling patterns all played their roles, the gross enlargement of the lower jaw and its co-option as a shovel-like feeding device, and possibly as an operculum, was an evolutionary novelty, indicating that at least some Mesozoic ammonites had strange life habits, a view reinforced by the bizarre coiling patterns of the heteromorphic forms.

As noted in Section 17.12.1, ammonites arose from bac-tritan nautiliforms during the Pridoli (Silurian) or Lower Devonian (depending on whether or not the bactritids are included in the Ammonitia) (Figure 17.54). During their his-tory from the Lower Devonian, ammonites underwent and survived three major near extinctions, the first at the end of the Devonian. They were abundant from the Pridoli, but only about 10% survived the extinction event at the end of

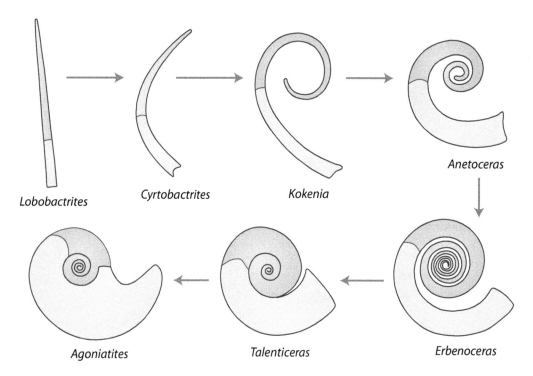

FIGURE 17.54 A transformation series showing the evolution of ammonite shell coiling. Redrawn and modified from Klug, C.A. and Korn, D., *Acta Paleont. Pol.*, 49, 235–242, 2004. The body chamber is shown in pale grey.

the Permian. These survivors resulted in a major diversification in the Triassic, only to suffer another near extinction at the end with the survivors spectacularly diversifying in the Jurassic and into the Cretaceous. They declined markedly towards the end of the Cretaceous during the Maastrichtian (Wiedmann 1988), and the remainder succumbed in the extinction event at the end of the Cretaceous (66 Mya) at the same time as the dinosaurs. There are over 1,500 ammonite genera recognised, most in the Mesozoic (e.g., Sepkoski et al. 2002).

Two ammonite taxa (orders) survived the End Permian extinction. One of them, the Ceratitida, underwent a major evolutionary radiation during the Triassic, evolving into about 80 families. A new group, the Ammonitida, evolved in the Lopingian (Permian) but remained insignificant. The extinction event at the end of the Triassic was survived by a few ammonites but saw the extinction of the ceratites and the only remaining orthoconic nautiliforms, the pseudorthoceratidans. The true ammonites (Ammonitida), with much more complex sutures than earlier groups, underwent an astounding burst of evolution in the Jurassic and later through the rest of the Mesozoic when they were the only ammonite group present. No ammonites survived the K–Pg extinction at the end of the Cretaceous (66 Mya), becoming extinct along with the dinosaurs and the belemnites. Why this happened is not known, but all three groups had declined before the end of the Cretaceous. Marine plankton production may have been severely reduced due to the reduction in sunlight for some time following the meteor impact that resulted in the K–Pg extinction event, so perhaps ammonite planktonic stages, which may have relied on feeding on plankton, succumbed while the lecithotrophic

juvenile stages of the ancestors of *Nautilus* that were feeding solely on yolk reserves survived.

17.12.2.6 The Major Groups of Ammonites

An outline of a classification of ammonites is given in the Appendix, although several alternative classifications have been proposed that differ in the detail and the ranks employed. For example, three major paraphyletic groups are sometimes recognised: the 'palaeoammonoids' in the Paleozoic, the 'mesoammonoids' in the Triassic, and the 'neoammonoids' in the rest of the Mesozoic. The number of ordinal-level taxa recognised varies considerably. Monks and Palmer (2002) recognised nine orders by elevating some suborders listed in the Appendix to orders, while other classifications recognise from three to ten (Shevyrev 2006). In the classification we have adopted (see Appendix), there are six orders. Some phylogenetic schemes that have appeared have assisted in clarifying relationships, for example, analyses of the 'Neoammonoidea' (Engeser & Keupp 2002), the hamitid ammonites (Monks 2002), and the Permian–Triassic ammonites (McGowan & Smith 2007).

17.12.2.6.1 *The Goniatitida, Clymeniida, and Ceratitida*

The 'palaeoammonoids' are usually either grouped in a paraphyletic 'order' Goniatitida or broken up into two or three separate groups which are often given ordinal status. They all have septa with round saddles and pointed lobes and together ranged from the Lower Devonian to the Lopingian (Permian). The Anarcestina contains the first ammonites, these differing in their coiled shells and ventral siphuncle from the Bactritidae from which they evolved. They arose in the Lower

Devonian and survived until the end of that period. The Clymeniida are found only in the Upper Devonian and differ from other ammonites in the siphuncle being located dorsally, although in the first chamber it is positioned ventrally as in other ammonites. Also, like the Anarcestina, but unlike most other ammonites, they had a hyponomic sinus, suggesting that a funnel may have been more developed than in most of the later ammonites. They were small (up to about 4 cm in diameter), and some had strange triangular coiling that has been interpreted as a means of shifting the centre of mass and the aperture downwards to optimise feeding on the seafloor (e.g., Korn 1992; Becker 1993). Despite their rather short history, they evolved into about 60 genera.

The Goniatitina (in the restricted sense) were the most diverse of the early ammonites, both in morphology and the number of species. Most were small, tightly coiled, and globose, and their suture lines ranged from simple 'goniatitic' to more complex with some resembling the ceratites; others had a weak hyponomic sinus, and, while most were smooth, a number of others had longitudinal or radial external sculpture. They arose in the Middle Devonian and survived until the end of the Permian.

The Ceratitida ranged from the Upper Devonian to the Triassic and included most of the Triassic ammonites. They are characterised by septa with round saddles and serrated lobes and fall into two main groups, the Prolecanitina and Ceratitina. Compared with the other early ammonites, some of the later members of this group had, for the first time, a laterally compressed shell which tapered to the narrow periphery, providing a more streamlined shape for efficient movement through the water. In other respects they were much more diverse in shell shape and ornamentation than earlier ammonites. Some were heteromorphs, with helical and uncoiled forms, and some other streamlined forms appear to have been oceanic.

17.12.2.6.2 Ammonitida

The Ammonitida first arose in the Triassic, and this highly diverse group (Figure 17.55) was a major component of the Mesozoic marine fauna until the end of the Cretaceous. Their septa are characterised by having finely folded saddles and lobes (the 'ammonite' pattern), and their shells are often sculptured. Many had extensions of the aperture (lappets and rostra) not seen in earlier ammonites. There have been many attempts at resolving the classification of the highly diverse Mesozoic ammonites (Donovan 1994). Four main groups (suborders or orders) are usually recognised. Ammonitida (as recognised here) is paraphyletic.

The Phylloceratina are the oldest group, first appearing in the Lower Triassic, and it persisted, along with the other ammonite groups, to the Upper Cretaceous. This small group of large-sized ammonites had involute, rounded shells lacking significant ornament, and apertural extensions, if present, were short. They presumably evolved from ceratitine ammonites but, unlike the rest of that group, survived the End Triassic extinction. They appear to have lived in oceanic

habitats, while most other ammonites preferred shallow-water environments.

The Lytoceratina probably evolved from the Phylloceratina in the Lower Jurassic and persisted to the Upper Cretaceous. Their shells, unlike those of many late Mesozoic ammonites, were smooth or only weakly ribbed. They differ from other ammonites in details of their ammonitella and suture lines and are thought to have inhabited open water. They may have also been able to live in deep water. The early members had evolute shells; later some became involute, and in a few the whorls were slightly dissociated.

The Ammonitina first appeared in the Lower Jurassic and along with the other groups of ammonites, survived to the end of the Upper Cretaceous, although they were in decline before their extinction. Despite this late decline, they were the most diverse ammonites, with up to 15 superfamilies recognised. Many species had restricted distributions, and numerous species persisted for only relatively short periods of time. As a group, they occupied a wide range of habitats (e.g., Westermann 1996) but were especially abundant in shallow coastal water environments. It is thought that many lived on or near the sea floor, but a few may have been oceanic. Most had planispiral shells, but a small number were heteromorphic. Putative sexual dimorphism (see Section 17.12.2) was common, and the majority were probably short-lived.

The Ancyloceratina includes most of the heteromorph ammonites. These appeared in the Lower Jurassic but were not common until the Cretaceous. Unlike the ammonitines, they remained common until the extinction event at the end of the Cretaceous. This group is characterised by shells that were either irregularly coiled, partly uncoiled, or uncoiled. Some reverted to planispiral coiling but tended to be sculptured and often spiny. Many were common and widespread, and this group includes some of the most useful index fossils. Their possible life habits and habitats are discussed in Section 17.12.2.3.

17.13 Coleoida

The coleoids comprise all living cephalopods other than nautilids, encompassing squid, octopuses, vampyromorphans, cuttlefish, and *Spirula*. In addition to the lack of an external shell, living coleoids differ from nautilids in having eight arms and, in living decabrachian taxa, two extra retractile tentacles. Unlike the arms in *Nautilus*, coleoid arms are typically lined with suckers and/or hooks and, sometimes, cirri. On the head, there is a pair of large lens-bearing eyes (in almost all taxa) that are markedly more complex than the pinhole eyes of *Nautilus*. Another significant difference from externally shelled cephalopods is the possession of a pair of fins, although these are (probably secondarily) absent in most octopods. The fins can both provide propulsion and generate lift. Nearly all coleoids are active carnivores, and all are capable of swimming. They differ from other cephalopods in that both living and extinct coleoids have an internal shell that has undergone varying degrees of reduction, including loss of calcification in many, and even complete loss in others. The reduced internal shells that retained chambers in many

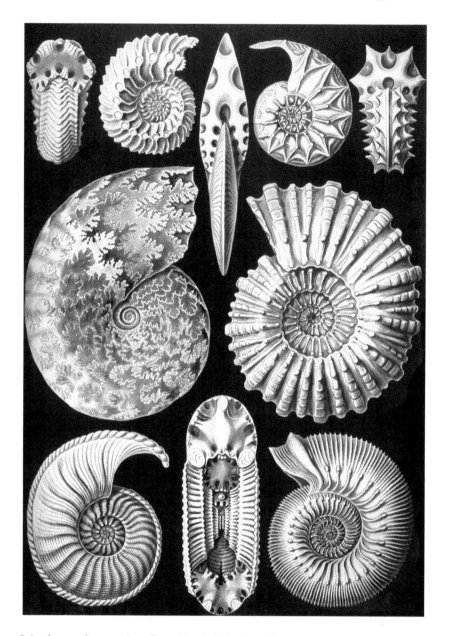

FIGURE 17.55 Some of the forms of ammonites. From Haeckel, E., *Kunstformen der Natur*, Verlag des Bibliographischen Instituts, Leipzig und Wien, 1904, plate 44.

early lineages and a few living taxa (sepiids, *Spirula*) still provided a means of achieving neutral buoyancy. Further reduction in squid and octopuses led to the necessity of constantly swimming, achieving buoyancy by other means (see Section 17.4.2.2) or becoming permanently benthic (as in many octopods). Their body is covered with thin skin providing a supplementary respiratory surface, essential in these mostly highly active animals.

The naked coleoids rely on camouflage, speed, or other behaviours to evade predators. They have been very successful in doing this and have evolved into a moderately large, diverse group that occupies most marine habitats and plays a significant role in the ecology of our oceans.

Unlike *Nautilus*, the thin skin usually contains chromatophores and may also contain other structures (leucophores, iridophores, photophores) involved in colour change (see

Section 17.4.4). In some, notably cuttlefish and many octopods, the skin is capable of being sculptured at will into complex projections, such as papillae or filaments.

A significant coleoid innovation is the thick muscular wall of the mantle cavity, the contractions of which facilitate both ventilation and jetting. The funnel is a fused tube (in contrast to the two flaps in *Nautilus*), and there are interlocking cartilages or fixed fusion points on the funnel and collar that stabilise the head and prevent water leaking from the edges of the mantle cavity during jetting. The mantle cavity contains a single pair of ctenidia and, associated with them, a single pair of auricles and kidneys as well as a pair of branchial hearts (see Section 17.5.3 and Chapter 6). Also, especially in decabrachians and unlike *Nautilus*, the circulatory system is effectively a closed one, with capillaries rather than haemocoelic blood spaces connecting the venous and arterial systems.

Besides the differences just noted, there are also many other structures present in coleoids that are not present in *Nautilus*, including marked differences in the head and tentacles, as well as the renopericardial, reproductive, and nervous systems (see relevant earlier sections in this chapter). Also, the beak is non-mineralised, and an ink sac is often present. The development of an ink sac may be unique to coleoids if the reports of possible ink in ammonites are erroneous, although this is currently unresolved (e.g., Nishiguchi & Mapes 2008).

In adults there is a single gonad, as in *Nautilus*. Nidamental glands are found in squid, cuttlefish, and *Nautilus*, but not in *Vampyroteuthis* or octopods where they have apparently been incorporated into the oviducal gland (Budelmann et al. 1997).

17.13.1 Phylogeny, Classification, Diversity, and Fossil History

The classification of coleoid fossils is, of necessity, mainly based on the shell or the uncalcified shell remnant, the gladius. Coleoid gladii are uncommon as fossils, as are the delicate phragmocones making up the chambered part of the calcified shell. Preservation of the body is much rarer than that of the shells, but those that are known give important clues about early coleoid evolution.

There are six major living ordinal-level taxa; the Sepiida, Spirulida, Myopsida, Oegopsida (these two previously combined as Teuthida), Vampyromorpha, and Octopoda (Nishiguchi & Mapes 2008; Donovan & Fuchs 2012; Allcock 2015). These fall into two main groups; the Decabrachia ('decapods' or Decapodiformes) and the Octobrachia ('octopods' or Octopodiformes), and as indicated by these names, the number and structure of the arms/tentacles are a key feature. The primitive number of arms appears to have been five pairs. These are identified in a dorsal to ventral paired series as I, II, III, IV, and V (see also below Figure 17.8). The extinct Belemnitidia, based on fossilised arm hooks, appears to have had ten equal-sized arms. Living decabrachians have modified one pair of arms (IV) as retractile tentacles while octopods have lost a pair (pair II). In the vampire squid (*Vampyroteuthis*) pair II is much reduced and used as a sensory filament. These groups also differ in their sucker morphology (see Section 17.3.2), with those of octopuses being attached by broad bases and lacking horny material. Decabrachians (squid and cuttlefish) have the suckers attached to the arm by a very narrow stalk and have a ring of horny material in the sucker, typically formed into teeth or hooks (Figure 17.10). *Vampyroteuthis* suckers are similar to those of octopods but have short, narrower bases.

The main coleoid groups recognised are briefly outlined below and in the Appendix. Attempts to examine the phylogeny of living coleoids have included morphological (e.g., Young & Vecchione 1996) and molecular analyses (e.g., Bonnaud et al. 1997; Carlini et al. 2000; Strugnell et al. 2005, 2006, 2014, 2017; Strugnell & Nishiguchi 2007; Lindgren 2010; Allcock et al. 2011, 2014; Lindgren et al. 2012; or both (Carlini et al. 2001). Various morphological characters have also been examined in detail with the aim to improve phylogenetic resolution, including, for example, the ctenidia (Young & Vecchione 2002) and shells and musculature (e.g., Bizikov 2008; Sutton et al. 2016). While the main groups are fairly robust, the resolution of their phylogenetic relationships remains somewhat ambiguous. Recent work incorporating fossil and molecular data to provide 'molecular clocks' suggested that several coleoid lineages may be much older than shown in the fossil record (Strugnell et al. 2006). In contrast to this interpretation, more recent divergence estimates suggest a Mesozoic origin of the crown group coleoids (Klug et al. 2016; Strugnell et al. 2017; Tanner et al. 2017), a finding congruent with the fossil record.

While about 800 species of living coleoids are recognised, the number of extinct species is much more difficult to assess.

The key innovations of coleoids were tied to the internalisation of the shell, this allowing the body to become at least partly unconfined and enabled the development of a muscular mantle wall and external fins. These changes led to more efficient jetting and, especially in octopods, increased flexibility. The external skin enabled the evolution of structures facilitating texture and colour changes. These and other innovations, such as the ink sac and modifications on the arms (cirri, hooks, suckers), not only enhanced prey capture but also greatly increased the chances of escape from predators.

As discussed above, coleoids appear to have evolved from the same group as the ammonites, the externally shelled Bactrita, but later, probably in the Mississippian (Lower Carboniferous), and from a different family, the Parabactritidae (see Section 17.12.1). It is possible that the coleoids may have originated earlier, perhaps in the Middle Mississippian or even the Upper Devonian, but Devonian records (Termier & Termier 1971; Bandel & von Boletzky 1988) have not been confirmed (Doyle et al. 1994). Nishiguchi and Mapes (2008) suggested the possibility that more than one group of bactritans gave rise to coleoids at different times in the Devonian and/or Carboniferous.

The shells of the first coleoids were similar in most respects to the shells of bactritans, being small to medium-sized, straight, narrow, external, with a long tubular living chamber, and the animal floated head downwards. They had a small spherical or oval protoconch with a caecum and prosiphon. None of the fossil coleoids from the late Paleozoic and Mesozoic had coiled shells as they had abandoned the protection afforded by the relatively cumbersome external shell in favour of a more active life style with internal shells. Thus, the (presumably gradual) development of an epithelial sac surrounding the shell would have been an important first step in the evolution of Coleoida. Along with shell reduction, this external mantle epithelium made it possible for muscles to become attached to the external surface of the shell or to the epithelial sac itself following strengthening with connective tissue and, later, muscle. These changes also enabled the development of fins and their cartilaginous support structures (e.g., Bizikov 2008) and greatly improved their swimming capabilities.

The first group of taxa to appear that are generally accepted as coleoids were the Aulacoceratia in the Mississippian. They had a small, cone-like rostrum (Figure 17.56 for terminology)

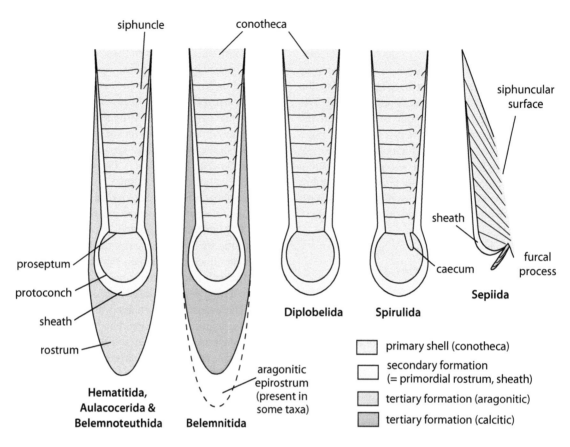

FIGURE 17.56 Schematic diagrams of the primary, secondary, and tertiary shell formations in the posterior part of selected coleoid shells. Note that the aragonitic thickening does not extend posteriorly as a rostrum in Hematitidae. Redrawn and modified from Fuchs, D., *Geobios*, 45, 29–39, 2012.

which surrounded the protoconch and the first part of the elongate phragmocone. The conotheca has a layer of nacre, and the body chamber was long and tubular, at least in juveniles (Bandel 1985; Doguzhaeva 2002). Some Aulacoceratida and all Hematitida have longitudinally ribbed rostra while xiphoteuthid aulacoceratoideans and belemnites have smooth rostra, calling into question the monophyly of aulacoceratoideans (Keupp & Fuchs 2014). In Aulacoceratia, the larval shell has the caecum and prosiphon reduced, and a closing membrane is developed that separates the protoconch from the rest of the phragmocone (Bizikov 2008). This situation differs from the arrangement of the caecum and prosiphon in Bactrita where the protoconch was not used as a part of the hydrostatic apparatus at the time of hatching, with that function taken over by septa formed before hatching (Bizikov 2008).

The fossilised remains of both bodies and hard parts have assisted in resolving the evolution of the group. Bizikov (2008) suggested that the ancestral coleoids probably had ten undifferentiated arms of about equal length, which were possibly thin and about half to three-quarters the body length, given their probable planktonic lifestyle. Like primitive living coleoids, they may have had an interbrachial web, and the arms may have had paired cirri and possibly uniserial suckers perhaps similar to those of *Vampyroteuthis*. The ctenidial axes were probably attached to the mantle wall by a suspensory membrane.

The internalisation of the shell did not immediately lead to the development of a muscular mantle because the earliest coleoids (Hematitida, Donovaniconida, Aulacoceratida)[24] retained a tubular living chamber. The enclosure of the shell by the mantle created a new outer shell layer, present in all coleoid shells. This was used by some early coleoids, notably in some early taxa, most famously including the belemnites, to deposit thick shell material on the outer surface posteriorly (the rostrum) (Figures 17.56 and 17.59) to balance the animal, enabling it to swim horizontally. Inside this layer, the original shell, the conotheca, remained thin. The development of fins enabled greater directional and orientation control and undulatory swimming.

For the mantle to be co-opted in jetting another evolutionary innovation had to occur – an anterior projection of the wall of the living chamber was needed to support the mantle, the pro-ostracum (Figure 17.59). The formation of the pro-ostracum may have occurred by the reduction of the ventral and lateral walls of the living chamber in the Phragmoteuthida, which lived from the Lopingian (Permian) to the Lower Jurassic, but the microstructure of the pro-ostracum in

---

[24] Supposed Devonian coleoids such as *Boletzkya* and *Naefiteuthis* (Bandel et al. 1983) have been shown to belong to other groups (e.g., Pignatti & Mariotti 1996).

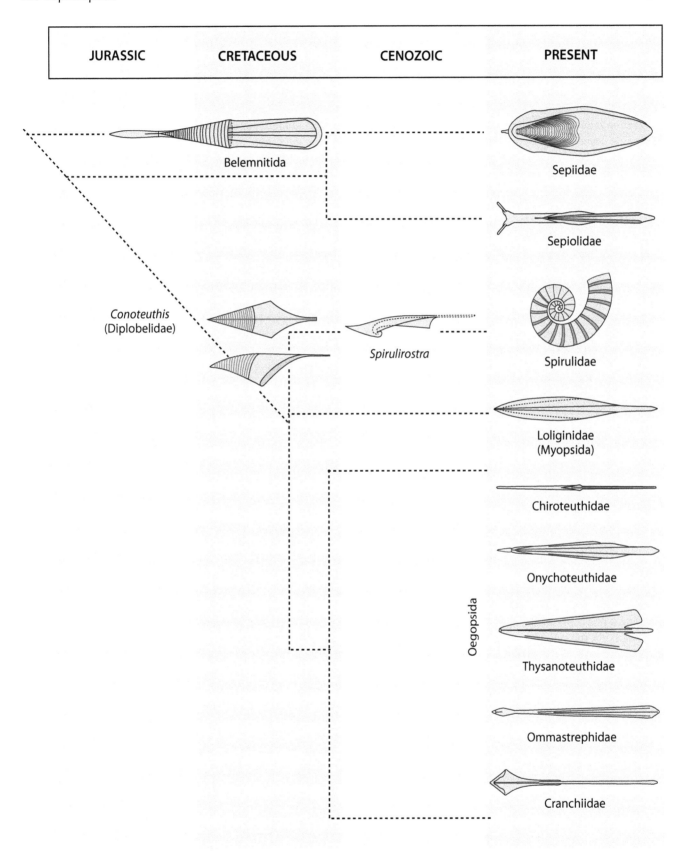

FIGURE 17.57 Some of the main lineages in coleoid evolution based on their shell morphology – part 1. Phylogeny done in consultation with Dr D. Fuchs. See also Figure 17.58. Shells and gladii redrawn and modified mainly from Bizikov, V.A., *Evolution of the Shell in Cephalopoda*, VNIRO Publishing, Moscow, Russia, 2008, *Conoteuthis* (D'Orbigny, A.D., *Mollusques vivants et fossiles*, Adolphe Delahays, Paris 1845 [1845–1847]).

belemnites suggests that it is a new structure (Doguzhaeva et al. 2002b). The radula of the Carboniferous *Saundersites* from the Pennsylvanian of the USA has 11 teeth in each row and, largely on this basis, a separate order, Donovaniconida, was created for it (Doguzhaeva et al. 2007a).

Largely freed from the confines of the shell, circular muscles in the mantle could develop and the mantle cavity could increase its capacity by elongating and expanding dorsolaterally. Importantly, these developments also required a straight living chamber (Bizikov 2008), not a curved one as in *Nautilus* and ammonites. Thus, although the shell no longer provided protection, its internalisation and subsequent evolution of the pro-ostracum provided support for the head, funnel retractor muscles, viscera, and muscular mantle and also stabilised the head and funnel during jetting. The shell was thus transformed into an inner skeleton providing support, not only for the body but for the muscles of fins, mantle, and funnel, resulting in a new combination of locomotory structures. These key innovations led to the success of the coleoids. With the mantle free to become muscular and create a respiratory current and a jet stream for locomotion, further efficiency was obtained by fixing the ctenidial axes with membranes along their length making it possible (unlike the situation in *Nautilus*) to continue breathing during fast jetting.

The evolution of fins created a new means of locomotion and led to major changes in the coleoid muscular system. The head retractors, so significant in locomotion in *Nautilus*, are much smaller in coleoids and do not have a locomotory function. Instead in most living coleoids they form a supporting envelope around the visceral sac, and the main locomotory muscles are those associated with the mantle and fins. The funnel retractors, weak in *Nautilus*, are much stronger in coleoids as they need to control the position of the funnel during fast jet swimming.

As noted above, the first coleoids were probably vertically oriented (head down), like their bactritan ancestors. Some evidence for this lies in cephalopod statocysts (Bizikov 2008). In cephalopods, the statoliths are attached to the morphologically anterior wall of the statocyst chamber, not ventrally as in other molluscs (Budelmann 1975, 1988) (see Chapter 7). In cephalopods, the anterior wall of the statocyst in a horizontally oriented animal is the equivalent of the ventral wall in a vertically oriented animal.

Various schemes of coleoid phylogeny that incorporate extinct taxa have been devised (e.g., Engeser & Bandel 1988; Pignatti & Mariotti 1996; Kluessendorf & Doyle 2000; Bizikov 2008; Nishiguchi & Mapes 2008; Donovan & Fuchs 2012; Kruta et al. 2016; Sutton et al. 2016; Strugnell et al. 2017; Tanner et al. 2017; Fuchs & Schweigert 2018); the one we have adopted is detailed in the Appendix. The relationships of the major groups have been clarified in these recent studies, in large part as a result of the study of high-quality fossils for some key taxa.

Of the nine or ten ordinal-level taxa usually recognised in coleoids, four or five are extinct (House 1988). The extinct groups all have internal phragmocones and

include the Aulacoceratida, Phragmoteuthida, Belemnitida, Belemnoteuthida, Donovaniconida, Diplobelida, and Hematitida, the latter sometimes being included within the Aulacoceratia (Jeletzky 1966b; Engeser & Reitner 1981; Teichert 1988). Of the living orders, the best-known fossil record is, unsurprisingly, for Sepiida and Spirulida.

The oldest undoubted coleoids include *Hematites*, *Paleoconus*, and *Bactritimimus* (Flower & Gordon 1959; Gordon 1964), from the Mississippian of the USA, for which the order Hematitida was created (Doguzhaeva et al. 2002a). These have a short living chamber apparently without a pro-ostracum, a key innovation in later coleoids (see above). Aulacoceratids are generally characterised by a long tubular body chamber, but *Hematites* has a short body chamber, one feature used by Doguzhaeva et al. (2002a) to place Hematitidae in a separate order; additionally the external aragonitic layer does not extend posteriorly as a rostrum. Typical aulacoceratids have longer chambers and a smaller apical angle than belemnites. The rather scant fossil record suggests these early coleoids underwent a considerable radiation. The early appearance of shell thickening resulted in several lineages (Hematitidae, Aulacoceratidae, Xiphoteuthidae, Donovaniconidae; one of which gave rise to the belemnites), which used this structure, including the development of a posterior rostrum, to achieve horizontal swimming.

Both belemnites and aulacoceratids have a rostrum, a ventral siphuncle in the phragmocone, a protoconch with a closing membrane, and an elongate tubular body. However, the belemnite innovations were to reduce the living chamber and form a pro-ostracum.

The Phragmoteuthida may be a stem group of the modern coleoid lineages (Doyle et al. 1994), other than Spirulida (Bizikov 2008). The gladius-bearing *Glochinomorpha* (Cisuralian, Permian) is somewhat enigmatic as it does not appear to have been derived from a phragmoteuthoid or belemnite ancestor (Doguzhaeva & Mapes 2015).

Even a reduced phragmocone, as in the phragmoteuthids, restricted the development of an efficient mantle–funnel jetting apparatus. This is because the mantle muscles terminate on the shell anterior to the phragmocone, so losing this structure; the major reduction of the conotheca enabled the muscular mantle to spread to the posterior end of the body, significantly increasing the volume of the mantle cavity (Bizikov 2008).

Coleoid shells were calcified as long as they retained a phragmocone, but with its loss in different coleoid lineages, the shells became decalcified. Fossil coleoids that appear to have a gladius but no phragmocone are mostly included in Vampyromorpha, but their relationships are often uncertain (e.g., Young et al. 1998).

Octobrachian fossils are common, but only *Proteroctopus* and *Vampyronassa* (both Middle Jurassic) and *Palaeoctopus* (Upper Cretaceous) are the only known fossils in which at least some details of the body have been preserved. *Vampyronassa* had some features similar to *Vampyroteuthis*, including a single row of suckers and cirri on the arms, an interbrachial web, fins, and two filaments. *Proteroctopus* had suckers without rings or stalks and otherwise generally resembled octobrachians, while *Palaeoctopus*, with a conchyloin gladius composed of two parts, is thought to be the sister of the incirrate octopods (Bizikov 2004, 2008; Sutton et al. 2016) or sister to the Octobrachia (Kruta et al. 2016) (Figure 17.58).

17.13.1.1 Belemnitidia (The Belemnites)

The bullet-like fossilised rostra of the shells (sometimes called 'guards' or 'lightning bolts') of belemnites (Lower Jurassic to end of Cretaceous) are by far the most conspicuous and common fossil remains of coleoids and are represented by the largest number of extinct coleoid taxa. It is generally assumed that all belemnites were horizontal swimmers and active carnivores.

Belemnite rostra are common in the Jurassic through to the end of the Cretaceous; belemnites coexisted with ammonites and, like them, became extinct at the End Cretaceous. They were squid-like in body shape but, based on fossils, had ten short arms about equal in length (i.e., they lacked long retractile tentacles). The arms had a row of suckers (Fuchs et al. 2010) which bore two (rarely one) rows of sharp hooks (*Onychites*) (Figure 17.9). The hooks are different in structure and origin from those found in some living squid (Engeser & Clarke 1988). In belemnites and other early coleoids (aulacoceratids, phragmoteuthids, and diplobelids) the hooks are probably homologues of the cirri seen in some octobrachians (Fuchs et al. 2013a) (Figure 17.9) rather than being formed from the chitinous ring in the suckers as they are in living hook-bearing squid (Figure 17.10). They also had an ink sac, lateral fins, a similar beak and radula to those of modern coleoids, and there is evidence that they also had similar eyes. The radula of only one belemnite is known, and that had nine elements (seven teeth and two marginal plates) in each row (Klug et al. 2016).

Complete belemnite shells are rarely preserved. The internal shell differed markedly from the gladius of modern squid in being calcified, robust, and chambered (Figure 17.59). Apically the shell was surrounded by the bullet-shaped, heavy calcite rostrum thought to have been a counterweight to the head and arms, assisting the animal to remain horizontal in the water. The typical belemnite rostrum, uniquely formed from radially deposited calcite, is usually the only part found as a fossil. At the front end of the rostrum is a conical hollow called the *alveolus*, which housed the more delicate phragmocone. The phragmocone is homologous with the external shells of other cephalopods and, like them, was chambered, conical, and aragonitic. It provided buoyancy, had a marginal (ventral) siphuncle, and the rostrum acted as a counterbalance. The phragmocone was enveloped by the *conotheca*, which extended anteriorly and dorsally as the leaf-shaped, mainly chitinous pro-ostracum (Figures 17.59 and 17.60). This latter structure represented the dorsal part of the living chamber which was bounded laterally by the muscular mantle. The rostrum occupied a third to a fifth of the length of the animal, including the arms. For example, the largest known belemnite, the Silurian *Megateuthis*

gigantea, had a rostrum up to 46 cm in length with the whole animal estimated to be around three metres long. In some Upper Cretaceous belemnites the rostrum was partly replaced by an organic capsule (Doguzhaeva & Bengtson 2011).

The belemnite shell differed from that of the aulacoceratidans and phragmoteuthidans in having a fully developed pro-ostracum and narrowly (not widely) spaced septa and a wider phragmocone. Occasionally a calcareous tube (the *epirostrum*) (Figure 17.56) was attached to and enclosed the rostrum and was up to several times its length. Belemnoteuthidae had a weakly developed rostrum while that of typical belemnites was large and well developed.

The taxonomy of belemnites is based on characters such as the shape and size of the rostrum, its cross-sectional shape, the different types of grooves on the surface, and impressions made by blood vessels. In some forms (e.g., the Jurassic *Hastites* and the Cretaceous *Hibolites*), the alveolar part of the rostrum was calcified only in its basal part while the frontal part was organic and decomposed after death, giving the remaining calcareous part (the *pseudoalveolus*) a weathered appearance in fossils.

Belemnites were probably agile, fast-swimming predators, using jetting as well as fin swimming. They are so common as fossils that, like some squid, they probably formed shoals in coastal waters.

Belemnites were undoubtedly carnivores which fed on various marine prey such as fish. They lived near the shore and on the continental shelf and were probably neutrally buoyant. Stomach contents of ichthyosaurs and plesiosaurs have shown that they fed on belemnites. They were also fed on by some octobrachians, as indicated by their arm hooks in stomach contents (Keupp et al. 2010).

The belemnites were thought to have first arisen in the lowermost Jurassic of Europe but are now known from Upper Triassic fossils from Japan and China (Iba et al. 2012, 2014a, 2014b). They were a very successful group through the Mesozoic and became extinct at the end of the Cretaceous.

Other Cretaceous taxa gave rise to the modern coleoids, although some details of the evolution of the non-belemnite coleoids remain poorly understood. Following the extinction of the belemnites, the modern coleoids underwent a major Cenozoic radiation, but this replacement had already started to occur in the north Pacific in the Late Cretaceous (Iba et al. 2011).

17.13.1.2 Decabrachia

The main distinguishing character of the Decabrachia (= Decapodiformes) is that in addition to eight arms they possess a pair of long, retractile tentacles. A molecular analysis of this group suggested that it may consist of only two orders rather than the four usually recognised, these being Sepioidea (containing Myopsida, Sepiidae, Sepiolidae, Idiosepiidae, *Spirula*) and Teuthoidea (Strugnell et al. 2006). In that analysis, *Spirula* grouped with Sepiidae, with which it shares some synapomorphies (Strugnell et al. 2005; Lindgren & Daly 2007), but Lindgren (2010) later found that *Spirula* was variously placed in a grouping containing Sepiida and Myopsida.

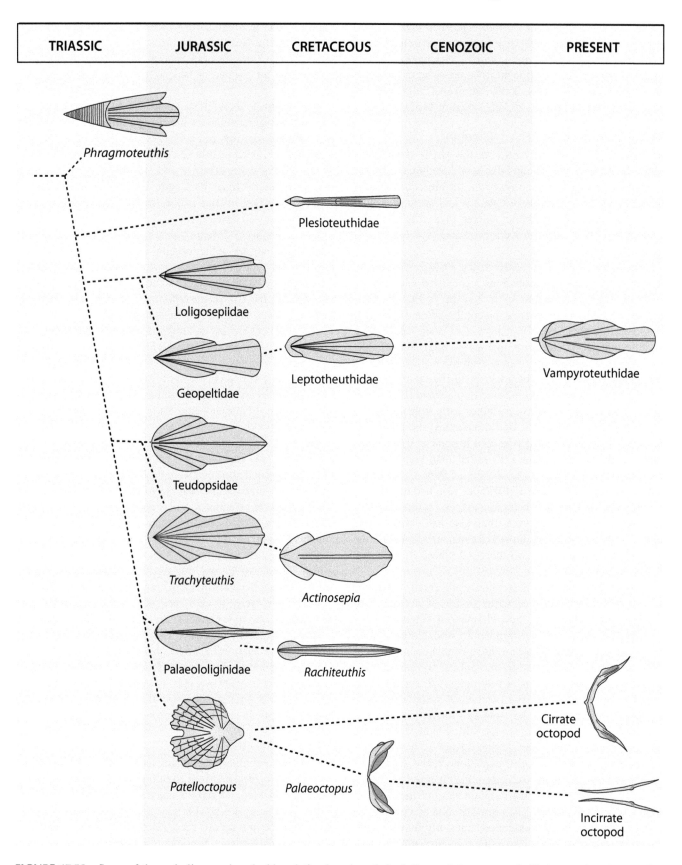

FIGURE 17.58 Some of the main lineages in coleoid evolution based on their shell morphology – part 2. Phylogeny done in consultation with Dr D. Fuchs. See also Figure 17.57. Shells and gladii redrawn and modified mainly from Bizikov, V.A., *Evolution of the Shell in Cephalopoda*, VNIRO Publishing, Moscow, Russia, 2008; *Patelloctopus* (Fuchs, D. and Schweigert, G., *Pal. Z*, 92, 203–217, 2018), *Actinosepia* (Fuchs, D. and Weis, R., *Neues Jahrb. Geol. Paläontol.*, 249, 93–112, 2009); *Rachiteuthis* (Fuchs, D. and Larson, N.L., *J. Paleont.*, 85, 234–239, 2011).

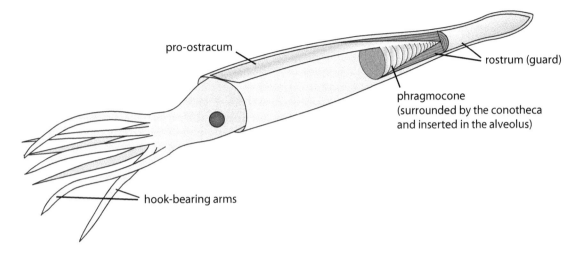

FIGURE 17.59   A generalised belemnite. Redrawn and modified from www.ucmp.berkeley.edu/taxa/inverts/mollusca/cephalopoda.php. Fins were present in the posterior part of the animal but are not shown here.

As shown by fossils, these latter two groups have had a long and separate evolutionary history. Using haemocyanin sequences, Warnke et al. (2011) estimated that the Spirulida–Sepiida split occurred about 150 Mya (latest Jurassic). In a review of molecular data concerning cephalopod phylogeny, Allcock et al. (2014) recognised five ordinal-level taxa: Spirulida, Sepiida, Sepiolida, Idiosepiida, and Myopsida with unresolved relationships within the Decabrachia. Two additional orders, Oegopsida and Bathyteuthida, formed a clade.

17.13.1.2.1   Sepiida

The traditional coleoid classification has this group containing Sepiidae (cuttlefish), Spirulidae, Idiosepiidae, Sepiadariidae, and Sepiolidae (dumpling squid) (e.g., Naef 1921, 1923). We include in this group only the cuttlefish and exclude the dumpling squid and *Spirula* following Allcock et al. (2014).

A number of early molecular studies and some morphological evidence suggested there is a closer relationship between Myopsida and Sepiida than with the other squid (Oegopsida) (e.g., Strugnell & Nishiguchi 2007; Nishiguchi & Mapes 2008), and this relationship was also suggested by Bizikov (2008) based on morphological data. However, recent molecular studies (Strugnell et al. 2017; Tanner et al. 2017) and morphological cladistic analyses (Kruta et al. 2016; Sutton et al. 2016) suggest that myopsidans are sister to the Oegopsida.

The position of Idiosepiidae is controversial, with molecular results not definitive. For example, Bonnaud et al. (2005) and Strugnell et al. (2017) suggested affinities with the oegopsids, Lindgren (2010) found it was grouped with sepiids, myopsidans, and sepiolids, and Tanner et al. (2017) found it was sister to *Euprymna* (Sepiolidae). It is now sometimes treated as a separate ordinal group (e.g., Akasaki et al. 2006; Allcock et al. 2014) and is treated as such here (see Appendix), a position also supported by recent morphological cladistic analyses (Kruta et al. 2016; Sutton et al. 2016).

The family Sepiidae (cuttlefish) is comprised of over 100 living species with the highest species richness (35) in Australian waters; the largest species can reach a metre in body length (mantle length up to 56 cm). They are benthic to benthopelagic, with most species found in shallow coastal water or on continental shelves and a few as deep as about 500 m. They are not found in cold waters; interestingly, they are absent from the Americas, and there is only one record from New Zealand. Cuttlefish are commercially fished in several parts of the world, and the cuttlebones are used for a calcium supplement (notably for caged birds) and as a mould for casting metal for jewellery.

The shell has ensured that there is a good fossil history with extinct ancestors including *Belosepia* and *Ceratosepia* (Haas 2003; Fuchs & Iba 2015), which give insights regarding the evolutionary transition to the cuttlebone. This was achieved by exponential growth of an endogastrically inrolled phragmocone (that lacked a pro-ostracum) to achieve a dorsally situated pro-ostracum-like structure by compressing and reducing the ventral shell wall and siphuncle (Haas 2003; Fuchs & Iba 2015).

A distinctive feature is the shell (sepion or cuttlebone) (Figure 17.61) which, because it floats, is frequently found washed ashore. The shell is very different from the simple phragmocones seen in some extinct coleoids. The dorsal wall is greatly expanded resulting in the closely spaced septa being strongly inclined. The chambers are subdivided by very thin sublamellae, and calcareous pillars act as supports in the chambers, keeping the thin septal walls apart (Figure 17.61). The septa cannot expand below so are tightly packed posteroventrally, and the viscera are accommodated beneath the broad shell and its posterior projection, the outer cone. In some parts of the shell there is no calcification with only organic pellicles secreted. Laterally, the septa converge to form the inner cone and posteriorly to form the ventral process. The extremely wide siphuncle forms much of the ventral surface. Because of these modifications to the shell, the centre of gravity in cuttlefish has moved to the mid-point of the animal, providing an efficient system for manoeuvrability and buoyancy.

The largest cuttlefish are *Sepia apama*, *S. latimanus*, and *S. officinalis* which all reach about half a metre in mantle

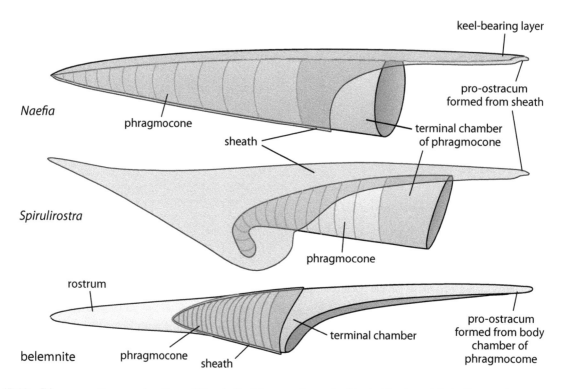

FIGURE 17.60 Diagrammatic reconstructions of the shell of three extinct coleoids. In *Naefia* and *Spirulirostra* the sheath is extended anteriorly to form the pro-ostracum and the terminal (body) chamber is tubular, whereas in belemnites the pro-ostracum is formed by the anterior extension of the dorsal part of the terminal chamber. Redrawn and modified after Fuchs, D. and Tanabe, K., Re-investigation of the shell morphology and ultrastructure of the Late Cretaceous spirulid coleoid *Naefia matsumotoi*, pp. 195–207, in Tanabe, K., Shigeta, Y., Sasaki, T., and Hirano, H. (eds.), *Cephalopods Present and Past, 8th International Symposium*, University of Bourgogne, France, 30 August–30 September 2010, Tokai University Press, Tokyo, 2010.

length, but the southern Australian *S. apama* has the largest body weight (Reid et al. 2005).

Although several generic names have been proposed, many classifications include only one genus (*Sepia*) while others recognise two or three additional genera (e.g., *Metasepia*, *Sepiella*).

Members of the superfamily Sepiolidea (variously called bobtail squid, dumpling squid, or stubby squid) have no shell, a more rounded mantle than cuttlefish, and are small (mantle length being between 1 and 10 cm). They are found in shallow coastal waters of the Indo-Pacific, and the two families contain over 70 species. A possible ancestor is the Eocene *Belosepiella* (Haas 2003).

The Idiosepiidae, with only about eight recognised species, contains the smallest Decabrachia, with males of some species having a mantle length of only 6 mm and females 8 mm. Their elongate body has small, posterior fins, and, uniquely, they have a dorsal attachment organ with which they cling to algae or seagrasses. They live in shallow water in the Indo-West Pacific and are often found in embayments and the lower parts of estuaries. Thus, unlike other cephalopods, they can tolerate moderate salinity fluctuations (Nugranad et al. 2005).

17.13.1.2.2   Spirulida

This group comprises only one living species and a few extinct taxa. *Spirula spirula* occurs almost world-wide and is oceanic, living in deep water where it lies vertically with its

head down. There is evidence of genetic differences between widely separated populations suggesting the possibility that some subdivision of this 'species' may be justified (Warnke 2007). Its rams-horn shell is commonly washed up on ocean beaches, but the living animal is rarely seen. The shell is coiled, endogastric, planispiral, and gyroconic. The animal has the decabrachian tentacle configuration, a vestigial radula, a photophore on the mantle edge, and, unlike *Sepia*, the eyes lack a cornea (the oegopsid type – Figure 17.42) (Warnke & Keupp 2005; Warnke 2007). The lifespan is around 20 months (Warnke 2007). The phylogenetic position of this group has stimulated much debate. Some workers support the traditional close relationship to sepiidans (e.g., Young et al. 1998), and in some recent classifications (e.g., Bizikov 2008) it is placed in its own order (as we do here), although some molecular analyses (Strugnell et al. 2006; Lindgren 2010) suggest a relationship with Sepiidae and Myopsida. A recent morphological cladistic analysis suggested that *Spirula* is closely related to the oegopsid *Gonatus* and the sepiolid *Rossia* and is part of a decabrach clade mainly containing Oegopsida (Sutton et al. 2016).

Typical spirulids are found only as far back as the Paleocene, but related orthoconic Cretaceous taxa are known (Fuchs et al. 2012a). Earlier reports that some Cretaceous forms were coiled are false, and coiling did not occur until the Cenozoic (Fuchs & Košťák 2015). Other supposed ancient

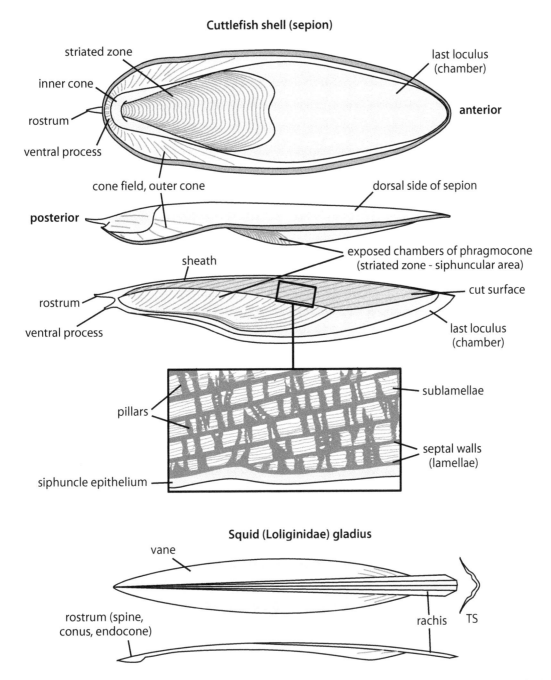

FIGURE 17.61 The shell (sepion or cuttlebone) of the cuttlefish *Sepia* (showing dorsal and ventral view and a longitudinal section with detail of the chambers) and a gladius of a loliginid squid. Redrawn and modified from the following sources: upper two figures (Bizikov, V.A., *Evolution of the Shell in Cephalopoda*, VNIRO Publishing, Moscow, Russia, 2008), longitudinal section (Haas, W., *Berliner Paläobiologische Abhandlungen*, 3, 113–129, 2003), detail in box (Denton, E.J. and Gilpin-Brown, J.B., *J. Mar. Biol. Assoc.* U.K., 41, 365–381, 1961b), and gladius (Okutani, T. et al., *Cephalopods from Continental Shelf and Slope around Japan*, Japan Fisheries Resource Conservation Association, Tokyo, 1987).

spirulids, *Naefia*[25] and *Adygeya*,[26] are superficially similar shells to those of some aulacoceratids in having a straight, calcified phragmocone and a rudimentary rostrum. Unlike members of the Aulacoceratia, spirulids lack a nacreous layer, although there is a nacre-like lamellar prismatic layer (Bandel & Stinnesbeck 2006), and their living chamber was reduced to a narrow pro-ostracum. Based on shell fragments from the Upper Carboniferous, *Shimanskya* has been considered as a possible early member of the *Spirula* lineage (Doguzhaeva et al. 1999), but this, like other Carboniferous records of Spirulida, is erroneous (D. Fuchs, pers. comm., Oct 2018).

[25] Bandel and Stinnesbeck (2006 p. 21) disputed this relationship and suggested that both *Naefia* and *Groenlandibelus* 'remain isolated regarding their taxonomic position among early coleoids'. Fuchs and Tanabe (2010) found *Naefia* shared similarities with belemnites (presence of tabular nacre) and Spirulida, while Fuchs et al. (2013b) treated them as Spirulida.

[26] D. Fuchs (pers. comm., Oct. 2018) indicated that this is based on a specimen that is not a coleoid.

The life history of *Spirula* is poorly known (Lukeneder et al. 2008). Shell isotope data suggest juveniles hatch below 1,000 m and migrate to shallower, warmer waters (400–600 m) as adults, later moving back down into deep, cold water (Lukeneder et al. 2008).

The *Spirula* shell superficially resembles that of some ammonites, but the siphuncle is marginal, on the inner side of the coil, and the sutures are simple. The shell is coiled endogastrically while ammonites are exogastrically coiled. Ammonite shells have four prismatic layers, but there are only two in *Spirula*. Nevertheless, Warnke and Keupp (2005) suggested that *Spirula* was a useful model for studying ammonite embryonic development, and there are similarities in the formation of the initial chambers (protoconch) of ammonites and *Spirula*. They have a similar mode of mineralisation (Warnke & Keupp 2005) and, probably, life history strategy (Lukeneder et al. 2008).

According to Bizikov (2008), the first supposed spirulids share similarities in their shell structure with aulacoceratians. The Cretaceous Adygeyidae and Groenlandibelidae, like *Spirula*, probably also had a head-down orientation and were planktonic with an elongated body. It is unlikely that the spirulids evolved from belemnites because the rostrum differed significantly and was more similar to that of aulacoceratids (Jeletzky 1966b). Also, importantly, the protoconch of spirulids had a caecum and a prosiphon (Figure 17.56), both of which are lacking in belemnites and aulacoceratids, but a caecum is found in *Cyrtobelus* (Groenlandibelidae), thought to be an early spirulid lineage (Fuchs et al. 2012a), which itself may have been derived from a diplobelidan (Fuchs et al. 2012b).

Some Paleogene coleoids that have been treated as spirulidans (Vasseuriidae, Belopteridae, and Spirulirostridae) had a massive rostrum and swam horizontally. Of these, the Vasseuridae had derived septal morphology, and their shell was straight except for an initial endogastric curve. Haas (2003) doubted the spirulid relationships of *Vasseuria* but instead suggested a loliginid relationship. In the families Belemnopseidae, Belopteridae, and Spirulirostridae endogastric growth of the shell increased together with a reduction of the pro-ostracum and the zone of mantle attachment shifting to the outer side of the phragmocone. Adult *Vasseuria* and *Spirulirostra* had a large rostrum that probably enabled horizontal swimming and presumably supported fins (Haas 2003), although juveniles had a weak or absent rostrum and presumably had head-down orientation. The larval shell in Belemnopseidae, Belopteridae, and Spirulirostridae was endogastrically coiled, and Bizikov (2008) suggested *Spirula* could be a paedomorphic form of Spirulirostridae, as it retains spiral growth through its life and there is no development of the rostrum.

17.13.1.3 Squid – A Grade Not a Clade – The Myopsida, Oegopsida, Bathyteuthoidea ('Teuthida')

Squid are diverse, with over 450 living species. These mostly streamlined decabrachians swim horizontally, have a distinct head and fins, and the shell is reduced to a horny gladius.

Both morphological (Bizikov 2008) and molecular phylogenetic analyses (e.g., Strugnell & Nishiguchi 2007; Nishiguchi & Mapes 2008; Lindgren 2010) have shown that squid (the 'Teuthida') are polyphyletic, and it is now apparent that the squid body form has evolved for pelagic life in several different decabrachian lineages (e.g., Bizikov 2008; Allcock et al. 2014).

The Myopsida includes two living families (see Appendix) that were traditionally grouped with Oegopsida in 'Teuthoidea' largely because of similarities in their gladii and tentacular clubs (e.g., Naef 1921, 1923). More recently it was suggested that Myopsida and Sepiida are related as they possess several shared characters, including a cornea, suckers with circular muscle, aspects of shell and beak morphology, a giant fibre node in the stellate ganglion, tentacle pockets, benthic eggs, and a vena cava ventral to the intestine (Haas 1997, 2002; Young et al. 1998; Bizikov 2008). This relationship has obtained equivocal support from molecular studies (e.g., Carlini et al. 2000; Lindgren et al. 2004; Strugnell et al. 2005; Strugnell & Nishiguchi 2007; Lindgren 2010).

Myopsida contains some of the most familiar squid, including *Loligo* and its relatives, these being diverse and abundant in coastal waters, with some the basis of important fisheries. Many are small, all less than 1 m, but members of the genus *Pickfordiateuthis* are tiny, reaching a maximum mantle length of only around 20 mm.

The Oegopsida have a gladius, but their eyes lack a cornea. While some earlier molecular studies suggested that Oegopsida may be polyphyletic, the more comprehensive analyses of Strugnell et al. (2005) and Strugnell and Nishiguchi (2007) supported their monophyly. These and some other studies show that the Bathyteuthidae, members of which possess some primitive characters, is the sister taxon to Oegopsida.

The fossil record for squid is poor, in part because, unlike octobrachians, their soft bodies do not fossilise except in exceptional circumstances and because many were presumably oceanic. The reason for the lack of fossilised remains is largely related to their use of ammonia to facilitate buoyancy, with the ammonia-rich tissues unable to generate a sufficiently low pH for calcium phosphate to replace soft tissue (Clements et al. 2017).

The replacement of the shell with the decalcified gladius provided a light, flexible internal skeleton for supporting the locomotory musculature. This innovation provided loligosepiids with a competitive edge over contemporary cephalopods (aulacoceratids, belemnites, and phragmoteuthids), and they diversified in the Lower Jurassic.

Arkhipkin et al. (2012) showed that organic layers in the gladius cone in squid are homologous with the organic component of the septa in the phragmocone of extinct coleoids, although the siphuncle and chambers have been lost. They also argued that the rostra of the gladius are homologous with the belemnite rostrum, perhaps indicating a phylogenetic relationship. In the Cranchiidae and Ommastrephidae the funnel retractor muscles lost their connection with the gladius and became attached to the mantle wall (Bizikov 2008). These two lineages have a very different type of gladius: wider in the slow-moving Cranchiidae and very narrow in the actively swimming Ommastrephidae.

Cranchiids are mainly slow-moving meso- and bathy-pelagic forms with adaptations to achieve neutral buoyancy. They are a very diverse group found in all oceans, and a few reach a very large size, notably the colossal squid (see Section 17.13.1.5). This large family has two groups: Tailless (all Cranchiinae and some Taoniinae) and tailed forms (most Taoniinae). The tailless cranchiids have retained the more primitive gladius with a wide cone, and they passively float horizontally or head down. Their large coelomic cavity is filled with ammonium chloride, giving them neutral buoyancy. The mantle has lost much of its musculature, and the reduced fins are used mainly for maintaining orientation. Because the gladius is no longer required for support it has become thin, narrow, and soft. The tailed cranchiids are more active swimmers, for which they use their long fins that require the support of the long, stiff gladius extending behind as a tail (*pseudocone*). The coelomic cavity is filled with ammonium chloride but is relatively smaller than that in Cranchiinae.

The mesopelagic 'needle-tailed' squid such as Chiroteuthidae, Mastigoteuthidae, and Joubiniteuthidae – the chiroteuthid group of families of Young and Vecchione (1996) – were thought to have originated from a spirulid ancestor by Bizikov (2008), although the analysis of Lindgren (2010) placed this group in Oegopsida.

The Onychoteuthoidea contains 15 families of modern oegopsid squid (see Appendix), including the Enoploteuthidae, Lycoteuthidae, Gonatidae, and Onychoteuthidae, and occurs in all oceans. All have a somewhat similar slender gladius, called the 'onychoteuthoid type' by Bizikov (2008), which provides support for funnel retractor muscles and the large, powerful fins which are the main means of locomotion. They live in all oceans where they mainly occupy meso- and bathy-pelagic habitats, and they are often abundant. They evolved much later than ommastrephids and may have been excluded by them from the epipelagic zone. Some meso- and bathy-pelagic onychoteuthoideans (Ancistrocheiridae, Octopoteuthidae) have a reduced funnel apparatus, enlarged fins, and attain neutral buoyancy with ammonium vacuoles in the mantle, head, or arms. Although most onychoteuthoideans are small (Enoploteuthidae, Lycoteuthidae, Pyroteuthidae, some Onychoteuthidae), the group includes some of the most successful squid families. A number (Neoteuthidae, Architeuthidae, some Onychoteuthidae) have become large, and a few very large, including the 'giant squid' (see Section 17.13.1.5).

The family Thysanoteuthidae is represented by a single living species, the 'diamond squid' *Thysanoteuthis rhombus*, a large (mantle a metre long), muscular, diamond-shaped squid found mainly in tropical and subtropical oceans around the world. It is distinct from other oegopsids in morphology, including gladius structure. Thysanoteuthids appear to reduce or avoid competitive interactions with other pelagic squid and fish by feeding during the daytime between 400 and 700 m, mainly on small fish. Ommastrephids feed at night near the surface, preying on fish and squid. Interestingly, *Thysanoteuthis* never shoals, and they pair for life – as far as known, a unique situation in living cephalopods (Nigmatullin & Arkhipkin 1998).

Bizikov (2008), on the basis of similarities in the gladius, suggested that *Thysanoteuthis* may represent a direct modified descendant of the Mesozoic Loligosepiidae, but those similarities are convergent as that group, and many others Bizikov treated as ancestral decabrachians, have subsequently been shown to belong to the Octobrachia (Donovan & Fuchs 2016).

Arkhipkin et al. (2012) showed that the organic layers of the gladius cone in modern squid are probably homologous with the organic part of the phragmocone in shell-bearing extinct coleoids. Similarly, the well-developed rostrum in the gladius of onychoteuthids and the small rostrum in the gladius of ommastrephids and gonatids are probably homologous with the belemnite rostrum, suggesting a relationship between belemnites and at least some squid. These authors argued that the reduction of the phragmocone and rostrum followed by decalcification supports the idea that squid evolved in the deep sea (e.g., Jeletzky 1966a; Donovan 1977) and colonised the continental shelves after the extinction of belemnites at the end of the Cretaceous while also remaining diverse in their original deep sea habitat.

17.13.1.4 Octobrachia (= Octopodiformes)

Included here are the Octopoda, comprised of the incirrate octopuses (including the paper argonauts), and the mainly pelagic cirrate octopods, as well as the Vampyromorpha. They all have four pairs of arms, and the internal shell is reduced to a gladius (Vampyromorpha), a U- or saddle-shaped, non-calcified structure (cirrate octopods), or a pair of stylets, or lost (Octopoda). They share similar sperm morphology (Healy 1989), embryological development (Naef 1928; Young & Vecchione 1996; von Boletzky 2003b), and radially symmetrical inner structure of the suckers, while decabrachian suckers are bilaterally symmetrical (Young & Vecchione 1996; Lindgren et al. 2004). The Vampyromorphina has some features that align it with decabrachians, for example, the gladius is similar to that of some squid (Bizikov 2008), although molecular studies have generally supported a monophyletic Octobrachia (e.g., Allcock et al. 2014).

The Jurassic Plesioteuthidae, possibly derived from the geopeltids, were squid-like, with a narrow gladius suggesting speed and manoeuvrability and a nektonic lifestyle. Narrowing of the gladius occurred in other members of this lineage in the Jurassic and Cretaceous. Bizikov (2008) suggested that another lineage from the geopeltids led to the Middle Jurassic family Mastigophoridae. *Mastigophora* is known from unusually well-preserved specimens with eight arms, uniserial ringless suckers, and fins (Fuchs 2014), while previous interpretations erroneously recognised a pair of tentacles, incorrectly placing it in the Decabrachia (Vecchione et al. 1999).

17.13.1.4.1 Vampyromorphina

This subordinal group comprises what is currently recognised as one living species (*Vampyroteuthis infernalis*) and several extinct taxa.

Vampyroteuthis lives in the oxygen minimum layer between about 500 and 1,500 m in oceans around the world. It is the only cephalopod known to spend its entire life cycle in this zone.

In some respects, the morphology of *Vampyroteuthis* is intermediate between octopuses and squid, and it probably has more features that resemble ancestral coleoids than any other living cephalopod. The body is similar in shape to a finned octopod, being short, with a pair of fins. There are four pairs of webbed arms (as in octopods) and a pair of velar filaments ('tentacles'). These unique, retractable, sensory filaments are the analogues of squid tentacles but are not homologues as they are derived from a different pair of arms (see Section 17.3.2 and Figure 17.8). The other four pairs of arms are webbed along most of their length and have, on their inner surfaces, a medial row of short suckers (with slightly narrower bases) and two rows of cirri. The gladius is large and uncalcified. The nervous system also shows some primitive features while other features are shared with octopods. The eyes are large relative to the rest of the body, and there is a pair of photoreceptors on top of the head. There are many photophores on the head, fins, arms, and mantle, and these are especially well developed on the arm tips and the fin bases. Gelatinous tissues rich in ammonium enable *Vampyroteuthis* to achieve neutral buoyancy, and, while fins provide most of the means of propulsion, it is also capable of jetting, although it has weak musculature. Its position in the water column is maintained using sophisticated statocysts that are intermediate in structure between those of octopods and decabrachians (see Chapter 7).

Vampyroteuthis is unlike any other octobrach in reproducing multiple times (Hoving et al. 2015). The eggs are about 8 mm in diameter and brooded in the arms. They hatch as squid-like paralarvae with a pair of tiny posterior fins (Young & Vecchione 1999).

There is no ink sac, but luminous fluid can be released from the arm tips when it is threatened (Robison et al. 2003). This cloud contains glowing blue orbs and can last several minutes, and the display may be accompanied by writhing of the glowing arms and a rapid escape. In another threat response, the webbed arms are held over the head to mask the photophores, displaying a seemingly larger body covered with spine-like cirri. The luminescent arm tips are held well above the head so that any attack aimed at them would be directed away from the vital organs.

They are reported to feed on detritus, including the remains of small animals such as crustaceans and cnidarians (Hoving & Robison 2012). In turn, they are fed on by deep-water fish and deep-diving whales and pinnipeds. When teleost fish appeared in the mid-Mesozoic, vampyromorphans probably survived by moving to an oxygen-poor bathy-pelagic refuge.

Loligosepiidae are vampyroteuthid fossils from the Lower Jurassic that had eight hooklet-free arms (Fuchs et al. 2013c).

Vampyronassa rhodanica from the Middle Jurassic (Callovian) of France (Fischer & Riou 2002) differs from *Vampyroteuthis* in the first (dorsal) pair of arms being longer than the others, the funnel more highly developed, and the body more elongate. Based on these records, the vampyromorphan lineage is perhaps 200 Ma old, but another study based on fossil and molecular data has estimated that the lineage may have evolved 252 Mya (Strugnell et al. 2006). Although the gladius of the squid-like ancestral octobrachians was chitinous but thick, the gladius of *Vampyroteuthis* is thin but retains the plesiomorphic shape and includes most of the muscle attachments and support for the fins (Bizikov 2008).

Fossilised remains of *Leptotheuthis* (Leptotheuthidae) and *Boreopeltis* (Plesioteuthidae) show that these were large, predatory, and squid-like, with a muscular mantle and large fins (Naef 1922; Jeletzky 1966b; Fuchs & Larson 2011; Fuchs 2014). Fuchs and Larson (2011) showed that *Leptotheuthis* was a member of the Loligosepiina, an extinct suborder of octobrachians. The classification of extinct octobrachians is detailed in the Appendix, and Sutton et al. (2016) carried out a cladistic analysis of these taxa.

17.13.1.4.2 Octopoda

There are two very distinct groups of octopods, the Incirrata or finless octopods and the Cirrata or finned octopods. The former comprises the familiar octopods including *Octopus* and similar taxa while the cirrate octopods are found in the deep sea. These groups were, however, not recovered as monophyletic in one molecular study (Carlini et al. 2001), and subsequent analyses have only included two (Strugnell et al. 2005, 2014) or three (Lindgren et al. 2012) cirrate taxa, although these form a monophyletic sister taxon to the incirrate octopods. Thus, the detail of their relationships currently remains unsettled.

The Cirrata are pelagic to benthopelagic and rely on swimming with fins. There are two lineages, one adapted to semi-benthic life with their head downwards and their body flattened (*Grimpoteuthis* and *Opisthoteuthis*). In these taxa the gladius is more primitive than in the other lineage which includes benthopelagic forms like *Cirroteuthis* and *Cirrothauma*, which rely more on fin swimming, and the gladius supporting the fins is wide, thick, and rigid (Bizikov 2008). The first incirrates were benthic, oriented with their mouth downwards, crawled using their arms, and adopted cryptic habits. They probably had a bipartite gladius vestige, whereas the common ancestor of Cirrata and Incirrata had a U-shaped gladius vestige (Fuchs 2016) and may have had fins, but once a benthic habitat was established, the fins were lost, and the flexible body became the key to their success. The gladius separated into paired rods, the stylets (Bizikov 2008). These changes are seen in one of the earliest (Late Cretaceous) fossil incirrates, *Palaeoctopus*, although it retains some cirrate features: fins (albeit reduced), and a thicker but bipartite gladius. While cirri have been reported on the arms (Haas 2002), they are apparently absent (Fuchs et al. 2009). The supposed cirrate palaeoctopod *Pohlsepia* from the Carboniferous (Kluessendorf & Doyle 2000), a sister to the Octobrachia in the cladistic analysis of Sutton et al. (2016), is doubtfully an octobrach (Fuchs et al. 2009). The divergence of the octopod lineage into Cirrata and Incirrata occurred before the Late Cretaceous, as shown by *Palaeoctopus*, a taxon from the Santonian and included with the incirrate octopods, as well as earlier (Cenomanian) fossil stylets (Fuchs et al. 2009).

The shell remnants (stylets) seen in many living octopods were derived from the gladius through reduction of the middle part of the pro-ostracum (Haas 2002; Bizikov 2004, 2008; Fuchs & Schweigert 2018). In the octopod lineage, the gladius was transformed into a U-shaped structure like that of modern cirrates (Bizikov 2004, 2008). Further evolution saw the reduction and loss of the basal connection between the two lateral elements and, independently in several lineages, the complete loss of the shell.

Members of the Incirrata are poor jet-swimmers, but their very plastic body allowed them to utilise small spaces for hiding or to seek prey.

At least two lineages of incirrates became secondarily pelagic – the Bolitenoidea and Argonautoidea. The gladius is completely lost in the former and the more advanced members of the latter.

Although the remnant of the gladius (the stylets) no longer supported the mantle musculature, it continued to support the fins. Because the mantle lost its anterior support, it ceased to have a fixed length, and body plasticity was substantially increased. This latter change was a significant exaptation for benthic life, as was the ability to explore and manipulate objects with the arms and suckers. These changes could only have happened in slow-moving forms in which the anterior dorsal mantle margin was fused with the head, and the fins were the main means of locomotion (Bizikov 2008). In incirrates the function of the stylets changed from fin support to supporting the funnel muscles. The shell was progressively reduced and lost in some lineages with the most developed stylets being found in some shallow-water benthic Octopodidae. Reduction and loss of the stylets in deep-water octopods may be related to their reduced activity levels (Voight 1997; Bizikov 2004, 2008). Also, some muscular shallow-water octopods (*Ameloctopus*, *Hapalochlaena*) lost the stylets and the ability to jet-swim (Voight 1997).

Incirrate octopods live in a wide variety of marine habitats and range from benthic to pelagic. They are very diverse in tropical areas, especially coral reefs.

There are about 300 species of incirrate octopods and 45 species of cirrates. The cirrate (finned) octopods are the most abundant octopods in the deep sea and are mostly smaller (10–130 cm in length) than many incirrates (Villanueva et al. 1997), and they all lack an ink sac. Their webbed arms have two rows of cirri and a single row of suckers (Figure 17.9). At least one species (*Stauroteuthis syrtensis*) is bioluminescent (Johnsen et al. 1999), and one species of *Cirrothauma* is effectively blind (Aldred et al. 1983). They live in water depths of about 100 to over 7,000 m, a record depth for cephalopods (Voss 1988), and feed on small invertebrates. Spawning appears to be continuous and may extend over several years (Villanueva 1992). The large eggs are usually laid in capsules on the seafloor, and at least some taxa brood them, as does the incirrate octopod *Graneledone boreopacifica*, which, at 53 months, has the longest-recorded egg-brooding of any animal (Robison et al. 2014).

The lineage containing deep-sea incirrate octopuses is thought to have evolved in Antarctica around 33 Mya and radiated into the deep sea with the initiation of the global thermohaline circulation (Strugnell et al. 2008). This finding is of interest as it was the first molecular evidence showing that a deep-sea fauna from other ocean basins originated from the Southern Ocean.

17.13.1.5 Giant Squid and Octopuses

Several cephalopods, particularly some in the deep sea, reach a very large size and are amongst the largest living invertebrates. The giant squid, *Architeuthis dux* (family Architeuthidae), is the longest, with a mantle length up to about three metres and a maximum total length of about 13 m. Reports of larger sizes are erroneous (e.g., O'Shea & Bolstad 2008). This cosmopolitan temperate-ocean species lives in depths of around 400–600 m and was thought to be rather slow moving and to feed on relatively small prey.

The colossal squid, *Mesonychoteuthis hamiltoni* (family Cranchiidae) also has a mantle length of up to about three metres and a confirmed total length of 9–10 m. It may attain much greater body weight (up to 494 kg [McClain et al. 2015], compared with ~300 kg) than *Architeuthis*. It is probably a slower moving predator than *Architeuthis* (Roper & Jereb 2010), with hooked arms, and feeds on fish. It has a circum-Antarctic distribution, and there are occasional records from more temperate waters, including one from Victoria, Australia (Reid 2016). Reports of another cranchid, *Galiteuthis phyllura*, with a mantle length of up to 2.7 m may be incorrect (Roper & Jereb 2010), this species only reaching about half a metre in mantle length.

The largest octopuses have a mantle length of about 0.5–0.7 m and total length of around four metres for *Haliphron atlanticus* (O'Shea 2004), with a total body length (including arms) of possibly up to 6 m for *Enteroctopus dofleini* which has been recorded as attacking divers (Anderson et al. 2007).

18 Gastropoda I – Introduction and the Stem Groups

18.1 INTRODUCTION TO THE GASTROPODA

Gastropods include limpets, snails and whelks, abalone, shell-less slugs, and there are even a few with bivalved shells. They are by far the largest class of molluscs and are one of the most diverse groups of animals, not only in sheer numbers of families and species but in their range of form and habits and in the variety of habitats they occupy. Gastropods live in nearly all biotic habitats; their greatest diversity is in marine environments, but there are also many non-marine species, particularly in the terrestrial realm. They are the only molluscs to have become terrestrial, a habit acquired independently in many groups. Estimates of the numbers of gastropod species are from around 40,000 to 90,000. In size, gastropods range from about 600 mm in shell length (the giant whelk *Syrinx aruanus*) to tiny gastropods a little less than a millimetre. One of the largest terrestrial snails (the giant African snail *Achatina achatina*) has a shell about 180 mm long.

Gastropods are crucial ecologically and are also important economically as sources of food, and their shells are used as ornaments (see Chapter 10). Gastropods have also been significant in many ecological, evolutionary, biomechanical, physiological, behavioural, and biological studies.

Gastropods differ from other molluscan classes in having undergone 'torsion', a 90–180° displacement during development (Box 18.2) that shifts the mantle cavity from its initially posterior position to an anterior one. Typically there is only a single shell, and there is an operculum in the veliger larvae and, often, in adults. The visceral mass is frequently extended dorsally as a spiral, inside a spiral shell.

In this section, we provide a brief introduction to, and overview of, the gastropods. Treatments of each of the main groups of gastropods are given in this and the following two chapters. Some details of extinct groups of gastropods are provided in Chapter 13 and the Appendix.

18.1.1 PHYLOGENY AND CLASSIFICATION

BOX 18.1   THE MAIN GROUPS OF LIVING GASTROPODS

(Subclass) **Eogastropoda**

This group contains only the true limpets and their coiled ancestors.

(Infraclass) **Patellogastropoda** (= Docoglossa) – includes only the true limpets. They are nearly all marine and have limpet-shaped shells. The shell is never nacreous, but some have foliate shell structure that can resemble nacre. The larval shell is tubular, and larvae are lecithotrophic. An operculum is present in larvae but is absent in juveniles and adults. The single ctenidium (when present) is bipectinate, and there are sometimes secondary gills. The radula is docoglossate.

(Subclass) **Orthogastropoda**

This group includes all the remaining gastropods.

(Infraclass) **Vetigastropoda** (= Archaeogastropoda in part) – contains the keyhole and slit-limpets (Fissurellidae), abalone (Haliotidae), slit shells (Pleurotomariidae), the top shells (Trochidae), and several other families (all Vetigastropoda) and also includes many of the 'hot-vent groups' (Neomphalina) and the cocculinoid limpets (Cocculinina). All are marine and have coiled to limpet-shaped shells, with only one shell-less (a fissurellid). The shell is often nacreous, with paucispiral larval shells, and their veliger larvae are lecithotrophic. An operculum, present in taxa with coiled shells, is circular with a central nucleus. The paired or single ctenidium is usually bipectinate, and the radula is nearly always rhipidoglossate.

(Infraclass) **Neritimorpha** (= Neritopsina) – contains marine, freshwater, and terrestrial members with coiled to limpet-shaped shells; there is one slug species. The shell is never nacreous and is, like the multispiral to paucispiral larval shell, resorbed internally in most living taxa. The veliger larvae are frequently planktotrophic. The operculum is asymmetrical, usually with a peg on the inner side. The single ctenidium is nearly always bipectinate, and the radula is rhipidoglossate.

(Infraclass) **Caenogastropoda** (covered in Chapter 19) – a very large, diverse group including taxa as different as periwinkles, volutes, cowries, tun shells, whelks, and slipper limpets. The shell is very rarely absent and is never nacreous. The larval shell is paucispiral or multispiral, and, except in direct developers, the veliger larva is lecithotrophic or planktotrophic. The operculum is often asymmetrical and usually present in adults. The single ctenidium is monopectinate and rarely absent. The radula has seven teeth in each transverse row (taenioglossate) or fewer (stenoglossate, toxoglossate), rarely more (e.g., ptenoglossate).

(Infraclass) **Heterobranchia** (sometimes called Heterostropha) (covered in Chapter 20) – a very large taxon which includes two groups previously given 'subclass' status: the 'opisthobranchs', containing most of the 'sea slugs' and their relatives, and the 'pulmonates', containing the majority of land snails and some marine and freshwater groups. The shell is never nacreous and can be coiled, reduced and internal, lost (slugs), or limpet-shaped or, rarely, bivalved. Unlike the other groups of gastropods, the usually multispiral larval shell is typically sinistrally coiled, or its coiling is offset to the direction of coiling of the adult shell (heterostrophic). The veliger larvae may be planktotrophic or lecithotrophic, and direct development is common. Most lack an operculum in the adult. They do not have a ctenidium, but some have one or more secondary gills. The radula is very diverse within this group, both in tooth morphology and in the numbers of teeth in each row.

A detailed classification for all gastropod groups is provided in the Appendix.

The main groups of gastropods are briefly summarised in Box 18.1. This classification differs considerably from that used until the late 1980s. Thiele (1925, 1929–1935) divided gastropods into three subclasses, the Prosobranchia, Opisthobranchia, and Pulmonata, and further divided the Prosobranchia into Archaeogastropoda, Mesogastropoda, and Stenoglossa (later called Neogastropoda) (Figure 18.1). Detailed reviews of the history of gastropod classification can be found in Bieler (1992) and Ponder and Lindberg (1997), and this information is not repeated here. Bouchet and Rocroi (2005) and Bouchet et al. (2017) have provided useful listings of the multitude of family-group and higher names proposed for gastropods and summaries of their classification.

Haszprunar (1988a), Salvini-Plawen and Steiner (1996), and Ponder and Lindberg (1997) provided detailed phylogenetic analyses of the Gastropoda based on morphology

(Figure 18.2) and Colgan et al. (2003), McArthur and Harasewych (2003), Aktipis et al. (2008), Aktipis and Giribet (2010, 2012), and Zapata et al. (2014) (Figure 18.3) covered all the main gastropod groups in their analyses using molecular data. There is considerable disagreement in the ordering of the stem groups between morphological and molecular studies, with the placements in some taxa in molecular analyses being unstable and often highly unlikely. The only combined morphological and molecular analysis of gastropods was carried out by Aktipis et al. (2008) (Figure 18.4).

The old concept of 'prosobranchs' comprises several very distinct groups – Patellogastropoda, Vetigastropoda, Neritimorpha (which all comprise the paraphyletic archaeogastropods), Caenogastropoda (most of the 'Mesogastropoda' and all Neogastropoda) while a few 'mesogastropod' groups are basal members of Heterobranchia. The opisthobranchs and pulmonates (together making up the Euthyneura) comprise the great majority of heterobranchs. The Cocculiniformia (particularly the Cocculinoidea) and sometimes the hotvent Neomphalina are often retained as separate groups, but we include them in Vetigastropoda for the reasons given in Section 18.5.2.

While the legitimacy of these groups has now been confirmed by several morphological, developmental, and molecular studies, there is no general agreement regarding the ranks applied to them. The two main clades (Eogastropoda and Orthogastropoda) have been used as subclasses as we have done, but others assign subclass rank to the next highest category (Patellogastropoda, Vetigastropoda, etc.).

18.1.2 Morphology

Gastropods typically move about on a large 'foot' (hence their name, which means 'stomach-foot') and in that way resemble chitons. They evolved from bilaterally symmetrical ancestors but are typically asymmetrical externally and always internally. During 'torsion' (see Box 18.2 and Chapter 8) the gut and nervous systems are twisted through 90–180°, and the mantle cavity is rotated.

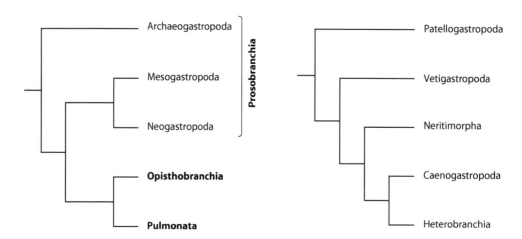

FIGURE 18.1   The relationships of the major groups of gastropods as previously (pre-1980 – left) and currently (right) understood.

Golikov & Starobogatov 1975

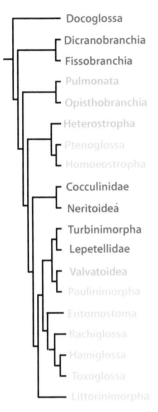

Barker 2001

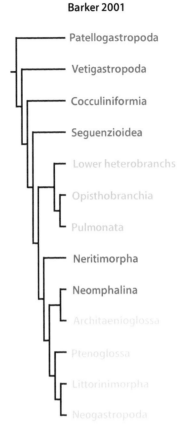

Key

Patellogastropoda
Vetigastropoda
Neritimorpha
Caenogastropoda
Heterobranchia

Haszprunar 1988

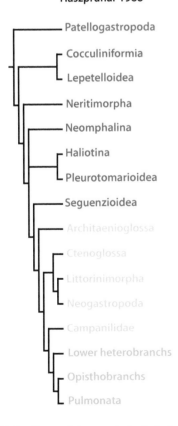

Ponder & Lindberg 1997

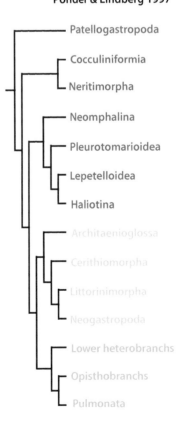

FIGURE 18.2   Some of the morphological phylogenies of Gastropoda.

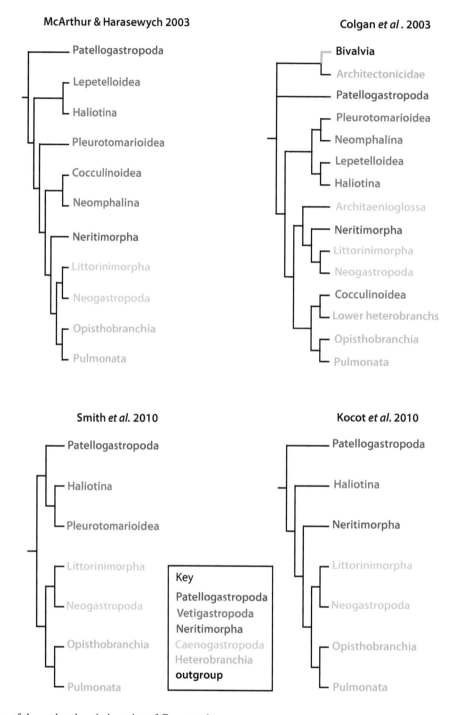

FIGURE 18.3   Some of the molecular phylogenies of Gastropoda.

BOX 18.2   TORSION

Torsion is the anti-clockwise rotation in the veliger larva of the visceral mass 90–180° relative to the anteroposterior axis of the head-foot complex, as shown in Figure 18.5.

See Chapter 8 for a more detailed discussion of this phenomenon.

The perceived advantages and disadvantages of torsion have been discussed by many workers, and various theories have been advanced as to its evolutionary advantages to either the larva, the adult or both (see Chapter 8 and Ponder and Lindberg 1997 for a summary). Despite the considerable differences between patellogastropods and orthogastropods, Ponder and Lindberg (1997) did not find any grounds for considering torsion as other than a single event that occurred at the commencement of gastropod evolutionary history.

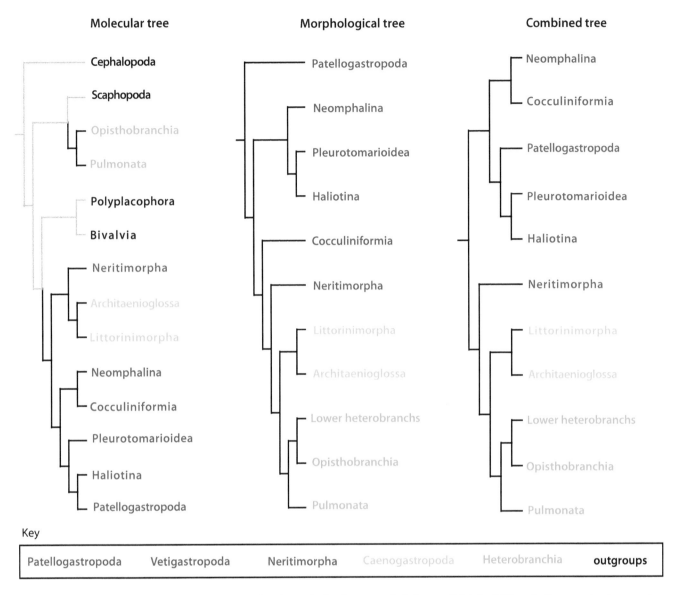

FIGURE 18.4 A combined morphological and molecular analysis of gastropod phylogeny (Aktipis, S.W. et al., Gastropoda: An overview and analysis, pp. 201–237, in Ponder, W.F. and Lindberg, D.R. (eds.), *Phylogeny and Evolution of the Mollusca*, University of California Press, Berkeley, CA, 2008).

Several accounts (Morton 1958a; Thompson 1967; Underwood 1972; Pennington & Chia 1985) refute the view that torsion bestows significant advantages on the larva, an idea initially advanced in a poem by Garstang (1929). Euthyneuran heterobranchs undergo at least some degree of 'detorsion' or, according to Page (1995), possibly never underwent the full 180° torsion, i.e., the 90° torsion they exhibit may be plesiomorphic.

Partly because of torsion and also because of coiling, a very different process, gastropods lose organs on the right side of their midline (or the left side in the rarely sinistrally coiled taxa). The results of torsion and coiling are the best-known asymmetries in gastropods, but numerous other asymmetries appear to be independent of the torsion process (Lindberg & Ponder 1996). With anopedal flexure (see Chapter 8), the mouth and anus are placed in juxtaposition. This is sometimes wrongly considered as part of torsion, but it is also seen in scaphopods, cephalopods, and, to a lesser extent, in bivalves (Lindberg 1985).

18.1.2.1 Shell, Operculum and External Body
18.1.2.1.1 Shell

Most gastropods have a single (often coiled) shell that is of considerable importance in protecting the animal from predation, desiccation, or other harmful effects and is often the main feature used in identification. Parallel trends in shell shape emphasise the plasticity of gross shell morphology – for example, in many lineages the shell has become limpet-like while in some the spire elongated, and in others the shell

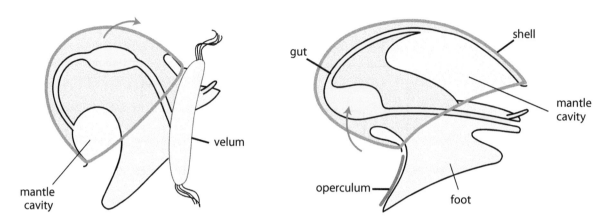

FIGURE 18.5 Diagrammatic illustration showing pre- and post-torsional states in a gastropod veliger with the latter having undergone 180° rotation. From several sources.

became globular. There are also convergent trends in features such as size, sculpture, and colour, all of which make the shell a useful tool in the hands of an expert, but they can provide many a trap for the inexperienced. The shell has been independently lost in several gastropod lineages, particularly within the Heterobranchia and typically with the adoption of a slug-like body.

The protoconch, also called the larval shell, is useful as an indicator of the group to which a gastropod belongs (Bandel 1982, 1997; Ponder & Lindberg 1997; Sasaki 1998; Bandel & Frýda 1999; Frýda et al. 2008a). It may consist of only the true larval shell laid down by the shell gland of the developing embryo, or an additional part is added during the swimming life of the veliger if that larva is feeding (i.e., is planktotrophic). This second phase of such protoconchs is usually called protoconch II or veloconch (Kesteven 1905), while the initial part is protoconch I. It is debatable whether or not protoconch II is homologous in the three major gastropod groups in which it is seen.

18.1.2.1.1.1 Shell Structure Gastropod shell structure (see Chapter 3 for details of shell structural types) is often rather diverse in the more basal taxa (Figure 18.6 and see the sections dealing with those taxa below), while crossed-lamellar shell structure is found throughout the higher gastropods.

18.1.2.2 Shell Muscles

Shell attachment muscles in other molluscs are typically paired, but most gastropods have a single 'columellar' muscle. Patellogastropods, neritimorphs, and many vetigastropods have two shell muscles. These are derived from both the left and right pretorsional mesodermal bands in patellogastropods (Smith 1935) and vetigastropod (Crofts 1955) (see Chapter 8 for details). Most caenogastropods and heterobranchs have a single shell muscle derived from the right pretorsional band, as is the single muscle in some 'higher' vetigastropods (Bandel 1982; Haszprunar 1985f; Ponder & Lindberg 1997), despite an earlier report (Crofts 1955) that

the muscle in vetigastropods was derived from the left pretorsional band. Paired shell muscles are known in very few caenogastropods, all of them members of the Velutinoidea, such as *Lamellaria*, *Trivia*, and *Velutina* (Fretter & Graham 1962). The paired muscles in velutinoideans may be heterochronous or due to a relatively minor change in gene expression, as it may also be in the heterobranch *Rissoella*, where a paired shell muscle has been reported (Fretter & Graham 1962; Haszprunar 1988a), although Simone (1995) described it as a bilobed muscle.

Members of the Haliotidae, like some other vetigastropods, have unequal shell muscles but differ from other gastropods in the right being larger.

18.1.2.3 Operculum

The operculum, a gastropod synapomorphy, is secreted by the posterior dorsal part of the foot and seals the shell opening (aperture) when the animal retracts into the shell. It is present in all gastropod veliger larvae, although it is secondarily absent in the embryos of some direct-developing land snails. An operculum is present in many adult gastropods but is lacking in juveniles and adults of many members of the Heterobranchia and all juvenile and adult patellogastropods.

The operculum is usually horny or, in some taxa, is calcareous as, for example, in turbinids. It is plesiomorphically made up of several spirals, with a central nucleus, but is modified in various lineages. Most caenogastropods, for example, have an operculum with a few rapidly increasing spirals (the paucispiral type) while in some taxa new material may be accreted around the entire edge resulting in concentric growth lines.

More details regarding the operculum are given in Chapter 3.

18.1.2.4 Head-Foot

There is a well-developed head with the mouth opening beneath a snout that is developed to varying degrees. In some taxa, the snout is introverted to form a proboscis (see Chapter 5).

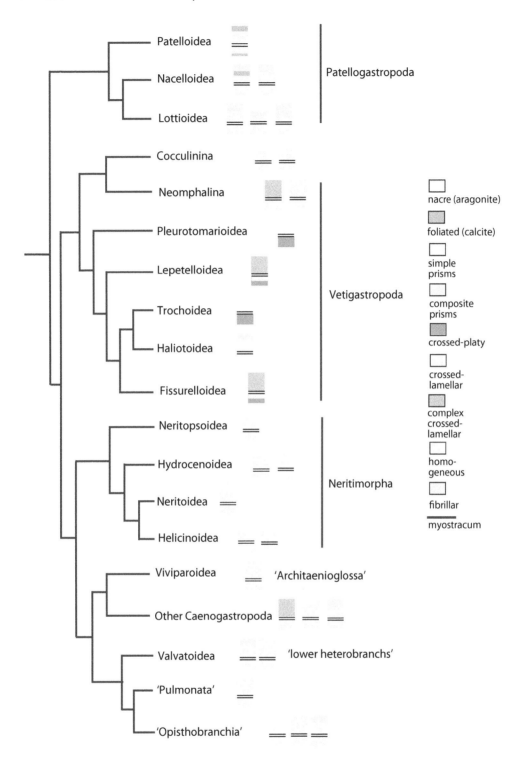

FIGURE 18.6 Gastropod shell structure and its phylogenetic distribution. The diagram represents shell layers with the outermost upper. Data from the following sources: Bøggild, O.B., *Det Kongelige Danske Videnskabernes Selskabs Skrifter, Naturvidenskabelig og Mathematisk Afdeling, ser. 9, 2*, 231–326, 1930, Carter, J.G. and Clark, G.R., Classification and phylogenetic significance of molluscan shell microstructure, pp. 50–71, in Broadhead, T.W. (ed.), *Mollusks: Notes for a short course organized by D.J. Bottjer, C.S. Hickman, and P.D. Ward. University of Tennessee Studies in Geology*, University of Tennessee Department of Geological Science, Knoxville, TN, 1985, Bandel, K., Shell structure of the Gastropoda excluding Archaeogastropoda, pp. 117–134, in Carter, J.G. (ed.), *Skeletal Biomineralization: Patterns, Processes and Evolutionary Trends*, Vol. 1, Van Nostrand Reinhold, New York, 1990, and Hedegaard, C., *Shell structures of the Recent Archaeogastropoda*, Science Thesis, University of Arhus, Denmark, 1990; Hedegaard, C., *Molluscs—Phylogeny and biomineralization*, Ph.D. dissertation, University of Aarhus, Denmark, 1996; Hedegaard, C., *J. Molluscan Stud.*, 63, 369–377, 1997.

The muscular foot is usually rather large and is used for 'creeping' in most species but in some is modified for burrowing, leaping (as in conchs – Strombidae), swimming (as in the planktonic heteropods), or clamping (as in limpets) (see Section 18.1.3.2 and Chapter 3 for more information regarding locomotion). In vetigastropods, an epipodium (epipodial skirt) may be present around the edges of the foot, and this may bear tentacles; epipodial tentacles are absent in other gastropods except for a few cerithiimorph caenogastropods. A pair of parapodial flaps arise from the edges of the foot in some 'opisthobranch' gastropods, for example, the sea hares (Aplysiidae), and, in some species, are used in swimming.

Details regarding the gastropod foot, its role in locomotion, and the crucial functions of cilia, mucus, and muscles are given in Chapter 3.

18.1.2.5   Mantle Cavity and Its Organs

An overview of the evolution of the gastropod mantle cavity and its critical role in gastropod evolution is given in Chapter 4 and is developed further below. As noted in Chapters 4 and 8, anopedal flexure and larval torsion uniquely resulted in the anterior to right side placement of the gastropod mantle cavity, a dramatic departure from its plesiomorphic posterior location. The first gastropods had a full complement of mantle cavity organs, with two pairs each of bipectinate ctenidia (and hence paired auricles), osphradia, and hypobranchial glands, the openings of the paired kidneys (with the right being the urinogenital aperture), and the anus. Most gastropod lineages have lost the right ctenidium (Figures 18.7 to 19.9) and the right component of the other paired mantle cavity structures. Some (including all heterobranchs) have lost the ctenidia altogether, although ctenidium-like gills have developed to replace them in some taxa (see Chapters 4 and 20). Some higher vetigastropods and some neritimorphs with only a left functional ctenidium also have a rudimentary right ctenidium.

The changes in the configuration of the mantle cavity organs came about independently in different gastropod lineages (Ponder & Lindberg 1996) (Figures 18.8 and 19.9), but the evolutionary forces driving those changes were similar in many of those lineages. Firstly, tighter shell coiling resulted in a narrowing of the aperture and a reduction of the space available on the right side of the cavity. Secondly, as shown in *Haliotis*, the development of the mantle cavity structures occurs rather late in ontogeny with the smaller right ctenidium developing later than the larger left. Thus, any taxon that is progenetically heterochronic could lose the right ctenidium for developmental and not for primarily functional reasons. A more detailed discussion is given below regarding the one major clade, the vetigastropods, which have retained some extant taxa still possessing both ctenidia (see Section 18.5.3.3).

The plesiomorphic gastropod ctenidium is like that of other molluscs in being composed of ciliated filaments on either side of an axis that contains blood vessels and nerves. In most vetigastropods, and in higher gastropods (apogastropods), the filaments have skeletal rods, but these rods are lacking in patellogastropods and neritimorphs. Efferent membranes are present in most vetigastropods and neritimorphs but not patellogastropods.

A pair of hypobranchial glands is present in many vetigastropods, including a number of those that have lost the right ctenidium and osphradium, although both are retained in some neritimorphs that have also lost the other right mantle cavity structures. A hypobranchial gland is absent in patellogastropods, and, if they are present in higher gastropods, there is only one (the left). While it is usually assumed to be homologous throughout the gastropods, this assumption has not been rigorously tested. In some gastropods, notably 'opisthobranchs', 'pallial' glandular areas occur that may not be homologous (Hyman 1967; Gosliner 1994). In the 'lower' heterobranchs the supposed hypobranchial gland is located anteriorly whereas in other gastropods it is typically posterior (Ponder & Lindberg 1997). The pigmented mantle organ (PMO) found in some lower heterobranchs is a modification of the hypobranchial gland (see Chapter 20).

The plesiomorphic condition is a pair of chemosensory osphradia, and this is seen in patellogastropods, where they are located on the floor of the mantle cavity; in some vetigastropods, as in other orthogastropods, they are on the mantle cavity roof. All other gastropods have only the left osphradium or, particularly in the case of some heterobranchs, have lost it. In many higher caenogastropods, the osphradium has become hypertrophied (Figure 18.8).

This presence of sense organs on the exterior of vetigastropods and heterobranchs (see Section 18.1.2.9 and Chapters 7 and 20) may be associated with a general lack of control over the inhalant flow of water into the mantle cavity (Ponder & Lindberg 1997; Lindberg & Ponder 2001). In contrast, sorbeoconch caenogastropods established inhalant control, which resulted in the hypertrophy of the chemosensory osphradium. These changes are often associated with the development of an anterior siphon which assists in the detection of prey, mates, or predators by adding directionality to the inhalant stream tested by the osphradium. These changes enabled caenogastropods to become food specialists to a much greater extent than vetigastropods, with several lineages independently becoming active carnivores. In contrast, vetigastropods rely largely on direct contact with the body to assess the direction of stimuli (Ponder & Lindberg 1997; Lindberg & Ponder 2001).

As detailed in Chapters 4 and 20, the euthyneuran heterobranchs show the greatest modifications to the mantle cavity. In 'opisthobranchs' (bubble shells and sea slugs) detorsion has resulted in the mantle cavity migrating down the right side and, in many sea slugs, it is lost altogether. Secondary gills, cryptic shapes and patterns, and/or noxious chemicals are developed in some taxa. In terrestrial gastropods the gill is lost, and the mantle cavity functions as a lung (pulmonary cavity) with a highly vascularised roof. In these 'pulmonates' the entrance to the lung is reduced to a small opening typically controlled by a sphincter muscle, which helps to prevent water loss.

18.1.2.6   Digestive System

The structure and function of the gastropod digestive system are described in Chapter 5, and we only present a brief overview below. It follows the usual molluscan pattern. There is an anterior mouth on the anteroventral side of the snout, although in some higher gastropods the snout is introverted as

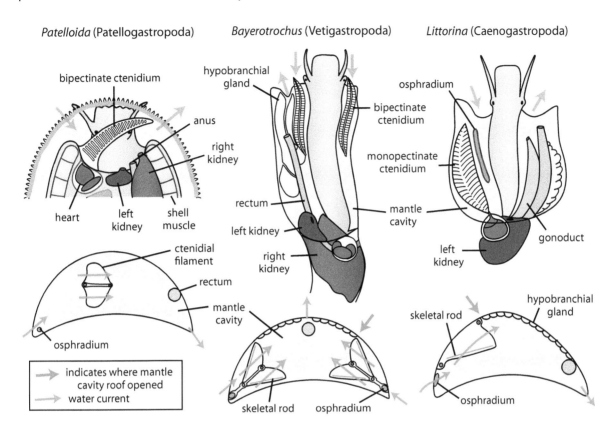

FIGURE 18.7   Examples of gastropod mantle cavities from three major clades. Redrawn and modified from the following sources: Top left (Fretter, V. and Graham, A.L., *British Prosobranch Molluscs: Their Functional Anatomy and Ecology*, Ray Society, London, 1962); top middle and right (Lindberg, D.R. and Ponder, W.F., *Am. Zool.*, 38, 98A, 1998); bottom row (Yonge, C.M., *Phil. Trans. R. Soc. B*, 232, 443–518, 1947).

a proboscis. Just inside the mouth lie the jaws. In patellogastropods, there is a single laterodorsal jaw of uniform composition, but when present in orthogastropods, the jaws are paired and composed of numerous small rods or even plates and lie laterally (see Chapter 5). The mouth opens into the buccal cavity, and protruding into this cavity is the muscular odontophore that bears the radula. The radula is typically supported by one or more pairs of odontophoral cartilages. Usually, a pair of salivary glands opens dorsally to the buccal cavity. In the more primitive taxa, and in many caenogastropods, the oesophagus has a large enzyme-secreting oesophageal gland and opens to the stomach. The stomach in lower orthogastropods typically has a ciliated sorting area, and part of its inner walls is cuticularised, with one thicker part forming a gastric shield against which a 'protostyle' rotates. A short style sac is lined with cilia that rotate the protostyle, a mucoid mass containing waste material. The paired digestive gland opens to the stomach and secretes enzymes involved in extracellular digestion. The intestine carries the faeces, primitively a string, or as pellets, to the rectum to be released at the anus, which in most gastropods lies within the mantle cavity. The intestine is looped to increase its length, with many loops in the more primitive taxa, notably in patellogastropods, and with shortening in the more apomorphic taxa.

Numerous modifications from the plesiomorphic conditions of the gut include the elongation of the snout or its introversion as a proboscis. The radula is modified from the assumed

plesiomorphic 'docoglossate' condition to various configurations through modification of the number of tooth rows (either increase or reduction), changes in tooth and cutting-edge shape, and relative size of the teeth. Some of the different arrangements are named rhipidoglossate, taenioglossate, ptenoglossate, or stenoglossate (including rachiglossate and toxoglossate), but many other distinctive radular configurations (mainly in the heterobranchs) remain unnamed, although a terminology for describing the various states is provided in Chapter 5. In some taxa, notably many suctorial feeders, the radula is lost. For a general account of the radula and its function see Chapter 5, while radular morphology for each major taxon is detailed in the following sections and Chapters 19 and 20.

18.1.2.7   Renopericardial System

As in other systems, the gastropod renal and vascular systems have been modified as a result of coiling of the viscera and torsion. Gastropods differ from other molluscs in having asymmetrical kidneys – either a pair comprised of two unequal sized kidneys carrying out different functions or a single (left) kidney. In caenogastropods and neritimorphs a remnant of the right kidney is incorporated in the reproductive tract. A nephridial gland is present in the left kidney of some vetigastropods, and a convergent structure is found in the single (left) kidney of caenogastropods. A detailed account of the structure and function of gastropod kidneys is given in Chapter 6.

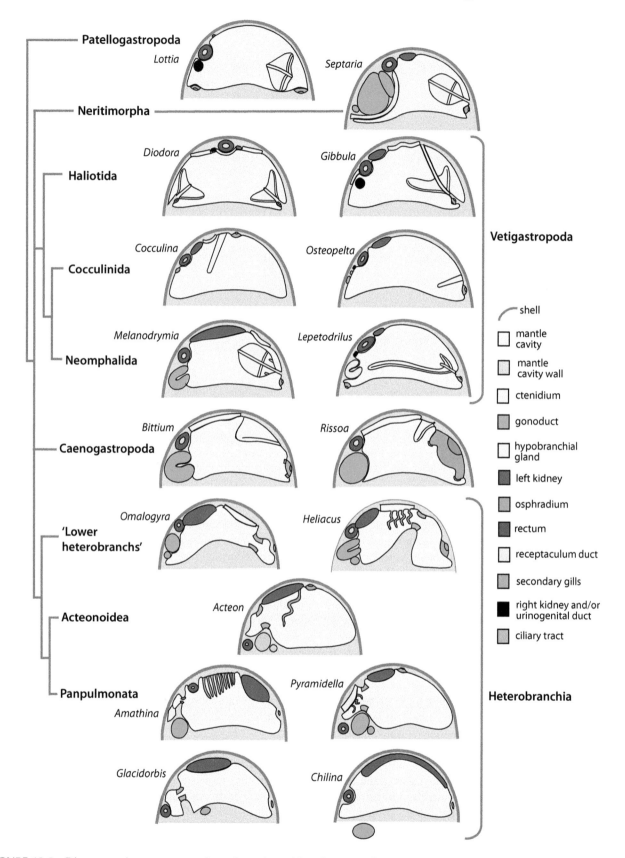

FIGURE 18.8 Diagrammatic transverse sections of mantle cavities of a range of gastropods to show the arrangement of the main structures. Redrawn and modified from Haszprunar, G., *Zool. Scr.*, 17, 161–179. 1988a.

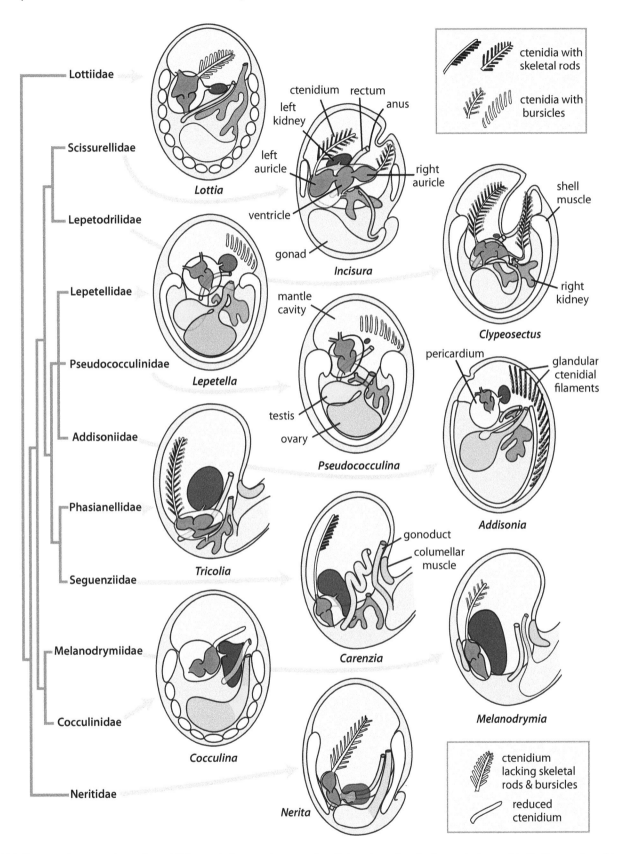

FIGURE 18.9 A comparison of ctenidial morphology and gonopericardial systems (seminal receptacles omitted) of examples of lower gastropods. Redrawn and modified from Haszprunar, G., *Zool. Scr.*, 17, 161–179, 1988a.

The chambered heart consists of a single ventricle and, plesiomorphically, two auricles that correspond to the presence of two ctenidia or one plus a rudiment. A single auricle is present in most gastropods, irrespective of whether a ctenidium is present or not. While arteries and veins are present, the circulatory system is mainly open with significant blood sinuses, notably in the head, viscera, and foot. Oxygenated blood returns to the auricle(s) via the gills and/or the mantle. Oxygenation can also occur via the kidney in those taxa where it lies in the mantle roof. Further details on the gastropod circulatory system are given in Chapter 6.

Filtration occurs by way of podocytes, usually on the auricle(s) but also on the ventricle in patellogastropods. In some, notably many heterobranchs, the podocytes are lost, and filtration occurs via slits (see Chapter 6 for details).

18.1.2.8 Reproductive System

All gastropods have a single gonad, the pretorsional left – i.e., post-torsional right, thus differing from the single gonad in scaphopods and cephalopods where it is the right. In primitive gastropods the gonad opens to the right kidney and is modified in different lineages, where it opens separately to the mantle cavity. In a few vetigastropods and all higher gastropods, ectodermal components are added to the terminal part of the mesodermal gonoduct, resulting in complex tubular and glandular structures being added that facilitate internal fertilisation and the production of encapsulated eggs. As noted above, in caenogastropods and neritimorphs a rudiment of the right kidney is incorporated in the genital duct, as evidenced by a pericardial connection in caenogastropods but not in neritimorphs.

Most gastropods have separate sexes, but a number, including almost all Heterobranchia, are hermaphroditic (see Chapter 8).

Accounts of reproduction, in general, are given in Chapter 8, along with the functional, and the main structural attributes of the reproductive systems, while the details of the morphological diversity of the reproductive systems are given in the sections dealing with each of the gastropod groups in this and the following two chapters. In copulating taxa, the gastropod pallial glandular system typically consists of a proximal albumen gland, a capsule (or equivalent) gland, both of which are in connection with the oviduct, a terminal vagina that receives the penis (or a spermatophore bursa), and sperm pouches. These latter structures are usually a sperm-receiving copulatory bursa (bursa copulatrix) and one or more seminal receptacles for sperm storage. Various accessory glands and structures are sometimes present, especially in euthyneuran heterobranchs.

Development and life history are summarised in Section 18.1.3.3.

18.1.2.9 Nervous System and Sense Organs

The nervous system is well-developed, with a concentration of nerve ganglia in all but the most primitive groups. Substantial pedal cords are present in patellogastropods, vetigastropods, and neritimorphs but not in heterobranchs and most caenogastropods. A streptoneurous visceral loop (the result of the body twisting during torsion) is present in most patellogastropods, vetigastropods, caenogastropods, and lower heterobranchs,

but most heterobranchs have undergone some detorsion resulting in euthyneury.

A cephalic eye is primitively situated at the outer bases of the paired cephalic tentacles, and in some taxa each eye is located on a stalk (ocular peduncle). The cephalic eyes are homologous in all gastropods and show an evolutionary sequence from open pits in patellogastropods, open but possessing a lens in many vetigastropods, and covered with a cornea in neritimorphs and higher gastropods. While in most gastropods the eyes maintain their plesiomorphic condition on the outer side of the cephalic tentacles, in most heterobranchs they lie in the middle or on the inner sides of those tentacles.

Specialised sense organs found in some patellogastropods include sensory strips and wart organs (see Section 18.4.3.7), and a suite of other sensory structures are only found in the vetigastropod Trochiformii, including the sensory papillae that cover the epipodial and cephalic tentacles, as well as some other parts of the head-foot, and the epipodial sense organs. These latter structures may be present with or without epipodial tentacles, but when the tentacles are present, they typically lie at their bases. The chemosensory bursicles that lie near the base of each gill filament and aid in predator detection are found in Trochiformii and some neomphalines. Osphradia (see Section 18.1.2.5) are insignificant in vetigastropods but are the main chemosensory organ in most caenogastropods. As noted above, in the latter group, they have become hypertrophied and they often develop ridges or filament-like folds to increase their sensory surface area. Also, in the heterobranchs sensory surfaces are usually on the exterior of the animal, and again the osphradium is small and of lesser importance or lacking altogether.

More details on gastropod sense organs are provided in Chapter 7 and the relevant section in the treatments of the main groups of gastropods.

18.1.3 Biology, Ecology, and Behaviour

Various aspects of gastropod ecology and biology are covered in Chapter 9 and in the sections and chapters dealing with each major gastropod group. We provide below an outline of some of these aspects.

18.1.3.1 Habits and Habitats

Because they are the only group of molluscs to have colonised the land, gastropods occupy a more extensive range of ecological niches than any other class. They range from the deepest ocean to high mountains and occupy virtually all freshwater habitats; some even live in salt lakes. By far the largest terrestrial group is the heterobranch stylommatophoran land snails and slugs, found in virtually all habitats ranging from rainforests to deserts and alpine regions, and they extend from the tropics to high latitudes. Most marine gastropods are benthic and mainly epifaunal. Their habitats include the deep sea, hot vents and seeps, the intertidal zone and supralittoral fringe, and estuaries; some inhabit anoxic habitats. A few such as the 'violet snails' (Epitoniidae) and the nudibranch *Glaucus* (Glaucidae) drift on the surface of the ocean, while another nudibranch *Phylliroe* (Phylliroidae), the caenogastropod heteropods, and the heterobranch Gymnosomata are active predators swimming in the plankton. Another plank-tonic group, the pteropod Thecosomata, gather tiny organisms in mucous nets (see Chapter 5).

18.1.3.2 Locomotion and Behaviour

An overview of gastropod locomotion is given in Chapter 3. Most gastropods crawl on their foot using cilia or muscular waves or a combination of both. Some, notably the patellogastropod limpets, use their disc-like foot and the special properties of mucus (see Chapter 3) to clamp firmly onto hard surfaces.

Shallow burrowing in soft sediments has evolved in some gastropod clades and is usually accompanied by modifications of the shell and the head-foot. The shell is typically stream-lined, but some have so-called ratchet sculpture (see Chapter 3), and the head-foot is usually modified to prevent sediment from entering the mantle cavity. In many burrowing higher caenogastropods the mantle cavity is on the right side of the animal. Burrowing taxa are less common in lower caeno-gastropods and vetigastropods, probably because the mantle cavity is situated anteriorly and requires special modifica-tions such as mantle tentacles or skirts to keep sediment out. Naticids have a substantial shield (formed from the enlarged propodium) that is convergent with the head-shield of burrow-ing cephalaspidean 'opisthobranchs'. Both act as a plough and protect the mantle cavity from being clogged with sediment.

Swimming gastropods range from those that take off from the bottom and swim awkwardly for a short distance to those that spend their entire lives in the water column. Swimming may be achieved by the thrashing movements of the foot, such as in some scissurellids and solariellids, or by parapodial flap-ping as in some species of *Aplysia*. At the other end of the spectrum, the shelled pteropods use parapodia to constantly maintain their position in the water column, while other naked pteropods and heteropods, or those with reduced shells, use their flattened bodies to swim in a fashion similar to fish.

Gastropods have developed a wide range of behaviours related to feeding, reproduction, predator avoidance, and defence, minimising exposure to unfavourable aspects of the environment by moving towards more favourable conditions (see major review by Chase [2002] and Chapter 7).

Apart from behavioural attributes, gastropods have evolved several lines of defence other than just the possession of a shell that the animal can withdraw into, with some of these being more important in different lineages. For example, in vari-ous slug groups, shell loss enables the animal to hide in small spaces not available to a similar-sized shelled species. Some slugs and a few shelled taxa secrete noxious chemicals, often metabolites derived from their food (see Chapter 9). Crypsis is a common method of predator avoidance and, in some sea slugs in particular, aposematism or mimicry (see Chapter 9). Another defence strategy, autotomy (Fleming et al. 2007), is uncommon but occurs in several groups (see Chapter 9).

18.1.3.3 Life History

Gastropod embryology and development is described in Chapter 8. Torsion (see Section 18.1.2 and Chapter 8) is under-gone during larval development. Primitive gastropods all have lecithotrophic development, but planktotrophic larvae

are found in many neritimorphs, caenogastropods, and het-erobranchs. Whether or not the evolution of planktotrophy in gastropods is a single event or evolved separately twice or three times remains an open question.

Lower gastropods practise external fertilisation, releasing their gametes into the water column. Others (neritimorphs, caenogastropods, and heterobranchs) fertilise internally using a penis to copulate, or they exchange spermatophores. Internal fertilisation enables the production of eggs surrounded by protective capsules or jelly, and their encapsulation (Figure 18.10) made the invasion of non-marine habitats possible.

In the basal groups, larval development usually occurs in the water column and, typically, a trochophore stage is followed by the veliger larva. The veliger then settles and metamorphoses into a juvenile snail. In those groups with encapsulated eggs, development within the capsule is under-gone to at least the veliger stage before release. While many marine species go through larval development, direct devel-opment is also common, and this is the normal mode for fresh-water and terrestrial taxa. Brooding of developing embryos in the mantle cavity, pallial oviduct, or other pouches is widely distributed in gastropods (see Chapter 8).

Eggs in direct-developing species are typically large, and this size is reflected in the initial size of the larval shell (proto-conch). Thus, the protoconch size can be useful in distinguish-ing feeding and non-feeding larvae in both living and extinct taxa. Also, the larval shells of neritimorphs and caenogastro-pods show a distinction between the initial protoconch (pro-toconch I) and the part that grows during planktonic feeding (protoconch II). While most gastropods have dextral larval shells, those of heterobranchs are sinistral, with the coiling direction changing to dextral at metamorphosis. This heteros-trophic condition is characteristic of heterobranchs and is rarely seen in other gastropods, although incipient heterostrophy can occur in some, such as observed in a few trochoideans (e.g., Hadfield & Strathmann 1990; Sasaki 1998 p. 91, figure 61a,b).

A few caenogastropods have an outer second larval shell and are known as echinospira larvae (see Chapter 19).

18.1.3.4 Feeding

Gastropod feeding habits are extremely varied and include grazing, browsing, suspension feeding, scavenging, detritus feeding, and carnivory. Plesiomorphic gastropod feeding is assumed to be grazing on surface films comprised of bacteria and other organisms including diatoms and algal sporlings. This feeding mode became more specialised in various lin-eages to become deposit feeding, grazing on colonial animals and macroalgae, or scraping rock surfaces. Grazing carnivory led to more specialised carnivores, including predation of mobile animals, while herbivory also extended to marine (and eventually) terrestrial angiosperms and fungi. While most gastropod feeding requires the use of the radula, some gas-tropods that feed suctorially have lost the radula. A few veti-gastropods and caenogastropods gather food as suspension feeders by modification of their ctenidium through extending the length of the filaments and developing food particle trans-port mechanisms. A few others use mucous nets to entangle their food (see Chapter 5).

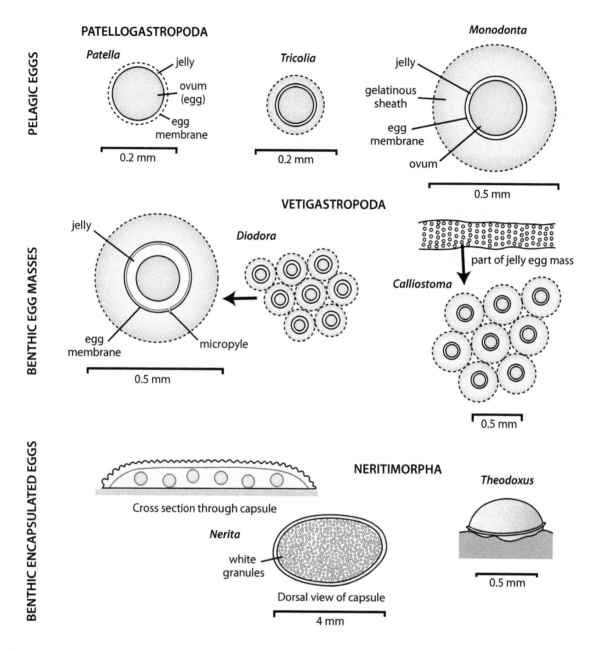

FIGURE 18.10 Examples of the spawn of patellogastropods, vetigastropods, and neritimorphs. Redrawn and modified from the following sources: Top and middle rows (Lebour, M.V., *J. Mar. Biol. Assoc.* U.K., 22, 105–166, 1937), bottom left (Anderson, D.T., *Proc. Linn. Soc. N.S.W.*, 87, 62–68, 1962), bottom middle (Lebour, M.V., *Proc. Zool. Soc. Lond.*, 114, 462–489, 1944), bottom right (Fretter, V. and Graham, A.L., *British Prosobranch Molluscs: Their Functional Anatomy and Ecology*, Ray Society, London, 1962).

Some gastropod carnivores drill holes in their shelled prey using a combination of radular scraping and chemical dissolution, this method of entry having been acquired independently in at least four groups of caenogastropods (Muricidae, Naticidae, Marginellidae, Cassidae) (see also Chapter 5) and a few stylommatophoran land snails.

18.1.3.5 Associations

Gastropods have numerous commensal and parasitic associations with other organisms including bacteria, 'protists', and many others (see Chapter 9). Apart from the important beneficial symbiotic bacteria in the gut, of particular note are the significant associations of many gastropods and trematodes, with a gastropod being a required intermediate host in the trematode life cycle (see Chapter 9). Some hot-vent gastropods have symbiotic bacteria in their gills or other parts of their bodies, although this phenomenon is not as significant as it is in some bivalves (see Chapter 9). A few gastropods are commensal in the tubes of burrowing animals, but, again, this is not as commonly seen as it is in bivalves (notably Galeommatoidea).

18.1.4 Economic Values

Various groups of gastropods are (and were) used as food and ornaments, including jewellery (see Chapter 10). Because of their abundance and diversity, especially in marine ecosystems, gastropods play a significant role in providing ecological services, particularly in the dynamics of shallow water and intertidal communities. They are also vital components of many freshwater and terrestrial ecosystems.

From a negative economic perspective, by far the most important gastropods are those that serve as intermediate hosts for human trematode parasites (see Chapter 10). Some terrestrial snails and slugs are also significant pests, notably in agriculture (e.g., Barker 2002).

18.1.5 Diversity and Fossil History

Gastropods are by far the largest group of molluscs, comprising about 73–78% of named living molluscs. Estimates of total extant species range from 40,000 to over 100,000 but may be as high as 150,000, with about 13,000 named genera for both living and extinct species (Bieler 1992). Gastropods not only show high taxonomic diversity but, as outlined above, considerable variety in size, shell, and body shape, as well as in habits. Many are of small size, and this vast, poorly known fauna of 'microgastropods' (0.5–5.0 mm) that live in marine, freshwater, and terrestrial environments comprise most of the many hundreds of undescribed species of gastropods.

There is a long, rich, gastropod fossil record with, as in most other animal groups, periodic extinctions of subclades, followed by diversification of new groups (Erwin & Signor 1991). As discussed in Chapter 13, several groups of tiny univalves range from the Early Cambrian, but these were not gastropods. Gastropods arose later in the Cambrian and gave rise to several major clades that flourished in the Paleozoic, several of which became extinct.

As briefly outlined in the sections above, numerous evolutionary pathways have been adopted by gastropods. Some common patterns emerge, and one of the most obvious is that there is, on the one hand, a reduction in the complexity of many characters and on the other an increase in complexity in some features (Haszprunar 1988a; Ponder & Lindberg 1997). Examples of simplification include:

- Simplification in shell microstructures, with most microstructures lost in 'higher' gastropods so their shells possess only crossed-lamellar microstructures.
- Reduction in the number of radular teeth.
- Changes that appear to result from shell coiling include:
  - Reduction and loss of the right ctenidium and loss of both ctenidia in some lineages.
  - Associated loss of the right auricle and associated circulatory and nervous system changes.
  - Loss of the right kidney.
- Simplification and reduction in odontophoral cartilages and muscles.

- Simplification of the oesophagus (loss or modification of the oesophageal gland) and loss of the ventral oesophageal fold.
- Simplification of the stomach and intestine.

In contrast, some characters show an increase in complexity, for example:

- Reproductive structures including:
  - Internal fertilisation and the associated development of copulatory organs and/or spermatophores and their associated structures.
  - Correlated production of egg capsules and the evolution of planktotrophic larvae.
  - Direct development.
- Increase in chromosome number.
- Greater complexity of some sensory structures (e.g., eyes, osphradium).

18.2 THE INITIAL RADIATION – EOGASTROPODS AND ORTHOGASTROPODS

The Cambrian radiation of gastropod-like taxa is described and discussed in Chapter 13. By definition, a gastropod has undergone torsion, but when working with the early fossil record, it is difficult to discern whether gastropod-like fossils contained torted animals. Interpreting such fossils has met with a wide divergence of views. For example, the Cambrian helcionellid 'limpets' are regarded as gastropods by some workers (see Parkhaev 2008 for a summary), while other workers have treated them as untorted 'monoplacophorans' or regarded them as a separate class (see Chapter 13 for details).

Rather than rely exclusively on the fossil data, here we work backwards by looking at the shared features of the most primitive living gastropods and try and reconstruct what the main features of an ancestral patellogastropod and orthogastropod might have been. From that base, we will attempt to determine the features possessed by both the eogastropod and the common gastropod ancestor.

18.2.1 Hypothetical Reconstructions

There have been very few attempts to construct a gastropod ancestor or 'archetype'. Our understanding of the ancestral gastropod has always focused around a torted animal, and that has usually been a coiled snail. In a marked departure, Haszprunar (1988a) argued for a limpet-shaped ancestor, influenced by his contention that 'Cocculiniformia' (see Section 18.5.2.2.2) were basal gastropods. While we generally agree with many features outlined by Haszprunar (1988a), here we focus on the phylogenetic basis behind ideas regarding a hypothetical gastropod ancestor and the precursors of the assumed two major gastropod lineages – the eogastropods and orthogastropods.

TABLE 18.1

Character States in the Outgroups (Polyplacophorans [Upper] and Monoplacophorans [Lower]) in the Second Column, with the States Probably Present in Ancestral Gastropods in the Third Column. The Two Right-Hand Columns Give the States Probably Present in Eogastropods [Based on Patellogastropods] and in Early Orthogastropods [Based Mainly on Vetigastropods].

Character	Outgroup Polyplacophora/ Monoplacophora	Ancestral Gastropod	Patellogastropoda	Orthogastropoda
Operculum in adult	Absent	? Absent	Absent	Present
Shell	Eight plates/ Limpet	Coiled or limpet-like	Limpet	Coiled
Protoconch	Absent/ Cap-like	Probably an elongated cap	Cap to caecum-like	Caecum-like to **coiled**
Protoconch offset	N/A/ None	None	Left	Right
Prismatic and nacre	Absent/ Present	Absent	Absent	**Present** (Vetigastropods)
Foliated shell structure	Absent/ N/A	Absent	**Present**	Absent
Shell aperture dilation in adult	N/A/ Posterior (untorted)	Anterior (torted)	**Posterior** (torted)	Anterior (torted)
Anterior end of shell muscle linked	N/A	?	Present	Absent
Cephalic tentacles	Absent/?rudimentary	Present	Present	Present
Cephalic eye structure	N/A	? Open pit	Open pit	? With lens
Mantle edge folds	Inner fold present	Inner fold present	Inner fold present	Inner fold present in vetigastropods, absent in caenogastropods
Retinal structure	N/A	?	? One type of retinal cell	Two or more types of retinal cells
Sensory retractile mantle edge tentacles	Absent	Absent	Present	Absent
Epipodial tentacles	Absent	Absent	Absent	**Present or absent**
Epipodial sense organs	Absent	Possibly present	Possibly present in late veliger	Present or absent
Head	Ligocephalic	Apocephalic?	Apocephalic	Apocephalic
Foot sole shape	Oval/ Disc-like	? Disc-like	Short-oval to disc-like	**Elongate** to secondarily? disc-like
Propodium and anterior pedal gland	Absent	Absent	Absent	**Present**
Anterior marginal groove with pedal glands	Absent	Absent	**Present**	Absent
Oral lappets	Present	Present	Present	**Usually absent** (present in cocculinoideans, some lepetelloideans, and in neomphalids; possibly in neritimorphs)
Oral hood	Present	Present	**Present**	**Absent** except in Cocculinoidea
Inner lips	Absent	? Absent	**Present**	Absent
Ctenidial skeletal rods	Absent	Absent	Absent	**Present** (absent in neritimorphs)
Gill axis structure (see Chapter 4, Figure 4.5)	Muscle bundles inner side of blood vessels/ Muscle bundles on outer side of blood vessels	? Muscle bundles on outer side of blood vessels	Muscle bundles on outer side of blood vessels	**Muscle bundles on inner side of blood vessels**
Ctenidial supporting membrane(s)	Absent	Absent	Absent	**Present**
Hypobranchial glands	Absent	Absent	Absent	**Present**
Wart organs	Absent	Absent	A pair or absent	Absent
Sensory streaks	Present/ Absent	Absent	Present	Absent
Pedal muscle fibres	Crossing/ Not crossing	Not crossing	Not crossing	**Crossing**

(Continued)

TABLE 18.1 (CONTINUED)

Character States in the Outgroups (Polyplacophorans [Upper] and Monoplacophorans [Lower]) in the Second Column, with the States Probably Present in Ancestral Gastropods in the Third Column. The Two Right-Hand Columns Give the States Probably Present in Eogastropods [Based on Patellogastropods] and in Early Orthogastropods [Based Mainly on Vetigastropods].

Character	Outgroup Polyplacophora/ Monoplacophora	Ancestral Gastropod	Patellogastropoda	Orthogastropoda
Statocyst position	Absent/ Lateral to pedal ganglia	Lateral to pedal ganglia	Lateral to pedal ganglia	**Dorsal to pedal ganglia**
Circum-mantle nerve (not lateral nerve cord)	Absent	?	Present	Absent
Labial ganglia	Absent	Absent?	**Present**	Fused with cerebral
Subradular ganglia	Paired/ Single (fused?)	?	**Paired**	?
Pedal cross connections	Present/ Absent	?	One or two	Well-developed (lost in later orthogastropods along with reduction in the pedal cords)
Buccal ganglia relative to labial ganglial	Anterior	?	Anterior and posterior	**Anterior**
Size of visceral loop	N/A	? Short	Short	**Long**
Ganglia of lateral nerve cords or visceral loop	Absent	Absent	**Supra- and suboesophageal ganglia present**	Absent (these form in later orthogastropods)
Innervation of ctenidia and osphradium	From lateral nerve cords	? From osphradial ganglia connected to visceral loop	From osphradial ganglia connected to supra- and suboesophageal ganglia on visceral loop	From osphradial ganglia connected to visceral loop
Innervation of shell muscles	From pedal ganglia	? From pedal and pleural ganglia	From pedal and pleural ganglia	From pleural ganglia or otherwise from visceral loop
Mantle cavity	Lateral groove	? Shallow anterior embayment + lateral groove	Shallow anterior embayment + lateral groove	**Deep anterior embayment**
Position of kidneys relative to pericardium	On either side	On either side	**Both on left**	On either side
Left kidney	Equal to right in size	Smaller than right	Much smaller than right	Much smaller than right. Becomes only kidney when right lost
Right kidney	Equal to left in size	Larger than left	Much larger than left	Much larger than left. Lost in more advanced orthogastropods or remnant incorporated in female genital system
Position of right kidney relative to digestive gland	Ventral/ Lateral	? Ventral	Ventral	**Dorsal**
Auricles	Two/ Two or more	Two	One	Two (reduced to one in derived orthogastropods)
Podocytes	On auricles only	On auricles only	On both auricle **and ventricle**	On auricles only
Ventricle attached to pericardium	Yes/ No	?	Yes	No
Circum-mantle blood vessel	Arterial sinus	? Absent	Present	Absent
Haemocyanin	Present/ ?	Present	**Absent**	Present
Rectum through pericardium	No/ Yes	Yes	**No**	Yes
Antero-lateral buccal cartilages	Present/ Absent	Absent	**Present**	Absent

(Continued)

TABLE 18.1 (CONTINUED)

Character States in the Outgroups (Polyplacophorans [Upper] and Monoplacophorans [Lower]) in the Second Column, with the States Probably Present in Ancestral Gastropods in the Third Column. The Two Right-Hand Columns Give the States Probably Present in Eogastropods [Based on Patellogastropods] and in Early Orthogastropods [Based Mainly on Vetigastropods].

Character	Outgroup Polyplacophora/ Monoplacophora	Ancestral Gastropod	Patellogastropoda	Orthogastropoda
Jaw	Absent/ Homogeneous, complex, dorsal	Homogeneous, complex, dorsal	Homogeneous, complex, dorsal	**Rods secreted by underlying epithelium; lateral plates – fused and dorsal in 'pulmonates'. Lost in some**
Jaw relative to buccal mass	N/A/ Attached to oral tube and free from buccal mass	Attached to oral tube and free from buccal mass	**Fixed to odontophore by muscle**	Attached to oral tube and free from buccal mass
Radular function	Partially flexoglossate	Partially flexoglossate	Stereoglossate	Flexoglossate
Radula	Docoglossate	Docoglossate	Docoglossate	**Rhipidoglossate**
Radular teeth mineralised with iron	Yes/ No	Yes	Yes	**No**
Basal plates of radular teeth	Absent	Absent	**Present**	Absent
Transverse labial muscles	Absent	Absent	**Present**	Absent
Ventral approximator muscles of odontophoral cartilages	One layer	One layer	**Two layers**	One layer
Dorsal protractor muscles of odontophore	Absent	Absent	**Present**	Absent
Post-dorsal buccal tensor muscle of odontophore	Absent	Absent	Absent	**Present**
Subradular organ (licker)	Present	Present	Well-developed	**Reduced; absent in 'higher' gastropods**
Salivary glands	Present, no ducts	Present, no ducts	Present, one or two pairs, with or without ducts	Present, originally no ducts
Oesophageal pouches/gland	Simple	? Simple	**Septate**	Simple; **papillate in higher vetigastropods**
Stomach	Simple	? Simple	**Very small, simple**	Large, complex
Gastric shield	Absent	? Absent	Absent	**Present**
Style sac	Absent/ Short, without long, stiff cilia	Short? without long, stiff cilia	**Very long, with short cilia** (sometimes this structure interpreted as part of the intestine)	Short, **with long, stiff cilia**
Intestine	Long, coiled	Long, coiled	Long, coiled	**Moderate length, one or two coils**
Anus	Mid-posterior	Mid-anterior	Anterior right	Mid-anterior, anterior right
Gonad number and position relative to digestive gland	Paired; ventral/ Paired; dorsal	Single; ventral	Single; ventral	Single; **dorsal**
Gonad structure	Tissue plates; open to haemocoel space/ ?	?	Internal trabeculae; open to haemocoel space.	Internal trabeculae present in some vetigastropods open to digestive gland
Chromosomes	8–16 pairs (m = 12)/ ?	?	Eight–ten pairs (m = 10)	13–18 pairs (m = 18)
Larva – size of velum	N/A	?	Small	Large
Larval shell muscles (in veliger)	N/A	?	Three	Two

Note: States in bold Indicate an apomorphic state for one of the two groups of gastropods; N/A = not applicable.

We determine the plesiomorphic features of gastropods by using chitons and monoplacophorans as outgroups. The distinctive characters of patellogastropods and orthogastropods are listed in Table 18.1. Also, based on the data in this table, the main features that an early eogastropod (represented in the extant fauna only by patellogastropods) and an ancestral orthogastropod were likely to have possessed can be estimated and serve to illustrate the distinctive characters of these two groups.

18.2.1.1 A Hypothetical Ancestral Gastropod

While we do not know for certain the features possessed by the first gastropods, we can piece together data from the fossil record and comparative anatomy to build up a picture of the characters they probably had. The fossil record for early gastropod-like molluscs is discussed in Chapter 13. Because it is not known if their shell was limpet-like or coiled, it is necessary to resort to comparative anatomy to look at the constraints imposed by the body of the first gastropods and then reconstruct possible ancestral stages. The first gastropod, by definition, must have been torted – that is, its larva underwent torsion – which has a profound effect on the adult body with the viscera rotated 180° relative to the head and foot. The possible causes of torsion and its selective advantages are discussed in Chapter 8. The post-torsional external and internal features of the animal and shell in these first gastropods were probably largely bilaterally symmetrical.

The larval shell was probably bowl-shaped and would soon become a short tube, and the adult shell was anchored to the foot by a pair of equal-sized shell muscles. It is unlikely that an operculum had developed at this early stage in either the adult or the larva (see Chapter 8 for discussion).

The apocephalic head had a ventral mouth, the sole of the foot was oval and the sides simple, and there was no propodium or anterior pedal gland. Some features associated with the head (cephalic tentacles with cephalic eyes at their outer bases and a short snout) may have developed soon after the appearance of the ancestral gastropod or were possibly present in the non-torted but long extinct gastropod sister taxon.

The shallow mantle groove contained one pair of small, bipectinate ctenidia that lacked skeletal rods, a pair of small, simple osphradia that lacked specialised sensory cells, and there were no hypobranchial glands.

The mouth had an oral lappet on either side of the head as in monoplacophorans and chitons, and it had outer lips, as in all molluscs.

Just inside the mouth was the jaw. A jaw is absent in chitons, but in monoplacophorans it is dorsal with lateral extensions and composed of uniform organic material. A similar jaw is seen in patellogastropods, but in basal vetigastropods the jaw is very different, being paired and made up of chitinous rods (see Chapter 5). By outgroup comparison, we assume that the plesiomorphic jaw in gastropods is like that seen in patellogastropods.

A ventral sensory subradular organ (licker) was present as it is in chitons and monoplacophorans.

The earliest gastropods had a radula like that of a chiton or monoplacophoran, probably with some of its lateral teeth mineralised with iron (a feature absent and possibly lost in

living monoplacophorans). The few marginal teeth were possibly capable of carrying out some sweeping movements – in other words, they may have had some limited flexoglossate capabilities (see Guralnick & Smith 1999 and Chapter 5). Thus, its feeding mode was scraping substrata, perhaps with some sweeping with the marginal teeth moving the dislodged particles towards the middle of the radula so they could be carried to the mouth.

A pair of large anterior (medial of Guralnick & Smith 1999) cartilages was present, as in the outgroups, and small, paired dorsolateral cartilages were probably also present. Sasaki (1998) considered that the dorsolateral cartilages (anterolaterals in his nomenclature) were apomorphies of patellogastropods and not homologues of the laterally placed cartilages in chitons or monoplacophorans. Because of their position, shape, and composition Guralnick and Smith (1999) argued that they are homologous, and it is on this supposition we assume that they were present in the HAG. In chitons and monoplacophorans, the paired anterior and dorsolateral cartilages are attached by a connective tissue sheath with the space between them forming a hollow space, the vesicle (Wingstrand 1985; Guralnick & Smith 1999). Besides the two pairs of cartilages mentioned above, small, paired, dorsal and ventral posterior cartilages were also probably present. These are found in some chitons where they are contained within the connective tissue making up the vesicle. In some patellogastropod lineages, similarly positioned but larger cartilages are thought to be homologous (Guralnick & Smith 1999). They are, however, absent in monoplacophorans and some patellogastropods (Lottiidae). Posterior cartilages are lacking in most orthogastropods and, when present, there is only a single pair which may be a result of the fusion of the two pairs (Guralnick & Smith 1999).

Based on their presence in monoplacophorans and polyplacophorans and their assumed homology with the salivary glands of gastropods, a pair of ductless salivary glands was present (Salvini-Plawen 1988).

As in chitons, the oesophagus was expanded laterally into lateral pouches that secreted carbohydrate-degrading enzymes but differed in being twisted by torsion before narrowing and entering the stomach. The U-shaped stomach has been rotated by torsion in an anteroposterior direction so the original posterior end is now anterior and the anterior is posterior, but the dorsal and ventral sides retained that orientation. Internally it had ciliated sorting surfaces, cuticular areas, and at least a rudimentary style sac with a pair of typhlosoles and epithelia bearing stiff cilia that rotated the contents. The stomach also received the pair of ducts (probably laterally as in the outgroups) from the large paired digestive glands. Digestion was mainly extracellular with enzymes secreted by the oesophageal gland and digestive gland.

As in chitons and monoplacophorans, in the early gastropods the intestine was long and coiled, and the rectum probably passed through the ventricle. The anus opened in the midline in the posterior part of the mantle cavity.

The heart had a ventricle and a pair of auricles, and the aorta that emerged from the ventricle posteriorly (the pretorsional anterior aorta) almost immediately bifurcated into two major aortic branches – one running anteriorly to the head-foot and the other posteriorly to the viscera. The ventricle is attached to the pericardium dorsally by a membrane (Fleure 1904), a condition also seen in chitons. Podocytes on the auricles were responsible for ultrafiltration, with filtrate being passed to the kidneys via the renopericardial ducts.

As in the outgroups, kidneys were probably symmetrical structures, and each had the same function. A renopericardial duct joined both kidneys to the pericardium. Such ducts are apparently absent in living monoplacophorans (Haszprunar & Schaefer 1997) but are present in chitons and bivalves. Each kidney removed nitrogenous waste from the blood of the head-foot and viscera before it went to the mantle skirt and then to the gills for oxygenation and from there to the heart. Each kidney was also presumably supplied with oxygenated blood by renal arteries arising from the posterior aorta so that the energetically expensive work of excretion could be performed (Fretter & Graham 1994).

Some very significant changes relating to the kidneys occurred in the 'proto-gastropods' before the eogastropod–orthogastropod split. Kidney asymmetry quickly developed, driven by shell coiling and possibly torsion, with the associated asymmetries in the circulatory system and the packing of organs in the visceral mass. Related to this asymmetry and a change in circulatory patterns the kidneys became specialised, with the right and left taking on separate roles that were previously shared. This separation of functions probably came about because, unlike most organs on the right side of the mantle cavity complex that became reduced in size or disappeared, the right kidney became much larger and retained its original position (at the base of the visceral mass) and its relationships with the vascular system. Its cytology became simplified, with the cells being predominantly excretory cells. The left kidney became smaller and specialised in a different way from the right. It originally lay in the visceral wall at the base of the mantle cavity, although it later migrated into the mantle roof in some taxa. Histologically and functionally it differed from the right kidney, and the original blood supply was disrupted. The new efferent connection was with a vessel or sinuses in the mantle skirt and, because the left afferent renal vein (which primitively would have carried deoxygenated blood from the viscera) was lost, it was replaced by a new connection with the left efferent ctenidial vein through which the afferent flow of oxygenated blood to the kidney was obtained. This new vessel also linked the left auricle and left kidney and thus also served as the major efferent renal vessel (Andrews 1985). The left kidney retained some of its excretory function but became specialised in absorption, regulating the ionic balance in the blood, in the same way the nephridial gland does in higher gastropods. Thus the essential conditions driving this change were (Andrews 1985; Fretter & Graham 1994): (1) migration of the left kidney into the mantle skirt, possibly mainly associated with coiling; (2) a supply of oxygenated blood by way of a new link with the efferent ctenidial vessel; and (3) the loss of a source of blood (containing waste metabolites) from the viscera. The arrangement depends on little or no venous blood reaching the left kidney from the afferent branchial vein and the establishment of a tidal flow in and out of the kidney from the efferent ctenidial vessel. Some resorptive and some excretory function persists in both, and each has renopericardial canals providing the same blood filtrate (primary urine) from the pericardium. The reduced nitrogenous excretion in the left kidney is of minor significance because the blood supplied to it has already passed through the right kidney (and the ctenidium) and thus had lost most of its waste material. Also, the lack of resorption in the right kidney is compensated for by the right renopericardial canal usually being significantly narrower than the left, thus reducing the flow of urine to the kidney. In addition, the greater volume of the right kidney enables the urine to accumulate there and stay longer allowing some resorption to occur, whereas the left kidney must deal with smaller volumes more quickly, requiring greater absorptive capacity. These conditions of the left and right kidneys persist in modern patellogastropods and in most vetigastropods.

Based on the outgroups, the position of the gonad in the first gastropods is uncertain – it is dorsal in chitons and *Nautilus* but ventral in monoplacophorans. It was single and opened directly to the right kidney in a similar fashion to that seen in monoplacophorans but in contrast to the situation in chitons where the gonad (a fused pair) opens by way of a pair of ducts directly to the exterior. The yolky eggs developed in the water column following broadcast spawning and passed through a trochophore and non-feeding veliger stage before metamorphosis.

These first gastropods had a nervous system with small, widely separated ganglia only slightly separated from the nerve trunks in which nerve cell bodies were present. The cerebral nerve loop lay at the anterior end of the buccal mass, and buccal ganglia and a labial commissure were present. The visceral loop had no ganglia besides the visceral, and the pleural ganglia were closely associated with the pedal ganglia from which long, thick pedal cords arose. There were a few cross connections between these cords. The statocysts, with statoconia, lay lateral to the pedal ganglia and were innervated directly from the cerebral ganglia. The left and right osphradial ganglia were innervated directly from the visceral loop.

Cephalic eyes and tentacles probably developed early in gastropod evolution (see above). The eyes were simple sensory pits, while the other sense organs were inherited from their untorted ancestors and included the licker, osphradia, and statocysts. The mantle edge and cephalic tentacles were also presumably well-endowed with scattered sensory cells.

18.2.1.2 Eogastropoda

The spirally coiled ancestors of Patellogastropoda present in the Paleozoic have not been definitively identified. Based on their fossil record, secondary flattening and the adoption of a limpet-shaped shell presumably occurred early in their history. These shells exhibit posterior aperture dilation as seen in monoplacophorans, but in all other gastropod limpets the dilation of the aperture is anteriorly directed. Living patellogastropods share several other plesiomorphic characters with

Polyplacophora and Monoplacophora which are not found in other gastropods (Table 18.1). These include their oval foot lacking an anterior pedal gland or propodium, a dorsal jaw, the basic docoglossate radular morphology and associated musculature, a subradular organ, lateral statocyst position, lack of skeletal support in the ctenidial filaments, the ventrally located (but unpaired) gonad, and, possibly plesiomorphic blood sinuses through the shell muscles. Oral lappets are present in patellogastropods, monoplacophorans and chitons, and a few basal orthogastropods. The shell muscle is symmetrical as in the outgroups but is reduced to a single pair, and their fibres do not cross in the ventral part of the foot as in the outgroups (Voltzow 1988).

There is also an impressive array of characters of patellogastropod limpets not seen in basal orthogastropods or the outgroups (Table 18.1). These include retractile mantle edge tentacles, a small anterior mantle cavity, a rotated pericardium, a muscular bulbous aorta (in some), a single auricle and ctenidium, nine pairs of chromosomes, posterior aperture dilation, foliated shell microstructure, caecum-like protoconch morphology[1] that is never partially coiled, three or more pairs of odontophoral cartilages (except Lepetidae, which have two pairs) and associated complex musculature, a docoglossate radula with stereoglossate function, a septate midoesophagus and a small, simple stomach with a very large style sac region, labial ganglia, and a short visceral loop. Also, a 'pallial line' which links the anterior ends of the shell muscle, the ventral gonad, labial ganglia, non-crossing pedal muscle fibres, and nine pairs of chromosomes are features not found in other gastropods.

Was the ancestor of patellogastropod limpets a limpet or a coiled snail? Other than their shared limpet-shape, many characters that unite patellogastropods are plesiomorphic (see above) based on outgroup comparison. Some anatomical features are difficult to explain if the common ancestor was a limpet (see below). Here we follow the arguments put forward by Lindberg (1981a) and Ponder and Lindberg (1997) in assuming that the ancestor was coiled. If this is the case, the question remains as to which patellogastropod characters are derived due to the assumed secondary limpet-shape and which are plesiomorphic?

The tubular protoconch seen in patellogastropods is straighter than the 'fish-hook'-shaped tubular protoconch seen in some Paleozoic taxa (e.g., Kaim 2004; Frýda et al. 2008a) but is not unlike the curved tubular protoconch of euomphaloideans, which may well be early eogastropods. The patellogastropod protoconch is separated from the teleoconch by a septum and is lost shortly after metamorphosis (Smith 1935), a process not seen in orthogastropods (Sasaki 1998). If this is also characteristic of the extinct coiled immediate ancestors of patellogastropods, it should be possible to detect this feature in fossils.

The protoconch in patellogastropods is offset to the left, an observation suggesting that the eogastropod ancestor may have been hypostrophically coiled (Lindberg 1981a).

The modern patellogastropods mainly have an oval to near-circular aperture, and we speculate that the coiled ancestors probably also had a circular or near-circular aperture that may not have been in contact with the parietal wall or was loosely attached. Also, we speculate that the teleoconch of the coiled ancestor may have only had one or a very few whorls.

It is not known whether the coiled patellogastropod ancestor had an operculum in the adult, but the presence of a larval operculum and the loss of the adult operculum in other limpet clades suggest that possibility. The operculum may have evolved after torsion as a protective measure for the larva (although this is unlikely given the results of Pennington and Chia 1985) or for the adult when the animal retracted into the shell. If the operculum was originally a response to the formation of a tubular larval shell and not an adult structure, it is possible that an adult operculum is an apomorphy of orthogastropods. Certainly, some ancient possible gastropod lineages such as the planispiral bellerophontids and the hypostrophic macluritoideans have apertural configurations that suggest that an adult operculum may not have been present. *Maclurites* does have a calcareous operculum, but it is unlike any found in modern orthogastropods.

This expanded mantle edge also has a major mantle nerve and blood vessel contained within it that enhance the sensory and respiratory functions of this structure. The inner side of the mantle edge is a respiratory surface, and, especially in larger species, outpocketings of the mantle blood vessel to produce secondary gill leaflets occur, probably independently, in Nacellidae, Patellidae, and some Lottiidae.

It is, of course, not known whether the expanded mantle edge, and its associated retractile tentacles, mantle nerve, and blood vessel, is a modification solely due to the adoption of a limpet-shape, but it seems probable these features may have been present in the coiled eogastropod ancestor. The shallow mantle cavity may have also been present, resulting in a gap in the mantle vein anteriorly, subsequently filled in a few larger species. The ctenidial vein on the left side becomes the left mantle vein when the ctenidium is absent.

The single left ctenidium in lottioideans is one of the main characters thought to indicate a coiled ancestor (Ponder & Lindberg 1997), with the loss of the right ctenidium being the result, in part at least, of the necessity to move waste material from left to right across the shallow embayment. This argument also suggests that the protoconch offset to the left mentioned above does not indicate a hypostrophically coiled ancestor. This gill was thought to be secondary by Thiele (1902) but is now generally accepted as homologous with the molluscan ctenidium on the basis of its position and details of its ciliation and structure (e.g., Yonge 1947).

There is no trace of the right ctenidium or its associated auricle in any living patellogastropod, but by outgroup comparison these structures must have been present in early eogastropods. The only remnant of paired mantle cavity organs is the pair of small osphradia which are retained in many patellogastropod taxa (Haszprunar 1985g), clearly a plesiomorphic condition, although the loss of the right osphradium, or both of them, has occurred in some taxa. At the ultrastructural

---

[1] A similar protoconch morphology is convergently present in some cocculiniforms.

level, the osphradium of patellogastropods is unique among gastropods in having its epithelium with free nerve ends only (i.e., lacking sensory cells) and in having cilia star cells (Haszprunar 1985g).

There are major differences between patellogastropods and orthogastropods in the structure and function of the feeding apparatus and the gut morphology. In the discussion below, and in Table 18.1, we attempt to tease out which of these structures may be plesiomorphic gastropod features and which may belong to the eogastropod lineage.

The change from a flexoglossate to a stereoglossate radula was a relatively simple one. The plesiomorphic condition is intermediate, as seen in the radula of chitons and monoplacophorans, which like patellogastropods are docoglossate, but they possibly have some flexoglossate features (Guralnick & Smith 1999). Thus, while the plesiomorphic condition is scraping plus some sweeping functions, in patellogastropods it is solely a scraping radula. The patellogastropod radula is also unique in having the teeth attached to expanded basal plates (Figure 18.13). Associated with the changes in radular function were changes in musculature and odontophoral cartilages (see above) (Guralnick & Smith 1999).

The first gastropods probably had ductless salivary glands (see Section 18.2.1.1), although salivary ducts are present in patellogastropods.

The early eogastropods probably moved slowly about, scraping rocks and/or stromatolites. Food particles were often coarse with substantial potential waste material. The plesiomorphic dorsal jaw became more complex to protect the dorsal anterior buccal cavity. The coarse nature of the food meant that gastric sorting areas could not function efficiently and the stomach proper became small, being only a vestibule to receive the oesophagus and the openings of the digestive glands. Digestion was extracellular, enabled by enzymes from the oesophageal glands and digestive glands. The style sac was greatly enlarged and elongated because a rotating protostyle was no longer needed and no gastric shield was necessary. The style sac and long, coiled intestine became the sites of digestion and resorption respectively. These changes are probably correlated with developing primary scraping and it is likely that a stereoglossate radular function occurred before the adoption of limpet-shaped shells. Uniquely the paired digestive glands open laterodorsally to the stomach (in all other gastropods they open ventrally).

The heart complex is uniquely configured in patellogastropods. Both the anterior and posterior aortae lie posteriorly (rather than on the right as in vetigastropods). This is probably not the primitive configuration because the aortae emerge from the posterior longitudinal face of the ventricle, not its posterior end (see Section 18.4.3.6). Unlike the outgroups and other gastropods, podocytes are on both the auricle and ventricle. Uniquely among gastropods, patellogastropods lack haemocyanin, but whether this is a feature shared with monoplacophorans is currently unknown.

In patellogastropods, the small left kidney uniquely lies on the right side of the pericardium in the visceral wall at the back of the mantle cavity, and there is an important difference in the location of the right kidney in patellogastropods and vetigastropods. The large right kidney of patellogastropods extends posteriorly under the digestive gland as in Polyplacophora, but in vetigastropods, the kidney lies on top of the digestive gland. The gonad lies ventrally as it does in monoplacophorans, unlike the dorsal position it occupies in chitons, *Nautilus*, and most other gastropods. It opens directly to the right kidney as it is assumed to do in the ancestral gastropod (see Section 18.2.1.1). In patellogastropods the renal epithelia in both kidneys differ in their histological and fine structural details from those of other gastropods, leading Andrews (1988) to indicate these differences lent support for the early divergence of this group from other gastropods. The wall of the left kidney that abuts the pericardium is thick and spongy, with tubules, and reminiscent of the structure of the nephridial glands seen in higher gastropods (Andrews 1985).

Reproduction and larval development are little changed from the ancestral condition.

The patellogastropod nervous system differs from the plesiomorphic condition seen in vetigastropods in having better-developed ganglia, including sub- and supraoesophageal ganglia on the visceral loop that respectively give off a nerve to the left and right osphradial ganglia. These ganglia, in *Patella* at least, innervate not only the osphradium (and ctenidium in Lottioidea) but also the pericardium, kidneys, and salivary glands.

Labial ganglia (absent in all other gastropods and outgroups) innervate the inner lips of the mouth. The pedal ganglia are closely connected to the pleural ganglia and give off large pedal cords with only one or two cross connections.

Unique sensory organs include the wart organs and sensory streaks (see Section 18.4.3.7), and the cephalic tentacle epithelium and osphradial ultrastructure differ from those of other gastropods.

The early fossil record of patellogastropods is poor, as discussed in Section 18.4.5 and in Chapter 13. Potential candidates for early eogastropods, including coiled members of the group, are considered in Chapter 13 and in more detail in the Appendix.

Some of the apomorphic anatomical characters listed in Table 18.1 are undoubtedly correlated – for example, the short visceral loop in the nervous system is related to the shallow mantle cavity.

18.2.1.3 Orthogastropoda

The lineage that gave rise to the orthogastropods was characterised by a shell with relatively tight coiling and associated deepening of the mantle cavity.

If the gastropod ancestor had a limpet-like shell, coiling might have occurred independently in the different groups of mainly hypostrophically coiled eogastropods. These gastropods also developed a coiled protoconch of about one whorl or less, a state indicative of lecithotrophic development in lower living gastropods. It is assumed that the protoconchs of the first orthogastropods were also like this, although some Paleozoic vetigastropod-like taxa had what appears to be a protoconch suggestive of planktotrophic larval development (see Section

18.5.5), indicating that dextral planktotrophic larval shells may have been present not only in early Neritimorpha and Caenogastropoda but perhaps also in an early lineage of vetigastropods. The Heterobranchia differ in having sinistral protoconch coiling. Because all Triassic to extant Vetigastropoda, as well as patellogastropods, have non-planktotrophic protoconchs of about one whorl or less, it is assumed that planktotrophic larvae were not basal in Gastropoda, and it remains unlikely that 'true' vetigastropods ever produced planktotrophic larvae (Nützel et al. 2007b).

The deepening of the mantle cavity was associated with an increase in the length of the ctenidia. This change increased the necessity for sanitation of the cavity, a need met in two main ways: (1) by moving the exhalant current to the right, reducing and eventually losing the right gill and, over time, the other right mantle cavity organs, or (2) by developing an embayment in the mantle, and hence in the underlying shell aperture, to provide an alternative exhalant route (Lindberg & Ponder 1998). In conjunction with these changes, mucus-secreting hypobranchial glands evolved to bind waste material, facilitating its rejection.

The enlargement of the ctenidia necessitated the formation of skeletal rods to assist in stiffening the larger and more numerous filaments. This stiffening increased the efficiency of the lateral cilia in driving the respiratory water currents through the mantle cavity. Also, afferent and efferent membranes evolved to help stabilise the gills. This change occurred in a common ancestor of vetigastropods and caenogastropods but may not have happened before the development of the neritimorph lineage, as they lack skeletal rods. Skeletal rods are also lacking in cocculinids, but that loss may be secondary as the gill in these limpets is much reduced.

Another key change was to the radula where we see a significant enhancement in the sweeping capabilities and less reliance on scraping, resulting in the intake of finer food particles with associated profound changes to the gut. This change in feeding enabled the early orthogastropods to utilise the rich biofilms covering all substratum surfaces without simultaneously taking in large amounts of sediment or other coarse mineral material. Hard mineralised radular teeth are not only unnecessary but a disadvantage in such feeding. Thus, with the increasing flexoglossate capabilities of the radula came an increase in the number of teeth in the marginal fields, increased flexibility of the teeth, and associated changes in the underlying musculature and cartilages. Orthogastropods retained the plesiomorphic pair of anterior (medial of Guralnick & Smith 1999) cartilages but lost the pair of dorsolateral cartilages (Guralnick & Smith 1999).

The salivary glands retained their plesiomorphic morphology in vetigastropods but developed ducts in higher gastropods. Sasaki (1998) argued that the vetigastropod salivary glands differed in being ramified and opening to the buccal cavity by long slits, and this suggested that, given their absence in cocculinids, neomphalines, and neritimorphs, they might be new structures. The salivary glands in apogastropods (caenogastropods and heterobranchs) are similar to the plesiomorphic condition, suggesting two equally plausible scenarios – either the loss of these glands is an apomorphy of neritimorphs and some vetigastropods, or the glands in vetigastropods and apogastropods evolved in parallel.

A subradular organ (licker) is retained in early orthogastropods although it is reduced in size, possibly due to a shift in food 'tasting' to the lips.

The oral lappet was retained in early orthogastropods but in living representatives is seen only in cocculinoideans, a few lepetelloideans, and neomphalines; there is a pair of similar structures (the cephalic lappets) in the neritimorph *Bathynerita* that may not be homologous (Warén & Bouchet 1993).

A heavy chitinous dorsal jaw was not required as the radula took on more of a sweeping rather than solely rasping function; instead thickenings were developed laterally and took on different roles such as nipping. Their structure changed from thick uniform material to a series of chitinous rods.

Unlike patellogastropods, the ventricle was not attached to the pericardium dorsally, and both auricles were retained. The ventricle had rotated so the bifurcated aorta, which maintained its primitive position, emerged on the left side. As in the outgroups, podocytes are present only on the auricle.

The different location and vascular supply of the two kidneys led to them maintaining different functions (as described in Section 18.2.1.1) and morphology. In many living vetigastropods, the right kidney is much larger than the left and is solely excretory. It lies in the visceral wall behind the mantle cavity and, unlike the situation in patellogastropods and chitons, it is on top of the digestive gland. The left kidney is small and located in the visceral wall in some (fissurellids, neritimorphs, caenogastropods) but lies in the mantle skirt in most vetigastropods (including neomphaloideans and cocculinoideans) and heterobranchs.

In contrast to its ventral position in patellogastropods, the gonad lies above the digestive glands (except in cocculinoideans), but, as noted above, it is unclear as to whether or not the dorsal position of the gonad is apomorphic. It opens to the right renopericardial duct rather than the kidney itself and shares a common duct to the exterior with the right kidney. Reproduction and larval development in early orthogastropods were little changed from the ancestral condition.

The nervous system initially retained its plesiomorphic configuration (see above) although more cross connections occurred between the pedal cords due to narrowing of the foot and increased muscular flexibility. In addition, as the mantle cavity deepened the visceral loop elongated by extending posteriorly.

The development of sense organs included the addition of a lens and, eventually, a cornea to the eyes and, in vetigastropods in particular, the addition of sensory structures on the head-foot and bursicles on the gills. The osphradium remained small in orthogastropods except for the caenogastropods where it markedly increased in size. In all orthogastropods, the osphradial epithelium, in contrast to patellogastropods, developed sensory cells with processes (Haszprunar 1985g).

18.3   THE LOWER GROUPS OF GASTROPODS

Several Paleozoic gastropod taxa do not appear to be related to living groups. These are discussed in Chapter 13, and their classification is outlined in the Appendix under Gastropoda. The living taxa and their relatives are summarised below.

18.4   PATELLOGASTROPODA (= DOCOGLOSSA, CYCLOBRANCHIA, EOGASTROPODA IN PART)

18.4.1   INTRODUCTION

Patellogastropod limpets are ubiquitous in hard shore intertidal communities throughout the world and include some of the best studied intertidal gastropods across disciplines such as ecology, physiology, behaviour, and reproduction. This group is of considerable biological importance (see Section 18.4.4.1 and Chapter 9), and many of the larger species are (or have been) used as food for humans.

Patellogastropods represent a subset of a group of gastropods previously referred to as 'archaeogastropods'. They are considered to be the sister taxon of all other living gastropods (Ponder & Lindberg 1997) (see Section 18.2.1.1.2), and while many of their characters are plesiomorphic (see Section 18.2.1.1.2 and Table 18.1), at the same time they are highly specialised and modified. Members of this taxon are characterised by their limpet-shaped shells that show no trace of coiling, and they have an uncoiled protoconch. They have the most diverse range of shell structures of any gastropods, and these include cone crossed-lamellar and, uniquely, foliated layers. They have lost one or both ctenidia and, in some, secondary mantle gills are present. Wart organs (= 'Spengel's organs') are present in some (see Section 18.4.3.7).

The characters of eogastropods listed above are almost entirely based on those of patellogastropods as the only living representatives of the group. Apomorphies of patellogastropods are also discussed in Section 18.2.1.1.2 and are listed in Table 18.1.

Some patellogastropod characters have evolved in parallel with other groups of gastropods (i.e., are homoplastic) including the loss of the right auricle and right (or both) ctenidia, simplification of the stomach (including the loss of the gastric shield), and the differentiation of the left and right kidneys. The displacement of both kidneys to the right of the pericardium is a unique patellogastropod character. Also, convergent with higher gastropods, the intestine lies outside the pericardium, and the heart has a single auricle. An operculum and one or two short, posterior, tentacle-like structures are present in the veliger, but they are absent in juveniles and adults (Smith 1935; Anderson 1965).

More subtle differences that set patellogastropods apart from other gastropods (see Table 18.1) include the ultrastructure of the cephalic tentacle epithelium (Künz & Haszprunar 2001), the osphradium (Haszprunar 1985g), and the kidney (Andrews 1985). Also, Daban et al. (1990) have shown that the protamines of patellogastropod sperm differ significantly from those of vetigastropods.

18.4.2   PHYLOGENY AND CLASSIFICATION

Despite Fleure (1904 p. 273) stating 'the Docoglossa diverged from a very primitive Prosobranch stock, and thus, though the divergence may be considerable, the group is an important one in connection with the problems of Gastropod phylogeny', the group was considered to be a superfamily within 'Archaeogastropoda' for most of the 20th century.

The recognition of the distinctiveness of patellogastropods is not a recent phenomenon as several 19th century malacologists proposed classifications recognising this. For example, Cuvier (1817) placed them with the polyplacophorans in the Cyclobranchia, a name most recently used by Golikov and Starobogatov (1975) at the subclass level for the docoglossate limpets alone. Troschel (1856–1879) proposed Docoglossa (in 1866) in recognition of the differences in the patellogastropod radula from that of other gastropods. Sars (1878) coined Onychoglossa on the basis of the claw-like morphology of the lateral teeth while Lankester (1883) proposed Phyllidiobranchia because of the structure of the secondary mantle gills in *Patella*, and Perrier (1889) proposed Heterocardia based on the patellogastropod heart. Despite this early recognition of their distinctiveness, Thiele (1925) placed them in his order Archaeogastropoda, ignoring the unique character combinations recognised by earlier workers. Their position as the sister group to other gastropods was again hypothesised by Golikov and Starobogatov (1975), and their distinctiveness has since been further clarified (Graham 1985; Haszprunar 1988a; Lindberg 1988c; Ponder & Lindberg 1997; Sasaki 1998; McArthur & Harasewych 2003; Lindberg 2008; Nakano & Sasaki 2011).

Patellogastropods were again subsumed in Archaeogastropoda in the classifications of Salvini-Plawen (1980) and Salvini-Plawen and Haszprunar (1987), although treated as a group of equal rank to Vetigastropoda and 'Neritopsina'. Wingstrand (1985) recognised patellogastropods as the most primitive living gastropods and Haszprunar (1988a), Lindberg (1988c), and Ponder and Lindberg (1997) subsequently recognised Patellogastropoda as sister to the rest of the gastropods. Currently this view is not held universally as molecular analyses tend to variably place the group within the lower parts of the gastropod tree, usually within the vetigastropods (e.g., Tillier et al. 1994; Harasewych & McArthur 2000; Colgan et al. 2003; McArthur & Harasewych 2003; Giribet et al. 2006; Aktipis & Giribet 2010; Uribe et al. 2016b). This result was also obtained in an analysis using morphology and several genes (Aktipis et al. 2008). These variable results are presumably caused by analytical problems with long branch attraction[2] or alignment (Nakano & Sasaki 2011; Uribe et al. 2019), with both patellogastropod and pleurotomarioidean sequences having long insertions (Harasewych 2002; Giribet 2003; Williams & Ozawa 2006; Nakano & Ozawa 2007; Aktipis & Giribet 2010). In some analyses patellogastropods are excluded because they are so divergent; for example Williams et al. (2014 p. 44) noted that

---

[2] Presumably because of the extinction of most of the related taxa.

'the single published patellogastropod genome was excluded from this analysis because it was deemed to be too divergent'.

The most recent gastropod classification (Bouchet et al. 2017 p. 370) recognised Patellogastropoda as a separate 'subclass' but at the same time did not recognise Eogastropoda because that hypothesis was 'rejected by recent molecular phylogenies'. The relationship of patellogastropods to other gastropods is discussed further in Section 18.2.1.1.

The classification of patellogastropods has changed markedly as just a few decades ago only a few genera in three families (Lepetidae, Acmaeidae, and Patellidae) were recognised. For example, most members of what we now know as Lottiidae were included in the genus *Acmaea*, a genus now regarded as a separate family (Acmaeidae) (Lindberg 1988c). Early classifications were based on features of the shell, radula, and gills, but later new characters involving shell microstructure, anatomy, sperm ultrastructure, and development, as well as data from molecules and fossils, suggested widespread convergence in the traditional characters (Lindberg 2008).

Ponder and Lindberg (1997) introduced Eogastropoda for Patellogastropoda plus their extinct coiled ancestors. Implicit in this proposal was a single derivation of the limpet form from the coiled lineage. Lindberg (2004) suggested that two patellogastropod groups may have independently evolved limpet-shaped shells (Lindberg 2008), one including the patellids and other the nacellids, lottiids, acmaeids, and lepetids. The patelloids are more nearly bilaterally symmetrical, as shown by some features of the nervous system and the possession of a pair of osphradia and wart organs, while the others show the loss and reduction of structures on the right side of the body typical of coiled snails. Whether the patellogastropods represent a single or multiple derivation remains equivocal (see Lindberg 2008 for additional discussion).

Recent investigations of relationships within Patellogastropoda have used both morphological and molecular data (see Nakano & Ozawa 2007; Lindberg 2008; Nakano & Sasaki 2011 for recent summaries). Some molecular analyses (Nakano & Ozawa 2007; Nakano & Sasaki 2011) show the family Eoacmaeidae at the base of the patellogastropod tree with the patellids the next branch above. From what is known, morphologically (radula, shell structure, ctenidium) the eoacmaeids are very like lottiids. Given the issues of molecular analyses using a limited set of genes, we prefer to give weight to the substantial morphological data that would place the patellids at the base of the tree assuming a single derivation of the limpet form. The split into two main clades identified by Lindberg (2008), i.e., Patelloidea + 'Nacelloidea' (Nacellidae), and the remaining patellogastropods, is supported by the analysis of Nakano and Ozawa (2004), although this split was not recovered by a later molecular analysis (Nakano & Ozawa 2007) where patellids were basal and the nacelloidean taxa the sister to the remaining patellogastropods. This configuration differs in other molecular analyses (see Nakano & Sasaki 2011 for a review), and as a result the phylogenetic arrangements of patellogastropods remain in a state of flux pending further study. A summary of the classification we adopt here is given in Box 18.3 and further details are provided in the Appendix.

18.4.2.1 Sister Group Relationships

All living patellogastropods are limpets with cap-shaped shells (Figure 18.11), but, as discussed above, it is likely that they are descended from coiled snails. This idea is in contrast to the view that gastropods were primarily limpets (Haszprunar 1988a), based on the shell and muscle morphology of monoplacophorans (see also Haszprunar 2008). The basal position of patellogastropods is not supported by the known fossil record because the earliest undoubted patellogastropod is from the Triassic of Italy (see Chapter 13), with their coiled ancestors not yet confidently identified (Frýda et al. 2008a and Chapter 13). As outlined above, based on a suite of morphological, developmental, and chromosomal evidence, patellogastropods are the most 'primitive' living gastropods (see also Section 18.2.1.1.2), although in molecular analyses their placement on the gastropod tree remains problematic with little agreement (see above).

18.4.2.2 Outline of Classification

Living patellogastropods are divisible into two main groups – the Patellida and Nacellida, and the Nacellida can be further divided into two suborders, Nacellina and Lottiina (see Box 18.3). Given the lack of agreement in recent phylogenetic analyses, this classification is adopted primarily on morphological differences for the reasons given in Section 18.4.2.

BOX 18.3   SUMMARY OF THE HIGHER CLASSIFICATION OF PATELLOGASTROPODA

(Infraclass) **Patellogastropoda**
(Order) **Patellida**
Superfamily Patelloidea
(Order) **Nacellida**
(Suborder) **Nacellina**
Superfamilies Nacelloidea and Lepetoidea
(Suborder) **Lottiina** (new suborder)
Superfamily Lottioidea

See the Appendix for a more detailed classification and the differentiating characters of the new taxon.

18.4.3   MORPHOLOGY

18.4.3.1   Shell and Protoconch

All living patellogastropods have cap-shaped shells (Figure 18.11) with the apex typically at the centre or anteriorly placed. Their adult shells range from about 5 to over 200 mm in length, but most intertidal species usually range between 20 and 40 mm. Both the smallest and largest species are typically found in the lowest intertidal zone or the subtidal zone.

The inner surface of the shell bears a horseshoe-shaped muscle scar opening anteriorly. The protoconch is a simple, caecum-like shell that may be bent but is never coiled, although it is often offset slightly to the left relative to the adult shell axis. The protoconch may be variously sculptured and is usually lost in the early juvenile stage.

FIGURE 18.11 Examples of living patellogastropod limpets. (a) *Acmaea mitra* (Acmaeidae), Monterey, California. © 2008 Gary MacDonald. (b) *Patelloida saccharina* (Lottiidae), Singapore. Courtesy of R. Tan. (c) *Lottia austrodigitalis* (Lottiidae), Cambria, California. Courtesy of D.R. Lindberg. (d) *Cellana radiata* (Nacellidae), Singapore. Courtesy of R. Tan. (e) *Cellana radiata* (ventral view), Singapore. Courtesy of R. Tan. (f) *Helcion pellucidus* (Patellidae), Dalebrook, South Africa. Courtesy of D.R. Lindberg. (g) *Patella vulgata* (Patellidae), Portaferry, Northern Ireland. Courtesy of D.R. Lindberg. (h) *Scutellastra cochlear* (Patellidae), Dalebrook, South Africa. Courtesy of D.R. Lindberg.

The adult shell aperture is generally oval, but some are modified for particular habitats. Elongated apertures with parallel sides are found in species that live on algae or seagrass. Species that occur on algae with round stipes have an elongated aperture with raised anterior and posterior edges so they fit the curved surface. Species with home scars often have highly crenulate edges or bear large rays that engage in a 'lock-and-key' arrangement with the contours of the scar. External sculpture, if present, is often of fine to strong concentric growth lines and/or radial ribs, or combinations of both, as in Pectinodontidae which have reticulate sculpture.

Subtidal species are typically white or pink while intertidal species are usually browns and greys with white spots and rays. Colour patterns can be genetically or environmentally induced, the latter as a result of incorporating pigments from food (see Chapter 3). Thus shell colour can be similar to the substratum on which the limpet occurs because of the incorporation of plant compounds into the shell (Lindberg & Pearse 1990); this colour can change when the limpet changes diet (Sorensen & Lindberg 1991).

The protoconch offset may be as much as 20° to the right of the anterior–posterior axis of the teleoconch and is in marked contrast to other secondarily flattened gastropod taxa such as various caenogastropod limpets (e.g., *Capulus*, *Crepidula*) where the protoconch is offset to the left. In those taxa, if the initial coiling were maintained, a dextral shell would result. In comparison, if coiling were initiated in the juvenile patellogastropod shell, the right offset would result in a sinistral shell. Given dextral anatomy and the counter-clockwise coiling of the larval operculum, it is reasonable to suggest that the coiled patellogastropods were hyperstrophic (Lindberg 1981b).

18.4.3.1.1 Shell Structure
The shell structure of patellogastropods distinguishes them from all other living gastropods (MacClintock 1967) and is typically comprised of four to six distinct shell layers. These microstructures include prismatic, foliated, and crossed-lamellar components, in contrast to vetigastropod shells, which are composed of combinations of intersected crossed-platy, nacreous, and prismatic structures (Hedegaard 1990, 1997).

Typically, the shell consists of outer layers of prismatic or foliated structures with the inner layers of either concentric or complex crossed-lamellar on either side of the myostracum (muscle scar). In different lineages, these basic types have been differentially emphasised (Lindberg 1988c; Fuchigami & Sasaki 2005) and, in most taxa, the exterior of the shell is prismatic rather than foliated. The youngest families have emphasised crossed-lamellar and prismatic structures, whereas most of the older groups have about equal foliated and crossed-lamellar components capped by a prismatic layer. Shell structure types are not associated with specific habitats, and, although geographic restrictions are apparent in some shell structure groups in living taxa, these patterns are not maintained when fossils are examined (Lindberg 1988c; Kase & Shigeta 1996; Fuchigami & Sasaki 2005). The hotvent neolepetopsids have a more simplified shell structure than some other patellogastropods, with prismatic complex

crossed-lamellar layers in all three taxa examined (Kiel 2004), and the addition of complex crossed-lamellar in one, and regularly foliated and simple prismatic in another.

18.4.3.2 Head-Foot, Mantle, and Operculum
The head has a pair of cephalic tentacles with eyes at their outer bases. There is a short, broad snout, and the mouth opens ventrally and is surrounded by a hood. The muscular foot is large, oval, and enables the limpet to adhere firmly to the substratum. The visceral mass is contained within a cavity delineated by the foot and the horseshoe-shaped attachment muscle (Figure 18.12). A marginal groove with pedal glands extends around the periphery of the foot, although there is no propodium (Grenon & Walker 1978; Lindberg & Dwyer 1983; Stützel 1984).

The mantle edge surrounds the shell opening and comprises two folds, and numerous retractile sensory tentacles emerge from the outer fold (Stützel 1984). The mantle extends past the shell attachment muscle to form the mantle groove between its ventral surface and the side of the foot. Anteriorly, the mantle cavity lies between the anterior portions of the shell attachment muscle forming a shallow embayment (this part is often called the 'nuchal cavity') and extends into the mantle groove, a structure not present in coiled gastropods. A narrow mantle muscle arises from the anterior ends of the shell muscle and runs through the anterior mantle well behind the edge.

Patellogastropods have two shell muscles joined posteriorly to form a U shape. Narrow gaps in muscles are caused by the passage of blood sinuses from the visceral mass to the mantle edge, a feature also seen in a few other gastropod taxa, including some Cocculinoidea and Neritimorpha (see Chapter 3).

18.4.3.3 Mantle Cavity
Patellogastropods have two main gill configurations: in the Lottiina (acmaeids and lottiids) a ctenidium is located over the head in a shallow nuchal (mantle) cavity, but is lacking in Patellida and Nacellina. Secondary gills occur in the mantle groove in Patellida, some Nacellina, and some larger Lottiina such as a few species of *Lottia* and *Scurria* (both Lottiidae). A few patellogastropods have no gills, including species of the seep taxon *Neolepetopsis* and the mantle cavity brooding taxon *Erginus*.

All patellogastropod gills lack skeletal support. The acmaeid and lottiid single (left) ctenidium is bipectinate with alternating filaments, and (as in chitons and monoplacophorans) they do not have afferent or efferent membranes. As in other ctenidia, there are ciliary bands on the filaments, but only irregular, unclumped cilia are found on the secondary gills (Yonge 1947; Nuwayhid et al. 1978). The lottiid *Rhodopetala* has a very reduced ctenidium that resembles a single filament of the patellid secondary gills (Lindberg 1981c).

In the Patellina, and in those Lottiina where secondary gill leaflets are present, they are located around the edge of the foot as 'mantle (or pallial) gills'. When present, these secondary gills arise from the circum-mantle blood vessel and are made up of numerous simple, flat, triangular leaflets. The

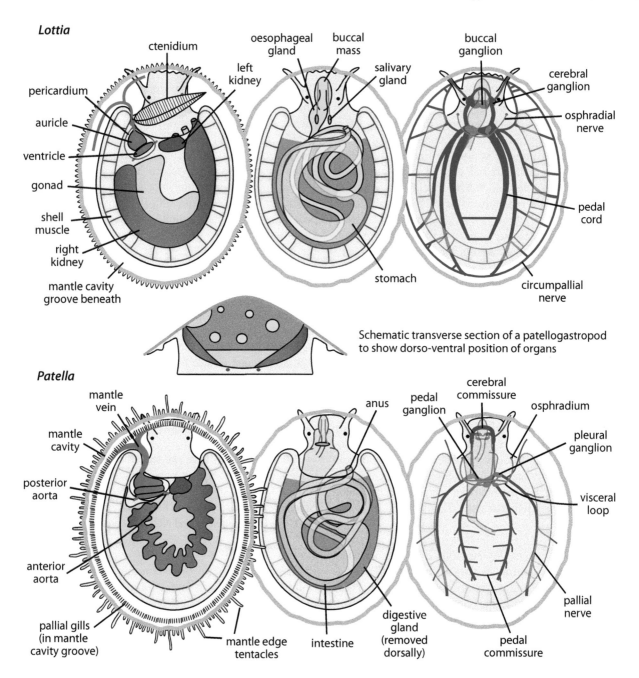

FIGURE 18.12 Comparison of the main anatomical features of a lottiid and a patellid patellogastropod. Redrawn and modified from the following sources: Top row left (Golikov, A.N. and Starobogatov, I., *Malacologia*, 15, 185–232, 1975), middle (Walker, C.G., *The Veliger*, 11, 88–97, 1968), right (Willcox, M.A., *Jenaische Zeitschrift fur Natürwissenschaft*, 32, 411–456, 1898). Bottom row left original based on various sources, middle (Sasaki, T., *Bull. Univ. Tokyo Mus.*, 38, i–vi, 1–223, 1998), right (Stützel, R., *Zoologica (Stuttgart)*, 46, 1–54, 1984), transverse section original.

presence and configuration of the secondary gills have been used in classification (Powell 1973; Lindberg 1998b). In some Patellina the gill ring is complete while in other taxa it is interrupted in front of the head (as is the circum-mantle blood vessel in those taxa). Some lottiids have secondary gills in addition to the ctenidium, and some large lottiids also have anterior secondary gills and thus converge on the condition in some Patellina. This development of the secondary gills to the anterior mantle skirt in some lottiine taxa is enabled by the secondary extension of the circum-mantle blood vessel across the anterior end of the mantle (Fisher 1904).

Unorganised cilia are found in the mantle groove and nuchal cavity (Yonge 1947; Fretter & Graham 1962) as well as over the body surface (Voltzow 1994). Sensory streaks that bear cilia are also found (see Section 18.4.3.7); these also produce mucoid secretions and may function in an analogous way to hypobranchial glands (Yonge 1947), as true hypobranchial glands are absent.

The mantle cavity of the Patellina often contains a pair of wart organs, which have been thought to represent paired gill rudiments by some authors (Stützel 1984; Ponder & Lindberg 1997), but here we treat them as possible sensory structures

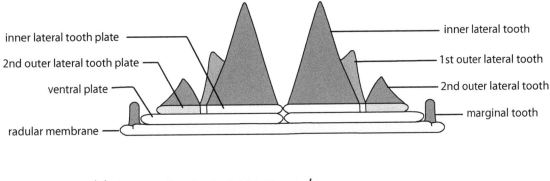

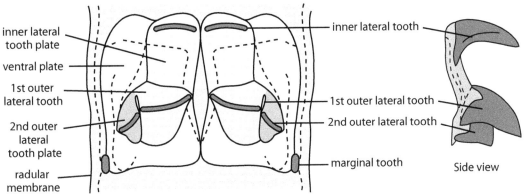

FIGURE 18.13   Structure of the radular teeth of *Lottia*. Redrawn and modified from Moskalev, L.I., *Trudy Instituta Okeanologii*, 88, 174–212, 1970.

as suggested by Haszprunar (1985g) (see Section 18.4.3.7). They are associated with the osphradia, which may be paired, single, or absent and are discussed in Section 18.4.3.7.

18.4.3.4   Digestive System

The alimentary system consists of a buccal cavity, with the odontophore usually containing more than one pair of cartilages, a docoglossan radula, a glandular oesophagus, a simple stomach, a long, coiled intestine, and a rectum. Unlike many vetigastropods, the patellogastropod stomach lacks a caecum and gastric shield, and the rectum does not penetrate the ventricle.

The patellogastropod mouth region is the most complex of any gastropod. It is typically composed of outer lips, inner lips (sometimes called palps), and oral lappets.

Variation in the mouth structures occurs, with various parts absent in some taxa, and sometimes there are elaborations such as papillae. For example, the lateral portions of the outer lips of *Nacella* and *Cellana* are uniquely fringed with clusters of small papillae. Nacellids and some other patellogastropods have a continuous outer lip, but in many others (acmaeids, lottiids, and patellids), they are bifurcated ventrally. The oral lappets are extensions of a thin flap of tissue that spreads dorsally over the outer lips and expands ventrally and laterally. They are found in lepetids, pectinodontids, acmaeids *Erginus* and *Rhodopetala*, lottiids, and patellids. Papillae of unknown function line the middle part of the oral lappets in the acmaeids *Erginus* and *Rhodopetala*, lottiids, and patellids.

The inner and outer lips are histologically similar, being composed of a combination of high epithelium, sensory,

and secretory cells. These lips are sometimes attached to each other and sometimes not (Thiem 1917b). A marked sensory area is developed on both inner lips in lepetids (see Section 18.4.3.7) and may be present in at least some other patellogastropods.

Just inside the mouth opening lie the jaws, licker, and radula, which are described below.

The cuticularised jaw lies dorsally and is present in all patellogastropods. It is composed of four distinct subunits that vary in size in different groups (Thiem 1917b; Sasaki 1998); a thick anterior jaw element that is rugose anteriorly, a pair of thin lateral extensions, a pair of large, thin, wing-like posterior lateral flanges, and a thick middle posterior jaw element which lies between the two posterior flanges.

The configuration of the patellogastropod radula (Figures 18.13 and 18.14) is docoglossate (see Chapter 5); there are relatively few teeth per row, and the general trend in the group is towards reduction and loss of teeth. There are three or fewer pairs of marginal teeth or plates and two to six pairs of lateral teeth impregnated with ferrous oxides that impart a brown or black colour. A 'central tooth' can be present or absent. Each tooth consists of a shaft and cusp and base; the base affixes to a lateral tooth plate that in turn rests upon a basal plate attached to the radular membrane (Figure 18.13). In some taxa (e.g., Lepetidae) the lateral teeth appear affixed directly to the basal plate. This configuration of the tooth and the basal plates is unique among gastropods. Cusps are usually distinguished by their differential mineralisation (ferrous oxides), while the shaft and bases mineralise mostly with silicon dioxides (Runham et al. 1969; Rinkevich 1993). The marginal teeth are always unmineralised.

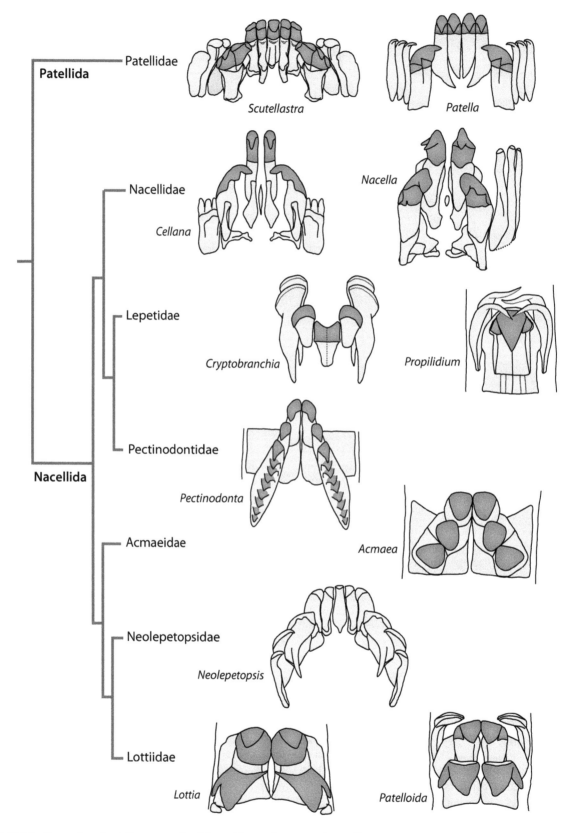

FIGURE 18.14 Examples of patellogastropod radulae with complete rows shown. The darker colouration indicates the mineralised parts of the teeth. Redrawn and modified from various sources. The radulae of the species illustrated do not necessarily reflect the range of variation in the genera they belong to or of their higher taxon.

'Central teeth' (a fusion of inner lateral teeth) occur in some patellids and neolepetopsids. When present, the tooth is usually bilaterally symmetrical and small; a relatively large tooth occurs in neolepetopsids. Central teeth are absent in all other patellogastropods (Figure 18.14).

In the Patelloidea there are four inner lateral teeth with a median lateral present in some taxa. The individual teeth usually rest on a single plate. A similar configuration is also (convergently?) found in the Neolepetopsidae. There are only two inner lateral teeth in the Lottiidae, Nacellidae, Eoacmaeidae, and Pectinodontidae. In the Lepetidae, the inner field is not expressed, and the two outer lateral teeth are typically fused and share a common basal plate. The basal plates underlying these teeth are often complex, especially in the Lottioidea and Nacelloidea. The outer lateral teeth are usually unicuspid but in some taxa are multicuspid (called *pluricuspid*) with two–four cusps in most taxa, but *Pectinodonta* has a unique outer lateral composed of up to ten cusps per tooth (Figure 18.14). All taxa, except lepetids and neolepetopsids, have mineralised outer lateral teeth, and the inner laterals are mineralised in all taxa except Neolepetopsidae. There may be three marginal teeth in Patelloidinae (a subfamily of Lottiidae) and Nacellidae, two in some Lottiidae, Lepetidae, and Eoacmaeidae, one pair in some Lottiidae, or none in some Lottiidae and Pectinodontidae. These teeth, also called *uncini*, are smaller than the other teeth, and, in *Lottia*, the single pair of marginal teeth is substantially reduced in size. Additional accessory plates are present in the marginal tooth fields of some Lottioidea and Nacelloidea.

Besides this variation in the number of teeth and development of the basal plates, the alignment of the teeth also differs among the taxa with the teeth arranged in stepped rows or in a posteriorly diverging V shape (alveate configuration).

The radular sac lies coiled behind the buccal mass; in patellids and nacellids it lies between the viscera and the foot muscle mass, but in other patellogastropods it coils within the visceral mass. The subradular organ (licker) is an unpaired structure on the anterodorsal wall of the sublingual pouch, anterior to the anterior limit of the radular membrane. It can be distended with blood from the buccal sinus, and, when retracted, concentric ridges are formed on the surface. Externally it is covered with cuticle or small cuticular scales.

Some patellogastropods have more odontophoral cartilages than other gastropods. There are three (Sasaki 1998) to five (Nowikoff 1912; Graham 1964) pairs of cartilages in patellids and six in nacellids – the anterior, posterior, anterolateral, anteroventral, and posterodorsal (Sasaki 1998; Guralnick & Smith 1999) in patellids, and with the addition of a small pair of anterior labial cartilages in *Cellana* (Sasaki 1998). Only some patellids and nacellids have posterior cartilages while acmaeids, lottiids, and lepetids have two pairs (anterior and anterolateral). Anterolateral cartilages are unique to patellogastropods.

The largest pair of cartilages, the anterior, are homologous with those found in other gastropods, but in patellogastropods, they differ by tending to become deepened anteriorly and by being fused anteriorly or at least closely approximated (Guralnick & Smith 1999). Anterolateral cartilages (dorsolateral of Guralnick & Smith 1999) are also probably plesiomorphic in Mollusca (Guralnick & Smith 1999), although Sasaki (1998) considered them to be apomorphic in patellogastropods and not homologous with the lateral cartilages in chitons or monoplacophorans. Guralnick and Smith (1999), however, argued that, based on position and morphological similarity, the dorsolateral cartilages of chitons, monoplacophorans, and patellogastropods are probably homologous. While their laterodorsal position is largely maintained in patellids, in lottioideans they are more reduced and have shifted further dorsally.

Guralnick and Smith (1999) pointed out some interesting modifications in the patellogastropod buccal apparatus. In outgroups such as polyplacophorans, the anterior part of the subradular membrane rests on a dorsal pair of cartilages that lie above the dorsolateral pair. Modifications in the Patellogastropoda include the replacement of the dorsal pair of cartilages by a pad of connective tissue (as in *Patella*), or in the basal lottioidean *Eoacmaea*, where the subradular membrane rests on the dorsolateral pair of cartilages. In those lottiids that have lost the dorsal pair of cartilages, the subradular membrane rests on the hypertrophied buccal mass muscles. When present, the dorsal cartilages are located only in the anterior part of the buccal mass. In Lottioidea, the radula rests on muscle while in patellids, it lies on a pair of dorsomedial cartilages (Guralnick & Smith 1999).

The musculature of the patellogastropod buccal mass is complex and has been described in general terms in Chapter 5; detailed accounts are given by Graham (1964), Fretter and Graham (1962), and Sasaki (1998). Patellogastropods differ from other gastropods in having transverse labial muscles associated with the paired inner lips, the dorsal protractor muscles of the odontophore, and two layered approximator muscles of the odontophoral cartilages (Sasaki 1998).

The configuration of patellogastropod salivary glands varies within the groups (Sasaki 1998). They may be large or small and sometimes lie behind the buccal mass to which they are connected by a pair of long ducts, as in *Cellana* and most Lottiidae. Patellids have a pair of large glands, with two ducts arising from each. A few patellogastropods, including the lottiid *Niveotectura* and some lepetids and pectinodontids, have ductless tubular salivary glands on the buccal mass (Sasaki 1998).

The oesophagus has been described in *Patella* (Graham 1932; Bush 1989). It is broad and complex with a deep ciliated ventral food channel and a trilobed ciliated dorsal fold in the short anterior oesophagus. In the midoesophagus, close, transverse, sheet-like partitions fill the pouch-like lateral parts and alternate on each side. The oesophagus is rotated because of torsion so the dorsal fold becomes ventral and the food channel dorsal before reaching the stomach, and the pouches are much reduced in size in the posterior part (the posterior oesophagus). The partitions that fill the lateral pouches are covered with ciliated and glandular cells, the latter producing amylase (Bush 1989). The oesophagus of other patellogastropods is similar (Sasaki 1998), although the number of lobes on the dorsal fold varies.

The posterior oesophagus enters the small stomach through a valve. The common digestive gland duct opens to the distal end of this short section, and then the stomach becomes the style sac. The style sac region is long and markedly broader than the stomach proper and the intestine and makes a large loop around the viscera. Its columnar epithelium contains pigment granules and has short cilia, thus differing from the style sac in other gastropods. The transition from style sac to the intestine is indicated by a marked narrowing and a change in histology.

The digestive gland is large and surrounds the gut. It is composed of two lobes fused into a single, lobulated organ. The digestive gland opens through two ducts that join into a common duct just before entering into the distal end of the stomach proper.

The patellogastropod intestine is much longer than that of other gastropods. Based on histology, Graham (1932) divided it into four regions, excluding the style sac region of the stomach. The pattern of gut looping in patellogastropods differs from other gastropods and the outgroups. As the stomach ends, the beginning of the intestine loops anteriorly before passing posteriorly and usually makes several coils before the final loop back anteriorly. From there it becomes the rectum, which is always on the right-hand side of the anterior mantle cavity. It then opens as the anus on the end of a short papilla, on either side of which are the left and right kidney openings.

The amount of gut looping has been used in patellogastropod systematics (e.g., Lindberg 1988c) although there is much variation within each group, and the basic pattern of looping does appear to differ in different groups.

18.4.3.5 Reproductive System

Patellogastropods have a single gonad that lies on the ventral surface of the visceral mass against the dorsal surface of the foot. The gonad opens into the right kidney, and the gametes are expelled from its opening in the mantle cavity. The structure of the ovary has not been well studied except in southern African Patellidae where Hodgson and Eckelbarger (2000) found it differs from studied vetigastropods in the trabeculae having large haemocoelic spaces and hypertrophic epithelial cells, in lacking jelly cells and smooth muscle, and in collagen fibres not being prominent. Most patellogastropods have separate sexes, although some are simultaneous or protandric hermaphrodites. Protandry is often correlated with territorial species that typically become female on acquiring a feeding territory (Branch 1974; Lindberg & Wright 1985).

Most discharge their gametes directly into the sea, where fertilisation and development take place. Thus there is no courtship or mating between individual limpets, although group spawning occurs in at least one Antarctic species of nacellid with individuals stacked on top of each other, increasing the probability of fertilisation (Picken & Allan 1983) (see Chapter 8, Figure 8.21).

There have been several studies involving the ultrastructure of patellogastropod sperm (e.g., Azevedo 1981; Hodgson & Bernard 1988; Hodgson & Chia 1993; Hodgson 1995; Hodgson et al. 1996). Patellogastropod sperm are typical

ectaquasperm, the plesiomorphic state in Gastropoda. The headpiece (nucleus and acrosome) grades from squat to elongate, and the midpiece consists of four to five mitochondria and two centrioles.

The larger species produce millions of eggs, but smaller species produce far fewer. While most patellids, nacellids, and many lottiids typically have yearly reproductive cycles, juveniles and some small lottiid species can spawn throughout the year (Fritchman 1962; Nicotri 1974). The eggs are typically small, about 90 μm in diameter, lack an outer vitelline layer, and contain sufficient yolk to last the developing limpet through settling and metamorphosis.

Only *Erginus rubella* uses the much-enlarged left kidney as a brood chamber, and in that species, the kidneys have been rotated 90° so that the left is dorsal and the right ventral (Lindberg 1988c). Other brooding taxa, *Erginus sybaritica* and *Rhodopetala*, use the enlarged nuchal cavity as a brood chamber (Golikov & Kussakin 1972; Lindberg 1983). *Erginus sybaritica* has a tentacle-like 'subcephalic tentacle' or 'penis' on the right side of the head (Golikov & Kussakin 1972; Lindberg 1987; Sasaki 1998).

18.4.3.6 Renopericardial System

The left and right kidneys lie on either side of the rectum, and the left is much smaller than the right, except in one species where it is enlarged and modified as a brood chamber (see the previous section). The histology of the left and right kidney is distinct, with the left kidney epithelium columnar and involved in resorption, while the right kidney is lined with ciliated vacuolated cells containing excretory spherules and these are involved in active transport and excretion (Andrews 1985) (see Chapter 6). In patellogastropods the large right kidney extends posteriorly under the digestive gland (as in chitons). Both kidneys open to the posterior part of the anterior mantle cavity on the right side and either side of the anal papilla. In some species, the right kidney opening, which also functions as the genital aperture, is extended as a long urinogenital papilla.

In patellogastropods, the larger right kidney lies behind the mantle cavity and extends over the dorsal surface of the foot, while the small left kidney lies at the base of the visceral hump behind the mantle cavity. In contrast, in most vetigastropods, the small left kidney (called the 'papillary sac') lies in the mantle cavity roof anterior to the pericardium and the right kidney posterior to the mantle cavity.

The circulatory system is open and, unlike other gastropods, the blood lacks haemocyanin. The heart consists of a ventricle and a single auricle, and there is often also a muscular 'aortic bulb' that assists in pumping blood through the anterior aorta to the head and the posterior aorta to the viscera. This bulb is derived from heart tissue, not from the aorta (see Chapter 6). The heart is rotated so that the left auricle lies anterior to the ventricle, and instead of the aorta emerging from the originally posterior end that now lies on the left side (as in vetigastropods), it has migrated so it arises from the middle of the ventral side (Fleure 1904). The ventricle is attached to the pericardium dorsally.

The rotation involving the heart is also evident by the post-torsional left and right kidneys being on the right of the pericardium and the renopericardial duct of the left kidney lying above that of the right. Thus, the left kidney is uniquely positioned on the right side of the pericardium, where it lies in the visceral wall at the back of the mantle cavity. The large right kidney extends posteriorly under the digestive gland. In patellogastropods (as in most vetigastropods), the right kidney receives deoxygenated blood primarily from the visceral sinuses while the resorptive left kidney receives partially oxygenated blood primarily from the head, mantle, and foot (see Chapter 6, Figure 6.10). The blood is oxygenated as it passes through the thin tissues of the mantle margin and, if present, the secondary gill leaflets. It is collected in the substantial circum-mantle vessel that reaches the auricle via a distinctive Y-shaped vessel. The secondary gill leaflets are outgrowths from the circum-mantle vessel (see Section 18.4.3.3). In those taxa with a ctenidium, some of the blood returning to the heart passes through the gill before reaching the auricle.

The number of blood sinuses through shell muscles is related to the available respiratory surfaces with the most numerous vessels in lepetids that lack a ctenidium or secondary gills (Thiem 1917b; Fretter & Graham 1962).

Although only investigated in *Patella*, filtration occurs via podocytes that are uniquely present on both the auricle and the ventricle (Økland 1982).

In all patellogastropods, the rectum lies outside the pericardium except in the assumed progenetic *Erginus apicina*, where it is surrounded by the pericardium, not the ventricle (Lindberg 1988c, d).

18.4.3.7 Nervous System and Sense Organs

The patellogastropod nervous system has some important features that differentiate it from that of vetigastropods. These include the presence of labial ganglia and a labial commissure. The former are absent in all other gastropods and the outgroups, but the commissure is present in the outgroups, neritimorphs and ampullariids (Sasaki 1998). Other unusual features include having a much smaller visceral loop, which in some, such as *Notoacmea fragilis* (Willcox 1898), is euthyneurous; a pair of ganglionic nerve cords that circle the mantle margin; and only one or a few commissures that connect the substantial pedal nerve cords. The supra- and suboesophageal ganglia each give off a nerve that runs to the osphradial ganglion, and the left osphradial nerve also supplies the ctenidium when that is present.

The cephalic eyes have no lens or cornea, being simple open pits. They are innervated by a short optic nerve that originates from each cerebral ganglion. Two osphradia and their ganglia exist in many patellogastropods including patellids, lottiids, and nacellids, where they lie on either side of the neck near the shell attachment muscles and near the base of the ctenidium on the left side in lottioids. A single osphradium occurs in some lottioids, such as *Testudinalia tessulata* (Yonge 1947), and osphradia are absent in several taxa such as lepetids, pectinodontids, *Niveotectura*, *Erginus* (Sasaki 1998), and *Bathyacmaea* (Sasaki et al. 2006a). A single osphradial ganglion was found in *Eulepetopsis vitrea* (Fretter 1990), but no osphradium.

Besides the osphradia, some patellogastropods have two unique sensory structures, the wart organs and sensory streaks (Thiele, 1893; Thiem, 1917a; Yonge, 1947; Stützel, 1984; Haszprunar, 1985g; Sasaki, 1998). The wart organs (or tubercles or tubercle organs), best studied in patellids, lie adjacent to the osphradia and are innervated by the osphradial nerves. They are present in patellids and nacellids but appear to be absent in most lottiids and acmaeids. Each consists of a small, often brightly coloured, lump of spongy tissue that contains nerve tissue, blood spaces, and connective tissue (Stützel 1984; Haszprunar 1985g; Sasaki 1998). These organs were thought to be ctenidial rudiments by some (see Section 18.4.3.3) but are most likely sensory organs (e.g., Haszprunar 1985g). Some workers (e.g., Fretter & Graham 1962; Walker 1968) have wrongly regarded these structures as osphradia.

When the wart organs and osphradia occur together they are called an osphradial complex, and these are known in some species of *Patella*, *Cellana*, *Nacella*, and *Nipponacmea* (Thiem 1917a; Sasaki 1998). Left and right wart organs may substantially differ morphologically, the left often, but not always, more lobate than the right (Thiem 1917a; Stützel 1984). Interpretations of their function are problematic (see above): a sense organ function is unproven because of the lack of nervous tissue compared to the adjacent osphradia, while as ctenidial rudiments the co-occurrence of a full ctenidium and wart organ in some *Nipponacmea* renders this unlikely. Possibly the wart organs enhance the sensory role of the osphradia, perhaps by facilitating the capture of molecules by changing boundary current conditions around the osphradium in a way analogous to that seen in some crustaceans, where manipulation of flow is used to enhance the concentration of odour plumes across antennules (Koehl et al. 2001). Also, the asymmetry of the wart organs could indicate that the left and right osphradia may be providing different information to the animal. Because the function of the patellogastropod wart organs remains open to conjecture, they are a subject worthy of much more detailed histological, biomechanical, and functional study.

Other unique sensory structures in patellogastropods include paired sensory strips innervated by nerves from the osphradial and pleural ganglia (Thiele 1893; Thiem 1917b; Haszprunar 1985g). These include the subpallial[3] strips that run around the anterior edge of the shell muscle but do not reach the osphradia. They have been described in detail in only a few patellid[4] and nacellid taxa. Some species of *Acmaea* and *Pectinodonta* have just the subpallial sensory streak (Thiele 1893; Haszprunar 1985g). The dorsal sensory streak is another type of sensory structure which has been described in some patellids, nacellids, and lottiids (Thiele 1893; Haszprunar 1985g) but may be present in most patellogastropods. It is a streak of sensory epithelium on the left part of the roof of the anterior mantle cavity beneath the pericardium and adjacent to the anterior end of the left shell muscle (Haszprunar 1985g).

---

[3] Subpallial refers to structures in the mantle groove.
[4] Some TEM details have been provided for the patellid *Patella* (*Helcion*) *pellucidum* (Haszprunar 1985g).

Osphradia, wart organs, and the subpallial and dorsal sensory streaks are all innervated by the osphradial ganglia or nerves arising from them. These structures are all absent in lepetids, which lack osphradial ganglia, but that group has a third type of sensory streak composed of 'neurolymphoid' tissue on the anterolateral mantle groove, which is innervated by the mantle nerve (Angerer & Haszprunar 1996). Lepetids also have an apparently unique paired sensory organ on the inner lips that also consists of 'neurolymphoid' tissue similar to that of wart organs and the sensory streak, but it is innervated by a nerve from the labial ganglion (Angerer & Haszprunar 1996).

While these different types of sensory streak, the wart organs and the osphradia are present in various patellogastropod taxa, knowledge of the details of their distribution within the different superfamilies remains poor.

Haszprunar et al. (2017) argued that one or two short, posterior, tentacle-like structures found in the late veligers of some patellogastropods were epipodial sense organs, although they had no evidence for this apart from them having a similar position to those seen in some vetigastropods and a similar (very simple) morphology. We are doubtful of the supposed homology of these structures, particularly in the absence of any ultrastructural evidence.

The statocysts are plesiomorphically located laterally to the pedal ganglia and contain statoconia.

18.4.4 Biology, Ecology, and Behaviour

18.4.4.1 Habits and Habitats

Patellogastropods are marine, being a predominately intertidal and shallow subtidal group where they form a conspicuous element of the fauna. Patellids and nacellids are restricted to intertidal and shallow subtidal habitats, but the remaining taxa range from the upper reaches of the intertidal to depths of over 4,000 m. A few live in brackish habitats, and one possibly extinct lottiid (*Potamacmaea fluviatilis*) may have lived in brackish and fresh water in the major rivers that drain into the Bay of Bengal in Southeast Asia (Lindberg 1990). Some taxa, like the Lepetidae and Acmaeidae, are mostly subtidal, while others such as the deep-water pectinodontid genera *Pectinodonta* and *Bathyacmaea* can live in depths of over 4,000 m (Sasaki et al. 2006a, b) where they feed on bacteria-rich waterlogged wood and detritus. *Eulepetopsis* (Neolepetopsidae, Lottioidea) and *Bathyacmaea* species are found living in deep-sea cold seeps and around hydrothermal vents (McLean 1990b; Nakano & Sasaki 2011). Some neolepetopsids also occur on whalebone in the deep sea (McLean 2008).

Most littoral species live on hard surfaces, with some restricted to calcareous substrata such as limestone or the shells of other molluscs, or on coralline algae, others on brown algae, and some live exclusively on seagrasses. The eastern Pacific lottiid '*Tectura*' *paleacea* and '*Tectura*' *depicta* live and feed on marine angiosperms (e.g., Vermeij 1992), and the Australian lottiid *Naccula parva* is associated with *Posidonia* and *Amphibolis* (Hickman 2005), but the exact composition of their food has not been studied.

Detailed ecological studies involving patellogastropods have been conducted in many parts of the world (Underwood 1979; Branch 1981 and references therein) but notably in the United Kingdom, southern Africa, North America, New Zealand, and Australia. Because they are primary consumers in the food chain, and predominately sessile and easily manipulated and marked, they have been called the 'guinea pigs' of rocky shore ecologists. They also characterise tidal zones and habitats and have been used in monitoring the ecological health of rocky shore communities (e.g., Navrot et al. 1974; Crump et al. 2003; Roy et al. 2003).

18.4.4.2 Locomotion and Behaviour

Because a wide range of predatory animals feed on patellogastropods (see Chapter 9), they are adapted to avoid predation by their shell morphology and clamping behaviour involving a powerful suction that requires considerable force to dislodge the animal. Intertidal species are also subjected to desiccation, temperature extremes, and osmotic stress, and these pressures, including those resulting from predation, have resulted in the evolution of some complex activity patterns. For example, many species are most active at night at low tide when visual predation is less effective and desiccation less likely. Recruitment of intertidal species generally occurs in the lower intertidal, with subsequent up-shore movement. Migration down-shore during winter has also been reported in colder climates.

At least three groups are territorial, with the best-known some southern African Patellidae. Some territorial patellids aggressively use their shells as battering rams to drive both conspecifics and other herbivorous species from their territory, and they are often larger than related non-territorial species (see also Chapter 9).

18.4.4.3 Life History

In most species eggs and sperm are liberated into the water column where fertilisation occurs, and the larvae pass through a trochophore and a non-feeding (lecithotrophic) veliger stage before settling and metamorphosis. Parental brood protection has evolved at least twice where the eggs are fertilised by sperm casting in the female mantle cavity, and the embryos develop into crawl-away young. A copulatory structure is known in only one species (see Section 18.4.3.5).

18.4.4.4 Feeding and Diet

Littoral patellogastropods indiscriminately graze on diatoms, blue-green algae, algal spores, or small pieces of plant material from the substratum. Some lottiid and patellid species utilise (as adults) specific host algae or marine angiosperms (e.g., seagrasses), and some of these feed directly on the host plant. Several subtidal taxa are found only on calcareous algae (e.g., *Acmaea*, *Erginus*) (Lindberg 1998b), and some others live only on the shells of oysters or other molluscs, but they feed on the biofilms and fungi on those surfaces. Experimentally removing the limpets demonstrates the marked effect of their

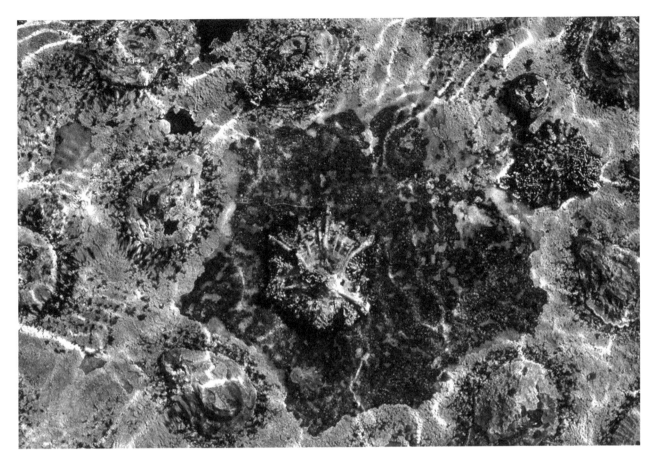

FIGURE 18.15   Southern Africa gardening limpets. The patch gardener *Scutellastra longicosta* surrounded by *Scutellastra cochlear* – a periphery gardener. Brookdale, South Africa. Courtesy of D. R. Lindberg.

feeding as it results in increased algal growth (Lodge 1948; Farrell 1988; Coleman et al. 2006).

Some large tropical and subtropical patellid species maintain small 'gardens' of particular algae adjacent to their home scars that are fertilised by the limpet (e.g., Branch 1981) (see below). These gardening limpets are mostly species of *Scutellastra* (Patellidae), including *S. mexicana* that reaches 258 mm in length, and *S. kermadecensis* at up to 174 mm, while other smaller gardening species range from about 60 mm to 130 mm (Lindberg 2007b). The southern African species *Scutellastra cochlear* lives on coralline algal substrata and keeps a garden around the periphery of the shell, but it is not the coralline algae that are maintained but a non-coralline red alga growing on its surface; other species tend more extensive algal patches over which they graze (Figure 18.15). In both cases, the limpets increase the productivity of their gardens through nutrient enhancement done either by way of nitrogenous excretions (Plagányi & Branch 2000), as with most periphery gardeners, or through pedal mucus left on the patch (Connor & Quinn 1984; Connor 1986).

Patellogastropod grazing is enhanced by powerful radular strokes as the radula rasps the substratum. Unlike other gastropods, there is no rotation of the teeth as they pass through the bending plane of the feeding stroke, and thus there is no sweeping by the few marginal teeth (see Section 8.4.3.4 and Chapter 5).

18.4.4.5   Associations

There is not much available information on associations in patellogastropods. A cestode and several trematodes are known from species of *Patella* in the United Kingdom, and parasite infestations are very low in South African taxa (Branch 1981). Ciliates have been found in various limpets, often attached to their gills, and mantle and commensal copepods, isopods, amphipods, flatworms, and polychaetes are also found in some species (see Chapter 9).

18.4.5   DIVERSITY AND FOSSIL HISTORY

Patellogastropods have low diversity relative to other major gastropod clades. This cosmopolitan group has about 350 living species and occurs in all major oceans. Some species are found associated with biogenic substrata, cold seeps, and hydrothermal vents in the deep sea (see also Section 18.4.4.1).

South Africa has the highest diversity of patellid limpets, with other significant faunas in Australasia and western Europe. Nacellidae are most speciose in the tropical and warm temperate Pacific Ocean (including Australia and New Zealand) while Lottiidae are around the Pacific Rim. Overall, Australasia, Japan, and southern Africa have diverse patellogastropod faunas including patellids, nacellids, and lottiids. Chile has only nacellids and lottiids and Europe only lottiids and patellids. In the western United States and the Caribbean, lottiids are the only patellogastropods (Lindberg 1988c), but nacellid taxa were present in the Eocene of western United States (Lindberg & Hickman 1986), and patellids occurred in the Caribbean during the Pliocene (Lindberg 2007b). Numerous co-existing taxa often occupy similar overlapping niches. For example, along the Californian coast, there may be as many as 10–15 co-occurring species, with most species living on exposed rock substratum, although they tend to be vertically segregated on the shore, and some species live on algae, seagrasses, or the shells of other gastropods (Lindberg 2007a).

The fossil history of patellogastropod limpets is rather poor with the earliest confirmed example (based on shell structure) from the Triassic of Italy (Hedegaard et al. 1997), but possible earlier patellogastropods exist. Horný (1961) described *Damilina* from the Silurian of Bohemia that has a patellogastropod-like morphology and muscle scar pattern, and Knight et al. (1960b) included several extinct taxa ranging from the Middle Silurian in their 'Patellacea' (see Section 18.2.1.1.2 and Chapter 13 for discussion).

Despite there being many limpet fossils in the Paleozoic, none convincingly have the unique shell characters possessed by patellogastropods (Frýda et al. 2008a; Lindberg 2008). For example, the Ordovician limpet *Floripatella rousseaui* was thought by Yochelson (1988) to be the earliest patellogastropod, but there is a Y-shaped impression on the posterior margin that may be an efferent mantle blood vessel, suggesting that the shell belonged to an untorted mollusc. This lack of fossil evidence for the existence of the clade in the Paleozoic is further frustrated by the failure to positively identify the putative coiled patellogastropod ancestors (Frýda et al. 2008a; Lindberg 2008). The only gastropods that share foliated shell microstructure with patellogastropods are the members of the extinct Paleozoic Platyceratoidea (Ponder & Lindberg 1997). Some workers (e.g., Bandel 2007) have suggested that, based on protoconch morphology, platyceratoideans are more closely related to Neritimorpha, but this evidence is not particularly convincing. Also, Sutton et al. (2006) have described some partially preserved internal anatomy in a Silurian gastropod that was identified as a platyceratid[5] and suggested that it has patellogastropod rather than neritimorph affinities. Given the highly specialised habits of platyceratoideans (they were external symbionts of crinoids) it is unlikely that they are the direct ancestors of modern patellogastropods but could, nonetheless, be a separate limpet group that evolved as part of the early eogastropod radiation.

By the Upper Cretaceous and Paleogene most of the modern patellogastropod groups had appeared in the fossil record (e.g., Akpan 1982; Lindberg & Marincovich 1988; Lindberg & Squires 1990; Kase & Shigeta 1996).

Some clades of patellogastropods have undergone extensive radiations within the last three million years (e.g., Lottiinae), but others appear to have maintained morphological stasis for tens of millions of years. For example, a Tethyan clade

(*Eoacmaea*) has a 99-million-year history in which shell morphology, habitat, and radular morphology appear unchanged (Lindberg & Vermeij 1985; Kirkendale & Meyer 2004), and this is the same period that saw more than half of the living caenogastropods evolve (Sepkoski et al. 2002).

18.5 VETIGASTROPODA

18.5.1 Introduction

This large and very diverse group of marine gastropods are found worldwide from the intertidal to the deep sea. They represent the largest subset of gastropods previously referred to as 'Archaeogastropoda'.

Vetigastropods have various plesiomorphic features. Scissurelloidea, Pleurotomarioidea, and Haliotoidea are the only gastropods that retain paired ctenidia, hypobranchial glands, and osphradia (Figure 18.16). Most vetigastropods have paired auricles and paired kidneys. The majority have coiled shells, but several groups have independently adopted a limpet shape, including the superficially bilaterally symmetrical Fissurellidae, the Cocculinoidea, Lepetelloidea, and a few Trochoidea. The shells of most of those taxa with paired ctenidia have slits or other secondary openings. While some shell reduction has occurred, notably in a few fissurellids (*Scutus* and *Fissurellidea*) where the shell has become embedded in the animal and even where loss has occurred in *Fissurellidea* (McLean 1984a), there are no other examples of shell loss in vetigastropods. Unlike all other gastropods, the shell is nacreous in many vetigastropods, but in some groups, all members lack nacre (see below).

A circular operculum is present in coiled taxa but is lost in many of those taxa that have secondarily adopted a limpet or abalone-shaped shell. The operculum is calcified and thick in some Trochoidea (Turbinidae, Phasianellidae) while in Liotiidae the horny operculum has rows of calcareous beads. The radula has many teeth in each row (rhipidoglossate) and is asymmetrical in many taxa, with this latter feature facilitating efficient packing in the radular sac (Hickman 1981).

Apomorphic structures include the elaborate sensory structures on the head-foot and the bursicles on the ctenidia (see Section 18.5.3.7). In addition, the oesophageal gland of higher vetigastropods has a unique internal papillate structure, and in some the stomach has a spiral caecum.

The group includes most of the gastropod taxa living in hydrothermal vents (see Section 18.5.4.1.1).

18.5.2 Phylogeny and Classification

Although vetigastropods had a Paleozoic origin, the details of their evolutionary history are difficult to interpret because the early fossil record contains diverse groups of gastropods that are unrelated in any obvious way to modern forms (see Chapter 13). Many of the early gastropods had shells with slits, holes, or indentations, and such features have sometimes been regarded as diagnostic of 'primitive' vetigastropods and indicative of paired ctenidia, but it is likely that slits and indentations were derived several times from ancestors that

---

[5] This identification was considered incorrect by Frýda et al. (2009), but no alternative taxon was suggested.

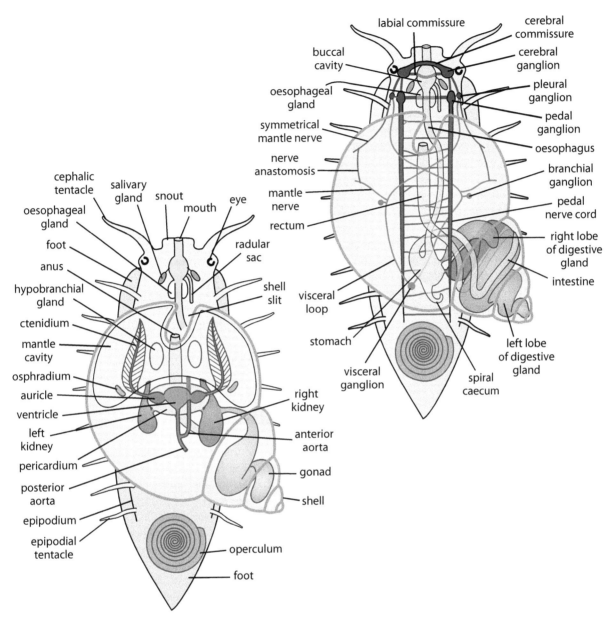

FIGURE 18.16 Generalised structure of a pleurotomarioid vetigastropod. Redrawn and modified from Ivanov, D.L., Class Gastropod Mollusks [in Russian], pp. 323–455 in Dogelya, V.A. and Zenkeyicha, L. A. (eds.), *Manual of Zoology*, Vol. 2, Academy of Science of the USSR, Moscow, 1940.

lacked them. Loss of these structures is also probably common, but it is not clear as to whether their absence in some extant vetigastropods is secondary or primary. An apertural indentation or slit is not necessarily an indication of paired ctenidia as these occur in the shells of some extant caenogastropods that possess a single gill (see Chapter 19).

By the 1980s evidence was accumulating to show that the 'Archaeogastropoda' comprised several distinct lineages deserving elevation to a higher rank.[6] The first significant, modern attempt to do this was that of Golikov and Starobogatov (1975) in which profound changes were made (Figure 18.2), several of which were not subsequently accepted. These authors divided Thiele's archaeogastropods into three 'subclasses', Cyclobranchia (containing only the patellogastropods), Scutibranchia (containing the orders Dicranobranchia [with superfamilies Fissurelloidea and the extinct Bellerophontoidea] and Fissobranchia [with superfamilies Haliotoidea, Pleurotomarioidea, and presumably the extinct Murchisonioidea], which retain paired ctenidia and are equivalent to the Zeugobranchia of Thiele [1925]), and Pectinibranchia, with a single ctenidium. Within their Pectinibranchia Golikov and Starobogatov (1975) grouped the vetigastropods with a single ctenidium (as Turbinimorpha) along with caenogastropods and several groups of lower heterobranchs. Salvini-Plawen (1980) first formulated the concept of the Vetigastropoda which comprises the groups we now recognise, with the exception of the hot-vent taxa which were described shortly after that paper appeared. Haszprunar (1988b) subsequently used the name Vetigastropoda for a subset of the taxa originally included in it and excluded the 'cocculiniform' taxa (Cocculinoidea + Lepetelloidea) and the Seguenziidae. The hot-vent neomphaline taxa related to *Neomphalus* were also treated as distinct. Ponder and Lindberg (1997) included Lepetelloidea and Seguenziidae in Vetigastropoda but excluded the neomphalines and Cocculinoidea. Sasaki (1998) provided a comprehensive morphological re-evaluation of the group and recognised that both cocculinids and *Neomphalus* were most closely related to Vetigastropoda, as did Geiger et al. (2008) (Figure 18.17).

The majority of Vetigastropoda form a well-demarcated group placed near the base of the orthogastropod tree (e.g., Geiger et al. 2008), with several apomorphic characters that define the living members of the group. These include the bursicles (see Chapter 7 and Section 18.5.3.7), small sensory organs at the base of each gill filament in most vetigastropods, although these are absent in some pleurotomariids, *Neomphalus* (but not the related *Melanodrymia*), and some (not all) species of *Lepetodrilus*. The reduced gill of cocculinids also lacks this feature. No other gastropods have bursicles. In many vetigastropods the glandular epithelium in the oesophageal glands is characteristically papillate, a condition unknown in other groups of gastropods. The papillate oesophageal gland is also lacking in cocculinoideans and neomphaloideans, as are other features characterising the majority of stem vetigastropods (Trochiformii), including the sensory papillae on their tentacles and, in some taxa, other parts of the head-foot (see Chapter 7 and Section 18.5.3.7), although they do possess epipodial sense organs (Haszprunar et al. 2017).

While the taxonomic concept of 'Archaeogastropoda' has usually been used to encompass the paraphyletic group recognised by Thiele (1925), some authors (e.g., Hickman 1988; Bandel 1990; Kiel & Bandel 2002) have used 'Archaeogastropoda' in a restricted sense in preference to Vetigastropoda.

Thus, even though some extant vetigastropods (Figures 18.19 and 18.20) show the most plesiomorphic configuration of any living gastropods regarding their mantle cavity organs and related structures, most living vetigastropods share several unique apomorphic characters as outlined above.

18.5.2.1 Outline of Classification

BOX 18.4 SUMMARY OF THE HIGHER CLASSIFICATION OF VETIGASTROPODA

(Infraclass) **Vetigastropoda**
(Superorder) **Bellerophontia**
(see Appendix for details and Chapter 13 for discussion)
(Superorder) **Trochia**
(Order) **Pleurotomariida**
Superfamily Pleurotomarioidea and some extinct superfamilies (see Appendix)

(Order) **Haliotida**
Superfamily Haliotoidea
(Order) **Seguenziida**
Superfamily Seguenzioidea
(Order) **Lepetellida**
(Suborder) **Lepetellina**
Superfamily Lepetelloidea
(Order) **Fissurellida**
(Suborder) **Fissurellina**
Superfamily Fissurelloidea
(Suborder) **Scissurellina**
Superfamilies Scissurelloidea, Lepetodriloidea
(Order) **Trochida**
Superfamilies Trochoidea (= Turbinoidea, Phasianelloidea, Angarioidea)
(Order) **Neomphalida**
(Suborder) **Cocculinina**
Superfamily Cocculinoidea
(Suborder) **Neomphalina**
Superfamily Neomphaloidea (the 'hot-vent taxa')

See the Appendix for details.

Some molecular phylogenies (see Geiger et al. 2008) indicate that there are four major branches of living vetigastropods. These four clades are treated here as separate orders as summarised in Box 18.4 and presented in more detail in the Appendix. Some other molecular phylogenies (Aktipis et al. 2008; Aktipis & Giribet 2010; Aktipis et al. 2011 [Figure 18.18]) have patellogastropods as a clade within the vetigastropod taxa, although the monophyletic groupings within vetigastropods are generally recovered in these analyses. As discussed above, and summarised in Table 18.1, the placement of patellogastropods within vetigastropods is not in agreement with most of the morphological evidence. Other analyses have recovered the patellogastropods and vetigastropods as sister taxa (Zapata et al. 2014; Cunha & Giribet 2019). The issue of long branch attraction in patellogastropods, which was suspected of being responsible for these groupings, was addressed by Uribe et al. (2019) with additional sampling within the patellogastropods as well as the application of site-heterogeneous evolutionary models and the removal of fast-evolving sites. Several of their subsequent trees placed the patellogastropod in a basal position as the outgroup of all other gastropods which they considered to be the best hypotheses for gastropod phylogeny, given the correspondence of their mitochondrial gene analysis with both nuclear and morphological data.

The possession of two ctenidia is a plesiomorphic feature so, not surprisingly, this condition is present in more than one clade. Thus, those vetigastropods with shell slits or holes and two ctenidia (Pleurotomarioidea, Scissurelloidea, Fissurelloidea, and Haliotoidea) are not recovered as a single clade in molecular analyses (Figure 18.19). Scissurelloideans are all small in size and include some of the smallest gastropods, while the other members of this 'group' are larger,

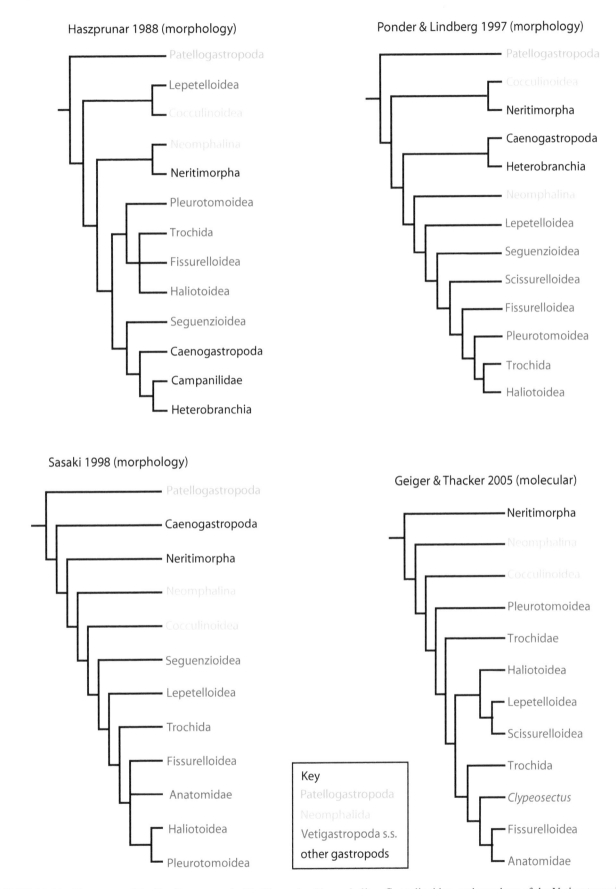

FIGURE 18.17 Placement of the Patellogastropoda, Neritimorpha, Neomphalina, Cocculinoidea, and members of the Vetigastropoda s.s. in some recent classifications of the Gastropoda using morphological and molecular data.

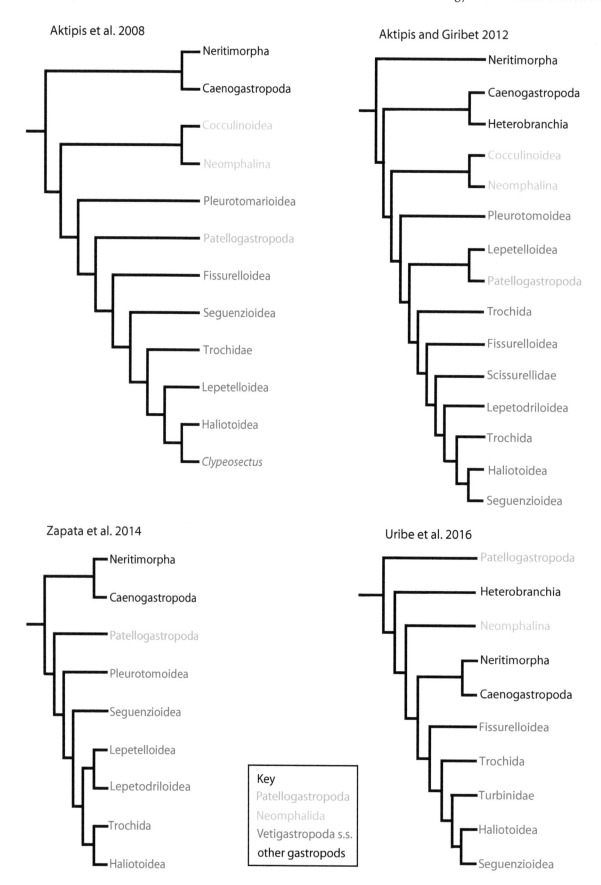

FIGURE 18.18 Placement of the Patellogastropoda, Neritimorpha, Neomphalina, Cocculinoidea, and members of the Vetigastropoda s.s. in some recent classifications of the Gastropoda using molecular data.

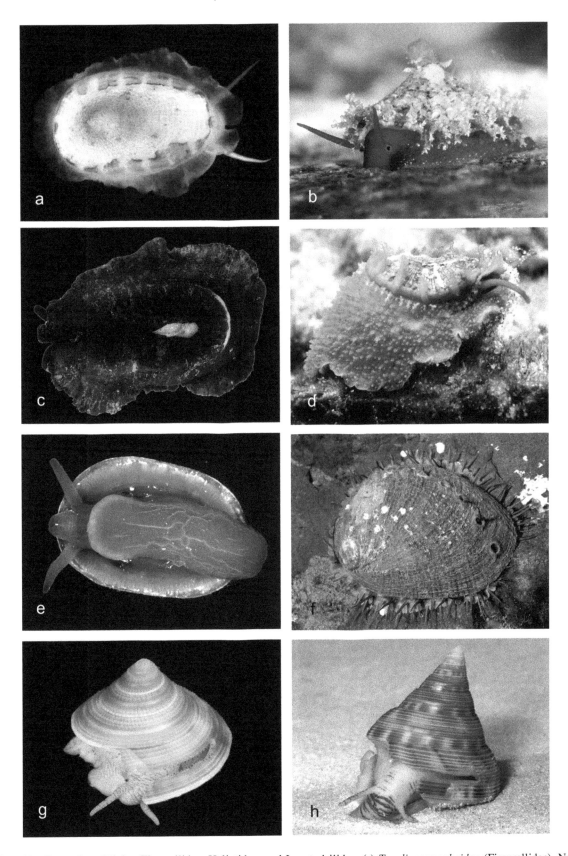

FIGURE 18.19 Examples of living Fissurellidae, Haliotidae, and Lepetodrilidae. (a) *Tugali parmophoidea* (Fissurellidae), New South Wales, Australia. Courtesy of D. Riek. (b) *Diodora* sp. (Fissurellidae), South Madagascar. Courtesy of B. Abela - MNHN. (c) *Scutus* sp. (Fissurellidae), South Madagascar. Courtesy of B. Abela - MNHN. (d) *Amblychilepas javanicensis* (Fissurellidae), New South Wales, Australia. Courtesy of D. Riek. (e) *Lepetodrilus* sp. (Lepetodrilidae), ventral view, 13°N, Eastern Pacific Rise. Courtesy of P. Batson and J, Voight. (f) *Haliotis rufescens* (Haliotidae), California. © 2008 Gary MacDonald. (g) *Bayerotrochus* sp. (Pleurotomariidae), Guadeloupe, French West Indies. Courtesy of L. Charles - MNHN. (h) *Perotrochus* sp. (Pleurotomariidae), Ile des Pins, New Caledonia. Courtesy of P. Maestrati - MNHN.

FIGURE 18.20 Examples of living Trochoidea. (a) *Calliostoma annulatum* (Calliostomatidae), Monterey Bay, California. © 2008 Gary MacDonald. (b) *Norrisia norrisii* (Tegulidae), San Nicolas Island, California. Courtesy of D.R. Lindberg. (c) *Euchelus* sp. (Chilodontaidae), Singapore. Courtesy of R. Tan. (d) *Munitiella* sp. (Skeneidae), Madang, Papua New Guinea. Courtesy of L. Charles - MNHN. (e) *Phasianella* sp. (Phasianellidae), Madang, Papua New Guinea. Courtesy of L. Charles - MNHN. (f) *Lithopoma phoebium* (Turbinidae) Guadeloupe, French West Indies. Courtesy of P. Maestrati - MNHN. (g) *Gena* sp. (Stomatellinae), Madang, Papua New Guinea. Courtesy of L. Charles - MNHN. (h) *Solariella*? sp. (Solariellidae), Madang, Papua New Guinea. Courtesy of L. Charles - MNHN.

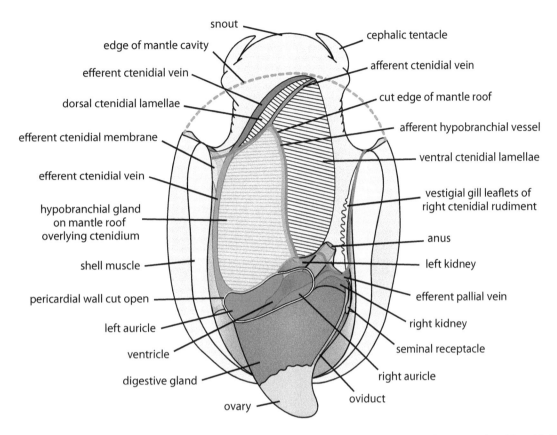

FIGURE 18.21 *Lepetodrilus* – a hot-vent vetigastropod. Redrawn and modified from Sasaki, T., *Bull. Univ. Tokyo Mus.*, 38, i–vi, 1–223, 1998.

with the shells of some pleurotomariids and haliotids reaching around 285 and 300 mm respectively in maximum dimensions.

In those taxa with shells lacking secondary shell openings, such as the Trochoidea (Figure 18.20), the turbinids were considered to belong to a separate superfamily (Bouchet & Rocroi 2005) based on molecular analyses (Williams, 2006, 2008), but Williams et al. (2008) reunited Turbinidae and Liotiidae in the superfamily Trochoidea (where they had been in most earlier classifications). Angariidae, Colloniidae, and Phasianellidae were elevated to three different superfamilies but were subsequently reunited under Trochoidea (Lee et al. 2016; Bouchet et al. 2017; Uribe et al. 2017).

The Lepetelloidea (including Pseudococculinidae) used to be included with the Cocculinoidea in the 'Cocculiniformia' (see below), both being small-sized, white limpets living on biogenic substrata in the deep sea. The enigmatic Seguenzioidea, the mainly hot-vent Lepetodriloidea (Figure 18.21), and Neomphaloidea are also vetigastropods. These are discussed in Section 18.5.2.2.

Several vetigastropod families live mostly in hot-vent and cold-seep habitats (see Chapter 9). There are two main groups: Those related to *Neomphalus* (Neomphaloidea; Neomphalidae, Melanodrymiidae, Peltospiridae, Cyathermiidae) (Figure 18.22) and those related to *Lepetodrilus* (Figures 18.19e and 18.21) and included in Lepetodriloidea (Lepetodrilidae, Sutilizonidae, and Gorgoleptidae). While the preference for chemosynthetic environments is seen in the great majority of these taxa, some are

also known from various non-chemosynthetic habitats (Kano 2008; Warén & Bouchet 2009; Sasaki et al. 2010).

Some authors (e.g., Hess et al. 2008; Sasaki et al. 2010; Bouchet et al. 2017), have treated the Neomphalina as a separate major clade, sometimes including the Cocculinoidea, equivalent in rank to Vetigastropoda and Neritimorpha. Others have included them, albeit sometimes tentatively, in the vetigastropods (e.g., Aktipis et al. 2008; Geiger et al. 2008). Given the possession of bursicles by some taxa we include neomphalines in the vetigastropods (Geiger et al. 2008; Hess et al. 2008) (see 18.5.3.7), although, as detailed in Table 18.2, they do not share many of the key anatomical vetigastropod apomorphies (e.g., Ponder & Lindberg 1997; Sasaki 1998; Geiger et al. 2008; Hess et al. 2008 table 3). Hess et al. (2008) pointed out that most shared characters are plesiomorphic for gastropods and that the neomphalines share several derived characters with cocculinoideans, neritimorphs, and the higher gastropod groups (apogastropods), although it is likely that these conditions are convergent. They include, for example, the loss of the right kidney and the possession of glandular gonoducts.

In some recent phylogenetic analyses, the neomphaloidean taxa are sister to the Cocculinoidea, as in the combined morphological and molecular analysis of Aktipis et al. (2008) and molecular analysis of Aktipis and Giribet (2012), or both taxa share a basal position (e.g., Geiger & Thacker 2006). Here we include both these taxa in Neomphalida, and these groups are discussed further in the next section.

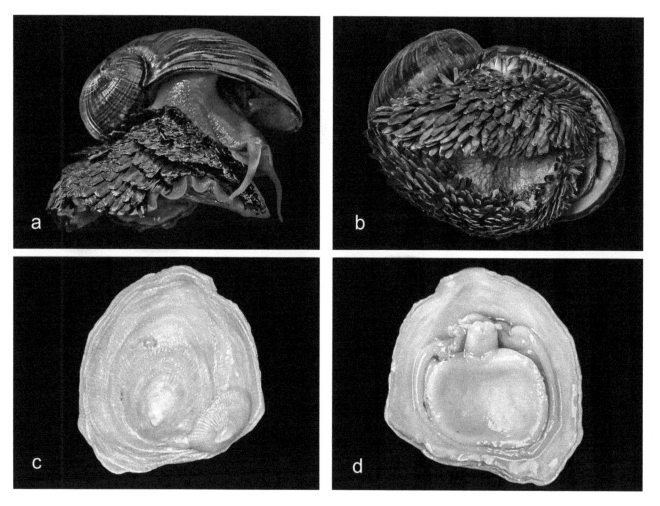

FIGURE 18.22 Neomphalida. (a, b) *Chrysomallon squamiferum* (Peltospiridae), showing the scales on the sides of the foot; (a) Kairei field, Central Indian Ridge. (b) Longqi field, Southwest Indian Ridge. (c, d) *Symmetromphalus regularis* (Neomphalidae), Alice Springs field, Mariana Trough. Courtesy of C. Chen.

18.5.2.2 Deep-Sea Enigmas; The Neomphaloidean and Cocculinoidean Taxa

The neomphaloideans and cocculinoideans share some derived characters that seemingly contradict the basal position they occupy in most phylogenetic analyses. They have lost the organs on the right side of the mantle cavity, the right auricle, and the right kidney. They also have simple anterior oesophageal pouches and a single pair of odontophoral cartilages. The gonad opens independently to the mantle cavity (i.e., not via the right kidney), and internal fertilisation is probably associated with this change. Also, the rectum does not penetrate the ventricle, and the statocysts each have a single statolith. In all these features both groups resemble those seen in higher gastropods, many of which were achieved by way of heterochronic processes, although presumably developed independently.

Many have adopted a limpet-shape, and a number (including all cocculinoideans) are hermaphroditic and live in either biogenic or vent or seep habitats in the deep sea. They have either lost, or never had, some key vetigastropod characters

– sensory papillae and papillate oesophageal gland – but *Melanodrymia* and some other small neomphaloideans have bursicles, another key character (Haszprunar 1989b; Hess et al. 2008), and most have epipodial sense organs (Haszprunar et al. 2017). Given these characters, and that they group as sister taxa (or as paraphyletic basal taxa) in some of the few molecular analyses available that deal with both these taxa, we treat them as belonging to the same major group but as separate taxa at the subordinal level. There are significant morphological differences between Neomphaloidea and Cocculinoidea; jaws are dorsal in cocculinoideans and lateral (as in other vetigastropods) in Neomphaloidea; a deep rather than shallow mantle cavity; cocculinids have oral lappets that are absent in other vetigastropods other than some lepetelloideans and neomphalines; and the shell muscles are traversed by blood sinuses in the same way as in patellogastropods (solid in other vetigastropods other than a few lepetelloideans). Both Cocculinoidea and Neomphaloidea are similar in having only the left kidney, a single auricle, and a single (left) ctenidium, although it is much modified in cocculinoideans.

18.5.2.2.1 The Neomphalina

Shell morphologies in the Neomphalina[7] (Figure 18.22) range from limpets such as *Neomphalus*, *Symmetromphalus*, and *Nodopelta* to coiled snails such as *Cyathermia*, *Melanodrymia* and *Chrysomallon*, and loosely coiled snails (e.g., *Pachydermia laevis*). These are grouped into three families, Neomphalidae, Peltospiridae, and Melanodrymiidae. Some are moderately large (e.g., *Neomphalus* shells are up to about 40 mm in maximum diameter, *Gigantopelta* up to 55 mm in maximum diameter), but most are small, with some being minute. The larvae are lecithotrophic (Sasaki et al. 2010).

The anatomical features of the Neomphalina are an odd mix of modified and, as detailed below, plesiomorphic characters shared with other vetigastropods. All known members of the group are restricted to deep-sea vent and seep habitats, apart from a few on deep-sea biogenic substrata (Sasaki et al. 2010).

18.5.2.2.2 The Cocculiniform Limpets

Small-sized white limpets occur on wood and other biogenic substrata in the deep sea and comprise several families based on differences in gill morphology, their radula, and other aspects of their digestive systems. They were grouped into two superfamilies, Cocculinoidea and Lepetelloidea, and collectively known as 'Cocculiniformia' (Haszprunar 1988a, e, 1998). Cocculinoidea contains only two families, Cocculinidae (Figure 18.23) and Bathysciadiidae, while the Lepetelloidea contains several families (see Appendix).

It was suggested that 'Cocculiniformia' should be placed near the base of the gastropods (Haszprunar 1988e, a, 1998), but other morphological (Ponder & Lindberg 1997) and molecular (Colgan et al. 2000; McArthur & Harasewych 2003) studies have shown that this grouping comprises two unrelated groups. The superfamily Lepetelloidea clusters with the majority of vetigastropods (Trochiformii) and provides an excellent example of how some modified groups of vetigastropods could lose many of the characteristic vetigastropod features. Most lepetelloideans have lost the sensory papillae and the epipodium but have retained the epipodial sense organs (Haszprunar et al. 2017). The gill is much modified, some have lost the bursicles, and the oesophageal gland is not papillate.

The placement of Cocculinoidea in or among the other major gastropods remains problematic, it being treated as either a sister to the Neritimorpha (Ponder & Lindberg 1997) or as a separate higher clade of gastropods (e.g., Bouchet & Rocroi 2005).[8] The most recent molecular studies usually show the cocculinids and Neomphaloidea as sister groups, with those

groups sisters to the rest of the vetigastropods. The pleurotomariids are often sister to the remaining vetigastropods.

Given the remarkable convergence in the two groups of small, white, deep-sea limpets, we list in Table 18.2 many of the shared features that characterise Cocculiniformia (*sensu* Haszprunar 1988a) and those that separate Cocculinoidea and Lepetelloidea. We compare these with Neomphaloidea, because, as already noted, this group clusters with cocculinids in many molecular analyses. In the table we also contrast the features seen in Haliotoidea, a 'typical' vetigastropod group, as they also share several characters.

Both groups of 'cocculiniform' limpets have white, cap-shaped shells that range in length from about 5 to 15 mm, are smooth or weakly sculptured and have the apex typically at the centre or nearer the posterior end. The inner surface of the shell has a horseshoe-shaped muscle scar that opens anteriorly at the location of the head. The foot is rounded and sucker-like with an anterior pedal gland, and adults lack an operculum.

The radula is rhipidoglossate but highly divergent in different groups within both superfamilies. Tooth reduction, particularly in the marginal field, occurs, especially in taxa that do not feed on wood.

Copulatory organs are present in both groups and are typically part of the right cephalic tentacle or distinct structures located anteriorly. Both groups are hermaphroditic, with cocculinoideans being simultaneous hermaphrodites with discrete regions of the gonad producing eggs and sperm while lepetelloideans have a separate ovary and testis. Their eggs differ, with those of cocculinoideans lacking a vitelline layer, but this is present in lepetelloideans, as in other typical vetigastropods. As in other basal gastropods, the eggs in both groups are rich in yolk, and development is presumably lecithotrophic, although, because of the ephemeral nature of the biogenic substrata used by members of both groups, their larvae must be able to swim and drift sufficiently far to locate suitable habitat on the deep-sea floor.

These limpets are distributed throughout the oceans of the world in the deep sea, including at bathyal and abyssal depths, where they inhabit a range of biogenic substrata including cephalopod beaks, waterlogged wood, and whale bones, with different genera and families specialising on different substrata. It is unlikely that they obtain nutrition directly from the biogenic substrata they live on but are more probably feeding on microbes and fungi, breaking them down. This has been demonstrated in *Pyropelta* where bacteria are the main food source (Warén et al. in Desbruyères et al. 2006 p. 93).

Given the vastness of the deep-sea floor the habitats these animals occupy would appear to be rare, but the fact that two unrelated but convergent groups of gastropod limpets have evolved to utilise these substrata suggests that this impression may need reconsideration.

18.5.3 Morphology

18.5.3.1 Shell and Protoconch

Vetigastropods range in size from some minute Scissurelloidea, Skeneidae (Trochoidea), and seguenzioideans, which may be

---

[7] This subordinal name Neomphalina (e.g., Sasaki 2010) has been used, as has the ordinal name Neomphalida (Hess et al. 2008), while Bouchet et al. (2017) elevated the group to subclass level (Neomphaliones), although including the Cocculinoidea. We use the ordinal name Neomphalida to include both Neomphalina and Cocculinina.

[8] They used Cocculiniformia in a restricted sense to include only the Cocculinoidea, a course of action we avoid because of potential confusion with the earlier broader concept.

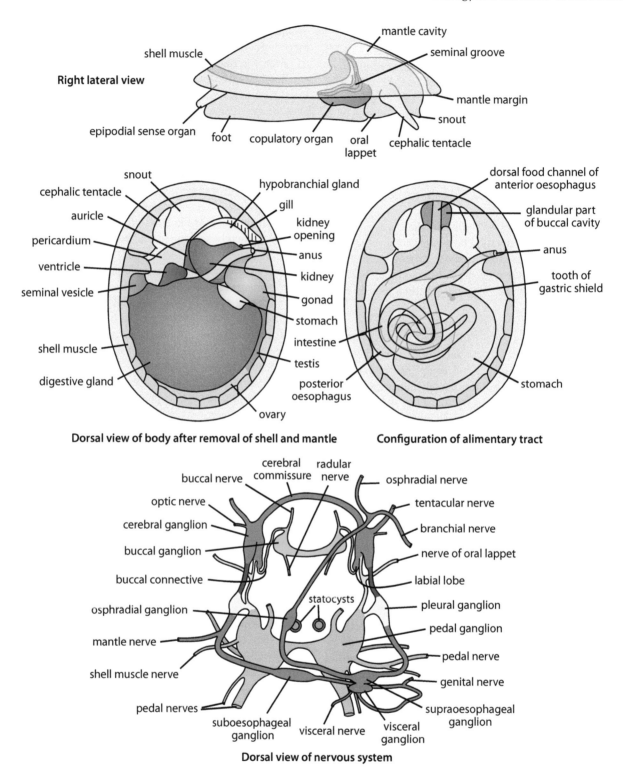

FIGURE 18.23 Morphology of *Cocculina*. Top three figures redrawn and modified from Sasaki, T., *Bull. Univ. Tokyo Mus.*, 38, i–vi, 1–223, 1998; nervous system redrawn and modified from Haszprunar, G., *J. Molluscan. Stud.*, 53, 46–51.

around 1 mm or even less, to members of the Haliotoidea, which may reach over 300 mm in length. Their shells can be coiled, ranging from squat and globose to elongate, or limpet-shaped. Vetigastropod limpets have evolved at least 15 times in living taxa (Vermeij 2016) although in many examples a coiled part is still present in the early shell (e.g., abalone, *Neomphalus*, and the trochid *Broderipia*), while

others (Fissurellidae, Cocculinidae, most Lepetelloidea, and some Peltospiridae) are symmetrical limpets. The shell aperture is usually circular to oval and never has a siphonal canal, although indentations or slits can occur or, in some fissurellids, haliotids, and scissurellids, there can be a hole behind the edge of the aperture. The holes and slits serve as exit points

TABLE 18.2

The Key Similarities and Differences between the Two Groups that Were Treated as Cocculiniformia[1], and Comparison with Neomphaloidea[2] and Haliotoidea (Haliotidae)[3]

	Cocculinoidea	Lepetelloidea	Neomphaloidea	Haliotoidea
Shell	Limpet-shaped Protoconch symmetrically coiled with free tip Nacre absent	Limpet-shaped Protoconch symmetrically folded with tip immersed Nacre absent	Coiled to limpet-like Protoconch coiled, with free tip Nacre absent	Coiled with greatly expanded aperture Protoconch coiled Nacre present
External body	Disc-shaped foot Pair of posterior epipodial sense organs usually present Mantle edge simple	Disc-shaped foot One or more pairs of epipodial sense organs and sometimes a pair of epipodial tentacles present Mantle edge simple or with tentacles	Disc-shaped to moderately elongate foot Several pairs of posterior epipodial sense organs often present Mantle edge simple or with small tentacles	Broad foot pointed posteriorly Numerous epipodial tentacles usually present, also an epipodial skirt and epipodial sense organs Mantle edge with tentacles
Mantle cavity	Mantle cavity shallow Left ctenidium only, reduced and simplified Skeletal rods absent Left hypobranchial gland present, often enclosed within a pouch Subpallial gland[4] sometimes present	Mantle cavity shallow Single monopectinate ctenidium on right side but innervated from left – reduced in some Skeletal rods sometimes present Hypobranchial glands absent Subpallial gland absent	Mantle cavity deep Single bipectinate ctenidium Skeletal rods sometimes present Left hypobranchial gland present and sometimes a small right hypobranchial gland; both absent in *Chrysomallon* Subpallial gland absent	Mantle cavity deep Left (larger) and right bipectinate ctenidium Skeletal rods present Left and right hypobranchial glands Subpallial gland absent
Shell muscles	Paired Symmetrical Blood sinuses through muscle ring	Paired Usually symmetrical Solid, blood sinuses uncommon	Paired or single Asymmetrical Solid	Paired Asymmetrical Solid
Alimentary canal	Mouth/snout with oral hood and oral lappets Hairs around mouth Single dorsal jaw, weak (except in *Teuthirostria* where it is well-developed) Single pair of odontophoral cartilages Salivary glands simple pouches at posterior end of buccal cavity except in *Fedikovella* and *Teuthirostria* which have prominent glands Midoesophagus with simple glandular pockets Gastric shield present. No style sac Intestine looped Single digestive gland opening in Cocculinidae; digestive gland absent in Bathysciadiidae but oesophageal gland much enlarged	Mouth/snout usually lacking oral hood and oral lappets Lacking hairs around mouth Jaws paired or lost One or two pairs of odontophoral cartilages Salivary glands simple Midoesophagus with simple glandular pockets Gastric shield present or absent; no style sac Intestine looped to simple. Expanded part of intestine functions as stomach in some One to several digestive gland openings to stomach	Mouth/snout lacking oral hood; oral lappets sometimes present Lacking hairs around mouth Jaws paired One pair of odontophoral cartilages Salivary glands simple or reduced Midoesophagus with simple glandular pockets Gastric shield present. Style sac region not well differentiated Intestine looped Two to five digestive gland openings to stomach Small gastric caecum present	Mouth/snout lacking oral hood and oral lappets Lacking hairs around mouth Jaws paired Two pairs of odontophoral cartilages Salivary glands simple Midoesophagus with papillate oesophageal glands Gastric shield present. Style sac region well differentiated Intestine looped Four digestive gland openings to stomach Spiral caecum present
Radula	Modified rhipidoglossate Marginal teeth very reduced and central teeth very broad in bathysciadiids	Rhipidoglossate Marginal field very reduced in some	Rhipidoglossate Marginal field well-developed	Rhipidoglossate Marginal field well-developed

(Continued)

TABLE 18.2   (CONTINUED)

The Key Similarities and Differences between the Two Groups that Were Treated as Cocculiniformia, Neomphaloidea, and Haliotoidea (Haliotidae), Representing a 'Typical' Vetigastropod

	Cocculinoidea	Lepetelloidea	Neomphaloidea	Haliotoidea
Food and habitat	Wood, bone, cephalopod beaks	Wood, bone, fish spines, seagrass stems, polychaete tubes, crab carapaces, cephalopod beaks, elasmobranch egg cases, hot vents	Browsers and/or suspension feeders. Some vent and seep taxa with endosymbionts	Algal feeders
Renopericardial system	Left kidney only. Right auricle lost	Left and right kidney. Right auricle lost	Left kidney only. Right auricle lost	Left and right kidney. Both auricles retained
Reproductive system	Hermaphroditic gonad, ventrally located. Internal fertilisation. Open seminal groove on the right side of neck. Copulation via modified right cephalic tentacle, oral lappet, or pedal organ. Glandular gonoduct. One or two seminal receptacles present. Eggs lack a vitelline layer	Separate testis and ovary, dorsal. Internal fertilisation. Open seminal groove on the right side of neck. Copulation via modified right cephalic tentacle. Genital ducts never glandular. Seminal receptacles sometimes present. Eggs with a vitelline layer	Separate sexes, gonad dorsal. Internal fertilisation. Open seminal groove on the right side of neck. Copulation via modified left tentacle or copulatory organ absent. Genital ducts glandular. Seminal receptacles sometimes present. Eggs lack a vitelline layer	Separate sexes, gonad dorsal. Internal fertilisation. No seminal groove and copulatory organ absent. Female genital system not modified. Seminal receptacles absent. Eggs with a vitelline layer
Nervous system	Pedal ganglia developed	True pedal ganglia rarely developed	Pedal ganglia developed	Pedal ganglia developed
Sense organs	Statocysts with a single statolith. Left osphradium present. Bursicles absent. Sensory papillae absent. Eyes often reduced or absent. Subradular organ absent	Statocysts with statoconia. Left osphradial ganglion present but osphradial epithelium absent. Bursicles usually in those that retain ctenidium. Sensory papillae present in Pseudococculinidae. Eyes often reduced or absent. Subradular organ absent	Statocysts with single statolith. Left osphradium present. Bursicles sometimes present. Sensory papillae absent. Eyes absent or reduced. Subradular organ absent	Statocysts with statoconia. Left and right osphradia present. Bursicles present. Sensory papillae present. Eyes functional. Subradular organ present

[1]   Based mainly on Haszprunar (1987b, 1988f, 1998); Strong et al. (2003).
[2]   Based mainly on Fretter et al. (1981); Haszprunar (1989b); Israelsson (1998); Hess et al. (2008).
[3]   Based mainly on Crofts (1929); Sasaki (1998).
[4]   A glandular epithelium in the mantle groove found in some of the cocculinid genera *Paracocculina* and *Coccopigya* (Haszprunar 1987b).

for the exhalant current and are associated with the presence of two ctenidia.

Most vetigastropods do not exhibit determinate growth (see Chapter 3), but some of those that do, such as the trochid *Clanculus*, develop adult apertural ornament. A terminal varix is produced in some living chilodontaids (Seguenzioidea) and some extinct members of that family had multiple varices (Webster & Vermeij 2017).

Most vetigastropods have lecithotrophic larvae, although a few develop directly after brooding. Thus all vetigastropods have a protoconch that consists of only the original larval shell (i.e., protoconch I) and is one whorl or less. The protoconch may be smooth or sculptured with netting-like sculpture, fine spirals, pits, pustules, or, rarely, heavy spiral ridges or axial ribs.

Several neomphalid taxa have been shown to have shell pores (e.g., Kiel 2004; Hess et al. 2008; Sasaki et al. 2010; Chen et al. 2017) (see Chapter 3).

18.5.3.1.1   Shell Structure

Vetigastropods have a diverse array of shell structures (Hedegaard 1990, 1997) and are the only gastropods with nacre and intersected crossed-platy shell structures. Nacre shell structures predominate in the Pleurotomariida, Trochida, Haliotida, and Seguenziida, and in some larger species their presence has led to the commercial harvesting of shells (e.g., some abalone, *Trochus*, and *Tectus*). The remaining taxa predominately have crossed-lamellar shell structures (e.g., Lepetellina, Fissurellina, Scissurellina, Cocculinina, Neomphalina). Intersected crossed-platy shell structure was first reported in the Pleurotomariidae (Erben & Krampitz 1972) but occurs broadly throughout the vetigastropods in both nacre and crossed-lamellar shell groups (Hedegaard 1997).

In the Pleurotomariida, Trochida, Haliotida, and Seguenziida columnar nacre is found in combination with prismatic, homogeneous, and intersected crossed-platy shell layers. Homogeneous layers are usually found on the outer surface of the shell, and in the Trochina the columnar nacre is typically 'sandwiched' between two prismatic layers. Intersected crossed-platy has not been reported in the Haliotidae, which have an outer prismatic layer and inner nacre layer. A lamellofibrillar layer grading into intersected crossed-platy structure has been reported in the trochid *Stomatella* by Hedegaard (1990), this being the only report of a lamellofibrillar layer in living vetigastropods. One major exception to the distribution of nacre in the Trochina is its absence in the Phasianellidae, which instead have shells consisting of prismatic, crossed-lamellar, and intersected crossed-platy layers.

Where nacre is absent, a wide variety of crossed-lamellar structures are found in the remaining taxa in Trochida and Neomphalida, including simple, co-marginal, radial, and cone complex crossed-lamellar structures. As in the groups with nacre, these structures co-occur with prismatic, homogeneous, and intersected crossed-platy shell layers. In Fissurellina, intersected crossed-platy structure is known only in *Macrochisma* (Hedegaard 1990), and it appears to be absent in the Cocculinina (Hedegaard 1990). In the Neomphalida intersected crossed-platy structure has only been reported in the Peltospiridae (Hedegaard 1990) and *Retiskenea* (Kiel 2004).

18.5.3.2 Head-Foot, Mantle, Operculum, and Locomotion

Many coiled species have a circular multispiral to paucispiral operculum with a central nucleus. Most taxa have a horny operculum but some (Turbinidae, Phasianellidae) have a thick calcareous operculum, and a few, as in many Lottiidae, have calcareous granules on the operculum. An operculum is usually absent in adult limpet-shaped taxa, but a small operculum is retained in subadults of *Gorgoleptis*, although lost in adults (Warén & Bouchet 2001). Spirally coiled neomphalids and peltospirids have a multispiral operculum, but in the limpet-like neomphalid *Symmetromphalus* (Figure 18.22) a small operculum persists in the subadult; in *Neomphalus* it is lost in juveniles. Similarly, in the weakly coiled peltospirids, the operculum is lost, except in *Hirtopelta*, where a small operculum is present on the end of the foot, as it is in *Peltospira operculata*.

In most vetigastropods, there is a distinct snout with a ventral mouth, but the snout is very short (almost non-existent) in the limpet-shaped Neomphalidae, which have a prominent oral lappet projecting from each side. In the related *Melanodrymia* (Melanodrymiidae) and the Peltospiridae, the snout is rather long, tapered, and simple distally. In cocculinids the snout is short, and a well-developed oral hood is present.

In probably the most bizarre case, the reduced operculum of the peltospirid *Chrysomallon squamiferum* is complemented by the development of proteinaceous scales[9] on the sides of the foot which fill the aperture when the animal retracts (Warén et al. 2003; Chen et al. 2015b) (Figure 18.22).

The cephalic tentacles have a pair of eyes at their outer bases, and these are usually raised on ocular peduncles, although eyes and peduncles are absent in some deep-sea taxa. In a few taxa the right cephalic tentacle serves as a copulatory organ (see Section 18.5.3.5). In some vetigastropods a smooth sub-optic tentacle is present on the right side, and, in Anatomidae (Scissurelloidea), there is a pair, one on the right and one on the left (Haszprunar et al. 2017).

In most typical vetigastropods unique, tiny, sensory papillae cover many of the head-foot structures, particularly the tentacles (see Section 18.5.3.7). Additional unique features seen on the epipodium in most vetigastropods are knob-like structures, the epipodial sense organs, which typically lie at the base of each epipodial tentacle when they are present; in trochids the epipodial sense organs are beneath the neck lobe (Haszprunar et al. 2017) (see Section 18.5.3.7).

Some other head-foot features of typical vetigastropods are very distinctive. Many have an epipodial skirt that often bears long epipodial tentacles (except in Fissurelloidea which have only epipodial sense organs).

Some peltospirids have well-developed epipodial sense organs and, in most accounts, these were wrongly identified as epipodial tentacles, as were those in neomphaloideans and cocculinoideans. In *Neomphalus* and some other neomphaloideans, there is a thin epipodial ridge around the foot and some posterior epipodial sense organs, while most cocculinids have two posterior epipodial sense organs (Haszprunar 1987b; McLean 1987).

Flap-like, usually tentaculate, neck lobes on the sides of the head/neck are found in many trochoideans (Figure 18.24) and those larger seguenzioideans (Eucyclidae) previously included in Trochoidea but are absent in the smaller

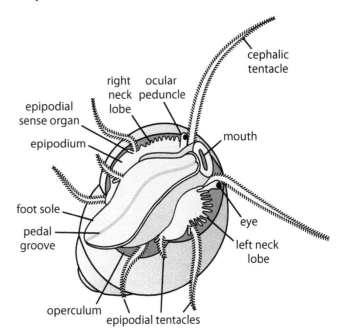

FIGURE 18.24 Ventral view of *Tricolia*, an example of a trochoidean vetigastropod. Redrawn and modified from Fretter, V. and Graham, A.L., *British Prosobranch Molluscs: Their Functional Anatomy and Ecology*, Ray Society, London.

---

[9] The scales are often covered with a layer of iron sulphide depending on its availability in the environment (Nakamura et al. 2012).

seguenzioideans (Quinn 1983), most peltospirids (McLean 1989), and at least some other neomphaloideans. Simple neck lobes that are modifications for suspension feeding are present in *Neomphalus* and *Symmetromphalus* (McLean 1981b, 1990a; Beck 1992a).

Cephalic lappets – a pair of flaps on the head between the cephalic tentacles – are found in Haliotoidea and many trochoideans. These are fringed with short tentacles and may be fused into a single structure.

The mantle edge may be smooth or bear short to moderately long tentacles or papillae. It can have a deep embayment that corresponds to a shell slit, as in scissurelloideans, pleurotomariids, and some fissurellids.

The typically long and mobile foot tapers behind, and the sole is often divided into left and right segments. In 'cocculiniforms' and other limpet-shaped taxa, the foot is usually disc-like. All have an anterior pedal gland opening to a slit anteriorly and sometimes anterolaterally.

18.5.3.3 Mantle Cavity

As noted above, some vetigastropods are the only gastropods that have retained paired organs in their mantle cavity, although a right hypobranchial gland and a vestigial right gill are seen in some neritimorphs (see Section 18.6.3.3). The right mantle cavity structures have been independently reduced or lost in several lineages (see Chapter 4). Reasons for this loss are outlined in Section 18.1.2.5 and in Chapter 4.

Members of the Pleurotomarioidea, Haliotoidea, and Fissurelloidea all retain a pair of ctenidia, as do most Scissurelloidea, but in that superfamily, the very small-sized species of *Larochea* (Larocheidae) have lost the shell slit and the right and left ctenidia are monopectinate (Marshall 1993). While most members of the Lepetodriloidea have a single ctenidium, *Clypeosectus* has two, and the Sutilizonidae have a pair of monopectinate ctenidia (Haszprunar 1989a). Single ctenidia (and osphradia) are found in all other living vetigastropods, including Cocculinoidea[10] and Neomphaloidea. Thus, based on our current understanding of vetigastropod phylogeny, the loss of the right ctenidium has occurred several times at different periods throughout their history. Loss of the right ctenidium is sometimes, but not always, associated with losing the right auricle and has occurred in Neomphaloidea and in at least some small-sized Seguenzioidea (Quinn 1983). A vestigial right ctenidium has been recognised in some trochoideans, Lepetodrilidae (Sasaki 1998), and *Neomphalus* (Fretter et al. 1981).

Most ctenidia in vetigastropods are bipectinate, and some are partially or completely monopectinate, the latter condition being seen in small-sized taxa such as several scissurelloideans (see above), seguenzioideans, skeneids, halistylines (Trochidae), and in at least one small neomphaloidean (Hess et al. 2008). In some tiny vetigastropods, such as species of scissurellids, the filaments are finger-shaped rather than lamellate (Fretter & Graham 1962). The larger, limpet-like Lepetodriloidea have a single left ctenidium that is bipectinate

anteriorly where it is free, but monopectinate posteriorly. The single (usually large) ctenidium in neomphaloids is bipectinate. Suspension-feeding trochoidean taxa also tend to have a monopectinate ctenidium (Hickman 1985; Hickman & McLean 1990; Hickman 2003).

The ctenidial filaments are strengthened with skeletal rods, although the rods are lost in cocculinoideans and in *Melanodrymia*, in which the gill is unusually small (Haszprunar 1989b). Some other small, coiled taxa related to *Melanodrymia* also lack skeletal rods, but others possess them (Hess et al. 2008). The gills are supported by a ventral (or efferent) and a dorsal (or afferent) membrane that extends for at least part of the gill length. The osphradia (or osphradium if only one gill is present) lie on the anterior edge of the ventral membrane.

In most typical vetigastropods the right hypobranchial gland is reduced, but surprisingly, in scissurelloideans it is the left hypobranchial gland that is reduced (Bourne 1910; Haszprunar 1989a) or lost, as in *Anatoma* (Anatomidae) (Sasaki 1998). Fissurellids lack hypobranchial glands (Sasaki 1998), as do peltospirid limpets (Neomphaloidea). In the lepetodrilid *Clypeosectus* there is a pair, as in males of *Pseudorimula* (females lack them), but in other lepetodrilids only the left is present. Several trochoideans have both hypobranchial glands, although only the left gill and osphradium are present. Cocculinoideans have a single hypobranchial gland, as do *Melanodrymia* and *Gigantopelta*; some other neomphaloideans have two hypobranchial glands, although the right is reduced (Israelsson 1998). In many cocculinids, the single hypobranchial gland is contained within a pouch enclosed by the kidney ventrally (Haszprunar 1987b).

18.5.3.4 Digestive System

The outer lip of the mouth may be papillate, as in pleurotomariids, haliotids, and calliostomatids, or smooth, as in fissurellids and trochoideans. Fissurellids have outer and inner lips, but most other vetigastropods have only outer lips. Most vetigastropods do not have oral lappets, but they are present in cocculinids, some lepetelloideans, and neomphalids. The lepetelloidean *Bathysciadium* has a pair of lateral 'suckerlike' pads on either side of the mouth (Warén 1996) that presumably represent expanded lips. Pleurotomariids and some trochoideans have the outer lip cleft ventrally and, in some trochoideans, there is a ventral projection or 'pseudoproboscis' (Hickman & McLean 1990).

Lateral jaws made up of rods are present in most vetigastropods while cocculinids have a vestigial dorsal jaw composed of 'cuticle' (Haszprunar 1987b). Jaws are lacking in some Lepetelloidea and the neomphaloidean *Melanodrymia* (Haszprunar 1989b), but other lepetelloideans, *Neomphalus*, and peltospirids have lateral jaws.

The vetigastropod radula (Figures 18.25 to 18.28) is nearly always rhipidoglossate. The few exceptions include several taxa of cocculinoidean and lepetelloidean limpets and some seguenzioideans (Figure 18.25), where the number of lateral and marginal teeth is reduced. In some seguenziids, the reduced radula parallels the taenioglossate radula of caenogastropods (e.g., Quinn 1983, 1991). In cocculiniforms the

---

[10] The left ctenidium is often thought to be a secondary gill (Haszprunar 1987b; Strong et al. 2003). Here we follow Ponder and Lindberg (1997) in treating it as a simplified ctenidium.

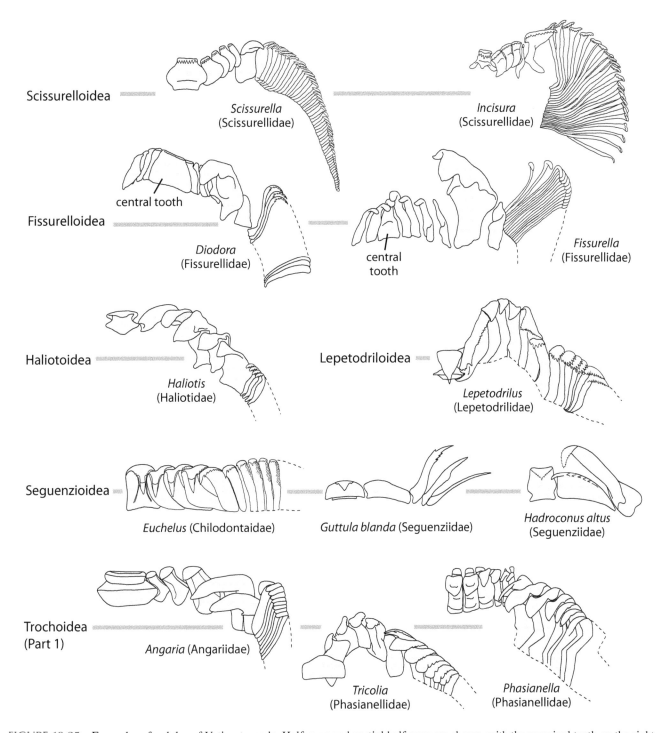

Scissurelloidea — *Scissurella* (Scissurellidae) — *Incisura* (Scissurellidae)

Fissurelloidea — central tooth — *Diodora* (Fissurellidae) — central tooth — *Fissurella* (Fissurellidae)

Haliotoidea — *Haliotis* (Haliotidae) — Lepetodriloidea — *Lepetodrilus* (Lepetodrilidae)

Seguenzioidea — *Euchelus* (Chilodontaidae) — *Guttula blanda* (Seguenziidae) — *Hadroconus altus* (Seguenziidae)

Trochoidea (Part 1) — *Angaria* (Angariidae) — *Tricolia* (Phasianellidae) — *Phasianella* (Phasianellidae)

FIGURE 18.25 Examples of radulae of Vetigastropoda. Half rows and partial half rows are shown, with the marginal teeth on the right. Redrawn and modified from various sources. The radulae of the species illustrated do not necessarily reflect the range of variation in the genera they belong to or of their higher taxon.

radula has been much modified for scraping substrata such as bone or chitin, and the lateral fields are greatly reduced (Figure 18.26). The pleurotomariid radula has a markedly increased number of teeth in the lateral and marginal fields, the hystrichoglossan condition (Hickman 1984) (Figure 18.27), an adaptation that may be related to their sponge-feeding habits. The sponge-feeding Trochaclididae also have highly modified radular teeth with the very long, marginal teeth having broom-like ends and there is a reduction of the lateral field (Marshall 1995). There is considerable variation in radular morphology in the large superfamily Trochoidea (Figure 18.28) including the strangely modified radula of *Thysanodonta* (Thysanodontinae, Calliostomatidae), which has 10–11 exceedingly narrow, elongate barbed teeth in each row of the latero-marginal field (Marshall 1988).

The radular sac is short to moderately long and usually has a bulbous bifid end. It may extend straight behind the odontophore or be bent upwards or coiled in some taxa (Sasaki 1998).

A ridged subradular organ (licker) is present in the sublingual pouch in peltospirids (Fretter 1989) and a smooth one in

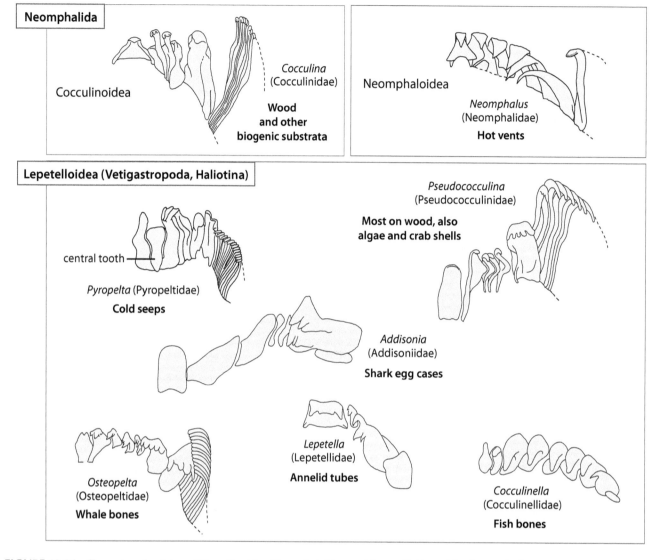

FIGURE 18.26 Examples of radulae of Cocculinoidea, Neomphaloidea, and Lepetelloidea with substrata indicated. Half rows and partial half rows are shown, with the marginal teeth on the right. Redrawn and modified from various sources. The radulae of the species illustrated do not necessarily reflect the range of variation in the genera they belong to or of their higher taxon.

pleurotomariids, haliotids, *Lepetodrilus*, and some trochids. A subradular organ is absent in turbinids, neomphaloideans, lepetelloideans, cocculinoideans, and Anatomidae (Sasaki 1998).

Large fused anterior and posterior cartilages (medial of Guralnick & Smith 1999) and separate small anterolateral cartilages are present in the fissurellid *Montfortula* (Golding et al. 2009a), and there are two pairs of cartilages in *Haliotis* (Sasaki 1998). The trochid *Austrocochlea* has a single pair of cartilages that consists of the fused anterior and posterior cartilages (Golding et al. 2009a).

Small posterior cartilages, abutting the posterior end of each anterior cartilage, are found in most 'typical' vetigastropods. However, cocculinids, neomphalines, and lepetodrilids are reported to have only a single pair of cartilages, although detailed studies have not been conducted to ascertain if the single cartilage element consists of two fused components.

The musculature of the buccal mass is complex. It is described in general terms in Chapter 5, and a detailed

account of the buccal musculature in vetigastropods is given by Sasaki (1998). Vetigastropods, with the apparent exception of *Neomphalus*, uniquely possess dorsal buccal tensor muscles, and most have postmedian retractor muscles of the radular sac (the latter absent in *Anatoma* and *Cocculina*) (Sasaki 1998).

The salivary glands of vetigastropods are 'unique in having [a] ramified lumen and longitudinally slit-like openings to the buccal cavity in contrast to the sac-like glands and small pore-like openings in others' (Sasaki 1998 p. 181). Sasaki argued that these unusual salivary glands are secondary because the original salivary glands were lost in neritimorphs, cocculin-oideans, neomphaloideans, and lepetodriloideans – groups he considered to be basal rhipidoglossate taxa. While this may be true, detailed histological comparisons are necessary to test this idea in conjunction with a robust phylogeny. Also, the vetigastropod taxa lacking salivary glands are found in deep-sea biogenic, seep, or vent habitats and may have lost

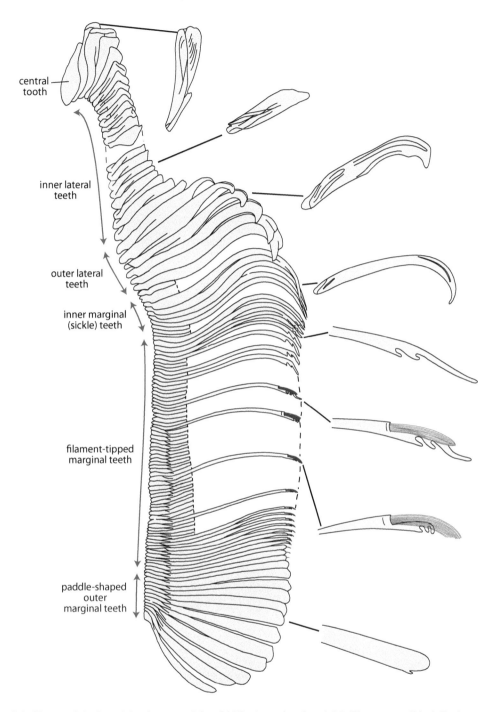

central
tooth

inner lateral
teeth

outer lateral
teeth

inner marginal
(sickle) teeth

filament-tipped
marginal teeth

paddle-shaped
outer
marginal teeth

FIGURE 18.27   A half row of the hystrichoglossan radula of *Mikadotrochus beyrichii* (Pleurotomariidae). Redrawn and modified from Woodward, M.F., *Q. J. Microsc. Sci.*, 44, 215–268, 1901.

the salivary glands independently, perhaps through feeding on bacterial films.

Dorsal and ventral folds run through the oesophagus in all vetigastropods. Lateral pouches are present in the anterior oesophagus of cocculinids and neomphaloideans, and these are expanded in other vetigastropods to cover the posterior part of the buccal mass (Sasaki 1998). These oesophageal pouches are very large in peltospirids.

In all vetigastropods the midoesophagus bears an oesophageal gland, and this is usually where the twist caused by torsion occurs. Outgrowths from the glandular walls result in the papillate condition of the oesophageal gland, one of the hallmarks of typical vetigastropods. In contrast, the walls of the oesophageal glands of cocculinids and neomphaloideans are relatively simple, lacking papillae, although in *Pachydermia* the gland cells of the midoesophagus form 'irregular stalked clumps' (Israelsson 1998 p. 99). In Bathysciadiidae there is no digestive gland but the expanded oesophageal gland occupies much the same position (Hartmann et al. 2011). *Neomphalus* and peltospirids have multiple openings to the digestive gland, and peltospirids have a somewhat similar stomach to *Pachydermia* but with a small sorting area.

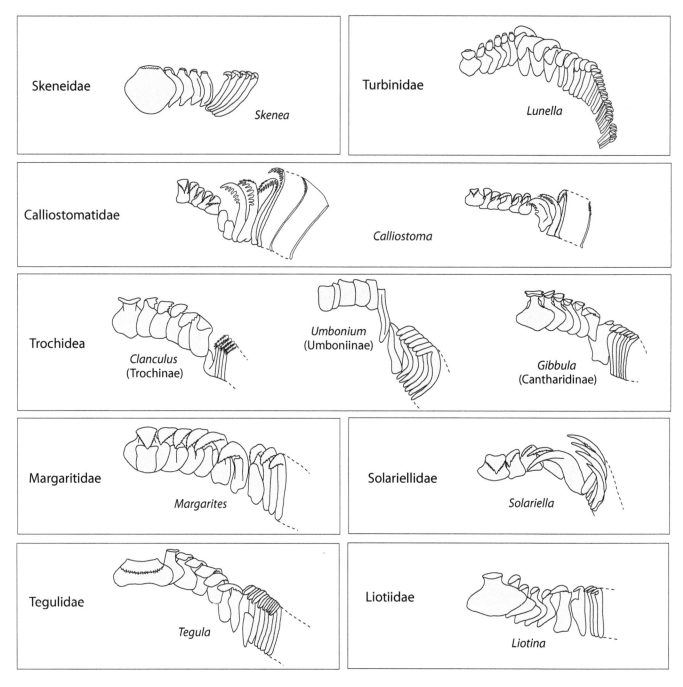

FIGURE 18.28 Examples of radulae of Trochina. Half rows and partial half rows are shown, with the marginal teeth on the right. Redrawn and modified from various sources. The radulae of the species illustrated do not necessarily reflect the range of variation in the genera they belong to or of their higher taxon.

There is a considerable range of stomach morphology in vetigastropods. Typical vetigastropods have often been thought to possess a primitive type of gastropod stomach with paired ventral openings to the digestive gland, a style sac with a protostyle, a gastric shield, and a sorting area extended into a caecum. This kind of stomach is found in most typical vetigastropods but with some variation. The caecum in pleurotomariids, haliotids, turbinids, and most trochids is greatly extended and forms a spiral, but in other vetigastropods, including some trochids (e.g., *Broderipia*), it is short and simple, while it is absent in the simplified stomachs of cocculinids and lepetelloideans. The stomach of the coiled

neomphaloidean *Pachydermia* is orientated as in other vetigastropods, but in the limpet-like, dorso-ventrally compressed *Neomphalus* it is orientated differently, perhaps, as argued by Fretter et al. (1981), the result of increased torsion – up to 270° – and a dorso-ventral flip so that the dorsal surface now lies ventrally. Both those stomachs have a gastric shield and short style sac region, a short, simple caecum, and lack a sorting area.

In the chemosynthetic vent taxa *Chrysomallon* and *Gigantopelta* the intestine is coiled only once (Chen et al. 2017), whereas in most vetigastropods the intestine is usually coiled at least twice. In some taxa, additional coils are present

(Sasaki 1998). The rectum penetrates the ventricle in most vetigastropods but does not in seguenziids, cocculinoideans, and neomphaloideans. The anus opens on the right side of the mantle cavity in taxa with a single ctenidium or towards the right when the right ctenidium is reduced and more centrally in those with two equal-sized ctenidia.

18.5.3.5 Reproductive System

Most vetigastropods are gonochoristic but cocculinoideans, lepetelloideans, and skeneids are hermaphroditic. Cocculinoideans are simultaneous hermaphrodites with distinct regions of the gonad producing eggs and sperm, and lepetelloideans are either simultaneous or protandric hermaphrodites with, as in the minute trochoidean *Skenea* (Kunze et al. 2008), a separate testis and ovary, although another hermaphroditic skeneid has a single gonad (Kunze et al. 2008). The small, coiled neomphaloidean *Leptogyra* and the scaly-footed *Chrysomallon squamiferum* are also simultaneous hermaphrodites with a single gonad (Hess et al. 2008; Chen et al. 2015a), but other related members of the latter group, such as *Leptogyropsis* and *Melanodrymia*, are gonochoristic. The Choristellidae, previously thought to be a gonochoristic member of the Lepetelloidea (e.g., Haszprunar 1988e), are now included in the Seguenzioidea (Y. Kano unpublished), all of which are gonochoristic.

In many vetigastropods, the gonad opens to the right kidney, and the gametes pass through part of the kidney before being released into the mantle cavity via the renal opening. Some, including seguenziids, cocculinids, neomphalids, and peltospirids, have a separate gonoduct, most of these taxa having internal fertilisation (Matabos & Thiebaut 2010). Copulatory organs are usually developed to enable internal fertilisation, either by modification of the right cephalic tentacle or another part of the anterior animal or as a separate, pedally derived, penial structure. These can be variable within a single group. For example, in Cocculinidae, a non-innervated copulatory structure is at the posterior end of the right oral lappet, and sperm travels to it by way of a ciliated groove. In *Paracocculina* it is on the right side of the foot, is innervated, and bears a flagellum, while in *Coccocrater* it is formed from the right cephalic tentacle (Haszprunar 1987b). The copulatory organ in Lepetelloidea is also the modified right cephalic tentacle to which sperm travels via a ciliated groove, as it generally does with other vetigastropod taxa having a copulatory structure. The peltospirids that apparently engage in internal fertilisation, possibly by way of spermatophores (see below), do not have a copulatory organ but have a well-developed prostate gland. A prostate gland is also found in neomphalids, and a copulatory organ derived from the left cephalic tentacle (McLean 1990a; Beck 1992a) is found in *Neomphalus* and *Symmetromphalus* but is typically absent in peltospirids. The other major group of vetigastropod vent taxa, the Lepetodriloidea, have a prostate gland, and some have a penis developed from the right cephalic tentacle, while in *Gorgoleptis* the penis is derived from the left side of the snout (Fretter 1988), and in some species of *Lepetodrilus*

and *Clypeosectus* a penis is present near the right cephalic tentacle. In Seguenzioidea a penis is developed from the right neck lobe in seguenziids (Quinn 1983), while the vent-living *Bathymargarites* has a copulatory organ developed from the right subocular peduncle (Warén & Bouchet 1989). Members of the Skeneidae, including *Skenea* and the vent genus *Protolira*, have a 'propodial penis' on the right anterior corner of the foot and consequently are assumed to have internal fertilisation (Warén & Bouchet 1989; Warén 1992; Kunze et al. 2008).

The spermatozoa of most vetigastropods are of the 'primitive' ectaquasperm type (e.g., Healy 1988; Hodgson & Foster 1992) (see Chapter 8), although those taxa with internal fertilisation produce modified sperm. Species of *Lepetodrilus* and *Addisonia* produce 'entaquasperm' suggesting the probability of fertilisation within the mantle cavity (Hodgson et al. 1997; Kano 2008) while clearly modified sperm involving internal fertilisation is seen, for example, in neomphalids (Healy 1988), peltospirids (Hodgson et al. 2009), and at least some seguenzioideans including the minute *Zalipais* (Healy 1990b), *Microcarina*, and *Brookula* (Healy 1990a). A larger seguenzioidean, a species of *Calliotropis*, has been shown to have ectaquasperm with an unusually short head (Healy 1989c).

Sperm storage sacs (seminal receptacles) are sometimes developed in taxa with copulatory structures and store the sperm before fertilisation. There are one or two seminal receptacles in Cocculinoidea that arise from the gonoduct (Strong et al. 2003), but these are absent in Lepetelloidea. Some Lepetodriloidea lack a seminal receptacle, but in *Lepetodrilus* and *Clypeosectus* one opens on the right side of the mantle cavity. There is also a seminal receptacle on the left side of the wall of the inner mantle cavity in at least some seguenziids (Sasaki 1998).

At least two hot-vent taxa, *Melanodrymia* and *Chrysomallon squamiferum,* use spermatophores to transfer sperm (Sasaki et al. 2010), but most vetigastropods do not (Sasaki 1998). Spermatophores may well be more widespread in the Neomphaloidea than currently recognised.

The eggs of many typical vetigastropods have a thick vitelline coat (gelatinous external envelope) which swells on contact with water, but this is absent in some, notably Neomphaloidea, Cocculinoidea, and lepetodrilids (Ponder & Lindberg 1997).

Most vetigastropods are broadcast spawners but in a few trochoideans (e.g., *Calliostoma, Cantharidus*) and the seguenzioidean *Euchelus* a jelly-like spawn is produced that is attached to the substratum (Duch 1969), and the young develop to an advanced veliger stage or crawling young. In *Margarites vorticiferus*, the young are brooded in the umbilicus of the shell until they hatch as crawl-away juveniles, and this mode of brooding is also seen in a species of *Clanculus* and some larocheids (Marshall 1993), solariellids (Marshall 1999), and liotiids (Lindberg & Dobberteen 1981). Brooding in 'cocculiniform' limpets has been reported on several occasions in the literature, but a review by Huys et al. (2002) indicated these records are erroneous, being based on the presence of eggs of a copepod (Nucellicolidae) associated with these limpets.

18.5.3.6 Renopericardial System

Two auricles are present in most vetigastropods, but only the left remains in Neomphaloidea and Cocculinoidea, as well as in some small seguenzioideans and skeneids. The heart is orientated transversely across the body in the coiled pleurotomarioideans, scissurelloideans, and trochoideans, including the limpet-like *Broderipia*, but is more obliquely orientated in the ear-shaped *Stomatia* and anteroventrally in Haliotidae (Sasaki 1998). The auricles are arranged on either side of the ventricle in all other taxa except for the fissurellids where they both lie anterolaterally to the transverse ventricle. In cocculinids, the single (left) auricle lies in front of the ventricle or anteriorly to the right (Haszprunar 1987b), while in lepetelloideans the auricle lies to the antero-right side of the ventricle. In *Neomphalus* the ventricle is oblique with the only (left) auricle on the left posterior side, suggesting a rotation of the viscera has occurred in this flattened species (Fretter et al. 1981). In other neomphaloideans such as *Pachydermia* and *Melanodrymia*, and in peltospirids, the auricle lies in front of the ventricle.

The rectum penetrates the ventricle in many vetigastropods but lies outside the pericardium in Cocculinoidea, some Lepetelloidea, Neomphaloidea, and at least some Seguenziidae and Skeneidae.

Many vetigastropods have two kidneys, the morphology and blood supply of which has been described by Andrews (1985) and Fretter and Graham (1994), and is further detailed in Chapter 6 and outlined in Section 18.2.1.1.1. The right kidney receives deoxygenated blood from the viscera and head-foot and, because this has a lower pH and is richer in metabolites than that reaching the left kidney, the right kidney is involved primarily in nitrogenous excretion. The left kidney, which lies in the mantle cavity roof, has modified vascular connections with the blood coming from the ctenidium being oxygenated and lower in metabolites (because the blood has passed through the right kidney), and the pH is higher. These are ideal conditions for absorption, and consequently this kidney has lost much of its excretory capacity and mainly resorbs soluble materials such as glucose from the urine in the lumen. Its epithelium also carries out pinocytosis, capturing larger molecules.

In vetigastropods other than fissurellids, the left kidney is also known as the papillary sac due to the many long papillae that project into the lumen. These vascularised papillae contain crystals of polymerised haemocyanin (Andrews 1985). The left kidney lies with its posterior wall adjacent to the anterior wall of the pericardium and its left side close to the efferent vessel from the left ctenidium.

In trochoideans, a spongy nephridial gland is located along the left kidney walls next to the pericardium and efferent branchial vessel, where tubular extensions of the kidney lumen open between the papillae. Both areas are concerned with absorption (see Chapter 6). In fissurellids, the left kidney is tiny but remains functional.

In those vetigastropods with only the left kidney (Cocculinoidea and Neomphaloidea), a condition convergent with higher gastropods, the kidney is enlarged; there is a nephridial gland in Neomphaloidea but not in Cocculinoidea. As already noted, a nephridial gland is also present in the left kidney of trochoideans, despite a right kidney still being present. The nephridial gland is rich in amoebocytes that phagocytose particulate matter in the blood.

In trochids, a little deoxygenated blood enters the left kidney from the nephridial gland vein, but most of the blood entering the kidney is oxygenated and from the link with the efferent branchial vessel, which, as in caenogastropods, interacts mainly with the nephridial gland. The nephridial gland vein connects with the auricle (with the branchial efferent vessel), and, in this vein, blood is shunted back and forth between the auricle and the nephridial gland (Andrews 1985). In peltospirids, which have only the left kidney, the nephridial gland is a protrusion from the kidney that lies between, and is fused to, the pericardial wall and the auricular wall. Perforations allow blood to flow back and forth between the auricle and outgrowth from the kidney (i.e., the nephridial gland) as the heart beats (Fretter 1989).

A transverse mantle vein is present in trochids, turbinids, and lepetodrilids but is absent in other groups of vetigastropods (Sasaki 1998). A basibranchial sinus, an expanded part of the efferent renal vessel, is found in vetigastropods with two ctenidia (Fretter & Graham 1962; Sasaki 1998).

18.5.3.7 Nervous System and Sense Organs

The nervous system is hypoathroid with rather poorly formed ganglia, with the exception of the peltospirid *Chrysomallon* which lacks true ganglia (Chen et al. 2015a). The pedal nerve cords range from being strongly developed with several cross connections in pleurotomariids, haliotids, and fissurellids, to being weaker with only one cross-connection in some lepetelloideans while some others lack any (Fretter & Graham 1962), including cocculinids (Haszprunar 1987b).

The visceral loop is streptoneurous and has supra- and suboesophageal ganglia that connect with the osphradial ganglia. According to Sasaki (1998 pp. 158, 184), a labial commissure is absent in vetigastropods but it is present in *Haliotis tuberculata* (see Crofts 1929 p. 109), although it is thin and inconspicuous but much exaggerated in the illustration of Fretter and Graham (1962 figure 158).

The cephalic eyes are, as in most gastropods, present at the outer bases of the cephalic tentacles where they are usually placed on the ends of short ocular peduncles. A lens is present but no cornea. In cocculinids, the eyes are degenerate and filled with mucous cells and have been termed the 'basitentacular gland' (Haszprunar 1987b).

As noted above, three important synapomorphic sensory structures are present in many typical vetigastropods, these being the sensory papillae, the epipodial sense organs, and the bursicles, which are all figured and described in Chapter 7. The sensory papillae adorn the cephalic and epipodial tentacles of most taxa and are often found on other parts of the head-foot as described by Crisp (1981). Each papilla is innervated and has a crown of immobile cilia, there being more cilia on the papillae on the cephalic tentacles than on the epipodial tentacles. In many vetigastropods, epipodial sense organs are also present (see Section 18.5.3.2) lying at the base of each epipodial tentacle when those are present. These small, knob-like organs are paler in colour than the surrounding epithelium, are innervated by a branch of the nerve supplying the epipodial tentacle, and have a central depression with a tuft of long immobile cilia. The structure of both the receptor structures of the sensory papillae and the epipodial sense organs are consistent with them being mechanoreceptors (Crisp 1981). Epipodial tentacles (see Section 18.5.3.2), often present on the sides of the foot, have some sensory capacity, even when sensory papillae are absent.

Bursicles are found in most vetigastropods and are unique to the group (Szal 1971; Haszprunar 1987a; Sasaki 1998; Geiger et al. 2008). These sensory pockets lie near the base of each ctenidial filament (see Chapter 7, Figure 7.27). Some Lepetelloidca, Pleurotomarioidea, several Neomphaloidea, and all Cocculinoidea lack them, suggesting that secondary loss has occurred in at least some of these taxa. Bursicles in the trochoidean *Tegula funebralis* have been experimentally shown to detect predatory starfish (Szal 1971).

The osphradium is on the anterior edge of the ventral (efferent) membrane of the gill(s), and its ganglion is within the membrane. A study of the osphradium of *Haliotis*, fissurellids, and at least some trochoideans showed that several cell types and simple sensory cells were present (Haszprunar 1985g) while in cocculinids and lepetelloideans the single osphradium apparently lacks sensory cells, and only nerve fibres reach the surface of the epithelium (Haszprunar 1987b, 1988d). These latter observations were made using light microscopy and need to be confirmed with TEM.[11] There is no published description of the ultrastructure of the sensory epithelium of the osphradium of any neomphaloidean.

All vetigastropods have statocysts that lie anterodorsally to the pedal ganglia (although they are innervated by the cerebral ganglia). They usually contain statoconia, but in a few (Cocculinoidea, Neomphaloidea) a single statolith is present, a condition that converges on many higher gastropods (Ponder & Lindberg 1997). The cocculinid *Macleaniella moskalevi*, based on the observation of two individuals, is exceptional in reportedly having either statoliths or statoconia (Strong & Harasewych 1999), but this requires confirmation.

18.5.4 Biology, Ecology, and Behaviour

18.5.4.1 Habits and Habitats

Vetigastropods have radiated and diversified throughout the oceans worldwide, in the tropics, temperate, and polar regions. They are found in all marine benthic habitats, from the intertidal to the deep sea, including estuarine environments, but have not successfully colonised the uppermost littoral or supralittoral zones. They are found on rocky substrata,

both on and in soft sediments, including living in sand in the surf zone. Some of the smallest taxa are found living interstitially among sand or gravel. Striking diversifications have occurred at deep-sea hydrothermal vents and hydrocarbon (or cold) seeps and on exotic deep-sea biogenic substrata such as waterlogged wood, bone, egg cases of sharks and skates, and plant debris (see next Section).

True swimming does not occur in vetigastropods, but brief excursions into the water column are achieved in a few, using a foot-thrashing mode of locomotion. Examples include the trochoidean Solariellinae (Herbert 1987), Umboniinae (Hickman 1985, 2003), and some scissurellids (Haszprunar 1988c; Hickman & Porter 2007).

18.5.4.1.1 Biogenic Deep-Sea Substrata, Cold Seeps, and Hydrothermal Vents

Biogenic substrata in the deep sea are an important habitat for some groups of vetigastropods (see also Chapter 9). For example, some members of the Cocculinidae (Cocculinoidea), Pseudococculinidae (Lepetelloidea), Melanodrymiidae (Neomphaloidea), as well as various 'skeneimorphs', live on wood (Kunze et al. 2008). Osteopeltidae, Cocculinellidae, and Pyropeltidae (all Lepetelloidea) are found on whalebone and the Lepetellidae (Lepetelloidea) live on polychaete tubes. Addisoniidae (Lepetelloidea) and Choristellidae (Seguenzioidea) live on elasmobranch egg cases, the lepetelloidean pseudococculinids on sunken algal holdfasts and carapaces of deep-sea crabs, Bathyphytophilidae on sunken seagrass rhizomes, and the cocculinoidean Bathysciadiidae lives on detrital cephalopod beaks.

In 1977 DSV *Alvin* was the first deep-sea submersible to examine hydrothermal (i.e., hot) vents, and the first vent mollusc was described in 1981 (Fretter et al. 1981; McLean 1981b). Since then many unique taxa have been described from vents and seeps, including the majority of the known Neomphaloidea, Lepetodriloidea, and the Pyropeltidae (Lepetelloidea). Additional family-group taxa found in vents and seeps are mostly represented in other marine habitats (see Sasaki et al. 2010 for review and Chapter 9). Those in this latter category include Fissurellidae, Pseudococculinidae, Seguenzioidea, and Trochoidea. Some additional families have a few species found only in seeps, including Cataegidae, Chilodontaidae, Calliostomatidae, Trochidae, and Sollariellidae (Sasaki et al. 2010).

18.5.4.2 Feeding and Diet

Most vetigastropods are browsers, sweeping the substratum with their rhipidoglossate radula and picking up mainly micro-organisms such as bacteria and detrital particles. Some deep-sea taxa ingest sediments and a few feed directly on plant material such as algae and marine angiosperms, a habit that has evolved in several groups, most notably the Haliotoidea and Trochoidea. Larger-sized haliotids capture and feed on drift algae by clamping their shell on pieces of algae.

Suspension feeding has evolved in several taxa in the Umboniinae in the Trochoidea, these having a monopectinate ctenidium with elongated filaments (Fretter 1975a; Hickman 1985, 2003; Hickman & McLean 1990). Some umboniine

---

[11] Transmission electron microscope.

taxa such as *Umbonium* and *Isanda* live in high densities in sand in the lower part of the intertidal zone while in parts of Australia the tall-spired *Bankivia* and *Leiopyrga* can be abundant in sand in the surf zone below low tide.

As noted above, taxa living on biogenic substrata are probably feeding on bacteria, other microbes, and fungi associated with the decomposition of the substratum rather than the substratum itself. Lepetodrilids combine grazing and suspension feeding, as well as having bacterial symbionts on the gill that provide nutrition (Bates 2007a, b). The intestine of *Neomphalus* contained grit, and radiolarian, foram, and crustacean skeletons (Fretter et al. 1981) while the gut of *Pachydermia* contained detritus, polychaete bristles, sponge spicules, grains of ion sulphide, and unicellular algae (Israelsson 1998).

Some vetigastropods are carnivorous grazers on encrusting invertebrate organisms such as hydroids, sponges, bryozoans, and tunicates and are therefore often associated with them. These include calliostomatids, pleurotomariids, and some fissurellids. There are no known active macro-predators among vetigastropods, but *Clypeosectus curvus* is a specialist 'predator' on folliculinid ciliates (Bergquist et al. 2007).

18.5.4.3 Life History

Most vetigastropods are dioecious, and some are sexually dimorphic, although hermaphroditism does occur, especially in some deep-sea taxa (see Section 18.5.3.5). Dwarf males have been reported living on the shells of females in the scissurelloidean *Larocheopsis* (Marshall 1993). Most discharge gametes directly into the water column where fertilisation and development take place. Larger species produce millions of eggs per reproductive season and typically have yearly cycles while smaller species produce fewer eggs, and some spawn throughout the year. Copulatory structures (often derived from the right cephalic tentacle – see Section 18.5.3.5) enabled internal fertilisation to evolve in a few groups, especially in some deep-sea taxa (see Section 18.5.3.5).

Most vetigastropods have small eggs that develop into non-feeding (lecithotrophic) larvae, passing through both a trochophore and veliger stage before settling and undergoing metamorphosis. Some internal fertilisers have glandular pallial structures that produce simple benthic jelly egg masses. In others, early development may occur within these eggs while they are contained in the spawn. Direct development may occur in some, with a few brooding species using shell features such as the umbilicus or surface sculpture to hold the developing young (see Section 18.5.3.5).

18.5.4.4 Associations

A few hot-vent taxa, notably *Lepetodrilus fucensis* (Lepetodrilidae) (Bates 2007a, b), *Cyathermia* (Neomphalidae) (Zbinden et al. 2015), and *Hirtopelta* (Peltospiridae) (Beck 2002) house epibiotic bacteria in their gills. It has also been suggested that the tubular spiny shell sculpture of *Ctenopelta porifera* (Peltospiridae) and the setae on the foot might be related to bacterial symbioses (Warén & Bouchet 1993). One of the most remarkable bacterial associations is found in the peltospirids

Chrysomallon squamiferum and in the genus *Gigantopelta*; symbiotic bacteria are maintained internally in a 'trophosome' ontogenetically derived from the oesophageal gland (Chen et al. 2017; Heywood et al. 2017; Chen et al. 2018). A network of bacteriocytes has been reported in the mantle groove of the lepetelloidean *Lepetella sierra*, and symbiotic bacterial associations are also suspected in *Addisonia* and *Bathyphytophilus* (Judge & Haszprunar 2014). As in other gastropods, numerous gut bacteria assist in digestion (see Chapters 5 and 9).

A few have commensal relationships with other invertebrates such as crustaceans and polychaetes that live in their mantle cavity (see Chapter 9). Parasites of vetigastropods have not been well studied, but trematodes have been recorded from several, mainly intertidal taxa (e.g., Fretter & Graham 1962).

18.5.4.5 Behaviour

Because most vetigastropods are broadcast spawners, in the great majority of taxa there is usually no courtship or mating behaviour. However, putative 'mating stacks' have been reported for two peltospirids – *Gigantopelta chessoia* and *Gigantopelta aegis* – at hydrothermal vents (Chen et al. 2018). Grange (1976) and Hickman and Porter (2007) have also reported mass synchronised spawning in vetigastropods, while in most other species the spawning period is more spread out (e.g., Bell 1992).

Several vetigastropods have been shown to have escape responses from predators (whelks, sea stars), which include swaying and tilting the shell to avoid tube feet as well as short, relatively rapid movement away from potential predators after contact. Similar shell manipulations are used by intertidal taxa that co-occur with aggressive territorial limpets (Stimson 1970) (see Section 18.4.4).

Most subtidal and intertidal species are active at night when visual predation is less effective, while intertidal species are seldom active during daytime low tides to avoid physiological stress due to drying.

Shell wiping with the foot is an interesting behavioural habit in at least some calliostomatids and is thought to remove fouling organisms and possibly supplement the diet (Holmes et al. 2001, 2011; Jones et al. 2006).

18.5.5 DIVERSITY AND FOSSIL HISTORY

Groups treated as vetigastropods were common in the Paleozoic where they are lumped mainly in Pleurotomarioidea or Trochoidea. Some pleurotomariid-like taxa extended back to the Upper Cambrian (e.g., Knight et al. 1960a) (see Chapter 13), although careful analysis based on shell structure and protoconch morphology suggests that the first undoubted vetigastropods arose in the Silurian (Frýda et al. 2008a) with the earlier taxa belonging to extinct groups of uncertain relationship. Molecular estimates of divergence times place the origin of the vetigastropods somewhere between the Ordovician and Carboniferous with the node in the Lower Devonian (Stöger et al. 2013 figure 3). Vetigastropods were diverse during the remainder of the Paleozoic and remained abundant during the Mesozoic but became less varied in the Late Mesozoic and

Cenozoic. There are many extinct families and some superfamilies, and these are listed in the Appendix.

A few vetigastropod-like Paleozoic fossils with slits and a turbiniform[12] shape have protoconchs of about two smooth whorls showing an abrupt transition to the teleoconch (Nützel & Mapes 2001; Kaim 2004), thus combining a possible planktotrophic protoconch with an apparently typical pleurotomarioid teleoconch. Whether these gastropods are vetigastropods with planktotrophic larval development, slit-bearing caenogastropods, or an entirely extinct group is uncertain. At present, the consensus is that they are not vetigastropods (Nützel et al. 2007b; Frýda et al. 2008a; Geiger et al. 2008), all of which have the typical protoconch described above.

Of the living taxa, the modern pleurotomariids are relicts of a once-diverse group that first appeared in the Jurassic (Bandel 2009). The trochoideans are the most diverse of the living taxa with over 500 genera (Hickman 1996), but their first appearance in the fossil record is unclear, given their lack of definitive shell features.

The hot-vent vetigastropod taxa are known from fossil vent habitats back to the Upper Jurassic (Kiel 2010; Kaim et al. 2014), but they presumably had a much longer history. The extinct ancestors in this lineage and that of the cocculiniform limpets have not been determined, but based on available molecular phylogenies, these two groups must have diverged from stem vetigastropods early in their evolution. One potential early peltospirid is the enigmatic gastropod *Elmira* found in large numbers at Upper Cretaceous hydrocarbon-seep deposits (Nobuhara et al. 2016).

18.6 NERITIMORPHA (= NERITOPSINA)

18.6.1 INTRODUCTION

This medium-sized group represents the last subset of gastropods previously referred to as the 'Archaeogastropoda'. Marine neritimorphs are found from the intertidal to the deep sea, including hydrothermal vent habitats, and, unlike the other groups dealt with in this chapter, they have also diversified into freshwater and terrestrial habitats.

Most neritimorphs have coiled shells (Figure 18.29), but some have adopted a limpet-shape, including the Phenacolepadidae and some freshwater taxa. In all neritimorphs except the slug-like, shell-less Titiscaniidae, the shell, which is never nacreous, covers the animal. Most species absorb the internal whorls of the shell as they grow, permitting the body of the snail to be more ovoid rather than coiled, irrespective of its shell morphology. An oval to D-shaped operculum is present in coiled taxa but is lost in some of those with limpet-like shells. The operculum is paucispiral, thick, and often partially calcified and has one, or rarely two, peg-like projections on the inner surface.

Neritimorphs retain some plesiomorphic features along with possessing a host of apomorphies. All neritimorphs have retained paired auricles but have a single (left) kidney, with the right represented by a remnant incorporated in the reproductive system. Only the left ctenidium is functional, but a tubercle-like structure seen in some taxa probably represents a vestigial right ctenidium. Besides the paired auricles, other plesiomorphic features include the rectum penetrating the ventricle, a rhipidoglossate radula, paired shell muscles, and a bipectinate ctenidium. The ctenidium lacks skeletal rods, but it is unclear whether this is a plesiomorphic feature or a secondary loss. Besides the suppression of the right mantle cavity organs, other apomorphic characters include lateral ciliated fields associated with the osphradium and the development of pallial genital structures facilitating internal fertilisation and the ability to produce encapsulated eggs. Also, unlike the other basal gastropod taxa, the veliger larva is planktotrophic like that of higher gastropods.

The mix of primitive and derived characters in neritimorphs has led to much speculation about their relationships. In a classic anatomical investigation of neritids, Bourne (1909) argued that the group had arisen from early gastropods, an observation supported by the fossil record which suggests the group originated before the Middle Devonian (see Section 18.6.5). An early derivation is also suggested by general morphology (Ponder & Lindberg 1997), the formation of the protoconch (Bandel 1982), and embryonic cleavage patterns (van den Biggelaar & Haszprunar 1996) and is supported by molecular data (e.g., Zapata et al. 2014).

Some Neritidae are found in both brackish conditions and fresh water, with some of the freshwater taxa retaining an estuarine or marine larval phase. The terrestrial neritimorphs include the Helicinoidea and Hydrocenidae, and a few members of the otherwise marine and freshwater Neritidae can survive in semiterrestrial conditions.

18.6.2 PHYLOGENY AND CLASSIFICATION

Early 20th century workers noted that neritids and their relatives differed greatly from other 'archaeogastropods', but this has only been reflected in gastropod classification in the last few decades.

Unlike other living neritimorphs, *Neritopsis* does not absorb the internal whorls of its shell. Based on this plesiomorphic feature and some other anatomical characters, and its long fossil history, *Neritopsis* is considered the most basal of living neritimorphs as reflected in recent molecular phylogenies (Kano et al. 2002; Uribe et al. 2016a). The closely related *Titiscania* has a slug-like body, having lost the shell.

The two terrestrial groups Helicinoidea and Hydrocenoidea have a lung instead of a gill and have intracapsular direct development. These two groups are not closely related and differ in their shells, radula, operculum, and reproductive systems. Both Bourne (1911) and Haszprunar (1988a) hypothesised that they independently colonised the land, an idea supported by the molecular phylogenies of Kano et al. (2002) and Uribe et al. (2016a).

Hydrocenoids are represented by only a single family while there are four living and two extinct families of helicinoideans recognised (Bouchet & Rocroi 2005), but neither group has had a modern review. The two groups are quite different

---

[12] Shaped like members of the genus *Turbo*.

FIGURE 18.29 Examples of living Neritimorpha. (a) *Neritopsis radula* (Neritopsidae), Marshall Islands. Courtesy of S. Johnson. (b) *Titiscania limacina* (Neritopsidae), Marshall Islands. Courtesy of J. Johnson. (c) *Theodoxus fluviatilis* (Neritidae), Europe. M. Mañas. Reproduced here under Creative Commons Attribution 4.0 International Licence [https://creativecommons.org/licenses/by/4.0/deed.en] (d) *Nerita undulata* (Neritidae), Singapore. Courtesy of R. Tan. (e) *Neripteron violaceum* (Neritidae), Australia. Courtesy of D. Riek. (f) *Smaragdia viridis* (Neritidae), Kavieng, Papua New Guinea. Courtesy of P. Maestrati - MNHN. (g) *Plesiothyreus cinnamomeus* (Phenacolepadidae), New South Wales, Australia. Courtesy of D. Riek. (h) *Alcadia pellucida* (Helicinidae), French Guiana. Courtesy of O. Gargominy - MNHN.

anatomically, indicating that they may well represent two independent terrestrial invasions (e.g., Haszprunar 1988a). Bandel (2001) argued that the hydrocenids were related to the neritiliids but molecular data place this latter marine and freshwater group as the sister to the helicinids while the hydrocenids are the next most basal taxon to Neritopsoidea (Kano et al. 2002; Fukumori & Kano 2014), in accordance with the conclusion reached by Bourne (1911) using anatomical characters.

18.6.2.1 Sister Group Relationships

It is possible that neritimorphs may have arisen as early as the Ordovician (see Section 18.6.5). If so, this would be around the same time that the first putative caenogastropods appeared (in the Upper Ordovician), making a sister group relationship with them possible.

The Paleozoic group Cyrtoneritimorpha is often included in Neritimorpha, but this placement is not universally accepted. Cyrtoneritimorphs are distinguished from the true Neritimorpha (separated as Cycloneritimorpha if both groups are included) by having an uncoiled to openly coiled tubular protoconch. We do not accept this arrangement and regard the cycloneritimorphs as belonging to an extinct group of uncertain relationships (see Chapter 13 and the Appendix).

A relationship between cocculinids and neritimorphs has been proposed (Thiele 1903; Ponder & Lindberg 1997) but has not received support from recent phylogenetic analyses.

18.6.2.2 Outline of Classification

In earlier classifications, neritimorphs were recognised as a 'superfamily' (Neritacea) within Archaeogastropoda, but Cox and Knight (in Knight et al. 1960b) introduced the 'suborder' Neritopsina for the group, emphasising its distinctiveness. That name was an amended form of the name Neritimorphi introduced by Koken in 1896, and Golikov and Starobogatov (1975) reintroduced that name as Neritimorpha, using it as a 'superorder' and including the cocculinids. Neritimorpha (excluding the cocculinids) has been used by most recent authors, although Neritopsina has sometimes been used (e.g., Salvini-Plawen 1980; Bieler 1992; Ponder & Lindberg 1997; Beesley et al. 1998b; Kano et al. 2002) for the group. Sasaki (1998) treated them as an order, also in the archaeogastropods, as did Haszprunar (1988a) within his 'superorder' Flexoglossata, but he used the earlier name Neritimorpha for the group. Ponder and Lindberg (1997), using the name Neritopsina, elevated the group to equal status to that of Vetigastropoda, Caenogastropoda, and Heterobranchia in their unranked classification, as did Bouchet and Rocroi (2005) and Lindberg et al. (2008). Bouchet et al. (2017) recognised the rank of these groups as that of subclass while here we use the rank infraclass.

Nine living families are generally recognised – the marine, freshwater, and semiterrestrial Neritidae, the freshwater and marine Neritiliidae, the marine Phenacolepadidae, Neritopsidae, and Titiscaniidae, and the terrestrial Helicinidae, Proserpinellidae, Proserpinidae (which make up the Helicinoidea), and Hydrocenidae.

The ancient Neritopsidae has a rich fossil history (Bandel 2000; Bandel & Kiel 2003; Bandel 2007; Bandel 2008). The shell-less slug *Titiscania* is usually included in its own family, but Kano et al. (2002) included it in the Neritopsidae, a course not adopted here because of the substantial morphological differences.

The Neritidae contains the great majority of living marine and freshwater neritimorphs. Three living subfamilies, Neritinae, Neritininae (= Septariini and Theodoxinae), and Smaragdiinae, are recognised by Bouchet et al. (2017), along with an extinct subfamily Velatinae. They range from globular snails to limpets. Two of the subfamilies (Neritininae, Smaragdiinae) are sometimes recognised as separate families (e.g., Bandel 2001). The Phenacolepadidae is a related family of limpets adapted to life in anoxic habitats. The marine and freshwater Neritiliidae contains two living genera (*Neritilia* and *Pisulina*) (Kano & Kase 2000; Kano et al. 2002; Kano & Kase 2002), although earlier classifications had members of this group in the neritids.

The Helicinidae and Neritiliidae are related, but we treat them below as belonging to separate superfamilies and place them in a separate suborder from Neritoidea and Neritopsoidea. The clade (Helicinoidea) is rather poorly supported in the molecular phylogeny of Kano et al. (2002), with no clear morphological synapomorphy, although their female reproductive systems are rather similar and differ markedly from other neritimorphs (Kano et al. 2002). In particular, the vaginal and oviduct openings are widely separated, and the opening of the vaginal cavity is deep inside the mantle cavity. Also, the two families have similar sperm ultrastructure (Kano et al. 2002). The Helicinidae is by far the largest of the terrestrial families and is split into six subfamilies (Bouchet et al. 2017), one of which is extinct.

The small terrestrial family Hydrocenidae comprises only four genera and differs markedly from the helicinids in anatomy and shell and opercular morphology.

The classification outlined in Box 18.5 is based largely on the molecular results of Kano et al. (2002) and Uribe et al. (2016a).

BOX 18.5 SUMMARY OF THE HIGHER CLASSIFICATION OF NERITIMORPHA

(Infraclass) **Neritimorpha** (= Neritomorpha, 'Neritopsina')
(Order) **Neritopsida** (= Cycloneritimorpha)
(Suborder) **Neritopsina**
Superfamily Neritopsoidea
(Suborder) **Neritina**
Superfamily Neritoidea
(Suborder) **Hydrocenina**
Superfamily Hydrocenoidea
(Suborder) **Helicinina**
Superfamilies Neritilioidea (new name) and Helicinoidea

See Appendix for details of classification.

18.6.3 MORPHOLOGY

18.6.3.1 Shell and Protoconch

Neritimorphs are generally small to medium-sized gastropods with a range of 2–40 mm, a relatively small size range compared with other groups of gastropods. Shell morphologies (Figures 18.29 and 18.33) range from coiled conical snails (Hydrocenidae) to limpets (Phenacolepadidae), but none have developed very high-spired shells. One taxon, *Titiscania*, is a slug. The rather globular shells of most neritids have an aperture with a flat, callus-covered inner lip, and the columellar edge often bears folds, teeth, or pustules. While most neritimorphs have coiled shells, a symmetrical limpet-like morphology has independently evolved in several lineages, including three living ones – Phenacolepadidae, Neritidae (*Septaria*), and in an undescribed genus of Neritiliidae (Kano et al. 2002) – while some limpet-like taxa with the early whorls coiled also occur in several clades. The freshwater genus *Septaria* (Figure 18.31) has an internal shelf converging on the caenogastropod slipper limpets (Calyptraeidae).

External colour patterns are variable, ranging from solid dark and light colours to bright greens. Shell markings are often geometrical in zig-zag patterns. Terrestrial neritimorphs can have bright colour markings and glossy shells, but others are more cryptic. Some cases of colour polymorphism have been documented in neritids (Huang 1995) (see also Chapter 9). Freshwater species are usually drab, and many of these are covered by a thick dark periostracum to protect the shell from erosion by acidic water. Some brackish and freshwater species of *Clithon* develop long spines (Haynes 2005), presumably as protection from predators, a feature not seen in living marine taxa, although short spines were present in a few extinct taxa.

The terrestrial Helicinidae have subconical, globose, or flattened shells usually less than 30 mm in diameter, and although some have projections within the aperture (Bishop 1980) that act as barriers, most lack such ornament. The apertural ornament consists of lamellae and other tooth-like barriers inside the inner lip and/or on the columella. In contrast, the other terrestrial group, the Hydrocenidae, mostly have conical shells usually less than 5 mm in length and lack apertural ornament.

Most living neritimorphs have a low-spired, rounded shape that may be linked with the resorption of the inner parts of the teleoconch and protoconch. A taller spired shell with deep sutures with the internal parts removed would probably be too vulnerable to mechanical pressures by environmental forces and predators (Nützel et al. 2007a).

Other than neritopsoids, all living neritimorphs progressively resorb the early whorls and columella of the shell as the animal grows. Thus, while the shell grows spirally, with the whorls visible externally, the interior forms a large cavity, and the visceral mass does not spirally twist but instead forms an ovoid mass that curves to the right (Figure 18.30). What advantages internal resorption, accompanied by a shortening of the viscera, afforded ancestral neritimorphs is uncertain. This feature has led to suggestions that neritids evolved

coiling independently of other gastropods (e.g., Thompson 1980; Haszprunar 1988a), but this is not supported by the fossil record (e.g., Ponder & Lindberg 1997) as the internal absorption probably first evolved in ancestors derived from Paleozoic naticopsoideans without this habit (Nützel et al. 2007a).

Microtubules from the mantle penetrate the shells of some vent-living phenacolepadids, and some of these 'shell pores' extend to the outer surface (Sasaki et al. 2003).

The protoconch features of neritimorphs are variable. The neritoideans have a smooth, convolute larval shell (Figure 18.32), which, like the teleoconch, has resorbed inner whorls. In contrast, the Paleozoic members of the group had both smooth and strongly sculptured spiral larval shells that were not internally resorbed (Nützel et al. 2007a), as in living *Neritopsis*. Two living species of *Neritopsis* have lecithotrophic development, but a third has planktotrophic development with a larval shell consisting of protoconch I and II (Bandel 2007) (see also Section 18.6.4.3).

Naticopsoidean protoconchs are generally similar to those of many caenogastropods, in being dextral and orthostrophic, with well-defined sutures and not internally resorbed, hinting that this type of protoconch may be plesiomorphic for both groups and that they may have possibly shared a common ancestor. If this was the case, they may have shared plesiomorphic larval planktotrophy (Nützel et al. 2007a).

18.6.3.1.1 Shell Structure

Shell structure in neritimorphs was treated in the thesis and dissertation of the late Claus Hedegaard (Hedegaard 1990, 1996) and is the basis for the information summarised in this section.

Most neritimorphs have solid shells with an outer calcite and an inner aragonite layer. They can have several types of crystalline structure in their shells. Cone complex crossed-lamellar structure is found in most neritimorphs (but not in *Neritilia*, *Georissa*, and phenacolepadids), although some variations occur. For example, in many species, this microstructure grades into one that closely resembles irregular complex crossed-lamellar structure. Aragonitic comarginal crossed-lamellar structure occurs in some helicinids, most neritids, and in phenacolepadids. Aragonitic radial crossed-lamellar structure is less common than the comarginal form, being known only in *Neritodryas* and one fossil *Theodoxus*.

Calcitic homogeneous shell structure forms the outer layer of the shell of nearly all neritimorphs, being absent in only a few helicinids. Aragonitic homogeneous shell structure comprises the middle shell layer in *Georissa* (Hydrocenidae) but is not present in other hydrocenids or other neritimorphs. In many neritimorphs examined, unidentified relics of other shell structures were found in the homogeneous layers of some taxa.

Simple prismatic structures are present in a few neritimorphs – including some hydrocenids, Helicinidae, *Neritilia* and the hot-vent phenacolepadid *Shinkailepas*.

There is a common sequence of shell structures in neritimorphs from the outside to the inside of the shell – calcitic homogeneous, commarginal crossed-lamellar, cone complex crossed-lamellar, and simple prismatic. In various taxa, a few

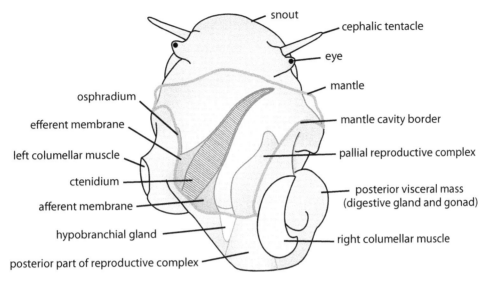

Dorsal view of neritid, removed from shell

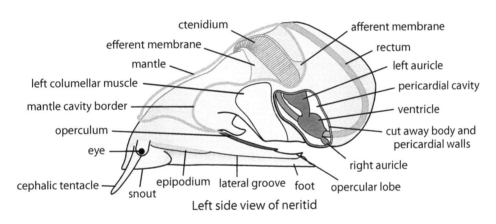

Left side view of neritid

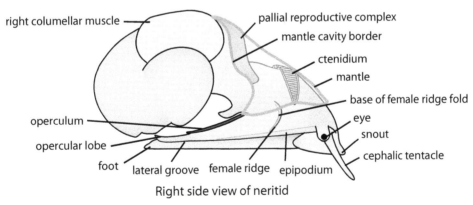

Right side view of neritid

FIGURE 18.30 External morphology of *Nerita*, with shell removed to show significant structures. Redrawn and modified from Holthuis, B.V., *Evolution between marine and freshwater habitats: A case study of the gastropod suborder Neritopsina*, Ph.D. dissertation, University of Washington, 1995.

deletions may occur, or one layer may be replaced by another microstructure.

A report of nacre in the hot-vent phenacolepadid *Shinkailepas tufari* (Beck 1992b) is a misinterpretation of scale-like deposits on the inside of the shell.

In many neritimorphs an obvious periostracum is present, although taxa such as *Nerita, Smaragdia, Neritopsis*, and some Helicinidae lack a distinct periostracum.

18.6.3.2 Head-Foot, Mantle, Operculum, and Locomotion

The head bears a short snout extended laterally to form oral lobes, and the neck is short, except in *Phenacolepas*. A pair of moderately long to very long (as in *Pisulina*, Kano & Kase 2002) (Figure 18.33) cephalic tentacles are present that lack the sensory papillae found in vetigastropods. The cytology of

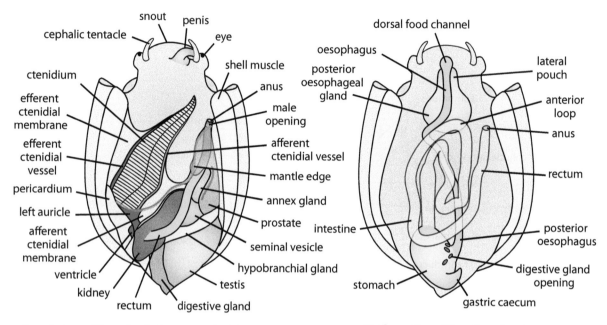

Dorsal view of the animal after removal of the shell and mantle

Configuration of the alimentary tract

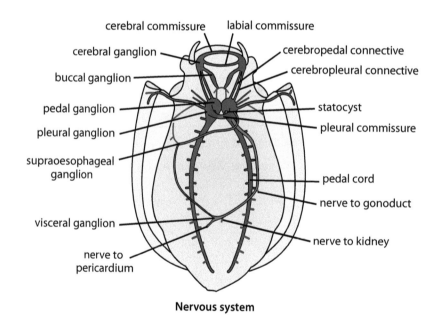

Nervous system

FIGURE 18.31 Anatomy of *Septaria*, a limpet-shaped neritid. Dorsal views with organs exposed to show the mantle cavity, gut, and nervous system. Redrawn and modified from Sasaki, T., *Bull. Univ. Tokyo Mus.*, 38, i–vi, 1–223, 1998.

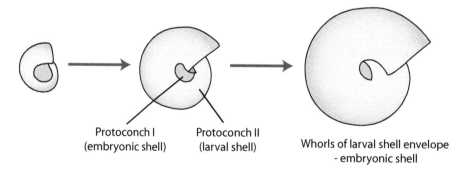

FIGURE 18.32 Growth stages of the larval shell (protoconch II) of the neritid *Smaragdia*. Redrawn and modified from Bandel, K., *Fazies*, 7, 1–198, 1982.

the tentacle epithelium differs from other groups of gastropods but lacks any clearly defining structures (Künz & Haszprunar 2001). A pair of ocular peduncles are present in *Neritopsis* and many neritids, but these are absent in phenacolepadids and Neritiliidae. Hydrocenids have very short lobes bearing eyes that probably represent the ocular peduncles, and they lack cephalic tentacles. Phenacolepadids are the only neritimorphs to have cephalic lappets – flaps on the snout on the inner sides of the tentacles. The mantle edge bears sensory tentacles in phenacolepadids, but in other neritimorphs it is simple. The foot may have a simple, narrow longitudinal epipodial ridge on each side; epipodial tentacles or sense organs are never present, but some sexual elaborations on the right are sometimes apparent. An epipodium is absent in *Neritilia*, *Neritopsis* (Holthuis 1995), and *Titiscania* (Bergh 1890; Marcus & Marcus 1967). A smooth epipodium is present in Helicinidae (Bourne 1911), Hydrocenidae (Thiele 1910), and Neritidae (Bourne 1909; Holthuis 1995), while an epipodium with short tentacles is present in some Phenacolepadidae (Okutani et al. 1989; Beck 1992b).

A distinct propodial pedal gland is present in Helicinidae (Bourne 1911; Baker 1925, 1926), *Neritopsis* (Holthuis 1995), Phenacolepadidae (Fretter 1984; Beck 1992b), and *Titiscania* (Marcus & Marcus 1967) but is absent in Neritidae (Bourne 1909).

The slug *Titiscania* has defensive glands on its dorsal surface that discharge white threads when the animal is disturbed (Bergh 1890; Marcus & Marcus 1967) (Figure 18.33).

Neritids all move using mucus, muscle, and cilia on the sole, as in most other gastropods (see Chapter 4). None are much modified for burrowing, and none can swim.

The terrestrial taxa show differences in the configuration of the foot sole, which is furrowed in hydrocenids but is usually uniform in helicinids. The sole in *Helicina delicatula* is tripartite with the broad central zone being mainly involved in locomotion, and numerous retrograde pedal waves are developed (Baker 1928). While species of *Helicina* are rather active, members of another helicinid genus, *Schasicheila*, are relatively slow and inactive (Baker 1928). This latter taxon has an elliptical, rather short sole divided into two halves by a longitudinal groove. There are irregular retrograde waves, with only one or two on the sole at any one time, usually moving together on both sides of the foot (Barker 2001). In contrast to the helicinids, in another helicinoidean family, Proserpinidae, the sole extends laterally up on to the sides of the foot where it is delimited by a parapodial groove and is thus similar to the aulacopodous condition seen in some terrestrial stylommatophoran pulmonates.

Most neritimorphs have a calcified oval to semi-circular operculum (Figure 18.34) with an internal apophysis (spur or peg) or ridge-like projection (as in *Bathynerita* or helicinids), the purpose of which is to provide a greater surface area for muscle attachment. The calcified layers are covered externally above and below by thin corneous layers except in *Nerita* where an extra calcareous layer is secreted externally. A few groups have lost the operculum, and in some it is not calcified. In most, complete withdrawal into the shell occurs

with the operculum sealing the aperture, but this cannot occur in the limpet-like taxa where the operculum is modified or reduced. In *Phenacolepas* it is minute, while in another phenacolepadid, the vent-living *Olgasolaris*, it is large and extends under the viscera (Beck 1992b). Opercular growth is spiral in most living neritimorphs but is concentric in *Neritopsis*, some helicinids, neritiliids, and neritids, and at least some fossil naticopsid taxa (Kano et al. 2002; Bandel 2008). In the limpet-like *Septaria*, the operculum commences growth spirally and then becomes concentric.

The operculum of helicinids has an inner corneous and an outer calcareous layer and lacks a true apophysis on the inner surface, having only a ridge. In the helicinoid Proserpinidae and Proserpinellidae the operculum is secondarily absent. The great diversity of opercular types in neritimorphs has been discussed in relation to their occurrence in the fossil record by Kaim and Sztajner (2005) and Bandel (2008).

In many neritimorphs, both columellar (i.e., shell) muscles are present, though the left is usually reduced in size but hydrocenids have only one (the right) muscle which is posteriorly divided into two strands (Thiele 1910).

18.6.3.3 Mantle Cavity

The single left ctenidium, present in aquatic taxa, is bipectinate along its entire length, and the filaments lack skeletal rods. Dorsal and ventral supporting membranes are present, the efferent extending along up to half the length of the ctenidium and the afferent much shorter. In aquatic taxa, the single left osphradium lies at the anterior base of the efferent membrane. In most aquatic taxa the ctenidium extends most of the length of the mantle cavity but is shorter in a few, especially in *Neritilia* in which the gill is only about a quarter of the length of the cavity. The ctenidium is also present in *Titiscania* (Bergh 1890; Marcus & Marcus 1967). In the terrestrial helicinoids and Hydrocenidae, there is no ctenidium or osphradium in the mantle cavity that functions as a lung. In other neritimorphs, the mantle cavity roof and floor are vascularised, and this may have been an exaptation to the colonisation of land (Barker 2001) enabling a simple transition to a lung.

The right side of the mantle cavity is largely occupied by the pallial part of the genital system and the rectum. In several neritoideans, a small tubercle or flap probably represents a vestigial right gill (Lenssen 1902; Bourne 1909; Fretter 1965), but this is not present in *Neritopsis* (Holthuis 1995) and is lacking in some other taxa such as phenacolepadids.

When present, the left hypobranchial gland is on the mantle roof above the gill. It is absent in Helicinidae, Hydrocenidae, *Neritina*, *Septaria*, *Theodoxus* (Bourne 1909; Thiele 1910; Bourne 1911; Holthuis 1995), *Phenacolepas* (Fretter 1984), and *Titiscania* (Bergh 1890). A right hypobranchial gland is also present in many taxa but is absent in *Bathynerita* and *Phenacolepas* (Bourne 1909, 1911; Yonge 1947; Fretter 1965; Holthuis 1995). When present, it is situated either to the right of the reproductive complex as in *Neritilia* and *Neritopsis* or projects posteriorly into the visceral mass as in most Neritidae and *Theodoxus* (Yonge 1947; Holthuis 1995) (Figure 18.37).

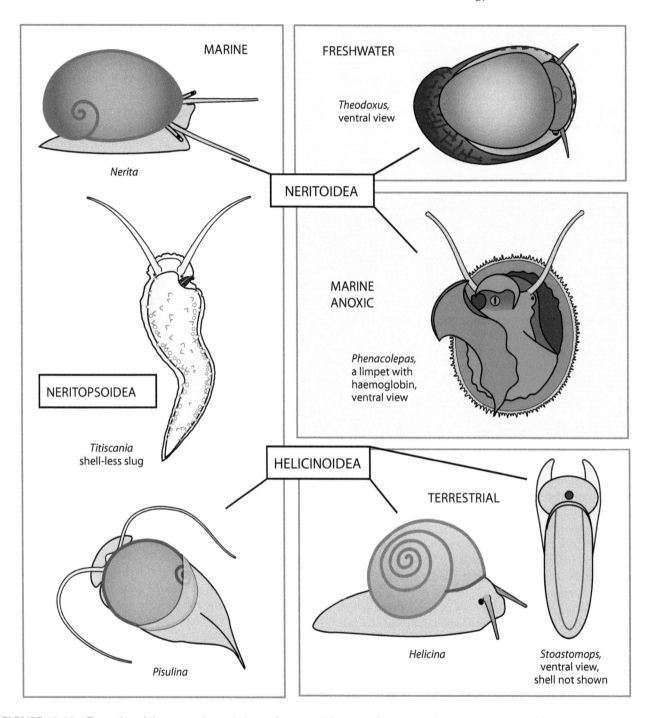

FIGURE 18.33 Examples of the external morphology of some neritimorphs. Redrawn and modified from the following sources: Upper left (Gray, M.-E., *Figures of Molluscous Animals Selected from Various Authors; Etched for the Use of Students. Volumes 1–4*, Longman & Company, London, 1842–1850), middle left (Templado, J. and Ortea, J., *The Veliger*, 44, 404–406, 2001), bottom left (Kano, Y. and Kase, T., *J. Molluscan. Stud.*, 68. 365–383, 2002), upper right original, middle right (Fretter, V., *J. Molluscan Stud.*, 50, 8–18, 1984), lower right (Baker, H.B., *Proc. Acad. Nat. Sci. Philadelphia*, 78, 35–56, 1926), lower middle original.

18.6.3.4 Digestive System

The oval to circular mouth is surrounded by an oral hood or veil and opens to the oral tube, which lacks jaws. The radula is rhipidoglossate and symmetrical. Detailed descriptions and figures of neritimorph radulae are available (Troschel 1856–1879; Thiele 1891–1893; Baker 1922, 1923; Unabia 1996), and some examples are shown in Figure 18.35. The central

teeth are absent in *Neritopsis*, *Titiscania*, Hydrocenidae, *Neritilia* and present in Helicinidae, the remaining Neritidae, and Phenacolepadidae. The inner lateral field may have zero to three teeth on each side. There are no teeth in the inner lateral field of *Titiscania* and *Georissa*, one in *Hydrocena* and *Neritilia*, two in *Neritopsis*, and three in Helicinidae, Neritidae, and Phenacolepadidae. The outer lateral field may

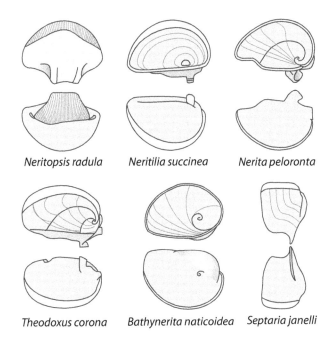

Neritopsis radula   Neritilia succinea   Nerita peloronta

Theodoxus corona   Bathynerita naticoidea   Septaria janelli

FIGURE 18.34 Examples of neritimorph opercula showing their inner and outer surfaces. Redrawn and modified from Holthuis, B.V., *Evolution between marine and freshwater habitats: A case study of the gastropod suborder Neritopsina*, Ph.D. dissertation, University of Washington, 1995.

have one, two, or several fused teeth or none. There are none in *Georissa*, one in *Hydrocena*, two in *Titiscania*, and several fused teeth in the Helicinidae, Neritidae, *Neritopsis*, and Phenacolepadidae. There are numerous marginal teeth, and the inner marginal teeth may have one or several cusps.

There are a pair of large anterior cartilages (called medial by Guralnick & Smith 1999), a pair of dorsolateral cartilages (median of Sasaki 1998), and a pair of small posterior cartilages (called posterior ventral cartilages by Guralnick & Smith 1999) are present (Sasaki 1998; Golding et al. 2009a) and a paired or single medial cartilage below the radula (Sasaki 1998; Golding et al. 2009a). The medial cartilage(s) are a synapomorphy of neritimorphs (Sasaki 1998).

The musculature of the buccal mass is complex and is described in general for gastropods in Chapter 5. For a detailed account of the buccal musculature in neritimorphs see Sasaki (1998). They possess three pairs of buccal muscles unique to the group, the dorsal levator muscles of the odontophore, the median levator muscles of the odontophore, the latter absent in the helicinid *Waldemaria*, and the tensor muscles of the anterior cartilages (Sasaki 1998). Unlike other basal gastropods, there is no subradular organ in neritimorphs.

Salivary glands are absent in neritimorphs, but Sasaki (1998) noted that the absence of these glands was functionally compensated for by the extensive glandular epithelium of the buccal cavity and by the sublingual glands, which are found in many (but possibly not all) neritimorphs. These glands are globular outgrowths from the sublingual pouch, which lies between the floor of the snout and the ventral part of the anterior buccal mass.

The oesophagus is rather long and, in most neritimorphs, oesophageal glands are septate glandular pouches separated from each other and the oesophagus posteriorly (Sasaki 1998).

The stomach is large, with a small posterior caecum (sometimes vestigial), a sorting area, a gastric shield, and a style sac, but a crystalline style is not developed.

The intestine is long, forming a single or double forward loop. The rectum extends anteriorly to the front of the mantle cavity. In neritids and other aquatic neritimorphs, the rectum passes through the ventricle, but in the terrestrial taxa, the rectum does not penetrate the pericardium. Faecal material is in the form of pellets.

18.6.3.5 Renopericardial System

The circulatory system is generally similar to that of vetigastropods. Right and left auricles are present in Neritidae (except *Bathynerita*), *Neritopsis*, *Titiscania*, Neritiliidae, and primitive helicinoideans (see below), but the right is absent in most Helicinidae, Hydrocenidae, *Bathynerita* (Warén & Bouchet 1993), and *Phenacolepas*. The anatomically left auricle in Helicinidae has rotated to the right due to rotation of the body (as it also is in Cocculinidae). In the most primitive helicinoideans, there are two auricles. In *Ceres* (Proserpinellidae), the right auricle is almost as large as the left (Thompson 1980), but in *Hendersonia* (Helicinidae, Hendersoninae) it is much reduced in size, although functional (Baker 1925).

The primary filtrate is formed by ultrafiltration by podocytes in the auricular wall (Andrews 1985; Estabrooks et al. 1999) and is moved to the kidney via a renopericardial duct.

The left kidney is the only functional one and possesses a bladder that opens by way of the renal opening at the posterior end of the mantle cavity (Andrews 1981; Holthuis 1995). The bladder wall usually has some lamellae dorsally but is smooth ventrally where the kidney wall is fused to the pericardium. A lobe-like tubular structure attached ventrally contains the renal epithelium and opens to the bladder posteriorly and the renopericardial duct anteriorly. There is no nephridial gland (Andrews 1988). In freshwater and terrestrial taxa the kidney is involved in ion resorption, both having the ability to produce dilute urine (Little 1972). The epithelium throughout the kidney contains mucoid cells whose secretions may assist in decreasing surface tension and, in terrestrial species, decreasing evaporation (Estabrooks et al. 1999). Marked rotation of the kidney has occurred in helicinids (Bourne 1911; Little 1972).

Phenacoplepadids are unusual in possessing haemoglobin contained within erythrocytes circulating in the blood which may be related to a potential association of symbiotic bacteria with this taxon (Warén & Bouchet 1993) and to their preference for anoxic environments. Other neritimorphs have haemocyanin as the respiratory pigment.

18.6.3.6 Reproductive System

All neritimorphs have separate sexes and practise internal fertilisation. The eggs lack an external vitelline layer. Oogenesis has been examined ultrastructurally in *Bathynerita naticoidea* (Eckelbarger & Young 1997) and is generally like

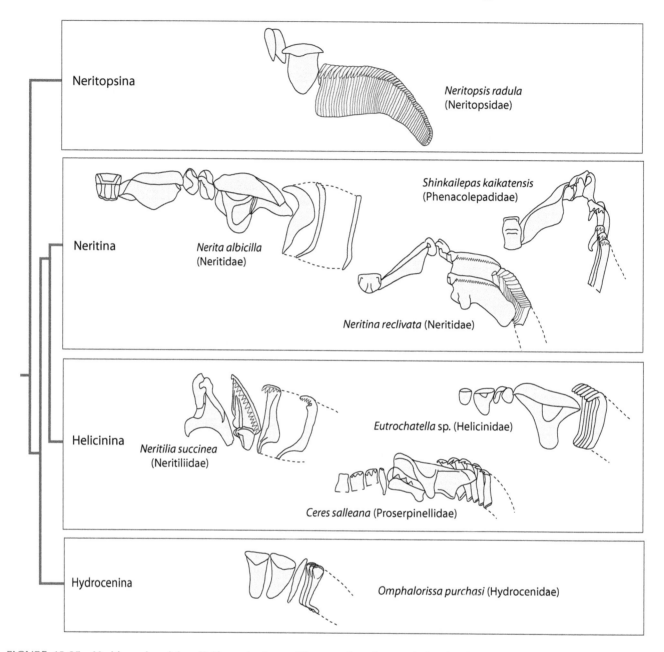

FIGURE 18.35 Neritimorph radulae. Half row is shown. The central teeth are missing in *Neritopsis*, *Neritilia*, and *Omphalorissa*. Redrawn and modified from various sources. The radulae of the species illustrated do not necessarily reflect the range of variation in the genera they belong to or of their higher taxon.

that in other gastropods. In an examination of neritimorph sperm, Koike (1985) found that sperm of Neritidae and Phenacolepadidae were similar and distinct from those of Helicinidae. In Neritidae and Phenacolepadidae, the midpiece has two elongate mitochondria parallel to the axoneme and a midpiece rod adjacent to the axoneme. The midpiece of the Helicinidae (Koike 1985) differs markedly, with four or five mitochondria arranged helically, and there is no midpiece rod. Neritid sperm differs from phenacolepadid sperm in the morphology of the tail. In the neritids, the tail is bent at the junction of the midpiece, and the tailpiece and the terminal part is expanded as a bag-like or fan-like structure. The bending correlates with the midpiece rod extending only to the junction of

the midpiece and tail. In the Phenacolepadidae, the tail is not expanded terminally, and the rod extends into the tailpiece, which is not bent. The sperm of the hot-vent *Bathynerita* is more similar to phenacolepadids than to neritids (Hodgson et al. 1998).

In both sexes the genital duct is divisible into gonadal, 'renal', and pallial sections. Regarding the 'renal' part of the reproductive system, there is no gonopericardial duct connecting the gonoduct and the pericardium, suggesting that the 'renal' component is not derived from the right kidney but is rather a *de novo* structure. Alternatively, the kidney–pericardial connection may have been lost with its incorporation into the reproductive tract and loss of kidney function. Currently,

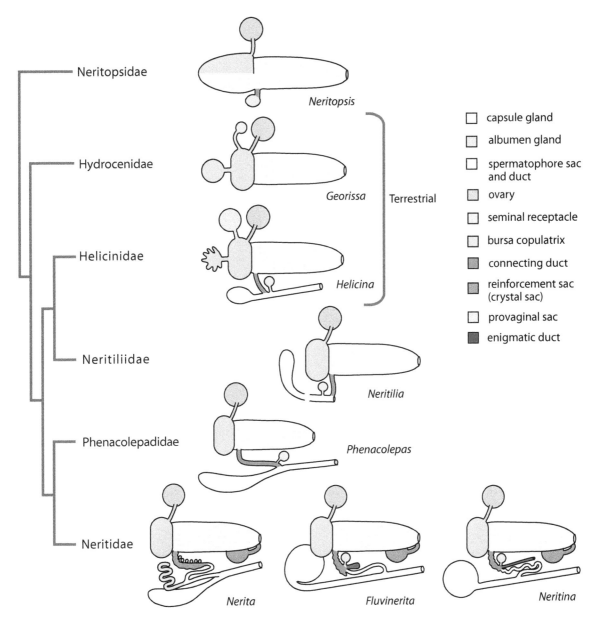

FIGURE 18.36 Diagrammatic representations of neritimorph female reproductive anatomy. Reproductive morphology redrawn and modified from Holthuis, B.V., *Evolution between marine and freshwater habitats: A case study of the gastropod suborder Neritopsina*, Ph.D. dissertation, University of Washington, 1995; tree after Kano, Y. et al., *Proc. R. Soc. B*, 269, 2457–2465, 2002.

there is no reliable cellular or developmental evidence to resolve the speculation regarding this question.

The bursa copulatrix of the helicinids and the spermatophore sac of the neritids were first thought to represent the metamorphosed right kidney and the vaginal opening its orifice (Thiele 1902). This scenario was disputed by Bourne (1909) who preferred the oviduct opening as the orifice. In yet another interpretation, Fretter (1965 p. 70) noted that in *Nerita* and *Neritina* a short duct 'leaves the gonoduct and opens to the mantle cavity alongside the vestigial right gill', the opening of which has 'the same relationship to the right gill as that of the left kidney to the left gill, and appears to be homologous with the opening of the right kidney of archaeogastropods'. This short duct that Fretter referred to is presumably the enigmatic duct (see below).

The female system is complex with one to three openings into the mantle cavity, usually with separate copulatory and oviduct orifices. The pallial gonoduct apparently develops by the glandular 'renal' gonoduct extending along the anterior mantle vein on the right side of the mantle cavity, and the rectum follows the same course (Fretter 1965; Berry et al. 1973; Barker 2001). The pallial oviduct (often called the 'ootype') may have single or paired albumen glands and a capsule gland (Sasaki 1998) (Figure 18.36).

As far as is known, in all neritimorphs the glandular oviduct produces gelatinous egg capsules, and these contain fertilised ova together with nutritive material from the albumen gland (Andrews 1937). The egg capsules are produced either singly or in an egg mass and may be further strengthened by impregnation with calcium carbonate spherulites or other

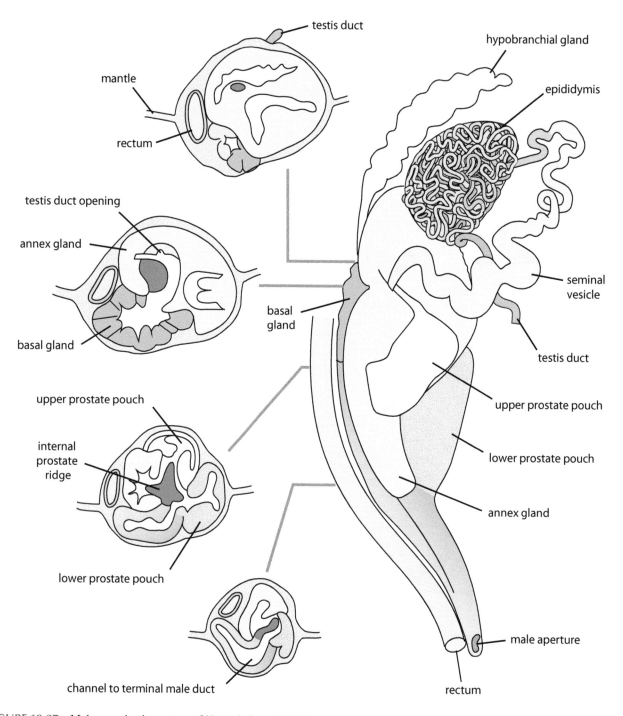

FIGURE 18.37   Male reproductive system of *Nerita balteata* (= *N. birmanica*). Redrawn and modified from Berry, A.J. et al., *J. Zool.*, 170, 189–200, 1973.

material from the crystal gland (see below). In most aquatic taxa the veliger larvae are planktotrophic, but direct development occurs in some freshwater and all terrestrial taxa.

In the Neritopsidae and Hydrocenidae there is a single female opening that functions as both the vaginal and oviduct opening, and this is presumably the plesiomorphic condition. In other neritimorphs, there are usually two, or sometimes three, openings, the third being the aperture of the 'ductus enigmaticus'. The two openings may be wide apart, as in Helicinidae and Neritiliidae, or abut, as in Neritidae and Phenacolepadidae.

Both male and female pallial reproductive organs include branching tubules of tissue equated with the hypobranchial gland by Bourne (1909), although this interpretation was questioned by Fretter and Graham (1962) and Berry et al. (1973). These tubules form lobules, a glandular pocket, or a diverticulum and are at the posterior end of the mantle cavity. They are of unknown function, although sperm has been found in the diverticulum in the helicinid *Hendersonia* (Baker 1925). In the helicinids, the single (left) glandular tubule opens independently of the reproductive system to the posterior end of the mantle cavity in the Proserpinellidae, but in the Helicinidae a secondary duct from the gland opens to the duct of the bursa copulatrix (Barker 2001). In the Phenacolepadidae, these glandular structures are incorporated into the proximal part of the pallial oviduct (Fretter 1984). In the Neritidae, the tubules lie in the mantle wall near the bursa copulatrix of females and the highly convoluted epididymis of males (Figure 18.37) and open to the posterior mantle cavity. The tubules are lined with unciliated gland cells containing secretory material that is not mucoid (Berry et al. 1973). These structures probably equate with the cluster of vesicles, opening separately to the mantle cavity, which were called vaginae by Marcus and Marcus (1967) in *Titiscania*. These were, however, not found in the *Titiscania* females examined by Houston (1990).

The female reproductive system (Figure 18.36) may have up to three openings to the exterior, the glandular oviduct, the vagina, and the enigmatic duct (ductus enigmaticus) (Bourne 1909), the latter of unknown function. In all taxa the glandular oviduct opens to the mantle cavity, but the enigmatic duct and the vagina may be either open or closed. The enigmatic duct is absent in most neritimorphs but is present and open in *Nerita* (*Puperita*), *Neritina*, *Septaria*, and *Smaragdia* (Bourne 1909; Andrews 1937; Starmühlner 1993; Holthuis 1995) and present and a blind tube in *Neritodryas* (Holthuis 1995). When present, the vagina may either have its own opening to the mantle cavity or share the opening of the glandular oviduct.

There is a vagina with a separate opening in Neritidae, except in *Nerita* (*Puperita*) (Holthuis 1995), Phenacolepadidae (Beck 1992b; Holthuis 1995), and Helicinidae (Sasaki 1998). A vagina shares the opening of the glandular oviduct in Hydrocenidae (Thiele 1910) and *Nerita* (*Puperita*) (Holthuis 1995). The vagina may be directly connected to the glandular oviduct or by way of a vaginal canal. There is a direct connection in Hydrocenidae (Thiele 1910; Bourne 1911), and a vaginal canal is present in other neritimorphs except those lacking a vagina (*Neritopsis* and *Titiscania*) where the sperm reception and egg conducting systems are not separated (Marcus & Marcus 1967; Holthuis 1995).

The bursa copulatrix and the seminal receptacle (if present) open into a fertilisation chamber that lies at the junction of the visceral and pallial portions of the gonoduct. The bursa receives allosperm or spermatophores by way of its opening to the mantle cavity. In those taxa with spermatophores, sperm are not found in the bursa, and they presumably pass through the spermatophore filament to the seminal receptacle (Berry et al. 1973) as these narrow filaments are often seen in the vaginal canal (Bourne 1909).

The seminal receptacle(s) that contains and stores orientated sperm is absent in *Neritopsis* (Holthuis 1995), but in most neritimorphs it is on the duct between the capsule gland and the vagina or on the glandular oviduct in Hydrocenidae and *Titiscania*. In *Neritilia* it is on the vagina (Holthuis 1995). While in most neritimorphs the seminal receptacle is a small sac, it is subdivided into 'accessory receptacles' in *Nerita* (except in the subgenus *Puperita*) (Holthuis 1995).

A reinforcement sac (crystal sac) is associated with the female reproductive system at the nearby openings of the female genital duct and the rectum in Neritidae but is absent in other neritimorphs. A small anterior empty sac was identified as a crystal sac in *Titiscania* by Marcus and Marcus (1967), but this was not found by Houston (1990), so we assume this structure is absent. In *Nerita*, the sac contains spherulites of calcium carbonate or in species of *Neritina*, *Theodoxus*, *Clithon*, *Dostia*, and *Septaria* externally derived sand and diatoms (Andrews 1937; Tan & Lee 2009), the material being used to coat and reinforce the egg capsules. The origin of the calcium carbonite spherulites in the crystal sac, which in some species can be very uniform, is unknown (Tan & Lee 2009). Neritiliid capsules are reinforced with sand or diatoms derived from the external environment (Kano & Kase 2002).

Hydrocena capsules are calcareous (Berry 1965), as they are also assumed to be in Helicinidae (Richling 2004), although, surprisingly, there do not appear to be published descriptions or figures of helicinid egg capsules. In another unexplained mystery, helicinids apparently release eggs into an egg sac at the upper end of the upper oviduct before fertilisation. Nothing is known about the source of nutrients to the developing embryo in any terrestrial neritimorph, indicating that the reproductive biology of terrestrial neritimorphs is very much in need of investigation. The egg capsules of *Neritopsis* are also unknown.

In males, the highly lobed testis opens to the upper visceral vas deferens, a narrow tube that runs anteriorly to the posterior end of the prostate gland and then bends back to become a very long, highly coiled duct that forms a tight mass behind the prostate gland. This duct then opens to a seminal vesicle, a broader lower part of the vas deferens in which sperm is stored, that narrows and enters the prostate gland. There is no gonopericardial duct. In the Neritopsidae, the male pallial glandular gonoduct (the prostate gland) is open ventrally, but the prostate is closed in other neritimorphs. The pallial prostate gland is where the spermatophores are formed in those taxa that possess them. It is made up of several types of prostatic tissues and pouches. One such pouch is identified as the 'auxiliary gland' in neritids, in addition to three others (see below), and in hydrocenids and helicinids there are two diverticulae (Baker 1925, 1926; Berry 1965; Haase & Schilthuizen 2007). In aphallate taxa the pallial gonoduct opens in front of the anus, but in phallate taxa a closed vas deferens or a seminal groove runs to the penis from the main pallial duct.

The complexity of the prostate and associated structures in neritids in particular is unusual and is described here in some detail, largely based on the thorough description by Berry et al. (1973) of a species of *Nerita* (Figure 18.37). The upper visceral vas deferens enters the main bulk of the prostate complex that is composed of fine glandular follicles that empty into the left side of the lumen (this has been variously called the 'annex' or 'auxiliary', or the prostate). Other than its spacious posterior part, the prostate lumen is U-shaped in section due to an internal dorsal glandular ridge, the cells being of different histology to those in the outer walls of the 'annex gland'. Another gland, again with different histology, the 'basal gland', opens dorsally near the point where the visceral vas deferens enters the prostate. Also, a glandular pouch with distinctive glandular histology opens ventrally in the posterior part while another glandular pouch opens more anteriorly.

Neritopsis radula   *Neritina taitensis*   *Nerita funiculata*

FIGURE 18.38   Three examples of neritimorph spermatophores. Redrawn and modified from Holthuis, B.V., *Evolution between marine and freshwater habitats: A case study of the gastropod suborder Neritopsina*, Ph.D. dissertation, University of Washington, 1995.

The distal part of the prostate complex narrows and is lined with mucoid-secreting and ciliated epithelium internally and is muscular externally. This channel often contains rudimentary spermatophores (Berry et al. 1973) (Figure 18.37).

A penis is absent in Helicinidae, Hydrocenidae, *Neritilia*, and *Neritopsis*. A cephalic lappet with a seminal groove forms a penis in Phenacolepadidae and possibly *Titiscania* (a similar structure is convergently present in some species of *Cocculina*). Neritidae (except *Neritilia*) have a separate, external, cerebrally innervated penis at the base of the right tentacle. These penes have a deep sperm groove and may be simple or more complex triangular flaps or may be cylindrical, with a pointed filament.

In some taxa spermatophores (Figure 18.38) are used to facilitate sperm transfer. When present, spermatophores may or may not have distinct filaments. Spermatophores are absent in Helicinidae and Hydrocenidae, *Bathynerita*, *Nerita* (*Puperita*), *Septaria*, *Theodoxus*, *Phenacolepas*, and *Titiscania*. Spermatophores without filaments are found in *Neritina* and *Neritopsis*, whereas *Nerita*, *Neritilia*, *Neritodryas*, and *Smaragdia* have spermatophores with filaments. The aphallate terrestrial taxa also lack spermatophores, and the details of sperm transfer are unknown.

18.6.3.7   Nervous System and Sense Organs

The nervous system is plesiomorphic in being hypoathroid with the pleural and pedal ganglia abutting or more or less fused and with the pedal ganglia extended posteriorly as cords connected by one to several commissures. Some apomorphic features include a well-developed labial commissure (but no distinct labial ganglia), and there is a thick connective between the pleural ganglia, the 'pleural commissure' (Sasaki 1998 figure 102C, D) (Figure 18.31), which is not found in other gastropods. Unusually, both limbs of the visceral loop originate from only the right pleural ganglion (instead of both left and right), a condition otherwise found only in Ampullariidae (Caenogastropoda) (Sasaki 1998). The suboesophageal ganglion lies close to the pleural ganglia, and, together with the pedal ganglia, these three ganglia form a tight ring.

The connective between the visceral nerve and the small supraoesophageal ganglion is very thin (e.g., Sasaki 1998) and could easily be missed, perhaps accounting for some reports in the literature of an incomplete visceral loop, as in Helicinidae (Bourne 1911), *Theodoxus* (Lenssen 1902; Bourne 1909), and *Titiscania* (Bergh 1890).

The cephalic eyes have a lens and a cornea except in *Pisulina* and *Neritilia*, both members of the Neritiliidae, which have 'open pit' eyes, having lost the cornea and vitreous body (Kano & Kase 2002), a condition probably related to the dark submarine caves and other cryptic habitats in which they occur. Eyes are absent in the hot-vent taxa *Olgasolaris* and *Shinkailepas* (Beck 1992b).

Cephalic eyes may be sessile at the outer bases of the tentacles, as in Helicinidae, *Bathynerita*, Neritiliidae, *Smaragdia*, *Phenacolepas*, and *Titiscania*, or on a small protrusion or a longer peduncle as in other Neritidae and *Neritopsis* and in Hydrocenidae (which lack cephalic tentacles).

Epidermal sensory cells are present in the cephalic tentacles (Künz & Haszprunar 2001), but these are not elaborated into specialised structures as they are in vetigastropods. Dermal light receptors involved in shadow responses have been reported in a species of *Nerita* (Gutiérrez & Womersley 2001).

Aquatic neritimorphs have an osphradium on the left side of the mantle cavity where it lies between the efferent membrane of the ctenidium and the anterior edge of the mantle cavity. There is no osphradium in the terrestrial helicinoideans and hydrocenids and supposedly in the deep-sea vent *Olgasolaris* (Beck 1992b). The osphradium is innervated by the osphradial nerve from the osphradial ganglion, which is in turn innervated from the supraoesophageal ganglion linked by zygosis to the left pleural ganglion (Haszprunar 1985g) (Chapter 7, Figure 7.7). The ultrastructure of the osphradium differs from that in vetigastropods and patellogastropods in having the sensory epithelium divided into central and lateral zones with ciliated cells. Although it resembles the osphradia of caenogastropods, the characteristic caenogastropod cell types are lacking, and, in addition, the epithelia are two layered, and the lateral zones are formed as grooves (Haszprunar 1985g).

The statocysts are located anterodorsally to the pedal ganglia and contain statoconia.

18.6.4   Biology and Ecology

18.6.4.1   Habits and Habitats

Marine neritimorphs are found worldwide with the exception of polar regions but are most diverse in tropical and subtropical environments. The Neritidae frequent mainly marine intertidal, brackish, and freshwater habitats with one genus being semiterrestrial. The marine taxa are often common on tropical to warm temperate intertidal and supratidal rocks while estuarine taxa frequent mangrove and other brackish habitats, and some also occur in fresh water (e.g., *Theodoxus*, *Clithon*, *Neritina*, and *Septaria*). These neritimorphs are typically on hard substrata, although members of the neritid subfamily Smaragdiinae are associated with angiosperms (mainly seagrasses) in shallow marine habitats. This latter taxon is the only non-cryptic inhabitant of shallow subtidal waters among living neritimorphs. Of the cryptic taxa, some, such as phenacolepadids, Neritopsidae, and the slug-like *Titiscania*, live deep under boulders, in subtidal coral rubble, or submarine caves.

The freshwater Neritidae are found in the lower courses of rivers and streams. There have been two separate radiations – one comprising *Septaria* and *Neritina* and related genera in the freshwater streams and rivers of the tropical Indo-West Pacific and the Americas, and the other is a radiation of *Theodoxus* in the rivers of Europe and central Asia.

Species of *Theodoxus* (Neritidae) live in streams, rivers, and springs. In Portugal, *T. fluviatilis* is found in warm spring habitats where the relatively constant temperature allows for year-round reproduction compared with much more variable streams (Graça et al. 2012). A few neritids are semiterrestrial with, for example, *Neritodryas cornea* recorded as living on *Pandanus* trees at the back of mangroves in Palau (Cowie & Smith 2000), and some supralittoral neritids have been described as 'almost terrestrial' (Vermeij & Frey 2008).

The Phenacolepadidae normally inhabit dysoxic, sulphide-rich environments in warm temperate to shallow tropical seas. They live under embedded stones or wood in mud (Fretter 1984; Warén & Bouchet 2001). A few phenacolepadids (*Bathynerita*, *Shinkailepas*, *Olgasolaris*) are found in deep-sea (500–2,500 m) vent and seep chemosynthetic communities. Modifications for these environments include long cephalic and mantle edge tentacles, presumably as a consequence of lack of light (Fretter 1984). They have red blood with haemoglobin in erythrocytes that enhances the oxygen-carrying capacity (Fretter 1984; Sasaki 1998).

The vent and seep neritimorphs are closely related to shallow water phenacolepadids and it has been suggested that the ability of their larvae to recognise dysoxic, sulphur-rich habitats and chemosynthetically nourished biotopes, as well as the use of haemoglobin as a respiratory pigment, were exaptations to life on vents and seeps (Warén & Bouchet 1993, 2001). *Bathynerita naticoidea*, from seeps in the Gulf of Mexico, is the only coiled member of Phenacolepadidae, and the presumably plesiomorphic shell is accompanied by the apparent lack of erythrocytes and a neritid-like radula (Kano et al. 2001; Warén & Bouchet 2001). Both coiled and limpet-like phenacolepadids have occupied chemosynthetic deep-sea communities since the Upper Cretaceous or Paleogene (Kano et al. 2002).

Related to the terrestrial Helicinidae, the Neritiliidae occupy a wide range of habitats with species of *Neritilia* living in freshwater streams, brackish estuaries, groundwaters, and

anchialine waters (bodies of haline water with subterranean connections to the sea overlain by a freshwater lens) (Stock et al. 1986; Kano et al. 2001; Kano & Kase 2002; Sasaki & Ishikawa 2002). Some species of the neritiliid genus *Pisulina* live in submarine sea caves (Kano & Kase 2000), and no marine species is known outside those habitats. The riverine species of *Neritilia* may have migrated into these cave habitats by way of underground water connections (Kano et al. 2002) as some neritiliids live in such habitats (Kano et al. 2001). *Neritilia* has an almost pan-tropical distribution (the Caribbean, western Africa, and islands of the tropical and subtropical Indo-Pacific, including Japan). One species of *Neritilia* has been found in a freshwater phreatic environment (12 m deep well) in Japan (Sasaki & Ishikawa 2002).

The two groups of living terrestrial neritimorphs, Helicinoidea and the Hydrocenoidea, are both typically associated with litter on moist forest floors, but some helicinids are arboreal, and a few hydrocenids live on tree trunks or limestone rocks, with both groups often associated with limestone. A few helicinids are found in xeric scrubland, and many can aestivate (Barker 2001). Based on their excretory physiology and kidney structure, helicinids were probably derived from freshwater ancestors (Little 1972).

18.6.4.2 Feeding and Diet

Marine neritids are mainly grazers on algal spores, diatoms, and detritus. In a study of the stomach contents of three species of *Nerita* from Kenya, D'Souza (1981) found a variety of algae and 'protists' that formed films on the rocks on which they lived. Phenacolepadids live in dysoxic sulphur rich environments and probably feed on chemosynthetic bacteria (Kano et al. 2002). Deep-sea and vent taxa are probably detritivores. *Bathynerita naticoidea* feeds on bacteria and the decomposing periostracum of the shells of *Bathymodiolus* with which it is associated (Dattagupta et al. 2007).

The Hawaiian neritid *Smaragdia bryanae* is a specialist herbivore on the seagrass *Halophila*, feeding on the epidermal cell contents (Unabia 2011). Other species of *Smaragdia* are also found on seagrasses, and members of the genus have been associated with these plants at least back to the Miocene (Unabia 2011). They have a modified radula and feed by the outer lateral teeth puncturing the epidermal cells; the contents are probably swept up by the long, thin cusps of the marginal teeth.

Theodoxus, and possibly some other freshwater neritimorphs, include aquatic insect larvae in their diets, probably fortuitously picked up when grazing on biofilms. Terrestrial neritimorphs are thought to feed on detritus, algal spores, moss, and lichens, although there is little detail available. The Malaysian limestone-associated hydrocenid *Hydrocena monterosatiana* has been shown to include algae, lichen, and moss in its diet (Berry 1961).

18.6.4.3 Life History

Neritids such as *Nerita* and *Neritina* produce dome-shaped capsules a day or two after mating. These are released from the oviduct, pass out of the mantle cavity, and are then pressed against the substratum by the foot. Each capsule contains a few tens to several hundred (depending on the species) fertile eggs that potentially all hatch as planktotrophic veligers (Andrews 1935; Bandel 2001; Tan & Lee 2009) or, sometimes, as post-metamorphic juveniles (Anderson 1962; Przeslawski 2011). In most neritimorphs, the capsule contents are released when it splits or ruptures.

In most marine neritimorphs the embryo passes through a trochophore stage within the capsule before hatching as feeding (planktotrophic) veliger larvae. A similar strategy is employed by brackish and freshwater taxa such as *Neritina*. In these, the veligers have an estuarine or marine period before returning to freshwater streams and rivers. Besides some Thiaridae (Glaubrecht 2006; Hidaka & Kano 2014), there are few other examples of freshwater gastropods with a marine larval phase (see also Chapter 9).

The nutrients utilised by neritimorph larvae are mainly the yolk within the eggs, although passing references to free yolk (Berry et al. 1973) or albumen (Andrews 1937) have been made.

All the aquatic neritimorph families possess planktotrophic species that have very characteristic protoconchs (see Section 18.1.2.1.1). The development of the embryonic shell in neritids has been studied by Bandel (1982, 1992). The primary shell is secreted by the shell gland and then detaches at less than half a whorl. At this stage, the body is partially connected to the inside of the shell, and the embryonic retractor muscle attaches later. The embryonic shell is mineralised by a layer of aragonitic biocrystallites, but further shell growth is by way of increments via the mantle edge. Before hatching, the embryonic shell 'resembles an egg with its slender upper part cut off obliquely' (Bandel 2001 p. 69), and the coiled larval shell is secreted during the planktotrophic larval stage, with overlap resulting in much of the early whorls being hidden below the last one. In addition to external growth, resorption of the inner walls of the larval shell results in little change in shape to the interior, other than an increase in size (Bandel 1992).

Interestingly the smallest larval shells are on those taxa with planktotrophic larvae that live in fresh water (families Neritiliidae and Neritidae – Theodoxinae). These larvae are transported downstream by the current and grow little before they reach the sea. Two Upper Cretaceous freshwater lineages (see Section 18.6.5) also have a similar protoconch size (Fukumori & Kano 2014). The evolutionary reasons for selection of a small larval settlement size are unclear but may be related to faster attainment of metamorphic competence because of their more variable and precarious environment (Fukumori & Kano 2014).

Some freshwater and all terrestrial species have direct development and hatch as juvenile snails. Thus taxa related to *Theodoxus* are direct developers, and these have one or a few embryos in each capsule that feed on nurse eggs to eventually hatch as juveniles (Blochmann 1882; Bandel 1982; Bandel 2001). Other freshwater species, including *Fluvinerita alticola* and the neritiliid *Neritilia succinea*, have independently adopted a similar strategy (Andrews 1937; Holthuis 1995;

Bandel 2001). These taxa have distinctly different protoconch morphology from planktotrophic neritids (Bandel 2001).

The terrestrial taxa presumably develop in the capsule and emerge as juveniles, although nothing is known about the details or the supply of food for the developing embryos. There are no records of parental care or brooding in neritimorphs.

Some freshwater neritimorphs are relatively long-lived with, for example, the neritid *Clithon retropictus* estimated to live for up to 20 years (Shigemiya & Kato 2001) (see Chapter 9).

18.6.4.4 Associations

Few associations have been recorded in neritimorphs. One is a sphaeromatid isopod that inhabits the mantle cavity of an intertidal neritid (Nishimura 1976), and various parasites have been recorded from neritimorphs. These include the sporozoan *Nematopsis gigas* in *Nerita ascensionis* (Azevedo & Padovan 2004) and several trematodes (see Chapter 9). A globular bacterium has been found on the ctenidia of the deep-sea vent phenacolepadid *Olgasolaris* (Beck 1992b), but the nature of the association is unknown. While neritimorphs undoubtedly have symbiotic gut bacteria, no symbiotic bacterial associations appear to have been reported in the group.

Intertidal neritids are preyed on by other predatory gastropods, and these and freshwater species by crabs, fish, and birds. Little is known about the biology of the terrestrial species, but presumably, they are preyed on by the same taxa that are predators of other terrestrial snails (see Chapter 9).

18.6.4.5 Behaviour

As with many intertidal organisms, the activity patterns of intertidal neritids are largely influenced by light and tide cycles; up-shore migration is common, with the largest individuals often located highest up the shore (Safriel 1969; Underwood 1977; Levings & Garrity 1983; Bovbjerg 1984; Chelazzi et al. 1988), and the use of visual cues (Chelazzi & Vannini 1980). Also, as with other intertidal snails, clustering is a common behaviour seen in some intertidal neritids as it reduces predation, desiccation, and overheating risks (e.g., Chelazzi et al. 1984). Neritids also have trail-following behaviours, following both their own and conspecific mucous trails on the substratum and facilitating the relocation of resting or feeding locations and building aggregations (e.g., Chelazzi et al. 1983c, 1985). Anti-predator responses (to predatory gastropods) in intertidal neritids involve raising the shell, its rotation, tentacle flailing, and rapid movement (Hoffman et al. 1978).

Mating behaviour has only rarely been described in neritimorphs. In *Nerita funiculata* mating has been documented as follows:

'The entire mating process takes anywhere from 10 minutes to one-half hour, depending on the individual pair. Initially, the male climbs onto the right side of the shell of the female and inserts the penis into the right side of the mantle cavity. During this time the pair makes back and forth movements and simultaneously rotate to and fro through a 90-degree arc. After pausing for a period of about one minute, they oscillate in the opposite direction, and then the spermatophores are transferred to the mantle cavity of the female. When copulation is completed the male either crawls down and away or withdraws into the shell and falls off' (Houston 1990 p. 106).

Freshwater taxa with an estuarine or marine larval phase have behaviours that cue their settling at the mouths of rivers and streams followed by their movement against the flow, literally crawling upstream into the freshwater habitat (Schneider & Lyons 1993; Blanco & Scatena 2005, 2007). In some cases, several species crawl upstream together in mass migrations and some 'hitchhiking' has been observed with the young of one species riding on the back of another (Kano 2009) (see Chapter 9).

18.6.5 DIVERSITY AND FOSSIL HISTORY

There are about 2,000 living species of neritimorphs (Fukumori & Kano 2014; Uribe et al. 2016a), with a large proportion (around 550 species) being in the Helicinidae (Richling 2004). Possible neritimorphs extend back to the Late Ordovician, although confirmed records (based on protoconch morphology) are much later, in the Mid-Devonian (Frýda et al. 2008a). These supposed neritimorphs of the Late Paleozoic have turbinate, sometimes sculptured adult shells in which the inner whorls are not resorbed, and their larval shells comprise about two smooth whorls (Nützel et al. 2007a). Taxa with the smooth involute larval shell of modern neritimorphs and resorbed inner whorls do not appear until the Triassic (Bandel & Frýda 1999; Bandel 2000) and by the Upper Triassic were well diversified with 36 species in seven families in the St Cassian Formation of Italy (Bandel 2007). The freshwater genus *Theodoxus*, with a distinctive development reflected in its protoconch (see Section 18.6.4.3), has a fossil history from the Paleocene (Bandel 2001).

Of the extant neritimorphs, the family Neritopsidae has the longest fossil history, dating to the Devonian or possibly Silurian (Knight et al. 1960b; Batten 1984; Bandel & Frýda 1999; Bandel 2000). The only living genus, *Neritopsis*, dates from the Middle Triassic and is well represented in the Jurassic and Cretaceous (e.g., Kase & Maeda 1980; Batten 1984). There are only two living species with the genus ranging through the Caribbean and tropical Indo-West Pacific where they live in deep rubble or submarine caves (Kase & Hayami 1992; Warén & Bouchet 1993; Holthuis 1995; Kano et al. 2002). Molecular analysis supports the neritopsids being the sister to the rest of the living neritimorphs (Kano et al. 2002), as does their anatomy (Holthuis 1995). The slug *Titiscania* is the sister group of *Neritopsis*, and, despite the lack of a shell, many features of the anatomy are similar although the accounts disagree in some important respects, particularly regarding aspects of the female reproductive system (e.g., Marcus & Marcus 1967; Houston 1990).

Some Mesozoic fossils have been included in Neritidae, even back to the Lower Triassic (Knight et al. 1960b; Tracey et al. 1993), but these relate to other neritimorphs (e.g., Bandel

2008) and fossils of true Neritidae range only from the Upper Cretaceous (e.g., Bandel 2007; Bandel 2008). The oldest fossil recognised as a member of the Neritiliidae is from the Eocene, but the family is probably older (Kano et al. 2002). The earliest record of a hydrocenid is Late Miocene (Tracey et al. 1993), but the fossil record for the group is poor, and the current wide distribution of the family suggests that the group has an ancient, possibly Pangean, origin (Bandel 2000). They are found from Europe and Asia to Australasia and southern Africa with the largest radiation in Borneo (Thompson & Dance 1983). The Senonian (Upper Cretaceous) *Schwardtina cretacea* from fresh water or brackish coal deposits in Hungary is similar in shell shape, size, and protoconch morphology to a hydrocenid and, like most neritimorphs, has a resorbed columella (Bandel & Riedel 1994; Bandel 2001). Thus this fossil may represent the earliest record of a neritimorph related to hydrocenids (Kano et al. 2002), although Bandel (2001) suggested it might have affinities with Neritiliidae.

Deianira and *Schwardtina* comprise two freshwater-brackish-water lineages in the Mesozoic, and another fossil, *Mesoneritina ajkaensis*, from the Upper Cretaceous of Hungary, is like a typical neritid and may represent a separate freshwater invasion. During the Cenozoic, both the Neritidae and Neritiliidae each gave rise to freshwater lineages, with one in Neritiliidae (see below) and at least two in Neritidae (Kano et al. 2002). To enable successful incursions into freshwater, an osmoregulatory kidney must have been independently developed in each lineage, initially in brackish water. Once this ability was achieved, the move to fresh water could occur and probably took place several times within neritids (Holthuis 1995).

Bunje and Lindberg (2007) demonstrated six deep phylogenetic subdivisions within European *Theodoxus* taxa, with each group occupying a major post-Tethyan marine basin. Their divergence was estimated to have occurred during the Miocene with the breakup of the Mediterranean and Paratethys seas. Subsequent diversification within the basins began during the Pliocene, primarily in the eastern and western Mediterranean Sea, the Pannonian Basin, and the Black Sea.

There are about 75 living genera of marine and freshwater neritimorphs. Most families have a rather low diversity in living taxa, with the marine Neritopsidae and Titiscaniidae having three and one species respectively, Neritidae, around 200 (mostly marine, but also brackish and freshwater), Septariidae, about 15 (brackish and freshwater), and the marine limpets of the Phenacolepadidae about 23 species. The Neritiliidae has only two living genera. Of the terrestrial families, Helicinidae is very diverse with hundreds of species, the related Proserpinidae has around 30, and the Hydrocenidae about 20 species (data mostly from Holthuis 1995).

The Upper Carboniferous *Dawsonella meeki*, possibly the oldest known land snail, is sometimes placed in its own family (Dawsonellidae) (e.g., Knight et al. 1960b). It was found in freshwater limestone in coal deposits in North America, so it is uncertain if it was of terrestrial or freshwater origin. The affinities of this fossil are of great interest given that it is sometimes attributed to the Helicinidae (Solem & Yochelson 1979) or Helicinoidea (e.g., Solem 1983). The inner whorls of the shell are resorbed (Solem 1983) as in most living neritimorphs. Bourne (1911), probably correctly, thought it was convergent with helicinids given the lack of fossils of similar taxa since the Carboniferous. He proposed that *Dawsonella* may have been derived from the contemporary marine naticopsids which he suggested were the ancestors to the neritids.[13] This concept of an early independent terrestrial incursion by neritimorphs is also in agreement with molecular data. Kano et al. (2002) tested the timing of the divergence of helicinids using a molecular clock (28S sequences) and came up with a result of a little more than 100 million years. This timing agrees with the first appearance of *Dimorphoptychia*, a fossil helicinid from the Upper Cretaceous of Europe (Tracey et al. 1993) and North America (Bishop 1980), and the fossil history is continuous from that time (Kano et al. 2002). The view that helicinids had a freshwater ancestry, based on

their renal physiology (Little 1972, 1990), is also in accordance with some fossil evidence. The oldest freshwater neritimorph, *Mesoneritina morrisonensis*, is Upper Jurassic, well before the first helicinids appear (Kano et al. 2002). Also, the Cretaceous Deianiridae, which resemble helicinids, are found in fresh or brackish coal deposits in Hungary (Bandel & Riedel 1994). Living helicinids have a tropical to warm temperate distribution in the tropical Americas and West Indies, Indo-Asia, Polynesia, and Australia.

Thus, neritimorphs became terrestrial at least three times, commencing with dawsonellids in the Late Paleozoic, then the hydrocenids probably in the Mesozoic, and the helicinids in the Cretaceous (Kano et al. 2002). In addition, one genus of Neritidae, *Neritodryas*, are also semiterrestrial including some that are arboreal (Little 1990; Cowie & Smith 2000), but these species are restricted to living near streams or estuaries as they still retain planktotrophic larvae (Holthuis 1995; Kano et al. 2002).

---

[13] In a departure from the normally accepted relationships of *Dawsonella*, Lieb et al. (2006) treated this fossil as the oldest record of a 'pulmonate' land snail.

19 Gastropoda II – The Caenogastropoda

19.1 INTRODUCTION

Caenogastropods are the most diverse group of living gastropods with about 157 extant families recognised (Bouchet et al. 2017). They display almost every shell form and include well-known groups such as periwinkles, cowries, whelks, cone shells, volutes, mitres, worm snails, and slipper shells. Despite their diverse shell morphologies, the shell itself has become internal and reduced in only a few families (see Section 19.3.1). Caenogastropods include some of the largest and smallest gastropods, with shells ranging in adult size from about 1 to over 900 mm in length. They include many of the most colourful shelled gastropods, some having elaborate markings and patterns on the shell and/or external animal. Examples of a range of living caenogastropods are shown in Figures 19.2–19.11.

They comprise about 60% of all gastropod species and include many ecologically, and some commercially, important marine families. The group is predominantly marine, but caenogastropods are found world-wide in every major habitat as several lineages have invaded and radiated in fresh water and a few have become terrestrial.

19.2 PHYLOGENY AND CLASSIFICATION

Until relatively recently, caenogastropods were treated as a subset of 'prosobranchs', comprising most of the 'Mesogastropoda' and all the 'Stenoglossa' (see Section 19.2.2). Some major groups of caenogastropods are recognised and supported in recent phylogenetic treatments, but the detailed relationships within those groups remain largely unresolved. This difficulty in the resolution within the main lineages may be, at least in part, due to heterochrony being a major factor in their evolution (Ponder & Lindberg 1997), as is the rapid radiation of several clades that occurred, particularly during the latter part of the Mesozoic and Paleogene (see Section 19.6 and Ponder et al. [2008] for more detail).

The main groups recognised here within caenogastropods are listed in Box 19.1 and are described in more detail in Section 19.7. As summarised in Box 19.1, the largely non-marine basal group known as Architaenioglossa are a grade rather than a clade (Simone 2004b; Ponder et al. 2008). The remaining caenogastropods form the Sorbeoconcha, with the Cerithiimorpha forming the basal group and the majority of caenogastropods making up the Hypsogastropoda. Despite ongoing morphological and molecular studies, the details of the relationships within this latter group remain somewhat unsettled.

19.2.1 SISTER GROUP RELATIONSHIPS

Recent morphological (including palaeontological), molecular, and combined analyses usually show the Heterobranchia as the sister group to caenogastropods, with these two clades together called the Apogastropoda (Ponder & Lindberg 1997). However, in a few molecular and combined analyses the Neritimorpha are the sister taxon (see Ponder et al. 2008 for details). The supposed sister relationships of caenogastropods and heterobranchs may well break down with a better understanding of the fossil record, as some other extinct non-heterobranch taxon may be the actual caenogastropod sister taxon.

19.2.2 OUTLINE OF CLASSIFICATION

The palaeontologist Klaus Bandel proposed several higher-category names (Bandel 1991a, b, 1993, 2002a). Some of these names, such as Palaeocaenogastropoda for those with Paleozoic origins, Metamesogastropoda for those first appearing in mid-Mesozoic, and Neomesogastropoda which appear in the Late Mesozoic (excluding the Neogastropoda) are not used here because they do not reasonably reflect what we know of the phylogeny of this group (see Ponder et al. 2008 for discussion) (Box 19.1 and Figure 19.1).

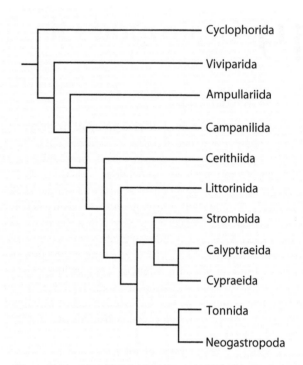

FIGURE 19.1 A tree illustrating the likely relationships of the main caenogastropod taxa (orders) based largely on recent morphological and molecular phylogenetic analyses.

Although the name Caenogastropoda was first introduced in 1960, it was largely ignored, and Mesogastropoda and Neogastropoda were included in the paraphyletic 'subclass' 'Prosobranchia' until the late 1980s (Haszprunar 1988f; Ponder & Warén 1988; Bieler 1992; Ponder & Lindberg 1997). Caenogastropods encompass most of the 'Mesogastropoda' (first proposed by Thiele 1925), with some of that group now known to be basal heterobranchs (see Chapter 20), and all of Thiele's Stenoglossa (= Neogastropoda).

Apart from architaenioglossans, another commonly used 'traditional' higher-category name, the Ptenoglossa, has also been recently abandoned because the grouping is polyphyletic. This grouping included three rather odd, unrelated carnivorous taxa (Eulimoidea, Triphoroidea, and Epitonioidea), many with divergent radulae and some having lost the radula. Another name, Heteropoda, included a group of highly modified permanently planktonic predatory caenogastropods which comprises a single superfamily (Pterotracheoidea) but is no longer formally used. The term 'littorinimorphs' is used here informally for the non-neogastropod hypsogastropods.

19.3 MORPHOLOGY

The general body plan of a generalised asiphonate 'littorinimorph' caenogastropod is shown in Figure 19.12.

FIGURE 19.2 Examples of living architaenioglossans and a pomatiid. (a) *Cyclophorus* sp. (Cyclophoridae), Thailand. Courtesy of Y. Tatara and H. Fukuda. (b) *Blaesospira echinus infernalis* (Pomatiidae), Cuba. Courtesy of S. Groves. (c) *Pomacea canaliculata* (Ampullariidae), South America. Courtesy of S. Ghesquiere. (d) *Cernina fluctuata* (Ampullinidae), Philippines. Courtesy of T. Kase. (e) *Viviparus viviparus* (Viviparidae), England. Courtesy of Prof. B. Eversham. (f) *Notopala* sp. (Viviparidae), Northern Territory, Australia. Courtesy of WFP. (g) *Campanile symbolicum* (Campanilidae), SW Australia. Courtesy of C. Bryce - Western Australian Museum. (h) *Plesiotrochus* sp. (Plesiotrochidae), Madang, Papua New Guinea. Courtesy of L. Charles - MNHN.

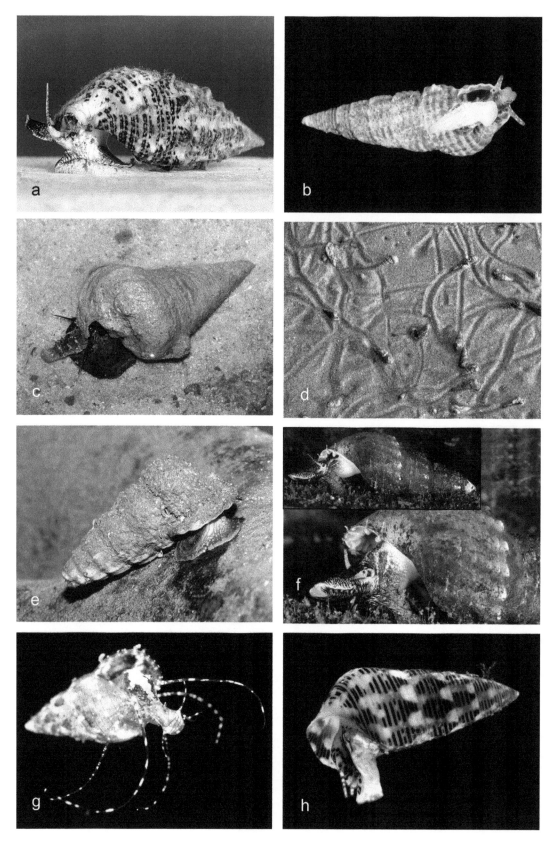

FIGURE 19.3 Examples of living Cerithiimorpha. (a) *Cerithium litteratum* (Cerithiidae), Guadeloupe, French West Indies. Courtesy of P. Maestrati - MNHN. (b) *Bittium reticulatum* (Cerithiidae), Wales, Great Britain. © 2016 I.F. Smith. (c) *Telescopium telescopium* (Potamididae), Singapore. Courtesy of R. Tan. (d) *Batillaria zonalis* (Batillariidae), Singapore. Courtesy of R. Tan. (e) *Cerithidea obtusa* (Potamididae), Singapore. Courtesy of R. Tan. (f) *Plotiopsis balonnensis* (Thiaridae), New South Wales, Australia. Courtesy of H. Jones. (g) *Styliferina* sp. (Litiopidae), Madang, Papua New Guinea. Courtesy of L. Charles - MNHN. (h) *Diala flammea* (Dialidae), Madang, Papua New Guinea. Courtesy of L. Charles - MNHN.

FIGURE 19.4   Examples of living heteropods, Provannidae, Littorinoidea, and Calyptraeidae. (a) *Carinaria japonica* (Carinariidae), Gulf of California, Mexico. © 2015 MBARI. (b) *Atlanta peroni* (Atlantidae), Atlantic coast, NE USA. Courtesy of B. Buge. (c) *Alviniconcha* sp. (Provannidae), Chamorro vent field, Marianas. Courtesy of K. Peijnenburg. (d) *Nodilittorina pyramidalis* (Littorinidae), New South Wales, Australia. Courtesy of D. Riek. (e) *Littorina obtusata* (Littorinidae), Europe. Courtesy of T. Gosliner. (f) *Pomatias elegans* (Pomatiidae), France. Courtesy of O. Gargominy - MNHN. (g) *Crepidula fornicata* (Calyptraeidae), England. Courtesy of K. Hiscock. (h) *Siphopatella walshi* (Calyptraeidae), Singapore, in *Tonna* shell. Courtesy of R. Tan.

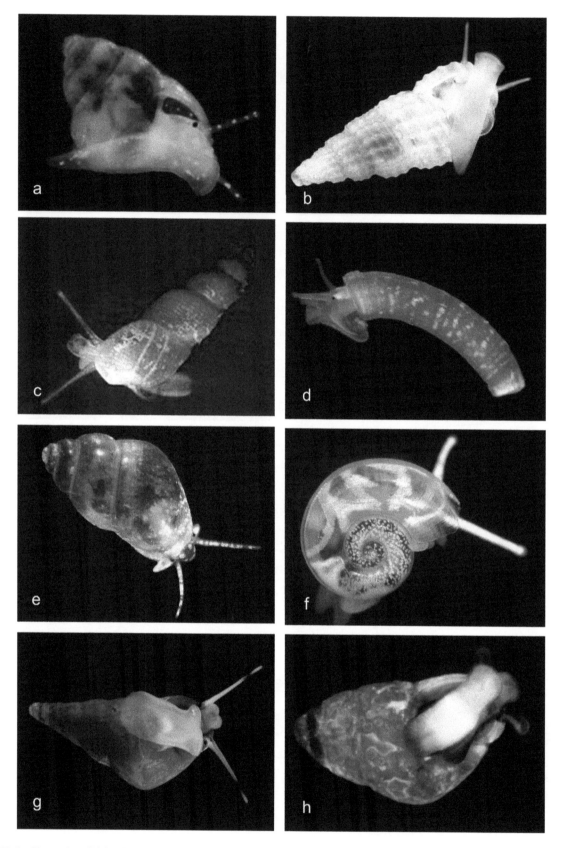

FIGURE 19.5 Examples of living Rissooidea and Truncatelloidea. (a) *Lucidestea nitens* (Rissoidae), New South Wales, Australia. Courtesy of D. Riek. (b) *Rissoina (Phosinella) allanae* (Rissoinidae), New South Wales, Australia. Courtesy of D. Riek. (c) *Ceratia nagashima* (Iravadiidae), Japan. Courtesy of H. Fukuda. (d) *Caecum* sp. (Caecidae), Guadeloupe, French West Indies. Courtesy of P. Maestrati - MNHN. (e) *Stenothyra basiangulata* (Stenothyridae), Japan. Courtesy of H. Fukuda. (f) *Vitrinella* sp. (Tornidae), Guadeloupe, French West Indies. Courtesy of P. Maestrati - MNHN. (g) *Tatea huonensis* (Tateidae), New South Wales, Australia. Courtesy of D. Riek. (h) *Amphithalamus incidatus* (Anabathridae), New South Wales, Australia. Courtesy of D. Riek.

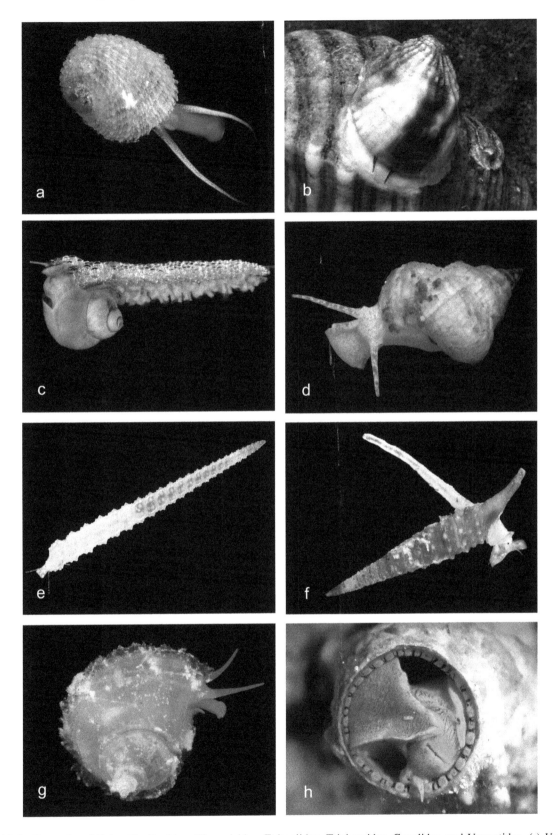

FIGURE 19.6 Examples of living Vanikoridae, Hipponicidae, Epitoniidae, Triphoroidea, Capulidae, and Vermetidae. (a) *Vanikoro* sp. (Vanikoridae), Papua New Guinea. Courtesy of L. Charles - MNHN. (b) *Sabia conicus* (Hipponicidae), New South Wales, Australia. Courtesy of D. Riek. (c) *Janthina exigua* (Epitoniidae), showing bubble raft with egg capsules attached, New South Wales, Australia. Courtesy of D. Riek. (d) *Epitonium* sp. (Epitoniidae), Papua New Guinea. Courtesy of L. Charles - MNHN. (e) *Inella* sp. (Triphoridae), Papua New Guinea. Courtesy of L. Charles - MNHN. (f) *Mastoniaeforis* sp. (Triphoridae), Papua New Guinea. Courtesy of L. Charles - MNHN. (g) *Trichotropis* sp. (Capulidae), Papua New Guinea. Courtesy of L. Charles - MNHN. (h) *Thylacodes decussatus* (Vermetidae), Florida, USA. Courtesy of R. Bieler.

FIGURE 19.7 Examples of living Naticidae, Xenophoridae, and Strombidae. (a) *Tanea undulata* (Naticidae), Philippines. Courtesy of T. Gosliner. (b) *Polinices* (*Conuber*) *sordidus* (Naticidae), New South Wales, Australia. Courtesy of D. Riek. (c) *Naticarius onca* (Naticidae), Singapore. Courtesy of R. Tan. (d) *Xenophora* sp. (Xenophoridae), South Madagascar. Courtesy of B. Abela - MNHN. (e) *Canarium microurceus* (Strombidae), New South Wales, Australia. Courtesy of D. Riek. (f) *Strombus pugilis* (Strombidae), Guadeloupe, French West Indies. Courtesy of P. Maestrati - MNHN.

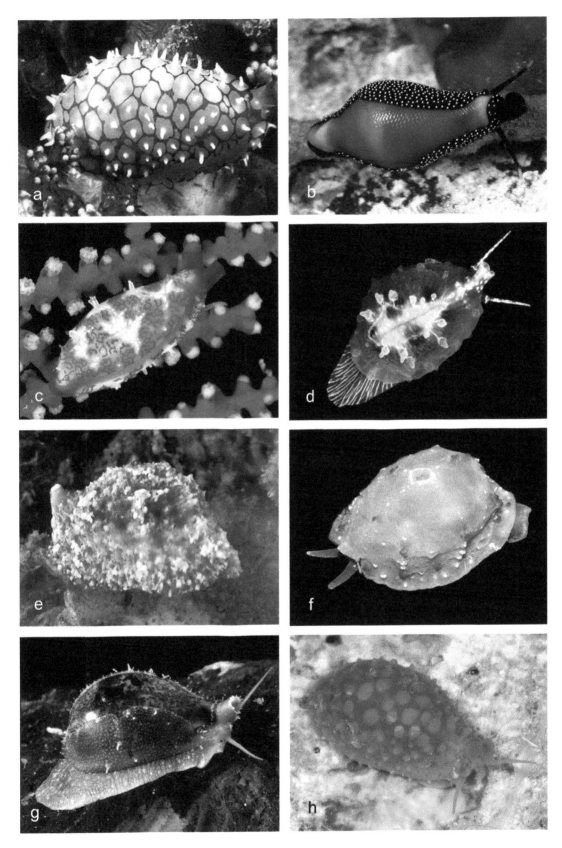

FIGURE 19.8   Examples of living Cypraeoidea and Velutinoidea. (a) *Prionovolva brevis* (Ovulidae), New South Wales, Australia. Courtesy of D. Riek. (b) *Crenovula aureola* (Ovulidae), Australia. Courtesy of T. Gosliner. (c) *Simnia* sp. (Ovulidae), French Guiana. Courtesy of L. Charles - MNHN. (d) *Trivia* sp. (Triviidae), French Guiana. Courtesy of L. Charles - MNHN. (e) Unidentified Velutinidae, New South Wales, Australia. Courtesy of D. Riek. (f) Unidentified Velutinidae, Ile des Pins, New Caledonia. Courtesy of P. Maestrati - MNHN. (g) *Purpuradusta gracilis* (Cypraeidae), New South Wales, Australia. Courtesy of D. Riek. (h) *Cribrarula cribraria* (Cypraeidae), N. W. Madagascar. Courtesy of T. Gosliner.

FIGURE 19.9 Examples of living Ficidae and Tonnoidea. (a) *Ficus variegata* (Ficidae), Singapore. Courtesy of R. Tan. (b) *Monoplex parthenopeus* (Cymatiidae), New South Wales, Australia. Courtesy of D. Riek. (c) *Eudolium crosseanum* (Tonnidae), Guadeloupe, French West Indies. Courtesy of L. Charles - MNHN. (d) *Tonna perdix* (Tonnidae), Papua New Guinea. Courtesy of L. Charles - MNHN. (e) *Turritriton labiosus* (Cymatiidae), New South Wales, Australia. Courtesy of D. Riek. (f) *Semicassis bisulcata* (Cassidae), Philippines. Courtesy of T. Gosliner. (g) *Personopsis grasi* (Personidae), Guadeloupe, French West Indies. Courtesy of L. Charles - MNHN. (h) *Charonia rubicunda* (Charoniidae), New South Wales, Australia. Courtesy of D. Riek.

FIGURE 19.10   Examples of living neogastropods. (a) *Cancellaria* sp. (Cancellariidae), Ile des Pins, New Caledonia. Courtesy of P. Maestrati - MNHN. (b) *Nassarius glans* (Nassariidae), Papua New Guinea. Courtesy of L. Charles - MNHN. (c) *Euplica scripta*, (Columbellidae), Singapore. Courtesy of R. Tan. (d) *Antillophos* sp. (Nassariidae), Guadeloupe, French West Indies. Courtesy of P. Maestrati - MNHN. (e) *Leucozonia nassa* (Fasciolariidae), Guadeloupe, French West Indies. Courtesy of P. Maestrati - MNHN. (f) *Babylonia spirata* (Babyloniidae), Singapore. Courtesy of R. Tan. (g) *Dentimargo* sp. (Marginellidae), Papua New Guinea. Courtesy of L. Charles - MNHN. (h) *Columbarium* sp. (Columbariidae), Ile des Pins, New Caledonia. Courtesy of P. Maestrati - MNHN.

FIGURE 19.11 Examples of living neogastropods – Volutoidea and Conoidea. (a) *Cymbiola vespertilio* (Volutidae), Philippines. Courtesy of T. Gosliner. (b) *Cronia amygdala* (Muricidae), New South Wales, Australia. Courtesy of D. Riek. (c) *Crithe* sp. (Cystiscidae), Philippines. Courtesy of T. Gosliner. (d) *Amalda* sp. (Ancillariidae), Ile des Pins, New Caledonia. Courtesy of P. Maestrati - MNHN. (e) *Harpa amouretta* (Harpidae), Madang, Papua New Guinea. Courtesy of L. Charles - MNHN. (f) *Bathytoma viabrunnea* (Borsoniidae), Guadeloupe, French West Indies. Courtesy of L. Charles - MNHN. (g) *Conus geographus* (Conidae), Papua New Guinea. Courtesy of L. Charles - MNHN. (h) *Lienardia rubida* (Clathurellidae), Papua New Guinea. Courtesy of L. Charles - MNHN.

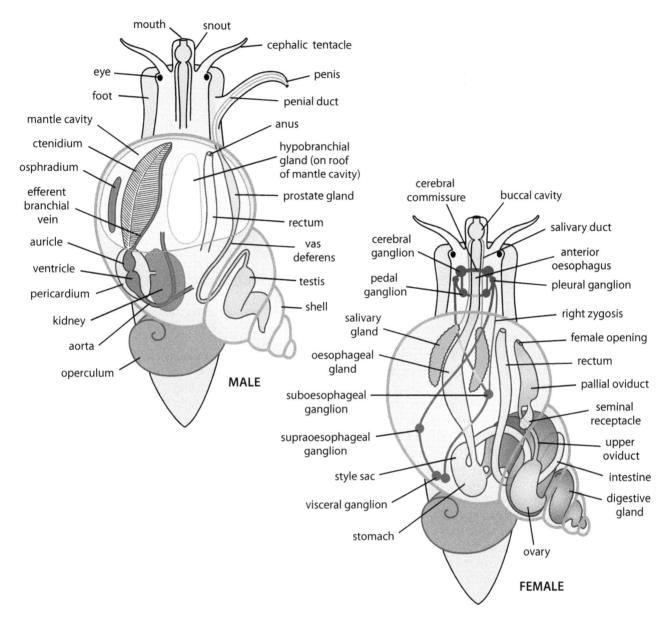

FIGURE 19.12 Body plan of a generalised asiphonate caenogastropod. The upper figure shows the renopericardial and pallial structures and the male reproductive system; the lower figure shows the nervous, digestive, and female reproductive systems. Redrawn and modified from Lang, A., *Text-book of Comparative Anatomy, Part 2*, Macmillan, London, 1896.

19.3.1 Shell and Protoconch

The shell (Figures 19.2 to 19.11) is typically dextrally coiled and is highly diverse, ranging from high spired to globose or flattened and even abalone-like or limpet-like (see below). Members of one family only (Triphoridae) are mostly sinistral, but sinistral species can occur sporadically in some families (Vermeij 1975; Robertson 1993).

A 'fusiform' shape is common, being determined by a feature unique to many caenogastropods – a canal in the anterior end of the aperture. This structure is encountered mostly in higher caenogastropods (e.g., 'whelks' of various kinds). Among the 'lower' caenogastropods, many cerithioideans also have a usually short, anterior canal in the aperture, but intermediate groups do not, notably those in the 'asiphonate clade'

of Colgan et al. (2007). The development of an anterior siphon in caenogastropods is explored further in Section 19.3.2.4.

The columella sometimes bears folds or plaits, a feature not uncommon in neogastropods but less common in other caenogastropods. The function of these folds remains controversial (e.g., Signor & Kat 1984; Price 2003).

The shells of the terrestrial cyclophoroideans are particularly diverse in shape and size (1–60 mm). Their shells range from flat spired to high spired, and while most are dextral, a few are sinistral. Some cyclophoroideans are partially or completely uncoiled with a few Pupillidae and Diplommatinidae undergoing dramatic changes in coiling direction as they grow. One of the most bizarre examples is the Malaysian limestone endemic *Opisthostoma vermiculum* (Diplommatinidae) which changes its coiling axis four times as it grows (Clements et al. 2008).

FIGURE 19.13   Varices in a cymatiid, *Monoplex vespaceus* (left) and a muricid, *Chicoreus denudatus* (right). Courtesy of D. Riek.

Shells showing varying degrees of uncoiling are seen in several caenogastropod families such as typical Siliquariidae and a few members of other families (e.g., Epitoniidae, 'Hydrobiidae' *s. l.*) (Rex & Boss 1976), but extreme examples resulting in tube-like shells are seen in some siliquariids, most vermetids, *Vermicularia* (Turritellidae), *Magilus* (Coralliophilinae), and a few minute truncatelloideans, notably Caecidae and a few subterranean taxa (e.g., Hershler & Longley 1986).

Some such as *Pedicularia* (Ovulidae), *Sinum* (Naticidae), and a few Muricidae (*Concholepas, Quoyula*) have a much-expanded aperture and reduced spire to give the shells a somewhat limpet-like or 'abalone'-like appearance. Even more limpet-like taxa have evolved in some families that normally have coiled shells, including Eulimidae (*Thyca*) and Capulidae (*Capulus*). Some Calyptraeidae have limpet-like shells with little or no coil, but others (e.g., *Calyptraea*) have coiled shells but have a limpet-like aperture. Most hipponicids are very limpet-like and secrete a shelly plate beneath their shell (Figure 19.6b), becoming effectively a cemented bivalve. The planktonic Carinariidae have reduced cap-shaped shells.

Internal, reduced shells are found only in the slug-like Velutinidae, and some eulimids are internal worm-like parasites (see Section 19.4.5.1) that have lost the adult shell, as have the pelagic 'heteropods' of the family Pterotracheidae.

The protoconch of caenogastropods is orthostrophic and multispiral except in some direct-developing taxa. The size and shape of the protoconch, including the number of larval whorls and the size of protoconch I, are very useful to infer the type of larval development (see Chapter 8). Protoconch II is often elaborately sculptured and the outer lip of the larval aperture often deeply notched.

Most caenogastropods are small (<10 mm), and these 'microgastropods' are often the most abundant and diverse components of gastropod marine faunas (e.g., Bouchet et al. 2002). Some superfamilies contain numerous species less than 5 mm in size with many species 1–3 mm in maximum shell dimension (e.g., most rissooideans, truncatelloideans, cingulopsoideans, eulimoideans, and triphoroideans). Also, many members of the families Marginellidae, Cystiscidae, Columbellidae, and many conoidean families are less than 10 mm in maximum shell size. In contrast, shells of some species in various hypsogastropod carnivorous families are more than 30 cm (e.g., Cymatiidae, Charoniidae, Bursidae, Cassidae, Volutidae, Fasciolariidae, Turbinellidae, and Melongenidae) while there are members of families such as Strombidae, Tonnidae, Buccinidae, and Muricidae which have shells more than 200 mm in length. Some non-hypsogastropods are also large, including some species in Ampullariidae, Campanilidae, Batillariidae, Potamididae, and Cerithiidae.

Another feature of many caenogastropods is *determinate growth* where the aperture is thickened at maturity, in contrast to those gastropods that retain a thin-edged, undifferentiated aperture throughout life (Vermeij & Signor 1992) (see also Chapter 3). Cessation of growth can be seen in various forms; typically it is a thickening of the aperture, the formation of internal thickening or teeth or a varix-like rib externally, and may also involve an expansion of the outer lip. Some taxa thicken the aperture periodically (*periodic growth*), leaving the remains of previous lip thickening events (*varices*) behind as they grow. These are sometimes arranged in a highly uniform way (e.g., many cymatiids and muricids) (Figure 19.13) or rather more erratically as in many cerithiids. Varices have evolved many times in different groups of gastropods, with the great majority of these in caenogastropods (Webster & Vermeij 2017).

The outer apertural edge may be simple, notched, or, in some neogastropods, may even bear a spine (Vermeij 2001) (also see Section 19.4.3).

Caenogastropod shell structure is typically aragonitic crossed lamellar, or rarely calcitic, and nacre is never present (see Bandel 1990 for a review) (see also Chapter 3).

19.3.2 Head-Foot, Mantle, and Operculum

In caenogastropods, unlike vetigastropods, the exposed parts of the animal are typically relatively simple. Plesiomorphically, the head has a prominent, protruding snout and bears a pair of rather long, narrow, tapering, smooth or ciliated tentacles with eyes at their outer bases, although both the head-foot and mantle show considerable modification throughout the group (Figures 19.2–19.11).

19.3.2.1 Head Structures

A narrow snout is characteristic of basal caenogastropods. It is sometimes very extensile (as in some cerithioideans, rissooideans, truncatelloideans, and stromboideans). The snout is often called a proboscis, but the term proboscis is used here when the snout is infolded to form an introvert, as has happened independently in several groups, and is thought to be associated with the adoption of a carnivorous diet (see Section 19.4.3).

The snout is used to assist in locomotion in Truncatellidae (see Chapter 3, Figure 3.36) and some Assimineidae and Pomatiopsidae (Davis 1967; Ponder 1988; Fretter & Graham 1994).

Long cephalic tentacles are plesiomorphic, but they show considerable variation in length throughout the group. They are short in some amphibious truncatelloidean taxa (e.g., *Blanfordia*) and are vestigial or lost in many Assimineidae. A number of small caenogastropods (notably Truncatelloidea) have complexly ciliated tentacles, and the tips of the tentacles of some taxa have long, stationary compound cilia.

The eyes are closed and have a lens. They are plesiomorphically located laterally at the base of the tentacles, in small bulges or on short peduncles. Eyes are reduced or even lost in some taxa, such as in naticids, where they are covered by the propodium, or unpigmented or lost in some deep-sea taxa and phreatic and stygobitic (freshwater) truncatelloideans. In contrast to basal heterobranchs, the eyes are always on the outer side of the tentacles although in some they are on short stalks (e.g., in most architaenioglossans, xenophoroideans, and some stromboideans), while in others (e.g., tonnoideans and neogastropods) they have migrated upwards along the outer side of the tentacle.

In most caenogastropods the neck region is simple, but in a few taxa (Viviparidae, Ampullariidae, Turritellidae, and Vanikoridae) there are flaps of cephalo-pedal origin on the sides of the neck or foot. In the architaenioglossan Viviparidae and Ampullariidae, these so-called nuchal lobes form paired siphons (left-inhalant, right-exhalant). These structures are not homologous to the inhalant and exhalant siphons developed in some other caenogastropods, as those are derived from the mantle.

In many caenogastropods, a penis is on the right side of the head behind the cephalic tentacle (see Section 19.3.6.2).

19.3.2.2 The Foot

The foot usually bears an oval to elongate creeping sole and an operculum. There are an anterior propodium and associated anterior pedal gland(s), and, in some taxa, there is an internal metapodial pedal mucous gland that opens to the sole (see below), besides the normal epithelial and subepithelial sole glands.

Plesiomorphically, the foot is a creeping disc that moves by way of muscular or ciliary action of the foot sole (see Chapter 3 and below). The foot is enlarged and broadened in many neogastropods, tonnoideans, and some other taxa or has become disc-like as in the sessile or near-sessile limpet-like Calyptraeidae and Hipponicidae that use their foot to clamp to the substratum. It is very reduced in sessile taxa (Vermetidae, Siliquariidae, and the coral-living muricid *Magilus*) that are attached by shell cementation or by some other means. In some the foot is elongate, to enable crawling on, for example, seagrass leaves or the narrow branches of sea fans (gorgonians). Lateral compression of the foot enables it to be used as a lever, and in stromboideans (e.g., *Strombus*, *Struthiolaria*, Aporrhaidae) and the related Xenophoridae, it is a muscular organ capable of leaping. The so-called heteropods (Pterotracheoidea) are permanent members of the plankton and swim using muscular actions of their body and a fin derived from the foot.

The shell can be covered or partially covered by lateral or anterior extensions of the foot and/or mantle in Naticidae, Cypraeoidea, Triviidae, Olividae, Marginellidae, Ficidae, some Volutidae, and particularly in Velutinidae.

The sides of the foot are simple in most taxa, but a few cerithiimorphs have epipodial tentacles (Litiopidae and some members of the cerithiid Bittiinae, and there are weak papillae in Plesiotrochidae), and one or a few short metapodial tentacles emerge from the posterior end of the foot in a small number of groups (some rissooideans and truncatelloideans, and Nassariidae).

The caenogastropod foot typically has a distinct propodium with the anterior pedal gland(s) opening to a groove along the anterior edge of the foot between this and the mesopodium. In some taxa, this groove extends further backwards and in a few cerithiimorphs extends around the whole foot. In a few terrestrial taxa, such as the Pomatiidae, the propodium and transverse groove are reduced or absent. The anterior pedal gland can consist of a series of glands opening to the edge of the foot, or the glands are compacted into a single compound gland with one opening (e.g., in pomatiids). In many taxa a posterior pedal gland is absent, including all non-marine species, although this structure is found beneath the middle of the sole of some, mainly small-sized, sorbeoconch caenogastropods such as litiopids, cingulopsoideans, some triphoroideans, rissooideans, abyssochrysoideans, and a few truncatelloideans and eulimoideans. The gland may open via a small slit-like aperture to the otherwise unmodified sole or to a long longitudinal groove in the sole that extends to the posterior end of the foot. The terrestrial pomatiids have a mass of tubular glands in the foot haemocoel which open at a pore in the middle of a longitudinal groove that divides the sole. This gland is not homologous with the posterior pedal gland and is involved in osmoregulation (Delhaye 1974).

Many small-sized caenogastropods such as rissooideans, cingulopsoideans, and truncatelloideans move by ciliary gliding, but in larger taxa movement is aided by muscular waves (see Chapter 3, Figures 3.32 to 3.35). In some taxa (e.g., most Littorinidae), the foot sole is divided longitudinally into two halves by a groove. This indicates a functional division of the underlying musculature and retrograde ditaxic locomotion (Miller 1974b) (see also Chapter 3, Figure 3.34), although some taxa exhibiting this locomotory pattern show no such groove. There are other exceptions, for example, the pomatiids have a longitudinally divided foot sole but have direct ditaxic waves, as do most other littorinoideans (Miller 1974a, b).

A so-called 'ventral pedal gland' is found in the foot sole of females of many neogastropods (see Chapter 8, Figure 8.17) and in velutinoideans, and is involved in egg capsule moulding, while muricids have a 'boring organ' on the anterior part of the sole (see Chapter 5).

Amphibious or terrestrial truncatelloidean taxa (notably Assimineidae and Pomatiopsidae) have a deep groove, the *omniphoric groove* (Davis 1967), running down each side of the neck; this carries mucus and waste to the foot. In some terrestrial assimineids, the mucus is pumped down almost tube-like grooves and streams from the grooves over the side of the foot, presumably enabling cleansing of the mantle cavity and moistening the foot (Ponder 1988). These grooves are modified, ciliated, often unpigmented strips found in many aquatic truncatelloideans (e.g., Johansson 1939; Marcus & Marcus 1963). The left groove assists water flow into the mantle cavity, and the other carries waste out.

19.3.2.3 The Operculum

Most 'littorinimorphs' have a paucispiral operculum, but in some, mainly cerithiimorphs, it is multispiral. Higher caenogastropods (tonnoideans and neogastropods) have either concentric opercula or the nucleus is terminal. The operculum sits on an opercular lobe which may be expanded, and in some Eatoniellidae and Lacuninae one or two tentacles emerge laterally from it. In several families an adult operculum is absent (e.g., the cowries and similar families, Velutinidae, the heteropods, Calyptraeidae, and Hipponicidae, and some neogastropod families such as Harpidae, Mitridae, Marginellidae, Cystiscidae, some conoideans, and many Volutidae).

19.3.2.4 Mantle Edge

The anterior mantle edge is extended as an inhalant siphon in many caenogastropods. Vermeij (2007) estimated that the siphonal canal had evolved and been lost numerous times during caenogastropod evolution. This condition occurs, probably independently, in living members of Ampullarioidea, Campaniloidea, Cerithioidea, Rissooidea, Stromboidea, Triphoroidea, and the higher Hypsogastropoda. It may extend as a long tube beyond the siphonal canal in the shell or be enclosed within it. Otherwise, except for those taxa where the shell can be covered or partially covered by lateral or anterior extensions of the foot and/or mantle, the mantle edge rarely extends much beyond the edge of the shell, but it can have papillae or even one or more tentacles that do.

19.3.3 Mantle Cavity

The plesiomorphic condition in caenogastropods is that only the left ctenidium (always monopectinate and with skeletal rods), osphradium (typically hypertrophied with a unique histology) (Haszprunar 1985g) (see Section 19.3.7), hypobranchial gland, and kidney are present, although elements of the right kidney are incorporated in the oviduct. The heart has a single (left) auricle, corresponding to the single (left) ctenidium.

Adoption of a semiterrestrial or terrestrial lifestyle resulted in the atrophy or loss of the ctenidium. It is retained in the amphibious Ampullariidae, but part of the right side of the mantle cavity is modified as a lung.

A critical innovation of the Sorbeoconcha was a change from exhalant to inhalant control of the water currents through the mantle cavity (Ponder & Lindberg 1997; Lindberg & Ponder 2001). In many groups of sorbeoconchs this change to inhalant control is reflected in the independent development of an anterior (or inhalant) notch or siphonal tube in the shell which houses an elaboration of the mantle edge, the siphon (see above). Some taxa also have a posterior notch in the shell facilitating the exhalant current, and in a few this is prolonged as a shelly tube. The development of a siphon enabled the more efficient use of the chemosensory osphradium which is considerably enlarged and elaborated in higher caenogastropods. This very important organ is discussed in Section 19.3.7. In the Sorbeoconcha, water flow into the mantle cavity is not only driven by the ctenidium as in many other gastropods but in part by the large, ciliated osphradium.

Unlike vetigastropods and heterobranchs, the kidney is usually located just behind the mantle cavity and, with few exceptions, opens through the posterior wall of the mantle cavity. It has extended into the roof of the mantle cavity in some non-marine taxa, including architaenioglossans and, for example, *Pomatias* (Littorinoidea) and bithyniids (Truncatelloidea), although the opening remains posterior except in viviparids and ampullariids where it has migrated forwards. In at least some terrestrial taxa water is secreted by the kidney to moisten the mantle cavity.

The gonoduct and rectum lie on the right side of the mantle cavity, and the hypobranchial gland occupies the mantle roof between these structures and the ctenidium. The mantle cavity structure is surprisingly uniform with modifications involving convergent elongation of the gill filaments and the development of a mucus-secreting endostyle and food groove in suspension-feeding taxa (see Section 19.4.3 and Chapter 5).

The ctenidium is retained in intertidal taxa, including many littorinids, but several members of that family which inhabit the supralittoral (e.g., *Tectarius* spp.) or the mangrove tree-living *Littoraria* (Reid 1986b) have a reduced ctenidium.

In some amphibious and terrestrial taxa the ctenidium and hypobranchial gland are lost, notably in the Cyclophoridae, where the osphradium is also very reduced or lost. In the terrestrial and supralittoral assimineids, the filaments are reduced to a few posterior stubs, and are essentially lost in the terrestrial *Pomatias*, being represented by only a few epithelial folds. In these snails, the mantle skirt provides the main respiratory surface (e.g., Delhaye 1974). Similarly, in the amphibious or aquatic Pomatiopsidae, the ctenidium is well-developed (e.g., *Pomatiopsis* and *Tomichia*) but is much reduced or absent in the semiterrestrial *Blanfordia*. Ctenidial reduction or even loss has also occurred in a few marine taxa (e.g., Rastodentidae) due to their very small body size. In these cases, the ciliated border of an enlarged osphradium produces a water current (Ponder 1966, 1968).

19.3.4 DIGESTIVE SYSTEM

The caenogastropod gut is a simplified version of that seen in many vetigastropods (see Chapters 5 and 18).

Plesiomorphically, there is a short to rather long snout with a pair of lateral jaws located just behind the anteroventral mouth. The jaws are typically made up of minute rods and a homogeneous layer that covers part of the outer surface (Strong 2003). They are particularly well-developed in the herbivorous freshwater ampullarioideans, in the related marine campaniloideans, and in some mainly carnivorous taxa such as Tonnoidea and Cypraeoidea. The jaws have been lost in a few lineages, including Pomatiidae, nearly all Littorinidae, some Assimineidae, and all Neogastropoda. In neogastropods, the cuticle lining of the oral tube is sometimes thickened to form a dorsal, jaw-like structure as in muricids or a funnel-shaped structure as in Volutomitridae and Cancellariidae (Figure 19.18).

In many carnivorous taxa, the snout has been introverted in various ways to form a proboscis (see Chapter 5 and below).

A subradular organ ('licker') is found in basal gastropods (Ponder & Lindberg 1997) (see Chapter 5), and it was thought that the only caenogastropods that have what is assumed to be this structure were some ampullariids (see Strong 2003 for a review). However, it is also present in cyclophoroideans (Strong 2003), and in both these groups of architaenioglossans it has a complexly folded, non-cuticularised epithelium rich in goblet cells and is not obviously innervated. It lies below the radula, and when the odontophore is retracted it projects into the mouth opening, suggesting that the subradular organ in caenogastropods has a lubricatory rather than a plesiomorphic sensory function (Strong 2003). Reduced but similar elaborations of the subradular epithelium were found in several other non-neogastropod sorbeoconchs by Strong (2003), who argued these were homologues of the subradular organ. She also observed this structure in two proboscidate taxa, a species of *Neverita* (Naticidae), and in one neogastropod, *Panarona* (Cancellariidae), but noted that it was absent in a cypraeid (with a short proboscis).

Caenogastropods have only a single pair of large buccal cartilages, but sometimes smaller cartilage structures are present that have only recently been identified (Golding et al. 2009b).

At times some unique structures are found associated with the snout or proboscis. For example, a *pseudoproboscis* is formed in the Capuloidea (e.g., Pernet & Kohn 1998), somewhat similar to that seen in some trochoideans. A unique epiproboscis occurs in the Mitridae (Ponder 1972; West 1990)

(see Chapter 5, Figure 5.23) and rhyncodeal outgrowths are seen in some conoidean taxa (see below).

The radular apparatus is simpler than in vetigastropods. The radular diverticulum has been lost in most caenogastropods, and the radular sac is rather short in most, although it can be long in some grazing taxa and is very long and coiled in Littorinidae, many of which graze on hard surfaces. The radular teeth, however, are not hardened with iron or silica as in some other surface-scraping molluscs such as patellogastropods and chitons.

As with other gastropods, the radula has proved to be important for caenogastropod taxonomy at all levels and forms the basis of names such as Architaenioglossa, Taenioglossa, Stenoglossa, Rachiglossa, and Ptenoglossa. Examples illustrating the range of radular morphology seen in caenogastropods are shown in Figures 19.14 to 19.18.

Within caenogastropods, the radula is plesiomorphically taenioglossate (seven teeth in each row – two marginal teeth and a lateral on each side of the central tooth), and this radula-type is highly conserved, being found in all architaenioglossans and the great majority of non-neogastropod sorbeoconchs, despite the wide diversity of feeding habits (see Chapter 5 and Section 19.4.3; Figures 19.14 and 19.16). Regardless of their relatively simple structure, determining the homology of tooth elements across all caenogastropods is problematic with our current level of knowledge (Ponder & Lindberg 1997), even with all the radular types presumably derived from a taenioglossate ancestral condition.

Significant modifications to the taenioglossate condition have occurred. In the Rissooidea where the great majority of taxa have a 'normal' taenioglossate radula, *Emblanda* (Emblandidae) has lost the marginal teeth (Figure 19.14). Some families, such as the Cingulopsidae (Cingulopsoidea), show a great variety of morphology of the radular teeth, but besides *Tubbreva* where the radula is lost, they have retained seven teeth in each row (Ponder & Yoo 1980). A surprising and significant departure in tooth morphology is also seen in the cingulopsoidean Rastodentidae (Figure 19.14), which have developed spiked teeth, similar to those seen in carnivorous taxa such as ovulids, triphorids, and neogastropods, but retain seven teeth in each row (Ponder 1966). These minute gastropods probably feed on bryozoans.

Triphoroideans stand out as having very diverse tooth morphology, with these small proboscidate gastropods, as far as known, all being obligate sponge feeders, and their radulae are presumably adapted to feeding on the great variety of food substrata this group offers. They exhibit an extraordinary diversity in tooth number, size, and shape (Figure 19.15) exceeded only by the 'opisthobranchs' (see Chapter 20). In this group, the conservative seven-tooth row constraint has broken down in some taxa, and many configurations can be found, including, in some triphorids, considerable multiplication of teeth (e.g., Marshall 1983; Nützel 1997), with some having numerous similar teeth (Figure 19.15). The related cerithiopsids retain a taenioglossan configuration, although the teeth are often highly modified (Marshall 1978; Nützel 1997) (Figure 19.15).

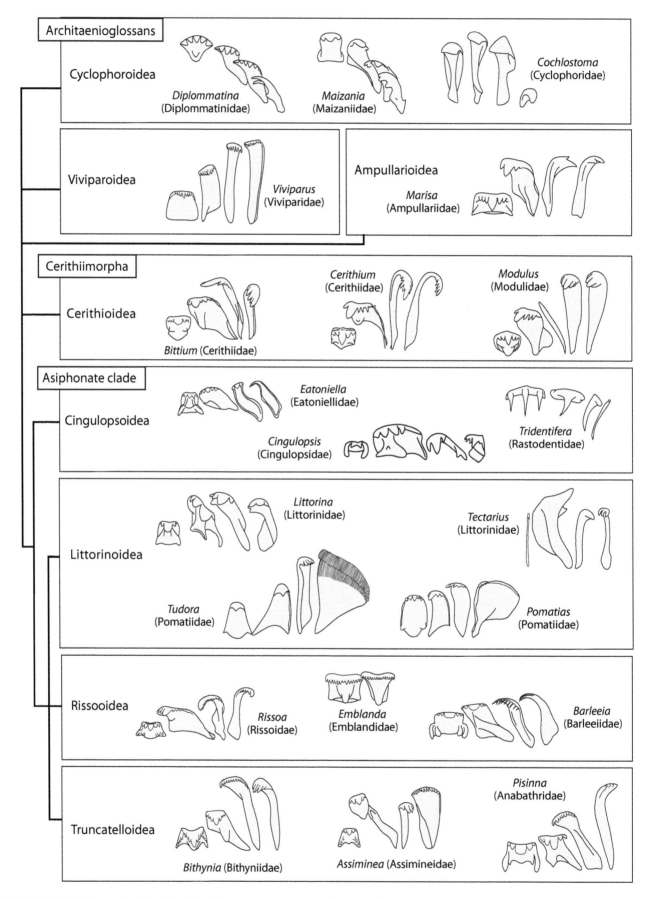

FIGURE 19.14 Examples of radulae from lower caenogastropods. Redrawn from various sources. Note that the radulae of the species illustrated do not necessarily reflect the range of variation in the genera they belong to or of their higher taxon.

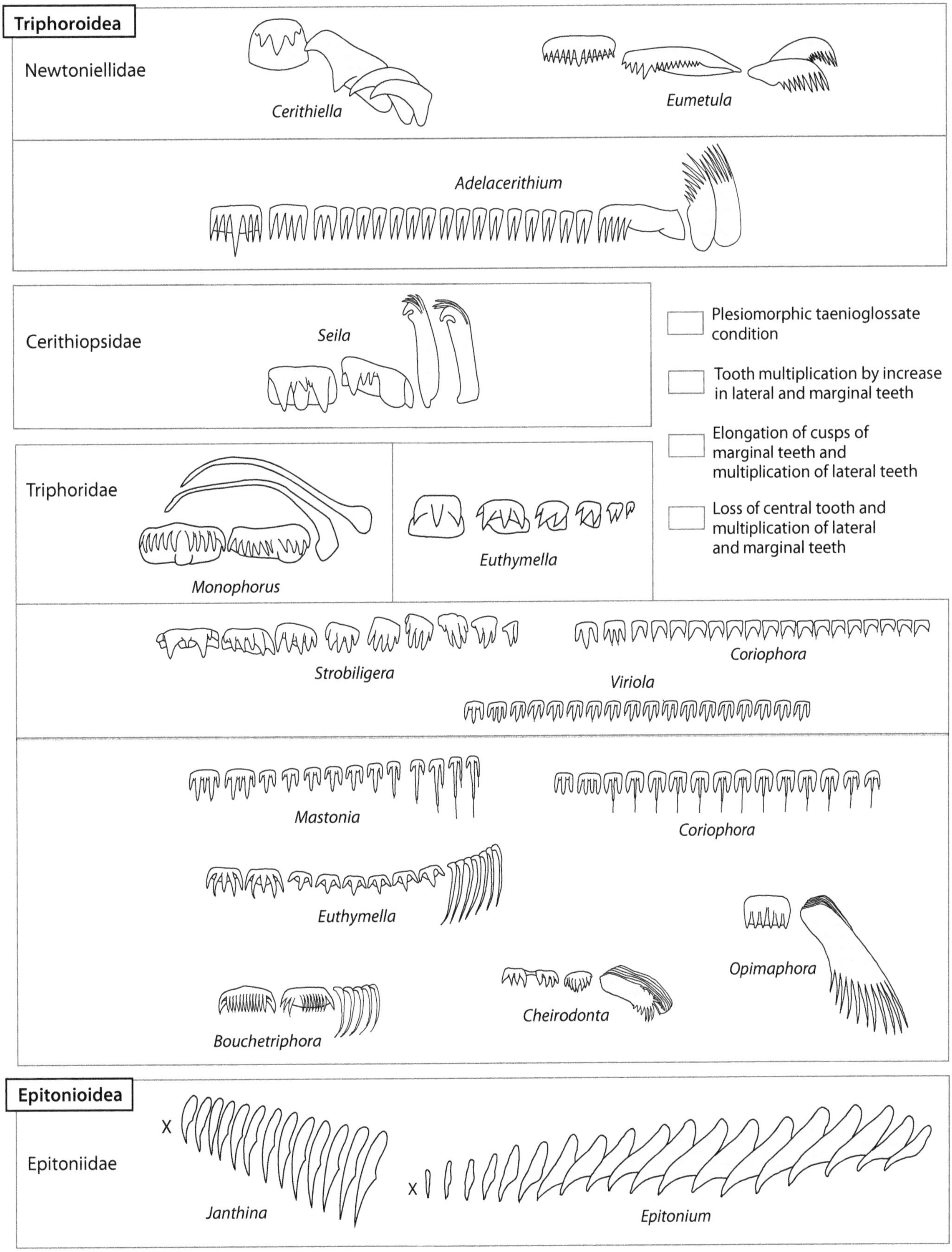

FIGURE 19.15 Examples of radular tooth multiplication in caenogastropods. X – central tooth absent. Redrawn from various sources. Note that the radulae of the species illustrated do not necessarily reflect the range of variation in the genera they belong to or of their higher taxon.

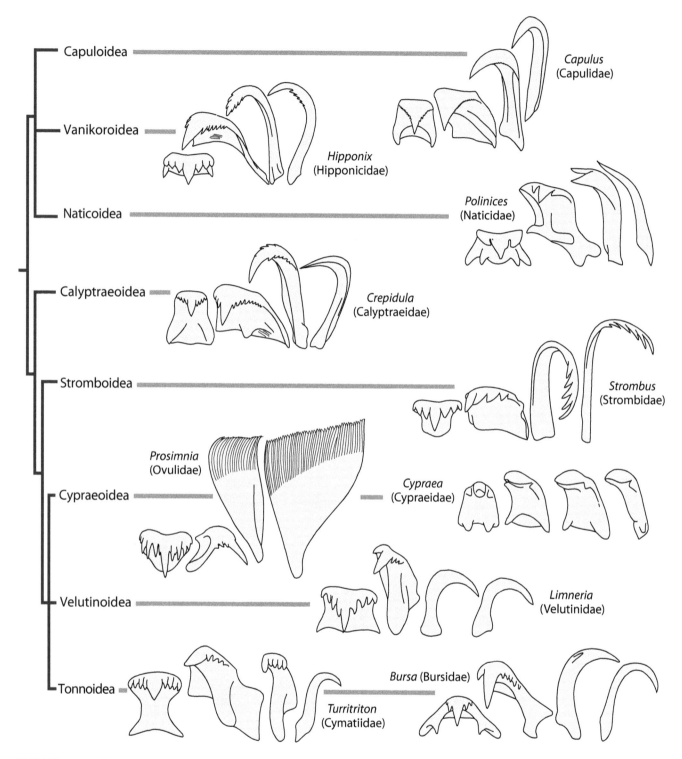

FIGURE 19.16 Examples of caenogastropod radulae. Redrawn from various sources. Note the radulae of the species illustrated do not necessarily reflect the range of variation in the genera they belong to or of their higher taxon.

The triphorid radula is small, rarely over two mm long and much longer than broad with numerous rows of very minute teeth. The number of teeth per row ranges from five to at least 63 in different species (Marshall 1983), with the multitoothed radular type called *rhinioglossate*. Similar tooth multiplication has also occurred in one group of newtoniellids,

the Adelacerithiinae (Marshall 1984), but most newtoniellids and cerithiopsids have a taenioglossate configuration (Figure 19.15).

The homologies of the teeth in triphorid radulae are not certain. Marshall (1983) considered that each transverse row consisted of a central tooth, one pair of lateral teeth, and one or more pairs of marginal teeth, with the inner and outer marginals often very different. Marshall (1983 p. 7) considered the lateral and marginal teeth 'almost certainly not homologous with those of mesogastropods, and certainly not with those of archaeogastropods'. While we do not agree with the lack of homology of the radular teeth with those in other lower caenogastropods, we accept that determining homology is difficult given their modification. Kosuge (1966) treated the central and the pair of lateral teeth of *Iniforis* and *Risbecia* as three central teeth, a position rightly disputed by Marshall (1983).

Some triphorids, such as *Tetraphora* and *Sagenotriphora*, have only five to seven teeth per row, and seven is presumably the plesiomorphic number for the family (e.g., Bandel 1984), although Kosuge (1966) and Marshall (1983) interpreted this as a reduction, which it may be in some cases.

Marshall (1983 p. 8) noted the tendency for abnormal teeth to be high in the group, with many exhibiting bilateral asymmetry in tooth shape and the cusps, and suggested this may be a factor in 'morphological changes [being] readily assimilated into the gene pool'.

In some triphorid taxa, the outer marginal teeth are very long and narrow, and this appears to have come about by the elongation of the middle cusp on the lateral teeth, as seen in several genera including *Mastonia* and *Bouchetriphora*. In *Cheirodonta* the whole outer marginal tooth has elongated and is multicuspid (Figure 19.15).

The unusual radula of triphorids has resulted in speculation regarding the relationships of the group, including a connection with neogastropods (Risbec 1955; Habe & Kosuge 1966; Golikov & Starobogatov 1975). The Triphoridae was grouped with Mathildidae, Architectonicidae, and Epitoniidae as 'Heterogastropoda' (Habe & Kosuge 1966), a group placed with the opisthobranchs by Climo (1975). Marshall (1984) noted that the arrangement of the teeth in each radular row was anterior to the central teeth in triphorids as in other non-neogastropod caenogastropods (including cerithiopsids) and vetigastropods, while in the newtoniellid *Adelacerithium*, the teeth run posteriorly from the central teeth as in neogastropods and many euthyneurans. Marshall (1984) used the term *sagittate* to describe the anteriorly pointing radular cross-row which he likened to an arrowhead (see Chapter 5).

Broad multicuspate outer marginal teeth are found in several groups; in some, the cusps are long and narrow so the tooth appears brush-like. Such teeth are found in Pelycidiidae, ovulids, and in some terrestrial littorinoideans and assimineids, notably the terrestrial Omphalotropinae (Figures 19.14 and 19.16). In some, the cusps are so deeply subdivided that in a few instances they have been mistaken for separate marginal teeth (e.g., Ponder & Hall 1983) and sometimes have been wrongly considered to represent fused marginal teeth.

Other than some triphoroideans, the main exceptions to the basic taenioglossate morphology in the non-neogastropod sorbeoconchs are seen in all epitonioideans and in eulimoideans that possess a radula (Warén 1983a). This highly derived type has been termed the ptenoglossate condition, where the radula possesses many rows of similar teeth, although this condition is convergent in the two groups. It is unclear whether the multiplication involves only the marginal teeth or both the marginal and lateral teeth (Ponder & Lindberg 1997).

Reduction in the number of tooth rows is seen in all neogastropods. Their main configurations are the rachiglossan type (one to three teeth per row) and the toxoglossan type (Conoidea only – one to five teeth), the extreme modification being the toxoglossan 'harpoon' teeth (see Chapter 5) referred to as *hypodermic* teeth. Coupled with the formation of harpoon teeth a sac is formed from the sublingual pouch, the *radular caecum*, that is analogous to the quiver of an archer in which the teeth are stored before use. The radular caecum is a key innovation in the major group of conoideans with toxoglossan teeth (i.e., the Conidae, Conorbidae, Borsoniidae, Clathurellidae, Mitromorphidae, Raphitomidae, and Mangeliidae). These 'families' all form a clade in one molecular analysis of Puillandre et al. (2011) but not in some others (Abdelkrim et al. 2018). Hypodermic teeth may have evolved independently at least three times in the Terebridae (Taylor et al. 1993; Castelin et al. 2012), with basal members lacking this tooth structure. The radular caecum is lacking in other conoideans; some of these have hypodermic teeth, and others have members that have lost the radula (see Table 19.1). Another type of marginal tooth is the so-called duplex type (Kantor & Taylor 2000) where the teeth are wishbone shaped.

An unusual very elongate tooth type is found in Cancellarioidea (the *nematoglossan* type) which is derived from the central teeth (Harasewych & Petit 1982; Taylor & Morris 1988). These teeth are used for piercing the skin of the host (e.g., rays, bivalves) on which the cancellariids feed by sucking their blood. The teeth are sheathed in a modified jaw-like structure (Figure 19.18) that forms a tube along which a tooth is thrust.

The radula has not only been lost in some conoideans (see Table 19.1) but also in a few other caenogastropods, including some neogastropods (e.g., the muricid subfamily Coralliophilinae and some Cancellariidae), many eulimids, and one genus of Cingulopsidae (see above), while it is much reduced in the suctorial Colubrariidae.

Introversion of the snout to form a proboscis has occurred independently in several groups of carnivorous caenogastropods (Golding et al. 2009a) (see also Chapter 5). The proboscis is often capable of considerable extension, and in some taxa can reach much more than the body length of the animal.

Golding et al. (2009a) demonstrated that there was considerable diversity in the muscular composition of the proboscis wall and the details of the attached retractor muscles. These data indicate that a proboscis evolved separately at least four times, and probably more, in caenogastropod evolution. Proboscis extension is by way of a hydrostatic skeleton

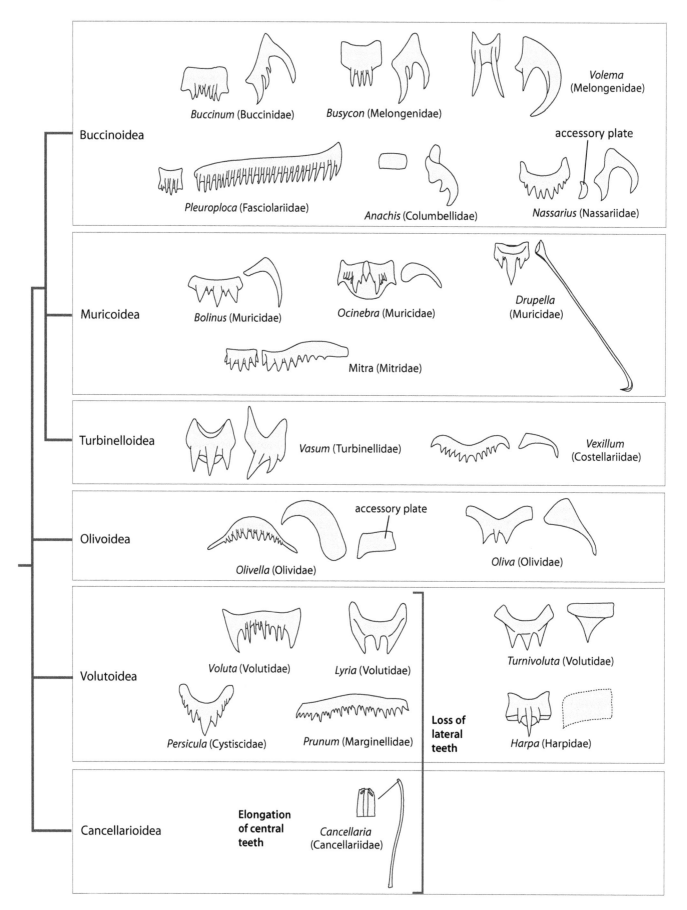

FIGURE 19.17   Examples of neogastropod radulae, with the exception of Conoidea. Redrawn from various sources. Note that radulae of the species illustrated do not necessarily reflect the range of variation in the genera they belong to or of their higher taxon.

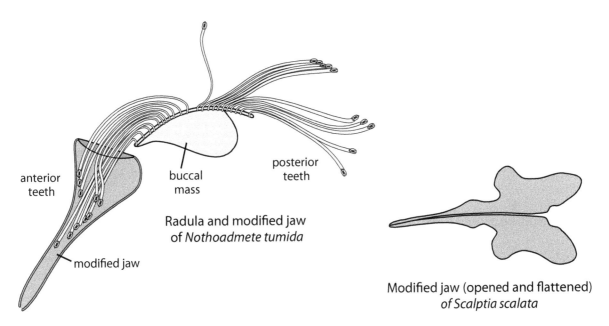

FIGURE 19.18   Radula and lateral view of the modified jaw-like structure of the cancellariid *Nothoadmete tumida* and a dorsal view of the flattened 'jaw' of *Scalptia scalata*. Top left figure redrawn and modified from Oliver, P.G., *B.A.S. Bull.*, 57, 15–20, 1982, bottom right from Olsson, A.A., *Palaeontogr. Am.*, 7, 19–32, 1970.

TABLE 19.1

Distribution of Some Anterior Gut Characters in Conoidea

Family	Marginal Teeth	Radular Loss	Radular Caecum	Central Tooth	Lateral Teeth	Venom Gland	Odontophore
Conidae	Hypodermic	No	Yes	Lost	Lost	Present	Absent
Terebridae	Solid to hypodermic	Some	Some	Lost	Lost	Present or absent	Present or absent
Turridae	Duplex	Some	No	Present in most	Fused to centrals if present; lost in some	Present	Present
Clavatulidae	Simple to hypodermic	No	No	In some	In some	Present	Present
Drilliidae	Simple to hypodermic	No	No	Present	Present	Present	Present
Pseudomelatomidae	Simple to inrolled	No	No	In some	In some	Present or absent	Present
Horaiclavidae	Duplex	Some	No	No	No	Present	Absent
Cochlespiridae	Duplex	No	No	Present		Present	Present
Borsoniidae	Hypodermic	Lost in one genus	Yes	Lost	Lost	Present or absent	Absent
Mitromorphidae	Hypodermic	No	Yes	Lost	Lost	Present	Absent
Clathurellidae	Hypodermic	No	Yes	Lost	Lost	Present	Absent
Conorbidae	Hypodermic	No	Yes	Lost	Lost	Present	Absent
Raphitomidae	Hypodermic	Several	Yes	Lost	Lost	Present or absent	Absent
Mangeliidae	Semi-inrolled to hypodermic	No	Yes	Lost	Lost	Present	Absent
Fusiturridae	Duplex	No	?	Lost?	Lost	?	?
Marshallenidae	Duplex	No	?	Present	Lost	?	?

Classification and radular data based on Bouchet et al. (2011) and venom gland and odontophore data are mainly from Taylor et al. (1993), Fedosov (2008), and Abdelkrim et al. (2018), with some modifications by Y. Kantor (pers. comm.)

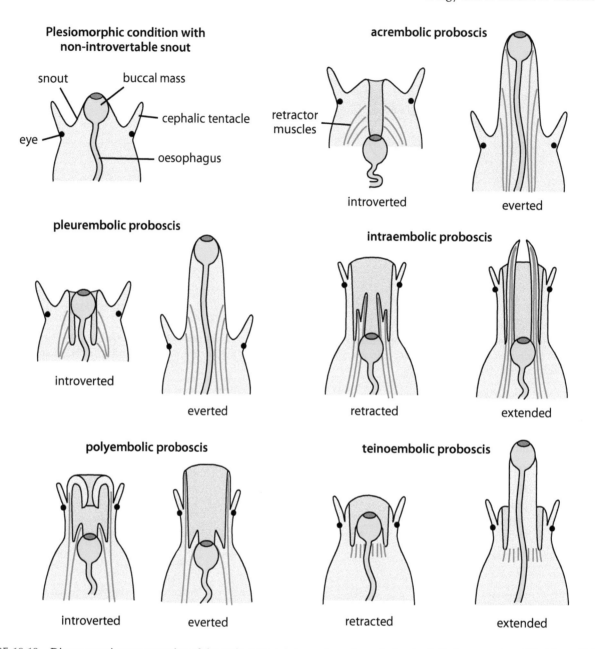

FIGURE 19.19 Diagrammatic representation of the main proboscis types in gastropods (see text). Redrawn and modified from Golding, R.E. et al., *J. Morphol.*, 270, 558–587, 2009a.

with the haemocoel in the proboscis isolated during eversion in some taxa, but, in the more derived groups, extension is brought about by way of a complex muscular hydrostat (Golding et al. 2009a).

Three of the main proboscis types recognised in caenogastropods are acrembolic, pleurembolic, and intraembolic, these differing mainly where the retractor muscles are attached (Figure 19.19). An *acrembolic* proboscis has the retractor muscles inserted distally in the proboscis wall so the entire proboscis is introverted and the buccal mass lies behind it in the haemocoel. In a *pleurembolic* proboscis the retractor muscles are inserted more proximally so only partial introversion occurs and the buccal mass remains within

the proboscis. While the acrembolic type appeared several times, it is possible that the pleurembolic proboscis appeared only once. An acrembolic proboscis is found in Epitonioidea, Triphoroidea, and Eulimoidea, while a pleurembolic proboscis is found in Cypraeoidea, Ficoidea, Tonnoidea, and Neogastropoda (Golding et al. 2009a). The Naticoidea have a unique type similar to the pleurembolic configuration but with different musculature, as the main muscles involved differ from those of other proboscidate caenogastropods (Golding et al. 2009a).

The *intraembolic* type of proboscis is non-introvertable and housed within an external sheath (*rhynchodeum*), with the buccal mass at the base of the proboscis (Figures 19.19 and 19.20).

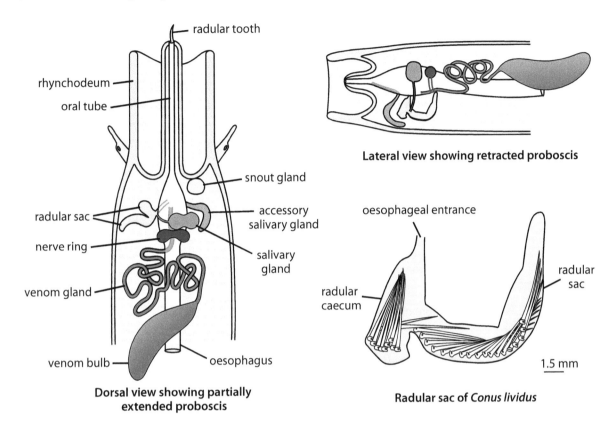

Lateral view showing retracted proboscis

Dorsal view showing partially extended proboscis

Radular sac of *Conus lividus*

FIGURE 19.20 Diagrammatic views of the anterior gut showing the intraembolic proboscis, rhynchodeum, and radular sac of a species of Conidae. Redrawn and modified from the following sources: dorsal view (Miller, J.A., *J. Molluscan Stud.*, 55, 167–181, 1989) lateral view original, radular sac (Marsh, H., *J. Molluscan Stud.*, 43, 1–11, 1977).

The intraembolic type occurs in conoideans and shows highly diverse configurations in different taxa (Figure 19.21). In some conoideans, the rhynchodeum can at least partially introvert, and this is called the *polyembolic* type (Figure 19.19), and the actual extendible part (the rhynchodeum) is not homologous with the proboscis of other caenogastropods.

In some conoideans an outgrowth from the rhynchodeum (the accessory rhyncodeal lobe) occurs which may be elaborate, as in the terebrids *Hastula bacillus* and *Myurella affinis* (Taylor 1990; Taylor & Miller 1990) (Figure 19.22), while in others, such as the horaiclavid *Horaiclavus* (Fedosov 2008) it is simpler (Figure 19.21). The origins of these structures and their functions are uncertain, and they are probably not homologous. Taylor and Miller (1990) suggested that the complex structure in *Hastula bacillus* was sensory and probably helped to locate the polychaete prey.

Another type of proboscis with an external rhynchodeum is found in some tonnoideans and has been called the '*Argobuccinum*' type (Day 1969). It is non-introvertable and is retracted only by the contraction of longitudinal muscles in its walls. This proboscis type has been named *teinoembolic* by Golding et al. (2009a). These authors suggest that it may be a derived pleurembolic proboscis with the retractor muscle function replaced by the muscular hydrostat in the proboscis wall.

Paired salivary glands open to the buccal cavity by way of short to long ducts. They are lacking in a few taxa, notably in those where the whole buccal apparatus is also lost.

In some taxa, the salivary glands and/or their ducts lie anterior to the nerve ring, but in most, they pass through the nerve ring. In neogastropods and a few other sorbeoconchs (notably Rissooidea and Truncatelloidea) they pass over the nerve ring (Ponder 1974; Ponder & Lindberg 1997). In a few acrembolic proboscidate taxa, such as epitoniids and triphoroideans, the salivary glands (and buccal mass) lie well behind the nerve ring although they are pulled through the nerve ring when the proboscis is extended during feeding. This is not always the case in taxa with an acrembolic proboscis, as in the less modified eulimids where the salivary glands (and buccal mass) are not lost (e.g., *Eulima*), they lie just behind the buccal mass and markedly anterior to the nerve ring when the proboscis is retracted or extended (Warén 1983b). In triphorids and cerithiopsids, which have long, narrow shells, the left and right salivary glands are staggered and of different sizes (Fretter 1951b).

Another kind of salivary gland, usually called *accessory salivary glands*, occur in some caenogastropods with true salivary glands. These accessory glands are not all homologous, so the different kinds are discussed separately below.

Tubular accessory salivary glands are found in some neogastropods (notably in families such as Muricidae [Figure 19.23], Costellariidae, Volutidae, and Olividae, but also in some marginellids, cancellariids, and conoideans) (e.g., Ponder 1974; Taylor et al. 1993; Strong 2003). These glands typically open by way of narrow ducts to the mid-ventral floor

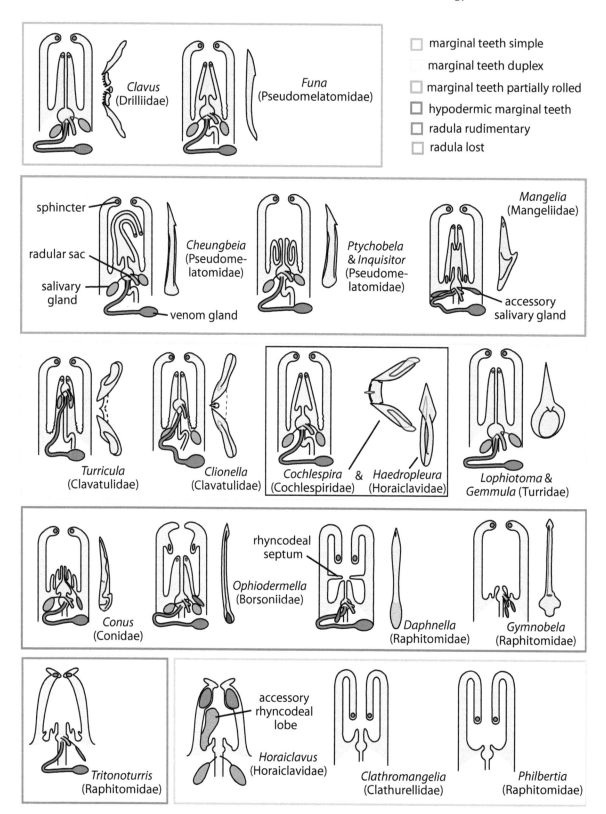

FIGURE 19.21 Foregut morphology in several of the conoidean families, with radulae shown where they are present. Figure 19.22 for Terebridae. Anterior gut redrawn and modified from Taylor, J.D. et al., *Bull. Br. Mus. Nat. Hist. Zool.*, 59, 125–170, 1993, Taylor, J.D., Foregut anatomy of the larger species of *Turrinae, Clavatulinae* and *Crassispirinae* (Gastropoda: Conoidea) from Hong Kong, pp. 185–213, in Morton, B. (ed.), *The Malacofauna of Hong Kong and Southern China, III: Proceedings of the Third International Workshop on the Malacofauna of Hong Kong and Southern China, Hong Kong*, Vol. 3, Hong Kong University Press, Hong Kong, 1994, and Fedosov, A.E., *Doklady Biol. Sci.*, 419, 136–138, 2008; radulae redrawn and modified from Kantor, Y.I. and Taylor, J.D., *J. Zool.*, 252, 251–262, 2000 and Powell, A.W.B., *Bull. Auckland Inst. Mus.*, 5, 1–184, 1966.

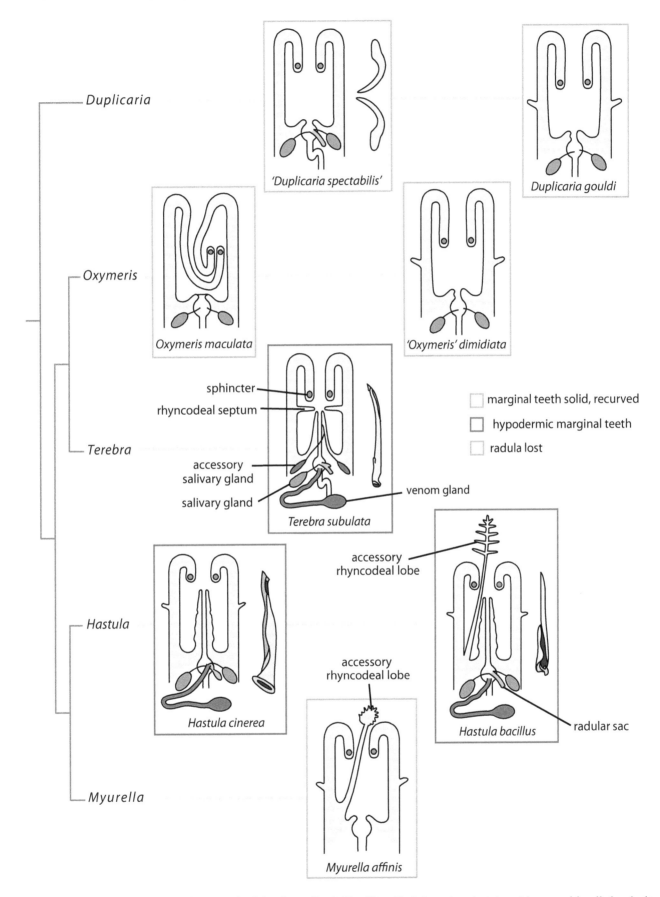

FIGURE 19.22 Variation in the anterior gut and radula of some Terebridae (Conoidea). Some taxa (not shown) have semi-inrolled or duplex marginal teeth (Castelin, M. et al., *Mol. Phylogenetics Evol.*, 64, 21–44, 2012). Redrawn and modified from Taylor, J.D., *Malacologia*, 32, 19–34, 1990.

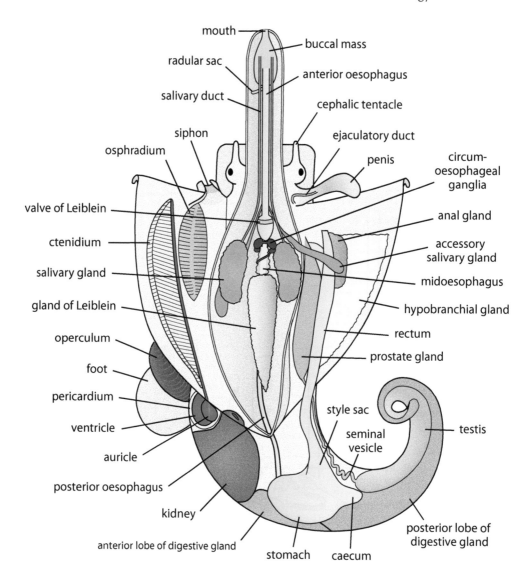

FIGURE 19.23 Generalised muricid (Neogastropoda) removed from its shell and viewed dorsally with the mantle cavity and anterior body cavity opened mid-dorsally and the proboscis extended. Redrawn and modified from Ponder, W.F., *Malacologia*, 12, 295–338, 1974.

of the buccal cavity. The rather haphazard distribution within neogastropods is not correlated with diet or mode of feeding (Andrews 1991).

Somewhat similar tubular glands are found in epitoniids (e.g., Strong 2003), and it has been suggested they are homologous with those in neogastropods (Andrews 1991) because they are similar in morphology and also open at the mouth, although Ball et al. (1997) rejected their homology, mainly because of histological differences.

Although very different in appearance to the tubular accessory salivary glands of neogastropods and epitoniids, in some tonnoideans modified sections of the salivary glands have been called accessory salivary glands (Ball et al. 1997) but are more correctly referred to as anterior and posterior lobes of the salivary glands (Barkalova et al. 2016). Ball et al. (1997) rejected homology of these so-called accessory glands of tonnoideans with those of neogastropods because of many differences, including their position and fine structure.

Oesophageal pouches (sometimes called buccal pouches, but see Chapter 5 for buccal pouches proper) are found in a few families of architaenioglossan and asiphonate caenogastropods. They are lateral pockets off the posterior-most part of the buccal cavity or, more usually, the most anterior part of the oesophagus. Such structures may be continuous with the oesophageal gland or separate from it (Ponder 1983; Sasaki 1998). In some asiphonate caenogastropods, the separate structures differ histologically from the oesophageal gland. Oesophageal pouches are typically a pair of small thin-walled or glandular sacs with narrow openings and are known in several families including Cyclophoridae and Ampullariidae (but not Viviparidae) (Amaudrut 1898; Strong 2003; Simone 2004b), *Campanile* (Houbrick 1981), at least one batillariid, *Lampanella* (Strong 2003), Littorinidae (Fretter & Graham 1962), and *Pellax* (Eatoniellidae) (Ponder 1968). In some lower caenogastropod taxa (e.g., *Eatoniella* [Eatoniellidae], *Macquariella* [Littorinidae], and *Provanna*

[Provannidae]) pouch-like structures are not, or are little, differentiated histologically from the anterior part of the oesophageal gland (Ponder 1983; Warén & Ponder 1991), or, as in *Barleeia*, these structures are present as anterior lateral pouches, and the oesophageal gland is lacking (Ponder 1983). Pouches on the anterior oesophagus have also been described in cypraeoideans (e.g., Kay 1960), and these may be homologous with oesophageal pouches (Strong 2003), but the unpaired 'mucus diverticulum' of *Erato* (Fretter 1951a) is considered a homologue of the oesophageal gland (Fretter & Graham 1962).

The oesophagus of caenogastropods, while possessing a pair of dorsal folds as found in lower gastropods, lacks conspicuous ventral folds, although reduced versions of these structures had been recognised as such in some taxa (e.g., Ponder & Lindberg 1997; Strong 2003). Ventral oesophageal folds are absent in neogastropods and some other carnivorous caenogastropods such as in the naticid and epitoniid species examined by Strong (2003). The structures interpreted by Strong as ventral folds in caenogastropods are composed of a glandular epithelium and, unlike the situation in lower gastropods, the ventral folds in caenogastropods do not extend into the midoesophagus.

An oesophageal gland (= midoesophageal gland) is plesiomorphically present, but it lacks the papillate structure seen in many vetigastropods and is often highly modified or lost altogether. The lateral pouches of the oesophageal gland are reduced or lost in some asiphonate caenogastropods, particularly in many of those with a crystalline style such as rissooideans and truncatelloideans (Graham 1939), so that in those taxa the midoesophagus is a simple tube containing only the dorsal folds in addition to irregular longitudinal ridges.

The different types of oesophageal gland are discussed in Chapter 5, with the greatest diversity seen in caenogastropods. To summarise, the main types are derived from the most basic type seen in some outgroup taxa (patellogastropods and some other basal gastropod taxa such as cocculinids and neritimorphs), which have simple lateral pouches and internal septae. This type differs considerably from the large, internally papillate gland found in many vetigastropods. Architaenioglossans lack an oesophageal gland, and it is lost in some other groups (see below).

The non-neogastropod caenogastropods have an oesophageal gland that is typically of the plesiomorphic caenogastropod type, but in neogastropods (Figure 19.23), the gland has been stripped from the oesophagus to form a gland of Leiblein or, in conoideans, a venom (= poison) gland (Ponder 1974) (Chapter 5, Figure 5.43). In one neogastropod group, the Cancellariidae, the midoesophageal gland is represented by a ventral glandular strip (Graham 1966; Ponder 1974; Strong 2003). Marginellids have a long, coiled duct that resembles the venom gland of conoideans. It is apparently derived from the stripped-off glandular dorsal folds of the oesophagus, and the gland of Leiblein is the muscular terminal sac (Ponder 1974; Ponder & Taylor 1992; Strong 2003), although this assumed homology has been questioned by Page (2011). The

marginellid and costellariid conditions provide insights into the derivation of the conoidean venom gland and may have evolved in parallel in a similar way (Ponder 1970a, 1974; Fedosov & Kantor 2010).

Blood in the sinus surrounding the gland of Leiblein is collected into the afferent renal vessel. A small terminal ampulla on the gland of Leiblein extends into the afferent renal vein in some neogastropod taxa (e.g., Muricidae, Buccinidae, Nassariidae) but not in others (Strong 2003).

The loss of the oesophageal gland has occurred in many caenogastropods where it has been correlated with the possession of a crystalline style (Graham 1939), a structure that, like the oesophageal gland, produces digestive enzymes. A crystalline style and an oesophageal gland do coexist in a few caenogastropods, as for example in some cerithioideans (Houbrick 1988).

In many neogastropods, a pyriform valve of Leiblein (Figure 19.23) lies at the junction of the anterior oesophagus and midoesophagus (Ponder 1974) and is only found in that group. A ciliated valve-like cone lies inside, with its walls comprised of a tall glandular epithelium. The function of this structure is discussed in Chapter 5. The homology of the valve with oesophageal structures in other caenogastropods has long been debated. Golding and Ponder (2010) showed that the histology of glandular dorsal folds in the anterior oesophagus in tonnoideans is very like that of cells lining the valve of Leiblein and suggested that they may be homologous.

There is a correlation between the presence or absence of the valve of Leiblein in neogastropods and the gland of Leiblein. In those having an unmodified gland of Leiblein, the valve of Leiblein is usually present, but in those taxa where the gland is lost, the valve is also absent. In taxa where the gland of Leiblein is modified as a venom gland or similar structure that opens near or into the buccal cavity, the valve is absent, as in most conoideans and a number of Marginellidae (e.g., Ponder 1970a; Strong 2003), although the latter sometimes have an oesophageal caecum which may be a rudiment of the valve of Leiblein (Strong 2003). It is present in Costellariidae, some of which have a very modified gland of Leiblein (Fedosov & Kantor 2010).

The general layout of the stomach is somewhat similar to that in vetigastropods in having a style sac, which primitively contains a mucoid protostyle, a gastric shield, and a sorting area, although the latter is simpler. The stomach is quite diverse in structure (Strong 2003) and sometimes simplified and reduced, especially in some neogastropods and eulimids.

In some vetigastropods, the gastric caecum is developed as a spiral caecum (Chapter 5, Figure 5.48), but such elaboration is absent in caenogastropods, although other caecum-like gastric structures are found (Strong 2003). In various cerithioideans, truncatelloideans, stromboideans, and calyptraeoideans a pocket lies above and behind the gastric shield near the posterior end of the stomach. In most, only crescentic ciliated folds and grooves extend from the oesophagus to enter this pocket, but in at least some calyptraeids, the truncatelloidean *Bithynia* (and in a similar convergent caecum seen in some neritoideans), the minor typhlosole (or a fold derived from it) enters (Strong 2003). Given the variation in this kind of caecum, Strong (2003 p. 510) suggested that they were probably not all homologous but comprised 'a heterogeneous assortment of structures that represent many independent modifications'. Other kinds of gastric caecum-like structures include the long posterior extension of the stomach that occurs in some littorinids (Johansson 1939; Graham 1949; Strong 2003) and the pouch-like extension of the stomach separated from the main chamber by a fold in some neogastropods (e.g., Smith 1967b; Medinskaya 1993; Strong 2003).

A well-developed, ventrolaterally positioned gastric shield is plesiomorphically present but is lost in many taxa, including naticoideans, cypraeoideans, epitoniids, and most neogastropods (Ponder & Lindberg 1997; Strong 2003). The gastric shield is supported and secreted by a 'glandular pad' which is unusually conspicuous in cerithioideans (Strong 2003).

A crystalline style has independently evolved in members of several groups of caenogastropods, including the littorinoidean Pomatiidae, some cerithioideans, stromboideans, vermetids, calyptraeids, rissooideans, truncatelloideans, and even in a few members of the neogastropod Nassariidae, such as *Cyclope* (Morton 1960) and *Tritia* (Strong 2003). It would appear that a crystalline style can evolve from a protostyle in response to more or less continuous microphagous or suspension feeding when proteolytic enzymes are otherwise lacking in the gut (Yonge 1930, 1932; Graham 1939). Irrespective of the presence or absence of a style, the style sac is usually a prominent feature of the caenogastropod stomach although it is reduced or absent in carnivorous taxa, including all cypraeoideans (Simone 2004a). Despite the diversity of feeding habits seen in cypraeoideans, their stomach morphology may indicate that carnivorous feeding is plesiomorphic in that taxon.

Pyloric caecae have been described in architaenioglossans, cypraeoideans, and possibly velutinoideans (Strong 2003). These lie beneath the style sac and open to the intestinal groove near the anterior (distal) end of the style sac.

While many caenogastropods have both digestive gland lobes, the anterior lobe is often reduced. The digestive gland opens by way of one or two, to many, openings to the stomach.

The caenogastropod intestine differs from that of vetigastropods in not penetrating the pericardium and not being markedly looped, except in some architaenioglossan taxa. It typically runs posteriorly along the style sac and then abruptly turns to run anteriorly but is straight in some carnivorous taxa. A typhlosole is found only in the proximal part of the intestine. The course of the rectum along the right mantle roof is usually straight, but in some taxa, it may form one or a few loops. The anus opens anteriorly in the mantle cavity. Faecal pellets rather than strings are produced.

Rectal (= anal) glands are found in a few caenogastropods, but the most studied are those in some neogastropods (see Chapter 5 for details).

19.3.5 RENOPERICARDIAL SYSTEM

Caenogastropods have a single (left) kidney and a single (left) auricle. Details of these systems are provided in Chapter 6. The possession of a single kidney necessitates it dealing with nitrogenous excretion in addition to its usual functions (Andrews 1985) (see Chapter 6). In non-marine taxa, the kidney is involved in resorption of water and solutes or even the storage of nitrogenous waste. A nephridial gland is also often present (see Chapter 6 for details).

19.3.6 REPRODUCTIVE SYSTEM

Apart from a few protandric taxa and a couple of simultaneous hermaphrodites, caenogastropods are dioecious, and all fertilise internally. In both sexes, the single proximal gonad (the post-torsional right as in other gastropods) is connected by a duct to the pallial gonoduct (a pallial, usually glandular, oviduct in females and a prostate gland usually present in males) which is formed from ectoderm of the mantle. The distal part of this duct is a short renal section, a remnant of the post-torsional right kidney. A gonopericardial duct may or may not be present. A penis is usually present, although a few groups, such as the cerithiimorphs, Eatoniellidae, epitonioideans, and triphoroideans, lack this structure, and some of these transfer sperm by way of spermatophores. A pallial sperm groove, or if this is closed over, a sperm duct (pallial vas deferens), transfers sperm from the prostate to the penis.

Protandry has been recorded in relatively few caenogastropods. These include *Campanile* (Houbrick 1981), epitonioideans, Calyptraeidae and some Eulimidae (Hoagland 1978), *Hipponix* (Yonge 1953c), the littorinid *Mainwaringia rhizophila* (Reid 1986a), the pelagic epitoniids *Janthina* and *Recluzia* (Lalli & Gilmer 1989), the stromboidean *Aporrhais pespelecani* (Johansson 1948), the tornid *Cyclostremiscus beauii* (Bieler & Mikkelsen 1988), the assimineid *Rugapedia* (Fukuda & Ponder 2004), some Triviidae (Gosliner & Liltved 1982, 1987b), and the mangeliid (Conoidea) *Propebela turricula* (Smith 1967a). Some vermetids are also protandrous while one species is known to change from male to female and then reverts to male (Calvo & Templado 2005). A species of the campaniloidean genus *Plesiotrochus* has been shown to be a simultaneous hermaphrodite (Houbrick 1990), as is *Janthina pallida* (Calabrò et al. 2018).

A few species are typically parthenogenic, including *Melanoides tuberculata* (Thiaridae), *Campeloma* spp. (Viviparidae), and *Potamopyrgus antipodarum* (Tateidae). At least some populations of all these taxa have occasional to very rare males (Schalie 1965; Berry & Kadri 1974; Wallace 1992).

In caenogastropods, internal fertilisation is correlated with the formation of complex pallial genital ducts which enable the encapsulation of eggs so that development can take place within the capsule, resulting in the release of veligers or crawl-away young. Nektonic larvae are either planktotrophic or lecithotrophic, with some of the former able to survive in the plankton for more than a year (Strathmann & Strathmann 2007) (see Section 19.4.4.1.2). Planktotrophic taxa produce a characteristic, and sometimes elaborately sculptured, secondary larval shell, protoconch II (see Section 19.3.1).

19.3.6.1 The Female System

The female system consists of an ovary in the distal part of the visceral mass, from which the eggs are transported by a thin-walled tube, the upper oviduct. This oviduct opens to an often short, thicker-walled section derived from the right kidney rudiment, sometimes connected to a renopericardial canal and thus termed the renal oviduct. The upper and renal oviducts lie in the visceral part of the animal, and the latter opens to the pallial oviduct which lies, as its name suggests, either entirely or mainly in the right side of the mantle cavity roof. Examples of the lower visceral and pallial parts of the female system are shown in Figures 19.25 and 19.26. The pallial oviduct typically has an upper albumen gland (which is often embedded, at least in part, in the viscera) and a lower capsule gland. The opening of the pallial oviduct, which is usually (but not always) anteriorly located, is where the encapsulated eggs are released and, in many caenogastropods, also serves for the reception of the penis (or spermatophore) during mating.

Typically, two kinds of sperm sacs are associated with the oviduct: The seminal receptacle and the bursa copulatrix. There is, however, considerable variation in the relationships of these sperm sacs to one another and the oviduct and, in a few taxa, one of the sperm sacs may be absent. Given their diverse locations, the sperm sacs, particularly the bursae, are not homologous in all caenogastropods.

During copulation, sperm are deposited in the sac-like copulatory bursa (bursa copulatrix) or, in those taxa using spermatophores, in the spermatophore bursa. The bursae can be located anteriorly or posteriorly. The posterior bursa may be derived from the distal renal oviduct and located posterior to the mantle cavity in the visceral mass. In such cases, the sperm may travel to the bursa via a ventral channel in the pallial oviduct or via separate duct(s) that open near the oviduct opening, as in the littorinids *Bembicium* and *Risellopsis* (Reid 1988). In some taxa, the duct leading to the bursa has become a separate tube that opens separately to the mantle cavity. Such modifications are evident in some Rissooidea (Barleeiidae) and Truncatelloidea where this tube may be derived from the ventral channel (Figure 19.24) (as seen in some Pomatiopsidae, some Cochliopidae and Geomelaniinae [Truncatellidae]) (Davis 1979; Ponder 1988; Hershler & Thompson 1992; Hershler & Ponder 1998), with this condition arising several times independently. The Pomatiopsidae have particularly diverse arrangements for sperm collection. In some the duct opens via a separate short tube to the posterior end of the mantle cavity (as in a number of Pomatiopsinae and Cochliopidae), but a few members of another pomatiopsid subfamily, Triculinae, have an opening in the pericardial wall or the kidney (also in Truncatellidae) or via a sperm duct that passes through the kidney wall (Davis et al. 1976; Davis 1979; Davis et al. 1983, 1984). In these latter two cases, the sperm enters the female system via a gonopericardial duct or a renogonadial duct respectively. Copulation via the kidney

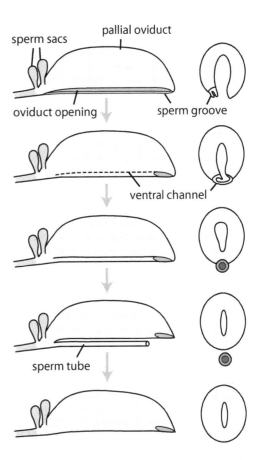

FIGURE 19.24 A generalised scenario for the evolution of the sperm tube and its loss in some truncatelloidean and rissooidean gastropods. Original.

opening occurs in some truncatelloideans (Ponder 1988) and the rissoid *Benthonella* (Ponder 1985). Copulation via the kidney is derived in caenogastropods, not a 'primitive feature' as sometimes suggested (e.g., Barker 2001a).

Typically, the bursa not only functions for the reception of sperm during mating but also resorbs excess gametes. Sperm is housed only temporarily in the bursa, from where it is transferred to one (sometimes two or several) seminal receptacle(s). Seminal receptacle(s) are typically thin-walled sacs either derived from, or closely associated with, the renal oviduct, although in some taxa they emerge from the albumen gland. These structures store orientated sperm long-term. In taxa which lack a distinct seminal receptacle, the renal oviduct itself may store sperm.

The eggs are fertilised by the sperm from the seminal receptacle(s) before they pass through the albumen and capsule gland where they receive nutrients and protective coatings before being laid as spawn (see Section 19.4.4.1.1).

During ontogeny, the pallial oviduct develops as an open groove. In most caenogastropods this then closes to form the glandular duct, but in some, notably most cerithioideans, vermetids, epitonioideans and eatoniellids (Cingulopsoidea), and some triphoroideans, it remains open when mature, and this is considered the plesiomorphic condition. Open pallial oviducts occur in a few taxa in groups that normally have a closed oviduct. For example, *Pomatias* (Pomatiasidae) is open, unlike the condition in the related Littorinidae, and some cyclophorids and rissooideans and the truncatelloidean *Ascorhis* also have open female genital ducts (Ponder & Clark 1988).

The closure of the originally open duct during evolution has occurred independently in many lineages and, not unexpectedly, this has happened in slightly different ways. In some, the closure has occurred so that the opening is located posteriorly, as seen in several groups such as in the truncatelloidean Tornidae and Iravadiidae, but in most caenogastropods the opening is terminal or subterminal and may be a small pore or a larger slit.

The capsule or jelly gland forms the anterior part of the glandular oviduct and is always on the right side of the mantle cavity. In those caenogastropods that produce jelly egg strings rather than capsules, the anterior region is a jelly gland, while in those that produce capsules it is known as the capsule gland and usually comprises more than one distinct glandular region. The relative importance of the different glandular regions is determined by the reproductive strategy (ovoviviparity or viviparity), structure of the capsule, and, if ovoviviparous, as in the great majority of caenogastropods, the nature of the spawn.

In nearly all caenogastropods the oviduct glands form a simple glandular tube, which is often laterally compressed, but in a few groups, it is more complex (Figures 19.25, 19.26). This complexity is particularly evident in the littorinids, where both the albumen and capsule glands can be complexly folded and coiled (Reid 1986b, 1989). The glandular epithelium comprising the oviduct glands is primitively epithelial but becomes more complex and subepithelial in the more derived taxa. The simple epithelial glandular tissue of the oviduct glands is composed of simple, elongate glandular cells while in those with more complex histology, the gland cells are in invaginated clusters. Only the simple glandular type is found in Rissooidea, and both kinds are found in the Truncatelloidea (Ponder 1988) and Littorinoidea (Reid 1989), but most other groups have not been sufficiently investigated.

The sperm sacs usually comprise one or more seminal receptacles and a bursa copulatrix. In some neogastropods there is also an ingesting gland that is probably derived from the posteriorly located bursa copulatrix seen in many 'lower' caenogastropods. The ingesting gland, as its name suggests, ingests excess sperm and yolk (Fretter & Graham 1962). In some neogastropods an anterior bursa has evolved to receive sperm during copulation (Figure 19.26).

A so-called 'ventral pedal gland' is found in the foot sole of females of many neogastropods and velutinoideans (e.g., Fretter & Graham 1962 figure 68) but is not homologous in the two groups. In neogastropods, it is used to assist in moulding and attaching the soft and partially formed egg capsule that is passed to it from the opening of the oviduct. The 'gland' lies mid-ventrally in the anterior part of the sole and is a deep cavity with folded walls which are ciliated but not glandular. The equivalent structure in the Velutinoidea receives the egg capsules from the oviduct which it forces into cavities premade (by the radula) in the compound ascidians on which they feed. It does this with the aid of a ramrod-like muscular central

core in the 'gland' (Fretter & Graham 1962), a structure not seen in neogastropods.

Spawn is described in Section 19.4.4.1.1 and larvae and larval development in Section 19.4.4.1.2, with more general information provided in Chapter 8.

19.3.6.2 The Male System

The male system in most caenogastropods is simple, the visceral parts consisting of the testis and a vas deferens which contains a sperm storage area, the seminal vesicle. The testis may consist of a simple sac or discrete lobes, and, in some taxa, the arrangement of the lobes can be quite complex. The pallial component of the male system consists of a prostate gland and, in many taxa, a penis.

The plesiomorphic condition for caenogastropods is aphally – that is, lacking a copulatory organ. This condition is still found in several groups (Campaniloidea, Cerithioidea, Cingulopsoidea, Triphoroidea, and Epitonioidea) most of which transfer sperm by way of spermatophores (see below), although this is not recorded in Cingulopsoidea. The primitive condition (Figure 19.27) is for an open groove and open prostate gland as seen in many cyclophoroideans, most littorinids, some rissooideans, and even some neogastropods. In some taxa with a closed duct and sperm tube, the line of closure is indicated by a strip of epithelial cells.

If a penis is present, a seminal groove or duct (sperm duct or pallial vas deferens) runs from the prostate to the penis, and similarly a groove or duct extends from the base of the penis to its tip. Closed sperm ducts have evolved independently in many caenogastropod lineages, including in some cyclophoroideans. In the more advanced state, the sperm tube, and in some the penial duct, are surrounded by muscles to form an ejaculatory duct. Where muscles are absent, the movement of sperm is aided by cilia.

In most hypsogastropods the penis (sometimes called a 'verge') is on or near the right side of the head, but in the architaenioglossan groups it is more variable, being derived from the right cephalic tentacle in Viviparoidea, the mantle in Ampullarioidea, and the head in Cyclophoroidea. The hypsogastropod penis typically has a broad basal portion and a slender distal part. This structure elongates during copulation, and its distal portion enters the female reproductive tract and, typically, deposits sperm in or near the bursa copulatrix of the female.

Campaniloids, cerithiimorphs, cingulopsoideans, vermetoideans, triphoroideans, and epitonioideans all lack a penis, as do some pupinids and diplommatinids (Cyclophoroidea). These taxa either have spermatophores (see below) or, in triphoroideans and epitonioideans, sperm is carried to the female on large transport parasperm (*spermatozeugma*).

Because all caenogastropods practise internal fertilisation, their sperm is modified accordingly (see Chapter 8), being structurally complex and usually dimorphic (Healy 1988, 1996b; Buckland-Nicks 1998). Thus, euspermatozoa often occur with non-fertile sperm (paraspermatozoa), except in a few groups including Eulimoidea, Naticoidea, Rissooidea, and Truncatelloidea. In some groups the parasperm are large

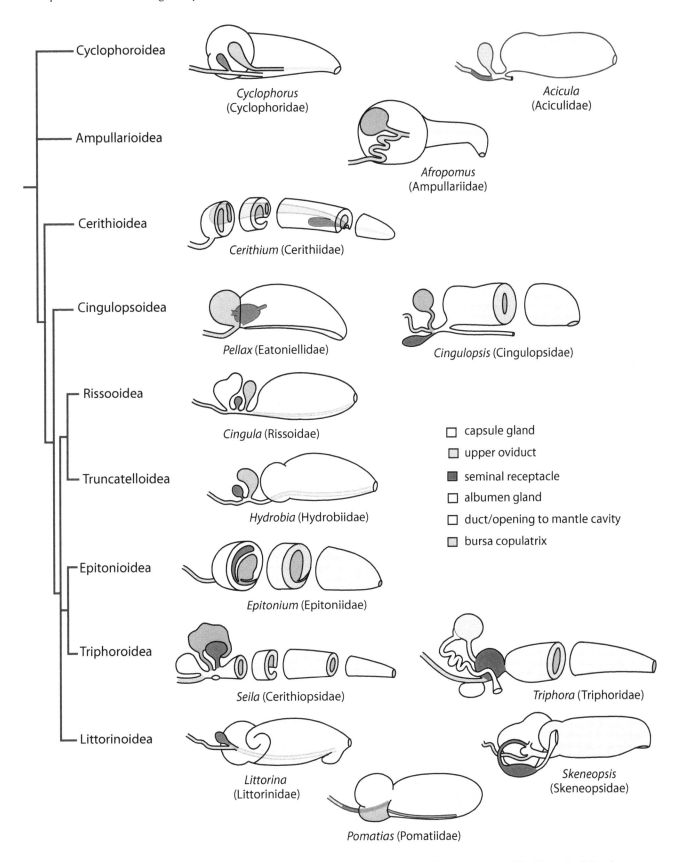

FIGURE 19.25 Some examples of lower caenogastropod female reproductive systems. Redrawn and modified from the following sources: *Cyclophorus* (Tielecke, H., *Archiv für Naturgeschichte*, 9, 317–371, 1940), *Afrodromus* (Berthold, T., *Zoomorphologie*, 108, 149–159, 1988), *Cerithium, Seila* and *Triphora* (Houston, R.S., *J. Molluscan Stud.*, 51, 183–189, 1985), *Pellax* (Ponder, W.F., *Records of Dominion Museum Wellington*, New Zealand, 6, 61–95, 1968), *Skeneopsis* and *Cingulopsis* (Fretter, V., *Proc. Linn. Soc. Lond.*, 164, 217–224, 1953), remainder (Fretter, V. and Graham, A.L., *British Prosobranch Molluscs: Their Functional Anatomy and Ecology*, Ray Society, London, 1962).

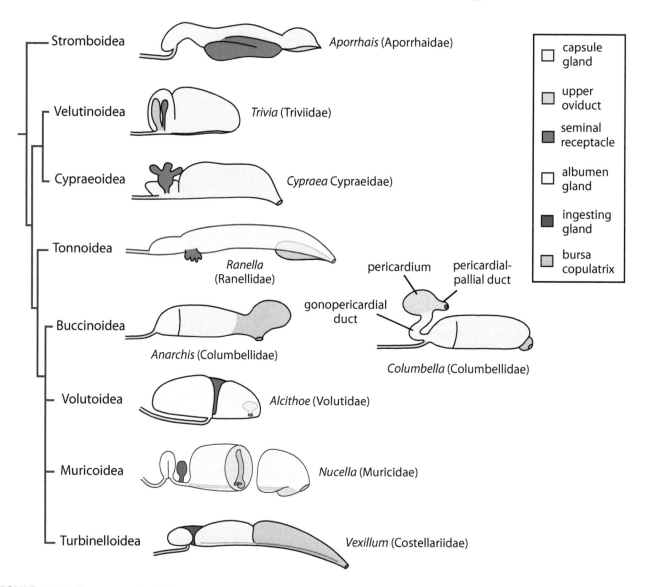

FIGURE 19.26 Some examples of higher caenogastropod female reproductive systems. Redrawn and modified from the following sources: *Aporrhais* (Morton, J.E., *Trans. R. Soc. N.Z.*, 78, 451–463, 1950), *Trivia* (Gosliner, T.M. and Liltved, W.R., *Zool. J. Linn. Soc.*, 74, 111–132, 1982), *Cypraea* (Kay, A.E., *International Revue der gesamten Hydrobiologie*, 45, 175–196, 1960), *Mayena* (Laxton, J.H., *Zool. J. Linn. Soc.*, 48, 237–253, 1969), *Columbella* and *Anarchis* (deMaintenon, M.J., *Invertebr. Biol.*, 118, 258–288, 1999), *Nucella* (Fretter, V. and Graham, A.L., *British Prosobranch Molluscs: Their Functional Anatomy and Ecology*, Ray Society, London, 1962), *Alcithoe* (Ponder, W.F., *J. Malacol. Soc. Aust.*, 2, 55–81, 1970b), *Vexillum* (Ponder, W.F., *Malacologia*, 11, 295–342, 1972).

and highly modified, while in others they differ only subtly from the eusperm. The parasperm can have distinctive morphologies which are discernible using a compound microscope (Nishiwaki 1964; Tochimoto 1967) (see Chapter 8, Figure 8.10 and Table 8.1).

There are two main types of eusperm in caenogastropods with one being found in the more primitive taxa (the architaenioglossan groups and Cerithiimorpha) and the other in the remaining caenogastropods, including the neogastropods (Healy 1988, 1996b, a, 2000). Structural details of euspermatozoa are obtained by ultrastructural investigations using TEM. The structure of the midpiece of the eusperm can be particularly informative, as it can differ in the number and arrangement of periaxonemal mitochondria and the structure of the mitochondrial cristae (plates) or their derivatives.

The ultrastructural data obtained can be very informative in determining the taxonomic position of a taxon, as some groups have very characteristic morphology (e.g., Healy 1988). For example, cerithioideans have four straight mitochondria in which the cristae are complex and parallel and lack a segmented sheath. In contrast, campaniloideans have seven to eight straight mitochondria with unmodified cristae, which are partly enclosed in a sheath. The architaenioglossan cyclophoroideans have structures intermediate between these two but, unlike campaniloideans, lack a sheath (Ponder et al. 2008).

Spermatophores (see Chapter 8) are known in Cerithioidea, Vermetoidea, Pterotracheoidea ('heteropods'), Triphoroidea (Robertson 1989), and some Cyclophoroidea (Barker 2001a) and have evolved independently in each group (Robertson

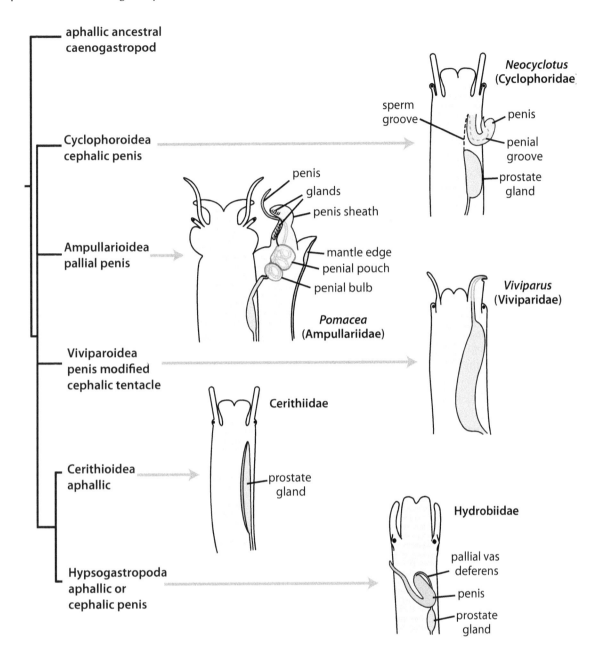

FIGURE 19.27   Some examples of lower caenogastropod male reproductive systems. Redrawn and modified from the following sources: *Neocyclotus* (Simone, L.R.L. de, *Morphology and Phylogeny of the Cypraeoidea (Mollusca, Caenogastropoda)*, Papel Virtual Editora, Rio de Janeiro, 2004b), *Pomacea* (Andrews, E.A., *Proc. Malacol. Soc. Lond.*, 36, 121–140, 1964; Gamarra-Luques, C. et al., *Biocell*, 30, 345–357, 2006), *Viviparus* (Vail, V.A., *Malacologia*, 16, 519–540, 1977), Cerithiidae original, Hydrobiidae (*Boetersiella*) (Arconada, B. and Ramos, M.A., *J. Nat. Hist.*, 35, 949–984, 2001).

1989). The spermatophore is secreted by the modified prostate gland and encases the eusperm. The whole structure is thus a sperm packet that can be passed to the female. Except for cyclophoroideans and Pterotracheoidea, the spermatophore-producing caenogastropod taxa lack a penis.

Apart from a few aphallic taxa (see above), most cyclophoroideans have a penis located behind the right cephalic tentacle or in the midline of the head (Figure 19.27), a configuration also seen in many hypsogastropods. Simone (2004b) considered the cyclophoroidean penis homologous with that of hypsogastropods, but, given the phylogenetic relationships within

caenogastropods, these structures are more probably convergent. An open penial groove is seen in many cyclophoroideans, but in some it is closed, and there may be an intromittent filament which, in a few, is retractile. While the features of cyclophoroidean penes are similar to those seen in many hypsogastropods, other architaenioglossans have evolved different copulatory organs, as outlined below.

In viviparids, the pallial vas deferens differs from that of other architaenioglossans in being a thick muscular duct, and the right cephalic tentacle is modified as a copulatory organ (Figure 19.27), with some having a retractile tip (Simone

2004b). Convergent modification of the right tentacle as a copulatory organ has occurred in cocculinoideans.

Ampullariids have a unique and complex copulatory structure of pallial origin (Figure 19.27). The narrow pallial vas deferens opens to a penis bulb near the anus, but there is no direct connection with the penis. The penial pouch contains a slender, convolute penis within which runs a penial duct in some genera or an open groove in others. When protruded, the penis is protected (and perhaps supported) by a penis shield which lies at the mantle edge. This latter structure can be very large, sometimes more than half the volume of the mantle cavity, and can contain one or more glands (Berthold 1989; Simone 2004b).

In most hypsogastropods the penis is behind the right cephalic tentacle or towards or in the midline of the head. Modifications of the penis include the development of lobes and/ or glands and the formation of a terminal copulatory papilla, stylet, or filament. The penial glands are sometimes large and aid in maintaining the penis in position during copulation.

Some evidence indicates that the penis in hypsogastropods has evolved several times from aphallate ancestors, as suggested by some basal clades being aphallate and also by the different innervation of the penis. For example, the penial nerve originates from the right pedal ganglion in littorinoideans, rissooideans, and truncatelloideans, except for Anabathridae and Emblandidae, where the penial nerve originates from the cerebral ganglion (Ponder 1988) (Figure 19.28).

Penial glands, if present, can be complex and may be internal or external. External glands range from glandular fields or ridges, globular 'apocrine' glands, and glandular papillae to mammiform glands, the latter being particularly complex and well-developed in some Littorinidae (Reid 1989). The internal glands can be tubular, and, in the truncatelloidean families Bithyniidae and Amnicolidae, they extend into the haemocoel of the head (Hershler & Ponder 1998). Penial glands are often used taxonomically, especially in littorinids and some truncatelloidean families (Hershler & Ponder 1998), particularly the Cochliopidae (Hershler & Thompson 1992).

19.3.7 Nervous System and Sense Organs

Compared to 'lower' gastropods, the nervous system of caenogastropods is concentrated, with prominent cerebral and pedal ganglia. Except for some architaenioglossans which are hypoathroid or partially so, caenogastropods are epiathroid (the pleural ganglia lie close to, or are fused with, the cerebral ganglia). Most caenogastropods are streptoneurous, although the visceral nerve ring is generally more compact than in vetigastropods.

The most plesiomorphic caenogastropod nervous system is seen in the Cyclophoroidea, which have a hypoathroid nervous system and a streptoneurous visceral loop, with the pedal ganglia having cords that extend through the foot and are cross-connected by many commissures. The Ampullariidae and the ampullinid *Cernina* are also hypoathroid (Kase 1990), but the Viviparidae are closer to epiathroid (Van Bocxlaer & Strong 2016), and all other caenogastropods are epiathroid.

The architaenioglossan central nervous system appears to be primitive in other respects – the cerebral ganglia are widely separated, as are most of the other ganglia (as in vetigastropods), and, at least in *Viviparus*, neurosecretory cells are dispersed through the ganglia and nerve cords (Gorf 1961).

Nearly all other caenogastropods (other than cyclophorids and cypraeids) lack pedal cords, these having become compacted as large ganglia usually connected by a single short commissure, although a second thin commissure is sometimes present. A unique feature is that the pedal ganglia give off numerous nerves anteriorly whereas other gastropods have only a few (Ponder & Lindberg 1997). In some taxa (e.g., Pomatiopsidae; Davis 1967), two pairs of small ganglia lie in front of the pedal ganglia, the upper *propodial* and lower *metapodial* ganglia, but in most caenogastropods these ganglia are fused with the pedal ganglia.

In many caenogastropods, cross-linked nerves presumably impart more efficient neural coordination. Many, even some cyclophoroideans, exhibit zygoneury (see Chapter 7).

Concentration of the nervous system has sometimes been related to small body size, but this does not necessarily follow. There are many variations in the relative length of major connectives, even in related taxa. For example, the right pleurosupraoesophageal connective is long in Littorinidae, some Hydrobiidae and Tateidae but shortened in Pomatiopsidae and is short in Bithyniidae, Assimineidae, and Truncatellidae. In many 'higher' caenogastropods both the 'oesophageal' ganglia lie close to, or are fused with, the pleural ganglia. Besides these changes, shortening of the visceral loop occurs in many caenogastropods, culminating in a euthyneurous condition in a few. This shortening is often associated with small body size, as in some rissoids, assimineids, cingulopsids, and eulimids, while in others, such as calyptraeids and most neogastropods, ganglionic fusion occurs, but a long, streptoneurous visceral loop is retained. Species of *Trivia* exhibit a euthyneurous visceral loop, despite this being elongate (Gosliner & Liltved 1982, 1987a, b).

A single nerve in each cephalic tentacle is found in cyclophoroideans, viviparoideans, and at least some ampullariids, but many other caenogastropods have a bifid tentacle nerve (Haszprunar 1988f; Ponder & Lindberg 1997), although this is by no means always the case (Strong 2003).

The eyes of all caenogastropods are well-developed and have a cornea covering the lens. They typically lie at the outer bases of the cephalic tentacles, but in neogastropods and a few other groups, notably Strombidae, they have migrated up the tentacle bases, although they always lie on the outer side of the tentacles.

In most caenogastropods, the paired statocysts each contain a single statolith but some primitive taxa show the plesiomorphic gastropod condition with multiple statoconia.

One of the most important sensory organs in most caenogastropods is the osphradium, the comparative morphology and ultrastructure of which has been investigated in some detail (Haszprunar 1985g; Taylor & Miller 1989).

The osphradia of architaenioglossans differ from those of other caenogastropods. In the Ampullarioidea it is short and

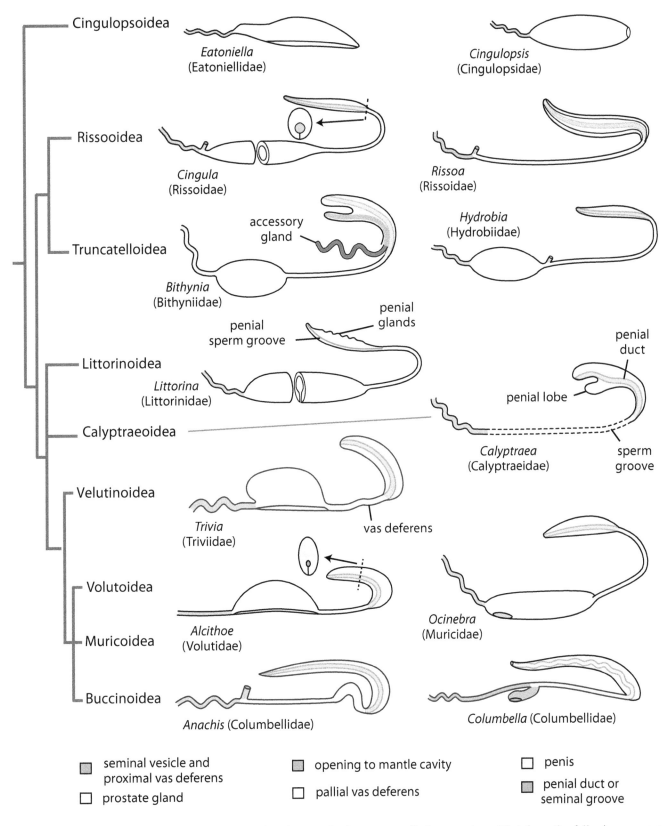

FIGURE 19.28 Some examples of hypsogastropod male reproductive systems. Redrawn and modified from the following sources: *Eatoniella* (Ponder, W.F., *Records of Dominion Museum Wellington, New Zealand*, 6, 61–95, 1968), *Trivia* (Gosliner, T.M. and Liltved, W.R., *Zool. J. Linn. Soc.*, 74, 111–132, 1982), *Alcithoe* (Ponder, W.F., *J. Malacol. Soc. Aust.*, 2, 55–81, 1970b), *Anachis* (Marcus, E., *Boletim, Faculdade se Filosofia, Ciencias e Letras da Universidade de Sao Paulo*, 260, 25–66, 1962), *Columbella* (Houston, R.S., *The Veliger*, 19, 27–46, 1976), remainder (Fretter, V. and Graham, A.L., *British Prosobranch Molluscs: Their Functional Anatomy and Ecology*, Ray Society, London, 1962).

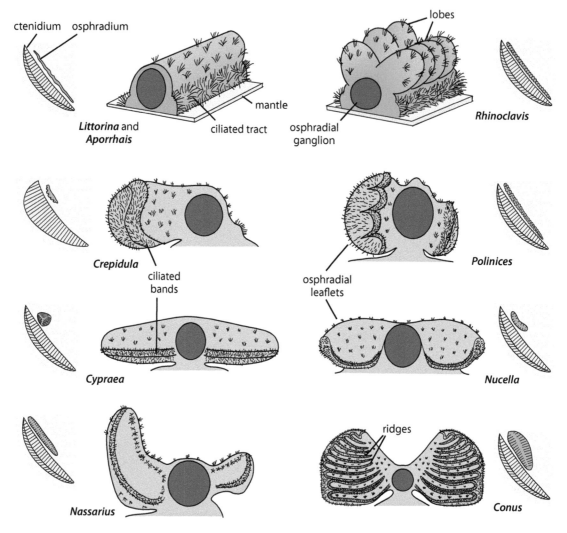

FIGURE 19.29 Some examples of caenogastropod osphradia showing the diversity of the lateral ciliated fields. Redrawn and modified from Taylor, J.D. and Miller, J.A., *J. Molluscan Stud.*, 55, 227–237.

broad while in Viviparoidea it forms a narrow ridge. In both these taxa, the osphradium is composed of a sensory epithelium overlying the osphradial nerve and ciliated lateral zones, with the lateral zones uniquely forming deep pits in viviparids and infoldings in ampullariids, the latter resulting in a gill-like structure (Haszprunar 1985g). Some cyclophoroideans have lost the osphradium, and in others it is small to rudimentary, but no detailed structural or ultrastructural information is available. In *Campanile* the osphradium is bipectinate, with a unique histology (Haszprunar 1992a).

Other caenogastropods exhibit considerable diversity of form in the shape and appearance of the osphradium (Figure 19.29), but all possess a unique ultrastructure (Haszprunar 1985g). The sensory epithelium overlays the osphradial nerve forming a raised central (= sensory) zone, and this is bordered by the ciliated lateral zone on each side. The lateral zone has two unique cell types, the microvillus Si1 cells and ciliated Si2 cells (terminology from Welsch & Storch 1969; Haszprunar 1985g, 1988f). The outer parts of the central zone have another unique type of cell, the Si4 cells. These, together

with the Si1 and Si2, are diagnostic of sorbeoconch caenogastropods (Ponder & Lindberg 1997).

An interesting transitional series linking the structure of the osphradium to habitat was described in European littorinids by Haszprunar (1985g). In lower shore species the ciliated ridges surrounding the osphradium are symmetrical, but in higher shore species the left ridge is reduced, and the enlarged right forms a groove. In the terrestrial littorinoidean *Pomatias*, the left ridge is lost and the right is enclosed in a deep groove containing large mucous cells, presumably to prevent drying.

Major elaborations of the osphradium occur in some caenogastropods (Figure 19.29). In many lower taxa, the osphradium forms a simple, narrow ridge, but in others such as the cerithiid *Rhinoclavis*, the central zone is elaborated into alternating lobes while the lateral zone remains simple (Taylor & Miller 1989). In others, such as many rissooideans and truncatelloideans, the osphradium has a wide central zone bordered by ciliated ridges which, in some small taxa at least, probably aid in producing respiratory currents. In several, mostly carnivorous, groups of hypsogastropods the osphradium develops lateral leaflets resulting in monopectinate, bipectinate, or even tripectinate osphradia. Unlike the situation in cerithioideans, these leaflets are composed of both the central and lateral zones. Cypraeids are unique in having a tripectinate osphradium, and despite their different diets this osphradial morphology is found in all species which range from grazing carnivory to herbivory. In neogastropods and other predatory caenogastropods (Naticoidea, Tonnoidea), the osphradium is bipectinate with numerous well-developed leaflets. Elaborate secondary folding on the large leaflets has occurred in the predatory Conidae. In the suspension-feeding 'slipper limpet' *Crepidula* the osphradium is asymmetrically developed and essentially monopectinate with short leaflets on the right side and the left side remains simple or has only two or three short leaflets (Taylor & Miller 1989).

Studies on the sensory functions of caenogastropod cephalic tentacles are long overdue. These tentacles are often long and in some taxa, highly mobile. Their epithelium undoubtedly contains at least simple sensory cells, and in some there are stationary cilia, sometimes long and compound, that may also have a sensory function.

19.4 BIOLOGY, ECOLOGY, AND BEHAVIOUR

19.4.1 HABITS AND HABITATS

Caenogastropods occur in virtually all marine, estuarine, freshwater, and terrestrial habitats. Most are marine, and their diversity is highest in the tropics and nearshore, being reduced towards the poles and in the deep sea. Several groups have independently radiated in fresh water, notably some families of the Truncatelloidea (Hydrobiidae and several related families including Bithyniidae and Pomatiopsidae), some Littorinidae, notably the genus *Cremnoconchus* in India (Reid et al. 2013), Cerithioidea (Thiaridae and a few other families), and the architaenioglossan Viviparidae and Ampullariidae. Freshwater truncatelloideans include a number of the smallest caenogastropods, with some living in caves or interstitially in groundwater while marine interstitial species live in sandy sediments. Freshwater caenogastropod taxa often include species having very narrow ranges with resulting conservation issues (see Chapter 10). Of the terrestrial taxa, the architaenioglossan cyclophoroideans and the littorinoidean Pomatiidae are the largest groups, but there are also significant numbers of terrestrial taxa in the truncatelloidean Omphalotropinae (Assimineidae) and Truncatellidae. These taxa typically live in forests on the ground in moisture-retaining microhabitats such as leaf litter, under or in dead wood, or in moss. Some are arboreal, living on tree trunks or foliage or epiphytes. Others are obligate calcicoles confined to limestone rock, and, sometimes in the Cyclophoroidea, there are spectacular radiations in these typically isolated habitats.

Most marine caenogastropods are epifaunal, but several groups are burrowers (e.g., Naticidae, Terebridae, Olividae, Struthiolariidae), and two groups are pelagic (see below). While there are some limpet-like families (notably Hipponicidae and Calyptraeidae), these are mainly sublittoral, as there are no intertidal caenogastropod limpets similar to patellogastropods or siphonariids. Nevertheless, some intertidal or shallow sublittoral snails have become effectively limpet-like by increasing their aperture size and developing a clamping foot (e.g., various muricid whelks, notably the Chilean 'loco' *Concholepas concholepas*). A few limpet-like taxa such as capulids, *Thyca* (Eulimidae), and Pediculariidae are attached to their hosts.

Eulimids are external or internal parasites on echinoderms, with some internal worm-like taxa having substantially reduced their anatomy and lost the adult shell (e.g., *Entoconcha*) (Figure 19.33), the only caenogastropods to do so besides some heteropods (Pterotracheidae). Reduced internal shells are seen in the only truly slug-like caenogastropods, the Velutinidae, although some Naticidae have somewhat slug-like bodies and reduced shells completely enclosed within the crawling animal. Several groups have the animal (mantle lobes or lobes from the foot) extended over the shell when they are active. Developing such lobes is particularly evident in the burrowing Naticidae and Olividae. Mantle lobes are also seen in some Volutidae and Marginellidae, not all of which burrow, and the non-burrowing Cypraeidae, Ovulidae, and Triviidae. Where the mantle covers the shell, as in cypraeids, some volutes, and others, there is a thin glaze secreted giving the shells of these taxa a shiny surface.

While most caenogastropods are mobile, some have adopted stationary lifestyles, being embedded in corals or sponges (e.g., some Coralliophilinae, some Siliquariidae, and the turritellid *Vermicularia*) or their echinoderm host (some parasitic Eulimidae). The tube worm-like vermetids have evolved the ability to cement their shell to the substratum. This ability is unknown in other caenogastropods except for a very poorly known small freshwater *Helicostoa* of uncertain affinities from China which cements its shell to the substratum (Pruvot-Fol 1937). The limpet-like hipponicids have a shelly plate, attached to the substratum, which they secrete with their foot (e.g., Yonge 1953c; Knudsen 1991).

Some benthic hypsogastropod caenogastropods gave rise to pelagic taxa. The surface-drifting epitoniid genera *Janthina* and *Recluzia* float on the surface of the oceans of the world grazing on drifting cnidarians. They are derived from benthic taxa (Churchill et al. 2011), as are the 'heteropod' pterotracheoideans. These active swimming predators appear to have been derived from lower hypsogastropods related to littorinoideans (Ponder et al. 2008).

19.4.2 LOCOMOTION

Most caenogastropods move by way of muscular or ciliary action of the foot sole (see Chapter 3). Stromboideans (e.g., *Strombus*, *Struthiolaria*, Aporrhaidae), and the related Xenophoridae, move using muscular leaping actions of the foot. Some terrestrial or semiterrestrial truncatelloideans (Truncatellidae and some Pomatiospidae and Assimineidae) move using the snout together with their short foot (see

Chapter 3). The heteropods (Pterotracheoidea) are permanent members of the plankton and swim using muscular actions of their body and a fin derived from the foot. A few caenogastropods are sessile, some of which have worm-like tubular shells (Vermetidae, Siliquariidae, and the coral-living *Magilus*) or are limpet-like (e.g., Calyptraeidae, Hipponicidae) (see Section 19.3.1).

19.4.3 Feeding and Diet

The majority of 'lower' caenogastropods are generalised browsers or deposit feeders, with many small-sized species selectively browsing on fine organic debris or films of micro-organisms.

The original browsing caenogastropods underwent major dietary specialisations, involving the multiple evolution of herbivory, suspension feeding and carnivory within the group.

Herbivory is uncommon in caenogastropods but occurs in the architaenioglossan Ampullariidae, the ampullinoidean *Cernina*, the campaniloideans *Plesiotrochus* and *Campanile*, and in the sorbeoconch Strombidae, as well as in some species of Cypraeidae, Littorinidae, and, surprisingly, the neogastropod family Columbellidae (deMaintenon 1999). This distribution indicates that herbivory has evolved independently several times in caenogastropods.

Some caenogastropods are suspension feeders, and this feeding mode has, like herbivory, been independently acquired in several groups (Declerck 1995) (see Chapter 5). The most specialised suspension feeders are the Viviparidae (Viviparoidea), Turritellidae and Siliquariidae (Cerithioidea), Calyptraeidae (Calyptraeoidea), and Struthiolariidae (Stromboidea). Juvenile suspension-feeding gastropods such as calyptraeids and capulids typically graze microalgae and are usually more active than the adults (Declerck 1995). Similar modifications occur convergently in all these groups – elongated gill filaments, the development of a mucus-secreting endostyle, and a food groove which leads to the mouth. This latter structure is very pronounced in viviparids, terminating at the right 'siphon' (nuchal lobe) which conducts the food to the mouth. Less specialised suspension feeding occurs in Bithyniidae (Truncatelloidea), which have relatively short, ctenidial filaments and no endostyle or food groove. Bithyniids use a mixed feeding strategy as they are also capable of normal feeding by browsing (e.g., Brendelberger & Juergens 1993). At least some species of *Capulus* and *Trichotropis* (Capuloidea) also suspension feed but have only slightly elongated ctenidial filaments and have no endostyle or food groove (Yonge 1938, 1962a). Suspension-feeding vermetids have moderately elongated ctenidial filaments and a food groove but no endostyle; many others employ a mucous net or string secreted by the foot to entangle drifting particles, or a combination of both (e.g., Yonge & Iles 1939; Schiaparelli & Cattaneo-Vietti 1999; Schiaparelli et al. 2006). Mucous net feeding also occurs in at least a few species of the sand-beach neogastropod *Olivella* (see Chapter 5).

Many caenogastropods are grazing carnivores, feeding on encrusting organisms such as tunicates, sponges, and bryozoans, while numerous others are active carnivores. Some hypsogastropod families show a range of feeding habits. Various cypraeids graze on algae or sponges and triviids on algae, sponges, and tunicates. In contrast, the related ovulids feed only on soft corals, and the slug-like velutinids feed on tunicates. The mainly small-sized, but very speciose, triphoroideans are specialised sponge feeders, and the small to moderately large Epitonioidea all feed on cnidarians. As noted above, a few species in this latter group, the 'violet snails' (Epitoniidae), are pelagic surface drifters, feeding on siphonophores such as the 'blue bottle' *Physalia* and *Velella*. The 'heteropods' (Pterotracheoidea), seek prey among the plankton such as other molluscs (including small cephalopods), crustaceans, and fish. The small-sized and highly diverse Eulimidae are ecto- or endoparasites in and on echinoderms (see below).

Terrestrial caenogastropods mainly feed on 'algae' and fungi associated with decaying vegetation or on the surface of leaves but not on living plant tissue (Barker 2001a). The small cyclophoroidean *Acicula* feeds on decaying leaves and fungi but also eats the eggs of other terrestrial gastropods (Barker & Efford 2002).

A few caenogastropods are faecal feeders (coprophages) – *Capulus ungaricus*, for example, lives on gastropods such as turritellids and is located at the exit point of the exhalant current. Other species of *Capulus* are attached to bivalves such as scallops, and some drill holes through the host shell and are kleptoparasites (see Chapter 5).

The adoption of active carnivory has been accompanied by the evolution of a variety of feeding methods. Most hypsogastropods bite or rasp their prey using the radula and, in some, the jaws are modified for grasping or nipping. Many benthic caenogastropods are active hunters seeking animal prey, with a few groups engulfing whole prey (see also Chapter 5).

Tonnoideans, as with their sister group the neogastropods, have a highly diverse range of prey. The tonnoidean families feed on a range of prey including echinoderms, molluscs, polychaetes, and tunicates. Some tonnoideans and the related Ficoidea are specialised feeders at the family level, with tonnids swallowing whole holothurians and others feeding on polychaetes (Personidae, Ficidae), echinoids (Cassidae), or a variety of prey (Cymatiidae). Many (but not all) cymatiids are more generalist feeders with some feeding on molluscs, ascidians, or polychaetes. Bursids also utilise a range of prey including echinoderms (ophiuroids) in some species (e.g., Taylor 1978). Hughes and Hughes (1981) provided a comprehensive list of prey consumed by cassids, and Morton (1990a), and Andrews et al. (1999) reviewed tonnoidean feeding in general. While Tonnoidea includes the only gastropods that are predators of echinoderms, the Eulimidae are echinoderm ecto- and endoparasites.

Many tonnoideans have modified salivary glands (see Section 19.3.4) that secrete acid used to assist with boring through the tests of their echinoderm prey. Cassids preying on echinoids make a hole in the test using sulphuric acid secretions (pH 0.13 in *Cassidaria echinophora*) from the salivary glands (Fänge & Lidman 1976). *Cassis tuberosa* can even cope with the long, sharp spines of *Diadema* on which it feeds (Hughes & Hughes 1971).

Naticids actively drill holes in their mainly bivalve prey. Shell drilling is a habit acquired independently by the neogastropod family Muricidae and a few species of Cassidae, Marginellidae, Buccinidae, and Capulidae (Carriker 1961). Naticids and muricids have independently evolved accessory boring organs (Carriker & Gruber 1999) (see Chapter 5), but the drilling mechanisms employed by the other families have not been investigated. Some predatory gastropods have devised other ways of entering shelled prey such as developing a strong lever-like spine on the shell aperture to force entry to bivalves, a feature independently evolved in several neogastropod families (Vermeij & Kool 1994; Marko & Vermeij 1996, 1999), as have shell chipping methods and even the use of toxins.

Members of some neogastropod families ingest a variety of prey (e.g., Conidae, Muricidae, Marginellidae), while some other family-group taxa are highly specialised, including the Mitridae, which feed only on sipunculids, and coralliophiline muricids which are associated with, and feed on, hard or soft corals. The very diverse conoideans mainly feed on polychaetes, although some species of the Conidae feed on gastropods or small fish, and a raphitomid, *Phymorhynchus buccinoides*, feeds on *Bathymodiolus* mussels in a deep-sea seep (Fujikura et al. 2009).

Some neogastropods are scavengers, notably the Nassariidae (e.g., Morton 2003a, 2011; Morton & Britton 2003; Morton & Jones 2003), although a few nassariids are selective deposit feeders (Connor & Edgar 1982; Morton 2011).

Significant changes in the digestive system are associated with carnivory (see Chapter 5), such as the development of a proboscis (introvert), which has occurred independently in several groups (see Section 19.3.4). Of the other major modifications to the anterior gut of carnivorous caenogastropods, the most derived is the conoidean venom glands from which, in many conoideans, the harpoon-like hypodermic radular teeth are charged and held at the end of their slender proboscis enabling members of this group to actively hunt worms, molluscs, and even fish (see Chapter 5, Figures 5.8 and 5.9).

Some neogastropods have become suctorial feeders and could be classed as parasites. A number of marginellids and colubrariids feed suctorially on parrot fish 'sleeping' in mucous cocoons on coral reefs (Bouchet & Perrine 1996), while some cancellariids feed on bivalves, and others feed on rays resting on the bottom of sandy bays (O'Sullivan et al. 1987).

19.4.4 Life History and Growth

19.4.4.1 Spawn, Development, and Larvae

19.4.4.1.1 Spawn

All caenogastropods differ from most vetigastropods in producing spawn where the eggs are wrapped in jelly or enclosed in a capsule. The nature of the spawn differs greatly between and among different groups, with differences including the shape and size of capsules (Figure 19.30) and the thickness of their walls (e.g., Thorson 1946; Fretter & Graham 1962; D'Asaro 1970; Bandel 1975).

The capsule is usually manipulated by the foot after it leaves the genital opening, and this often provides a final moulding of the capsule as well as enabling its fixation to the substratum. In some neogastropods, moulding of the egg capsules is aided by special glandular structures in the foot sole (the 'pedal gland'), for example as seen in some muricids. Female *Trivia* have a protrusible structure on the foot sole that may be used in egg capsule formation or manipulation (Fretter 1946).

Sometimes, particles such as sand or mud are applied by the foot to the capsule in order to provide camouflage. In naticids, most species have flat, coiled, jelly egg masses (Figure 19.30) in which sand is embedded with the aid of the foot. Females of the estuarine truncatelloidean *Ascorhis tasmanica* store sand in pouches in the snout which is then applied to the capsules (Ponder & Clark 1988).

Capsules can enhance embryo survival as they protect the developing embryos from predation and environmental hazards such as salinity changes, desiccation, ultraviolet light (Rawlings 1994; Przeslawski 2004; Przeslawski et al. 2004), and microbial infection (Benkendorff et al. 2001). The egg capsules are energetically expensive to produce, with the energy expended on capsule wall material sometimes approaching that used in the production of ova (Perron 1981).

Capsules can be attached to a firm substratum such as rock or, in many small species, algal fronds. In some carnivorous direct-developing species the spawn is laid on, or embedded in, the colonial organism on which they feed, for example in velutinids and triviids on compound ascidians (Fretter 1946), ovulids on soft corals (e.g., Knudsen 1997), and the costellariid *Austromitra rubiginosa* in compound ascidians (Ponder 1972). Various cypraeids continue to care for their spawn after laying while other caenogastropods brood capsules in the mantle cavity, oviduct, or, as in some cerithiimorphs (Planaxidae, Thiaridae, and Paludomidae [*Tanganyicia*]), in chambers in the haemocoel of the head-foot (see Chapter 8) and, in the provaniid *Ifremeria*, in the posterior pedal gland (Reynolds et al. 2010).

Some, for example ampullariids, campanilids, strombids, many cerithioideans, and some littorinids, produce eggs embedded in jelly. The spawn may be a short, often coiled mass or an elongate string and contain numerous eggs, but in most groups, the thin outer 'skin' of the jelly has been elaborated to form a capsule typically composed of more than one firm layer. Such capsules are sometimes species-specific in form, while others may have a rather uniform morphology, even at the family level. Many caenogastropods lay one or more eggs in solitary lens-shaped, hemispherical or globular capsules attached to various substrata. These relatively simple capsules are elaborated in different lineages. Pelagic (i.e., unattached) capsules are produced by many littorinids and heteropods, with those of the latter being single or in a mucoid string (Lalli & Gilmer 1989), although the egg string

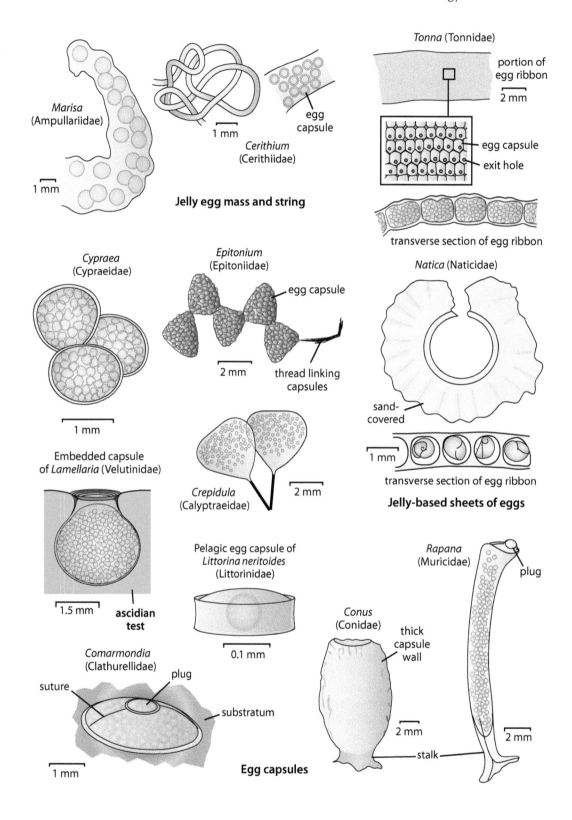

FIGURE 19.30 Spawn morphology. Redrawn and modified from the following sources: *Marisa* and *Conus* (Bandel, K., *The Veliger*, 18, 249–271, 1976), *Cerithium* (Lebour, M.V., *Proc. Zool. Soc. Lond.*, 114, 462–489, 1944), *Cypraea* (Knudsen, J., Further observations on the egg capsules and reproduction of some marine prosobranch molluscs from Hong Kong, pp. 283–306, in Morton, B. (ed.), *The Malacofauna of Hong Kong and Southern China, III: Proceedings of the Third International Workshop on the Malacofauna of Hong Kong and Southern China, Hong Kong*, Hong Kong University Press, Hong Kong, 1994), *Littorina*, *Lamellaria*, *Epitonium*, and *Comarmondia* (Fretter, V. and Graham, A.L., *British Prosobranch Molluscs: Their Functional Anatomy and Ecology*, Ray Society, London, 1962), *Crepidula* (Lebour, M.V., *J. Mar. Biol. Assoc. U.K.*, 22, 105–166, 1937), *Natica*, *Tonna*, *Rapana* (Thorson, G., *Danish Scientific Investigations in Iran*, 2, 159–238, 1940).

is permanently attached to the female of the pterotracheid *Firoloida desmaresti* (Owre 1964).

In some sessile vermetids and hipponicids, sac-like capsules are attached by narrow stalks within the anterior mantle cavity, as they are also in a few turritellids. In a number of species of *Dendropoma* the egg capsules lie free in the mantle cavity (Golding et al. 2014). In calyptraeids, they are attached to the substratum but are concealed and protected beneath the limpet-like shell, and in vanikorids they are brooded beneath the foot. The epitoniid *Janthina janthina* is viviparous, but other species of *Janthina* lay their sac-like capsules on the bubble raft used by these snails to float on the surface (Figure 19.6) (Churchill et al. 2011). Other epitoniids produce egg capsules attached to strings, and each contains multiple eggs (Figure 19.30).

More complex capsules are produced by many higher caenogastropods, and these are often laid in clusters or, in some taxa, in columns or large ball-like masses.

Considerable variation in the form of the spawn occurs in some groups. For example, within Tonnoidea, *Tonna* lays long flat gelatinous sheets in which hundreds of eggs are embedded (Figures 8.16 and 19.30). Other tonnoideans, such as *Fusitriton*, *Argobuccinum*, and *Charonia*, lay flat sheets of rectangular capsules attached to the substratum but in cymatiids (e.g., *Cabestana*, *Monoplex*, *Mayena*, and *Linatella*) (Laxton 1969; Muthiah & Sampath 2000) and at least some bursids, the egg masses are modified to form a cup-like egg case with the outer surface of the cup covered with a stiff material. The formation of this cup-like egg case has been described in the cymatiid *Cabestana spengleri* by Laxton (1969). The first capsule is laid in the centre, and the other capsules are laid in a spiral around it for an area the size of the shell aperture. Subsequent capsules are then tilted towards the centre with their bases forming the wall of the egg case. The size of the egg case corresponds to the diameter of the shell aperture, but the number of capsules laid depends on the height of the case, so the number of capsules laid is not necessarily a function of the size of the female. Capsules at the base of the egg case are shorter than the ones on the sides. Thus, in the taxa that produce these 'half orange' egg masses, the apertural lips match the size of the egg mass. The multi-layered outer surface of the egg case may be formed by the periostracal mantle groove (A. G. Beu, pers. comm.), and, if this is the case, the outer edge of the egg mass must be attached in some way to the mantle edge in the aperture. The female sits on the case for a considerable time while forming it. Possibly correlated with this behaviour is an expanded aperture in female bursids and some other tonnoideans, while those of males are narrower.

The egg masses of cassids differ from those of cymatiids and bursids with some species laying a tall, spiral column of egg cases, sometimes communally. The related ficids have flattened, almost square pouch-like capsules containing numerous eggs (Riedel 1995; Knudsen 2000).

The egg capsules of neogastropods exhibit considerable diversity of form. Muricids produce egg capsules that range from lens-shaped to very elongate vase-shaped (Figure 19.30),

and the muricid subfamily Coralliophilinae retain their sac-like capsules within the mantle cavity. Some other neogastropods such as buccinids and columbellids also exhibit a wide range of capsule form.

Also, within some non-neogastropod families, the spawn exhibits a considerable range of morphology. For example, in the Littorinidae, a number of genera such as *Bembicium* and *Lacuna* have gelatinous spawn, and some species hatch as planktotrophic veligers. Members of the littorinid subfamilies Laevilitorininae and Lacuninae are direct developers, with the eggs undergoing their development in the capsules embedded in their gelatinous spawn. In contrast, many littorinines have pelagic egg capsules, a useful adaptation to life in the high intertidal zone. These capsules release planktotrophic larvae, while in some the capsules are attached to the substratum with direct development, and in others, the eggs are brooded in the pallial oviduct before releasing the young as crawl-away juveniles. As would be expected, the terrestrial littorinoidean *Pomatias* is a direct developer, with the young developing in spherical capsules deposited in the soil.

19.4.4.1.2 Larvae and Larval Development

In all caenogastropods, the trochophore stage is suppressed or passed within the egg capsule, and the veliger is the only free-swimming larval form except for the evenly ciliated Warén's larva in the provannid *Ifremeria* (Reynolds et al. 2010) (see Chapter 8). The planktonic veligers can either feed in the water column (planktotrophic) or may carry their food (yolk) with them (lecithotrophic) (see Chapter 8 for general discussion). While many caenogastropods are lecithotrophic, as in vetigastropods, this developmental mode appears to be secondary in the group, as planktotrophy is probably plesiomorphic. Planktotrophic development (which is reflected in the protoconch and is therefore discernible in fossils) is diagnostic of caenogastropods. The initial veliger has the primary larval shell (protoconch I), but this is extended by growth during the feeding (planktotrophic) phase to form what is usually a distinctly different part of the protoconch, protoconch II. Planktotrophic development also occurs in heterobranchs and neritimorphs, but their secondary larval shell is not markedly different from the initial protoconch I. Whether or not planktotrophy evolved just once, or independently in all those groups (Ponder 1991), or twice (neritimorphs and apogastropods) (Haszprunar et al. 1995a) is still a matter of conjecture (see also Chapter 8).

Planktotrophic and direct developers can occur within the same family and even the same genus. In many caenogastropods, the entire larval life and metamorphosis occur within the egg capsule, a developmental mode that has evolved many times through caenogastropod history. While direct development is common in marine taxa, especially those in temperate to polar regions, it was also a necessary step before successful colonisation of the land or fresh water. Thus, in the Hydrobiidae, Tateidae, and Assimineidae, for example, planktotrophic larvae occur in some estuarine species, but freshwater and terrestrial (in the case of Assimineidae) taxa have only direct development.

All non-marine architaenioglossans have direct development. The Ampullariidae and Cyclophoroidea produce spherical capsules containing a single embryo. Cyclophoroideans are ovoviviparous or viviparous, and all viviparids are ovoviviparous. In the marine architaenioglossan taxa, *Campanile* is lecithotrophic, as is *Cernina*, with a protoconch of only two whorls (Houbrick 1981), but at least one extinct species of *Campanile* has a multi-whorled protoconch suggestive of planktotrophic development, as do some species of *Plesiotrochus* (Houbrick 1990; Bandel 1993).

Egg size is determined by the amount of yolk the egg contains and is reflected in the initial size of the juvenile shell (protoconch I) which, combined with the number of protoconch whorls, is a useful guide to distinguishing feeding and non-feeding larvae in both living and extinct taxa. In caenogastropods, the egg develops into a veliger larva within the egg capsule. The veliger may either be released to eventually settle and undergo metamorphosis to form a juvenile snail or is retained within the capsule where they develop into crawl-away juveniles. There are various ways the developing embryos are nourished within the capsules, including yolk, albumen, infertile 'nurse' eggs, and by adelphophagy (cannibalism) (see Chapter 8). Direct development is common in cold-water marine species and is the norm in freshwater and terrestrial taxa.

Veliger larvae show considerable diversification in caenogastropods (Figure 19.31) and have been the subject of many studies. Lower caenogastropods have larvae with broad, short velar lobes which tend to have a relatively short larval life. While similar larvae are found among higher caenogastropods, there are others, typically with long velar lobes, that have become significantly modified to enhance their ability to live in the plankton for longer periods and, as a consequence, potentially travel long distances before metamorphosis (see below).

In the Capuloidea and Velutinoidea, the larvae uniquely have their true larval shell surrounded by a pseudo-shell – the *echinospira* (Lebour 1935) (Figure 19.31). A 'more advanced' version of this in *Marseniopsis* (Velutinidae) has been called a *limacosphaera* larva (Bandel et al. 1993). Limacosphaera larvae differ from echinospira larvae in covering the larval shell with a 'lacunous muscular mantle that can change its volume by interaction with the body fluid and muscle activity' (Bandel et al. 1993 p. 1). In both these larval types, the pseudo-shell (echinospira or scaphoconcha) is much larger than the larval shell and, because it is very thin, probably assists with flotation. In some veligers, such as those of *Lamellaria* (Velutinidae) (Figure 19.31), the velar lobes have subdivided into three lobes on each side, a modification also seen in some heteropods (Lalli & Gilmer 1989), and stromboideans such as *Strombus* (Thiriot-Quiévreux 1983) (Figure 19.31) and *Aporrhais* (Lebour 1937). There are only two lobes in *Struthiolaria* (Stromboidea) (Morton 1950), several heteropod genera (Lalli & Gilmer 1989), and *Marseniopsis* (Velutinoidea) (Bandel et al. 1993), indicating that these differences are not uniform within higher taxa. Most other gastropods have a single, broad lobe, or this is subdivided into two lobes. Hypertrophy of the paired velar lobes occurs in some higher caenogastropods, such as in some naticids, tonnoideans (Figure 19.31), ovulids, and neogastropods, which in some cases can be extremely long. These modifications have not occurred in the Neritimorpha or heterobranchs, except architectonicids (lower Heterobranchia) which have long, bilobed velar lobes (Robertson et al. 1970) convergent with those seen in many higher caenogastropods.

Some larvae are large and can survive for long periods of time (sometimes over a year), enabling them to be transported for long distances by ocean currents (see Chapter 9). These larvae, termed *teleplanic* (Scheltema 1971; Scheltema 1988b), occur in just a few families. The largest larval shells are found in tonnoideans (Figure 19.31) where veligers can reach over 5 mm in length and up to 6.7 whorls (Bandel et al. 1994), a size seldom reached or exceeded by most adult living neogastropods. It is perhaps not surprising these larval shells have occasionally been described as adult species. The velar lobes have greatly increased in length, presumably to assist in the support of these large shells in the water column. While the teleplanic larvae of *Fusitriton oregonensis* only reach about 3.9 mm in length, they can spend up to 4.5 years in the nekton (Strathmann & Strathmann 2007). This longevity is probably reflected in its extensive range which includes both Asian and North American coasts as well as occurrences on seamounts and islands in the North Pacific (Smith 1970; Strathmann & Strathmann 2007).

19.4.4.1.3 Growth and Age

An overview of growth rates and the longevity of molluscs is given in Chapter 9. The information on caenogastropods is very scattered and has not been reviewed since the overview of Heller (1990). Most small-sized caenogastropods reach maturity in a year or less, but in others, growth rates and longevity vary considerably, and some littorinids live for several years and up to 15 in one species (Heller 1990). Of those studied, the longest-lived caenogastropod is the large Argentinean volute *Adelomelon beckii* that is estimated to live for around 29 years (Arrighetti et al. 2011), while other cold-water volutes reach ages of 17 to 20 years (Giménez et al. 2004; Cledón et al. 2005; Bigatti et al. 2007). Other medium to large gastropod species studied do not live as long; for example, the cold-water buccinids *Neptunea antiqua* (Powell & Stanton 1985) and *Buccinum undatum* (Gendron 1992) live for around ten and 12 years respectively. Long life is not restricted to cold water. A tropical muricid, *Morula musiva*, lives for around nine years (Tong 1986), while temperate muricids have been recorded to live for 15–20 years (Heller 1990) and the colder water *Concholepas concholepas* for ten years (Stotz 2000). Some tropical representatives of various other neogastropod families have also been shown to live for ten years or more (Heller 1990). Other caenogastropods have long lives; for example, *Cerithium caeruleum* lives for up to ten years (Ayal 1978), *Cerithidea decollata* for nine (Powell

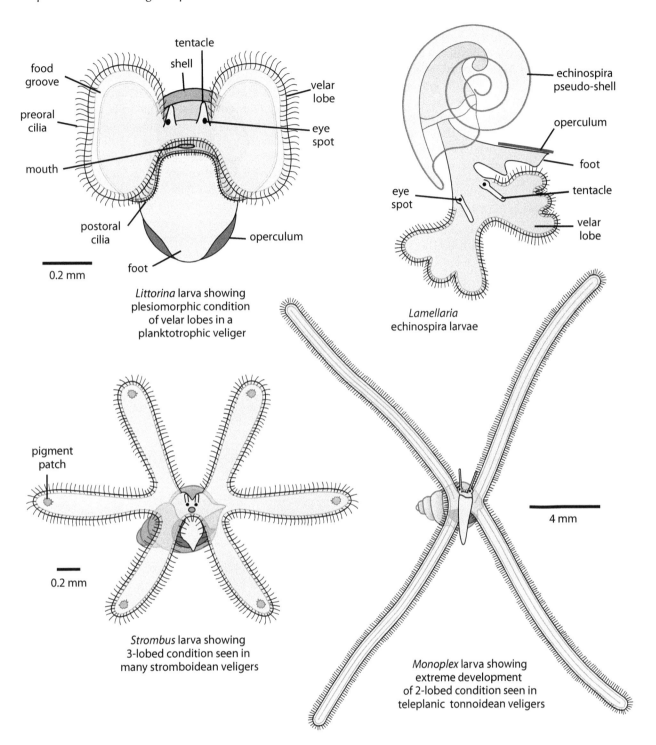

FIGURE 19.31 Veliger morphology, including the echinospira larva of *Lamellaria*. Redrawn and modified from the following sources: *Littorina* (Fretter, V. and Graham, A.L., *British Prosobranch Molluscs: Their Functional Anatomy and Ecology*, Ray Society, London, 1962), *Strombus* (Thiriot-Quiévreux, C., Estuaries, 6, 387–398, 1983), *Lamellaria* (Lebour, M.V., *Proc. Zool. Soc. Lond.*, 1935, 163–174, 1935), *Monoplex* (Lebour, M.V., *Proc. Zool. Soc. Lond.*, 114, 462–489, 1944).

& Stanton 1985), the turritellid *Gazameda gunnii* for seven years (Carrick 1980), and *Strombus costatus* and *S. gigas* for five and seven years respectively (Wefer & Killingley 1980).

In species with varices, the intervariceal growth usually occurs rapidly until the varix is formed, and this is followed by a growth pause (e.g., Laxton 1970).

In those caenogastropods with terminal growth, this may be reached long before death. For example, the queen conch, *Strombus gigas*, forms the mature shell aperture in a little over three years and reproduces shortly after that (Appeldoorn 1988), but this species lives to around seven years (Wefer & Killingley 1980).

19.4.5 ASSOCIATIONS

The great majority of caenogastropods do not form close associations with other animals. Most associations with other biota are with their food, in particular with the grazing carnivores or parasites (see Section 19.4.3 and next section). Some tornids are commensals, living in host-specific association with tube-dwelling worms or crustaceans (Bieler & Mikkelsen 1988; Morton 1988; Ponder 1994). Some associations with bacteria, protists, and other organisms with caenogastropods are discussed in Chapter 9.

19.4.5.1 The Evolution of Parasitism in Eulimidae

All Eulimidae appear to be parasitic on echinoderms (see Warén 1983a for a review of the shelled genera of this family), some have reduced shells, and a few have lost the shell altogether and become worm-like. Some shelled eulimids (Figures 19.32 and 19.33) crawl about freely and attach to their hosts only when feeding, but many others are permanently attached, either externally or internally. Some of the shell-less endoparasitic eulimoideans are so highly modified that

their systematic position was difficult to determine because most organ systems are absent in adults, although they all possess shelled veliger larvae, indicating their gastropod relationship. A few workers (Fischer 1880–1887; Mandahl-Barth 1941; Tikasingh & Pratt 1961) included the shell-less taxa in the Opisthobranchia (in an order Parasita), but they have otherwise long been recognised as related to eulimids (e.g., Schiemenz 1889; Vaney 1913). The opisthobranch affinities were suggested in part because of the mistaken belief that some of the endoparasitic forms were simultaneous hermaphrodites, but this is not so as the so-called testis in these forms is a tiny parasitic male (Mandahl-Barth 1941; Lützen 1968). The parasitic, shell-less species have been classified in a separate family, Entoconchidae, or sometimes (e.g., Mandahl-Barth 1941; Tikasingh & Pratt 1961), with an additional family, Enteroxenidae, recognised. It is now generally accepted that they all represent a single family, Eulimidae (Ponder & Gooding 1978; Warén 1983a). This idea is reinforced by there being a progression of modification to parasitism seen within eulimids (Vaney 1913) (Figure 19.33). Some genera such as *Megadenus* and *Thyca* are permanently

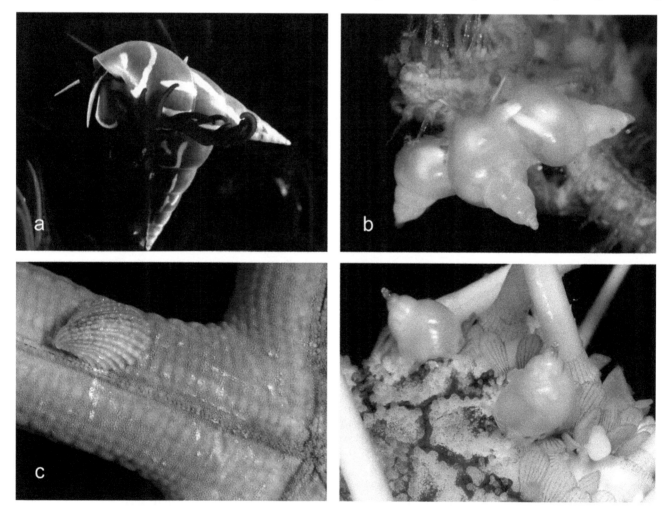

FIGURE 19.32 Some examples of ectoparasitic eulimids on their hosts. (a) *Annulobalcis yamamotoi* on a crinoid, New South Wales, Australia. Courtesy of D. Riek. (b) *Stilapex* sp. on an ophuroid, Papua New Guinea. Courtesy of P. Maestrati - MNHN. (c) *Thyca* sp. on the starfish *Linkia*, Panglao, Philippines. Courtesy of S. Tagaro - MNHN. (d) *Monogamus minibulla* on the urchin *Eucidaris tribuloides*, Guadeloupe, French West Indies. Courtesy of L. Charles - MNHN.

ECTOPARASITES

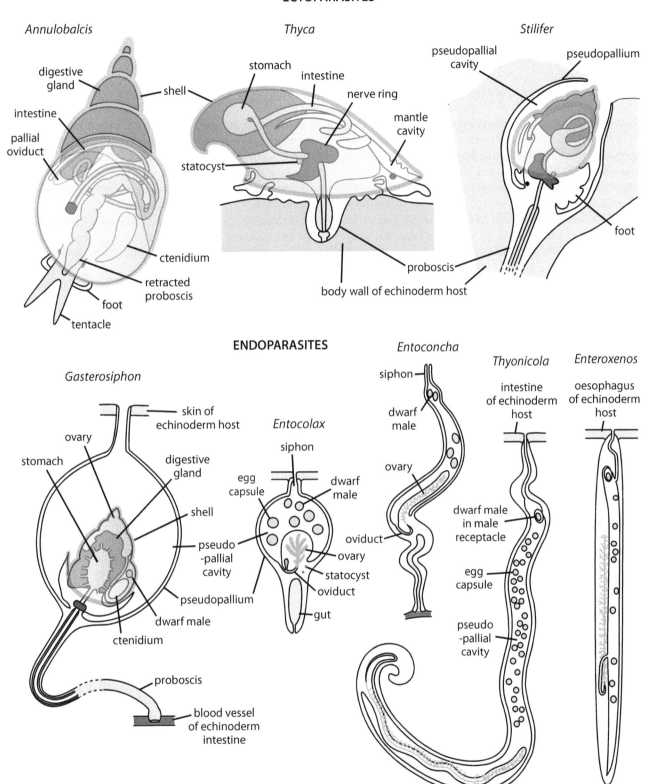

FIGURE 19.33 Various stages in the evolution of parasitism in eulimids. Redrawn and modified from Ivanov, A.V., *Comptes rendus (Doklady) de l'Academie des sciences de l'URSS*, 48, 450–452, 1945, except for *Annulobalcis* (Simone, L.R.L. de and Martins C.M., *J. Conchol.*, 35, 223–235, 1995), and *Thyca*, *Stilifer*, and *Gasterosiphon* (Vaney, C., *Bull. sci. de la France et Belg.*, 47, 1–87, 1913).

attached externally with some partially or, as in some species of *Stilifer* for example, almost completely embedded in the integument of the host echinoderm. *Gasterosiphon* attaches to the inside body wall of its holothurian host but has retained its shell, is relatively unmodified, and maintains contact with the exterior. *Paedophoropus* is found in the polian vesicle of the holothurian *Eupyrgus*. It lacks a shell as an adult but retains a visceral hump and most organs. *Entocolax* lacks a shell, is entirely internal, and has one end attached to a blood vessel in the body wall of its holothurian host but retains contact with the outside through a small opening (Figure 19.33).

Enteroxenos, *Thyonicola*, and *Entoconcha* are more modified and worm-like and usually attached to the intestine (although *Enteroxenos* may sometimes lie free in the body cavity) of the holothurian host. *Entoconcha* and *Entocolax* both have a short intestine but lack an anus, and the intestine is lost in *Enteroxenos*, *Thyonicola*, and *Comenteroxenos*. *Thyonicola dogieli* (Ivanov 1945) from the intestinal connective tissue of the holothurian *Cucumaria frondosa japonica* has a tube-like body that can reach 128 cm in length, and *Thyonicola mortenseni* in *Cucumaria miniata* reaches 303 cm, making it the longest caenogastropod by far (Figure 19.33).

How the worm-like endoparasites infest their holothurian hosts is poorly understood, but the larval parasite may be ingested, or it enters via the cloaca. In some, the larva may settle on the outside of the host and burrow inside. Egg capsules of species attached to the intestine are released via the cloaca or when the holothurian eviscerates, and in others, the eggs are liberated into the coelom to be released during evisceration (Tikasingh 1962; Lützen 1979; Byrne 1985).

19.4.6 PREDATORS AND DEFENCE

Predators of caenogastropods include most marine, estuarine, freshwater, and terrestrial carnivores such as fish, decapod crustaceans (particularly crabs), sea stars, other molluscs, birds, and mammals (see Chapter 9). Caenogastropods have responded to predation pressure in many ways including avoidance of visual predation by adopting cryptic habits or seeking crevices and other hiding places during daylight. Others have thickened their shells, and many have developed thickened apertures (so-called terminal growth) or formed varices and/or spines (e.g., Vermeij & Signor 1992).

Escape responses have evolved in some, where the presence of predators evokes a powerful sideways movement of the foot, as seen in strombids and nassariids whose muscular foot rapidly twists and turns so the animal jerks about and has a good chance of moving away from the predator.

Foot autotomy, where the posterior end of the foot is detached and may continue to move, is seen in some cypraeids (Cernohorsky 1964), *Harpa* (Stasek 1967), and the olivid *Agaronia* (Rupert & Peters 2011), while *Ficus* detaches some of its expanded mantle when threatened (Liu & Wang 1996; Liu & Wang 2002) (see also Chapter 9).

Acidic secretions from the epidermis occur in at least some cypraeids and velutinids (Kniffen 1968; Thompson 1988). In the velutinid slugs noxious chemicals are probably obtained

from their ascidian prey (e.g., Thompson 1960b; Carroll & Scheuer 1990; McClintock et al. 1992) while the velutinid *Marseniopsis mollis* probably deters predators with the secondary metabolite homarine derived from an epizoite that lives on its cnidarian prey (McClintock et al. 1994).

19.4.7 OTHER BEHAVIOURAL ATTRIBUTES

Caenogastropod behaviour is rather poorly documented but appears to be mainly related to feeding, the avoidance of predators, and mating, most of which is determined by chemical signals. Responses to visual cues are limited (with some exceptions) to shadow responses and phototaxis, with a corresponding lack of sexual, antagonistic, or other display behaviours. At least one *Littorina* species uses vision to discriminate shade intensity and shape orientation (Messenger 1991) (see also Chapter 7), and some male *Strombus pugilis* sequester and guard females. Fighting between males has been noted (Bradshaw-Hawkins & Sander 1981), but reproductive behaviours have apparently not been carefully observed in other species of *Strombus*.

As with other intertidal organisms, intertidal caenogastropod activity and feeding behaviours vary with the tidal cycle. Most are inactive at low tide (except at night) with activity increasing as the tide rises. In the subtidal, diurnal/nocturnal behaviours are important in avoiding predation. In the water column, vertical migrations of heteropods[1] have been documented with them moving to deeper water during daylight but coming to within 100 m of the surface at night.

Size-assortative mating (see Chapter 8) has been demonstrated in a few taxa, mainly species of *Littorina*.

19.5 ECONOMIC AND ECOLOGICAL VALUES

Some caenogastropods are used as food, and many groups are important in the shell trade, although over-collecting and habitat destruction have rendered many species vulnerable (see Chapter 10). In many societies, some species are important culturally, for example some charoniids used as trumpets, and the 'sacred chank' *Turbinella pyrum* in parts of Asia (see Chapter 10), have religious significance. Tyrian purple, a dye made by ancient Phoenicians, is made from the hypobranchial glands of muricids (see Chapter 10).

The large majority of caenogastropods have not been screened for possibly useful compounds although their potential is great, given the findings from extensive studies on the muricid *Dicathais orbita* (Benkendorff 2013) which found compounds with antibacterial and anticancer properties. Members of the Conidae are another well-known source of useful compounds, with some peptides from the venom glands being useful in pain regulation, and many others are under investigation (e.g., Kumar et al. 2015). Some of these compounds found in velutinids also appear to be potential anticancer agents (Bailly 2004; Cironi et al. 2005) (see also Chapter 10).

---

[1] Vertical migrations also occur in pteropods (Heterobranchia).

A few freshwater taxa are important as intermediate hosts of trematodes responsible for human diseases, and some are significant invasive species (see Chapter 10).

19.6 DIVERSITY AND FOSSIL HISTORY

There are about 157 extant, mostly marine, families recognised (Bouchet & Rocroi 2005; Bouchet et al. 2017), with many additional extinct families (see Appendix). Caenogastropods arose before the Carboniferous with their ancestors in the Early Paleozoic (Bandel 1993; Ponder et al. 2008), but the ancestral relationships of caenogastropods are uncertain, with various Paleozoic taxa having been suggested (e.g., Bandel 2002a; Frýda et al. 2008; Ponder et al. 2008). They undoubtedly lie within one of the extinct Early Paleozoic taxa assigned to vetigastropods. The first caenogastropod-like fossils appear in the Lower Ordovician about 490 Mya although their identity as members of this group remains tentative due to the absence of shell structure and, particularly, protoconch data, essential before an accurate assignment to a higher group can be made (Frýda et al. 2008; Ponder et al. 2008). Murchisonioids may have given rise to these caenogastropod-like taxa at least twice in the Ordovician, these being members of the families Loxonematidae and Subulitidae (Frýda et al. 2008). One group suggested as either the ancestral or sister group of caenogastropods is the Peruneloidea (Ordovician to Devonian), members of which had a peculiar uncoiled protoconch (Frýda et al. 2008). These have been allied to the Subulitoidea which arose in the Lower Ordovician (Nützel et al. 2000). The first fossil gastropods known to have caenogastropod-like protoconchs are limpet-like taxa (Pragoscutulidae) in the Devonian. These taxa are either aberrant offshoots of as yet unidentified coiled taxa (Ponder et al. 2008) or a different group with convergent protoconch morphology.

Some early tall-spired slit-bearing gastropods such as the Paleozoic vetigastropod Murchisonioidea had typical 'archaeogastropod' protoconchs and nacre while others with somewhat similar shell morphology, including the late Paleozoic Goniasmatidae (Orthonematoidea), had a protoconch and shell structure typical of caenogastropods. A slit in the middle of the outer lip does not necessarily indicate the possession of a pair of ctenidia as slits also occur in various living caenogastropod taxa such as some turritellids and siliquariids with only a single ctenidium.

Caenogastropods underwent significant separate radiations during the Jurassic, Cretaceous, and Paleogene and during their evolution have undergone extraordinary adaptive radiations, resulting in not only considerable morphological and ecological diversity but also a wide range of physiologies and behaviours.

The early caenogastropods were marine and benthic, probably fed on biofilms, and presumably had similar habits to some living small, shallow-water cerithioideans. Minor modifications from this way of life occurred along many lineages and included a move to algal habitats, sheltering beneath rocks and burrowing in sediment. Decrease in size enabled tiny snails to live interstitially while an increase in size enabled the development of a more robust shell and thus greater protection from predators.

Early offshoots from marine stem taxa gave rise to the enigmatic non-marine architaenioglossan taxa which are the most primitive of the living caenogastropods.

Caenogastropod diversification was driven largely by dietary specialisation – a divergence from the ancestral deposit feeding and surface grazing, which is still largely the norm in most architaenioglossans, cerithioideans, and lower hypsogastropods. Thus, much of the success of caenogastropod evolution resulted from multiple adoptions of herbivory, grazing carnivory or active predation, and even parasitism (Eulimidae and a few neogastropods). Of particular importance was the radiation of predatory gastropods, notably the neogastropods, which first appeared in the Cretaceous and have exhibited steady diversification since, which has given rise to one of the most diverse groups of living gastropods (Ponder 1974; Taylor & Morris 1988; Taylor 1998).

Most caenogastropod families are exclusively marine. While many are diverse in both temperate and tropical seas, some families are largely tropical (e.g., Strombidae, Cypraeidae, Ovulidae, Cerithiopsidae, Triphoridae, Epitoniidae, Olividae, Mitridae, Costellariidae, Terebridae, and Conidae), but some are more diversified in cooler waters (e.g., Buccinidae, Eatoniellidae, Rissoidae, Anabathridae, and Struthiolariidae).

Two attributes, internal fertilisation and larval planktotrophy, played a major role in the evolution of the group. Internal fertilisation enabled larvae to complete their early development in the protected environment of a capsule. Capsular development also enabled some taxa to undergo non-pelagic development, either within an external capsule or in capsules or eggs retained in a brood pouch within the animal (see also above). Capsular development also facilitated the caenogastropod invasions of marginal marine and non-marine habitats.

Extensive freshwater radiations have occurred independently in several caenogastropod lineages. These include the architaenioglossan Viviparidae and Ampullariidae and six families of cerithioideans (Strong et al. 2011; Bouchet et al. 2017). Truncatelloidean gastropods (Hydrobiidae, Cochliopidae, Tateidae, and several related families, including Pomatiopsidae and Bithyniidae) have undergone major freshwater radiations. Some other families also have freshwater members, including a single genus in Marginellidae (Strong et al. 2008) and two closely related nassariid genera (Strong et al. 2017).

Terrestrial taxa have also evolved several times but only within the architaenioglossan and lower hypsogastropod taxa. The Cyclophoroidea and two littorinoidean families (Pomatiidae and Annulariidae) are terrestrial, as are many species in the truncatelloidean Assimineidae and Truncatellidae. Members of both these latter families mainly occupy marine upper intertidal to supralittoral habitats, and various taxa have independently become terrestrial several times (e.g., Rosenberg 1996). The truncatelloidean taxon Pomatiopsidae, which includes freshwater, hyper-saline, and amphibious taxa, has given rise to two terrestrial taxa (*Blanfordia* and some species of *Fukuia*) in Japan.

Numerous examples of convergence occur within caenogastropods. Some are particularly notable such as that seen in the tubeworm-like shells of vermetids and siliquariids. These were included in the same family group until the 1950s – the latter is closely related to the cerithioidean turritellids, the former is of uncertain relationship within the Asiphonata. While there is considerable evidence of their lack of relationship, they are still sometimes incorrectly treated as closely related taxa (e.g., Simone 2001). These two families are not only similar in shell morphology, but both are also suspension feeders, with the siliquariids employing ctenidial feeding, as do some vermetids, while other vermetids use mucous strings to gather food. Both groups first appear in the Cretaceous (Bandel & Kowalke 1997; Bieler & Petit 2011). Similar 'worm snails' also evolved from turritellids (Vermiculariinae).

19.7　MAJOR GROUPS WITHIN CAENOGASTROPODA

The Caenogastropoda are defined by relatively few unique characters, but the suite of characters they possess readily distinguishes them from other gastropods.

Most caenogastropods have a simple foot, lacking an epipodium or epipodial tentacles, and all lack sensory papillae. The cephalic tentacles are often long, and the eyes (which have a lens and a cornea) are on the outer side of their bases.

The single (left) ctenidium has skeletal rods and is monopectinate. The single (left) osphradium is conspicuous, and there is a single (left) hypobranchial gland.

The mouth is simple, lacking a hood, and is at the distal end of a snout that is typically longer than broad. The radula is plesiomorphically taenioglossate, and there is a pair of lateral jaws (modified in some groups). The oesophagus plesiomorphically has an oesophageal gland and dorsal folds, but the ventral fold(s) are absent or rudimentary. The stomach typically features a protostyle and a gastric shield; a crystalline style developed in some; the intestine is short, and the rectum lies on the right side of the mantle cavity with the anus anteriorly located.

The heart has a single (left) auricle, and the rectum does not penetrate the ventricle or pericardium. There is a single typically viscerally located (left) kidney with a well-developed nephridial gland. A rudiment of the right kidney is incorporated in the genital system and is often connected to the pericardium via a renopericardial duct.

19.7.1　THE ARCHITAENIOGLOSSANS

This interesting and enigmatic mainly non-marine group has been thought to represent a grade, not a clade at the base of the caenogastropod tree (Ponder & Lindberg 1997; Ponder et al. 2008),[2] but in a molecular analysis by McArthur and Harasewych (2003), the non-marine architaenioglossans formed a monophyletic group. It traditionally comprised one terrestrial (Cyclophoroidea) and two freshwater groups

(Ampullarioidea and Viviparoidea), with the latter two having been treated as a single group in the past. We have added some marine taxa to this group (see below).

The first fossils attributed to (non-marine) architaenioglossans are Lower Carboniferous (Bandel 1993, 2002a), but according to Frýda et al. (2008), there are no unequivocal (non-marine) representatives in the Paleozoic. Their nervous system is hypoathroid to epiathroid and, while most possess a penis, it is formed from different anterior structures and probably independently in the three groups (see below).

Architaenioglossans share some plesiomorphic characters with lower gastropod taxa, including a partially or fully hypoathroid nervous system and subradular organ, but the three groups share few synapomorphies. Considerable modification of the pallial organs has occurred in all three groups, especially the loss of the gill and osphradium in the terrestrial cyclophorids. The viviparids are modified for suspension feeding and pallial brooding, while the ampullariids have a lung as a separate pocket of the mantle cavity.

Viviparids are distributed throughout most continents but are absent from Central and South America except as introduced species. They have a gill modified for suspension feeding and large collar-like flaps arising from the posterior part of the head. Their name is derived from the fact that they all brood their young in a brood pouch formed from the pallial oviduct.

Ampullariids are a tropical and subtropical group of large, mostly globose snails that can be amphibious. They have developed a separate lung in the highly vascularised anterior part of the mantle cavity with the relative development of the lung varying according to how amphibious the species is (Andrews 1965). Compared to those seen in viviparids, ampullariids have similar, but longer, flaps ('siphons') arising from the posterior part of the head. Ampullariids are naturally found in Africa, Madagascar, South-East Asia, and Central and South America. A few species, notably the invasive pest *Pomacea canaliculata*, originally from South America, have been introduced to many parts of South East Asia and elsewhere. They lay egg masses and have direct-developing young. Besides possessing a lung and a gill, ampullariids differ from viviparids, which were originally included with them in the same superfamily, in all being oviparous, having a hypoathroid, instead of a partially epiathroid, nervous system, and in having a penis derived from the mantle, while in viviparids it is formed from the modified right cephalic tentacle. Viviparids and ampullariids both have a concentric operculum which is calcified in some palaeotropical species of ampullariids, and they share similar sperm and osphradial ultrastructure.

The largest and most morphologically diverse architaenioglossan group are the exclusively terrestrial Cyclophoroidea, comprising several families (see Appendix). Members of this group have a hypoathroid nervous system, and the penis, when present, is cephalic, as in many other caenogastropods. Their shell is highly variable in size (1–60 mm), shape, and coiling. The operculum is circular and multispiral. Despite statements

---

[2] Barker (2001) conducted a morphological analysis on mainly terrestrial gastropods in which architaenioglossans were more closely related to Neritopsina and Neomphaloidea.

in the literature to the contrary, an osphradium is found in the mantle cavity of at least some taxa (Creek 1953; Kasinathan 1975), but a ctenidium is absent.

Cyclophorids are found in southern Africa, Madagascar, Asia, Australasia, and many islands of the Indo-Pacific, while members of the Neocyclotidae are found in South America, Central America, the West Indies, and the South Pacific. Diplommatinids occur in Europe, Madagascar, East Asia, and the western Pacific, and pupinids are found throughout most of the Indo-Pacific. Two small families, the craspedopomatids and maizaniids, are African while another small family, the Megalomastomatidae, is restricted to Jamaica. A small European family, Aciculidae, was previously included in the Littorinoidea.

19.7.1.1 Campanilida

Campanile was considered a member of the Cerithioidea by Houbrick (1981) and intermediate between caenogastropods and heterobranchs by Haszprunar (1988f). In the molecular analysis of Colgan et al. (2007), *Campanile* clustered with Ampullariidae. In an earlier analysis by Harasewych et al. (1998, Figure 4), *Campanile* clustered with the ampullariids and cyclophorids in a basal subclade with cerithiimorphs forming another subclade, with the single viviparid sister to both or the remaining caenogastropods. Hayes et al. (2009) included *Campanile* in their analysis investigating ampullariid phylogeny.

While the shell features of *Campanile* have little resemblance to those of ampullariids, the only living ampullinid *Cernina fluctuata* does. Anatomically *Cernina* also has features in common with ampullariids (Kase 1990; Kase & Ishikawa 2003), including the very similar hypoathroid nervous system, except that the pleural ganglia are separated from the pedal ganglia by shorter connectives. *Campanile* and *Plesiotrochus* have an epiathroid nervous system. Unlike other caenogastropods, both ampullariids and *Cernina* have a long intestine.

In *Cernina,* the snout is very short and bilobed and the cephalic tentacles long; they are shorter in *Campanile* and *Plesiotrochus* which also have a longer, narrower snout, although still shorter than that in cerithiimorphs. The eyes in *Cernina* and ampullariids are on peduncles while in *Campanile* and *Plesiotrochus* they are on the outer sides of the tentacle bases.

Like *Campanile* and *Plesiotrochus*, and unlike ampullariids, *Cernina* lacks a penis and has an open pallial gonoduct in both sexes. All have a seminal receptacle opening to the mantle cavity. Both *Cernina* and *Campanile* are protandrous hermaphrodites, while *Plesiotrochus* is a simultaneous hermaphrodite.

Cernina, Campanile, Plesiotrochus, and ampullariids have a small bipectinate osphradium that lacks the diagnostic cytology of sorbeoconchs (Haszprunar 1985g, 1992). Haszprunar (1992) did not directly contrast the osphradia of *Campanile* and ampullariids, but based on his figures and descriptions there are important differences, including the lack of two cell layers in *Campanile*. Interestingly, the

osphradial morphology and ultrastructure of ampullariids and viviparids are essentially the same (Haszprunar 1985g). The osphradial ultrastructure of *Campanile* (Haszprunar 1992) differs from that of any other caenogastropod and possesses a specialised cell with reticulate endoplasm that Haszprunar tentatively identified as an Si4 cell although the reticulate, rather than the finely concentric, contents of Si4 cells suggest that the specialised cell in *Campanile* is not an Si4 cell.

Eusperm ultrastructural data of ampullariids (and other architaenioglossans including *Campanile* and *Plesiotrochus*) are similar to Cerithioidea and similar parasperm are also present in these groups (Healy & Jamieson 1981; Healy 1982; Healy 1986, 1988, 1993a; Winik et al. 2001). Campaniloidea eusperm is characterised by having seven to eight straight mitochondria with unmodified cristae partly enclosed in a dense, segmented sheath (Healy 1993a). Apparently the eusperm of *Cernina* (Ampullinidae) is very similar (Kase & Ishikawa 2003).

Based on the above discussion of both molecular and morphological data we include Campaniloidea in the Architaenioglossa, bringing the total architaenioglossan superfamilies to five: Cyclophoroidea, Ampullarioidea, Viviparoidea, Ampullinoidea, and Campaniloidea. Similar arguments were made by Kase & Ishikawa (2003). This differs from the taxonomic allocation of Bouchet et al. (2017) who placed the 'subcohort' Campanilimorpha in the Sorbeoconcha outside Cerithiimorpha.

19.7.2 Sorbeoconcha

This grouping includes all caenogastropods other than the architaenioglossan taxa. Of the living taxa, the Cerithioidea are basal. Sorbeoconchs differ from most[3] members of the Architaenioglossa (and all other gastropods) in having control over the inhalant water flow into the mantle cavity instead of the exhalant flow, enabling the chemosensory function of the osphradium to be emphasised (Ponder & Lindberg 1997; Lindberg & Ponder 2001). There is a corresponding increase in osphradial size and complexity and the presence of lateral ciliated fields, and distinctive Si4 cells are unique sorbeoconch characters (Haszprunar 1985g, 1988; Ponder & Lindberg 1997).

Ponder and Lindberg (1997) noted that sorbeoconchs are also distinguished by the possession of an epiathroid nervous system (the pleural ganglia lie close to the cerebral ganglia), a seminal vesicle, a coiled radular sac, and the formation of a polar lobe in early development (Freeman & Lundelius 1992).

19.7.2.1 Cerithiimorpha

The first cerithiimorphs appeared in the Paleozoic, and modern members are grouped in a single superfamily, Cerithioidea, and divided into several families (see Appendix). While most families in this group are marine, several, including Thiaridae and Pleuroceridae, are found in fresh water (e.g., Lydeard et al. 2002; Strong et al. 2011). While it is not entirely clear how many

---

[3] Campaniloideans are an exception.

separate invasions of fresh water have occurred, one analysis suggests there have been at least three (Strong et al. 2011).

Cerithiimorphs have several distinctive morphological features including the lack of a penis, and they typically have open pallial genital ducts. Several have an anterior siphon that appears to be independently developed from that seen in the Hypsogastropoda. The ultrastructure of the sperm is also distinctive. Cerithioidean eusperm is characterised by having four straight, parallel mitochondria with complex cristae, and in lacking a segmented sheath (Healy 1983, 1988).

Cerithioideans have a well-developed snout with small jaws, and the foot is not very broad. A few cerithioideans have an epipodium and epipodial tentacles (Bittiinae, Litiopidae), but whether these structures are homologous with those in vetigastropods is unknown. The osphradium is simple and linear or in some develops weak lamellae.

19.7.2.2 Hypsogastropoda

Comprising all living caenogastropods other than Architaenioglossa and Cerithioidea, this group contains the great majority of extant caenogastropods. According to the classification of Bouchet et al. (2017) it contains one superorder, one order, and 31 superfamilies (see also Appendix).

Various higher taxa have been proposed to encompass higher hypsogastropods. These include the Pleurembolica and Vermivora, while another group, the Latrogastropoda, includes the Naticoidea + Pleurembolica (see Ponder et al. 2008; Golding et al. 2009a for further details and discussion).

Although relationships within the hypsogastropods remain largely uncertain, recent analyses have indicated there are two main groups, termed the asiphonate and siphonate clades (Colgan et al. 2007; Ponder et al. 2008), the latter corresponding to our usage of Latrogastropoda.

Hypsogastropoda is often supported in molecular analyses and is characterised by some ultrastructural characters (sperm, osphradial, and larval) (Ponder et al. 2008), but there are no known unique morphological characters. Members of the group have a single statolith in each statocyst (rather than several statoconia), and many have a cephalic penis.

19.7.2.2.1 Asiphonate Clade

This grouping is part of the Littorinimorpha as used by Bouchet and Rocroi (2005) and corresponds to the 'asiphonate group' of Ponder et al. (2008). The Littorinimorpha itself, as used by Bouchet and Rocroi (2005), is not monophyletic.

Its members have several features seen in cerithiimorphs; the osphradium is often a simple ridge, and the tentacular nerves are typically bifurcate, but males are phallate, and spermatozoa are sometimes dimorphic. Spawn may be laid in jelly masses, or eggs are encapsulated in globular to lens-shaped capsules. Plesiomorphically an oesophageal gland is present and the stomach contains a protostyle, although in groups that possess a crystalline style (notably Rissooidea and Truncatelloidea), the oesophageal gland is lost.

Some asiphonate superfamilies, the Littorinoidea, Cingulopsoidea, Rissooidea, and Truncatelloidea, have mainly conical shells and usually have paucispiral opercula.

The Littorinoidea comprises a few small marine families (see Appendix), but by far the most significant is the mainly intertidal Littorinidae with 13 extant genera and about 180 species grouped in three subfamilies (Reid 1998). While no littorinids have adopted a terrestrial life, some live in the semiterrestrial conditions of the supralittoral fringe, where most are confined due to their reproductive reliance on pelagic egg capsules (Reid 1989). Also included in this superfamily are the moderately diverse terrestrial Pomatiidae and Annulariidae (see below). Species in the amphibious Indian genus *Cremnoconchus* (Lacuninae) are the only freshwater littorinids.

Littorinids (winkles or periwinkles) dominate the mid to upper parts of intertidal shores around the world. They are small to moderate-sized taxa, often abundant and have diverse reproductive strategies. While some retain benthic jelly-coated spawn, others have pelagic egg capsules, and a number are viviparous. While many have pelagic larvae, some are direct developers, including a few examples of ovoviviparity in *Echininus*, *Tectarius*, and *Littoraria*, but these have evolved recently (Reid 1989). Indeed, typical littorinids appear to have a relatively recent origin, with the first fossils appearing in the Paleocene (Bandel 1993; Reid 1998), but the group is likely to have had an earlier origin with its members difficult to assign because of their rather featureless shells.

The terrestrial littorinoideans (Pomatiidae and Annulariidae) have an Upper Cretaceous origin and are found in Madagascar and the Americas, where they are particularly diverse, and also in the West Indies, Europe, the Middle East, India, and Africa. The Madagascan *Tropidophora cuveriana* is the largest living terrestrial operculate gastropod at around 50 mm in diameter. They are mostly ground-dwelling, but some species of the annulariid genus *Tudora* can be arboreal. Most are found in moist forests, but some (e.g., the Central American *Choanopoma*) are found in dry forests (Barker 2001b).

The highly modified pelagic 'heteropods' are, perhaps surprisingly, related to the Littorinidae (e.g., Colgan et al. 2007).

The largest groups of asiphonates are the marine Rissooidea and largely non-marine Truncatelloidea, which are mostly of small size (most <10 mm), and, as currently classified, together comprise 29 living families (see Appendix). Of these families, 18 are marine or estuarine, ten (all truncatelloideans) primarily or entirely non-marine, seven occur exclusively in fresh water, and three (Assimineidae, Truncatellidae, and Pomatiopsidae) have some terrestrial members. Assimineids and stenothyrids are primarily marine or estuarine, and both families contain some freshwater taxa.

While the shells of many rissooideans and truncatelloideans are basically conical, a great deal of variation in shell morphology is found in the group, including low-spired to planispiral shells, to those with very elongate shells. Some are partially to completely uncoiled, culminating in the tubular Caecidae. The operculum can be paucispiral to multispiral; in some species it bears one or more pegs on the inner surface to which muscles are attached.

The large marine family Rissoidae has been recently subdivided (Criscione et al. 2016), with the Rissoinidae and

Zebinidae now recognised as separate families, both having a Jurassic origin (Ponder 1985; Bandel 1993; Criscione et al. 2016). These three families, along with a few small marine families (see Appendix), together comprise the Rissooidea.

Some mainly freshwater truncatelloideans (Hydrobiidae, Tateidae, and Cochliopidae) include some estuarine and amphibious representatives, but others such as Amnicolidae, Bithyniidae, Moitessieriidae, and Lithoglyphidae do not. Most freshwater truncatelloideans have poor dispersal capabilities, and hence there are many narrow-range species, some confined to single streams, springs, or caves. Significant radiations of freshwater truncatelloideans have occurred in North America, Mexico, Australia, New Zealand, New Caledonia, Sulawesi, and southern Europe in both surface and subterranean waters (e.g., Ponder et al. 1993; Colgan & Ponder 1994; Haase & Bouchet 1998; Hershler 1999; Ponder & Colgan 2002; Hershler & Liu 2004; Haase 2008; Haase et al. 2010; Falniowski & Szarowska 2011; Zielske et al. 2011; Zielske & Haase 2014).

The truncatelloidean Pomatiopsidae have been comprehensively studied in a series of publications by G. M. Davis and his collaborators. A few species are important as significant intermediate hosts of medically important trematodes (see Chapter 10). The family comprises two subfamilies, the Asian freshwater Triculinae, and the much more widespread and ecologically diverse Pomatiopsinae (South America, northwest North America, Japan, Manchuria, China, Taiwan, the Philippines, Sulawesi, Australia, and South Africa). Members of this subfamily include freshwater aquatic and amphibious taxa and a few that are estuarine or terrestrial (Davis 1981). Most other non-marine truncatelloidean families are aquatic in fresh water, but the Truncatellidae are supralittoral to terrestrial, and the Assimineidae are mostly amphibious with marine, freshwater, and terrestrial members, including a few arboreal taxa such as *Pseudocyclotus* in Papua New Guinea.

Another superfamily of small, mainly conical marine species is the Cingulopsoidea, which comprises three divergent families (Eatoniellidae, Cingulopsidae, and Rastodentidae) that appear to form a monophyletic group (Criscione & Ponder 2013). While eatoniellids and cingulopsids are microphagous feeders like the great majority of rissooideans and truncatelloideans, the rastodentids may feed on bryozoans (Ponder 1966), and at least some species of Rissoinidae swallow whole foraminiferans (Ponder 1968).

Molecular analyses have shown that the Eulimoidea is related to the Rissooidea – these mostly elongate small snails are parasites of echinoderms and range from external limpets to internal shell-less 'worms'.

The asiphonate taxa also include the tall-spired, mostly small-sized but very diverse Triphoroidea that feed on sponges, and the Epitonioidea, a medium-sized group of cnidarian associates. Both these taxa were included in the polyphyletic Ptenoglossa, along with the Eulimidae.

Other asiphonate marine groups include the deep-sea Abyssochrysoidea, the limpet-like Hipponicoidea, the worm tube-like Vermetidae, the Vanikoridae, and the predatory moon snails (Naticoidea).

19.7.2.2.2   Latrogastropoda

This grouping contains the neogastropods and the remaining taxa that were included in the Littorinimorpha. While the taenioglossate latrogastropods may not form a monophyletic group, we recognise some ordinal taxa that reflect the relationships of these entirely marine gastropods (see Box 19.1 and the Appendix).

The Strombidae ('strombs' or 'conchs') and the Xenophoridae ('carrier shells') comprise the Strombida while the Calyptraeida includes the Capulidae, which contains both snails and limpets, and the Calyptraeidae ('slipper limpets'). The Cypraeida includes species of mostly medium to large size, such as the Cypraeidae (cowries), Ovulidae, and Velutinoidea. Velutinoidea differ from other caenogastropods in possessing a pair of shell muscles and are the only group to include members that have become slug-like. Whether or not these two attributes are related remains to be seen.

19.7.2.2.3   Peogastropoda

This grouping of mainly carnivorous gastropods was proposed by Simone (2011) and includes two important orders, the Tonnida and the Neogastropoda. Another name, Pleurembolica, proposed by Riedel (2000), is not used here as it also included the Calyptraeida.

19.7.2.2.4   Tonnida

Included here are the Ficoidea (Ficidae) and Tonnoidea (including the often large-sized tonnids, cassids, cymatiids, and charoniids). These gastropods are all proboscidate predators feeding on a range of invertebrates (see Section 19.4.3).

19.7.2.2.5   Neogastropoda

The Neogastropoda are almost exclusively marine and nearly all carnivorous. Many are large and conspicuous, and the group includes diverse families such as Muricidae, Buccinidae, Nassariidae, Mitridae, Costellariidae, Columbellidae, Marginellidae, Volutidae, Conidae, and several speciose families previously lumped together as 'Turridae'.

Neogastropods have long been recognised as a group supported by some novel gut characters, notably the nature of the radula, and some anatomical characters shared by many (but not all) of them including tubular accessory salivary glands, the valve and gland of Leiblein and the rectal gland (Ponder 1974). They also often have higher chromosome numbers and more cellular DNA than other gastropods (see Chapter 2). The group was, however, not well recovered in molecular analyses (see Ponder et al. 2008 for details), and the relationships within the group as a whole are still poorly resolved.

The identity of the sister taxon of neogastropods has been contentious, but among living gastropods it now appears to be the Tonnoidea. The origins of the group are possibly in the Triassic, as the Upper Triassic to Cretaceous Maturifusidae could either be a member or a sister taxon (Ponder et al. 2008). The Lower Triassic genus *Pseudotritonium* (Purpurinidae) has been suggested as the earliest member of this lineage (Nützel 2010).

20 Gastropoda III – The Heterobranchia

20.1 INTRODUCTION

This large, important group of gastropods shows the greatest morphological diversity of all the gastropod clades. They range from tiny worm-like interstitial slugs to large, colourful nudibranchs, and from intertidal limpets to terrestrial snails and slugs. They live from the deep sea to the intertidal, and in fresh water and virtually all terrestrial environments.

Heterobranchs include the sea slugs and bubble shells previously known as 'opisthobranchs', most of the terrestrial snails and all terrestrial slugs (previously known as 'pulmonates'), as well as some significant freshwater and other smaller marine groups. Many heterobranchs are well known – the nudibranch sea slugs for their beauty and diversity of form, land snails as food (escargot) and some, along with a number of slugs, as pests in crops and gardens. Less well known is the fact that a variety of heterobranchs, such as sea hares (*Aplysia*), pleurobranchs, the pond snail (*Lymnaea*), and some land snails, have served as important experimental animals from which we have learned much about nervous system function and even the basis of memory (see Chapter 10). Heterobranch land snails also include some of the most endangered animals on earth, accounting for a large proportion of known extinctions in the last 200 years (see Chapter 10).

In this chapter, we outline aspects of the diversity, natural history, biology, structure, evolution, and classification of the Heterobranchia.

20.2 PHYLOGENY AND CLASSIFICATION

The concept of the Heterobranchia is relatively new, being originally proposed by Gerhard Haszprunar (Haszprunar 1985b, 1988f). This significant insight markedly contrasted with the long-used classification of Thiele (1929–1935), in which three 'subclasses' – Prosobranchia, Opisthobranchia, and Pulmonata, were recognised. Instead, Heterobranchia incorporates Pulmonata, Opisthobranchia, several marine families, and one freshwater group (Valvatidae) that were included in Thiele's Mesogastropoda. Haszprunar (1985b) formally recognised this group as a clade, but the close relationships of 'opisthobranchs' and 'pulmonates' had long been suspected, as indicated by their recognition in a single group (Euthyneura) by some workers since the late 1800s and subsequently (e.g., Boettger 1954). Hyman (1967 p. 549), while maintaining the Opisthobranchia and Pulmonata as 'subclasses', stated that ' … no great difference exists between [them]; … hence the union of these two subclasses under the old group Euthyneura has some justification'.

Earlier, the planktonic pteropods were considered so different from other molluscs that they were made a separate class by Cuvier (1817), although their 'opisthobranch' affinities were later recognised by Blainville (1824) and subsequent workers (see Tesch 1904 for review).

The name Heterobranchia is now widely accepted, although an alternative name Heterostropha is used by a few palaeontologists. The name Heterostropha has also been used for just the basal heterobranchs (see Appendix), as has Allogastropoda. These taxa at the base of the heterobranch tree (the 'lower heterobranchs') are a paraphyletic collection of disparate non-euthyneuran families (Haszprunar 1988f; Ponder & Lindberg 1997; Dinapoli & Klussmann-Kolb 2010) that share some plesiomorphic characters, and most of them were previously thought to be 'prosobranchs'. Thus, no encompassing name for these basal taxa is needed except as a convenient term for discussion. Consequently, we refer to these taxa simply as the 'lower heterobranchs' in the text following. A detailed analysis involving all the extant family-level groups has not yet been carried out, although some recent molecular phylogenetic analyses (see text following) have included some basal groups, confirming that the situation is more complex than previously realised. For example, *Rhodope* (Rhodopemorpha), tiny, interstitial, marine turbellarian-like slugs that have lost both shell and radula, were originally thought to be highly modified 'opisthobranchs', but recently it has been shown that they are not euthyneurans but 'lower heterobranchs' (Schrödl et al. 2011).

Most phylogenetic treatments of heterobranchs focused on either 'pulmonates' or 'opisthobranchs' (for details see reviews in Mordan & Wade 2008; Wägele et al. 2008; Holznagel et al. 2010; Jörger et al. 2010; Dayrat et al. 2011; Wägele et al. 2014), so questions remained about relationships within the Euthyneura. Opisthobranchia and Pulmonata were recognised as 'subclasses' until recently, but these two groups are not well delineated. The most recent molecular analyses involving a reasonable number of taxa (Grande et al. 2004, 2008; Klussmann-Kolb et al. 2008; Dinapoli & Klussmann-Kolb 2010; Holznagel et al. 2010; Jörger et al. 2010; Dayrat et al. 2011; Göbbeler & Klussmann-Kolb 2011; Schrödl et al. 2011; Golding 2012; Kocot et al. 2013; Wägele et al. 2014; Kano et al. 2016) did not recover either Opisthobranchia or Pulmonata as monophyletic groups. Similarly, the Euthyneura (= Opisthobranchia + Pulmonata), which has usually been considered a well-supported monophyletic group, has the Pyramidellidae and Rissoellidae included within it, two groups previously grouped with the 'lower heterobranchs' (Dinapoli & Klussmann-Kolb 2010). These analyses differ substantially in the determination of euthyneuran basal taxa and generally dispute long-held ideas based on other sources of data (see Chapter 12 for discussion on issues with molecular analyses). The classification provided here (Box 20.1, Figure 20.1, and the Appendix) must be considered somewhat unstable; as with many other molluscan groups, there will continue

BOX 20.1   Summary of the Higher Classification of Living Heterobranchs

(Infraclass) **Heterobranchia**
   'Lower heterobranchs' (a paraphyletic grouping of convenience) – the primitive mostly marine groups (see Appendix for more detail), many of which were previously thought to be 'prosobranchs', and the detail of their relationships remains poorly resolved. It also includes the Rhodopoidea, that were included in 'Opisthobranchia'. Some taxa thought until recently to be 'lower heterobranchs' are now included in other groups.

(Order) Orbitestellida (Orbitestelloidea and Cimoidea)
(Order) Ectobranchia (Valvatoidea)
(Order) Allomorpha (Rhodopoidea and Murchisonelloidea)
(Order) Architectonicida
(Suborder) Architectonicina (Architectonicoidea and Mathildoidea)
(Suborder) Omalogyrina (Omalogyroidea)

(Cohort) **Euthyneura** (comprises most of the 'opisthobranchs' and all the 'pulmonates', as well as the Pyramidellidae and Rissoellidae previously included in the 'lower heterobranchs', and the glacidorbids, sometimes included in the 'lower heterobranchs')
(Subcohort) **Acteonimorpha**

(Order) Acteonida (Acteonoidea)
(Order) Rissoellida (Rissoelloidea)

(Subcohort) **Ringipleura**
(Megaorder) **Ringiculata**
(Order) Ringiculida (Ringiculoidea)
(Superorder) **Nudipleura** (the pleurobranchs and nudibranchs)
(Order) Pleurobranchida (Pleurobranchoidea)
(Order) Nudibranchia
(Suborder) Doridina
(Infraorder) Bathydoridoidei (Bathydoridoidea)
(Infraorder) Doridoidei (= Doridacea) (Doridoidea, Chromodoridoidea, Phyllidioidea, Onchidoridoidea, Polyceroidea)
(Suborder) Cladobranchia (Arminoidea, Dendronotoidea, Doridoxoidea, Proctonotoidea, Tritonioidea)
(Suborder) Aeolidiina (Flabellinoidea, Fionoidea, Aeolidioidea)
(Subcohort) **Tectipleura** (remaining 'opisthobranchs', including Umbraculoidea, and all 'pulmonates')
(Megaorder) **Euopisthobranchia** (includes the 'opisthobranchs' other than the nudibranchs, Acochlidia, Sacoglossa, and Rhodopemorpha)
(Order) Umbraculida (Umbraculoidea)
(Clade) **Pleurocoela**
(Order) Cephalaspidea (Bulloidea, Haminoeoidea, Philinoidea, Cylichnoidea, Newnesioidea)
(Suborder) Runcinina (Runcinoidea)
(Order) Aplysiida (= Anaspidea) (Aplysioidea, Akeroidea)
(Order) Pteropoda
(Suborder) Thecosomata
(Minorder) Euthecosomata (Cavolinioidea, Limacoidea)
(Minorder) Pseudothecosomata (Cymbulioidea)
(Suborder) Gymnosomata (Clionoidea, Hydromyloidea)
(Subcohort) **Panpulmonata** (includes the Sacoglossa and Acochlidia [previously 'Opisthobranchia'], two groups previously included in the 'lower heterobranchs' – Pyramidellidae and Glacidorbidae, as well as the classic 'pulmonates')
(Superorder) **Pylopulmonata**
(Order) Amphibolida
(Suborder) Amphibolina (Amphiboloidea)
(Suborder) Glacidorbina (Glacidorboidea)
(Order) Pyramidellida (Pyramidelloidea)
(Superorder) **Siphonariia**
(Order) Siphonariida (Siphonarioidea)

(Superorder) **Acochlidia**
(Order) Acochlida (Acochlidioidea, Parhedyloidea)
(Superorder) **Sacoglossa** (Oxynooidea, Plakobranchoidea, Platyhedyloidea)
(Superorder) **Hygrophila** (Chilinoidea, Lymnaeoidea)
(Superorder) **Eupulmonata**
(Order) Ellobiida (Ellobioidea)
(Clade) **Geophila**
(Megaorder) **Systellommatophora**
(Order) Onchidida (Onchidioidea)
(Order) Veronicellida (Veronicelloidea)
(Megaorder) **Stylommatophora**
(Order) Helicida
(Suborder) Achatinina (Achatinoidea, Streptaxoidea)
(Suborder?) Scolodontina (Scolodontoidea)
(Suborder) Helicina (Coelociontoidea, Papillodermatoidea, Plectopyloidea, Punctoidea, Testacelloidea, Urocoptoidea, Succineoidea, Athoracophoroidea, Oleacinoidea, Halotrematoidea, Pupilloidea, Clausilioidea, Othalicoidea, Aillyoidea, Rhytidoidea, Limacoidea, Gastrodontoidea, Trochomorphoidea, Parmacelloidea, Zonitoidea, Helicarionoidea, Arionoidea, Sagdoidea, Helicoidea)

The classification to family level of living and extinct heterobranchs is summarised in the Appendix, and further discussion on the phylogeny and classification of this group is given next and in the latter part of this chapter (see Section 20.5.1).

to be debate about the details of heterobranch phylogeny and classification for some time.

In a morphological cladistic analysis by Wägele and Willan (2000) Pleurobranchida and Nudibranchia formed a clade for which they proposed the name Nudipleura. It was distinguished from other euthyneurans in having an androdiaulic or triaulic reproductive system, loss of the osphradium in adults, and possession of a blood gland (see Section 20.3.5). This grouping has since been found in most subsequent morphological and molecular analyses. For example, the molecular analysis by Jörger et al. (2010) recovered the Nudipleura as sister to a group comprising the remaining euthyneurans (excluding Acteonoidea, which grouped with the 'lower heterobranchs'), and pyramidelloideans, previously thought to be 'lower heterobranchs'. They introduced two new groups, the Euopisthobranchia which encompassed the Umbraculoidea, Aplysiida (= Aplysiomorpha, Anaspidea), Runcinacea, Pteropoda, and Cephalaspidea s.s., and the Panpulmonata, which included Siphonarioidea, Sacoglossa, Glacidorboidea, Pyramidelloidea, Amphiboloidea, Hygrophila, Acochlidia, and Eupulmonata. A similar topology has been obtained in other recent molecular analyses (Dinapoli & Klussmann-Kolb 2010; Göbbeler & Klussmann-Kolb 2011; Kano et al. 2016) but not by White et al. (2011). In the Kano et al. (2016) analysis, Ringiculidae was sister to the Nudipleura, and the clade comprising those two taxa was named Ringipleura.

Examples of living heterobranchs are shown in Figures 20.2 to 20.12.

20.2.1 Sister Group Relationships

The sister group relationships of the major heterobranch groups, and of the Heterobranchia as a whole, has been, and

remains, controversial. It has long been recognised by some workers that the similarities between basal 'opisthobranchs' and 'pulmonates' are so marked that they probably had a common origin, although others preferred separate origins from 'archaeogastropods' (Hubendick 1945; Morton 1955a) or 'mesogastropods'. Thus, 'opisthobranchs' were usually thought to be derived from 'mesogastropods' such as Rissooidea, Cerithioidea, or Littorinoidea (e.g., Boettger 1954; Fretter & Graham 1962; Ghiselin 1966; Fretter 1975b; Brace 1977b; Gosliner 1981; Visser 1988). This latter idea was attractive because it neatly explained the lack of a ctenidium (lost by virtue of small size) and right kidney, the lack of an oesophageal gland (lost in all rissooideans), and the basic similarity of their reproductive tract. 'Pulmonates' were often thought to be derived from shelled 'opisthobranchs' such as the Acteonidae (e.g., Pelseneer 1894; Boettger 1954; Gosliner 1981).

The origin of heterobranchs is probably located deep in gastropod evolution and separate from, although possibly sister to, the caenogastropods (e.g., Ponder & Lindberg 1997). There is continued uncertainty about the actual sister taxon, and because it is almost certainly not an extant taxon, molecular results will probably continue to be somewhat misleading. Certainly long-held ideas suggesting that heterobranchs arose from caenogastropods have been rejected. Molecular results suggest caenogastropods are sister to Heterobranchia (Colgan et al. 2003; McArthur & Harasewych 2003; Castro & Colgan 2010), with neritimorphs sister to Caenogastropoda + Heterobranchia (= Apogastropoda) (see Chapter 18). The molecular analyses obviously exclude extinct taxa, and thus a sister group relationship with an extinct lineage outside Caenogastropoda cannot be rejected. Some clues may be found in the basal heterobranchs *Hyalogyra*, *Hyalogyrina*, and *Xenoskenea*, which have a rhipidoglossate-like radula. Two of

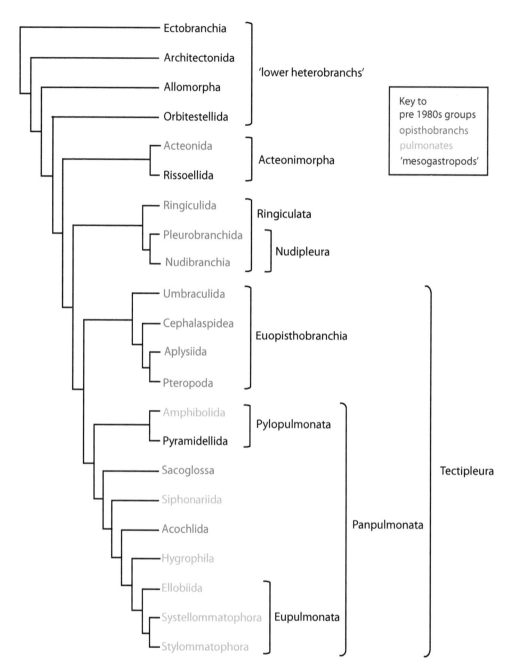

Key to
pre 1980s groups
opisthobranchs
pulmonates
'mesogastropods'

FIGURE 20.1   A summary tree showing likely relationships of the major heterobranch taxa based on recent phylogenetic studies.

these genera have crossed-platy shell structure (Hedegaard 1990; Kiel 2004), a feature previously thought to be restricted to the vetigastropods (e.g., Ponder & Lindberg 1997). These taxa have not, to date, been incorporated in molecular analyses.

In recent molecular analyses of heterobranchs (e.g., Jörger et al. 2010; Schrödl et al. 2011; Wägele et al. 2014; Kano et al. 2016), the sister to the Euthyneura is the unresolved 'lower heterobranchs' while the sister taxon to the Hygrophila + Acochlidia + Eupulmonata is a clade containing Glacidorboidea, Pyramidelloidea, and Amphiboloidea, and the siphonariids are at the base of the Panpulmonata with the Sacoglossa. These latter two taxa formed a clade in the analysis of Medina et al. (2011) which they called Siphoglossa, but

this has not been recovered as a monophyletic group in more recent analyses (e.g., Kano et al. 2016). It is likely that the details of some of the currently accepted relationships will change with more intensive gene and taxon sampling.

20.2.2   CHROMOSOME NUMBERS

The median number of chromosome pairs (n) in heterobranchs is 16, a value very close to the ancestral molluscan chromosome number (n = 17) calculated by Hallinan and Lindberg (2011). Thus, the tremendous anatomical, physiological, and ecological diversification of this group has not been correlated with a significant increase in chromosome number, except the Stylommatophora which have chromosome numbers ranging

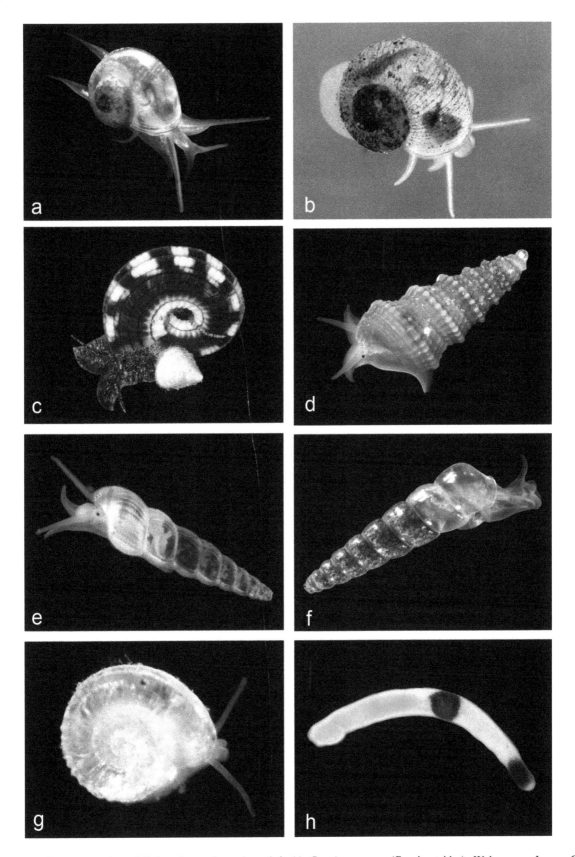

FIGURE 20.2 Some examples of living 'lower heterobranchs'. (a) *Cornirostra* sp. (Cornirostridae), Wakayama, Japan. Courtesy of H. Fukuda. (b) *Cincinna japonica* (Valvatidae), Lake Ogawara, Japan. Courtesy of H. Fukuda and M. Taru. (c) *Heliacus variagatus* (Architectonicidae), New South Wales, Australia. Courtesy of D. Riek. (d) *Mathilda* sp. (Mathildidae), Guadeloupe, French West Indies. Courtesy of L. Charles - MNHN. (e) *Graphis* sp. (Cimidae), Martinique, French West Indies. Courtesy of P. Maestrati - MNHN. (f) *Murchisonella* sp. (Murchisonellidae), Vanuatu. Courtesy of D. Brabant - MNHN. (g) *Orbitestella* sp. (Orbitestellidae), French Polynesia. Courtesy of G. Paulay. (h) *Rhodope* sp. (Rhodopidae), Guadeloupe, French West Indies. Courtesy of P. Maestrati - MNHN.

FIGURE 20.3   A rissoellid and some examples of living acteonoideans and cephalaspideans. (a) *Rissoella* cf. *diaphana* (Rissoellidae), Southern France. Courtesy of B. Brenzinger. (b) *Pupa strigosa* (Acteonidae) Vanuatu. Courtesy of D. Brabant - MNHN. (c) *Hydatina physis* (Aplustridae), Philippines. Courtesy of T. Gosliner. (d) *Acteocina* sp. (Acteocinidae), Philippines. Courtesy of P. Lozouet - MNHN. (e) *Scaphander japonicus* (Scaphandridae), Philippines. Courtesy of T. Gosliner. (f) *Haminoea cymbalum* (Haminoeidae), Philippines. Courtesy of T. Gosliner. (g) *Mariaglaja alexisi* (Aglajidae), Philippines. Courtesy of T. Gosliner. (h) *Gastropteron* sp. (Gastropteridae), Kavieng, Papua New Guinea. Courtesy of P. Maestrati - MNHN.

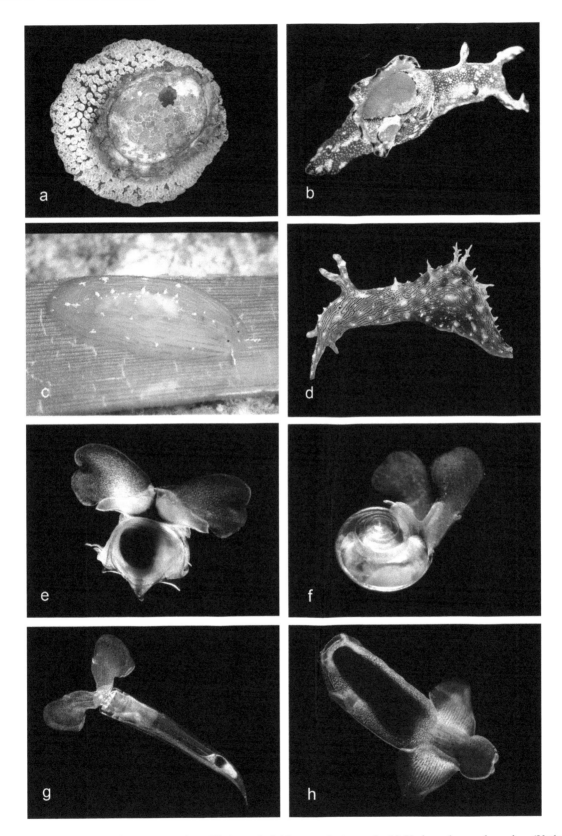

FIGURE 20.4 *Umbraculum* and some examples of living aplysioideans and pteropods. (a) *Umbraculum umbraculum* (Umbraculidae), South Madagascar. Courtesy of S. Schiaparelli - MNHN. (b) *Aplysia* sp. (Aplysiidae), Guadeloupe, French West Indies. Courtesy of P. Maestrati - MNHN. (c) *Phyllaplysia* sp. (Aplysiidae), Palau. Courtesy of T. Gosliner. (d) *Stylocheilus striatus* (Aplysiidae), New South Wales, Australia. Courtesy of M. Nimbs. (e) *Cavolinia uncinata* (Cavoliniidae), Northern subtropical Atlantic gyre. Courtesy of K. Peijnenburg. (f) *Limacina helicina antarctica* (Limacinidae), Southern subtropical convergence zone. Courtesy of K. Peijnenburg. (g) *Creseis virgula* (Creseidae), Equatorial Atlantic Ocean. Courtesy of K. Peijnenburg. (h) *Spongiobranchaea* sp. (Pneumodermatidae), Northern subtropical Atlantic gyre. Courtesy of K. Peijnenburg.

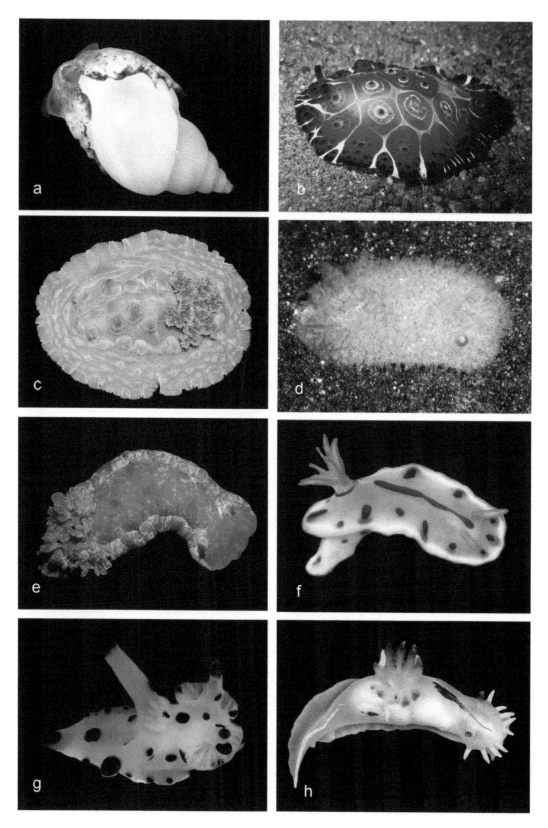

FIGURE 20.5 *Ringicula* and some examples of living nudipleurans. (a) *Ringicula* sp. (Ringiculidae), Kavieng, Papua New Guinea. Courtesy of P. Maestrati - MNHN. (b) *Pleurobranchus weberi* (Pleurobranchidae), Philippines. Courtesy of T. Gosliner. (c) *Asteronotus* sp. (Discodorididae), Madang, Papua New Guinea. Courtesy of A. Berberian - MNHN. (d) *Acanthodoris planca* (Onchidorididae), South Africa. Courtesy of T. Gosliner. (e) *Hexabranchus sanguineus* (Hexabranchidae), Bali. C. Watanabe, reproduced here under Creative Commons Attribution 2.0 Generic license [https://creativecommons.org/licenses/by/2.0/] (f) *Hypselodoris perii* (Chromodorididae), Philippines. Courtesy of T. Gosliner. (g) *Thecacera* sp. (Polyceridae), Philippines. Courtesy of T. Gosliner. (h) *Polycera* sp. (Polyceridae), South Africa. Courtesy of T. Gosliner.

FIGURE 20.6 Some examples of living nudibranchs. (a) *Aeolidia loui* (Aeolidiidae), California. Courtesy of T. Gosliner. (b) *Tenellia* sp. (Trinchesiidae), Philippines. Courtesy of T. Gosliner. (c) *Glaucus bennettae* (Glaucidae), New South Wales, Australia. Courtesy of D. Riek. (d) *Armina scotti* (Arminidae), Philippines. Courtesy of T. Gosliner. (e) *Bornella* sp. (Bornellidae), South Madagascar. Courtesy of M. Poddubetskaia - MNHN. (f) *Melibe leonina* (Tethydidae), California. © Gerald and Buff Corsi/Focus on Nature, Inc. (g) *Reticulidia fungia* (Phyllidiidae), Galapagos. Courtesy of T. Gosliner. (h) *Phyllidia* sp. (Phyllidiidae), Madang, Papua New Guinea. Courtesy of A. Berberian - MNHN.

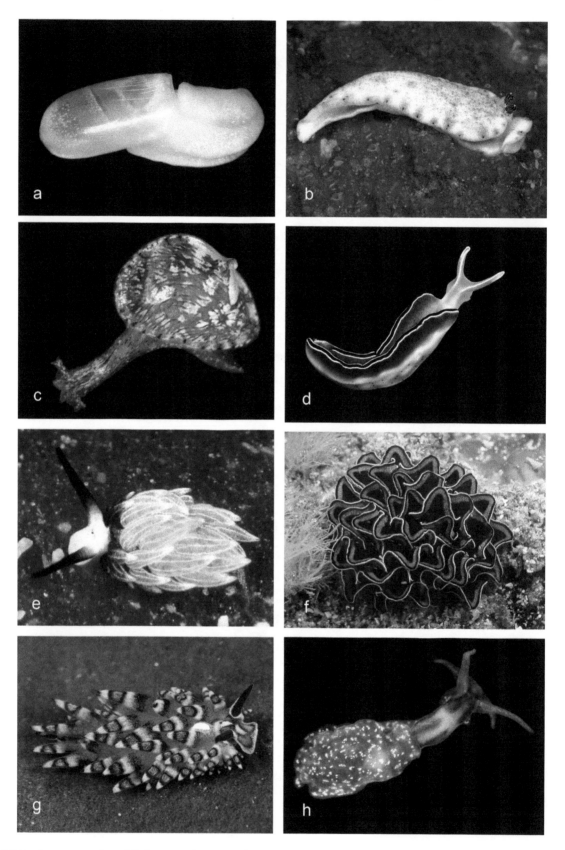

FIGURE 20.7 Some examples of living sacoglossans and an acochlidian. (a) *Ascobulla* sp. (Ascobullidae), Kavieng, Papua New Guinea. Courtesy of L. Charles - MNHN. (b) *Dermatobranchus caeruleomaculatus* (Arminidae), Galapagos. Courtesy of T. Gosliner. (c) *Berthelinia* cf. *caribbea*, French West Indies, Guadeloupe. Courtesy of P. Maestrati - MNHN. (d) *Elysia* sp. (Plakobranchidae), Papua New Guinea. Courtesy of A. Berberian - MNHN. (e) *Costasiella usagi* (Costasiellidae), Galapagos. Courtesy of T. Gosliner. (f) *Cyerce nigricans* (Hermaeidae), South Madagascar. Courtesy of P. Maestrati - MNHN. (g) *Stiliger ornatus* (Limapontiidae), tropical Indo-West Pacific. Courtesy of T. Gosliner. (h) *Acochlidium* sp. (Acochlidiidae), Panglao, Philippines. Courtesy of S. Tagaro - MNHN.

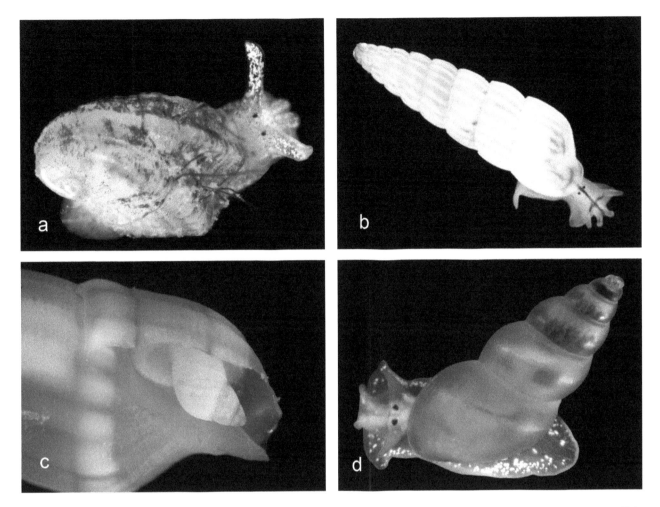

FIGURE 20.8   Some examples of living pyramidelloideans. (a) *Amathina* cf. *bicarinata* (Amathinidae), Kavieng, Papua New Guinea. Courtesy of L. Charles - MNHN. (b) *Turbonilla gravicosta* (Pyramidellidae), New South Wales, Australia. Courtesy of D. Riek. (c) *Odostomia*? sp. (Pyramidellidae) on a terebrid, Panglao, Philippines. Courtesy of P. Lozouet - MNHN. (d) Unknown pyramidellid, Vanuatu. Courtesy of D. Brabant - MNHN.

from five to 34 pairs. Higher numbers are seen in some heterobranchs shown to be polyploids (see Chapter 2). Taxon sampling is poor in the 'lower heterobranchs' with a single report of n = 10 in *Valvata tricarinata*.

Chromosome numbers in the euthyneurans are generally well sampled (>500 species) and range from ten to 18 pairs (Hallinan & Lindberg 2011), but several important groups have yet to be sampled, including the Umbraculoidea, Acochlidia, Glacidorboidea, and Trimusculoidea. Amphiboloidea and Pyramidelloidea are poorly sampled, with 18 and 17 pairs respectively. In the well-sampled Nudipleura, the chromosome numbers range from 12 to 16 pairs, with 13 being the usual number in nudibranchs and 12–13 in pleurobranchs. Only a single species of runcinoidean is known (n = 17), and Umbraculoidea is unsampled, but the remaining Euopisthobranchia are mainly well sampled with values between 10 and 17 pairs, with Cephalaspidea and Aplysiida usually having a haploid number of 17. Except for the Stylommatophora, chromosome numbers in the Panpulmonata cluster between 13 and 18 pairs. Sampling is good in Siphonarioidea and Sacoglossa where there are between 14 and 17 pairs, with 17 the usual number

for sacoglossans. Hygrophila range between 16 and 19 pairs. In the Eupulmonata, the Stylommatophora range between five and 44 pairs, with Succineidae 5–25, Athoracophoridae 44, this probably the result of polyploidy (Burch & Patterson 1971), with the remainder ranging between 20 and 34. The Stylommatophora have been identified by Hallinan and Lindberg (2011) as possible products of whole genome duplication. Besides the increased number of chromosomes, chromosome size is also elevated in the Stylommatophora (Gregory 2011). In the remaining eupulmonates (Systellommatophora and Ellobioidea), the chromosome numbers range between 16 and 18 pairs.

20.3   MORPHOLOGY

Useful reviews of the morphology of 'opisthobranchs' and 'pulmonates' can be found in Hyman (1967) and Franc (1968), for 'pulmonates' Luchtel et al. (1997), and 'opisthobranchs' Thompson and Brown (1976) and Gosliner (1994).

While most of the 'lower heterobranchs' closely resemble other shelled gastropods, at least in external features, the

FIGURE 20.9   Some examples of marine and freshwater panpulmonates. (a) *Salinator tecta* (Amphibolidae), New South Wales, Australia. Courtesy of R. Golding. (b) *Siphonaria funiculata* (Siphonariidae), ventral view, New South Wales, Australia. Courtesy of D. Riek. (c) *Smeagol climoi* (Otinidae), Wellington, New Zealand. Courtesy of WFP. (d) *Ophicardelus sulcatus* (Ellobiidae), New South Wales, Australia. Courtesy of D. Riek. (e) *Pythia* sp. (Ellobiidae), Papua New Guinea. Courtesy of P. Maestrati - MNHN. (f) *Cassidula* sp. (Ellobiidae), Papua New Guinea. Courtesy of P. Lozouet - MNHN. (g) *Lymnaea stagnalis* (Lymnaeidae), Europe. Courtesy of L. Peters. (h) *Physa acuta* (Physidae), Japan. Courtesy of N. Yotarou and reproduced here under Creative Commons Attribution 2.5 Generic Licence [https://creative-commons.org/licenses/by/2.5/deed.en]

FIGURE 20.10   Some examples of living systellommatophorans, stylommatophoran slugs, and a succineid. (a) *Peronia* sp. (Onchidiidae), ventral view, New South Wales, Australia. Courtesy of D. Riek. (b) *Onchidium daemellii* (Onchidiidae), New South Wales, Australia. Courtesy of D. Riek. (c) *Veronicella sloanei* (Veronicelliidae), Caribbean. Courtesy of D.G. Robinson [public domain]. (d) *Triboniophorus graeffei* (Athoracophoridae), New South Wales, Australia. Courtesy of D. Loughlin. (e) *Oxyloma elegans* (Succineidae), France. Courtesy of O. Gargominy - MNHN. (f) *Arion buttoni* (Arionidae), California. Courtesy of B. Roth. (g) *Arion rufus* (Arionidae), France. Courtesy of O. Gargominy - MNHN. (h) *Testacella haliotidea* (Testacellidae), France. Courtesy of O. Gargominy - MNHN.

FIGURE 20.11   Stylommatophorans. (a) *Xanthomelon* sp. (Camaenidae), Northern Territory, Australia. Courtesy of V. Kessner. (b) *Achatina fulica* (Achatinidae), French Polynesia. Courtesy of O. Gargominy - MNHN. (c) *Clausilia dubia geretica* (Clausiliidae), Europe. Courtesy of O. Gargominy - MNHN. (d) *Oleacina oleacea* (Oleacinidae), Cuba. Courtesy of S. Groves. (e) *Marmorana sepentina* (Helicidae), Corsica. Courtesy of O. Gargominy - MNHN. (f) *Amphidromus* sp. (Camaenidae), East Timor. Courtesy of V. Kessner. (g) *Ouagapia inaequalis* (Rhytididae), New Caledonia. Courtesy of O. Gargominy - MNHN. (h) *Insulivitrina* sp. (Vitrinidae), a 'semislug' from Tenerife. Courtesy of Prof. B. Eversham.

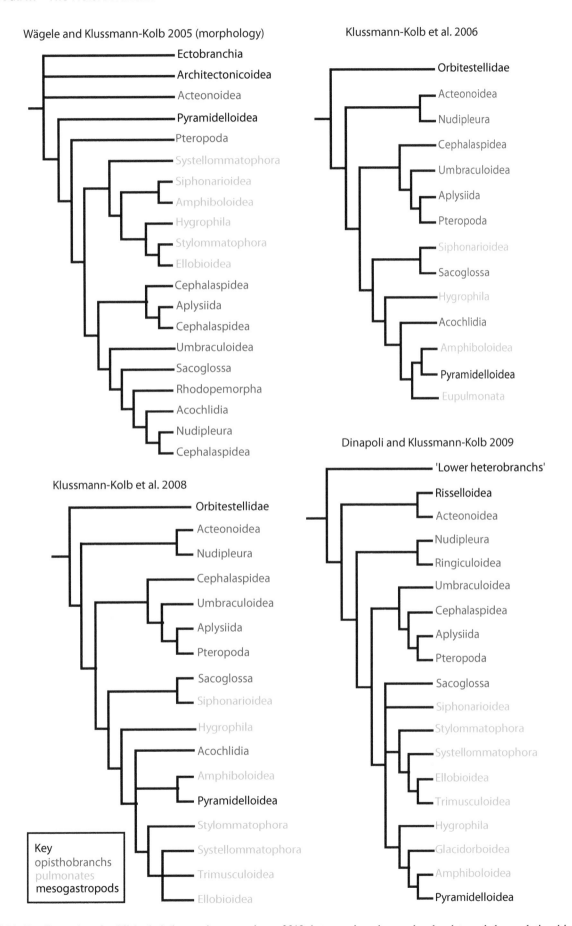

FIGURE 20.12 Examples of published phylogenetic trees prior to 2010 that were based on molecular data and show relationships within heterobranchs. The taxonomic placement of these taxa prior to the 1980s is indicated by colour coding. See also Figure 20.13.

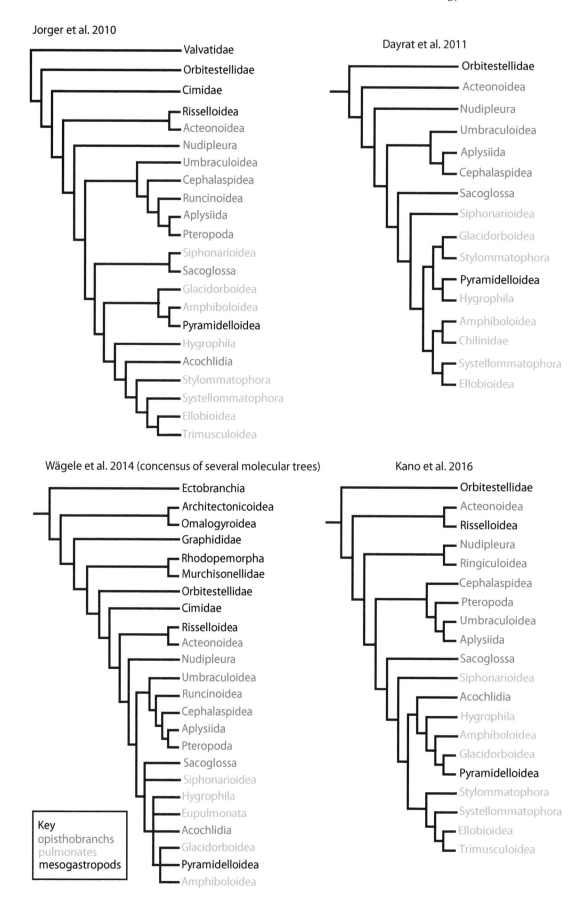

FIGURE 20.13 Examples of published phylogenetic trees since 2009 that were based on molecular data and show relationships within heterobranchs. The taxonomic placement of these taxa prior to the 1980s is indicated by colour coding. See also Figure 20.12.

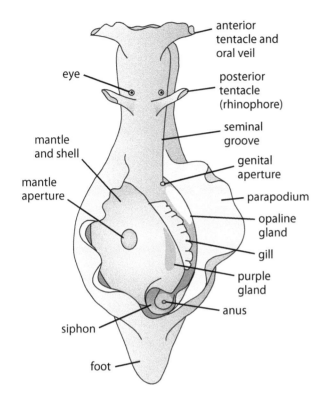

FIGURE 20.14 Dorsal view of the aplysiid *Aplysia punctata* showing the main external features. Redrawn and modified from Eales, N.B., *Bull. Br. Mus. Nat. Hist. Zool.*, 5, 267–404, 1960.

euthyneuran heterobranchs have undergone more departures from the original gastropod ground plan than any other group. Unlike other gastropod clades, shell-less slugs have evolved multiple times in the Euthyneura (e.g., Gosliner 1981; Medina et al. 2011).

Heterobranchs range in size from less than a millimetre to aplysiids (sea hares) several tens of centimetres in length, with the largest, the Californian *Aplysia vaccaria*, being about 75 cm long and weighing around 14 kg (Behrens 1981), making it one of the largest gastropods.

Marine euthyneurans in particular show a great range of body morphologies reflected in both their external (Figures 20.3 to 20.11) and internal anatomy (Figures 20.15 and 20.16), a diagrammatic representation of the internal anatomy of a primitive eupulmonate, an ellobioidean, is shown in Figure 20.17, and a stylommatophoran snail in Figure 20.18. Comparisons of the basic internal anatomy of a stylommatophoran snail and slug are provided in Figures 20.27 and 20.58.

20.3.1 SHELL AND OPERCULUM

Heterobranch shells are very diverse, ranging from large, rather heavy shells to internal rudiments, from tall-spired to planispiral and tightly coiled to limpet-shaped and even to tusk-shaped pteropods. Most are dextral, but all Planorbidae are sinistral, as are some members of a few other families, notably many species of the stylommatophoran land snail family Clausiliidae and some species in several other families including Camaenidae and Partulidae. The genetic and other

factors involved in the coiling direction (chirality) have been extensively studied in some land snails (see Chapter 2).

A well-developed shell and operculum are present in 'lower heterobranchs' except Rhodopoidea, and acteonoideans and rissoelloideans also have a coiled shell. In euopisthobranchs an adult shell is found in many Cephalaspidea and the thecosome pteropods, reduced shells in Aplysiida and all Umbraculoidea, but the shell is lost in gymnosome pteropods and lost or much reduced in some other taxa including runcinids and aglajids. In Ringipleura all Ringiculidae have an external shell; an internal shell is present in most pleurobranchs, but all adult Nudibranchia lack a shell. Some Sacoglossa have an external shell as do all pyramidelloideans, but it is lost in many sacoglossans and in all acochlidians. Thus shell reduction or loss, and associated loss of the adductor muscle and adult opercular loss, has occurred independently in nearly all clades (see Section 20.3.1.3). The shell may cover the animal completely or only partially, or it may be partially or completely covered by the mantle. If internal, the shell may be rudimentary, as in many stylommatophoran slugs where it can be reduced to a few granules. The shells may be calcareous and thick or thin and even lack calcium, being primarily organic.

The most bizarre range of shell morphologies is seen in the thecosome pteropods which were considered very different from other molluscs by early workers (see above). Their shells range from sinistrally coiled as in *Spiratella* to tusk-shaped as in *Creseis* or boat-shaped as in *Cavolinia* (Figure 20.4e-h).

In one sacoglossan lineage, the Juliidae, the shell is bivalved (Figure 20.7c). These strange gastropods were thought to be bivalves until the early 1960s when living animals were found. The left shell represents the true shell (including a coiled protoconch) while the right shell is an accessory structure (Baba 1961; Boettger 1963).

Many panpulmonates have a well-developed shell into which the animal can contract, but, perhaps surprisingly, most do not possess an operculum (see Section 20.3.1.5). The only operculate taxa that were included in 'pulmonates' are the amphiboloideans and glacidorboideans, with both these taxa also having other plesiomorphic features.

A high-spired shell, perhaps with columellar and parietal ornament (such as denticles, folds, lamellae), is considered plesiomorphic in Stylommatophora (Schileyko 1979; Nordsieck 1986). Apertural projections are probably adaptations for reducing predation. While many families have a reasonably consistent shell shape (Cain 1977), others are very heterogeneous, to some extent reflecting the wide range of habitats they occupy. Notable are the Streptaxidae and helicoideans, which exhibit most shell morphologies seen in stylommatophoran land snails.

Resorption of the internal parts of the shell is typical of Ellobiidae, except some members of the 'subfamily' Pedipedinae. Probably because of this resorption, the main part of the columellar muscle is free of attachment. This detachment results in the animal retracting into the shell by inversion of the head-foot rather than simple retraction (Barker 2001a).

The limpet shell form has evolved independently in several heterobranch clades, including the umbraculoideans, the

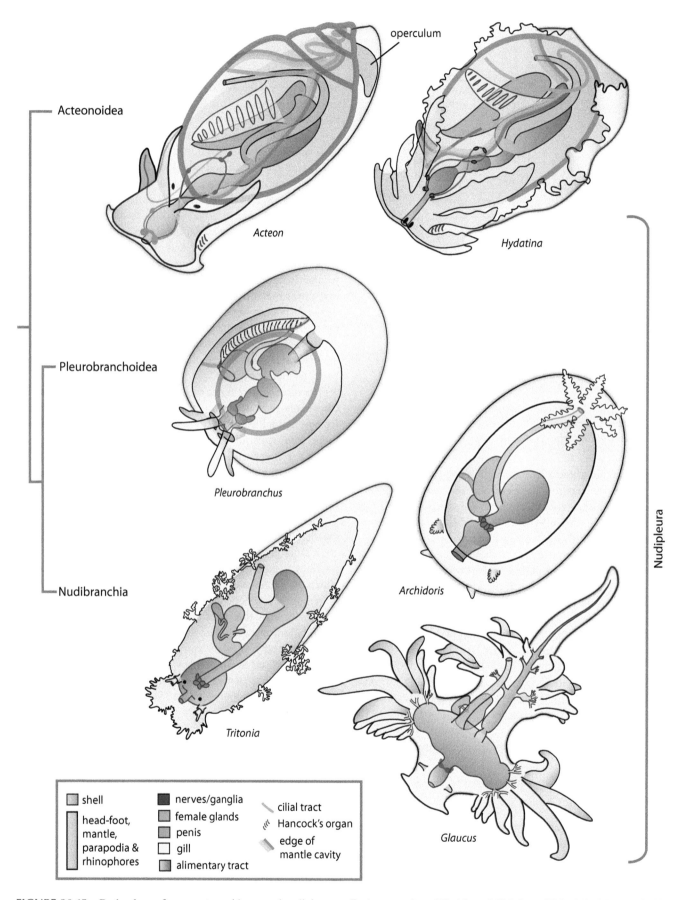

FIGURE 20.15 Body plans of some acteonoideans and nudipleurans. Redrawn and modified from Mikkelsen, P.M., *Adv. Mar. Biol.*, 42, 67–136, 2002.

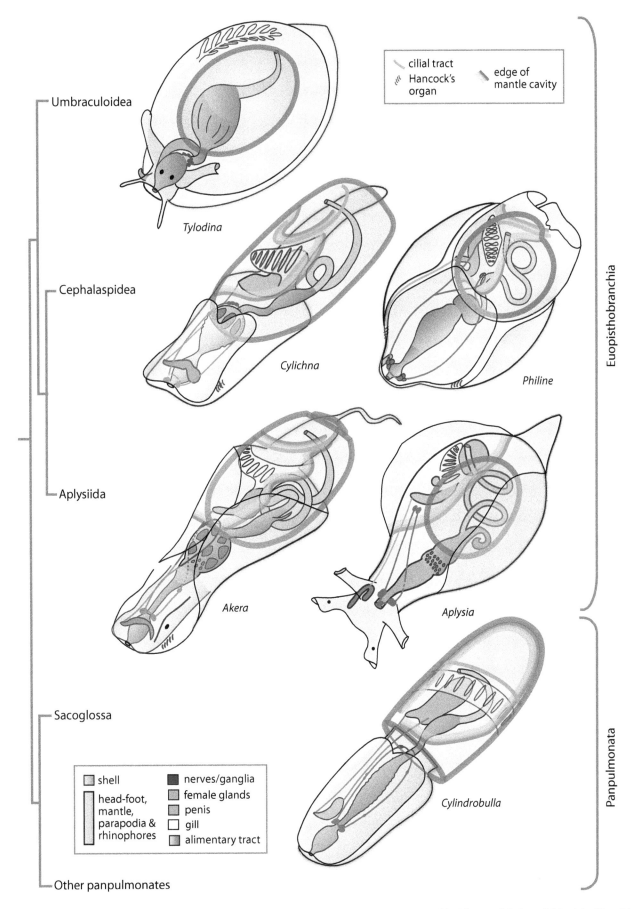

FIGURE 20.16 Body plans of euopisthobranchs and a shelled sacoglossan. Redrawn and modified from Mikkelsen, P.M., *Adv. Mar. Biol.*, 42, 67–136, 2002.

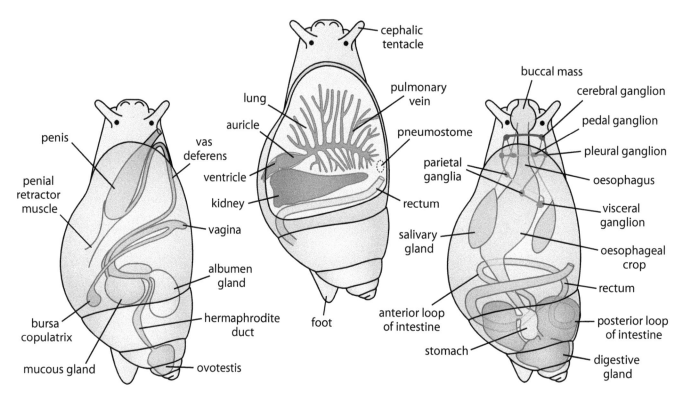

FIGURE 20.17   A generalised ellobioidean (Eupulmonata) showing some of the main anatomical features. Redrawn and modified from Barker, G.M., Gastropods on land: Phylogeny, diversity and adaptive morphology, pp. 1–146, in Barker, G.M. (ed.), *The Biology of Terrestrial Molluscs*, CABI Publishing, Wallingford, UK, 2001a.

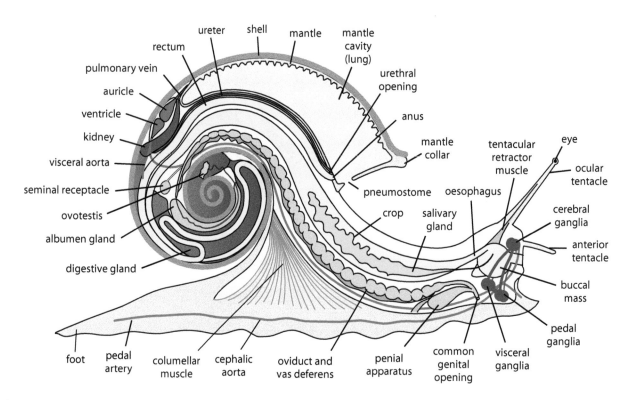

FIGURE 20.18   Stylised diagram showing the basic anatomy of a helicid snail (Stylommatophora). Redrawn and modified from Jammes, *Zoologie Pratique Basée sur la Dissection des Animaux les plus Répandus*, Masson, Paris, 1904.

marine siphonariids and trimusculids, and in the freshwater acroloxids and latiids as well as some planorbids (some of which were previously treated as a separate family, the 'ancylids') and lymnaeids (Lancinae). A few terrestrial taxa have reduced, abalone-like shells (many helicarionid semislugs, Testacellidae, the rhytidid *Schizoglossa*, and Boettgerilldae), and this shell form is also seen in the marine Otinidae.

'Limacisation' (the adoption of a slug form) has occurred independently in several lineages in geophilans (systellommatophorans and stylommatophorans). Similarly, a slender or elongated shell form is found in many stylommatophoran taxa. Such changes involve elongation of the body, with the shell and mantle becoming more posteriorly located, and are, in some cases, linked to the adoption of carnivory, although it is also possible these morphologies could have been exaptations for carnivory (Barker & Efford 2002). The slug body form has evolved independently multiple times and is mostly not associated with carnivory.

20.3.1.1 Shell Structure

Crossed lamellar shell structures predominate in the Heterobranchia (Bøggild 1930; Cox 1960b; Bandel 1990; Gosliner 1994). Bøggild (1930) surveyed a range of extinct and living heterobranch taxa and found that their shells typically consisted of three to four alternating, aragonitic lamellar layers and noted the shared similarity of 'opisthobranch' and 'pulmonate' shell structures and their similarity to those seen in caenogastropods. Cox (1960b) made similar comparisons, and Bandel (1990) recognised numerous aragonitic, lamellar shell structures in the heterobranchs, which either formed crossed lamellar structure, graded into it, or were closely similar. Luchtel et al. (1997) provided additional updates for pulmonate taxa.

Besides the lamellar layers, aragonitic spherulitic (prismatic), homogeneous, helical, intersected crossed-platy and scaly structures are also present in heterobranchs. Hedegaard (1990) first reported intersected crossed-platy in the lower heterobranchs *Bathyxylophila*, *Hyalogyrina*, and *Xyloskenea*, this being the first occurrence of this shell structure outside of the Vetigastropoda. Kiel (2004) later reported and illustrated its presence in *Xylodiscula* and an additional *Hyalogyrina* species. Spherulitic prismatic and homogeneous layers have been reported as both outer and inner layers in several heterobranch taxa (Bandel 1990; Hedegaard 1990). For example, the internal shell of *Aplysia* consists of an outer spherulitic prismatic layer that grades into a crossed lamellar layer (Bandel 1990), and homogeneous layers have been reported in several lower heterobranch groups (Hedegaard 1990), pteropods (Bøggild 1930), and pulmonates (Bandel 1990). Cavoliniid pteropods also have a unique helical structure (Bé et al. 1972), which, according to Bandel (1977), is derived from crossed acicular microstructure which also occurs unaltered in pteropod taxa. This structure is closely related to crossed lamellar structures, and transitional states between helical, crossed acicular, and crossed lamellar structures have been documented by Bandel (1990). Pteropod shell structure is receiving more attention of late because of increasing interest and monitoring of ocean

acidification (Roger et al. 2012). Scaly structures are found on the apertures of both terrestrial and freshwater pulmonates (e.g., Solem 1974).

Nacre shell structure was reported by Saleuddin (1971) and repeated by Luchtel et al. (1997) as being present as the innermost layer of the pulmonate *Helix pomatia*. This layer was first described in *H. pomatia* by Biedermann (1902), and Bøggild (1930) demonstrated that it was widespread in the helicoideans and was actually a longitudinal lamellar layer with the lamellae oriented parallel to the growth (mantle) surface. Inner layers of parallel lamellae or fibres were also noted in *Valvata* species by Bøggild (1930). Falniowski (1989, 1990) described and illustrated additional *Valvata* species, referring to the upper layers as 'palisade-like' (= crossed lamellar) and the innermost layer as an 'endostracum', which appeared as a thick, flat layer composed of long, slender fibres. Bandel (1990) also noted a similar layer in the planorbid *Ancylus* and referred to it as a fibrous prismatic structure. Carter and Clark (1985) referred to this shell structure as 'lamello-fibrillar'.

Calcitic shell structures are found in some stylommatophoran land snails and the internal shells of the terrestrial slugs *Deroceras* and *Arion* (Bandel 1990) and consist of fibrous and foliated shell structures. Calcitic structures are also found in the shell wall of some stylommatophoran eggs (Bandel 1990). Besides calcitic eggshells, the reproductive system of helicoidean stylommatophoran taxa also produces aragonitic copulatory darts composed of fine needles organised as spherulites (Bandel 1990). These and other structures not produced by mantle tissue, such as the calcareous opercula seen in some non-heterobranch gastropods, tend to be simple and lack the structural elements that characterise crossed lamellar structures (Bandel 1990).

20.3.1.2 Larval or Embryonic Shell

The protoconch of heterobranchs differs from that of most other gastropods in often being coiled in a different direction from the teleoconch – i.e., it is heterostrophic (see Chapter 3, Figure 3.11). The larval shells are sinistral, but the adult shell and anatomy are typically dextral. Those dextral taxa with no adult shell also have a sinistral larval shell. In sinistral adults, the larval shell is also heterostrophic as, for example, in *Limacina*.

Heterobranch larvae are often planktotrophic, with growth of the larval shell during feeding, but unlike many caenogastropods, protoconch II is not distinctly different.

20.3.1.3 Shell Reduction and Loss

As noted previously, shell reduction (and often loss) has occurred independently in several major euthyneuran lineages, with more plesiomorphic taxa having relatively well-developed shells compared with more derived taxa. For example, in sacoglossans, the shelled *Cylindrobulla* is sister to all other sacoglossans, while *Akera*, with its well-developed shell, is sister to the aplysiids (sea hares). As noted earlier, one sacoglossan lineage has produced a two-valved shell, the so-called 'bivalved gastropods', which has adductors and looks very much like a bivalve. In several lineages

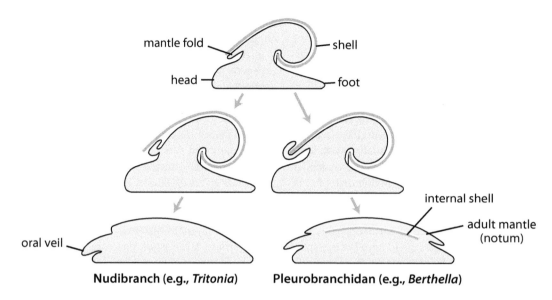

FIGURE 20.19 A diagrammatic scenario for the development of the embryonic mantle fold in nudipleuran slugs and its role in shell loss (on the left) in nudibranchs and internalisation (on the right) of the shell in pleurobranchs. Redrawn and modified from Thompson, T.E., *Phil. Trans. R. Soc. B*, 245, 171–218, 1962.

(e.g., pleurobranchs, aplysioideans, several groups of stylommatophoran slugs) the shell, or shell remnants (see Gosliner 1994 for details), have become internal, permanently covered by the mantle (Figure 20.19).

Shell reduction is often associated with an increase in shell protein and carbohydrate, rendering the shell more flexible (Poulicek et al. 1991), although this is far from universal in euopisthobranchs (T. Gosliner, pers. comm. Feb. 2019).

Reduction or loss of the shell results in a slug or semislug form which has evolved multiple times in distinct lineages. The lack of a shell makes terrestrial slugs more vulnerable to desiccation and predation but gives greater flexibility of the body so they can squeeze into narrow crevices and holes that would be impossible for a shelled snail of similar size. Terrestrial slugs are also limited by the availability of calcium essential for building shells. Marine eupulmonate slugs include most of the onchidiids and the maggot-like smeagolids, the latter living interstitially in the upper littoral in coarse sediments or pebbles. A few onchidiids are terrestrial as are all members of the related Veronicelloidea, but these slugs lack a secondary ureter, a structure extremely important in water conservation. Stylommatophoran slugs have a secondary ureter with either the sigmurethran or heterurethran condition (see Section 20.3.5).

Many shell-less marine euthyneurans compensate for the lack of a shell with calcareous spicules in their dermis (see Section 20.3.2.3), the incorporation of toxic or noxious chemicals (see Section 20.4.6.3.1), co-opting of nematocysts (see Section 20.4.6.4), or armour-like modifications to the epithelium (see Section 20.3.2.3).

20.3.1.4   Shell Variation

As with other molluscs, heterobranch shells show variation and convergence, and in some taxa in the Philinidae, in freshwater hygrophilans, and in the terrestrial stylommatophorans,

shell variation has been investigated more than in most other taxa. Such variation and convergence makes taxonomy based entirely on shell characters problematic, because it can be caused by genetic or environmental factors or a combination of both, although often the precise causes are not known. For example, a smaller aperture size to reduce water loss can be related to drier conditions, but such conditions presumably also have a genetic component (e.g., Goodfriend 1986). Snails living on calcareous substrata sometimes have thicker shells, but shell thickness is not closely related to moisture levels in the environment (Goodfriend 1986).

20.3.1.5   Operculum

A horny operculum is present in the shelled 'lower heterobranchs' and some euthyneurans (Rissoellidae, Pyramidellidae, Amphiboloidea, Glacidorbidae, Limacinidae, most Acteonidae, and some Retusidae) and in heterobranch larvae. The adult operculum varies in morphology from circular to ovate, and multispiral to paucispiral to concentric, the latter condition seen in Rissoellidae.

Some operculum-like structures are developed in stylommatophoran land snails. The most common way of sealing the aperture is with mucus, sealing the aperture against a hard surface (Figure 20.20b) which is also seen in some non-stylommatophorans, as for example in the aestivating planorbid *Isidorella* (WFP, pers. observ.). As noted in Chapter 9, an epiphragm, a sheet of hardened mucus that covers the aperture and is sometimes calcified, is also found in some taxa – the so-called 'free sealers' (e.g., Solem 1974) (Figure 20.20a).

In Clausiliidae, a curving calcareous structure with a widened end (the clausilium) closes the aperture. The narrow upper end of the clausilium slides in a groove in the columella and is controlled by muscles attached to it. Another type of calcareous operculum-like structure is seen in *Thyrophorella* (Punctoidea, Thyrophorellidae). It is formed

FIGURE 20.20   Aestivating camaenid snails. (a) *Torresitrachia* sp. showing free-sealing method with an epiphragm. (b) *Parachloritis* sp. showing the snails sealing against the bark of a tree. Note the mucus rings where snails have been sealed. Courtesy of V. Kessner.

from a prolongation of the upper half of the peristome beyond the aperture, and a hinge allows this projection to close over the aperture, thus acting like an operculum when the animal retracts.

20.3.2   EXTERNAL BODY MORPHOLOGY

Besides the shell, there have been other major changes to the external parts of the body, as in this respect, heterobranchs, and especially the marine euthyneurans, are much more varied than other gastropods.

20.3.2.1   Head Tentacles

A pair of head tentacles, homologous with the cephalic tentacles in other gastropods, is found in most 'lower heterobranchs'. These are very reduced in some Omalogyridae and lost in the Rhodopidae and the shelled burrowing euopisthobranchs with a prominent head shield, such as most cephalaspideans and Akeridae, as well as in some slug-like taxa such as Philinoglossidae and Runcinoidea. Two pairs of tentacles are found in Aplysiidae, Rissoellidae, some pleurobranchs, and in most nudibranchs. The anterior pair are the anterior or oral tentacles and, in rissoellids, the second pair is the cephalic tentacles. In nudibranchs, the homology of the second (posterior) pair of tentacles, the rhinophores, has been debated (see also Section 20.3.6.1 and text following), but detailed studies on their innervation have demonstrated their probable homology with the cephalic tentacles of other gastropods, including those of 'lower heterobranchs', and the optic tentacles of stylommatophorans (Huber 1993). Pyramidelloideans have 'rabbit-eared' triangular tentacles folded longitudinally so that the antero-ventral surface is an open gutter. Thecosomes have no head tentacles, but some of the naked gymnosomes have a pair of cephalic tentacles.

Rhinophores, with eyes at their bases, are found in most nudipleurans and euopisthobranchs, except cephalaspideans. They are chemosensory and formed from a rolled-up sheet of tissue in all but the nudibranchs, where they are solid. They

are often considered secondary structures (e.g., Hoffmann 1932–1939; Gosliner 1994) and whether or not they are homologous in all euthyneuran clades is uncertain, with independent evolution in several groups suggested (Gosliner 1994). The rhinophores in pleurobranchs and nudibranchs appear to be homologous based on phylogeny and because they are innervated by the same nerve, the rhinophoral nerve, although they are rolled in the former group and solid and circular in cross-section in the latter (Wägele & Willan 2000). The rhinophores have also been homologised with the Hancock's organs of cephalaspideans (Hoffmann 1932–1939; Gosliner 1994). Huber (1993) considered them homologous because the rhinophores and Hancock's organs are innervated by the same nerve (the rhinophoral nerve). Dayrat and Tillier (2002) doubted this interpretation, but more recent studies support this homology (see Chapter 7). The rhinophore-like tentacles of sacoglossans may not be homologous because the more plesiomorphic taxa (Cylindrobullidae and Ascobullidae) lack them, and they are innervated with three nerves, suggesting a fusion of anterior and posterior sense organs (Huber 1993).

Nudibranchs exhibit many kinds of rhinophoral elaboration, the most common being the lamellate type, but many different elaborations are found (Hoffmann 1932–1939; Gosliner 1994; Wägele & Willan 2000). A few nudibranchs secondarily lack rhinophores (e.g., some species of the interstitial *Pseudovermis*).

Two groups of nudibranchs have evolved *rhinophoral sheaths* into which the rhinophores can be withdrawn. These sheaths can be formed in two ways – by the front margin of the notum growing around the rhinophores (as in Tritonioidea and Dendronotoidea) or by the rim of the rhinophoral pocket growing around the rhinophore, as seen in some Doridoidea. The sheaths are absent in most Nudibranchia (many doridids, Zephyrinidae, Arminidae, Arminoidea, and Aeolidioidea) and all other euthyneurans.

Basal lung-bearing panpulmonates have a broad snout and short, non-contractile cephalic tentacles with eyes at their inner bases. In other lung-bearing panpulmonates the cephalic

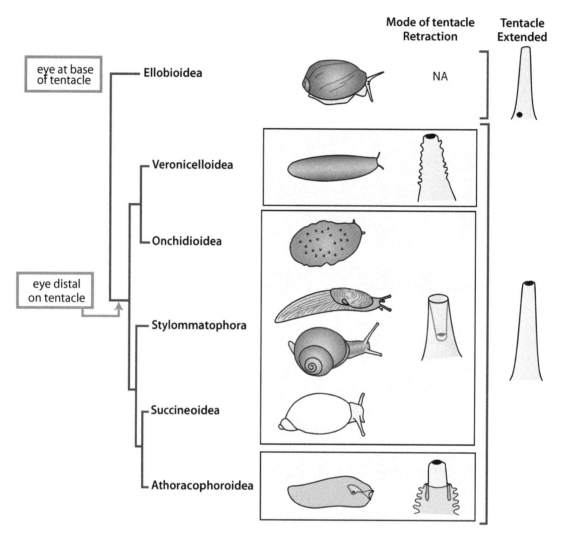

FIGURE 20.21 Differing methods of tentacle retraction in eupulmonates. Details of tentacle retraction redrawn and modified from Burch, J.B., *J. Malacol. Soc. Aust.*, 11, 62–67, 1968; animals redrawn from various sources.

tentacles range from stumpy lobes or triangular, to long and slender. The 'basommatophoran' taxa all have eyes at the bases of the tentacles while the members of the Geophila have eyes at the distal tips of the tentacles. These ocular tentacles are probably homologous with the cephalic tentacles of other gastropods and also with the rhinophores of nudibranchs (see Chapter 7). Within the Systellommatophora, the Onchidiidae have a single pair of retractile tentacles with eyes at their tips while the Veronicellidae and Rathouisiidae have two pairs of tentacles, the posterior pair with eyes at their tips. The distal eyes in systellommatophorans (see Section 20.3.6.1) and stylommatophorans may be convergent. Their tentacles are shortened by contraction while those of the stylommatophorans retract by introverting (Figure 20.21). The different modes of tentacle retraction in stylommatophorans were described by Burch (1968), with an autapomorphic method in athoracophorid slugs.

20.3.2.2 Cephalic (Head) Shield

A cephalic shield is formed by the dorsal head thickening and expanding laterally to form a triangular structure

in Acteonoidea, *Ringicula*, most cephalaspideans, *Akera* (Akeridae, Aplysiida, but not in aplysiids), and in some shelled sacoglossans. A somewhat similar structure is seen in some murchisonellids (e.g., Wise 1999). The head shield probably originally evolved in burrowing forms where it acts as a plough but is retained in surface-living acteonoideans and cephalaspideans; the sensory Hancock's organ (Section 20.3.6.1) lies beneath its lateral edges. In *Ringicula*, the posterior margin of the head shield is extended into an exhalant siphon, while in some Acteonoidea (Aplustridae) and Bullidae the lateral edges are extended to form rhinophore-like structures. A unique, siphon-like sensory structure is present in Gastropteridae (Ong et al. 2017).

20.3.2.3 Epithelial Modifications

Vacuolated epithelium, a specialised epithelial cell type found only in some nudibranchs (e.g., Wägele 1998; Wägele & Willan 2000), consists of elongate cells with numerous 'vacuoles' in the apical part of the cell. These large epithelial cells (Figure 20.22) make up the 'skin' and the lining of the stomach, and

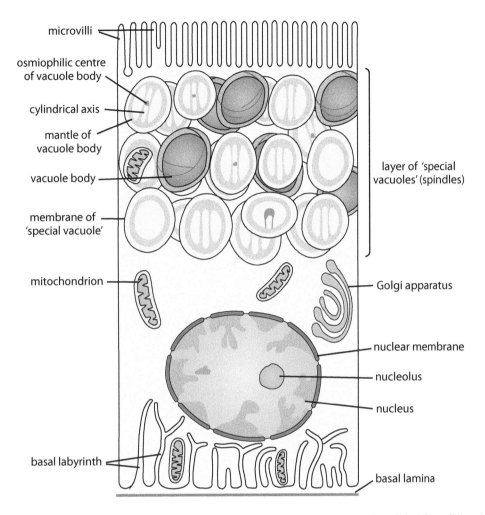

microvilli

osmiophilic centre
of vacuole body

cylindrical axis

mantle of
vacuole body

vacuole body

membrane of
'special vacuole'

mitochondrion

basal labyrinth

layer of 'special
vacuoles' (spindles)

Golgi apparatus

nuclear membrane

nucleolus

nucleus

basal lamina

FIGURE 20.22 An epidermal cell with spindles of an aeolidioidean nudibranch. Redrawn and modified from Schmekel, L., Aspects of evolution within the opisthobranchs, pp. 221–267, in Trueman, E.R. and Clarke, M.R. (eds.), Evolution. *The Mollusca*, Vol. 10, Academic Press, New York, 1985.

the large vesicles they contain are filled with ovoid granules. These granules are biconcave discs (termed spindles), about 5 µm across and contain a mesh of filaments internally (Martin & Walther 2003; Martin et al. 2007b), composed of chitin (Martin et al. 2007a). A study by Martin and Walther (2003) on two aeolidiids showed that when nematocysts of their hydrozoan prey were discharged, the impact resulted in the skin of the slug folding and portions containing epidermal vesicles being sloughed off. The spindles were released from the vesicles entangling the larger nematocysts. Small nematocysts were unable to penetrate the mucoid and microvillus layer on the skin surface.

These granule-like spindles, unique to nudibranchs, could be likened to a coating of sandbags or even flexible chain mail armour, in marked contrast to the shell or the external skeleton of arthropods. The evolution of this defence mechanism may be correlated with cnidarian feeding, developed as insulation against being stung by the nematocysts of their prey as the granules (spindles) occur in epithelial cells on parts of the body that would encounter them, such as the cerata, head and lips, non-retractile gills, and the stomach lining. They

are particularly dense in the cnidarian feeding aeolidiids and arminoideans which have relatively few epidermal gland cells, but are less dense in tritonioideans, dendronotoideans, and doridoideans in which glandular cells predominate (Martin et al. 2007b). The mixture of gland cells and cells with spindles enables two strategies to be used by nudibranchs – thick coats of mucus and a structural defence involving the chitinous spindles. In some aeolidiid nudibranchs the mucous coat has been shown to inhibit the discharge of nematocysts, even changing with prey type (Mauch & Elliott 1997; Greenwood et al. 2004).

The body of many doridoidean nudibranchs contains large numbers of calcium carbonate spicules that in some can impart a sandpaper-like texture. They can comprise a significant part of the dry body mass and may ramify through the entire body, including the foot, although they are in higher concentrations in the dorsum. They also play a supporting role in the gills and rhinophores. The large number of spicules in some nudibranchs could reduce their nutritional value as prey, perhaps offering another avenue of anti-predator deterrence (Penney 2002; Penney et al. 2018). The spicules are composed mainly of calcite ($CaCO_3$), but brucite ($Mg(OH)_2$) and smaller amounts of fluorite (CaF_2) are also found (Cattaneo-Vietti et al. 1995).

Their role in defence is not clear, but the spicules provide some rigidity to the dorsum so their primary function may be in providing support, including in structures such as papillae (Penney 2006, 2008). In some species the spicules, which are embedded in connective tissue, ramify through the body forming complex patterns that differ in different groups of doridoidean nudibranchs. Penney (2008) found three basic types of spicule network – in Discodorididae there is a cobweb-like unbraced framework, while in some Dorididae and Dendrodorididae there is a ramifying (dendritic) system of thick, spiculated tracts. In contrast, a lattice-like arrangement of distinct radial tracts is seen in a number of Dorididae and Phyllidiidae (Penney 2008). While these patterns have some phylogenetic utility, there is variation. For example, Penney (2008) found that although dendrodoridids he examined had a dendritic arrangement of spicules, most species of *Dendrodoris* lack them (as do all Chromodorididae). Within Doridoidea, *Cadlina* (Cadlinidae) and *Aldisa* (Dorididae) had a dendritic pattern, while *Doris* (Dorididae) had a lattice pattern.

There may be several kinds of spicules in a species, with variations including spindle-shaped, cruciform and spurred, and some with multiple spikes (e.g., Kress 1981; Chang et al. 2010, 2013). Spicule form has some taxonomic utility as, for example, doridoideans and most onchidoridids have only spindle-shaped spicules, but spicule morphology is much more varied among other doridine families.

Because spicules are absent among some basal nudibranchs, including *Bathydoris*, Penney (2008) argued that spicules were probably developed in nudibranchs independently from those found in other heterobranchs (see text following). Also, the spicule patterns in some taxa are simple, as in polycerids which have a few haphazardly arranged spicules, and aegirids, which have more spicules, but they are still disorganised. Some organisation is seen in goniodoridid spicules, with some arranged in two rows down the notum.

Calcareous spicules have been found in a few non-nudibranchs including some derived pleurobranchs, where they are usually stellate and thus different from the spindle-shaped spicules normally found in nudibranchs (Hyman 1967; Cervera et al. 2000). In several genera of discodoridids, dorsal sensory tubercles have a ring of protruding calcareous spicules and are termed caryophyllidia (Gosliner 1994). Similar structures are also seen in some species of *Onchidoris*, but these are probably not homologous. Spicules are also found in the body wall of the tiny 'lower heterobranch' slug *Rhodope* and some Acochlidia.

20.3.2.4 Oral Veil

Lateral extensions from the mouth called an oral veil are found in some euopisthobranchs, namely *Tylodina* (but not *Umbraculum*), pleurobranchs, and some nudibranchs, namely Armenidae and Zephyrinidae (Wägele & Willan 2000). Short tentacle-like processes are present on the anterior edge of the

oral veil in some Tritonioidea and Dendronotoidea. When present, the oral tentacles arise from the lateral parts of the oral veil, and, in many nudibranchs, these tentacles are all that remains of the oral veil (Wägele & Willan 2000). It is not clear whether the large cephalic cowl in Tethydidae used to capture crustaceans is an expansion of the head (Gosliner & Smith 2003; Gosliner & Pola 2012).

20.3.2.5 The Foot, Pedal Glands, and Parapodia

The heterobranch foot differs from most other gastropods in lacking a distinct propodium, with an exception being the basal heterobranch family Orbitestellidae, where it is narrow and on the anterior dorsal part of the foot. A propodium-like structure in *Otina*, *Smeagol*, and at least some ellobiids (Tillier & Ponder 1992) is analogous at best. In other respects the heterobranch foot is like that of most other gastropods, although it is very variable, especially in euopisthobranchs. In most groups of 'lower heterobranchs' the foot is usually rather simple. Omalogyrids have a small anterior mucous gland opening to the dorsal surface of the anterior foot below the mouth. An anterior mucous gland is present in most heterobranchs, and it is large and modified in many stylommatophorans (see below).

Most heterobranchs lack a discrete posterior pedal mucous gland, but such a gland is present in some 'lower heterobranchs' where it opens to a pore or slit in the middle of the sole. In hyalogyrids (Haszprunar et al. 2011), rissoellids and omalogyrids (Fretter 1948), and in orbitestellids (Ponder 1990a) it comprises two long lobes which, in the latter two taxa, extend back into the body cavity. A posterior pedal gland is also present in Architectonicidae and Mathildidae, and it discharges by way of a small opening in the middle of the foot sole. The gland itself extends through the ventral ciliated ridge in the mantle cavity (Haszprunar 1985c, d). A posterior pedal gland is also present in Gastropteridae (see below), and there is a 'metapodial gland' in pleurobranchs (Thompson 1976).

In marine euthyneurans the foot can range from being long and slender, as in aeolidiid nudibranchs, to short and broad as in doridids, and in some, such as gastropterids, it is not clearly demarcated from the rest of the body. In cephalaspideans like *Haminoea*, the posterior 'foot' is an extension of the mantle while the true foot is anterior and small. In pteropods, the foot is rudimentary or lost.

In most marine euthyneurans the anterior edge of the foot is smooth, simple, and often rounded, but in a number of species distinct, sometimes long, propodial tentacles are present, notably in many Aeolidioidea. In some, such as discodoridids, the anterior end of the foot is distinctly notched (Valdés & Gosliner 2001).

In cephalaspideans, most aplysiidans, and some sacoglossans, there is a pair of lateral outgrowths, the *parapodia*. These may partly cover the animal and/or the shell dorsally, and in some sacoglossans they can be deployed to expose or shade the chloroplast-containing tissue. In a few taxa, such as Gastropteridae, Akeridae, and Aplysiidae, the parapodia are employed in swimming. This ability is particularly developed in pteropods where swimming 'wings' are present, and the foot sole is rudimentary. In thecosomes the bases of these wings join around the mouth – they are equal and distinct in euthecosomes, but there is only one wing in Pseudothecosomata. In gymnosomes, the two parapodial wings abut the rudimentary foot ventrally.

In most heterobranchs, the whole ventral surface of the foot is richly supplied with glands which open to the foot sole. These may be epithelial or subepithelial, unicellular or multicellular. While most euthyneurans lack distinct pedal glands, in Gastropteridae a posterior pedal gland produces a yellowish secretion that may be defensive (Gosliner 1989) while a non-homologous posterior pedal gland is also found in many Pleurobranchidae (Willan 1987). In this latter case, the gland may produce pheromones to attract mates as it only appears as the animals reach sexual maturity (Gosliner 1994).

Sticky mucus is produced at the posterior end of the foot of some aplysiids, such as *Aplysia juliana*, which have developed this part of the foot into an effective adhesive organ. In many terrestrial slugs, the dorsal surface of the foot forms much of the dorsal surface of the body, and the mantle is much reduced.

In many heterobranchs, an anterior mucous gland is similar to, and presumably homologous with, the anterior mucous gland of other gastropods. This large gland is in the anterior part of the foot in most groups of heterobranchs, extending beyond the odontophore in orbitestellids (Ponder 1990a). It is the largest gland associated with the foot in stylommatophorans and systellommatophorans where it is tongue-shaped and called the suprapedal gland. It extends along much of the length of the animal, on the floor of the visceral cavity, and opens in the groove between the head (mouth) and the foot. This groove is continuous with the peripedal groove that runs around the lateral edges of the foot and contains mucoid cells. In stylommatophorans, the suprapedal gland is separated from the rest of the visceral cavity by a membrane, but in veronicellids and onchidiids it lies free in the visceral cavity.

A study by Cook and Shirbhate (1983) of the glands associated with the foot in the stylommatophoran slug *Limax pseudoflavus* showed that the pedal mucus is a mixture of mucopolysaccharide secreted by the suprapedal gland and mucoproteins secreted by glands in the anterior edge of the foot. The mucus on the dorsal part of the body is separated from the pedal mucus by the peripedal groove, which also secretes mucus. The foot sole, the ventral part of the peripedal groove, and the surrounds of the pneumostome (where there are rejection currents) are ciliated, while the rest of the body is unciliated, so mucus on the surface is not moved. The epithelium on the head, pneumostome, sole, and anterior edge of the foot can produce both a neutral or weakly acidic fluid mucus and a viscous acidic mucus, thus allowing for both adhesion and locomotion (see Chapter 3).

A posterior, dorsal, caudal gland occurs in some stylommatophorans (e.g., Arionidae, Zonitidae, Helicarionidae). The gland is apparently involved in mating, with copulating individuals eating the mainly mucus contents of the gland of the mating partner (Hyman 1967).

20.3.2.6 The Mantle

As in other molluscs, the mantle secretes the shell if one is present, but when the shell is reduced or lost the naked mantle may cover much (e.g., as in most doridoidean nudibranchs) or nearly all (e.g., as in onchidiids) of the dorsal surface (the notum). The mantle also secretes the calcareous spicules found in some sea slugs. In some taxa, notably in many stylommatophoran slugs, the mantle is greatly reduced, and the dorsal foot makes up much of the dorsal surface of the animal.

The distinction between mantle and foot is usually clearcut in marine euthyneurans, as in many the notum is distinctly separated from the foot by a rim (notal margin). In some cephalaspideans, such as *Haminoea*, the posterior foot-like structure is formed from the mantle, and in many aeolidiid and polycerid nudibranchs the notum and foot are not very distinctly separated.

The head is usually distinct in nudibranchs, but a fusion of the anterior part of the notum and head can occur, and in some the notal margin extends over the head, surrounding the rhinophores, as for example in *Ceratosoma* (Rudman 1988).

The notum of nudibranchs often bears papillae (including tubercles or pustules) or other projections, including the long, finger-shaped projections (cerata) seen in some nudibranchs and sacoglossans (see Section 20.3.2.6.2), or various gill-like or branched structures.

The notum is covered with cuticle in Onchidiidae and a few nudibranch taxa such as Corambidae and the polycerid *Vayssierea* (Wägele & Willan 2000).

In some thecosome pteropods, such as *Diacria* and *Cavolinia*, the external mantle is complex and may be used for the processing of food particles or perhaps to aid flotation, while other cavoliniid thecosomes have reduced external mantles (Gilmer & Harbison 1986).

The lack of a shell makes a slug vulnerable to predation, and the mantle surface may be modified for a new role. Some have special mantle epithelial adaptations including vesicles filled with granules and/or spicules (see Section 20.3.2.3) embedded in the mantle and/or glands containing noxious substances, as described next. It is sometimes cryptically or brightly coloured and can be ornamented with tubercles, papillae, or cerata. In some taxa the cerata have terminal sacs containing nematocysts (see Section 20.4.6.4).

20.3.2.6.1 Mantle Glands and Acid Glands

The mantle has mucus-secreting cells that serve various functions, from cleansing to antibacterial. In terrestrial eupulmonates the mucus-secreting gland cells serve to keep the surface moist. For example, the stylommatophoran slug *Limax pseudoflavus* produces a sulphated acid mucopolysaccharide/protein mixture secreted by five types of gland cell (Cook & Shirbhate 1983). Such mucoid secretions can also be distasteful, thus aiding in deterring predators.

Many naked marine euthyneurans have special mantle glands that produce secretions used in defence. These glands are especially developed in chromodoridid nudibranchs (Rudman 1984; Gosliner & Johnson 1999; Epstein et al.

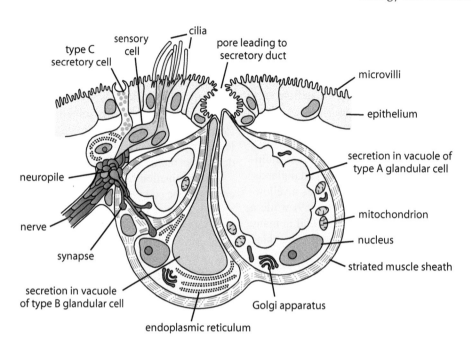

FIGURE 20.23 The repugnatorial gland on the dorsal surface of *Melibe leonina*. Redrawn and modified from Bickell-Page, L.R., *Zoomorphologie*, 110, 281–291, 1991.

2018). These 'repugnatorial glands' (Figure 20.23) can produce repugnant and/or toxic secretions of substances that are extracted from the food and either modified or synthesised *de novo* (e.g., Cimino & Ghiselin 1999, 2009) (see Chapter 9 for details).

Mantle glands have been investigated in sacoglossans (e.g., Edmunds 1966b), doridoideans (e.g., Edmunds 1968), including chromodoridids (Rudman 1984), aeolidiids (Edmunds 1966a), and in sea slugs generally (Wägele et al. 2006). Wägele et al. (2006) recognised three main types of defensive mantle glands in shell-less marine euthyneurans, commonly called sea slugs, as follows:

1. *Accumulating glands* – large glands with a restricted distribution in the mantle – contain metabolites from food.
2. *Biosynthetic cells* – small, widely distributed in the mantle – contain compounds created *de novo*.
3. *Acid glands* – produce inorganic acids.

There are other glands associated with the mantle cavity, and these are described next (see Sections 20.3.3 and 20.4.6.3.1).

The shelled euthyneurans *Acteon* and *Ringicula* have repugnatorial glands on the mantle and head shield (Gosliner 1994) prompting Faulkner and Ghiselin (1983) to suggest that their development preceded shell reduction and loss, rather than being acquired subsequently.

Mantle dermal formations are usually globular and composed of one to many cells, each with a single vacuole, and the whole structure is coated with a thin to thick layer of muscle. They are found most famously on chromodoridid nudibranchs where they accumulate noxious compounds from their sponge food in large, sac-like cells, often incorrectly called 'glands',

for which the term mantle dermal formations (MDF) was originally coined. The MDFs are branched in some chromodoridoideans (Gosliner & Behrens 1998; Wägele & Willan 2000). MDFs have also been described in non-chromodoridids, such as the triophid *Limacia clavigera* (Wägele 1998) as well as species of *Rostanga*, *Jorunna*, *Cadlina*, and many other doridoideans. Some do not possess a pore and can only liberate their contents when ruptured.

Other dermal glands include *saccular marginal glands* unique to the Arminidae which are, as their name suggests, arranged around the margin of the notum in members of that family. Wrongly thought to be cnidosacs by some workers (e.g., MacFarland & MacFarland 1966), and unlike the MDFs of chromodoridids, they comprise large cells that contain acid mucopolysaccharides. Each of these glands opens to the exterior via a pore (Wägele & Willan 2000). In doridoideans, there are also glandular follicles, or '*gill glands*', within the gill rachis or, mainly in Doridoidea, between the bases of the gill rachises (Wägele & Willan 2000). Also, in a few doridoideans (e.g., *Jorunna*), aggregations of cells with translucent cytoplasm near the rim of the notum have been called *mantle rim organs* (Foale & Willan 1987).

Dermal glands also occur in some other euthyneurans, including a number that do not feed on sponges, with records from a few cephalaspideans and even the sacoglossans *Plakobranchus* and *Elysia*. These glands differ in structural details in the different groups and are not homologous, having evolved separately several times (Wägele et al. 2006). Similar glands have also been described from the sides of the foot in *Siphonaria* (Pinchuck & Hodgson 2009). These glands comprise two cell types and secrete a sticky defensive secretion (composed mainly of mucopolysaccharides) if stimulated mechanically.

Other types of mantle glands in sea slugs include:

Spongy mantle glands on the mantle rim in acteonoideans, cephalaspideans, aplysiidans, and thecosome pteropods are assumed by Wägele et al. (2006) to have evolved early in euthyneuran evolution. These large sponge-like vacuolated epithelial cells have a minute nucleus, a non-staining vacuole, and open by way of an invaginated pore. These glands are not to be confused with the vacuolated epithelium found in many nudibranchs (see Section 20.3.2.3).

The *dorsal mantle gland* is found only in Umbraculoidea (*Tylodina* and *Umbraculum*) and lies in the anterior mantle with ducts opening to the anterior mantle margin.

In Pleurobranchidae, *acid glands* in the mantle and foot liberate mainly sulphuric acid (pH 1–2), and some hydrochloric acid is also present (Thompson & Slinn 1959; Thompson 1988; Gillette et al. 1991). Acid glands are also present in the cephalaspidean *Philine*, which likewise produce sulphuric acid (Thompson 1960b, 1976), and in some Doridina including a few onchidoridids, a doridid, and a number of discodoridids (Edmunds 1968). The acid is secreted by cells in the epidermis and, before release, is stored as precursors in vesicles. These acid secretions are thought to be distasteful to fish which reject them as food (e.g., Thompson 1960b; Marbach & Tsurnamal 1973), as does the nudibranch-eating cephalaspidean *Navanax* (Paine 1963). Pleurobranchs also have a buccal gland that produces sulphuric acid (see Section 20.3.4).

A *glandular strip* occurs on the right side of various opisthobranch taxa including many cladobranchs, some aplysiidans, and the sacoglossan *Elysia* (Wägele 1998). Given the distribution of these glandular strips in disparate euthyneuran taxa, they are probably independently derived. Other nudibranchs have similar glandular tissue on their papillae, as in some aeolidioideans, which has been considered possibly homologous (Wägele 1998; Wägele & Willan 2000). Comparable glandular epithelium also forms the gill glands (see above) in the doridoideans and some zephyrinids which lack the strip, as do pleurobranchs (Wägele 1998; Wägele & Willan 2000).

The *prebranchial pocket* or *prebranchial gland* is a gland of unknown function that opens to the exterior in front of the anterior end of the pinnate gill of some pleurobranchs (Pelseneer 1894; Dayrat & Tillier 2002).

20.3.2.6.2 Cerata

Cerata are long, hollow, finger-like extensions of the mantle and have evolved independently in the Aeolidioidea, Proctonotoidea (Proctonotidae), and Tritonioidea (*Marianina* in the Tritoniidae) where they are penetrated by a branch of the digestive gland. The degree of digestive gland branching is often correlated with the extent of symbiotic development with zooxanthellae or plastids in *Melibe* (Gosliner & Smith 2003) and *Phyllodesmium* (Rudman 1991a; Moore & Gosliner 2011) (see Section 20.4.7). In some, notably aeolidioideans, nematocysts containing cnidosacs are found at their distal ends (see Section 20.4.6.4).

Other cerata-like extensions found within the Doridina, and in the Limapontiidae and Hermaeidae within the Sacoglossa, are solid and not homologous. Both cerata and cerata-like processes increase the surface area for respiration.

20.3.3 MANTLE CAVITY AND ASSOCIATED STRUCTURES

Primitively, the heterobranch mantle cavity contains the same structures as in other gastropods except for the typical ctenidium. Thus, it contains the left osphradium and hypobranchial gland, the rectum and anus, pallial genital duct and its opening, and the kidney and its aperture. A mantle cavity with an anterior opening is typical of vetigastropods, neritimorphs, caenogastropods, and some less-derived heterobranchs, but there is great variation in euthyneurans. The mantle cavity of euthyneurans, and especially cephalaspideans, is highly variable and is absent in some taxa. In its less derived form, it is also the main location of the respiratory surface, which is often a highly modified gill referred to here as a plicatidium (see Section 20.3.3.1 and Chapter 4). Respiration in euthyneuran slugs occurs mainly or entirely via the body surface, and in some it may be enhanced by an increase in surface area by way of papillae, cerata, and other outgrowths, including gills such as those encircling the anus in many nudibranchs. Secondary lateral notal gills are found in cladobranchs such as the Tritoniidae, Bornellidae, and Dendronotidae, and other secondary gills are situated between the notum and foot in Phyllidiidae and *Armina* (Gosliner 1994).

A less-derived mantle cavity similar to that seen in many non-heterobranch gastropods is found in all shelled 'lower heterobranchs' and, in the Euthyneura, in rissoellids and the panpulmonate glacidorbids and pyramidelloideans. Many 'lower heterobranchs' have a modified, largely anteriorly placed, hypobranchial gland that often includes a pigmented mantle organ (PMO) (see Section 20.3.3.4). In most of the original 'pulmonates' (all of which are now in the Panpulmonata) the mantle cavity, which is located mainly on the right side, is transformed into a lung with a restricted opening (*pneumostome*) on the right and a highly vascularised roof.

The pteropod thecosome Limacinidae possess a coiled shell, and being located dorsally, the mantle cavity is oriented as in other coiled gastropods and the organs have the normal gastropod arrangement. With the uncoiling of the shell in other thecosomes the mantle cavity moved to a ventral position, and the anal opening moved from right to left and osphradium from left to right. The naked gymnosome pteropods have lost the mantle cavity.

In cephalaspideans, the opening of the mantle cavity varies from anterior to right posterior in orientation (Mikkelsen 1996), while in pleurobranchs and aplysiidans it is on the right. In the shell-less euthyneurans, reduction and loss of the mantle cavity have occurred in several lineages. Most have either a reduced mantle cavity or, as in nudibranchs, Acochlidia, and many sacoglossans, it has been lost. The reduction and loss were initiated by the mantle cavity (and its opening) being displaced on the right side of the body. In acteonids and ringiculids, the mantle cavity is large and the opening only slightly to moderately displaced to the right. *Aplysia* has a shallow, widely open mantle cavity in the middle

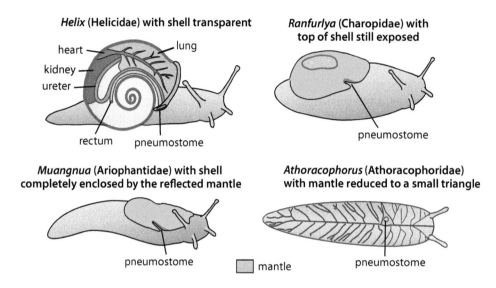

FIGURE 20.24 Comparison of body forms in a stylommatophoran snail, a semislug, and two slugs. All but lower right redrawn and modified from Solem, A.C., *The Shell Makers. Introducing Mollusks*, John Wiley & Sons, New York, 1974; lower right redrawn and modified from Burton, D.W., *N.Z. J. Zool.*, 8, 391–402, 1981.

of the right side whereas in pleurobranchids it is situated on the right side below the notal margin. This is by no means a simple linear progression; for example, in cephalaspideans (Mikkelsen 1996), the orientation of the cavity opening varies from anterior to right posterior in Aglajidae (Rudman 1974) and *Phanerophthalmus* (Austin et al. 2019).

Thus, through so-called detorsion, this body reorganisation has occurred independently in several lineages. With the migration of the mantle cavity along the right side and the associated shallowing, in the more apomorphic euthyneurans the head and foot can no longer be withdrawn into the mantle cavity. Associated with these changes are a reduction of the shell and the coiling of the visceral mass, which now rests on the foot. Both the mantle cavity and external shell are lost at metamorphosis in nudibranch sea slugs, the small to tiny acochlidian slugs, many sacoglossans, runcinoideans, a few cephalaspideans, and the terrestrial veronicelloidean slugs. Other slugs, including stylommatophoran slugs and onchidiids, retain at least a rudimentary mantle cavity. The structure identified by Climo (1980) as a lung in the marine slug *Smeagol* (Smeagolidae) is a gland, the pallial gland. There is a small posterior mantle cavity into which the kidney, anus, and pallial gland open, and it has a slit-like aperture. It is the homologue of the lung, but it is not the site of respiration, which occurs over the entire body surface, as shown by a net of blood sinuses below the epithelium (Ruthensteiner 1997).

Onchidiids were thought to differ from other panpulmonates because they had a posterior lung opening separate from the mantle cavity, and both these structures were thought to have developed separately (Fretter 1943). During its development, the lung in onchidiids forms a branching system of spaces somewhat reminiscent of the branching lung of athoracophorid slugs (Fretter 1943) (see below). Ruthensteiner (1997) showed that the lung developed from the anterior part

of the mantle cavity, with the rest of the cavity represented by the cloaca in the adult. Thus, unlike the situation in other panpulmonates with a lung, only part of the larval mantle cavity became the lung. The onchidiid lung is very small and only functions during activity, as otherwise cutaneous respiration suffices (Fretter 1975b). Like onchidiids, the terrestrial veronicellids have no shell, they lack a mantle cavity or lung and respire through their skin and possess a system of subcutaneous tubules.

The differences in lung placement and morphology in onchidiids and other lung-bearing panpulmonate slugs (see text following) may be largely the result of limacisation (Tillier 1984b).

Shelled panpulmonates mostly retain an anterior mantle cavity which allows the head and foot to be withdrawn into the shell. Although this cavity has been modified as a lung (Ruthensteiner 1997) (see Chapter 4, Figure 17), it has been markedly reduced and adapted in some groups, particularly in terrestrial slugs (Figure 20.24), giving rise to erroneous theories that the lung is a new structure and not homologous with the mantle cavity. The homology of the mantle cavity with the lung in less-derived panpulmonates has not been doubted (Hubendick 1947; Yonge 1952a; Fretter 1975b; Brace 1983) although typical mantle cavity structures, such as the hypobranchial gland, gill, and, in terrestrial taxa, the osphradium, are lost. The lung differs from a typical mantle cavity in that the only opening to the exterior is the small pneumostome. In terrestrial taxa and some others, it is contractile, being surrounded by a sphincter muscle. Thus, the pneumostome is contractile in Geophila and ellobiids but not in the estuarine Amphibolidae and the freshwater Hygrophila. There are, however, both simple and contractile pneumostomes in the marine Siphonariidae, suggesting that the origin of the pneumostome sphincter is independently derived (Barker 2001a) or has been lost in some aquatic clades.

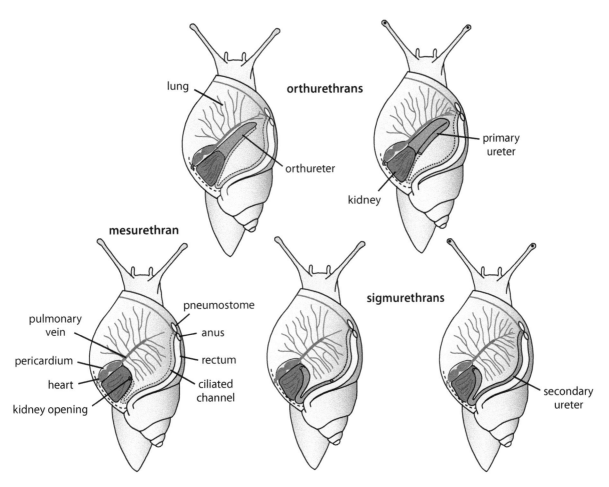

FIGURE 20.25 Different configurations of the arrangement of the ureter in stylommatophoran snails. Redrawn and modified from Barker, G.M., *Naturalised Terrestrial Stylommatophora (Mollusca: Gastropoda)*, Vol. 38, Manaaki Whenua Press, Otago, New Zealand, 1999.

The roof of the lung has a network of blood vessels where respiration occurs but is more poorly developed in ellobiids and the less-derived panpulmonates. The oxygenated blood drains into the auricle, and the volume of the lung cavity is controlled by muscles in its floor. The lung of many other marine and freshwater panpulmonates may be filled with water, and many can function in both air and water (as in Amphiboloidea, Hygrophila, and Siphonariidae). In hygrophilans, the lung is often used to trap a bubble of air while permanently submerged taxa do not use air to respire.

Respiration in air only occurs in terrestrial and some amphibious lung-bearing eupulmonates (Ellobioidea, Onchidiidae, Veronicelloidea, Stylommatophora). The marine ellobiids are typically amphibious, and a few are terrestrial. All have a well-developed lung, and none develop a secondary gill.

The homology of the narrow pneumostomal opening of siphonariids with the pneumostome of other panpulmonates is likely, although it has been questioned (Dayrat & Tillier 2002). Siphonariids possess secondary gills on the roof of the 'lung' and can respire in water or air. A secondary gill, the *pseudobranch*, is located just outside the mantle cavity (lung) in planorbids.

The panpulmonate lung has the heart and kidney located in the mantle cavity roof, unlike the posterior location of the renopericardial complex in caenogastropods. In most non-stylommatophoran lung-bearing panpulmonates the kidney opens by way of a papilla that lies inside the lung and distant from the pneumostome. In stylommatophorans, a long ureter opens alongside the anus. The orthurethran condition has a long straight ureter opening near the pneumostome, while in the mesurethran condition the short ureter is near the side of the kidney. Most stylommatophorans have the sigmurethran condition with the ascending ureter alongside the rectum (Figure 20.25). These different configurations were traditionally used in stylommatophoran taxonomy (see Section 20.5.1.2.4.2.10).

In stylommatophoran slugs, the marked reduction or loss of the shell is accompanied by the visceral mass being sunk into the elongated head-foot and a marked reduction of the size of the lung. Lung reduction is compensated for by cutaneous respiration facilitated by the body surface being kept moist by the secretion of mucus.

In athoracophorid slugs the mantle cavity is reduced to a small, highly specialised lung composed of many thin-walled diverticulae that radiate within a blood sinus, and the lung abuts the atrium (Figure 20.26). Its relative importance in

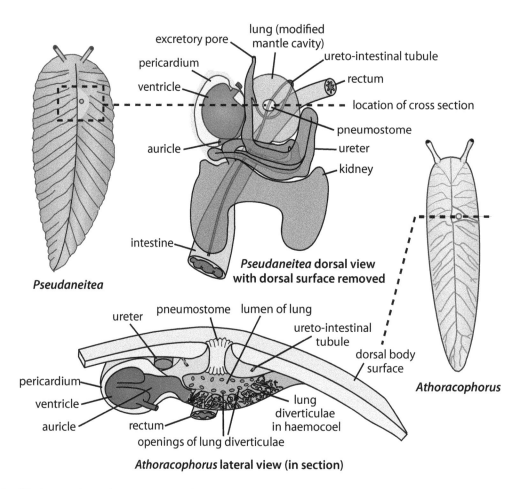

FIGURE 20.26 The lung and associated structures of an athoracophorid slug. Redrawn and modified from Burton, D.W., *N.Z. J. Zool.*, 8, 391–402, 1981.

respiration is not known because, as in most slugs, respiration can take place through the moist skin covering the dorsal body. The panpulmonate lung typically also contains some water and functions as a reservoir. In some athoracophorids, the ureter is much expanded and may have taken over this function (Burton 1981).

In stylommatophoran snails, the lung is well-developed and secondary gills are never present; in many stylommatophoran slugs the lung is modified, as outlined above. The pneumostomal area in stylommatophorans consists of the pneumostome itself, as well as the terminal ducts and openings of the ureter and rectum. The different arrangements of these structures in stylommatophorans were examined by Suvorov (2000). The 'primitive' arrangement has the excretory routes of the rectum and secondary ureter to the exterior open, while in the more derived conditions these are progressively closed over to form enclosed ducts in dissimilar and often homoplastic (Roth 2001) configurations in different groups. The pneumostome itself has pectinate folds that form a valve in some more derived taxa.

A pneumostome is usually regarded as a synapomorphy of 'pulmonates' (i.e., lung-bearing panpulmonates), but, according to Ruthensteiner (1997), this idea should be treated with caution. He argued that *Acteon* has a narrow opening to the mantle cavity (Fretter & Graham 1954), making it somewhat similar to the situation in *Siphonaria*.

20.3.3.1 Heterobranch Gills

No heterobranch has a typical ctenidium, but some have gills (Haszprunar 1985b, 1988; Gosliner 1994; Mikkelsen 1996; Ponder & Lindberg 1997) which may somewhat resemble a ctenidium (see Chapter 4 for discussion). In many heterobranchs that possess such a gill, it maintains the same position relative to the anus, heart, and renal opening as a ctenidium as well as its innervation and circulatory configuration. For these reasons, this type of gill has often been considered to be derived from a ctenidial homologue or, possibly, a rudiment of the ctenidium (e.g. Gosliner 1981, 1994; Schmekel 1982) and is termed a *plicatidium* (Perrier & Fischer 1911; Morton 1972) and is probably an example of atavism (see Chapter 4). These unpaired gills can superficially resemble the ctenidium of other gastropods, but they differ in lacking skeletal support, not having distinct lamellae like those of a true ctenidium, in lacking discrete cilial bands, in being only weakly ciliated, and in only the efferent vessel running in the mantle skirt, the afferent lying at the edge of the gill fold. Most heterobranch gills are usually located wholly or partially within the mantle cavity or are closely associated with it or its remnant.

Plicate gills (plicatidia) are found in Acteonoidea, many cephalaspideans, aplysiidans, pleurobranchs, and some shelled sacoglossans. They are also found in the thecosome pteropod *Cavolinia* and the 'lower heterobranch' Valvatoidea (see Chapter 4). They are formed from a fold of the mantle wall that is folded into two parallel rows of laminae (leaflets or plicae) or, in sacoglossans, a single row. Afferent and efferent blood vessels extend from the blood space formed inside the fold from which the lamellae protrude. Such gills occur in several heterobranch taxa, but based on structure and occurrence, they may have been derived and lost multiple times.

Various kinds of secondary gills are present in some cephalaspideans, pleurobranchs, aplysiidans and some nudibranchs, sacoglossans, pyramidelloideans (Haszprunar 1985b, 1988; Gosliner 1994), architectonicids (Haszprunar 1985c), thecosomes and gymnosomes (Lalli & Gilmer 1989), siphonariids (Yonge 1952a; de Villiers & Hodgson 1987), and planorbids (Hubendick 1978).

As the mantle cavity was reduced in the various lineages of marine euthyneurans, the plicatidium (when present) became more exposed. It lies anteriorly in the mantle cavity in acteonids, beneath the right edge of the mantle in taxa such as *Aplysia*, and is exposed on the right side of the body in pleurobranchs and gastropterids. In doridoidean nudibranchs, there is no mantle cavity, and the gill, when present, lies mid-dorsally surrounding the anus. It comprises a few to many branches, with each branch called a *rachis*. This type of gill is commonly called a *branchial plume* (MacFarland & MacFarland 1966), or an *anal gill* (Wägele & Willan 2000) or *branchial circlet* (Dayrat & Tillier 2002). The gill rachises contain muscle fibres which enable them to contract. Most doridoideans possess a *branchial pocket* into which the gill can be retracted, giving them greater protection. The Corambidae (Onchidoridoidea) have posterior gills between the notum and the foot that were probably derived from anal gills (Wägele & Willan 2000).

Some of the other types of gills seen in nudibranchs include the secondarily dichotomously branched or bushy structures in tritoniids which are on or near the edge of the notum, and the numerous branchial lamellae that lie between the notum and the foot in Phyllidiidae (Phyllidioidea) and most *Armina* (Arminoidea). These gills are considered convergent in the Phyllidiidae and Arminidae because they are differently arranged and very different in appearance from gills in other nudibranchs (Wägele & Willan 2000).

The name 'opisthobranch' refers to the position of the gill in relation to the auricle, with it lying behind the auricle. Acteonids and ringiculids are exceptions within Euthyneura because they possess an almost anterior mantle cavity and the gill (a plicatidium) is anterior to the auricle, as in vetigastropods and caenogastropods.

Siphonariids have a well-developed gill, consisting of many weakly ciliated leaflets with folded walls, which lie on the mantle roof. The gill is bordered by both the afferent and efferent vessels in a superficially similar way to a caenogastropod ctenidium (Fretter 1975b). Planorbids and acroloxids

have a secondary gill, the *pseudobranch*, that lies just external to the lung.

20.3.3.2 Opposed Ciliary Ridges (Raphes) and Mantle Tentacles

In shelled heterobranchs the weakly ciliated gill is not able to maintain a flow of water through the mantle cavity. Instead this is achieved by a long posterior ciliated pallial tentacle, as in the orbitestellids and cimids, a short posterior tentacle as in graphids, by two short tentacles as in Murchisonellidae (Warén 2013), or a dorsal and ventral opposed ciliated fold or band extending along the exhalant side of the cavity. These bands (or raphes) (Perrier & Fischer 1909; Dayrat & Tillier 2002) may also continue into a posterior extension of the mantle cavity which forms a caecum through which water circulates (Hubendick 1945, 1978; Yonge 1952a; Fretter 1975b; Brace 1977a; Mikkelsen 1996). In acteonids, the caecum is long and coils with the viscera, but it is short in cephalaspideans with reduced shells and a shallow mantle cavity, such as in *Scaphander*, Aglajidae, *Phanerophthalmus*, and *Philine*. The hygrophilan *Chilina* also has the bands extending into a pallial caecum (Harry 1964; Brace 1983), and, while ciliated bands are found in some other hygrophilans (i.e., Planorbidae), they lack a caecum. The weak ciliation on the gills of siphonariids is compensated for by the two ciliated ridges which provide the respiratory current in the water-filled lung (Haszprunar 1985b). While it is often assumed that the ciliated ridges are homologous in all heterobranchs (e.g., Haszprunar 1985b, 1988; Robertson 1985), their homology has been questioned (Ponder 1991), and there are very few studies describing their structure. The study by Pilkington et al. (1984) on *Amphibola* is an exception, and in that taxon, the ciliated tracts are a site of ionic exchange. The ciliated tracts are usually on the right side of the mantle cavity but are on the left in Architectonicidae, Mathildidae, and Omalogyridae (Haszprunar 1988f) (Figure 18.8). In architectonicids and mathildids, the ventral ridge contains the posterior pedal gland (Haszprunar 1985c, d). Opposed ciliated ridges are lacking in valvatoideans and weakly developed in Orbitestellidae (Ponder 1990a, b).

Because opposed ciliated ridges are absent in adults of other gastropods, they are considered a significant character in defining the heterobranchs (Haszprunar 1985b; Robertson 1985; Ponder & Lindberg 1997).

20.3.3.3 The Hypobranchial Gland

A hypobranchial gland is present in 'lower heterobranchs' and some shelled euthyneuran taxa (some acteonoideans, cephalaspideans, aplysiidans, thecosomes, and sacoglossans) that retain a mantle cavity and in some aquatic or semi-aquatic eupulmonates such as ellobiids. In the Thecosomata, the hypobranchial gland (also called the '*mantle gland*') secretes a mucous web to collect food (Gilmer & Harbison 1986). Its loss in other euthyneurans corresponds with the reduction and loss of the mantle cavity or its modification as a lung, except in the basal marine Amphibolidae (Golding et al. 2007). Other glands found in the mantle cavity of marine euthyneurans are discussed below in Section 20.3.3.5.

The hypobranchial gland is composed of elongate gland (typically mucous) cells with interspersed supporting cells. The gland may be large and able to secrete large quantities of mucus as in *Haminoea*, but in some species, it is reduced, consisting of only a few cells (Wägele et al. 2006).

20.3.3.4   The Pigmented Mantle Organ (PMO)

The heterobranch PMO (= pigmented larval organ or anal gland) has been considered an important character in diagnosing heterobranchs (Haszprunar 1985b, 1988f; Robertson 1985; Ponder 1991; Ruthensteiner & Schaefer 1991; Schaefer 1996; Ponder & Lindberg 1997). It is a vacuolated, pigmented organ on the roof of the mantle cavity close to the anus. In many 'lower heterobranchs' and a few euthyneurans, namely rissoellids and pyramidellids, it is retained in adults (Robertson 1985; Schander 1997), but in most euthyneurans it appears in the veliger or postveliger stage and disappears at metamorphosis. Thus, in nudibranchs it has been considered absent (Hurst 1967) or 'probably absent' (Robertson 1985), although Dayrat and Tillier (2002) pointed out that a PMO was said to be present in some nudibranch genera by Pelseneer (1901). The homology of the pigmented anal gland in Euthecosomata with the PMO is uncertain (Dayrat & Tillier 2002).

The significance and function of the PMO are still unresolved, and it has been variously homologised with the larval kidney, the anal gland, and hypobranchial gland. Schaefer (1996) considered it homologous with structures in various marine heterobranchs called anal glands, larval kidneys, and black larval kidneys but not with the hypobranchial gland. Dayrat and Tillier (2002) considered that the hypobranchial gland and PMO were different structures because they develop separately (Bickell & Chia 1979) and they produce different secretions, those of the hypobranchial gland differing from the pigmented and crystalline concretions of the PMO.

It is usually, but not always, pigmented and is sometimes incorrectly called a 'larval kidney'. Somewhat similar pigmented structures are also found in the larvae of a few other gastropods where they develop into the hypobranchial gland – for example, *Scissurella* and *Nassarius* (see Schaefer 1996). A suggestion (Robertson 1983, 1985) that the purple hypobranchial gland in adult epitoniids is a homologue of the PMO is not accepted.

20.3.3.5   Other Glands Associated with the Mantle Cavity

Gosliner (1994) listed several types of glands associated with the mantle cavity of marine euthyneurans. These include two glands in aplysiidans – the *opaline gland* (Bohadsch gland) that lies beneath the mantle floor and is thought to be defensive and the *purple gland* on the roof of the mantle cavity (see Chapter 9). There are glands similar to the opaline gland in the gymnosome pteropod *Clione* (Wägele et al. 2006).

A pallial (or 'opaline') gland in some Runcinoidea is not a homologue of the opaline gland in aplysiidans. It empties

a white fluid near the anus and may be excretory (Gosliner 1994) as it does not appear to contain defensive chemicals.

A 'yellow gland' is found in Aglajidae that may be defensive or excretory and empties to the posterior mantle cavity (Rudman 1974). It differs from the purple gland of aplysiidans in its yellow secretion and its position. Its structure has been described by Rudman (1972g).

Blochmann's glands occur in the mantle cavity of some cephalaspideans (e.g., *Haminoea*, *Bulla*, *Scaphander*, *Cylichna*). Each gland is a very large single subepithelial gland cell with a characteristic large nucleus and poorly staining contents. It is surrounded by muscle fibres, and has a multicellular duct (Wägele et al. 2006).

The aplysiidan purple or ink gland may be homologous to Blochmann's glands (Dayrat & Tillier 2002; Wägele et al. 2006), as may the aplysiidan opaline glands (Cimino & Ghiselin 2009).

Globular dorsal glands present in polycerid nudibranchs such as *Kaloplocamus*, *Plocamopherus*, and *Kalinga* are responsible for producing bioluminescence (Vallès & Gosliner 2006).

20.3.4   Digestive System

The general configuration of the digestive tract is described in Chapter 5. The mouth leads into a buccal tube, which is frequently lined with cuticle and opens by way of a sphincter to the often-muscular buccal cavity that contains the radula. There is usually a pair of dorsolateral jaws, but these are modified in many panpulmonates to form a single dorsal jaw. The buccal mass and radula show modifications not seen in other gastropods, and simplification of the rest of the digestive tract has also occurred (see Chapter 5).

Becoming a slug (limacisation) results in major bodily reorganisation necessitating the relocation of most of the midgut (including the digestive glands) and hindgut from the visceral coil to the pedal haemocoel (Figure 20.27).

Those heterobranchs feeding on encrusting organisms or surface films often have weak jaws and a wide, scraping radula whereas in others that feed on emergent organisms (plants or animals) the jaws tend to be well developed and the radular rows are reduced, in some to a single row. Where the radula is reduced or even absent, trituration occurs in a gizzard (see below).

A glandular area once thought to be sensory and named *Semper's organ*, is found inside the mouth area of stylommatophorans and produces mucus for the head and mouth area.

With a few exceptions, most heterobranchs lack a distinct snout or proboscis. The most notable exceptions are the long acrembolic proboscis (introvert) in pyramidelloideans, architectonicids, and mathildids (see also Chapter 5), with the pyramidelloidean proboscis being anatomically distinct. Many carnivorous taxa can evert their buccal mass to form a short proboscis-like structure – this includes most carnivorous stylommatophorans (Figure 20.61), Rathouisiidae, pseudothecosomatous pteropods, and some carnivorous cephalaspideans, pleurobranchs, and nudibranchs (Figure 20.28).

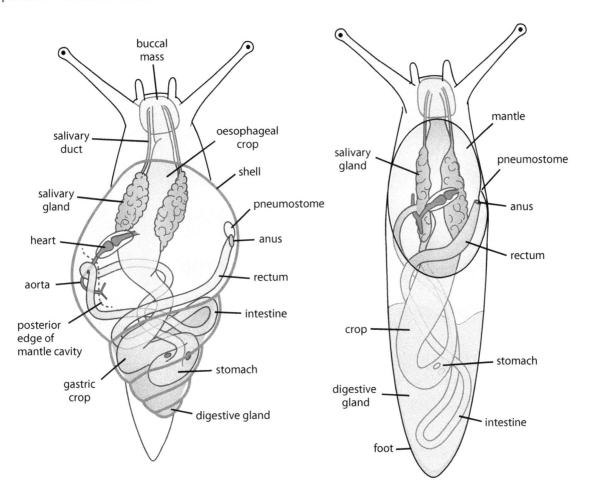

FIGURE 20.27 Comparison of the digestive systems in a generalised stylommatophoran snail and slug. Redrawn and modified from Barker, G.M., *Naturalised Terrestrial Stylommatophora (Mollusca: Gastropoda)*, Vol. 38, Manaaki Whenua Press, Otago, New Zealand, 1999.

The oral tube extends from the mouth to the buccal cavity and in acteonoideans, nudipleurans, and euopisthobranchs is lined with glandular epithelium. Several kinds of oral glands with ducts that open to the oral tube have been described in these taxa.

The acteonoideans *Bullina* and *Hydatina* (Aplustridae) have a single tubular *oral gland* that enters the posterior end of the oral tube (Rudman 1972e, b) and which is probably not homologous with the *labial glands* seen in certain cephalaspideans. These latter glands are found in some Philinidae and most Aglajidae (e.g., Rudman 1974, 1978) and consist of ventral and dorsal lobulate glands that may be paired or single.

Oral glands that are probably not homologous with those in cephalaspideans occur in some nudibranchs, including arminoideans and aeolidiids (Gosliner 1994). They are usually paired with either a pair of ducts or a single duct entering the oral tube ventrally, but a range of different forms is known (Wägele & Willan 2000). For example, in the genera *Dendrodoris* and *Ceratophyllidia* a pair of oral glands called *ptyaline glands* may be present. These glands have narrow ducts surrounded by muscle that open to the posterior end of the buccal mass where they join to form a single duct which runs ventrally along the buccal mass to open into the oral tube (Gosliner 1994; Wägele et al. 1999).

A large dorsal acid-producing buccal gland in pleurobranchids, called a *dendritic gland* or *acid gland*, comprises large cells with a non-staining vacuole (Thompson 1976; Willan 1987; Gosliner 1994; Wägele & Willan 2000). It opens medially to the anterior oral tube and produces concentrated sulphuric acid (pH ~ 1), which is used in defence. It may be tubular or ramifying and can fill much of the lumen of the visceral cavity; it is found in all pleurobranchoideans except *Tomthompsonia antarctica*.

A median buccal gland that secretes acid is also found in the polycerid *Plocamopherus ceylonicus*, but it seems not to be homologous with the acid gland of pleurobranchs (Wägele et al. 2006). The plesiomorphic condition of the jaws in heterobranchs is like that of most other gastropods – paired jaws made up of a series of rodlets that lie on either side of the entrance to the buccal cavity (see Chapter 5). The rissoellids and marine valvatoideans have large lateral paired jaws made up of numerous units with serrated edges while the freshwater *Valvata* has jaws with simple units. Omalogyrids have a thickened cuticle representing the jaws. In pyramidellids, the jaw forms a hollow, piercing stylet, but in the proboscidate architectonicids and mathildids, jaws are lacking.

The great variety of jaws seen in cephalaspideans was reviewed by Gosliner (1994). In some Cephalaspidea

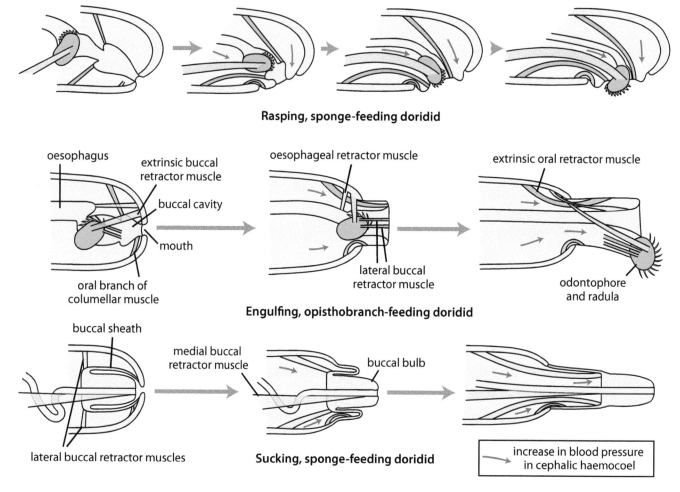

FIGURE 20.28 Protraction of the odontophore in dorid nudibranchs exhibiting different feeding strategies. Redrawn and modified from Young, D.K., *Malacologia*, 9, 421–445, 1969a.

(Retusidae, Aglajidae, and Philinidae), the Acochlidia, the Sacoglossa, Umbraculoidea, most gymnosome pteropods, and a number of doridoidean nudibranchs, distinct jaws are lost and only a cuticular lining remains. In those euthyneurans that have well-developed paired jaws, like some 'lower heterobranchs', the simple rodlets comprising the jaws in other gastropods have developed to form often elaborate tooth-like or cusp-like structures, and the jaws may rival the radula in size. In orbitestellids they are greatly enlarged, being much larger than the radula, and each jaw is composed of a few rows of one to four separate cusped plates (Ponder 1990a). In at least one murchisonellid, each jaw is hook-like and may be everted during feeding (Wise 1999). While jaws are also found in most Thecosomata and some Gymnosomata, in all aplysiidans and pleurobranchids the elongated jaws are embedded in pockets in the anterior wall of the buccal cavity. The elements making up the jaws may bear denticles, as in some 'lower heterobranchs' (see above), and in many acteonoideans, cephalaspideans, aplysiidans, nearly all pleurobranchs, some tritoniids, chromodoridids, and actinocyclids the jaw elements are multidenticulate. Most other nudibranchs have a single denticle on each jaw element, while some have two denticles and others none (Wägele & Willan 2000).

Some doridoideans possess relatively simple jaws composed of rodlets, but in polycerids, tritonioideans, arminoideans, and aeolidiids the rodlets are more or less fused into a thick pair of chitinous jaws, a condition termed *aliform* by Wägele and Willan (2000). Some doridoidean nudibranchs lack distinct jaws, but the pharyngeal cuticle may give rise to structures such as plates or hooks. Umbraculoideans lack true jaws, but the pharyngeal cuticle is papillate.

In some panpulmonates the jaw differs from that in other heterobranchs in being composed of three elements – usually thin paired lateral structures and a large, single, dorsal plate that forms the major part of the jaw. The major dorsal element is plesiomorphically composed of numerous plates bound together by a membrane. In stylommatophorans, fusion of these jaw plates into a single unit (the *oxygnathic* condition) has occurred in some taxa, and there is a convergent condition in onchidiids (Barker 2001a). In ellobiids, the lateral structures of the jaws are often reduced or absent, and they are absent in stylommatophorans. The Succineoidea have an accessory median plate on the jaw to which buccal muscles are attached – this type of jaw is termed *elasmognathous*. There is no jaw in trimusculids and amphibolids, and it is rudimentary in *Glacidorbis*.

In herbivorous eupulmonates the radula presses plant material against the jaw to cut it. The jaw also has a role in most carnivorous eupulmonates that feed by tearing up their prey prior to ingestion, but it is commonly reduced or lost in Rathouisiidae and those stylommatophorans that ingest their prey whole.

Some of the most elaborate modifications of the buccal region are seen in gymnosome pteropods such as *Clione* and *Crucibranchaea*, which in addition to their radular teeth have cephalic organs for grasping and manipulating the prey, including hooks, spines, and prehensile tentacles. In the Pneumodermatidae there is a pair of sucker-bearing arms reminiscent of the arms of cephalopods. These are retracted into a pair of sacs on the side of the head when not in use. Some other members of the Clionoidea lack tentacles but have paired adhesive buccal cones which range from one to four pairs. Some other genera lack either of these structures, although nearly all have paired hook sacs (Lalli & Gilmer 1989) (Figure 20.29).

The buccal cones are tentacle-like and hold the thecosome prey while the hooks in the hook sacs, which are protruded by hydrostatic pressure, remove the body of the prey from its shell (Lalli 1970). The adhesive material on the cones is secreted by subepithelial cells.

The hook sacs comprise several to many chitinous hooks that can be extruded when feeding and were thought to represent modified jaw elements (e.g., Pelseneer 1888; Lalli & Gilmer 1989). This interpretation was disputed by Morton (1958b) but confirmed in a detailed study by Vortsepneva and Tzetlin (2014) on the hook sacs of *Clione*.

Modification of the buccal mass occurs in different groups of euthyneurans to facilitate different feeding modes. In many, the radula is the primary food-collecting device while in some the jaw (see above) plays an important role. Others are suctorial, employing a muscular buccal pump (see below).

The buccal mass is relatively small and rounded in eupulmonate herbivores and detritivores, but in carnivorous taxa it is more elongate and muscular to facilitate capturing and swallowing large prey. In some carnivorous taxa it fills much of the body cavity, displacing other organs.

The buccal cavity is lined with cuticle and ends where it narrows to open to the oesophagus. The salivary ducts open to this cavity.

Some heterobranch taxa are suctorial, feeding on soft animal or plant tissue. Sacoglossans and omalogyrids pierce algal cells with a specialised uniseriate radula and suck out the cell contents with their muscular buccal bulb that acts as a pump (e.g., Gascoigne 1956; Jensen 1993b). Pyramidellids use their long proboscis to pierce the tissues of the invertebrates on which they feed and then suck out body fluid with a complex buccal pump (see Chapter 5, Figure 5.26). In some nudibranchs the buccal wall is modified as a muscular buccal pump with the families Onchidorididae, Corambidae, and Goniodorididae having a buccal pouch which lies on the dorsal side of the buccal mass (see Chapter 5, Figure 5.25). Sacoglossans have a muscular buccal mass modified as a pump, and some have an additional pair of muscular pouches (e.g., Jensen 1991).

Porostome doridans (Mandeliidae, Dendrodorididae, and Phyllidiidae) lack a radula and utilise a combination of oral and oesophageal glands, the buccal mass, and oesophageal musculature to dissolve sponge tissue and ingest it (Valdés & Gosliner 1999).

Unlike many other gastropods, the radular sac in most heterobranchs is at best only a little longer than the odontophore.[1] The radular teeth are non-mineralised and very diverse, providing an important taxonomic character suite in many groups of heterobranchs. A radula is present in all but a few suctorial feeders, namely pyramidelloideans, *Rhodope*, some Philinoidea (Retusidae and most Aglajidae), and some nudibranchs (dendrodoridids, mandeliids, and phyllidiids, which feed suctorially on sponges). The tritonioidean tethydids also lack a radula and use their oral hood to capture crustaceans. Among the eupulmonates, only one stylommatophoran is known to have lost the radula (see below).

The cells that produce the radular teeth, the odontoblasts, are larger (wider) in euthyneurans than in other gastropods (Franc 1968; Salvini-Plawen & Steiner 1996; Dayrat & Tillier 2002). The large odontoblasts are subterminal on the underside of the radular sac in most euthyneurans but are terminal at the end of the sac in cephalaspideans (Gabe & Prenant 1952a, b).

The radulae of 'lower heterobranchs' (Figure 20.30) are quite diverse but most have relatively few teeth in each row. Architectonicoideans have five teeth in each row, but in Architectonicidae there is a central tooth and one pair of lateral teeth and one pair of marginals while in Mathildidae the laterals are lost, and there are two pairs of marginal teeth (Bieler 1988). Orbitestellids have seven cuspate teeth in each row while valvatoideans have seven or eight and rissoellids have three to seven. *Omalogyra* has one row of teeth, the centrals, but another omalogyrid genus, *Ammonicera*, has no central and, on each side, a row of dagger-like lateral teeth and small lateral teeth (Sleurs 1985). Cimids are reported to have four teeth in each row of their minute radula – the outermost narrow with curved tips, the inner broader and cuspate (Warén 1993) (although only one side of the radula was figured and described – A. Warén pers. comm. 2012). This 'radula' is, however, a highly modified jaw. The poorly known murchisonellids have highly modified radulae (Warén 2013), but their diet is unknown. The Tjaemoeiidae also have four teeth in each row, the outermost being membranous plates, the inner hook-shaped (Warén 1991). Most unusual of all are the rhipidoglossate-like radulae of *Hyalogyrina* (n-1-1-1-n), *Hyalogyra* (n-6-1-6-n), and *Xenoskenea* (n-3-1-3-n).

It is usually thought that the broad rows of similar, simple teeth seen in basal euthyneurans are plesiomorphic (e.g., Hoffmann 1932–1939; Fretter 1975b; Gosliner 1994; Wägele & Willan 2000; Barker 2001a), with specialisation and reduction occurring independently in different lineages. This concept of the primitive euthyneuran

---

[1] There are a few exceptions such as in some species of *Glossodoris* (Rudman 1986) (now *Doriprismatica*) and *Halgerda*, where the radular sac approaches the buccal mass in length.

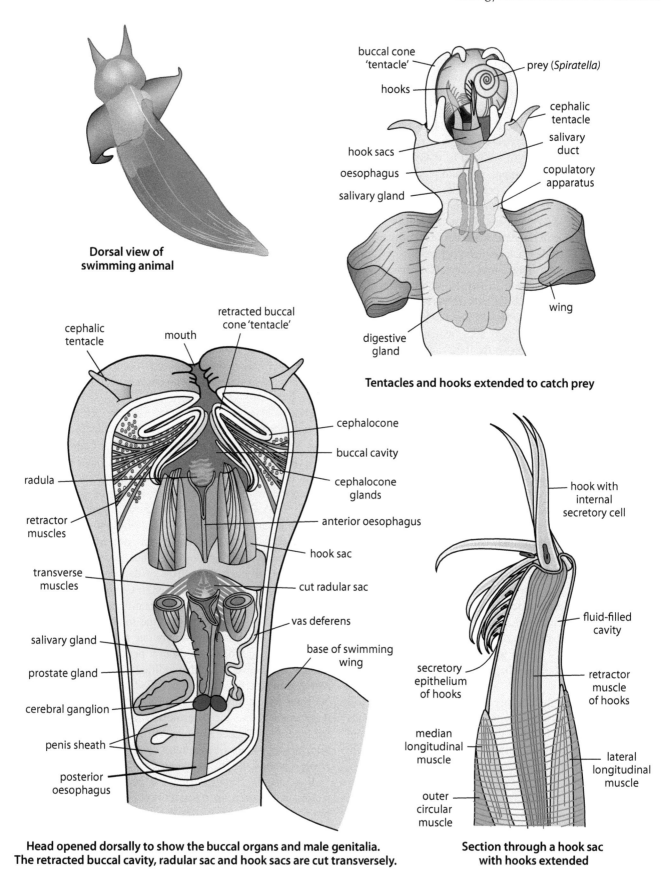

Dorsal view of swimming animal

Tentacles and hooks extended to catch prey

Head opened dorsally to show the buccal organs and male genitalia. The retracted buccal cavity, radular sac and hook sacs are cut transversely.

Section through a hook sac with hooks extended

FIGURE 20.29 Prey capture and buccal apparatus of the gymnosome, *Clione limacina*. Redrawn and modified from the following sources: Top and bottom right (Lalli, C.M., *J. Exp. Mar. Biol. Ecol.*, 4, 101–118, 1970), bottom left (Morton, J.E., *J. Mar. Biol. Assoc.* U.K., 37, 287–297, 1958b).

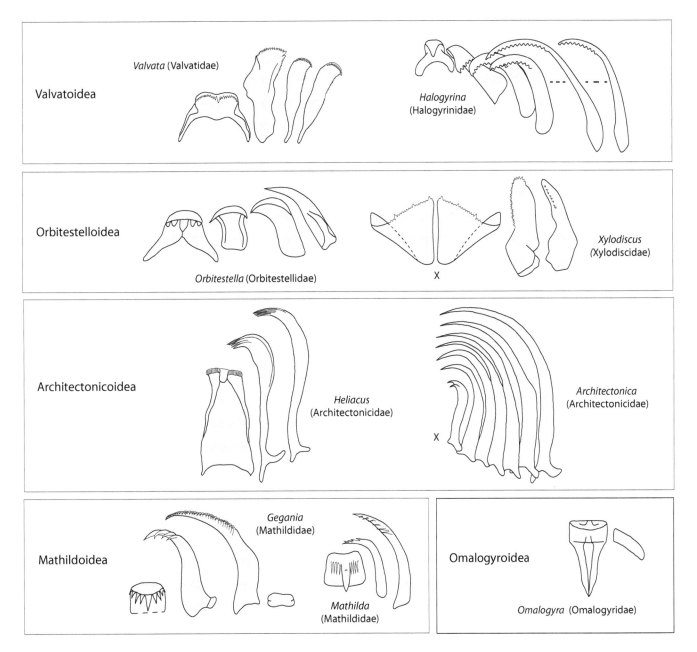

FIGURE 20.30 Examples of radulae of 'lower heterobranchs'. Redrawn and modified from various sources. X - central tooth absent. Note that the radulae of the species illustrated do not necessarily reflect the range of variation in the genera they belong to or of their higher taxon.

radula led to speculation on their origin from the rhipido-glossate vetigastropods (Hubendick 1945; Morton 1955a). For example, Morton (1955a) suggested that basal 'pulmonates' must have arisen before the evolution of the taenioglossate radula in monotocardian gastropods, and this proposition was accepted by Barker (2001a) but rejected by others (e.g., Fretter 1975b). Ponder (1991) and Ponder and Lindberg (1997) argued that by outgroup comparison a small number of radular teeth was the primitive configuration for Heterobranchia. Based on the configurations seen in the 'lower heterobranchs' this was thought to be probably around five to nine teeth in each row. The more recent recognition of the rhipidoglossate-like radulae in Halogyrinidae (Valvatoidea) (see above and Figure 20.30)

may give credence to a rhipidoglossate ancestor but one with a radula very different from the multiple lateral field teeth seen in many euthyneurans.

A well-illustrated review of radulae of marine euthyneurans is given by Gosliner (1994). They vary from many uniform, simple, hooked teeth to a marked reduction in the number of teeth in each row. Tooth reduction may proceed with retention or even increase in the significance of the central tooth or with the loss of that tooth row. In most aeolidioideans, all sacoglossans, and some tritonioideans (e.g., *Doto*) only the central teeth are present. Depending on the group, variation in tooth number is combined with tooth specialisation in either the centrals or laterals. Given this great variation, an overview is provided for each major clade.

Acteonoideans feed on polychaetes, and their radula (Figure 20.31) shows considerable diversity in the size, shape, and arrangement of the teeth. Most have many teeth in weakly sagittate to alveate rows with sharp cusps, and the central row of teeth is sometimes absent. *Pupa* has only five teeth on each side, no central teeth, and the main cusp on the outermost teeth is very long and slender (Rudman 1972f). The radula of *Acteon* is perhaps the most atypical with each row consisting of numerous (up to ~170 in a half row), very minute, denticulate teeth, and central teeth are absent.

The radulae of cephalaspideans are particularly diverse (Figure 20.32). In the carnivorous *Cylichna* and *Acteocina* (Cylichnidae, Philinoidea) the tooth rows are weakly sagittate to lineate, and the central teeth are broad with two cuspate bulges. The lateral teeth are few (3–11) and strongly curved, with the innermost much larger than the others. Radular evolution in the remaining Philinoidea involves the loss of the outer lateral teeth and the central teeth in various lineages and, in some, the loss of the entire radula.

The other major group of cephalaspideans, the Haminoeoidea, plesiomorphically have a broad radula with alveate rows and a row of central teeth. Most have simple, curved laterals, but the number of laterals can vary from about 60 to only one in *Cylichnatys*. The central teeth may be squarish, usually with three cusps as in *Haminoea* or broad and multicuspid as in *Bulla*, the latter genus only having two cuspate lateral teeth and an outermost smooth tooth. Some runcinids have central teeth and two pairs of lateral teeth, but the radula is reduced to vestigial in some.

The aplysiids, another herbivorous group, have a broad radula (Figure 20.32) with alveate rows and cuspate teeth, including a prominent central row. In this group, as in the herbivorous cephalaspideans, the radula rakes algal material into the crop from where it is passed to the grinding gizzard. In carnivorous cephalaspideans the gizzard crushes the prey.

The radulae of some groups traditionally placed with the cephalaspideans are much reduced. In the Ringiculidae there is a single pair of lateral teeth and small central teeth may be present or absent, while in the 'diaphanids' the central is present along with one to three laterals, or the laterals are absent (in *Newnesia*) (Figure 20.32). Umbraculids have numerous rather similar teeth (Figure 20.32).

Thecosome pteropods have a narrow radula (Figure 20.33) with only one or two laterals and a central row of teeth that each bear a prominent cusp. In contrast, gymnosomes have three to seven hook-shaped lateral teeth in each alveate row and a prominent central tooth (Figure 20.33).

The plesiomorphic radular configuration in nudipleurans is numerous, rather similar, hooked teeth in each row, including a central tooth. Similar radulae are found in most cephalaspideans, aplysiidans, umbraculoideans, pleurobranchoideans and many nudibranchs (e.g., Doridoidea, Tritoniidae, Arminidae) (Hoffmann 1932–1939; Gosliner 1994; Wägele & Willan 2000), and many panpulmonates (with the notable exceptions of sacoglossans, pyramidelloideans, and glacidorbids).

Some notaspideans have lost the central teeth but their radula mainly conforms to the plesiomorphic pattern, and

most have smooth, hook-shaped lateral teeth in alveate rows (Figures 20.34 and 20.35). Similarly, nudibranchs such as *Bathydoris* have a rather broad radula with a row of well-developed central teeth and hook-shaped laterals (Figure 20.34), but their tooth rows are usually sagittate, and there has been considerable modification within the various nudibranch clades. The cryptobranch nudibranchs have hook-shaped teeth, primitively with a central tooth row in some chromodoridids, but this is lost in the others.

In some groups, the inner laterals differ from the outer laterals, with the latter often being called 'marginal' teeth although some authors (e.g., Gosliner 1994; Ponder & Lindberg 1997; Wägele & Willan 2000) have indicated that they are not homologues of the marginal teeth in caenogastropods or vetigastropods. The number of lateral teeth is reduced in some nudibranch taxa, varying from a few on each side to just a single lateral, or the loss of all the lateral teeth as in most Aeolidioidea.

In some chromodoridids, the innermost lateral is enlarged, but in others, the numerous teeth are more uniform (Rudman 1984; Gosliner 1994). Many doridian nudibranch radulae lack central teeth, but these are present in some polycerids such as *Limacia* and *Crimora*. Some goniodoridids have vestigial central teeth, but they are absent in others (Gosliner 1994). The plesiomorphic configuration is seen in a few such as *Aegires*. In many others, the innermost laterals are markedly larger, for example as in most polycerids where they are cuspate and the remaining laterals are plate-like or elongate and hooked.

The tooth rows in tritonioideans and dendronotoideans vary from sagittate in most to strongly alveate (in Scyllaeidae), and in some, there is a reduction in the number of lateral teeth from the condition seen in the more plesiomorphic genera such as *Tritonia* (Figure 20.35), and the central teeth may also be lost. In *Hancockia* there is only a single lateral present (Figure 20.35), and in Dotidae only the central tooth remains. The teeth of the sponge-feeding doridid *Aldisa* are unusually elongate (Figure 20.34).

Most arminids have the plesiomorphic radular condition, with weakly sagittate tooth rows including a well-developed row of dentate central teeth and, usually, dentate inner laterals (Figure 20.35). Other cladobranchs have the laterals reduced to two (Dironidae) or one (Madrellidae) pairs. The central teeth are large in some and reduced in others.

The cnidarian feeding Aeolidioidea have large, cuspate central teeth with the most plesiomorphic members (Notaeolidiidae) having four to five lateral teeth in sagittate rows. The most apomorphic retain only one pair of laterals (i.e., are triserrate; e.g., *Flabellina* and *Eubranchus*), or, as in many, they are monoserrate (Figure 20.35). In some taxa, the central tooth is broad with two cuspate bulges, and the two segments suggest fused lateral elements (see Gosliner 1994 figures 37C–E). This so-called pectinate central is characteristic of the Aeolidiidae and is unique within the Nudibranchia. Also, the base of the central teeth of aeolidioideans is unusual in nearly always being strongly arched (Figure 20.35).

The radulae of two groups traditionally included in 'opisthobranchs', but now in the panpulmonates, the Acochlidioidea

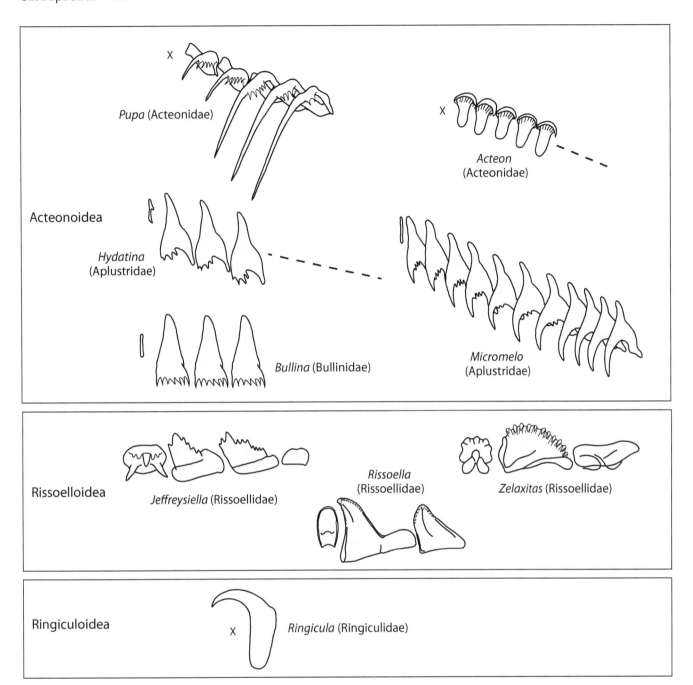

FIGURE 20.31 Examples of radulae of Acteonoidea, Ringiculoidea, and Rissoelloidea. X – central tooth absent. Redrawn and modified from various sources. Note that the radulae of the species illustrated do not necessarily reflect the range of variation in the genera they belong to or of their higher taxon.

and the Sacoglossa, share a unique character in having a U-shaped radular sac made up of 'ascending' and 'descending' limbs (Schrödl & Neusser 2010). In sacoglossans, the descending limb leads to a pouch beneath the odontophore called the ascus in which the worn, discarded teeth are collected (Jensen 1991, 1996a, b) (Figure 20.36). The radulae of sacoglossans are also simplified (Figure 20.37) compared with most euthyneuran taxa, being uniseriate with the elongate, pointed centrals interlocking and articulated and used for piercing algal cells or, in some, eggs of other marine euthyneurans. Details

of their morphology are discussed by Gascoigne (1977) and Jensen (1991, 1993a, b).

Gascoigne (1977) noted that there were two main types of sacoglossan teeth, one in which the slot to house the adjacent tooth commences near the apex and the other in which it is much lower. Bivalved sacoglossans have radular teeth with flexible sides and lateral processes. A few sacoglossans eat the eggs of other marine euthyneurans, and in these, the radula may be much reduced or highly specialised. In one species, *Calliopaea oophaga*, the radula produces three types of teeth

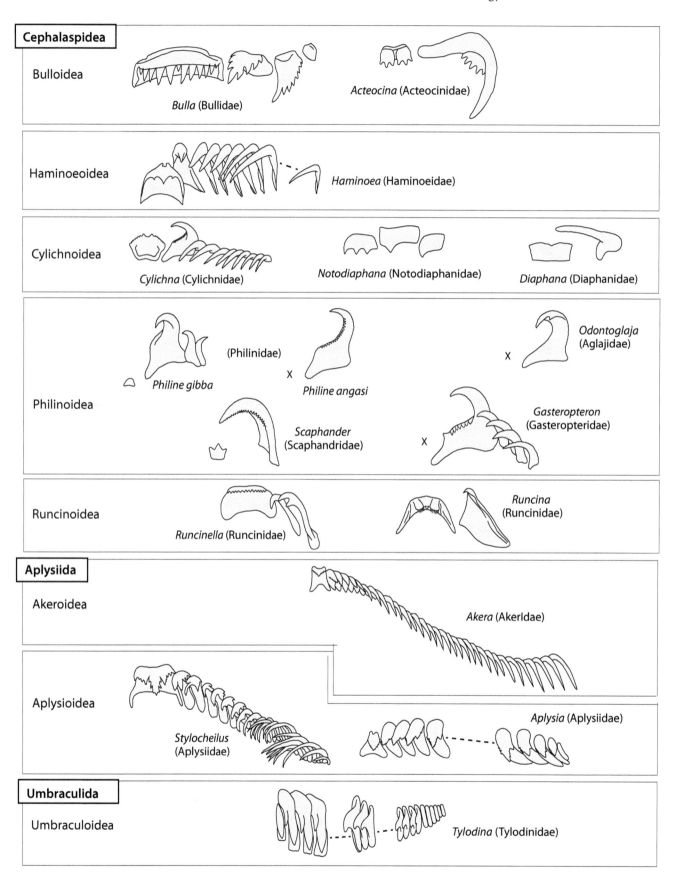

FIGURE 20.32 Examples of cephalaspidean, runcinoidean, and aplysioidean radulae. X – central tooth absent. Redrawn and modified from various sources. Note that the radulae of the species illustrated do not necessarily reflect the range of variation in the genera they belong to or of their higher taxon.

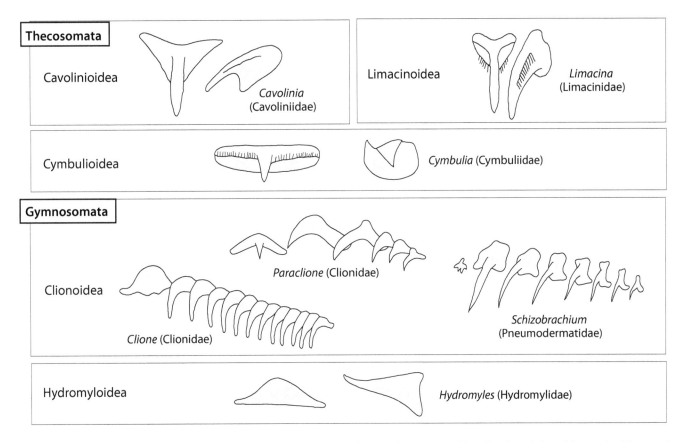

FIGURE 20.33 Examples of pteropod radulae. Redrawn and modified from various sources. Note that the radulae of the species illustrated do not necessarily reflect the range of variation in the genera they belong to or of their higher taxon.

over the lifetime of an individual, with the early ones used in post-larval feeding, the next in algal feeding, and the most mature are broader and used in egg feeding (Gascoigne & Sartory 1974). Probably all sacoglossans have juvenile teeth used for a short time in post-larval feeding before sap-sucking commences (Gascoigne 1977).

The Acochlidioidean radula (Figure 20.37) consists of a large, denticulate central plate, usually with one or two pairs of lateral plates, with the laterals sometimes single on one side and double on the other; some have only dagger-like central teeth somewhat reminiscent of the radula seen in sacoglossans. A few have asymmetric radulae with an additional tooth row on one side (Schrödl & Neusser 2010).

The radulae of most lung-bearing panpulmonates are more uniform than those of other euthyneurans. Most are broad and relatively short, and each row is comprised of many small, similar teeth, including the central teeth. Amphiboloidean radulae have been described by Golding et al. (2007). These all have alveate tooth rows, but the teeth show some diversity, with species of *Lactiforis* lacking central teeth while the others, including *Amphibola* (Figure 20.37), have prominent centrals. Species of *Salinator* have two distinctly different inner lateral teeth, the innermost small and simple with one or two cusps or rudimentary cusps and the other large and multicuspate. The remaining lateral teeth are sickle-like, as they are in other amphiboloideans, and in *Lactiforis* these

are needle-like. The small freshwater family Glacidorbidae are related to the estuarine amphiboloideans and, apparently, the marine pyramidellids (e.g., Golding 2012). Glacidorbids have a much-reduced radula which has one or three teeth in each row, with the central teeth being large and serrate and the outer pair, if present, being rudimentary (Ponder & Avern 2000) (Figure 20.37). As noted previously, pyramidellids lack a radula.

Siphonariids have a plesiomorphic configuration with the rows approximately linear to weakly sagittate in the outermost part (Harbeck 1996) (Figure 20.37), while onchidiids have markedly alveate tooth rows with well-developed central teeth (Figure 20.38). Veronicellids have more or less linear tooth rows, and the carnivorous rathouisiids have dagger-like teeth arranged in strongly sagittate rows (Figure 20.38).

In the basal Hygrophila (Figure 20.38), *Acroloxus* has a relatively narrow radula with up to about 39 cuspate teeth in each full alveate row (Hubendick 1962) while *Chilina* has about 121 cuspate teeth in each strongly alveate row (Harry 1964). The related *Latia* also has alveate tooth rows (Harbeck 1996).

Ellobiids (Figure 20.38) show considerable variation in their radular teeth (Martins 2007), but all have a broad radula, the tooth rows of which range from alveate to sagittate, and in most the rows change direction in the outer half

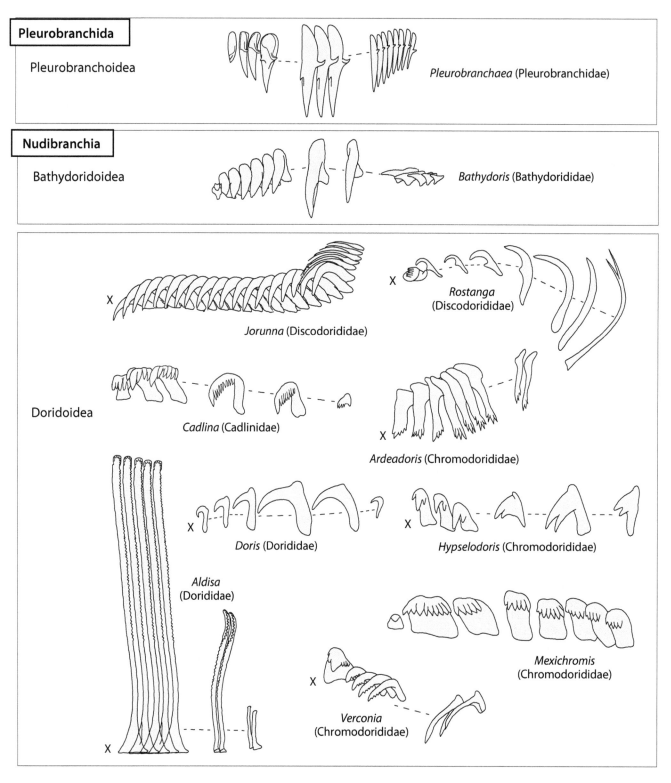

FIGURE 20.34 Examples of radulae of Nudipleura. X – central tooth absent. Redrawn and modified from various sources. Note that the radulae of the species illustrated do not necessarily reflect the range of variation in the genera they belong to or of their higher taxon.

(Harbeck 1996). Genera related to *Pedipes* (Pedipedinae) have rather narrow, hook-shaped radular teeth while those of *Otina* are broader and very different from those of *Smeagol* (Figure 20.38). The related *Trimusculus* is markedly sagittate (Harbeck 1996), with cuspate teeth. *Smeagol* has numerous simple, hooked teeth in each row which are linear through most of the radula but alveate in the outermost section. In contrast, *Otina* has lateral teeth with two or three cusps, and the outermost ones are plate-like with the rows alveate in the inner part of the radula but becoming linear in the outer part (Tillier & Ponder 1992). Both genera have unicuspid central teeth (Figure 20.38).

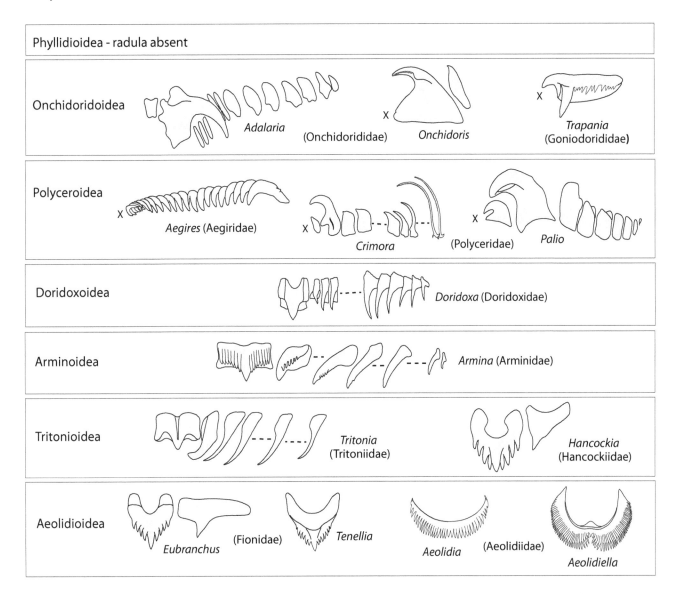

FIGURE 20.35 Additional examples of nudibranch radulae. X – central tooth absent. Redrawn and modified from various sources. Note that the radulae of the species illustrated do not necessarily reflect the range of variation in the genera they belong to or of their higher taxon.

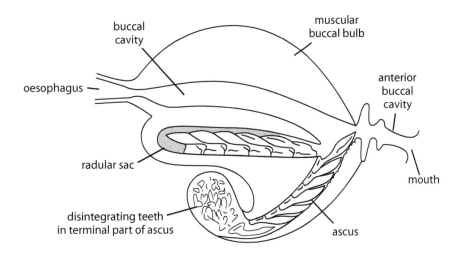

FIGURE 20.36 Longitudinal section of the buccal mass of the sacoglossan *Alderia modesta* showing the radula and the ascus. Redrawn and modified from Evans, *J. Molluscan Stud.*, 29, 249–258, 1953.

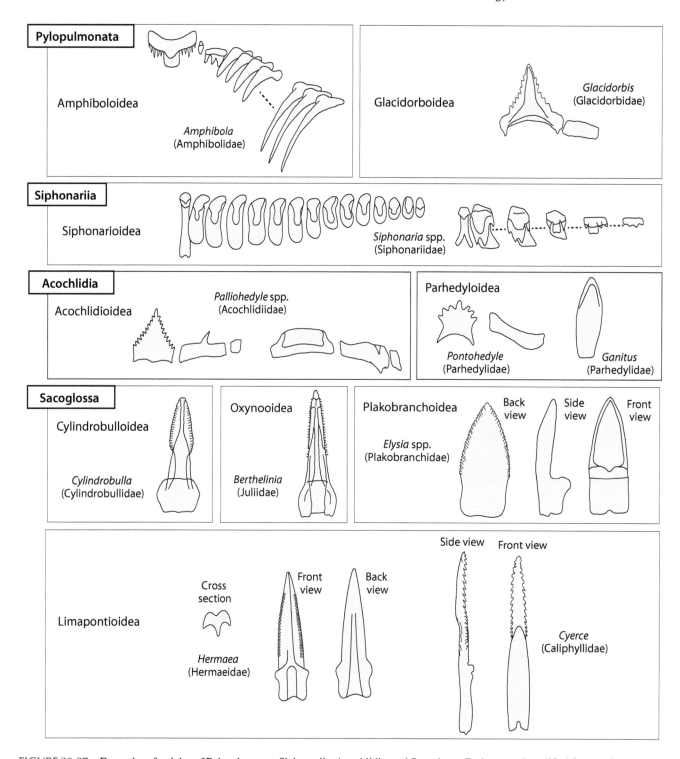

FIGURE 20.37 Examples of radulae of Pylopulmonata, Siphonariia, Acochlidia, and Sacoglossa. Redrawn and modified from various sources. Note that the radulae of the species illustrated do not necessarily reflect the range of variation in the genera they belong to or of their higher taxon.

Stylommatophoran radulae (Figure 20.39) show some diversity, although most conform to the plesiomorphic pattern with the tooth rows weakly alveate to linear and strongly sagittate in carnivorous groups, such as Testacellidae, Rhytididae, Streptaxidae, and Spiraxidae, all convergently having dagger-like teeth. In fact, there are considerable diet-related convergences seen generally in stylommatophoran

radulae (e.g., Solem 1973, 1978; Breure & Gittenberger 1982; Barker & Efford 2002).

Only one stylommatophoran has been shown to lack a radula, the streptaxid *Imperturbatia perelegans*, which has lost the entire buccal apparatus and is a carrion feeder that sucks on soft tissue (Gerlach & van Bruggen 1998).

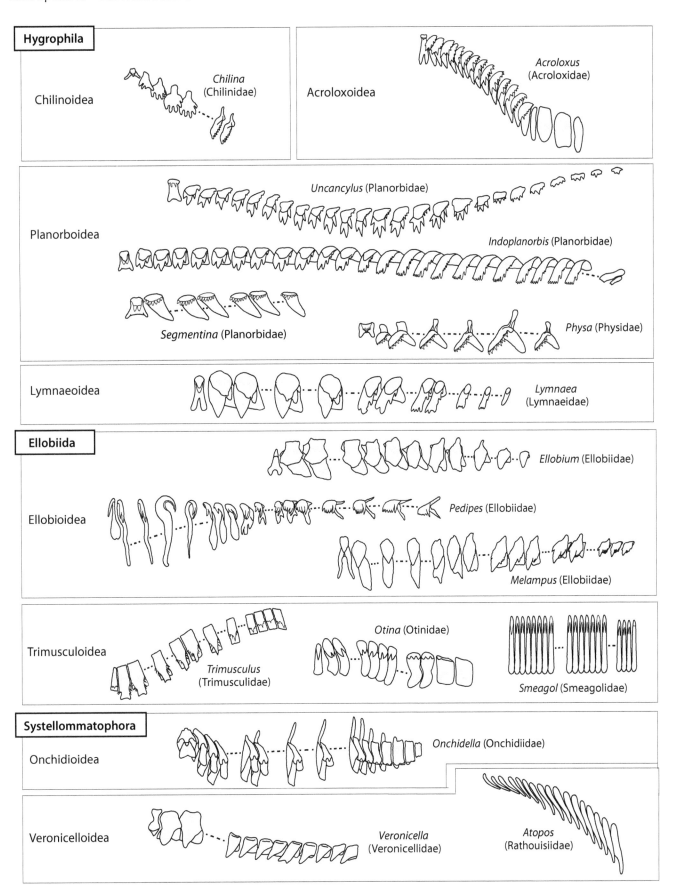

FIGURE 20.38 Lower eupulmonate radulae. Redrawn and modified from various sources. Note that the radulae of the species illustrated do not necessarily reflect the range of variation in the genera they belong to or of their higher taxon.

Stylommatophora

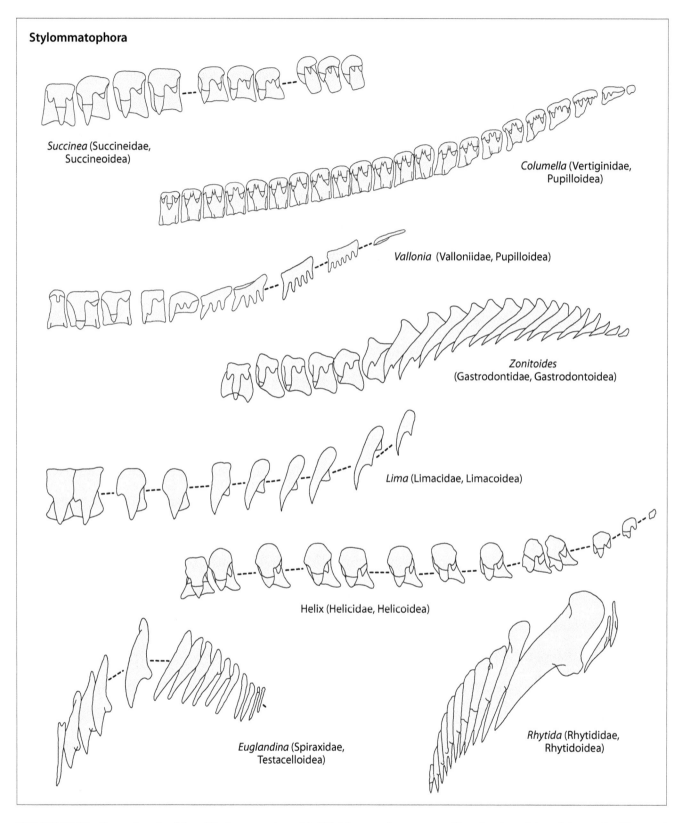

FIGURE 20.39 Examples of radulae of Stylommatophora. Modified from various sources. Note that the radulae of the species illustrated do not necessarily reflect the range of variation in the genera they belong to or of their higher taxon.

There is some controversy as to whether the buccal carti-lages seen in other gastropods are present in Heterobranchia. Homologous cartilages were thought to be absent by some workers (e.g., Haszprunar 1985b, 1988; Ponder & Lindberg 1997), while others argued that they had been retained, even if in a modified form (e.g., Barker 2001a). Haszprunar argued that the cartilage-like structures seen in some hetero-branchs differed in their structure and were hence 'second-ary structures' (Haszprunar 1988f p. 392), but some accounts described well-developed cartilages (e.g., Morton 1955b; Hubendick 1962; Ghose 1963a). The cartilage-like structures seen in many eupulmonates and some other euthyneurans are not well studied but apparently differ from those of other gastropods because they are a mixture of cartilage cells and muscle cells and not embedded in a matrix (Person & Philpott 1969; Curtis & Cowden 1977; Mackenstedt & Märkel 2001), but further investigation is required.

Regarding function, the broad euthyneuran radula operates more like a scoop than the broad sweeping action of the rhipi-doglossate radula, and because the odontophoral chondroid tissue is largely replaced by muscle, it has more flexibility and greater control (Fretter 1975b). Blood pressure plays a greater role in radular feeding movements than in caenogastropods and vetigastropods, resulting in more flexibility. The radular membrane is not folded away, as it is in vetigastropods when the odontophore is retracted, but covers the surface of the scoop (Fretter 1975b). Also, the membrane can only move forward (no lateral movement) over the bending plane. At this stage, the teeth become erect to rasp or to grip the food, which may also be sliced by the jaw. These feeding move-ments enable large pieces of food to be bitten off, which are then swallowed by a method apparently unique to eupulmo-nates (Fretter 1975b). The *collostylar hood* is a tongue of tis-sue that projects into the space left by the odontophore when it is protracted during feeding and which then directs the food into the oesophagus when the odontophore is retracted. The collostylar hood lies at the anterior end of the collostyle, a rod of connective tissue that lies in the trough-shaped concavity of the short radular sac (Fretter 1975b; Mackenstedt & Märkel 2001). The distribution of the collostylar hood in panpulmo-nates is not clear, but it certainly occurs in Stylommatophora and in Hygrophila. Heterobranchs usually have a pair of sali-vary glands that open to the buccal cavity and extend pos-teriorly along the oesophagus. In cephalaspideans they are usually simple, long, and tubular, but in acteonoids, they are more elongate and convolute (Gosliner 1994). Pleurobranchs have very long salivary ducts, and the salivary gland is cushion-like and lies next to the digestive gland (Wägele & Willan 2000). The salivary gland ducts of some nudipleurans (pleurobranchs and *Bathydoris*) have a small salivary bulb at the point where they enter the pharyngeal wall (Wägele & Willan 2000). In *Melibe*, the salivary glands are considerably reduced and, in the suctorial dendrodoridid and phyllidiid nudibranchs, the salivary glands are also reduced or absent. In *Dendrodoris nigra*, the paired salivary glands are small and globular and are somewhat similar in histology to the oral glands (Wägele et al. 1999).

The often large, paired, salivary glands of lung-bearing pan-pulmonates can be fused posteriorly. Histologically they are simple in *Otina* but more complex in Hygrophila and stylomma-tophorans which secrete mucus and amylase (Luchtel et al. 1997).

Dorsal folds are present in the oesophagus of some val-vatoideans (Cleland 1954; Ponder 1990b) but are absent in other heterobranchs. Ponder and Lindberg (1997) argued that because oesophageal glands develop late in ontogeny, the often simple heterobranch oesophagus reflects a paedomor-phic condition.

While in most heterobranchs the oesophagus is a simple tube, in some euthyneurans it expands to form a crop, although its delimitation may not be clear as it often bears no special-ised structure or epithelia.

In some euopisthobranchs (many cephalaspideans, theco-some pteropods, and aplysiidans) there is a thick muscular oesophageal gizzard that lies close to, or is contiguous with, the stomach and contains calcareous or chitinous plates or spines (Figure 20.40). The gizzard triturates the food which, depending on the group, may be algae, protists (foraminifer-ans), or animals such as gastropods, bivalves, or polychaetes. The gizzard contains hard plates of cuticle which may be calcified, with several in aplysiidans and only three in most cephalaspideans. The gizzard of thecosome pteropods is more similar to that of aplysiidans than those of many cephalaspi-deans which have few (usually 3) large, well-developed plates (Ghiselin 1966) (Chapter 5, Figure 5.41), but runcinids have four. The aplysiidan gizzard consists of an anterior portion with a few larger teeth and a posterior part with many small teeth or spines (Chapter 5, Figure 5.41).

The panpulmonate Sacoglossa pierce algal cells with their specialised radula and suck out the cell contents so there is no need for a gizzard, although many have a crop or an oesophageal pouch that may be muscular, glandular, or both (Jensen 1991). Their method of feeding differs from that seen in other algal- and plant-feeding panpulmonates, which ingest their food as plant fragments. In contrast to many euopistho-branchs, in lower panpulmonates where a gizzard is present it is part of the stomach, rather than the oesophagus, that takes on that role (see below).

A cuticle with plates lines the posterior oesophagus of Umbraculoidea (not the stomach as sometimes stated) and this area can also be considered an oesophageal gizzard (Wägele & Willan 2000). *Bathydoris* is the only nudibranch with a cuticular lining of the whole oesophagus, with the cuti-cle thin anteriorly and thickened posteriorly with longitudinal ridges bearing small spines (Wägele 1989b), this region pos-sibly being a remnant of an oesophageal gizzard (Hoffmann 1932–1939). In cladobranch nudibranchs, there is a cuticular-ised epithelium lining the anterior end of the oesophagus, and this is followed by a vacuolated non-cuticularised epithelium (Wägele & Willan 2000).

In the herbivorous cephalaspidean Bullidae and some spe-cies of Haminoeidae, as well as the Runcinidae, the oesopha-gus forms a paired or single oesophageal caecum or pouch of unknown function that lies just in front of the gizzard (e.g., Rudman 1971). Similar diverticula are present in Diaphanidae,

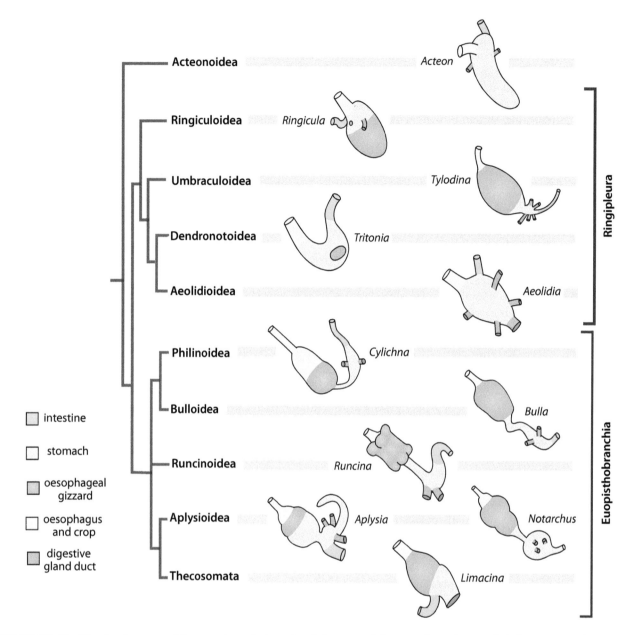

FIGURE 20.40   The development of the oesophageal gizzard and the stomach in acteonimorphs, euopisthobranchs and ringipleurans. Original, with figures redrawn and modified from several sources.

Colpodaspis, thecosome pteropods (Dayrat & Tillier 2002), and some Sacoglossa (Jensen 1996b, a; Mikkelsen 1996). Ghiselin (1966) suggested this diverticulum might be a homologue of the oesophageal gland seen in many vetigastropods and caenogastropods, but this is unlikely (Dayrat & Tillier 2002). In onchidiids, the oesophagus sometimes forms a crop, as it does in some stylommatophorans, although most lack this modification (Tillier 1989).

As in other gastropods, in most heterobranchs the stomach is a sac with the oesophagus on the right and the intestine anteriorly on the left. In many marine euthyneurans the stomach is a small, simple sac with a variable number of openings to the digestive gland. In cephalaspideans such as Bullidae, Haminoeidae, and Philinoidea the stomach is a one-way system with the oesophagus entering anteriorly and

the intestine exiting posteriorly, and some have ciliated tracts (Rudman 1971). Runcinids, acteonids, and aplustrids all have a simple sac-like stomach. The stomach in acteonids and at least some aplustrids is lined with cuticle (Rudman 1972c, d), and the broad, blind posterior caecum in *Acteon* is mostly cuticle-lined (Fretter 1939). In a few taxa, the stomach is muscular and may be lined in part with chitin. In *Ringicula* and many Tritonioidea, there are chitinous stomach plates present (Gosliner 1994), while other nudibranchs lack a cuticular gastric lining. In some 'lower pulmonates' part of the stomach may be heavily muscularised to form a gastric gizzard (Figure 20.41), but in other euthyneurans it is rather simple and in some is reduced to a small space to which the digestive gland ducts open. In pteropods, the lumena of the digestive gland and stomach merge so they are not readily separable. In some

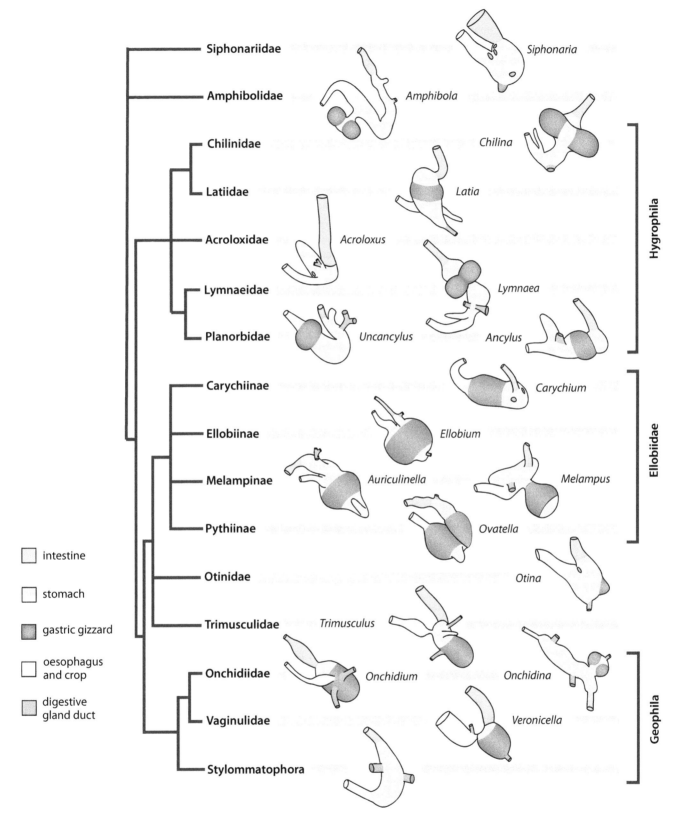

FIGURE 20.41 Comparative diagram showing the stomachs of lung-bearing panpulmonates, some with gastric gizzards. Original; figures redrawn and modified mainly from Hubendick, B. Systematics and comparative morphology of the Basommatophora, pp. 1–47 in V. Fretter, V. & Peake, *J. Pulmonates: Systematics, Evolution and Ecology*, 2A. Academic Press, London, 1978.

marine euthyneurans (aplysiidans, thecosomes, doridoideans, and tritonioideans) there is a blind, non-cuticularised stomach caecum (see below) which has sometimes been incorrectly referred to as a style sac. Some heterobranch stomachs, such as that of *Otina* (Morton 1955c), have a style sac region, but crystalline styles have not been reported except in *Acroloxus* (Hubendick 1962) and *Orbitestella* (Ponder 1990a).

In those taxa where the stomach is not clearly differentiated it can be defined as the part of the midgut where the intestine starts and the main ducts of the digestive glands open, although it can be difficult to distinguish between the stomach and the expanded digestive gland ducts other than histologically.

The stomachs of the more plesiomorphic lung-bearing panpulmonates, including ellobiids, have some plesiomorphic features such as ciliary currents, paired typhlosoles linking the digestive gland openings to the intestine, and a posterior caecum behind the entrance of the posterior digestive gland opening (Barker 2001a). The sorting area has become vestigial, the interior of the stomach largely cuticularised, and the gastric shield lost. A strong muscle coat surrounds the stomach to form a gizzard which becomes more or less isolated from the rest of the stomach in more apomorphic taxa (Figure 20.41). In stylommatophorans, the gizzard has been lost, and there is sometimes little distinction between an oesophageal crop and the stomach. Usually, the oesophagus enters the stomach and opens to an expanded area, the so-called 'gastric crop' which opens posteriorly to the 'gastric pouch' that often bears ciliated ridges (Barker 2001a), but both are simply the stomach lumen, and can be referred to as simply the stomach.

The stomach of onchidiids and most of the more plesiomorphic lung-bearing panpulmonates usually has an anterior chamber, a middle muscular gizzard, and a posterior caecum (Figure 20.41). This gizzard is typically a single muscular mass but in some taxa is partially (as in *Lymnaea*) or completely divided into two chambers (as in *Amphibola*). The gastric gizzard is reduced in *Siphonaria*, *Otina*, and *Smeagol* and absent (lost?) in *Acroloxus* and in all stylommatophorans.

The stomach wall of *Umbraculum* and *Tylodina* has rows of small teeth, an apomorphic condition for that group. Even more unusually, some tritoniids and tethydids have a posterior gizzard containing denticles, spines, or plates. It uniquely lies behind the stomach at the beginning of the intestine.

Barker (2001a) suggested that the oesophageal and gastric gizzards may be homologous, but this is unlikely and not generally accepted.

The digestive gland is usually the largest gland in the body. There are plesiomorphically two digestive gland lobes, each with a duct to the stomach, but often one lobe (the right) is much reduced or absent, and the number of digestive gland openings can vary from group to group. Histologically, as in other gastropods the digestive gland contains triangular excretory cells and tall digestive cells (see Chapter 5). The gland is a compact mass (*holohepatic*) as in other gastropods and in 'lower heterobranchs', lung-bearing panpulmonates, and in Cephalaspidea, Aplysiida, Pleurobranchida, Umbraculida, and doridoidean nudibranchs. In derived members of some

other groups of nudibranchs, it is branched (*cladohepatic*), as described below.

In some cladobranchs the separation between the holohepatic and cladohepatic conditions is not entirely clear-cut, and they show a range from simple, compact digestive glands to taxa with branched ducts that enter the cerata (Gosliner 1994). For example, the digestive gland diverticula extend into the notum in Curnonidae, and in the proctonotid genus *Janolus* some have digestive diverticula extending into the cerata, but others do not (Miller & Willan 1986).

The aeolidioidean nudibranchs and sacoglossan panpulmonates show the largest range of digestive gland morphology, although the more plesiomorphic members of both groups have a compact (holohepatic) digestive gland (Gosliner 1994). The branching of the digestive gland may be an adaptation to increase the digestive surface area and is also typically correlated with dietary specialisation (namely storage of cnidocysts, zooxanthellae, or chloroplasts) (Wägele & Willan 2000; Burghardt et al. 2008b; Wägele et al. 2010; Burghardt & Wägele 2014).

During larval development, the digestive gland develops from the midgut as two evaginations, with the right smaller than the left (Hamatani 1960, 1961; Bickell et al. 1981). In nudibranchs, the right portion is much smaller than the left (Schmekel & Portmann 1982), and in some doridoideans the small right lobe can develop as a 'stomach caecum' (see below). In the Sacoglossa the two lobes are equal in size (Schmekel 1985), and, unlike the aeolidiids, the digestive gland of sacoglossans attains an internal bilateral symmetry even though the animals of both groups appear externally to be bilaterally symmetrical. In tritonioideans such as *Tritonia* and *Marionia*, both diverticula are retained after metamorphosis. In aeolidioidean and tritonioidean adults, the left lobe is by far the larger, forming symmetrical branches along both sides of the body while the right lobe remains small, anterior, and visceral.

In the veliger larvae of *Amphibola*, the left lobe of the digestive gland develops while the right remains insignificant. In hygrophilans and higher pulmonates a larval 'albumen sac' is formed from the gut endoderm. This larval organ disappears in at least some Hygrophila (*Planorbis*, *Physa*, and *Lymnaea*), while the adult digestive gland is formed separately by the proliferation of cells from the stomach (Bloch 1938). This separate derivation may also be the case in some stylommatophorans (e.g., *Achatina*), but in others, the albumen sac appears to be incorporated in the adult digestive gland (Raven 1975).

As in other gastropods, there are plesiomorphically two openings of the digestive gland to the stomach, although most cladobranch nudibranchs have more than two (usually three) openings (Wägele & Willan 2000). When there are three openings, two open to the stomach anteriorly and the third opens mid-posteriorly. There are also three openings in onchidoideans, but the two anterior openings lie above the third.

In some Aeolidida and the plakobranchoidean sacoglossans, the left digestive gland may branch so extensively that it can extend through most parts of the body, including the rhinophores or their sheaths (as in some tritonioideans) and the oral veil and/or foot (some charcotiids) (Wägele & Willan 2000). In *Melibe engeli* (Burghardt & Wägele 2014) the fine digestive gland ducts were even found in muscular tissue and in the ovotestis. The digestive surface area is markedly extended in those taxa that have cerata (see Section 20.3.2.6.2) or other processes, including the ventral lamellae of *Armina*, into which the digestive gland penetrates. In cladobranchs with such processes, the only digestive gland tissue is in the lobes in the extensions, while the digestive gland tubes in the body itself lack glandular tissue (Wägele & Willan 2000). Such elaborations are usually, but not always, associated with photosynthesis (see Section 20.4.7 and Chapters 5 and 9), the exceptions being related to camouflage. In the Aeolidida the branches of the left digestive gland enter the cerata and, in many, end in a special sac, the cnidosac (see Section 20.4.6.4).

In sacoglossans (Fischer 1892; Tardy 1969), the digestive gland commences in the larva with the left lobe being larger than the right, but the lobes become equal in size as development proceeds. The two lobes are extended along the body (see Chapter 5 Figure 5.51) and in some taxa are extensively branched (e.g., in *Elysia*). In those sacoglossans that have cerata, they contain branches derived from both digestive glands, and in some of those taxa they may house chlorophyll-containing plastids obtained from their algal food (see Section 20.4.7).

There is a gastric caecum in the stomach of aplysiidans, most non-geophilid lung-bearing panpulmonates (Dayrat & Tillier 2002), the Bathydoridoidea, Doridoxoidea, and many other Doridina (Wägele & Willan 2000). In the nudibranch groups it appears to be formed from the right digestive gland (Schmekel & Portmann 1982) and is a small bulb lined with ciliated cells and lacks mucous cells (histology not investigated in Doridoxoidea). Its function is uncertain, but it may store sponge spicules (Hoffmann 1932–1939) or possibly has a sorting role. In those doridoidean taxa that lack a caecum, the right digestive gland has presumably been entirely lost. A caecum-like structure in *Dendrodoris* is not homologous with the stomach caecum but is formed from the intestine (Wägele et al. 1999).

In some arminoideans and aeolidioideans, *terminal sacs* are found at the distal ends of the papillae or cerata that contain branches of the digestive gland. They are comprised of large cells with extensive vacuoles, do not contain cnidocytes, and may have an excretory function (Wägele & Willan 2000). These can best be seen in the coral (*Porites*) feeding aeolidiid *Phestilla* (Rudman 1981b).

In many other aeolidioideans, there are sharply differentiated *cnidosacs* at the ends of the cerata that are demarcated with a short duct and a sphincter. The very large cells contain vacuoles (*kleptocnids*) that enclose functional cnidocytes (see Section 20.4.6.4 and Figure 20.68).

The terminal sacs were considered by Wägele and Willan (2000) to be the forerunners of the cnidosacs in aeolidioideans. A few aeolidioideans have cnidosac-like structures that do not contain cnidocytes, even though the animals feed on cnidarians, and some lack them altogether (Wägele & Willan 2000; Burghardt et al. 2008a).

Cnidosacs also occur in the papillae of the dendronotoidean *Hancockia* (Thompson 1972; Martin et al. 2009), but they differ from those in aeolidioideans in there being several small cnidosacs in each ceras. Also, cnidocytes are accumulated in the digestive cells as well as in the cells in the cnidosacs. Martin et al. (2009) suggested that in *Hancockia* the cnidosacs might function as a means of disposing of the cnidocytes rather than retaining them for defence. *Embletonia* (Embletoniidae) has cnidophages on the tips of the cerata, but they are not organised into cnidosacs (Martin et al. 2010).

In most heterobranchs the intestine is a slender tube, and, although there is some coiling in a few taxa, it is generally short compared with polyplacophorans and some vetigastropods. In herbivorous taxa the intestine is generally longer, and with some coiling, than in carnivorous taxa.

As in many other gastropods, the intestine usually originates from the anterior edge of the stomach, but in a few nudibranchs, this is not the case. In most cephalaspideans the oesophagus opens anteriorly but the intestine opens posteriorly, while in Bathydoridoidea the intestine arises ventrally on the left side of the stomach and opposite the point of entry of the oesophagus (Wägele 1989c). In *Tritoniella* the intestine emerges from the right side of the stomach (Wägele 1989a).

In some nudibranchs and many herbivorous cephalaspideans a distinct fold, the typhlosole, runs some of the length of the intestine from the stomach. This typhlosole is lacking in aplysiidans, many pleurobranchs, bathydoridoideans, and many doridoideans (Wägele & Willan 2000).

In many euthyneurans the anus opens on the anterior right side of the body beneath the edge of the notum or in the mantle cavity, this being the case in Cephalaspidea, Aplysiida, *Tylodina*, Pleurobranchoidea, shelled and naked pteropods, the pelagic phylliroid nudibranchs, and many nudibranchs and sacoglossans. In some nudibranchs the anus has shifted posteriorly along the right side of the body (e.g., *Doridoxa*, *Heterodoris*) or mid-posteriorly (e.g., *Fryeria*, *Corambe*), lying beneath the notum, a condition also seen in *Umbraculum* (Wägele & Willan 2000). The anus also opens mid-posteriorly, or nearly so, in Onchidiidae, Veronicellidae, Philinoidea, and Runcinidae. In some taxa, the anus migrates dorsally during ontogeny as in some Aeolidioidea (Schmekel & Portmann 1982), Cephalaspidea (Mikkelsen 1996), and Sacoglossa (Jensen 1996a). In doridoidean nudibranchs, the anus opens posterodorsally in the middle of the anal gill plume.

In shelled lung-bearing panpulmonates, the anus opens at or near the pneumostome on the right side, as it does in stylommatophoran slugs and Rathouisiidae.

The rectum and pericardium are closely associated in nearly all marine heterobranchs, even when there is a posterior shift (Wägele & Willan 2000). A few taxa such as *Janolus* diverge from this general condition and show a separation of the pericardium and the rectum, where the anus has migrated posteriorly but the heart has not.

A variety of structures loosely called anal or rectal glands occur in some marine euthyneurans such as runcinids, the gymnosome *Anopsia*, the nudibranchs *Okadaia*, and some species of *Janolus* (Hoffmann 1932–1939; Hyman 1967; Gosliner 1982). A caecum of various lengths is given off from the beginning of the rectum in terrestrial slugs of the families Agriolimacidae and Limacidae (Hyman 1967; Barker 2001a). In Veronicellidae and some Athoracophoridae, a narrow tubule connects the rectum to the ureter (Barker 2001a).

The faeces form rods in most heterobranchs, but faecal pellets are formed in a few, for example, *Limax* (Arakawa 1965) and *Gastropteron* (Arakawa 1968).

20.3.5 THE RENOPERICARDIAL SYSTEM

The vascular system is reviewed by Hyman (1967) and outlined in Chapter 6. The heart, as in caenogastropods, consists of a ventricle and an auricle (the left) and lies in the pericardium, which is not penetrated by the rectum. The auricle receives blood from afferent vessels and sinuses via the posterior aorta and shunts it to the ventricle, which pumps it into the anterior aorta, the efferent vessels, and lacunae.

In most euthyneurans the heart lies in the anterior half of the body, but in Bathydoridoidea and Doridina it has moved posteriorly. The heart is oriented plesiomorphically with the auricle anterior to the ventricle in most 'lower heterobranchs', acteonids, and many lung-bearing panpulmonates. The auricle has moved to the right along with the migration of the mantle cavity down the right side in aplysiids, *Haminoea*, acteonids, pleurobranchs, and lung-bearing panpulmonates, while most cephalaspideans have the auricle displaced further posteriorly. A posterior auricle is found in nudibranchs, acochlidians, some sacoglossans, runcinids, smeagolids, onchidiids, and a few stylommatophorans such as *Testacella* and gymnosome pteropods. In this last group, all but the gymnosomes have the pallial complex (or its remnants) displaced posteriorly.

The tiny nudibranch *Vayssierea caledonica* is reported to have a heart but lacks a pericardium (Baba 1937), and *Rhodope*, the sacoglossan *Alderia modesta*, and some small acochlidians lack a heart entirely (e.g., Brenzinger et al. 2013).

Major departures from other gastropods are seen in nudibranchs due to their secondary bilateral symmetry and loss of the mantle cavity.

While there are good descriptions of the circulatory system of some lung-bearing panpulmonates, there is relatively sparse information about this system in most other heterobranchs. In taxa with gills next to the anus, an efferent branchial vessel carries oxygenated blood from the gills to the auricle; a lateral sinus also carries oxygenated blood from both sides of the body into the auricle. In taxa that lack a gill, the oxygenated blood is carried only from the lateral sinuses.

As with other molluscs, heterobranch blood contains several kinds of blood cells. A useful review is provided by Luchtel et al. (1997) for pulmonates. As with other gastropods, two main kinds, amoebocytes and hyalinocytes, are present (see also Chapter 6).

Haemocyanin is the respiratory pigment found in the blood of most heterobranchs, an exception being the Planorbidae, which have haemoglobin (see Chapter 6).

The *blood gland* is a thin-walled, leaf-shaped organ found dorsally in the visceral cavity that lies dorsally anterior to the ventricle or more anteriorly in the anterior part of the visceral cavity (Figure 20.42) in Pleurobranchoidea, Bathydoridoidea, and Doridoidea but not in 'cladobranch' nudibranchs (Wägele & Willan 2000). It contains ramifying blood lacunae that open to a branch of the aorta and specialised blood gland cells loosely bound with connective tissue that contains amoebocytes. The blood gland cells contain many granules and have deep invaginations which form vacuoles containing what appears to be globular and polymerised haemocyanin. This observation, and evidence of a higher copper concentration than elsewhere in the body, strongly suggest that the blood gland concentrates or synthesises haemocyanin (Schmekel & Weischer 1973). The *crista aorta* in the pericardium of some cephalaspideans such as *Cylichna* (e.g., Lemche 1956) and *Aplysia* has been regarded as homologous (or analogous?) with the blood gland (Dayrat & Tillier 2002).

Surprisingly few detailed studies have been carried out on the excretory system of heterobranchs, and most of those have been undertaken on the specialised terrestrial and freshwater panpulmonate taxa (see below).

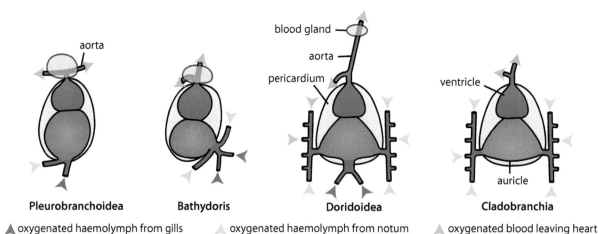

FIGURE 20.42 Pericardial complex and blood gland of doridians. Redrawn and modified from Wägele, H. and Willan, R.C., *Zool. J. Linn. Soc.*, 130, 83–181, 2000.

Plesiomorphically, podocytes, with numerous basal processes and filtration slits bridged by fine diaphragms, are present in the epithelium of the auricle, a condition found in the Thecosomata, Gymnosomata, Sacoglossa, and Acochlidia (Fahrner 2002). Auricular podocytes are presumably also present in the 'lower heterobranchs', as they are also the norm in non-heterobranch gastropods, but in the Nudibranchia there are numerous podocytes in the outer pericardial epithelium. In the cladobranch nudibranchs, there are also podocytes in the ventricular wall. Cephalaspideans lack true podocytes, but instead podocyte-like cells without diaphragms, so-called 'slashed cells', are involved in ultrafiltration and comprise the entire epicardium (Fahrner 2002). In *Runcina* they are also found in the outer pericardial epithelium. These specialised 'slashed cells' are thought to be an autapomorphy of Cephalaspidea (Fahrner 2002) and not the plesiomorphic condition in 'opisthobranchs' as suggested by Andrews (1988). Andrews based this idea on her observations on two cephalaspideans and, noting the absence of podocytes, together with the lack of basal infoldings of the kidney cells[2] and the available data on lung-bearing panpulmonates, she concluded that auricular ultrafiltration had probably been lost in the euthyneuran ancestor. The work of Fahrner (2002) has clearly shown, however, that podocytes are plesiomorphically present in most marine euthyneurans.

In several (but not all) doridine nudibranchs (including both bathydoridoideans and doridoideans), there are folds on the inner dorsal pericardial wall, usually in front of the ventricle, that have been termed 'pericardial glands'. Despite being termed glands, the folds are not glandular (Wägele & Willan 2000) but probably increase the surface area involved in filtration and presumably contain podocytes (Fahrner & Haszprunar 2002b), as do the similarly named 'pericardial glands' of bivalves (see Chapter 6).

Podocytes associated with the pericardium or heart have not yet been found in any lung-bearing panpulmonates, a major change which is possibly related to their colonisation of supratidal, freshwater, and terrestrial habitats (Luchtel et al. 1997). Ultrafiltration has been shown to occur in the auricular epicardium of *Lymnaea* (Andrews 1976) and in part of the ventricular epicardium of *Cornu aspersum* (Andrews 1988), but it is by way of slits. In both Hygrophila and Stylommatophora ultrafiltration usually occurs in the kidney, apparently by either paracellular or transcellular transport, or (in *Biomphalaria* at least) is restricted to a small specialised area of the kidney with arterial haemolymph supply (Luchtel et al. 1997; Fahrner 2002) (see also Chapter 6). In the latter case, podocyte-like cells line those parts of the kidney supplied with arterial blood (Matricon-Gondran 1990), and these cells may be homologous to podocytes (Fahrner 2002). However, these data are sketchy and, surprisingly, no marine lung-bearing panpulmonates have yet been studied.

Solitary rhogocytes (pore cells) are found in the connective tissue and haemocoel of investigated euthyneurans (see Chapter 6). These cells are structurally like podocytes, and it has been suggested that they may be involved in ultrafiltration (e.g., Fahrner 2002), although this is unlikely.

There are no podocytes and no evidence of ultrafiltration in *Alderia modesta*, a tiny sacoglossan that lacks a heart and pericardium, and the kidney shows none of the modifications needed for ultrafiltration (Fahrner & Haszprunar 2001). Another heterobranch lacking a heart, *Rhodope transtrosa*, also lacks podocytes but shows a unique pseudoprotonephridial system of ultrafiltration (Fahrner & Haszprunar 2002a).

The pericardium is connected with the kidney by a ciliated renopericardial duct in most euthyneurans, but in a few (e.g., *Creseis virgula* and *Cuthona caerulea*) the duct is absent, and the kidney opens directly to the pericardium (Fahrner 2002). The strongly ciliated funnel- to tube-shaped initial part of the renopericardial duct from the pericardium in nudibranchs is often called the *syrinx* (Wägele & Willan 2000), and its cytology markedly contrasts with the non-ciliated epithelium of the rest of the renopericardial duct (Fahrner & Haszprunar 2002b).

When a mantle cavity is present the single kidney (the post-torsional left) of heterobranchs lies in the mantle roof, into which it opens directly in most marine taxa. In those where the mantle cavity is much reduced or lost, the kidney lies in the same relative position in the dorsal body wall.

The kidney forms from the same primordium as the heart and gonad with the kidney developing on the pretorsional right side, and the corresponding cells on the left side either do not develop or disappear (Ghose 1963b; Raven 1975). Generally, descriptions of the formation of the adult kidney refer only to a single mass giving rise to the kidney (e.g., Thompson 1958; Raven 1975). Thus, unlike the situation in caenogastropods and neritimorphs, there is no evidence that a post-torsional right kidney rudiment is incorporated in the reproductive system. In stylommatophorans and some hygrophilans, the developing kidney bud reaches the mantle cavity and is joined by an ectodermal invagination that forms the ureter (Raven 1966).

In marine heterobranchs the kidney epithelium is unciliated except around the kidney opening. The epithelium consists of a single cell type with the cytology (large vacuoles, basal infoldings, and microvillus border) indicating that these cells are involved in both secretory activity and reabsorption. In contrast, the caenogastropod kidney has two main cell types, pigmented ciliated cells involved in reabsorption of organic solutes in the proximal region, and vacuolated excretory cells in the distal region (Andrews 1981, 1988). In euthyneurans the pigmented, ciliated cells are absent (Luchtel et al. 1997; Fahrner 2002), and Andrews (1988) considered that the histology of the kidney cells in two cephalaspideans she examined suggested that they had reduced activity because they lacked basal infoldings, glycogen-deposits, and only had weakly developed apical microvilli. She concluded that excretory activity might have been adopted by the digestive gland cells. The taxa examined by Fahrner (2002), which included

---

[2] Fahrner (2002) has cast doubt on this observation, noting that juvenile *Runcina* have weakly infolded bases but those of adults show deep infolding.

a runcinid, pteropods, sacoglossans, an acochlidian, and nudibranchs, all possessed ultrastructural features indicating excretory activity, although a juvenile of one species showed similar histological features to those described by Andrews. Fahrner (2002) suggested that the simplification of the kidney in marine euthyneurans was associated with the reduction and loss of the shell enabling diffusion of ammonia over much of the body surface. We consider this unlikely given the shell in several basal members of euthyneuran lineages. Regrettably, details of the renal epithelium of members of 'lower heterobranch' lineages are largely lacking, although Hawe et al. (2013) noted that the epithelium in the valvatid *Borysthenia* is uniform and nonciliated suggesting it is similar to that of euthyneurans. It will be of interest to see what detailed ultrastructural examination of the renal epithelium of marine 'lower heterobranchs' reveals.

Unlike the situation in many other heterobranchs (and most other molluscs), in lung-bearing panpulmonates the heart is not the only location of filtration where the primary urine is formed (Luchtel et al. 1997) (see Chapter 6). The heart lacks podocytes, but filtration probably occurs through tiny gaps in the wall of the auricle and/or ventricle. There is either a renopericardial canal present to carry the primary urine to the kidney lumen, or the pericardium and the kidney abut and interconnect. Filtration may also occur through pores in the walls of the kidney, and this has been demonstrated especially in some stylommatophorans and planorbids. Also, a small specialised area of the kidney in many lung-bearing panpulmonates is supplied with arterial blood, and podocyte-like cells have been described from the epithelium in that area in at least a planorbid (a species of *Biomphalaria*) (Matricon-Gondran 1990). Another freshwater hygrophilan, *Lymnaea stagnalis*, has been shown to conduct ultrafiltration through pores in the simple auricular epicardium that lacks podocytes (Andrews 1976).

The kidney of the freshwater 'lower heterobranch' *Valvata*, the development of which has been studied by Rath (1986), consists of three parts (Cleland 1954), a 'visceral kidney' which is the 'true' kidney with walls lined with renal epithelium; a 'pallial kidney', an anterior extension of the kidney in the roof of the mantle cavity in which the walls are folded and lined with renal epithelium; and a ureter[3] (Bernard 1890; Andrews 1988). The pallial and visceral sections of the kidney are largely separated by a septum, and the renopericardial duct enters the anterior end of the visceral kidney. According to Cleland (1954), the digestive gland of *Valvata* is probably also involved in excretion and osmoregulation.

The kidney opening (nephroproct) is usually near the anus even when the anus migrates posteriorly or dorsally, although in a single taxon, the Notaeolidiidae, the kidney opens in front of the genital aperture and is disconnected from the anus (Wägele 1990b; Wägele & Willan 2000).

The ureter is involved in osmoregulation of both ions and water (see Chapter 6). In stylommatophorans, the relationship of the ureter to the kidney and other pallial organs was the main character used to separate the major groups (orders) (see Section 20.5.1.2.4.2.10). In that group, a tubular ectodermal secondary ureter may be formed, or this is represented by a ciliated groove. The internal epithelium of the ureter contains two cell types (Bouillon 1960; Delhaye & Bouillon 1972a, b, c) while in Hygrophila and the orthurethran Stylommatophora, a tubular ectodermal ureter exhibits only one cell type (Delhaye & Bouillon 1972c). This latter type of ureter is sometimes called a primary ureter. In some ellobiids, a simple, short ureter-like structure forms a pouch-like extension of the kidney, termed an orthureter (Barker 2001a). The orthurethran condition is also seen in some stylommatophorans (pupilloideans). Both primary and secondary ureters in stylommatophorans are of ectodermal origin. The secondary ureter is only found in some stylommatophorans and can be a gutter or a closed tube along which urine is moved to the proximity of the pneumostome from where it can be passed to the exterior. It is probable that the sigmurethran condition seen in the great majority of stylommatophorans was derived from a stylommatophoran without a closed ureter and after losing the orthureter (i.e., the mesurethran condition) (Barker 2001a). Because the orthurethran condition is seen in ellobiids and some Hygrophila, it is usually thought to be plesiomorphic in Stylommatophora (e.g., Delhaye & Bouillon 1972b, c; Tillier 1989; Barker 2001a), although Delhaye and Bouillon (1972a) considered the orthureter to be a secondary structure derived from the anterior part of the kidney.

In the marine onchidiid *Onchidella* the kidney opens via a very short ureter to the reduced mantle cavity, as does the rectum (Fretter 1943). In a terrestrial onchidiid, *Semperoncis montana*, the ureter is, uniquely, a convoluted tube that runs from the posterior end of the kidney and passes through the floor of the lung to open into the rectal ampulla (Tillier 1983),[4] a structure that probably represents part of the reduced mantle cavity. The ureter opening to the anus is probably paedomorphic, reflecting the larval condition (Tillier 1983). The veronicellids (Delhaye & Bouillon 1972c) and rathouisiids (Tillier 1984a) also possess a well-developed ureter, which in the former group, is ectodermal in origin (Sarasin & Sarasin 1899). In veronicellids, a secondary ureter has one or more connections with the rectum (Barker 2001a). The elongate, tubular ureter seen in Onchidiidae, Veronicellidae, and Rathouisiidae has similar histology to that of the ureter in Succineidae (Delhaye & Bouillon 1972a, b), and while Barker (2001a) considered this might indicate homology, it may equally well indicate a homoplastic condition.

20.3.6 Nervous System and Sense Organs

The main nerve ring lies either at the front of the buccal mass (i.e., is *prepharyngeal*) or is at its posterior end (i.e., is *postpharyngeal*). The former condition is seen in acteonoideans and

---

[3] The histology and function of the ureter in valvatids have not been thoroughly compared with that of hygrophilans or stylommatophorans, but it is very unlikely that it is homologous.

[4] As *Platevindex apoikistes*.

in some of the following taxa: Cephalaspideans (e.g., philinoideans, pleurobranchoideans) and panpulmonates (including glacidorbids, and some hygrophilans and ellobiids). The post-pharyngeal condition is seen in some cephalaspideans such as *Haminoea*, pteropods, aplysiidans, umbraculoideans, all nudibranchs, and many panpulmonates including sacoglossans and amphiboloideans, some hygrophilans and ellobiids, systellommatophorans, and stylommatophorans. The anterior position is, by outgroup comparison, plesiomorphic (Gosliner 1994; Ponder & Lindberg 1997), although Mikkelsen (1996) judged the posterior position to be primitive in cephalaspideans. The migration of the nerve ring to behind the buccal mass clearly occurred independently in several groups. During development in *Adalaria*, the nerve ring moves from in front of the buccal mass to behind it (Thompson 1958).

As with other gastropods, the heterobranch circumoesophageal nerve ring includes three main pairs of ganglia – cerebral, pleural, and pedal. The cerebral and pedal ganglia are each joined by commissures, while the pedal and cerebral, pleural and pedal, and cerebral and pleural are each joined by pairs of connectives. An additional commissure, the *subcerebral commissure*, connects the cerebral ganglia in some panpulmonates (e.g., Bouvier 1887; Mol 1967) and a *parapedal commissure* may connect the pedal ganglia in addition to the pedal commissure (Dayrat & Tillier 2002).

The pleuropedal connective is usually long (epiathroid condition) but is short in some such as aplysiidans, ellobiids, *Trimusculus*, and some stylommatophorans where it approximates the hypoathroid condition (see also Chapter 7). During development, the nervous system of *Achatina* is initially hypoathroid and later epiathroid (Ghose 1962). In many euthyneurans, the pleural ganglia are about an equal distance from both the pleural and pedal ganglia (the 'hypo-epiathroid' condition) (Dayrat & Tillier 2002), and a few stylommatophorans show asymmetry in this condition (Tillier 1989).

A pair of buccal ganglia is usually present, and they are connected to the cerebral ganglia by cerebrobuccal connectives.

The euthyneuran nervous system has been discussed by Dayrat and Tillier (2000), and changes in its configuration have been a hallmark of the group, the most obvious of which was the untwisting of the visceral loop from its plesiomorphic figure of eight or *streptoneurous* state (the result of torsion) (see Chapter 7). A streptoneurous condition is limited to the lower heterobranchs and, in euthyneurans, to the Acteonidae, many cephalaspideans, *Akera*, and some less-derived Sacoglossa. The untwisted condition is seen in most Euthyneura (the euthyneurous condition).

The visceral loop has a normally single visceral (abdominal) ganglion, in addition to supra- and suboesophageal ganglia and, in most euthyneurans, two additional *parietal ganglia* lie between the visceral and sub- and supraoesophageal ganglion (sometimes called intestinal ganglia). These parietal ganglia are considered a synapomorphy of 'Pentaganglionata' (i.e., Euthyneura) by Haszprunar (1985b) (see below). The terminology, hypotheses of fusion, and homology of the parietal ganglia have been discussed by Dayrat and Tillier (2000). The suboesophageal and supraoesophageal ganglia can be identified by the nerves that arise from them, including when they are fused to other ganglia (usually the pleurals) or are sometimes vestigial (Haszprunar 1988f; Dayrat & Tillier 2000, 2002). As far as is known, they arise during ontogeny, with the exception of many nudibranchs (Thompson 1958, 1962; Carroll & Kempf 1994) where they form after settlement.

One or both of the oesophageal ganglia often lie close to, abut, or are fused with the pleural ganglia. The visceral ganglion plesiomorphically lies posteriorly among the viscera, and the visceral loop is long. In many euthyneurans, the visceral loop shortens, the visceral ganglion can lie close to the main nerve ring, and it may fuse with the suboesophageal ganglion, which might itself be fused with the left pleural ganglion.

Ganglion fusion is a hallmark of many euthyneuran taxa, but the plesiomorphic condition is unfused ganglia. The cerebral and pleural ganglia are separated in Cephalaspidea (except *Phanerophthalmus*), Aplysiida, Umbraculoidea, and Bathydoridoidea, and in a few Doridina (Wägele & Willan 2000). In most other nudibranchs, all pleurobranchoideans, and most sacoglossans, nearly all the ganglia are fused, except the visceral ganglion in some taxa.

The process of the so-called *detorsion* that results in euthyneury has long been the subject of speculation (e.g., Pelseneer 1901; Naef 1911; Krull 1934; Régondaud et al. 1974; Haszprunar 1985b, 1988; Gosliner 1994). This process undoubtedly occurred independently in several heterobranch lineages because of detorsion and/or shortening of the visceral commissure and concentration of the nervous system (Haszprunar 1985b, 1988; Dayrat & Tillier 2002).

According to Schmekel (1985) nervous system concentration was achieved in three independent ways in marine euthyneurans:

1. Shortening and detorsion of the visceral loop independently in Cephalaspidea, Sacoglossa, Aplysiida, and Nudibranchia.
2. The pleural ganglia are aligned with either the cerebral or pedal ganglion. In the former situation, the cerebral and pleural ganglia often fuse to form cerebropleural ganglia, and other visceral loop ganglia may also cluster with the pleural ganglia. The result is four ganglia in the nerve ring and two cerebropleural connectives on each side (seen in Sacoglossa, Nudibranchia, Pleurobranchida). In some taxa, the pleural ganglia are nearer to the pedal ganglia than the cerebral ganglia (thus with six ganglia in the nerve ring – as in Aplysiida, Umbraculida), and again the visceral loop ganglia may be more or less incorporated in the pleural ganglia (except the Aplysiida).
3. The oesophageal ganglia may fuse with the visceral ganglion (as in some Aplysiida) or with the pleural ganglia (as in Sacoglossa, Nudibranchia, Pleurobranchida, Umbraculida).

The most concentrated conditions in all three of these trends were independently achieved in some sacoglossans and aeolidiid nudibranchs (Schmekel 1985).

The nervous system in the more plesiomorphic euthyneurans has a long visceral loop with separate ganglia, but the concentration of the ganglia has occurred independently in several lineages (Haszprunar & Huber 1990).

The visceral ganglion usually lies, as its name suggests, within the viscera behind the anterior part of the body. The exact position of the visceral ganglion is, however, variable even within some species, as has been shown in some doridoideans and *Onchidella* (Wägele 1990a; Weiss & Wägele 1998).

Plesiomorphically in many marine euthyneurans (Umbraculoidea, Pleurobranchoidea) and in several Doridoidei (e.g., *Hexabranchus* and most Cladobranchia), a rhinophoral ganglion lies at or in the base of each rhinophore and is connected by a rather long rhinophoral nerve to the cerebral ganglion. In other taxa, the rhinophoral ganglion lies close to the cerebral ganglion, as in Doridoidea, Bathydoridoidea, and a few Aeolidioidea (Wägele & Willan 2000). Some tritonioidean nudibranchs lack rhinophoral ganglia.

A pair of buccal ganglia lies ventral to the oesophagus, emerging from the buccal area. They may have a short connective, as in many marine euthyneurans, or they may be well separated. The radula is innervated from the buccal ganglia by a nerve from each ganglion or a single nerve from the connective (Wägele & Willan 2000).

Whether or not the pentaganglionic condition is found in primitive lung-bearing panpulmonates such as siphonariids and amphibolids, as hypothesised by Haszprunar (1985b, 1988), remains uncertain. Haszprunar and Huber (1990) argued that the lack of these ganglia resulted from fusion and claimed the ancestral euthyneuran visceral loop had five ganglia (i.e., the 'Pentaganglionata' concept), the additional two being the pair of 'parietal' ganglia located close to the pleural ganglia, and the sub- and supraoesophageal and visceral ganglia. This concept was queried by Ponder and Lindberg (1997) and dismissed by Dayrat and Tillier (2000) who argued against parietal ganglia being present in ancestral euthyneurans, and Ruthensteiner (2006) found no evidence of fusion in pleural ganglia of siphonariids. Golding et al. (2007) incorrectly called the suboesophageal ganglion the parietal ganglion, there being no evidence of parietal ganglia in the amphibolid nervous system.

The most important apomorphic diagnostic characters of the lung-bearing panpulmonate central nervous system are the *cerebral lobes* of the cerebral ganglia, consisting of the *procerebrum*, *cerebral gland*, and *mediodorsal bodies*. They are considered a synapomorphy of that 'group' (previously 'Pulmonata') (e.g., Dayrat & Tillier 2002) and were the subject of an intensive study by Mol (1967). There are two kinds, the procerebrum (or lateral lobes) and the mediodorsal bodies (or dorsal lobes). The procerebrum contains a follicle gland (Geraerts & Goosse 1975), and it lies in the anterodorsal portion of the cerebral ganglion which usually gives rise to the tentacular nerves. Thus, in dissection, it appears to be a kind

of basal tentacular ganglion (Dayrat & Tillier 2002), and it probably processes olfactory information in terrestrial taxa. The procerebrum may contain either large cells, small cells (globineurons), or both (Mol 1967, 1974). The plesiomorphic condition is only large cells (Haszprunar & Huber 1990), and these taxa have an osphradium, are at least semi-aquatic, and include amphibolids, siphonariids, and hygrophilans. The small-celled (globineurons) procerebrum is found exclusively in air-breathing taxa with a contractile pneumostome (Barker 2001a), being found in all Eupulmonata, except the marine Trimusculidae. The procerebrum forms independently from the cerebral ganglion during ontogeny, although it is connected to it. According to Ruthensteiner (1997, 1999), the procerebrum is homologous with the rhinophoral ganglion of other euthyneurans and the tentacular ganglion of caenogastropods.

The probably neurosecretory cerebral gland lies laterally on the procerebrum and shows variation in size and organisation in different pulmonate taxa. It is formed from a remnant of the invagination trough of the cerebral ganglion (Ruthensteiner 1999). In ellobiids, trimusculids, succineids, and siphonariids, there is a tube-like extension from the cerebral gland to the base of each tentacle which connects the gland to the external environment (Mol 1967; Ruthensteiner 1999, 2006; Barker 2001a). This tube is lost during ontogeny in 'higher pulmonates' but is probably more common than currently known among 'lower pulmonates' as it is difficult to detect (Ruthensteiner 2006). Like the procerebrum, the cerebral gland forms independently from the cerebral ganglion during ontogeny and, according to Barker (2001a), may be homologous with the tiny subtentacular ganglion of some cephalaspideans (Lemche 1955).

Medio-dorsal bodies are endocrine organs in all lung-bearing panpulmonates that have been investigated, including amphibolids and siphonariids (e.g., Mol 1967). They lie dorsally on the cerebral ganglia near the origin of the cerebral commissure. A few neurosecretory cells occur in the cerebral (and in the pleural, parietal, and visceral) ganglia in some marine euthyneurans (e.g., Franc 1968), but they are not homologous with the dorsal bodies (Dayrat & Tillier 2002). The mediodorsal bodies are stimulated by day length and are involved in gamete development (Gomot de Vaufleury 2001).

Giant neurons are large nerve cells found in the central ganglia of euthyneurans, and their presence helps to define this grouping (Haszprunar 1988f). They achieve their large size by enlarging as the animal grows, resulting in neuronal gigantism (Gillette 1991). This characteristic enables individual cells to be targeted in experiments, most famously in *Aplysia* where significant behavioural experiments have been conducted by E. R. Kandel, A. O. Willows, D. L. Alkon, and others (see also Chapters 7 and 10). Other euthyneurans often used in experiments involving the nervous system include some nudibranchs and, especially, the freshwater hygrophilan *Lymnaea*. Terrestrial snails such as *Achatina* and *Helix* have also been used, along with other stylommatophoran snails and slugs.

20.3.6.1 Sense Organs

As with other molluscs, there is general sensitivity to chemical and mechanical stimuli over the entire body surface by way of receptor cells beneath the epithelium. Organs are developed where these receptors are concentrated.

A pair of cephalic eyes (Figure 20.43) is usually present on the head, but, unlike those in other gastropods, they characteristically lie in the middle of the bases of the cephalic tentacles or on their inner sides rather than on the outer sides. The eyes in 'lower heterobranchs' are typically rather well developed, but detailed studies of their morphology are lacking.

Eyes are present in all heterobranchs (except some deep-sea and interstitial taxa) and consist of a cornea made up of transparent flattened cells on the outer side, a more or less spherical lens, and a thick, pigmented retina made up of receptor cells, or a mixture of those cells and supporting cells. In aplysiids, the receptor cells have both a microvillus border and cilia, but in *Berthella*, *Pleurobranchus*, and nudibranchs they only have microvilli (Hughes 1970). Basally the receptor cells are connected to the optic nerve which originates from the cerebral ganglion. Heterobranch eyes are probably not capable of forming images but may be

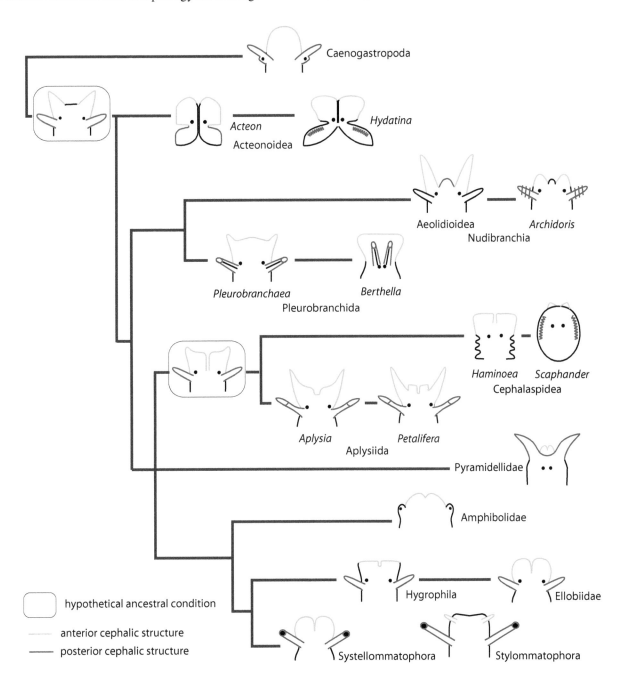

FIGURE 20.43 A diagrammatic representation of the cephalic sensory organs in euthyneurans. Redrawn and modified mainly from Staubach, S. *The evolution of the cephalic sensory organs within the Opisthobranchia*, Ph.D. dissertation, Johann Wolfang Goethe Universitat, 2008.

able to assess the direction and intensity of light (Luchtel et al. 1997).

In most marine euthyneurans the eyes are small and usually located at the base of the rhinophores (if these are present). The eyes typically lie beneath the epidermis, and in various taxa have sunk to lie close to the cerebral ganglia. In the latter cases, the optic nerve becomes very short, and in extreme cases, the eyes abut the cerebral ganglion (Hoffmann 1932–1939; Wägele & Willan 2000). In *Bulla* and *Aplysia* the eyes are better developed, rather like those of stylommatophorans and hygrophilans, with the retina composed of many small cells. In contrast, those of pleurobranchs and nudibranchs are smaller and more sunken and have fewer, larger cells in the retina (Willem 1892).

Pseudothecosome pteropods have reduced eyes that lack pigment and a lens and are located in the tips of the cephalic tentacles. Thecosomes lack eyes but have one or two tentacles at the junction of the wings that are probably light sensitive (Lalli & Gilmer 1989). Despite being efficient predators that are active in daylight, the gymnosome pteropods also lack eyes but have a second pair of probably light-sensitive short tentacles on the neck just behind the head (Lalli & Gilmer 1989).

The eyes of eupulmonates are relatively well developed. In stylommatophorans, they are at the tips of the posterior head tentacles and in systellommatophorans at the tips of the single pair of tentacles. In other lung-bearing panpulmonates they are located at the inner bases of the cephalic tentacles. Stylommatophoran eyes and those of *Lymnaea* are the most studied. Their retina usually consists of four types of cells: visual sensory cells containing photic vesicles and long microvilli, a second type of sensory cell with short microvilli and lacking photic vesicles, pigmented support cells, and a ganglion cell type (Luchtel et al. 1997). The eye of at least some eupulmonates can fully regenerate (Luchtel et al. 1997).

Besides the normal pair of cephalic eyes in the distal tips of the cephalic tentacles, the dorsal papillae of many onchidiid slugs usually have three to five eyes (although they range from one to seven) on a single papilla, and there may be up to 100 eyes on an individual (Hyman 1967). These are 'inverse' eyes (Chapter 7, Figure 7.40) and thus are structurally more similar to vertebrate eyes than the cephalic eyes of gastropods. The lens is unusual in comprising several large cells, and it fills the interior of the eye.[5] The inverted retina has the nervous layer next to the lens, followed by the layer of cell bodies and then a pigment layer. Innervation is from the pleural ganglia via branches from the mantle nerves. As in vertebrates, the nerve penetrates the retinal cells and spreads across the retina.

General epithelial chemosensory receptors are concentrated in areas such as around the mouth or on the head tentacles. For example, in the few panpulmonates investigated, concentration of the epidermal receptors has occurred in the lips (Luchtel et al. 1997). Such concentrations are particularly pronounced in some cephalaspideans, notably *Scaphander*, Philinidae, and Aglajidae, which have brown patches on either side of the mouth, the 'lip organs' of Edlinger (1980a). These

have a rich nerve supply from the cerebral ganglion and, in Aglajidae, bear long retractile bristles. They may be utilised in following the mucous trails of the prey of these carnivorous taxa (Rudman & Willan 1998).

The main olfactory organs are briefly described next. Further information can be found in a detailed review by Emery (1992). Interestingly, in *Achatina* and possibly other stylommatophorans, there are more neurons devoted to olfactory organs than to any other sense (Chase & Tolloczko 1993).

In hygrophilans, there are clusters of chemosensory cells at the bases of the cephalic tentacles or on the tentacles (Emery 1992). The main olfactory organ of stylommatophorans is in each optic tentacle just below the eye, with nervous connections to a tentacle ganglion lying distally in the tentacle, and the nerves from that ganglion run to the cerebropleural ganglion.

The Architectonicidae and Mathildidae both possess a well-developed, laminate osphradium (Haszprunar 1985a, c, d), but these are the only examples of a relatively large osphradium in the Heterobranchia. A small osphradium is present in the mantle cavity of at least several 'lower heterobranch' families and pyramidellids (Haszprunar 1985a), cephalaspideans (Edlinger 1980a), aplysiids (Theler et al. 1987), thecosome pteropods, shelled sacoglossans (Jensen 2011), and in most aquatic lung-bearing panpulmonates (Emery 1992; Luchtel et al. 1997). In gymnosome pteropods, which all lack a mantle cavity (and shell), the osphradium is exposed on the surface of the animal near the anus (Hyman 1967).

An osphradium is absent in adult Nudipleura (LaForge & Page 2007) although it is present in some larvae. It is also absent in Ellobiidae and those eupulmonate taxa with a contractile pneumostome (Harry 1964). Although reported from stylommatophoran embryos (Pelseneer 1901), an osphradium is of no significance in stylommatophorans because the cephalic tentacles have a chemosensory epithelium (Wondrak 1981) and because the mantle cavity has been modified as a lung (Haszprunar 1985b).

In some marine taxa such as *Aplysia* and *Tylodina*, and in at least some hygrophilans, the osphradium is sunk in a pit for protection (Benjamin 1971; Benjamin & Peat 1971). In *Amphibola*, *Siphonaria*, and *Chilina*, the osphradium is on the anterior side of the pneumostome, and, in higher Hygrophila, it is on the mantle edge and has been called the organ of Lacaze. Despite variation in the size of the osphradium, the histology is similar in all heterobranchs, consisting of mucous cells and ciliated effector and sensory cells (Haszprunar 1985a; Theler et al. 1987; Emery 1992).

Generally, the function of the osphradium in heterobranchs is poorly known. It has been suggested that it assists in regulating respiration in *Aplysia* by detecting changes in pH of the seawater (Croll 1985), and it may play a role in egg laying in *Lymnaea* (Nezlin 1997).

The Hancock's organs are cerebrally innervated, often pigmented, paired chemosensory organs found on either side of the head. If a head shield is present, they are located beneath it in the groove between that structure and the foot. These organs consist of some parallel, leaf-like epidermal folds and can be

---

[5] According to Labbé (1933) the lens in these eyes is silica (i.e., a glass lens).

very elaborate, as in the acteonoidean *Hydatina* (Rudman 1972b). They are found in cephalaspideans (Edlinger 1980a, b) but are absent in Ringiculidae and Runcinoidea. While previously thought to be absent in acteonoideans, these organs are actually present in several taxa and well developed in some (Rudman & Willan 1998). Similar sensory structures identified as Hancock's organs have also been reported from at least some Acochlidia (Brenzinger et al. 2011a), where they are innervated by a branch of the rhinophoral nerve, and in primitive shelled Sacoglossa. Careful comparative work needs to be done to see if these structures are homologous, as mapping them on current phylogenic trees suggests there may have been at least three different origins. Due to their innervation, Huber (1993) suggested that the rhinophores and anterior tentacles of the Acochlidia and many Sacoglossa, and both pairs of tentacles of gymnosomes, could all be derived from Hancock's organ.

In some acteonoids (Aplustridae) and the cephalaspidean Bullidae, the anterior corners of the head shield are expanded laterally and folded to form a pair of siphon-like extensions that direct water on to Hancock's organs. In acteonoideans, the water is carried posteriorly to be ejected via a posterior siphon formed from the mantle (Burn & Thompson 1998). These modifications parallel the funnelling of water over the osphradium seen in higher caenogastropods.

A pair of anterior (oral) tentacles and a second pair of posterior, sometimes complex, rhinophores are found in nudibranchs, aplysiids, umbraculoideans, and pleurobranchoideans. In some taxa the rhinophores are retractile, sometimes retracting into a sheath. The rhinophores may be simple tentacles or, as in acochlidians, aplysiids, umbraculoideans, and pleurobranchoideans, are folded longitudinally into a gutter, but in many nudibranchs they are conspicuous, erect, and highly complex.

One interpretation (Hoffmann 1932–1939) is that the Hancock's organs migrated posteriorly and the posterior corners of the cephalic shield extended to became rhinophores, presumably in a similar way to that seen in some acteonoideans and bullids. While the homology of Hancock's organs and rhinophores is indicated by their innervation (see Chapter 7), the basal position of cephalaspideans is disputed in recent phylogenies, suggesting that rhinophores may have evolved in an ancestor unlike a cephalaspidean and which lacked Hancock's organs.

Rhinophore-like tentacles are developed in many sacoglossans, but based on their innervation (Huber 1993) these appear to be derived from fused oral and cephalic tentacles so are not strictly homologous with the rhinophores of nudibranchs and aplysioideans. They are longitudinally folded or simple and are lost in some secondarily simplified sacoglossans such as Platyhedylidae and Gascoignellidae. Their often relatively complex structure provides important characters for taxonomy in some groups.

A pair of anterior (oral) tentacles is found in nudipleurans and all Geophila other than Onchidiidae. These are tactile and olfactory and used to test the ground in front of the animal, and are important in trail-following (Cook 2001) (see also Section 20.4.5). Other lung-bearing panpulmonates lack these additional tentacles although a pair of papillae or 'knobs' on the distal snout of ellobiids and onchidiids are possibly homologous (e.g., Barker 2001a).

The pair of small, spherical statocysts is, as in other ortho gastropods, situated near the pedal ganglia and function in geotaxis and orientation. They are lined with supporting cells and sensory hair cells, the latter with a nervous connection to the cerebral ganglia. Most heterobranch statocysts contain many minute calcareous statoliths, but sacoglossans and a few nudibranchs have a single otolith. There are at least two examples in nudibranchs of statocysts containing both a larger otolith and smaller statoconia (Wägele & Willan 2000).

Omalogyra atomus has a 'small tuft of tall ciliated cells', innervated by fibres from the pedal nerves, that project immediately below the head lobe at the junction of the neck and foot (Fretter 1948 p. 608). Fretter speculated these might be epipodial sense organs, but it is highly unlikcly that they are in any way related to the epipodial sense organs of vetigastropods.

20.3.7 Reproductive System

The heterobranch reproductive system consists of a mesodermal ovotestis, a hermaphroditic duct with sacs (seminal vesicles, ampulla) for storing endogenous sperm, and a fertilisation area followed by the gonoduct and its associated glandular structures and exogenous sperm sacs, all of which are ectodermal (Thompson 1962; Tardy 1970).

Except for some aphallic Acochlida (*Microhedyle*, *Strubellia*) (Schrödl & Neusser 2010) and *Thalassopterus* (Gymnosomata) which are gonochoric, all heterobranchs are hermaphrodites, and in most the gonad is an ovotestis with sperm and eggs being produced together. It is not uncommon for the male system to develop earlier, but most are effectively simultaneous hermaphrodites when mature. Some are protandrous, as for example in some stylommatophorans (e.g., Boato & Rasotto 1987; Tomiyama 1996), some hygrophilans (e.g., Russell-Hunter & McMahon 1976), and the gymnosome and thecosome pteropods (e.g., Morton 1958b; Lalli & Wells 1978).

20.3.7.1 Terminology

The study of heterobranch, and particularly euthyneuran, reproductive systems has been made more complicated by the terminology not being standardised, other than an attempt by Ghiselin (1966). Recent accounts include Dayrat and Tillier (2002) who use Thompson and Bebbington (1969) and Duncan (1960a, b) for euthyneuran terminology, and detailed accounts are given by Gosliner (1994) and Luchtel et al. (1997). A summary of terms commonly used for various reproductive structures in 'pulmonates' and 'opisthobranchs' is given in Table 20.1.

A general description of the ovotestis and sperm morphology in heterobranchs is given next along with the main features of the reproductive systems in each major group.

TABLE 20.1

Comparison of Common Terms Used in the Description of Reproductive Systems in Heterobranchs and Other Gastropods (Mainly Caenogastropods)

Other Gastropods	'Opisthobranchs'	'Pulmonates'	Used Here
Gonad, testis, or ovary	Ovotestis	Ovotestis	Ovotestis
Fertilisation chamber	Carrefour; fertilisation chamber	Carrefour; fertilisation chamber or pouch; talon	Fertilisation chamber, carrefour
Eggs, oocytes, ova	Eggs, oocytes, ova	Eggs, oocytes, ova	Eggs, oocytes, ova
Vas deferens or upper oviduct	Small (or little) hermaphrodite duct	Hermaphrodite duct	Hermaphrodite duct
Seminal receptacle (receptaculum seminalis)	Seminal receptacle; exogenous sperm sac; spermatocyst	Seminal receptacle; accessory bursa; talon; caecum	Seminal receptacle
Albumen gland	Albumen gland	Albumen gland	Albumen gland
	Mucous gland; nidamental gland; uterus (in part)	Muciparous gland; anterior mucous gland	Mucous gland
Capsule gland	Capsule gland (in part); membrane gland + mucous gland	Oothecal gland; oviduct gland	Oviducal gland
	Common genital opening, vestibule	Atrium, vestibule	Vestibule
Oviduct with ventral channel	Large hermaphrodite duct; spermoviduct, oviduct	Spermoviduct, oviduct	Spermoviduct
Vagina	Vagina	Vagina	Vagina
Oviduct	Uterus	Oviduct, female channel	Oviduct
(Part of capsule gland)	Winding gland; posterior mucous gland	Membrane gland; posterior mucous gland	Membrane gland
Bursa copulatrix[1] (copulatory bursa)	Gametolytic sac (or gland), bursa copulatrix	Spermatheca; spermathecal sac; bursa copulatrix	Bursa copulatrix
Male or female genital opening	Gonopore; male or female genital opening; common genital opening	Gonopore; male or female genital opening; common genital opening	Male or female genital opening; common genital opening
Received sperm; spermatozoa	Sperm; exosperm	Sperm; exosperm	Exosperm
Produced sperm; spermatozoa	Sperm; autosperm; endosperm	Endosperm; autosperm	Endogenous sperm, autosperm[2]
Seminal vesicle (temporary storage of autosperm)	Ampulla	Seminal vesicle	Seminal vesicle
	Spermiduct	Spermiduct	Spermiduct
Sperm groove; seminal groove	Seminal groove; seminal tract; external ciliated groove; ciliated sperm groove	Seminal groove; external ciliated groove	Sperm groove
Prostate (gland)	Prostate (gland)	Prostate (gland)	Prostate gland
-	-	Dart sac	Dart sac
Penis, verge, intromittent organ, copulatory structure	Penis	Penis (part of penis complex)	Penis
Penial duct, ejaculatory duct	Penial duct, ejaculatory duct	Penial duct, ejaculatory duct	Penial duct
-	-	Preputium	Preputium
		Prepuce (a cavity anterior to penial sheath). A preputial gland is sometimes present	Prepuce
	Penis sheath	Penial sheath; penis sac	Penis sheath
	Penial bulb		Penial bulb
		Penial gland	Penial gland
Pallial vas deferens; sperm duct; ejaculatory duct	Sperm duct; ejaculatory duct	Sperm duct; vas deferens	Sperm duct; vas deferens

(Continued)

TABLE 20.1 (CONTINUED)

TABLE 20.1 (CONTINUED)

Comparison of Common Terms Used in the Description of Reproductive Systems in Heterobranchs and Other Gastropods (Mainly Caenogastropods)

Other Gastropods	'Opisthobranchs'	'Pulmonates'	Used Here
-	-	Epiphallus (a muscular part of the sperm duct before the penial sheath; involved in spermatophore production – found in siphonariids and stylommatophorans)	Epiphallus
-	Penis retractor muscle(s)	Penis retractor muscles	Penis retractor muscles
-	-	Stylet; penial stylet	Stylet
-	-	Flagellum	Flagellum
-	-	Digitiform glands (specialised penial glands in some stylommatophorans)	Digitiform glands

Note: This list is not exhaustive but covers many of the terms used in several reviews.

[1] Mikkelsen (1996) suggested that the bursa copulatrix in 'opisthobranchs' is not homologous to the structure of the same name in Caenogastropoda.

[2] The term endosperm is not used because it is a major term in botany.

20.3.7.2 The Ovotestis and Gametes

The ovotestis is separated from the digestive gland in most heterobranchs, but in many (not all) nudibranchs and sacoglossans, the gonad spreads dorsally over the digestive gland and is sometimes intermingled with it and sometimes with the kidney.

The arrangement of the spermatogonia and oogonia differs somewhat in different heterobranch taxa. For example, although some have the male and female gametes in the same follicle, the spermatogonia can be concentrated in the medullary part of the follicle while the oogonia are cortical, an arrangement considered plesiomorphic (Mikkelsen 1996; Wägele & Willan 2000). Such separation of the sperm- and egg-producing areas occurs in most nudibranchs, but the separation is never complete. In one apomorphic arrangement in the Fionidae, oogonia are in groups of follicles which each surrounds one male follicle, but these follicles are all connected, and the eggs must pass the male part on their passage to the hermaphrodite duct (Gosliner 1994; Wägele & Willan 2000).

The ovotestis of lung-bearing panpulmonates is composed of acini containing both ova and sperm. Although functionally hermaphroditic, sperm usually mature first, and a few species are protandrous (Luchtel et al. 1997). Each acinus contains both male and, around the edges, female germ cells and follicle and Sertoli (nurse or supporting) cells. The follicle cells surround the developing ova while the Sertoli cells have developing sperm attached to them.

The gametes are transported from the ovotestis by a hermaphroditic duct into the fertilisation chamber. The median part of the hermaphroditic duct is swollen into an endogenous (autosperm) storage area (the seminal vesicle(s), as in many lung-bearing panpulmonates) or may be sac-like and is often called an *ampulla* (in many other euthyneurans). Sperm absorption in the seminal vesicle has been reported in some taxa, including *Acteon*, *Helix*, and *Limax* (Dayrat & Tillier 2002).

As already noted, in some euthyneurans, sperm mature before the ova, but usually they still function as simultaneous hermaphrodites. Some taxa, such as many (but not all) marine lung-bearing panpulmonates, exhibit various degrees of protandry (Berry 1977). Protandry may be the ancestral condition in euthyneurans, as this condition is common in 'lower heterobranchs'. Among euthyneurans, protandry is particularly pronounced in pteropods (Lalli & Gilmer 1989). It has also been recognised in a few panpulmonates, such as the freshwater limpet *Laevapex fuscus* (Russell-Hunter & McMahon 1976).

Sperm morphology in euthyneurans has been investigated in several studies, including some classical light-microscope studies (Franzén 1955; Thompson 1973). Ultrastructural studies show that heterobranch sperm differs from that of other gastropods in possessing a unique set of characters. These include a rounded acrosomal vesicle, a usually helical nucleus giving the sperm a screw-like appearance, a highly modified mitochondrial derivative, a spermatid acrosome associated with a dense nuclear plaque, and formation of the mitochondrial derivative through fusion of numerous small mitochondria along the length of the axoneme (Healy 1993).

Healy (1993) recognised three distinct groups of heterobranch gastropods based on their sperm ultrastructural and spermiogenic characters. These are (1) the Valvatoidea, (2) the Architectonicoidea, and (3) the Rissoelloidea, Omalogyroidea, Pyramidelloidea, and the euthyneurans. The distinct autapomorphic characters of the sperm of the Valvatoidea and Architectonicoidea suggest that they were early branches from the ancestral heterobranch lineage, a finding generally in accord with other morphological and molecular evidence.

Heterobranch sperm characters have also been used in more general phylogenetic analyses (e.g., Ponder & Lindberg 1997; Dayrat & Tillier 2002; Aktipis et al. 2008). Some of the more informative ultrastructural characters recognised by Dayrat and Tillier (2002) include the long glycogen piece in the sperm tail which may have granules in nine tracts or as a continuous sheath; the acrosomal vesicle (at the tip of the acrosome) which is rounded in most heterobranchs but irregular or absent in *Valvata* (it is often conical in other gastropods); subacrosomal material may form a curved dish shape but is usually columnar; paracrystalline material in the sperm midpiece may be absent or present; and cristae in mitochondria may be obvious or reduced or lost. Characters found in all heterobranchs include the sperm midpiece being long with the mitochondria forming a continuous sheath, and periaxonemal coarse fibres are present as are intra-axonemal dense granules. Developmental characters found in all apogastropods include temporary support cylinders associated with spermatids and a microtubular sheath being present in the development of the spermatid midpiece and nucleus. Heterobranchs do not have paraspermatozoa.

Some heterobranchs, notably eupulmonates, produce spermatophores (see Chapter 8) in a modified part of the ectodermal reproductive tract. These vary considerably in morphology, sometimes being rather elaborate (Figure 20.44).

Some, such as runcinids, philinoglossids, acochlidians, and rhodopids, have convergently become worm-like with parallel organ reduction and other changes, mostly as adaptations to interstitial life. Their reproductive systems have become simplified due to reduction and other modifications. Some acochlidiodeans (Microhedylidae) show gonochorism, loss of the penis, and spermatophore formation, and philinoglossids lack some typical cephalaspidean characters, presumably due to their minute size (Ghiselin 1966). Such significant modifications do not always occur with, for example, tiny punctids such as *Paralaoma* having similar reproductive systems to larger Punctidae (e.g., Climo 1973).

20.3.7.3   The Reproductive System of 'Lower Heterobranchs'

The simplest heterobranch reproductive system so far described is in Orbitestellidae (Ponder 1990a; Hawe & Haszprunar 2014) (Figure 20.45) and may approximate the plesiomorphic heterobranch reproductive system. The ovotestis produces eggs and sperm simultaneously in some individuals, but there is evidence that they may be protandric. There is a large seminal vesicle located proximally on the hermaphroditic duct, a simple glandular pallial oviduct with no sperm sacs and a ventral sperm groove lined with prostatic cells. Its opening is located ventrally at about two-thirds of the length of the duct, and a ciliated tract carries sperm to the external penis on the right side of the head just inside the mantle cavity. A similar reproductive tract was described in *Henrya* (Murchisonellidae) by Wise (1999) but with a small seminal receptacle at the posterior end of the pallial oviduct.

Other 'lower heterobranchs' that have been described are more complex and differ considerably from one another and euthyneurans, so are described individually next as an illustration of the remarkable diversity of the reproductive system in heterobranchs.

Architectonicoidea and Mathildoidea. Apart from one rare mathildid shown to have male and female individuals, but which could be protandric (Haszprunar 1985c), architectonicids and mathildids are protandrous or simultaneous hermaphrodites with a separate ovary and testis, and they also lack a penis and use spermatophores (Haszprunar 1985c, d). There is a large seminal vesicle, and a prostate gland is usually present and is proximal to the closed vas deferens that extends forward to open in the anterior section of the mantle cavity. The pallial oviduct forms an open glandular groove distally and, at least in one mathildid, posteriorly (proximal). A large seminal receptacle lies at the extreme posterior end of the mantle cavity and, attached to it, an equally large 'spermatolytic gland'. The seminal receptacle opens to the mantle cavity by way of a rather long muscular duct. These structures have no direct connection with the oviduct.

Very elongate spermatophores are moved via a spermatophore groove running from the mantle cavity to the side of the neck and then deposited in the mantle cavity of the partner. They then enter the opening of the muscular duct to the seminal receptacle located at the posterior end of the mantle cavity (Haszprunar 1985c; Robertson 1989). Sperm is released from the spermatophore when it is dissolved in the spermatolytic gland attached to the seminal receptacle, and they are stored in the seminal receptacle. When eggs are released, the sperm travels along the duct to the seminal receptacle, into the posterior mantle cavity where it enters the long slit-like opening of the oviduct and travels along the oviduct posteriorly to fertilise the eggs.

Omalogyroidea. As far as can be ascertained from the description of a single species of *Omalogyra* (Fretter 1948), omalogyrids are the only other 'lower heterobranchs' with a separate ovary and testis. This very minute gastropod is protandrous, and the separate oviduct and testicular duct (which functions as a seminal vesicle) join at the muscular pouch-like fertilisation chamber from which the separate oviduct and spermiduct depart (Figure 20.45). The latter is surrounded laterally and dorsally by a thick prostate gland, and a small sperm sac opens into its proximal part. The oviduct (see below) narrows and joins with the male system, and the common spermoviduct is a rather narrow tube that opens near the edge of the mantle cavity. It contains a unique, narrow, tubular, very muscular structure with an anterior origin that protrudes backwards into the lumen of the prostate (Figure 20.45). This structure is apparently a copulatory organ, and it passes through the dorsal wall of the spermoviduct into a sac-like structure interpreted as a bursa copulatrix. The copulatory organ represents the elongation and expansion of the bursal opening and is only present during spring when the male phase is dominant. The sperm sac is another structure present only during those months when it, as its name suggests, contains sperm and prostatic fluid, which show signs of being digested. Fretter (1948) suggested that the copulatory organ probably sucked up sperm and prostatic secretion from the region of the prostate and stored them in the bursa, but she did not actually observe the bursa with such contents. It was assumed that the copulatory organ would then reverse its orientation and be extruded from the common genital opening, but again this was not observed. It was suggested that the

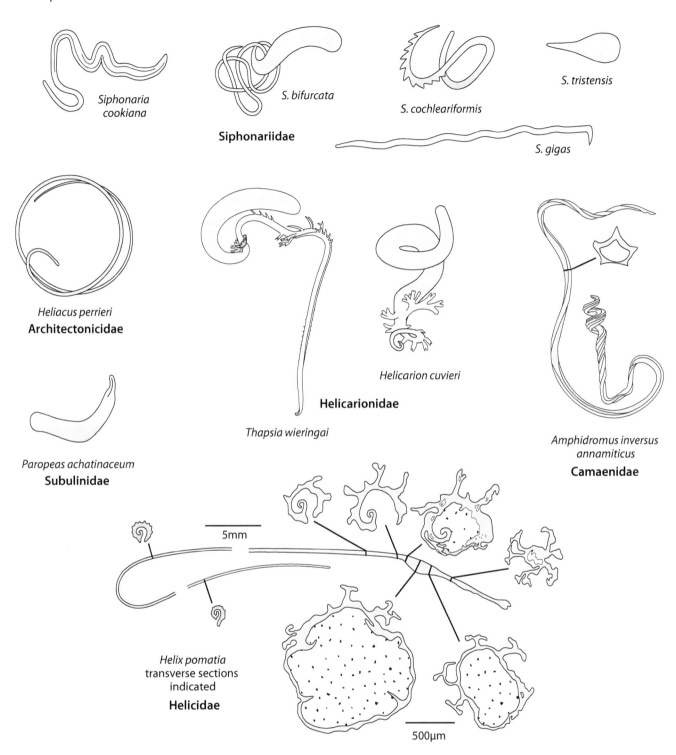

FIGURE 20.44 Spermatophores of various panpulmonates and an architechtonicid. Redrawn and modified from the following sources: *Heliacus* (Robertson, R., *Malacologia*, 30, 341–364, 1989), siphonariids (Berry, A.J., Gastropoda: Pulmonata, pp. 181–226, in Giese, A.C. and Pearse, J. S. (eds.), *Reproduction of Marine Invertebrates. Molluscs: Gastropods and Cephalopods*, Vol. 4, Academic Press, New York, 1977), *Thapsia* (Winter, A.J. de, *Zoologische Mededelingen (Leiden)*, 82, 441–477, 2008), *Helicarion* (Hyman, I.T. and Köhler, F., *Zool. J. Linn. Soc.*, 184, 933–968, 2018), *Amphidromus* (Sutcharit, C. and Panha, S., *J. Molluscan Stud.*, 72, 1–30, 2006), *Paropeas* (Naggs, F., *J. Molluscan Stud.*, 60, 175–191, 1994), *Helix* (Lind, H., *J. Zool.*, 169, 39–64, 1973).

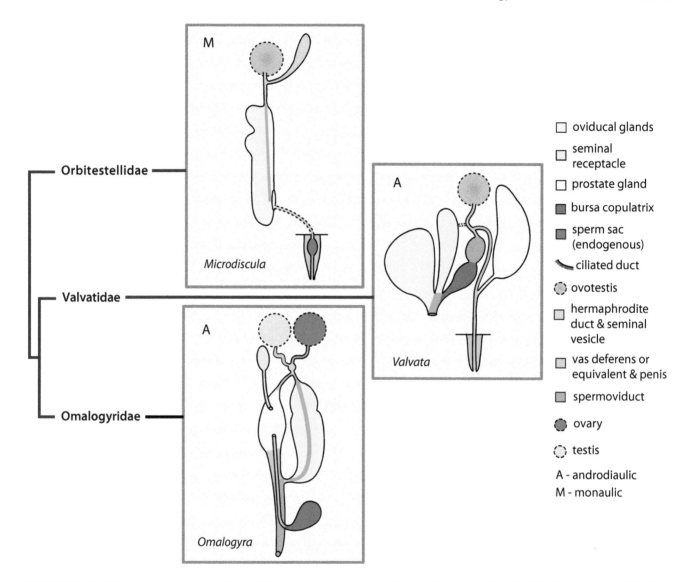

FIGURE 20.45 Diagrammatic representations of the reproductive systems of some 'lower heterobranchs'. Redrawn and modified from the following sources: *Microdiscula* (Ponder, W.F., *J. Molluscan Stud.*, 56, 515–532, 1990a), *Valvata* (Ponder, W.F., *J. Molluscan Stud.*, 56, 533–555, 1990b), *Omalogyra* (Fretter, V. and Graham, A.L., *British Prosobranch Molluscs: Their Functional Anatomy and Ecology*, Ray Society, London, 1962).

sperm sac functioned to remove excess sperm and prostatic fluid from the system.

The female system develops during the summer months. The ovary contains relatively very large eggs. The proximal part of the oviduct has the albumen gland followed by the mucous gland and then the capsule gland. The copulatory organ and sperm sac disappear, and the bursa develops a long duct and moves posteriorly. At this stage, it appears to function in digesting sperm in the same way as the sperm sac. Interestingly, individuals that develop in the summer mature directly into females with the male system only partially developed.

Valvatoidea. In the freshwater Valvatidae, the reproductive system is androdiaulic with an external cephalic penis anterior to the opening of the spermoviduct which lies near the opening to the mantle cavity (Garnault 1899; Cleland 1954) (Figure 20.45). The ovotestis contains both ova and sperm,

and the hermaphroditic duct acts as a seminal vesicle and bifurcates to form the narrow vas deferens and spermoviduct (as in rissoellids). In at least one species, the large prostate gland lies partly in the haemocoel and opens proximally to the vas deferens by way of a narrow duct, and beyond that point, the long vas deferens runs to the penis where it forms an ejaculatory duct. The female part of the system is initially a narrow tube which swells at the fertilisation chamber. This chamber opens directly into a second swelling that functions as a bursa copulatrix. The separated albumen gland opens narrowly to the oviduct just anterior to the bursa, and the separated capsule gland opens anteriorly close to the genital opening.

The reproductive system of the marine Cornirostridae (Ponder 1990b) is similar to that of the valvatids in most respects but differs in the penial duct being markedly muscular and swollen in the base of the penis forming an ejaculatory duct which also has a sperm storage function. Differences in

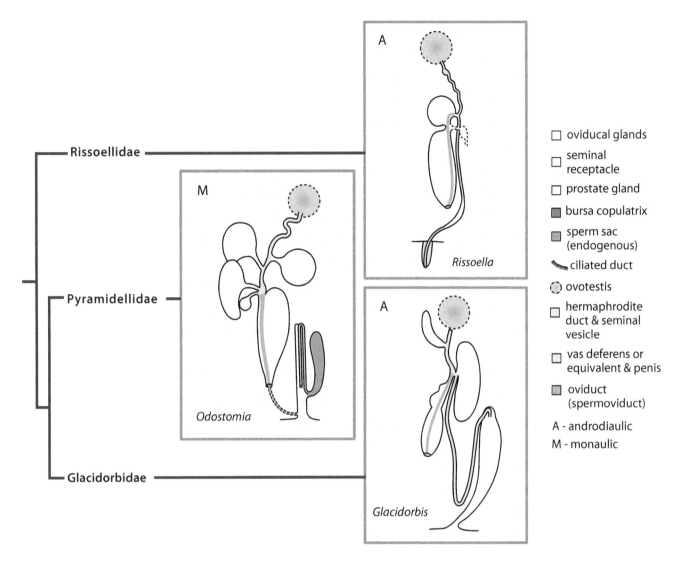

FIGURE 20.46 Diagrammatic representations of the reproductive systems of three heterobranch taxa of small size, two of which were, until recently, treated as 'lower heterobranchs'. Redrawn and modified from the following sources: *Rissoella* and *Odostomia* (Fretter, V. and Graham, A.L., *British Prosobranch Molluscs: Their Functional Anatomy and Ecology*, Ray Society, London, 1962), *Glacidorbis* (Ponder, W.F., *Zool. J. Linn. Soc.*, 87, 53–83, 1986).

the female organs include having a seminal receptacle opening to the albumen gland, the bursa opening by way of two short ducts to the spermoviduct, and a muscular pocket in the distal part of the oviduct just behind the short vagina that may receive sperm during copulation. There is a glandular pocket, not present in valvatids, which is anterior to the genital opening and may help mould the egg string.

The reproductive system of *Xenoskenea* and *Hyalogyrina* (Hyalogyrinidae) has a small copulatory organ which, during copulation, possibly functions together with a mantle lappet in transferring spermatophores to a receptacle that opens independently of the gonoduct, as in architectonicids and mathildids (Haszprunar et al. 2011). A large receptacle opening to the mantle cavity was also described in *Xylodiscula* (Xylodisculidae) which lack a copulatory organ, and sperm transfer by way of spermatophores is considered likely (Hawe et al. 2014).

The penis in valvatoideans and orbitestellids appears to have evolved independently as it is cerebrally innervated in

the former group and pedally innervated in the latter (Hawe & Haszprunar 2014).

20.3.7.4 Some Small Euthyneurans Previously Thought to Be Lower Heterobranchs

Three other small-sized taxa, two of which were, until recently, treated as 'lower heterobranchs', the other sometimes so, are the Rissoelloidea, a group now thought to be related to the Acteonoidea, and the Pyramidelloidea and Glacidorboidea, both now known to be panpulmonates. These groups have rather modified reproductive systems, probably due in part at least to their small size, and are illustrated together in Figure 20.46.

Rissoelloidea. The reproductive system in rissoellids has been described in some detail by Fretter (1948). An ovotestis opens into a hermaphroditic duct, the proximal part of which acts as a seminal vesicle. Rissoellids are simultaneous hermaphrodites, with the male system slightly more precocious. The reproductive system is androdiaulic, with a complete division of the pallial part of the endogenous male, and a pallial female duct which is a spermoviduct that opens near the outer edge of the mantle cavity. The vas deferens becomes glandular (prostate) distally towards the edge of the mantle cavity and then leads to a short, tubular penis that, at rest, lies in a groove in the body wall. The female system has a large proximal albumen gland, which also acts as a fertilisation chamber and is separated from the distal capsule gland by a short narrow section of the oviduct. In one species this receives a duct from a sac which opens to the posterior mantle cavity. At first glance this sac might appear to be a bursa with a separate pallial opening, as seen in some other marine euthyneurans (i.e., a triaulic system), but this is not the case as, according to Fretter (1948), the sac only ever contained 'waste' material from the oviduct which was from time to time discharged into the mantle cavity. Also, a second species investigated by Fretter lacked this sac and duct. The 'oviduct' is a spermoviduct as it contained a seminal groove that was observed to contain spermatozoa. Rissoellids lay hemispherical egg capsules, each containing one or two eggs, that are attached to algae.

Pyramidelloidea. Pyramidellids (Fretter & Graham 1949, 1962; Fretter 1953) are simultaneous hermaphrodites with sperm and ova produced in the same tubules in the ovotestis, although protandry can occur. The hermaphroditic duct (Figure 20.46) opens to a thin-walled section of the spermoviduct into which mucous and albumen glands open by way of narrow ducts. The genital aperture lies at the anterior end of the mantle cavity in some pyramidellids but on the propodium in *Turbonilla*. The walls of the lower section of the spermoviduct form a prostate gland when the male part of the system is mature. Exogenous sperm run along a ventral groove to the only sperm pouch, a seminal receptacle, which is behind the prostate gland. Autosperm travel down the spermoviduct and, after mixing with prostatic secretion, leave the genital aperture to move along a ciliated groove which runs along the right side of the head to the opening of the penis sheath. The retractile penis is positioned uniquely in gastropods. It lies within a long, narrow sheath which opens beneath the mentum and passes back through the nerve ring, and a sperm sac is attached to the penial sheath. The prostate gland is pallial, and in some taxa the seminal groove is open while, in others, it is closed to form a vas deferens or muscular ejaculatory duct (Fretter 1953). A sac-like structure that contains sperm is associated with the copulatory organ, which is invaginated and opens separately anteriorly. In the female system, separate 'albumen,' membrane, and mucous glands are somewhat isolated from the rest of the gonoduct and communicate with it by narrow ducts. There is no bursa copulatrix, and the seminal receptacle is located posteriorly. The oviducal glands open by way of narrow ducts to the upper part of the spermoviduct, the lower part of which is occupied by the prostate gland (Fretter & Graham 1949). Spermatophores are utilised to transfer sperm (e.g., Robertson 1978).

The glacidorbid reproductive system is briefly discussed in the next Section.

20.3.7.5 Euthyneuran Reproductive Systems

The euthyneuran reproductive system, as with most other systems, is not only highly modified, but there are considerable convergences and parallelisms. While aspects of the evolution of the reproductive systems of 'pulmonates' (e.g., Berry 1977; Visser 1977, 1981, 1988) and 'opisthobranchs' (e.g., Ghiselin 1966; Gosliner 1981, 1994) have been investigated separately, there has been little attempt to overview the whole group. In particular, the reproductive systems of the basal members of Euthyneura have been largely ignored with the focus being on the more derived members. Until recently ideas about the evolution of the reproductive system of both 'opisthobranchs' and 'pulmonates' were based on the erroneous idea that they were (usually independently) derived from dioecious caenogastropods. Instead, the ancestral euthyneurans already had a hermaphroditic reproductive system, and the search for the origins of the heterobranch hermaphroditic system must extend much further back in their evolution.

Early in euthyneuran evolution the pallial gonoduct and associated structures, including the penis, moved from the confines of the pallial floor into the cavity of the haemocoel (e.g., Ghiselin 1966), with this liberation profoundly enhancing evolutionary opportunities. What was the pallial section of the gonoduct now lay free in the haemocoel, connected to the hermaphroditic duct proximally and the body wall distally where it opened to the exterior. This previously pallial section is where, primitively, the sperm (exogenous and endogenous) and egg transport functions are combined in the spermoviduct.

Determination of the details of the ancestral conditions in the reproductive systems of the major clades of euthyneurans depends on their phylogenetic relationships which are, to some extent, still somewhat controversial. Recent molecular and morphological (e.g., Wägele et al. 2008) phylogenies have confirmed that the most basal euthyneuran group is the Acteonoidea (together with the apparently related, probably paedomorphic, rissoellids). The acteonoidean (Acteonidae and Aplustridae) reproductive system is, however, derived (Ghiselin 1966) and not ancestral to most other euthyneurans, and also differs markedly from the rissoellid reproductive system (Figures 20.46 and 20.47).

A 'typical' euthyneuran reproductive system consists of an ovotestis from which a hermaphrodite duct transports the gametes. The seminal receptacle opens to the 'carrefour' which may function as a fertilisation chamber, or, in some taxa, fertilisation may occur in a separate fertilisation pouch that opens to the 'carrefour'. The so-called 'talon' of lung-bearing panpulmonates is the fertilisation chamber.

A seminal vesicle (ampulla) lies at the distal section of the hermaphrodite duct and receives autosperm. In nudibranchs, it is typically sausage-shaped, but sometimes it appears as a bulge on one side of the hermaphrodite duct. In lung-bearing panpulmonates there are usually several small pouch-like seminal vesicles attached to the upper hermaphrodite duct. The upper hermaphrodite duct opens to the area where fertilisation occurs, this usually being a pouch (carrefour or talon) into which the seminal receptacle opens. In many Hygrophila, a sphincter enables the separation of male and female gametes

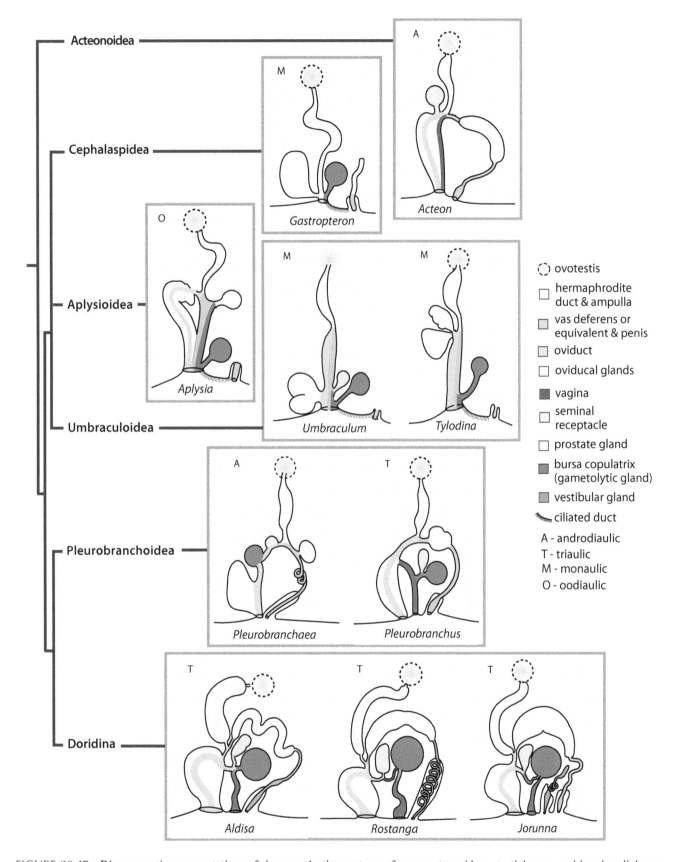

FIGURE 20.47 Diagrammatic representations of the reproductive systems of some acteonoidean, tectipleuran, and basal nudipleuran taxa. Redrawn and modified from Schmekel, L., Aspects of evolution within the opisthobranchs, pp. 221–267, in Trueman, E.R. and Clarke, M.R. (eds.), *Evolution. The Mollusca*, Vol. 10, Academic Press, New York, 1985, bottom group from Schmekel, L. and Portmann, A., *Opisthobranchia des Mittelmeeres, Nudibranchia und Saccoglossa*, pp. 20–327, Springer-Verlag, Berlin, 1982.

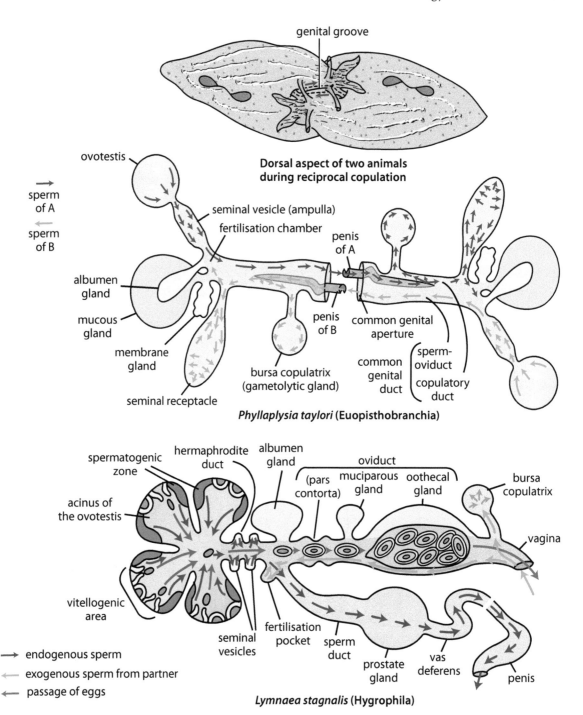

FIGURE 20.48 Two highly stylised euthyneuran reproductive systems showing the passage of endo- and exosperm during copulation above and sperm and eggs below. *Phyllaplysia* redrawn and modified from Beeman, R.D., *The Veliger*, 13, 1–31, 1970 and *Lymnaea* redrawn and modified from Geraerts, W.P.M. and Joosse, J., Freshwater snails (Basommatophora), pp. 141–207, in Tompa, A.S., Verdonk, N.H. and van den Biggelaar, J.A.M. (eds.), *Reproduction. The Mollusca*, Vol. 7, Academic Press, New York, 1984.

in the carrefour (Duncan 1958) while a sphincter that functions as a valve is present in the carrefour of many nudibranchs (Dayrat & Tillier 2002) and may control the flow of eggs into a separate fertilisation chamber (Thompson 1976).

The passage of both endosperm and exosperm in a copulating aplysiid is shown in Figure 20.48, as is the passage of sperm and eggs in a hygrophilan (*Lymnaea*) and, in Figure 20.49, a nudipleuran (*Tritonia*).

The large ectodermal spermoviduct carries gametes from the fertilisation chamber to the vestibule which opens at the cloaca-like common genital aperture or 'genital vestibule'. It is divided into three channels which separately carry the encapsulated eggs, autosperm, and exosperm. These 'channels' may be grooves or separate ducts. If there is a single tube the condition is called monaulic; with two separated tubes, it is diaulic; and with three separated tubes, it is triaulic (Figure

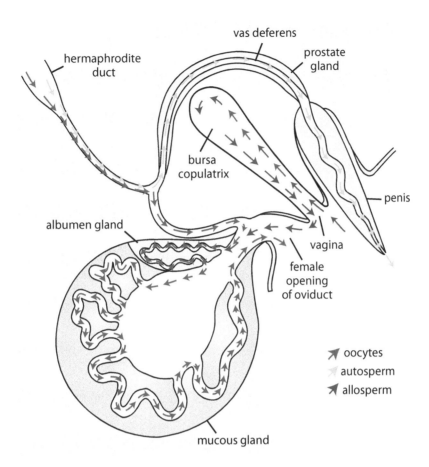

FIGURE 20.49 A diagrammatic representation of the reproductive system of *Tritonia hombergi* showing the passage of the gametes. Redrawn and modified from Thompson, T.E., *Q. J. Microsc. Sci.*, 102, 1–14, 1961b.

20.50). There has been confusion caused by these terms sometimes being applied to the number of genital openings rather than the number of separate ducts. A single genital opening is a condition known as *monotrematic* while two openings is known as *ditrematic* (e.g., Visser 1988). The number of openings is less significant for understanding phylogeny than the number of tubes (Schmekel 1985).

When there are three separate ducts, endogenous sperm (autosperm) is transported by one (the spermiduct), while the fertilised eggs are transported by the oviduct, and the third tube, the vaginal duct, transports the exogenous sperm after copulation. The fertilised eggs then pass down the oviduct where they are surrounded by albumen and then the egg membranes. These coatings are provided by the oviducal glands, of which there are usually three kinds recognised, the albumen, membrane, and mucous (or jelly) glands, the latter two comprising the 'capsule gland' (Ghiselin 1966; Beeman 1977; Hadfield & Switzer-Dunlap 1984; Dayrat & Tillier 2002). The diaulic condition can result from the separation of either the spermiduct or the oviduct. When the spermiduct is separated, and the oviduct and vaginal duct are united (i.e., the autosperm and exosperm are in two separate ducts), the condition is called androdiaulic while the oodiaulic condition is when the eggs and sperm are separated in two ducts. These terms have routinely been used for opisthobranchs (e.g., Gosliner 1994), but

Mikkelsen (1996) and Dayrat and Tillier (2002) have been critical of their use. For example, the gonoduct may be monaulic but functionally triaulic due to sperm-carrying grooves, as in *Aplysia*. Most cephalaspideans are monaulic and resemble pteropods in having a prostate associated with the penis. These groups also have similar spermoviducts but vary considerably in the structure of the copulatory apparatus (see below).

Visser (1988) studied the lung-bearing panpulmonate reproductive systems based on their ontogeny and argued that the more primitive members ('basommatophorans') were paedomorphic in their ditrematic (two openings) reproductive system, with the male system (penis + anterior vas deferens) developing later than the female duct. They are either semidiaulic (as in Trimusculidae, Chilinidae, Otinidae, and some ellobiids), or diaulic (as in some ellobiids, systellommatophorans, and 'higher' Hygrophila) or monotrematic monaulic as in the protandric Glacidorbidae (Ponder 1986) and the simultaneous hermaphroditic Siphonariidae and Amphiboloidea (Golding et al. 2010).

Stylommatophorans pass through a ditrematic condition early in the ontogeny of the reproductive system, and in their later development (and adult morphology) a monotrematic (and semidiaulic) system is established (Visser 1988).

The glandular parts of the reproductive tract that lie in the mantle cavity (i.e., are pallial) in 'lower heterobranchs' as

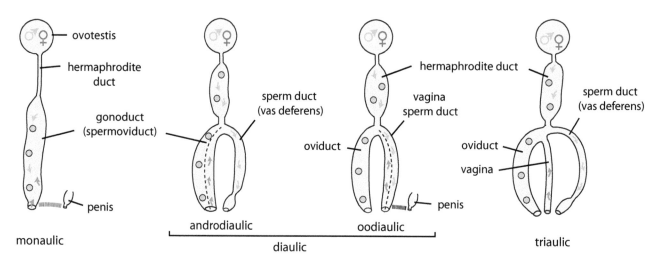

Stylised diagrams of the different configurations of reproductive ducts in heterobranchs

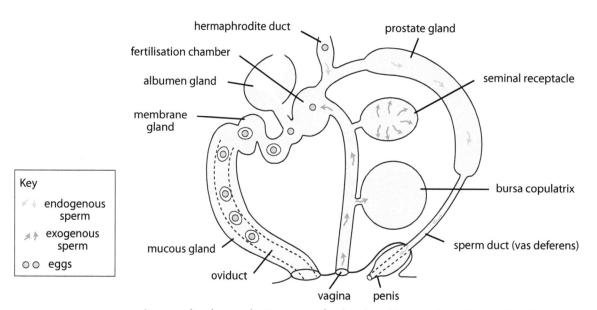

A generalised reproductive tract of a dorid nudibranch (triaulic condition)

FIGURE 20.50 Terminology of the different conditions of euthyneuran reproductive systems, with a more detailed diagram of a triaulic system of a dorid nudibranch. Redrawn and modified from Schmekel, L., Aspects of evolution within the opisthobranchs, pp. 221–267, in Trueman, E.R. and Clarke, M.R. (eds.), *Evolution. The Mollusca*, Vol. 10, Academic Press, New York, 1985.

well as in pyramidellids, rissoellids, and glacidorbids, have sunk into the haemocoel in nearly all other euthyneurans. The pallial spermoviduct opens in the mantle cavity in glacidorbids and pyramidellids (and most 'lower heterobranchs') and carries fertilised eggs as well as the exogenous and endogenous sperm. In other euthyneurans, it opens to the exterior. Similarly, the penis is external and attached behind the right tentacle in valvatoideans and acteonoideans, as it is in many caenogastropods. The penis is retracted into a penial sheath in the haemocoel in euthyneurans including glacidorbids and pyramidellids, and rissoellids also have a semi-internalised penis. In primitive euthyneurans, a copulatory apparatus is located further anteriorly than the opening to the pallial gonoduct, which is connected to the copulatory apparatus via a ciliated groove. In acteonoids, there is an internal sperm duct to the external penis, but in cephalaspideans, a sperm groove leads to the opening of the internal penis on the right of the head. The sperm groove is closed to form a tube (vas deferens or sperm duct) in many euthyneurans and in some runs directly from the carrefour. Architectonicids (see above) and siphonariids lack a penis and use spermatophores, which are also common in stylommatophorans, although a penis is present. The only instances of spermatophores in nudibranchs are in the Aeolidiidae (Tardy 1965).

The shape and position of the prostate gland vary in different groups. It can be elongated and may surround the autosperm channel (i.e., the vas deferens), or the spermoviduct, or the prostatic channel of the spermoviduct. Alternatively, it may be a separate gland that opens to either the carrefour (as in *Pleurobranchus* and *Achatinella*), the penis as a penial gland, as in *Glacidorbis*, cephalaspideans, and thecosomes, or to the pallial gonoduct as in some lung-bearing panpulmonates. In some sacoglossans and hygrophilans, the prostate may be extended along the spermiduct as many short evaginations. In some systellommatophorans, it may open into the proximal portion of the spermiduct or, as in some sacoglossans and some species of *Umbraculum* (Dayrat & Tillier 2002), it may even be absent.

Two pouches store exosperm, the seminal receptacle, and bursa copulatrix. The seminal receptacle is proximal, opens into the carrefour, and contains oriented sperm. The bursa copulatrix opens into the distal part of the gonoduct or vestibule near the common (or female) genital aperture and the sperm it contains are not oriented. Sperm absorption can occur in the bursa but not the seminal receptacle, although the distinctions are not always clear-cut with, for example, the distal sperm pouch in some heterobranch taxa having oriented sperm (Dayrat & Tillier 2002) (see below).

A seminal receptacle (exosperm storage) near the hermaphrodite duct is surrounded by some muscle, and the exosperm it stores are oriented at right angles to the usually folded wall (Schmekel 1971). The seminal receptacle is proximal in *Umbraculum* (Figure 20.47), but in many nudipleurans (pleurobranchs and nudibranchs), other than the Doridoidei, it is distally located, often with a separate duct and not entering the distal oviduct. Additional proximal seminal receptacles are found in some Aeolidioidea (especially Flabellinidae).

The bursa copulatrix is thin-walled, its epithelium made up of apocrine-secreting cells, and the lumen contains the degraded remains of gametes and secretions (Medina et al. 1988). In some nudibranchs such as *Armina*, oriented sperm suggests sperm storage also occurs in this structure (Wägele & Willan 2000). The bursa or its duct originates near the common genital opening. A proximal seminal receptacle and a distal bursa copulatrix are plesiomorphic in marine euthyneurans (Ghiselin 1966; Gosliner 1981; Schmekel 1985; Mikkelsen 1996). Both these sperm sacs are present in the Doridoidei, where they are usually attached to the vagina. In many other nudibranch taxa, there is only one sperm sac which functions mainly as a seminal receptacle (Wägele & Willan 2000).

Loss of either the bursa copulatrix and/or seminal receptacle can occur and, because these two structures are functionally similar (see Chapter 8), one can take over the function of the other. Thus, acteonids lack a seminal receptacle, but the bursa copulatrix occupies the position that is normal for the former structure. Similarly, in aeolidiids the bursa is missing, and the seminal receptacle is often located in the typical bursal location. There are also instances where the bursa and seminal receptacle appear to have fused into a single structure (Ghiselin 1966).

Ghiselin (1966) argued that the first 'opisthobranchs' probably had a monaulic system with the spermoviduct having albumen, membrane and mucous glands, a distal seminal receptacle, and a proximal bursa copulatrix. This arrangement of the sperm sacs makes functional sense with the bursa near the common genital opening so it can receive the incoming sperm, and the seminal receptacle is in a position enabling the sperm to fertilise the eggs before their passage through the oviducal glands (Ghiselin 1966). In taxa with an internally divided spermoviduct

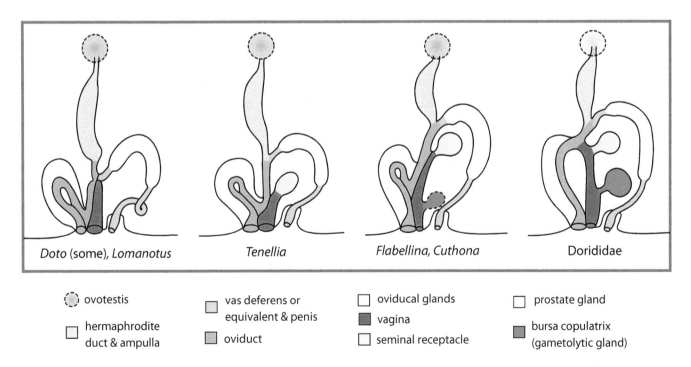

FIGURE 20.51 Diagrammatic representations of examples of nudibranch reproductive systems, all of which are androdiaulic. Redrawn and modified from Schmekel, L., Aspects of evolution within the opisthobranchs, pp. 221–267, in Trueman, E.R. and Clarke, M.R. (eds.), *Evolution. The Mollusca*, Vol. 10, Academic Press, New York, 1985.

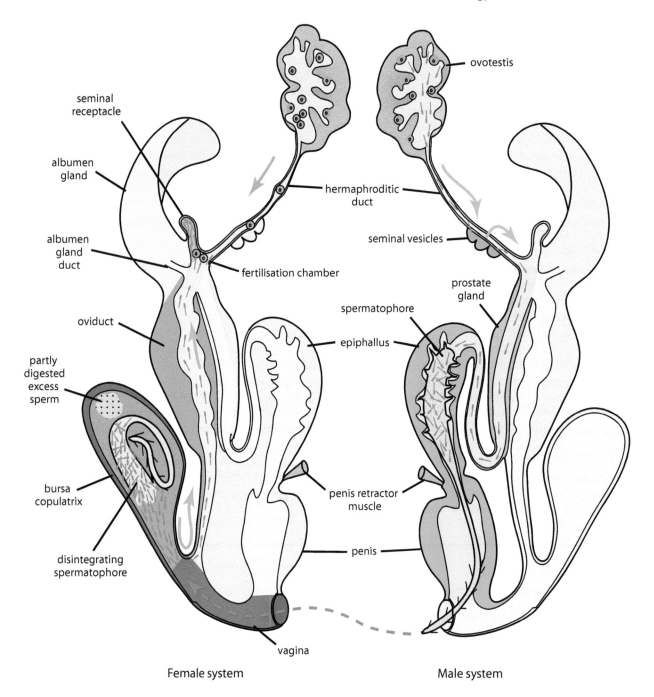

FIGURE 20.52 A diagrammatic representation of the passage of sperm in donor ('male') and recipient ('female') stylommatophorans. Redrawn and modified from Wiktor, A., *Malakol. Abh.*, 12, 85–100, 1987.

(such as in the Aplysiida and the sacoglossan *Cylindrobulla*), the seminal receptacle is nearer the genital opening.

There is great variation of the configuration of the sperm sacs in marine euthyneurans. Single sperm sacs are found in taxa such as acteonids, aeolidiids, and pteropods, as well as in pyramidellids. This appears to be either the result of the fusion of two functionally different sperm sacs in some lineages, as probably in the pteropods (Ghiselin 1966) or the loss of one of them in others, as for example the loss of the bursa in aeolidiids, while other nudibranchs possess a bursa. As demonstrated by Ghiselin (1966), sacoglossans show considerable

diversity in the position, function, and form of the sperm sacs with intermediates between the different configurations (Figure 20.53).

Nearly all nudibranchs are triaulic, although the basal *Bathydoris* is diaulic. Triauly has also arisen at least twice in cladobranch nudibranchs (*Dendronotus* and Arminoidea) and at least twice in aeolidioideans (Ghiselin 1966). It is also seen in many sacoglossans.

The oodiaulic condition in which the oviducal glands (and therefore the passage of the eggs) become separated from the spermoviduct is found in aplysiidans, the primitive

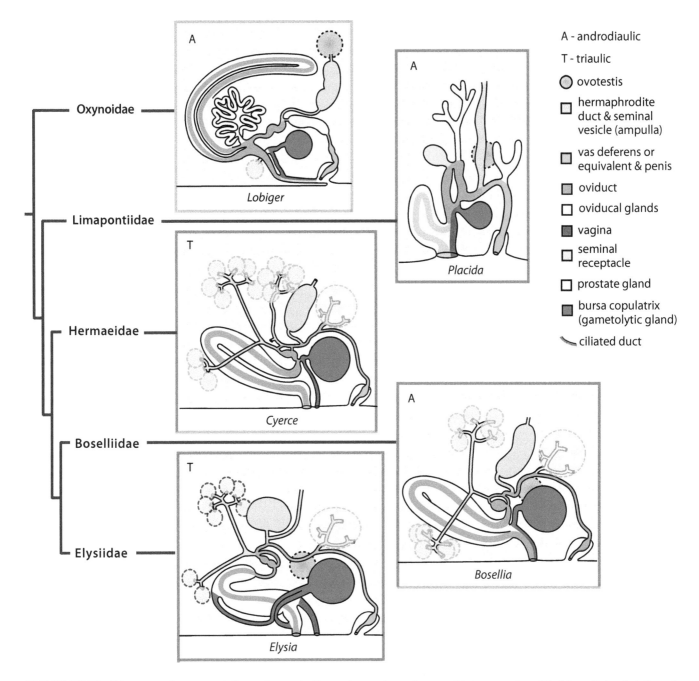

FIGURE 20.53 Diagrammatic representations of reproductive systems of sacoglossans. Redrawn and modified from Schmekel, L. and Portmann, A., *Opisthobranchia des Mittelmeeres, Nudibranchia und Saccoglossa*, pp. 20–327, Springer-Verlag, Berlin, 1982.

sacoglossan *Cylindrobulla*, diaphanids, and probably pteropods (Ghiselin 1966). The two groups of pteropods have often been considered monaulic but may actually be oodiaulic (Ghiselin 1966).

Only androdiaulic and triaulic conditions are found in the Nudipleura. Thus a triaulic system is seen in some pleurobranchoideans and nearly all Doridoidei, while almost all Cladobranchia are diaulic (Figure 20.51).

The albumen gland is usually pouch-like except when it surrounds the oviduct or the large hermaphrodite duct. This gland is usually the uppermost and unpaired but is paired in some sacoglossans. In some it is medial, and it opens to the vestibule of a common genital aperture in some cephalaspidean taxa and near the genital opening in the pteropod *Peraclis*. During the passage of the eggs, the secretion from the albumen gland surrounds the egg either in the lumen of the gland or, if the eggs do not pass through the gland, the secretion is discharged into either the carrefour or the oviduct.

The albumen gland is sometimes lost, but this source of nutrition for the embryos is readily compensated for by the addition of more yolk in the egg. The membrane gland can be combined with the mucous gland to form a glandular area in the capsule gland or may be separate from the mucous gland. In some cephalaspideans, a separate membrane gland opens to the gonoduct via the duct of the albumen gland, while in other cephalaspideans and hygrophilans, the membrane gland surrounds the oviduct. In certain other cephalaspideans the membrane gland opens into the vestibule, while in some ellobiids and *Otina*, it opens to the carrefour (Dayrat & Tillier 2002). According to Ghiselin (1966), the winding gland in aplysiidans is homologous to the membrane gland in other marine euthyneurans. The mucous gland is usually pyriform unless it surrounds the pallial oviduct or large hermaphrodite duct. In Stylommatophora (except Succineidae), the probable mucous gland homologue is associated with the uterus and secretes jelly and a calcareous layer around the egg capsules (Bayne 1968; Tompa 1984). In other euthyneurans, the mucous gland may consist of a single pouch, or a pair of pouches, and opens to the carrefour or, rarely, the vestibule, while in various cephalaspideans it is a J-shaped gland opening to the vestibule. In some it forms part of a gland that, in aplysiidans and *Tylodina*, opens proximally into the gonoduct while in others it opens distally or in the middle (Dayrat & Tillier 2002).

Siphonariids, amphiboloideans, glacidorboideans, and the Systellommatophora and Stylommatophora have a single genital opening (i.e., genital atrium). In Stylommatophora the genital opening is either anterior near the base of the optic or oral tentacles or is located more posteriorly near the pneumostome.

The plesiomorphic condition in euthyneurans is that the vas deferens conducts autosperm to the penis by way of an open groove (or tract), while in the more apomorphic condition this groove is closed to form a sperm duct. The vas deferens runs from the genital opening to the base of the penis where it enters a penial groove, as seen in most cephalaspideans, some pteropods, *Umbraculum* and *Tylodina*, aplysiidans, and a few ellobiids, but in most acteonoideans and in 'higher' euthyneurans the groove is closed to form a sperm (or ejaculatory) duct. In some shelled sacoglossans an external ciliated groove and an embedded duct occur together.

In many euthyneurans, the posterior and anterior apertures are fused into an anterior or posterior vestibule (Dayrat & Tillier 2002). In siphonariids and amphibolids, the female and male openings have fused by the anterior migration of the former, with a superficially similar condition in some sacoglossans and stylommatophorans where the vas deferens is in the visceral cavity (Dayrat & Tillier 2002). In many nudibranchs and *Otina*, a sperm groove or duct is absent because the male genital aperture has migrated posteriorly close to the female aperture, while in some nudibranchs the female and penial apertures form a common genital aperture. The three systellommatophoran families have separate male and female genital openings, but all stylommatophorans have a single opening (Figures 20.52 and 20.59).

The vaginal opening or genital atrium of some nudibranchs may have an accessory gland, the *vestibular gland* (e.g., many Doridoidei; absent in Pleurobranchoidea and Cladobranchia – Gosliner 1994; Wägele & Willan 2000). Also, the vagina of some nudibranchs is lined with cuticle or with cuticular armature (Gosliner 1994).

Some acochlidians have been shown to possess unusual strategies for sperm transfer, including by spermatophores and by hypodermic injection (see Section 20.4.4.1.2), while some are aphallic and a few have been shown, uniquely among euthyneurans, to be gonochoristic (see review by Schrödl & Neusser 2010).

Stylommatophoran reproductive systems have some special features (Figures 20.52, 20.57, and 20.59) including an epiphallus which contains endogenous sperm and builds the spermatophore. The spermatophore tail is formed by the flagellum. A diverticulum off the bursal duct is present in some taxa, and this receives the spermatophore; otherwise it is received by the bursal duct and bursa copulatrix, and this is where the spermatophore and excess sperm are digested. Sperm swim out via the spermatophore tail and enter the spermoviduct to make their way to the seminal receptacle(s) associated with the carrefour.

The sperm is mixed with secretions from the prostate gland which is tubular or in some species more bulbous; it is lined with secretory epithelium and forms part of the vas deferens. In many marine euthyneurans the prostate is often not markedly thicker than the rest of the vas deferens and is sometimes overlooked (Wägele & Willan 2000). The polycerid nudibranch *Plocamopherus* and some sacoglossans have a branched prostate gland.

Whether or not the prostate gland is homologous in all euthyneurans is unclear, but it typically contains similar secretory cells throughout the group. These cells contain corpuscular proteinaceous, strongly eosinophilic secretions, although other secretions may also be present. In the Sacoglossa, Aplysiida, and Diaphanidae the prostate gland tends to be pallial, and the gonoduct is oodiaulic.

As noted previously, in many cephalaspideans, pteropods, and some nudipleurans, the prostate gland differs from that of other heterobranchs in forming a separate diverticulum associated with the copulatory apparatus. In these taxa, the most primitive arrangement (seen in some cephalaspideans and gymnosomes) is one in which an open seminal groove extends to the tip of the penis, with the prostate gland liberating its secretions into this groove. Modifications to this arrangement include the closure of the seminal groove to form a sperm duct (vas deferens, or ejaculatory duct if it is muscular) and the formation of a spermatic bulb for sperm storage.

The penis is plesiomorphically simple, but within the nudipleurans, some ornamentation is generally present, this often taking the form of rows of cuticularised hooks arranged in lines or spirals. Cuticularised hooks on the penis are also found in some aplysiidans, but they are not present in umbraculoideans or pleurobranchoideans (Wägele & Willan 2000). A single distal cuticular spine that may be hollow or solid is present in some aeolidioideans (Gosliner 1994).

In acteonoideans, the penis is external and housed in the mantle cavity when at rest, and in *Umbraculum* and *Tylodina* the penis is permanently protracted and lies in a cleft in the middle of the anterior end of the large foot (Willan 1987). In most other euthyneurans the penis can be retracted, using attached retractor muscles, into a penial sheath that lies in the

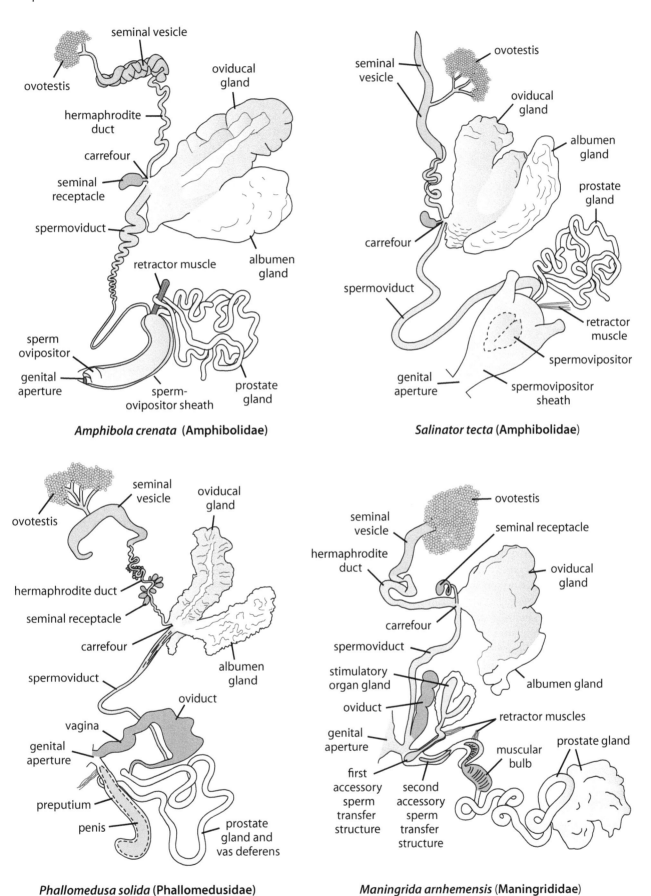

FIGURE 20.54 Reproductive systems of Amphiboloidea. Redrawn and modified from Golding, R.E. et al., *Zootaxa*, 50, 1–50, 2007 and Golding R.E. et al., *J. Morphol.*, 270, 558–587, 2010.

FIGURE 20.55   A scanning electron micrograph of the complex penis of *Phallomedusa solida*. Courtesy of R. Golding.

haemocoel. *Siphonaria* lacks a penis, and, in amphiboloideans, the male copulatory apparatus is remarkably diverse (see below).

In plesiomorphic euthyneuran taxa the copulatory apparatus consists of a retractile penis located on the right side of the head with a ciliated groove which extends to the opening of the spermoviduct on the right side of the body from where the sperm from the ovotestis enter the groove. In various more apomorphic lineages the groove has been transformed into a closed duct (vas deferens or sperm duct), and this duct can become muscular. The part of the vas deferens that carries the sperm through the penis is called the penial duct or, if highly muscular, the ejaculatory duct. In some taxa the penial apparatus has a sperm-storing '*spermatic bulb*', and in some of these, it forms spermatophores.

In most euthyneurans the penis is retractile, being enclosed in a penial sheath and attached to one or more retractor muscles. It can undergo considerable modification in size, shape, and ornament with features such as spines or a stylet sometimes being present. The penial stylet may also bear spines or hooks, as in the onchidid *Platyvindex mortoni* (Zhang et al. 2017). The most bizarre molluscan penis is that of the amphiboloidean *Phallomedusa* (Golding et al. 2007, 2010) (Figure 20.55), with the members of that superfamily having a much wider range of copulatory apparatus than seen in other panpulmonates.

In some pleurobranchs and nudibranchs a bulb-like *penial gland* is connected to the penial sheath and opens into the penis, but little is known about the function of this structure (Wägele & Willan 2000). Superficially similar *penial bulbs* are known from systellommatophorans (Tillier 1984a).

In an unusual elaboration, the aeolidioidean *Pruvotfolia*, which has a simple penis, has three sets of papillae (modified cerata) around the genital aperture which apparently assist in sperm transfer (Tardy 1969).

In euopisthobranchs, the Retusidae, Philinoglossidae, Bullidae, Atyidae, and Runcinidae all have a copulatory apparatus that stores sperm and has a prostate gland attached. In some cephalaspideans the prostate gland produces spermatophores, the prostate duct functions as an ejaculatory duct, and a penis may or may not be present (e.g., some Haminoeidae, Bullidae, Runcinidae, Retusidae) (Ghiselin 1966; Robertson 1989). Spermatophores are also produced in pyramidellids, architectonicids, thecosome and possibly the gymnosome pteropods, Acochlidia, and *Aeolidiella* (Aeolidioidea), Siphonarioidea, Chilinidae, Lymnaeidae, and many stylommatophorans (Robertson 1989, 2007) (Figure 20.53).

The female system of sacoglossans is particularly apomorphic. Many have a modified series of often multiple seminal receptacles. It appears to be primitively oodiaulic (as in *Cylindrobulla*), with most of the more derived sacoglossans being triaulic.

In some sacoglossans, when the ova leave the hermaphrodite duct and enter the oviduct, instead of entering the membrane gland, they move through a loop to the opposite end of the membrane gland, where they are fertilised by sperm from modified sperm sacs. They are then coated with the secretions from the albumen gland, covered by a membrane, and then the egg mass is completed in the mucous gland. Ghiselin (1966) suggested this rather unusual and complex configuration is due to 'historical accident' rather than being an optimal one and suggested how it could be derived from a primitive 'opisthobranch' configuration.

The panpulmonate Amphiboloidea are an estuarine group allied to the Glacidorboidea, although the placement of glacidorbids has been controversial (see Section 20.2). As shown in Figures 20.54 and 20.56, the reproductive systems of amphibolids and those of the siphonariids differ markedly from those of other lung-bearing panpulmonates so their main features are briefly described next.

Amphiboloideans have very diverse reproductive systems (Figure 20.54) and radulae (Golding et al. 2007, 2010). Of the three families now recognised, each has a very different reproductive anatomy, but all have a single genital opening, and, like the Glacidorbidae (Figure 20.45), none have a bursa copulatrix. The Amphibolidae have a monaulic system with a simple, narrow spermoviduct. A long tubular prostate gland separate from the spermoviduct joins the genital atrium where there is a unique spoon- or funnel-shaped structure called the spermovipositor which can retract into a sheath. Copulation is non-reciprocal, and sperm bundles are transferred, but these do not have a firm coating like that seen in spermatophores. In contrast, the other amphiboloidean families, Phallomedusidae and Maningrididae, are diaulic. Phallomedusids have a massive, spiral, tentaculate penis (Figure 20.55) unlike that of any other gastropod. *Maningrida* has two structures referred to as 'sperm transfer structures' (Golding et al. 2007, 2010). One of these structures probably acts as a penis and has a prostate gland attached, and the other possibly acts as a stimulatory organ and is covered with glandular bulbs. Copulation in the phallomedusids is reciprocal but has not been observed in *Maningrida*.

The condition in siphonariids is of interest as the copulatory organ is absent and its function taken over by the muscular vagina which is capable of some evagination. The basal part of the bursal duct has widened, and an epiphallus gland that produces spermatophores is present, as is a bursa copulatrix.

The protandric glacidorbids (Figure 20.46) have a diaulic system that lacks a bursa copulatrix and has two reproductive openings. The anterior one opens to the large praeputium which, in one species at least, is covered in large sucker-like structures and contains a small penis. Once copulation occurs, the female system develops, the penis is lost, and embryos are brooded in the pallial oviduct (Ponder 1986).

In hygrophilans and other so-called basommatophorans (Figure 20.56), the mainly non-glandular spermoviduct is largely undifferentiated, this contrasting with the situation in stylommatophorans where there has been extensive modification and functional diversification.

The layout of the stylommatophoran reproductive system is markedly different from other taxa and has led to some highly speculative scenarios to explain their evolution. For example, Visser (1977) suggested that the non-stylommatophorans were derived from 'male ancestors' and stylommatophorans from 'female ancestors' and that both independently acquired hermaphroditism by the elimination of either the male or the female from their gonochoristic ancestor. He argued that the vas deferens and prostate (and their equivalents) in the non-stylommatophorans (i.e., 'basommatophorans') and

stylommatophorans were not homologous. At a later time, after a study involving the ontogeny of their reproductive systems, Visser (1988) concluded that 'pulmonate' reproductive systems are developed from 'intermediate stages in the life cycle of a sequential protandrous prosobranch ancestor' with the 'basommatophoran' ditrematic condition (having two openings) being neotenous.

The basic plan of the stylommatophoran genital duct is shown in Figures 20.52 and 20.57. In some (not all) stylommatophorans exosperm is stored in one or more sperm sacs (the 'spermatheca') associated with the carrefour located at the junction of the hermaphrodite duct and the spermoviduct. In clausiliids and the 'orthurethran' taxa, the carrefour consists of only the fertilisation pouch while in most helicoideans there are multiple sperm pouches (Beese et al. 2008). Such seminal receptacle complexes, as well as epiphallus[6]-penis complexes (see text following), are present in stylommatophorans but are lacking in other lung-bearing panpulmonates. The general layout of the viscera, including the reproductive system, is dramatically modified with the adoption of a slug form (limacisation) (Figure 20.58).

Stylommatophorans have a genital atrium that acts as an oviducal opening and a vagina, and it is also the opening into which the penis sheath opens and through which the penis is protruded (the monotrematic condition). Even in the least modified stylommatophorans, the vas deferens runs from the penis as a separate duct and enters the spermoviduct posterior to the opening, a condition that can be considered semidiaulic.

There are several accessory structures associated with the penis complex in stylommatophorans, for example the dart sac and associated digitate glands in helicoideans and some other groups (Figure 20.57).

The spermoviduct has a sperm groove carrying endogenous sperm, and in many a separate groove, the seminal groove, that carries exogenous sperm. In many stylommatophorans, various degrees of separation occur so these grooves are separated incompletely or completely as ducts from the spermoviduct. Thus, in stylommatophorans, the terms monaulic, diaulic, and triaulic are usually not readily applied, in contrast to many marine euthyneurans and non-stylommatophoran lung-bearing panpulmonates, and the conditions found are not necessarily derived in a similar fashion. Thus, stylommatophorans with a three-duct system differ from similar modifications in non-stylommatophoran lung-bearing panpulmonates in that it is the middle part of the seminal groove that becomes separated from the spermoviduct, not the terminal part.

Large prostatic follicles embedded in the wall of the spermoviduct open into the sperm groove in many stylommatophorans, while in others the prostate gland is a separate structure. In some, such as the slugs *Deroceras*, *Arion*, and *Philomycus* (Visser 1977), there are, besides the prostatic follicles, flask-shaped gland cells that discharge into the sperm duct posteriorly.

---

[6] As noted above, an epiphallus is also present in siphonariids, but that group lacks a penis.

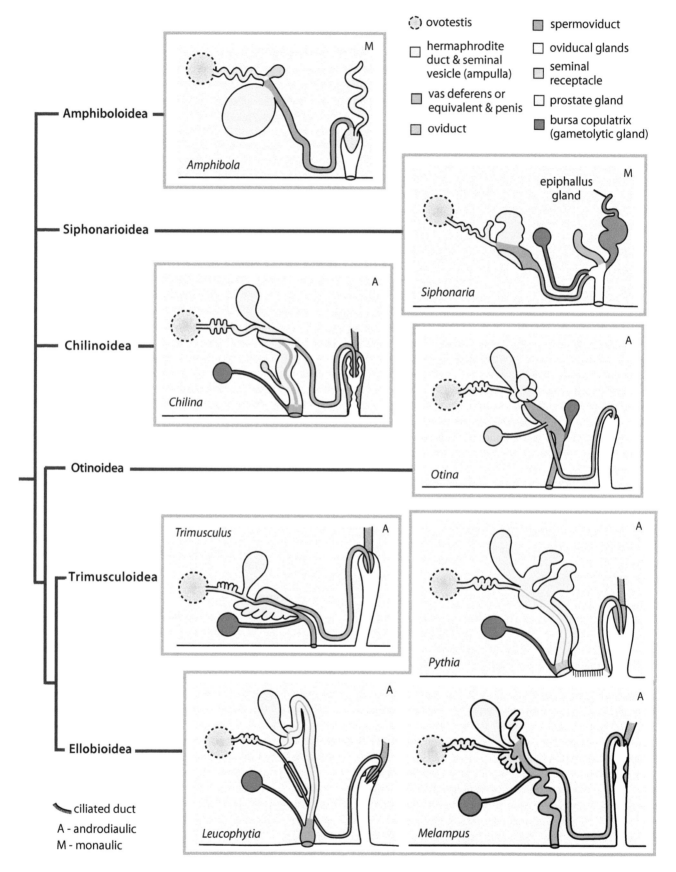

FIGURE 20.56 Diagrammatic representations of the reproductive systems of lower panpulmonates and basal eupulmonates. Redrawn and modified from Berry, A.J., Gastropoda: Pulmonata, pp. 181–226, in Giese, A.C. and Pearse, J. S. (eds.), *Reproduction of Marine Invertebrates. Molluscs: Gastropods and Cephalopods*, Vol. 4, Academic Press, New York, 1977.

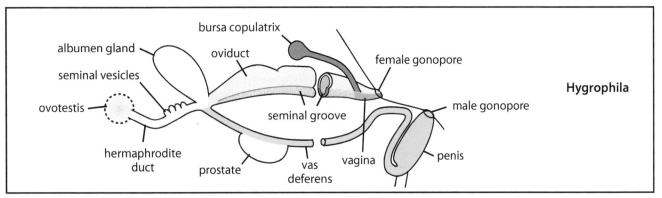

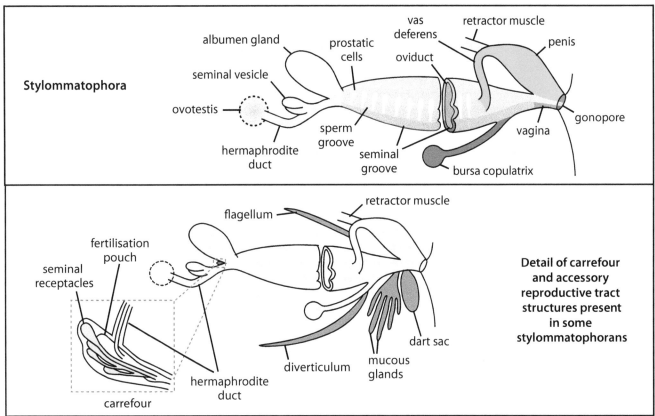

FIGURE 20.57 Comparison of a hygrophilan and a stylommatophoran reproductive system and schematically showing the accessory structures in some stylommatophorans. Upper two figures redrawn and modified from Visser, M.H.C., *Zool. Scr.*, 6, 43–54, 1977, bottom figure redrawn and modified from Beese, K. et al., *J. Zool. Syst. Evol. Res.*, 47, 49–60, 2008.

In some so-called triaulic forms, such as in species of Clausiliidae and *Zonitoides* (Figure 20.59), there is a connection between the duct of the bursa copulatrix and the seminal groove.

In some stylommatophoran slugs, such as *Agriolimax* and *Deroceras*, the spermoviduct lacks a seminal groove and has only the oviduct and sperm groove, the latter forming the vas deferens distally. This configuration probably resulted from a condition where the seminal groove had become separated as a diverticulum which was subsequently lost (Visser 1977). Where the seminal groove or duct is lost, the oviduct provides a passage for exogenous sperm to the carrefour and thus it has secondarily reverted to becoming a spermoviduct.

Some stylommatophorans such as Athoracophoridae and Succineidae can be described as diaulic, and this is also the predominant condition in Achatinellidae. Some achatinellids are viviparous, resulting in modifications to the oviduct, although not all modifications are in the viviparous forms. For example, in oviparous achatinellids, there is a distinct prostate gland with few to many large digitate follicles, and it is usually larger than the albumen gland. In the viviparous achatinellids, the prostate gland is conspicuously smaller and, sometimes, consists solely of small inconspicuous follicles on the vas deferens. The bursal duct in viviparous taxa enters the oviduct in the distal half while in oviparous forms it is proximal (Cook & Kondo 1960).

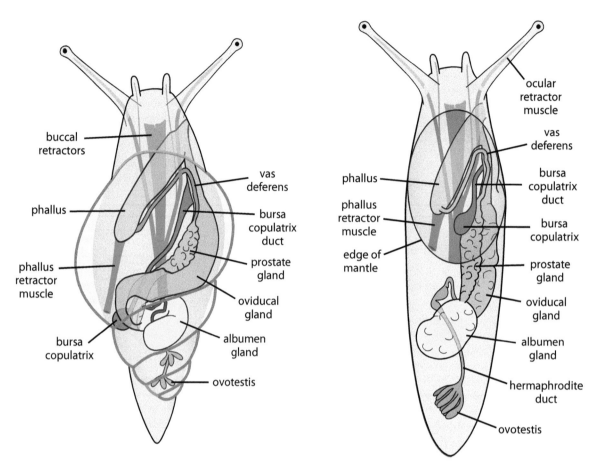

FIGURE 20.58 Comparison of the layout of the reproductive systems in a generalised stylommatophoran snail and slug. Redrawn and modified from Barker, G.M., *Naturalised Terrestrial Stylommatophora (Mollusca: Gastropoda)*, Vol. 38, Manaaki Whenua Press, Otago, New Zealand, 1999.

Some Succineidae have the sperm duct nearly separated from the oviduct + seminal duct[7] with only the most proximal part a spermoviduct. Thus, while the condition seen in these taxa is effectively diaulic, it is not fully so. They also differ from other stylommatophorans in having a discrete prostate gland made up of branching tubules which open, via a single duct controlled by a sphincter muscle, into the vas deferens (Rigby 1965; Visser 1977).

Beese et al. (2008) studied the evolution of the carrefour in stylommatophorans and showed that seminal receptacles associated with that structure originated more than once and were secondarily lost in several lineages. The fertilisation pouch was also lost in some lineages.

20.3.7.5.1 Accessory Mating Structures

Chitinous spines are associated with the vagina of a few cryptobranch nudibranchs, and many other nudibranchs have *vaginal* or *vestibular glands*. These glands are commonly stalked and sac-like, but some may be branched. In a few they are elongate. In some taxa, these glands contain a stylet located near the opening of the male aperture (Gosliner 1996). Other stimulatory organs are found associated with the male system and include the so-called 'love darts' in higher stylommatophorans as well as special glands or other structures associated with the copulatory apparatus. Mating in terrestrial euthyneurans may involve elaborate courtship, and some stylommatophoran land snails (e.g., Helicidae) exchange 'love darts' (see below), and some Camaenidae have a head-wart, an eversible glandular structure on the top of their head (see below) that is everted during courtship.

Examples of accessory structures seen in some stylommatophoran reproductive systems are shown in Figure 20.57, and some are discussed next. These include various modifications to the copulatory apparatus frequently referred to collectively as stimulatory structures (e.g., Tompa 1984), although that term may not accurately reflect the real function of at least several of them. They fall into two main types: (1) the *sarcobelum*, a fleshy club-like appendage arising from the vicinity of the penis, and (2) the *gypsobelum*, a hard, sharp calcified or chitinous structure usually arising from the female side (male side in zonitids), which notably includes the dart sac (see below).

The sarcobelum is a protuberance associated with the male genitalia found in many stylommatophorans (and also

---

[7] Visser (1977) argued that the seminal duct formed the vas deferens in succineids, but given the relationship with the prostate gland, for example, we consider that its derivation from the sperm duct is more likely.

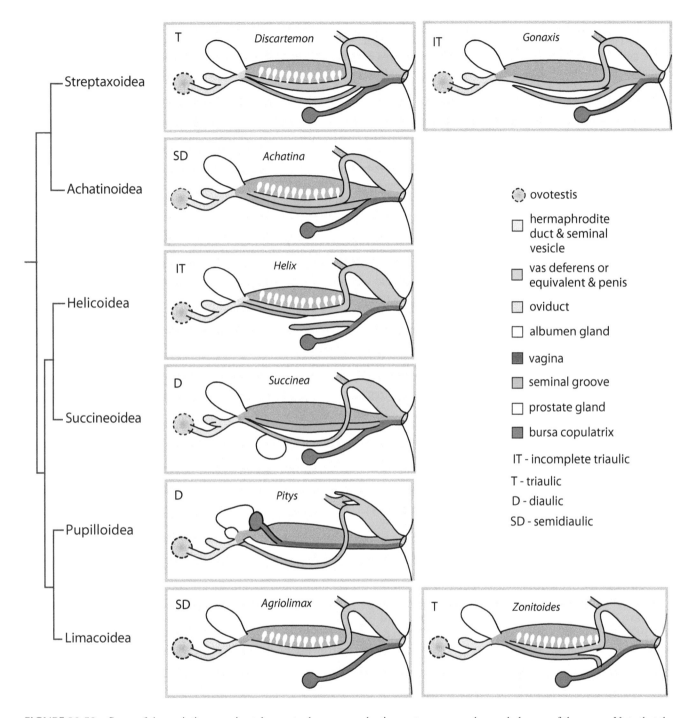

FIGURE 20.59 Some of the variation seen in stylommatophoran reproductive systems mapped on a phylogeny of the group. Note that the illustrations are not intended to convey the range of morphologies found in each of the higher taxa represented on the tree as several configurations are found in most of these. For example, in Clausiliidae, similar conditions can be found to those exhibited by *Gonaxis*, *Helix*, and *Agriolimax*. Redrawn and modified from Visser, M.H.C., *Zool. Scr.*, 6, 43–54, 1977.

Nudipleura). This structure plays a role in the positioning of the snail before copulation and stimulates the partner (Chase et al. 2010).

Other sarcobelum-like structures are more elaborate. Finger-like glands can be everted to stroke the dorsal surface of the mate during courtship in some slugs such as *Agriolimax* (Sirgel 1973) and *Deroceras*. In *D. panormitanum* (and presumably in other species) this structure is everted and applied

to the back of each partner after reciprocal sperm exchange. This application may produce a chemical signal to discourage other potential suitors, or it may have other effects such as inhibiting sperm digestion or even discouraging the partner from further mating to give the sperm a greater chance of fertilisation success (Benke et al. 2010).

In some taxa, complex structures with possibly similar functions to sarcobela are associated with the genital atrium rather than the copulatory organ. In *Milax gagates* there is a coiled muscular 'coniform organ' associated with the genital atrium, and there are similar (non-homologous) structures in veronicellids. In the slug *Arion hortensis*, the club-shaped end of the oviduct is everted during courtship and caresses the body of the partner (Quick 1960).

The gypsobelum includes the dart sac and similar structures. The so-called 'love dart', a sharp calcareous structure used to pierce the skin of the partner during courtship, is found in only a few stylommatophoran families (Tompa 1984; Davison et al. 2005; Koene & Schulenburg 2005). There is much diversity between taxa, even within closely related groups, and it has been lost in several lineages. The sacs show considerable structural diversity, and the darts they contain range from simple cones to elaborately bladed structures (Koene & Schulenburg 2005).

The dart sac consists of a pair of digitate glands and a muscular dart sac (*stylophore*) in which the dart is secreted and housed. The dart is forcefully everted from the dart sac before which the digitate glands have coated it with a secretion ('mucus' plus active agents).

The function of the 'love dart' has been long disputed (see Tompa 1984 for historical summary). Several hypotheses have been tested and found wanting. These were that dart shooting:

- Provided a 'nuptial gift' of calcium, but this was shown to be very unlikely by Koene and Chase (1998b), and, in some taxa, the dart is retained by the shooter and reused on the next mate (Koene & Schulenburg 2005).
- Provision of sexual stimulation – the dart is shot late in courtship, and the partner fires its dart shortly after it receives one with no difference in the duration of courtship as a result of either dart successfully penetrating (Chase & Vaga 2006).
- Aids species recognition – however, successful mating occurs even if no dart is shot (Chase 2007).
- Gave a signal of intention to copulate (Leonard 1992), but this idea has been falsified (see Adamo & Chase 1996; Chase 2007).

Instead, it appears that it is used to influence the success of the donated sperm (Chung 1987; Adamo & Chase 1996; Rogers &

Chase 2001; Chase 2007) by way of a chemical substance (an allohormone – see Koene & Maat 2001) in mucus covering the dart (Koene & Chase 1998a). The mucus and the allohormone are produced by the digitiform mucous glands associated with the dart sac. The allohormone is released into the blood system when the dart enters the body. The allohormone in *Cornu aspersum* causes the bursal duct diverticulum to become more accessible to the spermatophore and results in the closure of the duct of the bursa copulatrix, a sperm-digesting organ (Koene & Chase 1998a; Chase & Blanchard 2006).

The majority of studies on 'love darts' have been conducted on the common garden snail *Cornu aspersum*. During courtship, the dart sac (*stylophore*) is everted, and the calcareous dart is thrust into the side of the partner. Normally a dart is shot from both partners before the eversion of their penes. When this occurs, the penis of both partners is inserted into the partner simultaneously, and spermatophores are exchanged. The spermatophore is transferred either directly into the bursa copulatrix or an associated diverticulum. Sperm actively swim out of the tail of the spermatophore into the vagina and to the spermathecae where they are stored (Lind 1973), but before they get there enzymes usually break down about 99.9% of the exosperm (Lind 1973; Rogers & Chase 2001).

Some interesting questions remain regarding the evolution and function of the 'love dart' (see Chase 2007), not the least being how and why this structure first evolved. In some taxa from a few families, the dart is hollow and clearly acts as a hypodermic syringe (Tompa 1984).

Some stylommatophorans, notably camaenids (*sensu lato*) (Helicoidea) and some Urocyclidae (Helicarionoidea), have a cephalic head-wart (also called a frontal organ or cephalic accessory organ) located dorsally and commonly between the optic tentacles and is often erectile. The development of the head-wart is under hormonal control, and it is thought to produce sex pheromones that play a role in stimulation of the partner (Miles 1961; Binder 1969; Takeda & Tsuroka 1979; Takeda 1980, 1982; Winter 2008). There is also a 'cephalic dimple' in *Achatina* which has a similar function (Takeda 1982).

The enlargement in the buccal mass and often associated narrowing of the head and neck of carnivorous stylommatophorans (Figure 20.60) led to suggestions that there was little room for the development of accessory structures (e.g., notably the dart sac) associated with the male genitalia and

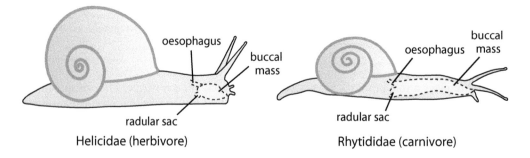

Helicidae (herbivore)   **Rhytididae (carnivore)**

FIGURE 20.60 Change in buccal mass size and anterior part of the body with the adoption of carnivory. Redrawn and modified from Beesley, P.L., Ross, G.J.B., and Wells, A. (eds.), *Mollusca: The Southern Synthesis, Part B*, CSIRO Publishing, Melbourne, VIC, 1998b.

this enlargment is responsible for the convergent loss of such structures (Tillier 1989). A later analysis by Barker and Efford (2002) showed no evidence these structural changes were linked.

As seen from this overview, the evolution of the genital systems of euthyneurans has generally included increasing complexity in the more derived taxa. This increase in complexity has involved in part the addition of accessory structures in some lineages but more generally there is a trend to separate the parts of the gonoduct carrying out different functions (see Chapter 8 for a general introduction to this topic).

20.4 BIOLOGY, ECOLOGY, AND BEHAVIOUR

20.4.1 HABITS AND HABITATS

Heterobranchs occupy the largest range of habitats of any gastropod group. The 'lower heterobranchs' are marine except for one group, the freshwater Valvatidae. The nudipleurans and euopisthobranchs are very successful, and almost exclusively marine, with only a few brackish-water taxa. Some are minute, and a few have become interstitial with highly modified (simplified), minute bodies. A few marine heterobranchs live in the deep sea, but the majority live in relatively shallow water. The pteropods are pelagic, as are a few nudibranchs (see next Section).

Despite the wide diversity of freshwater lung-bearing panpulmonate snails and limpets, the only known freshwater slugs are a few Acochlidia.[8] The ancestral Acochlidia were marine and interstitial (mesopsammic). There are two lineages which independently colonised fresh water (Schrödl & Neusser 2010); one is in the Caribbean (where it is represented by the small interstitial *Tantulum elegans* that live in mountain spring-fed swamps) and the other in the Indo-Pacific, where relatively large-sized benthic freshwater acochlidians are found living in some coastal streams. One small acochlidian group, Aitengidae, has amphibious and a terrestrial species (Kano et al. 2015), representing an independent acquisition of a terrestrial habit.

Most lower lung-bearing panpulmonates occupy estuarine or intertidal marine habitats. All amphibolids are estuarine, and siphonariids range from marine to estuarine habitats. Hygrophilans have radiated extensively in freshwater and, in the eupulmonates, ellobioideans are mostly estuarine, many inhabiting subtropical to tropical mangrove and saltmarsh. Some temperate species live in the upper littoral zone on exposed rocky shores, and one northern-hemisphere group (Carychiidae) is terrestrial. The related Trimusculidae, Otinidae, and Smeagolidae are marine intertidal. Other eupulmonates have undergone substantial terrestrial radiations, where they occupy all but the most extreme environments. The systellommatophoran slugs range from intertidal to supralittoral marine and estuarine to terrestrial. The Rathouisiidae and Veronicellidae are terrestrial, but most

onchidiids are marine and intertidal, with many living in mangrove and saltmarsh habitats and on rocky coasts. Two brackish water taxa can live in fresh water (*Onchidium typhae* from India and *Labbella ajuthiae* from Thailand), and two terrestrial species live in high-elevation rainforests (*Platevindex ponsonbyi* in Borneo and *Semperoncis montana* in the Philippines) (Dayrat 2009).

The great majority of terrestrial eupulmonates, the stylommatophorans, range from sub-Antarctic and sub-Arctic to tropical savannah, grasslands, and deserts but the greatest diversity is seen in moist habitats in temperate and tropical regions. Most are ground-dwellers, but there are also many arboreal species, and some are subterranean burrowers.

The physiological and other adaptations necessary for intertidal, freshwater, and terrestrial life are mainly discussed in Chapters 6 and 9. Some specific adaptations in different heterobranch groups are briefly discussed below.

20.4.2 LOCOMOTION

Heterobranchs mostly use their foot to glide on cilia and/or pedal waves. Movement is usually by cilia in heterobranchs, although muscular waves are employed in some of the larger taxa, such as several species of *Aplysia* and most terrestrial taxa. Modifications include an inch-worm mode of locomotion in a few taxa such as *Aplysia badistes*, *Phyllaplysia taylori*, and *Tethys*, and a 'galloping' or 'looping' gait in some land snails (see Chapter 3). Actual rates of progression for some 'opisthobranchs' and 'pulmonates' are summarised by Hyman (1967), and Thompson (1976) gave the fastest rate for an 'opisthobranch' as between 40 and 50 cm/min in large *Aplysia*.

Some, such as *Philine*, *Acteon*, *Cylichna*, *Scaphander*, *Cylindrobulla*, and *Ringicula*, actively plough through sediments, using their cephalic shield, but a few, including *Philine* and *Scaphander*, also use their whole-body movements to burrow deeper in the sediment (W. Rudman, pers. comm., 2019). Others, such as the sacoglossan *Plakobranchus ocellatus* and the pleurobranchs *Euselenops* and *Pleurobranchaea*, have a head shield-like structure but do not burrow or plough.

A few marine heterobranchs have become swimmers (reviewed by Farmer 1970) or floaters. Some smaller benthic species can temporarily move upside down by cilial action attached to the surface film, in a similar way to many small vetigastropods and caenogastropods. *Glaucus*, an aeolidiid nudibranch of the otherwise benthic family Glaucidae, floats permanently on the surface of the sea feeding on pelagic cnidarians such as *Velella*, *Porpita*, and *Physalia*. It stays afloat by taking a bubble of air into its gut where it is maintained by strong sphincter muscles (Thompson & Bennett 1970). Another nudibranch, *Phylliroe* (Phylliroidae), is also permanently planktonic and feeds on hydromedusae and larvaceans; some species are bioluminescent. *Phylliroe* is leaf-like, has lost the foot sole, and swims using lateral waves of contraction that pass along the whole body. Some other marine heterobranchs can swim by flapping their parapodia as seen in Gastropteridae, some species of the sacoglossan *Elysia*, and some aplysiidans

---

[8] Acochlidia were previously included in the 'opisthobranchs', but recent molecular studies have shown them to be nested in the 'pulmonates'.

(e.g., *Aplysia*, *Akera*). The aplysiidan *Notarchus* has dorsally fused parapodia except for a small slit-like opening anteriorly. It employs jet propulsion by water being taken in through the anterior aperture and forced out through the same aperture by strong muscular contractions. The edges of the aperture can become siphon-like and can be directed ventrally causing the sea hare to rapidly rise off the substratum, and successive expulsions of water cause it to somersault its way through the water (Martin 1966; Schuhmacher 1973).

Some marine heterobranchs that lack parapodia can swim briefly by vigorously undulating their body, examples being the pleurobranch *Pleurobranchaea*, the cladobranch nudibranchs *Melibe*, *Tethys*, *Scyllaea*, *Bornella*, *Tritonia*, and *Dendronotus*, the doridian nudibranchs *Nembrotha*, *Plocamopherus*, and *Hexabranchus*, and the aeolidiid *Flabellina*. *Pleurobranchus* swims clumsily by flapping the foot while some aeolidiids, such as *Cumanotus*, swim with the aid of beating cerata, as is also the case with the sacoglossan *Cyerce nigra* (Thompson 1976). Besides body flexing, *Hexabranchus* also employs synchronous waves that pass along the wide mantle skirt (Thompson 1976). These short-term swimming activities are usually related to predator avoidance and, in the case of *Hexabranchus*, also result in a spectacular aposematic warning display. Sometimes, however, the reason for this activity is not known, as with the European *Pleurobranchus membranaceus* which periodically swarm in non-feeding and non-reproducing aggregations (Thompson 1976).

The thecosome and gymnosome pteropods are permanent members of the plankton. Their foot sole is rudimentary or lost. The thecosomes swim upside down flapping their parapodia (often called 'wings' or 'fins') and undergo diurnal migrations, rising to the surface at night. They are so common in the plankton that their shells form a pteropod ooze on some parts of the deep ocean floor. The gymnosome pteropods are found in the near-surface waters during the day. They have streamlined bodies, and they only use their parapodia in swimming. The parapodia are smaller than in thecosomes, but the movements of these structures are more controlled, and with narrower basal attachments they can make a 'small twist after each stroke so that the leading anterior edge is directed relatively more strongly downwards in a down-stroke and upwards on the return' (Morton 1958b p. 289). The result is an upward and forward thrust resulting in efficient and fast swimming, enabling gymnosomes to outswim their thecosome prey. Morton (1958b) likened the swimming style in thecosomes to 'rowing' and in gymnosomes to 'sculling'.

Locomotion on land presents its particular problems, and mucus is a critical ingredient. Large pedal glands are present in Geophila and also in the terrestrial ellobioidean, *Carychium*, although the development of these structures in the two groups may be convergent.

Stimulation of the anterior part of *Aplysia* using strong touch, a weak electric shock, contact with a predator, or the application of a chemical to the skin will cause the animal to turn and crawl rapidly away, while pinching the tail will cause it to contract the tail and undertake rapid galloping locomotion (Leonard & Lukowiak 1986). *Melibe* can be induced to swim

with similar stimulants to those applied to *Aplysia* (Lawrence & Watson 2002), with swimming lasting for a few seconds to up to 25 min. After long periods of swimming, *Melibe* will rest with its broad cerata spread over the surface (Lawrence & Watson 2002). Chemosensory detection of a predator will also initiate swimming in some taxa. For example, detection of a predatory starfish (*Pycnopodia helianthoides*) (by the rhinophores) will cause *Tritonia diomedea* to move away from the source (Wyeth & Willows 2006b; Wyeth et al. 2006) and sometimes swim (Wyeth & Willows 2006a).

Some land snails and slugs use writhing movements of their foot to deter predators, but one snail, the helicarionid *Ovachlamys fulgens*, uses vigorous movements of the foot to leap several body lengths when disturbed.

20.4.3 Food and Feeding

'Lower heterobranchs' show a range of diets from general surface browsing (e.g., Rissoellidae), algal cell piercing (Omalogyridae), to specialised carnivory (Architectonicoidea). The diets of members of several small-sized, poorly known families (e.g., Orbitestellidae) remain unknown, but the tofanellid *Graphis albida* has been reported feeding on sabellariid polychaetes (Killeen & Light 2000). Pyramidellids are ectoparasitic, mainly on other molluscs and on polychaetes.

Marine heterobranchs, apart from the herbivorous aplysiidans and sacoglossans and the suspension-feeding Thecosomata, are mainly specialised carnivores. Comprehensive information about nudibranch food items has been compiled (Thompson 1976; Nybakken & McDonald 1981; Todd 1981). While carnivory is plesiomorphic in nudipleurans, it is less certainly so in euopisthobranchs, with most workers arguing for carnivory (e.g., Haszprunar 1985b; Rudman & Willan 1998; Cimino & Ghiselin 1999) and a few for herbivory (e.g., Mikkelsen 1996). In a few cases, some marine heterobranchs have been shown to feed on symbiotic organisms in the cnidarians or sponges on which they live, rather than the tissues of the host (e.g., Rudman 1991a; Becerro et al. 2003).

Sacoglossans are suctorial herbivores, sucking out contents of the cells of the green algae on which they feed (see Section 20.4.3.1). Nothing is known of the diet or feeding behaviour of Acochlidia, other than Aitengidae which are insectivorous, and one observation of a freshwater species, *Strubellia wawrai*, feeding on *Neritina* egg capsules in captivity (Brenzinger et al. 2011a).

In contrast to most marine euthyneurans, lung-bearing panpulmonates predominantly feed on decaying vegetable matter, fungi, or are true herbivores with marine taxa scraping algae from rock surfaces (siphonariids) or feeding on muddy surface deposits (Amphiboloidea, Ellobioidea, Onchidiidae). Members of the systellommatophoran Veronicellidae are herbivores while the Rathouisiidae are carnivores. Many stylommatophorans are fairly generalist in their diets, a useful strategy as it helps to conserve energy (Speiser 2001). Some feed on decaying vegetable and/or fungal material, others are omnivores, but some preferentially eat living plants (herbivores), fungi (fungivores), or animal tissue (carnivores).

Those 'sea slugs' which carry symbiotic zooxanthellae or plastids in their tissues (see Section 20.4.7) may be mixotroph – obtaining nutrients from their symbionts and supplementing it with normal feeding, while some appear to rely entirely, or almost so, on the symbionts.

Many stylommatophoran snails and slugs have broad diets compared with most other heterobranchs, feeding on various organic materials, including green or dead plants, decomposing leaves, wood and bark, fungi, and algae as well as animal scats and rotting animal remains. Carnivorous species consume other gastropods and organisms such as earthworms and nematodes. Land snails commonly rasp empty shells, limestone or cement to obtain calcium.

20.4.3.1 Herbivory

Aplysiidans and some cephalaspideans, such as *Bulla* and *Haminoea*, are herbivorous algal feeders. *Aplysia punctata* prefers some algae over others (Carefoot 1967); such preferential feeding may not be uncommon but remains largely unstudied.

The tiny omalogyrids feed by piercing filamentous algal cells with their dagger-like radula teeth. A somewhat similar mode of feeding is seen in sacoglossans, which are all herbivorous except for a few species that feed on egg masses (see below). Links between their algal food and their evolution have been well studied (e.g., Jensen 1981, 1993b, 1994, 1997; Händeler & Wägele 2007; Maeda et al. 2012; Christa et al. 2014a). They are all specialised suctorial herbivores using their single row of sharp, dagger-like teeth to pierce the algal cell walls. The most primitive members, the shelled taxa, all feed on species of *Caulerpa*. The shell-less taxa have moved to other algae, mainly siphonalean or septate green algae (Ulvophyceae). Jensen has argued that co-speciation and host switching occurred during their evolution, but these findings need to be tested with molecular phylogenies.

Siphonariids feed primarily on algae grazed from intertidal rocks. The terrestrial veronicellids and many stylommatophorans are voracious herbivores, and some have become agricultural pests.

As with sponges and some other colonial animals, plants also use chemical defences. Some of the most potent plant toxins are various alkaloids. Experiments with a slug (*Arion lusitanicus*) show it has a high tolerance of plant toxins, and tolerance and detoxification can be further induced by feeding slugs with non-lethal doses of toxic alkaloids (Aguiar & Wink 2005). These abilities enable slugs to feed on plants that would be toxic to other animals.

20.4.3.2 Fungivory

Many stylommatophorans feed partly, and often preferentially, on fungi but little detailed information is available. Many detritivores are also ingesting many fungal hyphae, and these presumably form an important part of their diet.

20.4.3.3 Active Carnivory

Among the 'lower heterobranchs', the architectonicoideans are specialised cnidarian feeders. Most nudipleurans,

acteonoideans, and gymnosomes are carnivorous, as are many cephalaspideans. Some feed by grazing on sponges, cnidarians, bryozoans, or tunicates, and some are active carnivores, feeding on polychaetes (e.g., Acteonoidea), small bivalves (Philinidae), crustaceans, or other heterobranchs (Aglajidae).

Many cephalaspideans (Cylichnidae, Retusidae, Scaphandridae) feed on forams, molluscs, or polychaetes. Some have lost the radula and swallow large active prey whole, this being particularly apparent in aglajids which swallow prey of large size. For example, the aglajid *Melanochlamys* eats large nemerteans and polychaetes while some species of *Philinopsis* eat whole cephalaspideans (Rudman 1972a), as do some species of *Aglaja*, as well as consuming other heterobranchs. Species of *Gymnodoris* also feed on other heterobranchs (Knutson & Gosliner 2014). Most species of the polycerid genus *Roboastra* feed exclusively on the closely related genus *Tambja* (Pola et al. 2003). The tritonioidean *Melibe* also lacks a radula and captures and eats small crustaceans and other small animals using its large fringed cephalic hood like a net.

The primitive food source in Doridoidea (Holohepatica) was sponges, with a shift in some groups to other kinds of food such as bryozoans, ascidians, and even other heterobranchs (e.g., Cimino & Ghiselin 1999). Doridoideans have extensively radiated, but the mainly cnidarian-feeding Cladobranchia (see Goodheart 2017 for review) are, except the Aeolidida, less speciose.

Nudibranch feeding strategies differ considerably. In *Tritonia*, food is sliced off by the jaws, and the broad radula is mainly used to move food to the oesophagus. Similarly, cropping of hydroids and tubularians by facelinids, eubranchids, and other nudibranchs occurs using the jaws (Thompson 1976). In contrast, in many doridoidean nudibranchs (e.g., *Archidoris* and *Jorunna*), jaws are absent, and the prey is gathered, broken up, and passed into the oesophagus by radular action and associated musculature (Thompson 1976). Others (e.g., *Onchidoris*, *Adalaria*, and *Corambella*) feed on the soft parts of barnacles or bryozoans using their buccal mass musculature to break them up, and the contents are sucked out using a buccal pump (Thompson 1976).

While some families are specialist predators on a particular group, such as Dorididae, Discodorididae, and Chromodorididae on sponges and Arminidae and Tritoniidae on alcyonarian corals, many others are more versatile. Aeolidiids are primarily cnidarian feeders, but a few species feed on other aeolidiids or egg masses (see below). Most are hydrozoan feeders, but many aeolidiids feed on Actiniaria (sea anemones) and the pelagic *Glaucus* feeds on chondrophores and siphonophores, as does the pelagic *Fiona* (Fionidae), although the main food of *Fiona* is stalked barnacles (*Lepas*) growing on floating objects. Other families that include hydrozoan feeders are the aeolidioid Flabellinidae, Notaeolidiidae, most Fionidae, and Facelinidae, the Dendronotidae, Lomanotidae, Dotidae, and Heroidae. Many Proctonotidae and Onchidorididae feed on bryozoans (Ectoprocta), and a few nudibranchs, notably species of *Trapania* (Goniodorididae), feed on entoprocts

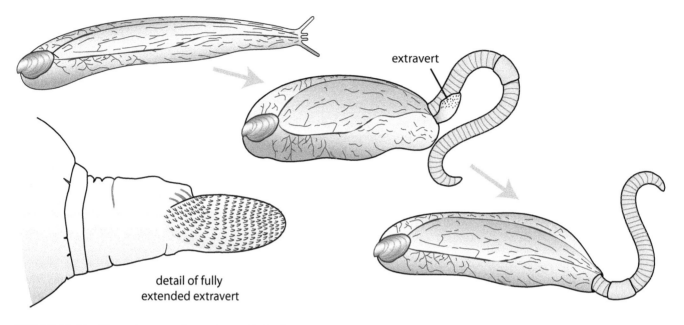

extravert

detail of fully
extended extravert

FIGURE 20.61   The stylommatophoran *Testacella* feeding on an earthworm using its everted buccal mass and hook-like teeth. Redrawn and modified from Webb, W.M., *The Zoologist*, 17, 281–289, 1893.

(kamptozoans). At least one onchidoridid (*Onchidoris bilamellata*) feeds on fixed barnacles (e.g., *Elminius, Balanus*). Members of the glaucid genus *Phyllodesmium* feed on octocorals, and several have symbiotic zooxanthellae (see Section 20.4.7 and Chapter 9).

While most nudibranchs feed on their favoured encrusting organism by radular scraping, dendrodoridids, phyllidiids, and mandeliids, which lack a radula, feed on sponges by sucking the sponge tissue through a proboscis-like modification of the buccal area. Pleurobranchs use an extendible proboscis-like buccal apparatus to penetrate ascidian tunics and suck out the contents.

The cladobranch *Dirona albolineata* feeds on small-shelled caenogastropods in shallow water but in deeper water is less selective, feeding on a range of invertebrates as well as coralline algae and detritus (Robilliard 1971). Arminids feed on pennatulaceans (sea pens) and soft corals while the related *Doridomorpha* (Doridomorphidae) feeds on *Heliopora* (blue coral), a massive octocoral.

The majority of doridoideans specialise in sponge feeding with some being very specific (e.g., some chromodoridids) (Rudman & Bergquist 2007). Thompson (1976) noted that those nudibranchs grazing on encrusting animals tend to be large and have a broad, flattened body with a wide foot while those feeding on more erect organisms, such as hydroids, actiniarians, and algae, are smaller and more elongate with a long, narrow foot. Many chromodoridids also have rather slender bodies.

Eating eggs is a specialised form of carnivory and has been adopted by members of several heterobranch families. A few aeolidiids have adopted this highly nutritious mode of feeding. Species of *Calma* (Calmidae) feed on cephalopod and fish eggs and are so adapted to this rich, nutritious diet that their anal opening is closed. Anal closure has not occurred in another aeolidiid, *Favorinus*, that eats the eggs of other marine euthyneurans. Egg feeding is also seen in a few sacoglossans (some species of *Stiliger, Calliopea,* and *Olea*) where they use their piercing radula to penetrate the egg capsule and suck out the contents.

Some terrestrial eupulmonate taxa (the systellommatophoran Rathouisiidae and several stylommatophoran groups) are predators, and those groups have convergent modifications to their gut (see below).

Barker and Efford (2002) extensively reviewed carnivorous terrestrial gastropods. Snail families that contain at least some carnivorous species include Subulinidae (e.g., *Rumina decollata* used in the control of helicid pests) and zonitids (notably *Oxychilus*). Some stylommatophorans are predators of other snails. Some of the better known include some members of the families Oleacinidae (the notorious *Euglandina* [see Chapter 10] and its relatives) and the closely related Spiraxidae, the American Haplotrematidae, Steptaxidae (and sometimes other invertebrates), and the Australasian and southern African Rhytididae (also worms and arthropods). Stylommatophorans which feed on other animals include the southern African Chlamydephoridae, the European and Middle Eastern Daudebardiinae, and the Haplotrematidae from the Americas, Testacellidae (*Testacella* – mainly earthworms – Figure 20.61), and Trigonochlamydidae. The Rathouisiidae feed on other gastropods but also include fungi and plant matter in their diet (Barker & Efford 2002). *Atopos* drills holes with its radula in the shells of the small land snails it feeds on (Schilthuizen et al. 2006). Rathouisiids can extend their buccal mass into a proboscis-like structure that can reach into the shell of the prey, and this ability is shared by some carnivorous stylommatophorans.

Even though many eupulmonates have become specialist carnivores, none have evolved an introvert proboscis and, unlike the situation in some other carnivorous euthyneurans, the gut is generally little modified.

Several groups of terrestrial snails and slugs, while being primarily herbivorous, contain species that may sometimes consume or scavenge insects, molluscs, or other animals. These include a few acavids, camaenids, helicarionids, helicids, partulids, polygyrids, sagdids, and vitrinids among the snails and the Agriolimacidae (*Deroceras*), arionids, limacids, milacids, and philomycids among the slugs (Barker & Efford 2002). One species of *Glacidorbis* is known to feed on injured invertebrates (Ponder 1986).

The animal or plant fed on often has defensive devices that need to be overcome. These include hard parts, such as an external shell, or internal or external spicules or spines, hard concretions, or other structures such as lignin in plants, noxious chemicals, or, with cnidarians, nematocysts. Heterobranchs have devised various ways of overcoming such defences. Some marine heterobranchs, for example, secrete copious mucus to surround the food to help protect against nematocysts, abrasion from spicules, and so on. Chemical deterrents are counteracted by species specialising in a particular food source and evolving the chemical pathways needed to detoxify the protective compounds (see Section 20.4.6.3.1 and Chapter 9). It is well known that some caenogastropods drill holes in the shells of prey (see Chapter 5), and this tactic has also evolved in a few heterobranchs. The polyceroidean nudibranch, *Vayssierea elegans,* drills into calcareous spirorbid and serpulid polychaete worm tubes using a secretion and the radula (Young 1969b). A few predatory stylommatophoran land snails, such as the North American *Haplotrema concavum* (Haplotrematidae), can drill holes in the shells of their prey (Pearce & Gaertner 1996).

Another form of carnivory is cannibalism which may be between young stages prior or post-hatching, or developing young eating eggs, and occurs in some marine and terrestrial (see Barker & Efford 2002) heterobranchs. Cannibalism also occurs between subadult or adult individuals in some aquatic and terrestrial taxa, with a detailed summary of records in terrestrial gastropods given by Barker and Efford (2002). Cannibalism has also been found in some 'opisthobranchs', including polycerids of the genus *Gymnodoris* (Knutson & Gosliner 2014).

Barker and Efford (2002) listed various modifications often associated with carnivorous terrestrial stylommatophorans involved in active prey capture. These include modification of the radular teeth which typically become longer and sharp, the buccal mass enlarges and becomes protrusible, the jaw may be lost, and there is an enlarged gastric crop. These changes are usually associated with elongation of the anterior body (Figure 20.60), the shell or limacisation. Other changes to the external body include enlargement of the suprapedal gland and elaboration of the locomotory foot sole. Other anatomical changes shown by many carnivorous stylommatophorans are a reduction in the terminal genitalia, retention of a long anterior nerve ring, the concentration of the cerebral ganglia, development of a sigmurethrous excretory system (see Section 20.3.3), and elaboration of the olfactory organs of the ocular peduncles.

While elongate radular teeth (Figure 20.39) are typical of carnivorous taxa, they are also seen in facultative carnivores and some omnivorous taxa. Some facultative carnivorous taxa can have short multicuspid teeth.

20.4.3.4 Parasitism

Pyramidelloideans, unlike any other euthyneurans, have a long introvert proboscis. The small pyramidellids are all suctorial ectoparasites on polychaetes and molluscs and are typically host specialists although some are more generalist, for example feeding on a variety of either molluscs or polychaetes. The prey of the larger tropical taxa such as *Pyramidella* is not known, but their anatomy shows they are also suctorial feeders.

A species of *Gymnodoris* has been observed feeding on the skin and mucus of resting shrimp gobies (*Ctenogobiops pomastictus*) (Williams & Williams 1986).

20.4.3.5 Suspension Feeding

Thecosome pteropods were thought to collect phytoplankton by way of cilia on their parapodia in an analogous fashion to the velar lobes of feeding veliger larvae, but they are now known to be mucous net feeders, using their extensive nets to entangle food as they drift in the plankton (Lalli & Gilmer 1989) (see Chapter 5). The marine limpet *Trimusculus* lives in intertidal caves and crevices and catches floating particles in a mucous net secreted by the anterior mantle edge. When a wave surge occurs the shell is lifted, water flows around the edges of the foot and then passes through the net, which is ingested from time to time by the odontophore (Walsby et al. 1973).

20.4.3.6 Coprophagy

Some terrestrial gastropods are attracted to, and feed on, faecal material (Cain 1983; Speiser 2001; Garvon & Bird 2005) because it is often a source of moisture as well as food, being rich in bacteria and other microorganisms, nutrients, and (if from a herbivore) plant particles. This feeding strategy has existed since at least the Upper Cretaceous where snails fed on dinosaur dung, as shown by their association with coprolites (Chin et al. 2009). Faecal feeding can also enhance transmission of parasites such as nematodes (e.g., Garvon & Bird 2005).

20.4.4 Reproduction, Development, and Growth

For a general account of the reproductive system and its basic terminology see Chapter 8 while a more detailed account is given in Section 20.3.7 above.

Mating strategies in marine euthyneurans have been well studied (e.g., Hadfield & Switzer-Dunlap 1984) with particularly detailed investigations in aplysiidans (e.g., Angeloni & Bradbury 1999; Angeloni et al. 2003; Yusa 2008), in some cephalaspideans (e.g., Leonard & Lukowiak 1985; Chaine & Angeloni 2005; Anthes & Michiels 2007a, b; Anthes et al. 2008), some nudibranchs (e.g., Karlsson & Haase 2002), and a few sacoglossans (Angeloni 2003; Gianguzza et al. 2004; Schmitt et al. 2007). Some mate when they are small, for example, the aeolidiid nudibranch *Tenellia lugubris* often mates when it is juvenile and stores sperm until it can produce eggs (Todd et al. 1997, as *Phestilla sibogae*), and small-sized individuals of the sacoglossan *Oxynoe olivacea* also often copulate (Gianguzza et al. 2004).

Mating between hermaphroditic species can be *unilateral* (one partner acts as the male, the other the female) or *reciprocal* (both partners act as both male and female).

All nudibranchs have a single reproductive event in their lifetime (i.e., are semelparous) (Wägele & Willan 2000). Unusual copulatory behaviour includes chain copulation involving several individuals forming a line, or even a ring, notably in aplysiidans (see also Chapter 8) and hypodermic insemination with penial stylets has evolved multiple times in marine heterobranchs (see Section 20.4.4.2).

Mating behaviour in stylommatophorans has been reviewed (Davison et al. 2005; Davison & Mordan 2007; Jordaens et al. 2009). In species with unilateral sperm transfer, the 'male' mounts the shell of its 'female' partner which will receive the sperm, but none of these 11 families have 'love darts'. Most simultaneous reciprocal-mating species (15 out of 18 families) mate 'face to face' within which about nine families have 'love darts' (see below) (Davison et al. 2005). Interestingly, there is also a correlation between mating behaviour and shell shape. The groups that mate 'face to face' are predominantly low-spired shells or slugs while those that shell-mount tend to be high-spired (Davison et al. 2005).

In some species of reciprocally mating land snails, one individual can be the initiator of copulation by mounting the shell of the comparatively passive partner. These behavioural differences can sometimes be related to body size (e.g., Dillen et al. 2010).

20.4.4.1 Self-Fertilisation

Self-fertilisation, or 'selfing', is the ability in some hermaphrodites to use their sperm to fertilise their eggs when a mate is hard to come by. This characteristic, along with direct development, particularly ovoviviparity, is an exaptation for the successful invasion of new areas as only one juvenile (or a fertilised egg in oviparous species) needs to be transported.

Self-fertilisation is common in stylommatophorans and has been demonstrated in all hygrophilans tested (Jarne et al. 1993). It is not reported in other lung-bearing panpulmonates and does not appear to occur in most marine euthyneurans, although selfing has been reported from two species of the sacoglossan genus *Berthelinia* (Kawaguti & Yamasu 1961; Grahame 1969). Pola and González Duarte (2008) recorded an instance of possible self-fertilisation in the nudibranch *Nembrotha kubaryana*, but otherwise, no other instances of selfing are known in nudibranchs, and there are very few reports in other marine euthyneurans, with most apparently self-incompatible (Hadfield & Switzer-Dunlap 1984).

It is not at all clear why some hermaphrodite taxa commonly self-fertilise, and others do not. Some avoid selfing by active copulatory behaviour, and it is also probable that the complex reproductive systems of nudipleurans, with the physical separation of both endogenous and exogenous sperm in

separate ducts, prevents selfing. Self-incompatibility is known among stylommatophorans (Tompa 1984). In marine euthyneurans, it has been suggested that sperm is activated only when the animal receives exogenous sperm (e.g., Thompson 1976; Beeman 1977), and in some a chemical process may be required to activate the capacitation of the sperm, differentiating exosperm and autosperm (Jarne et al. 1993).

In taxa where it occurs, the incidence of selfing can be negatively related to mate density but is usually delayed so it is a last resort. For example, *Physa acuta* delays selfing for two weeks (Tsitrone et al. 2003), and the minute *Alderia* delays for five to seven days. Thus there appears to be a balance between the costs of inbreeding and the costs of a delay in reproduction (Smolensky et al. 2009).

Despite the ability to self-fertilise, most studied natural populations outcross, presumably to avoid inbreeding depression although selfing or partial selfing may be the dominant mechanism in a few populations (e.g., Jarne et al. 1992; Armbruster & Schlegel 1994).

20.4.4.2 Hypodermic Insemination

Hypodermic insemination occurs in some acochlidians (Haase & Wawra 1996; Schrödl & Neusser 2010), cephalaspideans (e.g., Anthes & Michiels 2007b), sacoglossans (e.g., Trowbridge 1995; Angeloni 2003; Schmitt et al. 2007), and a few nudibranchs (e.g., Rivest 1984) may show hypodermal injection via hollow penial stylets. Sperm can be injected into the vaginal duct, into the genital system, or elsewhere into the body, with the location of the injection site affecting fertilisation success (Angeloni 2003).

The ability to carry out hypodermic insemination could allow an individual to inject sperm into its own body, thus facilitating selfing as has been observed in *Alderia willowi* (Smolensky et al. 2009).

Hedylopsid acochlidians impregnate hypodermically (Schrödl & Neusser 2010), and within that group there is a transitional series from a simple copulatory system to a complex one involving hypodermal injection, culminating in a large, spiny rapto-penis as seen in some of the freshwater Acochlidiidae. These latter systems involve a stylet for sperm injection and an additional injection system with an accessory gland. The marine *Hedylopsis ballantinei* is parthenogenic and possesses a penial stylet in the male phase and is aphallate in the female phase (Kohnert et al. 2011).

Hypodermic insemination with penial stylets has evolved multiple times in sacoglossans, many of which have a long penis, and most species that practise hypodermic insemination have a sharp penial stylet capable of piercing the body of the partner (Jensen 2001).

20.4.4.3 Aphally

Some euthyneurans lack a copulatory organ (i.e., are aphallic – in contrast to the possession of a copulatory organ – euphallic). While hermaphrodite euphallic individuals can potentially outcross as male or female, or self-fertilise, aphallic individuals can outcross only as females but are also capable of selfing. Aphally occurs in *Rhodope* (Brenzinger et al. 2011b), some

Hygrophila, and Stylommatophora but is not common. Some of the most studied aphallate taxa are certain species of the African planorbid genus *Bulinus* (e.g., Jarne et al. 1992).

Aphally has been reported in a few stylommatophorans including vertiginids (Pokryszko 1987) and clausiliids (Köhler & Burg Mayer 2016). In the latter case at least, it may have resulted from hybridisation.

As already noted, some acochlidians lack any penial apparatus, and some become aphallic during their ontogeny (Kohnert et al. 2011). Several of the aphallic taxa are gonochoristic, a unique condition among heterobranchs (Schrödl & Neusser 2010).

Dermal insemination using spermatophores occurs in the acochlidian *Pontohedyle* (Jörger et al. 2009), and Brenzinger et al. (2011b) suggested a similar process in the aphallic *Rhodope rousei*.

Using labelled sperm, Paraense (1976) showed that most exosperm are destroyed in the bursa copulatrix, with only a small proportion migrating along the female tract, and although there was no evidence of sperm build-up in the carrefour, exosperm accumulated in the ovotestis. He concluded that outcrossing occurred in the ovotestis but that selfing occurred in the seminal vesicles. Thus the exosperm must pass by the autosperm stored in the seminal vesicles. The mechanisms preventing their mixing are unknown.

20.4.4.4 Asexual Reproduction

Although apomictic parthenogenesis has been reported in the stylommatophoran slug *Deroceras laeve* (Nicklas & Hoffmann 1981), other cases of parthenogenesis in heterobranchs do not appear to have been recorded.

In the Pteropoda, members of the Cavoliniidae have long been known to have aberrant forms often referred to as 'skinny' or 'minute' (Spoel 1967). Some of these morphological oddities were shown to be preservation artefacts (Gilmer 1986; Lalli & Gilmer 1989). Spoel (1973) reported that some of these aberrant forms represented individuals that had reproduced asexually by splitting into two and referred to this form of reproduction as strobilation. Subsequent studies of both preserved and living individuals (Pafort-van Iersel & Spoel 1986; Pokora 1989) have revealed a complex form of reproduction in which asexual reproduction is part of an environmentally mediated reproductive strategy that also includes both protandry and self-fertilisation. When environmental conditions are favourable, the reproductive strategy of members of the Cavoliniidae consists of the outcrossing protandry, but when oceanographic conditions are unfavourable, the females of some taxa undergo fission, producing two individuals – the primary individual which once again returns to the male stage of the protandric cycle and the hermaphroditic, metamorphosed individual that self-fertilises (see Chapter 8, Figure 8.13). This form of asexual reproduction in the molluscs is only known in cavoliniid taxa although it resembles schizogamy, which occurs in some annelid worms. It is likely that the origins of the fission component of this complex reproductive strategy have some association with selection for autotomy in these pelagic molluscs (see Chapter 3). Janssen (1985) reported similar 'skinny' and minute forms in the

Miocene of the North Sea and Aquitaine Basins which suggested this reproductive strategy may have occurred throughout much of the Neogene.

20.4.4.5 Life Cycles

Most nudibranchs are semelparous – they have a single spawning period and then die. Some, mainly small cryptic species, tend to be opportunists and have several spawning periods within the same year (e.g., many small aeolidiids). Some larger species, such as most of those that live on encrusting animals, live for a year while others (e.g., as shown in some British species of *Tritonia*, *Archidoris*, and *Jorunna*) may live for two years before spawning (Todd 1981).

Mass migrations of marine heterobranchs are sometimes reported, and despite suspicions that these may not actually occur (Todd 1981), subtidal spawning aggregations and mass movements in *Onchidoris bilamellata* have been observed (Claverie & Kamenos 2008).

Lung-bearing panpulmonates have more variable lifespans than most marine euthyneuran groups, with siphonariids living for up to six years (Powell & Cummins 1985), hygrophilans all having short life spans, and stylommatophoran land snails and slugs varying from short-lived to 15 or more years (Heller 1990, 2001). Longer-lived species tend to be shelled, larger in size, and often live in variable and unpredictable environments (Heller 2001) (see also Chapter 9).

20.4.4.6 Reproductive Behaviour

Virtually nothing is known of reproductive behaviour in 'lower heterobranchs', but aplysiids, some nudibranchs, a few Hygrophila, and some stylommatophoran snails and slugs have been investigated. Mating is short or long in stylommatophorans but is short in hygrophilans. Complex courtship behaviour is seen in some terrestrial snails but is more common in slugs. These behaviours range from a pair circling each other to aerial mating seen in *Limax maximus* where the entwined pair is suspended from a tree branch for many hours.

Because heterobranchs are simultaneous hermaphrodites, self-fertilisation can occur in the absence of a mate in many lung-bearing panpulmonates (see Section 20.4.4.1) but is apparently rare in most marine euthyneurans (Thompson 1976). Behavioural and morphological adaptations co-evolved to efficiently keep incoming (exosperm) and outgoing autosperm separated, resulting in the highly complex diaulic and triaulic ducts seen in some euthyneurans (see Section 20.3.7.5 and Chapter 8).

Accessory spines and glands are found on, or associated with, the penis of many euthyneurans, including the 'love darts' in helicoidean snails (see Section 20.3.7.5.1). Hypodermic insemination, using a hollow penial stylet to pierce the side of the partner, is practised in a few taxa (see Section 20.4.4.1.2).

20.4.4.7 Development

Many 'lower heterobranchs', and marine euthyneurans, including the panpulmonate amphibolids, siphonariids, many onchidiids, and some ellobiids, retain a larval swimming stage, but most eupulmonates and some marine euthyneurans have direct development. Those with a free-swimming veliger larva may be lecithotrophic or planktotrophic. The larval shells in those heterobranchs with veliger larvae are typically heterostrophic whereas the heterostrophy is usually obscured in those taxa undergoing direct development. The eggs of freshwater and terrestrial lung-bearing panpulmonates are relatively large, their development modified by yolk-filled cells derived from the egg, as well as the formation of the podocyst and cephalic vesicle (see Chapter 8).

Unlike caenogastropods, heterobranch embryos lack polar lobes during the first stages of development.

20.4.4.8 Spawn

The spawn of 'lower heterobranchs' (Figure 20.62) is either capsules or gelatinous egg strings. The spherical egg capsules of valvatids contain several eggs connected by chalazae, but the spawn of the marine valvatoidean cornirostrids differs markedly in being a long, tightly coiled spiral mucoid string, and the eggs are apparently not connected by chalazae. Omalogyrids lay ovoid egg capsules attached to algae, and each contains one, or sometimes two, eggs (Fretter 1948). The apparently related *Architectonica* produces a long gelatinous string packed with minute capsules connected by chalazae and each containing a single egg (Bandel 1976) while at least some species of another architectonicid genus, *Philippia*, lay a gelatinous egg mass in the wide umbilicus of the shell which consists of several hundred capsules, each containing a single egg (Robertson 1970).

Capsules are also produced by a few small-sized euthyneuran taxa. Rissoellids lay hemispherical capsules containing one or two eggs attached to the substratum (Fretter 1948) and the egg masses of some pyramidellids are lens-shaped and also contain numerous egg capsules connected by chalazae (Figure 20.62).

Many marine euthyneuran egg masses have been rather well studied with their morphology reviewed by Hurst (1967) and Soliman (1987), and their ultrastructure has also been investigated (Eyster 1986; Klussmann-Kolb & Wägele 2001). These egg masses are not protected by the parent, so they have built-in protection against predators and environmental damage (see below and Chapter 8). They have a rather uniform ultrastructure with the embryos embedded in albumen in some Cephalaspidea *s. l.*, all Aplysiida and Sacoglossa, but in some other Cephalaspidea *s. l.*, and Nudibranchia the albumen forms an additional compact layer which dissolves during development and is utilised by the developing veligers in the capsules. There is only one embryo per capsule in most marine euthyneurans, but in the majority of aplysioideans, there are two to several. The embryos (and the albumen) are surrounded by a mucoid membrane, which forms the capsule, and these are embedded in a mucous matrix (Figure 20.63).

Capsules can be constructed with a single-layered membrane or with a double layer composed of a thin outer membrane and an inner albuminous layer (Figure 20.63). The capsules are usually surrounded by inner mucous layers which may be thick or thin and may or may not connect adjacent egg capsules. When the capsules are connected by the inner

mucus layer, these form the so-called chalazae (e.g., Dayrat & Tillier 2002), although fine strands between the eggs (not the capsules) have also been called chalazae (e.g., Haszprunar 1985b). Some egg ribbons (e.g., *Acanthodoris*, *Alderia*) have fine tube-like mucoid structures within them (Klussmann-Kolb & Wägele 2001). The egg capsules may be arranged haphazardly or, in some egg masses, in tube-like structures with the tubes folded within the egg mass (e.g., Figure 20.64).

The egg mass is covered with a multi-layered outer mucous sheath. The structure of this outer sheath differs in density and the thickness of the mucous layers in different taxa. It is probably the chemical and structural properties of this sheath that protect the egg capsules from desiccation and attack by, for example, algae, nematodes, and bacteria as well as the impact of ultraviolet light. There are also differences in the thickness and fine structure of the capsules and the detail of the mucous matrix in which they are embedded (Klussmann-Kolb & Wägele 2001).

Morphologically, benthic marine euthyneuran egg masses fall into four main types (Hurst 1967; Thompson 1976; Soliman 1987) (Figure 20.64):

1. Ovoid to globular sac attached by a stalk (as in *Acteocina*, *Philinopsis*, and many other cephalaspideans).
2. Egg strings, often attached to the substratum along one side (*Aplysia*, many aeolidiids [e.g., *Flabellina*], tritonioideans, and sacoglossans such as *Elysia*).
3. Ovoid or kidney-shaped jelly mass attached on one side (some small aeolidiids such as *Eubranchus*, the arminid *Dermatobranchus*, the aplysiid *Phyllaplysia*, the cephalaspidean *Haminoea*, etc.).
4. A ribbon attached by one edge (most doridoid nudibranchs, including *Chromodoris*, and *Polycera*).

The pelagic *Glaucus* has simple, short egg strings that are released into the sea and float. The egg string of another pelagic nudibranch, *Fiona*, is a spiral ribbon attached to a variety of floating objects (Lalli & Gilmer 1989). The gymnosome 'pteropods' have more or less spherical floating egg masses while thecosomes produce floating egg ribbons.

The spawn of non-stylommatophoran lung-bearing panpulmonates (Figure 20.65) is varied in form but not as markedly as in other marine euthyneurans. Amphiboloidean spawn varies from short, curved jelly masses attached to the substratum in *Salinator*, to near-circular jelly egg masses laid on the surface of the mud in *Amphibola*, to *Phallomedusa* which has long, narrow egg strings covered with sand (Golding et al. 2007) (Figure 20.65). Siphonariids lay jelly egg masses attached by one side to the substratum. They may be irregular and thick in section or coiled and sometimes narrow in section, somewhat resembling the ribbon spawn of many doridoideans (Chambers & McQuaid 1994). At least one Australian species, *Siphonaria virgulata*, has sausage-shaped pelagic egg masses (Creese 1980) (Figure 20.65).

Ellobiids produce jelly egg masses (e.g., *Cassidula*, *Leucophytia*) or strings (e.g., *Auricula*, *Ovatella*, *Ellobium*)

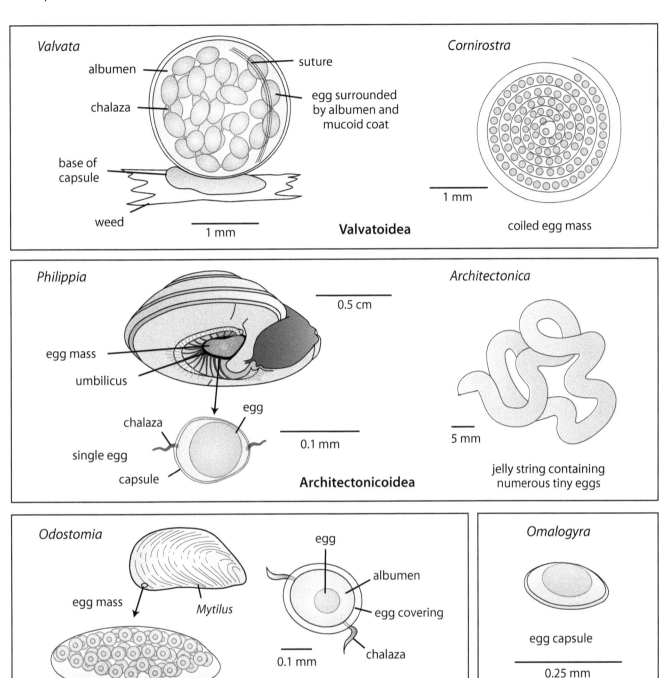

FIGURE 20.62 Spawn of 'lower heterobranchs' and the pyramidelloidean *Odostomia*. Redrawn and modified from the following sources: *Valvata*, *Omalogyra*, and *Odostomia* (Fretter, V. and Graham, A.L., *British Prosobranch Molluscs: Their Functional Anatomy and Ecology*, Ray Society, London, 1962), *Cornirostra* (Ponder, W.F., *J. Molluscan Stud.*, 56, 533–555, 1990b), *Architectonica* original, *Philippia* (Robertson, R., *Pac. Sci.*, 24, 66–83, 1970).

while *Carychium* lays batches of single capsules. The capsules within the egg masses or strings are connected by chalazae and contain a single egg (Morton 1955b; Berry et al. 1967; Duncan 1975).

Hygrophilans have simple (rounded to curved) egg masses attached to the substratum, and each egg mass contains a

relatively small number of eggs not connected by chalazae (Figure 20.65).

Onchidiids produce ovoid to irregular egg masses containing many eggs (a few tens to several thousand depending on the species) (Smith & Kenny 1987) connected by chalazae. In contrast, the eggs of the terrestrial veronicellids are laid

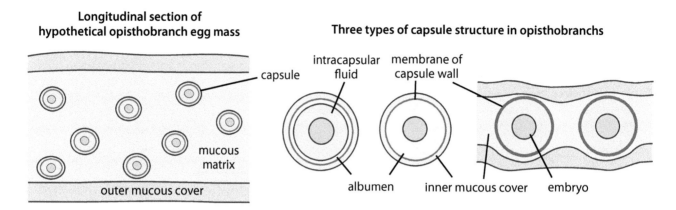

Longitudinal section of hypothetical opisthobranch egg mass

capsule
mucous matrix
outer mucous cover

Three types of capsule structure in opisthobranchs

intracapsular fluid
membrane of capsule wall
albumen
inner mucous cover
embryo

FIGURE 20.63 A section of a generalised egg mass and a comparison of three types of capsule structure in euopisthobranchs, nudipleurans, and sacoglossans. Redrawn and modified from Klussmann-Kolb, A. and Wägele, H., *Zool. Anz.*, 240, 101–118, 2001.

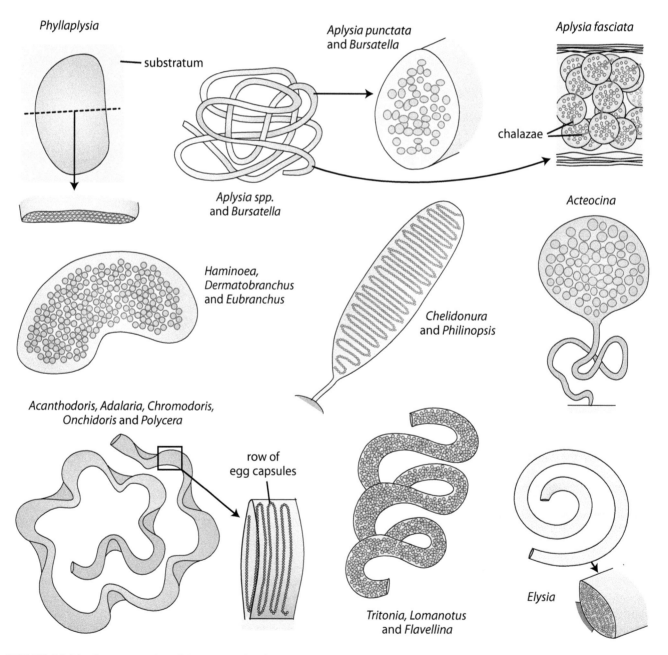

Phyllaplysia

substratum

Aplysia punctata and *Bursatella*

Aplysia fasciata

chalazae

Aplysia spp. and *Bursatella*

Acteocina

Haminoea, Dermatobranchus and *Eubranchus*

Chelidonura and *Philinopsis*

Acanthodoris, Adalaria, Chromodoris, Onchidoris and *Polycera*

row of egg capsules

Elysia

Tritonia, Lomanotus and *Flavellina*

FIGURE 20.64 Some examples of the spawn of various marine euthyneurans. Redrawn and modified from Klussmann-Kolb, A. and Wägele, H., *Zool. Anz.*, 240, 101–118, 2001.

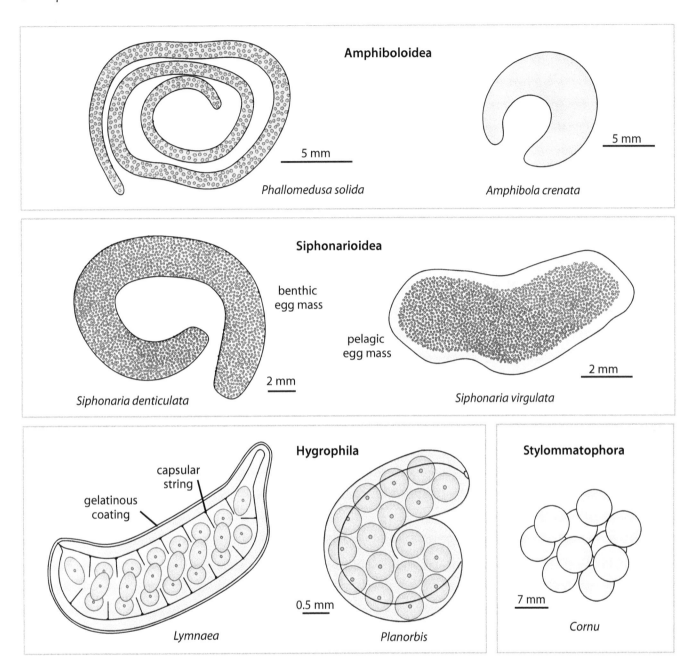

FIGURE 20.65 Some examples of lung-bearing panpulmonate spawn. Redrawn and modified from the following sources: *Phallomedusa* (Golding, R.E. et al., *Zootaxa*, 50, 1–50, 2007), *Amphibola* and *Cornu* original, Siphonaria (Creese, R.G., *Aust. J. Mar. Freshwater Res.*, 31, 37–47, 1980), hygrophilans (Bondesen, *P., Nat. Jutl.*, 3, 1–209, 1950).

in rounded clutches with the eggs connected by a chalazae-like string, but Dayrat and Tillier (2002) considered that the 'chalazae' of veronicellids, which are made up of filaments of mucus, are not homologous with other chalazae.

Stylommatophorans are much more conservative in the nature of their spawn than other euthyneurans, as most lay individual egg capsules containing single eggs ranging from 0.5 to 5.0 mm in diameter, and they are often calcified (Tompa 1984). The uncalcified eggs of succineids differ from other stylommatophorans in being laid in a continuous string thus resembling those of 'lower' lung-bearing panpulmonates. The outer covering of stylommatophoran eggs (Figure 20.66) may be uncalcified, with just a jelly-like outer layer (as in

Agriolimax carvanae and *Succinea*), or partly calcified with calcium carbonate crystals in a jelly matrix as in *Helix*, or well calcified with the outer layer being hard and brittle as in *Arion ater* (Tompa 1984). The calcified egg shells are comprised of calcium carbonate crystals in an organic matrix, although the calcified layer offers little protection from desiccation – it is the gelatinous layers that also make up the egg shell that serve this function. This gelatinous layer is outside the calcareous shell in many of the most calcified eggs (Tompa 1976), as in the punctoidean *Anguispira alternata*, and is hygroscopic, presumably assisting in reducing desiccation. Many stylommatophoran eggs (36 of 65 families) (Tompa 1976) have calcified shells, although some families, such as Helicidae, contain

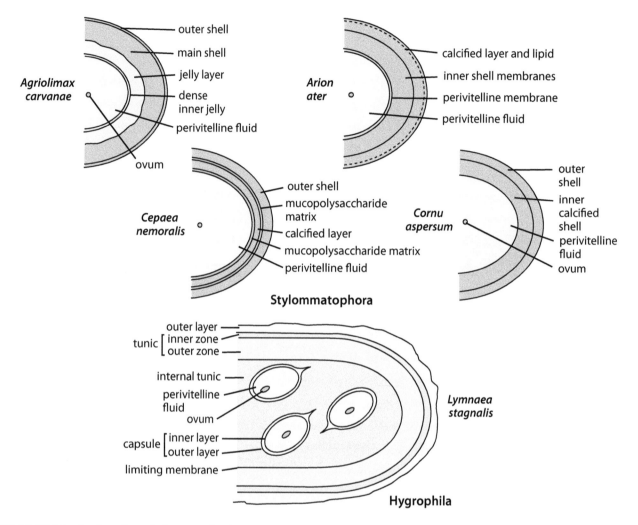

FIGURE 20.66 Sections through the eggs of stylommatophorans and lymnaeid spawn. Redrawn and modified from Tompa, A.S., Land snails (Stylommatophora), pp. 47–140, in Tompa, A.S., Verdonk, N.H., and van den Biggelaar, J.A.M. (eds.), *Reproduction. The Mollusca*, Vol. 7, Academic Press, New York, 1984 (Stylommatophora), and Geraerts, W.P.M. and Joosse, J., Freshwater snails (Basommatophora), pp. 141–207, in Tompa, A.S., Verdonk, N.H., and van den Biggelaar, J. A. M. (eds.), *Reproduction. The Mollusca*, Vol. 7, Academic Press, New York, 1984.

species with either calcified or partially calcified eggshells. Stylommatophoran slug eggs are usually uncalcified or partially calcified with some exceptions being *Testacella* and *Arion ater*. Other arionids have uncalcified or partially calcified eggs (Tompa 1976).

The calcified egg shells are not wasted. The embryos absorb much of the eggshell calcium during their development, and the newly hatched juvenile eats the remaining egg shell (Tompa 1976).

The eggs of stylommatophorans are typically laid in clutches, frequently in shallow depressions in soil or litter, and are then covered by the parent snail, or, in a few species, in rotten wood. A few arboreal species lay eggs among litter accumulated in epiphytes or on branches, and others lay them on the surface of leaves (Tompa 1976). *Testacella* lays its eggs in tunnels a metre or more below the surface.

Development of stylommatophoran eggs usually takes a month or two but can range from about two weeks in some taxa to over two years in *Testacella* (Tompa 1976).

Various forms of parental care occur sporadically in stylommatophorans, but is generally uncommon (Baur 1994), although several families brood young (see Chapter 8), usually in the distal part of the oviduct. Most of these are ovoviviparous, but a few cases of likely euviviparity are known in the Achatinellidae, where calcium transfer through a podocyst occurs (Tompa 1984). A few charopids and endodontids lay their eggs in the umbilicus of the shell. Parental care in hygrophilans is very rare, with the only known case the freshwater limpet *Protancylus* from Sulawesi (Albrecht & Glaubrecht 2006).

20.4.4.9 Larvae and Larval Shells

Unlike all other gastropods, besides a few sinistral taxa, the larval shells of heterobranchs are sinistral, even if there is a dextrally coiled adult. A sinistral protoconch on a dextral teleoconch often results in the protoconch being inclined at a different angle from the adult shell (teleoconch) – the heterostrophic condition.

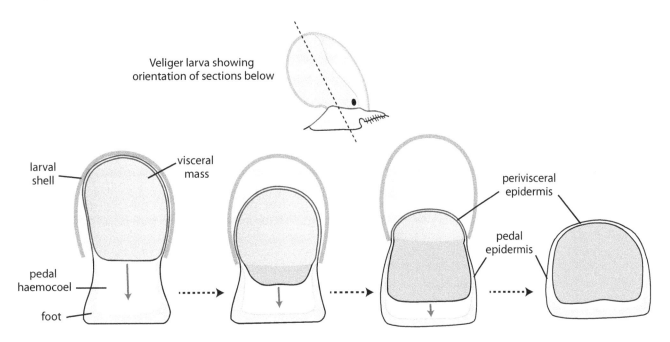

FIGURE 20.67 Diagrammatic representation of the metamorphosis of a nudibranch larva showing the movement of the visceral mass into the pedal haemocoel and the loss of the larval shell. Redrawn and modified from Bonar, D.B., *Am. Zool.*, 16, 573–591, 1976.

The veliger larva has a shell and an operculum. In marine heterobranchs, three kinds of larval shells can be distinguished (Vestergaard & Thorson 1938; Thorson 1946; Todd 1981). Most heterobranch larval shells are spiral, of three-quarters of a whorl to one whorl. Some have a simple cup-like shell with no spiral and others an inflated, egg-shaped shell, non-spiral with a small aperture. Cap-shaped larval shells are seen in very few, non-pelagic larval nudibranchs (Todd 1981), and this kind of larval shell is sometimes produced abnormally, prompting Thompson (1961a) to dismiss it as a separate category. The egg-shaped larval shells are produced by a few taxa with swimming larvae (e.g., *Onchidoris* and *Cadlina*).

In taxa that lack an adult shell the larval shell is cast off at metamorphosis (Figure 20.67), notably in nudibranchs and most sacoglossans.

The veliger larvae of heterobranchs possess a pair of simple rounded velar lobes, except architectonicids where the lobes are subdivided into two long lobes on either side (Robertson et al. 1970) and are thus similar to the veligers of some higher caenogastropods (see Chapter 19). Some architectonicids have teleplanic larvae capable of long-distance dispersal (Scheltema 1988).

20.4.5 BEHAVIOUR

Aspects of the sensory abilities of euthyneurans are covered in Chapter 7, and their sense organs are reviewed in Section 20.3.6.1. Their behaviour and the underlying neuronal and hormonal mechanisms involved have been reviewed by Chase (2002).

Some taxa have been shown to engage in trail-following of prey, conspecifics, or to use their trail to return to their home shelters (Ng et al. 2013). Examples include nudibranchs (Nakashima 1995), *Siphonaria* (Thomas 1973; Cook & Cook

1975), *Onchidium* (McFaruume 1980), and various stylommatophorans, for example, the slug *Limax* (Cook 1977). Gelperin (1974) has proposed that *Limax* can also detect air-borne olfactory signals. There are numerous examples of carnivorous heterobranch taxa tracking the slime trails of prey (e.g., Carté & Faulkner 1986; Clifford et al. 2003; Ng et al. 2013).

Behavioural adaptations can be vital in reducing water loss in terrestrial eupulmonates. Many show a strong preference for moist, shaded environments, but those that live in harsher conditions, such as deserts, survive by aestivating. Some species bury themselves in the soil while others shelter in crevices or beneath rocks. Some preferentially find crevices above the general ground level to take advantage of cooler air. Many arid-zone snails can remain in aestivation for long periods of time, even years, to emerge when it rains, producing conditions favourable for feeding and reproducing (Schmidt-Nielsen et al. 1971; Yom-Tov 1971). Some snails attach to vertical rock walls, tree trunks, or upside down on the ceilings of overhangs. All these behaviours not only assist in surviving desiccation but are also effective in hiding from predators.

20.4.6 PREDATION AND DEFENCE

20.4.6.1 The Shell as a Means of Defence

For terrestrial snails, the shell reduces one of their greatest threats, water loss, and is also the main means of defence against predators. As with other gastropods, there are numerous ways the shell of heterobranchs can be modified to resist predation, whether it be strengthening to avoid crushing, or developing ribs or other ornament, or by adopting cryptic colouration or shape. Shells can also be covered in periostracal processes that trap dirt, aiding in camouflage. A few may attach dirt, fragments of leaves, or lichens to their shells (e.g., Allgaier 2007).

Shells damaged in unsuccessful attacks by a predator are readily repaired by the mantle resulting in 'repair' scars on the shell. Similarly, minor damage to parts of the body, such as the foot and tentacles, can usually be repaired by regeneration (see Chapter 2).

Some of the strangest land snail shells are the distorted diplommatinid genera such as *Opisthostoma* and *Plectostoma*. Their unusual shell morphologies may be a response to predators such as the rathousiid slug *Atopos* and predatory beetle larvae (*Pteroptyx*) (Schilthuizen et al. 2006; Liew & Schilthuizen 2014). The last whorl of the adult shell is greatly distorted, and the shell also bears axial flanges, both these developments being thought to assist in deterring predators.

The loss of the operculum in most shelled euthyneurans is puzzling given that, for terrestrial snails in particular, an operculum would be a useful aid in preventing the entry of small predators and in reducing desiccation. It is likely that the stylommatophoran ancestor (which perhaps resembled an ellobiid) had lost the operculum. In its absence, mucus is used to seal the aperture to reduce desiccation and predation, particularly in aestivating snails. When this mucous seal forms a distinct covering, it is called an epiphragm, and it is sometimes hardened with calcium carbonate (Figure 20.20). Reduction in the size of the aperture or the formation of apertural barriers, such as folds or teeth that can sometimes almost fill the aperture, also assist in preventing predators gaining access. Such barriers do not prevent predators adapting, as seen for example in a carabid beetle that evolved a narrower head to allow it to gain access to restricted apertures (Symondson 2004). Apertural barriers are also found in ellobiids but otherwise are usually not as well developed in marine and estuarine taxa as they are in some terrestrial snails. While they are usually thought to be defensive, other functions suggested include trapping air when immersed in water (Emberton 1995) or perhaps assisting with balance (Suvorov 1993, 1999) and being a potentially useful calcium reserve.

20.4.6.2 Mucus

Heterobranch snails and slugs use mucus for locomotion (see Chapter 3) and, in terrestrial taxa, keeps the skin moist. When attacked they can also produce copious quantities of mucus which can be an effective deterrent (e.g., Eisner & Wilson 1970), this being particularly the case with terrestrial slugs (e.g., Pakarinen 1994a). The mucus may contain repellent chemicals, as it does in many marine euthyneuran slugs (see next section, Section 20.4.6.3.1, and Chapter 9).

20.4.6.3 Defence without a Shell

While the shell is the main defence mechanism in many molluscs, new defence mechanisms evolved along with its loss in many euthyneuran lineages. Some marine slugs in particular developed cryptic colouration, and some have very effective camouflage through employing cryptic shape, colour, and behaviour (Gosliner & Behrens 1990; Gosliner 2001).

The incorporation of various defensive glands and other mechanisms (Edmunds 1966b), including chemicals obtained from their food (Cimino & Ghiselin 1999, 2009) in the body of many slugs (see Section 20.4.6.3.1, next section, and Chapter 9), has been a major driver in their evolution. Some terrestrial slugs may also have defensive chemicals released in their mucus (see Chapter 3), or the tissues themselves are toxic (e.g., Symondson 2004). Terrestrial slugs also have behavioural strategies to avoid predation; some are very cryptic, and a few practise autotomy of their posterior body (see Section 20.4.6.6).

A variety of chemical defence systems have not only evolved in various sea slugs, but some have also developed spectacular warning colouration (see Section 20.4.6.5). The noxious chemicals are produced in special glands often associated with the mantle (see Section 20.3.2.6.1) (Wägele et al. 2006), as is the purple secretion produced by some aplysiidans (see Section 20.3.3.5 and Chapter 9). Some groups (Pleurobranchoidea, Doridoidea, and Acochlidioidea) have evolved dermal spicules (see Section 20.3.2.6 and Chapter 3). Also, some nudibranchs sequester nematocysts to act as a deterrent to predators (see Section 20.4.6.4).

Some gymnosome 'pteropods' have stellate chromatophores in their dermis which are similar to those in coleoid cephalopods although they are very slow-acting (Lalli & Gilmer 1989).

20.4.6.3.1 Chemical Deterrents

It is well known that some marine slugs defend themselves by secreting toxins and noxious chemicals, an adaptation that may have come about because of the necessity to dispose of chemical by-products from food or from the metabolism of that food. Storing them in the digestive gland or skin has led to the incorporation of toxic substances derived from food as a means of protection. This ability of some gastropods to utilise metabolites from food as defensive chemicals has been taken to a new level with shell loss (e.g., Cimino & Ghiselin 1999, 2009) (see Chapter 9). Some secretions produced by sea slugs can be very toxic and capable of repelling predatory sea stars and killing sea anemones, amphipods, and sacoglossans (e.g., Ajeska & Nybakken 1976; Jensen 1984).

In some shell-less marine euthyneurans the noxious chemicals derived from food are enhanced or replaced by toxic compounds synthesised *de novo* by the animal itself (e.g., Cimino & Ghiselin 2009) (see Chapter 9). While most of the noxious chemicals employed by these sea slugs are organic compounds, some (Philioidea, Pleurobranchoidea, and some Doridoidea) use acid secretions (see Section 20.3.2.6.1), notably sulphuric acid.

There have been many more studies on the chemical products of heterobranchs than on other gastropods and, within heterobranchs, studies on nudibranchs comprise the majority of those investigated (Benkendorff 2010) with several recent reviews (see Cimino & Gavagnin 2006; Cimino & Ghiselin 2009; Carbone et al. 2013; Wang et al. 2013). By 2010, 386 compounds had been isolated from 102 nudibranch species, and the Aplysiida yielded 247 compounds from 18 species (Benkendorff 2010). Some smaller or difficult to obtain euthyneuran groups including Acochlidia, Rhodopemorpha, and Thecosomata have not been investigated. For a more detailed

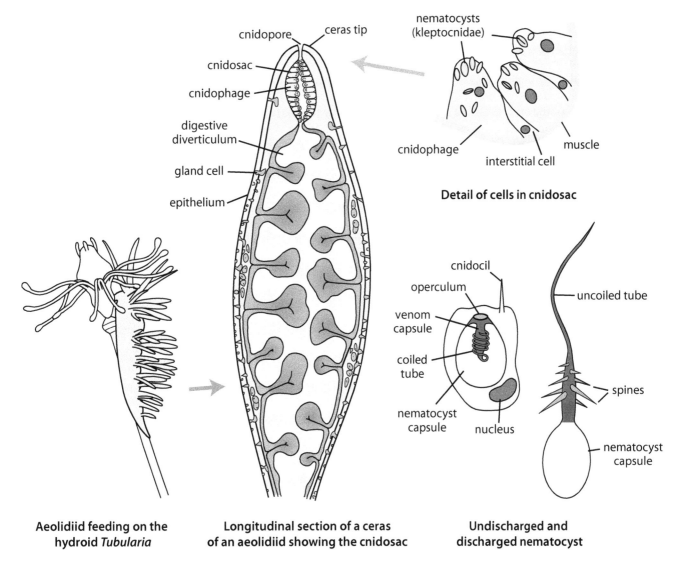

cnidopore | ceras tip
cnidosac
cnidophage
digestive diverticulum
gland cell
epithelium

nematocysts (kleptocnidae)

cnidophage | muscle
interstitial cell

Detail of cells in cnidosac

cnidocil
operculum
venom capsule
coiled tube
nematocyst capsule | nucleus

uncoiled tube
spines
nematocyst capsule

Aeolidiid feeding on the hydroid *Tubularia*   **Longitudinal section of a ceras of an aeolidiid showing the cnidosac***   **Undischarged and discharged nematocyst**

FIGURE 20.68 Cnidosacs of aeolidiid nudibranchs, with details of cnidophages and undischarged and discharged nematocysts. Redrawn and modified from the following sources: Longitudinal section of ceras (Edmunds, M., *J. Linn. Soc. A*, 47, 27–71, 1966a), cnidosac cells (Glaser, O.C., *J. Exp. Zool.*, 9, 117–142, 1910), nematocysts and aeolidiid feeding from various sources.

account of the secondary metabolites of 'opisthobranchs' in particular, see Chapter 9.

An entirely different strategy is employed by some physids using environmental chemicals to escape potential predators by living in a sulphide-rich stream (Covich 1981). Terrestrial slugs and snails often live on distasteful plants, but there is little evidence that the chemicals involved are used in defence against predators.

20.4.6.3.2 *Alarm Signals*

Like some other gastropods, various heterobranchs, when disturbed or attacked by predators, discharge chemical 'alarm signals'. These may be released from the injured part of the body, the urine, mucus, or, in aplysiids, the ink secretion (e.g., Fiorito & Gherardi 1990; Nolen et al. 1995). These alarm signals have been shown, in gastropods generally, to warn nearby members of the same species to initiate predator avoidance behaviour (e.g., changing direction and speed

of movement; burrowing) (e.g., Jacobsen & Sabell 2004) (see Chapter 9).

20.4.6.4 Use of Nematocysts

Cnidarian nematocysts (stinging cells) are co-opted by aeolidiid nudibranchs that feed on hydrozoans or anthozoans. They use a novel strategy involving taking the undischarged nematocysts from their prey and storing them in special *cnidosacs* at the ends of their cerata (Figure 20.68). Cnidosacs are not restricted to aeolidioideans as similar structures are also found in the cnidarian-feeding cladobranch nudibranchs *Doto*, *Hancockia*, and *Embletonia* (Martin et al. 2009, 2010), but the discussion below mainly relates to work done on aeolidioideans as they have been studied in more detail (see also a recent review by Goodheart et al. 2018).

As the sea slug feeds, it devours masses of cnidocysts, with large numbers (both discharged and intact) in the gut. They are also found in the faecal material, and some are in vacuoles in digestive gland cells but are apparently not digested (Martin 2003). Some observers have found intact, mature nematocysts in the gut and faeces (Martin 2003; Schlesinger et al. 2009) as mucus may well inhibit nematocyst discharge during feeding and in the gut (Mauch & Elliott 1997; Greenwood et al. 2004). However, it remains unclear as to whether mature nematocysts can be incorporated in the cnidosacs. Just how many of the nematocysts remain undischarged during feeding and passing through the gut (with their discharge possibly inhibited by mucus) is not well understood. There are also questions as to why the nematocysts do not discharge when the nudibranch feeds and how they are transported through the gut into the cnidosacs on the tips of the cerata without discharging. These matters were recently investigated by Obermann et al. (2012) who concluded that the most likely explanation was that immature, non-functional nematocysts are ingested while feeding (e.g., Greenwood & Mariscal 1984b) which mature and become functional within the cnidosacs due to a process involving acidification similar to that seen in the cnidarians themselves.

Thus, it seems probable that most mature nematocysts discharge and are voided in the faeces while the immature ones are moved via the stomach to the digestive diverticula. While there is evidence that immature cnidocysts are selectively incorporated in some species and that they complete their maturation in the cnidosac (Greenwood & Mariscal 1984b), this is not the case with all species (see below).

The nematocysts enter the cnidosac via a narrow canal and are then phagocytosed by the specialised epithelial cells (*cnidophages*) lining the cnidosac and stored in vacuoles in those cells which are called *kleptocnids* (Figure 20.68). There are many kleptocnids contained in each cnidophage. After completing their development, the kleptocnids can be used for defence. The kleptocnids are retained in the cnidophage cells for several days or even weeks, and they may obtain metabolic support during this period but following which they are slowly digested (Greenwood 2009). Experiments have shown that the transfer of the nematocysts can be quite rapid, being received in the cnidosacs within about two to three hours following feeding (e.g., Martin 2003).

Each cnidosac lies at the tip of a ceras and, below the external epithelium, the ceras is lined with muscle and, internally, is lined with a single-layered epithelium composed mainly of the cnidophages. It opens to the digestive gland by a narrow canal and opens at the tip of each ceras by a small pore (*cnidopore*) surrounded by muscle. If pressure is applied to the cerata, or they are autotomised, muscles surrounding the cnidosac force cells and nematocysts from ruptured cells out at the cnidopore, providing a defensive mechanism for the sea slug. The kleptocnids also discharge when experimentally squeezed from the cnidosac through the tip of the ceras (e.g., Greenwood & Mariscal 1984a; Greenwood & Garrity 1991; Greenwood 2009). A typical aeolidiid cnidosac contains about 3,000 nematocysts (Greenwood & Mariscal 1984b).

Some species show variation in the kinds of cnidocysts they utilise, reflecting differences in their food and hence the varieties of nematocysts available (e.g., Frick 2005). There are usually several kinds of nematocysts in the cnidarian prey, but the kind that is dominant (known as the *microbasic eurytele*) is usually not the most abundant in the cnidosacs (Day & Harris 1978), suggesting selection. There is also some evidence that those cnidocysts that would be most effective in defence (e.g., Thompson & Bennett 1969) are preferentially selected. While in some aeolidiids such selective incorporation in the cnidosacs of not only immature cnidocysts but of different types (Greenwood 1988) may occur, this has not been observed in some taxa (e.g., Martin 2003).

When aeolidiids that have cnidosacs are disturbed by a potential predator their body curls and their cerata 'bristle', and the tips may point towards the predator (e.g., Millen & Hamann 1992). The muscles around the cnidosacs contract, forcing the nematocysts from their cnidophage cells and through the cnidopore. The nematocysts discharge on contact with seawater, striking anything in close contact.

The tips of the cerata are often conspicuously coloured. The cerata are sometimes autotomised (see Section 20.4.6.6), and when this occurs, they may writhe about, distracting the predator.

The effectiveness of the defence provided by the nematocysts (kleptocnids) in the cnidosacs has been the subject of some debate (see Greenwood 2009 for review; Martin et al. 2009), although some are effective. For example, the oceanic-drifting *Glaucus* is, like other aeolidiids, able to utilise the large, highly toxic nematocysts from the siphonophores on which they feed, notably *Physalia*, the 'blue bottle' or 'Portuguese man of war'. Contact with this slug has resulted in stings to humans (Thompson & Bennett 1969; Bebbington 1986).

A few aeolidioideans, notably species of *Phyllodesmium*, have cnidosacs which never contain functional nematocysts (Burghardt et al. 2008a).

20.4.6.5 Aposematic (Warning) Colouration and Batesian and Müllerian Mimicry

Nudibranchs are commonly brightly coloured, often with vivid contrasting patterns. Such conspicuous colouring is assumed to be aposematic, alerting predators of chemical (or other) defences (e.g., Edmunds 1991; Gosliner 2001) (see also Chapter 9). For aposematic colouration to be effective, predators must have good colour vision and be daytime hunters (fish, birds, and some crustaceans), but regrettably, little is known about the predators of most marine heterobranchs and the extent to which vision is involved in their predation (see Section 20.4.6.7).

Aposematism in some non-cryptic nudibranchs has long been assumed. It has been shown that conspicuousness (contrast against the background and/or between two colours in a pattern – assuming that colour vision is not necessarily required) is correlated with toxicity (Cortesi & Cheney 2010). They also found little evidence of phylogenetic signal in their results, although Gosliner (2001) found that basal taxa tended to lack, or have less well-developed, aposematic colouration. Studies on fish behaviour have found they use a variety of cues for finding and testing prey, and experiments by Miller and Pawlik (2013) showed that fish could learn to avoid unpalatable prey by colour alone. These clues are not necessarily entirely visual as fish and crabs have also been shown to quickly learn to reject food with extracts from a nudibranch (Long & Hay 2006).

Aeolidiids also exhibit apparent aposematism with brightly coloured banding on the cerata (e.g., Bürgin 1965; Aguado & Marin 2007), all of which bear a distal cnidosac, and the cerata may be waved about to make them more conspicuous (see previous Section). Aeolidiids also have gland cells that may produce noxious secretions (Edmunds 1966a), and there is some question as to whether the nematocysts or the secretions are mainly responsible for deterring predators (e.g., Edmunds 1966a; Penney 2009).

Batesian mimicry occurs when there are one or more palatable species among toxic species sharing the same geographic space and which have similar colouration (see also Chapter 9). Examples include non-molluscan taxa such as crustaceans, juvenile holothurians, and flatworms that mimic toxic nudibranchs (e.g., Gosliner & Behrens 1990).

Within Chromodorididae in particular, there are 'colour groups' which are very similar in appearance and geographic area (e.g., Rudman 1991b). These colour groups appear to be an example of Müllerian mimicry (see Chapter 9) because several species have the same general colour and markings and all are protected by toxic compounds, although sometimes camouflage may also be involved. Recently, Layton et al. (2018) have shown that some species of *Chromodoris* can have geographically dependent colour patterns, and a few species can alter their colour patterns to mimic different congeners in different geographical regions. In the genus *Hypselodoris*, Epstein et al. (2018) demonstrated convergence in colour and mimicry patterns between members of distinct lineages in situations of sympatry.

Terrestrial slugs have not attracted the same attention as their colourful marine counterparts, but their often drab colours suggest crypsis while some, such as several athoracophorid taxa, have leaf-like veins and a leaf-shaped body.

20.4.6.6 Autotomy

For an introduction and a general discussion of autotomy see Chapter 3. In heterobranchs, examples include the autotomy of cerata in many aeolidiid nudibranchs, particularly in those without cnidosacs, when they make contact with predators, or when they are stressed by chemicals or fresh water (Rudman 1991a; Miller & Byrne 2000). The cerata may continue to writhe after autotomy and thus act as a decoy, and they regenerate readily.

The cerata of *Melibe leonina* also undergo autotomy, and this has been studied by Bickell-Page (1989). Autotomy is initiated by contact with crabs which stimulates two rings of nerves, one inside the other, which branch from the ceratal nerve.[9] The nerves cause the contraction of two bands of sphincter muscles at the base of the cerata – this zone is called

the *autotomy plane*. The contracted muscles seal off both sides of the detached cera. Unlike some aeolidiid cerata, those detached from *Melibe* show little movement. Another possible cause for autotomy of cerata (in solar-powered nudibranchs generally) could be the active regulation of symbiont density (Burghardt & Wägele 2014).

Parts of the mantle of some doridids and pleurobranchs can also be autotomised (Rudman 1998).

In the sacoglossan *Lobiger*, the large parapodia can be autotomised when the animal is attacked (Gonor 1961), while in another sacoglossan, *Oxynoe*, the end of the foot can be autotomised and regenerate later (Lewin 1970; Warmke & Almodóvar 1972). Autotomy of the posterior part of the foot is also seen in some terrestrial stylommatophoran slugs (arionids, agriolimacids, and limacids) when they are attacked by predators such as carabid beetles (e.g., Deyrup-Olsen et al. 1986; Pakarinen 1994b; Foltan 2004; Symondson 2004). In an arionid, autotomy of the foot is caused by the rapid contraction of a ring of muscles (Luchtel & Deyrup-Olsen 2001).

20.4.6.7 Predators of Heterobranchs

The predators of aquatic heterobranchs are generally the same as those that target other molluscs, and these are reviewed in Chapter 9. The predators of land snails and slugs were also reviewed in Chapter 9 and in Barker (2004). They include, besides predatory gastropods, many reptiles, birds, and mammals, several families of beetles, some specialised fly predators, and parasitoids, ants, bugs, centipedes, and harvestmen, some mites, and some planarians. Spiders will occasionally feed on terrestrial gastropods, but none are specialists.

Despite having evolved a range of defence mechanisms, nudibranchs and aplysioideans are preyed on by anemones (e.g., Meij & Reijnen 2012), fish, other sea slugs, and crabs (Rudman 2000; Meij & Reijnen 2012), and there has recently been a report of predation of the nudibranch *Triopha* by the toxoglossan caenogastropod *Californiconus californicus* (Valdés et al. 2013).

Some predators of particular interest include those animals that have become specialised heterobranch feeders. Examples include the southern Californian large (>20 cm in length) cephalaspidean *Navanax inermis*, that feeds on a variety of other marine heterobranch slugs. It tracks down its prey by following their mucous trails (Paine 1963). As noted previously, some stylommatophoran carnivores specialise in feeding on other snails, including *Euglandina rosea* that has been responsible for the extinction of hundreds of species on islands in the Pacific (see Chapter 10).

Vertebrate predators of gastropods are discussed in Chapter 9. Fish are predators on many marine euthyneurans while birds, reptiles, and small mammals are significant predators of terrestrial snails, and birds and fish feed on marine and freshwater euthyneurans. Green turtles (*Chelonia mydas*) have been recorded eating *Aplysia* but perhaps accidentally (Seminoff et al. 2002). Over 20 species of colubrid snakes feed exclusively on slugs and snails, including species in Asia and Brazil (Agudo-Padrón 2012, 2013). One such snake (*Pareas iwasakii*) has asymmetrical jaws ideal for prising

---

[9] The ceratal nerves are branches of the pleural nerves which arise from each cerebropleural ganglion.

snails out of a dextral shell. In Southeast Asia where these snakes occur, sinistral snails are unusually common, and the dextrally adapted snake has little success in consuming those snails (Hoso et al. 2007) (see also Chapter 9).

Some hygrophilans (Physidae, Lymnaeidae) initiate avoidance behaviour in response to freshwater crayfish by crawling above the water line, but planorbids do not (Alexander & Covich 1991; Covich et al. 1994).

20.4.7 Associations

Some sea slugs have utilised the ability of plants to convert solar energy into sugars (i.e., photosynthesis) and have done this in two different ways, one involving the single-celled zooxanthellae that are symbiotic with their cnidarian food and the other by utilising the plastids from plant cells. In both these associations metabolites from the photosynthetic symbionts or chloroplasts are used by the sea slug host, and they have been described as 'solar-powered' slugs or 'crawling leaves'.

Sequestering dinoflagellate (zooxanthellae: *Symbiodinium*) symbionts from their cnidarian food has evolved in at least three groups within the Nudibranchia, occurring in several aeolidiids, notably *Phyllodesmium*, some species of which are highly modified (Rudman 1981a, 1991a; Burghardt et al. 2008a; Burghardt et al. 2008b; Moore & Gosliner 2011) and *Pteraeolidia* (Burghardt et al. 2008b), the arminoideans *Pinufius* and *Doridomorpha* (Rudman 1981b, 1982), and the dendronotoidean *Melibe* (e.g., Burghardt & Wägele 2014). Gosliner and Smith (2003) showed that evolution of zooxanthellae with *Melibe* was limited to more derived, Indo-Pacific taxa and was not evident in less derived lineages found in temperate regions.

Many sacoglossans can utilise algal chloroplasts in their tissues, a habit supposedly acquired independently several times (Christa et al. 2014a). These are retrieved from their algal food and are maintained intracellularly in the epithelium of the digestive gland where they remain photosynthetically active (see Chapter 5 for more details). Experiments have shown that at least some can survive in the dark indicating that they are not totally reliant on photosynthesis (e.g., Christa et al. 2014b; Cartaxana et al. 2017).

Besides the usual gut bacteria, two unidentified symbiotic bacteria are known from the nudibranch *Dendrodoris nigra*, one in the epithelial cells of the notum and mantle (Zhukova & Eliseikina 2012) and the other in the vestibular gland associated with the female reproductive system and in the egg masses (Klussmann-Kolb & Brodie 1999). These bacteria may be involved in producing some secondary metabolites such as defensive chemicals or nutritive fatty acids. Other endosymbiotic bacteria have been found associated with the cerata of various nudibranchs. These are obtained from their cnidarian prey, as unidentified bacteria found in cells in tentacles of various cnidarians have also been found in cells in the tips of the cerata of the nudibranchs that feed on them (Doepke et al. 2012; Schuett & Doepke 2013).

Other associations with nudibranchs include some with crustaceans such as parasitic copepods and a commensal shrimp, as well as with scaleworms (Polychaeta) and even juvenile gobies (various articles at www.seaslugforum.net).

An overview of parasites of gastropods (and other molluscs) is provided in Chapter 9. They include trematodes (e.g., Blair et al. 2001), nematodes, ciliates, microsporidians, bacteria, and viruses. Terrestrial gastropods have parasitic mites (Fain 2004), nematodes (Morand et al. 2004), ciliates (van As & Basson 2004), and microsporidians (Selman & Jones 2004). Diseases caused by bacteria and viruses occur but are poorly understood (Raut 2004).

20.4.8 Economic and Ecological Values

There are serious conservation concerns for many terrestrial eupulmonates as a result of habitat destruction and other human activities such as the introduction of predatory pests (see Chapter 10).

Intensive studies on secondary metabolites in shell-less marine euthyneurans have located some potentially useful compounds. For example, the peptide Kahalalide F (Hamann et al. 1996), from the sacoglossan *Elysia rufescens*, has antitumour properties and has been undergoing clinical tests (e.g., Serova et al. 2013).

A few species of stylommatophorans are used for food with some, especially some helicids, farmed (see Chapter 10).

Some euthyneurans are important laboratory animals largely because of their large neurons (see Chapter 10).

Helicid snails have been used for medical purposes since antiquity (Bonnemain 2005). Land snail mucus is known for antibacterial properties (Cilia & Fratini 2018), with some constituents used to treat a range of skin conditions. Tissue adhesives for wound dressing and tissue repair have also been inspired by slug mucus (Li et al. 2017a).

As outlined in Chapter 10, various freshwater taxa (Hygrophila) are invasive, and some are intermediate hosts of significant parasites of stock or humans. Various terrestrial snails and slugs are also invasive, and a number are agricultural pests.

20.5 DIVERSITY AND FOSSIL HISTORY

The first heterobranchs were marine, probably tiny, and perhaps not unlike modern *Orbitestella* or *Cima*. They may have arisen in the Devonian about 370 Mya (Frýda et al. 2008) or even earlier in the Ordovician, but the earliest undoubted fossil heterobranch is the Triassic *Cylindrobullina* (240 Mya) (Frýda et al. 2008) which is considered by some to be a basal euthyneuran lineage. The 'lower heterobranchs' first appear a little later in the fossil record (Mathildidae 230 Mya, Architectonicidae 210 Mya) (Tracey et al. 1993; Kiel et al. 2002). Recognisable pyramidellids appear much later at 70 Mya (Kiel et al. 2002). Jörger et al. (2010) used a 'relaxed molecular clock approach' to estimate that the initial radiation of euthyneurans occurred in the late Paleozoic, and the major diversification occurred from the Mesozoic.

The taxa previously known as 'opisthobranchs' comprise over 100 living families and around 6,000 species (Wägele et al. 2008). The tendency for shell loss in this group means there is no fossil record for several groups, notably the largest clade, the Nudibranchia (ca. 3,000 species) (Wägele 2004), with the Doridina alone having about 2,000 (Valdés 2004; Penney 2008). These, with the pleurobranchs, comprise the Nudipleura which, judging from molecular phylogenies, diverged rather early from other euthyneurans. The only shelled nudipleurans are some pleurobranchs that have thin internal shells with a wide aperture and a reduced apical coil. While this shell morphology may provide clues, it is likely that it was highly modified through internalisation. Thus, while we can only guess at what the shelled ancestors of this clade were like, recently it was shown that the Ringiculidae are sister to the Nudipleura and that group has a fossil history back to the Middle Jurassic (Kano et al. 2016).

Despite this, and because even those taxa with shells show few diagnostic characters, Wägele et al. (2008) noted that the timing of the 'opisthobranch' clades in the fossil record mostly agreed with their phylogeny, with the earliest 'true opisthobranchs' (questionably identified as Hydatinidae – i.e., Aplustridae) first appearing about 190 Mya, although Triassic (about 220 Mya) 'opisthobranchs' have been reported (Bandel 1994, 2002b). Wägele et al. (2008) gave the appearance of the other groups as: Acteonoidea (about 160 Mya), true cephalaspideans (Bullidae) (180 Mya), and Aplysiida (*Akera*: 190 Mya); other shelled groups do not appear until 60 Mya or less although some of these are undoubtedly much older than indicated by the fossil record. It is interesting to note that the basal members of each major group in the Nudipleura tend to be found in Antarctic waters (Wägele et al. 2008).

Shell reduction and loss in many heterobranch lineages had consequences as well as opportunities. This trend was a prerequisite for the evolution of many of the unique biological features of euthyneuran marine slugs, such as the utilisation of secondary metabolites in chemical defence, and the incorporation of chloroplasts by sacoglossans and zooxanthellae and cnidocysts by aeolidioideans.

Other factors drove the diversification of some clades, for example, the single row of piercing radular teeth in sacoglossans and subsequent changes in tooth shape were major factors in the evolution of that group (Jensen 1997). In groups such as chromodorid nudibranchs, the adoption of sponge feeding and incorporating metabolites in their dermis were very significant (see Section 20.4.6.3.1) while some other nudibranch clades developed protective spicules (see Section 20.3.2.3) in their dermis, often in addition to chemical deterrents.

The lung-bearing panpulmonates do not have a particularly good fossil history. Marine taxa are all found in the mid to high intertidal, and these faunas are not well represented in the fossil record. Similarly, terrestrial conditions are often not conducive to fossil preservation because of lack of opportunity and often acidic conditions. As with 'opisthobranchs', the slug groups lack a fossil history. Stylommatophoran land snails have highly convergent shells, so the reliable identification of fossils is often very difficult or impossible. Bandel (1997) suggested that the earliest 'pulmonate' fossils were Late Carboniferous (<310

Mya) and were ellobiids, but other ellobioideans, siphonariids, otinids, and Hygrophila did not appear in the fossil record until the Middle to Upper Jurassic (~180 Mya) (Bandel 1994, 2002b). Stylommatophoran land snails first appear at the Jurassic–Cretaceous boundary but underwent most of their diversification in the Cenozoic. Fossils of some marine groups (e.g., Trimusculidae) do not appear until the Cenozoic, but the fossil record for some taxa is unreliable. For example, the record for Amphiboloidea is very poor, with the first known fossils being from the Pliocene, but the group must be much older (Golding 2012).

20.5.1 Major Groups within the Heterobranchia

20.5.1.1 'Lower Heterobranchs'

This paraphyletic or polyphyletic grouping contains the primitive, mostly marine groups (see Appendix for details of the classification), many of which were treated as 'prosobranchs' in earlier classifications (Figures 20.12 and 20.13). It also now includes the Rhodopemorpha. Until recently these tiny slugs were thought to be 'opisthobranchs', but molecular data show that they are located among the 'lower heterobranchs' (Schrödl et al. 2011). These are the only shell-less 'lower heterobranchs' and are marine, tiny, interstitial, and turbellarian-like slugs that lack a radula (e.g., Haszprunar & Hess 2005). Pyramidelloideans and rissoellids were thought to be 'lower heterobranchs' but are now included in Euthyneura based on recent molecular phylogenetic studies.

The living basal heterobranchs comprise more than a dozen extant families (see Appendix). Most are shelled and operculate and include the 'sundial shells' (Architectonicidae), the related Mathildidae, and the smaller-sized freshwater Valvatidae. Other families are small-sized and generally poorly known. They include the marine sister taxa of the Valvatidae, the Cornirostridae, and (probably) Xylodisculidae and the discoidal, tiny omalogyrids, one of which, *Ammonicera minortalis*, is reputedly the smallest adult mollusc with a shell at only 0.32–0.46 mm maximum diameter (Bieler & Mikkelsen 1998). These taxa are often the sister to the much larger architectonicids in molecular analyses. Other particularly poorly known shelled groups include the Cimidae and Tjaemoeiidae and some extinct taxa (see Appendix). The phylogenetic relationships of all these groups are unknown or poorly understood.

20.5.1.2 Euthyneura

This grouping encompasses the majority of what used to be called Opisthobranchia and Pulmonata, but recent molecular analyses show it also includes the Pyramidellidae and Rissoellidae previously treated as 'lower heterobranchs', and the glacidorbids, that were sometimes included there.

Except for some changes to the cephalaspideans, such as the removal of the acteonoideans and ringiculids, the other traditionally recognised 'opisthobranch' groups are fairly stable, despite the considerable parallel evolution in many features (e.g., shell loss, reproductive and nervous systems).

20.5.1.2.1 *Acteonimorpha*

The Acteonoidea (and Ringiculoidea) were excluded from Cephalaspidea by Mikkelsen (1996) based on morphology, and this has been confirmed in most subsequent studies using molecular data (e.g., Kano et al. 2016). They were instead included in the 'lower heterobranchs' by Bouchet and Rocroi (2005), and Acteonoidea was placed there in earlier molecular phylogenies such as Malaquias et al. (2009). The acteonids, now treated as the most basal euthyneurans, have an operculum, lack a gizzard, and the cerebral and pleural ganglia are fused.

Although ringiculids and acteonids have been regarded as the most basal 'opisthobranchs' (e.g., Gosliner 1981), a recent molecular analysis placed Ringiculoidea as sister to the Nudipleura (Kano et al. 2016). While *Ringicula* has a monaulic reproductive system (Gosliner 1981), that of nudipleurans and acteonids is diaulic (Wägele et al. 2008). Convergence is possible as similar systems have also evolved in sacoglossans and lung-bearing panpulmonates.

Acteonimorpha contains two main groups (orders), the Acteonida (Acteonoidea) and the Rissoellida (Rissoellidae). The latter family is included here because Rissoellidae is sister to Acteonoidea in recent molecular analyses (Figures 20.12 and 20.13).

20.5.1.2.2 *Ringipleura*

This grouping, indicated in the phylogenetic analysis of Kano et al. (2016), includes the Ringiculidae and the Nudipleura. *Ringicula* is rather poorly known, with only a few fairly detailed studies of its morphology (Fretter 1960; Gosliner 1981; Kano et al. 2016).

20.5.1.2.3 *Nudipleura*

Nudipleura contains Nudibranchia and Pleurobranchoidea (the pleurobranchs or side-gilled slugs) but does not include Umbraculoidea, which was thought to be related to pleurobranchoideans, and is now included in Euopisthobranchia. Nudipleura is sister to the euopisthobranchs in recent molecular phylogenies of heterobranchs that did not include Ringicula.

The pleurobranchs are a small group comprising three families, the pleurobranchids (seven genera), the Pleurobranchaeidae (three genera), and Quijotidae which contains a single species. The nudibranchs in comparison are a large group comprising numerous families (see Appendix for details). They are divided into two main groups, the Doridina and Cladobranchia.

The cladobranchs include the Dendronotoidea which, according to recent phylogenetic analyses (Pola & Gosliner 2010; Goodheart 2017), is not monophyletic, and contains, with the related Tritonioidea, the most primitive cladobranch nudibranchs, some with an only partially divided digestive gland and some with an anterior velum. More advanced cladobranchs show considerable ramification of the digestive gland, in addition to the development of cerata and loss of the velum and its transformation into tentacles (Ghiselin 1966).

Nevertheless, the Dendronotoidea, with the Tritonioidea, have many anatomical features, including the structure of the reproductive system, linking them to the Arminoidea and Aeolidioidea, as does their cnidarian diet. Pola and Gosliner (2010) have shown that the Arminoidea, as it was previously recognised, was not monophyletic. The remaining nudibranchs, the Doridina (= Euctenidiacea, Holohepatica, Anthobranchia), comprise the bathydoridids, and the doridids and their relatives, all of which have an unbranched digestive gland (see Appendix for details).

20.5.1.2.4 *Tectipleura*

In the classification proposed by Schrödl et al. (2011) based on their molecular phylogeny, this grouping included all the non-nudipleurans, including Umbraculoidea, sacoglossans, and acochlidians, as well as all 'pulmonates' and pyramidelloideans.

20.5.1.2.4.1 Euopisthobranchia
This grouping includes the taxa previously included in Opisthobranchia other than the nudipleurans, Acochlidia, Sacoglossa, and Rhodopemorpha.

Cephalaspidea was previously a catch-all for shelled 'opisthobranchs'. Most possess a few well-developed gizzard plates and have some common features of the reproductive system.

Two families (Gastropteridae and Aglajidae) lack gizzards, a condition probably the result of secondary loss, and although Wägele et al. (2008) considered these groups to be basal, this idea is contradicted by molecular phylogenies (Malaquias et al. 2009; Oskars et al. 2015). The position of the Diaphanidae has been problematic, and the group has often been placed near the acteonids. They have a shell but no operculum, and, like acteonoideans, they lack gizzard plates. *Diaphana* is placed at the base of Cephalaspidea *s.s.* in the Malaquias et al. (2009) and Oskars et al. (2015) molecular analyses of cephalaspideans.

The gizzard-bearing taxa – the cephalaspideans, aplysidans, and the pteropods – are a natural group confirmed in morphological (Dayrat & Tillier 2003) and molecular analyses (e.g., Jörger et al. 2010).

The pteropods (Thecosomata and Gymnosomata) are highly modified for swimming. Most recent studies (e.g., Ghiselin 1966; Klussmann-Kolb & Dinapoli 2006; Klussmann-Kolb et al. 2008) have found pteropods to be monophyletic, although two separate groups are indicated in the analysis of Dayrat and Tillier (2002). Their precise relationships with cephalaspideans and aplysiidans are also in doubt, although they are clearly related. Huber (1993) considered them to be neotenous 'bullomorphs' based on their nervous systems.

20.5.1.2.4.2 Panpulmonata
This name was introduced by Jörger et al. (2010) to include Siphonarioidea, Sacoglossa, Glacidorboidea (a group sometimes included in the 'lower heterobranchs' but that was originally considered to be a 'pulmonate'), Pyramidelloidea, Amphiboloidea, Hygrophila, Acochlidia, and Eupulmonata. This group thus includes highly divergent taxa and is very diverse. Each group is

briefly outlined next, and more details regarding their taxonomic composition are given in the Appendix.

20.5.1.2.4.2.1 **Acochlidia** This small group (~30 species) of small-size taxa live mainly interstitially in marine coastal sand but there are also a few freshwater species. Morphological analyses placed them with convergent tiny, interstitial worm-like taxa, such as Rhodopidae, Runcinidae, and Philinoglossidae (Wägele & Klussmann-Kolb 2005; Schrödl & Neusser 2010), but more recent molecular analyses show that the group is nested in the panpulmonates (Klussmann-Kolb et al. 2008; Schrödl et al. 2011), as are another group of marine slugs, the Sacoglossa. While ancestral acochlidians were marine and probably interstitial (mesopsammic), two lineages have independently colonised fresh water (Schrödl & Neusser 2010) (see Section 20.4.1). Interestingly, some taxa show convergent features with sacoglossans, a few of which, such as *Platyhedyle*, were previously thought to be acochlidians, leading Gosliner (1994) to suggest that at least some acochlidians were derived from sacoglossans. More recently it has been shown that the enigmatic amphibious/terrestrial and insectivorous family Aitengidae is also acochlidian (Neusser et al. 2011).

20.5.1.2.4.2.2 **Sacoglossa** In many recent molecular phylogenies, sacoglossans are sister to, or paraphyletic with, the siphonariids. Shell-less sacoglossans may have evolved only once from shelled ancestors (e.g., Jensen 1996a; Händeler & Wägele 2007). Cimino and Ghiselin (2009 p. 282) argued that 'the defensive use of caulerpenyne and its derivatives from the algae allowed [the shelled sacoglossans] to move above the substrate' on the assumption they were derived from 'burrowing herbivores' like cephalaspideans. This is probably not so if recent phylogenetic findings as to the relationships of sacoglossans are upheld – their ancestors may well have been shallow-water (possibly intertidal), surface-dwelling gastropods that shared a common ancestor with some of the lower marine pulmonates. Early in their evolution they specialised in exploiting siphonaceous green algae and evolved physiological (coping with the chemical defences of the algae) and morphological (the unique radular apparatus) adaptations leading to their successful radiation, subsequently enhanced by their use of plastids to photosynthesise (see Section 20.4.7).

20.5.1.2.4.2.3 **Siphonariida** These marine limpets are common on intertidal shores in many parts of the world. They lack tentacles and several key 'pulmonate' characters such as pulmonary blood vessels and a contractile pneumostome (although they have a pneumostome-like opening on the right side that corresponds with a shell groove). The mantle cavity has a small osphradium and secondary gill leaflets on the mantle roof. While most siphonariids are intertidal, most species of *Williamia* are subtidal. Siphonariids have long been considered to be basal or near basal 'pulmonates'. This position is supported in most recent molecular phylogenies where siphonariids are included basally in the panpulmonates as sister to, or paraphyletic with, sacoglossans (Klussmann-Kolb

et al. 2008; Dinapoli & Klussmann-Kolb 2010; Dayrat et al. 2011; Göbbeler & Klussmann-Kolb 2011; Schrödl et al. 2011; Golding 2012) (Figures 20.12 and 20.13).

Siphonariidae and Amphibolidae have sometimes been grouped – either as Amphiboloidea (including also Trimusculidae) (Tillier 1984a) or Thalassophila (e.g., Nordsieck 1992) – but these groupings are not supported in recent molecular analyses.

20.5.1.2.4.2.4 **Pylogastropoda** This grouping was used by Bouchet et al. (2017) to include the estuarine amphibolids, the freshwater glacidorbids, and the marine pyramidellids.

Glacidorbidae and Amphiboloidea form a monophyletic group in some recent molecular analyses (e.g., Holznagel et al. 2010) but not in others (Dayrat et al. 2011; Schrödl et al. 2011), although both are within the panpulmonates. In some analyses, both groups form a clade with the pyramidellids (Dinapoli & Klussmann-Kolb 2010; Schrödl et al. 2011; Golding 2012). Interestingly, all three of these groups, uniquely in the Panpulmonata, possess opercula.

Amphiboloideans are globose, operculate estuarine intertidal snails found in the Indo-West Pacific and notably in Australasia where they have recently been shown to be much more diverse than previously recognised (Golding et al. 2007; Golding 2012). These deposit feeders lack a jaw, have very broad, short tentacles with eyes at the base, and the mantle cavity has a pneumostome and contains a small osphradium (Golding et al. 2007). Despite nervous system characters (Ruthensteiner 1997) and early molecular results that appeared to exclude amphibolids from the 'pulmonates' (Tillier et al. 1996), recent molecular analyses include them in the broader concept of Panpulmonata. Similarly, the Glacidorbidae, a family of tiny near planispiral freshwater operculate snails found only in Australia and South America, were placed in the 'lower heterobranchs', mainly on the basis of the nervous system lacking some key pulmonate characters (Haszprunar & Huber 1990; Huber 1993; Dayrat & Tillier 2002), although some (Ponder 1986; Smith & Stanisic 1998; Ponder & Avern 2000) considered them to be paedomorphic basal 'pulmonates'. This latter position has been upheld in recent molecular analyses (e.g., Holznagel et al. 2010; Golding 2012), usually as sister to amphiboloideans. Glacidorbids have a mantle cavity lacking a pneumostome, and the pallial part of the genital system is not internal, unlike the situation in most other panpulmonates.

The pyramidellids differ greatly from amphibolids and glacidorbids, making it difficult to see how their relationship can be justified on morphological grounds. They comprise a speciose group of mostly tall-spired shelled heterobranchs that are ectoparasites on polychaetes and molluscs. They lack most characters seen in typical euthyneurans and have been included previously in the 'mesogastropods' or lower 'opisthobranchs' or, more recently, in the 'lower heterobranchs' (with which they share several morphological features). Recent molecular analyses have placed pyramidelloideans in the panpulmonates, but details of this surprising result differ in different analyses. For example, pyramidellids are sister to the amphibolids and glacidorbids in Jörger et al. (2010) and

Dinapoli and Klussmann-Kolb (2010) while in some other analyses (Dayrat et al. 2011; Medina et al. 2011) they are found nested among other pulmonate groups. It will be interesting to see how their relationships are determined in future studies.

The protoconch and teleoconch features of the Mesozoic Nerineoidea have similarities to 'typical' pyramidellids, like the large tropical *Pyramidella*, perhaps indicating a relationship.

The name Basommatophora has been variously used for basal 'pulmonates' (e.g., Mol 1974; Hubendick 1978; Haszprunar & Huber 1990; Nordsieck 1992; Salvini-Plawen & Steiner 1996). It is, however, a paraphyletic or polyphyletic group comprising all lung-bearing panpulmonates other than Geophila, with the name referring to the plesiomorphic position of the eyes at the base of the cephalic tentacles. Archaeopulmonata, in the sense used by Harbeck (1996), is also polyphyletic as it contained all the same groups as Basommatophora other than Hygrophila but also included onchidiids, because they possess a veliger larva.

A key innovation in non-marine panpulmonates was the ability to regulate the composition of body fluids in freshwater or terrestrial environments (see Chapter 6). For example, modifications to the kidney in the terrestrial taxa involved the development of a ureter for the resorption of ions. The ability to extract the oxygen they needed from the air is facilitated by the modification of their mantle cavity as a vascularised lung. Air flow could be controlled by way of the small, contractile opening, the pneumostome, a unique 'pulmonate' adaptation.

20.5.1.2.4.2.5 **Hygrophila** This important group of panpulmonates are nearly all found in fresh water but are otherwise not very well defined morphologically. The primitive South American Chilinidae lives in both fresh and brackish water. Hygrophila includes several well-known families such as the Planorbidae (cosmopolitan) and Physidae (Americas, Europe), both of which are sinistral, and the dextral cosmopolitan Lymnaeidae. Several groups of limpets have evolved – Acroloxidae (Eurasia), Latiidae (New Zealand; related to the South American Chilinidae), Lancinae (Lymnaeidae) (North America), and some planorbid limpets including those previously known as Ancylidae. The molecular phylogeny of Dayrat et al. (2011) has *Chilina* and *Latia* as sister to Amphiboloidea and the remainder of the group sister to the pyramidellids, but this relationship has not yet been confirmed in other molecular analyses which recovered a monophyletic Hygrophila (Klussmann-Kolb et al. 2008; Dinapoli & Klussmann-Kolb 2010; Holznagel et al. 2010; Göbbeler & Klussmann-Kolb 2011; Kocot et al. 2013). Anatomical synapomorphies have proved difficult to identify (Dayrat & Tillier 2002, 2003).

20.5.1.2.4.2.6 **Eupulmonata** The eupulmonates comprise the marine intertidal minute abalone-shaped Otinidae, the slug-like Smeagolidae and the limpet-shaped Trimusculidae, the mostly estuarine Ellobiidae, the marine and terrestrial Systellommatophora, and the entirely terrestrial Stylommatophora.

Many terrestrial eupulmonates have the ability to tolerate desiccation and temperature extremes. They lost the operculum in their early evolutionary history, but many compensate using mucus to either clamp or seal their aperture against hard surfaces or to cover the aperture with an epiphragm to protect themselves against desiccation and predators. Thus the eupulmonates became the most successful non-marine molluscs by adopting the necessary structural, physiological, and behavioural adaptations to conquer the land.

Eupulmonata is monophyletic within the Panpulmonata, and significant characters of the nervous, digestive, and reproductive systems support their monophyly (e.g., Haszprunar 1985b; Nordsieck 1992; Ruthensteiner 1997). It is also recovered in recent molecular phylogenies (Klussmann-Kolb et al. 2008; Mordan & Wade 2008; Dinapoli & Klussmann-Kolb 2010; Holznagel et al. 2010; Göbbeler & Klussmann-Kolb 2011), although it was not recovered by Dayrat et al. (2011).

20.5.1.2.4.2.7 **Ellobiida** This diverse assemblage of marine, estuarine, and semiterrestrial taxa are usually classified in five 'subfamilies' (Martins 1996), one of which is the northern-hemisphere terrestrial Carychiinae (or Carychiidae) (Weigand et al. 2014). It is likely that some of the other subfamilies will be raised to family level. Five independent terrestrial invasions are recognised in the group (Romero et al. 2016). Molecular data have shown that the marine limpet-shaped Trimusculidae, the small abalone-shaped intertidal crevice-dwelling Otinidae (Europe), and interstitial intertidal slugs (Smeagolidae) (Australasia and Japan) are related to ellobiids (Dayrat et al. 2011), and the authors of a recent revision (Romero et al. 2016) have even suggested that a single family (Ellobiidae) can encompass the whole group. Given the marked morphological disparity, we retain the families Trimusculidae, Otinidae, and Smeagolidae with the ellobiids in the Ellobioidea (see Appendix) as such an arrangement is more in keeping with the morphological data and is also largely concordant with molecular results.

20.5.1.2.4.2.8 **'Geophila'** The terrestrial stylommatophorans and marine and terrestrial systellommatophorans make up the remaining eupulmonates, with the former group being by far the largest. This grouping, named Geophila, was monophyletic in some morphological analyses (Barker 2001a; Dayrat & Tillier 2002), but Nordsieck (1992) found that ellobiids were sister to stylommatophorans. Geophila has been recovered in some molecular analyses (see review in Mordan & Wade 2008) but not the most recent ones (Dinapoli & Klussmann-Kolb 2010; Holznagel et al. 2010; Dayrat et al. 2011; Göbbeler & Klussmann-Kolb 2011). Thus, Geophila may not be a monophyletic group.

Taxa contained within 'Geophila' have eyes at the tip of the cephalic tentacles, the pedal gland is long and lies on the floor of the visceral cavity, they lack gizzard plates in the stomach, have a concentrated central nervous system and an unpaired dorsal jaw. If the Systellommatophora and Stylommatophora are not sister taxa, as suggested in molecular analyses (Dayrat et al. 2011; Göbbeler & Klussmann-Kolb 2011), characters such as the distal eyes on the cephalic tentacles must have been acquired independently.

20.5.1.2.4.2.9 Systellommatophora This monophyletic group (Dayrat et al. 2011), comprises slugs with one or two pairs of tentacles with distal eyes on the main ones. The Onchidiidae is a well-defined group of mainly marine intertidal slugs, with some possessing dorsal 'gills' and many with dorsal eyes. The family contains a few terrestrial taxa (Dayrat 2009), and the related tropical Veronicellidae and Rathouisiidae are entirely terrestrial.

The placement of onchidiids has been contentious in the past (see Britton 1984). They were transferred to the Opisthobranchia by Fretter (1943), a placement followed by some others (e.g., Hyman 1967), but it is now agreed that they are eupulmonates (e.g., Ruthensteiner 1997; Mordan & Wade 2008).

20.5.1.2.4.2.10 Stylommatophora The Stylommatophora contains the eupulmonate terrestrial snails and most of the slugs and is by far the largest group of terrestrial molluscs. It is very diverse, with about 25 superfamilies recognised (Bouchet et al. 2017) and around 30,000 species. Details of their classification are given in the Appendix. Unlike some other groups of heterobranchs, they form a rather uniform group anatomically, but they range in adult body size from just over a millimetre (some punctids) to giant achatinids with shells over 150 mm long. The relatively uniform anatomy and body form have frustrated attempts to classify this very large group. Many higher taxa have been erected that are probably not equivalent to those given the same rank in other groups of gastropods. Their basic classification, first outlined by Pilsbry (1900a), recognised three main groups based on the nature of the secondary ureter (Figure 20.25). The most plesiomorphic groups were included in the Orthurethra, with a straight ureter; the Heterurethra (Succineidae only) with a primitively open ureter running along the kidney and then at right angles along the rectum; and the Sigmurethra, including all the other families, with an S-shaped ureter. Baker (1955) later added Mesurethra for four families that lacked a ureter, and Solem (1959) added Tracheopulmonata as an additional group with a multi-looped ureter opening into a respiratory pore or directly to the exterior. The Pilsbry-Baker classification has been long used as a basis for classification (e.g., Nordsieck 1985; Tillier 1989; Barker 2001a), although it has been somewhat modified in recent years as a result of morphological and molecular phylogenetic studies (see discussion in Barker 2001a, and Mordan & Wade 2008 for a summary, and the Appendix for an outline of their classification).

21 Molluscan Research – Present and Future Directions

21.1 MOLLUSCAN RESEARCH PUBLICATIONS

Solem (1974) provided an early estimate of the number of papers published per year on molluscs. We have updated Solem's survey using publication data from the Clarivate Analytics [Thomson Reuters] Web of Science database®. Comparing estimates by Solem (1974) of publication productivity to the data available from the Web of Science for the same period shows similar publication patterns. The major difference between the two estimates is that Solem's publication data is on average 35% lower than the Web of Science data, except for the 1911–1920 period (12% lower).

The Web of Science data reported here resulted from an 'all databases' search using the string 'mollus*'. The 'all databases' option includes: The Web of Science Core Collection; the BIOSIS Citation Index; BIOSIS Previews; CABI: CAB Abstracts® and Global Health® databases; Chinese Science Citation Database℠; Current Contents Connect; the Data Citation Index; Derwent Innovations Index; Inspec®; KCI-Korean Journal Database; MEDLINE®; The U.S. National Library of Medicine® (NLM®) Premier Life Sciences database; the Russian Science Citation Index; SciELO Citation Index; and the Zoological Record. The general search string was also combined with taxon names and some keywords that highlighted some of the major areas of molluscan research. The string 'mollus*' retrieved almost 560,000 items.

We are well aware that the Web of Science and associated databases do not include all publications on molluscs. For example, a Google Scholar search using the same search string recovered an additional 90,000 items. Moreover, most citation databases probably underestimate the number of taxonomic and systematic works because most taxonomic journals were not indexed in early amalgamations of data (Wägele et al. 2014). This also substantially underestimates citation indices and can also affect the impact factor of a journal (Bouchet et al. 2016). Citation indices provide an important and readily available insight into the publication record of our research community, its breadth, and its contributions to our understanding of the biodiversity and processes of the world. Lastly, the reduction in the numbers of publications as the 2016 cutoff is approached in most categories is an artefact due to sampling error similar to the Signor-Lipps effect (Signor & Lipps 1982) but with missing publications rather than taxa. Unlike missing fossil taxa, many of these 'missed' publications are subsequently entered into the databases, and after a while, the number of publications typically rises above preceding years.

As also shown in Chapter 11, five arbitrary changes in publication rates can be recognised: Two corresponding to world wars and three to global recessions/depressions. Recoveries from these events appear to lag behind event cessation by about one year and are typically followed by substantial, sustained increases in publication rates. Of the major molluscan classes, publications featuring gastropods were dominant until 2002 when they were supplanted by bivalve publications. Gastropod research productivity appears to have been more affected by the global recession of the 1990s than that of either bivalves or cephalopods (Figure 21.1A). In contrast, publications in all three classes showed similar responses to the Great Depression and World War II. The overall patterns in publication productivity in the lesser classes shows much more of a 'boom and bust' pattern with frequent spikes in research output followed by a return to more typical levels of productivity (Figure 21.1B). The 2013 spike in both polyplacophoran and aplacophoran 'publications' is due to data in the Dryad Digital Repository (http://datadryad.org/) associated with the publication of Smith (2013). The actual publication numbers for this period are about half of the values reported here, erasing the most recent spike in aculiferan publications in Figure 21.1B.

21.1.1 PROVISION OF BASIC INFORMATION

Basic studies include biological systematics, an essential component that underlies all of modern biology. Many molluscs (and other invertebrates) are not even named, let alone had their relationships tested. Basic biological (feeding, reproduction, life history, habitat requirements, etc.) and distributional information is essential to make informed decisions for resource management, ecological studies, assessment of communities and their interactions, commercial shellfisheries, biomedical studies, etc., but this information is often lacking, even for many common species. Unfortunately, the gathering of basic biological (natural history) information is not given a high priority in universities and other research institutions, and even publishing such information is typically not encouraged by the editorial boards of many journals. Consequently, much of the cited basic biological information relies on data gathered from mainly European species in the 1800s to the mid-1900s. The reliability of using such information from other species as a surrogate depends on how well known the relationships between the taxa in question are. The more reliable the phylogenetic relationships, the more reliable the predictions are likely to be. Further, construction of phylogenies, especially those based on morphological data, are invaluable for the identification of missing data for critical taxa and many other purposes. We also need to find a way to continue educating graduate students in natural history and find incentives for journals to publish such studies.

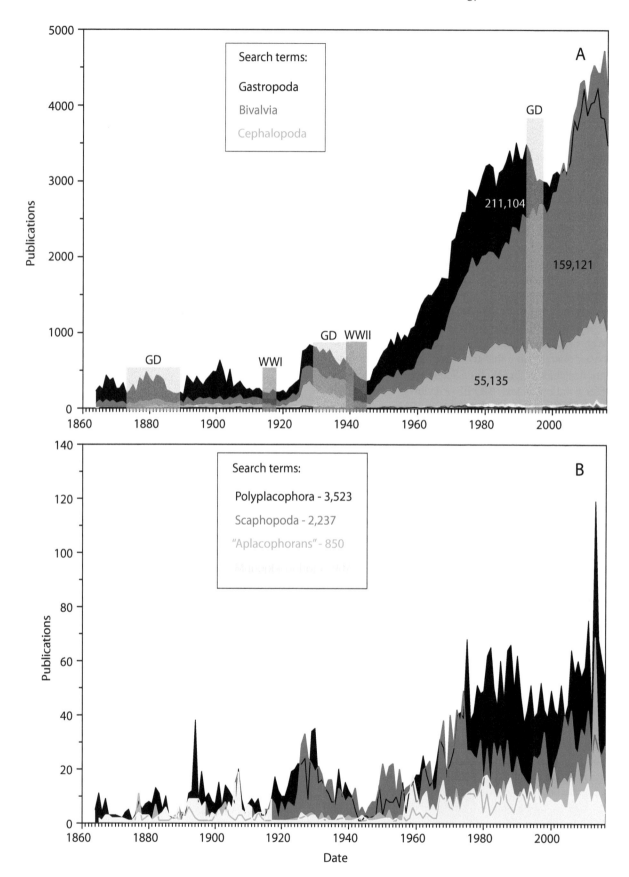

FIGURE 21.1 Area plots of molluscan publications by class between 1864 and 2016. Searches were based on each class name and its common synonyms. (A) Major classes, (B) Lesser classes. Five reductions in publication rates can be seen: Two coinciding with world wars (WW) and three with global recession/depressions (GD). The number of citations is also given in A.

Access to basic biological and ecological information is essential in management and the assessment of threatened species. It is also vital if the values of a particular species are to be assessed. Such assessment would usually include aspects such as economic or social importance, biological or ecological roles, and phylogenetic uniqueness. Unfortunately, for the great majority of species, only generalised data are typically available. How can this situation be rectified? Ways to achieve rapid increases in knowledge that have been successfully used in recent times include the use of museum collections and research workshops, as outlined below.

21.1.1.1 Use of Museum Collections

The largest single resource for distributional and general diversity information relating to molluscs is not the literature but the accumulated information in museum collections (Ponder 1999; Ponder et al. 2010; Marshall et al. 2018). By digitising the information in these collections, it can be placed on the web and accessed in various ways. In the relatively near future, these data should be available as combined information that will provide a powerful tool to give not only good distributional information but also other basic data. Such information might include what taxa are present in a particular area, what their accepted names are, and, because there is an historical component to these data, what their past and present distributions were.

In addition to distributional information, identification aids, checklists, and integrated systems will be available on the web. Following important data cleaning steps including taxonomic name resolution, integrated access to these data with links to literature and other resources will provide information about nomenclature, ecology and biology, molecular, phylogenetic relationships, stratigraphic occurrences, etc. Some resources, including the World Register of Marine Species (WoRMS) (nomenclature and classification), FossilWorks (palaeontological occurrences), GenBank (molecular data), and Morphobank (morphological data) are already online.

21.1.1.2 Research Workshops

Short-term gatherings of researchers in various parts of the world have been highly successful in quickly getting together much useful information about taxa and communities. The goal of these research workshops has been to pull together scientists and students to work on projects involving the local fauna within their areas of expertise. Molluscan workers have been particularly active in this area, and some examples with published results include the Tropical Marine Mollusc Programme (TMMP) funded by the Danish Ministry of Foreign Affairs and held in Thailand, India, Indonesia, Cambodia, and Vietnam (Hylleberg 2010). A series of important marine workshops held in Hong Kong (see Morton 2003d for review) and in Western Australia (Wells 2002) have been particularly successful. Others have been held in various areas including the 1988–2006 Azores Workshops (Martins 2009) and Lizard Island, Australia (Ponder 1975).

Other workshops have been taxon, habitat, or topic based. Many taxon-specific workshops focus on commercially important species such as cephalopods (Hatfield & Hochberg 2002), pectinids (Anonymous 2009), and abalones (Anonymous 2003). Other taxon-based workshops have featured invasive species such as *Dreissena* (zebra mussels), or threatened or listed taxa such as the alpine snails of Europe, while others address understudied or difficult groups (hydrobiids, Unionoidea, Ampullariidae, Conidae) (Coan & Kabat 2018). Workshops have also featured higher taxa such as scaphopods (Jong 2014), bivalves (Mikkelsen & Bieler 2004), 'Opisthobranchia' (Troncoso et al. 2011), and aplacophorans (Schander & Scheltema 2003). Topic-related workshops have included: genomic tools for molluscan research, archaeomalacology, habitat restoration, conservation, medical malacology, and sclerochronology (Coan & Kabat 2018).

On a much finer scale, and with different objectives, are the major workshops organised by Philippe Bouchet in New Caledonia and other parts of the western Pacific. These involve many workers contributing to the common goal to collect and identify the fauna in a particular area. With one exception (Bouchet et al. 2002), the results from these massive exercises have yet to be published but, when they are, they will undoubtedly rewrite much of what we know about Indo-West Pacific molluscan diversity.

21.1.1.3 DNA Barcoding

Using short segments of genes, particularly COI, to identify taxa was advocated by Hebert et al. (2003) and has generated substantial discussion of this approach (e.g., Darling 2006; Hajibabaei et al. 2007; Köhler 2007; Miller 2007; Waugh 2007; Wiens 2007). It is now widely accepted as a useful, but not necessarily infallible, tool (e.g., Schander & Willassen 2005; Hickerson et al. 2006; Elias et al. 2007; Siddall et al. 2009) that is generally known as DNA barcoding.

There are important differences, however, between barcoding as a reference system and the concept of a DNA taxonomy based only on limited sequence data (Köhler 2007). Barcoding as a tool for distinguishing species is conceptually and empirically problematic (Will et al. 2005) and should only be used cautiously when conventional morphology is difficult or impossible to apply (e.g., fragmentary material, larvae, etc.) for applications in molluscs (e.g., Kelly et al. 2007; Mikkelsen et al. 2007; Johnson et al. 2008; Layton et al. 2014).

21.1.1.4 Genetics

The term genetics, like ecology, is broadly used and can cover studies involving heredity, genes, chromosomes, variation, etc. Genetics had its roots in the selective breeding of plants and animals, but modern genetic research is very sophisticated and includes gene functions and interactions. It is not our intention to provide a synopsis of genetic theories, methods, or terminology here as these are readily available from many sources (e.g., Silva & Russo 2000; Sunnucks 2000; Deyoung et al. 2005; Karr 2007), but clearly the study of genetic variation is essential in understanding the structure and connections between populations, how speciation occurs, sources of variation, and in development. Not only have genetic studies provided much valuable and important scientific data, but there are also many practical applications, especially in commercial shellfisheries, sports fisheries, and the control of pest and invasive species. Genetic studies also provide a view of population structure unknowable from other observable traits, thereby providing vital information for management and conservation decision making.

Various techniques have been widely employed, including allozymes (Johnson 1976; Johnson et al. 1977; Turner 1977), DNA-RNA hybridisation (Karp & Whiteley 1973), DNA-DNA hybridisation (Milyutina & Petrov 1989), and radio-immuno-assay (RIA) (Harte 1992); today most genetic studies use sequences of amino and nucleic acids (e.g., RNA and DNA). Early sequencing of molluscs included the amino acid sequence of histone H2A from the cuttlefish *Sepia officinalis* (Wouters-Tyrou et al. 1982) followed later by base pair sequencing from C-DNA of *Octopus* rhodopsin (Ovchinnikov et al. 1988), and 18S rRNA sequences representing four molluscan taxa were included in an early molecular estimate of metazoan relationships (Field et al. 1988).

Suffice to note that molluscs have played an important role in our understanding of the principles and properties of genetics including insights into various evolutionary processes such as speciation, expatriation, vicariance, dispersal, epigenetics, developmental constraints, and species selection. Because of their low mobility and varied dispersal strategies, many molluscs are good subjects for studies on population genetics (e.g., Backeljau et al. 2001; Azuma et al. 2017; LaBella et al. 2017; Próckow et al. 2017; Richling et al. 2017; Rico et al. 2017). This branch of genetics is critical to our understanding of basic evolutionary processes as it involves the investigation of changes in the frequencies of alleles, genotypes, and phenotypes within populations, and hence microevolution within species. These changes are driven by selection, mutation, migration, and random genetic drift as well as non-random mating and recombination, some of which are influenced by changes in the abiotic realm.

21.2 PALAEOBIOLOGY

Since the inception of modern palaeobiology in the early 1900s, molluscs have been prominent foci in numerous research areas and time periods; some studies even predated the formal recognition of the field. For example, macroevolutionary trends, driven by heterochronic change in ammonite morphology, were characterised by Alpheus Hyatt (1838–1902) (Gould 1977), while the malacologist William H. Dall (1877) considered stasis in the fossil record and proposed an early model of punctuated equilibrium (Lindberg 1998a). Many of the major questions addressed by palaeobiology also appeared earlier, including ideas of catastrophism, transformation, and uniformitarianism (Cuvier 1796; Lyell & Deshayes 1830) and ultimately natural selection (Darwin 1859). The earlier Age of Exploration (see Chapter 1) had also 'primed

the pump' by stocking the museums of Europe, and later North America, with collections of living and fossil molluscs from around the world (Bentham et al. 1858; Dance 1966). Initially, these collections were primarily the subject of taxonomic work, and as workers dealt with an ever-increasing and unfamiliar diversity from foreign regions, many expanded their understanding of more familiar local taxa and described new higher taxa to account for the increasing varied character mosaics they saw. Curation of new fossil material also revealed rare specimens with better-preserved characters which recalled similarities with living taxa. For example, in Germany Wilhelm Wenz (1886–1945) hypothesised that fossils with muscle scars could provide evidence of the presence or absence of torsion in Early Paleozoic univalve shells (Wenz 1940).

In North America J. Brookes Knight (1952) not only further championed the merging of neontological and palaeontological data to understand the relationships of Paleozoic univalve shells but also ventured into early molluscan phylogeny, diversification, and functional morphology. From these beginnings, molluscan palaeobiology thrived in the second half of the 20th century, especially in the United States. Leaders in the field include James Valentine, Ellis Yochelson (1929–2006), Robert Linsley (1930–2006), John Pojeta (1935–2017), David Raup (1933–2015), Stephen Gould (1941–2002), Steven Stanley, Geerat Vermeij, and David Jablonski, all of whom used molluscs to examine questions of macroevolution, background and mass extinction, speciation, evolutionary morphology, and distributions in space and time.

In Australia, Bruce Runnegar expanded our understanding of Cambrian molluscan diversity, especially their shell microstructure, and together with John Pojeta reconsidered molluscan phylogeny from a more palaeontological perspective and described a new class of extinct molluscs – the Rostroconchia (Pojeta et al. 1972; Runnegar & Pojeta 1974). Working from Moscow, Russia, Pavel Parkhaev made major contributions to the enigmatic Cambrian 'small shelly fossils'. In Europe, John Peel further developed the lines of evidence first laid out by Knight for the diversity and functional morphology of Paleozoic molluscs. Lastly, Derek Briggs and Simon Conway Morris, who are best known for their analyses and re-interpretation of the mollusc-poor Burgess Shale fauna (Gould 1989), would later make substantial contributions to our understanding of Early Paleozoic aculiferans and halkieriids, respectively.

The contributions and advancements made by these and other palaeobiologists were and remain largely driven by an expansive view of the potential of fossils and the patterns they provide integrated with a fundamental knowledge of living molluscan anatomy, biology, and ecology. Such a perspective is essential as fossils are points in a historical continuum that ends in the living biota, which is a consequence of this history. To quote James Valentine from his seminal book *Evolutionary Paleoecology of the Marine Biosphere* (1973, p. 14) 'uniformitarianism must work in both ways, from the past to the present as well as from the present to the past'. We continue to encourage a breakdown in the distinction between neontology and palaeontology, and we remain convinced that molluscan hard parts contain more information about their

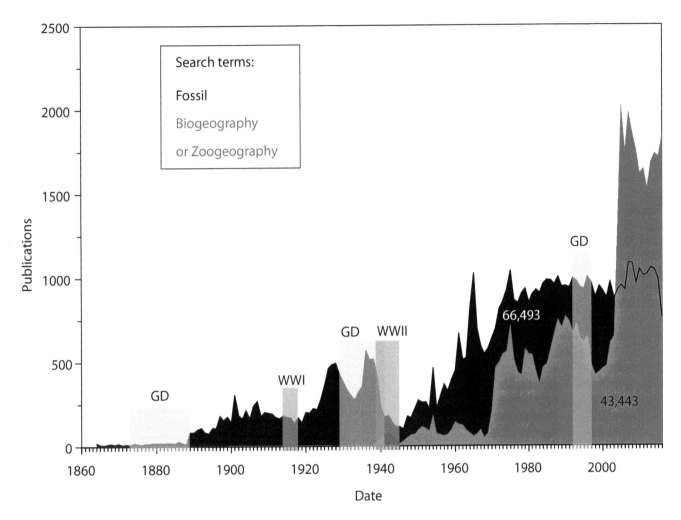

FIGURE 21.2 Area plots of molluscan publications by the topics of fossil and biogeography or zoogeography between 1864 and 2016. Five reductions in publication rates can be recognised: Two coinciding with world wars (WW) and three with economic downturns (GD). The number of citations is also given.

anatomy, biology, and ecology than is often credited and that the long and rich fossil record remains the only direct source of evidence of its evolutionary history.

Molluscan publications featuring fossils generally follow the track of other molluscan investigations (Figure 21.2). Paleontological publications experienced sustained increases following the Long Depression (1873–1886) and World War I. The Great Depression (1929–1941) corresponds to a substantial drop in publications, while the nascent recovery that began in 1934 was terminated by World War II. Sustained increases in publications followed World War II but plateaued between the mid-1970s and 2007 after which publication rates again increased (Figure 21.2). Unlike some other molluscan research topics, publications on fossils showed relatively little response to the global depression of the 1990s.

21.3 TAPHONOMY

To exploit the fossil record to the full, one must recognise its limitations (Kidwell & Holland 2002). The accumulation of remains of the living fauna or death assemblage is the raw

material for the fossil record, but taphonomic processes alter both biodiversity and abundances in the death assemblage and may bias reconstructions of abundances, biodiversity, and environments. Taphonomy is the study of what happens to organisms following their death and includes decomposition, post-mortem transport, burial, compaction, and other chemical, biological, or physical events. The remains of different molluscan taxa have different taphonomic outcomes which are determined by both anatomy and habitat. For example, shelled taxa (e.g., bivalves) living in low-energy environments have high probabilities of making it through the gauntlet of events associated with becoming part of the fossil record, while others such as nudibranchs living along exposed rocky shores rapidly become non-existent following decomposition. Other taphonomic outcomes are not as severe but still can affect abundances or representatives of specific habitats (Kidwell 2001; Kidwell 2002).

Death assemblages and the fossil record provide important estimates of biodiversity and community composition and surprisingly strong signals of rank-order abundances of living species (Kidwell 2001, 2002, 2008) and therefore are potentially important in estimating biodiversity and community composition. When combined with past environmental records, they can be used to estimate species responses to climate change. Smith (2008) surveyed death assemblages at ten headlands in northern New South Wales, Australia, and then compared the resulting biodiversity indices to species lists for the living fauna. Smith found that data from two or more sites fully represented both nearshore diversity and regional diversity of bivalves, but regional diversity of gastropods was poorly represented. Death assemblages can also identify past habitat associations. Russell (1991) compared four subtidal death assemblages from two habitats (rocky shore and sandy beach) with four fossil assemblages on San Nicolas Island, California. While death assemblages were found accumulating throughout the rocky shore habitat, sandy beach death assemblages were only found in localised sediment traps, typically adjacent to sandstone benches or boulders. Comparison of the Pleistocene fossil assemblages with the modern death assemblages reveals that the habitat could be confidently inferred from relative abundance measures of the molluscan species present due to the shorter time-scale and better environmental and anthropogenic resolution. For example, Edgar and Samson (2004) examined nearshore death assemblages from 13 dated sediment cores ranging in age from 120 years to the present in an inlet in southeastern Tasmania. They found declines in mollusc species richness at all sites during the last century. They correlated these declines with the presence of a scallop fishery which also ultimately declined and failed. In their study, the molluscan death assemblages provided previously undocumented evidence of major recent losses in mollusc biodiversity.

21.4 SYSTEMATICS

Today the practise of systematics (see Chapter 1 for an overview) is very different from what it was for the previous two centuries. Today's classifications are based on phylogenetic hypotheses, subject to recurrent testing on a scale never imagined or experienced by the monographers of the 19th and 20th centuries. In the past, type material and allocated specimens were only occasionally examined to test an earlier classification. Today, classifications can be readily tested with additional taxa or sequences, and the rate of addition of new character sets, especially molecular ones, as well as new taxa, continues to increase every year. Phylogenetic classifications of today are also valued for their predictive value in disciplines as disparate as ecology, physiology, biogeography, medicine, and conservation biology. Because this is all beyond the capability of any single researcher, we now see teams of specialists, often from around the world, working together. In addition, the research breadth of many of the current systematists has broadened and deepened. Not only is extraordinary knowledge of a specific taxon still required, but additional knowledge of anatomy, developmental biology, molecular biology, and palaeontology, etc. (all at multiple scales) is also often required, as well as a working understanding of the methods of character analysis and phylogenetic hypothesis testing.

21.5 SYSTEMATIC, PHYLOGENETIC, AND EVOLUTIONARY STUDIES

Molluscan publications in biological systematics have been a mainstay of molluscan research from its very origins through to today (Bouchet 1997; Bouchet et al. 2016). Publication rates in molluscan taxonomy and systematics were relatively low and flat until about 1960. Slowly increasing numbers of these publications continued until 1990 which saw a 300% increase in publication rates (Figure 21.3). This yearly increase has moderated slightly but continues today. Unlike many of the other trends seen in molluscan publications, there is little effect of financial downturns and global conflicts on taxonomic and systematic and phylogenetic publication rates. Papers reflecting phylogenetic studies first occurred in the late 1890s, and like taxonomy and systematics output, phylogenetic papers began to slowly increase in the early 1960s; in 2004 there was a second notable increase in publication rate which is sustained today (Figure 21.3).

21.5.1 DIFFERENT DATASETS

Early molluscan taxonomic work primarily focused on shell characters, including that of Linnaeus (1758), the Sowerby family (Cleevely 1974), Reeve (1843–1878), Carpenter (1857, 1864), and sometimes on molluscan anatomy (e.g., Cuvier 1795). In Europe, molluscan phylogeny was investigated by H. G. Bronn (Bronn 1862, 1862–1866), H. von Ihering (von Ihering 1876b), and beginning in the 1880s, R. Lankester, J. Thiele, P. Pelseneer, and H. Simroth. In the United States, taxon focus was more directed at individual classes or orders. For example, bivalve phylogeny was investigated by the embryologist W. K. Brooks (1876), the invertebrate zoologist B. Sharp (1888), and the neo-Lamarckian R. T. Jackson (1890). Other neo-Lamarckians proposing molluscan phylogenies in the 1890s included the palaeontologist A. Hyatt (ammonites) and the malacologist W. H. Dall (gastropods and bivalves). Other American malacologists such as H. A. Pilsbry, documented internal anatomy (reproductive systems) in eupulmonate systematics (Pilsbry 1948), but anatomical characters were sparse in the marine taxa monographed in *Manual of Conchology* by G. W. Tryon where shell characters were the standard.

Shell characters have continued to be refined, and with the advent of geometric morphometrics (Bookstein 1997; Zelditch et al. 2012; Adams et al. 2013) and high throughput image capture and processing (Boyer et al. 2015; Hsiang et al. 2016; Kahanamoku et al. 2017) dense, highly resolved, quantitative shell character data are now obtainable and subject to robust analysis. Anatomical character sets have also benefitted from new imaging and reconstruction techniques (e.g., Metscher 2009a, b; Kerbl et al. 2013; Candás et al. 2016), which were refined for molluscs by the group in Munich, Germany, led by Gerhard Haszprunar (e.g., Bäumler et al. 2008 and subsequent papers). The ability to process large numbers of individuals or taxa remains elusive but will probably be greatly improved because of advances in machine learning which

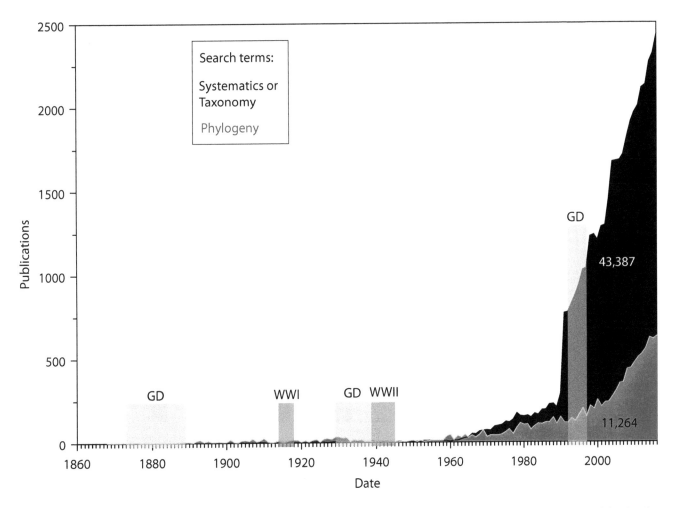

FIGURE 21.3 Area plots of molluscan publications in systematics or taxonomy and phylogeny between 1864 and 2016. Minimal reductions in publication rates occur during World Wars I and II and during the Great Depressions. The number of citations is also presented.

can potentially automate accurate image capture. At present a dense sampling of vitally important ultrastructure characters also remains elusive given the time-consuming protocols and procedures associated with transmission electron microscopy.

21.5.1.1 Molecular Datasets

Modern molecular datasets have proliferated over the last 30 years; in early 2019 nucleotide data for molluscs in GenBank exceeded 3,800,000 items (Figure 21.4), but there were only 26 genome assemblies and 384 mitochondrial genomes. Moreover, the diversity of molecular data is concentrated in a small fraction of total molluscan diversity. The ten molluscs with the most GenBank entries are: *Aplysia californica*, *Biomphalaria glabrata*, *Octopus bimaculoides*, *Bithynia siamensis goniomphalos*, *Mytilus galloprovincialis*, *Elliptio complanata*, *Arion vulgaris*, *Mizuhopecten yessoensis*, *Lymnaea stagnalis*, and *Crassostrea angulata*. Six of these species are also part of genome projects, and their ranking is not surprising given their roles in neurology, parasitism, behaviour, and management. Clearly, the totality of molluscan molecular diversity, and its potential across the breadth of biology, have yet to be sampled. Currently, 99% of our nucleotide sequence data comes from three classes – gastropods,

bivalves, and cephalopods, and 65% from gastropods alone (Figure 21.4). Half of the top ten taxa are heterobranch gastropods; *A. californica* alone accounts for almost one-third of the nucleotide items. The remaining taxa include one cephalopod, three pteriomorphian bivalves, and a unionid bivalve.

Not surprisingly, the number of nucleotide items available in each class is positively correlated ($r^2 = 0.9516$) with the number of taxa sequenced (Figure 21.4). However, we suspect that taxon numbers from GenBank are inflated because of nomenclatural issues with many entries. For example, although 15 monoplacophoran taxa are reported as having nucleotide data, only five (one-third) are identified to species rank. If this ratio is present in other classes it is likely that molecular data in GenBank represent only about 9.2 to 12.2% of the total estimated molluscan species diversity (see Table 21.1).

The number of nuclear genomes should be increased in the near future as sequencing costs drop and assembly and informatic approaches improve.

New sources of molecular data are also likely. While most data for molecular work have traditionally come from soft tissues, there is also active research to recover both genomic and non-genomic molecules from the shells of fossils and sub-fossil molluscs (Geist et al. 2008). In a study of the ampullariid

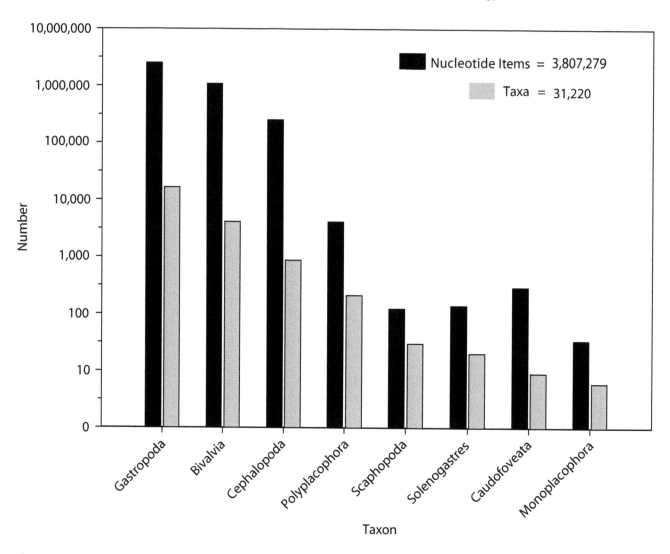

FIGURE 21.4 Grouped bar chart of the number of molluscan nucleotide items and the number of molluscan taxa in the National Centre for Biotechnology Information database GenBank (14 Feb 2019) by class.

TABLE 21.1

Percentages of Molluscan Nucleotide Items, Sampled Taxa, and Estimated Class Species Diversity Represented in the NCBI GenBank Database (14 February 2019)

	Gastropoda	Bivalvia	Cephalopoda	Polyplacophora	Scaphopoda	Solenogastres	Caudofoveata	Monoplacophora
% of molluscan spp. diversity	73.81%	24.09%	0.78%	0.62%	0.40%	0.19%	0.09%	0.02%
% of GenBank molluscan items	65.13%	28.18%	6.58%	0.11%	0.003%	0.004%	0.008%	0.0009%
% of GenBank molluscan taxa	75.11%	19.15%	3.88%	1.09%	0.20%	0.31%	0.21%	0.05%
Estimate of molluscan spp. diversity in GenBank	9.5–12.5%	7.5–9.9%	39.8–52.6%	9.3–12.2%	2.5–3.4%	10.2–13.5%	13.8–18.2%	11.7–15.5%

Pomacea, Andree and Lopez (2013) were able to extract and sequence the COI gene from ten-year-old empty shells, and their sequences were identical to tissue-based samples from *P. canaliculata*. In another recent study Hawk and Geller (2019) used decades-old shells of the abalone *Haliotis sorenseni* to estimate historical mitochondrial molecular diversity of this endangered taxon. The use of both fossil and recent shells as sources for molecular data opens the vast resources of museum collection holdings and provides both a window into deeper time and potentially even access to molecules from extinct taxa.

21.6 BIOGEOGRAPHY

Biogeographic or zoogeographic research underwent its first growth phase during the final years of the Great Depression (Figure 21.2). This growth was reduced by World War II but rebounded quickly in 1945, and research productivity remained variable year to year until 1970, when a second productivity increase occurred, primarily under the term biogeography. Most early molluscan zoogeographic and biogeographic studies often concluded that dispersal was the most likely model of distributional history (Briggs 1974; Darlington 1982). Dispersal was consistent with the presence of a pelagic larval stage in many marine species but was less applicable to terrestrial taxa. There were also obvious examples of vicariance in the marine environment, such as the isolation of tropical eastern Pacific and Caribbean molluscan faunas by the emergence of Panama (Vermeij 1978; Vermeij & Petuch 1986; Coates et al. 1992; Jackson et al. 1993). Glaciations in both the North and South Pacific and North Atlantic (Cox et al. 2014) also divided once continuous faunas. In molluscan biogeography, the age of the formation of barriers, and the timing of their removal, are important in hypothesis testing. Examples include the aforementioned emergence of the Panamanian Isthmus approximately 3.0 Mya (Jackson & O'Dea 2013), or the opening of the Bering Strait and the establishment of a Pacific–Atlantic Seaway about 5.32 Mya (Gladenkov et al. 2002).

The late 1970s and early 1980s were an especially rich period of development of biogeographical theory and methodology, following the introduction of the idea of vicariance into biogeographic studies (Croizat et al. 1974; Nelson 1974; Platnick 1976). The early model of Croizat was called panbiogeography and was used by Climo (1989) to interpret the biogeographical patterns of New Zealand using terrestrial gastropods. Terrestrial snails, with their apparent limited dispersal potential, have long been favoured as models for testing for vicariance in biogeographical studies (e.g., Hausdorf 1995; Gittenberger 1999; Emberton 2001; Naegele & Hausdorf 2015). In aquatic environments vicariance is often hypothesised in freshwater taxa (e.g., Davis 1982; Altaba 1998; Liu & Hershler 2007; Rintelen et al. 2014) while in marine taxa dispersal is often argued (e.g., Gonzalez-Wevar et al. 2017; Yahagi et al. 2017). Vicariance has also been implicated in the geographical structure of the thiotrophic endosymbiont bacteria hosted by *Bathymodiolus* mussels at deep-sea hydrothermal vents and hydrocarbon seeps (Ho et al. 2017), but usually both vicariance and dispersal are likely to have been operating in parts of the distributional history of most taxa (e.g., Hershler et al. 2015; Ho et al. 2015; Bolotov et al. 2016). With the advent of molecular techniques and phylogeography, finer scale dispersion and dispersal patterns are now being resolved and are providing additional evidence of the presence of both vicariance and dispersal events.

21.7 DISCOVERY

Intensive sampling over about the last hundred years has given us major insights into molluscan diversity. For example, highlights of modern deep-sea exploration were the discovery

in the 1950s of the first living monoplacophoran (see Chapter 14) and the sampling of deep-sea hydrothermal vents and cold seeps (see Chapter 9) in the late 1970s and 1980s, which brought to light a vast array of new animals including many molluscs. Studies on these faunas have enriched our knowledge and views on molluscan biology and evolution.

New taxa will continue to be discovered. Some will come from remote habitats which remain difficult to access and sample (e.g., the deep-sea, interstitial, groundwater) as well as ones which are thought to be well sampled, at least regarding larger taxa. This later source of diversity harbours mostly smaller, micro-taxa (e.g., Albano et al. 2011) and has been well demonstrated at tropical Pacific locales such as New Caledonia (Bouchet 1979; Bouchet et al. 2002) and in the Philippines (Bouchet 2009; Bouchet et al. 2009). While the New Caledonia sampling effort revealed over 2,700 species of marine molluscs (several times the number previously reported for other regions), this diversity was subsequently eclipsed by the Philippine sampling project which suggested that species diversity there was 40 to 50% greater than in New Caledonia. Many of these taxa were represented by few specimens (i.e., they are rare), and many are small (<5 mm) in size. Few nearshore regions, within or outside the tropics, have been sampled with this kind of effort. Thus, when shelf, slope, and deeper abyssal habitats are added for consideration, one can only assume that marine molluscan diversity is underestimated.

Similar discrepancies probably exist in terrestrial and freshwater faunas, and with increasing collecting effort associated with surveys and monitoring for global change, estimates will probably go up and the undescribed species diversity will also increase. Bouchet et al. (2016) have estimated that at the current pace of description it will take 300 years to name the ~150,000 undescribed species currently known. Enhanced use of molecular data will be required if any understanding of the distribution and the real loss of biodiversity and ecological services is to be gained, especially in tropical land and freshwater faunas (Bouchet 1997).

While the discovery of new taxa will continue, the discovery of significant new molluscan habitats probably concluded with the discovery of vent and seep faunas near the end of the last century. Certainly, more diversity will undoubtedly be found in marginal habitats, and our understanding of the breadth of molluscan niches is limited, but this does not preclude the continuance of discoveries associated with these habitats and niches.

21.8 THE FUTURE

In 1882 the American malacologist William Healy Dall, in his vice-presidential address to the Biology section of the American Association for the Advancement of Science, reviewed potential research programs in malacology (Lindberg 1998a). In a time of primarily descriptive malacology, the study areas highlighted by Dall included biogeography, deep-sea faunas, molluscan development, behaviour, and evolution – all still worthy and rich topics today. While progress and knowledge in these areas have been gained in the subsequent 136 years, much remains to be done.

In this volume, we have tried to incorporate and integrate data from both fossil and living molluscs to provide a historical dimension to many of the topics covered. Over 90% of the molluscs that have ever lived are now extinct. Some of these provide interesting and often baffling morphologies, palaeoecologies, and palaeobiological attributes, which complicates the understanding of their relationships to living taxa as well as interpreting their roles and interactions in past communities. The deep molluscan lineages have survived regional and global climate change, mass extinction events, ocean acidification, atmospheric changes, and no doubt other catastrophic events. The descendants of these taxa have survived and speciated to be one of the most physiologically, morphologically, and ecologically diverse groups on our planet.

Understanding the relationships of living taxa and their extinct ancestors allows the study and elucidation of many important biological traits through time. Such an investigation can require the involvement of disciplines as disparate as ecology, physiology, behaviour, biogeography, mineralogy, biochemistry, and conservation biology. This illustrates an important change in molluscan research that is already happening – the development of research groups or teams comprising scientists with a wide variety of expertise. Such collaborative research is essential in the future if complex questions that cut across traditional disciplines are to be addressed. Too often, we have focused on a specific group or processes of interest and ignored the incredible diversity afforded to us by the phylum. Molluscs are model organisms for evolutionary and developmental studies and investigations of neurobiology, endocrinology, ecology, extreme environments, and so on. Now is the time to start connecting the dots digitally and work to integrate our existing knowledge so we can quickly and robustly access and build on the hundreds of thousands of molluscan papers containing text and images from investigations across the entire breadth of biology and the history of life. To make progress, there must be a shared vocabulary and a network of malacologists coordinating this enterprise. We discuss below some possible future tools and some of the exciting lines of current research directions and possibilities.

21.8.1 Text Mining

The molluscan literature is vast and deep, and much of it, both old and new, is now digital. Google Scholar reports 327,000 items and there are 458,973 in the Web of Science. Molluscan research would be substantially enhanced if these items were turned into data (entities, concepts, relationships) which were subject to text analytics and other techniques which would ultimately classify, link, and establish meaning and relationships in our digital resources. Text mining is also applicable to museum records, geological and geographic data, and other collection and supplemental curatorial records. Text mining is already being applied to oceanographic, geological, and palaeobiological data, including the EarthCube building block project of the U.S. National Science Foundation (Richard et al. 2014), which is building a cyber infrastructure to support text mining and the construction and augmentation of knowledge

bases in the geosciences and biosciences. PaleoDeepDive (Peters et al. 2014) extracts and repackages data from text, tables, and figures in palaeobiological publications. A molluscan text mining effort will require oversight and coordination (e.g., a research coordination group) and a molluscan ontology, as well as text mining software.

21.8.2 New Technology

The introduction of new technology will play a large role in a better understanding of molluscan morphology. The increasing use of non-destructive three-dimensional imaging techniques such as confocal laser scanning microscopy (cLSM), optical projection tomography (OPT), magnetic resonance imaging (MRI), and micro-computed tomography (micro-CT) is revolutionising the acquisition of 3D anatomical data (Boistel et al. 2011; Laforsch et al. 2012; Faulwetter et al. 2013) and will only become more sophisticated and 'user-friendly' in the future.

While computed tomography and other imaging technologies made huge advances in the medical sciences, their usefulness in small-size objects was limited in early applications because of poor resolution. Small (micro-CT) scanners were invented in the early 2000s, and we are now seeing their use in resolving aspects of three-dimensional molluscan anatomy previously difficult to visualise. Recent developments in this equipment can now resolve down to 100 nm, and specimens scanned may be up to 200 mm in diameter. Several workers have already applied these methods to molluscan research (Golding & Jones 2007; Golding et al. 2009a; Metscher 2009b; Monnet et al. 2009; Alba-Tercedor & Sánchez-Tocino 2011; Handschuh et al. 2013; Kerbl et al. 2013; Candás et al. 2016).

Besides the aforementioned high-end technologies, software applications like *AutoMorph* (Automated Morphometrics: Image segmentation, 2D and 3D shape extraction, classification, and analysis) (Hsiang et al. 2018) make possible high throughput image analysis using digital image stacks of multiple specimens at either micro- or macroscopic size levels, thereby enabling community-scale morphometrics. Using such rapid throughput methods on museum specimens could provide a database with immediate utility in aiding ecological studies, monitoring community composition, and the detection of invasive species. The gathering of museum anatomical data is also likely to accelerate as well with advances in 3D microscopic imaging techniques and machine learning applications in 3D reconstruction.

At the other end of the spectrum are new advanced light source (ALS) technologies which enable high-resolution imaging within shells and other hard structures. ALS uses synchrotron-produced bright beams of x-rays to penetrate small objects and produce highly resolved images of internal structure (Figure 21.5).

21.8.3 Phylogenetic and Evolutionary Research

Ponder and Lindberg (2008) and Sigwart and Lindberg (2015) discussed two major outstanding issues in molluscan

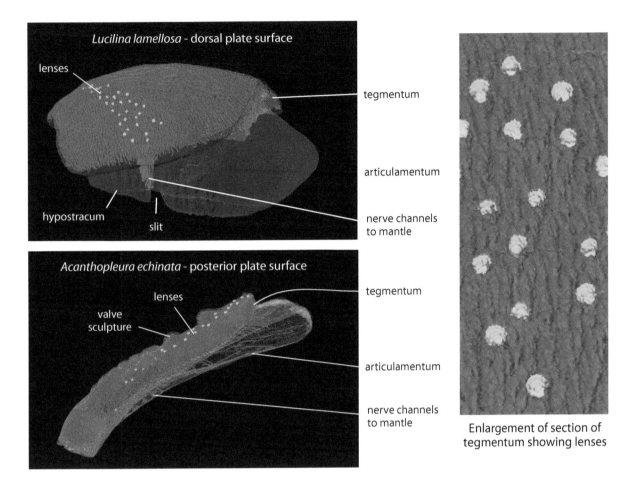

FIGURE 21.5 Advanced light source false colour images of polyplacophoran valves. Pink = aesthete canals; green = lens. Courtesy of J. Sigwart.

phylogenetic research: (1) the resolution of the sister taxon to Mollusca, and (2) the deep branches within the molluscan tree, both living and fossil. Regarding issue (1), available genomic data now support Brachiozoa as the sister taxon of living Mollusca (Stechmann & Schlegel 1999; Luo et al. 2015) (see Chapter 12), but the second issue – the connections of the deep molluscan branches – remains obscure. Addressing these deep branches necessitates a more systematic approach to filling in missing data, and high on the list of priorities is the incorporation of early fossil taxa. Datasets will also need to include developmental and physiological data, digital morphological descriptions, complete genomes, and other broadscale molecular and ultrastructural data. The use of total evidence analyses and reconstructions will undoubtedly assist in elucidating some of these deep branches.

The need for a more systematic approach for gathering data became readily apparent to us in the preparation of these volumes. While on the one hand we were often amazed by the considerable data that are available, we were also often reminded of just how few taxa some data are based on. For example, Sigwart and Sumner-Rooney (2015) tabulated the number of species for which neural descriptions were available in the 'minor' classes (Table 21.2). Except for the monoplacophorans, the number with neural data is less than 10% of estimated species diversity in each taxon. Surprisingly,

TABLE 21.2

Neural Anatomy Coverage in the Five Minor Classes

Taxon	Species Examined	Total Diversity
Monoplacophora	7 (21.9%)	32
Solenogastres	25 (10.4%)	240
Caudofoveata	10 (8.3%)	120
Polyplacophora	60 (7.5%)	800
Scaphopoda	20 (3.9%)	517

Neural studies from Sigwart and Sumner-Rooney (2015) and diversity estimates from Lindberg et al. (2004). Percentages represent the percentage of the total diversity of each taxon.

the monoplacophorans are over 20% of estimated diversity, which is probably due to their unique history of pseudo-extinction and supposed 'living fossil' status. While the major classes (Gastropoda, Bivalvia, and Cephalopoda) have been studied to a greater extent, their substantially larger diversities probably keep the percentage of species examined low (see Table 21.2).

In addition to a low sampling of molluscan species diversity, our knowledge is also limited by repeated sampling of certain taxa often determined by economic (aquaculture, fisheries) or husbandry considerations. In gastropods, heterobranchs account for over 60% of the taxa cited more than ten times in the *Microscopic Anatomy of Invertebrates* chapters on molluscs (Figure 21.6). Common taxa include the pond snail *Lymnaea stagnalis*, the land snails *Cornu aspersum* and *Helix pomatia*, and the schistosomiasis vector *Biomphalaria glabrata*. Caenogastropods with more than ten index entries are also well represented (29%) and include *Littorina littorea*, *Buccinum undatum*, *Conus* spp., and *Busycotypus canaliculatus*. Vetigastropods are represented by the commercially important *Haliotis* spp., and patellogastropods by the intertidal 'Guinea pig' *Patella vulgata*. There were no neritimorphs with more than ten entries, however four *Nerita* spp. and the euryhaline *Neritina reclivata* and *Theodoxus fluviatilis* are referenced.

In cephalopods, taxa supporting large-scale commercial fisheries such as *Sepia officinalis*, *Octopus vulgaris*, and loliginid squids are well represented, while the remaining taxa include the curiosities *Nautilus* spp., *Spirula*, and *Vampyroteuthis*. Most bivalves with over ten entries are commercially important pteriomorphians and venerids (Figure 21.6).

In the minor classes polyplacophorans and aplacophorans are represented by a handful of species, while in comparison, the monoplacophorans and scaphopods are well represented by referenced taxa (Figure 21.6). As discussed above, monoplacophorans have been a focus of interest since the 'discovery' of extant taxa in 1956 (Lemche & Wingstrand 1959) and have continued to be well documented when specimens are procured (e.g., McLean 1979; Wingstrand 1985; Haszprunar & Schaefer 1996; Schaefer & Haszprunar 1996; Ruthensteiner et al. 2010). Scaphopods have also received more attention recently (e.g., Reynolds 1990a,b; Steiner 1990, 1992a,b, 1998b, 1998a; Ruthensteiner 1991; Reynolds 1992, 2002).

Under-sampling limits our understanding of molluscan systems and biology at all levels – from our estimates of molluscan diversity to our understanding of variation in biochemical pathways and our polarity hypotheses for character transformations at all taxonomic levels. In systematics, under-sampling results from various limitations associated with taxonomy, including specimen procurement and availability, time-consuming techniques or limited observational time, and the lack of practitioners (Bouchet et al. 2016). All studies are further confounded by the normal distributions of most biological features, which require an adequate level of sampling to observe rarer, outlying states.

Thus, our knowledge of most molluscan characters used in analyses is probably limited to a few of the commonest states because of the under-sampling of most taxa. While every attempt is typically made to describe and account for the range of variation in populations and taxa, most taxonomic work is limited to less than 10–20 specimens. An exception to this generalisation was the work of Pelseneer (1920, 1929) who set out to address the question of heritable variation in

molluscan characters. Because of the large sample sizes, often over 3,000 individuals, Pelseneer recorded both substantial variations and probable teratologies in the gastropods he observed. For example, in the nassariid *Tritia reticulata* he observed variation in the number of cusps on the central teeth of the radula (Figure 21.7) and 11 different morphologies for the posterior pedal tentacles (Figure 21.8). The most common was a single symmetrical pair (85%), followed by a single median tentacle (10%). The remaining nine morphologies were found in only 5% of the individuals. Pelseneer's data can be used to estimate on average how many individuals need be observed to discover alternative character states. For example, in *Tritia reticulata* ten specimens would record two character states, but the next three states would probably require over 70 individuals and the final five states more than 400 (Figure 21.8).

Most population, reproductive, and experimental ecological studies are adequately replicated and sampled to ensure statistical rigour, while other large sample initiatives are frequently applied to fisheries (Tirado et al. 2017) or conservation concerns. Museum collections also provide major resources of shell morphology and, often, anatomical material. However, the data collected in many relatively well-sampled studies are sometimes limited (e.g., size, sex, reproductive state, dry weight), with little or no recording of other visible traits or features.

21.8.4 BIODIVERSITY

Biodiversity has veracity and is the product of lineages subjected to natural selection and stochastic events through time. Diversity is thought to provide communities with resilience to perturbation through functional redundancy. Lastly, there is also a human aesthetic for rich, diverse natural settings.

Understating biodiversity, both molluscan and in general, is important for several reasons. One common argument is the incompleteness of our knowledge of it, especially in marine tropical regions and in non-marine habitats (Bouchet 1997, 2006, 2009; Ponder & Lunney 1999; Bouchet et al. 2002; Lydeard et al. 2004; Strong et al. 2008; Albano et al. 2011). Incompleteness hinders an accurate perception of current community composition, organisation, and potential interactions, as well as our ability to make informed decisions regarding the status and potential conservation concerns for various taxa and their habitats. At larger scales, underestimated biodiversity can affect the testing of macroevolutionary hypotheses such as latitudinal diversity gradients, tempo, and mode in molluscan evolution, extinction dynamics, etc. (Smith et al. 1993; Roy et al. 1998, 2009a; Jablonski & Roy 2003; Régnier et al. 2009, 2015). More recently arguments have been constructed that emphasise the importance of biodiversity for maintaining ecological services that molluscs and other taxa provide (Balvanera et al. 2006; Dudgeon et al. 2006; Worm et al. 2006; Coen et al. 2007; Vaughn 2010; Mace et al. 2012). A consensus regarding the relationship between diversity and services has been difficult to obtain (Balvanera et al. 2006), often because of limited knowledge of the ecological role of a particular

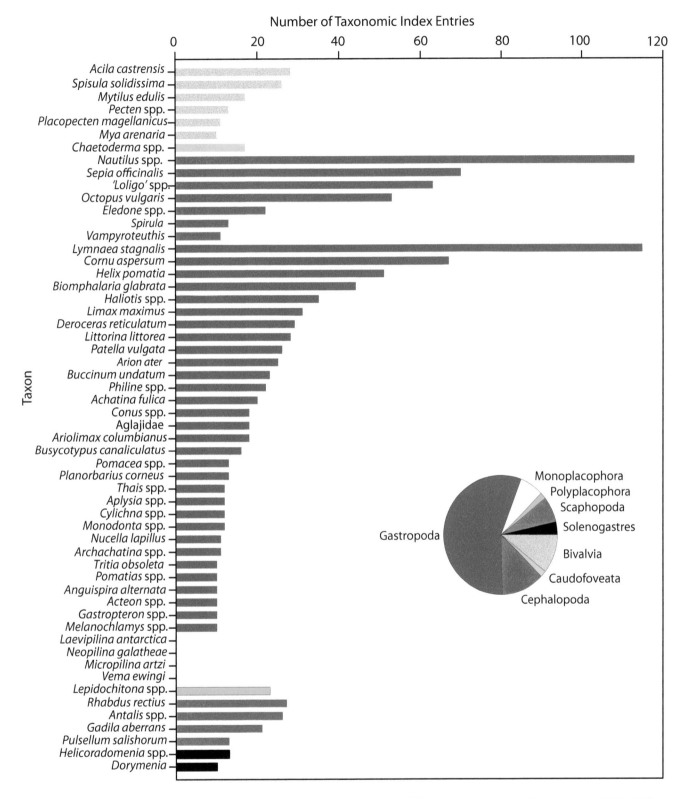

FIGURE 21.6 Taxa with more than ten index entries in the molluscan volumes of *Microscopic Anatomy of Invertebrates* (1994; 1997a,b) grouped by class.

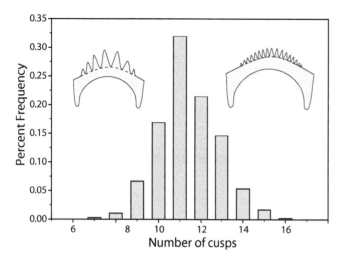

FIGURE 21.7 Cusp variation in the central radular tooth of *Tritia reticulata*; n = 3,028 specimens. Data from Pelseneer (1929).

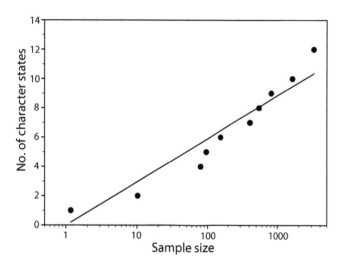

FIGURE 21.8 Relationship between sample size and the number of character states of the posterior pedal tentacles in *Tritia reticulata*; n = 3,254. Data reanalysed from Pelseneer (1929).

taxon and interactions in the community or ecosystem (i.e., ecological surprises) (Doak et al. 2008). Human-induced well-intentioned changes in species abundances (e.g., restoration initiatives) can also have unanticipated effects on biodiversity. For example, some models focusing on the restoration of the oyster *Crassostrea virginica* to reduce eutrophication in the Chesapeake Bay predicted negative effects on pelagic taxa and highlighted the trade-offs that, when known, must be considered before altering biodiversity schemes (Mann & Powell 2007). Further complicating restoration, and even general biodiversity studies, are shifting baselines of our perception of past and current biodiversity and the interactions among the taxa (Rothschild et al. 1994; Jackson et al. 2001; Kirby & Miller 2005; Alleway & Connell 2015).

As outlined in Chapter 10, global human-induced changes, notably climate change and habitat modification and destruction, are the major threats to molluscan biodiversity and are producing non-reversible changes to both numbers of species

and the ecological services these species provide. Molluscs, because of their extensive and deep fossil record, are among the best taxa to study to observe the impact of such changes during significant climatic and other past events. Molluscan lineages have survived numerous past global change events including hypoxia, acidification, heating or cooling, and hydrologic events. While past responses may provide some guidance and predictive insights into the current situation, rates of these past changes were mostly slower than current rates of change (Lawton & May 2002). Loss of habitat due to human modification and climate change will have the greatest effect on molluscan species, especially in terrestrial and freshwater taxa (Lydeard et al. 2004). Habitat loss and fragmentation also reduces organisms' resilience to change. While economic exploitation is unlikely to be a driver of extinction it will probably contribute to the functional loss and the possibility of extinction of many ecological interactions (Valiente-Banuet et al. 2015). As reviewed in Chapter 9, molluscs are important participants in numerous ecological interactions. Loss of ecological function is especially critical when top predators, which are often of economic interest, undergo major declines (Jackson et al. 2001; Estes et al. 2011). Loss of ecological function at any level often produces changes in competitive hierarchies and interactions. For example, the over-harvesting of marine mammals, sharks, and fish is thought to have been responsible for an increase in the numbers of cephalopods in the oceans of today (Caddy & Rodhouse 1998; Baum & Worm 2009).

Much past molluscan biodiversity research has focused on species numbers and the plethora of undescribed species, but future biodiversity research needs to be more expansive and integrative, especially given the anticipated perturbations brought about by climate change and other human-induced modifications. Recent work on biodiversity assessment has emphasised that biodiversity is best considered in an evolutionary manner by using the whole phylogenetic tree as a measuring device rather than just counts of species. While the taxonomic description of backlogged and newly discovered taxa is needed and will continue, substantial effort also needs to be made in building phylogenies and using these to understand biogeography, evolution, and conservation (Thornhill et al. 2016) as well as molluscan roles in ecological services beyond their usually cited contributions via filtration and ecosystem engineering (Gutiérrez et al. 2003; Coen et al. 2007) including their roles in ecological cascades.

21.8.5 CITIZEN SCIENCE

Almost all institutional molluscan research in recent decades has been supported by conventional government and academic funding. This relationship is especially apparent in the response of molluscan research productivity illustrated in the area plots which accompany this chapter and those in Chapter 11. These often show correlations between substantial decreases in research productivity and world wars and global economic slowdowns and turndowns (e.g., Figure 21.1). Many major contributions to molluscan research were, in the past, made by unfunded private researchers including specialist amateur researchers and retired scientists who found, like many citizen scientists, that the satisfaction of contributing to increasing knowledge was sufficient reward.

Bouchet et al. (2016) have called for the involvement of non-professionals in the description of new molluscan species as a way of reducing the estimated 300 years it will take to describe most of the known extant, undescribed species. Non-professional contributions to molluscan systematics have a long tradition in malacology (Kilburn 1999; Sturm et al. 2006; Coan & Kabat 2018; Mikkelsen 2010), sometimes with mixed results. The Bouchet et al. (2016) estimate of 300 years probably assumed a traditional timeline for species description, but molecular research might soon provide tools that provide unique molecular signatures for species far surpassing the utility of COI barcoding. While this will not alleviate the need for morphological description and quantification of fossil taxa and those taxa lacking available suitable tissue, it will greatly increase our knowledge of molluscan diversity even if it only exists as a cloud-based collection of 'signatures' and localities. While such advances will increase our abilities to assess biodiversity, they will probably result in further consolidation of such research endeavours in well-funded laboratories and further isolate the interested amateurs and lone scientists in small institutions.

Nevertheless, numerous other important venues for citizen science in malacology remain. The exhaustive efforts of Phillippe Bouchet and his team in conducting molluscan inventories in the tropical western Pacific have involved using groups of professionals and some specialist amateurs (Bouchet 1997; Bouchet et al. 2002, 2009). Other inventory efforts involving citizen scientists are also available. For example, BioBlitz surveys are intense periods of taxon inventories within a designated area. BioBlitz teams consist of scientists, naturalists, and citizen scientists and typically spend up to 24 hours identifying and photographing species within the designated area. BioBlitz events are global in scope and have been conducted in more than 15 countries in Africa, Asia, Australasia, Central and South America, Europe, and North America. Molluscan BioBlitz events with ample social media and cloud-based results are obvious investments for our community, and staging them with annual meetings of societies would ensure the participation of numerous professionals and highlight malacology.

The recent acquisition of the ability to document molluscan biology with high-resolution images has been made possible by revolutionary advances in cameras, including those in phones, and software, and has provided micro- and macro-imaging capability for countless amateur naturalists and potential citizen scientists. Besides documenting living individuals, their shells, and habitats, their behaviours are also recorded on digital videos. As an example, YouTube includes videos which document previously debated aspects of limpet behaviour. In one time-lapse video of nocturnal grazing by *Patella vulgata* on the southern coast of the United Kingdom aggressive territorial interactions between limpets were documented supporting an early report by Jones and Baxter (1985). Additional YouTube videos document the ingestion of algal thalli and blades by patellogastropods, which traditionally have been thought to be limited to substratum grazing because of their stereoglossate radular morphology (Guralnick & Smith 1999). Cephalopod videos are especially abundant, and in the recent reports of 'octopus cities' in Australia readily available, GoPro™ cameras were used to document numerous social interactions among the densely packed inhabitants (Scheel et al. 2017). One has only to search for images of any molluscan taxon on an internet browser to see the potential of a cadre of additional observers documenting molluscan biology. If these were connected to specimens, taxa, and localities, and made available, what a resource that would be! Numbers are important, but we must go beyond diversity estimates and work towards a more informatic and quantitative approach to malacology.

Appendix

HIGHER CLASSIFICATION OF THE EXTANT CLASSES OF MOLLUSCA

This appendix provides a classification of all extinct and living members of the extant classes of molluscs. The classification provided is mostly based on publications to the end of 2018, and a few from early 2019. Details of the major sources of information are given where appropriate.

The classification includes the higher taxonomic units, superfamilies, and families recognised as valid. Authorship of taxon names is not given. Those details, if needed, are readily available in the cited literature for each section or on the internet via several reliable sources.

One or more examples in each (usually superfamily) unit are illustrated. In some cases, notes and brief diagnostic statements are given, but time and space constraints precluded this for many taxa. Several colleagues have provided input (see acknowledgements in the foreword), but we accept responsibility for the content including any spelling or other factual errors. In the Cephalopoda, the level of input necessary was such that four experts in fossil cephalopod taxa have been listed as coauthors to those sections (see the introductions to nautiliform, coleoid, and ammonite sections for details).

Stratigraphic ranges for taxa have been primarily determined from the Paleobiology Data Base (PBDB) (https://paleobiodb.org/navigator/). Other sources included the Sepkoski Compendium (Sepkoski et al. 2002), the *Treatise on Invertebrate Paleontology* (Moore 1950), and primary literature as needed. The ranges are assigned to the appropriate System/Period and Series/Epoch category using the International Chronostratigraphic Chart v 2018/08 (Cohen et al. 2013; updated). In a few cases where there were highly discontinuous occurrence records reported by the PBDB, we do not report the extreme outliers. For example, if a taxon had a continuous fossil occurrence from the Devonian to the Lower Jurassic and a single, reported occurrence in the Miocene, we do not report the Miocene record. We consider this to be a conservative approach, and we assume that the outlying record is likely a mis-identification of the taxon.

Because of the coarse resolution of stratigraphic ranges presented here, the first occurrences and fossil ages reported should not be used in the construction of time-measured phylogenies. Instead, we encourage workers to delve into the primary paleontological literature, which is constantly undergoing detailed nomenclatural and stratigraphic resolution.

The following conventions apply throughout the appendix.

Extinct taxa are indicated by the † symbol. If the taxon under which families are listed is extinct, then the † symbol is not used at the family level.

The ranks given to each taxon higher than family level are given in brackets as an indication that these are highly subjective and likely to change in different classifications.

Where possible, the following endings are used for higher taxon names:

Superorder – ia
Order – ida
Suborder – ina
Hyporder – oidei
Superfamily – oidea

Family names are listed with the first name given being the one from which the superfamily name is derived, and the others are listed alphabetically. Where both extinct and living taxa are present in the same superfamily, the living families are listed first and the extinct families (distinguished by the † symbol) are listed last.

Generic names are given for each illustration. Where these are the type genus for a listed family name (e.g., *Pyramidella* for Pyramidellidae), no family name is given. Where the genus is not the source of a listed family name, the family to which it is allocated is also given.

In this appendix we use many figures from two main sources. The first is the two-volume work on Mollusca comprising Volume 5 of the *Fauna of Australia* series (© Copyright, Commonwealth of Australia) (Beesley et al. 1998a, b), which remains an important contribution to molluscan biology. The second significant source of figures for the appendix are those from the molluscan volumes of the *Treatise on Invertebrate Paleontology* (© Copyright, The University of Kansas Press and the Geological Society of America, Inc,). We thank the Australian Biological Resources Study for permission to use the *Fauna of Australia* images, and Dr Michael Cormack of the Paleontological Institute of the University of Kansas for permission to use the *Treatise* images. The references to these copyrighted figures have been abbreviated in the appendix to save space. These are the two *Fauna of Australia* volumes – **FA1** Beesley et al. 1998a, **FA2** Beesley et al. 1998b, and the *Treatise of Invertebrate Paleontology* volumes as follows: **TI** Moore 1964b (minor classes, lower gastropods), **TL** Moore 1957; Moore 1960; Moore 1996) (ammonites), **TK** Moore, 1964a (nautiliforms), **TN** Moore & Teichert, 1969 and 1971 (bivalves). We also wish to acknowledge the Paleobiology Database community for providing online access to an invaluable stratigraphic, taxonomic, and bibliographic resource in the preparation of the appendix.

The remaining appendix figures include original material, images from sources out of copyright, and the following images used by permission of the publisher and/or author.
Acavus, Ariophanta, Cerion, Corneosagda, Dextroformosana, Diaphera, Dyakia, Eua, Euglandina, Formosana, Gastrocopta, Granaria, Hedleyoconcha, Hypselostoma, Insulivitrina, Mastoides, Perrottetia, Ptychodon, Pukunia, Pupilla, Pyramidula, Sagda,

(Class) **POLYPLACOPHORA** (= Polyplaxiphora, Loricata, Placophora)
The chitons – with (usually) eight shell valves surrounded by a fleshy girdle covered with scales, spicules, or hairs.
The classification presented here largely follows that of Sirenko (2006). A more dated, but well-illustrated classification can be found in Moore (1964).
See Chapter 14 for more information on this group.

(Subclass) **Paleoloricata** † (= Palaeoloricata)
Cambrian (Miaolingian) to Silurian (Llandovery) (501–443.7 Ma)
Thick, massive valves that lack insertion plates, an articulamentum layer with associated sutural laminae. This group contains the earliest chitons with all its members extinct. Some taxa (*Phthipodochiton*, *Kulindroplax*) may have had greatly reduced ventral apertures – the foot-less chitons.

(Order) **Chelodida** †

(Superfamily) **Matthevioidea** † Cambrian (Miaolingian) to Devonian (Lower) (501–416 Ma) Elongate bodies with arrowhead-shaped intermediate valves. Family Mattheviidae (= Chelodidae, Preacanthochitonidae) and Echinochitonidae	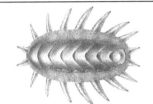 *Echinochiton*, from Pojeta and Dufoe (2008)	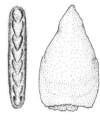 *Chelodes*, from TI (reconstruction) and FA1 (valves)

(Suborder) **Septemchitonina** † (= Heptaplacota)		
(Superfamily) **Septemchitonoidea**† Cambrian (Furongian) to Devonian (Lower) (489.5–416 Ma) Originally described as having only seven, but was shown later to have eight plates. Families Septemchitonidae, Alastegiidae, Gotlandochitonidae, Helminthochitonidae, and Solenocarididae?	 *Septemchiton*, from TI	
'Foot-less' chitons † Ordovician (Upper) to Silurian (Pridoli) (453–422.9 Ma) Seven dorsal plates enclosed by a cylindrical girdle of spicules joined by a ventral ridge. Higher taxa have yet to be diagnosed. Included genera *Phthipodochiton* and *Kulindroplax* (originally described as an aplacophoran).	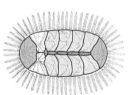 *Kulindroplax* – original	

(Subclass) **Neoloricata**
Devonian (Lower) to present (418–0 Ma)
Valves consist of three distinct shell layers – tegmentum, articulamentum, and hypostracum; articulamentum often forms insertion plates and sutural laminae.
Unplaced families:
Sirenko (2006) does not place the following four Cretaceous families: Haeggochitonidae, Ivoechitonidae, Olingechitonidae, and Scanochitonidae (and the 'order' Scanochitonida).

(Order) **Multiplacophorida** † Devonian (Middle) to Permian (Guadalupian) (383.7–265 Ma) Shell of 17 dorsal plates consisting of head and tail valves and five triplets of intermediate values (two lateral and one central plate); surrounded by spines. Family Strobilepidae	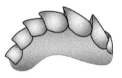 *Polysacos* (Strobilepidae), from Vendrasco et al. (2004)

(Order) **Lepidopleurida**
Devonian (Lower) to present (393.3–0 Ma)

(Suborder) **Cymatochitonina** †
Devonian (Lower) to Permian (Lopingian) (393.3–254 Ma)
The valves have a weakly developed articulamentum and narrow sutural laminae.

(*Continued*)

(Superfamily) **Cymatochitonoidea** † Devonian (Lower) to Permian (Lopingian) (393.3–254 Ma) Families Cymatochitonidae, Acutichitonidae, Gryphochitonidae, Lekiskochitonidae, and Permochitonidae	*Cymatochiton* valves, from TI 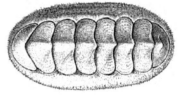
(Suborder) **Lepidopleurina** Devonian (Lower) to present (393.3–0 Ma) The valves of members of this group have well-developed sutural laminae and slit-less insertion plates.	
(Superfamily) **Lepidochitonoidea** Devonian (Lower) to present (393.3–0 Ma) Families Leptochitonidae (= Lepidopleuridae), Abyssochitonidae (= Ferreiraellidae, Xylochitonidae), Hanleyidae, Nierstraszellidae, Camptochitonidae †, Glyptochitonidae †, Mesochitonidae †, and Protochitonidae †	*Hanleya*, from Pilsbry (1892)  *Parachiton* valves (Leptochitonidae), from FA1 (ventral view on left, dorsal on right)
(Order) **Chitonida** Permian (Cisuralian) to present (299–0 Ma)	
(Suborder) **Chitonina** Permian (Cisuralian) to present (299–0 Ma) Members of this group have adanal gills and narrow-based egg hull projections.	
(Superfamily) **Chitonoidea** Permian (Cisuralian) to present (299–0 Ma) Families Chitonidae, Callistoplacidae, Callochitonidae, Chaetopleuridae, Ischnochitonidae, Loricidae, and Ochmazochitonidae †	*Ischnochiton*, from FA1 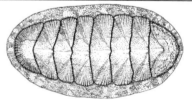  *Ornithochiton* (Chitonidae) valves, from FA1 (ventral view on left, dorsal on right)
(Superfamily) **Schizochitonoidea** Paleogene (Paleocene) to present (65.5–0 Ma) Family Schizochitonidae	*Schizochiton*, from Iredale and Hull (1926) 

(Continued)

(Suborder) **Acanthochitonina**	
Jurassic (Lower) to present (174.1–0 Ma) Members of this group have abanal gills and wide-based egg hull projections.	

(Superfamily) **Mopalioidea**
Jurassic (Lower) to present (174.1–0 Ma)
Families Mopaliidae, Choriplacidae,
  Lepidochitonidae, Schizoplacidae, and
  Tonicellidae
	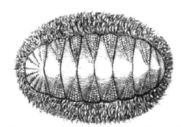
	Mopalia, from Pilsbry (1892)

(Superfamily) **Cryptoplacoidea** Paleogene (Oligocene) to present (33.9–0 Ma) Families Cryptoplacidae, Acanthochitonidae, Hemiarthridae, and Makarenkoplacidae †	*Cryptoplax*, from FA1 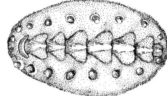  *Acanthochitona*, from FA1

(Class) **CAUDOFOVEATA** (= Chaetodermatomorpha, Chaetodermomorpha, Scutopoda)	
No fossil record Worm-like – apparently less diverse than the solenogasters. See Chapter 14 for general information on this group.	
(Order) **Chaetodermatida**	
No pedal groove.	

(Superfamily) **Chaetodermatoidea**
Radula bipartite or with a single pair of teeth
  (Chaetodermatidae), with or without
  supports. Mouth shield disc U-shaped or
  paired. Body often extended as a 'tail'
  posteriorly.
Families Chaetodermatidae, Limifossoridae,
  and Prochaetodermatidae
	Rhabdoderma (Prochaetodermatidae),
	from FA1 

	 *Falcidens* (Chaetodermatidae), from FA1

(Class) **SOLENOGASTRES** (= Neomeniomorpha)
No unequivocal fossil record
Worm-like, spiculate, and with a ventral pedal groove. See Chapter 14 for general information on this group.
Note: Four superfamily names are introduced below based on the earliest family name within the group, with the exception of Heteroherpioidea in which   all the family names were erected in the same publication.
Two Silurian (425–420 Ma) taxa (*Acaenoplax*, *Kulindroplax*) have been reported as aplacophorans and are discussed in Chapter 13.
(Subclass) **Aplotegmentaria**
Members of this group have a thin cuticle and flat spicules.

(Continued)

(Order) **Neomeniida** (= Neomenamorpha)	
(Superfamily) **Neomenioidea** Families Neomeniidae and Hemimeniidae	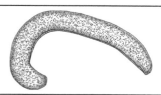 *Neomenia*, from Hansen (1888)

(Order) **Pholidoskepia**	

(Superfamily) **Dondersioidea**
Families Dondersiidae, Gymnomeniidae,
  Lepidomeniidae, Macellomeniidae,
  Meiomeniidae, and Sandalomeniidae

	Nematomenia (Dondersiidae), from FA1

(Subclass) **Pachytegmentaria**	
With thick cuticle with multi-layered, mostly needle-like spicules.	

(Order) **Sterrofustia**	
(Superfamily) **Heteroherpioidea** Families Heteroherpiidae, Imeroherpiidae, Phyllomeniidae, and Rhabdoherpia	 *Phyllonemia*, from FA1

(Order) **Cavibelonia**	

(Superfamily) **Proneomenioidea**
Families Proneomeniidae, Acanthomeniidae,
  Amphimeniidae, Drepanomeniidae,
  Epimeniidae, Notomeniidae, Perimeniidae
  (= Pruvotinidae), Rhipidoherpiidae,
  Rhopalomeniidae, Simrothiellidae,
  Strophomeniidae, and Syngenoherpiidae
	Epimenia, from FA1

	Eleutheromenia (Perimeniidae), from FA1

(Class) **MONOPLACOPHORA** (= Triblidia, Tryblidiacea)
Cambrian (Miaolingian) to present (509–0 Ma)
Limpet-shaped shell with five–eight pairs of shell attachment muscle clusters.
Note that several additional Paleozoic taxa have been included in Monoplacophora by Bouchet et al. (2017) and others, but we have assigned those taxa to different groups, mainly Brachiopoda. See Chapter 13 for details.
General information on this class can be found in Chapter 14.
Unplaced families: Ladamarekiidae † and Peelipilinidae †

(Order) **Tryblidiida** (= Pilinea) †
Cambrian (Miaolingian) to Silurian (Pridoli) (509–421 Ma)
Shell with multiple paired shell muscle scars; radular attachment and diaphragm scars also typically present. Oblique and medial pedal muscles fused in shell muscle clusters.

| (Superfamily) **Tryblidioidea** †
Ordovician (Lower) to Silurian (Pridoli)
  (485.4–421 Ma)
Apex not raised and at anterior end of shell.
Families Tryblidiidae, Bipulvinidae, and
  Drahomiridae | 
Tryblidium, from Lindström (1884)

Pilina (Tryblidiidae), from Lindström (1884) |

(Continued)

(Superfamily) **Propilinoidea** new name †. Cambrian (Miaolingian) to Ordovician (Lower) (501.0–470 Ma) This group is distinguished by its elevated apex, which is anterior to the aperture, and concentric sculpture. Family Propilinidae. New name.	*Propilina*, from TI
(Superfamily) **Archaeophialoidea** † Ordovician (Upper) (458.4–443.8 Ma) Shell elevated, aperture circular, and apex subcentral. Family Archaeophialidae	*Archaeophiala*, from TI
(Superfamily) **Pygmaeoconoidea** † Ordovician (Middle) to Ordovician (Upper) (466.0–445.6 Ma) Shell elevated and often laterally compressed; apex subcentral. Posterior shell muscle clusters sometimes fused. Individuals have been found occurring as epizoics on hyolith conchs. Family Pygmaeoconidae	*Pygmaeoconus*, from Horný (2006)
(Order) **Neopilinida** Quaternary (Pleistocene) to present (1.8–0 Ma) Shell thin, oblique, and medial pedal muscles distinct, not fused in shell muscle clusters; muscle attachment areas poorly defined, diaphragm scars absent.	
(Superfamily) **Neopilinoidea** Quaternary (Pleistocene) to present (1.8–0 Ma) Family Neopilinidae (= Vemidae, Laevipilinidae, Micropilinidae, and Monoplacophoridae)	*Neopilina*, from Lindberg et al. (2004)

(Class) **BIVALVIA** (= Lamellibranchiata, Elatobranchiata, Elatocephala, Acephala (in part), Pelecypoda, Cormopoda – see Bouchet & Rocroi 2010 for details)

The classification used here is mainly based on Bieler and Mikkelsen (2006), Nevesskaja (2009), Bieler et al. (in Bouchet and Rocroi) (2010), Carter et al. (2011), and Bieler et al. (2014). In particular, we acknowledge our reliance on the detailed classifications produced by Cox et al. (1969), Bieler et al. (in Bouchet and Rocroi) (2010), and Carter et al. (2011) which have been invaluable in compiling this appendix.

Note: *Lamellodonta* (Lamellodontidae), described as a Paleozoic bivalve, is a brachiopod.

(Grade) **Euprotobranchia** †

Cambrian (Series 2) to Cambrian (Miaolingian) (520.0–501.0 Ma)

The following two Cambrian groups are presumably the ancestors of other bivalves and as such do not constitute a monophyletic group. Carter et al. (2011) used this basal grade of Bivalvia for two orders, Fordillida and Tuarangiida.

(Order) **Fordillida** †

Cambrian (Series 2) to Cambrian (Miaolingian) (520.0–501.0 Ma)

(Superfamily) **Fordilloidea** † Cambrian (Series 2) to Cambrian (Miaolingian) (520.0–501.0 Ma) *Fordilla* is the earliest bivalve from upper Early Cambrian (Jell 1980). They are thought to have evolved from helcionelloidean 'monoplacophorans'. They are very small in size, lack a well-defined hinge, and possess foliated shell microstructure (Vendrasco et al. 2011). Runnegar and Bentley (1983) placed Fordillidae in the Mytiloida while *Pojetaia* was included in Nuculoida. Yochelson (1981) doubted that *Fordilla* was a bivalve. Families Fordillidae, and Camyidae	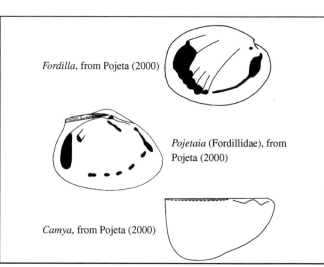 *Fordilla*, from Pojeta (2000) *Pojetaia* (Fordillidae), from Pojeta (2000) *Camya*, from Pojeta (2000)

(Clade) **Eubivalvia** †

Ordovician (Lower) to present (478.6–0 Ma)

This name was coined by Carter in Carter et al. (2011) to encompass both Protobranchia and Autobranchia.

Unplaced families: Bieler et al. (in Bouchet and Rocroi) (2010) listed these two families as 'order uncertain'; they were not included in Carter et al. (2011). Cirravidae †, and Laurskiidae †

(Order) **Tuarangiida** †

Cambrian (Miaolingian) (504.0–501.0 Ma)

MacKinnon (1982) tentatively regarded *Tuarangia* as a pteriomorphian while Runnegar (1983) and Carter (1990) regarded it as a bivalved monoplacophoran. Carter et al. (2000) treated this taxon as a basal bivalve.

(Superfamily) **Tuarangioidea** † Cambrian (Miaolingian) (504.0–501.0 Ma) Small, elongate, subquadrate to trapezoidal shells with long straight hinge lines, umbones near middle. Hinge teeth taxodont, with several oblique teeth on both sides of umbo. Ligament narrow, erect, amphidetic interposed between teeth. Foliated shell microstructure. Family Tuarangiidae	*Tuarangia*, from Pojeta (2000) 

(*Continued*)

(Subclass) **Protobranchia** (= Palaeotaxodonta in part)

Ordovician (Lower) to present (478.6–0 Ma)

Gill typically small, with short, simple, triangular filaments (protobranch condition). Shells sometimes nacreous, often with a (pseudo) taxodont hinge.

Living protobranchs fall into two distinct, sometimes paraphyletic, clades based on morphology and molecular data (see Chapter 15).

The classification below mainly follows Carter et al. (2011) which differs from that in Bieler et al. (in Bouchet and Rocroi) (2010).

A grouping (variously called Nuculiformii,[1] Foliobranchia, or Opponobranchia) comprising the Nuclida and Solemyida (including the Nucinellidae), has generally been recognised in recent classifications and was defined by an open mantle, with inhalant water entering anteriorly, the shells usually having nacre and the gill filaments opposite (not alternating) along ctenidial axis. This arrangement has, however, not been supported in recent multigene analyses (e.g., Combosch et al. 2017) with solemyids and nuculoideans forming separate clades. We thus recognise four orders, with Solemyidae and Nucinellidae separate from both the Nuculida and Nuculanida.

Unplaced family: Antactinodiontidae †

(Order) **Nuculida** (= Nuculoida)

Ordovician (Lower) to present (478.6–0 Ma)

Shells equivalve, isomyarian, with pseudotaxodont hinge, and the ligament is usually amphidetic. Shell microstructure is nacreous or crossed lamellar. The gills are small, and the labial palps large. There are no siphons, and water currents enter the animal anteriorly and leave posteriorly.

(Superfamily) **Nuculoidea**

Ordovician (Lower) to present (478.6–0 Ma)

Shells nacreous, with pseudotaxodont hinge and ligament within resilium. No pallial sinus.

Praenuculidae was removed from Nuculoidea by Carter et al. (2000) and Carter (2001), who identified it as basal to both the nuculid and solemyid lineages. However, Carter (2011) included it in Nuculoidea, which we follow here.

Families Nuculidae and Praenuculidae †

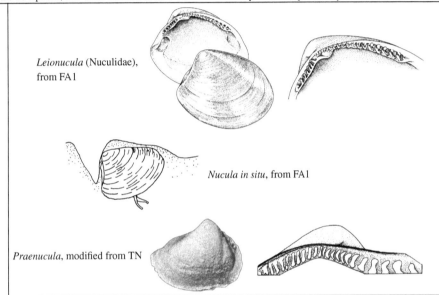

Leionucula (Nuculidae), from FA1

Nucula in situ, from FA1

Praenucula, modified from TN

(Order) **Solemyida** (= Solemyoida, Cryptodonta in part, Lipodonta)

Ordovician (Lower) to present (478.6–0 Ma)

This group was included in 'subclass' Cryptodonta in the *Treatise* (Cox et al. 1969) (which also included the Praecardioidea). Cope (1996b) used Lipodonta as a subclass for Solemyida alone.

The molecular analyses of Sharma et al. (2013) and Combosch et al. (2017) did not support the Manzanellidae being included in Solemyida (it is paraphyletic or polyphyletic) although those authors retained superfamily Manzanelloidea in their order Solemyida. Both superfamilies have opposite gill filaments (Allen & Sanders 1969) and we treat them here as separate orders.

(Continued)

---

[1] We prefer the use of Nuculiformii (first used by Starobogatov in 1992) following Carter (2000) in preference to Opponobranchia of Giribet (2008) or the much earlier Foliobranchia (see Bouchet & Rocroi 2010 for details). The latter name has not been used in recent literature, although it has been incorrectly applied to the Nuculaniformii rather than the Nuculiformii (Sharma et al. 2013).

(Superfamily) **Solemyoidea**
Ordovician (Lower) to present (478.6–0 Ma)
Shells are thin, non-nacreous, mostly
  periostracum with large periostracal frills and
  lacking a hinge. The ctenidia are large and
  house symbiotic bacteria, and the gut is reduced.
  They are dimyarian and some ancestral taxa had
  taxodont teeth (Carter 2001), being previously
  included in a separate superfamily
  (Ctenodontoidea).
Families Solemyidae, Clinopisthidae †,
  Ctenodontidae †, and Ovatoconchidae †.
	Solemya,
	from FA1 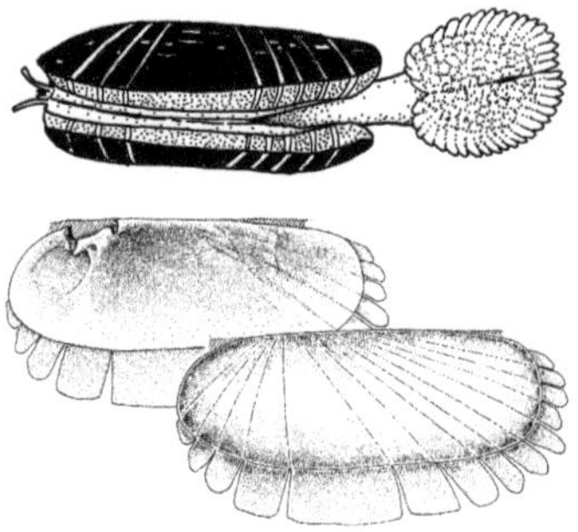
	
	Solemya
	in situ,
	from FA1 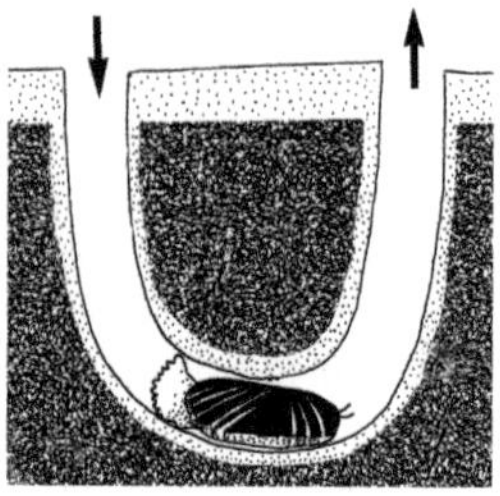	

(Order) **Nuculanida** (= Nuculanoida)

Ordovician (Lower) to present (478.6–0 Ma)

Shells elongate posteriorly, resilifer present or absent, with pseudotaxodont hinges.

(Suborder) **Nuculanina**

Ordovician (Lower) to present (478.6–0 Ma)

Members of this group have the mantle fused posteriorly with siphons developed, and a pallial sinus is usually present. Inhalant water enters posteriorly (instead of anteriorly as in other protobranchs), as in most autobranchs, and the shells lack nacre. Gill filaments alternate along the ctenidial axis.

(Continued)

(Superfamily) **Nuculanoidea**

Ordovician (Lower) to present (478.6–0 Ma)

Malletiidae, Tindariidae, Cucullellidae, Pseudocyrtodontidae, and Strabidae were included in a separate superfamily Malletioidea by Carter et al. (2011). However, this superfamily was not recognised by Bieler et al. (in Bouchet and Rocroi) (2010) or by Sharma et al. (2013), who included these families in Nuculanoidea.

Families Nuculanidae, Lametilidae, Malletiidae, Neilonellidae, Phaseolidae, Siliculidae, Tindariidae, Yoldiidae, Cucullellidae †, Isoarcidae †, Polidevciidae †. Pseudocyrtodontidae † and Strabidae †.

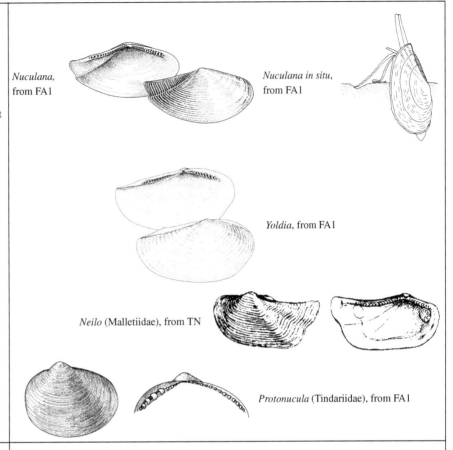

Nuculana, from FA1

Nuculana in situ, from FA1

Yoldia, from FA1

Neilo (Malletiidae), from TN

Protonucula (Tindariidae), from FA1

(Superfamily) **Sareptoidea**

Devonian (Lower) to present (409.1–0 Ma)

Sareptidae has previously been included in Nuculanoidea but was separated by Sharma et al. (2013) based mainly on molecular and morphological data on *Pristigloma*.

Pristiglomidae was treated as a subfamily of Sareptidae in Bieler et al. (in Bouchet and Rocroi) (2010) and as a superfamily and family by Carter et al. (2011), with the former classification being supported with molecular data of Sharma et al. (2013).

Family Sareptidae (= Pristiglomidae)

Pristigloma, from Saunders and Allen (1973). Image courtesy of Museum of Comparative Zoology, Harvard University.

(Suborder) **Afghanodesmatina** †

Ordovician (Lower) to present (478.6–0 Ma)

Carter et al. (2011) recognised this group as an order containing the following two superfamilies which were previously unassigned to superfamilies or, in the case of Nucularcidae, included in Nuculoidea, by Bieler et al. (in Bouchet and Rocroi) (2010).

(*Continued*)

(Superfamily) **Tironuculoidea** † Ordovician (Lower) to present (478.6–0 Ma) Families Tironuculidae, Nucularcidae, and Similodontidae *Trionucula*, from Morris and Fortey (1976)	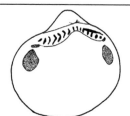
Superfamily **Afghanodesmatoidea** † Ordovician (Lower) to Devonian (Lower) (possibly Triassic [Upper]) (477.7–419.2 Ma [237. Ma]) (from Carter ms) Distinguished by the presence of anterior palaeotaxodont or pseudotaxodont teeth, which are abruptly enlarged relative to the posterior palaeotaxodont teeth, and the presence of an opisthodetic simple ligament without a resilium (Carter 2001). Carter (pers. comm., March 2018) suggested that Eritropidae is probably ancestral to Nuculanida. According to Cope (2000), cardiolariids (i.e., afghanodesmatids) are the earliest autobranch bivalves because they have the hinge modified for wider valve opening facilitating more effective disposal of pseudofaeces – this feature being linked to the development of a filibranch gill. This interpretation is not followed here. Families Afghanodesmatidae (= Cardiolariidae) and Eritropidae *Afghanodesma*, redrawn and modified from Desparmet et al. (1971)	  *Ekstadia* (Eritropidae), from TN

(Order) **Manzanellida**

Permian (Cisuralian) to present (284.4–0 Ma)

This grouping is introduced here to reflect the phylogenetic results of Sharma et al. (2013) and Combosch et al. (2017). In the Combosch et al. (2017) analysis Nucinellidae was placed at the base of the autobranch clade while in the Sharma et al. (2013) analysis it is sister to the Nuculida + Nuculanida, although no autobranchs were included in that analysis. Thus, there is a possibility that this group is a sister to the Autobranchia.

(Superfamily) **Manzanelloidea** Permian (Cisuralian) to present (284.4–0 Ma) This small group has been placed in the Solemyida based mainly on anatomical and molecular similarities (Allen & Sanders 1969; Allen 1985; Oliver & Taylor 2012). They are unusual amongst protobranch bivalves in being monomyarian. Like Solemyidae they have symbiotic bacteria in their gills (Oliver & Taylor 2012) and the gill filaments are opposite as in nuculids and solemyids. This group was previously classified amongst the taxodont pteriomorphians. Families Manzanellidae † and Nucinellidae *Nucinella*, from Hedley (1904)	

(Subclass) **Autobranchia** (= Autolamellibranchiata [original]; Autolamellibranchia)

Ordovician (Lower) to present (479–0 Ma)

Gills with elongate filaments extended into a W-shape (the autobranch condition).

(Infraclass) **Pteriomorphia** (= Filibranchia)

Ordovician (Lower) to present (478.6–0 Ma)

This is one of the largest groups of bivalves, members of which have filibranch, pseudoeulamellibranch, or eulamellibranch gills, often have nacre, and many are anisomyarian or monomyarian.

Carter (1990) recognised two major clades ('subclasses') Pteriomorphia and Isofilibranchia, the latter containing Modiomorphida and Mytilida. The classification used here largely follows Carter et al. (2011).

The Mytilida is now included in Pteriomorphia (e.g., Waller 1998; Giribet 2008; Bieler et al. 2014; Lemer et al. 2019) while the earlier (e.g., Carter 1990) concept of Modiomorphida is polyphyletic, with some members pteriomorphians while others are heteroconchs.

(Continued)

(Cohort) Mytilomorphi Ordovician (Lower) to present (478.6–0 Ma) A 'cohort' recognised by Carter et al. (2011) to include the orders Mytilida and Colpomyida.	
(Order) Mytilida (= Mytiloida) Ordovician (Lower) to present (478.6–0 Ma) Mytilida have filibranch to pseudoeulamellibranch gills and are heteromyarian. Their calcitic shells have an inner nacreous layer. The Mytiloida (and a polyphyletic concept of the extinct Modiomorphida) were included in a separate higher grouping Isofilibranchia by some workers (Pojeta 1978; Pojeta & Runnegar 1985; Carter 1990) but are now mainly included in the Pteriomorphia, although the earlier (e.g., Carter 1990) concept of Modiomorphida is polyphyletic, containing some pteriomorphians and some heteroconchs.	

(Superfamily) Modiolopsoidea † Ordovician (Middle) to Ordovician (Upper) (466–452 Ma A paraphyletic basal pteriomorphian group (Carter et al. 2000) characterised by a discontinuous fibrous ligament and an almost edentulous hinge with a variable number of anterior teeth beneath the hinge line (Fang & Morris 1997). This group includes some of the earliest lamellibranch bivalves such as the modiolopsid *Goniophorina* from the Lower Ordovician which were probably epifaunal or shallow infaunal byssate forms (e.g., Sánchez 2006). Families Modiolopsidae, Ischyrodontidae, and Modiolodontidae	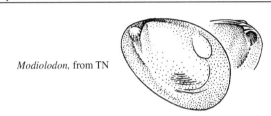 *Modiolodon*, from TN 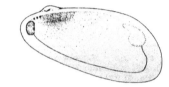 *Modiolopsis*, from Pojeta (1985)
(Superfamily) Mytiloidea Ordovician (Lower) to present (476.8–0 Ma) Mytiloideans have a dysodont hinge, an opisthodetic, parivincular ligament and are heteromyarian. The ventral mantle lobes are free, but posteriorly they are fused below the exhalant aperture. A few mytilids form siphons from mantle extensions but there is no mantle fusion involved. The gill is filibranch. The mytilids are a very large and ancient family with several (about eight) subfamilies recognised. Two of these (Crenellidae and Septiferidae) are elevated to family by Carter et al. (2011), and Morton (2015) recognised an additional two families, Modiolidae and Musculidae, as well as Mytilidae and Crenellidae, on the basis of anatomical differences (mainly musculature). Families Mytilidae and Mysidiellidae †	 *Mytilus*, from FA1  *Modiolus* (Mytilidae), from FA1 *Lithophaga* (Mytilidae), from FA1   *Musculus* (Mytilidae), original

(Order) Colpomyida † Ordovician (Lower) (478.6–473.9 Ma) This group is considered to be among the basal pteriomorphians (Malchus 2004).	
(Superfamily) Colpomyoidea † Families Colpomyidae and Evyanidae	*Colpomya*, redrawn and modified from Moore et al. (1952) 

(Continued)

'Cohort' Ostreomorphi

This grouping was devised by Carter et al. (2011) to encompass the remaining pteriomorphian taxa listed below (Arcida, Cyrtodontida, Myalinida, Ostreida, and Pectinida).

Unplaced families: Ischyrodontidae †, Matheriidae †, and Myodakryotidae †

(Order) Cyrtodontida †

Ordovician (Lower) to Devonian (Upper) (476.8–360.7 Ma)

Cope (1996a) defined this group as having ovoid shells, generally with prominent prosocline umbones, isomyarian or anisomyarian and with well-developed hinge teeth divided into an anterior and posterior group separated by an edentulous area. The posterior teeth are parallel or near parallel to the shell margin. The ligament is opisthodetic external, lamellar, duplivincular, or without a grooved ligament area.

Waller (1978) included cyrtodontoideans with the arcoideans in a 'superorder' Prionodata, a name also treated as a superorder by Healy et al. (2000) who included Arcoidea and Limopsoidea within it.

(Suborder) Cyrtodontina †

Ordovician (Lower) to Permian (Cisuralian) (478.6–290.1 Ma)

(Superfamily) Cyrtodontoidea † Ordovician (Lower) to Permian (Cisuralian) (478.6–290.1 Ma) The ligament of this group has fine longitudinal ridges and grooves as in a duplivincular ligament but may be just lamellar (Waller 1978). Family Cyrtodontidae	 *Cyrtodonta*, redrawn and modified from Desparmet et al. (1971)  *Ptychodesma* (Cyrtodontidae), modified from TN
(Superfamily) Falcatodontoidea † Ordovician (Lower) (478.6–471.8 Ma) Preduplivincular ligament simple (no grooves) (Fang & Cope 2008). Family Falcatodontidae.	*Falcatodonta*, modified from Cope (1996a) 
(Superfamily) Pichlerioidea † Triassic (Upper) (235–232 Ma) Of uncertain affinities. This group was included in Limopsidae in the *Treatise* (Cox et al. 1969), but Carter (1990) treated the group as a superfamily. Family Pichleriidae	*Pichleria*, modified from TN 

(Suborder) Praecardiidina †

Silurian (Llandovery) to Devonian (Upper) (436–360.7 Ma)

Shells circular to elongate, thin, isomyarian, weak or absent hinge plate edentulous or pseudotaxodont. Pallial line entire or slightly sinuate.

Kříž (2007) introduced Nepiomorphia, which was given equal rank to the Pteriomorphia, for the orders Praecardioida and Antipleuroida. They included epifaunal to infaunal taxa. Like Carter et al. (2011), we treat this grouping as a suborder.

(Superfamily) Praecardioidea † Silurian (Wenlock) to Devonian (Upper) (428.2–360.7 Ma) Families Praecardiidae and Buchiolidae	*Praecardium*, modified from TN 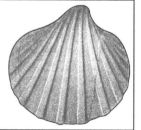

(Continued)

(Superfamily) **Cardioloidea** † Silurian (Wenlock) to Mississippian (Lower) (428.2–345.3 Ma) Families Cardiolidae and Slavidae	*Cardiola*, modified from TN
(Superfamily) **Dualinoidea** (= Antipleuroidea) † Silurian (Wenlock) to Triassic (Upper) (428.2–201.6 Ma) Families Dualinidae, Praelucinidae, Spanilidae, and Stolidotidae	*Antipleura* (Dualinidae), modified from TN
(Order) **Arcida** (= Arcoida) Ordovician (Lower) to present (476.8–0 Ma) Arcoideans are characterised by their duplivincular ligament and complex cross-lamellar shell structure similar to that found in heterodont bivalves. The mantle margin is completely open, and the gills are filibranch. Cope (1996b) proposed using the subclass Neotaxodonta for Arcoidea and Limopsoidea alone, excluding Glyptarcoidea, which he included in the Palaeoheterodonta (Order Actinodontoida).	
(Superfamily) **Glyptarcoidea** † Ordovician (Middle to Upper) (466–457.5 Ma) Cardinal teeth radiate out from beneath umbo but ventral to hinge plate and towards centre of valves (i.e., opposite direction to 'Actinodontoidea'). Ligament preduplivincular. Families Glyptarcidae and questionably Pucamyidae	*Glyptarca*, exterior of shell modified from Cope (1996a), hinge from Cope (1996b)

(Continued)

(Superfamily) **Arcoidea**

Ordovician (Upper) to present (457.5–0 Ma)

Arcoidea have a taxodont hinge and duplivincular ligament. Arcidae, Noetiidae, and Cucullaeidae have trapezoidal to quadrate shells, mostly with some radial ribbing, while Glycymerididae have rounded shells. Some arcoideans are byssate as adults.

Limopsidae is sometimes included in this superfamily, but this is not supported by recent analyses (Xue et al. 2012; Bieler et al. 2014).

A recent molecular analysis (Feng et al. 2015) has cast doubt on most of the extant familial groupings listed below. However, we maintain the current arrangement until confirmation of these findings.

Families Arcidae, Cucullaeidae, Glycymerididae, Limopsidae, Noetiidae (treated as a subfamily by Carter et al. 2011), Catamarcaiidae †, Frejidae †, and Parallelodontidae †

Arca, from FA1

Arcopsis (Noetiidae), from FA1

Cucullaea, from FA1

Limopsis, from FA1

Parallelodon, from TN

(Superfamily) **Philobryoidea**

Triassic (Middle) to present (247.2–0 Ma)

The separation of this group from Arcoidea requires testing.

Family Philobryidae

Philobrya, from Cotton (1961)

Lissarca (Philobryidae), from FA1

(Superfamily) **Plicatuloidea**

Triassic (Middle) to present (245.0–0 Ma)

Plicatulids have filibranch gills and are cemented by the right valve.

Waller (1978) separated the monomyarian plicatuloideans from the anisomyarian Dimyoidea and allied both groups with the oysters. They have strong interlocking secondary teeth in both valves.

Families Plicatulidae and Chondrodontidae †

Plicatula, from FA1

(*Continued*)

(Superfamily) **Dimyoidea** Triassic (Middle) to present (247.2–0 Ma) This group is anisomyarian, with the anterior adductor muscle much smaller than the posterior. They have filibranch gills and are cemented on the right side. The hinge lacks secondary teeth in the left valve, but these are sometimes present on the right valve. Family Dimyidae	 *Dimya*, from FA1
(Superfamily) **Prospondyloidea** † Permian (Lopingian) to Jurassic (Lower) (254–189.6 Ma) The byssal notch is completely closed in early ontogeny. Newell and Boyd (1970) included the type genus (*Prospondylus*) in Pseudomonotidae in 'Pectinacea', but in that family the byssal notch is open in early ontogeny. Family Prospondylidae	 *Prospondylus*, from Newell and Boyd (1970)

(Suborder) **Anomiina** Mississippian (Middle) to present (339.4–0 Ma) Carter et al. (2011) included both the 'hyporders' Aviculopectinodei and Anomioidei in this order – an arrangement we follow below. Unplaced family: Saharopteriidae †

(Superfamily) **Pseudomonotoidea** † Mississippian (Middle) to Cretaceous (Lower) (339.4–136.4 Ma) Pseudomonotids have an external alivincular ligament, are byssally attached in the younger stages at least, and often have distorted shells. The hinge is edentulous or has simple teeth, and the exterior shell is often radially ribbed, sometimes spinose. It has been suggested that this group gave rise to the oysters (Waller 1998; Waterhouse 2008). Family Pseudomonotidae †	*Pseudomonotis*, modified from TN
(Superfamily) **Anomioidea** Pennsylvanian (Lower) to present (318.1–0 Ma) The byssus is usually calcified (except in *Enigmonia*). The hinge lacks any teeth, and the ligament is a fibrous internal resilium supported in the right valve by a chondrophore or a stalked crurum (Yonge 1977). Families Anomiidae, Placunidae, and Permanomiidae †	*Monia* (Anomiidae), from FA1 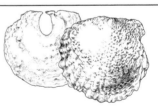  *Placuna*, from FA1

(Order) **Pectinida** Silurian (Llandovery) (439–0 Ma)
(Suborder) **Pectinina** Silurian (Llandovery) to present (439–0 Ma) Members of this group are monomyarian and lack nacre. The gills are filibranch or pseudoeulamellibranch. Several fossil groups included in the superfamily 'Pectinacea' in the *Treatise* (Cox et al. 1969) are now recognised as comprising several superfamilies and even suborders, but at least some of this may be a result of unnecessary taxonomic inflation. Waterhouse (2008) recognised three orders – Pterinopectinida, Limida (which included suborders Aviculopectinidina, Monotidina, Limidina), and Ostrida (suborder Etheripecininidia). In contrast, Nevesskaja (2009) used two suborders (Aviculopectinidina and Pectinina) and recognised only four superfamilies. The classification presented below largely follows that of Carter et al. (2011).

(Continued)

(Superfamily) **Pectinoidea**

Silurian (Wenlock) to present (428.2–0 Ma)

Pectinoideans are a large, diverse group that included byssally attached, cemented, or free-living forms. They lie on the right valve whether at rest, cemented, or byssally attached. The ligament is alivincular. The mantle lobes are completely unfused and tentaculate and may bear eyes.

Propeamussiidae was previously included here but is now placed in the superfamily Entolioidea.

Families Pectinidae (with several subfamilies), Cyclochlamydidae, Spondylidae, Neitheidae †, Pleuronectitidae †, and Tosapectinidae †

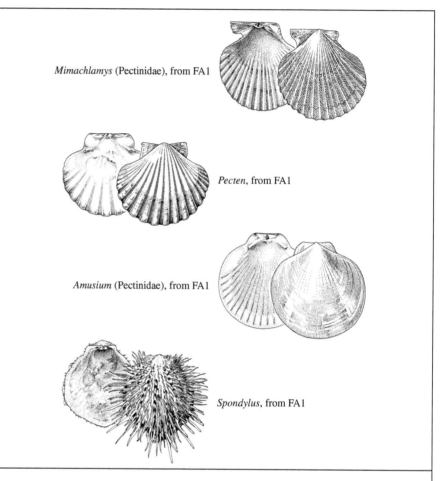

Mimachlamys (Pectinidae), from FA1

Pecten, from FA1

Amusium (Pectinidae), from FA1

Spondylus, from FA1

(Order) **Myalinida** †

Ordovician (Lower) to Cretaceous (Upper) (471.8–66 Ma)

This group is given subordinal status by Waterhouse (2008) (as Ambonychiidina) which he distinguished from pterioideans in their opisthodetic (usually duplivincular, platyvincular, or multivincular) ligament and (usually) lack of anterior wings. They are hetero- or monomyarian. Carter et al. (2011) recognised the group to a Megaorder.

(Superfamily) **Ambonychioidea** †

Ordovician (Middle) to Triassic (Upper) (468.0–216.5 Ma)

Mostly marine, but some Myalinidae lived in brackish and freshwater.

Families Ambonychiidae, Lunulacardiidae, Monopteriidae, Myalinidae, Mysidiellidae, and Ramonalinidae

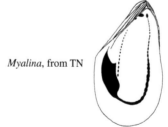

Myalina, from TN

(*Continued*)

(Superfamily) **Inoceramoidea** † Pennsylvanian (Lower) to Cretaceous (Upper) (318.1–66 Ma) Inoceramids sometimes attained a very large size (up to 1.8 m) and were epifaunal or semi-infaunal. They were included in the Pterioidea in the *Treatise* (Cox et al. 1969). Johnston and Collom (1998) argued on the basis of their ligament and hinge structure that inoceramids are similar to some Paleozoic praecardiodeans, and fossilised remains of the gills show they have a generally similar structure to that of Pteriidae (Knight et al. 2014). Families Inoceramidae, Atomodesmatidae, Kolymiidae, and Retroceramidae.	*Inoceramus*, modified from TN
(Superfamily) **Prokopievskioidea** † Permian (Cisuralian) (279.5–272.5 Ma) This group of freshwater bivalves was recently reviewed by Silantiev et al. (2015). Families Prokopievskiidae, Anadontellidae, and Naiaditidae	*Indonellina* (Prokopievskiidae), redrawn and modified from Silantiev et al. (2015)
(Superfamily) **Alatoconchoidea** † Permian (Guadalupian to Lopingian) (272.5–259 Ma) These were the largest Paleozoic bivalves, reaching up to a metre in size (Isozaki & Aljinović 2009). Families Alatoconchidae and Saikraconchidae	*Shikamaia* (Alatoconchidae), redrawn and modified from Kochansky-Devide (1978)

(Order) **Ostreida** (= **Ostreoida**)   Ordovician (Middle) to present (468–0 Ma)   This monomyarian group, the oysters, are cemented by their left valve. They have a prismatic shell layer and alivincular ligament. The gills are pseudoeulamellibranch.   This grouping as recognised by Carter et al. (2011) is equivalent to the superorder Pectinida recognised by Waterhouse (2008) which contained the orders Pterinopectinida, Limida, Ostreida, and Pectinida. Neither this group nor the 'Megaorder' Ostreata introduced by Carter et al. (2011) to include all the pteriomorphians below are recovered in the molecular analysis of Lemer et al. (2016).

(Superfamily) **Ostreoidea** Triassic (Lower) to present (249.7–0 Ma) Carter et al. (2011) split the rock oysters (*Crassostrea*, etc.) into a separate family, Flemingostreidae, but we continue to include these in Ostreidae as this was not supported in the analyses of Salvi et al. (2014, 2017). Families Ostreidae, Gryphaeidae, Arctostreidae †, Chondrodontidae †, Eligmidae †, Palaeolophidae †, and Rhombopteriidae †	*Ostrea*, from FA1 *Gryphaea*, from Fischer (1880–1887) *Hyotissa* (Gryphaeidae), from FA1

(Continued)

(Order) **Pteriida**	
Ordovician (Lower) to present (470.0–0 Ma)	
Members of this group are byssate, often inequivalve, and heteromyarian to monomyarian. The gills are filibranch or pseudoeulamellibranch.	
Treated as a superorder by Waterhouse (2008) who included Pteriidina, Pinnidina, and Ambonychiidina as separate suborders.	
(Suborder) **Pteriina**	
Ordovician (Lower) to present (470.0–0 Ma)	
This grouping is equivalent to the 'order' Pteriida in Carter et al. (2011).	
The family Pterineidae † is not allocated to a superfamily in the Carter et al. (2011) classification. It had previously been treated as a member of Pterioidea.	
(Superfamily) **Pterioidea** Ordovician (Lower) to present (470.0–0 Ma) Equivalve to subequivalve, dimyarian to monomyarian with anterior wings small, posterior wing large to absent, ligament usually alivincular or duplivincular, and hinge teeth are often present. The gills are filibranch or pseudoeulamellibranch, usually byssate with life position typically more or less upright. Pulvinitidae was previously treated as a separate family but was ranked as a subfamily of Malleidae in the Carter et al. (2011) classification. Families Pteriidae, Isognomonidae, Malleidae, Bakevelliidae †, Cassianellidae †, Kochiidae †, Pergamidiidae †, Plicatostylidae †, and Vlastidae †	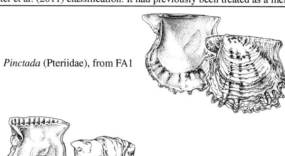 *Pinctada* (Pteriidae), from FA1  *Isognomon*, from FA1  *Foramelina* (Malleidae), from FA1  *Malleus*, from FA1
(Superfamily) **Posidonioidea** † Devonian (Middle) to Paleogene (Eocene) 388.1–48.6 Ma Posidoniidae is dimyarian, has no distinct byssal notch at maturity, the ligament is duplivincular, and the hinge edentulous. This group was included in the 'Pectinacea' in the *Treatise* (Cox et al. 1969). Posidoniidae is included in the Pterioidea by Waterhouse (2008). Families Posidoniidae, Aulacomyellidae, Daonellidae, and Halobiidae	 *Posidonia*, modified from TN
(Superfamily) **Rhombopterioidea** † Silurian (Wenlock) to Devonian (Middle) (428.2–391.9 Ma) This group was included in the 'Pectinacea' in the *Treatise* (Cox et al. 1969). Families Rhombopteriidae and Umburridae	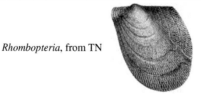 *Rhombopteria*, from TN

(Continued)

(Superfamily) **Pinnoidea** Silurian (Wenlock) to present (428.2–0 Ma) Triangular, monomyarian, byssate bivalves which live part buried in substratum. Their gills are pseudoeulamellibranch. Family Pinnidae	*Pinna*, from FA1
(Suborder) **Aviculopectinoidei** Silurian (Llandovery) to Cretaceous (Upper) (443.7–99.7 Ma). The taxa below were included in the 'Pectinacea' in the *Treatise* (Cox et al. 1969).	
(Superfamily) **Aviculopectinoidea** † Devonian (Middle) to Cretaceous (Upper) (391.9–99.7 Ma) The ligament is alivincular and external, and some nacre may be present. Families Aviculopectinidae, Deltopectinidae, and Limatulinidae	*Aviculopecten*, modified from TN
(Superfamily) **Pterinopectinoidea** † Silurian (Llandovery) to Triassic (Middle) (436–237 Ma) Distinguished on the basis of an external ligament which is amphidetic and chevron-duplivincular, the shell has moderate to large posterior wings which are usually not well delineated, and anterior right valve auricle usually with a byssal notch. Waterhouse (2008) recognised this monomyarian group as a new order (Pterinopectinida) in which he recognised three superfamilies and six families. He considered that this group was the 'root stock' of all other Pectinida. Families Pterinopectinidae, Claraiidae, and Natalissimidae	*Pterinopecten*, modified from Fischer (1880–1887)
(Superfamily) **Chaenocardioidea** † Silurian (Llandovery) to Triassic (Middle) (443.7–247.2 Ma) Ligament external and amphidetic. The right anterior auricle and byssal notch usually well developed. Families Chaenocardiidae and Streblochondriidae	*Streblocondria*, modified from TN
(Superfamily) **Heteropectinoidea** † Devonian (Upper) to Permian (Lopingian) (360.7–252.3 Ma) Families Heteropectinidae, Annuliconchidae, Antijaniridae, Hunanopectinidae, Limipectinidae, and Ornithopectinidae	*Annuliconcha*, modified from TN

(Continued)

(Order) Limida Devonian (Upper) to present (360.7–0 Ma) This monomyarian group is characterised by their equally convex shell valves which lack nacre and have an alivincular ligament. The gills are, uniquely for Pteriomorphia, eulamellibranch. Waterhouse (2008) interpreted this group more broadly than other workers and included within it the suborders Aviculopectinidina, Monotidina, and Limidina. Waller (1978), Bieler and Mikkelsen (2006), and Bieler et al. (in Bouchet & Rocroi) (2010) treated this group as an order, but Carter et al. (2011) raised it to 'hyporder'.	
(Superfamily) Limoidea Devonian (Upper) to present (360.7–0 Ma) Shell with weak hinge and characteristic long mantle edge tentacles are present. Families Limidae and Isolimeidae	 *Lima*, from FA1  *Limaria* (Limidae) animal, from FA1
(Suborder) Monotidina † Devonian (Middle) to Paleogene (Paleocene) (391.9–61.7 Ma)	
(Superfamily) Buchioidea † Triassic (Middle) to Cretaceous (Upper) (247.2–66 Ma) The monomyarian shell usually lacks a posterior wing, and the ligament is opisthodetic, external, platyvincular, or with resilifer. The right anterior auricle is well developed to small or modified and partly articulated. The hinge may have simple large teeth. Waterhouse (2008) included Buchiidae in his 'suborder' Monotidina as superfamily Eurydesmatoidea. He also included a superfamily Monotoidea in Monotidina. Families Buchiidae, Dolponellidae, and Monotidae	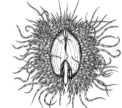 *Malayomaorica* (Buchiidae), from TN  *Monotis*, modified from TN
(Superfamily) Eurydesmatoidea † Permian (Cisuralian) (295–290.1 Ma) Families Eurydesmatidae and Manticulidae	 *Eurydesma*, redrawn and modified from TN
(Superfamily) Oxytomoidea † Devonian (Middle) to Paleogene (Paleocene) (391.9–61.7 Ma) Family Oxytomidae	 *Oxytoma*, from TN

(*Continued*)

(Suborder) Entoliidina Devonian (Upper) to present (360.7–0 Ma)	
(Superfamily) Entolioidea Devonian (Upper) to present (360.7–0 Ma) Propeamussiidae is included here by Carter et al. (2011), rather than in Pectinida where it is usually placed. Families Entoliidae, Propeamussiidae, Entolioidesidae †, and Pernopectinidae †	*Entolium*, from TN
(Superfamily) Euchondrioidea † Devonian (Upper) to Permian (Lopingian) (360.7–252.3 Ma) This group has the right valve less inflated than the left, and a series of denticles form a distinctive hinge. Family Euchondriidae	*Euchondria*, modified from TN

(Infraclass) Heteroconchia Ordovician (Lower) to present (478.6–0 Ma) This grouping includes the Palaeoheterodonta and the Heterodonta. Both are groups that have often been considered separate subclasses. The gills are filibranch in Trigonoidea only and eulamellibranch in the remainder.
(Cohort) Palaeoheterodonta (= Uniomorphi of Carter et al. 2011) Ordovician (Upper) to present (478.6–0 Ma) This grouping, sometimes used as a subclass, includes the Trigoniida and Unionida. The shells are prismato-nacreous, with trigonioideans having filibranch gills and unionidans having eulamellibranch gills. These two groups are probably not closely related (e.g., Morton 1987) despite both sharing similar shell structure and calcareous gill spicules (Taylor et al. 1973). Unplaced family: Thoraliidae †
(Subcohort) Unioni Ordovician (Upper) to present (478.6–0 Ma)

? (Superfamily) Lyrodesmatoidea † Ordovician (Middle) to Devonian (Middle) (467.3–383.7 Ma) This group was unplaced within Unioni by Carter et al. (2011) and under Heterodonta by Bieler et al. (in Bouchet & Rocroi) (2010). Families Lyrodesmatidae and Pseudarcidae	*Lyrodesma*, redrawn and modified from Desparmet et al. (1971) 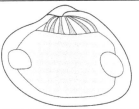

(Megaorder) Unionata Ordovician (Upper) to present (455.8–0 Ma)
(Order) Trigoniida (= Trigonioida) Ordovician (Upper) to present (455.8–0 Ma) Several different superfamilies are now recognised (Cooper 1991). In comparison, the *Treatise* (Cox et al. 1969) only recognises one ('Trigonacea'), and the list below may well be a case of unnecessary taxonomic inflation. For example, *Megatrigonia* and *Myophorella* were both included in the Trigoniidae in the *Treatise* (Cox et al. 1969). Members of this group have nacreous equivalve shells, trigonal to quadrate in shape, with a posterior area that is often demarcated from the rest of the shell and may be differently sculptured. The ligament is external, short, and opisthodetic. The characteristic hinge consists of large cardinal teeth which are typically transversely ridged. This type of hinge first appeared in the Triassic members of the group (Cooper 1991). Uniquely in the heteroconch bivalves, the gill is filibranch.

(Continued)

(Superfamily) **Trigonioidea** Ordovician (Upper) to present (455.8–0 Ma) Cooper (1991) recognised a separate superfamily for Myophoriidae, but this is not upheld by Boyd and Newell (1997) or Bieler et al. in Bouchet et al. (2010), although it is recognised by Carter et al. (2011). Families Trigoniidae, Eoschizodidae †, Groeberellidae †, Myophoriidae †, Prosogyrotrigoniidae †, Scaphellinidae †, and Schizodidae †	 *Neotrigonia* (Trigoniidae), from FA1 *Myophoria*, from Fischer (1880–1887)
(Superfamily) **Myophorelloidea** † Triassic (Middle) to Cretaceous (Upper) (247.2–66.0 Ma) Members of this group are elongated posteriorly, with subterminal umbones, nodulose sculpture in adults, and a distinctly differentiated and differently sculptured posterior area. Cooper recognised this group as a separate suborder, Myophorellina, in which he included this group and the Megatrigonioidea. Families Myophorellidae, Buchotrigoniidae, and Laevitrigoniidae	 *Myophorella*, modified from TN
(Superfamily) **Pseudocardinioidea** † Jurassic (Middle to Upper) (164.7–150.8 Ma) Families Pseudocardiniidae, and Utschamiellidae	 *Pseudocardinia*, from TN
(Superfamily) **Megatrigonioidea** † Jurassic (Middle) to Cretaceous (Upper) (164.7–66 Ma) This group differs from the myophorelloideans in lacking a distinct posterior area and in having strong radial sculpture with nodules usually poorly developed or absent. Families Megatrigoniidae, Iotrigoniidae, and Rutitrigoniidae	 *Megatrigonia*, modified from TN
(Superfamily) **?Trigonioidoidea** (or in Unionida?) † Jurassic (Upper) to Cretaceous (Upper) (145.5–94.3 Ma) Members of this non-marine group are mainly from Asia and are derived from ancient unionoideans according to Guo (1998) who first recognised this group as a separate superfamily. Families Trigonioididae, Nakamuranaiadidae, Nippononaiidae, Plicatounionidae, and Pseudohyriidae	 *Trigonioides*, modifed from TN

(Continued)

(Superfamily) **Trigonodoidea** † Permian (Guadalupian) to Jurassic (Lower) (265.0–199.3 Ma) Families Trigonodidae and Desertellidae	*Unionites* (Trigonodidae), from TN  

(Order) **Unionida** (= Unionoida, Unioniformes)   Triassic (Middle) to present (247–0 Ma)   Living unionidans have eulamellibranch gills, but based on impressions left in fossils at least some Triassic unionidans had filibranch gills (Whyte 1992). Skawina and Dzik (2011 p. 874) defined this group as 'Freshwater bivalves characterized by linearly arranged pedal elevator attachments in the beak region, but having plesiomorphic nacreous internal shell layer, and, in underived forms, transversely ribbed cardinal teeth of the hinge'.

(Superfamily) **Unionoidea** Triassic (Middle) to present (245.0–0 Ma) Freshwater mussels with a glochidium larva. Families Unionidae, Margaritiferidae, and Sancticarolitidae †	*Unio*, modified from TN  *Cyclonaias* (Unionidae), from TN  *Margaritifera*, from TN 

(*Continued*)

(Superfamily) **Etherioidea** (= Mullerioidea, Muteloidea) Cretaceous (Upper) to present (70.6–0 Ma) Freshwater mussels with a lasidium larva; some are cemented. Families Etheriidae (= Mulleriidae), Iridinidae (= Mutelidae), and Mycetopodidae	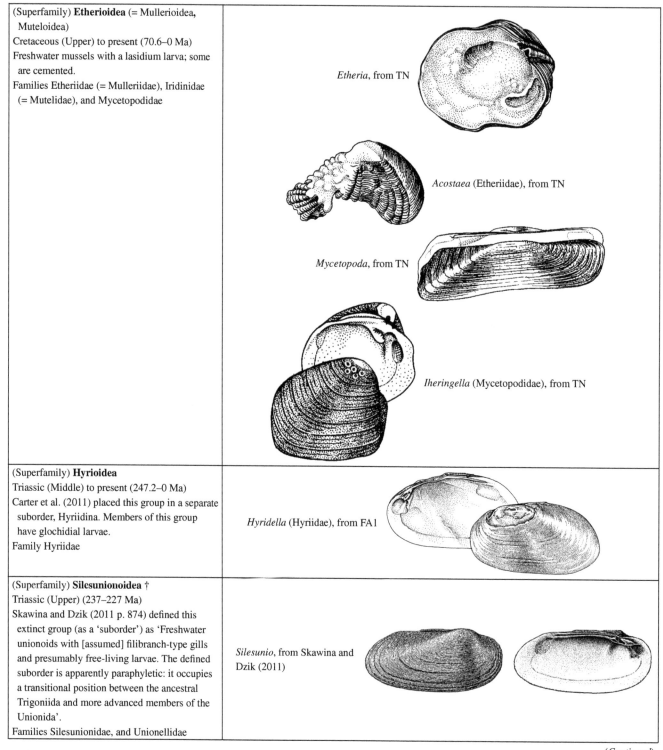 *Etheria*, from TN *Acostaea* (Etheriidae), from TN *Mycetopoda*, from TN *Iheringella* (Mycetopodidae), from TN
(Superfamily) **Hyrioidea** Triassic (Middle) to present (247.2–0 Ma) Carter et al. (2011) placed this group in a separate suborder, Hyriidina. Members of this group have glochidial larvae. Family Hyriidae	*Hyridella* (Hyriidae), from FA1
(Superfamily) **Silesunionoidea** † Triassic (Upper) (237–227 Ma) Skawina and Dzik (2011 p. 874) defined this extinct group (as a 'suborder') as 'Freshwater unionoids with [assumed] filibranch-type gills and presumably free-living larvae. The defined suborder is apparently paraphyletic: it occupies a transitional position between the ancestral Trigoniida and more advanced members of the Unionida'. Families Silesunionidae, and Unionellidae	*Silesunio*, from Skawina and Dzik (2011)

(Continued)

(Cohort) Heterodonta (= Cardiomorphi of Carter et al. 2011) Ordovician (Lower) to present (478.6–0 Ma) This group was divided into two major clades (Giribet 2008) largely on the basis of molecular studies. We use the same rank given by Carter et al. (2011).	
(Subcohort) Archiheterodonta (= Carditioni of Carter et al. 2011, which they rank as 'subcohort') Ordovician (Lower) to present (478.6–0 Ma) In most classifications and phylogenies, the archiheterodonts are sister to the euheterodonts, although in Sharma et al. (2011) they formed a clade with the palaeoheterodonts.	
(Order) Actinodontida (= Actinodontoida) † Ordovician (Lower) to Permian (Lopingian) (478.6–254 Ma) Members mostly marine but include some brackish and freshwater taxa.	
(Superfamily) Anodontopsoidea † (= Actinodontoidea) Ordovician (Lower) to Permian (Guadalupian) (478.6–272.5 Ma) Cycloconchidae was previously placed in its own superfamily and was included in Modiomorphoidea in the *Treatise* (Cox et al. 1969). Families Anodontopsidae, Actinodontidae, Baidiostracidae, Cycloconchidae, Intihuarellidae, and Redoniidae	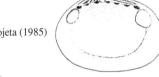 *Cycloconcha*, from Pojeta (1985)  *Actinodonta*, redrawn and modified from Desparmet et al. (1971)
(Superfamily) Nyassoidea † Ordovician (Lower) to Permian (Guadalupian) (478.6–272.5 Ma) Hinge with numerous pseudocardinal and pseudolateral teeth and an opisthodetic ligament. Family Nyassidae (= Zadimerodiidae)	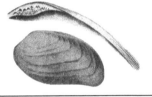 *Nyassa*, from TN
(Superfamily) Palaeomuteloidea † Devonian (Upper) to Permian (Lopingian) (388.1–254 Ma) A brackish to freshwater Permian group that is shaped like a modern freshwater mussel and has a reduced taxodont-like hinge with numerous very small teeth (Cox, 1969; Silantiev, 1998). Palaeanodontidae is placed in its own superfamily in Silantiev et al. (2015). Families Palaeomutelidae (= Palaeanodontidae), Amnigeniidae, Montanariidae, and Zadimerodiidae	*Palaeomutela*, redrawn and modified from Urazaeva et al. (2015)   *Montanaria*, modified from Desbiens (1994)
(Order) Cardiitida (= Carditoida) Ordovician (Lower) to present (478.6–0 Ma) Unplaced families: Archaeocardiidae † and Eodonidae †	

(*Continued*)

(Superfamily) **Crassatelloidea**

Ordovician (Lower) to present (478.6–0 Ma)

Carditids and condylocardiids were included in a separate superfamily (Carditioidea) in earlier classifications.

Families Crassatellidae, Astartidae, Carditidae, Condylocardiidae, ?Aenigmoconchidae †, Cardiniidae †, and Myophoricardiidae †

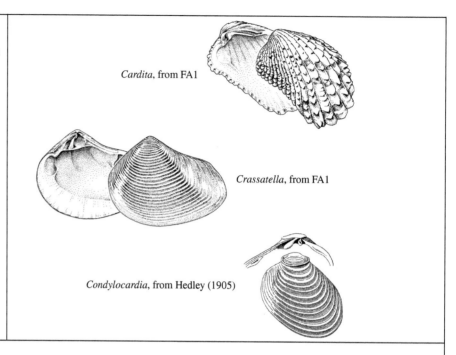

Cardita, from FA1

Crassatella, from FA1

Condylocardia, from Hedley (1905)

(Subcohort) **Euheterodonta**

Ordovician (Lower) to present (478.6–0 Ma)

This clade includes the remaining bivalves, encompassing the Anomalodesmata and the Imparidentia.

(Megaorder) **Anomalodesmata** (= Poromyata)

Ordovician (Lower) to present (478.6–0 Ma)

The classification of this group was considerably inflated by Carter et al. (2011) from a single order (Pholadomyida) in Bieler et al. (in Bouchet & Rocroi) (2010) to a 'megaorder' containing four orders. We follow the Carter et al. (2011) classification below, although it needs testing.

(Order) **Poromyida**

Triassic (Upper) to present (237–0 Ma)

Approximately equivalent to 'Septibranchia'.

(Superfamily) **Poromyoidea**

Cretaceous (Lower) to present (112–0 Ma)

Families Poromyidae and Cetoconchidae

Poromya, from FA1

(Superfamily) **Cuspidarioidea**

Triassic (Upper) to present (237–0 Ma)

Families Cuspidariidae, Halonymphidae, Protocuspidariidae, and possibly Spheniopsidae

Cuspidaria, from FA1

(Superfamily) **Parilimyoidea**

Jurassic (Upper) to Cretaceous (Upper) (161.2–66 Ma)

Family Parilimyidae

Parilimya, from FA1

(Continued)

(Superfamily) **Verticordioidea** Cretaceous (Upper) to present (72.1–0 Ma) Families Verticordiidae, Euciroidae, and Lyonsiellidae	*Verticordia*, from FA1 *Euciroa*, from FA1
(Order) **Pholadomyida** (= Pholadomyoida) Ordovician (Middle) to present (468.0–0 Ma)	
(Superfamily) **Pholadomyoidea** Ordovician (Middle) to present (468.0–0 Ma) Families Pholadomyidae, Arenigomyidae †, Margaritariidae †, and Ucumariidae †	*Pholadomya*, original
(Order) **Pandorida** Triassic (Middle) to present (2.47.2–0 Ma)	
(Superfamily) **Pandoroidea** Triassic (Middle) to present (247.2–0 Ma) Families Pandoridae, Laternulidae, and Lyonsiidae	*Pandora*, from FA1 *Laternula*, from FA1

(Superfamily) **Cercomyoidea** †
Triassic (Upper) to Cretaceous (Upper)
  208.5–66 Ma.
Cercomya was included in the Laternulidae by
  Keen and Cox (1968) but was moved, somewhat
  tentatively, by Carter (1990) to the Pterioida.
  Aberhan (2004) includes this group within
  Laternulidae. The present location follows
  Carter in ms.
Family Cercomyidae
	Cercomya, modified from TN

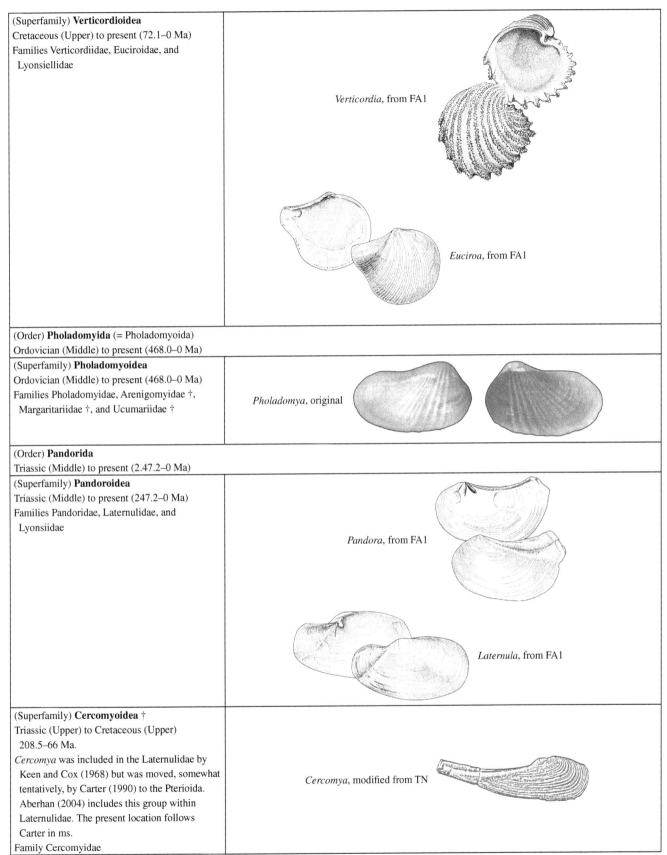

(Continued)

(Superfamily) **Clavagelloidea**
Cretaceous (Upper) to present (83.6–0 Ma)
Families Clavagellidae and Penicillidae

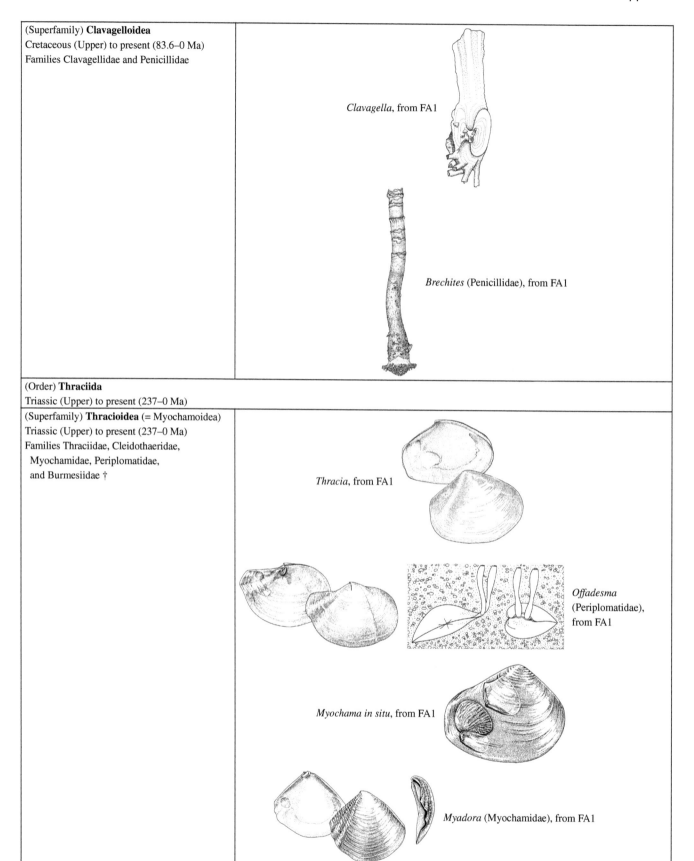

Clavagella, from FA1

Brechites (Penicillidae), from FA1

(Order) **Thraciida**
Triassic (Upper) to present (237–0 Ma)

(Superfamily) **Thracioidea** (= Myochamoidea)
Triassic (Upper) to present (237–0 Ma)
Families Thraciidae, Cleidothaeridae,
  Myochamidae, Periplomatidae,
  and Burmesiidae †

Thracia, from FA1

Offadesma
(Periplomatidae),
from FA1

Myochama in situ, from FA1

Myadora (Myochamidae), from FA1

(*Continued*)

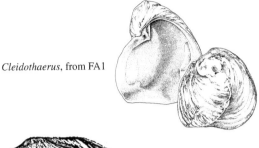

Cleidothaerus, from FA1

Burmesia, modified from TN

(Megaorder) **Imparidentia** (= 'subcohort' Cardioni of Carter et al. 2011)

Ordovician (Lower) to present (478.6–0 Ma)

This grouping was introduced by Bieler et al. (2014) to include the euheterodonts excluding the Anomalodesmata.

Imparidentia of uncertain relationships

The extinct taxa listed below are not assigned to an order and only tentatively placed in Imparidentia.

(Superfamily) **Grammysioidea** † Ordovician (Lower) to Paleogene (Paleocene) (471.8–61.6 Ma) Families Grammysiidae and Sanguinolitidae	 *Grammysia*, from TN

(Superfamily) **Modiomorphoidea** †

Ordovician (Lower) to Jurassic (Upper)
  (478.6–155.7 Ma)

The analysis in Carter et al. (2000) showed
  Modiomorphoida as recognised in the *Treatise*
  (Cox et al. 1969) was paraphyletic which, as
  also acknowledged by Newell (1969), contained
  early members of various lamellibranch groups.
  Thus, members of this group have been
  variously treated – for example the superfamily
  has been included in the Palaeoheterodonta by
  Newell (1969), Isofilibranchia along with the
  mytilids by Carter (1990) and Campbell et al.
  (1998), but placed in the Anomalodesmata by
  Carter et al. (2000). Even in the restricted sense
  used here it is highly likely that this group is
  paraphyletic.

Families Modiomorphidae, Cypricardiniidae,
  Hippopodiumidae, Palaeopharidae, and
  Tusayanidae

Cypricardinia, from TN, lower figure modified

Modiomorpha, from TN

Guerangeria (Modiomorphidae), from
Fischer (1880–1887)

(Suborder?) **Anthracosiidina** †

Pennsylvanian (Lower) to Permian (Lopingian) (318.1–252.3 Ma)

These extinct non-marine bivalves are given subordinal status by Carter et al. (2011).

(Superfamily) **Anthracosioidea** †
Pennsylvanian (Lower) to Permian (Lopingian)
  (318.1–252.3 Ma)
An extinct, non-marine group. Some reached
  several cm in length.
Families Anthracosiidae and possibly
  Ferganoconchidae, and Shaanxiconchidae
	Anthracosia, redrawn and
	modified from TN  

(*Continued*)

(Superfamily) **Prilukielloidea** † Permian (Guadalupian) (268–265 Ma) A freshwater group. Families Prilukiellidae and Senderzoniellidae	*Bakulia* (Senderzoniellidae), from Silantiev et al. (2015) 

(Order) **Hippurtida** Devonian (Lower) to Cretaceous (Upper) (412.3–66 Ma) This order contains the extinct rudists, comprised of four superfamilies. This group has often been treated as a single superfamily (Hippuritacea) (e.g., Dechaseaux 1969), but here we slightly modifiy the classification of Carter et al. (2011) who recognised four superfamilies in two orders. A phylogenetic analysis of the group was undertaken by Skelton and Smith (2000).
(Suborder) **Megalodontina** † Silurian (Llandovery) to Cretaceous (Upper) (436–66 Ma)

(Superfamily) **Megalodontoidea** † Devonian (Lower) to Cretaceous (Upper) (412.3–94.3 Ma) Families Megalodontidae, Ceratomyopsidae, Dicerocardiidae, Pachyrismatidae, and Wallowaconchidae	*Megalodon*, inside valve modified from TN, end view from Fischer (1880–1887) 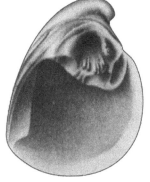 
(Superfamily) **Mecynodontoidea** † Devonian (Middle-Upper) (393.3–358.9 Ma) Taxa included here were assigned in the *Treatise* (Cox et al. 1969) to very diverse groups – namely 'Arcticacea' (Mecynodontidae), 'Crassatellacea' (*Prosocoelus*), and the pterioidan 'Ambonychiacea' (*Congeriomorpha*). Families Mecynodontidae, Beichuaniidae, Congeriomorphidae, Plethocardiidae, and Prosocoelidae	*Mecynodonta*, from TN  

(Suborder) **Hippuritina** (= Hippuritoida, Rudistes, Rudista) † Jurassic (Upper) to Cretaceous (Upper) (163.5–66 Ma)

(Superfamily) **Requienioidea** † Jurassic (Upper) to Cretaceous (Upper) (155.7–66 Ma) These include the so-called spirogyrate rudists, some of which form gastropod-like shells with one valve much smaller and operculum-like. Families Requieniidae and Epidiceratidae	*Requienia*, from Fischer (1880–1887) 

(Continued)

(Superfamily) **Radiolitoidea** (= Caprotinoidea, Hippuritoidea) † Jurassic (Upper) to Cretaceous (Upper) (163.5–66 Ma) Families Radiolitidae, Antillocaprinidae, Caprinidae, Caprinulidae, Caprotinidae, Diceratidae, Hippuritidae, Ichthyosarcolitidae, Monopleuridae, Plagioptychidae, Polyconitidae, and possibly Trechmannellidae	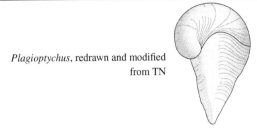 *Plagioptychus*, redrawn and modified from TN 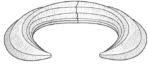 *Titanosarcolites* (Caprinidae), redrawn and modified from Mitchell (2002) *Diceras*, from Fischer (1880–1887), internal views of valves redrawn and modified. 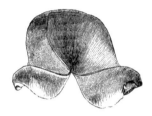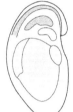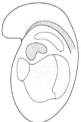  *Hippurites*, from Fischer (1880–1887)

(Order) **Lucinida** (= Lucinoida)   Ordovician (Lower) to present (478.6–0 Ma)   The phylogenetic position of this group has been controversial. They have been considered to be basal heterodonts (e.g., Newell 1969), but the cladistic analysis of Carter et al. (2000) had lucinoids outside the heterodonts. However, molecular analyses (Giribet & Distel 2003; Taylor et al. 2007; Giribet 2008; Lemer et al. 2019) have shown that this group are the most basal living euheterodonts.

(Superfamily) **Babinkoidea** † Ordovician (Lower to Middle) (478.6–457.5 Ma) Families Babinkidae and Coxiconchiidae	*Babinka*, modified from TN 

(*Continued*)

(Superfamily) **Lucinoidea**

Ordovician (Middle) to present (470–0 Ma)

Lucinids are a distinctive group with an anterior
inhalant and posterior exhalant water flow. Their
gills consist of a single demibranch and are
packed with chemosymbiotic bacteria
necessitating gill-like mantle modifications to
assist with respiration (Taylor & Glover 2006).

Taylor and Glover (2006) considered the
Mactomyidae (Devonian–Cretaceous) an
uncertain member of this superfamily.

Families Lucinidae (= Fimbriidae),
?Mactomyidae †, and Paracyclidae †

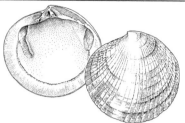

Codakia (Lucinidae), from FA1

Fimbria (Lucinidae), from FA1

(Superfamily) **Thyasiroidea**

Jurassic (Upper) to present (145.5–0 Ma)

This group was until relatively recently included
in Lucinoidea. It differs in having both pairs of
demibranchs (instead of just the inner pair) but
has a similar life style in that it has
chemosymbiotic bacteria in its gills.

Family Thyasiridae

Thyasira, from FA1

(Order) **Gastrochaenida**

Triassic (Upper) to present (237–0 Ma)

Carter (2011) provided a suborder for Gastrochaenidae, but the ordinal rank was introduced by Lemer et al. (2019) based on the results of their molecular
phylogeny where it is sister to all Imparidentia other than Lucinida.

(Superfamily) **Gastrochaenoidea**

Triassic (Upper) to present (237–0 Ma)

Gastrochaenidae was placed outside Myoida in
the analyses of Taylor et al. (2007; 2009) but
was poorly resolved in the analysis by Bieler et
al. (2014). It was grouped with thyasirids and
lucinids in the Combosch et al. (2017) analysis,
although poorly supported.

Family Gastrochaenidae

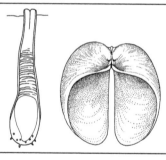

Gastrochaena, from FA1

(Order) **Solenida** (= Solenata, Adapedonta in part)

Ordovician (Middle) to present (460.9–0 Ma)

In recent molecular phylogenies the solenids and related taxa were shown to be distinctly different from the myoideans, where they had been placed in
older classifications. Carter et al. (2011) provided Solenata (a megaorder) and Solenida for the group while Bieler et al. (2014) used Adapedonta, a name
introduced in 1909 that also included 'Edentulacea' (Gastrochaenidae) and Myacea (Bouchet et al. 2010). The molecular analysis of Combosch et al.
(2017) supported the grouping of galeommatoideans, hiatellids, and solenoideans but excluded myids and gastrochaenids, so the use of 'Adapedonta' by
Combosch et al. (2017) is not appropriate. The placement of galeommatoideans in this grouping in the Combosch et al. (2017) analysis was not
maintained in the Lemer et al. (2019) analysis, with the latter recognising the composition below (extant taxa only) in their Adapedonta.

(Superfamily) **Orthonotoidea** †

Ordovician (Middle)–Permian (Lopingian)
(460.9–259 Ma)

Bieler et al. (in Bouchet & Rocroi) (2010) treated
this group as an order (Orthonotida) in
Anomalodesmata, but Carter et al. (2011)
included it in Solenata.

Families Orthonotidae, Konduriidae, Prothyridae,
and Solenomorphidae

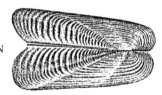

Orthonota, from TN

(Superfamily) **Solenoidea** Mississippian (Upper) to present (326.4–0 Ma) Families Solenidae and Pharidae (= Cultellidae)	*Solen*, from FA1 *Cultellus* (Pharidae), from FA1 *Pharus*, from FA1
(Superfamily) **Hiatelloidea** Jurassic (Lower) to present (199.6–0 Ma) Families Hiatellidae and Saxicavellidae	*Hiatella*, from FA1
(Superfamily) **Edmondioidea** † Ordovician (Upper) to Jurassic (Lower) (449.5–199.6 Ma) Families Edmondiidae and Pachydomidae	*Edmondia*, redrawn and modified from TN

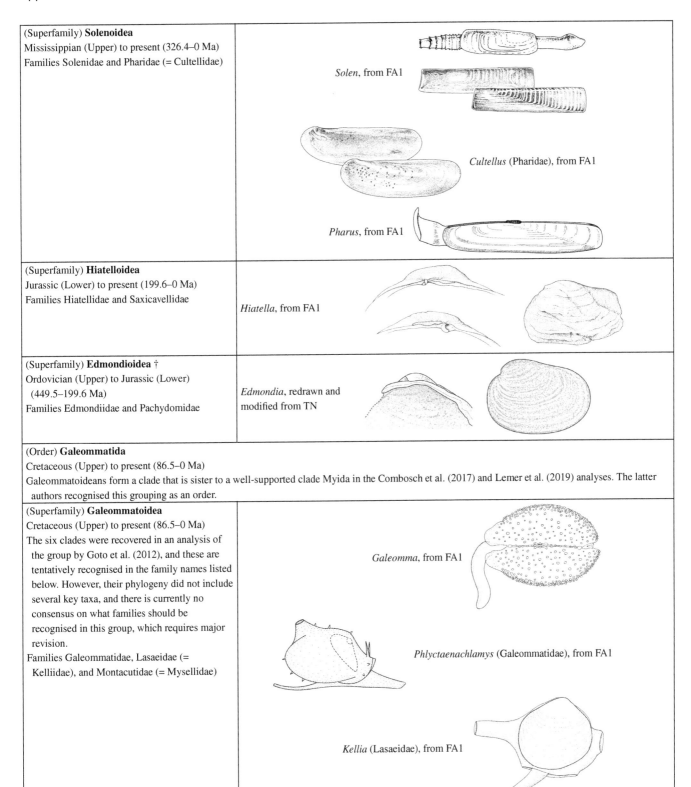

(Order) **Galeommatida**
Cretaceous (Upper) to present (86.5–0 Ma)
Galeommatoideans form a clade that is sister to a well-supported clade Myida in the Combosch et al. (2017) and Lemer et al. (2019) analyses. The latter
  authors recognised this grouping as an order.

(Superfamily) **Galeommatoidea**
Cretaceous (Upper) to present (86.5–0 Ma)
The six clades were recovered in an analysis of
  the group by Goto et al. (2012), and these are
  tentatively recognised in the family names listed
  below. However, their phylogeny did not include
  several key taxa, and there is currently no
  consensus on what families should be
  recognised in this group, which requires major
  revision.
Families Galeommatidae, Lasaeidae (=
  Kelliidae), and Montacutidae (= Mysellidae)
	Galeomma, from FA1
	
	Phlyctaenachlamys (Galeommatidae), from FA1
	
	Kellia (Lasaeidae), from FA1

(Order) **Cardiida**
Ordovician (Upper) to present (445.6–0 Ma)
Unplaced family: Palaeocarditidae †

(*Continued*)

(Superfamily) **Kalenteroidea** † Devonian (Middle) to Cretaceous (Upper) (391.9–66 Ma) Family Kalenteridae	*Myoconcha* (Kalenteridae), from TN
(Superfamily) **Cardioidea** Ordovician (Upper) to present (445.6–0 Ma) Families Cardiidae (including Tridacninae), and Pterocardiidae †	*Vepricardium* (Cardiidae), from FA1 *Cerastoderma* (Cardiidae), from TN *Fragum* (Cardiidae), from FA1 *Tridacna* (Cardiidae), from FA1

(Megaorder) **Neoheterodontei** (= Cardiata)
Ordovician (Lower) to present (478.6–0 Ma)
This large group contains many of the familiar heterodont families. It was first introduced by Taylor et al. (2007) as an unranked group to include Chamoidea, Cyrenoidea, Gaimardioidea, Mactroidea, Myida, Sphaerioidea, Ungulinoidea, and Veneroidea.
(Order) **Sphaeriida**
Jurassic (Upper) to present (145.5–0 Ma)
This is the most basal extant member of the Neoheterodontei and was given ordinal level by Lemer et al. (2019).

(Continued)

(Superfamily) **Sphaerioidea** Jurassic (Upper) to present (145.5–0 Ma) Families Sphaeriidae and Neomiodontidae †	*Sphaerium*, from FA1 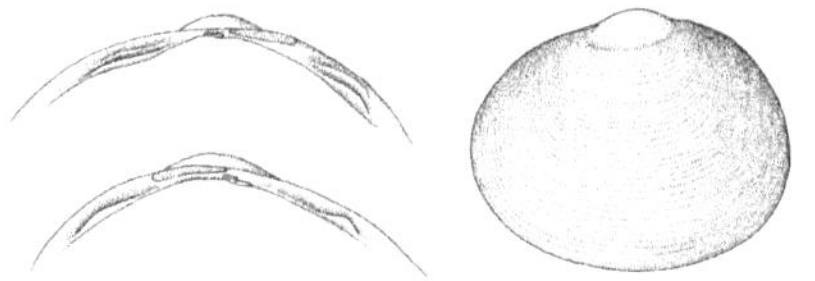
(Superfamily) **Tellinoidea** Triassic (Upper) to present (237–0 Ma) This is the most speciose bivalve superfamily. The shells are compressed laterally and are often slightly inequilateral. The ligament is external and seated on a shell plate (nymph). They have a powerful digging foot, separate siphons that are narrow and often long; the labial palps are large and, like the stomach, modified for deposit feeding. A characteristic feature of the group is the cruciform muscle located at the ventro-posterior margin. Families Tellinidae, Donacidae, Psammobiidae, Semelidae Solecurtidae, Icanotiidae †, Quenstedtiidae †, Sowerbyidae †, Tancrediidae †, and Unicardiopsidae †	*Tellina*, from FA1 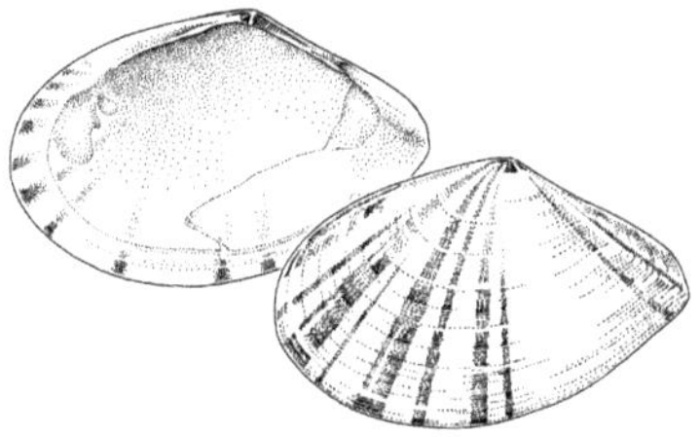  *Semele*, from FA1 *Gari* (Psammobiidae), from FA1 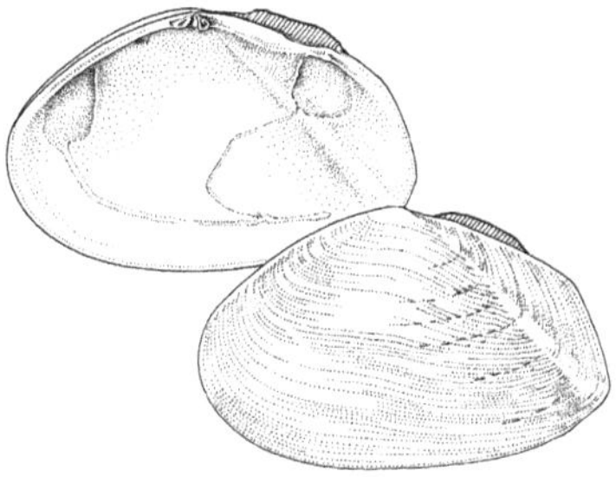 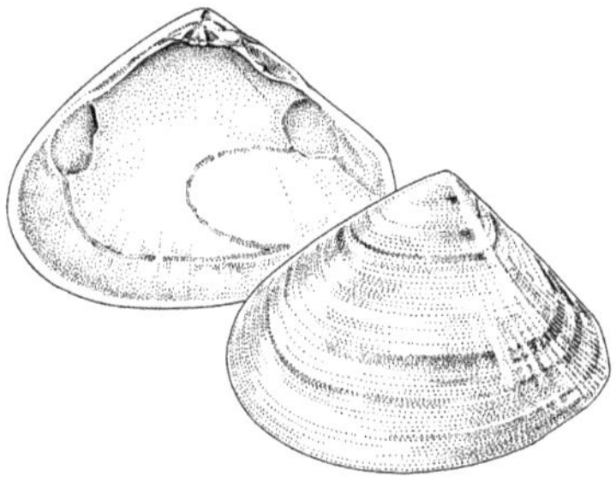 *Plebidonax* (Donacidae), from FA1 *Solecurtus*, from FA1 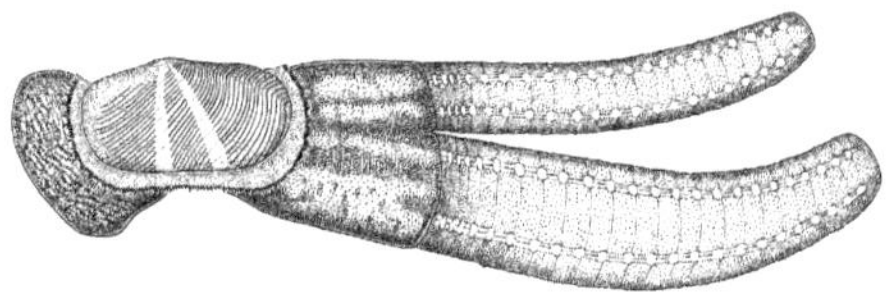

(Order) **Venerida**

Ordovician (Middle) to present (468–0 Ma)

This grouping contains the majority of neoheterodonts. There are a number of groupings within it that were recovered in recent molecular analyses, but the groupings vary to some extent.

(*Continued*)

(Superfamily) **Mactroidea**
Triassic (Upper) to present (203.6–0 Ma)
Families Mactridae, Anatinellidae, Cardiliidae, and Mesodesmatidae

Mactra, from FA1 (left), *Spisula* (Mactridae), from Fischer (1880–1887) (right)

Zenatina (Mactridae), from FA1

Paphies (Mesodesmatidae), from FA1

Cardilia, from FA1

(Superfamily) **Ungulinoidea**
Cretaceous (Lower) to present (140.2–0 Ma)
The hinge is characterised by a pair of cardinal teeth, the posterior one being bifid. This group was previously thought to be related to Lucinidae, but unlike that family it has two demibranchs and lacks bacterial symbionts in the gills. Members of this family have a narrow, flattened foot and lack siphons.
Ungulinidae grouped with the Cyamioidea and Galeommatoidea in the analyses of Taylor et al. (2007, 2009) and Bieler et al. (2014) but is sister to Chamidae in the Lemer et al. (2019) analysis.
Family Ungulinidae (= Diplodontidae)

Numella (Ungulinidae), from FA1

(Superfamily) **Chamoidea**
Cretaceous (Lower) to present (129.4–0 Ma)
Chamids are cementing bivalves that have been thought (incorrectly) to be related to rudists. They cement by either the left or right valve.
The relationships of Chamidae to other Venerida have been unclear, but its placement in this grouping is confirmed by Lemer et al. (2019).
Family Chamidae

Pseudochama (Chamidae), from FA1

(Superfamily) **Hemidonacoidea**
Neogene (Pleistocene) to present (3.6–0. Ma)
This family clustered with the Arcticidae in the Combosch et al. (2017) analysis and with Glossidae in Lemer et al. (2019).
Family Hemidonacidae

Hemidonax, from FA1

(*Continued*)

(Superfamily) **Glossoidea** (= Trapezoidea) Triassic (Upper) to present (221.5–0 Ma) This group may not be monophyletic (Taylor et al. 2007). Families Glossidae and Trapezidae	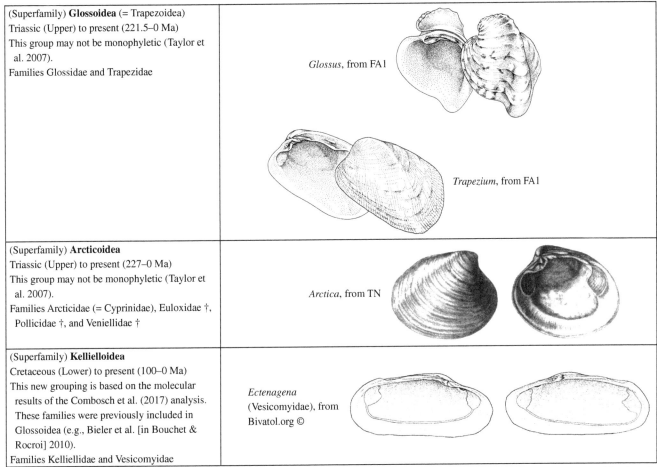 *Glossus*, from FA1 *Trapezium*, from FA1
(Superfamily) **Arcticoidea** Triassic (Upper) to present (227–0 Ma) This group may not be monophyletic (Taylor et al. 2007). Families Arcticidae (= Cyprinidae), Euloxidae †, Pollicidae †, and Veniellidae †	*Arctica*, from TN
(Superfamily) **Kellielloidea** Cretaceous (Lower) to present (100–0 Ma) This new grouping is based on the molecular results of the Combosch et al. (2017) analysis. These families were previously included in Glossoidea (e.g., Bieler et al. [in Bouchet & Rocroi] 2010). Families Kelliellidae and Vesicomyidae	*Ectenagena* (Vesicomyidae), from Bivatol.org ©

(Continued)

(Superfamily) **Veneroidea**
Mississippian (Middle) to present (339–0 Ma)
Representatives of the previously recognised
families Turtoniidae and Petricolidae are nested
within Veneridae based on molecular results
(Mikkelsen et al. 2006), and these family group
taxa are treated as subfamilies of Veneridae by
Bieler et al. (in Bouchet & Rocroi) (2010), in
addition to another 12 subfamilies.
Families Veneridae, Neoleptonidae, and
Isocyprinidae †

(Superfamily) **Cyrenoidea** (= Corbiculoidea)
Jurassic (Lower) to present (201.6–0 Ma)
Cyrenoididae was located in the Lucinoidea until
recently, but Taylor et al. (2009) showed that it
should be included here rather than with
Veneroidea where it was previously located.
Families Cyrenidae (= Corbiculidae),
Cyrenoididae, and Glauconomidae

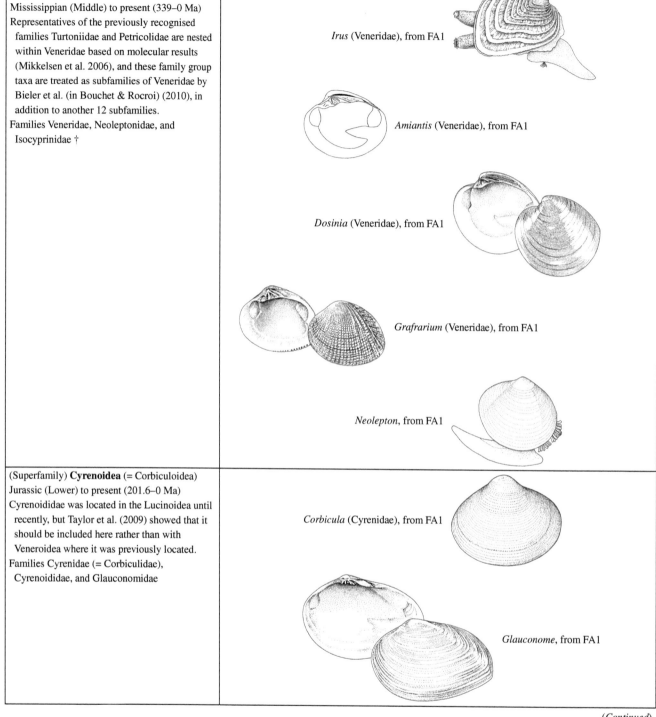

Irus (Veneridae), from FA1

Amiantis (Veneridae), from FA1

Dosinia (Veneridae), from FA1

Grafrarium (Veneridae), from FA1

Neolepton, from FA1

Corbicula (Cyrenidae), from FA1

Glauconome, from FA1

(Continued)

(Superfamily) **Cyamioidea**

Jurassic (Middle)? to present (174.1–0 Ma)

The Jurassic date is questionable as the next record is Miocene.

Basterotia groups with galeommatoideans in the molecular analysis of Taylor et al. (2007) and Bieler et al. (2014) suggesting that the composition of this group may well need to change.

In the molecular analysis of Taylor et al. (2007), *Gaimardia* was shown to be distinct from Cyamiidae, where it had sometimes been placed. However, Bieler et al. (2014) and Combosch et al. (2014) found it grouped with Cyamiidae and also Ungulinidae in their analyses.

The analysis of Goto et al. (2012) placed *Basterotia* in the Galeommatoidea. The relationships of other genera usually included in the Sportellidae have not been investigated. However, anatomically their relationship with galeommatoideans is unlikely because they do not have an anterior–posterior water flow typical of that group, while the sportellids that have been examined have both incurrent and excurrent flows at the posterior end as in most bivalves. Because the molecular analyses of Goto et al. (2012, 2018) did not include any (other) cyamioideans, and there were long branches separating the clades containing *Basterotia*, we prefer to maintain the current arrangement until more comprehensive analyses are conducted.

Families Cyamiidae, Gaimardiidae, Galatheavalvidae, and Sportellidae (?= Basterotiidae)

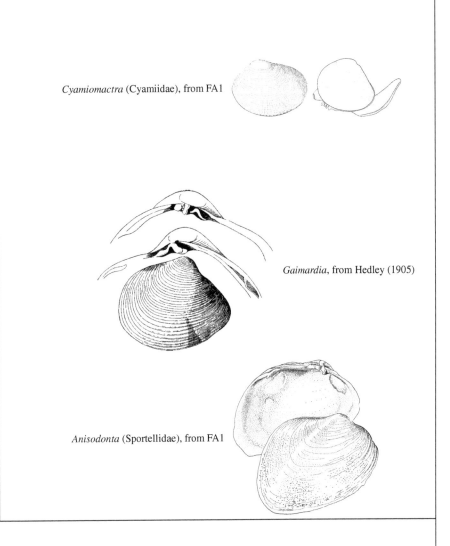

Cyamiomactra (Cyamiidae), from FA1

Gaimardia, from Hedley (1905)

Anisodonta (Sportellidae), from FA1

(Order) **Myida**

Ordovician to present (468–0 Ma)

This well-supported (e.g., Lemer et al. 2019) order includes a diverse assemblage of families.

(*Continued*)

(Superfamily) **Dreissenoidea**
Jurassic (Lower) to present (196.5–0 Ma)
This superfamily contains both marine and
  freshwater taxa.
Dreissena nested between Myoidea and
  Pholadoidea in the molecular analyses of Taylor
  et al. (2007, 2009).
Family Dreissenidae
	Dreissena, modified from Fischer (1880–1887)
	
	Mytilopsis (Dreissenidae), from TN
	
	Congeria, (Dreissenidae) modified from Fischer
	(1880–1887)

(Superfamily) **Pholadoidea**
Jurassic (Lower) to present (199.3–0 Ma)
This superfamily contains the classic wood- and
  rock-boring bivalves.
Families Pholadidae, Teredinidae, and
  Xylophagaidae
	Martesia (Pholadidae), from FA1
	
	Pholas, from FA1
	
	Teredora (Teredinidae), from FA1

(Continued)

(Superfamily) **Pleuromyoidea** † Ordovician (Middle) to Neogene (Miocene) (468–23 Ma) Families Pleuromyidae, Ceratomyidae, and Vacunellidae	 *Pleuromya*, from Fischer (1880–1887)  *Ceratomya*, from Fischer (1880–1887)
(Superfamily) **Myoidea** Triassic (Upper) to present (216.5–0 Ma) A few myoideans live in freshwater, but the great majority are marine. Families Myidae, Corbulidae (= Erodonidae), Pleurodesmatidae †, and Raetomyidae †	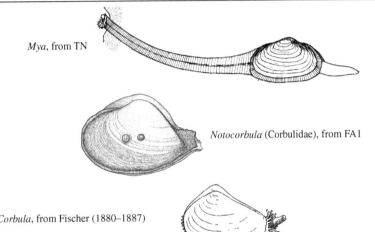 *Mya*, from TN *Notocorbula* (Corbulidae), from FA1 *Corbula*, from Fischer (1880–1887)

(Class) **SCAPHOPODA** (= Solenoconchia, Prosopocephala)   Silurian (Ludlow) to present (427.4–0 Ma)   Scaphopods are bilaterally symmetrical and have tubular tapering shells that are open at both ends. All scaphopods are marine and burrow into soft sediments.   The classification below is based on Steiner and Kabat (2001) and Reynolds and Steiner (2008). A more dated but well-illustrated classification can be found in Moore (1964).   See Chapter 16 for more information on this group.
Order **Dentaliida**   Silurian (Ludlow) to present (427.4–0 Ma)   The shell is longitudinally ribbed or smooth, rarely with concentric ridges. The widest part of the shell is at the anterior aperture. The foot has a pair of flanges.

(Superfamily) **Dentalioidea** Families Dentaliidae, Anulidentaliidae, Baltodentaliidae, Calliodentaliidae, Fustiariidae, Gadilinidae, Laevidentaliidae, Omniglyptidae, Prodentaliidae, and Rhabdidae	*Dentalium*, from FA1 *Lentigodentalium* (Dentaliidae), from FA1 *Episiphon* (Gadilinidae), from FA1

(Order) **Gadilida**   Permian (Cisuralian) to present (290.1–0 Ma)   The shell is similar to those of Dentaliida or is smooth, polished, and with the widest portion some distance behind the aperture. The foot terminates in a fringed disk and retracts by introverting within itself.

(Continued)

(Suborder) **Entalimorpha** Paleogene (Eocene) to present (48.6–0 Ma) The shell is similar to those of Dentaliida in being widest at the anterior aperture, smooth or longitudinally ribbed. The foot retracts by bending into the shell.		
(Superfamily) **Entalinoidea** Family Entalinidae	*Entalina*, from FA1	
(Suborder) **Gadilimorpha** Permian (Cisuralian) to present (290.1–0 Ma) The shell is usually small, smooth, and polished with the widest point typically about the middle.		
(Superfamily) **Gadiloidea** Families Gadilidae, Pulsellidae, and Wemersoniellidae	*Gadila*, from FA1 *Polyschides* (Gadilidae), from FA1	*Cadulus* (Gadilidae), from FA1

(Class) **CEPHALOPODA** See Chapter 17 for more information on this group. The superfamily rank is not used in some cephalopod groups because most available cephalopod classifications do not use this level, only orders and suborders.

(*Continued*)

(Subclass) **Palcephalopoda**

Cambrian (Furongian) to present (497–0 Ma)

Cephalopods can be divided into two major groups – the paraphyletic Palcephalopoda and the probably monophyletic Neocephalopoda (see Chapter 17 for details).

Palcephalopoda closely equates with the use of 'Nautiloidea' in the traditional three subclass classification (Nautiloidea, Ammonoidea, Coleoidea). However, some 'nautiloids' are included as stem groups in Neocephalopoda.

The rank used for this grouping varies, and it has also been used as an informal name. We refer to members of this grouping informally as nautiliforms.

The major groups below are often treated as separate subclasses, but we treat them as 'cohorts' as the use of subclass rank would seem to be unwarranted taxonomic inflation. Similarly, we follow the common practise of using ordinal rank for many of the subgroups.

Mutvei (2015) divided the 'Nautiloidea' into two 'superorders', Nautilosiphonata and Calciosiphonata, based on differences in their siphuncular structure, the first with '*Nautilus*-type' connecting rings, the second with 'calcified-perforate-type' connecting rings.

Note that where terms dorsal and ventral are used in descriptions they refer to the actual anterior and posterior sides of the shell.

Since this text was written, King and Evans (2019) have recognised five 'subclasses' (Plectronoceratia, Multiceratia, Orthoceratia, Tarphyceratia, Nautilia) on the basis of muscle-scar types.

When referencing this section, it should be cited as Ponder, W. F., Lindberg, D. R. and King, A. H. (2019), Appendix. Classification of the living molluscan classes. Palcephalopoda, in Ponder, W. F., Lindberg, D. R., and Ponder, J. M., *Biology and Evolution of the Mollusca*, Volume Two.

Some specialised terms used in this section include:

Connecting rings – the calcified edges of the siphuncle between the septa – may be thin or thick.

Septal neck – the edge of the septum at the siphuncle. There is a complex terminology including the following terms:

Cyrtochoanitic – the neck is curved outwards

Hemichoanitic – the end curves downwards for a short distance

Holochoanitic – the neck extends downwards and meets the next ring

Macrochoanitic – the neck extends over the next ring

Orthochoanitic – a short, cylindrical neck

Venter – the outer part of the shell whorl (equivalent to the peripheral region in gastropod shells)

Shell shape terms are illustrated in Chapter 17 (Figure 17.7).

(Cohort) **Plectronoceratia** † (= Plectroceratoidea)

Cambrian (Furongian) (497–485.4 Ma)

The shell is very small, compressed laterally, curved (usually endogastric, a few exogastric) or straight, and the septa are closely spaced, with short septal necks, and the ventral siphuncle is rather large, and the connecting rings are sometimes expanded and are poorly calcified. Muscle scars are unknown. Diaphragms in the siphuncle are often present in some plectronoceratidans and yanheceratidans and in all protactinoceratidans and yanheceratidans.

This taxon is sometimes treated as an order within Ellesmeroceratia.

(Order) **Plectronoceratida** (= Plectronocerida) † Cambrian (Furongian) (497–485.4 Ma) Shells small, weakly curved endogastrically and moderately expanded to nearly straight, slender; where known diaphragms simple, nearly straight, or weakly concave. Families Plectronoceratidae and Balkoceratidae	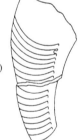 *Plectronoceras*, from Kobayashi (1935)
(Order) **Protactinoceratida** (= Protactinocerida) † Cambrian (Furongian) (497–485.4 Ma) The shells are small, rapidly expanding, and are slightly curved endogastrically. The septa are densely spaced, the siphuncle is wide and ventral, has distinctly inflated segments and is subdivided by W-shaped diaphragms. Family Protactinoceratidae	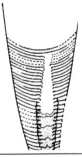 *Protactinoceras*, from Chen and Teichert (1983) (www.schweizerbart.de/journals/pala)

(*Continued*)

(Order) **Yanheceratida** (= Yanhecerida) † Cambrian (Furongian) (497–485.4 Ma) The shells are small, endogastrically curved or straight, and with a narrow ventral siphuncle, which is subdivided by conical diaphragms and is with or without calcareous deposits. Family Yanheceratidae	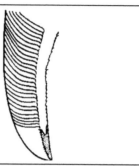 *Yanheceras*, from Chen and Teichert (1983) (www. schweizerbart.de/journals/pala)

(Cohort) **Ellesmeroceratia** (= Ellesmeroceratoidea, Multiceratoidea) †
Cambrian (Furongian) to Mississippian (497–330.9 Ma) (possibly to Permian [272.95 Ma])   The shells are small (typically 50–60 mm long) to medium-sized, elongated to breviconic, endogastrically curved or (rarely) exogastrically curved, or sometimes straight. They usually have a smooth surface and numerous closely spaced septa. The septal necks are short and orthochoanitic or holochoanitic, and the connecting rings vary but are often thick and layered. Multiple muscle scars are present in at least some Ordovician taxa (unknown in Cambrian taxa). Cameral deposits are absent.   Mutvei (2013) introduced a new 'superorder', Multiceratoidea, to include taxa with multiple muscle scars and the '*Nautilus* type' of siphuncular connecting ring.

(Order) **Ellesmeroceratida** (= Ellesmerocerida, Ecdyceratida) † Cambrian (Furongian) to Ordovician (Lower) (497–470 Ma) Shells are small, slightly curved endogastrically or (rarely) straight. The siphuncle is ventral. Septal necks are short, orthochoanitic, and connecting rings thick. This group is derived from the Plectronoceratida (e.g., Wade 1988). The illustrated *Eburoceras* is one of the most strongly curved members of this group. The close relationship between the Ellesmeroceratida, basslleroceratids, and earliest Tarphyceratida (in Nautilia) has long been recognised – the basslleroceratids are essentially 'morphological intermediates' between the two orders. Some workers currently place them (as the Family Basslleroceratidae) within the order Ellesmeroceratida (as we have done) as the constituent genera are weakly cyrtoconic (like some ellesmeroceratids) rather than weakly coiled (like tarphyceratids). Families Ellesmeroceratidae, Acaroceratidae, Baltoceratidae, Basslleroceratidae, Cyclostomiceratidae, Huaiheceratidae, Oneotoceratidae, Phthanoncoceratidae, Xiaoshanoceratidae, and questionably Apocrinoceratidae, and Shideleroceratidae	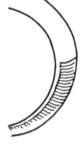 *Eburoceras* (Ellesmeroceratidae), from Chen and Teichert (1983)  *Bassleroceras*, from Teichert (1967)

(Continued)

(Order) **Cyrtocerinida** (= Bathmoceratiformes) †
Ordovician (Lower) (485.4–470 Ma)
The shells are larger than those of
  ellesmeroceratidans and typically longiconic to
  cyrtoconic. The siphuncle is ventral and wide
  with thick connecting rings inside which, in
  longitudinal section, are 'lingua-shaped'
  outgrowths which expand into the siphuncle.
This grouping was first recognised by Flower
  (1964) as the suborder Cyrtocerinina to include
  the three families below (these were previously
  included in the Ellesmeroceratida). See Mutvei
  (2015) for a recent review.
Families Cyrtocerinidae, Bathmoceratidae, and
  Eothinoceratidae
	Bathmoceras, from Mutvei (2015)

(Order) **Ascoceratida** (= Ascocerida) †
Ordovician (Upper) to Silurian (Pridoli)
  (457.5–418.7 Ma)
The shells are elongated, with the apical part
  slightly exogastrically curved, and the body
  chamber is inflated and constricted dorsally by a
  strongly curved septum. The siphuncle is ventral
  and ontogenetically changes from narrow to
  wide. The early shell is deciduous.
This group is included here following Mutvei
  (2013). Shevyrev (2006) had it as a subgroup of
  Orthoceratoidea, and it has also sometimes been
  treated as a group of equal rank to
  Orthoceratoidea.
Families Ascoceratidae, Choanoceratidae, and
  Hebetoceratidae
	Aphragmites (Ascoceratidae), from TK
	
	Ascoceras, from Zittel and Eastman (1913) and from TK

(Continued)

(Order) **Bisonoceratida** (= Bisonocerida) †
Ordovician (Lower) to Silurian (Wenlock)
(485.4–427.4 Ma)

The shells in early forms are curved cones that
are compressed in section, while later forms are
orthocones which are depressed or subcircular
in section. The siphuncle is broad, ventral with
long holochoanitic or macrochoanitic septal
necks. Endosiphuncular deposits are complex,
often comprising conchiolin crests, multiple
endosiphuncular 'blades', and infula with
numerous endosiphuncular tubes. The apical
end of the siphuncle is swollen, with a
'pre-septal cone' in some forms (see Evans &
King 2012 for further details).

This group was split off from the Endoceratida by
Evans and King (2012) to resolve polyphyly in
the latter group.

Families Bisonoceratidae, Allotrioceratidae,
Botryceratidae, Chihlioceratidae,
Coreanoceratidae, Emmonsoceratidae,
Humeroceratidae, Manchuroceratidae,
Najaceratidae, Piloceratidae,
Proterovaginoceratidae, Sinoendoceratidae, and
possibly Padunoceratidae

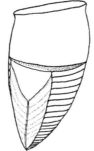

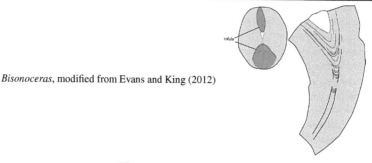

Bisonoceras, modified from Evans and King (2012)

Cassinoceras (Piloceratidae), from Flower (1947).

Manchuroceras, from TK

(Order) **Oncoceratida** (= Oncocerida) †
Ordovician (Middle) to Pennsylvanian (Lower)
(470–315.2 Ma)

The shells are exogastrically curved, or they have
straight breviconical shells, often with a closed
aperture. The siphuncle in this group is usually
either marginal and narrow, lacking lamellae, or
is wide, with longitudinally radial lamellae, with
cyrtochoanitic or suborthochoanitic septal
necks, and the thin connecting rings are either
convex or concave.

This group is included here, following Mutvei
(2013), rather than in Nautilida where it is often
assigned.

Families Oncoceratidae, Acleistoceratidae,
Aktjuboceratidae, Archiacoceratidae,
Bolloceratidae, Crytoceratidae, Diestoceratidae,
Guangyuanoceratidae, Hemiphragmoceratidae,
Jovellaniidae, Karoceratidae, Nothoceratidae,
Polyelasmoceratidae, Poterioceratidae,
Trimeroceratidae, Tripleuroceratidae,
Tripteroceratidae, and Valcourocratidae

Gomphoceras (Trimeroceratidae), from Zittel and
Eastman (1913)

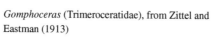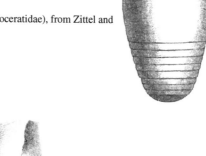

Amphicyrtoceras (Acleistoceratidae), from TK

(*Continued*)

(Order) **Discosorida** (= Ruedemannoceratina) †
Ordovician (Middle) to Mississippian (Middle)
  (470–339.4 Ma)

The shells are breviconic, endo- or exogastrically curved, straight, or, less commonly, coiled, with a closed or narrowed aperture. The siphuncle is usually marginal, with thick connecting rings inflated on the adapical end; sometimes intrasiphonal outgrowths in the form of longitudinal lamellae are produced.

The classification of this group has a complex history, particularly regarding the similarity of some genera with oncoceratidans and nautilidans (see Shevyrev 2006). Assignment of some of the following families to the order Discosorida should therefore be regarded as tentative.

Families Discosoridae, Brevicoceratidae, Cyrtogomphoceratidae, Devonocheilidae, Entimoceratidae, Lowoceratidae, Mandaloceratidae, Mecynoceratidae, Naedyceratidae, Phragmoceratidae, Reudemannoceratidae, Taxyceratidae, Ukhtoceratidae, Westonoceratidae, and questionably Gouldoceratidae and Mesoceratidae

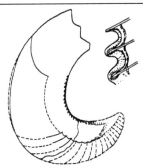

Konglungenoceras (Cyrtogomphoceratidae), from Teichert (1967)

Phragmoceras, from Zittel and Eastman (1913)

Phragmoceras, from Mutvei (2012) © Geologiska Föreningen, reprinted by permission of Taylor & Francis Ltd, www.tandfonline.com on behalf of Geologiska Föreningen.

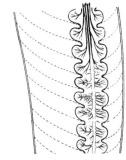

Gouldoceras, a sagittal section of the beaded siphuncle showing the siphuncular deposits, from FA2

(Cohort) **Orthoceratia** (= Orthoceratoidea, Orthocerida) †

Ordovician (Lower) to Triassic (Upper) (485.4–227 Ma), possibly Cretaceous (Lower) (112.6 Ma)

This group has shells that are long, orthoconic or weakly cyrtoconic. The siphuncle is central to ventral and narrow to broad. The septal necks are short orthochoanitic to long macrochoanitic. Connecting rings, siphonal deposits, and/or cameral deposits are present in most of the constituent orders. The muscle scars are dorsal in position.

Mutvei (2002b, 2002a) included (as orders) both Orthoceratida and Actinoceratida in his 'Orthoceratomorphi' on account of their dorsal muscle scars and an inner calcified-perforate layer of the connecting rings. Kröger and Evans (2011) recognised the orders Orthocerida, Dissidocerida, and Endocerida within their 'Subclass Orthoceratoidea'.

The record of an orthoceratid from the Lower Cretaceous (Upper Aptian) of the north-western Caucasus by Doguzhaeva (1995) requires confirmation.

(*Continued*)

(Order) **Orthoceratida** (= Michelinoceratida, Orthoceratoidea, Mixosiphonata?) †
Ordovician (Lower) to Triassic (Upper) (485.4–227 Ma), possibly Cretaceous (Lower) (112.6 Ma)

The shells are straight orthoconic to weakly cyrtoconic, circular in section, with a spherical apex which is bowl-shaped or blunt, and without a cicatrix. The external shell surface is smooth or with spiral or longitudinal ornamentation. The siphuncle is mainly narrow, central or subcentral (but broader and ventral in early forms), with short septal necks which are achoanitic to orthochoanitic, and the connecting rings are thin and tubular. Cameral deposits are present and often accompanied by siphonal annuli which coalesce apically to form a siphuncular lining; the deposits may be restricted to the apical-most portion of shell. The muscle scars are dorsal in position.

In the majority of orthoceratids, the ultrastructure of the connecting ring comprises an outer spherulitic-prismatic layer and an inner, thick chitinous layer. Mutvei (2017) erected the order Mixosiphonata on the basis of orthoceratids whose connecting rings consist of a 'single layer of calcareous spherulites, prisms and granules that are embedded in a chitinous substance'.

There are considerable problems with interpreting members of this group as there are relatively few characters to work with (e.g., Kröger & Mapes 2004). As amended by Kröger and Isakar (2006), this group only includes taxa with a spherical protoconch and simple siphuncle.

Two superfamilies previously recognised within the order Orthoceratida, the Orthoceratoidea and Pseudorthoceratoidea (King 1993), are now generally regarded as representing distinct orders. Doguzhaeva (1995) reported a fragmentary orthoceratid from the Lower Cretaceous of the north-western Caucasus.

Families Orthoceratidae, Arionoceratidae, Baltoceratidae, Boggyoceratidae, Clinoceratidae, Dawsonoceratidae, Geisonoceratidae, Michelinoceratidae, Paraphragmitidae, Sactorthoceratidae, Sichuanoceratidae, Sphaerorthoceratidae, Sphooceratidae and (questionably) Brachycycloceratidae, and Offleyoceratidae

Lyecoceras (Paraphragmitidae), modified from TK

Geisonoceras, from Teichert (1933)

Virgoceras (Geisonoceratidae), from TK

(Continued)

(Order) **Rioceratida** † Ordovician (Lower) (485.4 Ma) The shells are slender orthocones or weakly cyrtoconic with an empty (vacuosiphonate) ventral siphuncle, orthochoanitic to hemichoanitic septal necks, and the connecting rings are thin to only moderately thickened. Cameral deposits are absent. These are the earliest representatives of the cohort Orthoceratia and can be distinguished from all other orders by the combination of their vacuosiphonate marginal siphuncle and lack of cameral deposits (King & Evans 2019). Families Rioceratidae and Bactroceratidae	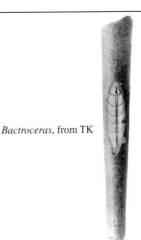 *Bactroceras*, from TK
(Order) **Pseudorthoceratida** † Ordovician (Lower) to Triassic (Lower) (478.6–252.3 Ma) The shells are straight orthocones to weakly cyrtoconic, circular in section, and the apex is bullet-shaped with a cicatrix. The external shell surface is smooth or annulate, with spiral or longitudinal ornamentation. The siphuncle is subcentral to subventral, and the septal necks are curved, with the connecting rings cyrtochoanitic, thin, and slightly to moderately concave producing a 'beaded siphuncle'. Cameral deposits are present and frequently extensive, especially on the ventral side of the shell. Siphonal deposits comprise discrete siphonal annuli that begin at the septal foramina and grow forward, backward, or both against the connecting rings, ultimately fusing with adjacent deposits to form a more-or-less continuous lining. Families Pseudorthoceratidae, Carbactinoceratidae, Cayutoceratidae, Proteoceratidae, Spyroceratidae, Trematoceratidae, and questionably Stereoplasmoceratidae	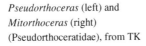   *Pseudorthoceras* (left) and *Mitorthoceras* (right) (Pseudorthoceratidae), from TK 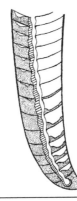 *Pseudocyrtoceras* (Pseudorthoceratidae), from TK

(Order) **Dissidoceratida** (= Dissidocerida) †

Ordovician (Lower) to Silurian (Llandovery) (485.4–433.4 Ma)

Mainly slender, smooth or annulate orthoconic shells with ventral to central, broad, tubular siphuncle, and short, orthochoanitic septal necks; cameral deposits present; siphonal deposits either rod-like and concentrated ventrally or slender, endoconic, and present around the whole circumference of the siphuncle.

This order was discussed by Zhuravleva (1994) and extended by Evans (2005) and Kröger (2008). Some genera were previously included within the Endoceratida on account of their endoconic-like siphonal linings.

(Continued)

(Suborder) **Dissidoceratina** (= Superfamily Dissidoceratoidea) † Ordovician (Lower) to Silurian (Middle) (485.4–427.4 Ma) Dissidoceratidans have intrasiphonal deposits (often in the form of a tapering 'rod') concentrated on the ventral side of the siphuncle. Families Dissidoceratidae, Cyptendoceratidae, Orthodochmioceratidae, Polymeridae, Protocycloceratidae (= Rhabdocycloceratidae?), and Rangeroceratidae	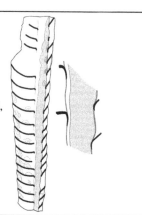 *Cyclorangeroceras* (Rangeroceratidae), modified from Evans (2005)
(Suborder) **Troedssonellina** (= Superfamily Troedssonelloidea) † Ordovician (Lower) to Silurian (Middle) (485.4–427.4 Ma) Dissidoceratidans have intrasiphonal deposits that extend over the whole circumference of the siphonal wall as thin, long endoconic-like lining; the intrasiphonal tube often contains diaphragms which are sometimes complex. Families Troedssonellidae and Narthecoceratidae	 *Troedssonella* from TK
(Order) **Intejoceratida** (= Bajkalocerida) † Ordovician (Lower to Middle Ordovician) (485.4–458.4 Ma) Members of this group are distinguished by their intrasiphonal deposits of numerous longitudinally radial lamellae.	
(Suborder) **Intejoceratina** † Ordovician (Lower to Middle Ordovician) (485.4–458.4 Ma) The straight shells have narrow camerae and a moderately wide, subcentral siphuncle, very short septal necks and thick, convex inward connecting rings. The intrasiphonal deposits have numerous, thick, longitudinal, radial lamellae and extensive cameral deposits. Given the form of their siphonal and cameral deposits, the intejoceratids are probably more closely related to the Dissidoceratida (or possibly Actinoceratida) rather than the Endoceratida (e.g., Flower 1976a). Bajkaloceratids exhibit some features in common with the intejoceratids but are poorly known, and any relationship with the Intejoceratida is not well established. Shevyrev (2006) regarded the Bajkaloceratida as a subgroup of the Orthoceratia. Families Intejoceratidae and Bajkaloceratidae?	 *Intejoceras*, from TK

(*Continued*)

(Order) **Lituitida** †

Ordovician (Lower) to Devonian (Lower)
(471.8–402.5 Ma)

The shells are slender orthocones or rapidly
expanding lituiticones, either weakly cyrtoconic
(Family Sinoceratidae) or with the apical part
spirally coiled (Family Lituitidae). The
siphuncle is subcentral, and the septal necks
orthochoanitic to hemichoanitic. Cameral
deposits are extensive, often fusing with the
siphonal deposits through broken connecting
rings. The shells are typically ornamented with
coarse growth lines or lirae which trace out a
ventral hyponomic sinus. At maturity the
aperture sometimes has paired lappets.

Although now considered to be related to the
cohort Orthoceratia (due to the presence of
cameral deposits and dorsal muscle scars), the
lituitids were formerly treated as part of the
order Tarphyceratida. Zhuravleva and
Doguzhaeva (2002) placed the order Litiuitida
within their new 'superorder' Astrovioidea
along with the order Pallioceratida.

Families Lituitidae and Sinoceratidae

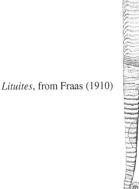

Lituites, from Fraas (1910)

Ancistroceras (Lituitidae), from Noetling (1884)

(Order) **Pallioceratida** †

Silurian to Devonian (443.8–358.9 Ma)

The shells are longiconic orthocones or are
slightly curved cyrtocones. The siphuncle is
displaced ventrally from the centre, and the
septal necks are short orthochoanitic, and the
connecting rings are thin and prone to complete
destruction apically by extensive cameral
deposits (including metacameral deposits) that
fuse with the ectosiphonal deposits.

Zhuravleva and Doguzhaeva (2002) provided a
detailed account of the 'superorder'
Astrovioidea which they created to encompass
the orders Lituitida and Pallioceratida on
account of their specialist siphuncular and
cameral deposits. They interpreted the formation
of the extensive cameral and siphonal deposits
and destruction of the connecting rings (during
life) as being related to the presence of 'cameral
mantle' and 'siphonal tissue'. This is not
accepted by all workers, and the validity of the
Pallioceratida as a natural grouping is debatable.

Families Astroviidae, Flowerinidae,
Lamellorthoceratidae, Ostreioceratidae,
Pallioceratidae, and possibly
Leurocycloceratidae

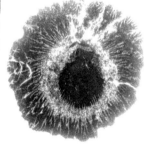

Lamellorthoceras, from TK

Esopoceras (Lamellorthoceratidae), showing the double-
layered cameral lamellae, from Stanley and Teichert (1976)

(Continued)

(Cohort) **Endoceratia** (= Endoceratoidea) †

Ordovician (Lower) to Silurian (Wenlock) (485.4–427.4 Ma)

The shells are large (up to 8 m in length), usually straight or occasionally endogastrically curved, and their outer surface is smooth or annulated. The siphuncle is wide, marginal to subcentral, and opens into a calyx-shaped protoconch and contains simple endocone-type deposits. Septal necks are short orthochoanitic to long macrochoanitic. The connecting rings vary in thickness but are composed of a single spherulite-prismatic layer (Mutvei 2012). A pair of dorsal muscle scars is present. Cameral deposits are absent.

The Endoceratia is usually considered to have evolved from the Family Baltoceratidae (formerly assigned to the order Ellesmeroceratida) via genera such as *Pachendoceras* which possessed broad ventral siphuncles with conical diaphragms. More recently, a possible origin for Endoceratia via the Cambrian Yanheceratida (which also possess conical diaphragms) has been proposed. The order Intejoceratida has previously been assigned to the 'Endoceratoidea', but the presence of extensive cameral deposits in the former suggests any relationship is unlikely.

(Order) **Endoceratida** (= Endocerida) † Ordovician (Lower) to Silurian (Wenlock) (485.4–419.2 Ma) These large shells are longiconic orthocones to weakly cyrtononic, and where known, the shell apex is slender and not swollen. The siphuncle is typically broad and ventral (occasionally subcentral) in position, and siphuncular deposits consist of simple 'funnel-shaped cones' (endocones) with a bifid or trifid endosiphuncular 'blade' structure radiating from a single central endosiphuncular tube. (This contrasts with the more complex endosiphuncular deposits seen in the order Bisonoceratida which also exhibit a different form of muscle scar pattern and a swollen shell tip, a 'pre-septal cone'.) The presence of endoconic-type deposits within the siphuncle is not confined to the order Endoceratida – endocones of various forms are also encountered in the unrelated orders Bisonoceratida, Dissidoceratida, and Discosorida. The order Endoceratida includes the largest known Paleozoic invertebrates, with some shells attaining lengths of several metres.	Structural elements in siphuncle of Endoceratida, from TK 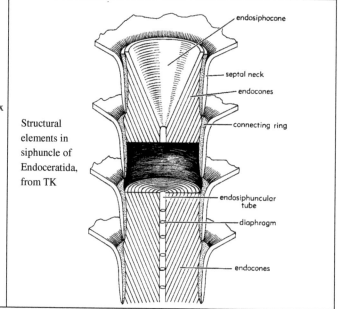

(Continued)

(Suborder) **Endoceratina** (= Endocerina) † Ordovician (Lower) to Silurian (Pridoli) (485.4–419.2 Ma) Shells are smooth or annulate and straight orthoconic to weakly cyrtoconic. Camerae are narrow to wide, and the siphuncle is broad and ventral, occasionally subcentral, with long holochoanitic to macrochoanitic septal necks. Families Endoceratidae, Cyrtendoceratidae, and Suecoceratidae	*Cameroceras* (right) (Endoceratidae), from TK. Septal necks and connecting rings (holochoanitic – *Paleocyclendoceras* (Endoceratidae) (left), macrochoanitic and *Dideroceras* (Endoceratidae) (middle), from Mutvei (1997) [CC by 4.0] 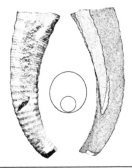 *Cyclocyrtendoceras* (Cyrtendoceratidae), from TK
(Suborder) **Proterocameroceratina** (= Proterocamerocerina) † Ordovician (Lower to Middle Ordovician) (485.4–458.4 Ma) Shells are smooth or annulate and straight orthoconic to weakly cyrtoconic, with camerae typically narrow. The siphuncle is ventral, and the septal necks typically short orthochoanitic in early forms, becoming hemichoanitic in later taxa. Families Proterocameroceratidae, Thylacoceratidae, and Yorkoceratidae	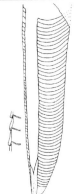 *Lamottoceras* (Proterocameroceratidae), from TK *Proterocameroceras*, from TK

(Cohort) **Actinoceratia** (= Actinoceratoidea) †

Ordovician (Lower) to Pennsylvanian (Upper) (485.4–298.9 Ma)

Shells are straight (or sometimes slightly curved), with the relatively wide siphuncle subcentral to subventral, with beaded edges and extensive deposits and a complex siphonal-vascular system composed of a longitudinal canal and radial tubes. The septal necks are recurved and cyrtochoanitic. Cameral deposits are present. Connecting rings are composed of two calcareous layers, an outer thin, spherulite-prismatic layer and an internal, thick, perforated layer (Mutvei 1997).

Usually a single order is recognised although this is not always the case (e.g., Starobogatov 1983).

(*Continued*)

(Order) **Actinoceratida** (= Actinocerida, Allotrioceratiformes) †

Ordovician (Lower) to Pennsylvanian (Upper) (485.4–298.9 Ma)

Details as above.

Families Actinoceratidae, Armenoceratidae, Discoactinoceratidae, Gonioceratidae, Huroniidae, Lambeoceratide, Meitanoceratidae, Ormoceratidae, Polydesmiidae, Wademidae (= Georginidae), and Wutinoceratidae

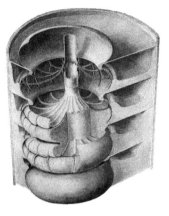

Actinoceratid siphuncular system, from TK

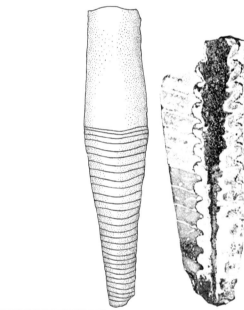

Actinoceras, from TK

(Cohort) **Nautilia** (= Nautiloidea in part)

Ordovician (Lower) to present (485.4–0 Ma)

Shell spirally coiled or, less commonly, curved. The protoconch is calyx-shaped with a cicatrix representing the centre from which the embryonic shell begins to grow in living *Nautilus* (Arnold 1988). The siphuncle is narrow, from central to marginal, with a caecum. The connecting rings consist of a thick spherulite-prismatic layer, which in extant *Nautilus* is lined from inside by the conchiolin layer (Mutvei 2002b). One or two pairs of ventral muscle scars (Order Tarphyceratida) or ventrolateral muscle scars (Order Nautilida) are present.

Most recent treatments use Nautilia in a more restricted sense than many older classifications in which all palcephalopod taxa were included in 'Nautiloidea'. An intermediate interpretation was adopted by Teichert & Moore (1964) where the Endoceratoidea and Actinoceratoidea were treated as subclasses and the subclass Nautiloidea encompassed the remaining 'nautiloids'. Wade and Stait (1998) recognised three orders in their subclass 'Nautiloidea'; here we use Nautilia in a more restricted sense.

(*Continued*)

(Order) **Tarphyceratida** (= Tarphycerida, Barrandeoceratida, Barrandeocerida) †

Ordovician (Lower) to Devonian (Middle) (478.6–383.7 Ma)

Shells are weakly curved to spirally coiled (often gyroconic, occasionally torticonic), some forms with weakly coiled inner whorls and a divergent last whorl. The surface is smooth or with coarse growth lines or ribs which trace out a broad ventral sinus. The siphuncle varies in position, the septal necks are short, and the connecting rings are thin to thick. Cameral and siphonal deposits are absent.

The tarphyceratids and barrandeoceratids were formerly regarded as distinct orders (or suborders) mainly due to differences in overall shell form and the thickness of the connecting rings; these differences are now considered to have arisen independently in various families. Consequently, the Barrandeoceratida was polyphyletic (Flower 1984), and the majority of its constituent genera have been reassigned within an expanded Tarphyceratida.

Families Tarphyceratidae, Apsidoceratidae, Barrandeoceratidae (= Bickmoritidae), Estonioceratidae, Lechritrochoceratidae, Nephriticeratidae, Ophidioceratidae, Plectoceratidae, Trocholitidae, and Uranoceratidae

Lechritrochoceras, from Teichert (1967)

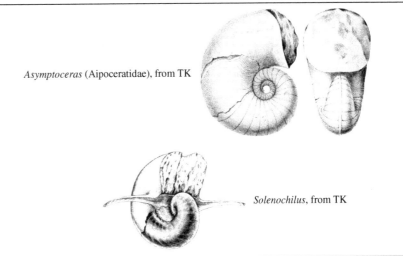

Estonioceras, from TK

(Order) **Nautilida**

Devonian (Lower) to present (410.8.4–0 Ma)

The shells are usually spirally coiled or, less commonly, curved, varying from evolute to involute, smooth or ornamented, with a narrow cylindrical siphuncle which occupies various positions. The septal necks are short and straight. Sutures are simple or complex, and internal siphonal or cameral deposits are lacking.

Nautilida appeared in the Lower Devonian (Pragian) and survived to the present day where they are represented by the genera *Nautilus* and *Allonautilus*. Some Devonian and Lower Carboniferous nautilidan genera require revision and may be better assigned to the Oncoceratida or Tarphyceratida.

Classification of the order Nautilida has a complex history, and several taxonomic schemes have been proposed including the recognition of various suborders (e.g., Flower 1988; Teichert 1988). Below we follow the classification of the Nautilida at superfamily level as employed by Teichert and Moore (1964).

(Superfamily) **Aipoceratoidea** †

Carboniferous (Mississippian) to Permian (Lopingian) (346.7–251.9 Ma)

Shells are typically rapidly expanding, cyrtoconic to loosely coiled, with the whorls rounded to flattened in section. The siphuncle is marginal and ventral and the shell surface smooth or ribbed. The shell aperture is modified and spinose in some forms. The septal necks are orthochoanitic, becoming cyrtochoanitic on the dorsal side, and the connecting rings are often weakly to strongly expanded.

Families Aipoceratidae, Scyphoceratidae, and Solenochilidae

Asymptoceras (Aipoceratidae), from TK

Solenochilus, from TK

(Continued)

(Superfamily) **Clydonautiloidea** †

Carboniferous (Mississippian) to Triassic (Upper) (346.7–227 Ma)

Shells are generally smooth, globular to compressed with a very small or occluded umbilicus. The siphuncle is typically small and subcentral. The shell sutures are straight in the earliest forms, but the superfamily is characterised by the suture developing prominent lobes and saddles ('goniatitic'-like) eventually becoming deeply sinuous in late Triassic forms.

Families Clydonautilidae, Ephippioceratidae, Gonionautlidae, Liroceratidae, and Siberionautilidae

Sibyllonautilus (Liroceratidae), modified from TK

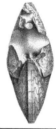

Gonionautilus, from TK

(Superfamily) **Trigonoceratoidea** †

Devonian (Upper) to Triassic (Upper) (382.7–227 Ma)

The shell is variable, being gyroconic to nautiliconic, with a quadrate whorl section in early forms. The venter is narrow to acute with a broad dorsum; some later forms have a broad, rounded or concave venter; siphuncle position highly variable, sutures weakly sinuous only becoming 'goniatitic'-like in late Carboniferous and Permian forms; shell form either smooth or strongly ornamented with nodules or spiral ridges.

Families Trigonoceratidae, Centroceratidae, Grypoceratidae, Permoceratidae, and Syringonautilidae

Chouteauoceras (Trigonoceratidae), from TK

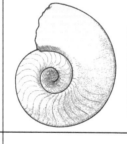

Domatoceras (Grypoceratidae), from TK (ex Dumble 1889)

(Superfamily) **Tainoceratoidea** †

Devonian (Lower) to Triassic (Upper) (410.8.4–201.6 Ma)

Shells with mainly depressed to quadrate whorl sections, rarely compressed; early forms with weakly cyrtoconic (nearly straight) to loosely coiled shells bearing spines, ribs, nodules, flanges or frills, and near-ventral siphuncle; later forms with more tightly coiled shells exhibiting nodules, ribs, or both and often with central siphuncle; some forms essentially smooth shelled. Sutures with very shallow lobes on venter and flanks.

Families Tainoceratidae, Koninckioceratidae, Rhiphaeoceratidae, Rutoceratidae, and Tetragonoceratidae

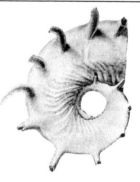

Cooperoceras (Tainoceratidae), from TK

Pleuronautilus (Tainoceratidae), from TK

(Continued)

(Superfamily) **Nautiloidea**

Triassic (Upper) to present day (201.6–0 Ma)

Shells are involute and generally smooth, with sinuous plications or ribs in some forms. The whorl section is compressed to depressed, sometimes with a flattened venter. Sutures are straight to strongly sinuous, and the siphuncle is central or dorsal in position.

Families Nautilidae, Aturiidae †, Cenoceratidae †, Hercoglossidae †, Paracenoceratidae †, Pseudonautilidae †, and (questionably) Cymatoceratidae †

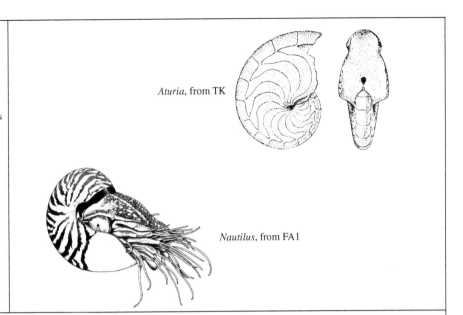

Aturia, from TK

Nautilus, from FA1

(Subclass) **Neocephalopoda**

This grouping encompasses the 'nautiloid' Bactrita, Ammonitida, and Coleoida. Some orthoceratidans are sometimes also included (e.g., Engeser 1996 – see above). The taxonomy given below largely follows that in Moore (1996), Furnish et al. (2009), and Howarth (2013).

The radula has few (about seven) teeth per row, the protoconch is small (1.5–3 mm long), approximately spherical and is separated from the rest of the phragmocone by a constriction; conical body chamber modified into a spiral in some groups (Engeser 1996). The details of mineralisation of the protoconch also differ from nautiliforms.

(Cohort) **Bactrita** †

Devonian (Lower) to Triassic (Upper) (412.3–221.5 Ma)

A small, probably paraphyletic group of cephalopods with straight or, less commonly, slightly exogastrically curved shells, with a semispherical protoconch, sometimes possessing a dome on the top instead of a cicatrix (Doguzhaeva 1996) and with a narrow ventral siphuncle. The suture is straight or wavy but always with a typically narrow and short ventral lobe. Siphonal or cameral deposits are absent.

In contrast to orthoceratians, bactritans lack a cicatrix and nacreous layer in the protoconch wall, and their siphuncle is always central. In contrast to ammonites, the primary varix is absent; the layers of the shell wall do not interrupt near the primary constriction, while the protoconch sometimes has a primordial dome on the top (Doguzhaeva 2002).

Although we have ranked this group as equivalent to Ammonitia, it is sometimes treated as an order within Ammonitia or a group within Orthoceratida.

(Order) **Bactritida** †

Devonian (Lower) to Triassic (Upper) (412.3–221.5 Ma)

(Superfamily) **Bactritoidea** †

Devonian (Lower) to Triassic (Upper) (412.3–221.5 Ma)

Families Bactritidae and Parabactritidae

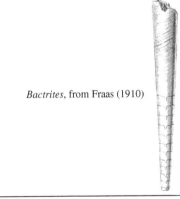

Bactrites, from Fraas (1910)

(Continued)

(Cohort) **Ammonitia** (= Ammonoidea, etc.) †

Devonian (Lower) to Cretaceous (Upper) (407.6–66 Ma)

Mainly with spirally coiled shells, with a nacreous layer, and a spherical protoconch without a cicatrix on the top. Members have a thin ventral siphuncle possessing a caecum and complexly incised sutures.

Over 2,300 generic names of ammonites have been proposed in this enormous group. They arose in the Lower Devonian and were dominant animals in the world's oceans until they became extinct along with the dinosaurs at the end of the Cretaceous. Major subdivisions in ammonites differ in the rank employed: In the *Treatise on Invertebrate Paleontology* (Moore 1957), the Ammonoidea was regarded as an order, while in the *Osnovy Paleontologii* (Orlov 1962) it was regarded as a superorder. Consequently, ranks of other units differed; for example the goniatitids were included in the suborder Goniatitina or the order Goniatitida, respectively. Today there is wide acceptance of the taxonomic scheme for the Paleozoic ammonites in *Osnovy* where five orders (Agoniatitida, Goniatitida, Clymeniida, Prolecanitida and Ceratitida.) are recognised.

When citing this section, the Paleozoic ammonites should be cited as Ponder, W. F., Lindberg, D. R., and Korn, D. (2019), Appendix. Classification of the living molluscan classes. Ammonitia (Paleozoic taxa), in Ponder, W. F., Lindberg, D. R., and Ponder, J. M., *Biology and Evolution of the Mollusca*, Volume Two and the Mesozoic ammonites as Ponder, W. F., Lindberg, D. R., and Hoffmann, R. (2019), Appendix. Classification of the living molluscan classes. Ammonitia (Mesozoic taxa), in Ponder, W. F., Lindberg, D. R., and Ponder, J. M., *Biology and Evolution of the Mollusca*, Volume Two.

Some specialised terms used in the descriptive text for the Paleozoic ammonites are listed below:

Terms for conchs (shells):

Pachyconic–conch fairly wide but not globular

Serpenticonic–conch snake-like formed by slow growing.

Platyconic– conch with flattened, parallel flanks.

Descriptive terms for septa:

Synclastically folded septa–septa simple, dome-shaped

Anticlastically folded septa–septa multiply folded

Septal pillars–septal folds connecting two major lobes

Descriptive terms for suture lines:

External lobe (ventral lobe in American terminology)–E lobe

Adventive lobe (lateral lobe in American terminology)–A lobe

Lateral lobe (umbilical lobe in American terminology)–L lobe

Umbilical lobe (internal lobe in American terminology)–U lobe

Internal lobe (dorsal lobe in American terminology)–I lobe

Thus sutural formulae using this terminology can be used. For example (E1 Em E1) A L U I - translates as uture line with a subdivided E lobe and the other lobes are simple.

Descriptive terms for ammonite ornament are:

Concavo-convex – growth lines extending with one projection across the flanks, forming a pronounced ventrolateral salient and a ventral sinus.

Biconvex – growth lines extending with two projections across the flanks, forming a pronounced ventrolateral salient and a ventral sinus.

(Order) **Agoniatitida** †

Devonian (Lower to Upper, Emsian to Frasnian) (407.6–372.2 Ma)

Ammonites with ventral siphuncle. Conchs gyroconic, advolute, evolute or involute, thinly discoidal to pachyconic; umbilicus very wide to closed. Very low to very high whorl expansion rate. Umbilical window present in the oldest representatives. Basic sutural formula E [L], many modifications during phylogeny (subdivision of the external lobe, development of umbilical and internal lobes). Septa in early forms simple-domed without inflexions; derived forms with anticlastically folded septa. Ornament with convex, concavo-convex or biconvex growth lines, ribs present in many species.

The Agoniatitida were the first ammonites, evolving from the Bactritina (in the Lower Devonian) from which they differ in their coiled shells and ventral siphuncle. They are distinguished from most other ammonites in possessing a hyponomic sinus (and hence presumably possessed a mobile funnel) and otherwise simple shell morphology and 'goniatitic' suture lines.

(*Continued*)

(Suborder) Agoniatitina † Devonian (Lower to Middle, Emsian to Frasnian) (407.6–372.2 Ma) Agoniatitidans with gyroconic, advolute, evolute or involute, thinly discoidal to pachyconic conchs; umbilicus very wide to closed. Expansion rate of the whorls low in the earliest forms but high or extremely high in derived species. Whorl overlap in the evolute, subinvolute and involute species low or moderate. Basic sutural formula E [L], in derived lineages with umbilical lobes, an internal lobe, and with a subdivided external lobe. Septa at the beginning simple-domed, later with weak lateral inflexions. Ribs present in many species, in more derived taxa restricted to the inner whorls or absent.	
(Superfamily) Mimosphinctoidea † Devonian (Lower to Middle, Emsian) (407.6–393.3 Ma) Families Mimosphinctidae, Mimoceratidae, and Teicherticeratidae	 *Erbenoceras* (Mimosphinctidae), from Klug (2001)
(Superfamily) Mimagoniatitoidea † Devonian (Lower to Middle, Emsian to Eifelian) (407.6–387.7 Ma) Families Mimagoniatitidae, Auguritidae, and Latanarcestidae	 *Mimagoniatites*, from Barrande (1865–1877)
(Superfamily) Agoniatitoidea † Devonian (Lower to Middle, Emsian to Givetian) (407.6–382.7 Ma) Families Agoniatitidae, Atlantoceratidae, Pinacitidae, and Tamaritidae	 *Agoniatites*, from Holzapfel (1895)
(Suborder) Gephuroceratina † Devonian (Lower to Upper, Givetian to Frasnian) (387.7–372.2 Ma) Agoniatitidans with thinly discoidal to pachyconic conch, narrow to wide umbilicus, low whorl overlap, and moderately high to high whorl expansion rate. Basic sutural formula (E2 El E2) L I, many modifications during phylogeny (development of external and umbilical lobes). Septa anticlastically folded; septal pillars parallel-arranged. Ornament usually with biconvex growth lines. Weak ribs and ventrolateral grooves often present in juveniles.	
(Superfamily) Gephuroceratoidea † Devonian (Middle to Upper, Givetian to Frasnian) (387.7–372.2 Ma) Families Gephuroceratidae, Devonopronoritidae, Ponticeratidae, and Taouzitidae	 *Manticoceras* (Gephuroceratidae), from Archiac and Verneuil (1842)
(Superfamily) Beloceratoidea Devonian (Middle to Upper, Givetian to Frasnian) (387.7–372.2 Ma) Families Beloceratidae, Acanthoclymeniidae, and Nordiceratidae	 *Beloceras*, from Sandberger and Sandberger (1850)

(*Continued*)

(Suborder) Anarcestina †	
Devonian (Lower to Middler, Givetian to Frasnian) (407.6–382.7 Ma)	
Agoniatitidans with subinvolute to involute and rarely evolute, thinly discoidal to pachyconic conchs; umbilicus very wide to closed. Expansion rate of the whorls usually low, rarely moderately high in adult specimens. Whorl overlap in most species moderate to high. Basic sutural formula E L [I], in derived lineages with an adventive or umbilical lobe and sometimes with a subdivided internal lobe. The lateral lobe has a subumbilical position in early growth stages. Septa synclastically folded with weak lateral inflexions. Riblets present in many species.	

(Superfamily) Anarcestoidea † Devonian (Lower to Middle, Emsian to Givetian) (407.6–382.7 Ma) Families Anarcestidae, Cabrieroceratidae, Sobolewiidae, and Werneroceratidae	*Anarcestes*, from Barrande (1865) 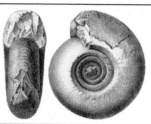

(Suborder) Pharciceratina †	
Devonian (Middle to Upper, Givetian to Frasnian) (387.7–372.2 Ma)	
Agoniatitidans with thinly discoidal to globose conch, narrow to wide umbilicus, deep whorl overlap, and very low whorl expansion rate. Basic sutural formula (E2 El E2) L U I, development of supplementary umbilical lobes during phylogeny. Septa anticlastically folded; septal pillars central-arranged. Ornament usually with biconvex growth lines, in some advanced forms with strong ribs and ventrolateral grooves.	

(Superfamily) Pharciceratoidea † Devonian (Middle to Upper, Givetian to Frasnian) (387.7–372.2 Ma) Families Pharciceratidae, and Petteroceratidae	*Lunupharciceras* (Pharciceratidae), from Sandberger and Sandberger (1850) 
(Superfamily) Triainoceratoidea † Devonian (Upper, Frasnian) (382.7–372.2 Ma) Family Triainoceratidae	*Triainoceras*, from Archiac and Verneuil (1842) 

(Order) Goniatitida †	
Devonian (Lower) to Permian (Upper) (Eifelian to Changhsingian) (407.6–251.9 Ma)	
Ammonites with an extremely wide variety of conch morphologies, ranging from globose and involute forms to thinly discoidal and evolute shapes. Suture line in almost all species with an adventive lobe, phylogenetically beginning with the elements E A L I. Multiple developments toward supplementary lobes (subdivision of the E lobe, secondary A, L, and U lobes, subdivision of the I lobe). Lobes in the early members of the order broadly rounded, becoming acute, denticulate, or digitate during phylogeny. Siphuncle almost always in ventral, only rarely in subventral position. Goniatitida were diverse with a wide range of shell forms and ornamentation, although most were thickly discoidal to globose with a narrow or closed umbilicus.	
(Suborder) Tornoceratina †	
Devonian (Middle) to Permian (Lower) (Eifelian to Changhsingian) (407.6–251.9 Ma)	
Goniatitidans with a wide variety of conch forms and ornaments. Usually thickly discoidal to globose with narrow or closed umbilicus. The conch shape is often modified during ontogeny; in many cases, the inner whorls are serpenticonic. Usually weakly ornamented, growth lines convex or biconvex. Basic sutural formula E A L I; adventive lobe in early forms widely rounded and acute in advanced lineages. External lobe simple, sometimes subdivided. During phylogeny, an umbilical lobe is produced and some lineages have supplementary adventive or lateral lobes. Septa synclastically folded in early forms. Advanced forms with anticlastically folded septa; septal pillars central-arranged.	

(Continued)

(Superfamily) **Tornoceratoidea** † Devonian (Middle to Upper, Givetian to Famennian) (387.7–358.9 Ma) Families Tornoceratidae, Kirsoceratidae, Parodoceratidae, Posttornoceratidae, and Pseudoclymeniidae	 *Tornoceras*, from Sandberger and Sandberger (1850–1856) 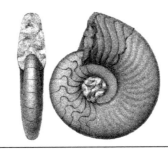 *Pseudoclymenia*, from Sandberger (1853)
(Superfamily) **Prionoceratoidea** † Devonian (Upper) to Permian (Upper) (Famennian to Changhsingian) (372.2–251.9 Ma) Families Prionoceratidae, Cheiloceratidae, Dimeroceratidae, Gattendorfiidae, Maximitidae, Praeglyphioceratidae, Prodromitidae, Pseudohaloritidae, and Sporadoceratidae	 *Cheiloceras*, from Sandberger and Sandberger (1850–1856) 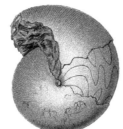 *Sporadoceras*, from Sandberger and Sandberger (1850–1856)
(Superfamily) **Dimeroceratoidea** † Devonian (Upper, Famennian) (372.2–358.9 Ma) Family Dimeroceratidae	*Dimeroceras*, from Sandberger and Sandberger (1850–1856) 

(Continued)

(Superfamily) **Prolobitoidea** † Devonian (Upper, Famennian) (372.2–358.9 Ma) Families Prolobitidae and Phenacoceratidae	*Prolobites*, from Sandberger and Sandberger (1850–1856)  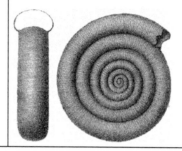 *Cycloclymenia* (Phenacoceratidae), from Sandberger and Sandberger (1850–1856)

(Suborder) Goniatitina †

Mississippian (Lower) to Permian (Upper) (Tournaisian to Changhsingian) (358.9–251.9 Ma)

Goniatitidans with a wide variety of conch forms and ornaments. Usually thickly discoidal to globose with narrow or closed umbilicus. The conch shape is often modified during ontogeny; in many cases, the inner whorls are serpenticonic. Usually weakly ornamented, growth lines convex or biconvex, often with spiral ornament. Basic sutural formula (E1 Em E1) A L U I; adventive nearly always acute. External lobe subdivided. During phylogeny, various modifications of the suture line occur including increase in the number of elements and secondary serration of lobes and notching of lobes and saddles.

The Goniatitina were the most diverse early ammonites. Most were small, tightly coiled, and globose, and their suture lines ranged from simple 'goniatitic' to more complex. Some had a weak hyponomic sinus, and, while most were smooth, some had external sculpture. They arose in the Lower Devonian and survived until the Upper Permian. Sometimes treated as an order.

(Superfamily) **Adrianitoidea** †
Pennsylvanian to Permian (Kasimovian to
    Wuchiapingian)
    (307.0–254.1 Ma)
Family Adrianitidae
	Emilites (Adrianitidae), from TL
	
	Adrianites, from TL

(Superfamily) **Girtyoceratoidea** †
Mississippian to Pennsylvanian (Tournaisian to
    Bashkirian)
    (358.9–315.2 Ma)
Families Girtyoceratidae and Baschkiritidae
	Girtyoceras, from TL

(Superfamily) **Dimorphoceratoidea** †
Mississippian to Pennsylvanian (Tournaisian to
    Bashkirian)
    (358.9–315.2 Ma)
Families Dimorphoceratidae and Berkhoceratidae
	Dimorphoceras, from Miller et al. (1952)

(Continued)

(Superfamily) **Gastrioceratoidea** †
 Pennsylvanian (Bashkirian to Moscovian)
 (323.2–307.0 Ma)
 Families Gastrioceratidae, Decoritidae,
  Homoceratidae, Reticuloceratidae, and
  Surenitidae
	Gastrioceras, from TL
	
	 Isohomoceras (Homoceratidae), from Korn and Penkert (2008)
	
	 Phillipsoceras (Reticuloceratidae), from Hodson (1957)

(Superfamily) **Goniatitoidea** † Mississippian (Lower) to Permian (358.9–259.1 Ma) Families Goniatitidae, Agathiceratidae, and Delepinoceratidae	*Goniatites*, courtesy of D. Korn *Progoniatites* (Goniatitidae), from Korn et al. (2010)

(Superfamily) **Gonioloboceratoidea** †
 Pennsylvanian to Permian (Bashkirian to
  Asselian) (323.2–295.0 Ma)
 Families Gonioloboceratidae,
  Gonioglyphioceratidae, and Wiedeyoceratidae
	Gonioloboceras, from TL

(Superfamily) **Marathonitoidea** † Pennsylvanian to Permian (Lopingian) (Kasimovian to Capitanian) (307.0–259.1 Ma) Family Marathonitidae	*Cardiella* (Marathonitidae), from Miller and Downs (1950)

(Continued)

(Superfamily) **Neodimorphoceratoidea** † Pennsylvanian (Serpukhovian to Virgilian) (330.9-286 Ma) Families Neodimorphoceratidae and Ramositidae	*Neodimorphoceras*, from TL

(Superfamily) **Neoglyphioceratoidea** †
Mississippian (Viséan to Serpukhovian)
(346.7–323.2 Ma)
Families Neoglyphioceratidae, Cravenoceratidae,
  Fayettevilleidae, Ferganoceratidae,
  Nuculoceratidae, and Rhymmoceratidae
	Lyrogoniatites (Cravenoceratidae),
	from Miller et al. (1952)
	
	Cravenoceras, from TL

(Superfamily) **Neoicoceratoidea** †
Pennsylvanian (Middle) to Permian (Moscovian
  to Changhsingian) (307.0–251.9 Ma)
Families Neoicoceratidae, Atsabitidae,
  Eothinitidae Metalegoceratidae, and
  Paragastrioceratidae
	Paragastrioceras, from TL

(Superfamily) **Nomismoceratoidea** † Mississippian to Pennsylvanian (Tournaisian to Serpukhovian) (358.9–323.2 Ma) Families Nomismoceratidae and Entogonitidae	*Entogonites*, from Korn and Sudar (2016)

(Superfamily) **Pericycloidea** †
Mississippian to Pennsylvanian (Tournaisian to
  Serpukhovian) (358.9–323.2 Ma)
Families Pericyclidae, Eogonioloboceratidae,
  Furnishoceratidae, Intoceratidae, Kozhimitidae,
  Maxigoniatitidae, and Muensteroceratidae
	Muensteroceras, from TL
	
	Pericyclus, from Korn et al. (2010)
	
	Semibollandites (Maxigoniatitidae), Bockwinkel
	et al. (2010)

(Continued)

(Superfamily) **Popanoceratoidea** † Permian (Asselian to Wuchiapingian) (298.9– 254.1 Ma) Families Popanoceratidae and Mongoloceratidae	*Neopopanoceras* (Popanoceratidae) from Zittel and Eastman (1913) *Popanoceras*, from Miller and Furnish (1940a)
(Superfamily) **Schistoceratoidea** † Pennsylvanian (Moscovian to Gzhelian) (315.2–298.9 Ma) Families Schistoceratidae, Axinolobidae, Christioceratidae, Orulganitidae, and Welleritidae	*Schistoceras*, from TL
(Superfamily) **Shumarditoidea** † Pennsylvanian to Middle Permian (Kasimovian to Roadian) (307.0–268.8 Ma Families Shumarditidae, Parashumarditidae, and Perrinitidae	*Perrinites*, from TL
(Superfamily) **Cycloloboidea** † Pennsylvanian to Permian (Kasimovian to Changhsingian) (307.0–251.9 Ma) Families Cyclolobidae, Hyattoceratidae, Kufengoceratidae, Neostacheoceratidae, and Vidrioceratidae	*Cyclolobus*, from TL *Martoceras* (Vibrioceratidae), from Haniel (1915)
(Superfamily) **Thalassoceratoidea** † Pennsylvanian to Permian (Bashkirian to Wordian) (323.2–265.1 Ma) Families Thalassoceratidae and Bisatoceratidae	*Eothalassoceras* (Thalassoceratidae), from TL

(*Continued*)

(Superfamily) **Somoholitoidea** † Pennsylvanian to Permian (Serpukhovian to Kungurian) (330.9–272.9 Ma) Families Somoholitidae, Clistoceratidae, Dunbaritidae, Glaphyritidae, Pseudoparalegoceratidae, and Stenolagphyritidae	*Neoshumardites* (Somoholitidae), from TL  

(Order) **Prolecanitida**
Mississippian to Triassic (Lower) (Tournaisian to Triassic) (358.9–?247.2 Ma)
Ammonites with a rather narrow variety of conch morphologies, ranging from serpenticone to platycone shapes. Suture line in all species with an adventive lobe, phylogenetically beginning with the elements (E1 Em E1) A L U I. Developments toward supplementary lobes, particularly of the U lobes. Siphuncle always in ventral position.
These ammonites were not very diverse in shape, and all had a ventral siphuncle.

(Suborder) **Prolecanitina** †
Mississippian to Triassic (Lower) (358.9–?247.2 Ma)
Prolecanitidans with a rather narrow variety of conch morphologies, ranging from serpenticone to platycone shapes. Suture line in all species with an adventive lobe, phylogenetically beginning with the elements (E1 Em E1) A L U I. Developments toward supplementary lobes, particularly of the U lobes, but also serration of lobes and ventrolateral saddle. Siphuncle always in ventral position.
This group is distinctive in that its young stages had sutures of the 'goniatitic' type and developed more complex suture patterns as they grew.

(Superfamily) **Medlicottioidea** † Mississippian to Triassic (Lower) (Viséan to Triassic) (346.7–?247.2 Ma) Families Medlicottiidae, Episageceratidae, Pronoritidae, Shikhanitidae, and Sundaitidae	*Artinskia* (Medlicottiidae), from TL 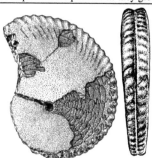

(Superfamily) **Prolecanitoidea** † Mississippian to Permian (Tournaisian to Wordian) (358.9–265.1 Ma) Families Prolecanitidae and Daraelitidae	*Epicanites* (Daraelitidae), from Miller and Furnish (1940b) 

(Order) **Clymeniida** †
Devonian (Upper; Famennian) (372.2–358.9 Ma)
Ammonites with dorsal siphuncle. Conchs thinly discoidal to pachyconic, umbilicus very wide to closed. Low to deep imprint zone, and low to moderately high whorl expansion rate. Basic sutural formula [L] I, many modifications during phylogeny (development of external, adventive, and umbilical lobes). Septa in early forms simple-domed without inflexion; derived forms with anticlastically folded septa. Ornament rarely with convex, mostly with concavo-convex or biconvex growth lines. Weak ribs present in many species.
Differ from other ammonites by having a dorsal siphuncle, although in the first chamber it is positioned ventrally as in other ammonites. Like Anarcestina, but unlike most other ammonites, they had a hyponomic sinus. Small (up to about 4 cm in diameter). Some had triangular coiling.

(Suborder) **Cyrtoclymeniina** †
Devonian (Upper, Famennian) (372.2–358.9 Ma)
Clymeniidans with thinly discoidal to pachyconic, usually 'goniatitoid' conch, narrow to moderately wide umbilicus. Deep imprint zone and moderately high whorl expansion rate. Basic sutural formula [L] I, many modifications during phylogeny (development of external, adventive, and umbilical lobes). Septa in early forms simple-domed without inflexion; derived forms with strong marginal inflexions. Ornament with concavo-convex or slightly biconvex growth lines, weak ribs present in many species.

(Continued)

(Superfamily) **Cyrtoclymenioidea** † Devonian (Upper, Famennian) (372.2–358.9 Ma) Families Cyrtoclymeniidae, Carinoclymeniidae, Cymaclymeniidae, Hexaclymeniidae, and Rectoclymeniidae	 *Cymaclymenia*, from Gümbel (1863) 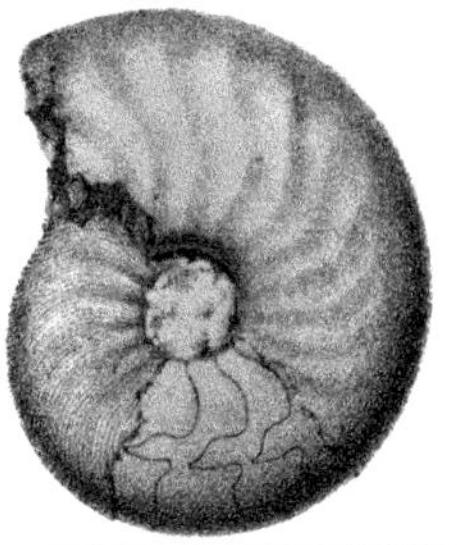
(Superfamily) **Biloclymenioidea** † Devonian (Upper, Famennian) (372.2–358.9 Ma) Families Biloclymeniidae and Pachyclymeniidae	 *Biloclymenia*, from Korn and Klug (2002) 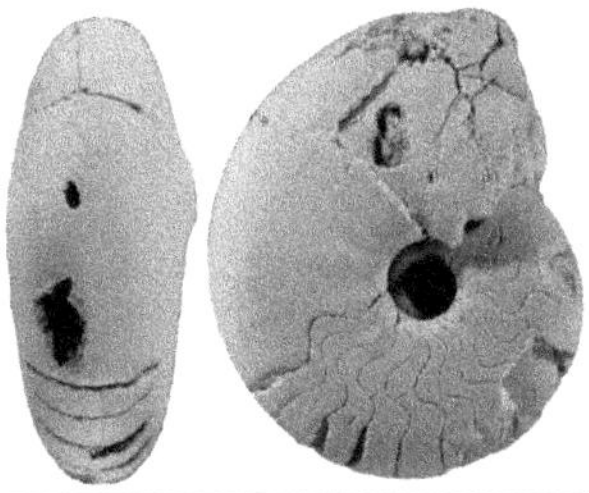
(Suborder) **Clymeniina** † Devonian (Upper, Famennian)	
(Superfamily) **Clymenioidea** (= Platyclymenioidea) † Devonian (Upper, Famennian) (372.2–358.9 Ma) Families Clymeniidae, Glatziellidae, Kosmoclymeniidae, Piriclymeniidae, and Platyclymeniidae	 *Clymenia*, from Gümbel (1863) 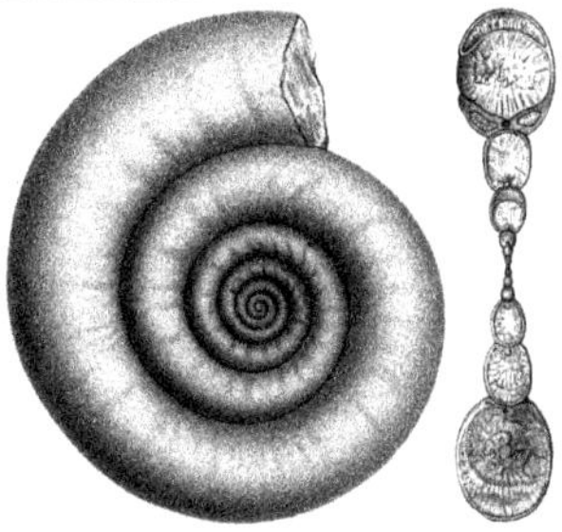
(Superfamily) **Wocklumerioidea** † Devonian (Upper, Famennian) (372.2–358.9 Ma) Families Wocklumeriidae and Parawocklumeriidae	 *Wocklumeria*, from Schindewolf (1937) 
(Superfamily) Gonioclymenioidea † Devonian (Upper, Famennian) (372.2–358.9 Ma) Families Gonioclymeniidae, Costalymeniidae, Sellaclymeniidae, and Sphenoclymeniidae	 *Gonioclymenia*, from Gümbel (1863) 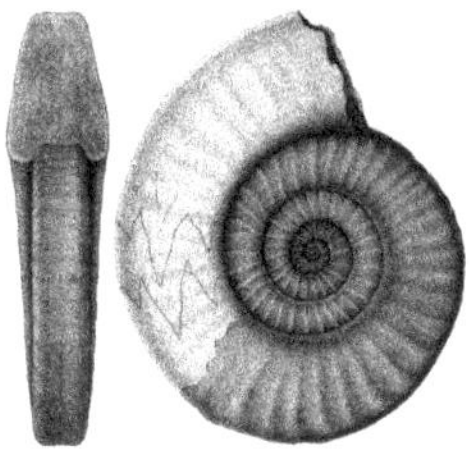

(Order) **Ceratitida** †

Lower Permian (Cisuralian) to Jurassic (Middle) (298.9–163.5 Ma)

Ammonites with a wide variety of conch morphologies, ranging from thinly discoidal and evolute shapes to globose and involute forms. Suture line in all species with an adventive lobe, primary suture line E A L I. Suture line development phylogenetically beginning with the elements E A L U I. Multiple developments toward supplementary lobes (subdivision of the E lobe, secondary L and U lobes, subdivision of the I lobe). Lobes usually denticulated, rarely unserrated; saddles usually rounded, rarely notched.

Septa have rounded saddles and serrated lobes. Compared with the other early ammonites some of the later members of this group had a flattened shell, and for the first time some taxa had a tapered edge to the shell possibly adding to their ability to move through the water efficiently.

(*Continued*)

(Suborder) Ceratitina † Permian (Cisuralian) to Triassic (Middle) (298.9–237 Ma) This group retained the 'ceratite' suture pattern throughout their life. Some were heteromorphs, with helical and uncoiled forms, and there were also streamlined forms.	
(Superfamily) Ceratitoidea † Triassic (Middle) to Jurassic (Middle) (247.2–201.3 Ma) Families Ceratitidae (= Beyrichitidae), Acrochordiceratidae, Badiotitidae, Balatonitidae, Celtitidae, Hungaritidae, Rimkinitidae, and Sibiritidae	*Ceratites*, from Zittel and Eastman (1913) 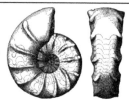  *Hungarites*, from Hyatt and Smith (1905) *Sibirites*, from Hyatt and Smith (1905) 
(Superfamily) Choristoceratoidea † Triassic (Upper) to Jurassic (Lower) (221.5–196.5 Ma) Families Choristoceratidae, Cochloceratidae, Cycloceltitidae, and Rhabdoceratidae	*Choristoceras*, from TL   *Cochloceras*, from TL
(Superfamily) Clydonitoidea † Triassic (Upper) (235–205.6 Ma) Families Clydonitidae, Clionitidae, Distichitidae, Heraclitidae, Sandlingitidae, Steinmannitidae, Thetiditidae, and Tibetitidae	*Leconteia* (Clionitidae), from Hyatt and Smith (1905)  *Clionites*, from Hyatt and Smith (1905)

(Continued)

(Superfamily) **Danubitoidea** † Triassic (Lower to Upper) (251.3–221.5 Ma) Families Danubitidae, Aplococeratidae, Lecanitidae, Longobarditidae, and Nannitidae	*Danubites*, from Hyatt and Smith (1905) *Lecanites*, from Hyatt and Smith (1905) *Nannites*, from Hyatt and Smith (1905)
(Superfamily) **Dinaritoidea** † Triassic (Lower to Middle) (251.3–242 Ma) Families Dinaritidae, Columbitidae, Helenitidae, and Tirolitidae	*Columbites*, from Hyatt and Smith (1905) *Dinarites*, from Hyatt and Smith (1905) *Tirolites*, from Fraas (1910)
(Superfamily) **Lobitoidea** † Triassic (Upper) (235–221.5 Ma) Family Lobitidae	*Lobites*, from TL

(*Continued*)

(Superfamily) **Meekoceratoidea** † Permian (Lopingian) to Triassic (Middle) (254–242 Ma) Families Meekoceratidae, Albanitidae, Arctoceratidae, Dieneroceratidae, Flemingitidae, Gyronitidae, Melagathiceratidae, Mullericeratidae, Ophiceratidae, Prionitidae, Proptychitidae, and Ussuriidae	 *Flemingites*, from Hyatt and Smith (1905) *Ussuria*, from Smith (1932) *Ophiceras*, from Hyatt and Smith (1905) *Meekoceras* (Prionitidae), from Hyatt and Smith (1905)
(Superfamily) **Megaphyllitoidea** † Triassic (Middle)–Jurassic (Lower) (247.2–201 Ma) Families Megaphyllitidae, Parapopanoceratidae, Procarnitidae, and Prosphingitidae	 *Prosphingites*, from Hyatt and Smith (1905)
(Superfamily) **Nathorstitoidea** † Triassic (Middle to Upper) (242–232 Ma) Families Nathorstitidae, Proteusitidae, and Thanamitidae	 *Nathorstites*, from TL

(*Continued*)

(Superfamily) **Noritoidea** † Triassic (Middle to Upper) (247.2–235 Ma) Families Noritidae, Inyoitidae, Lanceolitidae, and Stephanitidae	*Inyoites*, from Hyatt and Smith (1905) *Bosnites* (Noritidae), from TL
(Suborder) **Otoceratina** † Permian (Lopingian) to Triassic (Lower) (254–247.2 Ma)	

(Superfamily) **Otoceratoidea** †
  Permian (Lopingian) to Triassic (Lower)
  (254–247.2 Ma)
Families Otoceratidae, Anderssonoceratidae, and
  Araxoceratidae
	Otoceras, from TL

(Superfamily) **Pinacoceratoidea** †
Triassic (Middle to Upper) (247.2–201.6 Ma)
Families Pinacoceratidae, Carnitidae,
  Gymnitidae, Isculitidae, Klamathitidae, and
  Sagenitidae
	Sagenites, from Hyatt and Smith (1905)
	
	Placites (Gymnitidae), from Hyatt and Smith (1905)

(Superfamily) **Ptychitoidea** † Triassic (Middle to Upper) (247.2–232 Ma) Families Ptychitidae, Eosagenitidae, and Sturiidae	*Ptychites*, from Zittel and Eastman (1913)

(*Continued*)

(Superfamily) **Sageceratoidea** †
Permian (Lopingian) to Triassic (Upper)
  (252.3–232 Ma)
Families Sageceratidae, Aspenitidae,
  Beneckeiidae, and Hedenstroemiidae

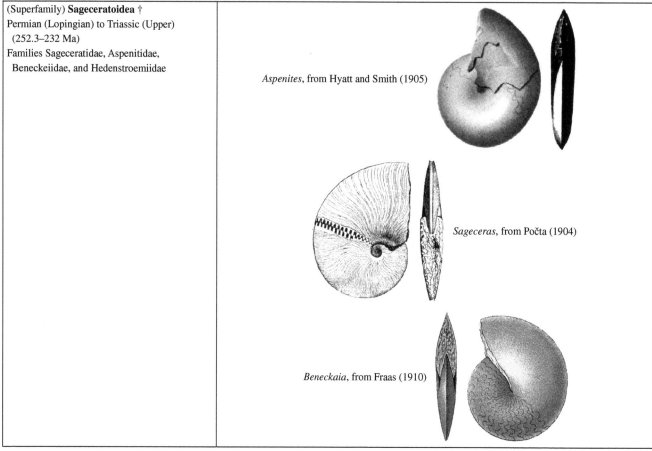

Aspenites, from Hyatt and Smith (1905)

Sageceras, from Počta (1904)

Beneckaia, from Fraas (1910)

(*Continued*)

| (Superfamily) **Tropitoidea** †
Triassic (Upper) to (235–201.6 Ma)
Families Tropitidae, Didymitidae, Episculitidae,
  Haloritidae, Juvavitidae, Parathisbitidae,
  Thisbitidae, Tropiceltidae, and Tropiceltitidae | 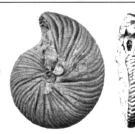 *Juvavites*, from Hyatt and Smith (1905)

 *Paratropites* (Tropitidae), from Hyatt and Smith (1905)

 *Tornquistites* (Tropiceltidae), from Hyatt and Smith (1905)

 *Discotropites* (Tropitidae), from Hyatt and Smith (1905) |
| (Superfamily) **Trachyceratoidea** †
Triassic (Middle to Upper) (242–205.6 Ma)
Families Trachyceratidae, Arpaditidae,
  Buchitidae, Cyrtopleuritidae, Dronovitidae, and
  Noridiscitidae |  *Trachyceras*, from Fraas (1910)

 *Buchites*, from TL |

(*Continued*)

(Suborder) **Paraceltitina** † Permian (Roadian) to Triassic (Middle) (272.9–242 Ma) Ceratitidans with discoidal conch with wide umbilicus. Ornamentation weak or composed of transverse ribs, often with ventrolateral nodules. Basic sutural formula (E1 E1) A L U I; more umbilical lobes may be introduced. Secondary serration of lobes common.	
(Superfamily) **Xenodiscoidea** † Permian (Roadian) to Triassic (Middle) (272.9–242 Ma) Families Xenodiscidae, Dzhulfitidae, Paraceltitidae, Pseudotirolitidae, and Xenoceltitidae	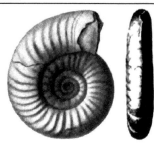 *Xenodiscus*, from Hyatt and Smith (1905) *Paraceltites*, from Miller and Furnish (1940a)

(Order) **Ammonitida** † Triassic (Upper) to Cretaceous (Upper) (201.6–66 Ma) The sutures of members of this group had folded saddles and lobes and fractal patterns. This group includes all the Mesozoic ammonites and there are four (or sometimes more) suborders recognised as well as many families and a number of superfamilies. An additional two suborders, Haploceratina and Perisphinctina, are sometimes separated from this group (e.g., Page 2008).	
(Suborder) **Ammonitina** † Triassic (Upper) to Cretaceous (Upper) (201.6–66 Ma) In this classificaiton we recognise this and two other suborders in the order Ammonitida.	
(Superfamily) **Acanthoceratoidea** † Cretaceous (Lower to Upper) (112.6–66 Ma) Families Acanthoceratidae, Brancoceratidae, Coilopoceratidae, Collignoniceratidae, Flickiidae, Forbesiceratidae, Leymeriellidae, Lyelliceratidae, Pseudotissotiidae, Sphenodiscidae, Tissotiidae, and Vascoceratidae	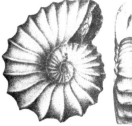 *Hysteroceras* (Brancoceratidae), from Stoliczka (1863–1866)  *Acanthoceras*, from Zittel and Eastman (1913)

(*Continued*)

(Superfamily) **Deshayesitoidea** † Cretaceous (Lower to Upper) (125.45–89.3 Ma) Families Deshayesitidae and Parahoplitidae	*Deshayesites*, from Fraas (1910)

(Superfamily) **Desmoceratoidea** †
 Cretaceous (Lower to Upper) (140.2–66 Ma)
 Families Desmoceratidae, Cleoniceratidae,
   Kossmaticeratidae, Munericeratidae,
   Pachydiscidae, and Silesitidae
	Pachydiscus, from Zittel and Eastman (1913)
	
	 Lewesiceras (Pachydiscidae), from Fraas (1910)
	
	 Kossmaticeras, from TL

(Superfamily) **Engonoceratoidea** † Cretaceous (Lower to Upper) (112.6–89.3 Ma) Families Engonoceratidae and Knemiceratidae	*Platiknemiceras* (Engonoceratidae), from Basse (1954)

(*Continued*)

(Superfamily) **Eoderoceratoidea** †
Jurassic (Lower to Middle) (196.5–168.4 Ma)
Families Eoderoceratidae, Amaltheidae,
  Coeloceratidae, Dactylioceratidae,
  Liparoceratidae, Phricodoceratidae, and
  Polymorphitidae

Phricodoceras, from Fraas (1910)

Amaltheus, from Zittel and Eastman (1913)

Liparoceras, from Fraas (1910)

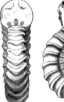

Aegoceras (Liparoceratidae), from Zittel and
  Eastman (1913)

Dactylioceras, from Zittel and Eastman (1913)

(Continued)

(Superfamily) **Haploceratoidea** †
Jurassic (Middle) to Cretaceous (Upper)
  (171.6–85.8 Ma)
Families Haploceratidae, Binneyitidae,
  Lissoceratidae, Oppeliidae, and Strigoceratidae

	Distichoceras (Oppeliidae), from Fraas (1910)

	Glochiceras (Haploceratidae), from Fraas (1910)
	
	Oppelia, from Počta (1904)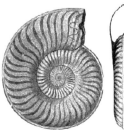

(Superfamily) **Hildoceratoidea** (=
  Hammatoceratoidea) †
Jurassic (Lower) to Cretaceous (Lower)
  (196.5–112.6 Ma)
Families Hildoceratidae, Graphoceratidae,
  Hammatoceratidae, Phymatoceratidae, and
  Sonniniidae
	Harpoceras (Hildoceratidae),
	  from Počta (1904)

	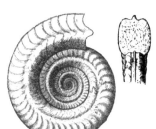*Hildoceras*, from Zittel and Eastman (1913)

(*Continued*)

(Superfamily) **Hoplitoidea** † Cretaceous (Lower to Upper) (145–66 Ma) Families Hoplitidae, Placenticeratidae, and Schloenbachiidae	 *Schloenbachia*, from Zittel and Eastman (1913)   *Hoplites*, from Zittel and Eastman (1913) *Placenticeras*, from Hyatt (1903) 

(Superfamily) **Lytoceratoidea** †
Jurassic (Lower) to Upper Cretaceous (200–66
  Ma)
This group was recently revised by Hoffmann
  (2015a).
Families Lytoceratidae and Tetragonitidae

	Lytoceras, from Zittel and Eastman (1913)

	Gabbioceras (Tetragonitidae), from TL

(Superfamily) **Perisphinctoidea** †
Jurassic (Middle) to Cretaceous (Upper)
  (171.6–66 Ma)
The Jurassic heteromorphs are now included in
  Perisphinctidae (Jain 2018). There is no clear
  connection between the Middle Jurassic
  heteromorphs and those in the Cretaceous.
Families Perisphinctidae, Aspidoceratidae,
  Ataxioceratidae, Aulacostephanidae,
  Dorsoplanitidae, Himalayitidae, Holcodiscidae,
  Morphoceratidae, Neocomitidae,
  Olcostephanidae, Oosterellidae, Pachyceratidae,
  Parkinsoniidae, Reineckeiidae, Simoceratidae,
  Tulitidae, and Virgatitidae

	Parkinsonia, from Počta (1904)

	Bullatimorphites (Tulitidae), from Vogt (1866)

(Superfamily) **Psiloceratoidea** † Jurassic (Lower to Upper) (201–161.2 M) This group is thought to have originated from the phylloceratids and gave rise to the Lytoceratoidea during the earliest Jurassic. Sometimes treated as an order or suborder. Families Psiloceratidae, Arietitidae, Echioceratidae, Oxynoticeratidae, Phyllytoceratidae, and Schlotheimiidae	 *Psiloceras*, from Fraas (1910) 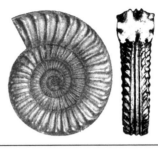 *Arietites*, from Zittel and Eastman (1913)
(Superfamily) **Pulchellioidea** † Lower Cretaceous (130–125.5 Ma) Family Pulchelliidae	 *Pulchellia*, from TL
(Superfamily) **Stephanoceratoidea** † Jurassic (Lower) to Cretaceous (Lower) (183.0–112.6 Ma) This group and the next were recently revised by Howarth (2017). Families Stephanoceratidae, Cardioceratidae, Kosmoceratidae, Macrocephalitidae, Otoitidae, and Sphaeroceratidae	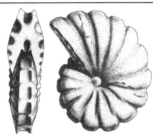 *Teloceras* (Stephanoceratidae), from Fraas (1910)  *Stephanoceras*, from Fraas (1910) *Otoites*, from Fraas (1910)  *Sinikosmoceras* (Kosmoceratidae), from Fraas (1910)

(Superfamily) **Spiroceratoidea** † Jurassic (Middle) (174.1–163.5 Ma) Family Spiroceratidae	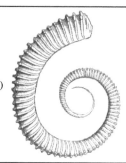 *Spiroceras*, from Zittel and Eastman (1913)

(Suborder) **Phylloceratina** †
Triassic (Middle) to Cretaceous (Upper) (247.2–66 Ma)
These large ammonites had involute shells that were weakly sculptured and thin shelled. The suture lines had phylloid-like saddles.

(Superfamily) **Phylloceratoidea** †
Triassic (Middle) to Cretaceous (Upper)
  (247.2–70.6 Ma)
Families Phylloceratidae, Discophyllitidae,
  Juraphyllitidae, and Ussuritidae
	Phylloceras, from Zittel and Eastman (1913) 

	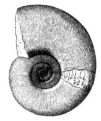 *Monophyllites* (Ussuritidae), from Zittel and Eastman (1913)

(Suborder) **Ancyloceratina** †
Jurassic (Middle) to Cretaceous (Upper) (171.6–66 Ma)
The connection between the Tithonian forms and the Cretaceous heteromorphs, which first appear in the Valanginian, is not clear.
The so-called Cretaceous heteromorph ammonites probably do not comprise a monophyletic group.

(Continued)

(Superfamily) **Ancyloceratoidea** † Jurassic (Middle) to Cretaceous (Upper) (171.6–70.6 Ma) Families Ancyloceratidae, Bochianitidae, Crioceratitidae, Hamulinidae, Hemihoplitidae, Heteroceratidae, Labeceratidae, Macroscaphitidae, and Ptychoceratidae	*Ancyloceras*, from Zittel and Eastman (1913) *Heteroceras*, from Zittel and Eastman (1913) *Crioceratites*, from TL *Macroscaphites*, from Zittel and Eastman (1913)
(Superfamily) **Douvilleiceratoidea** † Cretaceous (Lower to Upper) (125.45–89.3 Ma) Family Douvilleiceratidae, Astiericeratidae, and Trochleiceratidae	*Douvilleiceras*, from Zittel and Eastman (1913)
(Superfamily) **Scaphitoidea** † Cretaceous (Lower to Upper) (105.3–66 Ma) Family Scaphitidae	*Scaphites*, from Zittel and Eastman (1913).

(*Continued*)

(Superfamily) **Turrilitoidea** †
Cretaceous (Lower to Upper) (140.2–66 Ma)
Families Turrilitidae, Anisoceratidae, Baculitidae,
  Diplomoceratidae, Hamitidae, and
  Nostoceratidae

Baculites, from Smith (1901)

(a)   (b)

Turritelites, from Zittel and Eastman (1913)

Hamites (*Polyptychoceras*), from TL

Nipponites (Nostoceratidae), from Yabe (1904)

(*Continued*)

Unassigned to higher group (Order) **Colorthoceratida** † Pennsylvanian (Upper) (307–298.9 Ma) Recently described by Mutvei and Mapes (2018) the apparently internal shells of this group resemble orthoconic nautiliforms except for their shell structure which is more coleoid-like in lacking a nacreous layer and in having a high chitin content that made the shell 'semi-elastic'. However, there is no ink sac, and the authors could not exclude the possibility that the shell was external. In addition, the septum ultrastructure and the central position of the siphuncle differ from coleoids. Family Colorthoceratidae (= Colorthoceridae)	*Colorthoceras*, modified from Mutvei and Mapes (2018) 

(Cohort) **Coleoida** (= Coleoidea, Endocochlia)

Mississippian (Lower) to present (359.9–0 Ma)

Coleoida are distinguished from other cephalopods by their internalised shell which has undergone reduction or even loss. In addition, additional shell
layers (rostral layers) are deposited on the external surface of the primary shell (conotheca), septa composed or lamellofibrillar nacre, and there is usually
an ink sac.

An historical account of coleoid classification has been given by Donovan and Fuchs (2012).

Reports of Devonian coleoids cannot be substantiated, and most are likely bactritans while those reported by Bandel et al. (1983) (comprising the
'Paleoteuthomorpha') are thought to be a mix of bactritoids, orthoconic nautiliforms, trace fossils, and fish remains (Doyle et al. 1994).

This section should be cited as Ponder, W. F., Lindberg, D. R., and Fuchs, D. (2019), Appendix. Classification of the living molluscan classes. Coleoida, in
Ponder, W. F., Lindberg, D. R., and Ponder, J. M., *Biology and Evolution of the Mollusca*, Volume Two.

We base our classification for the living taxa largely on Nishiguchi and Mapes (2008) and Young et al. (2008). Fossil taxa are arranged according to Doyle
et al. (1994), the publications cited below and input from DF.

Uncertain relationships: The Pennsylvanian Shimanskyidae † (Doguzhaeva et al. 1999).

(Order) **Hematitida** † Mississippian (Middle) to Pennsylvanian (Lower) (346.7–315.2 Ma) Characterised by a thick, blunt, short rostrum which was partly calcified and partly organic. Conotheca with five–six shell layers. Siphuncle thin-walled. The young animal had a breviconic phragmocone, the chambers of which contained internal cameral deposits. The group is characterised by its short body chamber and probable lack of a pro-ostracum (Doguzhaeva et al. 2002). The aragonitic rostrum was secreted only as the animal approached maturity. Although this group is included in Aulacoceratida in older classifications and some modern ones (e.g., Bizikov 2008), it is now generally accepted that the two groups are distinct. One of the most important distinguishing features is that the post-alveolar rostrum is short in Hematitida and long in Aulacoceratida. Family Hematitidae	*Paleoconus* (Hematitidae), from Teichert (1967)  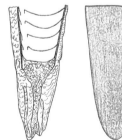 *Hematites* rostrum, right modified from Flower and Gordon (1959), longitudinal section redrawn and modified from Doguzhaeva et al. (2002)
(Order) **Donovaniconida** † Mississippian (Lower) to Pennsylvanian (Upper) (358.9–298.9 Ma) The radula of *Saundersites* has 11 teeth in each row, and, on this basis, this order was created (Doguzhaeva et al. 2007). Arm hooks and an ink sac have been found in *Gordoniconus* (Mapes et al. 2010). Families Donovaniconidae, Floweritidae, Gordoniconidae, Mutveiconitidae, Oklaconidae, and Rhiphaeoteuthidae	*Donovaniconus*, redrawn and modified from Doguzhaeva et al. (2010) 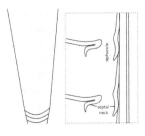

(Continued)

(Order) **Aulacoceratida** †	
Permian (Guadalupian) to Jurassic (Lower) (272.95–193.3 Ma)	
In this group the rostrum is aragonitic, long, and smooth or with incised grooves and sometimes also with ridges. The phragmocone has simple widely spaced sutures, high camerae, and a very narrow apical angle (usually < 12 degrees), and the siphuncle is thin-walled. The protoconch has a closing membrane, and the conotheca is thin, with two or three layers with the the pro-ostracum forming an entire body chamber. While the presence of arm hooks is assumed, they are still not confirmed in fossils.	
The superfamily classification below follows Pignatti and Mariotti (1996).	
Uncertain relationships: Palaeobelemnopseidae	
(Superfamily) **Aulacoceratoidea** † Pennsylvanian (Middle to Upper) (315.2–307 Ma) The members of this superfamily show a characteristic pattern of corrugated and folded lamellae in the rostrum that is absent in the Xiphoteuthidoidea (Mariotti & Pignatti 1999). Families Aulacoceratidae, Dictyoconitidae, and probably Palaeobelemnopseidae	 *Dictyoconites* reconstruction, from Bandel (1985) 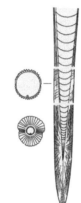 *Aulacoceras*, modified from Naef (1922)
(Superfamily) **Xiphoteuthidoidea** † Permian (Guadalupian) to Jurassic (Lower) (272.95–193.3 Ma) This superfamily was first recognised by Pignatti and Mariotti (1996). Previously the families in this group were included in Aulacoceratida. The smooth rostrum is highly varied in its development in different taxa. Families Xiphoteuthidae (= Chitinoteuthidae) and Palaeobelemnopseidae	 *Atractites* (Xiphoteuthidae), from Mariotti and Pignatti (1993)

(*Continued*)

(Order) **Phragmoteuthida** † Permian (Lopingian) to Jurassic (Upper) (259.1–150.8 Ma) The wide, three-lobed, fan-like pro-ostracum has long, wide median and lateral fields with convex ends. The phragmocone is breviconic with the septa closely to moderately spaced, and the thick-walled siphuncle differs from that of Belemnoidea. The shell wall consists of five to six layers (including three conothecal layers). The protoconch is unknown, and a rostrum is absent. The mantle was muscular, and arm hooks were present (Fuchs & Donovan 2018). This group has often been placed in the 'Belemnoida' (e.g., Engeser 1990; Bizikov 2008) or considered of uncertain status (Doyle et al. 1994). Doyle et al. (1994) and Donovan (2006) argued it is ancestral to all living coleoids and Bizikov (2008) to all living groups other than Spirulida. The hooks are long and weakly curved. The number of arms, while assumed to be ten, has not been confirmed in fossil material (Fuchs et al. 2013a). Family Phragmoteuthidae	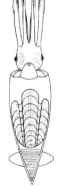 *Phragmoteuthis* reconstruction, from Naef (1922), modified by Bizikov (2008)

(Superorder) **Belemnitia** † Triassic (Upper) to Cretaceous (Upper) (337–66 Ma) Doyle and Shakides (2004) recognised four orders and Bizikov (2008) five. Belemnitians have a rostrum that is either calcitic or aragonitic and the ten equal-sized arms bear hooks. The morphological differentiation of the orders follows Doyle and Shakides (2004) and Donovan (2006). All have a small spherical protoconch. A list of the ca. 100 generic names (including synonyms) proposed for belemnites can be found at http://en.wikipedia.org/wiki/List_of_belemnites based on Sepkoski et al. (2002). See also Sepkoski's Online Genus Database, Cephalopoda entry. This group is treated as a superorder by Nishiguchi and Mapes (2008) and Bizikov (2008). Uncertain relationship: Lioteuthidae † *Lioteuthis*, the only member of the monotypic Lioteuthidae 'possibly represents a pro-ostracum of a belemnoid as the anterior margin is distinctly rounded and hyperbolar zones are absent' (Fuchs & Weis 2008 p. 96). It was previously considered to be an ancestral squid.

(Order) **Belemnitida** † Triassic (Upper) to Cretaceous (Upper) (237–66 Ma) Pro-ostracum narrower than Phragmoteuthida and broader than Diplobelida; rostrum calcitic. Arm hooks strongly curved, with obliquely truncated bases. Uncertain relationship: Chitinobelidae and Sinobelemnitidae

(*Continued*)

(Suborder) **Belemnitina** †

Cretaceous (Lower) (145–100.5 Ma)

The rostrum is composed of calcite deposited radially and the apical angle of the phragmocone is generally greater than 12 degrees. Mature conotheca are two-layered, although the nacreous layer is absent in some genera. The protoconch has a closing membrane, and the pro-ostracum is a single lobe.

Families Cylindroteuthidae, Hastitidae, Megateuthidae, Nipponoteuthidae, Oxyteuthidae, Passaloteuthidae, and Salpingoteuthidae

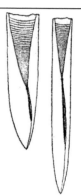

Belemnite shells – *Acroteuthis* (Cylindroteuthidae) (left), and *Oxytheuthis* (right), from Naef (1921)

Parapassaloteuthis (Passaloteuthidae), reconstruction, from Bandel and Boletzky (1988) (www.schweizerbart.de/9783510651337)

Megateuthis gigantea reconstruction, from Naef (1922)

(Suborder) **Belemnopseina** †

Jurassic (Lower) to Cretaceous (Lower) (201.3–139.8 Ma)

The massive calcitic rostrum has incised grooves and is conical in primitive members of the Belemnopseina but variable in shape in others. The phragmocone has simple sutures, narrow camerae, and a narrow apical angle. The pro-ostracum is spathulate, and the arm hooks have spurs.

Families Belemnopseidae, Belemnitellidae, Dicoelitidae, Dimitobelidae, Duvaliidae, Holcobelidae, Mesohibolitidae, Pseudobelidae, and Pseudodicoelitidae

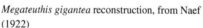

Hibolithes (Belemnopseidae), from Naef (1922) modified by Bizikov (2008)

(Continued)

(Suborder) **Belemnotheutidina** †

Cretaceous (Lower) (145–100.5 Ma)

The rostrum is poorly developed, aragonitic, sheath-like, conical, and with incised grooves and ridges. The phragmocone has simple sutures, low camerae, and a narrow apical angle. The pro-ostracum is spathulate, and the arm hooks are smooth with no spurs.

Families Belemnotheutidae, Chitinobelidae, and probably Ostenoteuthidae, and Sueviteuthidae

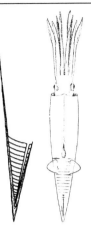

Belemnotheutis reconstruction, from Naef (1922) and longitudinal section of shell from Bather (1911)

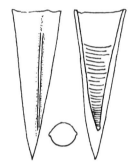

Antarctiteuthis (Belemnotheutidae), from Doyle and Shakides (2004)

(Suborder) **Diplobelida** †

Jurassic (Lower) to Cretaceous (Upper) (201.3–93.9 Ma)

The rostrum is calcitic, sheath-like or massive, conical, and with little ornament or smooth. The phragmocone sutures have a dorsal saddle, low camerae, and a narrow apical angle. The pro-ostracum is narrow and distinctively three lobed. The ten short arms bear hooks with nearly straight shanks and flat bases (Fuchs et al. 2013a).

This group is sometimes included in Belemnotheutidina.

Families Diplobelidae and Chondroteuthidae

Conoteuthis (Diplobelidae), redrawn from D'Orbigny (1845 [1845–1847])

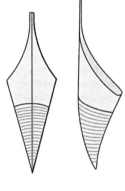

Diplobelus, from Zittel (1868)

(*Continued*)

(Superorder) **Decabrachia** (= Decembrachiata, Decembrachia, Decapoda, Decapodiformes)

Cretaceous (Lower) to present (145–0 Ma)

The classification of this group largely follows Nishiguchi and Mapes (2008) and Young et al. (2018). The diagnostic characters used largely follow Young et al. (2018). Recently Hoffmann (2015b) recommended using the name Decabrachia for this group, which we follow.

Decabrachia are characterised by the fourth pair of arms being modified as tentacles. The stalked suckers have horny rings, and a buccal crown is present. The statocyst has a single capsule, and the brain has the posterior buccal lobes widely separated, and the inferior frontal lobe system is absent. Fins are present, and the shell is a phragmocone, gladius, or is absent. The oesophagus lacks a crop. Nidamental glands are present, and the oviducal glands bilaterally symmetrical. The nephridial coeloms are joined to form a single coelom, and the visceropericardial coelom is extensive. Photosensitive vesicles are within the cephalic cartilage.

Uncertain relationships: Bayanoteuthidae † and Mississaepiidae †

The squids – previously lumped together as Teuthida – are now broken up into two main groups (Myopsida and Oegopsida).

The key distinguishing characters for each major decabrachian group are compared below in a tabular form (modified from Young et al. 2018).

	Shell	Cornea on Eye	Branchial Canal in Gill	Buccal Support Suckers	Sucker Circular Muscle	Tentacle Pocket in Head	Interstellate Connective	Oviduct	Egg Masses
Sepioidea	Flattened phragmocone	Yes	No	Yes/No	Yes	Yes	No	Left only	Benthic
Spiruloidea	Coiled phragmocone	No	No	No	No	Yes	No	Left only	?
Myopsida	Gladius	Yes	Yes	Yes/No	Yes	Yes	Yes	Left only	Benthic
Oegopsida	Gladius	No	Yes	No	No	No	Yes	Paired	Pelagic
Bathyteuthoidea	Gladius	No	Yes	Yes	No	Yes	Yes	Paired	Pelagic
Idiosepiidae	Gladius, reduced	Yes	No	No	No	Yes	?	Paired but only left functional	Benthic

(Order) **Spirulida** (= Spiruloidea)

Cretaceous (Upper) to present (83.6–0 Ma)

The placement of this group is uncertain – some recent analyses place it in the 'Teuthida' (see Nishiguchi & Mapes 2008), and it is often included in Sepiida, but we follow Bizikov (2008) in treating it as a separate higher taxon.

Calcareous shell a narrow phragmocone (apical angle 5–15 degrees), curved ventrally in open planispiral; round in cross section and possessing a ventral siphuncle and transverse septa. No nacreous layer. Pro-ostracum, if present narrow and rod-like; rostrum weak or absent.

Bizikov (2008) recognised three orders (here treated as superfamilies below).

(Superfamily) **Groenlandibeloidea** †

Cretaceous (Upper) (83.6–66 Ma)

These early members of the Spirulida had shells that lacked nacre, the rostrum was reduced to a rudiment, and the living chamber was reduced to a narrow, rod-like pro-ostracum, more like that in belemnites. Unlike belemnites, the protoconch had a caecum and a prosiphon and nacre was lacking.

The Cretaceous genera *Naefia* and *Cyrtobelus* are members of the Groenlandibelidae (Fuchs et al. 2012, 2013b).

Adygeya on which Adygeyidae is based has been previously included here but is based on an ammonite fragment (D. Fuchs pers. observ.).

Family Groenlandibelidae

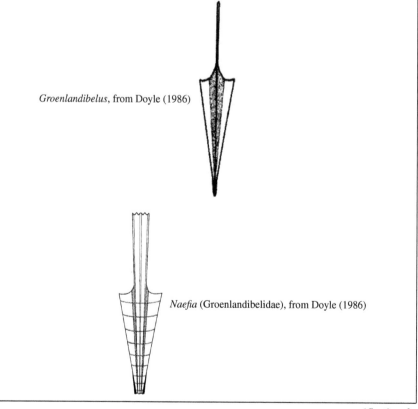

Groenlandibelus, from Doyle (1986)

Naefia (Groenlandibelidae), from Doyle (1986)

(Continued)

(Superfamily) **Belopteroidea** † Paleogene (Eocene) (56–33.9 Ma) Distinguished from groenlandibeloids in having an enrolled shell covered by a thick guard. Families Belopteridae and Belemnoseidae	*Belopterella* (?Belopteridae), from Naef (1922) *Belopterina* longitudinal section, from Dauphin (1986)
(Superfamily) **Spiruloidea** Neogene (Miocene) to present (23.03–0 Ma) *Spirula*, the only living member of this order, has some distinctive features which may or may not be characteristic of the other suborders. The animal is orientated vertically, and the eight arms have suckers in four rows while the clubs on the two tentacles have 16 rows. Both the fourth pair of arms are hectocotylised in males. The eyes lack a cornea, and the radula is absent. The fins are terminal and separate, with posterior lobes. A large photophore is at the posterior end of body. Females have accessory nidamental glands. A reduced ink sac is present but reduced but anal flaps are absent (Bizikov 2008). Families Spirulidae, Spirulirostridae †, and Spirulirostrinidae † Uncertain placement: Kostromateuthidae † (based on a shell fragment)	*Spirula* animal in life position and shell, from FA1 *Spirulirostra*, from Naef (1922)

(*Continued*)

(Order) Sepiida (= Sepioidea, Sepioidea) Cretaceous (Upper) to present (72.1–0 Ma) This group is characterised by the animal being orientated horizontally and the arms having suckers, which are usually in more than four series, with a circular muscle band and chitinous rings. The eyes are covered by a cornea. The shell is a cuttlebone (sepion), a gladius, or is absent. Accessory nidamental glands are present. Sepiida includes the cuttlefish (Sepiina) and dumpling squids (Sepiolioidea), both groups traditionally recognised as suborders.	
(Suborder) Sepiina Cretaceous (Upper) to present (72.1–0 Ma)	
(Superfamily) Vasseurioidea † Paleogene (Eocene) (56–33.9 Ma) The relationships of the two taxa in this group are obscure. Haas (2003) suggested that the Vasseuriidae were possibly ancestral to loliginids while the Belosepiellidae were ancestral sepiolids. Families Vasseuriidae and Belosepiellidae	*Belosepiella*, posterior part of shell, drawn from photograph in Haas (2003). 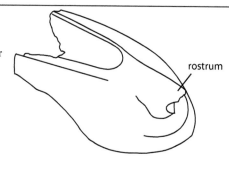 rostrum *Vasseuria*, pro-ostracum drawn from photograph of type specimen.
(Superfamily) Sepioidea Cretaceous (Upper) to present (72.1–0 Ma) The sepiids (cuttlefish) have a long, oval, dorso-ventrally flattened body. Narrow fins run along the body, and there is a substantial internal shell (sepion or cuttlebone) that lies dorsally beneath the skin and is involved in buoyancy control. There are two–four series of suckers on the arms, the ventral arms usually the longest and broadest with the left ventral arm hectocotylised in males. The tentacles are able to completely retract into pockets. *Belosepiella* has sometimes been considered a sepiolid. Families Sepiidae and Belosepiidae †	Animal of *Sepia* and shells (sepions), from FA1 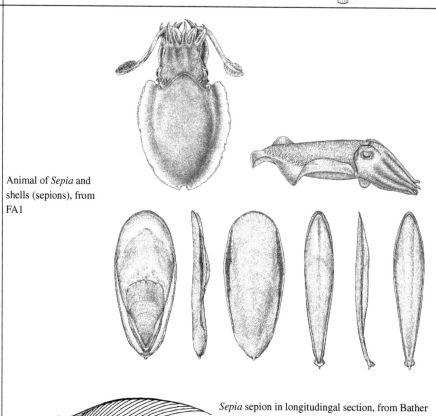  *Sepia* sepion in longitudingal section, from Bather (1911) *Belosepia* sepion in longitudingal section, from Bather (1911) 

(Continued)

(Suborder) **Sepiolina**	
Present (0 Ma)	
These small (up to 80 mm long) 'dumpling squids' have a short, rounded body and a pair of broad fins that are widely separated posteriorly. The shell is a rudimentary gladius.	
This group has variously been treated as a suborder or an order.	
(Superfamily) **Sepiolioidea** Present (0 Ma) In Sepiolidae any or all of the dorsal six arms may be hectocotylised but only the left fourth arm in Sepiadariidae. Shallow water and continental shelf. Families Sepiolidae and Sepiadariidae	*Sepiadarium*, from FA1 and *Rossia* (Sepiolidae) gladius (on right), from Bizikov (2008) 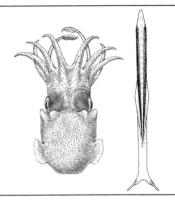

(Order) **Idiosepida**	
Present (0 Ma)	
Idiosepidans differ from sepiolioids in being smaller (up to about 20 mm long), in having only two rows of suckers on the arms, and the suckers lack circular muscles. In addition, a the gladius is vestigial, and both left and right ventral arms are hectocotylised.	
This group has variously been treated as a suborder or an order, and its relationships are uncertain.	
(Superfamily) **Idiosepioidea** Present (0 Ma) Members of the only family, Idiosepiidae (the pygmy squids), are the smallest decabrachs. They have an elongate body which is rounded posteriorly and short, separate posterior fins. The head has tentacle pockets. An attachment organ on the dorsal surface of the mantle is used to attach the animal to algae or seagrasses and is unique to this group. They are found in shallow water in the Indo-West Pacific and favour seagrass habitats. Family Idiosepiidae	*Idiosepius*, from FA1 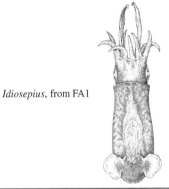

(Order) **Myopsida**
Paleogene (Eocene) to present (56–0 Ma)
Distinguished from Oegopsida by the eye being covered with a cornea. They have a well-developed gladius, and females have accessory nidamental glands, and the head has a tentacle pocket. The Eocene record is based on fossil statoliths (Neige et al. 2016).
Bizikov (2008) included this group in Sepiina, but it is treated as a separate order by Allcock et al. (2014).

(Continued)

(Superfamily) **Loliginoidea** Paleogene (Eocene) to present (56–0 Ma) Two extinct families included here by Bizikov (2008) (Teudopsidae and Palaeololiginidae) have been placed in Teudopseina (Vampyropoda) (Fuchs & Larson 2011a). Families Loliginidae (= Pickfordiateuthiidae) and Australiteuthidae	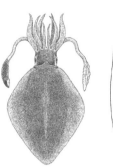 *Sepioteuthis* (Loliginidae), from FA1 and gladius, from Bizikov (2008) 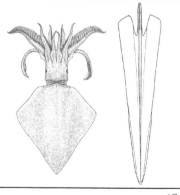 *Uroteuthis* (Loliginidae), from FA1

(Order) **Oegopsida** (= Teuthoidea in part)

Paleogene (Eocene) to present (56–33.9 Ma)

This large and very diverse grouping is certainly in need of refinement with recent analyses showing it is paraphyletic (Nishiguchi & Mapes 2008). The Middle Eocene record is based on statoliths (Neige et al. 2016).

Members of this group lack a cornea in the eye (thus the lens is in direct contact with the water), and the shell is a gladius. The buccal crown is suckerless, and this is the only decabrachian group where the head lacks tentacle pockets. The fins are usually joined posteriorly and usually lack posterior lobes. The females lack accessory nidamental glands. The funnel and head retractors are not fused where they attach to the gladius.

The relationships within this large group are uncertain. Below treat the list subgroups (orders) recognised by Bizikov (2008) on morphological grounds as superfamilies.

Uncertain placement: A superfamily Antarcticeroidea † with family Antarcticeridae from the Eocene of Antarctica were recently introduced by Doguzhaeva et al. (2017) as was a new higher taxon, Paracoleoidea. Fuchs et al. (2018) argued that this taxon was in all probability a coleoid and probably represented an early oegopsid.

(Superfamily) **Thysanoteuthoidea** new name Present (0 Ma) We introduce a superfamily name for this large squid (*Thysanoteuthis rhombus*) which is distinctive in its long fins that extend the entire length of the mantle, the rather short arms with two rows of suckers, and the tentacular clubs having four series of suckers. This taxon is the most basal member of the Oegopsida in the molecular phylogeny of Lindgren (2010). Family Thysanoteuthidae	 *Thysanoteuthis*, animal from FA1 and gladius, from Bizikov (2008)

(*Continued*)

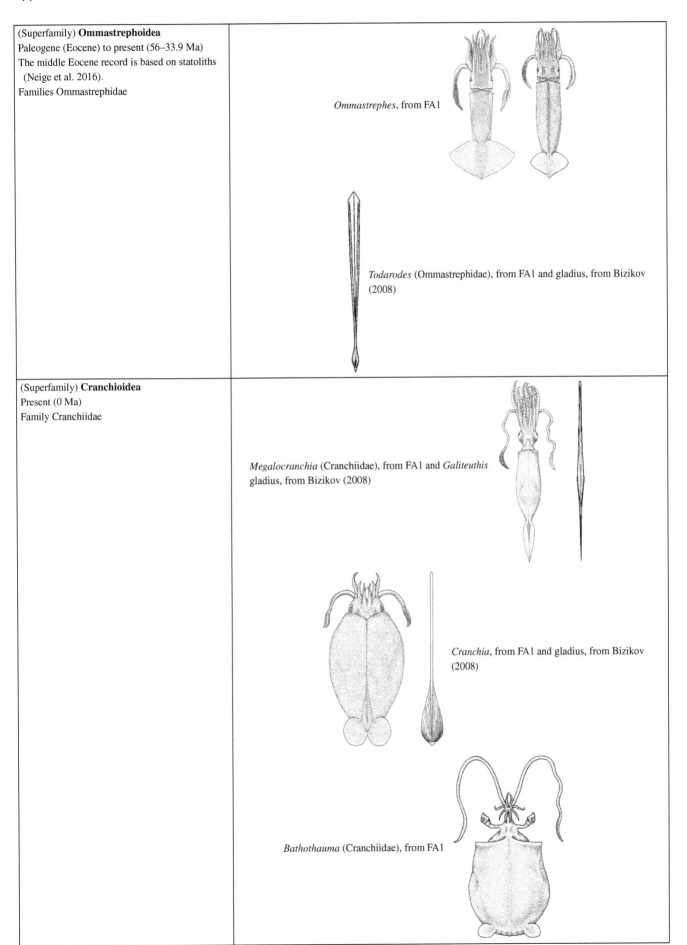

| (Superfamily) **Ommastrephoidea**
Paleogene (Eocene) to present (56–33.9 Ma)
The middle Eocene record is based on statoliths (Neige et al. 2016).
Families Ommastrephidae | *Ommastrephes*, from FA1

Todarodes (Ommastrephidae), from FA1 and gladius, from Bizikov (2008) |
| (Superfamily) **Cranchioidea**
Present (0 Ma)
Family Cranchiidae | *Megalocranchia* (Cranchiidae), from FA1 and *Galiteuthis* gladius, from Bizikov (2008)

Cranchia, from FA1 and gladius, from Bizikov (2008)

Bathothauma (Cranchiidae), from FA1 |

(*Continued*)

(Superfamily) **Onychoteuthoidea**

Neogene (Miocene) to present (23.03–0 Ma)

The Middle Miocene record is based on an arm hook (Fuchs & Donovan 2018). The fossil Eoteuthoidae has been attributed to this group, but it is based on a fragment and may not be a coleoid.

Families Onychoteuthidae (= Walvisteuthidae), Ancistrocheiridae, Architeuthidae, Brachioteuthidae, Cycloteuthidae, Enoploteuthidae, Gonatidae, Histioteuthidae, Lepidoteuthidae, Lycoteuthidae, Neoteuthidae, Octopoteuthidae, Pholidoteuthidae, Psychroteuthidae, and Pyroteuthidae

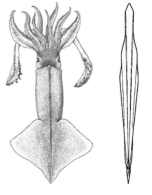

Onychoteuthis, from FA1 and *Kondakovia* gladius (Onychoteuthidae), from Bizikov (2008)

Architeuthis, from FA1

Histioteuthis, from FA1

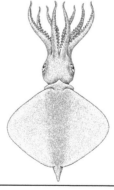

Octopoteuthis, from FA1

(Continued)

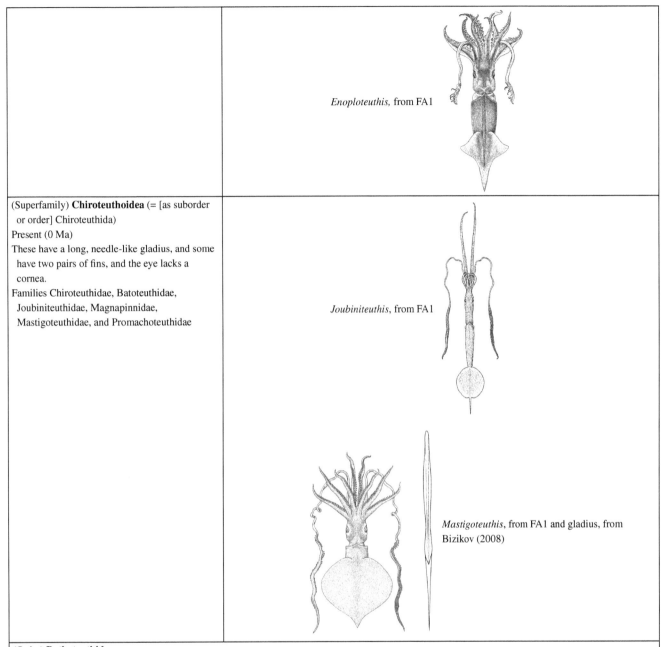

Enoploteuthis, from FA1

(Superfamily) **Chiroteuthoidea** (= [as suborder or order] Chiroteuthida) Present (0 Ma) These have a long, needle-like gladius, and some have two pairs of fins, and the eye lacks a cornea. Families Chiroteuthidae, Batoteuthidae, Joubiniteuthidae, Magnapinnidae, Mastigoteuthidae, and Promachoteuthidae	*Joubiniteuthis,* from FA1

Mastigoteuthis, from FA1 and gladius, from Bizikov (2008)

(Order) **Bathyteuthida**

Present (0 Ma)

In the molecular phylogeny of Lindgren (2010 p. 85), this group is sister to Oegopsida, and that author stated that it formed 'a well-supported clade distinct from Oegopsida, confirming its status as a separate order'.

(*Continued*)

(Superfamily) **Bathyteuthoidea** Present (0 Ma) This taxon differs from other Oegopsida in having tentacle pockets and suckers on the buccal membrane. The fins have posterior lobes. The gladius is expanded posteriorly, the females lack accessory nidamental glands, and the eyes lack a cornea. Donovan and Strugnell (2010) included the Jurassic–Cretaceous family Palaeololiginidae, but this is now included in Teudopseina (Fuchs & Larson 2011a). Families Bathyteuthidae and Ctenopterygidae

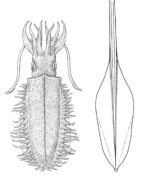

Ctenopteryx, from FA1 and gladius, from Bizikov (2008)

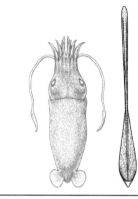

Bathyteuthis, from FA1 and gladius, from Bizikov (2008)

(Superorder) **Octobrachia** (= Octopodiformes, Vampyropoda)

Lower Triassic to present (252–0 Ma)

In Octobrachia the second pair of arms are modified or absent, and the fourth pair are unmodified. The suckers lack horny rings and have wide bases. A buccal crown is absent, and the head is fused to the mantle dorsally. An outer statocyst capsule is present, and the brain has the superior buccal lobes adjacent to or fused with the posterior buccal lobes, and an inferior frontal lobe system is present. The shell is reduced to a gladius, a pair of stylets, or to a cartilage fin support or is absent. Nidamental glands are absent, the oviducal glands are radially symmetrical, and the oesophagus usually has a crop. The nephridial coeloms are separate (Young et al. 2018).

We use the name Octobrachia in preference to Octopodiformes or Vampyropoda, following Hoffmann (2015b).

Fuchs (2009) suggested that Octopoda (i.e., Cirrata + Incirrata) may be diphyletic, based on his study of fossil taxa.

The Carboniferous fossil *Pohlsepia* is based on octopus-shaped stains and is probably not a valid record of an octobrach (D. Fuchs, pers. observ.), and Permian records are also doubtful.

(Order) **Vampyromorpha** (= Vampyroteuthoidea; Pseudoctobrachia)

Jurassic (Middle) to present (174.1–0 Ma)

All members but one are extinct.

(Suborder) **Prototeuthidina** (= Prototeuthida) †

Jurassic (Lower) to Cretaceous (Upper) (237–66 Ma)

The gladius of this group is comparatively slender, with the median field triangular, and the conus is closed ventrally. The lateral fields are shorter than half the length of the gladius. Both the median and lateral fields are reinforced, and the hyperbolar zones are reduced or absent (Fuchs et al. 2007; Schweigert & Fuchs 2012).

(Continued)

(Superfamily) **Plesioteuthoidea** new name †
Jurassic (Upper) to Cretaceous (Upper)
  (163.5–66 Ma)
Some taxa (e.g., *Plesioteuthis* and
  Senefelderiteuthis) from the Upper Jurassic of
  Solnhofen, Germany, were previously thought to
  be early 'teuthids' (e.g., Fischer & Riou 2002;
  Bizikov 2008). However, preserved remains
  show that these animals had eight arms and two
  pairs of short posterior fins (Klug et al. 2015).
  There is no evidence of cirri, hooks, or suckers
  in the impressions of *Dorateuthis* from the
  Cretaceous of Lebanon (Fuchs & Larson
  2011b), but there is evidence of cirri and radial
  suckers in *Plesioteuthis*, and there are no hooks
  (Fuchs & Hoffmann 2017).
Family Plesioteuthidae
	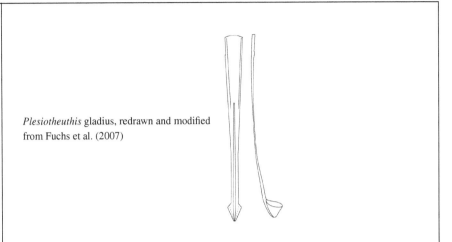
	
	Plesiotheuthis gladius, redrawn and modified
	from Fuchs et al. (2007)

(Suborder) **Loligosepiina** † Jurassic to Cretaceous (201.3–66 Ma) The gladius has a triangular median field, and the conus is cup-shaped and reduced ventrally. The hyperbolar zone is longer than half the length of the   gladius (Fuchs & Weis 2008).

(Superfamily) **Loligosepioidea** new name
Jurassic–Cretaceous (?Palogene)
The gladius of Necroteuthidae, which contains
  only *Necroteuthis* (Oligocene), appears to show
  'some loligosepiid characteristics' (Fuchs &
  Weis 2008 p. 96).
Families Loligosepiidae, Geopeltidae,
  Leptotheuthidae, Mastigophoridae, and possibly
  Necroteuthidae
	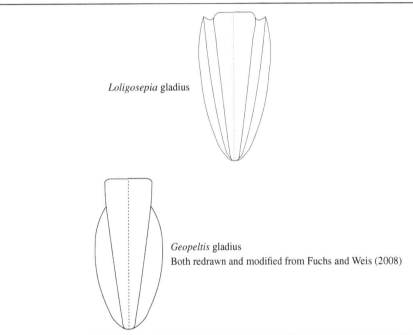
	Loligosepia gladius
	
	Geopeltis gladius
	Both redrawn and modified from Fuchs and Weis (2008)

(Suborder) **Vampyromorphina**
Present (0 Ma)
There are eight arms and a pair of thin filamentous filaments in pouches between arms i and ii. A single row of stalked suckers with chitinous rings is
  present on each arm, the dorsal mantle is joined to the head, and a pair of rounded fins is present. The internal shell (gladius) is well-developed and
  chitinous. Light organs are present in the mantle at the base of each fin and above the eyes.

(*Continued*)

(Superfamily) **Vampyroteuthoidea** Present (0 Ma) The vampire squid body is of medium-size, reaching 130 mm in mantle length. It possesses eight webbed arms and a pair of retractile filaments. They live in the deep sea in the oxygen minimum zone (600–900 m). Represented by a single species, *Vampyroteuthis infernalis*. Family Vampyroteuthidae	*Vampyroteuthis*, from FA1 and gladius, from Bizikov (2008) 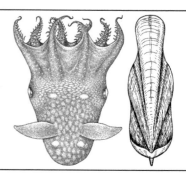

(Order) **Octopoda**
Jurassic (Lower) to present (201.3–0 Ma)
The suckers have a thin cuticular (not horny) lining and broad muscular bases. A nuchal cartilage and a funnel valve are absent, and the mantle cavity extends around the animal dorsally. The brain has the inferior frontal lobe system present and the superior buccal and posterior buccal lobes fused. The suprabrachial commissure is separate from the brain. The visceropericardial coelom is reduced, and the oviducal glands act, in part, as spermathecae and are subterminal on the oviducts. Photosensitive vesicles are located on the stellate ganglia.
Two taxa treated as suborders are recognised.

(Suborder) **Teudopseina** (= Mesoteuthina) †
Jurassic (Lower) to Cretaceous (182.7–66 Ma)
Most members of this group have a well-developed gladius, with an open spoon-shaped conus. The lateral fields and hyperbolar zones are 55% or less of the total gladius length, with the hyperbolar zones represented by broad furrows between the lateral and median fields. The anterior median field is more or less pointed (Fuchs & Weis 2009).
This group is polyphyletic in the cladistic analysis of Sutton et al. (2016), and it included *Vampyroteuthis*.

(Superfamily) **Teudopsoidea** † Jurassic (Lower) to Cretaceous (182.7–66 Ma) Families Teudopsidae, Palaeololiginidae, and Trachyteuthidae (= Actinosepiidae)	*Teudopsis* gladius, redrawn and modified from Fuchs and Weis (2010) 

(Superfamily) **Muensterelloidea** † Jurassic (Middle) to Cretaceous (Upper) (166.1–93.9 Ma) This superfamily was created by Fuchs and Schweigert (2018) for octobrachs with a short, almost patelliform gladius. Families Muensterellidae, Enchoteuthidae, and Patelloctopodidae	*Patelloctopus* gladius, redrawn and modified from Fuchs and Schweigert (2018) 

(Suborder) **Cirrata** (= Cirrina, Cirroctopoda)
Present (0 Ma)
Cirrates have fins, and the gladius, which has a cartilage-like consistency, is short and wide, being U-, V-, or saddle-shaped. Hectocotylisation is absent, but sexual dimorphism of suckers is present in some taxa. The arms have internal horizontal septa, cirri are present, and suckers are arranged in a single series. The eyes lack a cornea, and the gills lack branchial canals, and an ink sac is absent. Posterior salivary glands are often on or in the buccal mass, and the radula is reduced or absent. Anal flaps are absent. Spermatophores lack an ejaculatory apparatus, and the right oviduct is absent. The egg chorion lacks a stalk.

(Continued)

(Superfamily) **Cirroteuthoidea**
Present (0 Ma)
Families Cirroteuthidae, Cirroctopodidae,
    Opisthoteuthidae (= Grimpoteuthidae,
    Luteuthidae), and Stauroteuthidae
	Opisthoteuthis, from FA1 and gladius, from Bizikov (2008)
	
	Cirrothauma (Cirroteuthidae), from Abel (1916) and gladius, from Aldred et al. (1983)

(Suborder) **Incirrata** (= Incirrina, Octopoda s.s.)

Cretaceous (Upper) to present (100.5–0 Ma)

The incirrates are the typical octopuses and include both benthic or pelagic forms. Fins are absent, and the shell, if present, is represented by a pair of stylets or, in fossil taxa, blades. Their arms lack horizontal septa internally and cirri externally. Suckers are uniserial or biserial, and one of the two third arms forms the hectocotylus. The eyes have a cornea present, although it is reduced in some pelagic species, and the gills have branchial canals and are asymmetrical in cross-section. Posterior salivary glands are located behind the cephalic cartilage, and a radula and ink sac are usually present. Paired oviducts are present, and spermatophores are used. The chorion has a stalk.

(Continued)

(Superfamily) **Argonautoidea** Paleogene (Oligocene) to present (33.9–0 Ma) Families Argonautidae, Alloposidae, Ocythoidae, and Tremoctopodidae	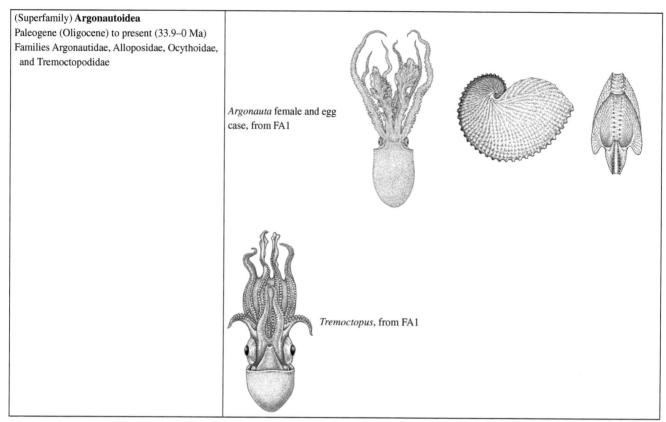 *Argonauta* female and egg case, from FA1 *Tremoctopus*, from FA1

(*Continued*)

(Superfamily) **Octopodoidea** (= Bolitaenoidea)
Cretaceous (Upper) to present (100.5–0 Ma)
The earliest known fossils of this group are
Keuppia and *Styletoctopus* from the Upper
Cretaceous (Fuchs et al. 2009).
The classification used here follows that of
Strugnell et al. (2014).
Families Octopodidae, Amphitretidae (=
Bolitaenidae, Idioctopodidae, and
Vitreledonellidae), Bathypolypodidae,
Eledonidae, Enteroctopodidae, and
Megaleledonidae.

Amphitretus, from FA1

Octopus, from FA1

Hapalochlaena (Octopodidae), from FA1

Enteroctopus gladius, from Bizikov (2008)

Eledone, from FA1

Class **Gastropoda**

Cambrian (Furongian) to present (488–0 Ma)

The torted animal has an operculum in the veliger larva and often in adults.

Note that we tentatively treat Paragastropoda as a separate class (see Chapter 13 for details).

The classification below in large part follows Bouchet et al. (2017) which should be consulted for details such as of authorship of taxa. We acknowledge the fact that our compilation has drawn very heavily on this valuable contribution and the preceding one (Bouchet & Rocroi, 2005), as well as the *Treatise* (Knight et al. 1960a; Knight et al. 1960b) and Wenz (1938–1944), without which putting together this appendix would have been vastly more difficult.

Paleozoic gastropod families of uncertain relationship

Codonocheilidae †, Craspedostomatidae †, Crassimarginatidae †, Discohelicidae †, Isospiridae †, Paraturbinidae †, Pragoserpulinidae †, Raphistomatidae †, Rhytidopilidae †, Scoliostomatidae †, Sinuopeidae †, and Yuopisthonematidae † (= Opisthonematidae)

We follow Bouchet et al. (2017) in listing the above families here. The Isospiridae was unplaced in the lower gastropods by Bouchet and Rocroi (2005) and Bouchet et al. (2017). Members of that family have also been considered to be untorted monoplacophorans by several workers, as for example, by Starobogatov and Moskalev (1987) who recognised Isospiroidea as a superfamily within Monoplacophora, with relationship to Crytonelloidea.

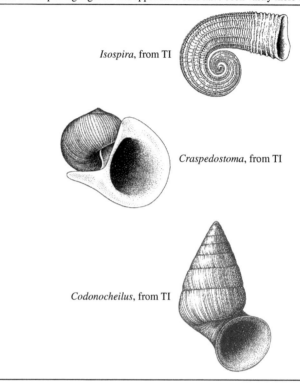

Isospira, from TI

Craspedostoma, from TI

Codonocheilus, from TI

(Subclass) **Eogastropoda**

Cambrian (Furongian) to present (488–0 Ma)

A grouping of early stem gastropods with a diversity of shell microstructures (crossed lamellar, foliated, prismatic, and nacre), protoconch morphologies, and shell morphologies. The differences between this group and the group comprising the great majority of gastropods, the orthogastropods, are discussed at length in Chapter 18.

(Infraclass) **Euomphaliformii** (= Euomphalina, Euomphalomorpha) †

Ordovician (Lower) to Triassic (Upper) (478.6–212 Ma)

This group has anisostrophically coiled shells with predominantly dextral coiling, but some members are sinistral, and some such as *Ecculiomphalus* are uncoiled. Protoconchs are loosely coiled and have a large initial chamber. The taxa included here are almost certainly a polyphyletic assemblage of probable and possible eogastropods. An ordinal-level classification for these taxa has not been established.

(*Continued*)

(Superfamily) **Euomphaloidea** † Ordovician (Lower) to Triassic (Lower) (478.6–251 Ma) Families Euomphalidae, Euomphalopteridae, Helicotomidae, Lesueurillidae, Omphalocirridae, Omphalotrochidae, and Straparollinidae	*Euomphalus*, from TI *Omphalocirrus*, from TI *Straparollina*, from TI
(Superfamily) **Macluritoidea** † Ordovician (Lower) to Triassic (Upper) (478.6–212 Ma) The operculum is calcareous and paucispiral. There are two retractor muscle scars in *Maclurites*, unknown in other genera. Family Macuritidae	*Maclurites*, from TI
(Superfamily) **Ophiletoidea** † Ordovician (Lower) to Devonian (Middle) (478.6–388 Ma) The whorls have a prominent lateral flange. Family Ophiletidae	*Ophileta*, from TI *Ecculiomphalus* (Ophiletidae), from TI *Lecanospira* (Ophiletidae), from TI
(Superfamily) **Palaeotrochoidea** † Devonian (Middle to Upper) (388–359 Ma) A poorly defined group most likely derived from euomphaliforms (Frýda et al. 2008a). Family Palaeotrochidae	*Palaeotrochus*, from TI

(Continued)

(Superfamily) **Platyceratoidea** †

Ordovician (Middle) to Jurassic (Upper) (461–161 Ma)

Platyceratoideans had calcitic shells. Early members are turbiniform
or trochiform while later taxa have expanded apertures and few
whorls. Some species developed spines.

Typical platyceratids were associates, possibly coprophages, on
crinoids and cystoids, or perhaps predators of those echinoderms
(Brett et al. 2004). Many species have been described.

Platyceratoideans have been thought to be related to
patellogastropods and neritimorphs. Evidence for an association
with patellogastropods includes shared calcitic foliated shell
structures (see Chapter 13). Alternatively, platyceratids formed a
separate clade that played no further part in gastropod evolution.

There are two types of protoconch that have been attributed to this
group, but it is not known from the type species of *Platyceras* (*P.
vetusta*, described from the Lower Carboniferous of Ireland). One
type is similar to protoconchs seen in some fossil taxa thought to be
Neritimorpha, and the other is the 'fish hook' type attributed to the
Cyrtoneritimorpha (= Orthonychioidea, see below). Bandel (1992)
found a neritimorph-like protoconch in the Triassic
Pseudorthonychia alata, the teleoconch of which resembled some
Platyceratidae, so he suggested that the Paleozoic Platyceratidae may
be neritimorphs. However other similar gastropods have 'fish-hook-
like' protoconchs and were included in the 'Cyrtoneritimorpha'.
Frýda et al. (2009) noted that the protoconchs of some Silurian and
Devonian species of platyceratids are orthostrophic and tightly
coiled. Others, such as the Lower Devonian *Praenatica* and the
Carboniferous–Permian *Orthonychia* have fish-hook protoconchs
like that of 'cyrtoneritimorphs' (i.e., Orthonychioidea) and are
probably stem members of that group. Both these latter taxa are often
treated as subgenera of *Platyceras*, but Frýda et al. (2008b) argued
that, on the basis of the protoconch morphology, the platyceratids are
diphyletic. They also argued that the development of a 'true larval
shell' also suggests that the platyceratids are not members of the
stem group of Patellogastropoda. Instead, these authors, like Bandel
(1992), favoured an ancestral relationship with neritimorphs.

Families Platyceratidae and Holopeidae

Platyceras, from TI

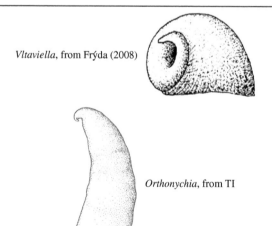

Holopea, from TI

(Superfamily) **Orthonychioidea** (= Cyrtoneritimorpha) †

Silurian (Wenlock) to Permian (Lopingian) (428.2–259.0 Ma).

Orthonychiidae have loosely coiled or limpet-like teleoconchs, and
the Vltaviellidae have naticiform to turbiniform teleoconchs (Bandel
& Frýda 1999; Frýda & Heidelberger 2003; Frýda et al. 2008b).
Their protoconchs are openly coiled and fish-hook-like. The
Carboniferous genus *Orthonychia* has an outer prismatic shell
structure with a middle calcitic semi-nacre layer and inner crossed
semi-foliate layer similar to that seen in the platyceratoideans
(Carter & Clark 1985).

Frýda (1999) thought that the Paleozoic 'Cyrtoneritimorpha' gave
rise to the modern neritimorphs ('Cycloneritimorpha') probably
during the Devonian. This idea was followed by Bandel and Frýda
(1999), who created the two orthonychioid families and later Frýda
and Heidelberger (2003) considered the Cyrtoneritimorpha to be of
'uncertain higher systematic position'. The relationships of this
group with Neritimorpha are, to say the least, questionable. We
tentatively include them here and note that this group has no
obvious relationships with other gastropod taxa except possibly with
other platyceratoideans with 'fish-hook' protoconchs.

Families Orthonychiidae and Vltaviellidae

Vltaviella, from Frýda (2008)

Orthonychia, from TI

Orthonychia protoconch, from Frýda et al. (2008b)

(Continued)

(Superfamily) **Archinacelloidea** † Cambrian (Furongian) to Triassic (Upper) (488–205.6 Ma) See Chapter 13 for discussion on this group. Families Archinacellidae, Archaeopragidae, and Metopomatidae	 *Archinocella*, from TI

(Infraclass) **Patellogastropoda** (= Docoglossa, Proteobranchiata, Onychoglossa) Mississippian (Lower) to present (358.9–0 Ma) The 'true limpets' (see Chapter 18). Unplaced family: Damilinidae †

(Order) **Patellida** Triassic (Middle) to present (247.2–0 Ma) Essentially central shell apex, nine pairs of chromosomes, calcitic foliate shell microstructures, no ctenidium but have a ring of mantle gills, three pairs of   marginal teeth (uncini), and a central lateral field with two pairs of teeth. Shell sculpture is primarily radial, often forming robust ribs.

(Superfamily) **Patelloidea** Triassic (Middle) to present (247.2–0 Ma) Primarily an intertidal and shallow subtidal (< 10 m) group. Contains the largest patellogastropods, reaching shell lengths > 35 cm. Family Patellidae	*Scutellastra* (Patellidae), from Pilsbry (1891)   *Cymbula* (Patellidae), from Pilsbry (1891) *Patella*, from Pilsbry (1891) 

(Continued)

(Order) Nacellida Cretaceous (Upper) to present (66–0 Ma) Distinguished by a reduced terminal chromosome, increased length of the radula, reduced length of the hindgut, the presence of an outer calcitic homogeneous shell layer, the reduction of the inner radular tooth field from two pairs to one pair of teeth, and reduction and increased fusion in the outer lateral teeth. Distinct segments for each tooth row and increasing development of supporting plate structures for the individual teeth across the group. A ctenidium is present or absent.	
(Suborder) Nacellina Cretaceous (Upper) to present (66–0 Ma) Ctenidium absent but secondary gills present in mantle groove.	
(Superfamily) Nacelloidea Cretaceous (Upper) to present (66–0 Ma) Shells composed primarily of calcitic foliate layers; crossed lamellar layer present only in *Cellana* below the myostracum. Family Nacellidae	*Cellana* (Nacellidae), from FA2  *Nacella*, from FA2
(Superfamily) Lepetoidea Paleogene (Oligocene) to present (23.03–0 Ma) Medium-sized (10–40 mm) limpets typically with anterior apices. A mostly deep-water group with white shells living on hard substrata in anoxic sediments, water-logged wood, vents, and at cold seeps. Radular morphology diverse given the broad array of habitats occupied by the group. Lepetoidei was established as a suborder by Golikov and Starobogatov (1989) for Lepetidae. Families Lepetidae and Pectinodontidae	*Lepeta*, from FA2   *Pectinodonta*, from FA2
(Suborder) Lottiina New suborder Mississippian (Lower) to present (358.9–0 Ma) Left ctenidium often present, secondary gills present in a few taxa.	

(Continued)

(Superfamily) **Lottioidea** (= Lepetopsina (suborder), Neolepetopsoidea, Eoacmaeoidea)

Mississippian (Lower) to present (358.9–0 Ma)

Small (5–10 mm) and medium-sized (10–40 mm) limpets typically with sub-central to anterior apices. Most basal clades (including Acmaeidae) are typically associated with calcium carbonate substrata, while the highly speciose Lottiidae occupy a great diversity of habitats from the highest intertidal to the deep sea.

Families Lottiidae, Acmaeidae, Eoacmaeidae, Neolepetopsidae, and Lepetopsidae †

Acmaea, from Pilsbry (1891)

Lottia, from FA2

Asteracmea (Lottiidae), from FA2

Patelloida (Lottiidae), from FA2

Lepetopsis, from Lesley (1889)

(*Continued*)

(Subclass) **Orthogastropoda**
Cambrian (Furongian) to present (488.3–0 Ma)
Defined by Ponder and Lindberg (1997), the orthogastropods include the remaining gastropods. The differences between this group and the Eogastropoda are discussed at length in Chapter 18.

(Infraclass) **Vetigastropoda**
Cambrian (Furongian) to present (488.3–0 Ma)
This grouping was originally used by Salvini-Plawen (1980) as including Macluritoidea, Pleurotomarioidea, Cocculinoidea, Trochoidea, and ?Murchisonioidea. Ponder and Lindberg (1997) extended the concept, using it to include Fissurelloidea, Seguenzioidea, Trochoidea, Lepetelloidea, Bellerophontoidea, Pleurotomarioidea, Haliotoidea, Scissurelloidea, and Lepetodriloidea (but not Peltospiridae, Neomphalidae, and *Melanodrymia*). With living taxa, Bouchet and Rocroi (2005) included the 'typical' vetigastropods (Fissurelloidea, Haliotoidea, Scissurelloidea, Pleurotomarioidea, Seguenzioidea, Trochoidea, Lepetelloidea) and the hot vent taxa (Lepetodriloidea and Neomphaloidea) but treated the Cocculiniformia (including only Cocculinoidea) as a group of equal rank to the vetigastropods. Bouchet et al. (2017) included in Vetigastropoda only the 'typical vetigastropods', including some hot vent groups, while Neomphaloidea and Cocculinoidea were placed in a separate 'subclass' Neomphaliones. Here we broaden Salvini-Plawen's (1980) original concept to include the hot vent groups (including Neomphalina), which were not known in 1980, thus extending the concept beyond that of Ponder and Lindberg (1997) but in agreement with that of Bouchet and Rocroi (2005), other than their exclusion of Cocculinoidea, which we include (see Chapter 18 for discussion). We exclude Macluritoidea, which we tentatively treat as eogastropods. Like Ponder and Lindberg (1997) we include bellerophontids (see Chapters 13 and 18 for justification and discussion).
We use Vetigastropoda in preference to Archaeogastropoda which has been sometimes used recently, especially in the palaeontological literature, but in different contexts. At times it includes the Eogastropoda/Patellogastropoda and the Neritimorpha (the original use of the name) or is redefined to exclude Neritimorpha and Patellogastropoda to equate with Vetigastropoda (Hickman 1988).
Our ordinal classification differs from that of Bouchet et al. (2017) and previous treatments. We base it on the topologies in recent molecular analyses (Kano 2008; Aktipis & Giribet 2010, 2012; Lee et al. 2016), as well as attempting to reflect the major morphological and molecular differences between these groups.

Paleozoic taxa of uncertain relationships	
Families Holopeidae † and Micromphalidae †	
(Superfamily) **Trochonematoidea** † Ordovician (Lower) to Cretaceous (Lower) (478.6–112.6 Ma) Families Trochonematidae † and Lophospiridae †	 *Trochonema*, from TI *Loxoplocus* (Lophospiridae), from TI
(Superfamily) **Amberleyoidea** † Triassic (Upper) to Cretaceous (Upper) (201.6–66.0 Ma) Families Amberleyidae and Nododelphinulidae	 *Amberleya*, from TI

(Continued)

(Superfamily) **Loxonematoidea** †
Ordovician (Lower) to Permian (Upper) (471.8–254.14 Ma)
Families Loxonematidae and Palaeozygopleuridae

Loxonema, from TI

Palaeozygopleura, from TI

(Superfamily) **Oriostomatoidea** †
Ordovician (Middle) to Jurassic (Lower) (449.5–183.0 Ma)
Oriostomatids have a calcareous, multispiral circular operculum
  (Yochelson & Linsley 1972), quite unlike the operculum of
  neritimorphs. They have been variously included in Trochina,
  Euomphaloidea, and Neritimorpha or as unassigned basal
  gastropods (Bouchet et al. 2017).
Families Oriostomatidae and ?Tubinidae

Oriostoma, from TI

Tubina, from TI

(Superorder) **Bellerophontia** (= Belleromorpha, Bellerophonina, Amphigastropoda, Galeroconcha [in part]) †
Cambrian (Furongian) to Triassic (Lower) (488.3–247.2 Ma)
Workers have used muscle scars to argue that bellerophontians were torted or untorted (see Chapter 13). Frýda et al. (2008a) reviewed their taxonomic
  position and considered that they were likely polyphyletic. Pending further research on this group, we tentatively treat them as a separate superorder
  within vetigastropods. Our justification for so doing is discussed in Chapter 13.
The name Belleromorpha was introduced by Naef (1911) to include Bellerophontidae, Tremanotidae, Zidoridae, and Cyrtolitidae. Bouchet and Rocroi
  (2005) used only the superfamily Bellerophontoidea for this group while Bouchet et al. (2017) included those bellerophontian taxa considered to be
  gastropods in the 'subclass' Amphigastropoda. Here we use it for all the bellerophon-like taxa which are readily recognised by having planispiral shells
  with median emargination or tremata, including some treated as monoplacophorans by Bouchet et al. (2017) or as vetigastropods (e.g., Ponder &
  Lindberg 1997). Their shell features suggest that these animals were torted and possessed a symmetrical anterior mantle cavity containing two ctenidia.

(Order) **Cyrtolitones** (= Sinuitopsida, Cyrtolitida) †
Cambrian (Furongian) to Permian (Guadalupian) (488.3–268.8 Ma)
Shell with about two whorls, and one pair of posterior muscle scars on periphery of last whorl. Starobogatov (1970), Starobogatov and Moskalev (1987),
  Peel (1991), and many others have treated this bellerophontiform group as monoplacophorans. However, the distinct ridge on the outer whorl of the shell
  appears to be an anal groove, and several taxa also have an indentation in the outer lip. These features suggest that members of the Cyrtolitones were
  probably torted.
Bouchet et al. (2017) used Cyrtolitiones as a subclass of Monoplacophora and Sinuitopsida as an order to include the superfamilies Cyrtolitoidea and
  Cyclocyrtonelloidea.

(Continued)

(Superfamily) **Cyrtolitoidea** † Cambrian (Furongian) to Devonian (Lower) (488.3–398 Ma) Families Cyrtolitidae and Carcassonnellidae	 *Cyrtolites*, from TI *Sinuitopsis* (Cyrtolitidae), from TI
(Superfamily) **Cyclocyrtonelloidea** † Cambrian (Furongian) to Permian (Guadalupian) (488.3–268.8 Ma) Families Cyclocyrtonellidae, Multifariitidae, Sinuellidae, and Sinuitinidae	*Sinuitina*, from TI
(Order) **Cyrtonellida** † Cambrian (Furongian) to Devonian (Middle) (488.3–383.7 Ma) Bouchet et al. (2017) used Cyrtonelliones (as a subclass of Monoplacophora) with order Cyrtonellida to include Cyrtonelloidea.	
Superfamily **Cyrtonelloidea** † Cambrian (Furongian) to Devonian (Middle) (488.3–383.7 Ma) Family Cyrtonellidae	*Cyrtonella*, from TI
(Order) **Bellerophontida** † Cambrian (Furongian) to Triassic (Middle) (488.3–245 Ma) Planispiral (isostrophically coiled), with sinus, slit, or in some, holes. Some loosely coiled, others limpet-like.	

(Superfamily) **Bellerophontoidea** †
Cambrian (Furongian) to Triassic (Middle) (488.3–245 Ma)
Members of this group may be stem vetigastropods (see Chapters 13
  and 18 for more details). They are also sometimes considered to be
  'monoplacophorans'.
Families Bellerophontidae, Bucanellidae, Bucaniidae, Euphemitidae,
  Pterothecidae, Sinuitidae, Tremanotidae, and Tropidodiscidae
	Bellerophon, from TI 

	Pterotheca, from TI
	
	Strepsodiscus (=
	Chalarostrepsis)
	(Tropidodiscidae),
	from TI 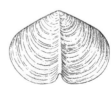 

(Superorder) Trochia Cambrian (Furongian) to present (488.3–0 Ma) This grouping comprises the 'typical' vetigastropods and differs from all other gastropods in that shells of many of its members possess nacre, which is lacking in Bellerophontia, and, unlike that group, members of Trochia are usually asymmetrically coiled.
(Order) Pleurotomariida Cambrian (Furongian) to present (488.3–0 Ma) This ancient group is characterised by dextrally coiled, nacreous shells that usually have a well-developed slit at the periphery of the last whorl. Pleurotomariidans were diverse in the Paleozoic, but the group is represented by only a single living family, Pleurotomariidae. Pleurotomariids are rather large, live on deep, hard substrata, and feed on sponges (Harasewych 2002).

(Superfamily) Eotomarioidea† Ordovician (Lower) to Jurassic (Lower) (471.8–199.3 Ma) Families Eotomariidae, Gosseletinidae, Luciellidae, Phanerotrematidae, Pseudoschizogoniidae, and Wortheniellidae	*Eotomaria*, from TI  *Ananias* (Gosseletinidae), from TI
(Superfamily) Pleurotomarioidea Ordovician (Lower) to present (471.8–0 Ma) Families Pleurotomariidae, Catantostomatidae †, Lancedellidae †, Phymatopleuridae †, Polytremariidae †, Portlockiellidae †, Rhaphischismatidae †, Stuorellidae †, Trochotomidae †, and Zygitidae †	*Bayerotrochus* (Pleurotomariidae), from FA2 
(Superfamily) Pseudophoroidea † Ordovician (Upper) to Devonian (Middle) (449.5–383.7 Ma) Shell depressed conical having a flat, concave, or weakly convex base within a surrounding frill which is sometimes spinose (Blodgett & Frýda 2003). Families Pseudophoridae and Planitrochidae	*Pseudophorus*, from TI 

(Superfamily) **Murchisonioidea** † Ordovician (Lower) to Triassic (Middle) (478.6–174.1 Ma) Families Murchisoniidae, Farewelliidae, Plethospiridae, and Ptychocaulidae	*Murchisonia*, from TI 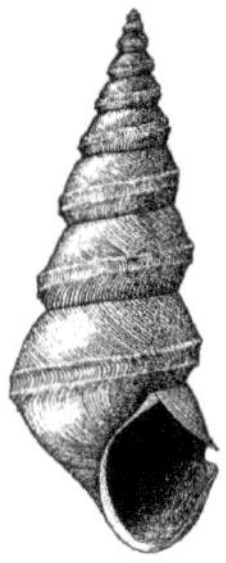 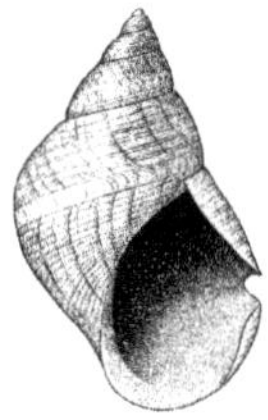 *Pithodea* (Plethospiridae), from TI
(Superfamily) **Porcellioidea** † Silurian (Wenlock) to Cretaceous (Lower) (430.5–125 Ma) Families Porcelliidae, Cirridae, and Pavlodiscidae	*Porcellia*, from TI 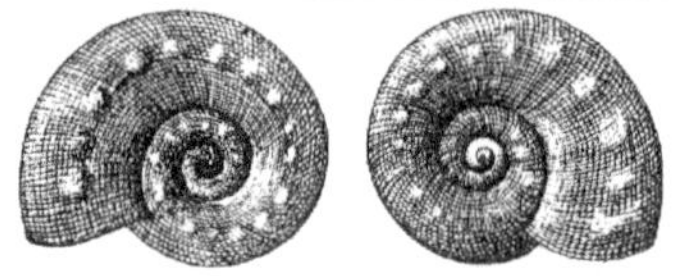
(Superfamily) **Ptychomphaloidea** † Jurassic (Lower) to Cretaceous (Lower) (428.2–182 Ma) Families Ptychomphalidae and Rhaphistomellidae	*Ptychomphalus*, from TI 
(Superfamily) **Schizogonioidea** † Triassic (Middle to Upper) (242–232 Ma) Families Schizogoniidae and Pseudowortheniellidae	*Schizogonium*, from TI 
(Superfamily) **Sinuspiroidea** † Pennsylvanian (Middle) (315.2–307 Ma) Family Sinuspiridae	*Sinuspira*, from TI 

(Order) **Haliotida**

Triassic (Middle) to present (247.2–0 Ma)

Members of the Haliotidae are distinctive in having nacreous shells with a row of small openings around the whorl angulation. There are two ctenidia with the right the largest, and epipodial tentacles and epipodial sense organs are both present. The aperture is widely expanded and the spire reduced.

(Continued)

(Superfamily) **Haliotoidea**	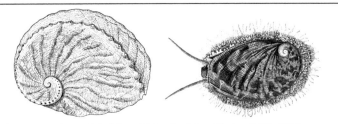
Triassic (Middle) to present (247.2–0 Ma)	
The relationships within haliotids have been recently investigated (Estes et al. 2005; Degnan et al. 2006; Streit et al. 2006).	
The family Temnotropidae has a nacreous shell (Bandel 1991) and is tentatively included here following Bouchet and Rocroi (2005). However, their broadly turbiniform shell has a simple slit and selenizone suggesting that they may actually be pleurotomariidans.	
Families Haliotidae and ?Temnotropidae †	*Haliotis*, upper figure of shell from FA2, animal from Fischer (1880–1887)

| (Order) **Seguenziida** |
| Cretaceous (Lower) to present (145.5–0 Ma) |
| The concept of Seguenzioidea was extended by Kano (2008) to include several taxa previously included in either Trochidae or Trochoidea. Given the inclusion of these taxa, there are no clear morphological synapomorphies for this grouping which is, however, well characterised in molecular phylogenies. |

(Superfamily) **Seguenzioidea**	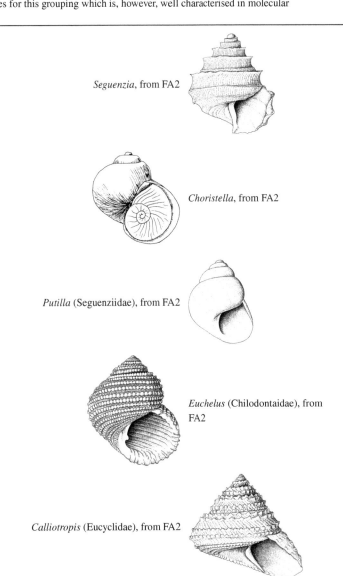
Cretaceous (Lower) to present (145.5–0 Ma)	
Families Seguenziidae, Cataegidae, Chilodontaidae, Choristellidae, Eucyclidae, Eudaroniidae, Pendromidae, Trochaclididae, Eucycloscalidae †, Eunemopsidae †, Lanascalidae †, Laubellidae †, Pseudoturcicidae †, and Sabrinellidae †	*Seguenzia*, from FA2
	Choristella, from FA2
	Putilla (Seguenziidae), from FA2
	Euchelus (Chilodontaidae), from FA2 
	Calliotropis (Eucyclidae), from FA2

(*Continued*)

(Order) **Lepetellida**

Neogene (Miocene) to present (23.03–0 Ma)

Bouchet et al. (2017) used Lepetellida to include Lepetelloidea, Fissurelloidea, Haliotoidea, Lepetodriloidea, and Scissurelloidea while admitting the likely paraphyly of this grouping. We restrict this order to Lepetelloidea alone.

(Superfamily) **Lepetelloidea** Neogene (Miocene) to present (23.03–0 Ma) Families Lepetellidae, Addisoniidae, Bathyphytophilidae, Caymanabyssiidae, Cocculinellidae, Osteopeltidae, Pseudocucculinidae, and Pyropeltidae	*Cocculinella*, from FA2  *Caymanabyssia*, from FA2  *Addisonia*, from FA2

(Order) **Fissurellida**

Pennsylvanian (Upper) to present (302.2–0 Ma)

(Suborder) **Fissurellina**

Pennsylvanian (Upper) to present (302.2–0 Ma)

The fissurellids are the only family recognised in this taxon. These externally bilaterally symmetrical gastropods have two equal-sized ctenidia, and the anus is located in the middle of the anterior mantle cavity, where it opens to either a small opening in the dorsal part of the shell, a slit, or more rarely, an internal groove in the shell. Epipodial tentacles are lacking, but epipodial sense organs are present.

(Superfamily) **Fissurelloidea** Pennsylvanian (Upper) to present (302.2–0 Ma) The phylogenetic relationships of the family and the taxa within it were investigated by Aktipis et al. (2011). Family Fissurellidae	*Macroschisma* (Fissurellidae) from FA2 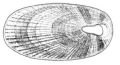  *Amblychilepas* (Fissurellidae), from Hedley (1916) *Emarginula* (Fissurellidae), left and middle figures from TI, right figure from Fretter and Graham (1962) 

(Suborder) **Scissurellina**

Triassic (Upper) to present (235–0 Ma)

(Superfamily) **Lepetodriloidea** Present (0 Ma) Families Lepetodrilidae (= Clypeosectidae) and Sutilizonidae	*Lepetodrilus*, shells original; animal from Fretter (1988) 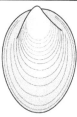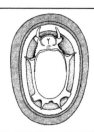

(*Continued*)

(Superfamily) **Scissurelloidea** Triassic (Upper) to present (235–0 Ma) Families Scissurellidae, Anatomidae, and Larocheidae	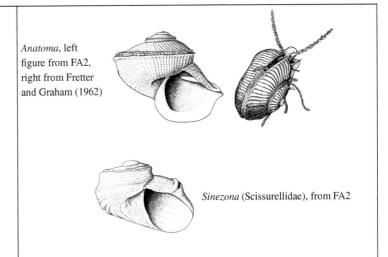 *Anatoma*, left figure from FA2, right from Fretter and Graham (1962) *Sinezona* (Scissurellidae), from FA2

(Order) **Trochida**	
Devonian (Middle) to present (388.1–0 Ma)	
As currently recognised this grouping includes a single superfamily, members of which have a single bipectinate ctenidium, an entire aperture, and well-developed epipodial tentacles and epipodial sense organs.	

(Superfamily) **Trochoidea** (= Turbinoidea, Phasianelloidea, Angarioidea) Devonian (Middle) to present (388.1–0 Ma) Several recently recognised superfamilies based largely on molecular work (Williams et al. 2008) were combined into Trochoidea by Bouchet et al. (2017) based on the most recent molecular study (Uribe et al. 2017). Families Trochidae, Angariidae, Areneidae, Calliostomatidae, Colloniidae, Conradiidae, Liotiidae, Margaritidae, Phasianellidae, Skeneidae, Solariellidae, Tegulidae, Turbinidae, Anomphalidae †, Araeonematidae †, Elasmonematidae †, Epulotrochidae †, Eucochlidae †, Metriomphalidae †, Microdomatidae †, Nododelphinulidae †, Proconulidae †, Sclarotrardidae †, Tychobraheidae †, and Velainellidae †	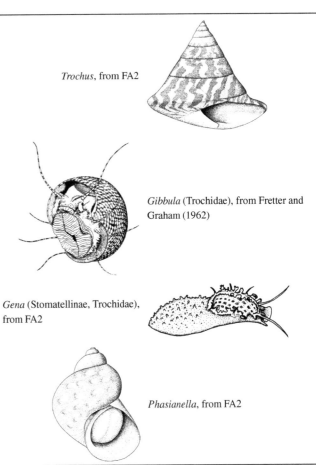 *Trochus*, from FA2 *Gibbula* (Trochidae), from Fretter and Graham (1962) *Gena* (Stomatellinae, Trochidae), from FA2 *Phasianella*, from FA2

(Continued)

(Order) **Neomphalida**	
Cretaceous (Lower) to present (133–0 Ma)	
Bouchet, et al. (2017) recognised Neomphaliones as a subclass. In contrast, we include this group within Vetigastropoda for the reasons discussed in Chapter 18. This grouping includes the Neomphalina ('hot vent' taxa) and Cocculinina.	

(Suborder) **Neomphalina**	
Cretaceous (Lower) to present (133–0 Ma)	
Members of this group contain both limpet-shaped and coiled taxa. They have a single, large bipectinate ctenidium (see Chapter 18 for more details).	

(Superfamily) **Neomphaloidea**[2]	
Cretaceous (Lower) to present (133–0 Ma) This group comprises the taxa previously known informally as the 'hot-vent taxa'. Families Neomphalidae, Melanodrymiidae, and Peltospiridae	 *Neomphalus*, modified from photograph by Rudo von Cosel  *Pachydermia* (Peltospiridae), original  *Peltospira*, original; animal from Fretter (1989)

(Suborder) **Cocculinina**	
Cretaceous (Lower) to present (112.6–0 Ma)	
The members of this group are all small white limpets that live on biogenic substrata. The gill is much modified as a single fold.	
Cocculinines are convergent with lepetelloideans, and the two groups were combined as cocculiniforms in earlier classifications (e.g., Haszprunar 1988b, 1998).	

(Superfamily) **Cocculinoidea**	
Cretaceous (Lower) to present (112.6–0 Ma) Families Cocculinidae, and Bathysciadiidae	 *Coccopigya* (Cocculinidae), from FA2  *Bathypelta* (Bathysciadiidae), from FA2

---

[2] As noted above, Frýda et al. (2009) considered Platyceratoidea to be diphyletic, containing members with uncoiled or with tighly coiled larval shells. We include platyceratids in the eogastropods.

(Infraclass) **Neritimorpha** (= Neritopsina, Neritomorpha, Cycloneritimorpha) Silurian (Lower) to present (440.8–0 Ma) Neritimorphs differ from vetigastropods in having a complex pallial genital system, a thickened paucispiral operculum, and a single (left) kidney. They differ from both vetigastropods and caenogastropods in possessing a bipectinate ctenidium lacking skeletal rods. Recent classifications have divided neritimorphs into Cyrtoneritimorpha and Cycloneritimorpha. We treat the former group as a synonym of Orthonychioidea, and thus the cycloneritimorphs are equivalent to Neritimorpha. Frýda et al. (2008b, 2009) argued that modern neritimorphs were either derived from Paleozoic 'platyceratids' with tightly coiled protoconchs, and possibly representing the oldest neritimorphs, or they evolved from the naticopsids. He pointed out that the strongly convolute protoconch that is an apomorphy of the crown-group neritimorphs probably originated after the Permian/Triassic mass extinction and before the Late Triassic.	
Possible neritimorphs of uncertain relationships	
(Superfamily) **Symmetrocapuloidea** † Jurassic (Middle) to Cretaceous (Lower) (164.7–112.6 Ma) The protoconch of this limpet-shaped taxon apparently consists of more than one whorl based on observations by Gründel (1998), who included the family in the patellogastropods following earlier classifications. Bouchet and Rocroi (2005), Bouchet et al. (2017) included this taxon in the neritimorphs. Family Symmetrocapulidae	*Symmetrocapulus*, from TI
(Superfamily) **Nerrhenoidea** † Devonian (Middle) (393.3–382.7 Ma) Described by Bandel and Heidelberger (2001), the internal whorls of the shell are not resorbed as they are in most other neritimorphs. The protoconch is low and of about two spirals, and the operculum is calcareous and paucispiral (Bandel & Heidelberger 2001). This group was included in the groups of uncertain position in Neritimorpha by Bouchet et al. (2017). Family Nerrhenidae	*Hessonia* (Nerrhenidae), redrawn and modified from Bandel (2008)
(Superfamily) **Dawsonelloidea** † Pennsylvanian (Middle) (315.2–307 Ma) These two families are considered possible sister taxa of the helicinids by Kano et al. (2002) but are much earlier than the first fossils of that family (see Chapter 18). Families Dawsonellidae and Deaniridae †	*Dawsonella*, from TI
(Superfamily) **Naticopsoidea** † Silurian (Wenlock) to Triassic (Early) (430.5–251 Ma) Families Naticopsidae, Scalaneritinidae, Trachyspiridae, and Tricolnaticopsidae	*Naticopsis*, from TI
(Order) **Neritopsida** We use this name in preference to Cycloneritimorpha.	
(Suborder) **Neritopsina** Silurian (Llandovery) to present (440.8–0 Ma) Neritopsines, like Nerrhenoidea, do not reabsorb the internal shell whorls, and the radula lacks central and inner lateral teeth.	

(Continued)

(Superfamily) **Neritopsoidea** (= Titiscanoidea) Silurian (Llandovery) to present (440.8–0 Ma) Titiscaniidae is slug-like, having lost the shell. We retain this taxon as a family separate from Neritopsidae with which it was synonymised by Bouchet et al. (2017). Families Neritopsidae, Titiscaniidae, Delphinulopsidae †, Fedaiellidae †, Palaeonaricidae †, Plagiothyridae †, and Pseudorthonychiidae †	*Neritopsis*, from FA2 *Titiscania*, from FA2
(Suborder) **Hydrocenina** new name Mesozoic to present (251.9–0 Ma) This terrestrial group was treated as a suborder by Golikov and Starobogatov (1989) and an order by Bandel (1992).	
(Superfamily) **Hydrocenoidea** Mesozoic to present (251.9–0 Ma) Family Hydrocenidae	*Hydrocena*, from FA2
(Suborder) **Helicinina** new name Silurian (Llandovery) to present (443.7–0 Ma) Bouchet et al. (2017 p. 373) included Neritiliidae in Helicinoidea but noted that 'the monophyletic nature of Helicinidae and Neritiliidae remains uncertain with insignificant nodal support in phylogenetic reconstructions (Kano, unpublished)'. Given the published topology in Kano et al. (2002), and the considerable morphological differences between the Neritiliidae and the families in Helicinoidea, we prefer to treat them as two separate superfamilies.	
(Superfamily) **Neritilioidea** new name Neogene (Miocene) to present (23.3–0 Ma) Marine and freshwater Family Neritiliidae	*Neritilia*, original
(Superfamily) **Helicinoidea** Cretaceous (Upper) to present (100.5–0 Ma) Cretaceous to recent. Terrestrial, ctenidium absent; respiration in the vascularised mantle cavity. This group is diverse in many tropical and subtropical environments. Families Helicinidae, Proserpinellidae, and Proserpinidae	*Helicina*, from FA2

(*Continued*)

(Suborder) **Neritina** new name

Silurian (Llandovery) to present (443.7–0 Ma)

(Superfamily) **Neritoidea**

Silurian (Llandovery) to present (443.7–0 Ma)

Phenacolepadids are shallow and deep-water marine limpets, but their anatomy is very similar to neritids which range from marine to freshwater and semiterrestrial.

Families Neritidae, Phenacolepadidae (= Shinkailepadidae), Cortinellidae †, Neridomidae †, Neritariidae †, Otostomidae †, Parvulatopsidae †, and Pileolidae †

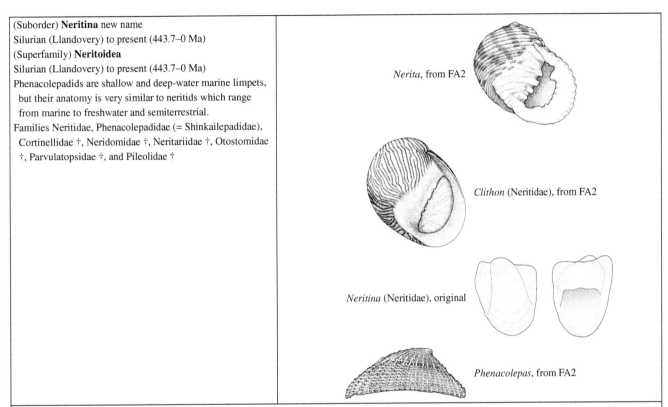

Nerita, from FA2

Clithon (Neritidae), from FA2

Neritina (Neritidae), original

Phenacolepas, from FA2

(Infraclass) **Caenogastropoda**

Ordovician (Lower) to present (478.6–0 Ma)

Caenogastropods are mostly operculate and differ from other gastropods in having only the left kidney, a single (left) hypobranchial gland, the left ctenidium with a single row of leaflets, and usually a well-developed ctenidium with the filaments containing skeletal rods. Sexes are nearly always separate, and fertilisation is internal.

(Continued)

Early, basal caenogastropod taxa of uncertain relationship
Not allocated to superfamily

The following families are 'unassigned to superfamily', in Bouchet et al. (2017).

Gigantocapulidae, a Late Cretaceous limpet-shaped taxon, was considered as possibly related to Vanikoridae by Beu (2007). A probable Early Cretaceous member of this latter family, *Brunonia* (Brunoniidae), was placed in the heteropod family Carinariidae by Kase (1988). Brunoniidae was suggested to be an earlier name for Gigantocapulidae by Beu (2007).

Globocornidae, Acanthonematidae[3] †, Ampezzanildidae[4] †, Brunoniidae (?=Gigantocapulidae) †, Coelostylinidae †, Kittlidiscidae[5] †, Plicatusidae †, Pragoscutulidae †, Spanionematidae †, and Spirostylidae †

Austroscutula (Pragoscutulidae), modified from Cook et al. (2008)

Brunonia, from TI

Globocornus, redrawn and modified from Espinosa and Ortea (2009)

Spanionema, from TI

(Superfamily) **Soleniscoidea** †

Silurian (Llandovery) to Triassic (Upper) (426.2–237 Ma)

According to Bandel (2002), the smooth protoconch is tightly coiled with an indistinct transition to the teleoconch which is also smooth.

Families Anozygidae, Meekospiridae, and Soleniscidae

Meekospira, from TI

Soleniscus, from TI

(Continued)

---

[3] Probably not a caenogastropod according to Bandel (2002).
[4] Included in Mathildoidea until Nützel and Kaim (2014) removed it to Caenogastropoda.
[5] Previously thought to be a pleurotomarioidean, this family is based on a caenogastropod according to Bandel (2009).

(Superfamily) **Subulitoidea** †
Ordovician (Lower) to Permian (Lopingian) (478.6–254 Ma)
The 'subulitoid' gastropods were previously considered to be
 a much larger group that has been shown to be polyphyletic
 (e.g., see Ponder et al. 2008). Nützel et al. (2000) concluded
 that most genera and families in this group became extinct at
 the end of the Permian. The earliest genus *Eroicaspira*
 ranged through the Ordovician, and three other genera also
 arose later in the Ordovician. However, the protoconchs of
 these taxa have not been studied to date. Their shells also
 differ in possessing a distinct anterior apertural notch that is
 not twisted.
Families Subulitidae and Ischnoptygmatidae

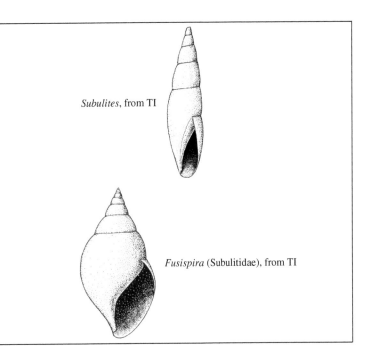

Subulites, from TI

Fusispira (Subulitidae), from TI

(Order) **Perunelomorpha** † Devonian (Lower) to Permian (Lopingian) (412–254 Ma) This grouping was established by Frýda (1999). Members are sometimes included in Caenogastropoda (Frýda et al. 2008a), or they are regarded as stem apogastropods (Frýda 1999). Bouchet et al. (2017 p. 373) treated this group as caenogastropods of uncertain position. The protoconch is smooth and openly coiled, and the teleoconch may be smooth or sculptured. Nützel et al. (2000) included this group in the Subulitoidea. This group and Soleniscoidea comprised Bandel's (2002) Procaenogastropoda.	
(Superfamily) **Peruneloidea** † Devonian (Lower) to Permian (Lopingian) (412–254 Ma) Families Chuchlinidae, Imoglobidae, Perunelidae, and Sphaerodomidae	 *Chuchlina* (Perunelidae), redrawn from Frýda et al. (2008a) and Frýda (2012)

(Paraphyletic group) **Architaenioglossa** Pennsylvanian (Lower) to present (318.1–0 Ma) We modify the concept of Haller's (1892) Architaenioglossa used by Thiele (1929–1935) so that it includes Campaniloidea. Thiele's concept is equivalent to Cyclophoroidei of Starobogatov and Sitnikova (1983). This grouping is usually considered to be a paraphyletic cluster of three non-marine superfamilies at the base of the Caenogastropoda. All have a taenioglossate radula and share a plesiomorphic eusperm morphology with Cerithiimorpha. All architaenioglossans lack Si4 cells in the osphradium and have a simple nervous system with a long cerebral commissure, and it is hypoathroid in some taxa.	
(Order) **Cyclophorida** (= Procyclophoroidea) Pennsylvanian (Lower) to present (318.1–0 Ma) Starobogatov and Sitnikova (1983) established Cyclophoroidei as a suborder containing Cyclophoroidea, Piloidea, and Aciculoidea. We redefine it as an order containing including Cyclophoroidea (= Aciculoidea) only, excluding Piloidea (= Ampullarioidea). Dendropupoidea is also tentatively included here, a group which Bandel (2000) placed in his order Procyclophoroida (see Bouchet et al. 2017 for details).	
(Superfamily) **Dendropupoidea** † Pennsylvanian (Lower to Middle) (318.1–314.6 Ma) This group may possibly be ancestral to Cyclophoroidea and was probably living in non-marine habitats (Bandel 2002). Families Anthracopupidae and Dendropupidae	 *Dendropupa*, from TI *Anthracopupa*, from TI

(Continued)

(Superfamily) **Cyclophoroidea**

Cretaceous (Upper) to present (70.6–0 Ma)

Cyclophoroideans possess a lung lacking a gill, most lack an osphradium, and all are terrestrial.

Families Cyclophoridae, Aciculidae, Craspedopomatidae, Diplommatinidae, Maizaniidae, Megalomastomatidae, Neocyclotidae, Pupinidae, and Ferussinidae †

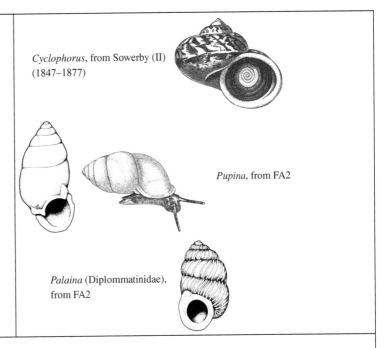

Cyclophorus, from Sowerby (II) (1847–1877)

Pupina, from FA2

Palaina (Diplommatinidae), from FA2

(Order) **Viviparida**

Triassic (Upper) to present (205.6–0 Ma)

Hydrogastropoda was a higher grouping named by Simone (2011) based on his morphological analysis. It included Ampullarioidea + his Epiathroidea (in which he included Viviparidae) and all sorbeoconch caenogastropods.

Viviparids share a unique osphradial morphology with ampullariids.

This clade was supported in the Colgan et al. (2007) molecular analysis.

(Superfamily) **Viviparoidea**

Triassic (Upper) to present (205.6–0 Ma)

These dextrally coiled, globose to conical snails all live in fresh water. They are suspension feeders with well-developed mantle cavity organs but also graze on surface films. The right cephalic tentacle is modified as a penis in males.

Families Viviparidae and Pliopholygidae (? = Amuropaludinidae)

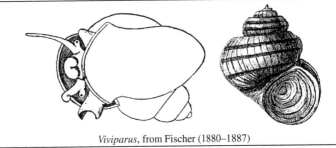

Viviparus, from Fischer (1880–1887)

(Order) **Ampullariida**

Triassic (Middle) to present (242–0 Ma)

The shell is typically globular with a simple aperture. The nervous system is hypoathroid, and the snout is very short and bilobed. The foot is broad, and the cephalic tentacles are long and slender. The eyes are on peduncles. See Chapter 19 for discussion of this taxon.

(Continued)

(Superfamily) **Ampullarioidea**

Jurassic (Middle) to present (167.7–0 Ma)

The apple snails all live in fresh water, and many are amphibious.
  The shell is usually dextral, rarely sinistral and they have a lung as
  a separate sac off the otherwise normal mantle cavity. They are
  usually herbivores but can also be omnivores.

Family Ampullariidae (= Pilidae)

Ampullaria, from Fischer
(1880–1887)

Pila (Ampullariidae), from Fischer
(1880–1887)

(Superfamily) **Ampullinoidea**

Triassic (Middle) to present (242–0 Ma)

This marine group is represented by a single living species, *Cernina
  fluctuata*.

Families Ampullinidae, Gyrodidae †, and Tylostomatidae †

Ampullina, original

Cernina (Ampullinidae), original

(Order) **Campanilida**

Jurassic (Middle) to present (164.7–0 Ma)

The shell is typically elongate, dxtral, with a siphonate aperture. The nervous system is epiathroid, and the snout is short. The foot is not very broad, and
  the cephalic tentacles have the sessile eyes on their outer bases.

This grouping was established as a suborder (as Campanilimorpha) for *Campanile* by Haszprunar (1988a). We have modified the classification and
  assignment of families by Bouchet et al. (2017 p. 373) by separating Ampullinoidea from Campanilida. See Chapter 19 for discussion.

(Superfamily) **Campaniloidea**

Jurassic (Middle) to present (164.7–0 Ma)

Families Campanilidae, Plesiotrochidae, Diozoptyxidae †,
  Metacerithiidae †, Settsassiidae †, Trypanaxidae †, and
  Vernediidae †

Campanile, from FA2

Plesiotrochus, from FA2

(*Continued*)

(Cohort) **Sorbeoconcha** Mississippian (Middle) to present (336–0 Ma) Includes the remaining caenogastropods. We here modify the definition of Sorbeoconcha to include all caenogastropods other than the Viviparoidea, Cyclophoroidea, Campaniloidea, and Ampullarioidea. Epiathroidea was introduced by Simone (2011) for Sorbeoconcha + Viviparidae which included taxa with an epiathroid central nervous system.	
Superfamilies of uncertain relationship	
(Superfamily) **Acteoninoidea** † Mississippian (Middle) to Cretaceous (Upper) (336–66 Ma) Family Acteoninidae	*Acteonina*, from TI
(Superfamily) **Palaeostyloidea** † Mississippian (Middle) to Permian (Guadalupian) (345.3–265 Ma) Family Palaeostylidae	*Palaeostylus*, from TI
(Superfamily) **Pseudozygopleuroidea** † (= Zygopleuroidea) Mississippian (Upper) to Paleogene (Eocene) (326.4–48.6 Ma) The Protorculidae were considered by Nützel (1997) to be a sister to Triphoroidea and Epitonioidea while Kaim (2004) suggested affinities with Rissooidea. Families Polygyrinidae, Pommerozygiidae, Protorculidae, Pseudozygopleuridae, and Zygopleuridae	*Zygopleura*, from TI *Pseudozygopleura*, modified from Nützel (1997)

(Continued)

(Megaorder) **Cerithiimorpha** (= Cerithimorpha, Cerithiomorpha, Cerithiiformes)

Devonian (Middle) to present (391.9–0 Ma)

Cerithiimorphs all possess a taenioglossate radula, and most have an elongate shell, with or without a siphonal canal. They lack a penis, and the pallial gonoducts are usually open.

As used here, Cerithiimorpha differs from the broader grouping originally envisaged by Golikov and Starobogatov (1975). Bandel (2006) further restricted the grouping to include Cerithioidea with the exclusion of Turritellidae, Siliquariidae, and Styliferinidae which he included in a separate group, Turritellimorpha, families usually included in Cerithioidea (Lydeard et al. 2002; Bouchet & Rocroi 2005; Strong et al. 2011), as they are here.

(Order) **Cerithiida**

Devonian (Middle) to present (391.9–0 Ma)

Cerithiidans have an epiathroid nervous system and a rather long snout. Most are surface scrapers, some are deposit feeders (e.g., Potamididae), while Turritellidae and Siliquariidae are suspension feeders.

In the molecular analysis of Takano and Kano (2014), Pickworthiidae, previously classified in Littorinoidea, were shown to belong in this grouping.

(Superfamily) **Orthonematoidea** †

Devonian (Middle) to Triassic (Upper) (391.9–232 Ma)

These mainly high-spired taxa have a slit or sinus in the outer lip of the aperture (Goniasmatidae), but this is absent in Orthonematidae, with intermediate situations found (Bandel 2002). The protoconch is almost smooth, heliciform in shape and of up to three rounded whorls (Bandel 2002).

Bouchet et al. (2017 p. 373) placed orthonematoids in *incertae sedis* under Sorbeoconcha, but we follow Nützel and Bandel (2000) by including these taxa in Cerithiimorpha. In contrast, Bandel (2006) considered members of the Orthonematidae to be possible ancestors of Campaniloidea.

Families Orthonematidae and Goniasmatidae

Orthonema, from TI

Goniasma, from TI

(*Continued*)

(Superfamily) **Cerithioidea** (= Turritelloidei, Turritellimorpha)
Pennsylvanian (Middle) to present (309–0 Ma)
Families – *marine* – Cerithiidae, Batillariidae, Dialidae,
  Diastomatidae, Litiopidae (= Styliferinidae), Modulidae,
  Pelycidiidae, Planaxidae (= Fossaridae), Pickworthiidae,
  Potamididae, Scaliolidae, Siliquariidae, Turritellidae; *fresh water*
  – Hemisinidae, Melanopsidae (? = Zemelanopsidae),
  Pachychilidae, Paludomidae, Pleuroceridae, Semisulcospiridae,
  Thiaridae, Brachytrematidae †, Canterburyellidae †, Cassiopidae
  †, Cryptaulacidae †, Eustomatidae †, Juramelanatriidae †,
  Ladinulidae †, Lanascalidae †, Lucmeriidae †, Maoraxidae †,
  Popenellidae †, Probittiidae †, Procerithiidae †, Prostyliferidae †,
  Propupaspiridae †, Prisciphoridae (= Prisciophoridae) †,
  Terebrellidae †, and Zardinellopsidae †.[6]
Note: The extinct Omalaxidae is treated as a subfamily of
  Turritellidae by Lozouet (2012).

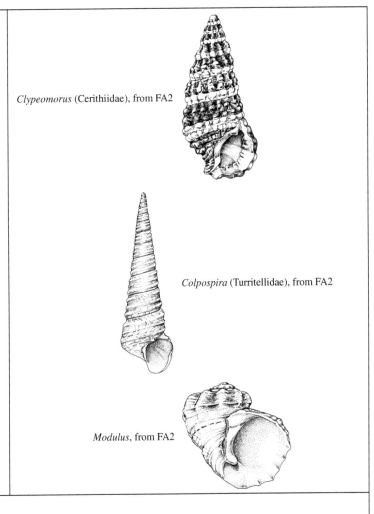

Clypeomorus (Cerithiidae), from FA2

Colpospira (Turritellidae), from FA2

Modulus, from FA2

(Megaorder) **Hypsogastropoda**
Silurian (Ludlow) to present (426.2–0 Ma)
This grouping contains the remaining caenogastropods. The detailed relationships within this group, which includes many of the old 'mesogastropods' and
  the neogastropods, remain largely unresolved. Most members possess a taenioglossan radula, but some have modified types including ptenoglossan and,
  in the neogastropods, stenoglossan radulae. Most have a snout, and, in the carnivorous groups, the snout is introverted to form a proboscis.
The classification below includes a few unranked taxa.

Asiphonate group
This grouping is the 'asiphonate group' of Colgan et al. (2007) and Ponder et al. (2008). It is equivalent in rank to superorder and will be formally named
  in another publication. This grouping is not equivalent to Littorinata named by Pchelintsev (1963) and including Littorinoidea, Rissooidea, and
  Calyptraeoidea, as it excludes the latter group. The name was emended to Littorinimorpha by Starobogatov and Sitnikova (1983). Bouchet et al. (2017 p.
  376) noted that 'molecular phylogenies have rejected the monophyly of Littorinimorpha and of Ptenoglossa (Takano & Kano, 2014) as construed in the
  classification of Bouchet and Rocroi (2005). The Latrogastropoda ('siphonate clade' of Ponder et al., 2008) are monophyletic, leaving the rest of the
  Hypsogastropoda paraphyletic or unresolved, with the exception of (Rissooidea + Truncatelloidea + Vanikoroidea) which form a monophyletic group'.

(Order) **Littorinida**
Devonian (Middle) to present (388.1–0 Ma)
Various higher taxa based on the name *Littorina* have been proposed including Littorinata, Littorinimorpha, and Littorinoidei, but they comprise different
  assortments of taxa (see Bouchet et al. 2017 for details) to the grouping used here.

(Continued)

---

[6] Several of these extinct families were placed in Cerithiimorpha 'taxa of uncertain position' in Bouchet et al. (2017).

Superfamilies of uncertain relationship

(Superfamily) **Cingulopsoidea**

Paleogene (Oligocene) to present (28.4–0 Ma)

These small-sized families include two surface film feeders and
Rastodentidae which are thought to feed on bryozoans. Most
cingulopsids and all rastodentids have a highly modified
taenioglossan radula – lost in the cingulopsid genus *Tubbreva.*

The superfamily was shown to be monophyletic in the molecular
analysis of Criscione and Ponder (2013), but the inclusion of more
taxa may well show that this is not the case.

Eatoniellidae is allied to triphoroideans in the molecular analysis of
Takano and Kano (2014).

Families Cingulopsidae, Eatoniellidae, and Rastodentidae

Cingulopsis, from Fretter and Graham
(1962)

Tubbreva (Cingulopsidae), from Ponder and
Yoo (1980)

Eatoniella, from Ponder and Yoo
(1978)

(Superfamily) **Abyssochrysoidea**

Jurassic (Upper) to present (145.5–0 Ma)

Both living families are found in the deep sea with the Provannidae
and the extinct Hokkaidoconchidae and Paskentanidae obligate hot
vent or seep taxa.

The two living families were included in Zygopleuroidea by
Bouchet and Rocroi (2005) but were separated as
Abyssochrysoidea by Kaim et al. (2008).

Both recent groups have a taenioglossate radula and are assumed to
feed on surface deposits. At least one provannid has symbiotic
autotrophic bacteria associated with it.

The placement of this group was thought to be close to
littorinoideans, but recent molecular analyses place it near
Triphoroidea, Eatoniellidae, and Naticidae (Takano & Kano 2014,
Osca et al. 2015).

Families Abyssochrysidae, Provannidae, Hokkaidoconchidae †, and
Paskentanidae †

Provanna, modified from Warén and
Ponder (1991)

(Superfamily) **Hipponicoidea**

Cretaceous (Upper) to present (125.5–0 Ma)

This limpet-like family was thought to be related to Vanikoridae, but
molecular data show this is not the case (Takano & Kano 2014)

Family Hipponicidae

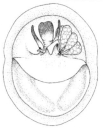

Sabia (Hipponicidae), shell and ventral view of animal, from FA2

(*Continued*)

(Superfamily) **Vermetoidea**

Mississippian (Middle) to present (342.8–0 Ma)

The Vermetidae (worm shells) are all marine and have the adult
shell variously coiled or uncoiled. Some are ctenidial suspension
feeders, and others use mucus secreted from the foot to trap
suspended material. The radula is taenioglossate. The relationships
of this family within Littorinida are at present unresolved. In one
recent molecular analysis using mitochondrial genome data (Osca
et al. 2015), it appears at the base of the caenogastropods, a result
clearly inconsistent with other data. Their vermetid data was
obtained from Rawlings et al. (2010 p. 18) who noted that their
'gene rearrangement may be a synapomorphy defining a clade
within the Littorinimorpha to which the Littorinidae (*Littorina*),
Pomatiopsidae (*Oncomelania*) and Vermetidae belong'. A
molecular analysis incorporating vermetids and a range of other
caenogastropods (Colgan et al. 2007) placed vermetids as sister to
Littorinidae + Pterotracheidae while in an analysis by Takano and
Kano (2014) the vermetid was in an unresolved polytomy with
hipponicids, epitoniids, and a clade including rissooideans,
truncatelloideans, and eulimoideans.

Families Vermetidae and Sakarahellidae †

Thylacodes (Vermetidae), from FA2

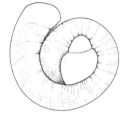

Novastoa (Vermetidae) protoconch and
early teleoconch, from FA2

(Superfamily) **Pseudomelanioidea** †

Permian (Guadalupian) to Neogene (Miocene) (272.5–23.03 Ma)

These tall-spired shells have a smooth protoconch of about three
whorls (Guzhov 2006).

Family Pseudomelaniidae (= Trajanellidae)

Pseudomelania, from Wenz
(1938–1944)

(Suborder) **Littorinina** new name

Permian (Lopingian) to present (259–0 Ma)

(Superfamily) **Littorinoidea**

Permian (Lopingian) to present (259–0 Ma)

The members of this superfamily (which may not be monophyletic)
include the littoral marine Littorinidae (periwinkles), the terrestrial
Pomatiidae and Annulariidae, and two small-size families, one of
which occurs in deep water (Zerotulidae). Most are surface film
scrapers, but some littorinids are herbivores. Their radula is
taenioglossate.

Families Littorinidae, Annulariidae, Pomatiidae, Skeneopsidae,
Zerotulidae, Bohaispiridae †, Leviathaniidae †, Purpuroideidae †,
and Tripartellidae †

Littorina, shell from Graham (1988),
animal from Fretter and Graham
(1962)

Skeneopsis, from Fretter and Graham
(1962)

(*Continued*)

(Superfamily) **Pterotracheoidea** (= Heteropoda)

Jurassic (Lower) to present (189.6–0 Ma)

This group, known as heteropods, are all pelagic and some genera
lack a shell. The foot is modified as a flattened fin. All are efficient
swimmers and are active carnivores feeding on zooplankton,
including crustaceans, molluscs, salps, small fish and fish eggs,
and larvae.

This group has been shown to be the sister group to Littorinoidea in
several molecular analyses.

Families Pterotracheidae, Atlantidae, Carinariidae, Bellerophinidae
† , and Coelodiscidae †

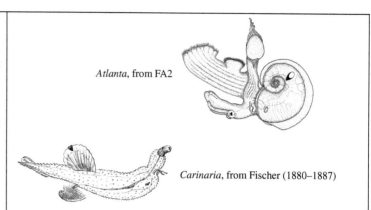

Atlanta, from FA2

Carinaria, from Fischer (1880–1887)

(Suborder) **Triphorina** new name

Jurassic (Middle) to present (164.7–0 Ma)

(Superfamily) **Triphoroidea**

Jurassic (Middle) to present (164.7–0 Ma)

These families are all sponge feeders. Their shell has a siphonal
canal and is usually tall-spired and strongly sculptured. They have
a very short to long proboscis and their radula has five to as many
as 63 teeth in each row (Marshall 1983).

Previously included in the polytypic Ptenoglossa (or Ctenoglossa or
Heteroglossa) along with Epitoniidae and Eulimidae.

Families Triphoridae (mostly sinistral), Cerithiopsidae,
Newtoniellidae (both dextral), and Berendinellidae †

Ataxocerithium (Newtoniellidae),
from FA2

Iniforis (Triphoridae), reproduced from Laseron (1958) with
permission from CSIRO Publishing.

(Suborder) **Rissoidina**

Permian (Lopingian) to present (259–0 Ma)

Introduced as a suborder (as Rissooidei) by Slavoshevskaja (1983) for superfamilies Rissooidea, Rissoinoidea, and Truncatelloidea. Here we extend the
concept to include the superfamilies Eulimoidea and Vanikoroidea.

Bouchet et al. (2017) used the informal group name 'Rissoiform clade' for the same grouping.

(*Continued*)

(Superfamily) **Rissooidea** (= Rissoinoidea)

Permian (Lopingian) to present (259–0 Ma)

This large, probably polyphyletic, grouping of small-sized snails contains several diverse marine families. Most are surface film scrapers, but *Rissoina* swallows whole forams. The radula has only five teeth in each row in Emblandidae.

Rissooideans are sister to Vanikoridae and Eulimidae in the molecular analysis of Takano and Kano (2014).

A number of families previously included in the Rissooidea are now placed in the Truncatelloidea (Criscione & Ponder 2013).

Families Rissoidae, Barleeiidae, Emblandidae, Lironobidae, Rissoinidae, Zebinidae, and Palaeorissoinidae †

Rissoa, animal from Fretter and Graham (1962), shell original

Rissoina, reproduced from Laseron (1956) with permission from CSIRO Publishing.

Zebina, reproduced from Laseron (1956) with permission from CSIRO Publishing.

Barleeia, from Graham (1988)

(*Continued*)

(Superfamily) **Truncatelloidea**

Jurassic (Lower) to present (189.6–0 Ma)

This large, possibly polyphyletic grouping of small-sized snails
  contains several diverse marine families (Anabathridae, Caecidae,
  Iravadiidae, Stenothyridae, and Tornidae) and several small
  families (Calopiidae, Elachisinidae, Epigridae, Falsicingulidae,
  and Hydrococcidae). It also contains several freshwater families,
  some of which are diverse (Hydrobiidae, Amnicolidae,
  Bithyniidae, Bythinellidae, Lithoglyphidae, Pomatiopsidae,
  Emmericiidae, and Cochliopidae) while Moitessieriidae is a small
  group and Helicostoidae is known from only one species.
  Assimineidae and Truncatellidae are high tidal to supratidal
  families, the former with some freshwater taxa, and there are some
  terrestrial taxa in Truncatellidae, Assimineidae, and
  Pomatiopsidae. Stenothyridae contains some freshwater taxa, but
  most, like the the majority of Clenchiellidae, live in brackish
  water. The extinct family Mesocochliopidae (Jurassic of China)
  may belong here.

Most are surface film scrapers, although bithyniids can both
  suspension feed and browse. Some tornids are commensals. The
  radula is taenioglossate, although highly modified in Epigridae.

Families Truncatellidae, Amnicolidae, Anabathridae, Assimineidae,
  Bithyniidae, Bythinellidae, Caecidae, Calopiidae, Clenchiellidae,
  Cochliopidae, Elachisinidae, Emmericiidae, Epigridae,
  Falsicingulidae, Helicostoidae, Hydrobiidae, Hydrococcidae,
  Iravadiidae, Lithoglyphidae, Moitessieriidae, Pomatiopsidae (=
  Tomichiidae), Stenothyridae, Tateidae, Tornidae (= Vitrinellidae),
  and Mesocochliopidae †

Two family names, Tomichiidae and Vitrinellidae, were recognised
  by Bouchet et al. (2017) but in our view with insufficient
  justification. Palaeorissoinidae was included in Truncatelloidea by
  those authors but we include that family in Rissooidea.

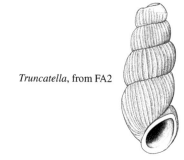

Truncatella, from FA2

Gabbia (Bithyniidae), from FA2

Ctiloceras (Caecidae), from FA2

Caecum, from Fretter and Graham (1962)

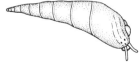

Circulus (Tornidae), from FA2

(Superfamily) **Eulimoidea**

Triassic (Upper) to present (235–0 Ma)

Eulimids are ecto- or endoparasites of echinoderms. The radula is
  ptenoglossan or, more commonly, absent.

Some of the internal parasites in this family are shell-less and were
  separated into separate families including Entoconchidae and
  several others (see Bouchet et al. 2017 for details).

This family was previously grouped with Triphoroidea and
  Epitonioidea (see remarks under Triphoroidea) and Bouchet et al.
  (2017) lumped this group with Vanikoroidea.

Aclididae was erroneously included in Heterobranchia by Dinapoli
  and Klussmann-Kolb (2010), but, as pointed out by Takano and
  Kano (2014), their material was a species of *Larochella*, a
  graphidid. Aclididae was lumped with Eulimidae in Bouchet et al.
  (2017) based on the results of Takano and Kano (2014), but those
  latter authors maintained Aclididae, which we follow pending
  more detailed investigation.

Families Eulimidae and Aclididae

Vitreolina (Eulimidae), from FA2

Echineulima (Eulimidae), from FA2

Aclis, from Graham (1988)

(Continued)

(Superfamily) **Vanikoroidea** Jurassic (Middle) to present (171.6–0 Ma) Vanikorids have coiled shells, probably feed on surface deposits, and have a taenioglossate radula. Vanikoroideans and eulimoideans are sister taxa in the molecular analysis of Takano and Kano (2014), and these two groups were united in the Vanikoroidea in Bouchet et al. (2017), a grouping that we do not accept due to their considerable morphological differences. Family Vanikoridae	*Vanikoro*, from FA2 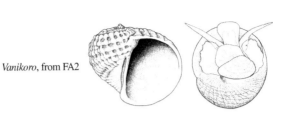
(Suborder) **Naticina** Devonian (Middle) to present (388.1–0 Ma) This name was introduced by Riedel (2000) for Naticoidea.	
(Superfamily) **Naticoidea** Devonian (Middle) to present (388.1–0 Ma) Naticids ('moon snails') are shallow burrowers with a large foot, sometimes capable of completely enclosing the shell. These carnivores have a proboscis with an accessory boring organ at its tip which aids in drilling the mainly bivalve prey. The radula is taenioglossan. Family Naticidae	*Euspira* (Naticidae), from FA2 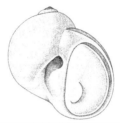 *Notocochlis* (Naticidae), from FA2 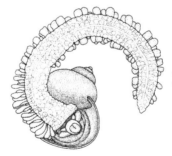
(Suborder) **Epitoniina** Triassic (Upper) to present (235–0 Ma) Originally introduced as an order (Epitoniida) by Minichev and Starobogatov (1979).	
(Superfamily) **Epitonioidea** Triassic (Upper) to present (235–0 Ma) Epitoniidae contains the 'wentletraps' and the pelagic violet snails. They are all cnidarian feeders with most epitoniids feeding on sea anemones and the pelagic *Janthina* feeding on siphonophores and chondrophores. The radula is ptenoglossate. Family Epitoniidae (= Janthinidae, Nystiellidae) While the relationships of the Devonian Spanionematidae are uncertain, some look remarkably like epitoniids.	*Epitonium*, from FA2  *Janthina* (Epitoniidae) and bubble raft with attached egg capsules, from FA2 

(Continued)

(Superorder) **Latrogastropoda**

Triassic (Upper) to present (235–0 Ma)

Following Bouchet et al. (2017) we use this name of Riedel (2000) for his grouping of Calyptraeoidea + Cypraeoidea + Naticoidea + Laubierinioidea + Ficoidea + Tonnoidea + Neogastropoda in preference to Rhynchogastropoda of Simone (2011) for Calyptraeoidea + his Adenogastropoda. Bouchet et al. (2017) modified Riedel's concept by excluding Naticoidea and including Strombida, a concept adopted here.

Strombogastropoda of Simone (2011) includes Stromboidea + Rhynchogastropoda (i.e., the remaining caenogastropods) and is identical to our concept of Latrogastropoda.

Unplaced family: Colombellinidae †. This family was previously included in Stromboidea.

(Order) **Strombida**

Triassic (Upper) to present (235–0 Ma)

This grouping was named Leptopoda by Gray (1857), but that name was used earlier in Bivalvia.

(Superfamily) **Stromboidea**

Triassic (Upper) to present (235–0 Ma)

Strombids ('strombs') are surface herbivores, aporrhaids are surface or subsurface deposit feeders, seraphsids are active burrowers, and struthiolariids burrowing suspension feeders. The snout is long and the radula taenioglossate.

Families Strombidae, Aporrhaidae, Rostellariidae, Seraphsidae, Struthiolariidae, Dilatilabridae †, Hippochrenidae †, Pereiraeidae †, and Thersiteidae †

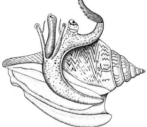

Laevistrombus (Strombidae), from FA2

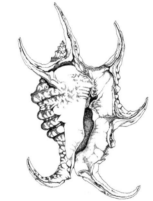

Lambis (Strombidae), from FA2

Terebellum (Seraphsidae), from FA2

Aporrhais, from Woodward (1851–1856)

(*Continued*)

(Superfamily) **Xenophoroidea**
Triassic (Upper) to present (212–0 Ma)
This deposit feeding family of 'carrier shells' have a taenioglossate
  radula, are closely related to stromboideans, and are sometimes
  included in the same superfamily.
Families Xenophoridae and Lamelliphoridae †
	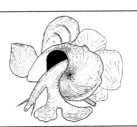
	Xenophora, from Fischer (1880–1887)

(Order) **Calyptraeida** (= Calyptraeiformii, Calyptraeiformes) Triassic (Upper) to present (237–0 Ma)

(Suborder) **Calyptraeina**(= Calyptraeoidei) Triassic (Upper) to present (237–0 Ma)

(Superfamily) **Calyptraeoidea**
Cretaceous (Lower) to present (112.6–0 Ma)
Calyptraeids (slipper limpets) are suspension feeders with highly
  modified mantle organs including a very large ctenidium. Some
  species can also feed on surface deposits. Their radula is
  taenioglossate.
Simone (2011) recognised this superfamily as encompassing not
  only Calyptraeidae but also Capulidae, Hipponicidae, and
  Vanikoridae. The latter family has been shown to be closely related
  to Eulimidae and the Capulidae and Hipponicidae are also
  unrelated.
Family Calyptraeidae

	Bostrycapulus (Calyptraeidae), from FA2

(Superfamily) **Capuloidea**
Triassic (Upper) to present (237–0 Ma)
Capulids have a coiled (e.g., *Trichotropis*) to limpet-shaped
  (*Capulus*) shell. Members of the small, deepwater family
  Haloceratidae also have coiled shells. Some *Capulus* species live
  associated with other molluscs (see Chapters 5 and 19). The mouth
  has a unique pseudoproboscis, and the radula is taenioglossate.
Family Capulidae (= Trichotropidae), Haloceratidae, and
  Gyrotropidae †

	Trichotropis (Capulidae), from FA2

Adenogastropoda A group introduced by Simone (2011) for Naticoidea + Siphonogastropoda.

Siphonogastropoda Jurassic (Upper) to present (150.8–0 Ma) Named by Simone (2011) for Cypraeoidea + Peogastropoda. Pleurombolica of Riedel (2000) is a similar grouping but included Calyptraeoidea. Unplaced family: Lyocyclidae. This poorly known group was included in the Vanikoridae, but Takano and Kano (2014) showed it was more closely related   to Cypraeidae.

(Order) **Cypraeida** Jurassic (Upper) to present (150.8–0 Ma)

(Continued)

(Superfamily) **Cypraeoidea** Jurassic (Upper) to present (150.8–0 Ma) This grouping contains the cypraeids ('cowries') and ovulids ('false cowries'). Ovulids feed on soft corals and gorgonians; some cypraeids feed on sponges, and others are herbivores. The radula is taenioglossate. The osphradium of cypraeids is uniquely triangular, with three sets of leaflets. Families Cypraeidae, Ovulidae, and Pediculariidae	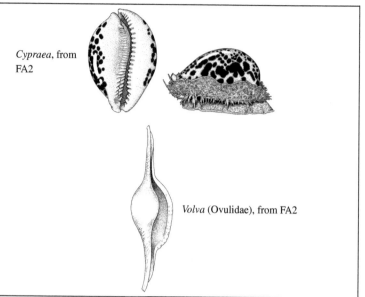 *Cypraea*, from FA2 *Volva* (Ovulidae), from FA2
(Superfamily) **Velutinoidea** Paleogene (Paleocene) to present (66–0 Ma) The osphradium is linear with two rows of leaflets. Velutinids have a reduced internal shell and are externally slug-like. All are ascidian feeders although one species of *Velutina* is known to feed on *Tubularia* (a hydroid) as well as compound ascidians. The radula is taenioglossate. This superfamily was combined with Cypraeoidea by Simone (2011) and Bouchet et al. (2017). Families Velutinidae (= Lamellariidae), Eratoidae, and Triviidae	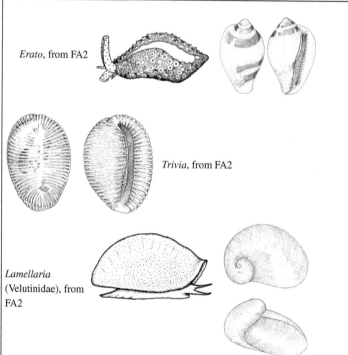 *Erato*, from FA2 *Trivia*, from FA2 *Lamellaria* (Velutinidae), from FA2

Peogastropoda

Cretaceous (Lower) to present (125.5–0 Ma)

Named by Simone (2011), this group includes the three superfamilies below which all have a taenioglossate radula and were previously called the 'higher mesogastropods' and neogastropods. It is equivalent to Vermivora of Riedel (2000), but that same name was proposed by Gray in 1860 for a group of stylommatophorans.

(Order) **Tonnida** new order.

Cretaceous (Lower) to present (125.5–0 Ma)

The two superfamilies below are here grouped in this order. It is equivalent to Cassina + Ficina + Troschelina (in part) of Riedel (2000).

(Continued)

(Superfamily) **Ficoidea** Cretaceous (Upper) to present (84.9–0 Ma) Ficids have a short-spired shell with a long anterior canal and siphon. The proboscis is very long, and they feed on polychaetes. Their radula is taenioglossate. Often included in the Tonnoidea, ficids are now considered to belong to their own superfamily (Riedel 2000), although this was not recognised by Simone (2011). Family Ficidae (= Pyrulidae)	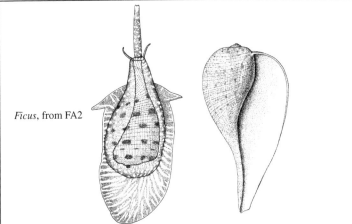 *Ficus*, from FA2
(Superfamily) **Tonnoidea** (= Cassoidea, Laubierinioidea) Cretaceous (Lower) to present (125.5–0 Ma) Tonnoideans have a relatively short (e.g., some ranellids) to very long (e.g., Personidae) proboscis. Cassids (helmet shells) can drill through an echinoid test, and tonnids (tuns) swallow their prey whole. Tonnids are holothurian feeders, cassids are echinoid feeders, ranellids and charoniids (tritons) are carnivores on echinoderms, tunicates, and molluscs while personids are polychaete feeders, and bursids (frog shells) feed on echinoderms and polychaetes. All have a taenioglossate radula. The recognition of the families below is based on the recent phylogeny of Strong et al. (2019). Families Tonnidae, Bursidae, Cassidae, Charoniidae, Cymatiidae, Laubierinidae (= Pisanianuridae), Personidae, Ranellidae, Thalassocyonidae, Eosassiidae †, and Mataxidae †	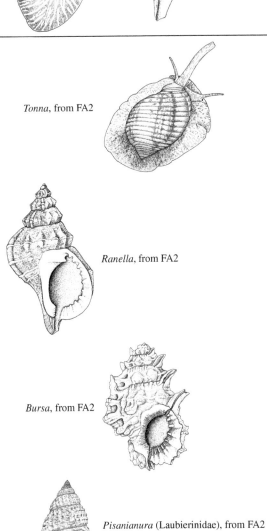 *Tonna*, from FA2 *Ranella*, from FA2 *Bursa*, from FA2 *Pisanianura* (Laubierinidae), from FA2

(Continued)

(Order) Neogastropoda Permian (Guadalupian) to present (272–0 Ma) Neogastropods are characterised particularly by modifications to their digestive system including the oesophageal gland being separated from the midoesophagus, the salivary ducts not passing through the nerve ring, a reduced stomach, and a radula with a reduced number of teeth in each row (usually a rachiglossate or toxoglossate type). They are generally assumed to be a monophyletic group, but attempts to recover a monophyletic Neogastropoda using molecular data have often failed. The relationships of the families included within neogastropods remain uncertain.	
Uncertain relationships Mesozoic families unassigned to a superfamily that are stem taxa may have given rise to neogastropods. The purpurinids date from the Triassic and are sometimes included in the Littorinoidea. Pseudotritoniidae (= Maturifusidae) † and Purpurinidae †	*Purpurina*, modified from www.bagniliggia.it/WMSD/HtmSpecies/6410000004.htm.
(Superfamily) Pholidotomoidea (= Sarganoidea) † Cretaceous (Upper) to Paleogene (Eocene) (99.7–48.6 Ma) This extinct Cretaceous group are thought to be basal neogastropods. Families Pholidotomidae (= Volutodermatidae), Sarganidae, Moreidae (= Pyropsidae), and Weeksiidae	*Sargana*, modified from www.fossilshells.nl/tencret12.html
(Superfamily) Cancellarioidea Cretaceous (Upper) to present (99.7–0 Ma) Located within Conoidea in one recent molecular analysis (Zou et al. 2011) and placed with a volutid in another analysis (Fedosov et al. 2015) which prompted Bouchet et al. (2017) to lump Cancellarioidea and Volutoidea together. However, given the differing molecular results, and the major anatomical differences between Volutidae and Cancellariidae, we maintain these two superfamilies. Family Cancellariidae	*Cancellaria*, from Fischer (1880–1887)   *Trigonostoma* (Cancellariidae), from FA2

(*Continued*)

(Superfamily) **Buccinoidea**

Jurassic (Upper) to present (161.2–0 Ma)

Buccinoideans include carnivores and scavengers as well as the suctorial Colubrariidae. They differ from other neogastropods in lacking accessory salivary glands. Some species of *Cominella* (a buccinid) have been reported to drill holes in their mollusc prey, but the details as to how this occurs have not been documented.

Families Buccinidae, Belomitridae, Colubrariidae, Columbellidae, Fasciolariidae, Melongenidae, Nassariidae, Pisaniidae, Echinofulguridae † and possibly Babyloniidae, Johnwyattiidae †, Taiomidae †, Strepsiduridae (= Melapiidae), and Perissityidae †

Buccinum, from Fischer (1880-1887)

Penion (Buccinidae), from FA2

Mitrella (Columbellidae), from FA2

Melongena, from Fischer (1880–1887)

Nassarius, from FA2

(Continued)

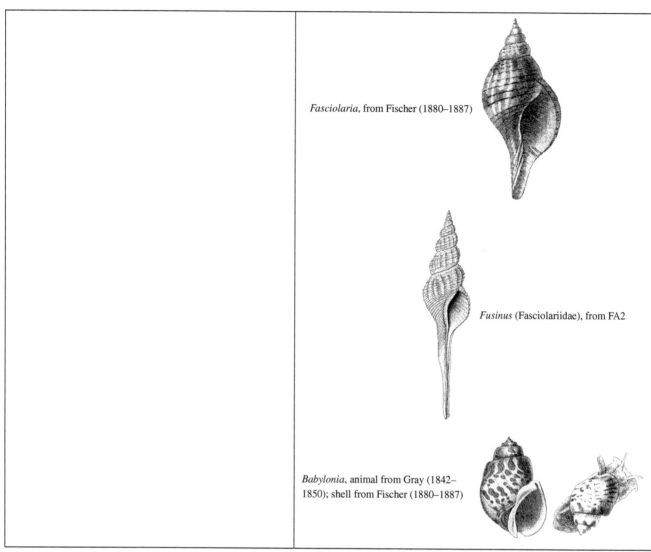

Fasciolaria, from Fischer (1880–1887)

Fusinus (Fasciolariidae), from FA2

Babylonia, animal from Gray (1842–1850); shell from Fischer (1880–1887)

(*Continued*)

(Superfamily) **Muricoidea**

Cretaceous (Lower) to present (140.2–0 Ma)

Muricidae (rock shells, oyster drills, etc.). Many species drill shells of molluscs and barnacles using an accessory boring organ on the sole of the foot.

Family Muricidae (with several subfamilies, including Coralliophilinae)

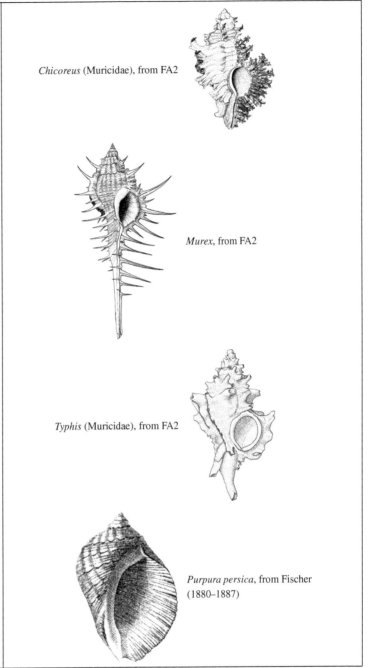

Chicoreus (Muricidae), from FA2

Murex, from FA2

Typhis (Muricidae), from FA2

Purpura persica, from Fischer (1880–1887)

(*Continued*)

(Superfamily) **Turbinelloidea**

Cretaceous (Upper) to present (94.3–0 Ma)

This grouping is based on the molecular analyses of Fedosov et al. (2015).

Families Turbinellidae, Columbariidae, Costellariidae, Ptychatrachidae, and Volutomitridae

Vexillum (Costellariidae), from FA2

Columbarium, from FA2

Vasum (Turbinellidae), from FA2

Exilia (Ptychatractidae), from FA2

(Continued)

(Superfamily) **Mitroidea** Cretaceous (Lower) to present (105.3–0 Ma) This group was split from Muricoidea on the basis of the molecular analysis of Fedosov et al. (2015), and the superfamily was further investigated by Fedosov et al. (2018). Families Mitridae (with Pleioptygmatinae as a subfamily), Charitodoronidae, and Pyramimitridae.	 *Domiporta* (Mitridae), from FA2  *Pterygia* (Mitridae), from FA2
(Superfamily) **Olivoidea** Paleogene (Paleocene) to present (66.0–0 Ma) Often included within Muricoidea. Pseudolividae has sometimes been included in its own superfamily while Olivoidea has been included in Volutoidea in older classifications. The molecular analysis of Fedosov et al. (2015) has these taxa as a monophyletic group. Families Olividae (= Olivellidae), Ancillariidae, Bellolividae, Benthobiidae, and Pseudolividae	*Oliva*, from FA2  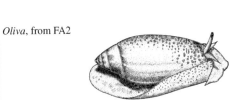  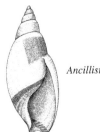 *Ancillista* (Ancillariidae), from FA2 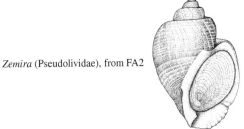 *Zemira* (Pseudolividae), from FA2

(*Continued*)

(Superfamily) **Volutoidea**

Cretaceous (Upper) to present (99.7–0 Ma)

This grouping is only tentatively recognised, partly on historical grounds and partly because the very small amount of published molecular data indicates that the volutes at least do not lie with the muricoideans where they are often placed.

Here we include the Volutidae (volutes, balers, etc.), Harpidae (harps), and the small-sized Marginellidae and Cystiscidae.

Bouchet et al. (2017) treated all but Volutidae as unassigned to superfamily. Volutidae groups with conoideans and cancellarioideans in the Zou et al. (2011) analysis, with Tonnoidea in Colgan et al. (2007) and with Cancellariidae in Fedosov et al. (2015).

Families Volutidae, Cystiscidae, Marginellidae, and possibly Harpidae

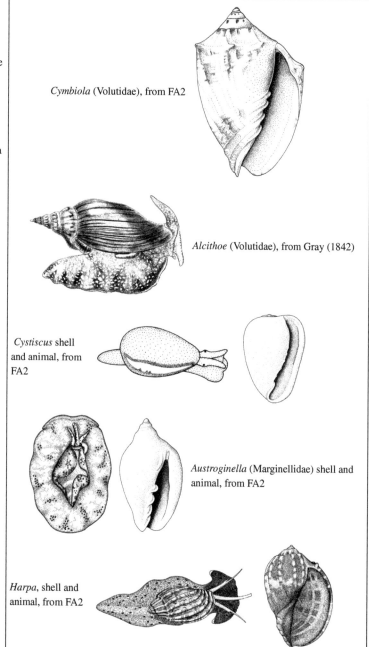

Cymbiola (Volutidae), from FA2

Alcithoe (Volutidae), from Gray (1842)

Cystiscus shell and animal, from FA2

Austroginella (Marginellidae) shell and animal, from FA2

Harpa, shell and animal, from FA2

(*Continued*)

(Superfamily) **Conoidea**

Permian (Guadalupian) to present (272.5–0 Ma)

This very large group includes the terebrids ('augers'), conids ('cones'), and a number of families now recognised that previously made up the huge family Turridae, initially as a result of morphological studies (e.g., Taylor et al. 1993; Kantor & Taylor 2002; Bouchet et al. 2011) and more recently based on molecular studies (e.g., Puillandre et al. 2009; Puillandre et al. 2011; Abdelkrim et al. 2018).

Families Conidae, Borsoniidae, Bouchetispiridae, Clathurellidae, Clavatulidae, Cochlespiridae, Conorbidae, Drilliidae, Fusiturridae, Horaiclavidae, Mangeliidae, Marshallenidae, Mitromorphidae, Pseudomelatomidae, Raphitomidae, Strictispiridae, Terebridae, Turridae, Cryptoconidae †, and possibly Speightiidae †

Lophiotoma (Turridae), from FA2

Eucithara (Mangeliidae), from FA2

Daphnella (Raphitomidae), from FA2

Terebra, from FA2

Conus, from FA2

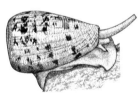

(Continued)

(Infraclass) **Heterobranchia**

See Chapter 20 for details of this group that largely consists of the previously recognised 'Opisthobranchia' and 'Pulmonata' together with some taxa that were previously treated as 'mesogastropods'.

Extinct taxa of uncertain relationship

(Superfamily) **Nerineoidea** † Jurassic (Lower) to Cretaceous (Upper) (196.5–66 Ma) These taxa were recently reviewed by Kollmann (2014). Families Nerineidae, Ceritellidae, Eunerineidae, Itieriidae, Nerinellidae, Pseudonerineidae (= Tubiferidae), and Ptygmatididae	 *Nerinea*, from TI  *Nerinella*, from TI

(Superfamily) **Streptacidoidea** †
 	
 Mississippian (Middle) to Cretaceous (Upper) (345.3–70.6 Ma)
 Streptacidoideans have been suggested as the common ancestor of
    Nerineoidea and Acteonelloidea and they may also be related to
    Murchisonelloidea.
 Families Streptacididae and Cassianebalidae

	Streptacis, from TI

	Donaldina (Streptacididae), from TI

(*Continued*)

(Superfamily) **Acteonelloidea** † Cretaceous (Lower to Upper) (112.6–66 Ma) Family Acteonellidae	*Acteonella* (left) and *Trochactaeon* (right) (Acteonellidae), modified from Wenz (1938–1944) 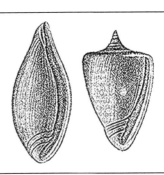

'Lower heterobranchs'
(= the paraphyletic Allogastropoda and Heterostropha *sensu* Ponder and Warén 1988, with some modifications).
Uncertain relationships: Tjaernoeiidae, Dolomitellidae †, Kuskokwimiidae †, and Misurinellidae †

(Order) **Ectobranchia**
Permian (Lopingian) to present (252.3–0 Ma)
Shell with moderate to low spire, protoconch not obviously heterostrophic, and PMO not present. Tentacles long, eyes in middle of bases; operculum present. Large secondary gill present. Browsers with well-formed radula with seven to many teeth in each row.

(Superfamily) **Valvatoidea** Permian (Lopingian) to present (252.3–0 Ma) Marine except for the freshwater Valvatidae. Families Valvatidae, Cornirostridae, Hyalogyrinidae, and Provalvatidae †	*Cornirostra*, from FA2  *Valvata*, shell, public domain 

(Order) **Architectonicida**
Triassic (Lower) to present (251–0 Ma)
We use this ordinal-level name introduced by Minichev and Starobogatov (1979) to include the three superfamilies below, with Omalogyridae included here on the basis of the molecular results of Dinapoli and Klussmann-Kolb (2010).
Originally proposed (as Architectonicoida) as a superorder for Architectonicida and Epitoniida by Minichev and Starobogatov (1979).

(Suborder) **Architectonicina**
Triassic (Lower) to present (251–0 Ma)
Medium-sized to rather large, feed on cnidarians.

(Superfamily) **Mathildoidea**
Triassic (Lower) to present (251–0 Ma)
Shell high-spired with protoconch distinctly heterostrophic.
  Operculum present. Tentacles long, eyes on outer bases. Poorly
  known biologically, but at least a few are coral feeders.
Families Mathildidae, Gordenellidae †, Schartiidae †, and
  Trachoecidae †
	Opimilda (Mathildidae), from FA2 

	Gegania (Mathildidae), from FA2 

(Continued)

(Superfamily) **Architectonicoidea**
Triassic (Upper) to present (228–0 Ma)
Shell with low spire, angulate periphery, and with operculum.
  Tentacles long, eyes on outer bases. Feed on various cnidarians.
Families Architectonicidae, Amphitomariidae †, and Cassianaxidae
  †
	Heliacus (Architectonicidae), from FA2 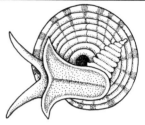
	
	Architectonica, from FA2 

(Suborder) **Omalogyrina** new name Paleogene (Oligocene) to present (28.4–0 Ma) Minute, suctorial on algal cells.	

(Superfamily) **Omalogyroidea**
Paleogene (Oligocene) to present (28.4–0 Ma)
Shell tiny, discoidal. Single pair of tentacles or lobes. Eyes in
  middle of tentacle bases and distinct PMO. Radula with single
  tooth row; suctorial feeders on algal cells.
Families Omalogyridae and Stuoraxidae †
	Omalogyra, from Graham (1988) 

	Ammonicera (Omalogyridae), from Bieler and Mikkelsen (1998)

(Order) **Orbitestellida**
Paleogene (Paleocene) to present (61.7–0 Ma)
This ordinal name is introduced here to include Orbitestellidae and Cimidae that are distinguished from other heterobranchs by the combination of a
  number of features including their very simple reproductive anatomy, lack of a secondary gill, prominent PMO, long pallial tentacle, and long cephalic
  tentacles with eyes in the middle of their bases.	

(Superfamily) **Orbitestelloidea** Paleogene (Oligocene) to present (33.9–0 Ma) Xylodisculidae is included here following Wilson et al. (2017) and Kano (unpublished molecular data – *fide* Bouchet et al. 2017). Families Orbitestellidae and Xylodisculidae	*Xylodiscula*, from FA2  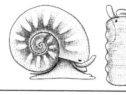 *Microdiscula* (Orbitestellidae), from FA2 *Orbitestella*, from Laseron (1954) 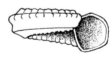 

(*Continued*)

(Superfamily) **Cimoidea** Paleogene (Paleocene) to present (61.7–0 Ma) The justification for the synonymy of Graphididae with Cimidae is given by Bouchet et al. (2017). Family Cimidae (= Graphididae and Tofanellidae)	 *Cima*, from Graham (1988) *Graphis* (Cimidae), original

(Order) **Allomorpha**
Jurassic (Lower) to present (189.6–0 Ma) This grouping was introduced by Wilson et al. (2017), on the basis of molecules and morphology, to include Rhodopemorpha and Murchisonellidae.

(Superfamily) **Rhodopoidea** Present (0 Ma) Rhodopoideans lack tentacles, radula, jaws or heart, and a well-defined foot. One species is known to feed on sponge larvae and another on placozoans. This highly modified group of tiny interstitial worm-like slugs has been variously assigned within the 'opisthobranchs', but recent molecular analyses show that it is a lower heterobranch. The name Rhodopemorpha has been used for this family alone, but its use in the current classification would seem unnecessarily inflated. Family Rhodopidae	 *Rhodope*, from FA2 *Helminthope* (Rhodopidae), from Salvini-Plawen (1991)

(Superfamily) **Murchisonelloidea** Jurassic (Lower) to present (189.6–0 Ma) Cephalic tentacles moderately long to absent, PMO present, head shield sometimes present. Probably related to Streptacidoidea. Murchisonellidae (= Ebalidae) and Donaldinidae †	 *Murchisonella*, from Hedley (1905) *Koloonella* (Murchisonellidae), from FA2

(Cohort) **Euthyneura**
Permian (Guadalupian) to present (272.5–0 Ma) Euthyneura (including Acteonimorpha) is recovered in some recent molecular analyses (Göbbeler & Klussmann-Kolb 2010; Zapata et al. 2014). In another analysis (Kano et al. 2016) euthyneurans form an unresolved trichotomy – Acteonimorpha, Ringipleura + Nudipleura and Tectipleura (remaining Euthyneura) – while in others (Dinapoli & Klussmann-Kolb 2010; Jörger et al. 2010) Acteonimorpha is sister to the remaining Euthyneura. The arrangement of Euthyneura here largely follows Bouchet et al. (2017) which in turn is based on the analyses of Dinapoli and Klussmann-Kolb (2010), Jörger et al. (2010), and Kano et al. (2016).

'Opisthobranchia'
'Opisthobranchia' was treated as a separate subclass in older classifications, but since the concept of Heterobranchia became established, it, along with 'Pulmonata', has slipped in rank (e.g., as a superorder in Beesley et al. [1998b] and as an 'informal group' in Bouchet and Rocroi [2005]). Recent phylogenetic analyses (e.g., Schrödl et al. 2011; Wägele et al. 2014; Zapata et al. 2014) found that this group is either paraphyletic or polyphyletic with two of the groups traditionally included in 'opisthobranchs', the Sacoglossa and Acochlidia, now grouping within the panpulmonates. Consequently, 'Opisthobranchia' is no longer used as a formal taxonomic name.

(Continued)

(Subcohort) **Acteonimorpha**

Jurrassic (Middle) to present (164.7–0 Ma)

This name is introduced by Bouchet et al. (2017) to include Acteonidae and probable relatives included in the Acteonoidea below, as well as Rissoellidae. The fossil groups included below may also belong here.

Other higher-level names based on *Acteon* exist but encompass quite different assortments of taxa (see Bouchet & Rocroi 2005 and Bouchet et al. 2017). The name Architectibranchia was introduced by Haszprunar (1985) as a superorder for Acteonoidea, Ringiculoidea, and Diaphanoidea, but the latter two families are now included in other groups within the Euthyneura (see below).

The relationships of this clade, which includes Acteonoidea and Rissoelloidea, are unresolved, being recovered as a sister group to Nudipleura (Göbbeler & Klussmann-Kolb 2010, Zapata et al. 2014 in part), sister to the other Euthyneura (e.g., Dinapoli & Klussmann-Kolb 2010; Jörger et al. 2010), or sister to Tectipleura (Zapata et al. 2014 in part).

(Order) **Acteonida**

Permian (Guadalupian) to present (272.5–0 Ma)

(Superfamily) **Acteonoidea**	
Permian (Guadalupian) to present (272.5–0 Ma) Shell and in some, an operculum present in some. Head-shield and external penis present. No parapodia but head-shield has two flaps that superficially resemble short parapodia. Secondary gill (plicatidium) present in mantle cavity. No gizzard. Feed on polychaetes. Families Acteonidae, Aplustridae (= Bullinidae Gray and Hydatinidae), Cylindrobullinidae †, Tubiferidae †, and Zardinellidae †	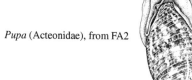 *Pupa* (Acteonidae), from FA2 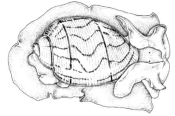 *Bullina* (Aplustridae), from FA2 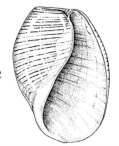 *Hydatina* (Aplustridae), from FA2

(Order) **Rissoellida**

Cretaceous (Lower) to present (112–0 Ma)

Rissoellina was introduced by Golikov and Starobogatov (1968) as a suborder including Skeneopsidae and Rissoellidae. As used here Rissoellida includes Rissoelloidea alone.

(Superfamily) **Rissoelloidea**	
Cretaceous (Lower) to present (112–0 Ma) Small, fragile shell with low to high spire, operculum concentric with lateral peg and two pairs of tentacles. Eyes in middle of bases of posterior tentacles and distinct PMO. Radula with five–seven teeth in each row. Browsers. Members of this group were treated as 'mesogastropods' until the 1980s and were then included in the 'lower heterobranchs'. More recent molecular data have shown them to be sister to the acteonoideans. Family Rissoellidae	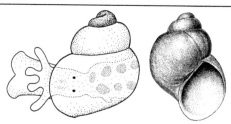 *Rissoella*, animal from FA2, shell from Graham (1988)

(*Continued*)

(Subcohort) **Ringipleura** Jurassic (Middle) to present (161.2–0 Ma) Introduced recently by Kano et al. (2016) for Ringiculoidea + Nudipleura recovered in their molecular analysis. It was previously included in or near the Acteonoidea or in the 'lower heterobranchs'. The subordinate higher taxa Ringiculimorpha and Ringiculida are also available.	

(Megaorder) **Ringiculata** (= Ringiculimorpha) Jurassic (Middle) to present (161.2–0 Ma)	

(Order) **Ringiculida** Jurassic (Middle) to present (161.2–0 Ma)	

(Superfamily) **Ringiculoidea** Jurassic (Middle) to present (161.2–0 Ma) Well-developed external shell with thick outer lip and bilobed cephalic shield. Possibly feed on small benthic prey as forams and crustaceans have been found in stomach contents. Family Ringiculidae	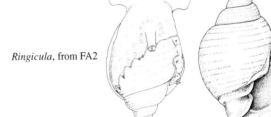 *Ringicula*, from FA2

(Superorder) **Nudipleura** Paleogene (Paleocene) to present (66–0 Ma) The Nudipleura include the pleurobranchs and the nudibranchs.	

(Order) **Pleurobranchida** (= Pleurobranchomorpha) Paleogene (Paleocene) to present (66–0 Ma) These carnivorous 'side-gilled' slugs have a well-developed plicatidium on the right side of the body, and most have a reduced internal shell.	

(Superfamily) **Pleurobranchoidea** Paleogene (Paleocene) to present (66–0 Ma) While members of this group feed on a range of animals, each genus has its own diet: for example, *Berthella* – sponges, *Berthellina* – sponges and cnidarians, *Pleurobranchus* – ascidians, *Pleurobranchaea* – opportunistic carnivore, hydroids, anemones, polychaetes, molluscs, carrion (Willan 1984). Families Pleurobranchidae, Pleurobranchaeidae, and Quijotidae	*Pleurobranchus*, from FA2 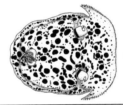  *Euselenops* (Pleurobranchaeidae), from FA2

(Order) **Nudibranchia** Present (0 Ma) Body fossils of nudibranchs are rare. Johnson and Richardson Jr (1966 p. 628) reported 'several specimens … complete with radula' from the Pennsylvanian Mazon Creek localities in Illinois, USA. In addition, sublethal predation and repair is present on Upper Cretaceous bryozoans suggesting the presence of nudibranch predation by this time (McKinney 1997; Berning 2007). As there is no definitive identifications of these remains we do not include these records in our stratigraphic ranges. Carnivorous sea slugs, many with a posterior circlet of secondary gills. Radula typically with numerous rows of teeth. Two 'suborders', Doridina and Cladobranchia, are recognised by Bouchet et al. (2017) based on the molecular results of Mahguib and Valdes (2015).	

(Suborder) **Doridina** (= Euctenidiacea, Holohepatica, Anthobranchia) This group was classically divided into four higher groups (suborders or superfamilies), namely Anadoridacea (or Phanerobranchia), Eudoridacea (or Cryptobranchia), Gnathodoridacea, and Porostomata, but they are mainly not well supported in the latest phylogenetic studies.	

(Infraorder) **Bathydoridoidei** (= Gnathodoridacea in part) Originally Gnathodoridacea included *Bathydoris* and *Doridoxa*, but the latter belongs in Cladobranchia (Mahguib & Valdés 2015).	

(Superfamily) **Bathydoridoidea** Carnivorous grazers on various colonial invertebrates. Polar regions and deep sea. *Bathydoris* is sister to the rest of the Doridina in the molecular phylogeny of Mahguib and Valdés (2015). Family Bathydorididae	*Bathydoris*, modified from Valdés (2002)

(Infraorder) **Doridoidei** (= Doridacea)	

(*Continued*)

(Superfamily) **Doridoidea** (= Cryptobranchia, Eudoridoidea) All Doridoidea are sponge feeders and have a posterior circlet of secondary gills. The Cryptobranchia is paraphyletic (Hallas et al. 2017). Families Dorididae and Discodorididae	*Halgerda* (Discodorididae), from FA2 *Hoplodoris* (Discodorididae), from FA2
(Superfamily) **Chromodoridoidea** This grouping was previously included in the Doridoidea. Hexabranchidae is included here following Hallas et al. (2017) who noted that this grouping and polycerids were closely related. Families Chromodorididae, Actinocyclidae, Cadlinidae, and Hexabranchidae	*Glossodoris* (Chromodorididae), from FA2
(Superfamily) **Polyceroidea** Most are bryozoan feeders, but *Nembrotha* feeds on ascidians, *Roboastra* feeds on nudibranchs. Those taxa that used to be included in Gymnodorididae feed on other nudibranchs, and one feeds on the fins of resting fish while those that were in Okadaiidae feed on tubicolous polychaetes. The current concept of the Polyceridae is based on the findings of Hallas et al. (2017). Family Polyceridae (= Gymnodorididae, Okadaiidae)	*Kalinga* (Polyceridae), from FA2 *Polycera*, from FA2 *Kaloplocamus* (Polyceridae), from FA2
(Superfamily) **Onchidoridoidea** Onchidorididae are mostly bryozoan feeders, but at least one species feeds on barnacles; Goniodorididae feed on bryozoans, kamptozoans, or ascidians and have a reduced number of teeth in each radular row. Corambidae are bryozoan feeders. Aegiridae feed on calcareous sponges. Families Onchidorididae, Aegiridae, Akiodorididae, Calycidorididae, Corambidae, and Goniodorididae	*Onchidoris*, from FA2 *Corambe*, from FA2 *Trapania* (Goniodorididae), from FA2 *Okenia* (Goniodorididae), from FA2

(*Continued*)

(Superfamily) **Phyllidioidea** (= Porostomata, Porodoridoidea). All phyllidioideans lack a radula and are probably all sponge feeders. Families Phyllidiidae, Dendrodorididae, and Mandeliidae	 *Phyllidia*, from FA2  *Dendrodoris*, from FA2

(Suborder) **Cladobranchia** (= Cladohepatica)
This major clade is recovered in the analysis of Goodheart (2017), but two of the three groups usually recognised within it, the Euarminida and the Dendronotida (= Dendronotoidea), were not supported. The classification below follows that of Bouchet et al. (2017), albeit tentatively, and is based on Pola and Gosliner (2010), Mahguib and Valdés (2015), and Goodheart et al. (2015, 2017). Most members of this grouping have cerata, a reduced mantle, and non-retractile rhinophores.

Families not assigned to superfamily: Bornellidae, Embletoniidae, Goniaeolididae, Heroidae, Lomanotidae, Madrellidae, Phylliroidae, and Pinufiidae The diets of these families are as follows (W. Rudman, pers. comm.): Bornellidae, Embletoniidae and Heroidae (hydroids), Goniaeolididae and Madrellidae (bryozoans), and Phylliroidae (pelagic, feeds on zooplankton).	 *Madrella*, from FA2  *Phylliroe*, from FA2 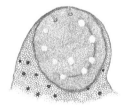 *Bornella*, from FA2

(Superfamily) **Arminoidea** (= Euarminida) Have lateral anus, oral veil and lack cerata. Gills, if present, beneath mantle edge. Arminids feed on pennatulaceans (sea-pens), and the illustrated doridomorphid is shown feeding on *Heliopora*, a type of octocoral. Families Arminidae and Doridomorphidae	 *Armina*, from FA2  *Doridomorpha* (on coral), from FA2

(Superfamily) **Dendronotoidea** Dotidae, Hancockiidae, Lomanotidae, and Scyllaeidae (hydroid feeders) and Dendronotidae (mostly hydroids, but one species of *Dendronotus* feeds on the tubicolous sea anemone *Cerianthus*), Tethydidae (including *Melibe*, small crustaceans). Rhinophores with basal sheaths and head usually has an oral veil. Cerata do not contain digestive gland or cnidosacs (except in *Hancockia*), sometimes elaborate, and may act as gills. The family Phylliroidae were previously included here, but we follow Bouchet et al. (2017) in not assigning that family to a superfamily. Families Dendronotidae, Dotidae, Hancockiidae, Scyllaeidae, and Tethydidae	 *Melibe* (Tethydidae), from FA2

(Continued)

(Superfamily) **Tritonioidea** Typical Tritoniidae feed on octocorals, and those that were classified in Aranucidae are hydroid feeders. Family Tritoniidae (= Aranucidae)	*Tritoniopsis* (Tritoniidae), from FA2

(Superfamily) **Doridoxoidea** (= Pseudoeuctenidiacea)
A small family containing a few species of *Doridoxa*. Distinctive in
  having the digestive gland transitional between the holohepatic
  (solid and simple) and cladohepatic (ramifying) states. Anus on
  right side of body and gills absent.
Family Doridoxidae
	Doridoxa, modified from http://divegall
	ery.com/French_Polynesia_1991.jpg

(Superfamily) **Proctonotoidea**
Diets are as follows (W. Rudman, pers. comm.). Proctonotidae
  (bryozoan feeders), Curnonidae (one known to feed on an
  octocoral), Dironidae (most bryozoan feeders, but a species of
  Dirona is an opportunistic predator (bryozoans, gastropods,
  crustaceans, hydroids, and ascidians).
Families Proctonotidae (= Zephyrinidae), Curnoidae (=
  Charcotiidae), Dironidae, and Lemindidae
	Caldukia (Proctonotidae), from FA2
	
	Pseudotritonia (Curnoidae), from FA2

(Suborder) **Aeolidiina**

Body with numerous cerata containing arms of digestive gland and terminal cnidosac. Oral tentacles present, and rhinophores lack sheath. No gills. Anus
  anterior on right. One or three teeth in each radular row. Most are cnidarian feeders.
Bouchet and Rocroi (2005) and Bouchet et al. (2017) recognised three superfamilies, but these are reduced to two to reflect the results of Korshunova et al.
  (2017).
Uncertain relationships: Pseudovermidae

| (Superfamily) **Fionoidea** (= Flabellinoidea, Pleuroprocta)
Diets are as follows: Pinufiidae (a single tropical species, with
  symbiotic zooxanthellae; feeds on the coral *Porites*);
  Embletoniidae, Eubranchidae, Flabellinidae, and Notaeolidiidae,
  (hydroids), Fionidae (goose barnacles), and Calmidae (cephalopod
  and fish eggs).
Using molecular and morphological data, Korshunova et al. (2017)
  recently split the Tergipedidae into three families and the
  Flabellinidae into several families.
Families Fionidae, Abronicidae, Apataidae, Calmidae,
  Coryphellidae, Cumanotidae, Cuthonellidae, Cuthonidae, Dotidae,
  Eubranchidae, Flabellinidae, Murmaniidae, Paracoryphellidae,
  Pinufiidae, Samlidae, Tergipedidae, Trinchesiidae, and
  Unidentiidae | *Pinufius*, from FA2

Fiona, from FA2

Eubranchus, from FA2

Cuthona, from FA2 |

(*Continued*)

(Superfamily) **Aeolidioidea**

Diets as follows: Aeolidiidae (anemones, one genus on hydroids), Glaucidae (benthic species feed on hydroids, except *Phyllodesmium* which feed on octocorals and many have symbiotic zooxanthellae; while the pelagic *Glaucus* feeds on siphonophores, Facelinidae (mostly hydroids, but *Favorinus* feeds on eggs of other opisthobranchs); Fionidae is also pelagic.

Pseudovermidae was previously included in this superfamily but is now unassigned under Aeolidiina.

Families Aeolidiidae, Babakinidae, Facelinidae (= Favorinidae), Flabellinidae, Flabellinopsidae, Glaucidae, Myrrhinidae, Notaeolidiidae, Piseinotecidae, and Pleurolidiidae (= Protaeolidiidae)

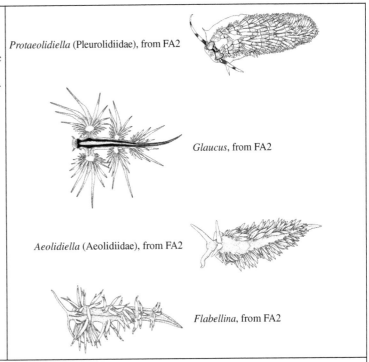

Protaeolidiella (Pleurolidiidae), from FA2

Glaucus, from FA2

Aeolidiella (Aeolidiidae), from FA2

Flabellina, from FA2

(Subcohort) **Tectipleura** Pennsylvanian (Middle) to present (315.2–0 Ma) This grouping is sister to Nudipleura and comprises Euopisthobranchia and Panpulmonata (Schrödl et al. 2011).	

(Megaorder) **Euopisthobranchia**

Jurassic (Middle) to present (164.7–0 Ma)

This grouping, named by Jörger et al. (2010), includes the Aplysiida, Cephalaspidea (without Acteonoidea), Pteropoda, Runcinacea, and Umbraculoidea, which form a monophyletic group in the molecular trees (e.g., Dinapoli & Klussmann-Kolb 2010 and Jörger et al. 2010).

(Order) **Umbraculida** (= Umbraculomorpha)

Paleogene (Paleocene) to present (61.7–0 Ma)

The umbrella slugs – with cap-shaped shells, two pairs of tentacles, and a plicatidium on the right side.

This group is the sister to all other euopisthobranchs (e.g., Jörger et al. 2010; Göbbeler & Klussmann-Kolb 2011; Zapata et al. 2014).

(Superfamily) **Umbraculoidea** (= Tylodinoidea)

Paleogene (Paleocene) to present (61.7–0 Ma)

This small group of sponge feeders have a rather large body size and are warty in appearance.

Families Umbraculidae and Tylodinidae

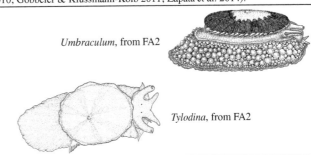

Umbraculum, from FA2

Tylodina, from FA2

(Clade) **Pleurocoela** (= Placoesophaga)

Jurassic (Middle) to present (164.7–0 Ma)

This grouping was introduced by Thiele (1926) for Cephalaspidea and the pteropods. Here we modify it to include the aplysioideans. These taxa form a monophyletic grouping in some phylogenetic analyses. This same group was named Placoesophaga by Medina et al. (2011).

(Order) **Cephalaspidea** (= Bulliformes)

Jurassic (Middle) to present (164.7–0 Ma)

This group includes most of the benthic shelled 'opisthobranchs', most of which have a gizzard having three–four plates, usually a mantle cavity, head-shield, and sometimes parapodia. They are carnivorous or herbivorous.

This taxon has in the past included the Acteonoidea and Ringiculoidea, but those taxa are now excluded (see above).

(Suborder) **Bulloidei**

Jurassic (Middle) to present (167.7–0 Ma)

We use this name established by Amitrov (1984) as a suborder (Bulloidei) of Bulliformes to avoid using a subordinal name that would be a homonym of the genus name *Bullina*.

(*Continued*)

(Superfamily) **Bulloidea**

Jurassic (Middle) to present (164.7–0 Ma)

External shell, head-shield with siphon-like folds; parapodia small.
  Herbivorous. Distinctive radula and with three gizzard plates.
  Mantle cavity with large plicatidium.

Retusids have lost the radula, and Rhizoridae have lost the gizzard
  plates and the radula.

Families Bullidae, Retusidae, Rhizoridae, and Tornatinidae (=
  Acteocinidae)

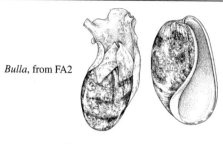

Bulla, from FA2

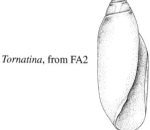

Retusa, from FA2

Tornatina, from FA2

(Superfamily) **Haminoeoidea**

Cretaceous (Lower) to present (130–0 Ma)

Have head shield and parapodia. Internal whorls of external shell
  dissolved. Mantle cavity with large plicatidium. Herbivorous.
  Three gizzard plates.

Families Haminoeidae (= Bullactidae and Smaragdinellidae)

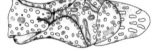

Haminoea, from FA2

Liloa (Haminoeidae), from FA2

Atys (Haminoeidae), from FA2

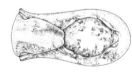

Smaragdinella (Haminoeidae),
  from FA2

(Continued)

(Superfamily) **Philinoidea**

Jurassic (Middle) to present (164.7–0 Ma)

Members of this grouping have three gizzard plates (except Gastropteridae and Aglajidae, the latter also lack a radula and jaws). Philinids and scaphandrids have external shells while Gastropteridae, Philinoglossidae, and Aglajidae have a reduced internal shell or no shell. Philinoglossidae are tiny interstitial slugs.

All members of this group are carnivorous, preying on a wide variety of invertebrates, although each tends to specialise.

Families Philinidae, Aglajidae, Alacuppidae, Colpodaspididae, Gastropteridae, Laonidae, Philinoglossidae (= Plusculidae), Philinorbidae, and Scaphandridae

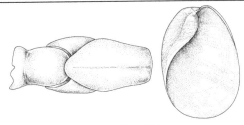

Philine, from FA2

Chelidonura (Aglajidae), from FA2

Colpodaspis, from FA2

(Superfamily) **Cylichnoidea** (= Diaphanoidea)

Jurassic (Middle) to present (167.7–0 Ma)

Shell external to internal. An almost certainly polyphyletic group based on recent molecular analyses.

Families Cylichnidae, Colinatydidae, Diaphanidae, Eoscaphandridae, and Mnestiidae

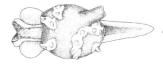

Diaphana, from FA2

Adamnestia (Cylichnidae), from FA2

Roxaniella (Cyclichnidae), from FA2

(Superfamily) **Newnesioidea**

Jurassic (Upper) to present (160–0 Ma)

According to Moles et al. (2017), this family is sister to all other cephalaspideans. It contains only three known species from the Antarctic.

Family Newnesiidae

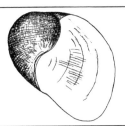

Newnesia, modified from Dell (1990)

(Suborder) **Runcinina**

Present (0 Ma)

Malaquias et al. (2009) excluded the Runcinoidea from the Cephalaspidea on the basis of their molecular phylogeny. They have been shown to be sister to Anaspidea + Pteropoda (Jörger et al. 2010; Göbbeler & Klussmann-Kolb 2011).

(Continued)

(Superfamily) **Runcinoidea**	
Present (0 Ma) Members of this group have four gizzard plates and no mantle cavity. A shell, if present, is tiny and posterior. Herbivorous, feeding on filamentous green algae. Families Runcinidae and Ilbiidae	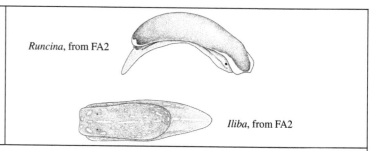 *Runcina*, from FA2 *Iliba*, from FA2

(Order) Aplysiida (= Aplysiomorpha, Anaspidea)

Jurassic (Middle) to present (164.7–0 Ma)

The sea hares have a gizzard with many plates, and a plicate gill in the mantle cavity. These herbivores may be a subgroup of Cephalaspidea. We follow Bouchet et al. (2017) in using Aplysiida in preference to Aplysiomorpha and Anaspidea.

(Superfamily) **Akeroidea**	
Jurassic (Middle) to present (164.7–0 Ma) Coiled external shell. Head shield small, no tentacles. Pallial caecum very long. Herbivores feeding on filamentous green and red algae. Family Akeridae	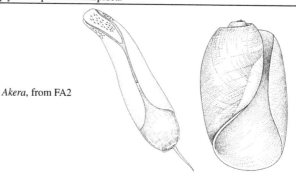 *Akera*, from FA2

(Superfamily) **Aplysioidea**	
Neogene (Miocene) to present (23–0 Ma) Reduced, partially internal shell, or shell in adult absent. Two pairs of tentacles and parapodia. Opaline and purple glands present. Aplysioideans are all herbivores. Family Aplysiidae	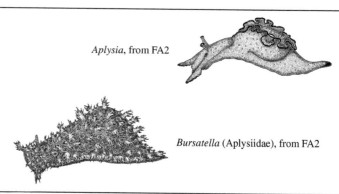 *Aplysia*, from FA2 *Bursatella* (Aplysiidae), from FA2

(Order) Pteropoda

Cretaceous (Lower) to present (136.4–0 Ma)

These two very different groups of pelagic 'opisthobranchs' have been variously treated as monophyletic or unrelated. However, the molecular analyses of Klussmann-Kolb and Dinapoli (2006) and Burridge et al. (2017) have shown the group to be monophyletic and sister to the Aplysiida.

(Suborder) Thecosomata

Paleogene (Eocene) to present (55.8–0 Ma)

The shelled planktonic 'pteropods'. They have a gizzard, and the radula has three–five teeth per row. Many are suspension feeders using a mucous web.

(Minorder) Euthecosomata

Paleogene (Eocene) to present (48.6–0 Ma)

Adult shell present.

(Continued)

(Superfamily) **Cavolinioidea** Paleogene (Eocene) to present (48.6–0 Ma) Families Cavoliniidae (= Cliidae, Cuvierinidae), Creseidae, Praecuvierinidae †, and Sphaerocinidae †	 *Creseis*, from FA2 *Cavolinia*, from Gray (1842) *Clio* (Cavoliniidae), from Gray (1842)
(Superfamily) **Limacinoidea** Paleogene (Eocene) to present (48.6–0 Ma) Limacinidae (= Spiratellidae)	 *Limacina*, from Gray (1842)
(Minorder) **Pseudothecosomata**	

(Superfamily) **Cymbulioidea**
Present (0 Ma)
Cymbuliidae have a 'pseudoconch' and no radula, Desmopteridae
  lack a shell or pseudoconch and gizzard, while Peraclidae have an
  adult shell.
Cymbulioideans feed on both phyto- and zooplankton.
Families Cymbuliidae, Desmopteridae, and Peraclidae
	
	Cymbulia, from FA2
	
	Corolla (Cymbuliidae), from FA2
	
	Desmopterus, from FA2

(Continued)

(Suborder) **Gymnosomata**

Cretaceous (Lower) to present (136.4–0 Ma)

The shell-less 'pteropods' which like the thecosomes spend their entire life in the plankton.

Active carnivores on zooplankton and other small pelagic animals and have grasping arms, a jawed proboscis, hook sacs, and suckered arms in
Pneumodermatidae. Like the thecosomes, they swim using a pair of 'wings' (parapodia). They lack a gizzard.

(Superfamily) **Clionoidea**
 	
Cretaceous (Lower) to present (136.4–0 Ma)
Members of this group have the body divided into a head and trunk.
Some are specialised feeders, for example *Clione limacina*
  apparently feeds exclusively on the thecosome *Spiratella*.
Families Clionidae, Cliopsidae, Notobranchaeidae, and
  Pneumodermatidae

	Paraclione (Clionidae), from FA2

(Superfamily) **Hydromyloidea** (= Gymnoptera) Present (0 Ma) Sometimes this group is treated as a separate suborder, Gymnoptera. The body is not divided into a separate head and trunk. Families Hydromylidae and Laginiopsidae	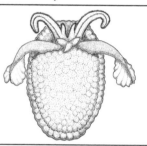 *Hydromyles*, from FA2

'Pulmonata'

This group was treated as a subclass in earlier classifications but has since been included within Heterobranchia. Recent molecular phylogenies have shown
that some groups previously treated as 'lower heterobranchs' and 'opisthobranchs' are included amongst the 'pulmonates'.

(Infrasubcohort) **Panpulmonata**

Pennsylvanian (Middle) to present (315.2–0 Ma)

Panpulmonata was introduced by Jörger et al. (2010) to include Acochlidia, Amphiboloidea, Eupulmonata, Glacidorboidea, Hygrophila, Pyramidelloidea,
Sacoglossa, and Siphonarioidea. This includes highly divergent basal taxa often included in Basommatophora (Amphiboloidea, Ellobioidea, Hygrophila,
and Siphonarioidea), a poly- or paraphyletic grouping, encompassing all non-stylommatophoran and systellommatophoran taxa (i.e., Geophila)
traditionally included in the pulmonates. Another grouping, Archaeopulmonata, was established to encompass all basommatophorans except Hygrophila.

Two groups, Acochlidia and Sacoglossa, have traditionally been associated with 'opisthobranchs' and pyramidelloideans with the 'lower heterobranchs',
but molecular analyses have shown that they are nested within the panpulmonate clade.

(Superorder) **Pylopulmonata**

Permian (Lopingian) to present (259–0 Ma)

This name, first introduced in the unpublished thesis of Teasdale (2017), was used by Bouchet et al. (2017). It includes the operculate Amphiboloidea,
Glacidorboidea, and Pyramidelloidea which form a clade in several molecular phylogenies (Dinapoli & Klussmann-Kolb 2010; Holznagel et al. 2010;
Jörger et al. 2010; Teasdale 2017).

(Order) **Amphibolida**

Neogene (Pliocene) to present (5.3–0 Ma)

This ordinal name was first used by Starobogatov (1970). The shell is typically globular, and these taxa are all intertidal in estuarine habitats.

(Suborder) **Amphibolina**

Neogene (Pliocene) to present (5.3–0 Ma)

This subordinal grouping was first used by Van Mol (1976) (as Amphibolacea).

(Continued)

(Superfamily) **Amphiboloidea** Neogene (Pliocene) to present (5.3–0 Ma) This operculate, estuarine group is particularly diverse in Australia (Golding et al. 2007; Golding 2012). In the analysis of Golding (2012), *Phallomedusa* nested in Amphibolidae, but, pending further investigation, we consider it a separate family given the major morphological differences, in particular in its reproductive anatomy. Families Amphibolidae, Maningrididae and Phallomedusidae	 *Amphibola*, from Cooke et al. (1927)  *Phallomedusa*, from FA2
(Suborder) **Glacidorbina** Neogene (Miocene) to present (23.03–0 Ma)	
(Superfamily) **Glacidorboidea** Neogene (Miocene) to present (23.03–0 Ma) Fresh water; this operculate group is found only in Australia and southern South America (Ponder & Avern 2000; Rumi et al. 2015). Family Glacidorbidae	*Glacidorbis*, from Ponder (1986) 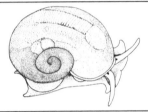

(Order) **Pyramidellida** (= Pyramidellimorpha) Permian (Lopingian) to present (259–0 Ma) The shell is typically elongate but some with short spire. A few (e.g., *Amathina*) with limpet-shaped shell. Cephalic tentacles are triangular, with the eyes on their inner bases, and there is often a distinct PMO. The proboscis is very long, a radula is absent, but there is a piecing stylet and a buccal pump. Ectoparasites on a wide range of invertebrates.

(Continued)

(Superfamily) **Pyramidelloidea**
Permian (Lopingian) to present (259–0 Ma)
Families Pyramidellidae, Amathinidae, and Heteroneritidae †

Pyramidella, from FA2

Otopleura (Pyramidellidae), from FA2

Chemnitzia (Pyramidellidae), from FA2

Linopyrga (Pyramidellidae), from Hedley (1916)

Odostomia (Pyramidellidae), from FA2

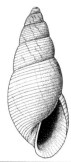

Leucotina (Amathinidae), from FA2

(*Continued*)

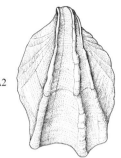

Amathina, from FA2

(Superorder) Siphonariia

Jurassic (Upper) to present (163.5–0 Ma)

The name Siphoglossa was introduced by Medina et al. (2011) for Siphonariidae + Sacoglossa, but in more recent analyses these taxa do not form a monophyletic group (e.g., Kano et al. 2016).

(Order) Siphonariida

Jurassic (Upper) to present (163.5–0 Ma)

(Superfamily) Siphonarioidea

Jurassic (Upper) to present (163.5–0 Ma)

This group of marine 'pulmonate' limpets contains the widespread intertidal Siphonariidae and the extinct family Acroreiidae based on an Eocene fossil.

Families Siphonariidae and Acroreiidae †

Siphonaria, shell from FA2, ventral view of animal from Yonge (1952)

(Superorder) Acochlidia (= Acochlidiacea, Acochlidimorpha)

Present (0 Ma)

Interstitial slugs and includes a few freshwater taxa, the only non-marine members of what were the 'opisthobranchs'. Members have no shell or mantle cavity. Some have dermal spicules. There are no jaws or gizzard and one to three teeth in each radular row.

(Order) Acochlida

(Superfamily) Acochlidioidea (= Hedylopsoidea, Palliohedyloidea, Strubellioidea)

Marine, fresh water (Acochlidiidae), and terrestrial (Aitengidae)

Families Acochlidiidae (= Palliohedylidae, Strubelliidae), Aitengidae, Bathyhedylidae, Hedylopsidae, Pseudunelidae, and Tantulidae

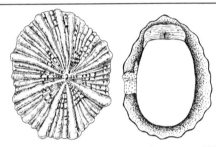

Acochlidium, from FA2

Hedylopsis (Hedylopsidae)

(Superfamily) Parhedyloidea

Marine only.

Families Parhedylidae and Asperspinidae

Asperspina, from FA2

Paragranitas (Parhedylidae), from FA2

(Superorder) Sacoglossa (= Ascoglossa)

Paleogene (Eocene) to present (47.8–0 Ma)

Shelled or shell-less slugs with a single row of radular teeth. Suctorial feeders on green algae.

(Continued)

(Superfamily) **Oxynooidea**
Paleogene (Eocene) to present (47.8–0 Ma)
Cylindrobullidae has been placed in its own superfamily, in some
  classifications either Cephalaspidea or Sacoglossa. It is sister to the
  rest of the Oxynooidea in several recent molecular phylogenies.
  Juliidae are the 'bivalve gastropods'.
Families Oxynoidae, Cylindrobullidae, Juliidae, and Volvatellidae
	Volvatella, from FA2
	
	Roburnella (Oxynoidae), from FA2
	
	Julia, from FA2

(Superfamily) **Plakobranchoidea** (= Limapontioidea)
Present (0 Ma)
Families Plakobranchidae (often spelt Placobranchidae; = Elysiidae,
  Boselliidae), Costasiellidae, Hermaeidae (= Caliphyllidae),
  Jenseneriidae, and Limapontiidae
	Plakobranchus, from FA2
	
	Costasiella, from FA2
	
	Cyerce (Hermaeidae), from FA2
	
	Hermaea, from FA2

(Superfamily) **Platyhedyloidea** Present (0 Ma) Family Platyhedylidae	*Platyhedyle*, from FA2

(Continued)

(Megaorder) Hygrophila Jurassic (Middle) to present (174.1–0 Ma) This is a mostly freshwater group. Bouchet et al. (2017) include Acroloxoidea and Planorboidea within Lymnaeoidea, but their justification for doing so is not clear, and some of it is based on unpublished work. We maintain the *status quo* pending further studies.	
(Order) Lymnaeida Jurassic (Middle) to present (174.1–0 Ma)	
(Superfamily) Chilinoidea Paleogene (Eocene) to present (54–0 Ma) Contains freshwater and estuarine taxa. Families Chilinidae and Latiidae	 *Chilina*, from Fischer (1880–1887)  *Latia*, from Powell (1979)
(Superfamily) Acroloxoidea Neogene (Pliocene) to present (5.332–0 Ma) A small group of freshwater limpets. Family Acroloxidae	*Acroloxus*, shells public domain; animal from Hubendick (1978) 
(Superfamily) Lymnaeoidea Jurassic (Middle) to present (167.7–0 Ma) Lymnaeids have triangular tentacles, are dextral and always found in fresh water. The North American genus *Lanx* is limpet-like. Family Lymnaeidae	 *Pseudosuccinea* (Lymnaeidae), from FA2  *Radix* (Lymnaeidae), from Cooke et al. (1927) *Lanx* (Lymnaeidae), original

(Continued)

(Superfamily) **Planorboidea**

Jurassic (Middle) to present (174.1–0 Ma)

Planorboideans are all in fresh water, are sinistral, and have long, slender tentacles and (except Physidae) a pseudobranch, and haemoglobin in their blood.

Bouchet et al. (2017) split the planorbids into three families based on the molecular results of Albrecht et al. (2007). We do not accept that the *Bulinus*-group should be elevated to family rank based on the results presented, although we do accept that the limpet-like *Burnupia* may be justified in being recognised as a separate family.

Physidae do not have haemoglobin or a pseudobranch and this family is only tentatively grouped with the planorbids as some molecular phylogenies suggest it is more closely related to Lymnaeidae.

Families Planorbidae (= Ancylidae and Bullinidae), Burnupiidae, Physidae, and Clivunellidae †

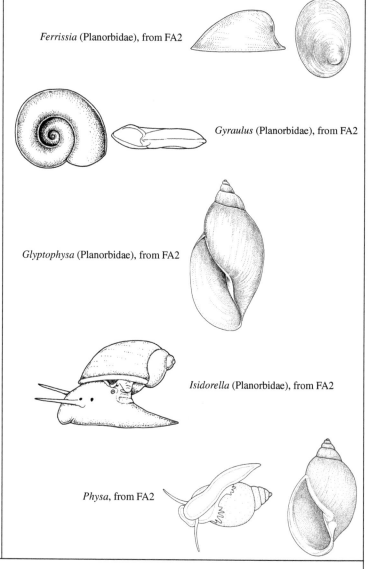

Ferrissia (Planorbidae), from FA2

Gyraulus (Planorbidae), from FA2

Glyptophysa (Planorbidae), from FA2

Isidorella (Planorbidae), from FA2

Physa, from FA2

(Megaorder) **Eupulmonata**

Jurassic (Upper) to present (157.3–0 Ma)

This clade encompasses the remaining 'pulmonates'.

(Order) **Ellobiida** (= Ellobiiformes, Ellobiacea, Actophila, Trimusculida)

Jurassic (Upper) to present (157.3–0 Ma)

Includes Ellobioidea and Trimusculoidea, both comprising intertidal taxa and, in the Ellobioidea, some that are supratidal and a few terrestrial.

(Continued)

(Superfamily) **Ellobioidea**

Jurassic (Upper) to present (157.3–0 Ma)

Family Ellobiidae (with five 'subfamilies'), includes mainly marine and a few terrestrial taxa.

Ophicardelus (Ellobiidae), shell from Powell (1979), animal from Hedley (1916)

Carychium (Ellobiidae), from Morton (1955b)

(Superfamily) **Trimusculoidea**

Paleogene (Oligocene) to present (33.9–0 Ma)

This entirely marine group was combined with Ellobioidea by Romero et al. (2016). However, their phylogeny does not preclude the existence of two clades, and, given the considerable morphological disparity between these two groups, we prefer to keep them separate until their relationships are investigated further.

Otinidae and Smeagolidae, both marine, are treated as one (Otinidae) by Bouchet et al. 2017 on the basis of a molecular analyses by Dayrat et al. (2011) and Romero et al. (2016). Given that their results were not clear-cut, and given the considerable morphological differences, we prefer to keep them as separate families.

Families Trimusculidae, Otinidae, and Smeagolidae

Trimusculus, shell from FA2, animal from Yonge (1958)

Otina, from Morton (1955a)

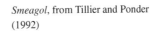

Smeagol, from Tillier and Ponder (1992)

(Infrasubcohort) **Geophila**

Pennsylvanian (Middle) to present (315.2–0 Ma)

A grouping that has sometimes been used to encompass all systellommatophoran and stylommatophoran pulmonates (e.g., Mordan & Wade 2008).

(Superorder) **Systellommatophora** (= Soleolifera)

Present (0 Ma)

This group contains only shell-less slugs.

(Order) **Onchidiida**

(Continued)

(Superfamily) **Onchidioidea** Littoral and supralittoral marine, a few terrestrial. Family Onchidiidae	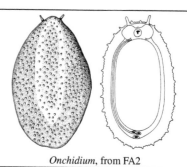 *Onchidium*, from FA2

(Order) **Veronicellida**

(Superfamily) **Veronicelloidea** An entirely terrestrial group. Families Veronicellidae (= Vaginulidae), and Rathouisiidae	*Atopos* (Rathousiidae), from FA2   *Vaginulus* (Veronicellidae), from FA2

(Superorder) **Stylommatophora**

Pennsylvanian (Middle) to present (315.2–0 Ma)

This entire taxon is terrestrial, containing the great majority of the land snails and land slugs.

(Order) **Helicida**

Pennsylvanian (Middle) to present (315.2–0 Ma)

Extinct families unassigned to superfamily by Bouchet et al. (2017): Anadromidae †, Anastomopsidae †, Cylindrellinidae †, Grandipatulidae †, Grangerellidae †, Palaeoxestinidae †, and Scalaxidae †

(Suborder) **Achatinina** (= 'Achatinoid Sigmurethra')

Paleogene (Eocene) to present (40.4–0 Ma)

This clade includes omnivorous and carnivorous taxa.

(Superfamily) **Achatinoidea** Paleogene (Eocene) to present (40.4–0 Ma) Families Achatinidae (= Subulinidae), Aillyidae, Ferussaciidae, and Micractaeonidae	*Achatina*, from FA2   *Ferussacia*, from FA2 *Eremopeas* (Achatinidae), from FA2 

(*Continued*)

(Superfamily) **Streptaxoidea** Paleogene (Eocene) to present (40.4–0 Ma) Families Streptaxidae and Diapheridae	 *Gulella* (Streptaxidae), from FA2 *Perrottetia* (Streptaxidae), from Schileyko (2000a)  *Diaphera*, from Schileyko (2000a)

(Suborder) **Scolodontina** Neogene (Miocene) to present (20.44–0 Ma)	

(Superfamily) **Scolodontoidea** Neogene (Miocene) to present (20.44–0 Ma) Scolodontidae was included in Rhytidoidea by Bouchet and Rocroi (2005) but was placed in its own suborder and superfamily by Bouchet et al. (2017) based on the molecular results of Ramírez et al. (2012). Family Scolodontidae	*Scolodonta*, redrawn and modified from Pilsbry (1900b)  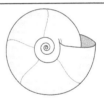

(Suborder) **Helicina** Pennsylvanian (Middle) to present (315.2–0 Ma) This suborder contains the remaining stylommatophorans. Superfamilies of uncertain relationship:

(*Continued*)

(Superfamily) **Coelociontoidea** Present (0 Ma) Family Coelociontidae	*Coelocion*, from FA2
(Superfamily) **Papillodermatoidea** Present (0 Ma) Family Papillodermatidae	*Papilloderma*, from Castillejo et al. (1995)
(Superfamily) **Plectopyloidea** Quaternary (Pleistocene) to present (0.126–0 Ma) Families Plectopylidae, Corillidae, and Sculptariidae	*Craterodiscus* (Corillidae), from FA2
(Superfamily) **Punctoidea** Paleogene (Paleocene) (59.2–0 Ma) The extinct family Anastomopsidae was included here by Bouchet and Rocroi (2005) but is unassigned to superfamily by Bouchet et al. (2017). Families Punctidae, Charopidae, Cystopeltidae, Discidae, Endodontidae, Helicodiscidae, Oopeltidae, and Oreohelicidae	*Anguispira* (Discidae), from Binney (1878) *Ptychodon* (Charopidae), from Schileyko (2001) *Otoconcha* (Charopidae), original, from F. Brooks photograph *Hedleyoconcha* (Charopidae), from Schileyko (2001) *Cystopelta*, from FA2
(Superfamily) **Testacelloidea** Paleogene (Eocene) to present (41.2–0 Ma) Taxa now placed in Oleacinoidea were previously included here (e.g., Bouchet & Rocroi 2005). Family Testacellidae	*Testacella*, from Griffith and Pidgeon (1834)

(Continued)

(Superfamily) Urocoptoidea Cretacous (Upper) to present (83.6–0 Ma) Families Eurocoptidae, Cerionidae, Epirobiidae, Eucalodiidae, and Holospiridae	*Cerion*, from Schileyko (1999a)
(Infraorder) Succineoidei (= Elasmognatha) Paleogene (Paleocene) to present (66–0 Ma)	
(Superfamily) Succineoidea Paleogene (Paleocene) to present (66–0 Ma) Family Succineidae	*Succinea*, from FA2 *Omalonyx* (Succineidae), from Cooke et al. (1927)
(Superfamily) Athoracophoroidea Present (0 Ma) Family Athoracophoridae	*Athoracophorus*, from Ponder and Warren (1965)
(Infraorder) Oleacinoidei Paleogene (Paleocene) to present (59.2–0 Ma)	
(Superfamily) Oleacinoidea Paleogene (Paleocene) to present (59.2–0 Ma) Families Oleacinidae, and Spiraxidae	*Spiraxis*, from Schileyko (2000a) *Euglandina* (Oleacinidae), shell from Schileyko (2000a); animal from Binney (1878)

(*Continued*)

(Superfamily) **Haplotrematoidea** Paleogene (Eocene) to present (42.5–0 Ma) Family Haplotrematidae	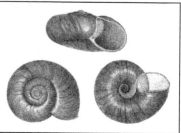 *Haplotrema*, from Binney (1878) 

(Infraorder) **Pupilloidei** Cretaceous (Lower) to present (140.2–0 Ma)	

(Superfamily) **Pupilloidea** Cretaceous (Lower) to present (140.2–0 Ma) Achatinellidae, Cochlicopidae (with Amastridae), Enidae (with Cerastidae), and Partulidae (with Draparnaudiidae) were included in their own superfamilies in Bouchet and Rocroi (2005). The extinct family Cylindrellinidae was included here by Bouchet and Rocroi (2005) but was unassigned to superfamily by Bouchet et al. (2017). Families Pupillidae, Achatinellidae, Agardhiellidae, Amastridae, Argnidae, Azecidae, Cerastidae, Chondrinidae, Cochlicopidae, Draparnaudiidae, Enidae, Fauxulidae, Gastrocoptidae (= Hypselostomatidae), Lauriidae, Odontocycladidae, Orculidae, Pagodulinidae, Partulidae, Pleurodiscidae, Pyramidulidae, Spelaeoconchidae, Spelaeodiscidae, Strobilopsidae, Truncatellinidae, Valloniidae, and Vertiginidae	*Pupilla*, from Schileyko (1998a) *Gastrocopta*, from Schileyko (1998b)  *Pleurodiscus*, from FA2 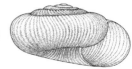 *Granaria* (Chondrinidae), from Schileyko (1998a)  *Mastoides* (Enidae), from Schileyko (1998b) 

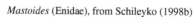

(Continued)

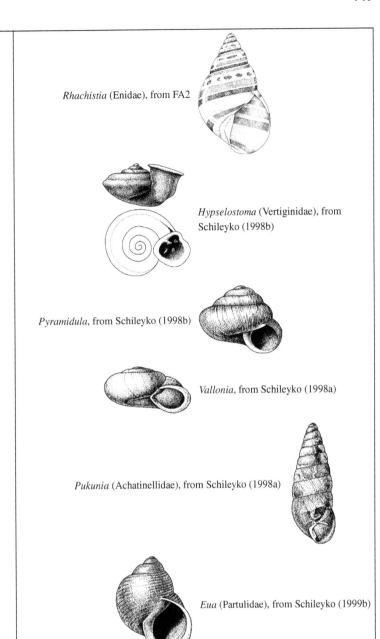

Rhachistia (Enidae), from FA2

Hypselostoma (Vertiginidae), from
Schileyko (1998b)

Pyramidula, from Schileyko (1998b)

Vallonia, from Schileyko (1998a)

Pukunia (Achatinellidae), from Schileyko (1998a)

Eua (Partulidae), from Schileyko (1999b)

(*Continued*)

	Samoana (Partulidae), from Schileyko (1999b) 
(Infraorder) Clausilioidei Paleogene (Paleocene) to present (66–0 Ma)	
(Superfamily) Clausilioidea Paleogene (Paleocene) to present (66–0 Ma) The extinct family Anadromidae was included here by Bouchet and Rocroi (2005) but was unassigned to superfamily by Bouchet et al. (2017). Families Clausiliidae, Filholiidae †, and Palaeostoidae †	*Formosana* (left) and *Dextroformosana* (right) (Clausiliidae), from Schileyko (2000b) 
(Infraorder) Orthalicoidei Paleogene (Eocene) to present (55–0 Ma)	
(Superfamily) Orthalicoidea Paleogene (Eocene) to present (55–0 Ma) The extinct family Grangerellidae was included here by Bouchet and Rocroi (2005) but was unassigned to superfamily by Bouchet et al. (2017). Families Orthalicidae, Amphibulimidae, Bothriembryontidae (=Placostylidae), Bulimulidae, Megaspiridae, Odontostomidae, Simpulopsidae, and Vidaliellidae †	*Bothriembryon*, from FA2  *Placostylus* (Bothyriembryontidae), from FA2 
(Infraorder) Rhytidoidei Paleogene (Eocene) to present (54–0 Ma)	

(*Continued*)

(Superfamily) **Rhytidoidea**
Paleogene (Eocene) to present (54–0 Ma)
Bouchet et al. (2017) combined Rhytidoidea and Acavoidea based
  on molecular work.
Families Rhytididae (= Chlamydephoridae), Acavidae, Caryodidae,
  Clavatoridae, Dorcasiidae, Macrocyclidae, Megomphicidae, and
  Strophocheilidae
	Strangesta (Rhytididae), from FA2
	
	Hedleyella (Caryodidae), from FA2
	
	Acavus, from Schileyko (1999a)

(Infraorder) **Limacoidei** Paleogene (Paleocene) to present (66–0 Ma)	
(Superfamily) **Limacoidea** Paleogene (Paleocene) to present (66–0 Ma) Families Limacidae, Agriolimacidae, Boettgerillidae, and Vitrinidae	*Limax*, from FA2 *Insulivitrina* (Vitrinidae), from Schileyko (2003b)
(Superfamily) **Gastrodontoidea** Paleogene (Paleocene) to present (61.7–0 Ma) Families Gastrodontidae, Oxychilidae, and Pristilomatidae	*Oxychilus*, from FA2 *Gastrodonta*, from FA2

(Continued)

(Superfamily) **Trochomorphoidea**
Neogene (Miocene) to present (13.82–0 Ma)
Dyakiidae and Staffordiidae were included in their own
  superfamilies in the Bouchet and Rocroi (2005) classification.
Families Trochomorphidae, Chronidae, Dyakiidae, Euconulidae,
  and Staffordiidae
	Trochomorpha, from FA2
	
	Sasakina (Dyakiidae), from Schileyko (2003a)
	
	Dyakia, from Schileyko (2003a)
	
	Staffordia, from Schileyko (2003a)

(Superfamily) **Parmacelloidea** Paleogene (Eocene) to present (40.4–0 Ma) Families Parmacellidae, Milacidae, and Trigonochlamydidae	*Milax*, from FA2 *Parmacella*, from Cooke et al. (1927)
(Superfamily) **Zonitoidea** Neogene (Miocene) to present (15.97–0 Ma) Family Zonitidae	*Zonitoides* (Zonitidae), from FA2
(Superfamily) **Helicarionoidea** Paleogene (Eocene) to present (37.2–0 Ma) Families Helicarionidae, Ariophantidae, and Urocyclidae	*Helicarion*, from FA2 *Nitor* (Helicarionidae), from FA2 *Ariophanta*, from Schileyko (2003a)
(Infraorder) **Arionoidei** Neogene (Miocene) to present (20.44–0 Ma)	

(Continued)

(Superfamily) **Arionoidea** Neogene (Miocene) to present (20.44–0 Ma) Families Arionidae, Anadenidae, Ariolimacidae, Binneyidae, and Philomycidae	*Arion*, from FA2
(Infraorder) **Helicoidei** Cretaceous (Upper) to present (100.5–0 Ma)	
(Superfamily) **Sagdoidea** Cretaceous (Upper) to present (100.5–0 Ma) Families Sagdidae, Solaropsidae, and Zachrysiidae	*Sagda*, from Schileyko (1998b) *Corneosagda* (Sagdidae), from Schileyko (1998b)
(Superfamily) **Helicoidea** Cretaceous (Upper) to present (100.5–0 Ma) Families Helicidae, Camaenidae (= Badybaenidae, Helicostylidae), Canariellidae, Cepolidae, Echinichidae, Elonidae, Geomitridae (= Cochlicellidae), Helicodontidae, Hygromiidae, Labyrinthidae, Pleurodontidae, Polygyridae, Sphincterochilidae, Thysanophoridae, Trichodiscinidae, Trissexodontidae, and Xanthonychidae	*Cornu aspersum* (Helicidae), from FA2 *Cochlicella* (Geomitridae), from FA2 *Bradybaena* (Camaenidae), from FA2 *Glyptorhagada* (Camaenidae), from FA2 *Ordtrachia* (Camaenidae), from FA2

References

Abdelkrim, J., Aznar-Cormano, L., Fedosov, A., Kantor, Y. I., Lozouet, P., Phuong, M., Zaharias, P. & Puillandre, N. (2018). Exon-capture based phylogeny and diversification of the venomous gastropods (Neogastropoda, Conoidea). *Molecular Biology and Evolution* 35: 2355–2374.

Abel, O. (1916). *Paläobiologie der Cephalopoden aus der gruppe der Dibranchiaten*. Jena, G. Fischer.

Aberhan, M. (2004). Early Jurassic Bivalvia of northern Chile. Part II. Subclass Anomalodesmata. *Beringeria* 34: 117–154.

Adamkewicz, S. L., Harasewych, M. G., Blake, J., Saudek, D. & Bult, C. J. (1997). A molecular phylogeny of the bivalve mollusks. *Molecular Biology and Evolution* 14: 619–629.

Adamo, S. A. & Chase, R. (1996). Dart shooting in helicid snails: An 'honest' signal or an instrument of manipulation? *Journal of Theoretical Biology* 180: 77–80.

Adamo, S. A., Ehgoetz, K., Sangster, C. & Whitehorne, I. (2006). Signaling to the enemy? Body pattern expression and its response to external cues during hunting in the cuttlefish *Sepia officinalis* (Cephalopoda). *Biological Bulletin* 210: 192–200.

Adams, D. C., Rohlf, F. J. & Slice, D. E. (2013). A field comes of age: Geometric morphometrics in the 21st century. *Hystrix* 24: 7–14.

Aguado, F. & Marin, A. (2007). Warning coloration associated with nematocyst-based defences in aeolidioidean nudibranchs. *Journal of Molluscan Studies* 73: 23–28.

Agudo-Padrón, A. I. (2012). Brazilian snail-eating snakes (Reptilia, Serpentes, Dipsadidae) and their alimentary preferences by terrestrial molluscs (Gastropoda, Gymnophila & Pulmonata): A preliminary overview. *Biological Evidence* 2: 2–3.

Agudo-Padrón, A. I. (2013). Snail-eating snakes ecology, diversity, distribution and alimentary preferences in Brazil. *Journal of Environmental Science and Water Resources* 2: 238–244.

Aguiar, R. & Wink, M. (2005). How do slugs cope with toxic alkaloids? *Chemoecology* 15: 167–177.

Ajeska, R. A. & Nybakken, J. (1976). Contributions to the biology of *Melibe leonina* (Gould, 1852) (Mollusca: Opisthobranchia). *The Veliger* 19: 19–26.

Akasaki, T., Nikaido, M., Tsuchiya, K., Segawa, S., Hasegawa, M. & Okada, N. (2006). Extensive mitochondrial gene arrangements in coleoid Cephalopoda and their phylogenetic implications. *Molecular Phylogenetics and Evolution* 38: 648–658.

Akpan, B. (1982). Limpet grazing on Cretaceous algal-bored ammonites. *Palaeontology* 25: 361–367.

Aksarina, N. A. (1968). Probivalvia – A new class of ancient molluscs [in Russian]. *Novye dannye po geologii i poleznym iskopaemum Zapadnoi Sibiri [New Data on Geology and Natural Resources of Western Siberia]* 3: 77–86.

Aktipis, S. W., Giribet, G., Lindberg, D. R. & Ponder, W. F. (2008). Gastropoda: An overview and analysis, pp. 201–237 *in* W. F. Ponder & Lindberg, D. R. (eds.), *Phylogeny and Evolution of the Mollusca*. Berkeley, CA, University of California Press.

Aktipis, S. W. & Giribet, G. (2010). A phylogeny of Vetigastropoda and other 'archaeogastropods': Reorganizing old gastropod clades. *Invertebrate Biology* 129: 220–240.

Aktipis, S. W., Boehm, E. & Giribet, G. (2011). Another step towards understanding the slit-limpets (Fissurellidae, Fissurelloidea, Vetigastropoda, Gastropoda): A combined five-gene molecular phylogeny. *Zoologica Scripta* 40: 238–259.

Aktipis, S. W. & Giribet, G. (2012). Testing relationships among the vetigastropod taxa: A molecular approach. *Journal of Molluscan Studies* 78: 12–27.

Alba-Tercedor, J. & Sánchez-Tocino, L. (2011). *The use of the SkyScan 1172 high-resolution micro-CT to elucidate if the spicules of the sea slugs (Mollusca: Nudibranchia, Opisthobranchia) have a structural or a defensive function*. SkyScan Users Meeting, Leuven, Belgium, Bruker microCT.

Albano, P. G., Sabelli, B. & Bouchet, P. (2011). The challenge of small and rare species in marine biodiversity surveys: Microgastropod diversity in a complex tropical coastal environment. *Biodiversity and Conservation* 20: 3223–3237.

Albrecht, C. & Glaubrecht, M. (2006). Brood care among basommatophorans: A unique reproductive strategy in the freshwater limpet snail *Protancylus* (Heterobranchia: Protancylidae), endemic to ancient lakes on Sulawesi, Indonesia. *Systematics and Biodiversity* 58: 49–58.

Albrecht, C., Kuhn, K. & Streit, B. (2007). A molecular phylogeny of Planorboidea (Gastropoda, Pulmonata): Insights from enhanced taxon sampling. *Zoologica Scripta* 36: 27–39.

Aldred, R. G., Nixon, M. & Young, J. Z. (1983). *Cirrothauma murrayi* Chun, a finned octopod. *Philosophical Transactions of the Royal Society B* 301: 1–54.

Alexander, J. E. & Covich, A. P. (1991). Predator avoidance by the freshwater snail *Physella virgata* in response to the crayfish *Procambarus simulans*. *Oecologia* 87: 435–442.

Algeo, T., Henderson, C. M., Ellwood, B., Rowe, H., Elswick, E., Bates, S., Lyons, T., Hower, J. C., Smith, C. & Maynard, B. (2012). Evidence for a diachronous Late Permian marine crisis from the Canadian Arctic region. *Geological Society of America Bulletin* 124: 1424–1448.

Allcock, A. L. & Piertney, S. B. (2002). Evolutionary relationships of Southern Ocean Octopodidae (Cephalopoda: Octopoda) and a new diagnosis of *Pareledone*. *Marine Biology* 140: 129–135.

Allcock, A. L., Cooke, I. R. C. & Strugnell, J. M. (2011). What can the mitochondrial genome reveal about higher-level phylogeny of the molluscan class Cephalopoda? *Zoological Journal of the Linnean Society* 161: 573–586.

Allcock, A. L., Lindgren, A. R. & Strugnell, J. M. (2014). The contribution of molecular data to our understanding of cephalopod evolution and systematics: A review. *Journal of Natural History* 48: 1–49.

Allcock, A. L. (2015). Systematics of cephalopods, pp. 1–16 *in* P. Gopalakrishnakone & Malhotra, A. (eds.), *Evolution of Venomous Animals and their Toxins*. Dordrecht, the Netherlands, Springer Science + Business Media.

Allen, J. A. (1958). On the basic form and adaptations to habitat in the Lucinacea (Eulamellibranchia). *Philosophical Transactions of the Royal Society B* 241: 421–484.

Allen, J. A. & Sanders, H. L. (1969). *Nucinella serrei* Lamy (Bivalvia: Protobranchia), a monomyarian solemyid and possible living actinodont. *Malacologia* 7: 381–396.

Allen, J. A. & Scheltema, R. S. (1972). The functional morphology and geographic distribution of *Planktomya henseni*, a supposed neotenic pelagic bivalve. *Journal of the Marine Biological Association of the United Kingdom* 52: 19–31.

Allen, J. A. & Turner, R. D. (1974). On the functional morphology of the family Verticordiidae (Bivalvia), with descriptions of new species from the abyssal Atlantic. *Philosophical Transactions of the Royal Society B* 268: 401–536.

Allen, J. A. & Morgan, R. E. (1981). The functional morphology of Atlantic deep water species of the families Cuspidariidae and Poromyidae (Bivalvia): An analysis of the evolution of the septibranch condition. *Philosophical Transactions of the Royal Society of London B* 294: 413–546.

Allen, J. A. (1985). Recent Bivalvia: Their form and evolution, pp. 337–403 *in* E. R. Trueman & Clarke, M. R. (eds.), *Evolution. The Mollusca.* Vol. 10. New York, Academic Press.

Allen, J. A. (2000). An unusual suctorial montacutid bivalve from the deep Atlantic. *Journal of the Marine Biological Association of the United Kingdom* 80: 827–834.

Alleway, H. K. & Connell, S. D. (2015). Loss of an ecological baseline through the eradication of oyster reefs from coastal ecosystems and human memory. *Conservation Biology* 29: 795–804.

Allgaier, C. (2007). Active camouflage with lichens in a terrestrial snail, *Napaeus* (*N.*) *barquini* Alonso and Ibáñez, 2006 (Gastropoda, Pulmonata, Enidae). *Zoological Science* 24: 869–876.

Allison, P. A. (1987). A new cephalopod with soft parts from the Upper Carboniferous Francis Creek Shale of Illinois, USA. *Lethaia* 20: 117–121.

Alroy, J., Aberhan, M., Bottjer, D. J., Foote, M., Fürsich, F. T., Harries, P. J., Hendy, A. J. W., Holland, S. M., Ivany, L. C., Kiessling, W., Kosnik, M. A., Marshall, C. R., McGowan, A. J., Miller, A. I., Olszewski, T. D., Patzkowsky, M. E., Peters, S. E., Villier, L., Wagner, P. J., Bonuso, N., Borkow, P. S., Brenneis, B., Clapham, M. E., Fall, L. M., Ferguson, C. A., Hanson, V. L., Krug, A. Z., Layou, K. M., Leckey, E. H., Nürnberg, S., Powers, C. M., Sessa, J. A., Simpson, C. T., Tomasovych, A. & Visaggi, C. C. (2008). Phanerozoic trends in the global diversity of marine invertebrates. *Science* 321: 97–100.

Altaba, C. R. (1998). Testing vicariance: Melanopsid snails and Neogene tectonics in the Western Mediterranean. *Journal of Biogeography* 25: 541–551.

Alyakrinskaya, I. O. (2002). Physiological and biochemical adaptations to respiration of hemoglobin-containing hydrobionts. *Biology Bulletin of the Russian Academy of Sciences* 29: 268–283.

Amaudrut, A. (1898). La partie antérieure du tube digestif et la torsion chez les mollusques gastéropodes. *Annales des sciences naturelles. Zoologie* 8: 1–291.

Amitrov, O. V. (1984). Gastropoda: taxa of orders and higher groups, pp. 36–41 *in* L. P. Tatarinov, Shimanskii, V. N. & Amitrov, O. V. (eds.), Spravochnik po sistematike iskopaemykh organizmov *[Handbook on the Taxonomy of Fossil Organisms].* Moscow, Paleontologicheskiĭ Institut (Akademiia nauk SSSR).

Amler, M. R. W. (1989). Die Gattung *Parallelodon* Meek & Worthen (Bivalvia, Arcoida) im mitteleuropäischen Unterkarbon. *Geologica et Palaeontologica* 23: 53–69.

Amler, M. R. W. (1999). Synoptical classification of fossil and Recent Bivalvia. *Geologica et Palaeontologica* 33: 237–248.

Anderson, D. T. (1962). The reproduction and early life histories of the gastropods, *Bembicium auratum* (Quoy and Gaimard) (Fam. Littorinidae), *Cellana tramoserica* (Sowerby) (Fam. Patellidae) and *Melanerita melanotragus* (Smith) (Fam. Neritidae). *Proceedings of the Linnean Society of New South Wales* 87: 62–68.

Anderson, D. T. (1965). The reproduction and early life histories of the gastropods, *Notoacmaea petterdi* (Ten. Woods), *Chlazacmaea flammea* (Quoy and Gaimard) and *Patelloida alticostata* (Angas) (Fam. Acmaeidae). *Proceedings of the Linnean Society of New South Wales* 90: 106–114.

Anderson, R. C., Shimek, R., Cosgrove, J. A. & Berthinier, S. (2007). Giant Pacific octopus, *Enteroctopus dofleini*, attacks on divers. *The Canadian Field Naturalist* 121: 423–425.

André, J., Grist, E. P. M., Semmens, J. M., Pecl, G. T. & Segawa, S. (2009). Effects of temperature on energetics and the growth pattern of benthic octopuses. *Marine Ecology Progress Series* 374: 167–179.

Andree, K. B. & López, M. A. (2013). Species identification from archived snail shells via genetic analysis: A method for DNA extraction from empty shells. *Molluscan Research* 33: 1–5.

Andrews, E. A. (1935). The egg capsules of certain Neritidae. *Journal of Morphology* 57: 31–59.

Andrews, E. A. (1937). Certain reproductive organs in the Neritidae. *Journal of Morphology* 61: 525–561.

Andrews, E. B. (1964). The functional anatomy and histology of the reproductive system of some pilid gastropod molluscs. *Proceedings of the Malacological Society of London* 36: 121–140.

Andrews, E. B. (1965). The functional anatomy of the mantle cavity, kidney and blood system of some pilid gastropods (Prosobranchia). *Journal of Zoology* 146: 70–94.

Andrews, E. B. (1976). The fine structure of the heart of some prosobranch and pulmonate gastropods in relation to filtration. *Journal of Molluscan Studies* 42: 199–216.

Andrews, E. B. (1981). Osmoregulation and excretion in prosobranch gastropods. Part 2: Structure in relation to function. *Journal of Molluscan Studies* 47: 248–289.

Andrews, E. B. (1985). Structure and function in the excretory system of archaeogastropods and their significance in the evolution of gastropods. *Philosophical Transactions of the Royal Society B* 310: 383–406.

Andrews, E. B. (1988). Excretory systems of molluscs, pp. 381–448 *in* E. R. Trueman & Clarke, M. R. (eds.), *Form and Function. The Mollusca.* Vol. 11. San Diego, CA, Academic Press.

Andrews, E. B. (1991). The fine structure and function of the salivary glands of *Nucella lapillus* (Gastropoda: Muricidae). *Journal of Molluscan Studies* 57: 111–126.

Andrews, E. B., Page, A. M. & Taylor, J. D. (1999). The fine structure and function of the anterior foregut glands of *Cymatium intermedius* (Cassoidea: Ranellidae). *Journal of Molluscan Studies* 65: 1–19.

Andrews, H. E. (1974). Morphometrics and functional morphology of *Turritella mortoni*. *Journal of Paleontology* 48: 1126–1140.

Angeloni, L. & Bradbury, J. (1999). Body size influences mating strategies in a simultaneously hermaphroditic sea slug, *Aplysia vaccaria*. *Ethology Ecology and Evolution* 11: 187–195.

Angeloni, L. (2003). Sexual selection in a simultaneous hermaphrodite with hypodermic insemination: Body size, allocation to sexual roles and paternity. *Animal Behaviour* 66: 417–426.

Angeloni, L., Bradbury, J. W. & Burton, R. S. (2003). Multiple mating, paternity, and body size in a simultaneous hermaphrodite, *Aplysia californica*. *Behavioral Ecology* 14: 554–560.

Angerer, G. & Haszprunar, G. (1996). Anatomy and affinities of lepetid limpets (Patellogastropoda = Docoglossa), pp. 171–175 *in* J. D. Taylor (ed.), *Origin and Evolutionary Radiation of the Mollusca.* Oxford, UK, Oxford University Press.

Anonymous (2003). Proceedings of the workshop on rebuilding techniques for abalone in British Columbia. *Canadian Technical Report of Fisheries and Aquatic Sciences* 2482: 1–106.

Anonymous (2009). 17th International Pectinid Workshop held in Santiago de Compostela, Spain. *National Shellfisheries Association Quarterly Newsletter* July 2009: 11.

Ansell, A. D. & Nair, N. B. (1969). A comparative study of bivalves which bore mainly by mechanical means. *American Zoologist* 9: 857–868.

Ansell, A. D. (1981). Functional morphology and feeding of *Donax serra* Röding and *Donax sordidus* Hanley (Bivalvia, Donacidae). *Journal of Molluscan Studies* 47: 59–72.

Anthes, N. & Michiels, N. K. (2007a). Reproductive morphology, mating behaviour, and spawning ecology of cephalaspid sea slugs (Aglajidae and Gastropteridae). *Invertebrate Biology* 126: 335–365.

Anthes, N. & Michiels, N. K. (2007b). Precopulatory stabbing, hypodermic injections and unilateral copulations in a hermaphroditic sea slug. *Biology Letters* 3: 121–124.

Anthes, N., Schulenburg, H. & Michiels, N. K. (2008). Evolutionary links between reproductive morphology, ecology and mating behavior in opisthobranch gastropods. *Evolution: International Journal of Organic Evolution* 62: 900–916.

Appeldoorn, R. S. (1988). Age determination, growth, mortality and age of first reproduction in adult queen conch, *Strombus gigas* L., off Puerto Rico. *Fisheries Research* 6: 363–378.

Arakawa, K. Y. (1965). Studies on the molluscan faeces (II). *Publications of the Seto Marine Biological Laboratory* 13: 1–21.

Arakawa, K. Y. (1968). Studies on the molluscan faeces (III). *Publications of the Seto Marine Biological Laboratory* 16: 127–139.

Archiac, V. d. & Verneuil, E. d. (1842). On the fossils of the older deposits in the Rhenish Provinces; preceded by a general survey of the fauna of the Palaeozoic rocks, and followed by a tabular list of the organic remains of the Devonian system in Europe. *Transactions of the Geological Society of London*, series 2 6: 303–410, plts 325–338.

Arconada, B. & Ramos, M. A. (2001). New data on Hydrobiidae systematics: Two new genera from the Iberian Peninsula. *Journal of Natural History* 35: 949–984.

Arkhipkin, A. I., Bizikov, V. A. & Fuchs, D. (2012). Vestigial phragmocone in the gladius points to a deepwater origin of squid (Mollusca: Cephalopoda). *Deep Sea Research Part I: Oceanographic Research Papers* 61: 109–122.

Arkhipkin, A. I. (2014). Getting hooked: The role of a U-shaped body chamber in the shell of adult heteromorph ammonites. *Journal of Molluscan Studies* 8: 354–364.

Armbruster, G. & Schlegel, M. (1994). The land-snail species of *Cochlicopa* (Gastropoda: Pulmonata: Cochlicopidae): Presentation of taxon-specific allozyme patterns, and evidence for a high level of self-fertilization. *Journal of Zoological Systematics and Evolutionary Research* 32: 282–296.

Arnold, J. M. & Young, R. E. (1974). Ultrastructure of a cephalopod photophore. I. Structure of the photogenic tissue. *Biological Bulletin* 147: 507–521.

Arnold, J. M., Young, R. E. & King, M. V. (1974). Ultrastructure of a cephalopod photophore. II. Iridophores as reflectors and transmitters. *Biological Bulletin* 147: 522–534.

Arnold, J. M. (1987). Reproduction and embryology of *Nautilus*, pp. 353–372 *in* W. B. Saunders & Landman, N. H. (eds.), *Nautilus: The Biology and Paleobiology of a Living Fossil. Topics in Geobiology*. Vol. 6. New York, Springer.

Arnold, J. M. (1988). Some observations on the cicatrix of *Nautilus* embryos, pp. 181–190 *in* J. Wiedmann & Kullmann, J. (ed.), *Cephalopods Present and Past. 2nd International Cephalopod Symposium*: O. H. Schindewolf Symposium, Tübingen 1985. Stuttgart, E. Schweizerbart'sche Verlagsbuchhandlung.

Arnold, J. M. (2010). Reproduction and embryology of *Nautilus*, pp. 353–372 *in* W. B. Saunders & Landman, N. H. (eds.), *Nautilus: The Biology and Paleobiology of a Living Fossil. Reprint with additions. Topics in Geobiology*. New York, Springer.

Arrighetti, F., Brey, T., Mackensen, A. & Penchaszadeh, P. E. (2011). Age, growth and mortality in the giant snail *Adelomelon beckii* (Broderip 1836) on the Argentinean shelf. *Journal of Sea Research* 65: 219–223.

van As, J. G. & Basson, L. (2004). Ciliophoran (Ciliophora) parasites of terrestrial gastropods, pp. 559–578 *in* G. M. Barker (ed.), *Natural Enemies of Terrestrial Molluscs*. Oxford, UK/ Cambridge, MA, CABI Publishing.

Atkins, D. G. (1936). On the ciliary mechanisms and interrelationships of lamellibranchs. Part I: New observations on sorting mechanisms. *Quarterly Journal of Microscopical Science* 79: 181–308.

Atkins, D. G. (1937a). On the ciliary mechanisms and interrelationships of lamellibranchs. Part IV: Cuticular fusion with special reference to the fourth aperture in certain lamellibranchs. *Quarterly Journal of Microscopical Science* 79: 423–445.

Atkins, D. G. (1937b). On the ciliary mechanisms and interrelationships of lamellibranchs. Part III: Types of lamellibranch gills and their food currents. *Quarterly Journal of Microscopical Science* 79: 375–421.

Atkins, D. G. (1937c). On the ciliary mechanisms and interrelationships of lamellibranchs. Part II: Sorting devices on the gills. *Quarterly Journal of Microscopical Science* 79: 339–373.

Atkins, D. G. (1938a). On the ciliary mechanisms and interrelationships of lamellibranchs. Part VI: The pattern of the lateral ciliated cells of the gill filaments of the Lamellibranchia. *Quarterly Journal of Microscopical Science* 80: 331–344.

Atkins, D. G. (1938b). On the ciliary mechanisms and interrelationships of lamellibranchs. Part VII: Latero-frontal cilia of the gill filaments and their phylogenetic value. *Quarterly Journal of Microscopical Science* 80: 345–436.

Atkins, D. G. (1938c). On the ciliary mechanisms and interrelationships of lamellibranchs. Part V: Note on the gills of *Amussium pleuronectes*. *Quarterly Journal of Microscopical Research* 80: 321–329.

Ausich, W. I. & Bottjer, D. J. (1982). Tiering in suspension-feeding communities on soft substrata throughout the Phanerozoic. *Science* 216: 173–174.

Austin, J., Gosliner, T. & Malaquias, M. A. E. (2019). Systematic revision, diversity patterns and trophic ecology of the tropical Indo-West Pacific sea slug genus *Phanerophthalmus* A. Adams, 1850 (Cephalaspidea, Haminoeidae). *Invertebrate Systematics* 32: 1336–1387.

Ax, P. (1999). *Das System der Metazoa. Ein Lehrbuch der phylogenetischen Systematik*. Vol. 2. Stuttgart/Jena/New York, Gustav Fischer.

Ax, P. (2000). *Multicellular Animals. The Phylogenetic System of the Metazoa*. Vol. 2. Berlin, Springer Science + Business Media.

Ayal, Y. (1978). *Geographical distribution, ecological niche and the strategy of reproduction of the colonizer Cerithium scabridum Phil. (Gastropoda: Prosobranchia) as compared with those of some other sympatric non-colonizing congeneric species* [in Hebrew]. Ph.D. dissertation, The Hebrew University of Jerusalem.

Azevedo, C. (1981). The fine structure of the spermatozoon in *Patella lusitanica* (Gastropoda: Prosobranchia), with special reference to acrosome formation. *Journal of Submicroscopic Cytology* 13: 47–56.

Azevedo, C. & Padovan, I. P. (2004). *Nematopsis gigas* n. sp. (Apicomplexa), a parasite of *Nerita ascencionis* (Gastropoda, Neritidae) from Brazil. *Journal of Eukaryotic Microbiology* 51: 214–219.

Azuma, N., Zaslavskaya, N. I., Yamazaki, T., Nobetsu, T. & Chiba, S. (2017). Phylogeography of *Littorina sitkana* in the northwestern Pacific Ocean: Evidence of eastward trans-Pacific colonization after the Last Glacial Maximum. *Genetica* 145: 139–149.

Baba, K. (1937). Contributions to the knowledge of a nudibranch, *Okadaia elegans* Baba. *Japanese Journal of Zoology* 7: 147–190.

Baba, K. (1938). The later development of a solenogastre, *Epimenia verrucosa* (Nierstrasz). *Journal of the Department of Agriculture of the Kyushu Imperial University* 6: 21–40.

Baba, K. (1940a). The early development of a solenogastre, *Epimenia verrucosa* (Nierstrasz). *Annotationes Zoologicae Japonenses* 19: 107–113.

Baba, K. (1940b). *Epimenia ohshimai* a new Solenogastre species from Amakusa, Japan. *Venus* 10: 91–96.

Baba, K. (1961). On the identification and the affinity of *Tamanovalva limax*, a bivalved sacoglossan mollusc in Japan. *Publications of the Seto Marine Biological Laboratory* 9: 37–62.

Babin, C., Delance, J.-H., Emig, C. & Racheboeuf, P. R. (1992). Brachiopodes et Mollusques Bivalves: Concurrence ou indifférence? *Geobios* 25: 35–44.

Babin, C. (2000). Ordovician to Devonian diversification of the Bivalvia. *American Malacological Bulletin* 15: 167–178.

Backeljau, T., Baur, A. & Baur, B. (2001). Population and conservation genetics, pp. 383–412 in G. M. Barker (ed.), *The Biology of Terrestrial Molluscs*. Wallingford, UK, CABI Publishing.

Baeumler, N., Haszprunar, G. & Ruthensteiner, B. (2011). Development of the excretory system in the polyplacophoran mollusc, *Lepidochitona corrugata*: The protonephridium. *Journal of Morphology* 272: 972–986.

Baeumler, N., Haszprunar, G. & Ruthensteiner, B. (2012). Development of the excretory system in a polyplacophoran mollusc: Stages in metanephridial system development. *Frontiers in Zoology* 9: 23.

Bailly, C. (2004). Lamellarins, from A to Z: A family of anticancer marine pyrrole alkaloids. *Current Medicinal Chemistry - Anti-Cancer Agents* 4: 363–378.

Baker, H. B. (1922). Notes on the radula of the Helicinidae. *Proceedings of the Academy of Natural Sciences of Philadelphia* 75: 29–67.

Baker, H. B. (1923). Notes on the radula of the Neritidae. *Proceedings of the Academy of Natural Sciences of Philadelphia* 75: 117–178.

Baker, H. B. (1925). Anatomy of *Hendersonia*: A primitive helicinid mollusk. *Proceedings of the Academy of Natural Sciences of Philadelphia* 77: 273–303.

Baker, H. B. (1926). Anatomical notes on American Helicinidae. *Proceedings of the Academy of Natural Sciences of Philadelphia* 78: 35–56.

Baker, H. B. (1928). Mexican mollusks collected for Dr. Bryant Walker in 1926, I. *Occasional papers of the Museum of Zoology, University of Michigan* 193: 1–65.

Baker, H. B. (1955). Heterurethrous and aulacopod. *The Nautilus* 68: 109–112.

Baldwin, J. (1987). Energy metabolism of *Nautilus* swimming muscles, pp. 325–330 in W. B. Saunders & Landman, N. H. (eds.), *Nautilus: The Biology and Paleobiology of a Living Fossil. Topics in Geobiology*. Vol. 6. New York, Springer.

Ball, A. D., Taylor, J. D. & Andrews, E. B. (1997). Development of the acinous and accessory salivary glands in *Nucella lapillus* (Neogastropoda: Muricoidea). *Journal of Molluscan Studies* 63: 245–260.

Balthasar, U. & Butterfield, N. J. (2009). Early Cambrian "soft-shelled" brachiopods as possible stem-group phoronids. *Acta Palaeontologica Polonica* 54: 307–314.

Balvanera, P., Pfisterer, A. B., Buchmann, N., He, J. S., Nakashizuka, T., Raffaelli, D. & Schmid, B. (2006). Quantifying the evidence for biodiversity effects on ecosystem functioning and services. *Ecology Letters* 9: 1146–1156.

Bambach, R. K., Knoll, A. H. & Sepkoski, J. J. (2002). Anatomical and ecological constraints on Phanerozoic animal diversity in the marine realm. *Proceedings of the National Academy of Sciences of the United States of America* 99: 6854–6859.

Bambach, R. K., Bush, A. M. & Erwin, D. H. (2007). Autecology and the filling of ecospace: Key Metazoan radiations. *Palaeontology* 50: 1–22.

Bandel, K. (1975). *Embryonalgehäuse karibischer Meso- und Neogastropoden (Mollusca)*. Mainz, Akademie der Wissenschaften und der Literatur Wiesbaden.

Bandel, K. (1976). Observations on spawn, embryonic development and ecology of some Caribbean lower Mesogastropoda (Mollusca). *The Veliger* 18: 249–271.

Bandel, K. (1977). Die Herausbildung der Schraubenschicht bei Pteropoden. *Biomineralisation* 9: 73–85.

Bandel, K. (1982). Morphologie und Bildung der frühontogenetischen Gehäuse bei conchiferen Mollusken [Morphology and formation of the early ontogenetic shells of conchiferan molluscs]. *Facies* 7: 1–198.

Bandel, K., Teiter, J. & Stürmer, W. (1983). Coleoids from the Lower Devonian Black Slate ('Hunsrück-Schiefer') of the Hunsrück (West Germany). *Neues Jahrbuch für Geologie und Paläontologie-Abhandlungen* 165: 397–417.

Bandel, K. (1984). The radulae of Caribbean and other Mesogastropoda and Neogastropoda. *Zoologische Verhandelingen (Leiden)* 214: 1–188.

Bandel, K. (1985). Composition and ontogeny of *Dictyoconites* (Aulacocerida, Cephalopoda). *Palaeontologische Zeitschrift* 59: 223–244.

Bandel, K. & Leich, H. (1986). Jurassic Vampyromorpha (dibranchiate cephalopods). *Neues Jahrbuch für Geologie und Paläontologie-Monatshefte* 1986: 129–148.

Bandel, K. & von Boletzky, S. (1988). Features of development and functional morphology required in the reconstruction of early coleoid cephalopods, pp. 229–246 in J. Wiedmann & Kullmann, J. (eds.), *Cephalopods Present and Past. 2nd International Cephalopod Symposium*: O. H. Schindewolf Symposium, Tübingen 1985. Stuttgart, E. Schweizerbart'sche Verlagsbuchhandlung.

Bandel, K. & Stanley, G. D. (1989). Reconstruction and biostratinomy of Devonian cephalopods (Lamellorthoceratidae) with unique cameral deposits. *Senckenbergiana lethaea* 69: 391–437.

Bandel, K. (1990). Shell structure of the Gastropoda excluding Archaeogastropoda, pp. 117–134 in J. G. Carter (ed.), *Skeletal Biomineralization: Patterns, Processes and Evolutionary Trends*. Vol. 1. New York, Van Nostrand Reinhold.

Bandel, K. (1991a). Schlitzbandschnecken mit perlmutteriger Schale aus den triassischen St. Cassian-Schichten der Dolomiten. *Annalen des Naturhistorischen Museums in Wien Serie* A 92: 1–53.

Bandel, K. (1991b). Über triassische 'Loxonematoidea' und ihre Beziehungen zu rezenten und paläozoischen Schnecken. *Palaeontologische Zeitschrift* 65: 239–268.

Bandel, K. (1992). Platyceratidae from the Triassic St. Cassian Formation and the evolutionary history of the Neritomorpha (Gastropoda). *Palaeontologische Zeitschrift* 66: 231–240.

Bandel, K. (1993). Caenogastropoda during Mesozoic times. *Scripta Geologica Special Issue* 2: 7–56.

Bandel, K., Hain, S., Riedel, F. & Tiemann, H. (1993). *Limacosphaera*, an unusual mesogastropod (Lamellariidae) larva of the Weddell Sea (Antarctica). *The Nautilus* 107: 1–8.

Bandel, K. (1994). Triassic Euthyneura (Gastropoda) from St. Cassian Formation (Italian Alps) with a discussion on the evolution of the Heterostropha. *Freiberger Forschungshefte-Reihe C-Geowissenschaften* 452: 79–100.

Bandel, K. & Riedel, F. (1994). The Late Cretaceous gastropod fauna from Ajka (Bakony Mountains, Hungary): A revision. *Annalen des Naturhistorischen Museums in Wien Serie* A 96: 1–65.

Bandel, K., Riedel, F. & Tiemann, H. (1994). A special adaptation to planktonic life in larvae of the Cassoidea (= Tonnoidea) (Gastropoda). *Marine Biology* 118: 101–108.

Bandel, K. (1997). Higher classification and pattern of evolution of the Gastropoda: A synthesis of biological and paleontological data. *Courier Forschungsinstitut Senckenberg* 201: 57–81.

Bandel, K. & Kowalke, T. (1997). Cretaceous *Laxispira* and a discussion on the monophyly of vermetids and turritellids (Caenogastropoda, Mollusca). *Geologica et Palaeontologica* 31: 257–274.

Bandel, K. & Frýda, J. (1998). The systematic position of the Euomphalidae (Gastropoda). *Senckenbergiana lethaea* 78: 103–131.

Bandel, K. & Frýda, J. (1999). Notes on the evolution and higher classification of the subclass Neritimorpha (Gastropoda) with the description of some new taxa. *Geologica et Palaeontologica* 33: 219–235.

Bandel, K. (2000). The new family Cortinellidae (Gastropoda, Mollusca) connected to a review of the evolutionary history of the subclass Neritimorpha. *Neues Jahrbuch für Geologie und Paläontologie-Abhandlungen* 217: 111–129.

Bandel, K. (2001). The history of *Theodoxus* and *Neritina* connected with description and systematic evaluation of related Neritimorpha (Gastropoda). *Mitteilungen aus dem Geologisch-Paläontologischen Institut der Universität Hamburg* 85: 65–164.

Bandel, K. & Heidelberger, D. (2001). The new family Nerrhenidae (Neritimorpha, Gastropoda) from the Givetian of Germany. *Neues Jahrbuch für Geologie und Palaontologie-Monatshefte* 12: 705–718.

Bandel, K. (2002a). Re-evaluation and classification of Carboniferous and Permian Gastropoda belonging to the Caenogastropoda and their relation. *Mitteilungen aus dem Geologisch-Paläontologischen Institut der Universität Hamburg* 86: 81–188.

Bandel, K. (2002b). About the Heterostropha (Gastropoda) from the Carboniferous and Permian. *Mitteilungen aus dem Geologisch-Paläontologischen Institut der Universität Hamburg* 86: 45–80.

Bandel, K. & Heidelberger, D. (2002). A Devonian member of the subclass Heterostropha (Gastropoda) with valvatoid shell shape. *Neues Jahrbuch für Geologie und Palaontologie-Monatshefte* 9: 533–550.

Bandel, K., Nützel, A. N. & Yancey, T. E. (2002). Larval shells and shell microstructures of exceptionally well-preserved Late Carboniferous gastropods from the Buckhorn Asphalt Deposit (Oklahoma, USA). *Senckenbergiana lethaea* 82: 639–689.

Bandel, K. & Kiel, S. (2003). Relationships of Cretaceous Neritimorpha (Gastropoda, Mollusca), with the description of seven new species. *Bulletin of the Czech Geological Survey* 78: 53–65.

Bandel, K. (2006). Families of the Cerithioidea and related superfamilies (Palaeo-Caenogastropoda; Mollusca) from the Triassic to the Recent characterized by protoconch morphology – including the description of new taxa. *Freiberger Forschungshefte-Reihe C-Geowissenschaften* 511: 59–138.

Bandel, K. & Stinnesbeck, W. (2006). *Naefia* Wetzel, 1930 from the Quiriquina Formation (Late Maastrichtian, Chile): Relationship to modern *Spirula* and ancient Coleoidea (Cephalopoda). *Acta Universitatis Carolinae Geologica* 49: 21–32.

Bandel, K. (2007). Description and classification of Late Triassic Neritimorpha (Gastropoda, Mollusca) from the St Cassian Formation, Italian Alps. *Bulletin of Geosciences* 82: 215–274.

Bandel, K. (2008). Operculum shape and construction of some fossil Neritimorpha (Gastropoda) compared to those of modern species of the subclass. *Vita Malacologica* 7: 19–36.

Bandel, K. (2009). The slit bearing nacreous Archaeogastropoda of the Triassic tropical reefs in the St. Cassian Formation with evaluation of the taxonomic value of the selenizone. *Berliner Paläobiologische Abhandlungen* 10: 5–47.

Barbosa, S. S., Byrne, M. & Kelaher, B. P. (2009). Reproductive periodicity of the tropical intertidal chiton *Acanthopleura gemmata* at One Tree Island, Great Barrier Reef, near its southern latitudinal limit. *Journal of the Marine Biological Association of the United Kingdom* 89: 405–411.

Barkalova, V. O., Fedosov, A. E. & Kantor, Y. I. (2016). Morphology of the anterior digestive system of tonnoideans (Gastropoda: Caenogastropoda) with an emphasis on the foregut glands. *Molluscan Research* 36: 54–73.

Barker, G. M. (1999). *Naturalised Terrestrial Stylommatophora (Mollusca: Gastropoda)*. Vol. 38. Otago, New Zealand, Manaaki Whenua Press, Manaaki Whenua-Landcare Research New Zealand Ltd.

Barker, G. M. (2001a). Gastropods on land: Phylogeny, diversity and adaptive morphology, pp. 1–146 *in* G. M. Barker (ed.), *The Biology of Terrestrial Molluscs*. Wallingford, UK, CABI Publishing.

Barker, G. M. (2001b). *The Biology of Terrestrial Molluscs*. Wallingford, UK, CABI Publishing.

Barker, G. M. (2002). *Molluscs as Crop Pests*. Wallingford, UK, CABI Publishing.

Barker, G. M. & Efford, M. G. (2002). Predatory gastropods as natural enemies of terrestrial gastropods and other invertebrates, pp. 279–404 *in* G. M. Barker (ed.), *Natural Enemies of Terrestrial Molluscs*. Oxford, UK/Cambridge, MA, CABI Publishing.

Barker, G. M. (2004). *Natural Enemies of Terrestrial Molluscs*. Oxford, UK/Cambridge, MA, CABI Publishing.

Barnard, F. R. (1974). *Septibranchs of the Eastern Pacific (Bivalvia: Anomalodesmata)*. Vol. 8. Los Angeles, CA, Allan Hancock Foundation, University of Southern California.

Barnes, R. D. (1968). *Invertebrate Zoology*, 2nd edition. Philadelphia, PA, W. B. Saunders Company.

Barrande, J. (1865–1877). *Systême silurien du Centre de la Bohême, Ière partie: Recherches Paléontologiques, vol. II, Céphalopodes*. Prague, Self-published.

Barratt, I. M. & Allcock, A. L. (2010). Ageing octopods from stylets: Development of a technique for permanent preparations. *ICES Journal of Marine Science: Journal du Conseil* 67: 1452–1457.

Barskov, I. S. (1996). Phosphatized blood vessels in the siphuncle of Jurassic ammonites. *Bulletin de l'Institut océanographique de Monaco* 14: 335–341.

Barskov, I. S., Boiko, M. S., Konovalova, V. A., Leonova, T. B. & Nikolaeva, S. V. (2008). Cephalopods in the marine ecosystems of the Paleozoic. *Paleontological Journal* 42: 1167–1284.

Bartol, I. K., Krueger, P. S., Thompson, J. T. & Stewart, W. J. (2008). Swimming dynamics and propulsive efficiency of squids throughout ontogeny. *Integrative and Comparative Biology* 48: 720–733.

Bartolomaeus, T. (1989). Larvale Nierenorgane bei *Lepidochiton cinereus* (Polyplacophora) und *Aeolidia papillosa* (Gastropoda). *Zoomorphologie* 108: 297–307.

Bartolomaeus, T. (1997). Ultrastructure of the renopericardial complex of the interstitial gastropod *Philinoglossa helgolandica* (Hertling), 1932 (Mollusca: Opisthobranchia). *Zoologischer Anzeiger* 235: 165–176.

Basil, J. A., Bahctinova, I., Kuroiwa, K., Lee, N., Mims, D., Preis, M. & Soucier, C. (2005). The function of the rhinophore and the tentacles of *Nautilus pompilius* L. (Cephalopoda, Nautiloidea) in orientation to odor. *Marine and Freshwater Behaviour and Physiology* 38: 209–221.

Basse, E. (1954). Sur une ammonite nouvelle du Turonien de Liban. *Notes et Mémoires sur le Moyen-Orient* 5: 200–204.

Bates, A. E. (2007a). Persistence, morphology, and nutritional state of a gastropod-hosted bacterial symbiosis in different levels of hydrothermal vent flux. *Marine Biology* 152: 557–568.

Bates, A. E. (2007b). Feeding strategy, morphological specialisation and presence of bacterial episymbionts in lepetodrilid gastropods from hydrothermal vents. *Marine Ecology Progress Series* 347: 87–99.

Bather, F. A. (1911). *A guide to the Fossil Invertebrate Animals in the Department of Geology and Palaeontology in the British Museum (Natural History)*, 2nd edition. London, British Museum (Natural History).

Batt, R. J. (1989). Ammonite shell morphotype distribution in the Western Interior Greenhorn Sea and some paleoecological implications. *PALAIOS* 4: 32–42.

Batten, R. L. (1972). The ultrastructure of five common Pennsylvanian pleurotomarian gastropod species of Eastern United States. *American Museum Novitates* 2501: 1–34.

Batten, R. L. (1984). *Neopilina, Neomphalus,* and *Neritopsis,* living fossil molluscs, pp. 218–224 *in* N. Eldredge & Stanley, S. M. (eds.), *Living Fossils*. New York, Springer.

Bauer, G. & Vogel, C. (1987). The parasitic stage of the freshwater pearl mussel (*Margaritifera margaritifera* L) [Part] I. Host response to Glochidiosis. *Archives of Hydrobiology* 76: 393–402.

Bauer, G. (1992). Variation in the life span and size of the freshwater pearl mussel. *Journal of Animal Ecology* 61: 425–436.

Bauer, G. (1994). The adaptive value of offspring size among freshwater mussels (Bivalvia: Unionoidea). *Journal of Animal Ecology* 63: 933–944.

Baum, J. K. & Worm, B. (2009). Cascading top-down effects of changing oceanic predator abundances. *Journal of Animal Ecology* 78: 699–714.

Baumiller, T. K. & Gahn, F. J. (2002). Fossil record of parasitism on marine invertebrates with special emphasis on the platyceratid-crinoid interaction. *Paleontological Society Papers* 8: 195–210.

Bäumler, N., Haszprunar, G. & Ruthensteiner, B. (2008). 3D interactive microanatomy of *Omalogyra atomus* (Philippi, 1841) (Gastropoda, Heterobranchia, Omalogyridae), pp. 108–116 *in* D. Geiger & Ruthensteiner, B. (eds.), *Micromolluscs: Methodological Challenges – Exciting Results. Proceedings from the Micromollusc Symposium of the 16th UNITAS Malacologica World Congress of Malacology July 15–20, 2007 in Antwerp, Belgium.* Vol. 1. Auckland, New Zealand, Magnolia Press.

Baur, B. (1994). Parental care in terrestrial gastropods. *Experientia* 50: 5–14.

Baxter, J. M., Jones, A. M. & Sturrock, M. G. (1987). The ultrastructure of aesthetes in *Tonicella marmorea* (Polyplacophora; Ischnochitonina) and a new functional hypothesis. *Journal of Zoology* 211: 589–604.

Baxter, J. M., Sturrock, M. G. & Jones, A. M. (1990). The structure of the intrapigmented aesthetes and the properiostracum layer in *Callochiton achatinus* (Mollusca: Polyplacophora). *Journal of Zoology* 220: 447–468.

Bayne, C. J. (1968). Histochemical studies on the egg capsule of eight gastropod molluscs. *Proceedings of the Malacological Society of London* 38: 199–212.

Bé, A. W. H., MacClintock, C. & Currie, D. C. (1972). Helical shell structure and growth of the pteropod *Cuvierina columnella* (Rang) (Mollusca, Gastropoda). *Biomineralization Research Reports* 4: 47–79.

Bebbington, A. (1986). Observations on a collection of *Glaucus atlanticus* (Gastropoda, Opisthobranchia). *Haliotis* 15: 73–81.

Becerro, M. A., Turon, X., Uriz, M. J. & Templado, J. (2003). Can a sponge feeder be a herbivore? *Tylodina perversa* (Gastropoda) feeding on *Aplysina aerophoba* (Demospongiae). *Biological Journal of the Linnean Society* 78: 429–438.

Beck, L. A. (1992a). Two new neritacean limpets (Gastropoda: Prosobranchia: Neritacea: Phenacolepadidae) from active hydrothermal vents at Hydrothermal Field 1 'Wienerwald' in the Manus Back-Arc Basin (Bismarck Sea, Papua New Guinea). *Annalen des Naturhistorischen Museums in Wien Serie B* 93: 259–275.

Beck, L. A. (1992b). *Symmetromphalus hageni* sp. n., a new neomphalid gastropod (Prosobranchia: Neomphalidae) from hydrothermal vents at the Manus Back-Arc Basin (Bismarck Sea, Papua New Guinea). *Annalen des Naturhistorischen Museums in Wien Serie B* 93: 243–257.

Beck, L. A. (2002). *Hirtopelta tufari* sp. n., a new gastropod species from hot vents at the East Pacific Rise (21 S) harbouring endocytosymbiotic bacteria in its gill (Gastropoda: Rhipidoglossa: Peltospiridae). *Archiv für Molluskenkunde* 130: 249–257.

Becker, R. T. (1993). Anoxia, eustatic changes, and Upper Devonian to lowermost Carboniferous global ammonoid diversity, pp. 115–163 *in* M. R. House (ed.), *The Ammonoidea: Environment, Ecology, and Evolutionary Change. Systematics Association Special Volume.* Oxford, UK, Oxford University Press.

Beeman, R. D. (1970). The anatomy and functional morphology of the reproductive system in the opisthobranch *Phyllaplysia taylori* Dall 1900. *The Veliger* 13: 1–31.

Beeman, R. D. (1977). Gastropoda: Opisthobranchia, pp. 115–179 *in* A. C. Giese & Pearse, J. S. (eds.), *Reproduction of Marine Invertebrates. Molluscs: Gastropods and Cephalopods.* Vol. 4. New York, Academic Press.

Beese, K., Armbruster, G. F. J., Beier, K. & Baur, B. (2008). Evolution of female sperm-storage organs in the carrefour of stylommatophoran gastropods. *Journal of Zoological Systematics and Evolutionary Research* 47: 49–60.

Beesley, P. L., Ross, G. J. B. & Wells, A. (eds.). (1998a). *Mollusca: The Southern Synthesis. Part A,* Melbourne, VIC, CSIRO Publishing.

Beesley, P. L., Ross, G. J. B. & Wells, A. (eds.). (1998b). *Mollusca: The Southern Synthesis. Part B,* Melbourne, VIC, CSIRO Publishing.

Behrens, D. W. (1981). *Pacific Coast Nudibranchs. A Guide to the Opisthobranchs of the Northeastern Pacific.* Los Osos, CA, Sea Challengers.

Bell, L. J. (1992). Reproduction and larval development of the West Indian top shell, *Cittarium pica* (Trochidae), in the Bahamas. *Bulletin of Marine Science* 51: 250–266.

Bengtson, S. & Missarzhevsky, V. (1981). Coeloscleritophora – a major group of enigmatic Cambrian metazoans. *US Geological Survey Open-File Report* 81: 19–21.

Bengtson, S. (1992). The cap-shaped Cambrian fossil *Maikhanella* and the relationship between coeloscleritophorans and molluscs. *Lethaia* 25: 401–420.

Bengtson, S. & Conway Morris, S. (1992). Early radiation of biomineralizing phyla, pp. 447–481 *in* J. H. Lipps & Signor, P. W. (eds.), *Origin and Early Evolution of the Metazoa*. New York, Plenum Press.

Bengtson, S. & Runnegar, B. N. (1992). Origins of biomineralization in metaphytes and metazoans, pp. 447–451 *in* J. W. Schopf & Klein, C. (ed.), *The Proterozoic Biosphere: A Multidisciplinary Study*. Cambridge, Cambridge University Press.

Bengtson, S. (1993). Molluscan affinity of coeloscleritophorans – Reply. *Lethaia* 26: 48.

Bengtson, S. (2002). Origins and early evolution of predation. *Paleontological Society Papers* 8: 289–318.

Bengtson, S. & Collins, D. (2015). Chancelloriids of the Cambrian Burgess Shale. *Palaeontologia Electronica* 18: 1–67.

Beninger, P. G. & Veniot, A. (1999). The oyster proves the rule: Mechanisms of pseudofeces transport and rejection on the mantle of *Crassostrea virginica* and *C. gigas*. *Marine Ecology Progress Series* 190: 179–188.

Beninger, P. G., Veniot, A. & Poussart, Y. (1999). Principles of pseudofeces rejection on the bivalve mantle: Integration in particle processing. *Marine Ecology Progress Series* 178: 259–269.

Beninger, P. G. & Dufour, S. C. (2000). Evolutionary trajectories of a redundant feature: Lessons from bivalve gill abfrontal cilia and mucocyte distributions, pp. 273–278 *in* E. M. Harper, Taylor, J. D. & Crame, J. A. (eds.), *Evolutionary Biology of the Bivalvia*. London, The Geological Society (Special Publication No. 177).

Benjamin, P. R. (1971). On the structure of the pulmonate osphradium. 1. Cell types and their organisation. *Zeitschrift für Zellforschung und Mikroskopische Anatomie* 117: 485–501.

Benjamin, P. R. & Peat, A. (1971). On the structure of the pulmonate osphradium. 2. Ultrastructure. *Zeitschrift für Zellforschung und Mikroskopische Anatomie* 118: 168–189.

Benke, M., Reise, H., Montagne-Wajer, K. & Koene, J. M. (2010). Cutaneous application of an accessory-gland secretion after sperm exchange in a terrestrial slug (Mollusca: Pulmonata). *Zoology* 113: 118–124.

Benkendorff, K., Davis, A. R. & Bremner, J. B. (2001). Chemical defense in the egg masses of benthic invertebrates: An assessment of antibacterial activity in 39 mollusks and 4 polychaetes. *Journal of Invertebrate Pathology* 78: 109–118.

Benkendorff, K. (2010). Molluscan biological and chemical diversity: Secondary metabolites and medicinal resources produced by marine molluscs. *Biological Reviews* 85: 757–775.

Benkendorff, K. (2013). Natural product research in the Australian marine invertebrate *Dicathais orbita*. *Marine Drugs* 11: 1370–1398.

Bentham, G., Busk, G., Carpenter, W. B., Darwin, C., Harvey, W. H., Henfrey, A., Henslow, J. S., Huxley, T. H. & Lindley, J. (1858). Public natural history collections. *Gardeners' Chronicle and Agricultural Gazette* 1858: 861.

Benton, M. J., Wills, M. A. & Hitchin, R. (2000). Quality of the fossil record through time. *Nature* 403: 534–537.

Benton, M. J. (2003). *When Life Nearly Died: The Greatest Mass Extinction of All Time*. London, Thames & Hudson.

Bergenhayn, J. R. M. (1955). Die fossilen schwedischen Loricaten nebst einer vorläufigen Revision des Systems der ganzen Klasse Loricata. *Lunds Universitet Arsskrift* 51: 1–42.

Bergenhayn, J. R. M. (1960). Cambrian and Ordovician loricates from North America. *Journal of Paleontology* 34: 168–178.

Bergh, L. S. R. (1890). Die Titiscanien, eine Familie der rhipidoglossen Gastropoden. *Gegenbaurs Morphologisches Jahrbuch* 16: 1–26.

Bergquist, D. C., Eckner, J. T., Urcuyo, I. A., Cordes, E. E., Hourdez, S., Macko, S. A. & Fisher, C. R. (2007). Using stable isotopes and quantitative community characteristics to determine a local hydrothermal vent food web. *Marine Ecology Progress Series* 330: 49–65.

Bernard, F. (1890). Recherches sur la *Valvata piscinalis*. *Bulletin scientifique de la France et de la Belgique* 22: 253–361.

Bernard, F. (1895). Sur le développement et la morphologie de la coquille chez les lamellibranches. *Bulletin de la Société géologique de France, ser. 3* 25: 104–154.

Bernard, F. R. (1979). New species of *Cuspidaria* from the Northeastern Pacific (Bivalvia, Anomalodesmata) with a proposed classification of septibranchs. *Venus* 38: 14–24.

Berning, B. (2007). Evidence for sublethal predation and regeneration among living and fossil ascophoran bryozoans. *Bryozoan Studies*: 1–7.

Berrie, A. D. & Boize, B. J. (1985). The fish hosts of *Unio* glochidia in the River Thames. *Verhandlungen der Internationalen Vereinigung für theoretische und angewandte Limnologie* 22: 2712–2716.

Berry, A. J. (1961). The habitats of some minute cyclophorids, hydrocenids and vertiginids on a Malayan limestone hill. *The Bulletin of the National Museum* 30: 101–105.

Berry, A. J. (1965). Reproduction and breeding fluctuations in *Hydrocena monterosatiana* a Malayan limestone archaeogastropod. *Journal of Zoology* 144: 219–228.

Berry, A. J., Loong, S. C. & Thum, H. H. (1967). Genital systems of *Pythia, Cassidula* and *Auricula* (Ellobiidae, Pulmonata) from Malayan mangrove swamps. *Proceedings of the Malacological Society of London* 37: 325–337.

Berry, A. J., Lim, R. & Sase, K. A. (1973). Reproductive systems and breeding condition in *Nerita birmanica* (Archaeogastropoda: Neritacea) from Malayan mangrove swamps. *Journal of Zoology* 170: 189–200.

Berry, A. J. & Kadri, A. B. H. (1974). Reproduction in the Malayan freshwater cerithiacean gastropod *Melanoides tuberculata*. *Journal of Zoology* 172: 369–381.

Berry, A. J. (1977). Gastropoda: Pulmonata, pp. 181–226 *in* A. C. Giese & Pearse, J. S. (eds.), *Reproduction of Marine Invertebrates. Molluscs: Gastropods and Cephalopods*. Vol. 4. New York, Academic Press.

Berthold, T. & Engeser, T. (1987). Phylogenetic analysis and systematization of the Cephalopoda (Mollusca). *Verhandlungen des Naturwissenschaftlichen Vereins in Hamburg* 29: 187–220.

Berthold, T. (1988). Anatomy of *Afropomus balanoideus* (Mollusca, Gastropoda, Ampullariidae) and its implications for phylogeny and ecology. *Zoomorphologie* 108: 149–159.

Berthold, T. (1989). Comparative conchology and functional morphology of the copulatory organ of the Ampullariidae (Gastropoda, Monotocardia) and their bearing upon phylogeny and paleontology, pp. 141–164 *in* N. Schmidt-Kittler & Willmann, R. (eds.), *Phylogeny and the Classification of Fossil and Recent Organisms*. Vol. 28. Hamburg, Verlag Paul Parey.

Beu, A. G. (2007). The "inoceramus limpet" *Gigantocapulus problematicus* (Nagao & Otatume, 1938) in New Zealand (Late Cretaceous Gastropoda or Monoplacophora, Gigantocapulidae n. fam.). *Palaeontologische Zeitschrift* 81: 267–282.

Biakov, A. (2015). Biogeography of the Permian marine Boreal basins based on bivalves. *Paleontological Journal* 49: 1184–1192.

Bickell-Page, L. R. (1989). Autotomy of cerata by the nudibranch *Melibe leonina* (Mollusca): Ultrastructure of the autotomy plane and neural correlate of the behaviour. *Philosophical Transactions of the Royal Society B* 324: 149–172.

Bickell-Page, L. R. (1991). Repugnatorial glands with associated striated muscle and sensory cells in *Melibe leonina* (Mollusca, Nudibranchia). *Zoomorphologie* 110: 281–291.

Bickell, L. R. & Chia, F.-S. (1979). Organogenesis and histogenesis in the planktotrophic veliger of *Doridella steinbergae* (Opisthobranchia: Nudibranchia). *Marine Biology* 52: 291–313.

Bickell, L. R., Chia, F.-S. & Crawford, B. J. (1981). Morphogenesis of the digestive system during metamorphosis of the nudibranch *Doridella steinbergae* (Gastropoda): Conversion from phytoplanktivore to carnivore. *Marine Biology* 62: 1–16.

Biedermann, W. (1902). Untersuchungen über Bau und Entstehung der Molluskenschalen. *Jenaische Zeitschrift für Naturwissenschaft. Neue folgen.* 36: 1–164.

Bieler, R. (1988). Phylogenetic relationships in the gastropod family Architectonicidae, with notes on the family Mathildidae (Allogastropoda), pp. 205–240 *in* W. F. Ponder, Eernisse, D. J. & Waterhouse, J. H. (eds.), *Prosobranch Phylogeny. Malacological Review Supplement.* Ann Arbor, MI, Malacological Review.

Bieler, R. & Mikkelsen, P. M. (1988). Anatomy and reproductive biology of two western Atlantic species of Vitrinellidae, with a case of protandrous hermaphroditism in the Rissoacea. *The Nautilus* 102: 1–29.

Bieler, R. (1992). Gastropod phylogeny and systematics. *Annual Review of Ecology and Systematics* 23: 311–338.

Bieler, R. & Mikkelsen, P. M. (1998). *Ammonicera* in Florida: Notes on the smallest living gastropod in the United States and comments on other species of Omalogyridae (Heterobranchia). *The Nautilus* 111: 1–12.

Bieler, R., Mikkelsen, P. M. & Prezant, R. S. (2005). Byssus-attachment by infaunal clams: Seagrass-nestling *Venerupis* in Esperance Bay, Western Australia (Bivalvia: Veneridae), pp. 177–197 *in* F. E. Wells, Walker, D. I. & Kendrick, G. A. (eds.), *Proceedings of the Twelfth International Marine Biological Workshop: The Marine Flora and Fauna of Esperance, Western Australia.* Vol. 1. Perth, WA, Western Australian Muscum.

Bieler, R. & Mikkelsen, P. M. (2006). Bivalvia – A look at the branches. *Zoological Journal of the Linnean Society* 148: 223–235.

Bieler, R. & Petit, R. E. (2011). Catalogue of Recent and fossil "worm-snail" taxa of the families Vermetidae, Siliquariidae, and Turritellidae (Mollusca: Caenogastropoda). *Zootaxa* 2948: 1–103.

Bieler, R., Mikkelsen, P. M., Collins, T. M., Glover, E. A., González, V. L., Graf, D. L., Harper, E. M., Healy, J., Kawauchi, G. Y., Sharma, P. P., Staubach, S., Strong, E. E., Taylor, J. D., Tëmkin, I., Zardus, J. D., Clark, S., Guzmán, A., McIntyre, E., Sharp, P. & Giribet, G. (2014). Investigating the Bivalve Tree of Life – An exemplar-based approach combining molecular and novel morphological characters. *Invertebrate Systematics* 28: 32–115.

Bigatti, G., Penchaszadeh, P. E. & Cledón, M. (2007). Age and growth in *Odontocymbiola magellanica* (Gastropoda: Volutidae) from Golfo Nuevo, Patagonia, Argentina. *Marine Biology* 150: 1199–1204.

Van den Biggelaar, J. A. M. & Haszprunar, G. (1996). Cleavage patterns and mesentoblast formation in the Gastropoda: An evolutionary perspective. *Evolution* 50: 1520–1540.

Van den Biggelaar, J. A. M. (1996). Cleavage pattern and mesentoblast formation in *Acanthochiton crinitus* (Polyplacophora, Mollusca). *Developmental Biology* 174: 423–430.

Billings, E. (1865). *Palaeozoic Fossils. Containing Descriptions and Figures of New or Little Known Species of Organic Remains from the Silurian Rocks, 1861–1865 [Volume 1].* Montreal, QC, Dawson Bros.

Bilyard, G. R. (1974). The feeding habits and ecology of *Dentalium entale stimpsoni* Henderson (Mollusca: Scaphopoda). *The Veliger* 17: 126–138.

Binder, E. (1969). Cephalic accessory sexual organ of *Gymnarion*: Speciation and phylogeny (Pulmonata, Helicarionidae). *Malacologia* 9: 59–64.

Binney, W. G. (1878). *The Terrestrial Air-Breathing Mollusks of the United States and Adjacent Territories of North America.* Vol. 5. Boston, MA, C. C. Little and J. Brown.

Bischoff, G. C. O. (1981). *Cobcrephora* n. g., representative of a new polyplacophoran order Phosphatoloricata with calciumphosphatic shells. *Senckenbergiana lethaea* 61: 173–215.

Bishop, M. J. (1980). Helicinid land snails with apertural barriers. *Journal of Molluscan Studies* 46: 241–246.

Bizikov, V. A. (2002). Reanalysis of functional design of *Nautilus* locomotory and respiratory systems. *American Malacological Bulletin* 17: 17–30.

Bizikov, V. A. (2004). *The Shell in Vampyropoda (Cephalopoda): Morphology, Functional Role and Evolution.* Moscow, Ruthenica.

Bizikov, V. A. (2008). *Evolution of the Shell in Cephalopoda.* Moscow, VNIRO Publishing.

de Blainville, H. M. D. (1825). *Manuel de malacologie et de conchyliologie.* Paris, F.-G. Levrault.

de Blainville, H. M. D. (1824). Mollusques, pp. 1–392. *Dictionnaire des Sciences Naturelles.* Vol. 32. Strasbourg, F-G. Levrault.

Blair, D., Davis, G. M. & Wu, B. (2001). Evolutionary relationships between trematodes and snails emphasizing schistosomes and paragonimids. *Parasitology* 123: S229–S243.

Blanco, J. F. & Scatena, F. N. (2005). Floods, habitat hydraulics and upstream migration of *Neritina virginea* (Gastropoda: Neritidae) in northeastern Puerto Rico. *Caribbean Journal of Science* 41: 55–74.

Blanco, J. F. & Scatena, F. N. (2007). The spatial arrangement of *Neritina virginea* (Gastropoda: Neritidae) during upstream migration in a split-channel reach. *River Research and Applications* 23: 235–245.

Blažek, R. & Gelnar, M. (2006). Temporal and spatial distribution of glochidial larval stages of European unionid mussels (Mollusca: Unionidae) on host fishes. *Folia Parasitologica (Prague)* 53: 98–106.

Blind, W. (1969). Die systematische Stellung der Tentakuliten. *Palaeontographica: Abteilung A* 133: 101–145.

Blind, W. (1976). Die ontogenetische Entwicklung von *Nautilus pompilius* (Linné). *Palaeontographica: Abteilung A* 153: 117–160.

Bloch, S. (1938). Beitrag zur Kenntnis der Ontogenese von Süsswasserpulmonaten mit besonderer Berücksichtigung der Mitteldanndrüse. *Revue suisse de Zoologie* 45: 157–219.

Blochmann, F. (1882). Über die Entwicklung der *Neritina fluviatilis* Müll. *Zeitschrift für wissenschaftliche Zoologie* 36: 125–174.

Blodgett, R. B. & Frýda, J. (2003). New Silurian-Devonian pseudophorid gastropods. *Bulletin of Geosciences* 78: 359–367.

Blumrich, J. (1891). Das integument der Chitonen. *Zeitschrift für wissenschaftliche Zoologie* 52: 404–476.

Boato, A. & Rasotto, M. B. (1987). Functional protandry and seasonal reproductive cycle in *Solatopupa similis* (Bruguière) (Pulmonata Chondrinidae). *Italian Journal of Zoology* 54: 119–125.

Bockwinkel, J., Korn, D. & Ebbighausen, V. (2010). The ammonoids from the Argiles de Timimoun of Timimoun (Early and Middle Viséan; Gourara, Algeria). *Fossil Record* 13: 215–278.

Boettger, C. R. (1954). Die Systematik der euthyneuren Schnecken. *Verhandlungen der Deutschen Zoologischen Gesellschaft Supplementband* 18: 253–280.

Boettger, C. R. (1963). Gastropoden mit zwei Schalenklappen. *Zoologischer Anzeiger.* Supplement 26: 403–439.

Bogan, A. & Bouchet, P. (1998). Cementation in the freshwater bivalve family Corbiculidae (Mollusca: Bivalvia): A new genus and species from Lake Poso, Indonesia. *Hydrobiologia* 389: 131–139.

Bogan, A. E. & Hoeh, W. R. (2000). On becoming cemented: Evolutionary relationships among the genera in the freshwater bivalve family Etheriidae (Bivalvia: Unionoida), pp. 159–168 *in* E. M. Harper, Taylor, J. D. & Crame, J. A. (eds.), *Evolutionary Biology of the Bivalvia.* London, The Geological Society (Special Publication No. 177).

Bogan, A. E. (2008). Global diversity of freshwater mussels (Mollusca, Bivalvia) in freshwater. *Hydrobiologia* 595: 139–147.

Bogan, A. E. & Roe, K. J. (2008). Freshwater bivalve (Unioniformes) diversity, systematics, and evolution: Status and future directions. *Journal of the North American Benthological Society* 27: 349–369.

Bøggild, O. B. (1930). The shell structure of the mollusks. *Det Kongelige Danske Videnskabernes Selskabs Skrifter. Naturvidenskabelig og Mathematisk Afdeling, ser. 9* 2: 231–326.

Boistel, R., Swoger, J., Kržič, U., Fernandez, V., Gillet, B. & Reynaud, E. G. (2011). The future of three-dimensional microscopic imaging in marine biology. *Marine Ecology* 32: 438–452.

von Boletzky, S. (2003a). A lower limit to adult size in coleoid cephalopods: Elements of a discussion. *Berliner Paläobiologische Abhandlungen* 3: 19–28.

von Boletzky, S. (2003b). Biology of early life stages in cephalopod molluscs. *Advances in Marine Biology* 44: 143–203.

von Boletzky, S. & Villanueva, R. (2014). Cephalopod biology, pp. 3–15 *in* J. Iglesias, Fuentes, L. & Villanueva, R. (eds.), *Cephalopod Culture.* New York, Springer.

Bolotov, I. N., Vikhrev, I. V., Bespalaya, Y. V., Gofarov, M. Y., Kondakov, A. V., Konopleva, E. S., Bolotov, N. N. & Lyubas, A. A. (2016). Multi-locus fossil-calibrated phylogeny, biogeography and a subgeneric revision of the Margaritiferidae (Mollusca: Bivalvia: Unionoida). *Molecular Phylogenetics and Evolution* 103: 104–121.

Bonar, D. B. (1976). Molluscan metamorphosis: A study in tissue transformation. *American Zoologist* 16: 573–591.

Bondesen, P. (1950). A comparative morphological-biological analysis of the egg-capsules of freshwater pulmonate gastropods: Hygrophila, Basommatophora, Pulmonata. *Natura Jutlandica* 3: 1–209.

Bone, Q., Brown, E. R. & Travers, G. (1994). On the respiratory flow in the cuttlefish *Sepia officinalis. Journal of Experimental Biology* 194: 153–165.

Bonnaud, L., Boucher-Rodoni, R. & Monnerot, M. (1997). Phylogeny of cephalopods inferred from mitochondrial DNA sequences. *Molecular Phylogenetics and Evolution* 7: 44–54.

Bonnaud, L., Pichon, D. & Boucher-Rodoni, R. (2005). Molecular approach of Decabrachia phylogeny: Is *Idiosepius* definitely not a sepiolid? *Phuket Marine Biological Center Research Bulletin* 66: 203–212.

Bonnemain, B. (2005). *Helix* and drugs: Snails for western health care from antiquity to the present. *Evidence-based Complementary and Alternative Medicine* 2: 25–28.

Bookstein, F. L. (1997). *Morphometric Tools for Landmark Data: Geometry and Biology.* Cambridge, UK, Cambridge University Press.

Boss, K. J. (1965). Symbiotic erycinacean bivalves. *Malacologia* 3: 183–195.

Boss, K. J. (1982). Mollusca, pp. 945–1166 *in* S. P. Parker (ed.), *Synopsis and Classification of Living Organisms.* Vol. 1. New York, McGraw Hill Book Company.

Botting, J. P. & Muir, L. A. (2008). Unravelling causal components of the Ordovician radiation: The Builth Inlier (central Wales) as a case study. *Lethaia* 41: 111–125.

Bottjer, D. J. & Carter, J. G. (1980). Functional and phylogenetic significance of projecting periostracal structures in the Bivalvia (Mollusca). *Journal of Paleontology* 54: 200–216.

Bottjer, D. J. & Ausich, W. I. (1986). Phanerozoic development of tiering in soft substrata suspension-feeding communities. *Paleobiology* 12: 400–420.

Bottjer, D. J. (2002). Enigmatic Ediacara fossils: Ancestors or aliens? Pp. 11–33 *in* D. J. Bottjer, Etter, W. J., Hagadorn, W. & Tang, C. M. *Exceptional Fossil Preservation: A Unique View on the Evolution of Marine Life.* New York, Columbia University Press.

Bottjer, D. J., Clapham, M. E., Fraiser, M. L. & Powers, C. M. (2008). Understanding mechanisms for the end-Permian mass extinction and the protracted Early Triassic aftermath and recovery. *GSA Today* 18: 4–10.

Boucaud-Camou, E. & Roper, C. F. E. (1995). Digestive enzymes in paralarval cephalopods. *Bulletin of Marine Science* 57: 313–327.

Bouček, B. (1964). *The Tentaculites of Bohemia: Their Morphology, Taxonomy, Ecology, Phylogeny and Biostratigraphy.* Prague: Czechoslovak Academy of Sciences.

Boucher-Rodoni, R. & Mangold, K. M. (1994). Ammonia production in cephalopods, physiological and evolutionary aspects, pp. 53–60 *in* H. O. Pörtner, O'Dor, R. K. & Macmillan, D. L. (eds.), *Physiology of Cephalopod Molluscs: Lifestyle and Performance Adaptations. Marine and Freshwater Behaviour and Physiology*, Special Issue. Basel, Gordon and Breach Science Publishers SA.

Bouchet, P. (1979). How many molluscan species in New Caledonia? *Hawaiian Shell News* 27: 10.

Bouchet, P. & Perrine, D. (1996). More gastropods feeding at night on parrotfishes. *Bulletin of Marine Science* 59: 224–228.

Bouchet, P. (1997). Inventorying the molluscan diversity of the world: What is our rate of progress? *The Veliger* 40: 1–11.

Bouchet, P., Lozouet, P., Maestrati, P. & Héros, V. (2002). Assessing the magnitude of species richness in tropical marine environments: Exceptionally high numbers of molluscs at a New Caledonia site. *Biological Journal of the Linnean Society* 75: 421–436.

Bouchet, P. & Rocroi, J.-P. (2005). Classification and nomenclator of gastropod families. *Malacologia* 47: 1–397.

Bouchet, P. (2006). The magnitude of marine biodiversity, pp. 31–62 *in* C. M. Duarte (ed.), *The Exploration of Marine Biodiversity: Scientific and Technological Challenges.* Bilbao, Spain, Fundación BBVA.

Bouchet, P. (2009). From specimens to data, and from seashells to molluscs: The Panglao Marine Biodiversity Project. *Vita Malacologica* 8: 1–8.

Bouchet, P., Ng, P. K. L., Largo, D. & Tan, S. H. (2009). Panglao 2004-investigations of the marine species richness in the Philippines. *Raffles Bulletin of Zoology* 20: 1–19.

Bouchet, P., Rocroi, J.-P., Bieler, R., Carter, J. G. & Coan, E. V. (2010). Nomenclator of bivalve families with a classification of bivalve families. *Malacologia* 52: 1–184.

Bouchet, P., Kantor, Y. I., Sysoev, A. V. & Puillandre, N. (2011). A new operational classification of the Conoidea (Gastropoda). *Journal of Molluscan Studies* 77: 273–308.

Bouchet, P., Bary, S., Héros, V. & Marani, G. (2016). How many species of molluscs are there in the world's oceans, and who is going to describe them? *Tropical Deep-Sea Benthos* 29: 9–24.

Bouchet, P., Rocroi, J.-P., Hausdorf, B., Kaim, A., Kano, Y., Nützel, A., Parkhaev, P., Schrödl, M. & Strong, E. E. (2017). Revised classification, nomenclator and typification of gastropod and monoplacophoran families. *Malacologia* 61: 1–526.

Bouillon, J. (1960). Ultrastructure des cellules rénales des mollusques. I. Gastéropodes pulmonés terrestres (*Helix pomatia*). *Annales des sciences naturelles. Zoologie* 2: 719–749.

Bourget, E. & Crisp, D. J. (1975). Factors affecting deposition of the shell in *Balanus balanoides* (L.). *Journal of the Marine Biological Association of the United Kingdom* 55: 231–249.

Bourne, G. C. (1909). Contributions to the morphology of the group Neritacea of the aspidobranch gastropods: Part 1. The Neritidae. *Proceedings of the Zoological Society of London* 8: 810–887.

Bourne, G. C. (1910). On the anatomy and systematic position of *Incisura* (*Scissurella*) *lytteltonensis*. *Quarterly Journal of Microscopical Science* 55: 1–47.

Bourne, G. C. (1911). Contributions to the morphology of the group Neritacea of the apsidobranch gastropods. Part II. The Helicinidae. *Proceedings of the Zoological Society of London* 11: 759–809 [plates: pp.730–742].

Bouvier, E. L. (1887). Système nerveux, morphologie générale et classification des gastéropodes prosobranches. *Annales des sciences naturelles. Zoologie* 3: 1–510.

Bovbjerg, R. V. (1984). Habitat selection in two intertidal snails, genus *Nerita*. *Bulletin of Marine Science* 34: 185–196.

Bowsher, A. L. (1955). Origin and adaptation of platyceratid gastropods. *University of Kansas: Paleontological Contributions* 5: 1–11.

Boyd, D. W. & Newell, N. D. (1997). A reappraisal of trigoniacean families (Bivalvia) and a description of two early Triassic species. *American Museum Novitates* 3216: 1–14.

Boyer, D. M., Puente, J., Gladman, J. T., Glynn, C., Mukherjee, S., Yapuncich, G. S. & Daubechies, I. (2015). A new fully automated approach for aligning and comparing shapes. *Anatomical Record* 298: 249–276.

Boyle, P. R. (1969). Fine structure of the eyes of *Onithochiton neglectus* (Mollusca, Polyplacophora). *Zeitschrift für Zellforschung und Mikroskopische Anatomie* 102: 313–332.

Boyle, P. R. (1974). The aesthetes of chitons. Part 2. Fine structure of *Lepidochitona cinereus* (L.). *Cell and Tissue Research* 153: 383–398.

Boyle, P. R. (1975). Fine structure of the subradular organ of *Lepidochitona cinereus* L. (Mollusca, Polyplacophora). *Cell and Tissue Research* 162: 411–417.

Boyle, P. R. (1977). The physiology and behaviour of chitons (Mollusca: Polyplacophora). *Oceanography and Marine Biology Annual Review* 15: 461–509.

Boyle, P. R., Mangold, K. & Froesch, D. (1979). The mandibular movements of *Octopus vulgaris*. *Journal of Zoology* 188: 53–67.

Boyle, P. R. (1999). Cephalopods, pp. 115–139 *in* T. Poole (ed.), *UFAW Handbook on the Care and Management of Laboratory Animals: Amphibious and Aquatic Vertebrates and Advanced Invertebrates*. Vol. 2. Oxford, UK, Wiley Blackwell.

Boyle, P. R. & Rodhouse, P. G. K. (2005). *Cephalopods: Ecology and Fisheries*. Oxford, UK, Wiley-Blackwell.

Brace, R. C. (1977a). Anatomical changes in nervous and vascular systems during the transition from prosobranch to opisthobranch organization. *Transactions of the Zoological Society of London* 34: 1–25.

Brace, R. C. (1977b). The functional anatomy of the mantle complex and columellar muscle of tectibranch molluscs (Gastropoda, Opisthobranchia). *Philosophical Transactions of the Royal Society B* 277: 1–56.

Brace, R. C. (1983). Observations on the morphology and behaviour of *Chilina fluctuosa* Gray (Chilinidae), with a discussion on the early evolution of pulmonate gastropods. *Philosophical Transactions of the Royal Society B* 300: 463–491.

Bradshaw-Hawkins, V. I. & Sander, F. (1981). Notes on the reproductive biology and behavior of the West Indian fighting conch, *Strombus pugilis* Linnaeus in Barbados, with evidence of male guarding. *The Veliger* 24: 159–164.

Branch, G. M. (1974). The ecology of *Patella* Linnaeus from the Cape Peninsula, South Africa. 2. Growth rates. *Transactions of the Royal Society of South Africa* 41: 161–193.

Branch, G. M. (1981). The biology of limpets: Physical factors, energy flow, and ecological interactions. *Oceanography and Marine Biology Annual Review* 19: 235–380.

Brasier, M. D. (1989). Towards a biostratigraphy of the earliest skeletal biotas, pp. 117–165 *in* J. W. Cowie & Brasier, M. D. (eds.), *The Precambrian–Cambrian Boundary. Oxford Monographs on Geology and Geophysics*. Oxford, UK, Clarendon Press.

Brayard, A., Bucher, H., Brühwiler, T., Galfetti, T., Goudemand, N., Guodun, K., Escarguel, G. & Jenks, J. (2007). *Proharpoceras* Chao: A new ammonoid lineage surviving the end-Permian mass extinction. *Lethaia* 40: 175–181.

Brayard, A., Escarguel, G., Bucher, H., Monnet, C., Brühwiler, T., Goudemand, N., Galfetti, T. & Guex, J. (2009). Good genes and good luck: Ammonoid diversity and the end-Permian mass extinction. *Science* 325: 1118–1121.

Brendelberger, H. & Juergens, S. (1993). Suspension feeding in *Bithynia tentaculata* (Prosobranchia, Bithyniidae), as affected by body size, food and temperature. *Oecologia* 94: 36–42.

Brenzinger, B., Neusser, T. P., Jörger, K. M. & Schrödl, M. (2011a). Integrating 3D microanatomy and molecules: Natural history of the Pacific freshwater slug *Strubellia* Odhner, 1937 (Heterobranchia: Acochlidia), with description of a new species. *Journal of Molluscan Studies* 77: 351–374.

Brenzinger, B., Wilson, N. G. & Schrödl, M. (2011b). 3D microanatomy of a gastropod 'worm', *Rhodope rousei* n. sp. (Heterobranchia) from southern Australia. *Journal of Molluscan Studies* 77: 375–387.

Brenzinger, B., Haszprunar, G. & Schrödl, M. (2013). At the limits of a successful body plan – 3D microanatomy, histology and evolution of *Helminthope* (Mollusca: Heterobranchia: Rhodopemorpha), the most worm-like gastropod. *Frontiers in Zoology* 10: 37.

Bretsky, P. W. (1973). Evolutionary patterns in the Paleozoic Bivalvia: Documentation and some theoretical considerations. *Geological Society of America Bulletin* 84: 2079–2096.

Brett, C. E., Gahn, F. J. & Baumiller, T. K. (2004). Platyceratid gastropods as parasites, predators, and prey and their possible effects on echinoderm hosts: Collateral damage and targeting. *Geological Society of America Abstracts with Programs* 36: 478.

Breure, A. S. H. & Gittenberger, E. (1982). The rock-scraping radula, a striking case of convergence (Mollusca). *Netherlands Journal of Zoology* 32: 307–312.

Briggs, D. E. G. (1977). Bivalved arthropods from the Cambrian Burgess Shale of British Columbia. *Palaeontology* 20: 595–621.

Briggs, D. E. G. & Fortey, R. A. (1989). The early radiation and relationships of the major arthropod groups. *Science* 246: 241–243.

Briggs, D. E. G., Fortey, R. A. & Wills, M. A. (1992). Morphological disparity in the Cambrian. *Science* 256: 1670–1673.

Briggs, D. E. G. & Kear, A. J. (1993). Decay and preservation of polychaetes: Taphonomic thresholds in soft-bodied organisms. *Paleobiology* 19: 107–135.

Briggs, D. E. G., Erwin, D. H., Collier, F. J. & Clark, C. (1994). *The Fossils of the Burgess Shale*. Vol. 238. Washington, DC, Smithsonian Institution Press.

Briggs, J. C. (1974). *Marine Zoogeography*. New York, McGraw Hill Book Company.

Bristow, G. A., Berland, B., Schander, C. & Vo, D. T. (2010). A new endosymbiotic bivalve (Heterodonta: Galeommatoidea), from Pacific holothurians. *Journal of Parasitology* 96: 532–534.

Britton, J. C. & Morton, B. (1979). *Corbicula* in North America: The evidence reviewed and evaluated, pp. 249–287 *in* J. C. Britton (ed.), *Proceedings of the First International Corbicula Symposium* October 13–15, 1977. Fort Worth, TX, Texas Christian University Research Foundation.

Britton, K. M. (1984). The Onchidiacea (Gastropoda, Pulmonata) of Hong Kong with a worldwide review of the genera. *Journal of Molluscan Studies* 50: 179–191.

Brocco, S. L. & Cloney, R. A. (1980). Reflector cells in the skin of *Octopus dofleini*. *Cell and Tissue Research* 205: 167–186.

Bromley, R. G. & Asgaard, U. (1990). *Solecurtus strigilatus*: A jet-propelled burrowing bivalve, pp. 313–320 *in* B. Morton (ed.), *The Bivalvia. Proceedings of a Memorial Symposium in Honour of Sir Charles Maurice Yonge (1899–1986), Edinburgh, 1986*. Hong Kong, Hong Kong University Press.

Bromley, R. G. & Heinberg, C. (2006). Attachment strategies of organisms on hard substrates: A palaeontological view. *Palaeogeography, Palaeoclimatology, Palaeoecology* 232: 429–453.

Bronn, H. G. (1862). Malacozoa, Abt. 1: Malacozoa Acephala [bivalves and brachiopods], pp. 1–706 *in* H. G. Bronn (ed.), *Die Klassen und Ordnungen des Thier-Reichs*. Vol. 3. Leipzig und Heidelberg, C. F. Winter.

Bronn, H. G. (1862–1866). *Malacozoa, Abt. 2: Malacozoa Cephalophora [gastropods, scaphopods, cephalopods]*. Vol. 3. Leipzig und Heidelberg, C. F. Winter.

Brooker, L. R., Lee, A. K., Macey, D. J., Van Bronswijk, W. & Webb, J. (2003). Multiple-front iron-mineralisation in chiton teeth (*Acanthopleura echinata*: Mollusca: Polyplacophora). *Marine Biology* 142: 447–454.

Brooks, W. K. (1876). The affinity of the Mollusca and the Molluscoida. *Proceedings of the Boston Society* 18: 225–236.

Brunn, A. F. (1943). The biology of *Spirula spirula* (L.). *Dana Report* 4: 1–46.

Brusca, R. C., Moore, W. & Shuster, S. M. (2016). *Invertebrates*. Sunderland, MA, Sinauer Associates.

Bucher, H., Landman, N. H., Klofak, S. M. & Guex, J. (1996). Mode and rate of growth in ammonoids, pp. 407–461 *in* N. H. Landman, Tanabe, K. & Davis, R. A. (eds.), *Ammonoid Paleobiology: From Anatomy to Ecology. Topics in Geobiology*. New York, Plenum Press.

Buckland-Nicks, J. & Howley, B. (1997). Spermiogenesis and sperm structure in relation to early events of fertilization in the limpet *Tectura testudinalis* (Muller, 1776). *Biological Bulletin* 193: 306–319.

Buckland-Nicks, J. A., Koss, R. & Chia, F.-S. (1988a). The elusive acrosome of chiton sperm. *International Journal of Invertebrate Reproduction & Development* 13: 193–198.

Buckland-Nicks, J. A., Koss, R. & Chia, F.-S. (1988b). Fertilization in a chiton: Acrosome-mediated sperm-egg fusion. *Gamete Research* 21: 199–212.

Buckland-Nicks, J. A., Chia, F.-S. & Koss, R. (1990). Spermiogenesis in Polyplacophora, with special reference to acrosome formation (Mollusca). *Zoomorphology (Berlin)* 109: 179–188.

Buckland-Nicks, J. A. (1993). Hull cupules of chiton eggs: Parachute structures and sperm focusing devices? *Biological Bulletin* 184: 269–276.

Buckland-Nicks, J. A. & Eernisse, D. J. (1993). Ultrastructure of mature sperm and eggs of the brooding hermaphroditic chiton, *Lepidochitona fernaldi* Eernisse 1986, with special reference to the mechanism of fertilization. *Journal of Experimental Zoology Part A* 265: 567–574.

Buckland-Nicks, J. A. (1995). Ultrastructure of sperm and sperm-egg interaction in *Aculifera*: Implications for molluscan phylogeny. *Mémoires du Muséum national d'Histoire naturelle* 166: 129–153.

Buckland-Nicks, J. A. & Scheltema, A. H. (1995). Was internal fertilization an innovation of early Bilateria? Evidence from sperm structure of a mollusc. *Proceedings of the Royal Society B* 261: 11–18.

Buckland-Nicks, J. A. (1998). Prosobranch parasperm: Sterile germ cells that promote paternity? *Micron* 29: 267–280.

Buckland-Nicks, J. A. & Hodgson, A. N. (2000). Fertilization in *Callochiton castaneus* (Mollusca). *Biological Bulletin* 199: 59–67.

Buckland-Nicks, J. A., Gibson, G. & Koss, R. (2002). Phylum Mollusca: Gastropoda, pp. 261–287 *in* C. M. Young (ed.), *Atlas of Marine Invertebrate Larvae*. San Diego, CA, Academic Press.

Buckland-Nicks, J. A. & Hadfield, M. G. (2005). Spermatogenesis in *Serpulorbis* (Mollusca: Vermetoidea) and its implications for phylogeny of gastropods. *Invertebrate Reproduction & Development* 48: 171–184.

Buckland-Nicks, J. A. (2006). Fertilization in chitons: Morphological clues to phylogeny. *Venus* 65: 51–70.

Buckland-Nicks, J. A. (2008). Fertilization biology and the evolution of chitons. *American Malacological Bulletin* 25: 97–111.

Buckland-Nicks, J. A. & Reunov, A. (2009). Ultrastructure of hull formation during oogenesis of *Rhyssoplax tulipa* (= *Chiton tulipa*) (Chitonidae: Chitoninae). *Invertebrate Reproduction & Development* 53: 165–174.

Buckland-Nicks, J. A. & Reunov, A. A. (2010). Egg hull formation in *Callochiton dentatus* (Mollusca, Polyplacophora): The contribution of microapocrine secretion. *Invertebrate Biology* 129: 319–327.

Buckland-Nicks, J. A. (2014). Apocrine secretion of the egg hull in oogenesis and exclusion of sperm organelles at fertilization make reproduction in Chitonida (Mollusca) unique. *Journal of Natural History* 48: 2885–2897.

Budd, G. E. & Jensen, S. (2000). A critical reappraisal of the fossil record of the bilaterian phyla. *Biological Reviews* 75: 253–295.

Budd, G. E. (2003). The Cambrian fossil record and the origin of the Phyla. *Integrative and Comparative Biology* 43: 157–165.

Budelmann, B.-U. (1975). Gravity receptor function in cephalopods with particular reference to *Sepia officinalis*. *Fortschritte der Zoologie* 23: 84–96.

Budelmann, B.-U. (1988). Morphological diversity of equilibrium receptor systems in aquatic invertebrates, pp. 757–782 *in* J. Atema, Fay, R. R., Popper, A. N. & Tavolga, W. N. (eds.), *Sensory Biology of Aquatic Animals*. New York, Springer-Verlag.

Budelmann, B.-U. & Bleckmann, H. (1988). A lateral line analogue in cephalopods: Water waves generate microphonic potentials in the epidermal head lines of *Sepia* and *Lolliguncula*. *Journal of Comparative Physiology A* 164: 1–5.

Budelmann, B.-U., Schipp, R. & von Boletzky, S. (1997). Cephalopoda, pp. 119–414 *in* F. W. Harrison & Kohn, A. J. (eds.), *Microscopic Anatomy of Invertebrates: Mollusca 2. Mollusca.* Vol. 6A. New York, Wiley-Liss.

Bullock, T. H. & Horridge, G. A. (1965). The Mollusca, pp. 1273–1515 *in* T. H. Bullock, Horridge, G. A. & Freeman, W. H. (eds.), *Structure and Function in the Nervous Systems of Invertebrates.* Vol. 2. San Francisco, CA/London, W. H. Freeman & Co.

Bulman, O. M. B. (1939). Muscle systems of some inarticulate brachiopods. *Geological Magazine* 76: 434–444.

Bunje, P. M. E. & Lindberg, D. R. (2007). Lineage divergence of a freshwater snail clade associated with post-Tethys marine basin development. *Molecular Phylogenetics and Evolution* 42: 373–387.

Burch, J. B. (1968). Tentacle retraction in Tracheopulmonata. *Journal of the Malacological Society of Australia* 11: 62–67.

Burch, J. B. & Patterson, C. M. (1971). The chromosome number of the athoracophorid slug, *Triboniophorus graeffei*, from Australia. *Malacological Review* 4: 25–26.

Burghardt, I., Schrödl, M. & Wägele, H. (2008a). Three new solar-powered species of the genus *Phyllodesmium* Ehrenberg, 1831 (Mollusca: Nudibranchia: Aeolidioidea) from the tropical Indo-Pacific, with analysis of their photosynthetic activity and notes on biology. *Journal of Molluscan Studies* 74: 277–292.

Burghardt, I., Stemmer, K. & Wägele, H. (2008b). Symbiosis between *Symbiodinium* (Dinophyceae) and various taxa of Nudibranchia (Mollusca: Gastropoda), with analyses of long-term retention. *Organisms Diversity & Evolution* 8: 66–76.

Burghardt, I. & Wägele, H. (2014). The symbiosis between the 'solar-powered' nudibranch *Melibe engeli* Risbec, 1937 (Dendronotoidea) and *Symbiodinium* sp. (Dinophyceae). *Journal of Molluscan Studies* 80: 508–517.

Bürgin, U. E. (1965). The color pattern of *Hermissenda crassicornis* (Eschscholtz, 1831). *The Veliger* 7: 205–215.

Burn, R. & Thompson, T. E. (1998). Order Cephalaspidea, pp. 943–959 *in* P. L. Beesley, Ross, G. J. B. & Wells, A. (eds.), *Mollusca: The Southern Synthesis. Part B. Fauna of Australia.* Melbourne, VIC, CSIRO Publishing.

Burton, D. W. (1981). Pallial systems in the Athoracophoridae (Gastropoda, Pulmonata). *New Zealand Journal of Zoology* 8: 391–402.

Bush, A. M. & Bambach, R. K. (2011). Paleoecologic megatrends in marine Metazoa. *Annual Review of Earth and Planetary Sciences* 39: 241–269.

Bush, M. S. (1989). The ultrastructure and function of the oesophagus of *Patella vulgata* Linnaeus. *Journal of Molluscan Studies* 55: 111–124.

Bush, S. L. & Robison, B. H. (2007). Ink utilization by mesopelagic squid. *Marine Biology* 152: 485–494.

Butterfield, N. J. (2006). Hooking some stem-group 'worms': Fossil lophotrochozoans in the Burgess Shale. *BioEssays* 28: 1161–1166.

Butterfield, N. J. (2008). An early Cambrian radula. *Journal of Paleontology* 82: 543–554.

Byrne, M. (1985). The life history of the gastropod *Thyonicola americana* Tikasingh, endoparasitic in a seasonally eviscerating holothurian host. *Ophelia* 24: 91–101.

Byrum, C. A. & Ruppert, E. E. (1994). The ultrastructure and functional morphology of a captaculum in *Graptacme calamus* (Mollusca, Scaphopoda). *Acta Zoologica* 75: 37–46.

Caddy, J. F. (1969). Development of mantle organs, feeding, and locomotion in postlarval *Macoma balthica* (L.) (Lamellibranchiata). *Canadian Journal of Zoology* 47: 609–617.

Caddy, J. F. & Rodhouse, P. G. (1998). Cephalopod and groundfish landings: Evidence for ecological change in global fisheries? *Reviews in Fish Biology and Fisheries* 8: 431–444.

Caetano, C. H. S. & Dos Santos, F. N. (2010). Mollusca, Scaphopoda, Gadilidae, *Stiocadulus magdalenensis* Gracia and Ardila, 2009: First record of the genus and species from Brazil. *Check List* 6: 687–689.

Cain, A. J. (1977). Variation in the spire index of some coiled gastropod shells, and its evolutionary significance. *Philosophical Transactions of the Royal Society B* 277: 377–428.

Cain, A. J. (1983). Ecology and ecogenetics of terrestrial molluscan populations, pp. 597–647 *in* W. D. Russell-Hunter (ed.), *Ecology. The Mollusca.* Vol. 6. New York, Academic Press.

Calabrò, C., Rindone, A., Bertuccio, C. & Giacobbe, S. (2018). Hermaphroditism in a violet snail, *Janthina pallida* (Gastropoda, Caenogastropoda): A contribution. *Biologia*: 74: 1–5.

Calvo, M. & Templado, J. (2005). Reproduction and sex reversal of the solitary vermetid gastropod *Serpulorbis arenarius*. *Marine Biology* 146: 963–973.

Campbell, D. C., Hoekstra, K. J. & Carter, J. G. (1998). 18S ribosomal DNA and evolutionary relationships within the Bivalvia, pp. 75–85 *in* P. A. Johnston & Haggart, J. W. (eds.), *Bivalves: An Eon of Evolution.* Calgary, AB, University of Calgary Press.

Campbell, D. C. (2000). Molecular evidence on the evolution of the Bivalvia, pp. 31–46 *in* E. M. Harper, Taylor, J. D. & Crame, J. A. (eds.), *Evolutionary Biology of the Bivalvia.* London, The Geological Society (Special Publication No. 177).

Candás, M., Díaz-Agras, G., Abad, M., Barrio, L., Cunha-Veira, X., Pedrouzo, L., Señarís, M., García-Álvarez, O. & Urgorri, V. (2016). Application of microCT in the study of the anatomy of small marine molluscs. *Microscopy and Analysis* 30: S8–S11.

Cao, C., Love, G. D., Hays, L. E., Wang, W., Shen, S. & Summons, R. E. (2009). Biogeochemical evidence for euxinic oceans and ecological disturbance presaging the end-Permian mass extinction event. *Earth and Planetary Science Letters* 281: 188–201.

Carbone, M., Gavagnin, M., Haber, M., Guo, Y. W., Fontana, A., Manzo, E., Genta-Jouve, G., Tsoukatou, M., Rudman, W. B., Cimino, G., Ghiselin, M. T. & Mollo, E. (2013). Packaging and delivery of chemical weapons: A defensive trojan horse stratagem in chromodorid nudibranchs. *PLoS ONE* 8: e62075 (62071–62079).

Carefoot, T. H. (1967). Growth and nutrition of *Aplysia punctata* feeding on a variety of marine algae. *Journal of the Marine Biological Association of the United Kingdom* 47: 565–589.

Carey, N., Galkin, A., Henriksson, P., Richards, J. G. & Sigwart, J. D. (2012). Variation in oxygen consumption among 'living fossils' (Mollusca: Polyplacophora). *Journal of the Marine Biological Association of the United Kingdom* 93: 197–207.

Carey, N., Sigwart, J. D. & Richards, J. G. (2013). Economies of scaling: More evidence that allometry of metabolism is linked to activity, metabolic rate and habitat. *Journal of Experimental Marine Biology and Ecology* 439: 7–14.

Carlini, D. B., Reece, K. S. & Graves, J. E. (2000). Actin gene family evolution and the phylogeny of coleoid cephalopods (Mollusca: Cephalopoda). *Molecular Biology and Evolution* 17: 1353–1370.

Carlini, D. B., Young, R. E. & Vecchione, M. (2001). A molecular phylogeny of the Octopoda (Mollusca: Cephalopoda) evaluated in light of morphological evidence. *Molecular Phylogenetics and Evolution* 21: 388–397.

Caron, J.-B., Scheltema, A. H., Schander, C. & Rudkin, D. (2006). A soft-bodied mollusc with radula from the Middle Cambrian Burgess Shale. *Nature* 442: 159–163.

Caron, J.-B., Scheltema, A. H., Schander, C. & Rudkin, D. (2007). Reply to Butterfield on stem-group 'worms': Fossil lophotrochozoans in the Burgess Shale. *BioEssays* 29: 200–202.

Carpenter, P. P. (1857). *Catalogue of the Collection of Matzatlan Shells in the British Museum*. Warrington, UK, Oberlin Press.

Carpenter, P. P. (1864). *Supplementary report on the present state of our knowledge with regard to the Mollusca of the west coast of North America. Report on the thirty-third meeting of the British Association for the Advancement of Science; held at Newcastle-upon-Tyne in August and September 1863.* Newcastle-upon-Tyne, UK, pp. 517–686.

Carrick, N. (1980). Aspects of the biology of *Gazameda gunni* (Reeve, 1849), a viviparous mesogastropod and potential 'indicator' of perturbation induced by sewage pollution. *Journal of the Malacological Society of Australia* 4: 254–255.

Carriker, M. R. (1961). Comparative functional morphology of boring mechanisms in gastropods. *American Zoologist* 1: 263–266.

Carriker, M. R. & Gruber, G. L. (1999). Uniqueness of the gastropod accessory boring organ (ABO): Comparative biology, an update. *Journal of Shellfish Research* 18: 579–595.

Carroll, A. R. & Scheuer, P. J. (1990). Kuanoniamines A, B, C, and D: Pentacyclic alkaloids from a tunicate and its prosobranch mollusk predator *Chelynotus semperi*. *The Journal of Organic Chemistry* 55: 4426–4431.

Carroll, D. J. & Kempf, S. C. (1994). Changes occur in the central nervous system of the nudibranch *Berghia verrucicornis* (Mollusca, Opisthobranchia) during metamorphosis. *Biological Bulletin* 186: 202–212.

Cartaxana, P., Trampe, E., Kühl, M. & Cruz, S. (2017). Kleptoplast photosynthesis is nutritionally relevant in the sea slug *Elysia viridis*. *Scientific Reports* 7: 7714.

Carté, B. & Faulkner, D. J. (1986). Role of secondary metabolites in feeding associations between a predatory nudibranch, two grazing nudibranchs, and a bryozoan. *Journal of Chemical Ecology* 12: 795–804.

Carter, J. G. & Aller, R. C. (1975). Calcification in the bivalve periostracum. *Lethaia* 8: 315–320.

Carter, J. G. & Michael, J. S. T. (1978). Shell microstructure of a Middle Devonian (Hamilton Group) bivalve fauna from central New York. *Journal of Paleontology* 52: 859–880.

Carter, J. G. & Clark, G. R. (1985). Classification and phylogenetic significance of molluscan shell microstructure, pp. 50–71 *in* T. W. Broadhead (ed.), *Mollusks: Notes for a short course organized by D.J. Bottjer, C.S. Hickman, and P.D. Ward. University of Tennessee Studies in Geology.* Knoxville, TN, University of Tennessee Department of Geological Science.

Carter, J. G. (1990a). Shell microstructural data for the Bivalvia: Part IV. Order Ostreoida, pp. 347–362 *in* J. G. Carter (ed.), *Skeletal Biomineralization: Patterns, Processes and Evolutionary Trends*. Vol. 1. New York, Van Nostrand Reinhold.

Carter, J. G. (1990b). Evolutionary significance of the shell microstructure in the Palaeotaxodonta, Pteriomorphia and Isofilibranchia (Bivalvia: Mollusca), pp. 135–296 *in* J. G. Carter (ed.), *Skeletal Biomineralization: Patterns, Processes and Evolutionary Trends*. Vol. 1. New York, Van Nostrand Reinhold.

Carter, J. G. (1990c). Shell microstructural data for the Bivalvia: Part II. Orders Nuculoida and Solemyoida, pp. 303–319 *in* J. G. Carter (ed.), *Skeletal Biomineralization: Patterns, Processes and Evolutionary Trends*. Vol. 1. New York, Van Nostrand Reinhold.

Carter, J. G. (1990d). Shell microstructural data for the Bivalvia: Part V. Order Pectinoida, pp. 363–389 *in* J. G. Carter (ed.), *Skeletal Biomineralization: Patterns, Processes and Evolutionary Trends*. Vol. 1. New York, Van Nostrand Reinhold.

Carter, J. G. & Hall, R. M. (1990). Polyplacophora, Scaphopoda, Archaeogastropoda and Paragastropoda (Mollusca), pp. 29–51 *in* J. G. Carter (ed.), *Skeletal Biomineralization: Patterns, Processes and Evolutionary Trends*. Vol. 2. New York, Van Nostrand Reinhold.

Carter, J. G. & Lutz, R. A. (1990). Bivalvia (Mollusca), pp. 5–28 *in* J. G. Carter (ed.), *Skeletal Biomineralization: Patterns, Processes and Evolutionary Trends*. New York, Van Nostrand Reinhold.

Carter, J. G. & Schneider, J. A. (1997). Condensing lenses and shell microstructure in *Corculum* (Mollusca: Bivalvia). *Journal of Paleontology* 71: 56–61.

Carter, J. G., Campbell, D. C. & Campbell, M. R. (2000). Cladistic perspectives on early bivalve evolution, pp. 47–79 *in* E. M. Harper, Taylor, J. D. & Crame, J. A. (eds.), *Evolutionary Biology of the Bivalvia*. London, The Geological Society (Special Publication No. 177).

Carter, J. G. (2001). Shell and ligament microstructure of selected Silurian and Recent palaeotaxodonts (Mollusca: Bivalvia). *American Malacological Bulletin* 16: 217–238.

Carter, J. G. (2004). Evolutionary implications of a duplivincular ligament in the Carboniferous pinnid *Pteronites* (Mollusca, Bivalvia, Pteriomorphia). *Journal of Paleontology* 78: 235–239.

Carter, J. G., Campbell, D. C. & Campbell, M. R. (2006). *Morphological Phylogenetics of the Early Bivalvia*. Barcelona, Spain, Universitat Autònoma de Barcelona. Abstracts and Posters of the International Congress on Bivalvia (Figures at www.senckenberg.de/odes/06-16/Carter_et_al_Phylogeny-EarlyBiv.pdf).

Carter, J. G., Altaba, C. R., Anderson, L. C., Araujo, R., Biakov, A. S., Bogan, A. E., Campbell, C., Campbell, M., Chen, J. H., Cope, J. C. W., Delvene, G., Dijkstra, H. H., Fang, Z. J., Gardner, R. N., Gavrilova, V. A., Goncharova, I. A., Harries, P. J., Hartman, J. H., Hautmann, M., Hoeh, W. R., Hylleberg, J., Jiang, B. Y., Johnston, P., Kirkendale, L. A., Kleemann, K. H. & Koppka, J. (2011). A synoptical classification of the Bivalvia (Mollusca). *Paleontological Contributions* 4: 1–47.

Casey, R. (1952). Some genera and subgenera, mainly new, of Mesozoic heterodont lamellibranchs. *Proceedings of the Malacological Society of London* 29: 121–176.

Castelin, M., Puillandre, N., Kantor, Y. I., Modica, M., Terryn, Y., Cruaud, C., Bouchet, P. & Holford, M. (2012). Macroevolution of venom apparatus innovations in auger snails (Gastropoda; Conoidea; Terebridae). *Molecular Phylogenetics and Evolution* 64: 21–44.

Castillejo, J., Garrido, C. & Santos, M. (1995). Las babosas de las familias Papillodermidae Wiktor, Martin et Castillejo, 1990, Parmacellidae, Gray, 1860 y Testacellidae Gray, 1840 en la Peninsula Iberica. Morfologia y distribucion (Gastropoda, Pulmonata, Terrestria nuda). *Revista Real Academia Galega de Ciencias* 14: 63–80.

Castro, L. R. & Colgan, D. J. (2010). The phylogenetic position of Neritimorpha based on the mitochondrial genome of *Nerita melanotragus* (Mollusca: Gastropoda). *Molecular Phylogenetics and Evolution* 57: 918–923.

Cattaneo-Vietti, R., Angelini, S., Gaggero, L. & Lucchetti, G. (1995). Mineral composition of nudibranch spicules. *Journal of Molluscan Studies* 61: 331–337.

Cavin, L. & Forey, P. L. (2007). Using ghost lineages to identify diversification events in the fossil record. *Biology Letters* 3: 201–204.

Cernohorsky, W. O. (1964). Autotomy in *Cypraea cribellum*. *The Cowry* 1: 96.

Cervera, J. L., García-Gómez, J. C. & Ortea, J. (2000). A new species of *Berthella* (Blainville, 1824) (Opisthobranchia: Notaspidea) from the Canary Islands (eastern Atlantic Ocean), with a re-examination of the phylogenetic relationships of the Notaspidea. *Journal of Molluscan Studies* 66: 301–311.

Chaine, A. & Angeloni, L. (2005). Size-dependent mating and gender choice in a simultaneous hermaphrodite, *Bulla gouldiana*. *Behavioral Ecology and Sociobiology* 59: 58–68.

Chamberlain, J., Friedman, G. M. & Chamberlain, R. B. (2002). Devonian archanodont unionoids from the Catskill Mountains of New York: Implications for the paleoecology and biogeography of the first freshwater bivalves. *Northeastern Geology and Environmental Sciences* 26: 211–229.

Chamberlain, J. A. (1980). Hydromechanical design of fossil cephalopods, pp. 289–336 in M. R. House & Senior, J. R. (eds.), *The Ammonoidea*. Vol. 18. Systematics Association Special Volume. New York, Academic Press.

Chamberlain, J. A. (1987). Locomotion of *Nautilus*, pp. 489–526 in W. B. Saunders & Landman, N. H. (eds.), *Nautilus: The Biology and Paleobiology of a Living Fossil. Topics in Geobiology*. New York, Springer.

Chambers, R. J. & McQuaid, C. D. (1994). A review of larval development in the intertidal limpet genus *Siphonaria* (Gastropoda: Pulmonata). *Journal of Molluscan Studies* 60: 415–423.

Chang, C. H., Mok, H. K., Huang, L. G. & Chang, Y. W. (2010). Differentiation according to body region and interspecific variation in the morphology of integumentary spicules of nudibranchs. *Molluscan Research* 30: 73–80.

Chang, Y. W., Willan, R. C. & Mok, H. K. (2013). Can the morphology of the integumentary spicules be used to distinguish genera and species of phyllidiid nudibranchs (Porostomata: Phyllidiidae)? *Molluscan Research* 33: 14–23.

Chanley, P. E. & Chanley, P. (1970). Larval development of the commensal clam *Montacuta percompressa* Dall. *Proceedings of the Malacological Society of London* 39: 59–67.

Chapman, G. & Newell, G. E. (1956). The role of the body fluid in the movement of soft-bodied invertebrates. II. The extension of the siphons of *Mya arenaria* L. and *Scrobicularia plana* (Da Costa). *Proceedings of the Royal Society B* 145: 564–580.

Chase, R. & Tolloczko, B. (1993). Tracing neural pathways in snail olfaction: From the tip of the tentacles to the brain and beyond. *Microscopy Research and Technique* 24: 214–230.

Chase, R. (2002). *Behavior and Its Neural Control in Gastropod Molluscs*. Oxford, UK, Oxford University Press.

Chase, R. & Blanchard, K. C. (2006). The snail's love-dart delivers mucus to increase paternity. *Proceedings of the Royal Society B* 273: 1471–1475.

Chase, R. & Vaga, K. (2006). Independence, not conflict, characterizes dart-shooting and sperm exchange in a hermaphroditic snail. *Behavioral Ecology and Sociobiology* 56: 732–739.

Chase, R. (2007). The function of dart shooting in helicid snails. *American Malacological Bulletin* 23: 183–189.

Chase, R., Darbyson, E., Horn, K. E. & Samarova, E. (2010). A mechanism aiding simultaneously reciprocal mating in snails. *Canadian Journal of Zoology* 88: 99–107.

Checa, A. G. & Garcia-Ruiz, J. M. (1996). Morphogenesis of the septum in ammonoids, pp. 253–296 in N. H. Landman, Tanabe, K. & Davis, R. A. (eds.), *Ammonoid Paleobiology: From Anatomy to Ecology. Topics in Geobiology*. New York, Plenum Press.

Checa, A. G. & Jiménez-Jiménez, A. P. (2003). Evolutionary morphology of oblique ribs of bivalves. *Palaeontology* 46: 709–724.

Checa, A. G., Ramirez-Rico, J., González-Segura, A. & Sánchez-Navas, A. (2009). Nacre and false nacre (foliated aragonite) in extant monoplacophorans (=Tryblidiida: Mollusca). *Die Naturwissenschaften* 96: 111–122.

Checa, A. G., Vendrasco, M. J. & Salas, C. (2017). Cuticle of Polyplacophora: Structure, secretion, and homology with the periostracum of conchiferans. *Marine Biology* 164: 64.

Chelazzi, G. & Vannini, M. (1980). Zonal orientation based on local visual cues in *Nerita plicata* L. (Mollusca: Gastropoda) at Aldabra Atoll. *Journal of Experimental Marine Biology and Ecology* 46: 147–156.

Chelazzi, G., Focardi, S. & Deneubourg, J. L. (1983a). A comparative study on the movement patterns of two sympatric tropical chitons (Mollusca: Polyplacophora). *Marine Biology* 74: 115–125.

Chelazzi, G., Focardi, S., Deneubourg, J. L. & Innocenti, R. (1983b). Competition for the home and aggressive behaviour in the chiton *Acanthopleura gemmata* (Blainville) (Mollusca: Polyplacophora). *Behavioral Ecology and Sociobiology* 14: 15–20.

Chelazzi, G., Innocenti, R. & Della Santina, P. (1983c). Zonal migration and trail-following of an intertidal gastropod analyzed by LED tracking in the field. *Marine Behaviour and Physiology* 10: 121–136.

Chelazzi, G., Deneubourg, J. L. & Focardi, S. (1984). Cooperative interactions and environmental control in the intertidal clustering of *Nerita textilis* (Gastropoda; Prosobranchia). *Behaviour* 90: 151–166.

Chelazzi, G., Della Santina, P. & Vannini, M. (1985). Long-lasting substrate marking in the collective homing of the gastropod *Nerita textilis*. *Biological Bulletin* 168: 214–221.

Chelazzi, G. & Parpagnoli, D. (1987). Behavioural responses to crowding modification and home intrusion in *Acanthopleura gemmata* (Mollusca, Polyplacophora). *Ethology* 75: 109–118.

Chelazzi, G., Focardi, S. & Deneubourg, J. L. (1988). Analysis of movement patterns and orientation mechanisms in intertidal chitons and gastropods, pp. 173–184 in G. Chelazzi & Vannini, M. (eds.), *Behavioral Adaptation to Intertidal Life*. New York, Plenum Press.

Chelazzi, G., Della Santina, P. & Parpagnoli, D. (1990). The role of trail following in the homing of intertidal chitons: A comparison between three *Acanthopleura* spp. *Marine Biology* 105: 445–450.

Chelazzi, G., Santini, G., Della Santina, P. & Focardi, S. (1993). Does the homing accuracy of intertidal chitons rely on active trail following? A simulation approach. *Journal of Theoretical Biology* 160: 165–178.

Chen, C., Copley, J. T., Linse, K., Rogers, A. D. & Sigwart, J. D. (2015a). The heart of a dragon: 3D anatomical reconstruction of the 'scaly-foot gastropod' (Mollusca: Gastropoda: Neomphalina) reveals its extraordinary circulatory system. *Frontiers in Zoology* 12: 1–16.

Chen, C., Linse, K., Copley, J. T. & Rogers, A. D. (2015b). The 'scaly-foot gastropod': A new genus and species of hydrothermal vent-endemic gastropod (Neomphalina: Peltospiridae) from the Indian Ocean. *Journal of Molluscan Studies* 81: 332–334.

Chen, C., Uematsu, K., Linse, K. & Sigwart, J. D. (2017). By more ways than one: Rapid convergence at hydrothermal vents shown by 3D anatomical reconstruction of *Gigantopelta* (Mollusca: Neomphalina). *BMC Evolutionary Biology* 17: 62.

Chen, C., Linse, K., Uematsu, K. & Sigwart, J. D. (2018). Cryptic niche switching in a chemosymbiotic gastropod. *Proceedings of the Royal Society B* 285: 20181099.

Chen, J.-Y. & Teichert, C. (1983a). Cambrian cephalopods. *Geology* 11: 647–650.

Chen, Y.-Y. & Teichert, C. (1983b). Cambrian cephalopods of China. *Palaeontographica. Abteilung A* 181: 1–102.

Cherns, L. (2004). Early Palaeozoic diversification of chitons (Polyplacophora, Mollusca) based on new data from the Silurian of Götland, Sweden. *Lethaia* 37: 445–456.

Cherns, L., Rohr, D. M. & Frýda, J. (2004). Polyplacophoran and symmetrical univalve mollusks, pp. 179–183 in B. D. Webby, Paris, F., Droser, M. L. & Percival, I. G. (eds.), *The Great Ordovician Biodiversification Event. Critical Moments and Perspectives in Earth History and Paleobiology*. New York, Columbia University Press.

Childress, J. J. & Seibel, B. A. (1998). Life at stable low oxygen levels: Adaptations of animals to oceanic oxygen minimum layers. *Journal of Experimental Biology* 201: 1223–1232.

Chin, K., Hartman, J. H. & Roth, B. (2009). Opportunistic exploitation of dinosaur dung: Fossil snails in coprolites from the Upper Cretaceous Two Medicine Formation of Montana. *Lethaia* 42: 185–198.

Chinzei, K., Savazzi, E. & Seilacher, A. (1982). Adaptational strategies of bivalves living as infaunal secondary soft bottom dwellers. *Neues Jahrbuch für Geologie und Paläontologie-Abhandlungen* 164: 229–244.

Choo, L. M., Choo, L. Q. & Tan, K. S. (2014). The origin and formation of hair on external valve surfaces of the tropical marine mussel *Modiolus traillii* (Reeve, 1857). *Journal of Molluscan Studies* 80: 111–116.

Christa, G., Händeler, K., Kück, P., Vleugels, M., Franken, J., Karmeinski, D. & Wägele, H. (2014a). Phylogenetic evidence for multiple independent origins of functional kleptoplasty in Sacoglossa (Heterobranchia, Gastropoda). *Organisms Diversity & Evolution* 15: 23–36.

Christa, G., de Vries, J., Jahns, P. & Gould, S. B. (2014b). Switching off photosynthesis: The dark side of sacoglossan slugs. *Communicative & Integrative Biology* 7: 20132493-3.

Chung, D. J. D. (1987). Courtship and dart shooting behavior of the land snail *Helix aspersa*. *The Veliger* 30: 24–39.

Churchill, C. K., ÓFoighil, D., Strong, E. E. & Gittenberger, A. (2011). Females floated first in bubble-rafting snails. *Current Biology* 21: R802–R803.

Cilia, G. & Fratini, F. (2018). Antimicrobial properties of terrestrial snail and slug mucus. *Journal of Complementary and Integrative Medicine* 15.

Cimino, G. & Ghiselin, M. T. (1999). Chemical defense and evolutionary trends in biosynthetic capacity among dorid nudibranchs (Mollusca: Gastropoda: Opisthobranchia). *Chemoecology* 9: 187–207.

Cimino, G. & Gavagnin, M., Eds. (2006). *Molluscs: From Chemo-Ecological Study to Biotechnological Application*. Berlin/Heidelberg, Springer-Verlag.

Cimino, G. & Ghiselin, M. T. (2009). Chemical defense and the evolution of opisthobranch gastropods. *Proceedings of the California Academy of Sciences* 60: 175–422.

Cironi, P., Albericio, F. & Alvarez, M. (2005). Lamellarins: Isolation, activity and synthesis. *Progress in Heterocyclic Chemistry* 16: 1–26.

Clapham, M. E. & James, N. P. (2008). Paleoecology of Early–Middle Permian marine communities in eastern Australia: Response to global climate change in the aftermath of the late Paleozoic Ice Age. *Palaios* 23: 738–750.

Clapham, M. E. & Payne, J. L. (2011). Acidification, anoxia, and extinction: A multiple logistic regression analysis of extinction selectivity during the Middle and Late Permian. *Geology* 39: 1059–1062.

Clark, R. B. (1960). The economics of Dentalium. *The Veliger* 6: 9–19.

Clarke, M. & Hart, M. (2018). Statoliths and coleoid evolution. *Treatise Online* 102, Part M, Chapter 11: 1–23.

Clarke, M. R. (1960). *Lepidoteuthis grimaldii* – A squid with scales. *Nature* 188: 955–956.

Clarke, M. R. (1962). Respiratory and swimming movements in the cephalopod *Cranchia scabra*. *Nature* 196: 351–352.

Clarke, M. R. (1986). *A Handbook for the Identification of Cephalopod Beaks*. London/Oxford, UK, Clarendon Press.

Clarke, M. R. & Maddock, L. (1988). Statoliths of fossil coleoid cephalopods, pp. 153–168 in M. R. Clarke & Trueman, E. R. (eds.), *Paleontology and Neontology of Cephalopods. The Mollusca*. Vol. 12. New York, Academic Press.

Clarke, M. R. (1996a). Cephalopods as prey. III. Cetaceans. *Philosophical Transactions of the Royal Society B* 351: 1053–1065.

Clarke, M. R. (1996b). The role of cephalopods in the world's oceans: An introduction. *Philosophical Transactions of the Royal Society of London: Part B* 351: 979–983.

Claverie, T. & Kamenos, N. A. (2008). Spawning aggregations and mass movements in subtidal *Onchidoris bilamellata* (Mollusca: Opisthobranchia). *Journal of the Marine Biological Association of the United Kingdom* 88: 157–159.

Cledón, M., Brey, T., Penchaszadeh, P. E. & Arntz, W. (2005). Individual growth and somatic production in *Adelomelon brasiliana* (Gastropoda: Volutidae) off Argentina. *Marine Biology* 147: 447–452.

Cleevely, R. (1974). A provisional bibliography of natural history works by the Sowerby family. *Journal of the Society for the Bibliography of Natural History* 6: 482–559.

Cleland, D. M. (1954). A study of the habits of *Valvata piscinalis* (Müller) and the structure and function of the alimentary canal and reproductive system. *Proceedings of the Malacological Society of London* 30: 167–203.

Clements, R., Liew, T. S., Vermeulen, J. J. & Schilthuizen, M. (2008). Further twists in gastropod shell evolution. *Biology Letters* 4: 179–182.

Clements, T., Colleary, C., De Baets, K. & Vinther, J. (2017). Buoyancy mechanisms limit preservation of coleoid cephalopod soft tissues in Mesozoic Lagerstätten. *Palaeontology* 60: 1–14.

Clifford, K. T., Gross, L., Johnson, K., Martin, K. J., Shaheen, N. & Harrington, M. A. (2003). Slime-trail tracking in the predatory snail, *Euglandina rosea*. *Behavioral Neuroscience* 117: 1086–1095.

Climo, F. M. (1973). The systematics, biology and zoogeography of the land snail fauna of Great Island, Three Kings Group, New Zealand. *Journal of the Royal Society of New Zealand* 3: 565–627.

Climo, F. M. (1975). The anatomy of *Gegania valkyrie* Powell (Mollusca: Heterogastropoda: Mathildidae) with notes on other heterogastropods. *Journal of the Royal Society of New Zealand* 5: 275–288.

Climo, F. M. (1980). Smeagolida, a new order of gymnomorph mollusc from New Zealand based on a new genus and species. *New Zealand Journal of Zoology* 7: 513–522.

Climo, F. M. (1989). The panbiogeography of New Zealand as illuminated by the genus *Fectola* Iredale, 1915 and subfamily Rotadiscinae Pilsbry, 1927 (Mollusca, Pulmonata, Punctoidea, Charopidae). *New Zealand Journal of Zoology* 16: 587–649.

Cloney, R. A. & Florey, E. (1968). Ultrastructure of cephalopod chromatophore organs. *Zeitschrift für Zellforschung und Mikroskopische Anatomie* 89: 250–280.

Cloney, R. A. & Brocco, S. L. (1983). Chromatophore organs, reflector cells, iridocytes and leucophores in cephalopods. *American Zoologist* 23: 581–592.

Cloud, P. E. (1948). Some problems and patterns of evolution exemplified by fossil invertebrates. *Evolution* 2: 322–350.

Coan, E. V., Valentich-Scott, P. & Bernard, F. R. (2000). *Bivalve Seashells of Western North America [including] Marine Bivalve Mollusks from Arctic Alaska to Baja California.* Santa Barbara, CA, Santa Barbara Museum of Natural History.

Coan, E. V. & Kabat, A. R. (2018) *2,400 Years of Malacology.* https://www.malacological.org/downloads/epubs/2400-years/2400yrs_of_Malacology_complete.pdf.

Coates, A. G., Jackson, J. B. C., Collins, L. S., Cronin, T. M., Dowsett, H. J., Bybell, L. M., Jung, P. & Obando, J. A. (1992). Closure of the isthmus of Panama – The near-shore marine record of Costa Rica and western Panama. *Geological Society of America Bulletin* 104: 814–828.

Coen, L. D., Brumbaugh, R. D., Bushek, D., Grizzle, R., Luckenbach, M. W., Posey, M. H., Powers, S. P. & Tolley, S. G. (2007). Ecosystem services related to oyster restoration. *Marine Ecology Progress Series* 341: 303–307.

Cohen, K. M., Finney, S. C., Gibbard, P. L. & Fan, J.-X. (2013; updated). The ICS international chronostratigraphic chart. *Episodes* 36: 199–204.

Coleman, R. A., Underwood, A. J., Benedetti-Cecchi, L., Åberg, P., Arenas, F., Arrontes, J., Castro, J., Hartnoll, R. G., Jenkins, S. R., Paula, J., Santina, P. D. & Hawkins, S. J. (2006). A continental scale evaluation of the role of limpet grazing on rocky shores. *Oecologia* 147: 556–564.

Colgan, D. J. & Ponder, W. F. (1994). The evolutionary consequences of restrictions on gene flow: Examples from hydrobiid snails. *The Nautilus* 108: 25–43.

Colgan, D. J., Ponder, W. F. & Eggler, P. E. (2000). Gastropod evolutionary rates and phylogenetic relationships assessed using partial 28S rDNA and histone H3 sequences. *Zoologica Scripta* 29: 29–63.

Colgan, D. J., Ponder, W. F., Beacham, E. & Macaranas, J. M. (2003). Gastropod phylogeny based on six segments from four genes representing coding or non-coding and mitochondrial or nuclear DNA. *Molluscan Research* 23: 123–148.

Colgan, D. J., Ponder, W. F., Beacham, E. & Macaranas, J. M. (2007). Molecular phylogenetics of Caenogastropoda (Gastropoda: Mollusca). *Molecular Phylogenetics and Evolution* 42: 717–737.

Collin, R. & Giribet, G. (2010). Report of a cohesive gelatinous egg mass produced by a tropical marine bivalve. *Invertebrate Biology* 129: 165–171.

Collins, A. J., LaBarre, B. A., Wong, B. S., Shah, M. V., Heng, S., Choudhury, M. H., Haydar, S. A., Santiago, J. & Nyholm, S. V. (2012). Diversity and partitioning of bacterial populations within the accessory nidamental gland of the squid *Euprymna scolopes.* *Applied and Environmental Microbiology* 78: 4200–4208.

Collins, M. & Villanueva, R. (2006). Taxonomy, ecology and behaviour of the cirrate octopods, pp. 277–322 in R. N. Gibson, Atkinson, R. J. A. & Gordon, J. D. M. (eds.), *Oceanography and Marine Biology Annual Review.* Vol. 44. London, Taylor & Francis.

Collins, M. A. & Rodhouse, P. G. K. (2006). Southern Ocean cephalopods. *Advances in Marine Biology* 50: 191–265.

Combosch, D. J., Collins, T. M., Glover, E. A., Graf, D. L., Harper, E. M., Healy, J. M., Kawauchi, G. Y., Lemer, S., McIntyre, E. & Strong, E. E. (2017). A family-level Tree of Life for bivalves based on a Sanger-sequencing approach. *Molecular Phylogenetics and Evolution* 107: 191–208.

Connor, M. S. & Edgar, R. K. (1982). Selective grazing by the mud snail *Ilyanassa obsoleta.* *Oecologia* 53: 271–275.

Connor, V. M. & Quinn, J. F. (1984). Stimulation of food species growth by limpet mucus. *Science* 225: 843–844.

Connor, V. M. (1986). The use of mucous trails by intertidal limpets to enhance food resources. *Biological Bulletin* 171: 548–564.

Connors, M. J., Ehrlich, H., Hog, M., Godeffroy, C., Araya, S., Kallai, I., Gazit, D., Boyce, M. & Ortiz, C. (2012). Three-dimensional structure of the shell plate assembly of the chiton *Tonicella marmorea* and its biomechanical consequences. *Journal of Structural Biology* 177: 314–328.

Conway Morris, S. (1976). A new Cambrian lophophorate from the Burgess Shale of British Columbia. *Palaeontology* 19: 199–222.

Conway Morris, S. & Peel, J. S. (1990). Articulated halkieriids from the Lower Cambrian of north Greenland. *Nature* 345: 802–805.

Conway Morris, S. & Peel, J. S. (1995). Articulated halkieriids from the Lower Cambrian of North Greenland and their role in early protostome evolution. *Philosophical Transactions of the Royal Society B* 347: 305–358.

Conway Morris, S. (2006). Darwin's dilemma: The realities of the Cambrian 'explosion'. *Philosophical Transactions of the Royal Society B* 361: 1069–1083.

Conway Morris, S. & Caron, J.-B. (2007). Halwaxiids and the early evolution of the lophotrochozoans. *Science* 315: 1255–1258.

Conway Morris, S. & Peel, J. S. (2009). New palaeoscolecidan worms from the Lower Cambrian: Sirius Passet, Latham Shale and Kinzers Shale. *Acta Palaeontologica Polonica* 55: 141–156.

Cook, A. (1977). Mucus trail following by the slug *Limax grossui* Lupu. *Animal Behaviour* 25: 774–781.

Cook, A. & Shirbhate, R. (1983). The mucus-producing glands and the distribution of the cilia of the pulmonate slug *Limax pseudoflavus.* *Journal of Zoology* 201: 97–116.

Cook, A. (2001). Behavioural ecology: On doing the right thing, in the right place, at the right time, pp. 447–487 in G. M. Barker (ed.), *The Biology of Terrestrial Molluscs.* Wallingford, UK, CABI Publishing.

Cook, A., Nützel, A. & Frýda, J. (2008). Two Mississippian Caenogastropod limpets from Australia and their meaning for the ancestry of the Caenogastropoda. *Journal of Paleontology* 82: 183–187.

Cook, A. G., Jell, P. A., Webb, G. E., Johnson, M. E. & Baarli, B. G. (2015). Septate gastropods from the Upper Devonian of the Canning Basin: Implications for palaeoecology. *Alcheringa* 39: 519–524.

Cook, C. M. & Kondo, Y. (1960). Revision of the Tornatellinidae and Achatinellidae (Gastropoda, Pulmonata). *Bernice Pauahi Bishop Museum Bulletin* 221: 1–303.

Cook, S. B. & Cook, C. B. (1975). Directionality in the trail-following response of the pulmonate limpet *Siphonaria alternata*. *Marine Behaviour and Physiology* 3: 147–155.

Cooke, A. H., Shipley, A. E. & Reed, F. R. C. (1927). *Molluscs, Brachiopods (Recent) and Brachiopods (Fossil)*. London, MacMillan & Co.

Cooke, I. R. C., Whitelaw, B., Norman, M., Caruana, N. & Strugnell, J. M. (2015). Toxicity in cephalopods, pp. 1–15 *in* P. Gopalakrishnakone & Malhotra, A. (eds.), *Evolution of Venomous Animals and their Toxins*. Dordrecht, the Netherlands, Springer Science + Business Media.

Cooper, K. M., Hanlon, R. T. & Budelmann, B.-U. (1990). Physiological color change in squid iridophores. II. Ultrastructural mechanisms in *Lolliguncula brevis*. *Cell and Tissue Research* 259: 15–24.

Cooper, M. R. (1991). Lower Cretaceous Trigonioida (Mollusca, Bivalvia) from the Algoa Basin, [South Africa] with a revised classification of the order. *Annals of the South African Museum* 100: 1–52.

Cope, J. C. W. (1996a). Early Ordovician (Arenig) bivalves from the Llangynog Inlier, south Wales [Part 4]. *Palaeontology* 39: 979–1025.

Cope, J. C. W. (1996b). The early evolution of the Bivalvia, pp. 361–370 *in* J. D. Taylor (ed.), *Origin and Evolutionary Radiation of the Mollusca*. Oxford, UK, Oxford University Press.

Cope, J. C. W. (1997a). Affinities of the early Ordovician bivalve *Catamarcaia* Sánchez & Babin, 1993 and its role in bivalve evolution. *Geobios* 30: 127–131.

Cope, J. C. W. (1997b). The early phylogeny of the class Bivalvia. *Palaeontology* 40: 713–746.

Cope, J. C. W. & Babin, C. (1999). Diversification of bivalves in the Ordovician. *Geobios* 32: 175–185.

Cope, J. C. W. (2000). A new look at early bivalve phylogeny, pp. 81–95 *in* E. M. Harper, Taylor, J. D. & Crame, J. A. (eds.), *Evolutionary Biology of the Bivalvia*. London, The Geological Society (Special Publication No. 177).

Cope, J. C. W. (2002). Diversification and biogeography of bivalves during the Ordovician Period. *Geological Society, London, Special Publications* 194: 35–52.

Cope, J. C. W. (2004). *Bivalve and rostroconch mollusks*. New York, Columbia University Press.

Cope, J. C. W. & Kříž, J. (2013). The Lower Palaeozoic palaeobiogeography of Bivalvia, pp. 221–241 *in* D. A. T. Harper & Servais, T. (eds.), *Early Palaeozoic Biogeography and Palaeogeography*. London, Geological Society of London, Memoirs.

Cortesi, F. & Cheney, K. L. (2010). Conspicuousness is correlated with toxicity in marine opisthobranchs. *Journal of Evolutionary Biology* 23: 1509–1518.

Cotton, B. C. (1961). *South Australian Mollusca. Pelecypoda*. Adelaide, SA, Government Printer.

Covich, A. P. (1981). Chemical refugia from predation for thin-shelled gastropods in a sulfide-enriched stream. *Verhandlungen der Internationalen Vereinigung für Theoretische und Angewandte Limnologie* 21: 1632–1636.

Covich, A. P., Crowl, T. A., Alexander, J. E. & Vaughn, C. C. (1994). Predator-avoidance responses in freshwater decapod-gastropod interactions mediated by chemical stimuli. *Journal of the North American Benthological Society* 13: 283–290.

Cowie, R. H. & Smith, B. D. (2000). Arboreal Neritidae. *The Veliger* 43: 98–99.

Cox, L. N., Zaslavskaya, N. I. & Marko, P. B. (2014). Phylogeography and trans-Pacific divergence of the rocky shore gastropod *Nucella lima*. *Journal of Biogeography* 41: 615–627.

Cox, L. R. (1960a). Thoughts on the classification of the Bivalvia. *Proceedings of the Malacological Society of London* 34: 60–88.

Cox, L. R. (1960b). Gastropoda. General characteristics of Gastropoda, pp. 84–169 *in* R. C. Moore (ed.), *Treatise on Invertebrate Paleontology Part I. Mollusca*. Vol. 1. Lawrence, KS, Geological Society of America and University of Kansas Press.

Cox, L. R. & Knight, J. B. (1960). Suborders of Archaeogastropoda. *Proceedings of the Malacological Society of London* 33: 262–264.

Cox, L. R., Newell, N. D., Boyd, D. W., Branson, C. C., Casey, R., Chavan, A., Coogan, A. H., Dechaseaux, C., Fleming, C. A., Haas, F., Hertlein, L. G., Kauffman, E. G., Keen, A. M., LaRocque, A., McAlester, A. L., Moore, R. C., Nuttall, C. P., Perkins, B. F., Puri, H. S., Smith, L. A., Soot-Ryen, T., Stenzel, H. B., Trueman, E. R., Turner, R. D. & Weir, J. (1969a). Bivalvia, pp. N1–N489; N491–N868 *in* R. C. Moore & Teichert, C. (eds.), *Treatise on Invertebrate Paleontology Part N [parts 1 and 2] Bivalvia*. Vol. 6 Mollusca. Lawrence, KS, Geological Society of America and University of Kansas Press.

Cox, L. R., Newell, N. D., Branson, C. C., Casey, R., Chavan, A., Coogan, A. H., Dechaseaux, C., Fleming, C. A., Haas, F., Hertlein, L. G., Keen, A. M., LaRocque, A., McAlester, A., Perkins, B. F., Puri, H. S., Smith, L. A., Soot-Ryen, T., Stenzel, H. B., Turner, R. D. & Weir, J. (1969b). Systematic descriptions, pp. N225-N869 *in* R. C. Moore (ed.), *Treatise on Invertebrate Paleontology Part N*. Vol. 6 Mollusca. Lawrence, KS, Geological Society of America and University of Kansas Press.

Coyne, K. J. & Waite, J. H. (2000). In search of molecular dovetails in mussel byssus: From the threads to the stem. *Journal of Experimental Biology* 203: 1425–1431.

Crame, J. A. (2000a). Evolution of taxonomic diversity gradients in the marine realm: Evidence from the composition of recent bivalve faunas. *Paleobiology* 26: 188–214.

Crame, J. A. (2000b). The nature and origin of taxonomic diversity gradients in marine bivalves, pp. 347–360 *in* E. M. Harper, Taylor, J. D. & Crame, J. A. (eds.), *Evolutionary Biology of the Bivalvia*. London, The Geological Society (Special Publication No. 177).

Crame, J. A. (2000c). Intrinsic and extrinsic controls on the diversification of the Bivalvia, pp. 135–148 *in* S. J. Culver & Rawson, P. F. (eds.), *Biotic Response to Global Change: The Last 145 Million Years*. Cambridge, UK, Cambridge University Press.

Crame, J. A. (2002). Evolution of taxonomic diversity gradients in the marine realm: A comparison of Late Jurassic and Recent bivalve faunas. *Paleobiology* 28: 184–207.

Creek, G. A. (1953). The morphology of *Acme fusca* (Montague) with special reference to the genital system. *Proceedings of the Malacological Society of London* 29: 228–240.

Creese, R. G. (1980). Reproductive cycles and fecundities of two species of *Siphonaria* (Mollusca, Pulmonata) in south-eastern Australia. *Australian Journal of Marine and Freshwater Research* 31: 37–47.

Creese, R. G. (1986). Brooding behaviour and larval development in the New Zealand chiton, *Onithochiton neglectus* de Rochebrune (Mollusca: Polyplacophora). *New Zealand Journal of Zoology* 13: 83–91.

Crichton, R., Killby, V. A. A. & Lafferty, K. J. (1973). The distribution and morphology of phagocytic cells in the chiton *Liolophura gaimardi*. *Australian Journal of Experimental Biology and Medical Science* 51: 357–372.

Crick, R. E. (1988). Buoyancy regulation and macroevolution in nautiloid cephalopods. *Senckenbergiana lethaea* 69: 13–42.

Criscione, F. & Ponder, W. F. (2013). A phylogenetic analysis of rissooidean and cingulopsoidean families (Gastropoda: Caenogastropoda). *Molecular Phylogenetics and Evolution* 66: 1075–1082.

Criscione, F., Ponder, W. F., Köhler, F., Takano, T. & Kano, Y. (2016). A molecular phylogeny of Rissoidae (Caenogastropoda: Rissooidea) allows testing the diagnostic utility of morphological traits. *Zoological Journal of the Linnean Society* 179: 23–40.

Crisp, M. (1981). Epithelial sensory structures of trocids. *Journal of the Marine Biological Association of the United Kingdom* 61: 95–106.

Crofts, D. R. (1929). *Haliotis. Liverpool Marine Biology Committee: Memoirs on Typical British Marine Plants and Animals.* Liverpool, UK, Liverpool Marine Biology Committee.

Crofts, D. R. (1955). Muscle morphogenesis in primitive gastropods and its relation to torsion. *Proceedings of the Zoological Society of London* 125: 711–750.

Croizat, L., Nelson, G. & Rosen, D. E. (1974). Centers of origin and related concepts. *Systematic Zoology* 23: 265–287.

Croll, R. P. (1985). Sensory control of respiratory pumping in *Aplysia californica. Journal of Experimental Biology* 117: 15–28.

Croxall, J. P. & Prince, P. (1996). Cephalopods as prey. I. Seabirds. *Philosophical Transactions of the Royal Society B* 351: 1023–1043.

Crump, R. G., Williams, A. D. & Crothers, J. H. (2003). West Angle Bay: A case study. The fate of limpets. *Field Studies* 10: 579–599.

Cruz, R., Lins, U. & Farina, M. (1998). Minerals of the radular apparatus of *Falcidens* sp. (Caudofoveata) and the evolutionary implications for the phylum Mollusca. *Biological Bulletin* 194: 224–230.

Cummins, S. F., Boal, J. G., Buresch, K. C., Kuanpradit, C., Sobhon, P., Holm, J. B., Degnan, B. M., Nagle, G. T. & Hanlon, R. T. (2011). Extreme aggression in male squid induced by a β-MSP-like pheromone. *Current Biology* 21: 322–327.

Cunha, T. J. & Giribet, G. (2019). A congruent topology for deep gastropod relationships. *Proceedings of the Royal Society B* 286: 20182776.

Currie, D. R. (1992). Aesthete channel morphology in three species of Australian chitons (Mollusca: Polyplacophora). *Journal of the Malacological Society of Australia* 13: 3–14.

Curtis, S. K. & Cowden, R. R. (1977). Ultrastructure and histochemistry of the supportive structures associated with the radula of the slug *Limax maximus. Journal of Morphology* 151: 187–212.

Cuvier, G. (1795). Deuxiéme Mémoire sur l'organisation et les rapports des animaux à sang blanc, dans lequel on traite de la structure des Mollusques et de leur division en ordre, lu à la Société d'Histoire naturelle de Paris, le 11 Prairial an troisième. *Magasin encyclopédique, ou Journal des sciences, des lettres et des arts* 2: 433–449.

Cuvier, G. (1796). Notice sur le squelette d'une très-grande espèce de quadrupède inconnue jusqu'à présent, trouvé au Paraguay, et déposé au cabinet d'Histoire naturelle de Madrid. *Magasin encyclopédique, ou Journal des sciences, des lettres et des arts* 7: 303–310.

Cuvier, G. (1817). *Le règne animal distribué d'après son organization, pour servir de base à l'histoire naturelle des animaux.* 1–4. Paris, Detérville.

Cuvier, G. L. (1798). *Tableau élémentaire de l'histoire naturelle des Animaux.* Paris, Baudouin.

D'Asaro, C. N. (1970). Egg capsules of prosobranch molluscs from south Florida and the Bahamas (and notes on spawning in the laboratory). *Bulletin of Marine Science* 20: 414–440.

D'Orbigny, A. D. (1845 (1845–1847)). *Mollusques vivants et fossiles.* Paris, Adolphe Delahays.

D'Souza, M. (1981). The food of some neritid prosobranch molluscs found at Mkomaini Mombasa, Kenya. *Kenya Aquatica* 3: 46–49.

Daban, M., Morriconi, E., Kasinsky, H. E. & Chiva, M. (1990). Characterization of the nuclear sperm basic proteins in one archaeogastropod: Comparison of protamines between species. *Comparative Biochemistry and Physiology Part B* 96: 123–127.

Daley, J. (2018) Never-Before-Seen Colony of 1,000 Brooding Octopuses Found Off California Coast. *Smithsonian.com*, pp. 1–2. https://www.smithsonianmag.com/smart-news/massive-colony-1000-brooding-octopuses-found-california-180970664/.

Dall, W. H. (1877). On a provisional hypothesis of saltatory evolution. *American Naturalist* 11: 135–137.

Dall, W. H. (1893). The phylogeny of the Docoglossa. *Proceedings of the Academy of Natural Sciences of Philadelphia* 45: 285–287.

Dame, R. F. (2012). *Ecology of Marine Bivalves: An Ecosystem Approach,* 2nd edition. Boca Raton, FL, CRC Press.

Dance, S. P. (1966). *Shell Collecting: An Illustrated History.* London, Faber and Faber.

Daniel, T. L., Helmuth, B. S. T., Saunders, W. B. & Ward, P. D. (1997). Septal complexity in ammonoid cephalopods increased mechanical risk and limited depth. *Paleobiology* 23: 470–481.

Darling, J. (2006). The value of barcoding. *BioScience* 56: 710–711.

Darlington, P. J. (1982). *Zoogeography: The Geographical Distribution of Animals.* Malabar, FL, Krieger.

Darmaillacq, A. S., Dickel, L. & Mather, J. A., Eds. (2014). *Cephalopod Cognition.* Cambridge, UK, Cambridge University Press.

Darragh, T. A. (1986). The Cainozoic Trigoniidae of Australia. *Alcheringa* 10: 1–34.

Darwin, C. (1838–1843). *The Zoology of the Voyage of H.M.S. Beagle, under the Command of Captain Fitzroy, R.N., during the Years 1832 to 1836.* London, Smith, Elder.

Darwin, C. (1859). *On the Origins of Species by Means of Natural Selection.* London, John Murray.

Dattagupta, S., Martin, J., Liao, S.-m., Carney, R. S. & Fisher, C. R. (2007). Deep-sea hydrocarbon seep gastropod *Bathynerita naticoidea* responds to cues from the habitat-providing mussel *Bathymodiolus childressi. Marine Ecology* 28: 193–198.

Dauphin, Y. (1986). Microstructure des coquilles de Céphalopodes: La partie apicale de *Belopterina* (Coleoidea). *Bulletin du Museum national d'Histoire naturelle de Paris (3èm sér.)* 8: 53–75.

Davis, G. M. (1967). The systematic relationship of *Pomatiopsis lapidaria* and *Oncomelania hupensis formosana* (Prosobranchia, Hydrobiidae). *Malacologia* 6: 1–143.

Davis, G. M., Kitikoon, V. & Temcharoen, P. (1976). Monograph on *Lithoglyphopsis aperta,* the snail host of Mekong River schistosomiasis. *Malacologia* 15: 241–287.

Davis, G. M. (1979). The origin and evolution of the gastropod family Pomatiopsidae, with emphasis on the Mekong River Triculinae. *Academy of Natural Sciences of Philadelphia* Monographs 20: 1–120.

Davis, G. M. (1981). Different modes of evolution and adaptive radiation in the Pomatiopsidae (Prosobranchia, Mesogastropoda). *Malacologia* 21: 209–262.

Davis, G. M. (1982). Historical and ecological factors in the evolution, adaptive radiation, and biogeography of freshwater mollusks. *American Zoologist* 22: 375–395.

Davis, G. M., Kuo, Y. H., Hoagland, K. E., Chen, P. L., Yang, H. M. & Chen, D. J. (1983). Advances in the systematics of the Triculinae (Gastropoda, Prosobranchia): The genus *Fenouila* of Yunnan, China. *Proceedings of the Academy of Natural Sciences of Philadelphia* 135: 177–199.

Davis, G. M., Kuo, Y. H., Hoagland, K. E., Chen, P. L., Yang, H. M. & Chen, D. J. (1984). *Kunwaingia*, a new genus of Triculinae (Gastropoda: Pomatiopsidae) from China: Phenetic and cladistic relationships. *Proceedings of the Academy of Natural Sciences of Philadelphia* 136: 165–193.

Davis, R. A., Landman, N. H., Dommergues, J. L., Marchand, D. & Bucher, H. (1996). Mature modifications and dimorphism in ammonoid cephalopods, pp. 463–539 *in* N. H. Landman, Tanabe, K. & Davis, R. A. (eds.), *Ammonoid Paleobiology: From Anatomy to Ecology. Topics in Geobiology*. New York, Plenum Press.

Davison, A., Wade, C. M., Mordan, P. B. & Chiba, S. (2005). Sex and darts in slugs and snails (Mollusca: Gastropoda: Stylommatophora). *Journal of Zoology* 267: 329–338.

Davison, A. & Mordan, P. B. (2007). A literature database on the mating behavior of stylommatophoran land snails and slugs. *American Malacological Bulletin* 23: 173–181.

Day, J. A. (1969). Feeding of the cymatiid gastropod, *Argobuccinum argus*, in relation to the structure of the proboscis and secretions of the proboscis gland. *American Zoologist* 9: 909–916.

Day, R. M. & Harris, L. G. (1978). Selection and turnover of coelenterate nematocysts in some aeolid nudibranchs. *The Veliger* 21: 104–109.

Dayrat, B. A. & Tillier, S. (2000). Taxon sampling, character sampling and systematics: How gradist presuppositions created additional ganglia in gastropod euthyneuran taxa. *Zoological Journal of the Linnean Society* 129: 403–418.

Dayrat, B. A. & Tillier, S. (2002). Evolutionary relationships of euthyneuran gastropods (Mollusca): A cladistic re-evaluation of morphological characters. *Zoological Journal of the Linnean Society* 135: 403–470.

Dayrat, B. A. & Tillier, S. (2003). Goals and limits of phylogenetics: The euthyneuran gastropods, pp. 161–184 *in* C. Lydeard & Lindberg, D. R. (eds.), *Molecular Systematics and Phylogeography of Mollusks. Smithsonian Series in Comparative Evolutionary Biology*. Washington, DC, Smithsonian Books.

Dayrat, B. A. (2009). Review of the current knowledge of the systematics of Onchidiidae (Mollusca: Gastropoda: Pulmonata) with a checklist of nominal species. *Zootaxa* 2068: 1–26.

Dayrat, B. A., Conrad, M., Balayan, S., White, T. R., Albrecht, C., Golding, R. E., Gomes, S. R., Harasewych, M. G. & de Frias Martins, A. M. (2011). Phylogenetic relationships and evolution of pulmonate gastropods (Mollusca): New insights from increased taxon sampling. *Molecular Phylogenetics and Evolution* 59: 425–437.

De Baets, K., Klug, C., Korn, D. & Landman, N. H. (2012). Early evolutionary trends in ammonoid embryonic development. *Evolution* 66: 1788–1806.

De Baets, K., Bert, D., Hoffmann, R., Monnet, C., Yacobucci, M. M. & Klug, C. A. (2015). Ammonoid intraspecific variability, pp. 359–426 *in* C. A. Klug, Korn, D., De Baets, K., Kruta, I. & Mapes, R. H. (eds.), *Ammonoid Paleobiology: From Anatomy to Ecology. Topics in Geobiology*. Vol. 13. Dordrecht, the Netherlands, Springer.

De Lisa, E., Salzano, A. M., Moccia, F., Scaloni, A. & Di Cosmo, A. (2013). Sperm-attractant peptide influences the spermatozoa swimming behavior in internal fertilization in *Octopus vulgaris*. *Journal of Experimental Biology* 216: 2229–2237.

Dean, C. D., Sutton, M. D., Siveter, D. J. & Siveter, D. J. (2015). A novel respiratory architecture in the Silurian mollusc *Acaenoplax*. *Palaeontology* 58: 839–847.

Dechaseaux, C. (1969). Classification, pp. N766 *in* R. C. Moore. *Mollusca 6. Bivalvia. Treatise on Invertebrate Paleontology*. Part N. Vols. 1 and 2. Boulder, CO/Lawrence, KS, The Geological Society of America, Inc. and the University of Kansas.

Declerck, C. H. (1995). The evolution of suspension feeding in gastropods. *Biological Reviews* 70: 549–569.

Degnan, S. A., Imron, B. J., Geiger, D. L. & Degnan, B. M. (2006). Evolution in temperate and tropical seas: Disparate patterns in southern hemisphere abalone (Mollusca: Vetigastropoda: Haliotidae). *Molecular Phylogenetics and Evolution* 41: 249–256.

Délage, Y. (1899). Études sur la mérogonie. *Archives de zoologie expérimentale et générale* 7: 383–417.

Delhaye, W. & Bouillon, J. (1972a). L'évolution et l'adaptation de l'organe excréteur chez les mollusques gastéropodes pulmonés: II: Histophysiologie comparée du rein chez les stylommatophores. *Bulletin biologique de France et de Belgique* 106: 123–142.

Delhaye, W. & Bouillon, J. (1972b). L'évolution et l'adaptation de l'organe excréteur chez les mollusques gastéropodes pulmonés: I: Introduction générale et histophysiologie comparée du rein chez les basommatophores. *Bulletin biologique de France et de Belgique* 106: 45–77.

Delhaye, W. & Bouillon, J. (1972c). L'évolution et l'adaptation de l'organe excréteur chez les mollusques gastéropodes pulmonés: III: Histophysiologie comparée du rein chez les soléolifêres et conclusions générales pour tous les pulmonés. *Bulletin biologique de France et de Belgique* 106: 295–314.

Delhaye, W. (1974). Recherches sur les glandes pédieuses des gastéropodes prosobranches, principalement les formes terrestres et leur rôle possible dans l'osmorégulation chez les Pomatiasidae et les Chondropomidae. *Forma et Functio* 7: 181–200.

Dell'Angelo, B. & Schwabe, E. (2010). Teratology in chitons (Mollusca, Polyplacophora): A brief summary. *Bollettino Malacologico* 46: 9–15.

Dell, R. K. (1990). Antarctic Mollusca, with special reference to the fauna of the Ross Sea. *Royal Society of New Zealand*. Bulletin 27: 1–311.

Dell'Angelo, B. & Tursi, A. (1990). Abnormalities in chitons shell-plates. *Oebalia* 14: 1–14.

deMaintenon, M. J. (1999). Phylogenetic analysis of the Columbellidae (Mollusca: Neogastropoda) and the evolution of herbivory from carnivory. *Invertebrate Biology* 118: 258–288.

Denton, E. J. & Gilpin-Brown, J. B. (1961a). The buoyancy of the cuttlefish, *Sepia officinalis* (L.). *Journal of the Marine Biological Association of the United Kingdom* 41: 319–342.

Denton, E. J. & Gilpin-Brown, J. B. (1961b). The distribution of gas and liquid within the cuttlebone. *Journal of the Marine Biological Association of the United Kingdom* 41: 365–381.

Denton, E. J. & Taylor, D. W. (1964). The composition of the gas in the chambers of the cuttlebone of *Sepia officialis*. *Journal of the Marine Biological Association of the United Kingdom* 44: 203–207.

Denton, E. J. & Gilpin-Brown, J. B. (1966). On the buoyancy of the pearly nautilus. *Journal of the Marine Biological Association of the United Kingdom* 46: 723–759.

Denton, E. J., Gilpin-Brown, J. B. & Howarth, J. V. (1967). On the buoyancy of *Spirula spirula*. *Journal of the Marine Biological Association of the United Kingdom* 47: 181–191.

Denton, E. J. (1971). Examples of the use of active transport of salts and water to give buoyancy in the sea. *Philosophical Transactions of the Royal Society B* 262: 277–287.

Denton, E. J. & Gilpin-Brown, J. B. (1973). Floatation mechanisms in modern and fossil cephalopods. *Advances in Marine Biology* 11: 197–268.

Denton, E. J. (1974). On buoyancy and the lives of modern and fossil cephalopods. *Proceedings of the Royal Society B* 185: 273–299.

Derby, C. D. (2007). Escape by inking and secreting: Marine molluscs avoid predators through a rich array of chemicals and mechanisms. *Biological Bulletin* 213: 274–289.

Derby, C. D., Kicklighter, C. E., Johnson, P. M. & Zhang, X. (2007). Chemical composition of inks of diverse marine molluscs suggests convergent chemical defenses. *Journal of Chemical Ecology* 33: 1105–1113.

Desbiens, S. (1994). Le bivalve *Montanaria* Spriestersbach, 1909: Habitat et morphologie d'une nouvelle espèce emsienne de la Formation de York River de Gaspé, Québec. *Canadian Journal of Earth Sciences* 31: 381–392.

Desbruyères, D., Segonzac, M. & Bright, M., Eds. (2006). *Handbook of Deep-sea Hydrothermal Vent Fauna*. Linz, Austria: Land Oberösterreich, Biologiezentrum der Oberösterreichische Landesmuseen.

Desparmet, R., Termier, G. & Termier, H. (1971). Sur un bivalve protobranche ante-arenigien trouve au nord de Wardak (Afghanistan). *Geobios* 4: 143–150.

Devaere, L., Clausen, S., Steiner, M., Álvaro, J. J. & Vachard, D. (2013). Chronostratigraphic and palaeogeographic significance of an early Cambrian microfauna from the Heraultia Limestone, northern Montagne Noire, France. *Palaeontologia Electronica* 16: 1–91.

Devaere, L., Clausen, S., Álvaro, J. J., Peel, J. S. & Vachard, D. (2014). Terreneuvian orthothecid (Hyolitha) digestive tracts from northern Montagne Noire, France; taphonomic, ontogenetic and phylogenetic implications. *PLoS ONE* 9: e88583.

DeVantier, L. M. & Endean, R. (1988). The scallop *Pedum spondyloideum* mitigates the effects of *Acanthaster planci* predation on the host coral *Porites*: Host defence facilitated by exaptation? *Marine Ecology Progress Series* 47: 293–301.

Deyoung, R. W., Honeycutt, R. L. & Brennan (2005). The molecular toolbox: Genetic techniques in wildlife ecology and management. *Journal of Wildlife Management* 69: 1362–1384.

Deyrup-Olsen, I., Martin, A. W. & Paine, R. T. (1986). The autotomy escape response of the terrestrial slug *Prophysaon foliolatum* (Pulmonata: Arionidae). *Malacologia* 27: 307–311.

Dillen, L., Jordaens, K., Van Dongen, S. & Backeljau, T. (2010). Effects of body size on courtship role, mating frequency and sperm transfer in the land snail *Succinea putris*. *Animal Behaviour* 79: 1125–1133.

Dilly, P. N. & Herring, P. J. (1974). The ocular light organ of *Bathothauma lyromma* (Mollusca: Cephalopoda). *Journal of Zoology* 172: 81–100.

Dilly, P. N. & Herring, P. J. (1978). The light organ and ink sac of *Heteroteuthis dispar* (Mollusca: Cephalopoda). *Journal of Zoology* 186: 47–59.

Dinamani, P. (1963). Feeding in *Dentalium conspicuum*. *Proceedings of the Malacological Society of London* 36: 1–5.

Dinamani, P. (1964). Burrowing behaviour of *Dentalium*. *Biological Bulletin* 126: 28–32.

Dinapoli, A. & Klussmann-Kolb, A. (2010). The long way to diversity – Phylogeny and evolution of the Heterobranchia (Mollusca: Gastropoda). *Molecular Phylogenetics and Evolution* 55: 60–76.

Distel, D. L. (2003). The biology of marine wood boring bivalves and their bacterial endosymbionts, pp. 253–271 *in* B. Goodel, Nicholas, D. D. & Schultz, T. P. (eds.), *Wood Deterioration and Preservation (Advances in Our Changing World). ACS Symposium Series*. Washington, DC, American Chemical Society.

Distel, D. L., Altamia, M. A., Lin, Z., Shipway, J. R., Han, A., Forteza, I., Antemano, R., Limbaco, M. G. J. P., Tebo, A. G. & Dechavez, R. (2017). Discovery of chemoautotrophic symbiosis in the giant shipworm *Kuphus polythalamia* (Bivalvia: Teredinidae) extends wooden-steps theory. *Proceedings of the National Academy of Sciences of the United States of America* 114: E3652–E3658.

Doak, D. F., Estes, J. A., Halpern, B. S., Jacob, U., Lindberg, D. R., Lovvorn, J., Monson, D. H., Tinker, M. T., Williams, T. M. & Wootton, J. T. (2008). Understanding and predicting ecological dynamics: Are major surprises inevitable. *Ecology* 89: 952–961.

Dodd, B. J., Barnhart, M. C., Rogers-Lowery, C. L., Fobian, T. B. & Dimock, R. V. (2006). Persistence of host response against glochidia larvae in *Micropterus salmoides*. *Fish and Shellfish Immunology* 21: 473–484.

Doepke, H., Herrmann, K. & Schuett, C. (2012). Endobacteria in the tentacles of selected cnidarian species and in the cerata of their nudibranch predators. *Helgoland Marine Research* 66: 43–50.

Doguzhaeva, L. A. & Mutvei, H. (1993). Structural features in Cretaceous ammonoids indicative of semi-internal or internal shells, pp. 99–114 *in* M. R. House (ed.), *The Ammonoidea: Environment, Ecology, and Evolutionary Change*. Oxford, UK, Clarendon Press.

Doguzhaeva, L. A. (1995). An Early Cretaceous orthocerid cephalopod from north-western Caucasus. *Palaeontology* 37: 889–900.

Doguzhaeva, L. A. (1996). Microstructure of juvenile shells of the Permian *Hemibactrites* sp. (Cephalopoda: Bactritoidea). *Doklady Akademii nauk SSSR* 349: 275–279.

Doguzhaeva, L. A. & Mutvei, H. (1996). Attachment of the body to the shell in ammonoids, pp. 43–63 *in* N. H. Landman, Tanabe, K. & Davis, R. A. (eds.), *Ammonoid Paleobiology: From Anatomy to Ecology. Topics in Geobiology*. New York, Plenum Press.

Doguzhaeva, L. A., Mapes, R. H. & Mutvei, H. (1997). Beaks and radulae of Early Carboniferous goniatites. *Lethaia* 30: 305–313.

Doguzhaeva, L. A., Mapes, R. H. & Mutvei, H. (1999). A Late Carboniferous spirulid coleoid from the southern mid-continent (USA). Shell wall ultrastructure and evolutionary implications, pp. 47–57 *in* F. Oloriz & Rodriguez-Tovar, F. J. (eds.), *Advancing Research on Living and Fossil Cephalopods*. New York, Kluwer Academic/Plenum Publishers.

Doguzhaeva, L. A. (2002). Adolescent bactritoid, orthoceroid, ammonoid and coleoid shells from the Upper Carboniferous and Lower Permian of South Urals. *Abhandlungen der Geologischen Bundesanstalt* 57: 9–56.

Doguzhaeva, L. A., Mapes, R. H. & Mutvei, H. (2002a). Early Carboniferous coleoid *Hematites* Flower and Gordon, 1959 (Hematitida ord. nov.) from midcontinent (USA), pp. 299–320 *in* H. Summesberger, Histon, K. & Daurer, A. (eds.), *Cephalopods Present and Past*. Vienna, Geologische Bundesanstalt.

Doguzhaeva, L. A., Mutvei, H. & Donovan, D. T. (2002b). Proostracum, muscular mantle and conotheca in the Middle Jurassic belemnite *Megateuthis*, pp. 321–329 *in* H. Summesberger, Histon, K. & Daurer, A. (eds.), *Cephalopods Present and Past*. Vienna, Geologische Bundesanstalt.

Doguzhaeva, L. A., Mapes, R. H. & Mutvei, H. (2003). The shell and ink sac morphology and ultrastructure of the Late Pennsylvanian cephalopod *Donovaniconus* and its phylogenetic significance. *Berliner Paläobiologische Abhandlungen* 3: 61–78.

Doguzhaeva, L. A., Mapes, R. H. & Mutvei, H. (2004). Occurrence of ink in Paleozoic and Mesozoic coleoids (Cephalopoda). *Mitteilungen aus dem Geologisch-Paläontologischen Institut der Universität Hamburg* 88: 145–155.

Doguzhaeva, L. A., Mapes, R. H. & Mutvei, H. (2007a). A Late Carboniferous coleoid cephalopod from the Mazon Creek Lagerstätte (USA), with a radula, arm hooks, mantle tissues, and ink, pp. 121–143 *in* N. H. Landman, Davis, R. A. & Mapes, R. H. (eds.), *Cephalopods Present and Past: New Insights and Fresh Perspectives. 6th International Symposium*, University of Arkansas, Fayetteville, 16–19 September 2004. Dordrecht, the Netherlands, Springer.

Doguzhaeva, L. A., Mapes, R. H., Summesberger, H. & Mutvei, H. (2007b). The preservation of body tissues, shell, and mandibles in the ceratitid ammonoid *Austrotrachyceras* (Late Triassic), Austria, pp. 221–238 *in* N. H. Landman, Davis, R. A. & Mapes, R. H. (eds.), *Cephalopods Present and Past. New Insights and Fresh Perspectives. 6th International Symposium*, University of Arkansas, Fayetteville, 16–19 September 2004. Dordrecht, the Netherlands, Springer.

Doguzhaeva, L. A., Mapes, R. H. & Mutvei, H. (2010). Evolutionary patterns of Carboniferous coleoid cephalopods based on their diversity and morphological plasticity, pp. 171–180 *in* K. Tanabe, Shigeta, Y., Sasaki, T. & Hirano, H. (eds.), *Cephalopods Present and Past. 8th International Symposium*, University of Bourgogne, France, 30 Aug–30 September 2010. Hadano, Japan, Tokai University Press.

Doguzhaeva, L. A. & Bengtson, S. (2011). The capsule: An organic skeletal structure in the Late Cretaceous belemnite *Gonioteuthis* from north-west Germany. *Palaeontology* 54: 397–415.

Doguzhaeva, L. A. & Mapes, R. H. (2015). Arm hooks and structural features in the Early Permian *Glochinomorpha* Gordon, indicative of its coleoid affiliation. *Lethaia* 48: 100–114.

Doguzhaeva, L. A., Bengtson, S., Reguero, M. A. & Mörs, T. (2017). An Eocene orthocone from Antarctica shows convergent evolution of internally shelled cephalopods. *PLoS ONE* 12: e0172169.

Donovan, D. A., Elias, J. P. & Baldwin, J. (2004). Swimming behavior and morphometry of the file shell *Limaria fragilis*. *Marine and Freshwater Behaviour and Physiology* 37: 7–16.

Donovan, D. T. (1964). Cephalopod phylogeny and classification. *Biological Reviews* 39: 259–287.

Donovan, D. T. (1977). Evolution of the dibranchiate Cephalopoda, pp. 15–48 *in* M. Nixon & Messenger, J. B. (eds.), *The Biology of Cephalopods. Symposia of The Zoological Society of London*. London, Academic Press.

Donovan, D. T. (1994). History of classification of Mesozoic ammonites. *Journal of the Geological Society* 151: 1035–1040.

Donovan, D. T. (2006). Phragmoteuthida (Cephalopoda: Coleoidea) from the Lower Jurassic of Dorset, England. *Palaeontology* 49: 673–684.

Donovan, D. T. & Strugnell, J. M. (2010). A redescription of the fossil coleoid cephalopod genus *Palaeololigo* Naef, 1921 (Decapodiformes: Palaeololiginidae) and its relationship to Recent squids. *Journal of Natural History* 44: 1475–1492.

Donovan, D. T. (2012). Ammonites and octopuses. *Treatise Online* No. 47, Part M, Chapter 18: 1–9.

Donovan, D. T. & Fuchs, D. (2012). History of higher classification of Coleoidea. *Treatise Online* No. 53, Part M, Chapter 14: 1–20.

Donovan, D. T. & Fuchs, D. (2016). Fossilized soft tissues in Coleoidea. *Treatise Online* 73, Part M, Chapter 13: 1–30.

Donovan, S. K. (1992). A plain man's guide to rudist bivalves. *Journal of Geological Education* 40: 313–320.

Dorgan, K. M. (2015). The biomechanics of burrowing and boring. *Journal of Experimental Biology* 218: 176–183.

Doubleday, Z. A., White, J., Pecl, G. T. & Semmens, J. M. (2011). Age determination in merobenthic octopuses using stylet increment analysis: Assessing future challenges using *Macroctopus maorum* as a model. *ICES Journal of Marine Science* 68: 2059–2063.

Downie, C., Fisher, D. W., Goldring, R. & Rhodes, F. H. T. (1967). Miscellanea, pp. 613–626 *in* W. B. Harland, Holland, C. H., House, M. R., Hughes, N. F., Reynolds, A. B. et al. *The Fossil Record*. London, Geological Society of London and the Palaeontological Association, Swansea.

Doyle, P. (1986). *Naefia* (Coleoidea) from late Cretaceous of southern India. *Bulletin of the British Museum of Natural History* 40: 133–139.

Doyle, P., Donovan, D. T. & Nixon, M. (1994). Phylogeny and systematics of the Coleoidea. *University of Kansas Paleontological Contributions. New Series* 5: 1–15.

Doyle, P. & Shakides, E. V. (2004). The Jurassic belemnite suborder Belemnotheutina. *Palaeontology* 47: 983–998.

Drapatz, H. (2010). *Paleobiology of the Tentaculitoids from the Lower Devonian Hunsruck Slate (Germany)*. Doctoral Thesis, Rheinische Friedrich-Wilhelms Universitat Bonn.

Dreier, A., Loh, W., Blumenberg, M., Thiel, V., Hause-Reitner, D. & Hoppert, M. (2014). The isotopic biosignatures of photo- vs. thiotrophic bivalves: Are they preserved in fossil shells? *Geobiology* 12: 406–423.

Drew, G. A. (1901). The life history of *Nucula delphinodonta* (Mighels). *Quarterly Journal of Microscopical Science* 44: 313–391.

Dreyer, H., Steiner, G. & Harper, E. M. (2003). Molecular phylogeny of Anomalodesmata (Mollusca: Bivalvia) inferred from 18S rRNA sequences. *Zoological Journal of the Linnean Society* 139: 229–246.

Driscoll, E. G. (1964). Accessory muscle scars, an aid to protobranch orientation. *Journal of Paleontology* 38: 61–66.

Droser, M. L. & Bottjer, D. J. (1989). Ordovician increase in extent and depth of bioturbation: Implications for understanding early Paleozoic ecospace utilization. *Geology* 17: 850–852.

Droser, M. L. & Sheehan, P. M. (1995). Paleoecological significance of the Ordovician radiation and end Ordovician extinction: Evidence from the Great Basin, pp. 64–106 *in* D. Cooper, Droser, M. L. & Finney, S. C. (eds.), *Ordovician Odyssey. Short papers for the 7th International Symposium on the Ordovician System held June 1995*. Fullerton, CA, Pacific Section Society for Sedimentary Geology (SEPM).

Droser, M. L., Bottjer, D. J. & Sheehan, P. M. (1997). Evaluating the ecological architecture of major events in the Phanerozoic history of marine invertebrate life. *Geology* 25: 167–170.

Droser, M. L. & Finnegan, S. (2003). The Ordovician radiation: A follow-up to the Cambrian explosion? *Integrative and Comparative Biology* 43: 178–184.

Dubin, L. S. & Jones, P. (1999). *North American Indian Jewelry and Adornment: From Prehistory to the Present*. New York, Harry N. Adams.

Duch, T. (1969). Spawning and development in the trochid gastropod *Euchelus gemmatus* (Gould, 1841) in the Hawaiian Islands. *The Veliger* 11: 415–417.

Dudgeon, D., Arthington, A. H., Gessner, M. O., Kawabata, Z.-I., Knowler, D. J., Lévêque, C., Naiman, R. J., Prieur-Richard, A.-H., Soto, D. & Stiassny, M. L. (2006). Freshwater biodiversity: Importance, threats, status and conservation challenges. *Biological Reviews* 81: 163–182.

Dufour, S. C. & Beninger, P. G. (2001). A functional interpretation of cilia and mucocyte distributions on the abfrontal surface of bivalve gills. *Marine Biology* 138: 295–309.

Dufour, S. C. (2005). Gill anatomy and the evolution of symbiosis in the bivalve family Thyasiridae. *Biological Bulletin* 208: 200–212.

Dufour, S. C. & Felbeck, H. (2006). Symbiont abundance in thyasirids (Bivalvia) is related to particulate food and sulphide availability. *Marine Ecology Progress Series* 320: 185–194.

Duméril, A. M. C. (1806). *Zoologie analytique; ou, Méthode naturelle de classification des animaux, rendue plus facile à l'aide de tableaux synoptiques.* Paris, Allais.

Duncan, C. J. (1958). The anatomy and physiology of the reproductive system of the freshwater snail *Physa fontinalis. Proceedings of the Zoological Society of London* 131: 55–84.

Duncan, C. J. (1960a). The genital systems of the freshwater Basommatophora. *Proceedings of the Zoological Society of London* 135: 339–356.

Duncan, C. J. (1960b). The evolution of the pulmonate genital system. *Proceedings of the Zoological Society of London* 134: 601–609.

Duncan, C. J. (1975). Reproduction, pp. 309–365 *in* V. Fretter & Peake, J. (eds.), *Pulmonates: Functional Anatomy and Physiology.* Vol. 1. London, Academic Press.

Dunn, C. W., Hejnol, A., Matus, D. Q., Pang, K., Browne, W. E., Smith, S. A., Seaver, E. C., Rouse, G. W., Obst, M., Edgecombe, G. D., Sørensen, M. V., Haddock, S. H. D., Schmidt-Rhaesa, A., Okusu, A., Kristensen, R. M., Wheeler, W. C., Martindale, M. Q. & Giribet, G. (2008). Broad phylogenomic sampling improves resolution of the animal tree of life. *Nature* 452: 745.

Dunn, C. W., Giribet, G., Edgecombe, G. D. & Hejnol, A. (2014). Animal phylogeny and its evolutionary implications. *Annual Review of Ecology, Evolution, and Systematics* 45: 371–395.

Dunstan, A. J., Ward, P. D. & Marshall, N. J. (2011). Vertical distribution and migration patterns of *Nautilus pompilius. PLoS ONE* 6: e16311.

Duperron, S. (2010). The diversity of deep-sea mussels and their bacterial symbioses, pp. 137–167 *in* S. Kiel (ed.), *The Vent and Seep Biota: Aspects from Microbes to Ecosystems. Topics in Geobiology.* Dordrecht, the Netherlands, Springer Netherlands.

Duperron, S., Pottier, M.-A., Léger, N., Gaudron, S. M., Puillandre, N., Le Prieur, S., Sigwart, J. D., Ravaux, J. & Zbinden, M. (2012). A tale of two chitons: Is habitat specialisation linked to distinct associated bacterial communities? *FEMS Microbiology Ecology* 83: 552–567.

Duval, D. M. (1963). The comparative anatomy of some lamellibranch siphons. *Proceedings of the Malacological Society of London* 35: 289–295.

Dzik, J. (1978). Larval development of hyolithids. *Lethaia* 11: 293–299.

Dzik, J. (1981a). Larval development, musculature, and relationships of *Sinuitopsis* and related Baltic bellerophonts. *Norsk Geologisk Tidsskrift* 61: 111–121.

Dzik, J. (1981b). Origin of the Cephalopoda. *Acta Palaeontologica Polonica* 26: 161–191.

Dzik, J. (1984). *Phylogeny of the Nautiloidea (Filogeneza łodzików).* Vol. 45. Warsaw, Państwowe Wydawn Naukowe (Polish Scientific Publishers).

Dzik, J. (1993). Early metazoan evolution and the meaning of its fossil record, pp. 339–386 *in* M. K. Hecht, R. J. MacIntyre, & M. T. Clegg (eds.), *Evolutionary Biology.* New York, Plenum Press.

Dzik, J. (1994). Evolution of 'small shelly fossils' assemblages of the Early Paleozoic. *Acta Palaeontologica Polonica* 39: 247–313.

Dzik, J. (2010). Brachiopod identity of the alleged monoplacophoran ancestors of cephalopods. *Malacologia* 52: 97–113.

Dzik, J. & Mazurek, D. (2013). Affinities of the alleged earliest Cambrian gastropod *Aldanella. Canadian Journal of Zoology* 91: 914–923.

Eales, N. B. (1960). Revision of the world species of *Aplysia* (Gastropoda, Opisthobranchia). *Bulletin of British Museum (Natural History), Zoology, London* 5: 267–404.

Ebel, K. (1990). Swimming abilities of ammonites and limitations. *Paläontologische Zeitschrift* 64: 25–37.

Ebel, K. (1992). Mode of life and soft body shape of heteromorph ammonites. *Lethaia* 25: 179–193.

Ebel, K. (1999). Hydrostatics of fossil ectocochleate cephalopods and its significance for the reconstruction of their lifestyle. *Paläontologische Zeitschrift* 73: 277–288.

Eckelbarger, K. J. & Young, C. M. (1997). Ultrastructure of the ovary and oogenesis in the methane-seep mollusc *Bathynerita naticoidea* (Gastropoda: Neritidae) from the Louisiana slope. *Invertebrate Biology* 116: 299–312.

Edgar, G. J. & Samson, C. R. (2004). Catastrophic decline in mollusc diversity in eastern Tasmania and its concurrence with shellfish fisheries. *Conservation Biology* 18: 1579–1588.

Edgecombe, G. D., Giribet, G., Dunn, C. W., Hejnol, A., Kristensen, R. M., Neves, R. C., Rouse, G. W., Worsaae, K. & Sørensen, M. V. (2011). Higher-level metazoan relationships: Recent progress and remaining questions. *Organisms Diversity & Evolution* 11: 151–172.

Edlinger, K. (1980a). Zur Phylogenie der chemischen Sinnesorgane einiger Cephalaspidea (Mollusca - Opisthobranchia). *Zeitschrift für Zoologische Systematik und Evolutionsforschung* 18: 241–256.

Edlinger, K. (1980b). Beiträge zur Anatomie, Histologie, Ultrastruktur und Physiologie der chemischen Sinnesorgane einiger Cephalaspidea (Mollusca, Opisthobranchia). *Zoologischer Anzeiger* 205: 90–112.

Edmunds, M. (1966a). Protective mechanisms in the Eolidacea (Mollusca, Nudibranchia). *Journal of the Linnean Society* 47: 27–71.

Edmunds, M. (1966b). Defense adaptation of *Stiliger vanellus* (Marcus), with a discussion on the evolution of 'nudibranch' molluscs. *Proceedings of the Malacological Society of London* 37: 37–81.

Edmunds, M. (1968). Acid secretion in some species of Doridacea (Mollusca, Nudibranchia). *Proceedings of the Malacological Society of London* 38: 121–133.

Edmunds, M. (1991). Does warning coloration occur in nudibranchs? *Malacologia* 32: 241–255.

Eernisse, D. J. (1988). Brooding in a chiton: Why synchronize to self-fertilization? *American Zoologist* 28: 170.

Eernisse, D. J. & Kerth, K. (1988). The initial stages of radular development in chitons (Mollusca: Polyplacophora). *Malacologia* 28: 95–103.

Eernisse, D. J., Terwilliger, N. B. & Terwilliger, R. C. (1988). The red foot of a lepidopleurid chiton: Evidence for tissue hemoglobins. *The Veliger* 30: 244–247.

Eernisse, D. J. & Reynolds, P. D. (1994). Polyplacophora, pp. 55–110 *in* F. W. Harrison & Kohn, A. J. (eds.), *Microscopic Anatomy of Invertebrates. Mollusca 1.* Vol. 5. New York, Wiley-Liss.

Eibye-Jacobsen, D. (2004). A re-evaluation of *Wiwaxia* and the polychaetes of the Burgess Shale. *Lethaia* 37: 317–335.

Eisner, T. & Wilson, E. O. (1970). Defensive liquid discharge in Florida tree snails (*Liguus fasciatus*). *The Nautilus* 84: 14–15.

Elias, M., Hill, R. I., Willmott, K. R., Dasmahapatra, K. K., Brower, A. V. Z., Mallet, J. & Jiggins, C. D. (2007). Limited performance of DNA barcoding in a diverse community of tropical butterflies. *Proceedings of the Royal Society of London B* 274: 2881–2889.

Elicki, O. & Gürsu, S. (2009). First record of *Pojetaia runnegari* [Jell, 1980] and *Fordilla* [Barrande, 1881] from the Middle East (Taurus Mountains, Turkey) and critical review of Cambrian bivalves. *Palaeontologische Zeitschrift* 83: 267–291.

Eltzholtz, J. R. & Birkedal, H. (2009). Architecture of the biomineralized byssus of the Saddle Oyster (*Anomia* sp.). *Journal of Adhesion* 85: 590–600.

Emberton, K. C. (1995). When shells do not tell: 145 million years of evolution in North America's polygyrid land snails, with a revision and conservation priorities. *Malacologia* 37: 69–109.

Emberton, K. C. (2001). Exploratory phylogenetic and biogeographic analyses within three land-snail families in southeastern-most Madagascar. *Biological Journal of the Linnean Society* 72: 567–584.

Emery, D. G. (1992). Fine structure of olfactory epithelia of gastropod molluscs. *Microscopy Research and Technique* 22: 307–324.

Engeser, T. & Reitner, J. (1981). Beiträge zur Systematik von phragmokontragenden Coleoiden aus dem Untertithonium (Malm zeta, Solnhofener Plattenkalk) von Solnhofen und Eichstätt (Bayern). *Neues Jahrbuch für Geologie und Paläontologie-Monatshefte* 9: 527–545.

Engeser, T. & Bandel, K. (1988). Phylogenetic classification of coleoid cephalopods, pp. 105–115 in J. Wiedmann & Kullmann, J. (eds.), *Cephalopods Present and Past. 2nd International Cephalopod Symposium*: O. H. Schindewolf Symposium, Tübingen 1985. Stuttgart, E. Schweizerbart'sche Verlagsbuchhandlung.

Engeser, T. (1990a). Major events in cephalopod evolution, pp. 119–138 in P. D. Taylor & Larwood, G. P. (eds.), *Major Evolutionary Radiations*. Vol. 42. Oxford, UK, Clarendon Press.

Engeser, T. (1990b). Phylogeny of the fossil coleoid Cephalopoda (Mollusca). *Berliner Geowissenschaftlichen Abhandlungen, A* 124: 123–191.

Engeser, T. S. & Clarke, M. R. (1988). Cephalopod hooks, both Recent and fossil, pp. 133–151 in M. R. Clarke & Trueman, E. R. (eds.), *Paleontology and Neontology of Cephalopods. The Mollusca*. Vol. 12. New York, Academic Press.

Engeser, T. S. (1996). The position of the Ammonoidea within the Cephalopoda, pp. 3–19 in N. H. Landman, Tanabe, K. & Davis, R. A. (eds.), *Ammonoid Paleobiology: From Anatomy to Ecology. Topics in Geobiology*. New York, Plenum Press.

Engeser, T. S. & Keupp, H. (2002). Phylogeny of the aptychi-possessing Neoammonoidea (Aptychophora nov., Cephalopoda): History of research. *Lethaia* 34: 79–96.

Epstein, H. E., Hallas, J. M., Johnson, R. F., Lopez, A. & Gosliner, T. M. (2018). Reading between the lines: Revealing cryptic species diversity and colour patterns in *Hypselodoris* nudibranchs (Mollusca: Heterobranchia: Chromodorididae). *Zoological Journal of the Linnean Society* 20: 1–74.

Erben, H. K. (1964). Die Evolution der ältesten Ammonoidea (Lieferung I). *Neues Jahrbuch für Geologie und Paläontologie-Monatshefte* 120: 107–212.

Erben, H. K., Flajs, G. & Siehl, A. (1968). *Über die Schalenstruktur von Monoplacophoren*. Mainz, Verlag der Akademie der Wissenschaften und der Literatur.

Erben, H. K. & Krampitz, G. (1972). Ultrastruktur und Aminosäuren-Verhältnisse in den Schalen der rezenten Pleurotomariidae (Gastropoda). *Biomineralisation* 6: 12–31.

Erlandson, J. M., Vellanoweth, R. L., Rick, T. C. & Reid, M. R. (2005). Coastal foraging at Otter Cave: A 6600-year-old shell midden on San Miguel Island, California. *Journal of California and Great Basin Anthropology* 25: 69–86.

Erwin, D. H. (1990a). Carboniferous-Triassic gastropod diversity patterns and the Permo-Triassic mass extinction. *Paleobiology* 16: 187–203.

Erwin, D. H. (1990b). The end-Permian mass extinction. *Annual Review of Ecology and Systematics* 21: 69–91.

Erwin, D. H. & Signor, P. W. (1991). Extinction in an extinction-resistant clade: The evolutionary history of the Gastropoda, pp. 152–160 in E. C. Dudley (ed.), *The Unity of Evolutionary Biology. Proceedings of the Fourth International Congress of Systematic and Evolutionary Biology, July 1990*. Portland, OR, Discordides Press (University of Maryland).

Erwin, D. H. (2006). *Extinction: How Life on Earth Nearly Ended 250 Million Years Ago*. Princeton, NJ, Princeton University Press.

Erwin, D. H., Laflamme, M., Tweedt, S. M., Sperling, E. A., Pisani, D. & Peterson, K. J. (2011). The Cambrian conundrum: Early divergence and later ecological success in the early history of animals. *Science* 334: 1091–1097.

Erwin, D. H. & Valentine, J. W. (2013). *The Cambrian Explosion: The Construction of Animal Biodiversity*. Greenwood Village, CO, Roberts and Company Publishers.

Espinosa, J. & Ortea, J. (2009). Nueva familia, género y especie de molusco gasterópodo (Mollusca: Gastropoda) de las cuevas submarinas de Cuba. *Revista de la Academia Canaria de Ciencias* 21: 93–98.

Estabrooks, W. A., Kay, E. A. & McCarthy, S. A. (1999). Structure of the excretory system of Hawaiian nerites (Gastropoda: Neritoidea). *Journal of Molluscan Studies* 65: 61–72.

Estes, J. A., Lindberg, D. R. & Wray, C. (2005). Evolution of large body size in abalones (*Haliotis*): Patterns and implications. *Paleobiology* 31: 591.

Estes, J. A., Terborgh, J., Brashares, J. S., Power, M. E., Berger, J., Bond, W. J., Carpenter, S. R., Essington, T. E., Holt, R. D. & Jackson, J. B. C. (2011). Trophic downgrading of planet Earth. *Science* 333: 301–306.

Etter, W. (2002a). La Voulte-sur-Rhône: Exquisite cephalopod preservation, pp. 293–305 in D. J. Bottjer, Etter, W., Hagadorn, J. W. & Tang, C. M. (eds.), *Exceptional Fossil Preservation: A Unique View on the Evolution of Marine Life*. New York, Columbia University Press.

Etter, W. (2002b). Solnhofen: Plattenkalk preservation with *Archaeopteryx*, pp. 327–352 in D. J. Bottjer, Etter, W., Hagadorn, J. W. & Tang, C. M. (eds.), *Exceptional Fossil Preservation: A Unique View on the Evolution of Marine Life*. New York, Columbia University Press.

Etter, W. & Tang, C. M. (2002). Posidonia shale: Germany's Jurassic Marine Park, pp. 265–291 in D. J. Bottjer, Etter, W., Hagadorn, J. W. & Tang, C. M. (eds.), *Exceptional Fossil Preservation: A Unique View on the Evolution of Marine Life*. New York, Columbia University Press.

Evans, D. H. (2005). The Lower and Middle Ordovician cephalopod faunas of England and Wales. *Monograph of the Palaeontographical Society* 628: 1–81.

Evans, D. H. & King, A. H. (2012). Resolving polyphyly within the Endocerida: The Bisonocerida nov., a new order of early palaeozoic nautiloids. *Geobios* 45: 19–28.

Evans, K. & Rowell, A. (1990). Small shelly fossils from Antarctica: An Early Cambrian faunal connection with Australia. *Journal of Paleontology* 64: 692–700.

Evans, T. J. (1953). The alimentary and vascular systems of *Alderia modesta* (Lovén) in relation to its ecology. *Journal of Molluscan Studies* 29: 249–258.

Eyster, L. S. (1986). The embryonic capsules of nudibranch molluscs: Literature review and new studies on albumen and capsule wall ultrastructure. *American Malacological Bulletin* 4: 205–216.

Fahrner, A. & Haszprunar, G. (2001). Anatomy and ultrastructure of the excretory system of a heart-bearing and a heartless sacoglossan gastropod (Opisthobranchia, Sacoglossa). *Zoomorphology* 121: 85–93.

Fahrner, A. (2002). *Comparative microanatomy and ultrastructure of the excretory systems of opisthobranch Gastropoda (Mollusca)*. Ph.D. dissertation, Ludwig Maximilians Universität.

Fahrner, A. & Haszprunar, G. (2002a). Ultrastructure of the renopericardial complex in *Hypselodoris tricolor* (Gastropoda, Nudibranchia). *Zoomorphology* 121: 183–194.

Fahrner, A. & Haszprunar, G. (2002b). Microanatomy, ultrastructure, and systematic significance of the excretory system and mantle cavity of an acochlidian gastropod (Opisthobranchia). *Journal of Molluscan Studies* 68: 87–94.

Fain, A. (2004). Mites (Acari) parasitic and predaceous on terrestrial gastropods, pp. 505–524 *in* G. M. Barker (ed.), *Natural Enemies of Terrestrial Molluscs*. Oxford, UK/Cambridge, MA, CABI Publishing.

Faller, S., Rothe, B. H., Todt, C., Schmidt-Rhaesa, A. & Loesel, R. (2012). Comparative neuroanatomy of Caudofoveata, Solenogastres, Polyplacophora, and Scaphopoda (Mollusca) and its phylogenetic implications. *Zoomorphology* 131: 149–170.

Falniowski, A. (1989). A critical review of some characters widely used in the systematics of higher taxa of freshwater prosobranchs (Gastropoda: Prosobranchia), and a proposal of some new, ultrastructural ones. *Folia Malacologica* 3: 73–93.

Falniowski, A. (1990). Anatomical characters and SEM structure of radula and shell in the species-level taxonomy of freshwater prosobranchs (Mollusca: Gastropoda: Prosobranchia): A comparative usefulness study. *Folia Malacologica* 4: 53–142.

Falniowski, A. & Szarowska, M. (2011). Radiation and phylogeography in a spring snail *Bythinella* (Mollusca: Gastropoda: Rissooidea) in continental Greece. *Annales Zoologici Fennici* 48: 67–90.

Fang, Z.-J. & Cope, J. C. W. (2008). Affinities and palaeobiogeographical significance of some Ordovician bivalves from East Yunnan, China. *Alcheringa* 32: 297–312.

Fang, Z.-Z. & Morris, N. J. (1997). The genus *Pseudosanguinolites* and some modioliform bivalves (mainly Palaeozoic) *Palaeoworld* 7: 50–74.

Fänge, R. & Lidman, U. (1976). Secretion of sulfuric acid in *Cassidaria echinophora* Lamarck (Mollusca: Mesogastropoda, marine carnivorous snail). *Comparative Biochemistry and Physiology Part A* 53: 101–103.

Farmer, W. M. (1970). Swimming gastropods (Opisthobranchia and Prosobranchia). *The Veliger* 13: 73–89.

Farrell, T. M. (1988). Community stability: Effects of limpet removal and reintroduction in a rocky intertidal community. *Oecologia* 75: 190–197.

Faulkner, D. J. & Ghiselin, M. T. (1983). Chemical defense and the evolutionary ecology of dorid nudibranchs and some other opisthobranch gastropods. *Marine Ecology Progress Series* 13: 295–301.

Faulwetter, S., Vasileiadou, A., Kouratoras, M., Dailianis, T. & Arvanitidis, C. (2013). Micro-computed tomography: Introducing new dimensions to taxonomy. *ZooKeys* 263: 1–45.

Fedonkin, M. A. & Waggoner, B. M. (1997). The Late Precambrian fossil *Kimberella* is a mollusc-like bilaterian organism. *Nature* 388: 868–871.

Fedonkin, M. A., Simonetta, A. & Ivantsov, A. Y. (2007). New data on *Kimberella*, the Vendian mollusc-like organism (White Sea region, Russia): Palaeoecological and evolutionary implications. *Geological Society, London, Special Publications* 286: 157–179.

Fedosov, A., Puillanre, N., Kantor, Y. I. & Bouchet, P. (2015). Phylogeny and systematics of mitriform gastropods (Mollusca: Gastropoda: Neogastropoda). *Zoological Journal of the Linnean Society* 175: 336–359.

Fedosov, A., Puillanre, N., Herrmann, M., Kantor, Y. I., Oliverio, M., Dgebuadze, P., Modica, M. V. & Bouchet, P. (2018). The collapse of *Mitra*: Molecular systematics and morphology of the Mitridae (Gastropoda: Neogastropoda). *Zoological Journal of the Linnean Society* 183: 253–337.

Fedosov, A. E. (2008). Reduction of the alimentary system structures in predatory gastropods of the superfamily Conoidea (Gastropoda: Neogastropoda). *Doklady Biological Sciences* 419: 136–138.

Fedosov, A. E. & Kantor, Y. I. (2010). Evolution of carnivorous gastropods of the family Costellariidae (Neogastropoda) in the framework of molecular phylogeny. *Ruthenica* 20: 117–139.

Feng, W., Chen, Z. & Sun, W. (2003). Diversification of skeletal microstructures of organisms through the interval from the latest Precambrian to the Early Cambrian. *Science in China. Series D: Earth Sciences* 46: 977–985.

Feng, W. M. & Sun, W. G. (2003). Phosphate replicated and replaced microstructure of molluscan shells from the earliest Cambrian of China. *Acta Palaeontologica Polonica* 48: 21–30.

Feng, W. M. & Sun, W. G. (2006). Monoplacophoran *Igorella*-type pore-channel structures from the Lower Cambrian in China. *Materials Science and Engineering: C* 26: 699–702.

Feng, Y., Li, Q. & Kong, L. (2015). Molecular phylogeny of Arcoidea with emphasis on Arcidae species (Bivalvia: Pteriomorphia) along the coast of China: Challenges to current classification of arcoids. *Molecular Phylogenetics and Evolution* 85: 189–196.

Feulner, G. (2017). Formation of most of our coal brought Earth close to global glaciation. *Proceedings of the National Academy of Sciences* 114: 11333–11337.

Field, K. G., Olsen, G. J., Lane, D. J., Giovannoni, S. J., Ghiselin, M. T., Raff, E. C., Pace, N. R. & Raff, R. A. (1988). Molecular phylogeny of the animal kingdom. *Science* 239: 748–753.

Filipiak, P. & Jarzynka, A. (2009). Organic remains of tentaculitids: New evidence from Upper Devonian of Poland. *Acta Palaeontologica Polonica* 54: 111–116.

Finn, J. K. & Norman, M. D. (2010). The argonaut shell: Gas-mediated buoyancy control in a pelagic octopus. *Proceedings of the Royal Society B* 277: 2967–2971.

Fiorito, G. & Gherardi, F. (1990). Behavioural changes induced by ink in *Aplysia fasciata* (Mollusca, Gastropoda): Evidence for a social signal role of inking. *Marine and Freshwater Behaviour and Physiology* 17: 129–135.

Fioroni, P. (1978). Cephalopoda, Tintenfische, pp. 1–181 *in* F. Seidel (ed.), *Morphogenese der Tiere, Erste Reihe: Deskriptive Morphologenese, Lieferung 2: G5–1*. New York, Gustav Fisher Verlag.

Fischer, F. P. (1880–1887). *Manuel de conchyliologie et de paléontologie conchyliologique où Histoire naturelle des mollusques vivants et fossiles*. Paris, Savy.

Fischer, F. P. & Renner, M. (1979). SEM – Observations on the shell plates of three polyplacophorans (Mollusca, Amphineura). *Spixiana* 2: 49–58.

Fischer, F. P. (1980). Fine structure of the larval eye of *Lepidochitona cinerea* L. (Mollusca, Polyplacophora). *Spixiana* 3: 53–57.

Fischer, F. P., Maile, W. & Renner, M. (1980). Die Mantelpapillen und Stacheln von *Acanthochiton fascicularis* L. (Mollusca, Polyplacophora). *Zoomorphology* 94: 121–132.

Fischer, F. P. (1988). The ultrastructure of the aesthetes in *Lepidopleurus cajetanus* (Polyplacophora: Lepidopleurina). *American Malacological Bulletin* 6: 153–159.

Fischer, F. P., Eisenhammer, B., Miltz, C. & Singer, I. (1988). Sense organs in the girdle of *Chiton olivaceus* (Mollusca, Polyplacophora). *American Malacological Union Bulletin* 6: 131–140.

Fischer, F. P., Alger, M., Cieslar, D. & Krafczyk, H. U. (1990). The chiton gill: Ultrastructure in *Chiton olivaceus* (Mollusca, Polyplacophora). *Journal of Morphology* 204: 75–87.

Fischer, H. (1892). Recherches sur la morphologie du foie des gastéropodes. *Bulletin des sciences Françaises et Belgiques* 24: 260–346.

Fischer, J. C. & Riou, B. (2002). *Vampyronassa rhodanica* nov. gen. nov. sp., vampyromorphe (Cephalopoda, Coleoidea) du Callovien inférieur de la Voulte-sur-Rhône (Ardèche, France). *Annales de Paléontologie* 88: 1–17.

Fisher, D. W. (1962). Small conoidal shells of uncertain affinities, pp. W98–W143 in R. C. Moore (ed.), *Treatise on Invertebrate Paleontology Part W*. Lawrence, KS, Geological Society of America & University of Kansas Press.

Fisher, W. K. (1904). The anatomy of *Lottia gigantea* Gray. *Zoologische Jahrbücher. Abteilung für Anatomie und Ontogenie der Thiere* 20: 1–66.

Fleming, K. (2009). The great pretenders. *Australian Geographic* 96: 84–93.

Fleming, P. A., Muller, D. & Bateman, P. W. (2007). Leave it all behind: A taxonomic perspective of autotomy in invertebrates. *Biological Reviews* 82: 481–510.

Fleure, H. J. (1904). On the evolution of topographical relations among the Docoglossa. *Transactions of the Linnean Society of London 2nd series: Zoology* 9: 269–290.

Flower, R. H. (1947). Holochoanites are endoceroids. *The Ohio Journal of Science* 47: 155–172.

Flower, R. H. (1954). *Cambrian Cephalopods*. Vol. 40. Socorro, NM, New Mexico Institute of Mining & Technology.

Flower, R. H. (1955a). Trails and tentacular impressions of orthoconic cephalopods. *Journal of Paleontology* 29: 857–867.

Flower, R. H. (1955b). Cameral deposits in orthoconic nautiloids. *Geological Magazine* 92: 89–103.

Flower, R. H. & Gordon, M. (1959). More Mississippian belemnites. *Journal of Paleontology* 33: 809–842.

Flower, R. H. (1964). The nautiloid order Ellesmeroceratida, Cephalopoda. *State Bureau of Mines and Mineral Resources, New Mexico Institute of Mining and Technology Memoir* 12: 1–234.

Flower, R. H. (1976a). Some Whiterock and Chazy endoceroids. *New Mexico Bureau of Mines & Mineral Resources Memoir* 25: 13–39.

Flower, R. H. (1976b). Ordovician cephalopod faunas and their role in correlation, pp. 523–552 in M. G. Bassett (ed.), *The Ordovician System: Proceedings of a Palaeontological Association Symposium, Birmingham, September 1974*. Cardiff, University of Wales Press and National Museum of Wales.

Flower, R. H. (1984). *Bodeiceras*, a new Mohawkian oxycone, with revision of the Order Barrandeoceratida and discussion of the status of the Order. *Journal of Paleontology* 58: 1372–1379.

Flower, R. H. (1988). Progress and changing concepts in cephalopod and particularly nautiloid phylogeny and distribution, pp. 17–24 in J. Wiedmann & Kullmann, J. (eds.), *Cephalopods Present and Past. 2nd International Cephalopod Symposium: O. H. Schindewolf Symposium, Tübingen 1985*. Stuttgart, E. Schweizerbart'sche Verlagsbuchhandlung.

Flügel, E. (1994). Pangean shelf carbonates: Controls and paleoclimatic significance of Permian and Triassic reefs. *Geological Society of America* Special Papers 288: 247–266.

Foale, S. J. & Willan, R. C. (1987). Scanning and transmission electron microscope study of specialized mantle structures in dorid nudibranchs (Gastropoda, Opisthobranchia, Anthobranchia). *Marine Biology* 95: 547–558.

Focardi, S. & Chelazzi, G. (1990). Ecological determinants of bioeconomics in three intertidal chitons (*Acanthopleura* spp.). *Journal of Animal Ecology* 59: 347–362.

Foltan, P. (2004). Influence of slug defence mechanisms on the prey preferences of the carabid predator *Pterostichus melanarius* (Coleoptera: Carabidae). *European Journal of Entomology* 101: 359–364.

Foote, M. (2003). Origination and extinction through the Phanerozoic: A new approach. *Journal of Geology* 111: 125–148.

Fort, G. (1937). Le spérmatophore des céphalopodes. Étude du spérmatophore d'*Eledone cirrhosa* Lam. *Bulletin biologique de France et de Belgique* 71: 357–373.

Fortey, R. A. (1989). There are Extinctions and Extinctions: Examples from the Lower Palaeozoic. *Philosophical Transactions of the Royal Society of London. B* 325: 327–355.

Fraas, E. (1910). *Der Petrefaktensammler. Ein Leitfaden zum Sammeln und bestimmen der Versteinerungen Deutschlands*. Stuttgart, K. G. Lutz.

Franc, A. (1960). Classe des bivalves, pp. 1845–2133 in P.-P. Grassé (ed.), *Traité de Zoologie*. Vol. 5, Fascicule 2. Paris, Masson.

Franc, A. (1968). Classe des gastéropodes (Gastropoda Cuvier 1798), pp. 1–986 in P.-P. Grassé (ed.), *Traité de Zoologie*. Vol. 5, Fascicule 3. Paris, Masson.

Frankenberg, D. & Smith, K. L. (1967). Coprophagy in marine animals. *Limnology and Oceanography* 12: 443–450.

Franzén, Å. (1955). Comparative morphological investigations into the spermiogenesis among Mollusca. *Zoologiska Bidrag från Uppsala* 30: 399–456.

Freeman, G. & Lundelius, J. W. (1992). Evolutionary implications of the mode of D quadrant specification in coelomates with spiral cleavage. *Journal of Evolutionary Biology* 5: 205–247.

Frenkiel, L. & Mouëza, M. (1980). Ciliated receptors in the cruciform muscle sense organ of *Scrobicularia plana* (Da Costa) (Mollusca Lamellibranchia Tellinacea). *Zeitschrift für mikroskopisch-anatomische Forschung* 94: 881–894.

Fretter, V. (1937). The structure and function of the alimentary canal of some species of Polyplacophora (Mollusca). *Transactions of the Royal Society of Edinburgh* 59: 119–164.

Fretter, V. (1939). The structure and function of the alimentary canal of some tectibranch molluscs, with a note on excretion. *Transactions of the Royal Society of Edinburgh* 59: 599–646.

Fretter, V. (1943). Studies on the functional morphology and embryology of *Onchidella celtica* (Forbes & Hanley) and their bearing on its relationships. *Journal of the Marine Biological Association of the United Kingdom* 25: 685–720.

Fretter, V. (1946). The genital ducts of *Theodoxus*, *Lamellaria* and *Trivia*, and a discussion on their evolution in the prosobranchs. *Journal of the Marine Biological Association of the United Kingdom* 26: 312–351.

Fretter, V. (1948). The structure and life history of some minute prosobranchs of rock pools: *Skeneopsis planorbis* (Fabricius), *Omalogyra atomus* (Philippi), *Rissoella diaphana* (Alder) and *Rissoella opalina* (Jeffreys). *Journal of the Marine Biological Association of the United Kingdom* 27: 597–632.

Fretter, V. & Graham, A. L. (1949). The structure and mode of life of the Pyramidellidae, parasitic opisthobranchs. *Journal of the Marine Biological Association of the United Kingdom* 28: 493–532.

Fretter, V. (1951a). Some observations on the British cypraeids. *Proceedings of the Malacological Society of London* 29: 14–20.

Fretter, V. (1951b). Observations on the life history and functional morphology of *Cerithiopsis tubercularis* (Montagu) and *Triphora perversa* (L.). *Journal of the Marine Biological Association of the United Kingdom* 29: 567–586.

Fretter, V. (1953). The transference of sperm from male to female prosobranch, with reference also to the pyramidellids. *Proceedings of the Linnean Society* 164: 217–224.

Fretter, V. & Graham, A. L. (1954). Observations on the opisthobranch mollusc *Acteon tornatilis* (L.). *Journal of the Marine Biological Association of the United Kingdom* 33: 565–585.

Fretter, V. (1960). Observations on the tectibranch *Ringicula buccinea* (Brocchi). *Proceedings of the Zoological Society of London* 135: 537–549.

Fretter, V. & Graham, A. L. (1962). *British Prosobranch Molluscs: Their Functional Anatomy and Ecology*. London, Ray Society.

Fretter, V. (1965). Functional studies of the anatomy of some neritid prosobranchs. *Journal of Zoology* 147: 46–74.

Fretter, V. (1975a). *Umbonium vestiarium*, a filter-feeding trochid. *Journal of Zoology* 177: 541–552.

Fretter, V. (1975b). Introduction, pp. xi–xxix *in* V. Fretter & Peake, J. (eds.), *Pulmonates: Functional Anatomy and Physiology*. Vol. 1. London/New York, Academic Press.

Fretter, V., Graham, A. L. & McLean, J. H. (1981). The anatomy of the Galápagos rift limpet, *Neomphalus fretterae*. *Malacologia* 21: 337–361.

Fretter, V. (1984). The functional anatomy of the neritacean limpet *Phenacolepas omanensis* (Biggs) and some comparisons with *Septaria*. *Journal of Molluscan Studies* 50: 8–18.

Fretter, V. (1988). New archaeogastropod limpets from hydrothermal vents; superfamily Lepetodrilacea. II. Anatomy. *Philosophical Transactions of the Royal Society B* 318: 33–82.

Fretter, V. (1989). The anatomy of some new archaeogastropod limpets (Superfamily Peltospiracea) from hydrothermal vents. *Journal of Zoology* 218: 123–169.

Fretter, V. (1990). The anatomy of some new archaeogastropod limpets (Order Patellogastropoda, Suborder Lepetopsina) from hydrothermal vents. *Journal of Zoology* 222: 529–555.

Fretter, V. & Graham, A. L. (1994). *British Prosobranch Molluscs: Their Functional Anatomy and Ecology*. London, Ray Society.

Frey, R. C., Beresi, M. S., Evans, D. H., King, A. H. & Percival, I. G. (2004). Nautiloid cephalopods, pp. 209–213 *in* B. D. Webby, Paris, F., Droser, M. L. & Percival, I. G. (eds.), *The Great Ordovician Biodoversification Event*. Vol. 3. New York, Columbia University Press.

Frick, K. E. (2005). Nematocyst complements of nudibranchs in the genus *Flabellina* in the Gulf of Maine and the effect of diet manipulations on the cnidom of *Flabellina verrucosa*. *Marine Biology* 147: 1313–1321.

Friedrich, S., Wanninger, A., Bruckner, M. & Haszprunar, G. (2002). Neurogenesis in the mossy chiton, *Mopalia muscosa* (Gould) (Polyplacophora): Evidence against molluscan metamerism. *Journal of Morphology* 253: 109–117.

Fritchman, H. K. (1962). A study of the reproductive cycle in the California Acmaeidae (Gastropoda). IV. *The Veliger* 4: 134–139.

Froesch, D. (1979). Antigen-induced secretion in the optic gland of *Octopus vulgaris*. *Proceedings of the Royal Society of London B* 205: 379–384.

Frýda, J. & Gutiérrez-Marco, J. C. (1996). An unusual new sinuitid mollusc (Bellerophontoidea, Gastropoda) from the Ordovician of Spain. *Journal of Paleontology* 70: 602–609.

Frýda, J. (1997). Oldest representatives of the superfamily Cirroidea (Vetigastropoda) with notes on early phylogeny. *Journal of Paleontology* 71: 839–847.

Frýda, J. & Bandel, K. (1997). New Early Devonian gastropods from the *Plectonotus* (*Boucotonotus*) – *Palaeozygopleura* community in the Prague Basin (Bohemia). *Mitteilungen aus dem Geologisch-Paläontologischen Institut der Universität Hamburg* 80: 1–57.

Frýda, J. (1999a). Shape convergence in gastropod shells: An example from the Early Devonian *Plectonotus* (*Boucotonotus*) – *Palaeozygopleura* community of the Prague Basin (Bohemia). *Mitteilungen aus dem Geologisch-Paläontologischen Institut der Universität Hamburg* 83: 179–190.

Frýda, J. (1999b). Suggestions for polyphyletism of Paleozoic bellerophontiform molluscs inferred from their protoconch morphology. *Abstracts, 65th Annual Meeting, American Malacological Society* (Pittsburgh): 30.

Frýda, J. (1999c). Higher classification of Paleozoic gastropods inferred from their early shell ontogeny. *Journal of the Czech Geological Society* 44: 137–154.

Frýda, J. & Blodgett, R. B. (2001). The oldest known heterobranch gastropod, *Kuskokwimia* gen. nov., from the Early Devonian of west-central Alaska, with notes on the early phylogeny of higher gastropods. *Bulletin of the Czech Geological Survey* 76: 39–54.

Frýda, J. & Heidelberger, D. (2003). Systematic position of Cyrtoneritimorpha within the class Gastropoda with description of two new genera from Siluro-Devonian strata of Central Europe. *Bulletin of the Czech Geological Survey* 78: 35–39.

Frýda, J. & Rohr, D. M. (2004). Gastropods, pp. 184–195 *in* B. D. Webby, Paris, F., Droser, M. L. & Percival, I. G. (eds.), *The Great Ordovician Biodiversification Event*. Vol. 3. New York, Columbia University Press.

Frýda, J., Heidelberger, D. & Blodgett, R. B. (2006). Odontomariinae, a new middle Paleozoic subfamily of slit-bearing euophaloidean gastropods (Euophalomorpha, Gastropoda). *Neues Jahrbuch für Geologie und Palaontologie-Monatshefte*: 225–248.

Frýda, J., Nützel, A. & Wagner, P. J. (2008a). Paleozoic Gastropoda, pp. 239–270 *in* W. F. Ponder & Lindberg, D. R. (eds.), *Phylogeny and Evolution of the Mollusca*. Berkeley, CA, University of California Press.

Frýda, J., Racheboeuf, P. R. & Frýdová, B. (2008b). Mode of life of Early Devonian *Orthonychia protei* (Neritimorpha, Gastropoda) inferred from its post-larval shell ontogeny and muscle scars. *Bulletin of Geosciences* 83: 491–502.

Frýda, J., Racheboeuf, P. R., Frýdová, B., Ferrová, L., Mergl, M. & Berkyová, S. (2009). Platyceratid gastropods – Stem group of patellogastropods, neritimorphs or something else? *Bulletin of Geosciences* 84: 107–120.

Frýda, J. (2012). Phylogeny of Palaeozoic gastropods inferred from their ontogeny, pp. 395–435 *in* J. A. Talent (ed.), *Earth and Life Global Biodiversity, Extinction Intervals and Biogeographic Perturbations through Time*. Dordrecht, the Netherlands, Springer Netherlands.

Frýda, J., Ebbestad, J. O. R. & Frýdová, B. (2019). The oldest members of Porcellioidea (Gastropoda): A new link between Baltica and Perunica. *Papers in Palaeontology* 5: 281–297.

Frye, C. J. & Feldmann, R. M. (1991). North American Late Devonian cephalopod aptychi. *Kirtlandia* 46: 49–71.

Fryer, G. (1959). Development of a mutelid lamellibranch. *Nature* 183: 1342–1343.

Fryer, G. (1961). The developmental history of *Mutela bourguignati* (Ancey) Bourguignat (Mollusca, Bivalvia). *Philosophical Transactions of the Royal Society B* 244: 259–298.

Fuchigami, T. & Sasaki, T. (2005). The shell structure of the Recent Patellogastropoda (Mollusca: Gastropoda). *Paleontological Research* 9: 143–168.

Fuchs, D., Klinghammer, A. & Keupp, H. (2007). Taxonomy, morphology and phylogeny of plesioteuthidid coleoids from the Upper Jurassic (Tithonian) Plattenkalks of Solnhofen. *Neues Jahrbuch für Geologie und Paläontologie-Abhandlungen* 245: 239–252.

Fuchs, D. & Weis, R. (2008). Taxonomy, morphology and phylogeny of Lower Jurassic logligosepiid coleoids (Cephalopoda). *Neues Jahrbuch für Geologie und Paläontologie-Abhandlungen* 249: 93–112.

Fuchs, D. (2009). Octobrachia – a diphyletic taxon. *Berliner Paläobiologische Abhandlungen* 10: 182–192.

Fuchs, D., Bracchi, G. & Weis, R. (2009). New octopods (Cephalopoda: Coleoidea) from the Late Cretaceous (Upper Cenomanian) of Hâkel and Hâdjoula, Lebanon. *Palaeontology* 52: 65–81.

Fuchs, D. & Weis, R. (2009). A new Cenomanian (Late Cretaceous) coleoid (Cephalopoda) from Hâdjoula, Lebanon. *Fossil Record* 12: 175–181.

Fuchs, D., Von Boletzky, S. & Tischlinger, H. (2010). New evidence of functional suckers in belemnoid coleoids (Cephalopoda) weakens support for the 'Neocoleoidea' concept. *Journal of Molluscan Studies* 76: 404–406.

Fuchs, D. & Tanabe, K. (2010). Re-investigation of the shell morphology and ultrastructure of the Late Cretaceous spirulid coleoid *Naefia matsumotoi*, pp. 195–207 in K. Tanabe, Shigeta, Y., Sasaki, T. & Hirano, H. (eds.), *Cephalopods Present and Past. 8th International Symposium, University of Bourgogne, France, 30 August–30 September 2010*. Tokyo, Tokai University Press.

Fuchs, D. & Weis, R. (2010). Taxonomy, morphology and phylogeny of Lower Jurassic teudopseid coleoids (Cephalopoda). *Neues Jahrbuch für Geologie und Paläontologie-Abhandlungen* 257: 351–366.

Fuchs, D. & Larson, N. L. (2011a). Diversity, morphology, and phylogeny of coleoid cephalopods from the Upper Cretaceous Plattenkalks of Lebanon-Part I: Prototeuthidina. *Journal of Paleontology* 85: 234–249.

Fuchs, D. & Larson, N. L. (2011b). Diversity, morphology, and phylogeny of coleoid cephalopods from the Upper Cretaceous Plattenkalks of Lebanon-Part II: Teudopseina. *Journal of Paleontology* 85: 815–834.

Fuchs, D. (2012). The 'rostrum' problem in coleoid terminology – an attempt to clarify inconsistencies. *Geobios* 45: 29–39.

Fuchs, D., Keupp, H., Trask, P. & Tanabe, K. (2012a). Taxonomy, morphology and phylogeny of Late Cretaceous spirulid coleoids (Cephalopoda) from Greenland and Canada. *Palaeontology* 55: 285–303.

Fuchs, D., Keupp, H. & Wiese, F. (2012b). Protoconch morphology of *Conoteuthis* (Diplobelida, Coleoidea) and its implications on the presumed origin of the Sepiida. *Cretaceous Research* 34: 200–207.

Fuchs, D., Donovan, D. T. & Keupp, H. (2013a). Taxonomic revision of '*Onychoteuthis*' *conocauda* Quenstedt, 1849 (Cephalopoda: Coleoidea). *Neues Jahrbuch für Geologie und Paläontologie-Abhandlungen* 270: 245–255.

Fuchs, D., Heyng, A. M. & Keupp, H. (2013b). *Acanthoteuthis problematica* Naef, 1922, an almost forgotten taxon and its role in the interpretation of cephalopod arm armatures. *Neues Jahrbuch für Geologie und Paläontologie-Abhandlungen* 269: 241–250.

Fuchs, D., Iba, Y., Ifrim, C., Nishimura, T., Kennedy, W. J., Keupp, H., Stinnesbeck, W. & Tanabe, K. (2013c). *Longibelus* gen. nov., a new Cretaceous coleoid genus linking Belemnoidea and early Decabrachia. *Palaeontology* 56: 1081–1106.

Fuchs, D. (2014). First evidence of Mastigophora (Cephalopoda: Coleoidea) from the early Callovian of La Voulte-sur-Rhône (France), pp. 21–27 in F. Wiese, Reich, M. & Arp, G. (eds.), *Spongy, Slimy, Cosy & More. Göttingen Contributions to Geosciences*. Göttingen, Germany, Universitätsverlag Göttingen.

Fuchs, D. & Iba, Y. (2015). The gladiuses in coleoid cephalopods: Homology, parallelism, or convergence? *Swiss Journal of Palaeontology* 134: 187–197.

Fuchs, D. & Košťák, M. (2015). *Amphispirula* gen. nov. from the Eocene of southern Moravia (Czech Republic): A new ancestor of the Recent deep-sea squid *Spirula*? *Journal of Systematic Palaeontology* 14: 91–98.

Fuchs, D. (2016). Part M, Chapter 9B: The gladius and gladius vestige in fossil Coleoidea. *Treatise Online* 83: 1–23.

Fuchs, D., Iba, Y., Tischlinger, H., Keupp, H. & Klug, C. (2016). The locomotion system of Mesozoic Coleoidea (Cephalopoda) and its phylogenetic significance. *Lethaia* 49: 433–454.

Fuchs, D. & Hoffmann, R. (2017). Arm armature in belemnoid coleoids. *Treatise Online* 91, Part M, Chapter 10: 1–20.

Fuchs, D. & Donovan, D. (2018). Systematic descriptions: Phragmoteuthida. *Treatise Online* 111, Part M, Chapter 23C: 1–7.

Fuchs, D., Keupp, H. & Klug, C. (2018). A critical review of *Antarcticeras* Doguzhaeva, 2017–teuthid affinities can explain the poorly mineralized phragmocone. *Historical Biology* 2018: 1–6.

Fuchs, D. & Schweigert, G. (2018). First Middle–Late Jurassic gladius vestiges provide new evidence on the detailed origin of incirrate and cirrate octopuses (Coleoidea). *PalZ* 92: 203–217.

Fujikura, K., Sasaki, T., Yamanaka, T. & Yoshida, T. (2009). Turrids whelk, *Phymorhynchus buccinoides* feeds on *Bathymodiolus* mussels at a seep site in Sagami Bay, Japan. *Plankton and Benthos Research* 4: 23–30.

Fukuda, H. & Ponder, W. F. (2004). A protandric assimineid gastropod: *Rugapedia androgyna* n. gen. and n. sp. (Mollusca: Caenogastropoda: Rissooidea) from Queensland, Australia. *Molluscan Research* 24: 75–88.

Fukuda, Y. (1987). Histology of the long digital tentacles, pp. 249–256 in W. B. Saunders & Landman, N. H. (eds.), *Nautilus: The Biology and Paleobiology of a Living Fossil. Topics in Geobiology*. New York, Springer.

Fukumori, H. & Kano, Y. (2014). Evolutionary ecology of settlement size in planktotrophic neritimorph gastropods. *Marine Biology* 161: 213–227.

Furnish, W. M. & Glenister, B. F. (1964). Nautiloidea - Ascocerida, pp. K261–K277 in R. C. Moore (ed.), *Treatise on Invertebrate Paleontology Part K. Mollusca*. Vol. 3. Lawrence, KS, Geological Society of America and University of Kansas Press.

Furnish, W. M., Glenister, B. F., Kullmann, J. & Zhou, Z. (2009). Carboniferous and Permian Ammonoidea (Goniatitida and Prolecanitida), pp. 1–258 in R. C. Moore (ed.), *Treatise on Invertebrate Paleontology Part L. Mollusca*. Vol. 2. Lawrence, KS, Geological Society of America & University of Kansas Press.

Furuya, H., Ota, M., Kimura, R. & Tsuneki, K. (2004). Renal organs of cephalopods: A habitat for dicyemids and chromidinids. *Journal of Morphology* 262: 629–643.

Furuya, H. (2006). Three new species of dicyemid mesozoans (phylum Dicyemida) from *Amphioctopus fangsiao* (Mollusca: Cephalopoda), with comments on the occurrence patterns of dicyemids. *Zoological Science* 23: 105–119.

Gabbott, S. E. (1999). Orthoconic cephalopods and associated fauna from the Late Ordovician Soom Shale Lagerstätte, South Africa. *Palaeontology* 42: 123–148.

Gabe, M. & Prenant, M. (1952a). Recherches sur la gaine radiculaire des mollusques. 4. L'appareil radiculaire d'*Acteon tornatilis* Linné. *Archives de zoologie expérimentale et générale* 89: 15–25.

Gabe, M. & Prenant, M. (1952b). Recherches sur la gaine radiculaire des mollusques. 5. L'appareil radiculaire de quelques opisthobranches céphalaspides. *Bulletin du Laboratoire maritime de Dinard* 37: 13–27.

Gainey, L. F. (1972). The use of the foot and the captacula in the feeding of *Dentalium* (Mollusca: Scaphopoda). *The Veliger* 15: 29–34.

Galle, A. & Parsley, R. L. (2005). Epibiont relationships on hyolithids demonstrated by Ordovician trepostomes (Bryozoa) and Devonian tabulates (Anthozoa). *Bulletin of Geosciences* 80: 125–138.

Gamarra-Luques, C., Winik, B., Vega, I., Albrecht, E., Catalan, N. & Castro-Vazquez, A. (2006). An integrative view to structure, function, ontogeny and phylogenetical significance of the male genital system in *Pomacea canaliculata* (Caenogastropoda, Ampullariidae). *Biocell* 30: 345–357.

Gardner, R. N. (2005). Middle-Late Jurassic bivalves of the superfamily Veneroidea from New Zealand and New Caledonia. *New Zealand Journal of Geology and Geophysics* 48: 325–376.

Garnault, P. (1899). Sur les organes reproducteurs de la *Valvata piscinalis*. *Zoologischer Anzeiger* 12: 266–269.

Garstang, W. (1929). The origin and evolution of larval forms. *Report of the British Association for the Advancement of Science* 96: 77–98.

Garvon, J. M. & Bird, J. (2005). Attraction of the land snail *Anguispira alternata* to fresh faeces of white-tailed deer: Implications in the transmission of *Parelaphostrongylus tenuis*. *Canadian Journal of Zoology* 83: 358–362.

Gascoigne, T. (1956). Feeding and reproduction in the Limapontiidae. *Transactions of the Royal Society of Edinburgh* 63: 129–151.

Gascoigne, T. & Sartory, P. K. (1974). The teeth of three bivalved gastropods and three other species of the order Sacoglossa, with an appendix on *Calliopaea oophaga* n. sp., a new sacoglossan. *Proceedings of the Malacological Society of London* 41: 109–124.

Gascoigne, T. (1977). Sacoglossan teeth. *Malacologia* 16: 101–105.

Gegenbauer, C. (1878). *Grundriss der vergleichenden Anatomie. Zweite verbesswerte Auflage*. Leipzig, W. Engelmann.

Geiger, D. L. & Thacker, C. E. (2006). Molecular phylogeny of basal gastropods (Vetigastropoda) shows stochastic colonization of chemosynthetic habitats at least from the mid Triassic. *Cahiers de Biologie Marine* 47: 343–346.

Geiger, D. L., Nützel, A. & Sasaki, T. (2008). Vetigastropoda, pp. 297–330 in W. F. Ponder & Lindberg, D. R. (eds.), *Phylogeny and Evolution of the Mollusca*. Berkeley, CA, University of California Press.

Geilenkirchen, W. L. M., Van, Timmermans, L. P. M., Dongen, C. A. M., Van & Arnolds, W. J. A. (1971). Symbiosis of bacteria with eggs of *Dentalium* at the vegetal pole. *Experimental Cell Research* 67: 477–479.

Geisler, J. H., Colbert, M. W. & Carew, J. L. (2014). A new fossil species supports an early origin for toothed whale echolocation. *Nature* 508: 383.

Geist, J., Wunderlich, H. & Kuehn, R. (2008). Use of mollusc shells for DNA-based molecular analyses. *Journal of Molluscan Studies* 74: 337–343.

Gelperin, A. (1974). Olfactory basis of homing behavior in the giant garden slug, *Limax maximus*. *Proceedings of the National Academy of Sciences of the United States of America* 71: 966–970.

Gendron, L. (1992). Determination of the size at sexual maturity of the waved whelk *Buccinum undatum* Linnaeus, 1758, in the Gulf of St. Lawrence, as a basis for the establishment of a minimum catchable size. *Journal of Shellfish Research* 11: 1–7.

Geraerts, W. P. M. & Goosse, J. (1975). The control of vitellogenesis and of growth of female accessory sex organs by the dorsal body hormone (DBH) in the hermaphrodite freshwater snail *Lymnaea stagnalis*. *General and Comparative Endocrinology* 27: 450–467.

Geraerts, W. P. M. & Joosse, J. (1984). Freshwater snails (Basommatophora), pp. 141–207 *in* A. S. Tompa, Verdonk, N. H. & van den Biggelaar, J. A. M. (eds.), *Reproduction. The Mollusca*. Vol. 7. New York, Academic Press.

Gerlach, J. & van Bruggen, A. C. (1998). A first record of a terrestrial mollusc without a radula. *Journal of Molluscan Studies* 64: 249–250.

Geyer, G. (1994). Middle Cambrian mollusks from Idaho and early conchiferan evolution. *Bulletin of the New York State Museum* 481: 69–86.

Geyer, G. & Streng, M. (1998). Middle Cambrian pelecypods from the Anti-Atlas, Morocco. *Revista Española de Paleontología, No. extraordinario, Homenajeal Prof. Gonzalo Vidal*: 83–86.

Ghiselin, M. T. (1966). Reproductive function and the phylogeny of opisthobranch gastropods. *Malacologia* 3: 327–378.

Ghiselin, M. T. (1988). The origin of molluscs in the light of molecular evidence, pp. 66–95 *in* P. H. Harvey & Partridge, L. (eds.), *Oxford Surveys of Evolutionary Biology*. Vol. 5. Oxford, UK, Oxford University Press.

Ghose, K. C. (1962). Morphogenesis of the nervous system of the giant land snail *Achatina fulica* Bowdich. *Zoologischer Anzeiger* 169: 467–475.

Ghose, K. C. (1963a). The alimentary system of *Achatina fulica*. *Transactions of the American Microscopical Society* 82: 149–167.

Ghose, K. C. (1963b). Morphogenesis of the pericardium and heart and kidney and ureter and gonad and gonoduct in the giant land snail *Achatina fulica* Bowdich. *Proceedings of the Zoological Society (Calcutta)* 16: 201–214.

Gianguzza, P., Badalamenti, F., Jensen, K. R., Chemello, R., Cannicci, S. & Riggio, S. (2004). Body size and mating strategies in the simultaneous hermaphrodite *Oxynoe olivacea* (Mollusca, Opisthobranchia, Sacoglossa). *Functional Ecology* 18: 899–906.

Gibbs, P. E. (1984). The population cycle of the bivalve *Abra tenuis* and its mode of reproduction. *Journal of the Marine Biological Association of the United Kingdom* 64: 791–800.

Gili, E., Masse, J. & Skelton, P. W. (1995). Rudists as gregarious sediment-dwellers, not reef-builders, on Cretaceous carbonate platforms. *Palaeogeography, Palaeoclimatology, Palaeoecology* 118: 245–267.

Gillette, R. (1991). On the significance of neuronal gigantism in gastropods. *Biological Bulletin* 180: 234–240.

Gillette, R., Saeki, M. & Huang, R. C. (1991). Defensive mechanisms in notaspid snails: Acid humor and evasiveness. *Journal of Experimental Biology* 156: 335–347.

Gilly, W. M. F. & Lucero, M. T. (1992). Behavioural responses to chemical stimulation of the olfactory organ in the squid *Loligo opalescens*. *Journal of Experimental Biology* 162: 209–229.

Gilmer, R. W. (1986). Preservation artifacts and their effects on the study of euthecosomatous pteropod mollusks. *The Veliger* 29: 48–52.

Gilmer, R. W. & Harbison, G. R. (1986). Morphology and field behavior of pteropod molluscs: Feeding methods in the families Cavoliniidae, Limacinidae and Peraclididae (Gastropoda: Thecosomata). *Marine Biology* 91: 47–57.

Gilmour, T. H. J. (1963). A note on the tentacles of *Lima hians* (Gmelin) (Bivalvia). *Proceedings of the Malacological Society of London* 35: 82–85.

Gilmour, T. H. J. (1967). The defensive adaptations of *Lima hians* (Mollusca, Bivalvia). *Journal of the Marine Biological Association of the United Kingdom* 47: 209–221.

Gilmour, T. H. J. (1990). The adaptive significance of foot reversal in the Limoida, pp. 249–263 *in* B. Morton (ed.), *The Bivalvia. Proceedings of a Memorial Symposium in Honour of Sir Charles Maurice Yonge (1899–1986), Edinburgh, 1986*. Hong Kong, Hong Kong University Press.

Giménez, J., Brey, T., Mackensen, A. & Penchaszadeh, P. E. (2004). Age, growth, and mortality of the prosobranch *Zidona dufresnei* (Donovan, 1823) in the Mar del Plata area, south-western Atlantic Ocean. *Marine Biology* 145: 707–712.

Giribet, G. (2016a). Phylum Nemertea: The ribbon worms, pp. 435–452 *in* R. C. Brusca, Moore, W. & Shuster, S. M. (eds.), *Invertebrates*. Sunderland, MA, Sinauer Associates.

Giribet, G. (2002). Current advances in the phylogenetic reconstruction of metazoan evolution. A new paradigm for the Cambrian explosion? *Molecular Phylogenetics and Evolution* 24: 345–357.

Giribet, G. & Wheeler, W. C. (2002). On bivalve phylogeny: A high-level analysis of the Bivalvia (Mollusca) based on combined morphology and DNA sequence data. *Invertebrate Biology* 121: 271–324.

Giribet, G. (2003). Stability in phylogenetic formulations and its relationship to nodal support. *Systematic Biology* 52: 554–564.

Giribet, G. & Distel, D. L. (2003). Bivalve phylogeny and molecular data, pp. 45–90 *in* C. Lydeard & Lindberg, D. R. (eds.), *Molecular Systematics and Phylogeography of Mollusks. Smithsonian Series in Comparative Evolutionary Biology*. Washington, DC, Smithsonian Books.

Giribet, G., Okusu, A., Lindgren, A. R., Huff, S. W., Schrödl, M. & Nishiguchi, M. K. (2006). Evidence for a clade composed of molluscs with serially repeated structures: Monoplacophorans are related to chitons. *Proceedings of the National Academy of Sciences of the United States of America* 103: 7723–7728.

Giribet, G. (2008a). Bivalvia, pp. 105–141 *in* W. F. Ponder & Lindberg, D. R. (eds.), *Phylogeny and Evolution of the Mollusca*. Berkeley, CA, University of California Press.

Giribet, G. (2008b). Assembling the lophotrochozoan (= spiralian) tree of life. *Philosophical Transactions of the Royal Society of London B* 363: 1513–1522.

Giribet, G., Dunn, C. W., Edgecombe, G. D., Hejnol, A., Martindale, M. Q. & Rouse, G. W. (2009). Assembling the spiralian tree of life, pp. 52–64 *in* M. J. Telford & Littlewood, D. T. J. (eds.), *Animal Evolution: Genes, Genomes, Fossils and Trees*. Oxford, UK, Oxford University Press.

Giribet, G. (2014). On Aculifera: A review of hypotheses in tribute to Christoffer Schander. *Journal of Natural History* 48: 2739–2749.

Giribet, G. (2016b). New animal phylogeny: Future challenges for animal phylogeny in the age of phylogenomics. *Organisms Diversity & Evolution* 16: 419–426.

Gittenberger, E. (1999). Dispersal, vicariance, and partial morphostasis in the evolutionary history of SE European Zonitini (Mollusca, Gastropoda, Pulmonata). *Zoologischer Anzeiger* 237: 243–258.

Gladenkov, A. Y., Oleinik, A. E., Marincovich, L. & Barinov, K. B. (2002). A refined age for the earliest opening of Bering Strait. *Palaeogeography Palaeoclimatology Palaeoecology* 183: 321–328.

Glaessner, M. F. & Daily, B. (1959). The geology and Late Precambrian fauna of the Ediacara fossil reserve. *Records of the South Australian Museum* 13: 369–401.

Glaessner, M. F. & Wade, M. (1966). The Late Precambrian fossils from Ediacara, South Australia. *Palaeontology* 9: 599–628.

Glaser, O. C. (1910). The nematocysts of eolids. *Journal of Experimental Zoology* 9: 117–142.

Glaubrecht, M. (2006). Independent evolution of reproductive modes in viviparous freshwater Cerithioidea (Gastropoda, Sorbeoconcha) – A brief review. *Basteria* 3: 23–28.

Glaubrecht, M., Fehér, Z. & von Rintelen, T. (2006). Brooding in *Corbicula madagascariensis* (Bivalvia, Corbiculidae) and the repeated evolution of viviparity in corbiculids. *Zoologica Scripta* 35: 641–654.

Glover, E. A. & Taylor, J. D. (2010). Needles and pins: Acicular crystalline periostracal calcification in venerid bivalves (Bivalvia: Veneridae). *Journal of Molluscan Studies* 76: 157–179.

Göbbeler, K. & Klussmann-Kolb, A. (2011). Molecular phylogeny of the Euthyneura (Mollusca, Gastropoda) with special focus on Opisthobranchia as a framework for reconstruction of evolution of diet. *Thalassas* 27: 121–154.

Gofas, S. (2000). Systematics of *Planktomya*, a bivalve genus with teleplanic larval dispersal. *Bulletin of Marine Science* 67: 1013–1023.

Golding, R. E. & Jones, A. S. (2007). Micro-CT as a novel technique for 3D reconstruction of molluscan anatomy. *Molluscan Research* 27: 123–128.

Golding, R. E., Ponder, W. F. & Byrne, M. (2007). Taxonomy and anatomy of Amphiboloidea (Gastropoda: Heterobranchia: Archaeopulmonata). *Zootaxa* 50: 1–50.

Golding, R. E., Ponder, W. F. & Byrne, M. (2009a). Three-dimensional reconstruction of the odontophoral cartilages of Caenogastropoda (Mollusca: Gastropoda) using micro-CT: Morphology and phylogenetic significance. *Journal of Morphology* 270: 558–587.

Golding, R. E., Ponder, W. F. & Byrne, M. (2009b). The evolutionary and biomechanical implications of snout and proboscis morphology in Caenogastropoda (Mollusca: Gastropoda). *Journal of Natural History* 43: 2723–2763.

Golding, R. E., Byrne, M. & Ponder, W. F. (2010). Novel copulatory structures and reproductive functions in Amphiboloidea. *Invertebrate Biology* 127: 168–180.

Golding, R. E. & Ponder, W. F. (2010). Homology and morphology of the neogastropod valve of Leiblein (Gastropoda: Caenogastropoda). *Zoomorphology* 129: 81–91.

Golding, R. E. (2012). Molecular phylogenetic analysis of mudflat snails (Gastropoda: Euthyneura: Amphiboloidea) supports an Australasian centre of origin. *Molecular Phylogenetics and Evolution* 63: 72–81.

Golding, R. E., Bieler, R., Rawlings, T. A. & Collins, T. M. (2014). Deconstructing *Dendropoma*: A systematic revision of a world-wide worm-snail group, with descriptions of new genera (Caenogastropoda: Vermetidae). *Malacologia* 57: 1–97.

Golikov, A. N. & Kussakin, O. G. (1972). Sur la biologie de la reproduction des Patellides de la familie Tecturidae (Gastropoda, Docoglossa) et sur la position systématique de ses subdivisions. *Malacologia* 11: 287–294.

Golikov, A. N. & Starobogatov, I. (1975). Systematics of prosobranch gastropods. *Malacologia* 15: 185–232.

Golikov, A. N. & Starobogatov, Y. I. (1989). Voprosy filogenii i sistemy perednezhabernykh briukhonogikh molliuskov. [Problems of phylogeny and system of the prosobranchiate gastropods] (in Russian). *Trudy Zoologicheskogo Instituta* 187: 4–77.

Gomot de Vaufleury, A. (2001). Regulation of growth and reproduction, pp. 331–355 *in* G. M. Barker (ed.), *The Biology of Terrestrial Molluscs*. Wallingford, UK, CABI Publishing.

Gonor, J. J. (1961). Observation on the biology of *Lobiger serrachifaldi*, a shelled saccoglossan opisthobranch from the Mediterranean. *Vie et Milieu Série* A 12: 381–403.

González-Tizón, A., Martínez-Lage, A., Ausio, J. & Méndez, J. (2000). Polyploidy in a natural population of mussel, *Mytilus trossulus*. *Genome* 43: 409–411.

Gonzalez-Wevar, C. A., Hune, M., Segovia, N. I., Nakano, T., Spencer, H. G., Chown, S. L., Saucede, T., Johnstone, G., Mansilla, A. & Poulin, E. (2017). Following the Antarctic Circumpolar Current: Patterns and processes in the biogeography of the limpet *Nacella* (Mollusca: Patellogastropoda) across the Southern Ocean. *Journal of Biogeography* 44: 861–874.

González, V. L., Andrade, S. C. S., Bieler, R., Collins, T. M., Dunn, C. W., Mikkelsen, P. M., Taylor, J. D. & Giribet, G. (2015). A phylogenetic backbone for Bivalvia: An RNA-seq approach. *Proceedings of the Royal Society. B* 282: 20142332.

González, V. L. & Giribet, G. (2015). A multilocus phylogeny of archiheterodont bivalves (Mollusca, Bivalvia, Archiheterodonta). *Zoologica Scripta* 44: 41–58.

Goodfriend, G. A. (1986). Variation in land-snail shell form and size and its causes: A review. *Systematic Zoology* 35: 204–223.

Goodheart, J. A. (2017). Insights into the systematics, phylogeny, and evolution of Cladobranchia (Gastropoda: Heterobranchia). *American Malacological Bulletin* 35: 73–81.

Goodheart, J. A., Bleidißel, S., Schillo, D., Strong, E. E., Ayres, D. L., Preisfeld, A., Collins, A. G., Cummings, M. P. & Wägele, H. (2018). Comparative morphology and evolution of the cnidosac in Cladobranchia (Gastropoda: Heterobranchia: Nudibranchia). *Frontiers in Zoology* 15: 43.

Gordon, M. E. (1964). Carboniferous cephalopods of Arkansas. *Geological Survey Professional Paper* 460: 1–322.

Goreau, T. F., Goreau, N. I., Yonge, C. M. & Yeumann, Y. (1970). On feeding and nutrition in *Fungiacava eilatensis* (Bivalvia, Mytilidae), a commensal living in fungiid corals. *Journal of Zoology* 160: 159–172.

Gorf, A. (1961). Untersuchungen über Neurosekretion bei der Sumpfdeckelschnecke *Vivipara vivipara* L. *Zoologische Jahrbucher. Abteilung für Allgemeine Zoologie und Physiologie der Tiere* 69: 379–404.

Gosliner, T. & Behrens, D. W. (1998). Five new species of *Chromodoris* (Mollusca: Nudibranchia: Chromodorididae) from the tropical Indo-Pacific Ocean. *Proceedings of the California Academy of Sciences* 50: 139–165.

Gosliner, T. M. (1981). Origins and relationships of primitive members of the Opisthobranchia (Mollusca: Gastropoda). *Biological Journal of the Linnean Society* 16: 197–225.

Gosliner, T. M. (1982). The genus *Janolus* (Nudibranchia, Arminacea) from the Pacific coast of North America, with a reinstatement of *Janolus fuscus* O'Donoghue, 1924. *The Veliger* 24: 219–226.

Gosliner, T. M. & Liltved, W. R. (1982). Comparative morphology of three South African Triviidae (Gastropoda: Prosobranchia) with the description of a new species. *Zoological Journal of the Linnean Society* 74: 111–132.

Gosliner, T. M. & Liltved, W. R. (1987a). Further studies on the morphology of the Triviidae (Gastropoda: Prosobranchia) with emphasis on species from southern Africa. *Zoological Journal of the Linnean Society* 90: 207–254.

Gosliner, T. M. & Liltved, W. R. (1987b). Comparative anatomy of the Triviidae and its bearing upon the phylogeny of velutinacean gastropods. *American Zoologist* 27: 62A.

Gosliner, T. M. (1989). Revision of *Gasteropteridae* (Opisthobranchia: Cephalaspidea) with descriptions of a new genus and six new species. *The Veliger* 32: 333–381.

Gosliner, T. M. & Behrens, D. W. (1990). Special resemblance, aposematic coloration and mimicry in opisthobranch gastropods, pp. 127–138 *in* M. Wicksten (ed.), *Adaptive Coloration in Invertebrates*. College Station, TX, Texas A&M University Sea Grant College Program.

Gosliner, T. M. (1994). Gastropoda: Opisthobranchia, pp. 253–355 *in* F. W. Harrison & Kohn, A. J. (eds.), *Microscopic Anatomy of Invertebrates. Mollusca 1*. Vol. 5. New York, Wiley-Liss.

Gosliner, T. M. (1996). The Opisthobranchia, pp. 161–213 *in* P. H. Scott, Blake, J. A. & Lissner, A. L. (eds.), *Taxonomic Atlas of the Benthic Fauna of the Santa Maria Basin and Western Santa Barbara Channel*. Vol. 9. San Diego, CA, Science Applications International Corporation.

Gosliner, T. M. & Johnson, R. F. (1999). Phylogeny of *Hypselodoris* (Nudibranchia: Chromodorididae) with a review of the monophyletic clade of Indo-Pacific species, including descriptions of twelve new species. *Zoological Journal of the Linnean Society* 125: 1–114.

Gosliner, T. M. (2001). Aposematic coloration and mimicry in opisthobranch mollusks: New phylogenetic and experimental data. *Bollettino Malacologico* 37: 163–170.

Gosliner, T. M. & Smith, V. G. (2003). Systematic review and phylogenetic analysis of the nudibranch genus *Melibe* (Opisthobranchia: Dendronotacea) with descriptions of three new species. *Proceedings of the California Academy of Sciences* 54: 302–355.

Gosliner, T. M. & Pola, M. (2012). Diversification of filter-feeding nudibranchs: Two remarkable new species of *Melibe* (Opisthobranchia: Tethyiidae) from the tropical western Pacific. *Systematics and Biodiversity* 10: 333–349.

Gosling, E. M. (2003). *Bivalve Molluscs: Biology, Ecology and Culture*. Oxford, UK, Fishing News Books.

Goto, R., Kawakita, A., Ishikawa, H., Hamamura, Y. & Kato, M. (2012). Molecular phylogeny of the bivalve superfamily Galeommatoidea (Heterodonta, Veneroida) reveals dynamic evolution of symbiotic lifestyle and interphylum host switching. *BMC Evolutionary Biology* 12: 172.

Goto, R., Fukumori, H., Kano, Y. & Kato, M. (2018). Evolutionary gain of red blood cells in a commensal bivalve (Galeommatoidea) as an adaptation to a hypoxic shrimp burrow. *Biological Journal of the Linnean Society* 125: 368–376.

Gould, S. J. (1977). *Ontogeny and Phylogeny*. Cambridge, MA, Harvard University Press.

Gould, S. J. & Calloway, C. B. (1980). Clams and brachiopods – Ships that pass in the night. *Paleobiology* 6: 383–396.

Gould, S. J. (1989). *Wonderful Life: The Burgess Shale and the Nature of History*. New York, W.W. Norton.

Gowlett-Holmes, K. L. (1987). The suborder Choriplacina Starobogatov & Sirenko, 1975 with a redescription of *Choriplax grayi* (H. Adams & Angas, 1864) (Mollusca: Polyplacophora). *Transactions of the Royal Society of South Australia* 111: 105–110.

Graça, M. A., Serra, S. R. & Ferreira, V. (2012). A stable temperature may favour continuous reproduction by *Theodoxus fluviatilis* and explain its high densities in some karstic springs. *Limnetica* 31: 0129–0140.

Gradstein, F. M., Ogg, J. G. & Smith, A. G. (2004). *A Geologic Time Scale 2004*. Vol. 86. Cambridge, UK, Cambridge University Press.

Graf, D. L. & Ó Foighil, D. (2000a). Molecular phylogenetic analysis of 28S rDNA supports a Gondwanan origin for Australasian Hyriidae (Mollusca: Bivalvia: Unionoida). *Vie et Milieu. Life and Environment* 50: 245–254.

Graf, D. L. & Ó Foighil, D. (2000b). The evolution of brooding characters among the freshwater pearly mussels (Bivalvia: Unionoidea) of North America. *Journal of Molluscan Studies* 66: 157–170.

Graf, D. L. & Cummings, K. S. (2006a). Freshwater mussels (Mollusca: Bivalvia: Unionoida) of Angola, with description of new species, *Mutela wistarmorrisi*. *Proceedings of the Academy of Natural Sciences of Philadelphia* 155: 163–194.

Graf, D. L. & Cummings, K. S. (2006b). Palaeoheterodont diversity (Mollusca: Trigonioida + Unionoida): What we know and what we wish we knew about freshwater mussel evolution. *Zoological Journal of the Linnean Society* 148: 343–394.

Graham, A. L. (1988). *Molluscs: Prosobranch and Pyramidellid Gastropods*. Second Edition. Leiden, E. J. Brill/Dr W. Backhuys.

Graham, A. L. (1932). On the structure and function of the alimentary system of the limpet. *Transactions of the Royal Society of Edinburgh* 57: 287–308.

Graham, A. L. (1934). The cruciform muscle of lamellibranchs. *Proceedings of the Royal Society of Edinburgh* 54: 17–30.

Graham, A. L. (1939). On the structure of the alimentary canal of style-bearing prosobranchs. *Proceedings of the Zoological Society of London* 109: 75–112.

Graham, A. L. (1949). The molluscan stomach. *Transactions of the Royal Society of Edinburgh* 61: 737–778.

Graham, A. L. (1959). The functional morphology of the prosobranch buccal mass, pp. 370–373 *in* H. R. Hewer & Riley, N. D. (eds.), *Proceedings of the 15th International Congress of Zoology London* 16–23 July 1958. London, Linnean Society of London.

Graham, A. L. (1964). The functional anatomy of the buccal mass of the limpet (*Patella vulgata*). *Proceedings of the Zoological Society of London* 143: 301–329.

Graham, A. L. (1966). The fore-gut of some marginellid and cancellariid prosobranchs. *Studies in Tropical Oceanography (Miami)* 4: 134–151.

Graham, A. L. (1973). The anatomical basis of function in the buccal mass of prosobranch and amphineuran molluscs. *Journal of Zoology* 169: 317–348.

Graham, A. L. (1985). Evolution within the Gastropoda: Prosobranchia, pp. 151–186 *in* E. R. Trueman & Clarke, M. R. (eds.), *Evolution. The Mollusca*. Vol. 10. New York, Academic Press.

Grahame, J. (1969). The biology of *Berthellina caribbea* Edmunds. *Bulletin of Marine Science* 19: 868–879.

Grande, C., Templado, J., Cervera, J. L. & Zardoya, R. (2004). Molecular phylogeny of Euthyneura (Mollusca: Gastropoda). *Molecular Biology and Evolution* 21: 303–313.

Grande, C., Templado, J. & Zardoya, R. (2008). Evolution of gastropod mitochondrial genome arrangements. *BMC Evolutionary Biology* 8: 1–15.

Grange, K. R. (1976). Larval development in *Lunella smaragda* (Gastropoda: Turbinidae). *New Zealand Journal of Marine and Freshwater Research* 10: 517–525.

Grave, B. H. (1932). Embryology and life history of *Chaetopleura apiculata*. *Journal of Morphology* 54: 153–160.

Gray, M.-E. (1842–1850). *Figures of Molluscous Animals Selected from Various Authors; Etched for the Use of Students. Volumes 1–4*. London, Longman & Company.

Greenfield, L. M. (1972). Feeding and gut physiology in *Acanthopleura spinigera* (Mollusca). *Journal of Zoology* 166: 37–47.

Greenwald, L., Cook, C. B. & Ward, P. D. (1982). The structure of the Chambered Nautilus siphuncle: The siphuncular epithelium. *Journal of Morphology* 172: 5–22.

Greenwald, L., Verderber, G. & Singley, C. (1984). Localization of Na–K ATPase activity in the *Nautilus* siphuncle. *Journal of Experimental Zoology Part A* 229: 481–484.

Greenwald, L. & Ward, P. D. (1987). Buoyancy in *Nautilus*, pp. 547–562 *in* W. B. Saunders & Landman, N. H. (eds.), *Nautilus: The Biology and Paleobiology of a Living Fossil. Topics in Geobiology*. New York, Springer.

Greenwood, P. G. & Mariscal, R. N. (1984a). The utilization of cnidarian nematocysts by aeolid nudibranchs: Nematocyst maintenance and release in *Spurilla*. *Tissue and Cell* 16: 719–730.

Greenwood, P. G. & Mariscal, R. N. (1984b). Immature nematocyst incorporation by the aeolid nudibranch *Spurilla neapolitana*. *Marine Biology* 80: 35–38.

Greenwood, P. G. (1988). Nudibranch nematocysts, pp. 445–462 *in* D. A. Hessinger & Lenhoff, H. M. *The Biology of Nematocysts*. San Diego, CA, Academic Press.

Greenwood, P. G. & Garrity, L. K. (1991). Discharge of nematocysts isolated from aeolid nudibranchs. *Hydrobiologia* 216/217: 671–677.

Greenwood, P. G., Garry, K., Hunter, A. & Jennings, M. (2004). Adaptable defense: A nudibranch mucus inhibits nematocyst discharge and changes with prey type. *Biological Bulletin* 206: 113–120.

Greenwood, P. G. (2009). Acquisition and use of nematocysts by cnidarian predators. *Toxicon* 54: 1065–1070.

Gregory, T. R., Ed. (2011). *The Evolution of the Genome*. Burlington, MA, Elsevier Academic Press.

Grenon, J.-F. & Walker, G. (1978). The histology and histochemistry of the pedal glandular system of two limpets, *Patella vulgata* and *Acmaea tessulata* (Gastropoda: Prosobranchia). *Journal of the Marine Biological Association of the United Kingdom* 58: 803–816.

Griffin, L. E. (1897). Notes on the anatomy of *Nautilus pompilius*. *Zoological Bulletin* 1: 147–161.

Griffin, L. E. (1900). The anatomy of *Nautilus pompilius*. *Memoirs of the National Academy of Sciences* 8: 101–230.

Griffith, E. & Pidgeon, E. (1834). *Mollusca and Radiata*. London, Whittaker and Co.

Grigioni, S., Boucher-Rodoni, R., Tonolla, M. & Peduzzi, R. (1999). Symbiotic relations between bacteria and cephalopods. *Bollettino della Società Ticinese di Scienze Naturali* 87: 53–55.

Gründel, J. (1998). Archaeo- und Caenogastropoden aus dem Dogger Deutschlands und Nordpolens. *Stuttgarter Beiträge zur Naturkunde*, Serie B 260: 1–39.

Gubanov, A. P., Peel, J. S. & Pianovskaya, I. A. (1995). Soft-sediment adaptations in a new Silurian gastropod from Central Asia. *Palaeontology* 38: 831–842.

Gubanov, A. P. (1998). The Early Cambrian molluscan evolution and its palaeogeographic and biostratigraphic implications. *Acta Universitatis Carolinae Geologica* 42: 419–422.

Gubanov, A. P., Kouchinsky, A. V. & Peel, J. S. (1999). The first evolutionary-adaptive lineage within fossil molluscs. *Lethaia* 32: 155–157.

Gubanov, A. P. & Peel, J. S. (2000). Cambrian monoplacophoran molluscs (Class Helcionelloida). *American Malacological Bulletin* 15: 139–145.

Gubanov, A. P. & Peel, J. S. (2001). Latest helcionelloid molluscs from the Lower Ordovician of Kazakhstan. *Palaeontology* 44: 681–694.

Gubanov, A. P., Kouchinsky, A. V., Peeland, J. S. & Bengtson, S. (2004). Middle Cambrian molluscs of 'Australian' aspect from northern Siberia. *Alcheringa* 28: 1–20.

Gude, G. K. (1905). On the occurrence of internal septa in *Glyptostoma newberryanum*. *Proceedings of the Malacological Society of London* 6: 283.

Guex, J. & Rakus, M. (1971). Sur la régulation bathymétrique des ammonites (Cephalopoda). *Bulletin de la Société vaudoise des sciences naturelles* 337: 1–8.

Gümbel, C. W. (1863). Über Clymenien in den übergangsgebilden des Fichtelgebirges. *Palaeontographica* 11: 85–165.

Guo, F. (1998). Origin and phylogeny of the Trigonioidoidea (non-marine Cretaceous bivalves), pp. 277–289 *in* P. A. Johnston & Haggart, J. W. (eds.), *Bivalves: An Eon of Evolution*. Calgary, AB, University of Calgary Press.

Guralnick, R. P. & Smith, K. (1999). Historical and biomechanical analysis of integration and dissociation in molluscan feeding, with special emphasis on the true limpets (Patellogastropoda: Gastropoda). *Journal of Morphology* 241: 175–195.

Guralnick, R. P. (2004). Life-history patterns in the brooding freshwater bivalve *Pisidium* (Sphaeriidae). *Journal of Molluscan Studies* 70: 341–351.

Gutiérrez, J. L., Jones, C. G., Strayer, D. L. & Iribarne, O. O. (2003). Mollusks as ecosystem engineers: The role of shell production in aquatic habitats. *Oikos* 101: 79–90.

Gutiérrez, L. M. & Womersley, C. Z. (2001). Shadow responses and the possible role of dermal photoreceptors in the Hawaiian black snail, *Nerita picea* (Gastropoda: Neritidae). *The Veliger* 44: 1–7.

Guzhov, A. V. (2006). Lower and Middle Callovian gastropod assemblages from Central European Russia. *Paleontological Journal* 40: 500–506.

Haag, W. R. & Rypel, A. L. (2011). Growth and longevity in freshwater mussels: Evolutionary and conservation implications. *Biological Reviews* 86: 225–247.

Haas, W. (1972a). Micro- und ultrastructure of Recent and fossil Scaphopoda, pp. 15–19. *General Proceedings of the 24th International Geological Congress*. Vol. 4. Montreal, QC, The Congress.

Haas, W. (1972b). Untersuchungen über die Mikro- und Ultrastruktur der Polyplacophorenschale. *Biomineralisation* 5: 3–52.

Haas, W. (1976). Observations on the shell and mantle of the Placophora. *Belle W. Baruch Library in Marine Science* 5: 389–402.

Haas, W. (1981). Evolution of calcareous hardparts in primitive molluscs. *Malacologia* 21: 403–418.

Haas, W. (1997). Der ablauf der Entwicklungsgeschichte der Decabrachia (Cephalopoda, Coleoidea) (The evolutionary history of the Decabrachia). *Palaeontographica: Abteilung A* 245: 63–81.

Haas, W. (2002). The evolutionary history of the eight-armed Coleoidea. *Abhandlungen der Geologischen Bundesanstalt* 57: 341–351.

Haas, W. (2003). Trends in the evolution of the Decabrachia. *Berliner Paläobiologische Abhandlungen* 3: 113–129.

Haase, M. & Wawra, E. (1996). The genital system of *Acochlidium fijiense* (Opisthobranchia: Acochlidioidea) and its inferred function. *Malacologia* 38: 143–151.

Haase, M. & Bouchet, P. (1998). Radiation of crenobiontic gastropods on an ancient continental island: The *Hemistomia*-clade in New Caledonia (Gastropoda: Hydrobiidae). *Hydrobiologia* 367: 43–129.

Haase, M. & Schilthuizen, M. (2007). A new *Georissa* (Gastropoda: Neritopsina: Hydrocenidae) from a limestone cave in Malaysian Borneo. *Journal of Molluscan Studies* 73: 215–221.

Haase, M. (2008). The radiation of hydrobiid gastropods in New Zealand: A revision including the description of new species based on morphology and mtDNA sequence information. *Systematics and Biodiversity* 6: 99–159.

Haase, M., Fontaine, B. & Gargominy, O. (2010). Rissooidean freshwater gastropods from the Vanuatu archipelago. *Hydrobiologia* 637: 53–71.

Habe, T. & Kosuge, S. (1966). *The Tropical Pacific*. Osaka, Japan, Hoikusha.

Hadfield, M. G. & Switzer-Dunlap, M. (1984). Opisthobranchs, pp. 209–350 *in* A. S. Tompa, Verdonk, N. H. & van den Biggelaar, J. A. M. (eds.), *Reproduction. The Mollusca*. Vol. 7. New York, Academic Press.

Hadfield, M. G. & Strathmann, M. F. (1990). Heterostrophic shells and pelagic development in trochoideans: Implications for classification, phylogeny and paleoecology. *Journal of Molluscan Studies* 56: 239–256.

Haeckel, E. (1904). *Kunstformen der Natur*. Leipzig und Wien, Verlag des Bibliographischen Instituts.

Hagadorn, J. W. (2002a). Burgess Shale: Cambrian explosion in full bloom, pp. 61–89 *in* D. J. Bottjer, Etter, W., Hagadorn, J. W. & Tang, C. M. (eds.), *Exceptional Fossil Preservation: A Unique View on the Evolution of Marine Life*. New York, Columbia University Press.

Hagadorn, J. W. (2002b). Bear Gulch: An exceptional upper Carboniferous plattenkalk, pp. 167–183 *in* D. J. Bottjer, Etter, W., Hagadorn, J. W. & Tang, C. M. (eds.), *Exceptional Fossil Preservation: A Unique View on the Evolution of Marine Life*. New York, Columbia University Press.

Hajibabaei, M., Singer, G. A. C., Hebert, P. D. N. & Hickey, D. A. (2007). DNA barcoding: How it complements taxonomy, molecular phylogenetics and population genetics. *Trends in Genetics* 23: 167–172.

Halanych, K. M., Bacheller, J., Aquinaldo, A. M., Liva, S., Hillis, D. M. & Lake, J. A. (1995). 18S rDNA evidence that the lophophorates are protostome animals. *Science* 267: 1641–1643.

Hall, J. & Clarke, J. M. (1888). *Trilobites and Other Crustacea of the Oriskany, Upper Helderberg, Hamilton, Portage, Chemung and Catskill Groups*. Vol. 7. Albany, NY, Charles Van Benthuysen.

Hallan, A., Colgan, D. J., Anderson, L. C., García, A. & Chivas, A. R. (2013). A single origin for the limnetic-euryhaline taxa in the Corbulidae (Bivalvia). *Zoologica Scripta* 42: 278–287.

Hallas, J. M., Chichvarkhin, A. & Gosliner, T. M. (2017). Aligning evidence: Concerns regarding multiple sequence alignments in estimating the phylogeny of the Nudibranchia suborder Doridina. *Royal Society Open Science* 4: 171095.

Haller, B. (1882). Die Organisation der Chitonen der Adria. *Arbeiten aus dem Zoologischen Institut der Universität Wien* 4: 1–74.

Hallinan, N. & Lindberg, D. R. (2011). Comparative analysis of chromosome counts infers three paleopolyploidies in the Mollusca. *Genome Biology and Evolution* 3: 1150–1163.

Hamada, N. & Matsukuma, A. (1995). Bivalve family Chamidae and evolutionary paleontology, with special reference to the shell mineralogy and transposition. *Science Reports: Department of Earth and Planetary Sciences, Kyushu University* 19: 93–102.

Hamann, M. T., Otto, C. S., Scheuer, P. J. & Dunbar, D. C. (1996). Kahalalides: Bioactive peptides from a marine mollusk *Elysia rufescens* and its algal diet *Bryopsis* sp. *Journal of Organic Chemistry* 61: 6594–6600.

Hamatani, I. (1960). Notes on veligers of Japanese opisthobranchs (2). *Publications of the Seto Marine Biological Laboratory* 8: 307–315.

Hamatani, I. (1961). Notes on veligers of Japanese opisthobranchs (3). *Publications of the Seto Marine Biological Laboratory* 9: 67–79.

Händeler, K. & Wägele, H. (2007). Preliminary study on molecular phylogeny of Sacoglossa and a compilation of their food organisms. *Bonner Zoologische Beiträge* 55: 231–254.

Handl, C. H. & Todt, C. (2005). Foregut glands of Solenogastres (Mollusca): Anatomy and revised terminology. *Journal of Morphology* 265: 28–42.

Handschuh, S., Baeumler, N., Schwaha, T. & Ruthensteiner, B. (2013). A correlative approach for combining microCT, light and transmission electron microscopy in a single 3D scenario. *Frontiers in Zoology* 10: 44.

Haniel, C. A. (1915). Die Cephalopoden der Dyas von Timor, pp. 1–153. *Paläontologie von Timor*. Vol. 3(6). Stuttgart, E. Schweizerbart, Nägele und Dr. Sprosser.

Hanlon, R. T. (1982). The functional organization of chromatophores and iridescent cells in the body patterning of *Loligo pealei* (Cephalopoda, Myopsida). *Malacologia* 23: 89–119.

Hanlon, R. T. & Messenger, J. B. (1988). Adaptive coloration in young cuttlefish (*Sepia officinalis* L.): The morphology and development of body patterns and their relation to behaviour. *Philosophical Transactions of the Royal Society B* 320: 437–487.

Hanlon, R. T., Cooper, K. M., Budelmann, B.-U. & Pappas, T. C. (1990). Physiological color change in squid iridophores. I. Behavior, morphology and pharmacology in *Lolliguncula brevis*. *Cell and Tissue Research* 259: 3–14.

Hanlon, R. T. & Messenger, J. B. (1996). *Cephalopod Behaviour*. Cambridge, UK, Cambridge University Press.

Hanlon, R. T., Smale, M. J. & Sauer, W. H. H. (2002). The mating system of the squid *Loligo vulgaris reynaudii* (Cephalopoda, Mollusca) off South Africa: Fighting, guarding, sneaking, mating and egg laying behavior. *Bulletin of Marine Science* 71: 331–345.

Hanlon, R. T., Conroy, L. A. & Forsythe, J. W. (2008). Mimicry and foraging behaviour of two tropical sand-flat octopus species off North Sulawesi, Indonesia. *Biological Journal of the Linnean Society* 93: 23–38.

Hansen, B. (1953). Brood protection and sex ratio of *Transennella tantilla* (Gould), a Pacific bivalve. *Videnskabelige Meddelelser fra Dansk naturhistorisk Forening* 115: 313–324.

Hansen, H. J. (1888). Malacostraca marina Groenlandiæ occidentalis. Oversigt over det vestlige Grønlands Fauna af malakostrake Havkrebsdyr. *Videnskabelige Meddelelser fra den Naturhistoriske Forening i Kjøbenhavn, Aaret 1887, Series 4* 9: 5–226.

Harasewych, M. G. & Petit, R. E. (1982). Notes on the morphology of *Cancellaria reticulata* (Gastropoda, Cancellariidae). *The Nautilus* 96: 104–113.

Harasewych, M. G., Adamkewicz, S. L., Plassmeyer, M. & Gillevet, P. M. (1998). Phylogenetic relationships of the lower Caenogastropoda (Mollusca, Gastropoda, Architaenioglossa, Campaniloidea, Cerithioidea) as determined by partial 18S rDNA sequences. *Zoologica Scripta* 27: 361–372.

Harasewych, M. G. & McArthur, A. G. (2000). A molecular phylogeny of the Patellogastropoda (Mollusca: Gastropoda). *Marine Biology* 137: 183–194.

Harasewych, M. G. (2002). Pleurotomarioidean gastropods. *Advances in Marine Biology* 42: 237–294.

Harbeck, K. (1996). Die Evolution der Archaeopulmonata. *Zoologische Verhandelingen (Leiden)* 305: 1–133.

Harnik, P. G. & Lockwood, R. (2011). Extinction in the Marine Bivalvia. *Treatise Online, Part N, Revised, Volume 1*. Lawrence, KS, KU Paleontological Institute. 1: 1–24.

Harper, D. A. T. (2006). The Ordovician biodiversification: Setting an agenda for marine life. *Palaeogeogaphy, Palaeoclimatology, Palaeoecology* 232: 148–166.

Harper, E. M. (1991). The role of predation in the evolution of cementation in bivalves. *Palaeontology* 34: 455–460.

Harper, E. M. (1992). Post-larval cementation in the Ostreidae and its implications for other cementing bivalves. *Journal of Molluscan Studies* 58: 37–47.

Harper, E. M. (1997). The molluscan periostracum: An important constraint in bivalve evolution. *Palaeontology* 40: 71–97.

Harper, E. M., Palmer, T. J. & Alphey, J. R. (1997). Evolutionary response by bivalves to changing Phanerozoic sea-water chemistry. *Geological Magazine* 134: 403–407.

Harper, E. M. (1998). The fossil record of bivalve molluscs, pp. 243–267 *in* S. K. Donovan & Paul, C. R. C. (eds.), *The Adequacy of the Fossil Record*. Chichester, NY, John Wiley.

Harper, E. M. (2000). Are calcitic layers an effective adaptation against shell dissolution in the Bivalvia? *Journal of Zoology* 251: 179–186.

Harper, E. M., Hide, E. A. & Morton, B. (2000a). Relationships between the extant Anomalodesmata: A cladistic test, pp. 129–143 *in* E. M. Harper, Taylor, J. D. & Crame, J. A. (eds.), *Evolutionary Biology of the Bivalvia*. London, Geological Society of London (Special Publication No. 177).

Harper, E. M., Taylor, J. D. & Crame, J. A., Eds. (2000b). *The Evolutionary Biology of the Bivalvia*. The Geological Society Special Publication. London, Geological Society of London.

Harper, E. M. & Morton, B. (2004). Tube construction in the watering pot shell *Brechites vaginiferus* (Bivalvia: Anomalodesmata: Clavagelloidea). *Acta Zoologica* 85: 149–161.

Harper, E. M., Dreyer, H. & Steiner, G. (2006). Reconstructing the Anomalodesmata (Mollusca: Bivalvia): Morphology and molecules. *Zoological Journal of the Linnean Society* 148: 395–420.

Harper, E. M. (2012). Cementing bivalves. *Treatise Online* Part N, Revised, Volume 1, Chapter 21: 1–12.

Harper, J. A. & Rollins, H. B. (1982). Recognition of Monoplacophora and Gastropoda in the fossil record: A functional morphological look at the bellerophont controversy. *Proceedings of the Third North American Paleontological Convention* 1: 227–232.

Harper, J. A. & Rollins, H. B. (2000). The bellerophont controversy revisited. *American Malacological Bulletin* 15: 147–156.

Harrington, M. J., Jehle, F. & Priemel, T. (2018). Mussel byssus structure-function and fabrication as inspiration for biotechnological production of advanced materials. *Biotechnology Journal* 13: e1800133.

Harrison, F. W. & Kohn, A. J., Eds. (1994). *Mollusca I. Microscopic Anatomy of Invertebrates*. Vol. 5. New York, Wiley-Liss.

Harrison, F. W. & Kohn, A. J., Eds. (1997a). *Mollusca 2A. Microscopic anatomy of invertebrates*. Vol. 6A. London, John Wiley and Sons.

Harrison, F. W. & Kohn, A. J., Eds. (1997b). *Mollusca 2B. Microscopic Anatomy of Invertebrates*. Vol. 6B. London, John Wiley and Sons.

Harry, H. W. (1964). The anatomy of *Chilina fluctuosa* Gray re-examined, with prolegomena on the phylogeny of the higher lymnic Basommatophora (Gastropoda: Pulmonata). *Malacologia* 1: 355–385.

Harte, M. E. (1992). A new approach to the study of bivalve evolution. *American Malacological Union Bulletin* 9: 199–206.

Hartmann, H., Hess, M. & Haszprunar, G. (2011). Interactive 3D anatomy and affinities of Bathysciadiidae (Gastropoda, Cocculinoidea): Deep-sea limpets feeding on decaying cephalopod beaks. *Journal of Morphology* 272: 259–279.

Hartwell, A. M., Voight, J. R. & Wheat, C. G. (2018). Clusters of deep-sea egg-brooding octopods associated with warm fluid discharge: An ill-fated fragment of a larger, discrete population? *Deep Sea Research Part I: Oceanographic Research Papers* 135: 1–8.

Harvey, A. W., Mooi, R. & Gosliner, T. M. (1999). Phylogenetic taxonomy and the status of *Allonautilus* Ward and Saunders, 1997. *Journal of Paleontology* 73: 1214–1217.

Haszprunar, G. (1985a). On the anatomy and systematic position of the Mathildidae (Mollusca, Allogastropoda). *Zoologica Scripta* 14: 201–213.

Haszprunar, G. (1985b). The Heterobranchia: A new concept of the phylogeny and evolution of the higher Gastropoda. *Zeitschrift für Zoologische Systematik und Evolutionsforschung* 23: 15–37.

Haszprunar, G. (1985c). On the anatomy and fine structure of a peculiar sense organ in *Nucula* (Bivalvia, Protobranchia). *The Veliger* 28: 52–62.

Haszprunar, G. (1985d). Zur Anatomie und systematischen Stellung der Architectonicidae (Mollusca, Allogastropoda). *Zoologica Scripta* 14: 25–43.

Haszprunar, G. (1985e). The fine morphology of the osphradial sense organs of the Mollusca. Part 2: Allogastropoda (Architectonicidae and Pyramidellidae). *Philosophical Transactions of the Royal Society B* 307: 497–505.

Haszprunar, G. (1985f). On the innervation of gastropod shell muscles. *Journal of Molluscan Studies* 51: 309–314.

Haszprunar, G. (1985g). The fine morphology of the osphradial sense organs of the Mollusca. Part 1: Gastropoda - Prosobranchia. *Philosophical Transactions of the Royal Society B* 307: 457–496.

Haszprunar, G. (1986). Fine morphological investigations on sensory structures of primitive Solenogastres (Mollusca). *Zoologischer Anzeiger* 217: 345–362.

Haszprunar, G. (1987a). Anatomy and affinities of cocculinid limpets (Mollusca: Archaeogastropoda). *Zoologica Scripta* 16: 305–324.

Haszprunar, G. (1987b). The fine structure of the ctenidial sense organs (bursicles) of Vetigastropoda (Zeugobranchia, Trochoidea) and their functional and phylogenetic significance. *Journal of Molluscan Studies* 53: 46–51.

Haszprunar, G. (1987c). The fine morphology of the osphradial sense organs of the Mollusca. Part 3. Placophora and Bivalvia. *Philosophical Transactions of the Royal Society B* 315: 37–61.

Haszprunar, G. (1987d). The fine morphology of the osphradial sense organs of the Mollusca. Part 4. Caudofoveata and Solenogastres. *Philosophical Transactions of the Royal Society B* 315: 63–73.

Haszprunar, G. (1988a). Anatomy and affinities of pseudococculinid limpets (Mollusca: Archaeogastropoda). *Zoologica Scripta* 17: 161–179.

Haszprunar, G. (1988b). *Sukashitrochus* sp., a scissurellid with heteropod-like locomotion (Mollusca, Archaeogastropoda). *Annalen des Naturhistorischen Museums in Wien Serie B* 90: 367–371.

Haszprunar, G. (1988c). A preliminary phylogenetic analysis of the streptoneurous gastropods, pp. 7–16 *in* W. F. Ponder, Eernisse, D. J. & Waterhouse, J. H. (eds.), *Prosobranch Phylogeny. Malacological Review Supplement*. Ann Arbor, MI, Malacological Review.

Haszprunar, G. (1988d). Anatomy and relationships of the bone-feeding limpets, *Cocculinella minutissima* (Smith) and *Osteopelta mirabilis* Marshall (Archaeogastropoda). *Journal of Molluscan Studies* 54: 1–20.

Haszprunar, G. (1988e). Comparative anatomy of cocculiniform gastropods and its bearing on archaeogastropod systematics, pp. 64–84 *in* W. F. Ponder, Eernisse, D. J. & Waterhouse, J. H. (eds.), *Prosobranch Phylogeny. Malacological Review Supplement*. Ann Arbor, MI, Malacological Review.

Haszprunar, G. (1988f). On the origin and evolution of major gastropod groups, with special reference to the Streptoneura (Mollusca). *Journal of Molluscan Studies* 54: 367–441.

Haszprunar, G. (1989a). New slit-limpets (Scissurellacea and Fissurellacea) from hydrothermal vents. Part 2: Anatomy and relationships. *Natural History Museum of Los Angeles County. Contributions in Science* 408: 1–17.

Haszprunar, G. (1989b). The anatomy of *Melanodrymia aurantiaca* Hickman, a coiled archaeogastropod from the East Pacific hydrothermal vents (Mollusca, Gastropoda). *Acta Zoologica* 70: 175–186.

Haszprunar, G. & Huber, G. (1990). On the central nervous system of Smeagolidae and Rhodopidae, two families questionably allied with the Gymnomorpha (Gastropoda: Euthyneura). *Journal of Zoology* 220: 185–199.

Haszprunar, G. (1992a). Ultrastructure of the osphradium of the Tertiary relict snail, *Campanile symbolicum* Iredale (Mollusca, Streptoneura). *Philosophical Transactions of the Royal Society B* 337: 457–469.

Haszprunar, G. (1992b). The first molluscs – small animals. *Bollettino di Zoologia* 59: 1–16.

Haszprunar, G. (1992c). On the anatomy and relationships of the Choristellidae (Archaeogastropoda: Lepetelloidea). *The Veliger* 35: 295–307.

Haszprunar, G. (1993). Sententia: The Archaeogastropoda – A clade, a grade or what else? *American Malacological Bulletin* 10: 165–177.

Haszprunar, G., von Salvini-Plawen, L. & Rieger, R. M. (1995a). Larval planktotrophy – A primitive trait in the Bilateria? *Acta Zoologica* 76: 141–154.

Haszprunar, G., Schaefer, K., Warén, A. & Hain, S. (1995b). Bacterial symbionts in the epidermis of an Antarctic neopilinid limpet (Mollusca, Monoplacophora). *Philosophical Transactions of the Royal Society B* 347: 181–185.

Haszprunar, G. (1996). The Mollusca: Coelomate turbellarians or mesenchymate annelids? pp. 1–28 *in* J. D. Taylor (ed.), *Origin and Evolutionary Radiation of the Mollusca*. Oxford, UK, Oxford University Press.

Haszprunar, G. & Schaefer, K. (1996). Anatomy and phylogenetic significance of *Micropilina arntzi* (Mollusca, Monoplacophora, Micropilinidae Fam. nov.). *Acta Zoologica* 77: 315–334.

Haszprunar, G. & Schaefer, K. (1997). Monoplacophora, pp. 415–457 *in* F. W. Harrison & Kohn, A. (eds.), *Microscopic Anatomy of Invertebrates. Mollusca 2*. Vol. 6B. New York, Wiley-Liss.

Haszprunar, G. (1998). Superorder Cocculiniformia, pp. 653–664 *in* P. L. Beesley, Ross, G. J. B. & Wells, A. (eds.), *Mollusca: The Southern Synthesis. Part B. Fauna of Australia*. Melbourne, VIC, CSIRO Publishing.

Haszprunar, G. (2000). Is the Aplacophora monophyletic? A cladistic point of view. *American Malacological Bulletin* 15: 115–130.

Haszprunar, G. & Wanninger, A. (2000). Molluscan muscle systems in development and evolution. *Journal of Zoological Systematics and Evolutionary Research* 38: 157–163.

Haszprunar, G., Friedrich, S., Wanninger, A. & Ruthensteiner, B. (2002). Fine structure and immunocytochemistry of a new chemosensory system in the *Chiton* larva (Mollusca: Polyplacophora). *Journal of Morphology* 251: 210–218.

Haszprunar, G. & Hess, M. (2005). A new *Rhodope* from the Roscoff area (Bretagne), with a review of *Rhodope* species. *Spixiana* 28: 193–197.

Haszprunar, G. (2008). Monoplacophora (Tryblidia), pp. 97–104 *in* W. F. Ponder & Lindberg, D. R. (eds.), *Phylogeny and Evolution of the Mollusca*. Berkeley, CA, University of California Press.

Haszprunar, G., Schander, C. & Halanych, K. M. (2008). Relationships of higher molluscan taxa, pp. 19–32 *in* W. F. Ponder & Lindberg, D. R. (eds.), *Phylogeny and Evolution of the Mollusca*. Berkeley, CA, University of California Press.

Haszprunar, G. & Wanninger, A. (2008). On the fine structure of the creeping larva of *Loxosomella murmanica*: Additional evidence for a clade of Kamptozoa (Entoprocta) and Mollusca. *Acta Zoologica* 89: 137–148.

Haszprunar, G., Speimann, E., Hawe, A. & Hess, M. (2011). Interactive 3D anatomy and affinities of the Hyalogyrinidae, basal Heterobranchia (Gastropoda) with a rhipidoglossate radula. *Organisms Diversity & Evolution* 11: 201–236.

Haszprunar, G., Kunze, T., Warén, A. & Hess, M. (2017). A reconsideration of epipodial and cephalic appendages in basal gastropods: Homologies, modules and evolutionary scenarios. *Journal of Molluscan Studies* 83: 363–383.

Hatfield, F. & Hochberg, E. (2002). A brief history of the Cephalopod International Advisory Council (CIAC). *Bulletin of Marine Science* 71: 17–30.

Hausdorf, B. (1995). A preliminary phylogenetic and biogeographic analysis of the Dyakiidae (Gastropoda: Stylommatophora) and a biogeographic analysis of other Sundaland taxa. *Cladistics- the International Journal of the Willi Hennig Society* 11: 359–376.

Hautmann, M. (2006). Shell mineralogical trends in epifaunal Mesozoic bivalves and their relationship to seawater chemistry and atmospheric carbon dioxide concentration. *Facies* 52: 417–433.

Hautmann, M., Ware, D. & Bucher, H. (2017). Geologically oldest oysters were epizoans on Early Triassic ammonoids. *Journal of Molluscan Studies* 83: 253–260.

Havlíček, V. & Kříž, J. (1978). Middle Cambrian *Lamellodonta simplex* Vogel: 'Bivalve' turned brachiopod *Trematobolus simplex* (Vogel). *Journal of Paleontology* 52: 972–975.

Hawe, A., Hess, M. & Haszprunar, G. (2013). 3D reconstruction of the anatomy of the ovoviviparous (?) freshwater gastropod *Borysthenia naticina* (Menke, 1845) (Ectobranchia: Valvatidae). *Journal of Molluscan Studies* 79: 191–204.

Hawe, A. & Haszprunar, G. (2014). 3D-microanatomy and histology of the hydrothermal vent gastropod *Lurifax vitreus* Warén & Bouchet, 2001 (Heterobranchia: Orbitestellidae) and comparisons with Ectobranchia. *Organisms Diversity & Evolution* 14: 43–55.

Hawe, A., Paroll, C. & Haszprunar, G. (2014). Interactive 3D-anatomical reconstruction and affinities of the hot-vent gastropod *Xylodiscula analoga* Warén & Bouchet, 2001 (Ectobranchia). *Journal of Molluscan Studies* 80: 315–325.

Hawk, H. L. & Geller, J. B. (2019). DNA entombed in archival seashells reveals low historical mitochondrial genetic diversity of endangered white abalone *Haliotis sorenseni*. *Marine and Freshwater Research* 70: 359–370.

Hayes, K. A., Cowie, R. H. & Thiengo, S. C. (2009). A global phylogeny of Apple Snails: Gondwanan origin, generic relationships, and the influence of outgroup choice (Caenogastropoda: Ampullariidae). *Biological Journal of the Linnean Society* 98: 61–76.

Haynes, A. (2005). An evaluation of members of the genera *Clithon* (Montfort, 1810) and *Neritina* (Lamarck 1816) (Gastropoda: Neritidae). *Molluscan Research* 25: 75–84.

Healy, J. M. & Jamieson, B. G. M. (1981). An ultrastructural examination of developing and mature paraspermatozoa in *Pyrazus ebeninus* (Mollusca, Gastropoda, Potamididae). *Zoomorphology* 98: 101–119.

Healy, J. M. (1982). An ultrastructural examination of developing and mature euspermatozoa in *Pyrazus ebeninus* (Mollusca, Gastropoda, Potamididae). *Zoomorphology* 100: 157–175.

Healy, J. M. (1983). Ultrastructure of euspermatozoa of cerithiacean gastropods (Prosobranchia: Mesogastropoda). *Journal of Morphology* 178: 57–76.

Healy, J. M. (1986). Euspermatozoa and paraspermatozoa of the relict cerithiacean gastropod *Campanile symbolicum* (Prosobranchia, Mesogastropoda). *Helgoländer Meeresuntersuchungen* 40: 201–218.

Healy, J. M. (1988). Sperm morphology and its systematic importance in the Gastropoda, pp. 251–266 *in* W. F. Ponder, Eernisse, D. J. & Waterhouse, J. H. (eds.), *Prosobranch Phylogeny. Malacological Review Supplement*. Ann Arbor, MI, Malacological Review.

Healy, J. M. (1989a). Spermiogenesis and spermatozoa in the relict bivalve genus *Neotrigonia*: Relevance to trigonioid relationships, particularly Unionoidea. *Marine Biology* 103: 75–85.

Healy, J. M. (1989b). Spermatozoa of the deep-sea cephalopod *Vampyroteuthis infernalis* Chun: Ultrastructure and possible phylogenetic significance. *Philosophical Transactions of the Royal Society B* 323: 589–608.

Healy, J. M. (1989c). Ultrastructure of spermiogenesis in the gastropod *Calliotropis glyptus* Watson (Prosobranchia: Trochidae) with special reference to the embedded acrosome. *Gamete Research* 24: 9–20.

Healy, J. M. (1990a). Sperm structure in the scissurellid gastropod *Sinezona* sp. (Prosobranchia, Pleurotomarioidea). *Zoologica Scripta* 19: 189–193.

Healy, J. M. (1990b). Euspermatozoa and paraspermatozoa in the trochid gastropod *Zalipais laseroni* (Trochoidea: Skeneidae). *Marine Biology* 105: 497–507.

Healy, J. M. (1993a). Transfer of the gastropod family Plesiotrochidae to the Campaniloidea based on sperm ultrastructural evidence. *Journal of Molluscan Studies* 59: 135–146.

Healy, J. M. (1993b). Comparative sperm ultrastructure and spermiogenesis in basal heterobranch gastropods (Valvatoidea, Architectonicoidea, Rissoelloidea, Omalogyroidea, Pyramidelloidea) (Mollusca). *Zoologica Scripta* 22: 263–276.

Healy, J. M. (1995). Sperm ultrastructure in the marine bivalve families Carditidae and Crassatellidae and its bearing on unification of the Crassatelloidea with the Carditoidea. *Zoologica Scripta* 24: 21–28.

Healy, J. M., Schaefer, K. & Haszprunar, G. (1995). Spermatozoa and spermatogenesis in a monoplacophoran mollusc, *Laevipilina antarctica*: Ultrastructure and comparison with other mollusca. *Marine Biology* 122: 53–65.

Healy, J. M. (1996a). Euspermatozoan ultrastructure in *Bembicium auratum* (Gastropoda): Comparison with other caenogastropods especially other Littorinidae. *Journal of Molluscan Studies* 62: 57–63.

Healy, J. M. (1996b). Molluscan sperm ultrastructure: Correlation with taxonomic units within the Gastropoda, Cephalopoda and Bivalvia, pp. 99–113 *in* J. D. Taylor (ed.), *Origin and Evolutionary Radiation of the Mollusca*. Oxford, UK, Oxford University Press.

Healy, J. M. (2000). Mollusca – relict taxa, pp. 21–79 *in* B. G. M. Jamieson (ed.), *Progress in Male Gamete Biology. Reproductive Biology of Invertebrates*. Vol. 9, B. New Delhi, India, Oxford and IBH Publishing.

Healy, J. M., Keys, J. L. & Daddow, L. Y. M. (2000). Comparative sperm ultrastructure in pteriomorphian bivalves with special reference to phylogenetic and taxonomic implications, pp. 169–190 *in* E. M. Harper, Taylor, J. D. & Crame, J. A. (eds.), *Evolutionary Biology of the Bivalvia*. London, The Geological Society (Special Publication No. 177).

Healy, J. M., Mikkelsen, P. M. & Bieler, R. (2006). Sperm ultrastructure in *Glauconome plankta* and its relevance to the affinities of the Glauconomidae (Bivalvia: Heterodonta). *Invertebrate Reproduction & Development* 49: 29–39.

Healy, J. M., Mikkelsen, P. M. & Bieler, R. (2008). Sperm ultrastructure in *Hemidonax pictus* (Hemidonacidae, Bivalvia, Mollusca): Comparison with other heterodonts, especially Cardiidae, Donacidae and Crassatelloidea. *Zoological Journal of the Linnean Society* 153: 325–347.

Heard, W. H. (1965). Comparative life histories of North American pill clams (Sphaeriidae: *Pisidium*). *Malacologia* 2: 381–411.

Heard, W. H. (1977). Reproduction of fingernail clams (Sphaeriidae: *Sphaerium* and *Musculium*). *Malacologia* 16: 421–455.

Heath, H. (1899). The development of *Ischnochiton*. *Zoologische Jahrbücher. Abteilung für Anatomie und Ontogenie der Tiere* 12: 567–656.

Heath, H. (1905a). The Morphology of a Solenogastre. *Zoologische Jahrbücher* 21: 701–734.

Heath, H. (1905b). The excretory and circulatory systems of *Cryptochiton stelleri* Midd. *Biological Bulletin* 9: 213–225.

Heath, H. (1911). The Solenogastres. *Memoirs of the Museum of Comparative Zoology* 45: 9–79.

Heath, H. (1918). Solenogastres from the Eastern coast of North America. *Museum of Comparative Zoology. Memoirs* 45: 185–263.

Heath, H. (1937). The anatomy of some protobranch mollusks. *Mémoires du Musée royal d'Histoire naturelle de Belgique* 10: 1–26.

Hebert, P. D., Cywinska, A. & Ball, S. L. (2003). Biological identifications through DNA barcodes. *Proceedings of the Royal Society B* 270: 313–321.

Hedegaard, C. (1990). *Shell structures of the Recent Archaeogastropoda*. Science Thesis, University of Århus, Denmark.

Hedegaard, C. (1996). *Molluscs – Phylogeny and biomineralization*. Ph.D. dissertation, University of Aarhus.

Hedegaard, C. (1997). Shell structures of the Recent Vetigastropoda. *Journal of Molluscan Studies* 63: 369–377.

Hedegaard, C., Lindberg, D. R. & Bandel, K. (1997). Shell microstructure of a Triassic patellogastropod limpet. *Lethaia* 30: 331–335.

Hedley, C. (1904). Additions to the marine molluscan fauna of New Zealand. *Records of the Australian Museum* 5: 86–97.

Hedley, C. (1905). Studies on Australian Mollusca. Part 9. *Proceedings of the Linnean Society of New South Wales* 30: 520–546.

Hedley, C. (1916). Studies on Australian Mollusca. Part XIII. *Proceedings of the Linnean Society of New South Wales* 41: 680–719 (+ 687 plts).

Heinberg, C. (1979). Evolutionary ecology of nine sympatric species of the pelecypod *Limopsis* in Cretaceous chalk. *Lethaia* 12: 325–340.

Heller, J. (1990). Longevity in molluscs. *Malacologia* 31: 259–295.

Heller, J. (2001). Life history strategies, pp. 413–445 *in* G. M. Barker (ed.), *The Biology of Terrestrial Molluscs*. Wallingford, UK, CABI Publishing.

Helmkampf, M., Bruchhaus, I. & Hausdorf, B. (2008). Phylogenomic analyses of lophophorates (brachiopods, phoronids and bryozoans) confirm the Lophotrochozoa concept. *Proceedings of the Royal Society B* 275: 1927–1933.

Hengsbach, R. (1996). Ammonoid pathology, pp. 581–605 *in* N. H. Landman, Tanabe, K. & Davis, R. A. (eds.), *Ammonoid Paleobiology: From Anatomy to Ecology. Topics in Geobiology*. New York, Plenum Press.

Hennig, W. (1979). *Taschenbuch der speziellen Zoologie: Wirbellose 1 (ausgenommen Gleidertiere)*. Vol. 1. Jena, Gustav Fischer.

Henry, J. Q., Okusu, A. & Martindale, M. Q. (2004). The cell lineage of the polyplacophoran, *Chaetopleura apiculata*: Variation in the spiralian program and implications for molluscan evolution. *Developmental Biology* 272: 145–160.

Herbert, D. G. (1987). Revision of the Solariellinae (Mollusca: Prosobranchia: Trochidae) in Southern Africa. *Annals of the Natal Museum* 28: 283–382.

Herring, P. J. (1988). Luminescent organs, pp. 449–489 *in* E. R. Trueman & Clarke, M. E. (eds.), *Form and Function. The Mollusca*. Vol. 11. New York, Academic Press.

Herring, P. J. (1994). Reflective systems in aquatic animals. *Comparative Biochemistry and Physiology Part A* 109: 513–546.

Herring, P. J., Dilly, P. N. & Cope, C. (2002). The photophores of the squid family Cranchiidae (Cephalopoda: Oegopsida). *Journal of Zoology* 258: 73–90.

Herringshaw, L. G., Thomas, A. T. & Smith, M. (2007). Systematics, shell structure and affinities of the Palaeozoic Problematicum *Cornulites*. *Zoological Journal of the Linnean Society* 150: 681–699.

Hershler, R. & Longley, G. (1986). Phreatic hydrobiids (Gastropoda: Prosobranchia) from the Edwards (Balcones Fault Zone) Aquifer Region, south-central Texas. *Malacologia* 27: 127–172.

Hershler, R. & Thompson, F. G. (1992). A review of the aquatic gastropod subfamily Cochliopinae (Prosobranchia: Hydrobiidae). *Malacological Review Supplement* 5: 1–140.

Hershler, R. & Ponder, W. F. (1998). A review of morphological characters of hydrobioid snails. *Smithsonian Contributions to Zoology* 600: 1–55.

Hershler, R. (1999). A systematic review of the hydrobiid snails (Gastropoda: Rissooidea) of the Great Basin, western United States, Part II: Genera *Colligyrus*, *Eremopyrgus*, *Fluminicola*, *Pristinicola* and *Tryonia*. *The Veliger* 42: 306–337.

Hershler, R. & Liu, H.-P. (2004). A molecular phylogeny of aquatic gastropods provides a new perspective on biogeographic history of the Snake River Region. *Molecular Phylogenetics and Evolution* 32: 927–937.

Hershler, R., Liu, H.-P. & Simpson, J. S. (2015). Assembly of a micro-hotspot of caenogastropod endemism in the southern Nevada desert, with a description of a new species of *Tryonia* (Truncatelloidea, Cochliopidae). *Zookeys* 492: 107–122.

Hess, M., Beck, F., Gensler, H., Kano, Y., Kiel, S. & Haszprunar, G. (2008). Microanatomy, shell structure and molecular phylogeny of *Leptogyra*, *Xyleptogyra* and *Leptogyropsis* (Gastropoda: Neomphalida: Melanodrymiidae) from sunken wood. *Journal of Molluscan Studies* 74: 383–401.

Hess, S. C. (1987). *Comparative morphology, variability and systematic applications of cephalopod spermatophores (Teuthoidea and Vampyromorpha)*. Ph.D. dissertation, University of Miami.

Hewitt, R. A. (1996). Architecture and strength of the ammonoid shell, pp. 297–339 *in* N. H. Landman, Tanabe, K. & Davis, R. A. (eds.), *Ammonoid Paleobiology: From Anatomy to Ecology. Topics in Geobiology*. New York, Plenum Press.

Hewitt, R. A. & Westermann, G. E. G. (1997). Mechanical significance of ammonoid septa with complex sutures. *Lethaia* 30: 205–212.

Hewitt, R. A. & Westermann, G. E. G. (2003). Recurrences of hypotheses about ammonites and *Argonauta*. *Journal of Paleontology* 77: 792–795.

Heywood, J. L., Chen, C., Pearce, D. A. & Linse, K. (2017). Bacterial communities associated with the Southern Ocean vent gastropod, *Gigantopelta chessoia*: Indication of horizontal symbiont transfer. *Polar Biology* 40: 2335–2342.

Hickerson, M. J., Meyer, C. P. & Moritz, C. (2006). DNA barcoding will often fail to discover new animal species over broad parameter space. *Systematic Biology* 55: 729–739.

Hickman, C. S. (1981). Implications of radular tooth-row functional integration for archaeogastropod systematics. *Malacologia* 25: 143–160.

Hickman, C. S. (1984). Form and function of the radulae of pleurotomariid gastropods. *The Veliger* 27: 29–36.

Hickman, C. S. (1985). Comparative morphology and ecology of free-living suspension-feeding gastropods from Hong Kong, pp. 217–234 *in* B. Morton & Dudgeon, D. (eds.), *Proceedings of the Second International Workshop on the Malacofauna of Hong Kong and of Southern China, Hong Kong, 6–24 April 1983*. Hong Kong, Hong Kong University Press.

Hickman, C. S. & Lindberg, D. R. (1985). Perspectives on molluscan phylogeny, pp. 13–18 *in* T. W. Broadhead. *Mollusks: Notes for a Short Course organized by D. J. Bottjer, C. S. Hickman and P. D. Ward. University of Tennessee Studies in Geology.* Vol. 13. Knoxville TN, University of Tennessee Department of Geological Sciences.

Hickman, C. S. (1988). Archaeogastropod evolution, phylogeny and systematics: A re-evaluation, pp. 17–34 *in* W. F. Ponder, Eernisse, D. J. & Waterhouse, J. H. (eds.), *Prosobranch Phylogeny. Malacological Review Supplement.* Ann Arbor, MI, Malacological Review.

Hickman, C. S. & McLean, J. H. (1990). Systematic revision and suprageneric classification of trochacean gastropods. *Science Series – Natural History Museum of Los Angeles County* 35: i–vi, 1–169.

Hickman, C. S. (1996). Phylogeny and patterns of evolutionary radiation in trochoidean gastropods, pp. 177–198 *in* J. D. Taylor (ed.), *Origin and Evolutionary Radiation of the Mollusca.* Oxford, UK, Oxford University Press.

Hickman, C. S. (1999). Larvae in invertebrate development and evolution, pp. 21–59 *in* B. K. Hall & Wake, M. H. (eds.), *The Origin and Evolution of Larval Forms.* San Diego, CA, Academic Press.

Hickman, C. S. (2003). Functional morphology and mode of life of *Isanda coronata* (Gastropoda: Trochidae) in an Australian macrotidal sandflat, pp. 69–88 *in* F. E. Wells, Walker, D. I. & Jones, D. S. (eds.), *The Marine Flora and Fauna of Dampier, Western Australia: Proceedings of the Twelfth International Marine Biological Workshop held in Dampier* 24 July–11 August 2000. Perth, WA, Western Australian Museum.

Hickman, C. S. & Porter, S. S. (2007). Nocturnal swimming, aggregation at light traps, and mass spawning of scissurellid gastropods (Mollusca: Vetigastropoda). *Invertebrate Biology* 126: 10–17.

Hidaka, H. & Kano, Y. (2014). Morphological and genetic variation between the Japanese populations of the amphidromous snail *Stenomelania crenulata* (Cerithioidea: Thiaridae). *Zoological Science* 31: 593–602.

Hikida, Y. (1996). Shell structure and its differentiation in the Veneridae (Bivalvia). *Journal of the Geological Society of Japan* 102: 847–865.

Hinz-Schallreuter, I. (1995). Muscheln (Pelecypoda) aus dem Mittelkambrium von Bornholm. *Geschiebekunde Aktuell* 11: 71–84.

Ho, P.-T., Kwan, Y.-S., Kim, B. & Won, Y.-J. (2015). Postglacial range shift and demographic expansion of the marine intertidal snail *Batillaria attramentaria. Ecology and Evolution* 5: 419–435.

Ho, P.-T., Park, E., Hong, S. G., Kim, E.-H., Kim, K., Jang, S.-J., Vrijenhoek, R. C. & Won, Y.-J. (2017). Geographical structure of endosymbiotic bacteria hosted by *Bathymodiolus* mussels at eastern Pacific hydrothermal vents. *BMC Evolutionary Biology* 17: 121.

Hoagland, K. E. (1978). Protandry and the evolution of environmentally mediated sex change: A study of the Mollusca. *Malacologia* 17: 365–391.

Hoagland, K. E. & Turner, R. D. (1981). Evolution and adaptive radiation of wood-boring bivalves (Pholadacea). *Malacologia* 21: 111–148.

Hoare, R. D. & Mapes, R. H. (1995). Relationships of the Devonian *Strobilepis* and related Pennsylvanian problematica. *Acta Palaeontologica Polonica* 40: 111–128.

Hoare, R. D. (2000). Considerations on Paleozoic Polyplacophora including the description of *Plasiochiton curiosus* n. gen. and sp. *American Malacological Bulletin* 15: 131–137.

Hochachka, P. W., Mommsen, T. P., Storey, J., Storey, K. B., Johansen, K. & French, C. J. (1983). The relationship between arginine and proline metabolism in cephalopods. *Marine Biology Letters* 4: 1–21.

Hochachka, P. W. (1985). Fuels and pathways as designed systems for support of muscle work. *Journal of Experimental Biology* 115: 149–164.

Hochberg, F. G., Norman, M. D. & Finn, J. K. (2006). *Wunderpus photogenicus* n. gen. and sp., a new octopus from the shallow waters of the Indo-Malayan Archipelago (Cephalopoda: Octopodidae). *Molluscan Research* 26: 128–140.

Hodgson, A. N., Baxter, J. M., Sturrock, M. G. & Bernard, R. T. F. (1988). Comparative spermatology of 11 species of Polyplacophora (Mollusca) from the suborders Lepidopleurina, Chitonina and Acanthochitonina. *Proceedings of the Royal Society B* 235: 161–178.

Hodgson, A. N. & Bernard, R. T. F. (1988). A comparison of the structure of the spermatozoa and spermatogenesis of 16 species of patellid limpet (Mollusca: Gastropoda: Archaeogastropoda). *Journal of Morphology* 195: 205–223.

Hodgson, A. N. & Foster, G. G. (1992). Structure of the sperm of some South African archaeogastropods (Mollusca) from the superfamilies Haliotoidea, Fissurelloidea and Trochoidea. *Marine Biology* 113: 89–97.

Hodgson, A. N. & Chia, F.-S. (1993). Spermatozoon structure of some North-American prosobranchs from the families Lottiidae (Patellogastropoda) and Fissurellidae (Archaeogastropoda). *Marine Biology* 116: 97–101.

Hodgson, A. N. (1995). Spermatozoal morphology of Patellogastropoda and Vetigastropoda (Mollusca: Prosobranchia). *Mémoires du Muséum national d'Histoire naturelle* 166: 167–177.

Hodgson, A. N., Ridgway, S., Branch, G. M. & Hawkins, S. J. (1996). Spermatozoan morphology of 19 species of prosobranch limpets (Patellogastropoda) with a discussion of patellid relationships. *Philosophical Transactions of the Royal Society B* 351: 339–347.

Hodgson, A. N., Healy, J. M. & Tunnicliffe, V. (1997). Spermatogenesis and sperm structure of the hydrothermal vent prosobranch gastropod *Lepetodrilus fucensis* (Lepetodrilidae, Mollusca). *Invertebrate Reproduction & Development* 31: 87–97.

Hodgson, A. N., Eckelbarger, K. J. & Young, C. M. (1998). Sperm morphology and spermiogenesis in the methane-seep mollusc *Bathynerita naticoidea* (Gastropoda: Neritacea) from the Louisiana slope. *Invertebrate Biology* 117: 199–207.

Hodgson, A. N. & Eckelbarger, K. J. (2000). Ultrastructure of the ovary and oogenesis in six species of patellid limpets (Gastropoda: Patellogastropoda) from South Africa. *Invertebrate Biology* 119: 265–277.

Hodgson, A. N., Eckelbarger, K. J. & Young, C. M. (2009). Sperm ultrastructure and spermatogenesis in the hydrothermal vent gastropod *Rhynchopelta concentrica* (Peltospiridae). *Journal of Molluscan Studies* 75: 159–165.

Hodson, F. (1957). Marker horizons in the Namurian of Britain, Ireland, Belgium and Western Germany. *Association pour l'Étude de la Paléontologie et de la Stratigraphie Houillères* 24: 1–26.

Hoeh, W. R., Bogan, A. E. & Heard, W. H. (2001). A phylogenetic perspective on the evolution of morphological and reproductive characteristics in the Unionoida, pp. 257–280 *in* G. Bauer & Wächtler, K. (eds.), *Ecology and Evolution of the Freshwater Mussels Unionoida. Ecological Studies.* Berlin, Springer.

Hoeh, W. R., Bogan, A. E., Heard, W. H. & Chapman, E. G. (2009). Palaeoheterodont phylogeny, character evolution, diversity and phylogenetic classification: A reflection on methods of analysis. *Malacologia* 51: 307–317.

Hoffman, D. L., Homan, W. C., Swanson, J. & Weldon, P. J. (1978). Flight responses of three congeneric species of intertidal gastropods. *The Veliger* 21: 293–296.

Hoffmann, H. (1929–1930a). Amphineura und Scaphopoda. Nachträge. *H. G. Bronn's Klassen und Ordnungen des Tier-Reichs wissenschaftlich dargestellt in Wort und Bild* 3, Abteilung 1–3: 1–511.

Hoffmann, H. (1929–1930b). Aplacophora, pp. 1–134 *in* H. G. Bronn (ed.), *H. G. Bronn's Klassen und Ordnungen des Tier-Reichs Wissenschaftlich Dargestellt in Wort und Bild*. Vol. 3, Abteilung 1. Leipzig, Akademische Verlagsgesellschaft.

Hoffmann, H. (1932–1939). Opisthobranchia. Teil 1, pp. i–xi, 1–1247 *in* H. G. Bronn (ed.), *H. G. Bronn's Klassen und Ordnungen des Tier-Reichs Wissenschaftlich Dargestellt in Wort und Bild*. Vol. 3, Mollusca. Abteilung 2, Gastropoda. Buch 3. Leipzig, Akademische Verlagsgesellschaft.

Hoffmann, R. & Warnke, K. (2014). *Spirula* - das unbekannte Wesen aus der Tiefsee. *Denisia* 32: 33–46.

Hoffmann, R. (2015a). The correct taxon name, authorship, and publication date of extant ten-armed coleoids. *Paleontological Contributions* 11: 1–4.

Hoffmann, R. (2015b). Lytoceratoidea. *Treatise Online* 57: 1–34.

Hoffmann, R. & Keupp, H. (2015). Ammonoid paleopathology, pp. 877–926 *in* C. Klug, Korn, D., De Baets, K., Kruta, I. & Mapes, R. H. (eds.), *Ammonoid Paleobiology: From Anatomy to Ecology. Topics in Geobiology*. Dordrecht, the Netherlands, Springer.

Hoffmann, R., Lemanis, R., Naglik, C. & Klug, C. A. (2015). Ammonoid buoyancy, pp. 613–648 *in* C. Klug, Korn, D., De Baets, K., Kruta, I. & Mapes, R. H. (eds.), *Ammonoid Paleobiology: From Anatomy to Ecology. Topics in Geobiology*. Dordrecht, the Netherlands, Springer.

Hoffmann, R., Weinkauf, M. F. G. & Fuchs, D. (2017). Grasping the shape of belemnoid arm hooks – A quantitative approach. *Paleobiology* 43: 304–320.

Hoffmann, S. (1949). Studien über das Integument der Solenogastres, nebst Bemerkungen über die Verwandtschaft zwischen den Solenogastres und Placophoren. *Zoologiska Bidrag från Uppsala* 27: 293–427.

Holland, C. H. (2003). Some observations on bactritid cephalopods. *Bulletin of Geosciences* 78: 369–372.

Holmes, S. P., Sturgess, C. J., Cherrill, A. & Davies, M. S. (2001). Shell wiping in *Calliostoma zizyphinum*: The use of pedal mucus as a provendering agent and its contribution to daily energetic requirements. *Marine Ecology Progress Series* 212: 171–181.

Holmes, S. P., Dekker, R. & Williams, I. D. (2004). Population dynamics and genetic differentiation in the bivalve mollusc *Abra tenuis*: Aplanic dispersal. *Marine Ecology Progress Series* 268: 131–140.

Holmes, S. P., Duncan, P. F. & Schnabel, K. E. (2011). Shell wiping in *Calliostoma alertae* Marshall, 1995 (Calliostomatidae; Trochoidea). *Molluscan Research* 31: 133–135.

Holthuis, B. V. (1995). *Evolution between marine and freshwater habitats: A case study of the gastropod suborder Neritopsina*. Ph.D. dissertation, University of Washington.

Holzapfel, E. (1895). Das Obere Miteldevon (Schichten mit *Stringocephalus Burtini* und *Maeneceras terebratum*) im Rheinischen Gebirge. *Abhandlungen der Königlich Preussischen Geologischen Landesanst* 16: 1–460.

Holznagel, W. E., Colgan, D. J. & Lydeard, C. (2010). Pulmonate phylogeny based on 28S rRNA gene sequences: A framework for discussing habitat transitions and character transformation. *Molecular Phylogenetics and Evolution* 57: 1017–1025.

Horný, R. J. (1961). New genera of Bohemian Monoplacophora and patellid Gastropoda. *Věstník Ustředního ústavu geologického* 36: 299–302.

Horný, R. J. (1965). *Cyrtolites* Conrad, 1838 and its position among the Monoplacophora (Mollusca). *Acta Musei Nacionalis Pragae* 21: 57–70.

Horný, R. J. (1992). Muscle scars in *Sinuites* (Mollusca, Gastropoda) from the Lower Ordovician of Bohemia. *Časopis Národního Muzea Rada Přírodovědná* 158: 79–100.

Horný, R. J. (1996). Retractor muscle scars in *Gamadiscus* (Mollusca, Tergomya). *Bulletin of the Czech Geological Survey* 71: 245–249.

Horný, R. J. (1998). Fossilized intestinal contents in the Lower Ordovician *Cyrtodiscus nitidus* (Mollusca, Tergomya) from the Barrandian Area (Bohemia). *Vestnik Ceského geologického ústavu* 73: 211–216.

Horný, R. J. (2005). Muscle scars, systematics and mode of life of the Silurian family Drahomiridae (Mollusca, Tergomya). *Acta Musei Nacionalis Pragae, Serie B, Historia Naturalis* 61: 53–76.

Horný, R. J. (2006). *Peelipilina*, a new tergomyan mollusc from the Middle Ordovician of Bohemia (Czech Republic). *Časopis Národního Muzea Rada Přírodovědná* 175: 97–108.

Hoso, M., Asami, T. & Hori, M. (2007). Right-handed snakes: Convergent evolution of asymmetry for functional specialization. *Biology Letters* 3: 169–172.

Houbrick, R. S. (1981). Anatomy, biology and systematics of *Campanile symbolicum* with reference to adaptive radiation of the Cerithiacea (Gastropoda: Prosobranchia). *Malacologia* 21: 263–289.

Houbrick, R. S. (1988). Cerithioidean phylogeny, pp. 88–128 *in* W. F. Ponder, Eernisse, D. J. & Waterhouse, J. H. (eds.), *Prosobranch Phylogeny. Malacological Review Supplement*. Ann Arbor, MI, Malacological Review.

Houbrick, R. S. (1990). Aspects of the anatomy of *Plesiotrochus* (Plesiotrochidae, fam. n.) and its systematic position in Cerithioidea (Prosobranchia, Caenogastropoda), pp. 237–250 *in* F. E. Wells, Walker, D. I., Kirkman, H. & Lethbridge, R. (eds.), *The Marine Flora and Fauna of Albany, Western Australia: Proceedings of the Third International Marine Biological Workshop*, Quaranup, Western Australia, January 11–28, 1988. Perth, WA, Western Australian Museum.

Houki, S., Yamada, M., Honda, T. & Komaru, A. (2011). Origin and possible role of males in hermaphroditic androgenetic *Corbicula* clams. *Zoological Science* 28: 526–531.

House, M. R. (1988). Major features of cephalopod evolution, pp. 1–16 *in* J. Wiedmann & Kullmann, J. (eds.), *Cephalopods Present and Past. 2nd International Cephalopod Symposium*: O. H. Schindewolf Symposium, Tübingen 1985. Stuttgart, E. Schweizerbart'sche Verlagsbuchhandlung.

Houston, R. S. (1976). The structure and function of neogastropod reproductive systems: With special reference to *Columbella fuscata* Sowerby, 1832. *The Veliger* 19: 27–46.

Houston, R. S. (1985). Genital ducts of the Cerithiacea (Gastropoda: Mesogastropoda) from the Gulf of California. *Journal of Molluscan Studies* 51: 183–189.

Houston, R. S. (1990). Reproductive systems of neritimorph archaeogastropods from the eastern Pacific, with special reference to *Nerita funiculata* Menke, 1851. *The Veliger* 33: 103–110.

Hoving, H. J. T., Roeleveld, M. A. C., Lipiński, M. R. & Melo, Y. (2004). Reproductive system of the giant squid *Architeuthis* in South African waters. *Journal of Zoology* 264: 153–169.

Hoving, H. J. T., Laptikhovsky, V. V., Piatkowski, U. & Önsoy, B. (2008). Reproduction in *Heteroteuthis dispar* (Rüppell, 1844) (Mollusca: Cephalopoda): A sepiolid reproductive adaptation to an oceanic lifestyle. *Marine Biology* 154: 219–230.

Hoving, H. J. T. & Robison, B. H. (2012). Vampire squid: Detritivores in the oxygen minimum zone. *Proceedings of the Royal Society B* 279: 4559–4567.

Hoving, H. J. T., Perez, J. A. A., Bolstad, K. S. R., Braid, H. E., Evans, A. B. E., Fuchs, D., Judkins, H., Kelly, J. T., Marian, J. E. A. R., Nakajima, R., Piatkowski, U., Reid, A., Vecchione, M. & Xavier, J. (2014). The study of deep-sea cephalopods. *Advances in Marine Biology* 67: 236–359.

Hoving, H. J. T., Laptikhovsky, V. V. & Robison, B. H. (2015). Vampire squid reproductive strategy is unique among coleoid cephalopods. *Current Biology* 25: R322–R323.

Howarth, M. K. (2013). Psiloceratoidea, Eodoceratoidea, Hildoceratoidea. *Treatise Online* 57: 1–139.

Howarth, M. K. (2017). Systematic descriptions of the Stephanoceratoidea and Spiroceratoidea. *Treatise Online* 57: 1–101.

Hoyal Cuthill, J. F. & Han, J. (2018). Cambrian petalonamid *Stromatoveris* phylogenetically links Ediacaran biota to later animals. *Palaeontology* 61: 813–823.

Hryniewicz, K., Jakubowicz, M., Belka, Z., Dopieralska, J. & Kaim, A. (2017). New bivalves from a Middle Devonian methane seep in Morocco: The oldest record of repetitive shell morphologies among some seep bivalve molluscs. *Journal of Systematic Palaeontology* 15: 19–41.

Hsiang, A. Y., Elder, L. E. & Hull, P. M. (2016). Towards a morphological metric of assemblage dynamics in the fossil record: A test case using planktonic foraminifera. *Philosophical Transactions of the Royal Society B* 371: 20150227.

Hsiang, A. Y., Nelson, K., Elder, L. E., Sibert, E. C., Kahanamoku, S. S., Burke, J. E., Kelly, A., Liu, Y. & Hull, P. M. (2018). AutoMorph: Accelerating morphometrics with automated 2D and 3D image processing and shape extraction. *Methods in Ecology and Evolution* 9: 605–612.

Huang, Q. (1995). *Polymorphism in twelve species of Neritidae: (Mollusca: Gastropoda: Prosobranchia) from Hong Kong*. Ph.D. dissertation, University of Hong Kong.

Hubendick, B. (1945). Phylogenie und Tiergeographie der Siphonariidae. Zur Kenntnis der Ordnung Basommatophora und des Ursprungs der Pulmonatengruppe. *Zoologiska Bidrag från Uppsala* 24: 1–216.

Hubendick, B. (1947). Phylogenetic relations between the higher limnic Basommatophora. *Zoologiska Bidrag från Uppsala* 25: 141–164.

Hubendick, B. (1962). Studies on *Acroloxus* (Mollusca: Basommatophora). *Göteborg Kungliga Vetenskaps och Vitterhets Samhälles Handlingar* B9: 1–68.

Hubendick, B. (1978). Systematics and comparative morphology of the Basommatophora, pp. 1–47 *in* V. Fretter & Peake, J. (ed.), *Pulmonates: Systematics, Evolution and Ecology*. Vol. 2A. London, Academic Press.

Huber, G. (1993). On the cerebral nervous system of marine Heterobranchia (Gastropoda). *Journal of Molluscan Studies* 59: 381–420.

Huber, M. (2010). *Compendium of Bivalves. A Full-Color Guide to 3,300 of the World's Marine Bivalves. A Status on Bivalvia after 250 years of Research*. Hackenheim, Germany, ConchBooks.

Huber, M., Langleit, A. & Kreipl, K. (2015). *Compendium of Bivalves 2. A Full-Color Guide to the Remaining Seven Families. A Systematic Listing of 8,500 Bivalve Species and 10,500 Synonyms*. Vol. 2. Hackenheim, Germany, ConchBooks.

Huffard, C. L. & Caldwell, R. L. (2002). Inking in a blue-ringed octopus, *Hapalochlaena lunulata*, with a vestigial ink sac. *Pacific Science* 56: 255–257.

Huffard, C. L. & Hochberg, F. G. (2005). Description of a new species of the genus *Amphioctopus* (Mollusca: Octopodidae) from the Hawai'ian Islands. *Molluscan Research* 25: 113–128.

Hughes, H. P. I. (1970). A light and electron microscope study of some opisthobranch eyes. *Zeitschrift für Zellforschung und mikroskopische Anatomie* 106: 79–98.

Hughes, R. N. & Hughes, H. P. (1971). A study of the gastropod *Cassis tuberosa* (L.) preying upon sea urchins. *Journal of Experimental Marine Biology and Ecology* 7: 305–314.

Hughes, R. N. & Hughes, H. P. I. (1981). Morphological and behavioural aspects of feeding in the Cassidae (Tonnacea, Mesogastropoda). *Malacologia* 20: 385–402.

Hunt, S. & Nixon, M. (1981). A comparative study of protein composition in the chitin-protein complexes of the beak, pen, sucker disc, radula and oesophageal cuticle of cephalopods. *Comparative Biochemistry and Physiology Part B* 68: 535–546.

Hurst, A. (1967). The egg masses and veligers of thirty northeast Pacific opisthobranchs. *The Veliger* 9: 255–288.

Huxley, T. H. (1853). On the morphology of the cephalous Mollusca, as illustrated by the anatomy of certain Heteropoda and Pteropoda collected during the voyage of H. M. S. Rattlesnake in 1846–1850. *Philosophical Transactions of the Royal Society* 143: 29–66.

Huys, R., Lopez-González, P. J., Roldan, E. & Luque, A. A. (2002). Brooding in cocculiniform limpets (Gastropoda) and familial distinctiveness of the Nucellicolidae (Copepoda): Misconceptions reviewed from a chitonophilid perspective. *Biological Journal of the Linnean Society* 75: 187–217.

Hwang, D. F., Arakawa, O., Saito, T., Noguchi, T., Simidu, U., Tsukamoto, K., Shida, Y. & Hashimoto, K. (1989). Tetrodotoxin-producing bacteria from the blue-ringed octopus *Octopus maculosus*. *Marine Biology* 100: 327–332.

Hyatt, A. (1903). Pseudoceratites of the Cretaceous. *Monographs of the United States Geological Survey* 44: 351.

Hyatt, A. & Smith, J. P. (1905). The Triassic cephalopod genera of North America. *U.S. Geological Survey, Professional Paper* 40: 1–394.

Hylander, B. L. & Summers, R. G. (1977). An ultrastructural analysis of the gametes and early fertilization in 2 bivalve mollusks *Chama macerophylla* and *Spisula solidissima* with special reference to gamete binding. *Cell and Tissue Research* 184: 469–490.

Hylleberg, J. & Nateewathana, A. (1991). Morphology, internal anatomy, and biometrics of the cephalopod *Idiosepius biserialis* Voss, 1962, a new record for the Andaman Sea. *Phuket Marine Biology Centre Research Bulletin* 56: 1–9.

Hylleberg, J. (2010). Russian contributions to the international Tropical Marine Mollusc Programme (TMMP), 1999–2003. *The Bulletin of the Russian Far East Malacological Society* 14: 119–130.

Hyman, I. T. & Köhler, F. (2018). Reconciling comparative anatomy and mitochondrial phylogenetics in revising species limits in the Australian semislug *Helicarion* Férussac, 1821 (Gastropoda: Stylommatophora). *Zoological Journal of the Linnean Society* 184: 933–968.

Hyman, L. H. (1967). *Mollusca I.* Vol. 6. New York, McGraw-Hill.

Iba, Y., Mutterlose, J., Tanabe, K., Sano, S.-i., Misaki, A. & Terabe, K. (2011). Belemnite extinction and the origin of modern cephalopods 35 my prior to the Cretaceous – Paleogene event. *Geology* 39: 483–486.

Iba, Y., Sano, S.-i., Mutterlose, J. & Kondo, Y. (2012). Belemnites originated in the Triassic – A new look at an old group. *Geology* 40: 911–914.

Iba, Y., Sano, S.-i. & Mutterlose, J. (2014a). The early evolutionary history of belemnites: New data from Japan. *PLoS ONE* 9: e95632.

Iba, Y., Sano, S.-i., Rao, X., Fuchs, D., Chen, T., Weis, R. & Sha, J. (2014b). Early Jurassic belemnites from the Gondwana margin of the Southern Hemisphere – Sinemurian record from South Tibet. *Gondwana Research* 28: 882–887.

von Ihering, H. (1876a). Beiträge zur Kenntnis des Nervensystems der Amphineuren und Arthrocochliden. *Morphologisches Jahrbuch* 3: 155–178.

von Ihering, H. (1876b). Versuch eines natürlichen Systemes der Mollusken. *Deutsche Malakozoologische Gesellschaft Jahrbücher* 3: 97–147.

Ikematsu, W. & Yamane, S. (1977). Ecological studies of *Corbicula leana* Prime. III: On spawning throughout the year and self-fertilization in the gonad. *Bulletin of the Japanese Society of Scientific Fisheries* 43: 1139–1146.

Isozaki, Y. & Aljinović, D. (2009). End-Guadalupian extinction of the Permian gigantic bivalve Alatoconchidae: End of gigantism in tropical seas by cooling. *Palaeogeography, Palaeoclimatology, Palaeoecology* 284: 11–21.

Israelsson, O. (1998). The anatomy of *Pachydermia laevis* (Archaeogastropoda: 'Peltospiridae'). *Journal of Molluscan Studies* 64: 93–109.

Ivanov, A. V. (1945). A new endoparasitic molluse *Parenteroxenos dogieli* nov. gen., nov. sp. *Comptes rendus (Doklady) de l'Académie des sciences de l'URSS* 48: 450–452.

Ivanov, D. L. (1940). Class Gastropod Mollusks [in Russian], pp. 323–455 *in* V. A. Dogelya & Zenkeyicha, L. A. (eds.), *Manual of Zoology.* Vol. 2. Moscow, Academy of Science of the USSR.

Ivanov, D. L. (1990). Evolutionary morphology of Mollusca. The radula in the class Aplacophora, pp. 159–198 *in* A. A. Shileyko (ed.), *Archives of the Zoological Museum of Moscow State University.* Vol. 28. Moscow.

Ivanov, D. L. (1996). Origin of Aculifera and problems of monophyly of higher taxa in molluscs, pp. 59–65 *in* J. D. Taylor (ed.), *Origin and Evolutionary Radiation of the Mollusca.* Oxford, UK, Oxford University Press.

Ivanov, D. L. & Moskalev, L. I. (2007). *Neopilina starobogatovi,* a new monoplacophoran species from the Bering Sea, with notes on the taxonomy of the family Neopilinidae (Mollusca: Monoplacophora). *Ruthenica* 17: 1–6.

Ivantsov, A. & Fedonkin, M. A. (2001). Locomotion trails of the Vendian invertebrates preserved with the producer's body fossils, White Sea, Russia. *PaleoBios* 21: 72.

Ivantsov, A. (2009). New reconstruction of *Kimberella*, problematic Vendian metazoan. *Paleontological Journal* 43: 601–611.

Ivantsov, A. (2010). Paleontological evidence for the supposed Precambrian occurrence of mollusks. *Paleontological Journal* 44: 1552–1559.

Ivantsov, A. (2012). Paleontological data on the possibility of Precambrian existence of mollusks, pp. 153–179 *in* A. Fyodorov & Yakovlev, H. (ed.), *Mollusks: Morphology, Behavior, and Ecology.* New York, Nova Science Publishing, Inc.

Iwata, Y., Shaw, P., Fujiwara, E., Shiba, K., Kakiuchi, Y. & Hirohashi, N. (2011). Why small males have big sperm: Dimorphic squid sperm linked to alternative mating behaviours. *BMC Evolutionary Biology* 11: 236.

Jablonski, D. & Lutz, R. A. (1983). Larval ecology of marine benthic invertebrates: Paleobiological implications. *Biological Reviews* 58: 21–89.

Jablonski, D. (1996). The rudists re-examined. *Nature* 383: 669–670.

Jablonski, D. (1998). Geographic variation in the molluscan recovery from the end-Cretaceous extinction. *Science* 279: 1327–1330.

Jablonski, D., Roy, K. & Valentine, J. W. (2000). Dissecting the latitudinal diversity gradient in marine bivalves, pp. 361–365 *in* E. M. Harper, Taylor, J. D. & Crame, J. A. (eds.), *Evolutionary Biology of the Bivalvia.* London, The Geological Society (Special Publication No. 177).

Jablonski, D. & Roy, K. (2003). Geographical range and speciation in fossil and living molluscs. *Proceedings of the Royal Society B* 270: 401–406.

Jablonski, D., Roy, K., Valentine, J. W., Price, R. M. & Anderson, P. S. (2003). The impact of the pull of the recent on the history of marine diversity. *Science* 300: 1133–1135.

Jabr, F. (2010). Fact or fiction: Can a squid fly out of the water? *Scientific American.* https://www.scientificamerican.com/article/can-squid-fly/

Jackson, J. B. C., Jung, P., Coates, A. G. & Collins, L. S. (1993). Diversity and extinction of tropical American mollusks and emergence of the isthmus of Panama. *Science* 260: 1624–1626.

Jackson, J. B. C., Kirby, M. X., Berger, W. H., Bjorndal, K. A., Botsford, L. W., Bourque, B. J., Bradbury, R. H., Cooke, R., Erlandson, J., Estes, J. A., Hughes, T. P., Kidwell, S. M., Lange, C. B., Lenihan, H. S., Pandolfi, J. M., Peterson, C. H., Steneck, R. S., Tegner, M. J. & Warner, R. R. (2001). Historical overfishing and the recent collapse of coastal ecosystems. *Science* 293: 629–638.

Jackson, J. B. C. & O'Dea, A. (2013). Timing of the oceanographic and biological isolation of the Caribbean Sea from the tropical eastern Pacific Ocean. *Bulletin of Marine Science* 89: 779–800.

Jackson, R. T. (1890). Phylogeny of the Pelecypoda, the Aviculidae and their allies. *Memoirs of the Boston Society of Natural History* 4: 277–400.

Jacobs, D. K. (1992). The support of hydrostatic load in cephalopod shells: Adaptive and ontogenetic explanations of shell form and evolution from Hooke 1695 to the present, pp. 287–349 *in* M. K. Hecht, Wallace, B. & MacIntyre, R. J. (eds.), *Evolutionary Biology.* Vol. 26. New York, Plenum Press.

Jacobs, D. K. & Landman, N. H. (1993). *Nautilus* – A poor model for the function and behavior of ammonoids? *Lethaia* 26: 101–111.

Jacobs, D. K. (1996). Chambered cephalopod shells, buoyancy, structure and decoupling: History and red herrings. *PALAIOS* 11: 610–614.

Jacobs, D. K. & Chamberlain, J. A. (1996). Buoyancy and hydrodynamics in ammonoids, pp. 169–224 *in* N. H. Landman, Tanabe, K. & Davis, R. A. (eds.), *Ammonoid Paleobiology: From Anatomy to Ecology. Topics in Geobiology.* New York, Plenum Press.

Jacobsen, H. P. & Sabell, O. B. (2004). Antipredator behaviour mediated by chemical cues: The role of conspecific alarm signalling and predator labelling in the avoidance response of a marine gastropod. *Oikos* 104: 43–50.

Jacquet, S. & Brock, G. (2015). New Lower Cambrian macromolluscs from South Australia and taphonomic constraints in the fossil record. *Molluscs 2015. Coffs Harbour, NSW, Malacological Society of Australasia*: 35.

Jain, S. (2018). Genus *Parapatoceras* Spath from Kachchh and the likely ancestor of *Epistrenoceras* Bentz (Ammonoidea, Middle Jurassic). *Neues Jahrbuch für Geologie und Paläontologie-Abhandlungen* 288: 255–272.

Jakubowicz, M., Hryniewicz, K. & Belka, Z. (2017). Mass occurrence of seep-specific bivalves in the oldest-known cold seep metazoan community. *Scientific Reports* 7: 14292.

Jammes, L. (1904). *Zoologie Pratique basée sur la Dissection des Animaux les plus Répandus*. Paris, Masson.

Jansen, W., Bauer, G. & Zahner-Meike, L. (2001). Glochidial mortality in freshwater mussels, pp. 187–205 *in* G. Bauer & Wächtler, K. (eds.), *Ecology and Evolution of the Freshwater Mussels Unionoida. Ecological Studies*. Berlin, Springer.

Janssen, A. W. (1985). Evidence for the occurrence of a 'skinny' or 'minute stage' in the ontogenetical development of Miocene *Vaginella* (Gastropoda, Euthecosomata) from the North Sea and Aquitaine Basins. *Mededelingen van de Werkgroep voor Tertiaire en Kwartaire Geologie* 21: 193–204.

Jarne, P., Finot, L., Bellec, C. & Delay, B. (1992). Aphally versus euphally in self-fertile hermaphrodite snails from the species *Bulinus truncatus* (Pulmonata: Planorbidae). *American Naturalist* 139: 424–432.

Jarne, P., Vianey-Liaud, M. & Delay, B. (1993). Selfing and outcrossing in hermaphrodite freshwater gastropods (Basommatophora): Where, when and why. *Biological Journal of the Linnean Society* 49: 99–125.

Jattiot, R., Brayard, A., Fara, E. & Charbonnier, S. (2015). Gladius-bearing coleoids from the Upper Cretaceous Lebanese Lagerstätten: Diversity, morphology, and phylogenetic implications. *Journal of Paleontology* 89: 148–167.

Jebram, B. D. (1986). The ontogenetical and supposed phylogenetical fate of the parietal muscles in the Ctenostomata (Bryozoa). *Journal of Zoological Systematics and Evolutionary Research* 24: 58–82.

Jeletzky, J. A. (1966a). Comparative morphology, phylogeny, and classification of fossil Coleoidea. *University of Kansas Paleontology Contributions and Articles* 30: 1–162.

Jeletzky, J. A., Ed. (1966b). *Mollusca: Comparative Morphology, Phylogeny, and Classification of Fossil Coleoidea [Article 7]*. Lawrence, KS, University of Kansas Publications.

Jell, P. A. (1978). Mollusca, pp. 269–271. *McGraw Hill Yearbook of Science and Technology, 1976*. New York, McGraw-Hill Book Company, Inc.

Jell, P. A. (1980). Earliest known pelecypod on Earth – A new Early Cambrian genus from South Australia. *Alcheringa* 4: 233–239.

Jensen, K. R. (1981). Observations on feeding methods in some Florida ascoglossans. *Journal of Molluscan Studies* 47: 190–199.

Jensen, K. R. (1984). Defensive behavior and toxicity of ascoglossan opisthobranch *Mourgona germineae* Marcus. *Journal of Chemical Ecology* 10: 475–486.

Jensen, K. R. (1991). Comparison of alimentary systems in shelled and non-shelled Sacoglossa (Mollusca, Opisthobranchia). *Acta Zoologica* 72: 143–150.

Jensen, K. R. (1993a). Morphological adaptations and plasticity of radular teeth of the Sacoglossa (= Ascoglossa) (Mollusca, Opisthobranchia) in relation to their food plants. *Biological Journal of the Linnean Society* 48: 135–155.

Jensen, K. R. (1993b). Evolution of buccal apparatus and diet radiation in the Sacoglossa (Opisthobranchia). *Bollettino Malacologico* 29: 147–172.

Jensen, K. R. (1996a). The Diaphanidae as a possible sister group of the Sacoglossa (Gastropoda, Opisthobranchia), pp. 231–248 *in* J. D. Taylor (ed.), *Origin and Evolutionary Radiation of the Mollusca*. Oxford, UK, Oxford University Press.

Jensen, K. R. (1996b). Phylogenetic systematics and classification of the Sacoglossa (Mollusca, Gastropoda, Opisthobranchia). *Philosophical Transactions of the Royal Society B* 351: 91–122.

Jensen, K. R. (1997). Evolution of the Sacoglossa (Mollusca, Opisthobranchia) and the ecological associations with their food plants. *Evolutionary Ecology* 11: 301–335.

Jensen, K. R. (2001). Review of reproduction in the Sacoglossa (Mollusca, Opisthobranchia). *Bollettino Malacologico* 37: 81–98.

Jensen, K. R. (2011). Comparative morphology of the mantle cavity organs of shelled Sacoglossa, with a discussion of relationships with other Heterobranchia. *Thalassas* 27: 169–192.

Jespersen, Å. & Lützen, J. (2006). Reproduction and sperm structure in Galeommatidae (Bivalvia, Galeommatoidea). *Zoomorphology* 125: 157–173.

Jin, Y. G., Wang, Y., Wang, W., Shang, Q. H., Cao, C. Q. & Erwin, D. H. (2000). Pattern of marine mass extinction near the Permian-Triassic boundary in South China. *Science* 289: 432–436.

Johansson, J. (1939). Anatomische Studien über die Gastropoden-familien Rissoidae und Littorinidae. *Zoologiska Bidrag från Uppsala* 18: 289–296.

Johansson, J. (1948). Über die Geschlechtsorgane von *Aporrhais pespelicani* nebst einigen Bemerkungen über die phylogenetische Bedeutung der Cerithiacea und Architaenioglossa. *Arkiv för Zoologi* 41A: 1–13.

Johnsen, S., Balser, E. J., Fisher, E. C. & Widder, E. A. (1999). Bioluminescence in the deep-sea cirrate octopod *Stauroteuthis syrtensis* Verrill (Mollusca: Cephalopoda). *Biological Bulletin* 197: 26–39.

Johnson, C. C. (2002). The rise and fall of rudist reefs: Reefs of the dinosaur era were dominated not by corals but by odd mollusks, which died off at the end of the Cretaceous from causes yet to be discovered. *American Scientist* 90: 148–153.

Johnson, M. S. (1976). Allozymes and area effects in *Cepaea nemoralis* on western Berkshire Downs. *Heredity* 36: 105–121.

Johnson, M. S., Clarke, B. & Murray, J. (1977). Genetic variation and reproductive isolation in *Partula. Evolution* 31: 116–126.

Johnson, R. G. & Richardson, E. S. (1968). Ten-armed fossil cephalopod from the Pennsylvanian of Illinois. *Science* 159: 526–528.

Johnson, S. B., Warén, A. & Vrijenhoek, R. C. (2008). DNA barcoding of *Lepetodrilus* limpets reveals cryptic species. *Journal of Shellfish Research* 27: 43–51.

Johnston, P. A. & Collom, C. J. (1998). The bivalve heresies – Inoceramidae are Cryptodonta, not Pteriomorphia, pp. 347–360 *in* P. A. Johnston & Haggart, J. W. (eds.), *Bivalves: An Eon of Evolution*. Calgary, AB, University of Calgary Press.

Johnston, P. A. & Haggart, J. W., Eds. (1998). *Bivalves: An Eon of Evolution*. Calgary, AB, University of Calgary Press.

Johnstone, J. (1899). *Cardium. Liverpool Marine Biology Committee: Memoirs* 2: i-vii, 1–84.

Jones, A. M. & Baxter, J. M. (1985). The use of *Patella vulgata* L. in rocky shore surveillance, pp. 265–273 *in* P. G. Moore & Seed, R. (eds.), *The Ecology of Rocky Coasts*. London, Hodder & Stoughton.

Jones, B. W. & Nishiguchi, M. K. (2004). Counterillumination in the Hawaiian bobtail squid, *Euprymna scolopes* Berry (Mollusca: Cephalopoda). *Marine Biology* 144: 1151–1155.

Jones, D. S. & Jacobs, D. K. (1992). Photosymbiosis in *Clinocardium nuttalli*: Implications for test of photosymbiosis in fossil molluscs. *PALAIOS* 7: 86–95.

Jones, H. D., Holmes, S. P., Sturgess, C. J., Cherrill, A. & Davies, M. S. (2006). Shell wiping in *Calliostoma granulatum* (Born, 1778). *Journal of Conchology* 39: 1–8.

Jong, J. G. de (2014). Report on the 2nd European workshop on scaphopod molluscs (Cismar, April 28th–May 2nd 2014). *Spirula Correspondentieblad van de Nederlandse Malacologische Vereniging* 339: 120.

Jordaens, K., Dillen, L. & Backeljau, T. (2009). Shell shape and mating behaviour in pulmonate gastropods (Mollusca). *Biological Journal of the Linnean Society* 96: 306–321.

Jörger, K. M., Hess, M., Neusser, T. P. & Schrödl, M. (2009). Sex in the beach: Spermatophores, dermal insemination and 3D sperm ultrastructure of the aphallic mesopsammic *Pontohedyle milaschewitchii* (Acochlidia, Opisthobranchia, Gastropoda). *Marine Biology* 156: 1159–1170.

Jörger, K. M., Stöger, I., Kano, Y., Fukuda, H., Knebelsberger, T. & Schrödl, M. (2010). On the origin of Acochlidia and other enigmatic euthyneuran gastropods, with implications for the systematics of Heterobranchia. *BMC Evolutionary Biology* 10: 323.

Jose, J., Krishnakumar, K. & Dineshkumar, B. (2018). Squid ink and its pharmacological activities. *GSC Biological and Pharmaceutical Sciences* 2: 17–22.

Judge, J. & Haszprunar, G. (2014). The anatomy of *Lepetella sierrai* (Vetigastropoda, Lepetelloidea): Implications for reproduction, feeding, and symbiosis in lepetellid limpets. *Invertebrate Biology* 133: 324–339.

Kaas, P. & Van Belle, R. A. (1985a). *Monograph of living Chitons. Volume 2. Sub-order Ischnochitonina; Ischnochitonidae; Schizoplacinae; Callochitoninae; Lepidochitoninae*. Vol. 2. Leiden, the Netherlands, Brill/Backhuys.

Kaas, P. & Van Belle, R. A. (1985b). *(Mollusca: Polyplacophora) Order Neoloricata: Lepidopleurida*. Leiden, the Netherlands, Brill/Backhuys.

Kaas, P. & Van Belle, R. A. (1985c). *Sub-order Ischnochitonina; Ischnochitonidae; Schizoplacinae; Callochitoninae; Lepidochitoninae*. Vol. 2. Leiden, the Netherlands, Brill/ Backhuys.

Kaas, P. & Van Belle, R. A. (1994). *(Mollusca, Polyplacophora). Sub-order Ischnochitonina; Ischnochitonidae; Ischnochitoninae (concluded) Callistoplacinae; Mopaliidae. [Additions to Volumes 1–4]*. Vol. 5. Leiden, the Netherlands, Brill/Backhuys.

Kaas, P., Jones, A. M. & Gowlett-Holmes, K. L. (1998). Class Polyplacophora – Introduction, pp. 161–177 *in* P. L. Beesley, Ross, G. J. B. & Wells, A. (eds.), *Mollusca: The Southern Synthesis. Part A. Fauna of Australia*. Melbourne, VIC, CSIRO Publishing.

Kahanamoku, S. S., Hull, P. M., Lindberg, D. R., Hsiang, A. Y., Finnegan, S. (2017). Twelve thousand Recent limpets (Mollusca, Patellogastropoda) from a northeastern Pacific latitudinal gradient. *Scientific Data* 5: 170197.

Kaim, A. (2004). The evolution of conch ontogeny in Mesozoic open sea gastropods. *Palaeontologia Polonica* 62: 1–183.

Kaim, A. & Sztajner, P. (2005). The opercula of neritopsid gastropods and their phylogenetic importance. *Journal of Molluscan Studies* 71: 211–219.

Kaim, A., Jenkins, R. G. & Warén, A. (2008). Provannid and provannid-like gastropods from the Late Cretaceous cold seeps of Hokkaido (Japan) and the fossil record of the Provannidae (Gastropoda: Abyssochrysoidea). *Zoological Journal of the Linnean Society* 154: 421–436.

Kaim, A., Jenkins, R. G., Tanabe, K. & Kiel, S. (2014). Mollusks from late Mesozoic seep deposits, chiefly in California. *Zootaxa* 3861: 401–440.

Kakabadzé, M. V. & Sharikadzé, M. Z. (1993). On the mode of life of heteromorph ammonites (heterocone, ancylocone, ptychocone). *Geobios* 26: 209–215.

Kano, Y. & Kase, T. (2000). Taxonomic revision of *Pisulina* (Gastropoda: Neritopsina) from submarine caves in the tropical Indo-Pacific. *Palaeontological Research* 4: 107–129.

Kano, Y., Sasaki, T. & Ishikawa, H. (2001). *Neritilia mimotoi*, a new neritiliid species from an anchialine lake and estuaries in southwestern Japan. *Venus* 60: 129–140.

Kano, Y., Chiba, S. & Kase, T. (2002). Major adaptive radiation in neritopsine gastropods estimated from 28S rRNA sequences and fossil records. *Proceedings of the Royal Society B* 269: 2457–2465.

Kano, Y. & Kase, T. (2002). Anatomy and systematics of the submarine-cave gastropod *Pisulina* (Neritopsina: Neritiliidae): Part 4. *Journal of Molluscan Studies* 68: 365–383.

Kano, Y. (2008). Vetigastropod phylogeny and a new concept of Seguenzioidea: Independent evolution of copulatory organs in the deep-sea habitats. *Zoologica Scripta* 37: 1–21.

Kano, Y. (2009). Hitchhiking behaviour in the obligatory upstream migration of amphidromous snails. *Biology Letters* 5: 465–468.

Kano, Y., Kimura, S., Kimura, T. & Warén, A. (2012). Living Monoplacophora: Morphological conservatism or recent diversification? *Zoologica Scripta* 41: 471–488.

Kano, Y., Neusser, T. P., Fukumori, H., Jörger, K. M. & Schrödl, M. (2015). Sea-slug invasion of the land. *Biological Journal of the Linnean Society* 116: 253–259.

Kano, Y., Brenzinger, B., Nützel, A., Wilson, N. G. & Schrödl, M. (2016). Ringiculid bubble snails recovered as the sister group to sea slugs (Nudipleura). *Scientific Reports* 6: 30908.

Kantor, Y. I. & Taylor, J. D. (2000). Formation of marginal radular teeth in Conoidea (Neogastropoda) and the evolution of the hypodermic envenomation mechanism. *Journal of Zoology* 252: 251–262.

Kantor, Y. I. & Taylor, J. D. (2002). Foregut anatomy and relationships of raphitomine gastropods (Gastropoda: Conoidea: Raphitominae). *Bollettino Malacologico*. 38: 83–110.

Karlsson, A. & Haase, M. (2002). The enigmatic mating behaviour and reproduction of a simultaneous hermaphrodite, the nudibranch *Aeolidiella glauca* (Gastropoda, Opisthobranchia). *Canadian Journal of Zoology* 80: 260–270.

Karp, G. C. & Whiteley, A. H. (1973). DNA-RNA hybridization studies of gene activity during the development of the gastropod, *Acmaea scutum*. *Experimental Cell Research* 78: 236–241.

Karr, T. L. (2007). Application of proteomics to ecology and population biology. *Heredity* 100: 200–206.

Kase, T. & Maeda, H. (1980). Early Cretaceous Gastropoda from the Choshi District, Chiba Prefecture, central Japan. *Transactions and Proceedings of the Palaeontological Society of Japan* 118: 291–324.

Kase, T. (1988). Reinterpretation of *Brunonia annulata* (Yokoyama) as an Early Cretaceous carinariid mesogastropod (Mollusca). *Journal of Paleontology* 62: 766–771.

Kase, T. (1990). Research report on ecology of a living fossil of extinct naticids, *Globularia fluctuata* (Sowerby) (Gastropoda, Mollusca) in Palawan, the Philippines – II. *Journal of Geology (Chigaku Zajjhi)* 99: 398–401 (in Japanese).

Kase, T. & Hayami, I. (1992). Unique submarine cave fauna: Composition, origin and adaptation. *Journal of Molluscan Studies* 58: 446–449.

Kase, T. & Shigeta, Y. (1996). New species of Patellogastropoda (Mollusca) from the Cretaceous of Hokkaido, Japan and Sakhalin, Russia. *Journal of Paleontology* 70: 762–771.

Kase, T. & Ishikawa, M. (2003). Mystery of naticid predation history solved: Evidence from a "living fossil" species. *Geology* 31: 403–406.

Kasinathan, R. (1975). Some studies of five species of cyclophorid snails from Peninsular India. *Proceedings of the Malacological Society of London* 41: 379–394.

Katz, S., Cavanaugh, C. M. & Bright, M. (2006). Symbiosis of epi- and endocuticular bacteria with *Helicoradomenia* spp. (Mollusca, Aplacophora, Solenogastres) from deep-sea hydrothermal vents. *Marine Ecology Progress Series* 320: 89–99.

Kauffman, E. G. (1969). Form, function and evolution, pp. N129–N204 *in* R. C. Moore (ed.), *Treatise on Invertebrate Paleontology Part N*. Vol. Mollusca 6. Lawrence, KS, Geological Society of America and University of Kansas Press.

Kauffman, E. G. & Johnson, C. C. (1988). The morphological and ecological evolution of Middle and Upper Cretaceous reef-building rudistids. *PALAIOS* 3: 194–216.

Kawaguti, S. & Yamasu, T. (1961). Self-fertilization in the bivalved gastropod with special reference to the reproductive organs. *Biological Journal of Okayama University* 7: 213–224.

Kawaguti, S. (1983). The third record of association between bivalve mollusks and zooxanthellae. *Proceedings of the Japan Academy Series B* 59: 17–20.

Kay, A. E. (1960). The functional morphology of *Cypraea caput serpentis* L. and an interpretation of the relationships among the Cypraeacea. *International Revue der gesamten Hydrobiologie* 45: 175–196.

Keferstein, W. M. (1862). *Klassen und Ordnungen der Weichthiere (Malacozoa): Wissenschaftlich dargestellt in Wort und Bild*. Vol. 3. Leipzig/Heidelberg, Germany, C. F. Winter'sche Verlagshandlung.

Kellogg, J. L. (1899). Special report on the life history of the common clam, *Mya arenaria*, pp. 78–95. *29th Annual Report, Commissioners of Inland Fisheries, Rhode Island, 1898*. Providence, RI, E.L. Freeman & Sons.

Kelly, R. P., Sarkar, I. N., Eernisse, D. J. & DeSalle, R. (2007). DNA barcoding using chitons (genus *Mopalia*). *Molecular Ecology Notes* 7: 177–183.

Kelly, S. R. A. (1988). Cretaceous wood-boring bivalves from western Antarctica with a review of the Mesozoic Pholadidae. *Palaeontology* 31: 341–372.

Kennedy, W. J., Morris, N. J. & Taylor, J. D. (1970). The shell structure, mineralogy and relationships of the Chamacea (Bivalvia). *Palaeontology* 13: 379–413.

Kennedy, W. J., Landman, N. H., Cobban, W. A. & Larson, N. L. (2002). Jaws and radulae in *Rhaeboceras*, a Late Cretaceous ammonite. *Abhandlungen der Geologischen Bundesanstalt* 57: 113–132.

Kerber, M. (1988). Mikrofossilien aus unterkambrischen Gesteinen der Montagne Noire, Frankreich. *Palaeontographica Abteilung A* 5: 127–203.

Kerbl, A., Handschuh, S., Nödl, M.-T., Metscher, B., Walzl, M. & Wanninger, A. (2013). Micro-CT in cephalopod research: Investigating the internal anatomy of a sepiolid squid using a non-destructive technique with special focus on the ganglionic system. *Journal of Experimental Marine Biology and Ecology* 447: 140–148.

Kesteven, H. L. (1905). The ontogenetic stages represented by the gastropod protoconch. *Journal of Cell Science* 2: 183–187.

Keupp, H. (2007). Complete ammonoid jaw apparatuses from the Solnhofen plattenkalks: Implications for aptychi function and microphagous feeding of ammonoids. *Neues Jahrbuch für Geologie und Paläontologie-Abhandlungen* 245: 93–101.

Keupp, H., Engeser, T., Fuchs, D. & Haeckel, W. (2010). Fossile Spermatophoren von *Trachyteuthis hastiformis* (Cephalopoda, Coleoidea) aus dem Oberkimmeridgium von Painten/Bayern. *Archaeopteryx* 28: 23–30.

Keupp, H. & Fuchs, D. (2014). Different regeneration mechanisms in the rostra of aulacocerids (Coleoidea) and their phylogenetic implications, pp. 13–20 *in* F. Wiese, Reich, M. & Arp, G. (eds.), *Spongy, Slimy, Cosy & More. Göttingen Contributions to Geosciences*. Göttingen, Germany, Universitätsverlag Göttingen.

Keupp, H., Hoffmann, R., Stevens, K. & Albersdörfer, R. (2016). Key innovations in Mesozoic ammonoids: The multicuspidate radula and the calcified aptychus. *Palaeontology* 59: 775–791.

Kidwell, S. M. (2001). Preservation of species abundance in marine death assemblages. *Science* 294: 1091–1094.

Kidwell, S. M. (2002). Time-averaged molluscan death assemblages: Palimpsests of richness, snapshots of abundance. *Geology* 30: 803–806.

Kidwell, S. M. & Holland, S. M. (2002). The quality of the fossil record: Implications for evolutionary analyses. *Annual Review of Ecology and Systematics* 33: 561–588.

Kidwell, S. M. (2008). Ecological fidelity of open marine molluscan death assemblages: Effects of post-mortem transportation, shelf health, and taphonomic inertia. *Lethaia* 41: 199–217.

Kiel, S. & Bandel, K. (2002). Further Archaeogastropoda from the Campanian of Torallola, northern Spain. *Acta Geologica Polonica* 52: 239–249.

Kiel, S., Bandel, K. & del Carmen Perrilliat, M. (2002). New gastropods from the Maastrichtian of the Mexcala Formation in Guerrero, southern Mexico, part II: Archaeogastropoda, Neritimorpha and Heterostropha. *Neues Jahrbuch für Geologie und Paläontologie-Abhandlungen* 226: 319–342.

Kiel, S. (2004). Shell structures of selected gastropods from hydrothermal vents and seeps. *Malacologia* 46: 169–183.

Kiel, S. (2010). The fossil record of vent and seep mollusks, pp. 255–277. *The Vent and Seep Biota: Aspects from Microbes to Ecosystems. Topics in Geobiology*. Dordrecht, the Netherlands, Springer Netherlands.

Kielan-Jaworowska, Z. (1962). New Ordovician genera of polychaete jaw apparatuses. *Acta Palaeontologica Polonica* 7: 291–332 + plates 291–213.

Kier, W. M. (1987). The functional morphology of the tentacle musculature of *Nautilus pompilius*, pp. 257–270 *in* W. B. Saunders & Landman, N. H. (eds.), *Nautilus: The Biology and Paleobiology of a Living Fossil. Topics in Geobiology*. New York, Springer.

Kier, W. M. (1988). The arrangement and function of molluscan muscle, pp. 211–252 *in* E. R. Trueman & Clarke, M. R. (eds.), *Form and Function. The Mollusca*. Vol. 11. New York: Academic Press.

Kier, W. M. & Smith, A. M. (1990). The morphology and mechanics of *Octopus* suckers. *Biological Bulletin* 178: 126–136.

Kier, W. M. & Stella, M. P. (2007). The arrangement and function of *Octopus* arm musculature and connective tissue. *Journal of Morphology* 268: 831–843.

Kilburn, R. N. (1999). A brief history of marine malacology in South Africa. *Transactions of the Royal Society of South Africa* 54: 31–41.

Killeen, I. J. & Light, J. M. O. (2000). *Sabellaria*, a polychaete host for the gastropods *Noemiamea dolioliformis* and *Graphis albida*. *Journal of the Marine Biological Association of the United Kingdom* 80: 571–573.

Kim, K. S., Webb, J., Macey, D. J. & Cohen, D. D. (1986). Compositional changes during biomineralization of the radula of the chiton *Clavarizona hirtosa*. *Journal of Inorganic Biochemistry* 28: 337–345.

King, A. H. (1993). Mollusca: Cephalopoda (Nautiloidea). *The Fossil Record*. 2: 169–188.

King, A. H. & Evans, D. H. (2019). High-level classification of the nautiloid cephalopods: A proposal for the revision of the Treatise Part K. *Swiss Journal of Palaeontology* 138: 65-85.

Kingston, A. C. N., Chappell, D. R. & Speiser, D. I. (2018). Evidence for spatial vision in *Chiton tuberculatus*, a chiton with eyespots. *Journal of Experimental Biology* 221: jeb183632.

Kirby, M. X. & Miller, H. M. (2005). Response of a benthic suspension feeder (*Crassostrea virginica* Gmelin) to three centuries of anthropogenic eutrophication in Chesapeake Bay. *Estuarine, Coastal and Shelf Science* 62: 679–689.

Kirkendale, L. A. & Meyer, C. P. (2004). Phylogeography of the *Patelloida profunda* group (Gastropoda: Lottidae): diversification in a dispersal-driven marine system. *Molecular Ecology* 13: 2749–2762.

Kirkendale, L. A. (2009). Their day in the sun: Molecular phylogenetics and origin of photosymbiosis in the 'other' group of photosymbiotic marine bivalves (Cardiidae: Fraginae). *Biological Journal of the Linnean Society* 97: 448–465.

Kirkendale, L. A. & Paulay, G. (2017). Photosynthesis in Bivalvia. *Treatise Online* 89: 31.

Kirschvink, J. L., Kobayashi-Kirschvink, A., Diaz-Ricci, J. C. & Kirschvink, S. J. (1992). Magnetite in human tissues: A mechanism for the biological effects of weak ELF magnetic fields. *Bioelectromagnetics* 13: 101–113.

Kisailus, D. & Nemoto, M. (2018). Structural and proteomic analyses of iron oxide biomineralization in chiton teeth, pp. 53–73 *in* T. Matsunaga, Tanaka, T. & Kisailus, D. (eds.), *Biological Magnetic Materials and Applications*. Singapore, Springer.

Klages, N. T. (1996). Cephalopods as prey. II. Seals. *Philosophical Transactions of the Royal Society B* 351: 1045–1052.

Kleemann, K. H. (1973). Der Gesteinsabbau durch Ätzmuscheln an Kalkküsten. *Oecologia* 13: 377–395.

Kleemann, K. H. (1990a). Evolution of chemically-boring Mytilidae (Bivalvia), pp. 111–124 *in* B. Morton (ed.), *The Bivalvia (Proceedings of a memorial symposium in honour of Sir Charles Maurice Yonge (1899–1986), Edinburgh, 1986.)*. Hong Kong, Hong Kong University Press.

Kleemann, K. H. (1990b). Coral associations, biocorrosion, and space competition in *Pedum spondyloideum* (Gmelin) (Pectinacea, Bivalvia). *Marine Ecology* 11: 77–94.

Kleemann, K. H. (1996). Biocorrosion by bivalves. *Marine Ecology* 17: 145–158.

Kluessendorf, J. & Doyle, P. (2000). *Pohlsepia mazonensis*, an early 'Octopus' from the Carboniferous of Illinois, USA. *Palaeontology* 43: 919–926.

Klug, C. A. (2001). Functional morphology and taphonomy of nautiloid beaks from the Middle Triassic of southern Germany. *Acta Palaeontologica Polonica* 46: 43–68.

Klug, C. A. & Korn, D. (2004). The origin of ammonoid locomotion. *Acta Palaeontologica Polonica* 49: 235–242.

Klug, C. A., Montenari, M., Schulz, H. & Urlichs, M. (2007). Soft-tissue attachment of Middle Triassic Ceratitida from Germany, pp. 205–220 *in* N. H. Landman, Davis, R. A. & Mapes, R. H. (eds.), *Cephalopods Present and Past. New Insights and Fresh Perspectives. 6th International Symposium*, University of Arkansas, Fayetteville, 16–19 September 2004. Dordrecht, the Netherlands, Springer.

Klug, C. A., Meyer, E. P., Richter, U. & Korn, D. (2008). Soft-tissue imprints in fossil and Recent cephalopod septa and septum formation. *Lethaia* 41: 477–492.

Klug, C., Kröger, B., Kiessling, W., Mullins, G. L., Servais, T., Frýda, J., Korn, D. & Turner, S. (2010). The Devonian nekton revolution. *Lethaia* 43: 465–477.

Klug, C. A., Riegraf, W. & Lehmann, J. (2012). Soft-part preservation in heteromorph ammonites from the Cenomanian-Turonian Boundary Event (OAE 2) in north-west Germany. *Palaeontology* 55: 1307–1331.

Klug, C. A. & Hoffmann, R. (2015). Ammonoid septa and sutures, pp. 45–90 *in* C. A. Klug, Korn, D., De Baets, K., Kruta, I. & Mapes, R. H. (eds.), *Ammonoid Paleobiology: From Anatomy to Ecology. Topics in Geobiology*. Dordrecht, the Netherlands, Springer.

Klug, C. & Lehmann, J. (2015). Soft part anatomy of ammonoids: Reconstructing the animal based on exceptionally preserved specimens and actualistic comparisons, pp. 507–529 *in* C. Klug, Korn, D., De Baets, K., Kruta, I. & Mapes, R. H. (eds.), *Ammonoid Paleobiology: From Anatomy to Ecology. Topics in Geobiology*. Dordrecht, the Netherlands, Springer Netherlands.

Klug, C., Kröger, B., Vinther, J., Fuchs, D. & De Baets, K. (2015a). Ancestry, origin and early evolution of ammonoids, pp. 3–24 *in* C. Klug, Korn, D., De Baets, K., Kruta, I. & Mapes, R. (eds.), *Ammonoid Paleobiology: From Macroevolution to Paleogeography. Topics in Geobiology*. Vol. 13. Dordrecht, the Netherlands, Springer.

Klug, C. A., De Baets, K., Kröger, B., Bell, M. A., Korn, D. & Payne, J. L. (2015b). Normal giants? Temporal and latitudinal shifts of Palaeozoic marine invertebrate gigantism and global change. *Lethaia* 48: 267–288.

Klug, C. A., Zatoń, M., Parent, H., Hostettler, B. & Tajika, A. (2015c). Mature modifications and sexual dimorphism, pp. 253–320 *in* C. A. Klug, Korn, D., De Baets, K., Kruta, I. & Mapes, R. H. (eds.), *Ammonoid Paleobiology: From Anatomy to Ecology. Topics in Geobiology*. Dordrecht, the Netherlands, Springer.

Klug, C., Schweigert, G., Fuchs, D., Kruta, I. & Tischlinger, H. (2016). Adaptations to squid-style high-speed swimming in Jurassic belemnitids. *Biology Letters* 12: 20150877.

Klunzinger, M. W., Thomson, G. J., Beatty, S. J., Morgan, D. L. & Lymbery, A. J. (2013). Morphological and morphometrical description of the glochidia of *Westralunio carteri* Iredale, 1934 (Bivalvia: Unionoida: Hyriidae). *Molluscan Research* 33: 104–109.

Klussmann-Kolb, A. & Brodie, G. D. (1999). Internal storage and production of symbiotic bacteria in the reproductive system of a tropical marine gastropod. *Marine Biology* 133: 443–447.

Klussmann-Kolb, A. & Wägele, H. (2001). On the fine structure of opisthobranch egg masses (Mollusca, Gastropoda). *Zoologischer Anzeiger* 240: 101–118.

Klussmann-Kolb, A. & Dinapoli, A. (2006). Systematic position of the pelagic Thecosomata and Gymnosomata within Opisthobranchia (Mollusca, Gastropoda) – Revival of the Pteropoda. *Journal of Zoological Systematics and Evolutionary Research* 44: 118–129.

Klussmann-Kolb, A., Dinapoli, A., Kuhn, K., Streit, B. & Albrecht, C. (2008). From sea to land and beyond – New insights into the evolution of euthyneuran Gastropoda (Mollusca). *BMC Evolutionary Biology* 16: 1–16.

Knauth, L. P. (2005). Temperature and salinity history of the Precambrian ocean: Implications for the course of microbial evolution. *Palaeogeography, Palaeoclimatology, Palaeoecology* 219: 53–69.

Kniffen, J. C. (1968). Acid secretion in *Cypraea*. *American Journal of Digestive Diseases* 13: 775–778.

Knight, J. B. (1947). Bellerophont muscle scars. *Journal of Paleontology* 21: 264–267.

Knight, J. B. (1952). Primitive fossil gastropods and their bearing on gastropod classification. *Smithsonian Miscellaneous Collections* 117: 1–56.

Knight, J. B. & Yochelson, E. L. (1958). A reconsideration of the relationships of the Monoplacophora and the primitive Gastropoda. *Proceedings of the Malacological Society of London* 33: 37–48.

Knight, J. B., Batten, R. L., Yochelson, E. L. & Cox, L. R. (1960a). Palaeozoic and some Mesozoic Caenogastropoda and Opisthobranchia, pp. (I)310–(I)331 *in* R. C. Moore (ed.), *Treatise on Invertebrate Paleontology Part I.* Mollusca 1. Lawrence, KS, Geological Society of America and University of Kansas Press.

Knight, J. B., Cox, L. R., Keen, A. M., Batten, R. L., Yochelson, E. L. & Robertson, R. (1960b). Systematic descriptions [Archaeogastropoda], pp. (I)169–(I)310 *in* R. C. Moore (ed.), *Treatise on Invertebrate Paleontology Part I.* Vol. Mollusca 1. Lawrence, KS: Geological Society of America and University of Kansas Press.

Knight, J. B. & Yochelson, E. L. (1960). Monoplacophora, pp. (I)77–(I)84 *in* R. C. Moore (ed.), *Treatise on Invertebrate Paleontology Part I.* Lawrence, KS: Geological Society of America and University of Kansas Press.

Knight, R. I., Morris, N. J., Todd, J. A., Howard, L. E. & Ball, A. D. (2014). Exceptional preservation of a novel gill grade in large Cretaceous inoceramids: Systematic and palaeobiological implications *Palaeontology* 57: 37–54.

Kniprath, E. (1980). Ontogenetic plate and plate field development in two chitons, *Middendorffia* and *Ischnochiton. Roux's Archives of Developmental Biology* 189: 97–106.

Knoll, A. H., Bambach, R. K., Payne, J. L., Pruss, S. & Fischer, W. W. (2007). Paleophysiology and end-Permian mass extinction. *Earth and Planetary Science Letters* 256: 295–313.

Knudsen, J. (1961). The bathyal and abyssal *Xylophaga* (Pholadidae, Bivalvia). *Galathea Report* 5: 163–209.

Knudsen, J. (1970). The systematics and biology of abyssal and hadal Bivalvia. *Galathea Report* 11: 2–241.

Knudsen, J. (1991). Observations on *Hipponix australis* (Lamarck, 1819) (Mollusca, Gastropoda, Prosobranchia) from the Albany area, Western Australia, pp. 641–660 *in* F. E. Wells, Walker, D. I., Kirkman, H. & Lethbridge, R. (eds.), *Proceedings of the Third International Marine Biological Workshop, January 1988: The Marine Flora and Fauna of Albany, Western Australia.* Vol. 2. Albany, WA, Western Australian Museum.

Knudsen, J. (1994). Further observations on the egg capsules and reproduction of some marine prosobranch molluscs from Hong Kong, pp. 283–306 *in* B. Morton (ed.), *The Malacofauna of Hong Kong and Southern China, III: Proceedings of the Third International Workshop on the Malacofauna of Hong Kong and Southern China, Hong Kong.* Hong Kong, Hong Kong University Press.

Knudsen, J. (1997). Observations on the egg capsules and reproduction of four species of Ovulidae and of *Nassarius (Zeuxis) siquijorensis* (A. Adams, 1852) (Gastropoda: Prosobranchia) from Hong Kong, pp. 361–370 *in* B. Morton (ed.), *The Marine Flora and Fauna of Hong Kong and Southern China IV: Proceedings of the Eighth International Marine Biological Workshop held in Hong Kong* 2–20 April 1995. Hong Kong, Hong Kong University Press.

Knudsen, J. (2000). Observations on egg capsules and protoconchs of some marine prosobranch Gastropoda from Hong Kong, pp. 183–201 *in* B. Morton (ed.), *The Marine Flora and Fauna of Hong Kong and Southern China V: Proceedings of the Tenth International Marine Biological Workshop held in Hong Kong* 6–26 April 1998. Hong Kong, Hong Kong University Press.

Knutson, V. L. & Gosliner, T. M. (2014). Three new species of *Gymnodoris* Stimpson, 1855 (Opisthobranchia, Nudibranchia) from the Philippines, pp. 129–143 *in* G. C. Williams & Gosliner, T. M. (eds.), *The Coral Triangle: The 2011 Hearst Philippine Biodiversity Expedition.* San Francisco, CA, California Academy of Sciences.

Kobayashi, T. (1933). Faunal study of the Wanwanian (basal Ordovician) series with special notes on the Ribeiridae and the ellesmereoceroids. *Journal of the Faculty of Science, Imperial University of Tokyo* 3: 249–328.

Kobayashi, T. (1935). On the phylogeny of the primitive nautiloids, with descriptions of *Plectronoceras liaotungense* and *Iddingsia shantungensis*, new species. *Japanese Journal of Geology and Geography* 12: 17–26.

Kobayashi, T. (1987). The ancestry of the Cephalopoda. From *Helcionella* to *Plectronoceras. Proceedings of the Japan Academy Series B* 63: 135–138.

Kochansky-Devide, V. (1978). *Tanchintongia ogulineci* new species: An aberrant Permian bivalve in Europe. *Palaeontologische Zeitschrift* 52: 213–218.

Kocot, K. M., Cannon, J. T., Todt, C., Citarella, M. R., Kohn, A. B., Meyer, A., Santos, S. R., Schander, C., Moroz, L. L., Lieb, B. & Halanych, K. M. (2011). Phylogenomics reveals deep molluscan relationships. *Nature* 477: 452–456.

Kocot, K. M. (2013). Recent advances and unanswered questions in deep molluscan phylogenetics. *American Malacological Bulletin* 31: 195–208.

Kocot, K. M., Halanych, K. M. & Krug, P. J. (2013). Phylogenomics supports Panpulmonata: Opisthobranch paraphyly and key evolutionary steps in a major radiation of gastropod molluscs. *Molecular Phylogenetics and Evolution* 69: 764–771.

Kocot, K. M., Struck, T. H., Merkel, J., Waits, D. S., Todt, C., Brannock, P. M., Weese, D. A., Cannon, J. T., Moroz, L. L., Lieb, B. & Halanych, K. M. (2017). Phylogenomics of Lophotrochozoa with consideration of systematic error. *Systematic Biology* 66: 256–282.

Koehl, M. A. R., Koseff, J. R., Crimaldi, J. P., McCay, M. G., Cooper, T., Wiley, M. B. & Moore, P. A. (2001). Lobster sniffing: Antennule design and hydrodynamic filtering of information in an odor plume. *Science* 294: 1948–1951.

Koene, J. M. & Chase, R. (1998a). The love dart of *Helix aspersa* Müller is not a gift of calcium. *Journal of Molluscan Studies* 64: 75–80.

Koene, J. M. & Chase, R. (1998b). Changes in the reproductive system of the snail *Helix aspersa* caused by mucus from the love dart. *Journal of Experimental Biology* 201: 2313–2319.

Koene, J. M. & ter Maat, A. (2001). 'Allohormones': A class of bioactive substances favoured by sexual selection. *Journal of Comparative Physiology A* 187: 323–326.

Koene, J. M. & Schulenburg, H. (2005). Shooting darts: Co-evolution and counter-adaptation in hermaphroditic snails. *BMC Evolutionary Biology* 5: 1–13.

Köhler, F. (2007). From DNA taxonomy to barcoding – How a vague idea evolved into a biosystematic tool. *Zoosystematics and Evolution* 83: 44–51.

Köhler, F. & Burg Mayer, G. (2016). Aphally in the stylommatophoran land snail *Phaedusa* (Clausiliidae: Phaedusinae) in Timor and its systematic implications. *Molluscan Research* 36: 239–246.

Kohnert, P., Neusser, T. P., Jörger, K. M. & Schrödl, M. (2011). Time for sex change! 3D-reconstruction of the copulatory system of the 'aphallic' *Hedylopsis ballantinei* (Gastropoda, Acochlidia). *Thalassas* 27: 113–119.

Koike, K. (1985). Comparative ultrastructural studies on the spermatozoa of the Prosobranchia (Mollusca: Gastropoda). *Science Report of the Faculty of Education, Gunma University* 34: 33–153.

Komaru, A., Kawagishi, T. & Konishi, K. (1998). Cytological evidence of spontaneous androgenesis in the freshwater clam *Corbicula leana* Prime. *Developmental Genes and Evolution* 208: 46–50.

Koninck, L. G. d. (1883). *Faune du calcaire carbonifère de la Belgique*. Vol. 4. Bruxelles, Belgium, Hayez.

Korn, D. (1992). Relationship between shell form, septal construction and suture line in clymeniid cephalopods (Ammonoidea; Upper Devonian). *Neues Jahrbuch für Geologie und Paläontologie-Abhandlungen* 185: 115–130.

Korn, D. & Penkert, P. (2008). Neue Ammonoideen-Funde aus den Namur-Grauwacken der Umgebung von Arnsberg (Westfalen). *Geologie und Palaeontologie in Westfalen* 70: 5–13.

Korn, D., Bockwinkel, J. & Ebbighausen, V. (2010). The ammonoids from the Argiles de Teguentour of Oued Temertasset (early Late Tournaisian; Mouydir, Algeria). *Fossil Record* 23: 35–152.

Korshunova, T., Martynov, A., Bakken, T., Evertsen, J., Fletcher, K., Mudianta, W., Saito, H., Lundin, K., Schrödl, M. & Picton, B. (2017). Polyphyly of the traditional family Flabellinidae affects a major group of Nudibranchia: Aeolidacean taxonomic reassessment with descriptions of several new families, genera, and species (Mollusca, Gastropoda). *ZooKeys* 717: 21885.

Kosnik, M. A., Alroy, J., Behrensmeyer, A. K., Fürsich, F. T., Gastaldo, R. A., Kidwell, S. M., Kowalewski, M., Plotnick, R. E., Rogers, R. R. & Wagner, P. J. (2011). Changes in shell durability of common marine taxa through the Phanerozoic: Evidence for biological rather than taphonomic drivers. *Paleobiology* 37: 303–331.

Kosuge, S. (1966). The family Triphoridae and its systematic position. *Malacologia* 4: 297–324.

Kouchinsky, A. V. (1999). Shell microstructures of the Early Cambrian *Anabarella* and *Watsonella* as new evidence on the origin of the Rostroconchia. *Lethaia* 32: 173–180.

Kouchinsky, A. V. (2000). Shell microstructures in Early Cambrian molluscs. *Acta Palaeontologica Polonica* 45: 119–150.

Kouchinsky, A. V. (2001). Mollusks, hyoliths, stenothecoids, and coeloscleritophorans, pp. 326–349 *in* A. Y. Zhuravlev & Riding, R. (eds.), *The Ecology of the Cambrian Radiation*. New York, Columbia University Press.

Kouchinsky, A. V., Bengtson, S., Runnegar, B. N., Skovsted, C. B., Steiner, M. & Vendrasco, M. J. (2012). Chronology of early Cambrian biomineralization. *Geological Magazine* 149: 221–251.

Kraemer, L. R., Swanson, C., Galloway, M. & Kraemer, R. (1986). Biological basis of behaviour in *Corbicula fluminea*: II Functional morphology of reproduction and development and review of evidence for self-fertilization. *American Malacological Bulletin Special Edition No. 2*: 193–202.

Kress, A. (1981). A scanning electron microscope of notum structures in some dorid nudibranchs (Gastropoda: Opisthobranchia). *Journal of the Marine Biological Association of the United Kingdom* 61: 177–191.

Kring, D. A. (2000). Impact events and their effect on the origin, evolution, and distribution of life. *GSA Today* 10: 1–7.

Kříž, J. (1979). Devonian Bivalvia, pp. 255–257 *in* M. R. House, Scrutton, C. T. & Bassett, M. G. (eds.), *The Devonian System. Special Papers in Palaeontology*. Vol. 23. London, Palaeontological Association.

Kříž, J. (1984). Autecology and ecogeny of Silurian Bivalvia, pp. 183–195 *in* M. G. Bassett & Lawson, J. D. (eds.), *Autecology of Silurian Organisms. Special Papers in Palaeontology*. Vol. 32. London, The Palaeontological Association.

Kříž, J. (2007). Origin, evolution and classification of the new superorder Nepiomorphia (Mollusca, Bivalvia, Lower Palaeozoic) *Palaeontology* 50: 1341–1365.

Kříž, J. (2011). Silurian *Tetinka* Barrande, 1881 (Bivalvia, Spanilidae) from Bohemia (Prague Basin) and Germany (Elbersreuth, Frankenwald). *Bulletin of Geosciences* 86: 29–48.

Kroger, B. & Mapes, R. (2007). Carboniferous actinoceratoid Nautiloidea (Cephalopoda) – A new perspective. *Journal of Paleontology* 81: 714–724.

Kröger, B. (2002). Antipredatory traits of the ammonoid shell: Indications from Jurassic ammonoids with sublethal injuries. *Paläontologische Zeitschrift* 76: 359–375.

Kröger, B. (2003). The size of the siphuncle in cephalopod evolution. *Senckenbergiana lethaea* 83: 39–52.

Kröger, B. & Mapes, R. H. (2004). Lower Carboniferous (Chesterian) embryonic orthoceratid nautiloids. *Journal of Paleontology* 78: 560–573.

Kröger, B. (2005). Adaptive evolution in Paleozoic coiled cephalopods. *Paleobiology* 31: 253–268.

Kröger, B. & Mapes, R. H. (2005). Revision of some common carboniferous genera of North American orthocerid nautiloids. *Journal of Paleontology* 79: 954–963.

Kröger, B. & Mutvei, H. (2005). Nautiloids with multiple paired muscle scars from Lower–Middle Ordovician of Baltoscandia. *Palaeontology* 48: 781–791.

Kröger, B. (2006). Early growth-stages and classification of orthoceridan cephalopods of the Darriwillian (Middle Ordovician) of Baltoscandia. *Lethaia* 39: 129–139.

Kröger, B. & Isakar, M. (2006). Revision of annulated orthoceridan cephalopods of the Baltoscandic Ordovician. *Fossil Record* 9: 139–165.

Kröger, B. (2007). Some lesser known features of the ancient cephalopod order Ellesmerocerida (Nautiloidea, Cephalopoda). *Palaeontology* 50: 565–572.

Kröger, B. & Mapes, R. H. (2007). On the origin of bactritoids (Cephalopoda). *Paläontologische Zeitschrift* 81: 316–327.

Kröger, B. (2008a). Nautiloids before and during the origin of ammonoids in a Siluro-Devonian section in the Tafilalt, Anti-Atlas, Morocco. *Special Papers in Palaeontology* 79: 1–110.

Kröger, B. (2008b). A new genus of middle Tremadocian orthoceratoids and the Early Ordovician origin of orthoceratoid cephalopods. *Acta Palaeontologica Polonica* 53: 745–749.

Kröger, B. & Zhang, Y.-B. (2009). Pulsed cephalopod diversification during the Ordovician. *Palaeogeography, Palaeoclimatology, Palaeoecology* 273: 174–183.

Kröger, B. & Evans, D. H. (2011). Review and palaeoecological analysis of the late Tremadocian–early Floian (Early Ordovician) cephalopod fauna of the Montagne Noire, France. *Fossil Record* 14: 5–34.

Kröger, B., Vinther, J. & Fuchs, D. (2011). Cephalopod origin and evolution: A congruent picture emerging from fossils, development and molecules. *BioEssays* 33: 602–613.

Kröger, B. (2013). Cambrian – Ordovician cephalopod palaeogeography and diversity, pp. 429–448 *in* D. A. T. Harper & Servais, T. (eds.), *Early Palaeozoic Biogeography and Palaeogeography*. Vol. 38. London, Geological Society of London.

Krull, H. (1934). Die Aufhebung der Chiastoneurie bei den Pulmonaten. *Zoologischer Anzeiger* 105: 173–183.

Kruta, I., Landman, N. H., Rouget, I., Cecca, F. & Tafforeau, P. (2011). The role of ammonites in the Mesozoic marine food web revealed by jaw preservation. *Science* 331: 70–72.

Kruta, I., Landman, N. H. & Cochran, J. K. (2014a). A new approach for the determination of ammonite and nautilid habitats. *PLoS ONE* 9: e87479.

Kruta, I., Landman, N. H., Mapes, R. H. & Pradel, A. (2014b). New insights into the buccal apparatus of the Goniatitina: Palaeobiological and phylogenetic implications. *Lethaia* 47: 38–48.

Kruta, I., Landman, N. H. & Tanabe, K. (2015). Ammonoid radula, pp. 485–505 in C. Klug, Korn, D., De Baets, K., Kruta, I. & Mapes, R. H. (eds.), *Ammonoid Paleobiology: From Anatomy to Ecology. Topics in Geobiology*. Dordrecht, the Netherlands/Heidelberg/New York/London, Springer.

Kruta, I., Rouget, I., Charbonnier, S., Bardin, J., Fernandez, V., Germain, D., Brayard, A. & Landman, N. (2016). *Proteroctopus ribeti* in coleoid evolution. *Palaeontology* 59: 767–773.

Krylova, E. M. & Sahling, H. (2006). Recent bivalve molluscs of the genus *Calyptogena* (Vesicomyidae). *Journal of Molluscan Studies* 72: 359–395.

Kubodera, T., Koyama, Y. & Mori, K. (2007). Observations of wild hunting behaviour and bioluminescence of a large deep-sea, eight-armed squid, *Taningia danae*. *Proceedings of the Royal Society B* 274: 1029–1034.

Kues, B. S., Mapes, R. H. & Yochelson, E. L. (2006). Nautiloid-scaphopod homeomorphy in the late Palaeozoic of the United States. *Lethaia* 39: 91–93.

Kulicki, C. (1979). The ammonite shell: Its structure, development and biological significance. *Paleontologia Polonica* 39: 97–142.

Kumar, P. S., Kumar, D. S. & Umamaheswari, S. (2015). A perspective on toxicology of *Conus* venom peptides. *Asian Pacific Journal of Tropical Medicine* 8: 337–351.

Künz, E. & Haszprunar, G. (2001). Comparative ultrastructure of gastropod cephalic tentacles: Patellogastropoda, Neritaemorphi and Vetigastropoda. *Zoologischer Anzeiger* 240: 137–165.

Kunze, T., Hess, M., Brückner, M., Beck, F. & Haszprunar, G. (2008). Skeneimorph gastropods in Neomphalina and Vetigastropoda – A preliminary report. *Zoosymposia* 1: 119–131.

Kuroda, S., Kunita, I., Tanaka, Y., Ishiguro, A., Kobayashi, R. & Nakagaki, T. (2014). Common mechanics of mode switching in locomotion of limbless and legged animals. *Journal of the Royal Society Interface* 11: 20140205.

Labandeira, C. C. (1998). Early history of arthropod and vascular plant associations 1. *Annual Review of Earth and Planetary Sciences* 26: 329–377.

Labandeira, C. C. (2002). The history of associations between plants and animals, pp. 248–261 in C. M. Herrera & Pellmyr, O. *Plant–Animal Interactions: An Evolutionary Approach*. Malden, MA, Wiley-Blackwell.

Labbé, A. (1933). La génèse des yeux dorsaux chez les Oncidiadés. *Comptes rendus de l'Académie des sciences* 114: 1002–1003.

LaBella, A. L., Dover, C. L., Van, Jollivet, D. & Cunningham, C. W. (2017). Gene flow between Atlantic and Pacific Ocean basins in three lineages of deep-sea clams (Bivalvia: Vesicomyidae: Pliocardiinae) and subsequent limited gene flow within the Atlantic. *Deep-Sea Research Part II – Topical Studies in Oceanography* 137: 307–317.

de Lacaze-Duthiers, H. (1856–1857). Histoire de l'organization et du développment du Dentale. I–II. *Annales des sciences naturelles. Zoologie* 6: 225–228, 319–385.

LaForge, N. L. & Page, L. R. (2007). Development in *Berthella californica* (Gastropoda: Opisthobranchia) with comparative observations on phylogenetically relevant larval characters among nudipleuran opisthobranchs. *Invertebrate Biology* 126: 318–334.

Laforsch, C., Imhof, H., Sigl, R., Settles, M., Hess, M. & Wanninger, A. (2012). Applications of computational 3D–modelling in organismal biology, pp. 117–142 in C. Alexandru. *Modeling and Simulation in Engineering*. Online, InTech.

Lalli, C. M. (1970). Structure and function of the buccal apparatus of *Clione limacina* (Phipps) with a review of feeding in gymnosomatous pteropods. *Journal of Experimental Marine Biology and Ecology* 4: 101–118.

Lalli, C. M. & Wells, F. E. (1978). Reproduction in the genus *Limacina* (Opisthobranchia: Thecosomata). *Journal of Zoology* 186: 95–108.

Lalli, C. M. & Gilmer, R. W. (1989). *Pelagic Snails: The Biology of Holoplanktonic Gastropod Mollusks*. Stanford, CA, Stanford University Press.

Lameere, A. (1936). Histoire de la classification des mollusques. *Mémoires du Musée royal d'Histoire naturelle de Belgique* 3: 1–12.

Lamprell, K. L. & Healy, J. M. (1998). A revision of the Scaphopoda from Australian waters (Mollusca). *Records of the Australian Museum, Supplement* 24: 1–189.

Landing, E. D. (1991). Upper Precambrian through Lower Cambrian of Cape Breton Island: Faunas, paleoenvironments, and stratigraphic revision. *Journal of Paleontology* 65: 570–595.

Landing, E. D. & Kröger, B. (2009). The oldest cephalopods from East Laurentia. *Journal of Paleontology* 83: 123–127.

Landing, E. D. & Kröger, B. (2012). Cephalopod ancestry and ecology of the hyolith 'Allatheca' degeeri s.l. in the Cambrian Evolutionary Radiation. *Palaeogeography, Palaeoclimatology, Palaeoecology* 353: 21–30.

Landman, N. H. & Cochran, J. K. (1987). Growth and longevity of *Nautilus*, pp. 401–420 in W. B. Saunders & Landman, N. H. *Nautilus: The Biology and Paleobiology of a Living Fossil. Topics in Geobiology*. New York, Springer.

Landman, N. H. (1988). Early ontogeny of Mesozoic ammonites and nautilids, pp. 215–228 in J. Wiedmann & Kullmann, J. *Cephalopods Present and Past. 2nd International Cephalopod Symposium*: O. H. Schindewolf Symposium, Tübingen 1985. Stuttgart, Germany, E. Schweizerbart'sche Verlagsbuchhandlung.

Landman, N. H. & Davis, R. A. (1988). Jaw and crop preserved in an orthoconic nautiloid cephalopod from the Bear Gulch Limestone (Mississippian, Montana). *New Mexico Bureau of Mines and Mineral Resources Memoir* 44: 103–107.

Landman, N. H. (1989). Iterative progenesis in Upper Cretaceous ammonites. *Paleobiology* 15: 95–117.

Landman, N. H., Cochran, J. K., Chamberlain, J. A. & Hirschberg, D. J. (1989). Timing of septal formation in two species of *Nautilus* based on radiometric and aquarium data. *Marine Biology* 102: 65–72.

Landman, N. H., Tanabe, K. & Shigeta, Y. (1996). Ammonoid embryonic development, pp. 343–405 in N. H. Landman, Tanabe, K. & Davis, R. A. *Ammonoid Paleobiology: From Anatomy to Ecology. Topics in Geobiology*. New York, Plenum Press.

Landman, N. H., Lane, J. A., Cobban, W. A., Jorgensen, S. D., Kennedy, W. J. & Larson, N. L. (1999). Impressions of the attachment of the soft body to the shell in late Cretaceous pachydiscid ammonites from the Western Interior of the United States. *American Museum Novitates* 3273: 1–31.

Lang, A. (1896). *Text-book of Comparative Anatomy. Part 2*. London, Macmillan.

Langer, M. R., Lipps, J. H. & Moreno, G. (1995). Predation on foraminifera by the dentaliid deep-sea scaphopod *Fissidentalium megathyris*. *Deep Sea Research Part I: Oceanographic Research Papers* 42: 849–857.

Lankester, E. R. (1883). Mollusca, pp. 632–695. Encyclopaedia Britannica. London, Encyclopaedia Britannica.

Laptikhovsky, V. V., Nikolaeva, S. & Rogov, M. (2018). Cephalopod embryonic shells as a tool to reconstruct reproductive strategies in extinct taxa. *Biological Reviews* 93: 270–283.

Laptikhovsky, V. V., Rogov, M. A., Nikolaeva, S. V. & Arkhipkin, A. I. (2013). Environmental impact on ectocochleate cephalopod reproductive strategies and the evolutionary significance of cephalopod egg size. *Bulletin of Geosciences* 88: 83–93.

Lardeux, H. (1969). *Les Tentaculites d'Europe Occidentale et d'Afrique du Nord*. Paris, Éditions du Centre national de la recherche scientifique.

Laseron, C. F. (1954). Revision of the Liotiidae of New South Wales. *Australian Zoologist* 12: 1–25.

Laseron, C. F. (1956). The families Rissoinidae and Rissoidae (Mollusca) from the Solanderian and Dampierian zoogeographical provinces. *Australian Journal of Marine and Freshwater Research* 7: 384–484.

Laseron, C. F. (1958). The family Triphoridae (Mollusca) from northern Australia; also Triphoridae from Christmas Island (Indian Ocean) *Marine and Freshwater Research* 9: 569–658.

Latyshev, N. A. (2004). A study on the feeding ecology of chitons using analysis of gut contents and fatty acid markers. *Journal of Molluscan Studies* 70: 225–230.

Lauckner, G. (1983). Diseases of Mollusca: Amphineura, pp. 963–977 in O. Kinne (ed.), *Diseases of Marine Animals. Introduction, Bivalvia to Scaphopoda*. Vol. 2. Hamburg, Germany, Biologische Anstalt Helgoland.

Laumer, C. E., Bekkouche, N., Kerbl, A., Goetz, F., Neves, R. C., Sørensen, M. V., Kristensen, R. M., Hejnol, A., Dunn, C. W. & Giribet, G. (2015). Spiralian phylogeny informs the evolution of microscopic lineages. *Current Biology* 25: 2000–2006.

Lawrence, K. A. & Watson, W. H. (2002). Swimming behavior of the nudibranch *Melibe leonina*. *Biological Bulletin* 203: 144–151.

Lawton, J. H. & May, R. M., eds. (2002). *Extinction Rates*. Oxford, UK, Oxford University Press.

Laxton, J. H. (1969). Reproduction in some New Zealand Cymatiidae (Gastropoda, Prosobranchia). *Zoological Journal of the Linnean Society* 48: 237–253.

Laxton, J. H. (1970). Shell growth in some New Zealand Cymatiidae (Gastropoda: Prosobranchia). *Journal of Experimental Marine Biology and Ecology* 4: 250–260.

Layton, K. K., Martel, A. L. & Hebert, P. D. (2014). Patterns of DNA barcode variation in Canadian marine molluscs. *PLoS ONE* 9: e95003.

Layton, K. K., Gosliner, T. M. & Wilson, N. G. (2018). Flexible colour patterns obscure identification and mimicry in Indo-Pacific *Chromodoris* nudibranchs (Gastropoda: Chromodorididae). *Molecular Phylogenetics and Evolution* 124: 27–36.

Leal, J. H. (2008). A remarkable new genus of carnivorous, sessile bivalves (Mollusca: Anomalodesmata: Poromyidae) with descriptions of two new species. *Zootaxa* 1764: 1–18.

Lebour, M. V. (1935). The echinospira larvae of Plymouth. *Proceedings of the Zoological Society of London* 1935: 163–174.

Lebour, M. V. (1937). The eggs and larvae of the British prosobranchs with special reference to those living in the plankton. *Journal of the Marine Biological Association of the United Kingdom* 22: 105–166.

Lebour, M. V. (1944). The eggs and larvae of some prosobranchs from Bermuda. *Proceedings of the Zoological Society of London* 114: 462–489.

Lee, A. P., Webb, J., Macey, D. J., Bronswijk, W., Van, Savarese, A. R. & De Witt, G. C. (1998). In situ Raman spectroscopic studies of the teeth of the chiton *Acanthopleura hirtosa*. *Journal of Biological Inorganic Chemistry* 3: 614–619.

Lee, H., Samadi, D. S., Puillandre, N., Tsai, M.-H., Dai, C.-F. & Chen, W.-J. (2016). Eight new mitogenomes for exploring the phylogeny and classification of Vetigastropoda. *Journal of Molluscan Studies* 82: 534–541.

Lee, T. & Ó Foighil, D. (2002). 6-phosphogluconate dehydrogenase (PGD) allele phylogeny is incongruent with a recent origin of polyploidization in some North American Sphaeriidae (Mollusca, Bivalvia). *Molecular Phylogenetics and Evolution* 25: 112–124.

Lee, T. & Ó Foighil, D. (2003). Phylogenetic structure of the Sphaeriinae, a global clade of freshwater bivalve molluscs, inferred from nuclear (ITS-1) and mitochondrial (16S) ribosomal gene sequences. *Zoological Journal of the Linnean Society* 137: 245–260.

Lefevre, G. & Curtis, W. C. (1910). Reproduction and parasitism in the Unionidae. *Journal of Experimental Zoology Part A* 9: 79–115.

Legg, D. A., Sutton, M. D. & Edgecombe, G. D. (2013). Arthropod fossil data increase congruence of morphological and molecular phylogenies. *Nature Communications* 4: 2485.

Lehmann, J., Klug, C. A. & Wild, F. (2015). Did ammonoids possess opercula? Reassessment of phosphatised soft tissues in *Glaphyrites* from the Carboniferous of Uruguay. *Paläontologische Zeitschrift* 89: 63–77.

Lehmann, U. (1967). Ammoniten mit Kieferapparat und Radula aus Lias-Geschieben. *Paläontologische Zeitschrift* 41: 38–45.

Lehmann, U. (1972). Aptychen als Kieferelemente der Ammoniten. *Paläontologische Zeitschrift* 46: 34–48.

Lehmann, U. & Hillmer, G. (1980). *Wirbellose Tiere der Vorzeit. Leitfaden der Systematischen Paläontologie*. Stuttgart, Ferdinand Enke Verlag.

Lehmann, U. (1981). *The Ammonites: Their Life and Their World*. Cambridge, UK, Cambridge University Press.

Lehmann, U. (1985). Zur anatomie der ammoniten: Tintenbeutel, kiemen, augen. *Paläontologische Zeitschrift* 59: 99–108.

Lehmann, U. (1988). On the dietary habits and locomotion of fossil cephalopods, pp. 633–640 in J. Wiedmann & Kullmann, J. (eds.), *Cephalopods Present and Past. 2nd International Cephalopod Symposium*: O. H. Schindewolf Symposium, Tübingen 1985. Stuttgart, Germany, E. Schweizerbart'sche Verlagsbuchhandlung.

Leise, E. M. (1984). Chiton integument: Metamorphic changes in *Mopalia mucosa* (Mollusca: Polyplacophora). *Zoomorphology* 104: 337–343.

Leise, E. M. (1988). Sensory organs in the hairy girdles of some mopaliid chitons. *American Malacological Bulletin* 6: 141–151.

Lemanis, R., Zachow, S., Fusseis, F. & Hoffmann, R. (2015). A new approach using high-resolution computed tomography to test the buoyant properties of chambered cephalopod shells. *Paleobiology* 41: 313–329.

Lemche, H. (1955). Neurosecretion and incretory glands in a tectibranch mollusc. *Experientia* 11: 320–322.

Lemche, H. (1956). *The Anatomy and Histology of Cylichna (Gastropoda; Tectibranchia)*. Vol. 16 of Skrifter udg. af Universitetets zoologiske Museum København. København, Denmark, E. Munksgaard.

Lemche, H. (1957). A new living deep-sea mollusc of the Cambrio-Devonian class Monoplacophora. *Nature* 179: 413–416.

Lemche, H. (1959). Molluscan phylogeny in the light of *Neopilina*, pp. 380–381 in H. R. Hewer & Riley, N. D. (eds.), *Proceedings of the 15th International Congress of Zoology*. London, International Congress of Zoology.

Lemche, H. & Wingstrand, K. G. (1959). The anatomy of *Neopilina galatheae* Lemche, 1957 (Mollusca, Tryblidiacea). *Galathea Report* 3: 9–71.

Lemer, S., González, V. L., Bieler, R. & Giribet, G. (2016). Cementing mussels to oysters in the pteriomorphian tree: A phylogenomic approach. *Proceedings of the Royal Society B* 283: 20160857.

Lemer, S., Bieler, R. & Giribet, G. (2019). Resolving the relationships of clams and cockles: Dense transcriptome sampling drastically improves the bivalve tree of life. *Proceedings of the Royal Society B* 286: 20182684.

Lenssen, J. (1902). Système nerveux, système circulatoire, système respiratoire et système excréteur de la *Neritina fluviatilis*. *La Cellule* 20: 289–339.

Leonard, J. L. & Lukowiak, K. (1985). Courtship, copulation, and sperm trading in the sea slug, *Navanax inermis* (Opisthobranchia: Cephalaspidea). *Canadian Journal of Zoology* 63: 2719–2729.

Leonard, J. L. & Lukowiak, K. (1986). The behavior of *Aplysia californica* Cooper (Gastropoda: Opisthobranchia). 1. Ethogram. *Behaviour* 98: 320–360.

Leonard, J. L. (1992). The 'Love-dart' in helicid snails: A gift of calcium or a firm commitment? *Journal of Theoretical Biology* 159: 513–521.

Leriche, M. (1891). Note sur le genre *Vasseuria* Munier-Chalmas. *Bulletin Société des sciences naturelles de l'Ouest de la France. Deuxieme Serie* 6: 185–187.

Lesley, J. P. (1889). *A Dictionary of the Fossils of Pennsylvania and Neighboring States Named in the Reports and Catalogues of the Survey*. Vol. 1. Harrisburg, PA, Edwin K. Meyers.

Levi-Kalisman, Y., Falini, G., Addadi, L. & Weiner, S. (2001). Structure of the nacreous organic matrix of a bivalve mollusk shell examined in the hydrated state using cryo-TEM. *Journal of Structural Biology* 135: 8–17.

Levings, S. C. & Garrity, S. D. (1983). Diel and tidal movement of two co-occurring neritid snails; differences in grazing patterns on a tropical rocky shore. *Journal of Experimental Marine Biology and Ecology* 67: 261–278.

Lewin, R. A. (1970). Toxin secretion and tail autotomy by irritated *Oxynoe panamensis* (Opisthobranchiata; Sacoglossa). *Pacific Science* 24: 356–358.

Lewy, Z. (1995). Hypothetical endosymbiontic zooxanthellae in rudists are not needed to explain their ecological niches and thick shells in comparison with hermatypic corals. *Cretaceous Research* 16: 25–37.

Lewy, Z. (1996). Octopods: Nude ammonoids that survived the Cretaceous–Tertiary boundary mass extinction. *Geology* 24: 627–630.

Lewy, Z. (2002a). The function of the ammonite fluted septal margins. *Journal of Paleontology* 76: 63–69.

Lewy, Z. (2002b). New aspects in ammonoid mode of life and their distribution. *Geobios* 35: 130–139.

Lewy, Z. (2003). Reply to Checa and to Hewitt and Westermann. *Journal of Paleontology* 77: 796–798.

Li, G., Zhang, Z., Hua, H. & Yang, H. (2014). Occurrence of the enigmatic bivalved fossil *Apistoconcha* in the lower Cambrian of southeast Shaanxi, North China Platform. *Journal of Paleontology* 88: 359–366.

Li, J., ÓFoighil, D. & Middelfart, P. (2012). The evolutionary ecology of biotic association in a megadiverse bivalve superfamily: Sponsorship required for permanent residency in sediment. *PloS ONE* 7: e42121.

Li, J., ÓFoighil, D. & Strong, E. E. (2016). Commensal associations and benthic habitats shape macroevolution of the bivalve clade Galeommatoidea. *Proceedings of the Royal Society B* 283.

Li, J., Celiz, A. D., Yang, J., Yang, Q., Wamala, I., Whyte, W., Seo, B. R., Vasilyev, N. V., Vlassak, J. J., Suo, Z. & Mooney, D. J. (2017a). Tough adhesives for diverse wet surfaces. *Science* 357: 378–381.

Li, L., Connors, M. J., Kolle, M., England, G. T., Speiser, D. I., Xiao, X., Aizenberg, J. & Ortiz, C. (2015). Multifunctionality of chiton biomineralized armor with an integrated visual system. *Science* 350: 952–956.

Li, L., Zhang, X., Yun, H. & Li, G. (2017b). Complex hierarchical microstructures of Cambrian mollusk *Pelagiella*: Insight into early biomineralization and evolution. *Scientific Reports* 7: 1935.

Lieb, B., Dimitrova, K., Kang, H. S., Braun, S., Gebauer, W., Martin, A., Hanelt, B., Saenz, S. A., Adema, C. M. & Markl, J. (2006). Red blood with blue-blood ancestry: Intriguing structure of a snail hemoglobin. *Proceedings of the National Academy of Sciences of the United States of America* 103: 12011–12016.

Lieb, B. & Todt, C. (2008). Hemocyanin in mollusks – A molecular survey and new data on hemocyanin genes in Solenogastres and Caudofoveata. *Molecular Phylogenetics and Evolution* 49: 382–385.

Liew, T.-S. & Schilthuizen, M. (2014). Association between shell morphology of micro-land snails (genus *Plectostoma*) and their predators' predatory behaviour. *PeerJ* 2: e329.

Liljedahl, L. (1984). *Janeia silurica*, a link between nuculoids and solemyoids (Bivalvia). *Palaeontology* 27: 693–698.

Liljedahl, L. (1992). The Silurian *Ilionia prisca*, oldest known deep-burrowing supension-feeding bivalve. *Journal of Paleontology* 66: 206–210.

Lind, H. (1973). The functional significance of the spermatophore and the fate of the spermatozoa in the genital tract of *Helix pomatia* (Gastropoda: Stylommatophora). *Journal of Zoology* 169: 39–64.

Lindberg, D. R. (1981a). Is there a coiled ancestor in the docoglossan phylogeny? *12th Annual Meeting of the Western Society of Malacologists*. Davis, CA. 12: 15.

Lindberg, D. R. (1981b). Acmaeidae: Gastropoda Mollusca, pp. 1–122 *in* W. L. Lee (ed.), *Invertebrates of the San Francisco Bay Estuary System*. Pacific Grove, CA, The Boxwood Press.

Lindberg, D. R. (1981c). Rhodopetalinae, a new subfamily of Acmaeidae from the boreal Pacific: Anatomy and systematics. *Malacologia* 20: 291–305.

Lindberg, D. R. & Dobberteen, R. A. (1981). Umbilical brood protection and sexual dimorphism in the boreal Pacific trochid gastropod *Margarites vorticiferus* Dall. *International Journal of Invertebrate Reproduction* 3: 347–355.

Lindberg, D. R. (1983). *Anatomy, systematics and evolution of brooding acmaeid limpets*. Ph.D. dissertation, University of California, Santa Cruz, CA.

Lindberg, D. R. & Dwyer, K. R. (1983). The topography, formation and role of the home depression of *Collisella scabra* (Gould) (Gastropoda: Acmaeidae). *The Veliger* 25: 229–234.

Lindberg, D. R. (1985). Aplacophorans, monoplacophorans, polyplacophorans, scaphopods: The lesser classes, pp. 230–247 *in* T. W. Broadhead. *Mollusks: Notes for a Short Course organized by D. J. Bottjer, C. S. Hickman and P. D. Ward. University of Tennessee Studies in Geology*. Vol. 13. Knoxville TN, University of Tennessee Department of Geological Sciences.

Lindberg, D. R. & Vermeij, G. J. (1985). *Patelloida chamorrorum*, new species: A new member of the Tethyan *Patelloida profunda* group (Gastropoda: Acmaeidae). *The Veliger* 27: 411–417.

Lindberg, D. R. & Wright, W. G. (1985). Patterns of sex change of the protandric patellacean limpet *Lottia gigantea* (Mollusca: Gastropoda). *The Veliger* 27: 261–265.

Lindberg, D. R. & Hickman, C. S. (1986). A new anomalous giant limpet from the Oregon Eocene (Mollusca: Patellida). *Journal of Paleontology* 60: 661–668.

Lindberg, D. R. (1987). Gastropod gut and radula morphology: Evolutionary implications of a micro-computer assisted study. *Annual Meeting of the Western Society of Malacologists*. Monterey, CA. 19: 26.

Lindberg, D. R. (1988a). Consilient data sets and the phylogeny of the Mollusca. *Geological Society of America, 1988 Centennial Celebration*. Denver, CO. 20: 186.

Lindberg, D. R. (1988b). Congruent data sets and the phylogeny of the Mollusca. *American Zoologist* 28: 11A.

Lindberg, D. R. (1988c). The Patellogastropoda, pp. 35–63 *in* W. F. Ponder, Eernisse, D. J. & Waterhouse, J. H. (eds.), *Prosobranch Phylogeny. Malacological Review Supplement*. Ann Arbor, MI, Malacological Review.

Lindberg, D. R. (1988d). Heterochrony in gastropods, a neontological view, pp. 197–216 *in* M. L. McKinney, Stehli, F. G. & Jones, D. S. (eds.), *Heterochrony in Evolution: A Multidisciplinary Approach. Topics in Geobiology*. New York, Plenum Publishing Corporation.

Lindberg, D. R. & Marincovich, L. N. (1988). New species of limpets (*Patelloida gradatus* sp. n.; *Niveotectura myrakeenae* sp. n.) from the Neogene of Alaska (Patellogastropoda: Mollusca). *Arctic* 41: 167–172.

Lindberg, D. R. (1990). Systematics of *Potamacmaea fluviatilis* (Blanford), a brackish water patellogastropod (Patelloidinae: Lottiidae). *Journal of Molluscan Studies* 56: 309–316.

Lindberg, D. R. & Pearse, J. S. (1990). Experimental manipulation of shell color and morphology of the limpets *Lottia asmi* (Middendorff) and *Lottia digitalis* (Rathke) (Mollusca, Patellogastropoda). *Journal of Experimental Marine Biology and Ecology* 140: 173–186.

Lindberg, D. R. & Squires, R. L. (1990). Patellogastropods (Mollusca) from the Eocene Tejon Formation of southern California. *Journal of Paleontology* 64: 578–587.

Lindberg, D. R. & Ponder, W. F. (1996). An evolutionary tree for the Mollusca: Branches or roots? pp. 67–75 *in* J. D. Taylor (ed.), *Origin and Evolutionary Radiation of the Mollusca*. Oxford, UK, Oxford University Press.

Lindberg, D. R. (1998a). William Healey Dall: A neo-Lamarckian view of molluscan evolution. *The Veliger* 41: 227–238.

Lindberg, D. R. (1998b). Order Patellogastropoda, pp. 639–652 *in* P. L. Beesley, Ross, G. J. B. & Wells, A. (eds.), *Mollusca: The Southern Synthesis. Part B. Fauna of Australia*. Melbourne, VIC, CSIRO Publishing.

Lindberg, D. R. & Ponder, W. F. (1998). A re-evaluation of the evolution of the pallial cavity of gastropod molluscs. *American Zoologist* 38: 98A.

Lindberg, D. R. & Ponder, W. F. (2001). The influence of classification on the evolutionary interpretation of structure: A re-evaluation of the evolution of the pallial cavity of gastropod molluscs. *Organisms Diversity & Evolution* 1: 273–299.

Lindberg, D. R. & Ghiselin, M. T. (2003). Fact, theory and tradition in the study of molluscan origins. *Proceedings of the California Academy of Sciences* 54: 663–686.

Lindberg, D. R. & Guralnick, R. P. (2003). Phyletic patterns of early development in gastropod molluscs. *Evolution & Development* 5: 494–507.

Lindberg, D. R. (2004). Are the living patellogastropods sister taxa? *Molluscan megadiversity: Sea, land, and freshwater, Perth, Western Australia, Sea, Land and Freshwater. World Congress of Malacology*, Perth, WA, 11–16 July 2004.

Lindberg, D. R., Ponder, W. F. & Haszprunar, G. (2004). The Mollusca: Relationships and patterns from their first half-billion years, pp. 252–278 *in* J. Cracraft & Donoghue, M. J. (eds.), *Assembling the Tree of Life*. New York, Oxford University Press.

Lindberg, D. R. & Pyenson, N. D. (2007). Things that go bump in the night: Evolutionary interactions between cephalopods and cetaceans in the Tertiary. *Lethaia* 40: 335–343.

Lindberg, D. R. (2007a). Patellogastropoda, pp. 753–761 *in* J. T. Carlton (ed.), *The Light & Smith Manual: Intertidal Invertebrates of the Central California Coast*. Vol. 18. Berkeley, CA, University of California Press.

Lindberg, D. R. (2007b). Reproduction, ecology, and evolution of the Indo-Pacific limpet *Scutellastra flexuosa*. *Bulletin of Marine Science* 81: 219–234.

Lindberg, D. R. (2008). Patellogastropoda, Neritimorpha and Cocculinoidea. The low diversity clades, pp. 271–296 *in* W. F. Ponder & Lindberg, D. R. (eds.), *Phylogeny and Evolution of the Mollusca*. Berkeley, CA, University of California Press.

Lindberg, D. R. (2009). Monoplacophorans and the origin and relationships of mollusks. *Evolution: Education and Outreach* 2: 191–203.

Lindberg, D. R. & Sigwart, J. D. (2015). What is the molluscan osphradium? A reconsideration of homology. *Zoologischer Anzeiger* 256: 14–21.

Lindgren, A. R., Giribet, G. & Nishiguchi, M. K. (2004). A combined approach to the phylogeny of Cephalopoda (Mollusca). *Cladistics* 20: 454–486.

Lindgren, A. R. & Daly, M. (2007). The impact of length-variable data and alignment criterion on the phylogeny of Decapodiformes (Mollusca: Cephalopoda). *Cladistics* 23: 464–476.

Lindgren, A. R. (2010). Molecular inference of phylogenetic relationships among Decapodiformes (Mollusca: Cephalopoda) with special focus on the squid order Oegopsida. *Molecular Phylogenetics and Evolution* 56: 77–90.

Lindgren, A. R., Pankey, M. S., Hochberg, F. G. & Oakley, T. H. (2012). A multi-gene phylogeny of Cephalopoda supports convergent morphological evolution in association with multiple habitat shifts in the marine environment. *BMC Evolutionary Biology* 12: 129.

Lindström, G. (1884). On the Silurian Gastropoda and Pteropoda of Gotland. *Kungliga Svenska vetenskaps-akademiens handlingar* 19: 1–250.

Lindström, G. (1890). The Ascoceratidae and the Lituitidae of the Upper Silurian formation of Gotland. *Kungliga Svenska vetenskaps-akademiens handlingar* 23: 1–54.

Linnaeus, C. (1758). *Systema Naturæ per Regna Tria Naturæ, Secundum Classes, Ordines, Genera, Species, cum Characteribus, Differentiis, Synonymis, Locis. Tomus I.* Holmiae (Stockholm), Laurentius Salvius.

Linsley, R. M. (1978). Locomotion rates and shell form in the Gastropoda. *Malacologia* 17: 193–206.

Linsley, R. M. & Kier, W. M. (1984). The Paragastropoda: A proposal for a new class of Paleozoic Mollusca. *Malacologia* 25: 241–254.

Linzmeier, B. J., Landman, N. H., Peters, S. E., Kozdon, R., Kitajima, K. & Valley, J. W. (2018). Ion microprobe–measured stable isotope evidence for ammonite habitat and life mode during early ontogeny. *Paleobiology* 44: 684–708.

Little, C. (1972). The evolution of kidney function in the Neritacea (Gastropoda, Prosobranchia). *Journal of Experimental Biology* 56: 249–261.

Little, C. (1990). *The Terrestrial Invasion*. Cambridge, UK, Cambridge University Press.

Liu, D.-Y. (1979). Earliest Cambrian brachiopods from southwest China. *Acta Palaeontologica Sinica* 18: 505–512.

Liu, H.-P. & Hershler, R. (2007). A test of the vicariance hypothesis of western North American freshwater biogeography. *Journal of Biogeography* 34: 534–548.

Liu, L.-L. & Wang, S.-P. (1996). Mantle autotomy of *Ficus ficus* (Gastropoda: Ficidae). *Journal of Molluscan Studies* 62: 390–392.

Liu, L.-L. & Wang, S.-P. (2002). Histology and biochemical composition of the autotomy mantle of *Ficus ficus* (Mesogastropoda: Ficidae). *Acta Zoologica* 83: 111–116.

Ljashenko, G. P. (1957). New families of Devonian Tentaculites. *Dokladi Akademii Nauk* 116: 141–144.

Llinás, R. R. (1999). *The Squid Giant Synapse: A Model for Chemical Transmission*. New York/Oxford, UK, Oxford University Press.

Lobo-da-Cunha, A., Batista, C. & Oliveira, E. (1994). The peroxisomes of the hepatopancreas in marine gastropods. *Biology of the Cell* 82: 67–74.

Lobo-da-Cunha, A. (1997). The peroxisomes of the hepatopancreas in two species of chitons. *Cell and Tissue Research* 290: 655–664.

Lodge, S. M. (1948). Algal growth in the absence of *Patella* on an experimental strip of foreshore, Port St. Mary, Isle of Man. *Proceedings of the Liverpool Biological Society* 56: 78–83.

Long, J. D. & Hay, M. E. (2006). Fishes learn aversions to a nudibranch's chemical defense. *Marine Ecology Progress Series* 307: 199–208.

Lord, J. P. (2011). Larval development, metamorphosis and early growth of the gumboot chiton *Cryptochiton stelleri* (Middendorff, 1847) (Polyplacophora: Mopaliidae) on the Oregon coast. *Journal of Molluscan Studies* 77: 182–188.

Lovén, S. (1844). *Chaetoderma*, ett nytt maskslägte n. g. *Öfversigt af Kongl. Vetenskaps-akademiens forhandlingar* 1: 116.

Lowenstam, H. A. (1978). Recovery, behaviour and evolutionary implications of live Monoplacophora. *Nature* 273: 231–232.

Lozouet, P. (2012). Position systématique de quelques gastéropodes de l'Éocène à dernier tour disjoint (Mollusca, Gastropoda, Caenogastropoda): *Delphinula conica, Omalaxis, Eoatlanta*. *Cossmanniana* 14: 57–66.

Lucas, J. M., Vaccaro, E. & Waite, J. H. (2002). A molecular, morphometric and mechanical comparison of the structural elements of byssus from *Mytilus edulis* and *Mytilus galloprovincialis*. *Journal of Experimental Biology* 205: 1807–1817.

Luchtel, D. L., Martin, A. W., Deyrup-Olsen, I. & Boer, H. H. (1997). Gastropoda: Pulmonata, pp. 459–718 *in* F. W. Harrison & Kohn, A. J. (eds.), *Microscopic Anatomy of Invertebrates. Mollusca 2*. New York, Wiley-Liss.

Luchtel, D. L. & Deyrup-Olsen, I. (2001). Body wall: Form and function, pp. 147–178 *in* G. M. Barker (ed.), *The Biology of Terrestrial Molluscs*. Wallingford, UK, CABI Publishing.

Lukeneder, A., Harzhauser, M., Müllegger, S. & Piller, W. E. (2008). Stable isotopes (δ18O and δ13C) in *Spirula spirula* shells from three major oceans indicate developmental changes paralleling depth distributions. *Marine Biology* 154: 175–182.

Lukeneder, A., Harzhauser, M., Müllegger, S. & Piller, W. E. (2010). Ontogeny and habitat change in Mesozoic cephalopods revealed by stable isotopes (δ18O, δ13C). *Earth and Planetary Science Letters* 296: 103–114.

Lummel, L. V. (1930). Untersuchungen über einige Solenogastres. *Zeitschrift für Morphologie und Okologie der Tiere* 18: 347–383.

Lundin, K. & Schander, C. (1999). Ultrastructure of gill cilia and ciliary rootlets of *Chaetoderma nitidulum* Loven 1844 (Mollusca, Chaetodermomorpha). *Acta Zoologica* 80: 185–191.

Lundin, K. & Schander, C. (2001a). Ciliary ultrastructure of polyplacophorans (Mollusca, Amphineura, Polyplacophora). *Journal of Submicroscopic Cytology and Pathology* 33: 93–98.

Lundin, K. & Schander, C. (2001b). Ciliary ultrastructure of neomeniomorphs (Mollusca: Neomeniomorpha = Solenogastres). *Invertebrate Biology* 120: 342–349.

Lundin, K. & Schander, C. (2003). Epidermal ciliary ultrastructure of adult and larval sipunculids (Sipunculida). *Acta Zoologica* 84: 113–119.

Lundin, K., Schander, C. & Todt, C. (2009). Ultrastructure of epidermal cilia and ciliary rootlets in Scaphopoda. *Journal of Molluscan Studies* 75: 69–73.

Luo, Y. J., Takeuchi, T., Koyanagi, R., Yamada, L., Kanda, M., Khalturina, M., Fujie, M., Yamasaki, S. i., Endo, K. & Satoh, N. (2015). The *Lingula* genome provides insights into brachiopod evolution and the origin of phosphate biomineralization. *Nature Communications* 6: 9301.

Lützen, J. (1968). Unisexuality in the parasitic family Entoconchidae (Gastropoda, Prosobranchia). *Malacologia* 7: 7–15.

Lützen, J. (1979). Studies on the life-history of *Enteroxenos* Bonnevie, a gastropod endoparasitic in aspidochirote holothurians. *Ophelia* 18: 1–51.

Lyashenko, G. P. (1955). New data on the systematics of tentaculitids, nowakiids and stylioinids [In Russian]. *Bulletin of the Moscow Society for the Investigation of Nature, Geology Division* 30: 94–95.

Lydeard, C., Holznagel, W. E., Glaubrecht, M. & Ponder, W. F. (2002). Molecular phylogeny of a circum-global, diverse gastropod superfamily (Cerithioidea: Mollusca: Caenogastropoda): Pushing the deepest phylogenetic limits of mitochondrial LSU rDNA sequences. *Molecular Phylogenetics and Evolution* 22: 399–406.

Lydeard, C., Cowie, R. H., Ponder, W. F., Bogan, A. E., Bouchet, P., Clark, S. A., Cummings, K. S., Frest, T. J., Gargominy, O., Herbert, D. G., Hershler, R., Perez, K. E., Roth, B., Seddon, M. B., Strong, E. E. & Thompson, F. G. (2004). The global decline of nonmarine mollusks. *BioScience* 54: 321–330.

Lyell, C. (1830). *Principles of Geology: Being an Attempt to Explain the Former Changes of the Earth's Surface, by Reference to Causes now in Operation*. Vol. 1. London, John Murray.

Lyell, K. M. (1881). *Life, Letters and Journals of Sir Charles Lyell, Bart*. Vol. 2. London, J. Murray.

MacClintock, C. (1967). Shell structure of patelloid and bellerophontoid gastropods (Mollusca). *Peabody Museum of Natural History Yale University Bulletin* 22: 1–140.

Mace, G. M., Norris, K. & Fitter, A. H. (2012). Biodiversity and ecosystem services: A multilayered relationship. *Trends in Ecology & Evolution* 27: 19–26.

Macey, D. J., Webb, J. & Brooker, L. R. (1994). The structure and synthesis of biominerals in chiton teeth. *Bulletin de l' Institut océanographique de Monaco* 14: 191–197.

Macey, D. J. & Brooker, L. R. (1996). The junction zone: Initial site of mineralization in radula teeth of the chiton *Cryptoplax striata* (Mollusca: Polyplacophora). *Journal of Morphology* 230: 33–42.

MacFarland, F. M. & MacFarland, O. H. (1966). *Studies of Opisthobranchiate Mollusks of the Pacific Coast of North America*. Vol. 6. San Francisco, CA, California Academy of Sciences.

Maciá, S., Robinson, M. P., Craze, P., Dalton, R. & Thomas, J. D. (2004). New observations on airborne jet propulsion (flight) in squid, with a review of previous reports. *Journal of Molluscan Studies* 70: 297–299.

Mackenstedt, U. & Märkel, K. (2001). Radular structure and function, pp. 213–236 *in* G. M. Barker. *The Biology of Terrestrial Molluscs*. Wallingford, UK, CAB Publishing.

Mackie, G. L., Qadri, S. U. & Clarke, A. H. (1974). Development of brood sacs in *Musculium securis* (Bivalvia: Sphaeriidae). *The Nautilus* 88: 109–111.

Mackie, G. L. (1984). Bivalves, pp. 351–418 *in* A. S. Tompa, Verdonk, N. H. & van den Biggelaar, J. A. M. (eds.), *Reproduction. The Mollusca*. Vol. 7. New York, Academic Press.

MacKinnon, D. I. (1982). *Tuarangia paparua* n. gen. and n. sp., a late Middle Cambrian pelecypod from New Zealand. *Journal of Paleontology* 56: 589–598.

Maeda, T., Hirose, E., Chikaraishi, Y., Kawato, M., Takishita, K., Yoshida, T., Verbruggen, H., Tanaka, J., Shimamura, S., Takaki, Y., Tsuchiya, M., Iwai, K. & Maruyama, T. (2012). Algivore or phototroph? *Plakobranchus ocellatus* (Gastropoda) continuously acquires kleptoplasts and nutrition from multiple algal species in nature. *PLoS ONE* 7: e42024.

Malakhovskaya, Y. E. (2008). Shell structure of *Kutorgina* Billings (Brachiopoda, Kutorginida). *Paleontological Journal* 42: 479.

Malaquias, M. A. E., Mackenzie-Dodds, J., Bouchet, P., Gosliner, T. M. & Reid, D. G. (2009). A molecular phylogeny of the Cephalaspidea *sensu lato* (Gastropoda: Euthyneura): Architectibranchia redefined and Runcinacea reinstated. *Zoologica Scripta* 38: 23–41.

Malchus, N. (2004). Constraints in the ligament ontogeny and evolution of pteriomorphian Bivalvia. *Palaeontology* 47: 1539–1574.

Malchus, N. (2008). Problems concerning early oyster evolution: A reply to Márquez-Aliaga and Hautmann. *Palaeogeography, Palaeoclimatology, Palaeoecology* 258: 130–134.

Maloof, A. C., Porter, S. M., Moore, J. L., Dudás, F. Ö., Bowring, S. A., Higgins, J. A., Fike, D. A. & Eddy, M. P. (2010). The earliest Cambrian record of animals and ocean geochemical change. *Bulletin of the Geological Society of America* 122: 1731–1774.

Mandahl-Barth, G. (1941). *Thyonicola mortenseni* n. gen., n. sp. eine neue parasitische Schnecke. *Videnskabelige Meddelelse fra Dansk naturhistorisk Forening* 104: 341–351.

Mangold-Wirz, K. M. & Fioroni, P. (1970). Die sonderstellung der Cephalopoden. *Zoologische Jahrbuch Anatomie Systematik* 97: 522–631.

Mangold, K. M., Bidder, A. M. & Portman, A. (1989). Organisation générale des céphalopodes, pp. 7–69 in P. P. Grassé (ed.), *Céphalopodes. Traité de Zoologie.* Vol. 5, Fascicule 4. Paris, Masson.

Mangum, C. P. & Towle, D. W. (1982). The *Nautilus* siphuncle as an ion pump. *Pacific Science* 36: 273–282.

Mann, R. & Powell, E. N. (2007). Why oyster restoration goals in the Chesapeake Bay are not and probably cannot be achieved. *Journal of Shellfish Research* 26: 905–917.

Mann, T. R. R., Martin, A. W. & Thiersch, J. B. (1970). Male reproductive tract, spermatophores and spermatophoric reaction in the giant octopus of the North Pacific, *Octopus dofleini martini. Proceedings of the Royal Society London B* 175: 31–61.

Mapes, R. H. (1987). Late Paleozoic cephalopod mandibles: Frequency of occurrence, modes of preservation, and paleoecological implications. *Journal of Paleontology* 61: 521–538.

Mapes, R. H. & Davis, R. A. (1996). Color patterns in ammonoids, pp. 103–127 in N. H. Landman, Tanabe, K. & Davis, R. A. (eds.), *Ammonoid Paleobiology: From Anatomy to Ecology. Topics in Geobiology.* New York, Plenum Press.

Mapes, R. H. & Dalton, R. B. (2002). Scavenging or predation? Mississippian ammonoid accumulations in carbonate concretion halos around *Rayonnoceras* (Actinoceratoidea - Nautiloidea) body chambers from Arkansas, pp. 407–422 *in* H. Summesberger, Histon, K. & Daurer, A. (eds.), *Cephalopods Present and Past.* Vienna, Geologische Bundesanstalt.

Mapes, R. H., Weller, E. A. & Doguzhaeva, L. A. (2007). An early Carboniferous coleoid cephalopod showing a tentacle with arm hooks and ink sac (Montana, USA), pp. 123–124. *Cephalopods Present and Past. Seventh International Symposium.* Abstracts. Sapporo, Japan, Hokkaido University.

Mapes, R. H., Weller, E. A. & Doguzhaeva, L. A. (2010). Early Carboniferous (Late Namurian) coleoid cephalopods showing a tentacle with arm hooks and an ink sac from Montana, USA, pp. 155–170 *in* K. Tanabe, Shigeta, Y., Sasaki, T. & Hirano, H. (eds.), *Cephalopods Present and Past. 8th International Symposium,* University of Bourgogne, France, 30 August–30 September 2010. Tokyo, Tokai University Press.

Mapes, R. H. & Larson, N. L. (2015). Ammonoid color patterns, pp. 25–44 *in* C. Klug, Korn, D., De Baets, K., Kruta, I. & Mapes, R. H. (eds.), *Ammonoid Paleobiology: From Anatomy to Ecology. Topics in Geobiology.* Dordrecht, the Netherlands/ Heidelberg/New York/London, Springer.

Marbach, A. & Tsurnamal, M. (1973). On the biology of *Berthellina citrina* (Gastropoda: Opisthobranchia) and its defensive acid secretion. *Marine Biology* 21: 331–339.

Marchand, W. (1907). Studien über Cephalopoden. I. Der männliche Leitungsapparat der Dibranchiaten *Zeitschrift für wissenschaftliche Zoologie* 86: 311–415.

Marchand, W. (1912). Studien über Cephalopoden. II. Uber die Spermatophoren *Zoologica* 67: 171–200.

Marcus, E. (1962). Studies on Columbellidae. *Boletim, Faculdade se Filosofia, Ciências e Letras da Universidade de São Paulo* 260: 25–66.

Marcus, E. d. B.-R. & Marcus, E. (1963). On Brazilian supralittoral and brackish water snails. *Boletim do Instituto Oceanográfico São Paulo* 13: 41–52.

Marcus, E. d. B.-R. & Marcus, E. (1967). American opisthobranch molluscs. 1. Tropical American Opisthobranchs 2. Opisthobranchs from the Gulf of California. *Studies in Tropical Oceanography* 6: 1–137, 141–248.

Marean, C. W., Bar-Matthews, M., Bernatchez, J., Fisher, E., Goldberg, P., Herries, A. I. R., Jacobs, Z., Jerardino, A., Karkanas, P., Minichillo, T., Nilssen, P. J., Thompson, E., Watts, I. & Williams, H. M. (2007). Early human use of marine resources and pigment in South Africa during the Middle Pleistocene. *Nature* 449: 905–908.

Marek, L. & Yochelson, E. L. (1976). Aspects of the biology of Hyolitha (Mollusca). *Lethaia* 9: 65–82.

Mariotti, N. & Pignatti, J. S. (1993). Remarks on the genus *Atractites* Gümbel, 1861 (Coleoidea: Aulacocerida). *Geologica Romana* 29: 355–379.

Mariotti, N. & Pignatti, J. S. (1999). The Xiphoteuthididae Bather, 1892 (Aulacocerida, Coleoidea), pp. 161–170. *Advancing Research on Living and Fossil Cephalopods.* Springer.

Marko, P. B. & Vermeij, G. J. (1996). Don't call these invertebrates spineless: Evolutionary history of labral spines in ocenebrine gastropods. *American Zoologist* 36: 133A.

Marko, P. B. & Vermeij, G. J. (1999). Molecular phylogenetics and the evolution of labral spines among eastern Pacific ocenebrine gastropods. *Molecular Phylogenetics and Evolution* 13: 275–288.

Marquez, F. & Re, M. E. (2009). Morphological and chemical description of the stylets of the red octopus, *Enteroctopus megalocyathus* (Mollusca: Cephalopoda). *Molluscan Research* 29: 27–32.

Marsh, H. (1977). The radular apparatus of *Conus. Journal of Molluscan Studies* 43: 1–11.

Marshall, B. A. (1978). Cerithiopsidae (Mollusca: Gastropoda) of New Zealand, and a provisional classification of the family. *New Zealand Journal of Zoology* 5: 47–120.

Marshall, B. A. (1983). A revison of the Recent Triphoridae of southern Australia (Mollusca: Gastropoda). *Records of the Australian Museum, Supplement* 2: 1–119.

Marshall, B. A. (1984). Adelacerithiinae: A new subfamily of the Triphoridae (Mollusca: Gastropoda). *Journal of Molluscan Studies* 50: 78–84.

Marshall, B. A. (1988). Thysanodontinae: A new subfamily of the Trochidae (Gastropoda). *Journal of Molluscan Studies* 54: 215–229.

Marshall, B. A. (1993). The systematic position of *Larochea* Finlay, 1927, and introduction of a new genus and two new species (Gastropoda: Scissurellidae). *Journal of Molluscan Studies* 59: 285–294.

Marshall, B. A. (1995). Recent and Tertiary Trochaclididae from the southwest Pacific (Mollusca: Gastropoda: Trochoidea). *The Veliger* 38: 92–115.

Marshall, B. A. (1999). A revision of the recent Solariellinae (Gastropoda: Trochoidea) of the New Zealand region. *The Nautilus* 113: 4–42.

Marshall, C., Finnegan, S., Clites, E., Holroyd, P., Bonuso, N., Cortez, C., Davis, E., Dietl, G., Druckenmiller, P. & Eng, R. (2018). Quantifying the dark data in museum fossil collections as palaeontology undergoes a second digital revolution. *Biology Letters* 14: 20180431.

Marshall, C. R. (2006). Explaining the Cambrian 'Explosion' of animals. *Annual Review of Earth and Planetary Sciences* 34: 355–384.

Marshall, C. R. & Jacobs, D. K. (2009). Flourishing after the end-Permian mass extinction. *Science* 325: 1079–1080.

Martí Mus, M. & Bergström, J. A. N. (2005). The morphology of hyolithids and its functional implications. *Palaeontology* 48: 1139–1167.

Martí Mus, M., Palacios, T. & Jensen, S. (2008). Size of the earliest mollusks: Did small helcionellids grow to become large adults? *Geology* 36: 175–178.

Martí Mus, M., Jeppsson, L. & Malinky, J. M. (2014). A complete reconstruction of the hyolithid skeleton. *Journal of Paleontology* 88: 160–170.

Martín-Durán, J. M., Passamaneck, Y. J., Martindale, M. Q. & Hejnol, A. (2016). The developmental basis for the recurrent evolution of deuterostomy and protostomy. *Nature Ecology & Evolution* 1: 0005.

Martin, A. W. (1983). Excretion, pp. 353–405 *in* A. S. M. Saleuddin & Wilbur, K. M. (eds.), *Physiology*, Part 2. *The Mollusca*. Vol. 5. New York, Academic Press.

Martin, R. (1966). On the swimming behaviour and biology of *Notarchus punctatus* Philippi (Gastropoda, Opisthobranchia). *Pubblicazioni della Stazione Zoologica di Napoli* 35: 61–75.

Martin, R. (2003). Management of nematocysts in the alimentary tract and in cnidosacs of the aeolid nudibranch gastropod *Cratena peregrina*. *Marine Biology* 143: 533–541.

Martin, R. & Walther, P. (2003). Protective mechanisms against the action of nematocysts in the epidermis of *Cratena peregrina* and *Flabellina affinis* (Gastropoda, Nudibranchia). *Zoomorphology* 122: 25–32.

Martin, R., Hild, S., Walther, P., Ploss, K., Boland, W. & Tomaschko, K.-H. (2007a). Granular chitin in the epidermis of nudibranch molluscs. *Biological Bulletin* 213: 307–315.

Martin, R., Walther, P. & Tomaschko, K.-H. (2007b). Protective skin structures in shell-less marine gastropods. *Marine Biology* 150: 807–817.

Martin, R., Hess, M., Schrödl, M. & Tomaschko, K.-H. (2009). Cnidosac morphology in dendronotacean and aeolidacean nudibranch molluscs: From expulsion of nematocysts to use in defense? *Marine Biology* 156: 261–268.

Martin, R., Tomaschko, K.-H., Hess, M. & Schrödl, M. (2010). Cnidosac-related structures in *Embletonia* (Mollusca, Nudibranchia) compared with dendronotacean and aeolidacean species. *Open Marine Biology Journal* 4: 96–100.

Martins, A. M. de Frias (1996). Relationships within the Ellobiidae, pp. 285–294 *in* J. D. Taylor (ed.), *Origin and Evolutionary Radiation of the Mollusca*. Oxford, Oxford University Press.

Martins, A. M. de Frias (2007). Morphological and anatomical diversity within the Ellobiidae (Gastropoda, Pulmonata, Archaeopulmonata). *Vita Malacologica* 4: 1–28.

Martins, A. M. de Frias (2009). The Azores Workshops. *Açoreana Supplement* 6: 9–13.

Massare, J. A. (1987). Tooth morphology and prey preference of Mesozoic marine reptiles. *Journal of Vertebrate Paleontology* 7: 121–137.

Matabos, M. & Thiebaut, E. (2010). Reproductive biology of three hydrothermal vent peltospirid gastropods (*Nodopelta heminoda*, *N. subnoda* and *Peltospira operculata*) associated with Pompeii worms on the East Pacific Rise. *Journal of Molluscan Studies* 76: 257–266.

Mather, J. A. (2011). Consciousness in cephalopods? *Journal of Cosmology* 14: 1–12.

Mather, J. A. (2012). Cephalopod intelligence, pp. 118–128 *in* T. K. Shackelford & Vonk, J. (eds.), *The Oxford Handbook of Comparative Evolutionary Psychology*. Oxford, UK/New York, Oxford University Press.

Mäthger, L. M. & Denton, E. J. (2001). Reflective properties of iridophores and fluorescent 'eyespots' in the loliginid squid *Alloteuthis subulata* and *Loligo vulgaris*. *Journal of Experimental Biology* 204: 2103–2118.

Mäthger, L. M., Collins, T. F. T. & Lima, P. A. (2004). The role of muscarinic receptors and intracellular Ca^{2+} in the spectral reflectivity changes of squid iridophores. *Journal of Experimental Biology* 207: 1759–1769.

Mäthger, L. M. & Hanlon, R. T. (2007). Malleable skin coloration in cephalopods: Selective reflectance, transmission and absorbance of light by chromatophores and iridophores. *Cell and Tissue Research* 329: 179–186.

Mäthger, L. M., Denton, E. J., Marshall, N. J. & Hanlon, R. T. (2009). Mechanisms and behavioural functions of structural coloration in cephalopods. *Journal of The Royal Society Interface* 6: S149–S163.

Matricon-Gondran, M. (1990). The site of ultrafiltration in the kidney sac of the pulmonate gastropod *Biomphalaria glabrata*. *Tissue and Cell* 22: 911–923.

Matsukawa, M. & Nakada, K. (2003). Adaptive strategy and evolution of corbiculoids based on the Japanese Mesozoic fossils *Bulletin of Tokyo Gakugei University Section* 4: 161–189.

Matsukuma, A. (1996). Transposed hinges: A polymorphism of bivalve shells. *Journal of Molluscan Studies* 62: 415–431.

Matsukuma, A., Hamada, N. & Scott, P. H. (1997). *Chama pulchella* (Bivalvia: Heterodonta) with transposed shell and normal dentition. *Venus* 56: 221–231.

Matthews, S. C. & Missarzhevsky, V. V. (1975). Small shelly fossils of late Precambrian and early Cambrian age: A review of recent work. *Journal of the Geological Society* 131: 289–303.

Mauch, S. & Elliott, J. (1997). Protection of the nudibranch *Aeolidia papillosa* from nematocyst discharge of the sea anemone *Anthopleura elegantissima*. *The Veliger* 40: 148–151.

Mazaev, A. V. (2012). Anetshelloida, a new Rostroconch order (Mollusca: Rostroconchia). *Paleontological Journal* 46: 121–131.

Mazaev, A. V. (2015). Middle Permian rostroconchs of the Kazanian Stage of the East European Platform. *Paleontological Journal* 49: 238–249.

Mazurek, D. & Zaton, M. (2011). Is *Nectocaris pteryx* a cephalopod? *Lethaia* 44: 2–4.

McArthur, A. G. & Harasewych, M. G. (2003). Molecular systematics and the major lineages of the Gastropoda, pp. 140–160 *in* C. Lydeard & Lindberg, D. R. (eds.), *Molecular*

Systematics and Phylogeography of Mollusks. Smithsonian Series in Comparative Evolutionary Biology. Washington, DC, Smithsonian Books.

McClain, C. R., Balk, M. A., Benfield, M. C., Branch, T. A., Chen, C., Cosgrove, J., Dove, A. D., Gaskins, L. C., Helm, R. R. & Hochberg, F. G. (2015). Sizing ocean giants: Patterns of intraspecific size variation in marine megafauna. *PeerJ* 3: e715.

McClintock, J. B., Slattery, M., Heine, J. & Weston, J. (1992). Chemical defense, biochemical composition and energy content of three shallow-water Antarctic gastropods. *Polar Biology* 11: 623–629.

McClintock, J. B., Baker, B. J., Hamann, M. T., Yoshida, W., Slattery, M., Heine, J. N., Bryan, P. J., Jayatilake, G. S. & Moon, B. H. (1994). Homarine as a feeding deterrent in common shallow-water Antarctic lamellarian gastropod *Marseniopsis mollis*: A rare example of chemical defense in a marine prosobranch. *Journal of Chemical Ecology* 20: 2539–2549.

McFadien-Carter, M. (1979). Scaphopoda, pp. 95–111 *in* A. C. Giese & Pearse, J. S. (eds.), *Reproduction of Marine Invertebrates. Molluscs: Pelecypods and Lesser Classes.* Vol. 5. New York, Academic Press.

McFaruume, I. D. (1980). Trail-following and trail-searching behaviour in homing of the intertidal gastropod mollusc, *Onchidium verruculatum*. *Marine and Freshwater Behaviour and Physiology* 7: 95–108.

McGhee, G. R. (1978). Analysis of shell torsion phenomenon in Bivalvia. *Lethaia* 11: 315–329.

McGowan, A. J. & Smith, A. B. (2007). Ammonoids across the Permian/Triassic boundary: A cladistic perspective. *Palaeontology* 50: 573–590.

McGowan, J. A. (1954). Observations on the sexual behavior and spawning of the squid, *Loligo opalescens*, at La Jolla, California. *California Fish and Game* 40: 47–54.

McKinney, M. L. (1997). Extinction, vulnerability and selectivity: Combining ecological and paleontological views. *Annual Review of Ecology and Systematics* 28: 495–516.

McLean, J. H. (1962). Feeding behavior of the chiton *Placiphorella*. *Proceedings of the Malacological Society of London* 35: 23–26.

McLean, J. H. (1979). A new monoplacophoran limpet from the continental shelf off southern California. *Natural History Museum of Los Angeles County. Contributions in Science* 306: 1–19.

McLean, J. H. (1981a). On the possible derivation of the Fissurellidae from the Bellerophontacea. *Bulletin of the American Malacological Union* 50: 41–42.

McLean, J. H. (1981b). The Galápagos rift limpet *Neomphalus*: Relevance to understanding the evolution of a major Paleozoic-Mesozoic radiation. *Malacologia* 21: 291–336.

McLean, J. H. (1984a). Shell reduction and loss in fissurellids: A review of genera and species in the *Fissurellidea* group. *American Malacological Bulletin* 2: 21–34.

McLean, J. H. (1984b). A case for derivation of the Fissurellidae from the Bellerophontacea. *Malacologia* 25: 3–20.

McLean, J. H. (1987). Taxonomic descriptions of cocculinid limpets (Mollusca, Archaeogastropoda): Two new species and three rediscovered species. *Zoologica Scripta* 16: 325–333.

McLean, J. H. (1989). New archaeogastropod limpets from hydrothermal vents: New family Peltospiridae, new superfamily Peltospiracea. *Zoologica Scripta* 18: 49–66.

McLean, J. H. (1990a). Neolepetopsidae, a new docoglossate limpet family from hydrothermal vents and its relevance to patellogastropod evolution. *Journal of Zoology* 222: 485–528.

McLean, J. H. (1990b). A new genus and species of neomphalid limpet from the Mariana Vents with a review of current understanding of relationships among Neomphalacea and Peltospiracea. *The Nautilus* 104: 77–86.

McLean, J. H. (2008). Three new species of the family Neolepetopsidae (Patellogastropoda) from hydrothermal vents and whale-falls in the Northeastern Pacific. *Journal of Shellfish Research* 27: 15–20.

McMahon, B. R., Burggren, W. W., Pinder, A. W. & Wheatly, M. G. (1991). Air exposure and physiological compensation in a tropical intertidal chiton, *Chiton stokesii* (Mollusca: Polyplacophora). *Physiological Zoology* 64: 728–747.

McMahon, R. F. (1991). Mollusca: Bivalvia, pp. 315–399 *in* J. H. Thorp & Covich, A. P. (eds.), *Ecology and Classification of North American Freshwater Invertebrates.* New York, Academic Press.

Medina, A., Griffond, B., Garcia-Gomez, J. C. & Garzia, F. J. (1988). Ultrastructure of the gametolytic gland and seminal receptacle in *Hypselidoris messinensis* (Gastropoda, Opisthobranchia). *Journal of Morphology* 195: 95–102.

Medina, M., Lal, S., Vallès, Y., Takaoka, T. L., Dayrat, B. A., Boore, J. L. & Gosliner, T. M. (2011). Crawling through time: Transition of snails to slugs dating back to the Paleozoic, based on mitochondrial phylogenomics. *Marine Genomics* 4: 51–59.

Medinskaya, A. I. (1993). Anatomy of the stomach of some Neogastropoda from the offshore zone of the Japan Sea. *Ruthenica* 3: 17–24.

Meeuse, B. J. D. & Fluegel, W. (1958). Carbohydrases in the sugar-gland juice of *Cryptochiton* (Polyplacophora, Mollusca). *Nature* 181: 699–700.

Meeuse, B. J. D. & Fluegel, W. (1959). Carbohydrate-digesting enzymes in the sugar gland juice of *Cryptochiton stelleri* Middendorff (Polyplacophora, Mollusca). *Archives Néerlandaises de Zoologie* 13: 301–313.

Mehl, J. (1984). Radula und Fangarme bei *Michelinoceras* sp. aus dem Silur von Bolivien. *Paläontologische Zeitschrift* 58: 211–229.

Mehl, J. (1990). Fossilerhaltung von Kiemen bei *Plesioteuthis prisca* (Rüppel 1829) aus untertithonen Plattenkalken der Altmühlalb. *Archaeopteryx* 8: 77–91.

Meij, S. E. T., Van der & Reijnen, B. T. (2012). First observations of attempted nudibranch predation by sea anemones. *Marine Biodiversity* 42: 281–283.

Menzies, R. J., Ewing, M., Worzel, J. L. & Clarke, A. H. (1959). Ecology of Recent Monoplacophora. *Oikos* 10: 168–182.

Menzies, R. J. (1968). New species of Neopilina of the Cambro-Devonian Class Monoplacophora from the Milne-Edwards Deep of the Peru-Chile Trench, R/V Anton Bruun. Proceedings of the Symposium on Mollusca, January 12–16, 1968 at Cochin. Cochin, India, *Marine Biological Association of India. Symposium Series* 3: 1–19.

Messenger, J. B. (1991). Photoreception and vision in molluscs, pp. 364–397 *in* J. R. Cronly-Dillon & Gregory, R. L. (eds.), *Vision and Visual Dysfunction.* Vol. 2. Houndmills, UK, Macmillan Press.

Messenger, J. B. (2001). Cephalopod chromatophores: Neurobiology and natural history. *Biological Reviews* 76: 473–528.

Metscher, B. D. (2009a). MicroCT for developmental biology: A versatile tool for high-contrast 3D imaging at histological resolutions. *Developmental Dynamics* 238: 632–640.

Metscher, B. D. (2009b). MicroCT for comparative morphology: Simple staining methods allow high-contrast 3D imaging of diverse non-mineralized animal tissues. *BMC Physiology* 9: 11.

Meyers, T. R., Millemann, R. E. & Fustish, C. A. (1980). Glochidiosis on salmonid fishes. IV. Humoral and tissue responses of coho and chinook salmon to experimental infection with *Margaritifera margaritifera* (L) (Pelecypoda: Margaritanidae). *Journal of Parasitology* 66: 274–281.

Middelfart, P. A. & Craig, M. (2004). Description of *Austrodevonia sharnae* n. gen. n. sp. (Galeommatidae: Bivalvia), an ecto-commensal of *Taeniogyrus australianus* (Stimpson, 1855) (Synaptidae: Holothuroidea). *Molluscan Research* 24: 211–219.

Mikami, S. & Okutani, T. (1981). A consideration on transfer of spermatophores in *Nautilus macromphalus*. *Venus* 40: 57–62.

Mikkelsen, N. T., Schander, C. & Willassen, E. (2007). Local scale DNA barcoding of bivalves (Mollusca): A case study. *Zoologica Scripta* 36: 455–463.

Mikkelsen, N. T., Todt, C., Kocot, K. M., Halanych, K. M. & Willassen, E. (2018). Molecular phylogeny of Caudofoveata (Mollusca) challenges traditional views. *Molecular Phylogenetics and Evolution* 132: 138–150.

Mikkelsen, P. M. (1996). The evolutionary relationships of Cephalaspidea s.l. (Gastropoda: Opisthobranchia): A phylogenetic analysis. *Malacologia* 37: 375–442.

Mikkelsen, P. M. (2002). Shelled opisthobranchs. *Advances in Marine Biology* 42: 67–136.

Mikkelsen, P. M. & Bieler, R. (2004). International Marine Bivalve Workshop 2002: Introduction and summary. *Malacologia* 46: 241–248.

Mikkelsen, P. M., Bieler, R., Kappner, I. & Rawlings, T. A. (2006). Phylogeny of Veneroidea (Mollusca: Bivalvia) based on morphology and molecules. *Zoological Journal of the Linnean Society* 148: 439–521.

Mikkelsen, P. M. & Bieler, R. (2007). Bivalves, pp. 284–287. *Seashells of Southern Florida: Living Marine Mollusks of the Florida Keys and Adjacent Regions*. Princeton, NJ/Oxford, UK, Princeton University Press.

Mikkelsen, P. M. (2010). Seventy-five years of molluscs: A history of the American Malacological Society on the occasion of its 75th annual meeting. *American Malacological Bulletin* 28: 191–213.

Miles, C. (1961). The occurrence of head-warts on the land snail *Rumina decollata* from Arizona. *Journal of Conchology* 101: 179–382.

Millen, S. V. & Hamann, J. C. (1992). A new genus and species of Facelinidae (Opisthobranchia: Aeolidacea) from the Caribbean Sea. *The Veliger* 35: 205–214.

Miller, A. K. & Downs, H. R. (1950). Ammonoids of the Pennsylvanian Finis shale of Texas. *Journal of Paleontology* 24: 185–218.

Miller, A. K. & Furnish, W. M. (1940a). Permian ammonoids of the Guadalupe Mountain Region and adjacent areas. *Special Papers, Geological Society of America* 26: 1–242.

Miller, A. K. & Furnish, W. M. (1940b). Studies on Carboniferous ammonoids: parts 1–4. *Journal of Paleontology* 14: 356–377.

Miller, A. K., Youngquist, W. & Nielsen, M. L. (1952). Mississippian cephalopods from western Utah. *Journal of Paleontology* 26: 148–161.

Miller, A. M. & Pawlik, J. R. (2013). Do coral reef fish learn to avoid unpalatable prey using visual cues? *Animal Behaviour* 85: 339–347.

Miller, J. A. (1989). The toxoglossan proboscis: Structure and function. *Journal of Molluscan Studies* 55: 167–181.

Miller, J. A. & Byrne, M. (2000). Ceratal autotomy and regeneration in the aeolid nudibranch *Phidiana crassicornis* and the role of predators. *Invertebrate Biology* 119: 167–176.

Miller, M. C. & Willan, R. C. (1986). A review of the New Zealand arminacean nudibranchs (Opisthobranchia: Arminacea). *New Zealand Journal of Zoology* 13: 377–408.

Miller, S. E. (2007). DNA barcoding and the renaissance of taxonomy. *Proceedings of the National Academy of Sciences of the United States of America* 104: 4775–4776.

Miller, S. L. (1974a). Adaptive design of locomotion and foot form in prosobranch gastropods. *Journal of Experimental Marine Biology and Ecology* 14: 99–156.

Miller, S. L. (1974b). The classification, taxonomic distribution, and evolution of locomotor types among prosobranch gastropods. *Journal of Molluscan Studies* 41: 233–261.

Milyutina, I. A. & Petrov, N. B. (1989). Divergence of unique DNA sequences in Mytilinae (Bivalvia Mytilidae). *Molekulyarnaya Biologiya (Moscow)* 23: 1373–1381.

Mironenko, A. A. & Rogov, M. A. (2016). First direct evidence of ammonoid ovoviviparity. *Lethaia* 49: 245–260.

Mironenko, A. A. (2018). Endocerids: Suspension feeding nautiloids? *Historical Biology* 2018: 1–9.

Mishler, B. D. (2010). Species are not uniquely real biological entities, pp. 110–122 in F. J. Ayala & Arp, R. (eds.), *Contemporary Debates in Philosophy of Biology*. Chichester, UK/Malden, MA, Wiley-Blackwell.

Mitchell, S. F. (2002). Field guide to the geological evolution of the Maastrichtian rocks of the Central Inlier, Jamaica *Caribbean Journal of Earth Science* 36: 27–38.

Mol, J.-J., Van (1967). *Étude morphologique et phylogénétique du ganglion cérébroïde des Gastéropodes pulmonés (Mollusques)*. Vol. 37. Bruxelles, Palais des Académies.

Mol, J.-J., Van (1974). Evolution phylogénétique du ganglion cérébroïde chez les Gastéropodes pulmonés. *Haliotis* 4: 77–86.

Moltschaniwskyj, N. A. (2004). Understanding the process of growth in cephalopods. *Marine and Freshwater Research* 55: 379–386.

Monks, N. & Young, J. R. (1998). Body position and the functional morphology of Cretaceous heteromorph ammonites. *Palaeontologia Electronica* 1: 1–15.

Monks, N. (1999). Cladistic analysis of Albian heteromorph ammonites. *Palaeontology* 42: 907–925.

Monks, N. (2002). Cladistic analysis of a problematic ammonite group: The Hamitidae (Cretaceous, Albian–Turonian) and proposals for new cladistic terms. *Palaeontology* 45: 689–707.

Monks, N. & Palmer, P. (2002). *Ammonites*. Washington, DC/London, Smithsonian Institution Press & Natural History Museum.

Monnet, C., Zollikofer, C., Bucher, H. & Goudemand, N. (2009). Three-dimensional morphometric ontogeny of mollusc shells by micro-computed tomography and geometric analysis. *Palaeontologia Electronica* 12: 1–13.

Moor, B. (1983). Organogenesis, pp. 123–177 in N. H. Verdonk, van den Biggelaar, J. A. M. & Tompa, A. S. (eds.), *Development. The Mollusca*. Vol. 3. New York, Academic Press.

Moore, E. J. & Gosliner, T. M. (2011). Molecular phylogeny and evolution of symbiosis in a clade of Indopacific nudibranchs. *Molecular Phylogenetics and Evolution* 58: 116–123.

Moore, H. B. (1957). Cephalopoda: Ammonoidea, pp. i–xxii, L1–L490 in R. C. Moore (ed.), *Treatise on Invertebrate Paleontology Part L*. Vol.L Mollusca 4. Boulder, CO/Lawrence, KS, Geological Society of America & University of Kansas Press.

Moore, H. B., Ed. (1996). Cretaceous Ammonoidea. *Treatise on Invertebrate Paleontology Part L*. Vol. Mollusca 4. Boulder, CO/Lawrence, KS, Geological Society of America and University of Kansas Press.

Moore, R. C. (1950). Treatise on Invertebrate Paleontology. *The Micropaleontologist* 4: 10–11.

Moore, R. C., Lalicker, C. G. & Fischer, A. G. (1952). *Invertebrate Fossils*. New York, McGraw-Hill.

Moore, R. C. (1960). Ammonoidea, pp. xxii+1–490 in R. C. Moore (ed.), *Cretaceous Ammonoidea. Treatise on Invertebrate Paleontology*. Part L. *Mollusca 4*. Lawrence, KS, Geological Society of America and University of Kansas Press.

Moore, R. C., Ed. (1964a). *Cephalopoda – General Features. Endoceratoidea - Actinoceratoidea - Nautiloidea.* Treatise on Invertebrate Paleontology Part K. Vol. Mollusca 3. Lawrence, KS, Geological Society of America and University of Kansas Press.

Moore, R. C., Ed. (1964b). *Mollusca – General features, Scaphopoda, Amphineura, Monoplacophora, Gastropoda – General features, Archaeogastropoda and some (mainly Paleozoic) Caenogastropoda and Opisthobranchia.* Treatise on Invertebrate Paleontology Part I. Vol. Mollusca 1. Lawrence, KS, Geological Society of America and University of Kansas Press.

Moore, R. C. & Teichert, C., Eds. (1969). *Bivalvia.* Treatise on Invertebrate Paleontology Part N [parts 1 and 2] Bivalvia. Vol. Mollusca 6. Lawrence, KS, Geological Society of America and University of Kansas Press.

Moore, R. C. & Teichert, C., Eds. (1971). *Bivalvia. Oysters.* Treatise on Invertebrate Paleontology Part N [part 3] Bivalvia. Vol. Mollusca 6. Lawrence, KS, Geological Society of America and University of Kansas Press.

Moore, R. C. & Sylvester-Bradley, P. C. (1996). Taxonomy and nomenclature of aptychi, pp. (L)465–(L)471 *in* R. C. Moore & Kaesler, R. L. (eds.), *Treatise on Invertebrate Paleontology Part L. Treatise on Invertebrate Paleontology.* Vol. Part L Mollusca 4 revised. Lawrence, KS, Geological Society of America and University of Kansas Press.

Morand, S., Wilson, M. J. & Glen, D. M. (2004). Nematodes (Nematoda) parasitic in terrestrial gastropods, pp. 525–558 *in* G. M. Barker. *Natural Enemies of Terrestrial Molluscs.* Oxford, UK/Cambridge, MA, CABI Publishing.

Mordan, P. B. & Wade, C. M. (2008). Heterobranchia II: The Pulmonata, pp. 409–426 *in* W. F. Ponder & Lindberg, D. R. (eds.), *Phylogeny and Evolution of the Mollusca.* Berkeley, CA, University of California Press.

Moroz, L., Nezlin, L., Elofsson, R. & Sakharov, D. A. (1994). Serotonin and FMRFamide-immunoreactive nerve elements in the chiton *Lepidopleurus asellus* (Mollusca, Polyplacophora). *Cell and Tissue Research* 275: 277–282.

Morris, N. J. & Fortey, R. A. (1976). The significance of *Tironucula* gen. nov. to the study of bivalve evolution. *Journal of Paleontology* 50: 701–709.

Morris, N. J. & Eagar, R. M. C. (1978). The infaunal descendants of the Cycloconchidae: An outline of the evolutionary history and taxonomy of the Heteroconchia, superfamilies Cycloconchacea to Chamacea. *Philosophical Transactions of the Royal Society B* 284: 259–275.

Morris, N. J., Dickens, J. M. & Astafieva-Urbaitis, K. (1991). Upper Palaeozoic anomalodesmatan Bivalvia. *Bulletin of the British Museum (Natural History). Geology* 47: 51–100.

Morris, P. J. (1991). *Functional morphology and phylogeny: An assessment of monophyly in the Kingdom Animalia and Paleozoic nearly-planispiral snail-like mollusks.* Ph.D. dissertation, Harvard University.

Morris, R. W. & Felton, S. H. (1993). Association of crinoids, symbiotic and platyceratid in the upper Ordovician cornulites of the Cincinnati, (Cincinnatian) Ohio region. *Palaios* 8: 465–476.

Morse, M. P. & Zardus, J. D. (1997). Bivalvia, pp. 7–118 *in* F. W. Harrison & Kohn, A. J. (eds.), *Microscopic Anatomy of Invertebrates. Mollusca 2.* Vol. 6A. New York, Wiley-Liss.

Morton, B. S. (1973). The biology and functional morphology of *Galeomma (Paralepida) takii* (Bivalvia, Leptonacea). *Journal of Zoology* 169: 135–150.

Morton, B. (1976). Secondary brooding of temporary dwarf males in *Ephippodonta (Ephippodontina) oedipus* sp. nov. (Bivalvia: Leptonacea). *Journal of Conchology* 29: 31–39.

Morton, B. (1977). The hypobranchial gland in Bivalvia. *Canadian Journal of Zoology* 55: 1225–1234.

Morton, B. (1979a). The biology and functional morphology of the coral-sand bivalve *Fimbria fimbriata* (Linnaeus 1758). *Records of the Australian Museum* 32: 389–420.

Morton, B. S. (1979b). A comparison of lip structure and function correlated with other aspects of the functional morphology of *Lima lima, Limaria (Platilimaria) fragilis,* and *Limaria (Platilimaria) hongkongensis* sp. nov. (Bivalvia: Limacea). *Canadian Journal of Zoology* 57: 728–742.

Morton, B. (1980). Anatomy of the 'living fossil' *Pholadomya candida* Sowerby 1823 (Bivalvia: Anomalodesmata: Pholadomyacea). *Videnskabelige Meddelelser fra Dansk naturhistorisk Forening* 142: 7–101.

Morton, B. & Scott, P. J. B. (1980). Morphological and functional specializations of the shell, musculature and pallial glands in the Lithophaginae (Mollusca: Bivalvia). *Journal of Zoology* 192: 179–203.

Morton, B. (1981a). The biology and functional morphology of *Chlamydoconcha orcutti* with a discussion on the taxonomic status of the Chlamydoconchacea (Mollusca: Bivalvia). *Journal of Zoology* 195: 81–121.

Morton, B. (1981b). The mode of life and function of the shell buttress in *Cucullaea concamerata* (Martini) (Bivalvia: Arcacea). *Journal of Conchology* 30: 295–301.

Morton, B. (1981c). Prey capture in the carnivorous septibranch *Poromya granulata* (Bivalvia: Anomalodesmata: Poromyacea). *Sarsia* 66: 241–256.

Morton, B. (1982a). The functional morphology of *Parilimya fragilis* (Bivalvia: Parilimyidae nov. fam.) with a discussion on the origin and evolution of the carnivorous septibranchs and a reclassification of the Anomalodesmata. *Transactions of the Zoological Society of London* 36: 153–216.

Morton, B. (1982b). The mode of life and functional morphology of *Gregariella coralliophaga* (Gmelin 1791) (Bivalvia: Mytilacea) with a discussion on the evolution of the boring Lithophaginae and adaptive radiation in the Mytilidae, pp. 875–895 *in* B. S. Morton & Tseng, C. K. (eds.), *The Marine Flora and Fauna of Hong Kong and Southern China I: Proceedings of the First International Marine Biological Workshop held in Hong Kong April 18*–May 10, 1980. Hong Kong, Hong Kong University Press.

Morton, B. S. (1982c). The biology, functional morphology and taxonomic status of *Fluviolanatus subtorta* (Bivalvia: Trapeziidae), a heteromyarian bivalve possessing 'zooxanthellae'. *Journal of the Malacological Society of Australia* 5: 113–140.

Morton, B. (1983a). The sexuality of *Corbicula fluminea* (Müller) in lentic and lotic waters in Hong Kong. *Journal of Molluscan Studies* 49: 81–83.

Morton, B. S. (1983b). The biology and functional morphology of the twisted ark *Trisidos semitorta* (Bivalvia: Arcacea) with a discussion on shell 'torsion' in the genus. *Malacologia* 23: 375–396.

Morton, B. (1985a). A pallial boring gland in *Barnea manilensis* (Bivalvia: Pholadidae)? pp. 191–197 *in* B. Morton & Dudgeon, D. (eds.), *Proceedings of the Second International Workshop on the Malacofauna of Hong Kong and of Southern China, Hong Kong, 6–24 April 1983.* Hong Kong, Hong Kong University Press.

Morton, B. (1985b). Adaptive radiation in the Anomalodesmata, pp. 405–459 *in* E. R. Trueman & Clarke, M. R. (eds.), *Evolution. The Mollusca.* Vol. 10. New York, Academic Press.

Morton, B. (1987a). The mantle margin and radial mantle glands of *Entodesma saxicola* and *Entodesma inflata* (Bivalvia: Anomalodesmata: Lyonsiidae). *Journal of Molluscan Studies* 53: 139–151.

Morton, B. (1987b). The functional morphology of *Neotrigonia margaritacea* (Bivalvia: Trigoniacea), with a discussion of phylogenetic affinities. *Records of the Australian Museum* 39: 339–354.

Morton, B. (1988). *Partnerships in the Sea: Hong Kong's Marine Symbioses*. Hong Kong, Hong Kong University Press.

Morton, B. S. & Scott, P. J. B. (1988). Evidence for chemical boring in *Petricola lapicida* (Gmelin, 1791) (Bivalvia: Petricolidae). *Journal of Molluscan Studies* 54: 231–237.

Morton, B. S. & Thurston, M. H. (1989). The functional morphology of *Propeamussium lucidum* (Bivalvia: Pectinacea), a deep-sea predatory scallop. *Journal of Zoology* 218: 471–496.

Morton, B. (1990a). Prey capture, preference and consumption by *Linatella caudata* (Gastropoda: Tonnoidea: Ranellidae). *Journal of Molluscan Studies* 56: 477–486.

Morton, B. (1990b). Corals and their bivalve borers – The evolution of a symbiosis, pp. 11–46 *in* B. Morton. *The Bivalvia. Proceedings of a Memorial Symposium in Honour of Sir Charles Maurice Yonge (1899–1986), Edinburgh, 1986*. Hong Kong, Hong Kong University Press.

Morton, B., Ed. (1990c). *The Bivalvia. Proceedings of a Memorial Symposium in Honour of Sir Charles Maurice Yonge (1899–1986) at the IXth International Malacological Congress, 1986* Edinburgh, Scotland, UK. Hong Kong, Hong Kong University Press.

Morton, B. (1996). The evolutionary history of the Bivalvia, pp. 337–359 *in* J. D. Taylor (ed.), *Origin and Evolutionary Radiation of the Mollusca*. Oxford, UK, Oxford University Press.

Morton, B. (2001). The evolution of eyes in the Bivalvia. *Oceanography and Marine Biology Annual Review* 39: 165–205.

Morton, B. (2002a). Biology and functional morphology of the watering pot shell *Brechites vaginiferus* (Bivalvia: Anomalodesmata: Clavagelloidea). *Journal of Zoology* 257: 545–562.

Morton, B. (2002b). The biology and functional morphology of *Humphreyia strangei* (Bivalvia: Anomalodesmata: Clavagellidae): An Australian cemented 'watering pot' shell. *Journal of Zoology* 258: 11–25.

Morton, B. (2003a). Observations on the feeding behaviour of *Nassarius clarus* (Gastropoda: Nassariidae) in Shark Bay, Western Australia. *Molluscan Research* 23: 239–249.

Morton, B. (2003b). The biology and functional morphology of *Dianadema* gen. nov *multangularis* (Tate, 1887) (Bivalvia: Anomalodesmata: Clavagellidae). *Journal of Zoology* 259: 389–401.

Morton, B. (2003c). The functional morphology of *Bentholyonsia teramachii* (Bivalvia: Lyonsiellidae): Clues to the origin of predation in the deep water Anomalodesmata. *Journal of Zoology* 261: 363–380.

Morton, B. (2003d). Hong Kong's international malacological, wetland and marine biological workshops (1977 – 1998): Changing local attitudes towards marine conservation. *Perspectives on marine environmental change in Hong Kong and southern China, 1977–2001. Proceedings of an International Workshop Reunion Conference, Hong Kong 21–26 October 2001*. Hong Kong, Hong Kong University Press.

Morton, B. & Britton, J. C. (2003). The behaviour and feeding ecology of a suite of gastropod scavengers at Watering Cove, Burrup Peninsula, Western Australia, pp. 147–171 *in* F. E. Wells, Walker, D. I. & Jones, D. S. (eds.), *The Marine Flora and Fauna of Dampier, Western Australia: Proceedings of the Twelfth International Marine Biological Workshop held in Dampier* 24 July–11 August 2000. Vol. 1. Perth, WA, Western Australian Museum.

Morton, B. & Jones, D. S. (2003). The dietary preferences of a suite of carrion-scavenging gastropods (Nassariidae, Buccinidae) in Princess Royal Harbour, Albany, Western Australia. *Journal of Molluscan Studies* 69: 151–156.

Morton, B. (2005). Biology and functional morphology of a new species of endolithic *Bryopa* (Bivalvia: Anomalodesmata: Clavagelloidea) from Japan and a comparison with fossil species of *Stirpulina* and other Clavagellidae. *Invertebrate Biology* 124: 202–219.

Morton, B. (2007). The evolution of the watering pot shells (Bivalvia: Anomalodesmata: Clavagellidae and Penicillidae). *Records of the Western Australian Museum* 24: 19–64.

Morton, B. & Peharda, M. (2008). The biology and functional morphology of *Arca noae* (Bivalvia: Arcidae) from the Adriatic Sea, Croatia, with a discussion on the evolution of the bivalve mantle margin. *Acta Zoologica* 89: 19–28.

Morton, B. (2011). Behaviour of *Nassarius bicallosus* (Caenogastropoda) on a north western Western Australian surf beach with a review of feeding in the Nassariidae. *Molluscan Research* 31: 90–94.

Morton, B. (2015). Evolution and adaptive radiation in the Mytiloidea (Bivalvia): Clues from the pericardial–posterior byssal retractor musculature complex. *Molluscan Research* 35: 227–245.

Morton, J. E. (1950). The Struthiolariidae: Reproduction, life history and relationships. *Transactions of the Royal Society of New Zealand* 78: 451–463.

Morton, J. E. (1955a). The evolution of the Ellobiidae with a discussion on the origin of the Pulmonata. *Proceedings of the Zoological Society of London* 125: 127–168.

Morton, J. E. (1955b). The functional morphology of *Otina otis*, a primitive marine pulmonate. *Journal of the Marine Biological Association of the United Kingdom* 34: 113–150.

Morton, J. E. (1955c). The functional morphology of the British Ellobiidae (Gastropoda, Pulmonata) with special reference to the digestive and reproductive systems. *Philosophical Transactions of the Royal Society B* 239: 89–160.

Morton, J. E. (1958a). Torsion and the adult snail: A re-evaluation. *Proceedings of the Malacological Society of London* 33: 2–10.

Morton, J. E. (1958b). Observations on the gymnosomatous pteropod *Clione limacina* (Phipps). *Journal of the Marine Biological Association of the United Kingdom* 37: 287–297.

Morton, J. E. (1959). The habits and feeding organs of *Dentalium entalis*. *Journal of the Marine Biological Association of the United Kingdom* 38: 225–238.

Morton, J. E. (1960). The habits of *Cyclope neritea*, a style-bearing stenoglossan gastropod. *Proceedings of the Malacological Society of London* 34: 96–105.

Morton, J. E. & Yonge, C. M. (1964). Classification and structure of the Mollusca, pp. 1–58 *in* K. M. Wilbur & Yonge, C. M. (eds.), *Physiology of Mollusca*. Vol. 1. New York, Academic Press.

Morton, J. E. & Miller, M. C. (1968). *The New Zealand Sea Shore*. London/Auckland, Collins.

Morton, J. E. (1972). The form and functioning of the pallial organs in the opisthobranch *Akera bullata*, with a discussion on the nature of the gill in Notaspidea and other tectibranchs. *The Veliger* 14: 337–349.

Morton, N. & Nixon, M. (1987). Size and function of ammonite aptychi in comparison with buccal masses of modern cephalopods. *Lethaia* 20: 231–238.

Moskalev, L. I. (1970). The gastropod mollusks of the genus *Collisella* (Prosobranchia, Acmaeidae) of the outlying Asiatic Seas of the Pacific Ocean (in Russian, English summary). *Trudy Instituta Okeanologii. Akademii Nauk SSSR* 88: 174–212.

Moskalev, L. I., Starobogatov, I. & Filatova, Z. A. (1983). New data on the Monoplacophora of the abyssal of the Pacific and the Southern Atlantic Ocean. *Zoologicheskii Zhurnal* 62: 981–995.

Moyne, S. & Neige, P. (2004). Cladistic analysis of the Middle Jurassic ammonite radiation. *Geological Magazine* 141: 115–123.

Moysiuk, J., Smith, M. R. & Caron, J.-B. (2017). Hyoliths are Palaeozoic lophophorates. *Nature* 541: 394–397.

Mukai, H., Terakado, K. & Reed, C. G. (1997). Bryozoa, pp. 45–206 *in* F. W. Harrison & Woollacott, R. M. (eds.), *Microscopic Anatomy of Invertebrates. Lophophorates, Entoprocta, and Cycliophora.* Vol. 13. New York, Wiley-Liss.

Müller, A. H. (1974). Über den Kieferapparat fossiler und rezenter Nautiliden (Cephalopoda) mit Bemerkungen zur Ökologie, Funktionsweise und Phylogenie. *Freiberger Forschungshefte C* 298: 7–17.

Muntz, W. R. A. (1986). The spectral sensitivity of *Nautilus pompilius*. *Journal of Experimental Biology* 126: 513–517.

Muntz, W. R. A. (1987a). Visual behaviour and visual sensitivity of *Nautilus pompilius*, pp. 231–244 *in* W. B. Saunders & Landman, N. H. (eds.), *Nautilus: The Biology and Paleobiology of a Living Fossil. Topics in Geobiology.* New York, Springer.

Muntz, W. R. A. (1987b). A possible function of the iris groove in *Nautilus*, pp. 245–248 *in* W. B. Saunders & Landman, N. H. *Nautilus: The Biology and Paleobiology of a Living Fossil. Topics in Geobiology.* New York, Springer.

Muthiah, P. & Sampath, K. (2000). Spawn and fecundity of *Cymatium* (*Monoplex*) *pileare* and *Cymatium* (*Linatella*) *cingulatum* (Gastropoda: Ranellidae). *Journal of Molluscan Studies* 66: 293–300.

Mutvei, H. (1964). Remarks on the anatomy of Recent and fossil Cephalopoda: With description of the minute shell structure of belemnoids. *Stockholm Contributions in Geology* 11: 79–102.

Mutvei, H., Arnold, J. M. & Landman, N. H. (1993). Muscles and attachment of the body to the shell in embryos and adults of *Nautilus belauensis* (Cephalopoda). *American Museum Novitates* 3059: 1–15.

Mutvei, H. (1997). Siphuncular structure in Ordovician endocerid cephalopods. *Acta Palaeontologica Polonica* 42: 375–390.

Mutvei, H. & Doguzhaeva, L. A. (1997). Shell ultrastructure and ontogenetic growth in *Nautilus pompilius* L. (Mollusca: Cephalopoda). *Palaeontographica: Abteilung A* 246: 33–52.

Mutvei, H. (2002a). Nautiloid systematics based on siphuncular structure and positon of muscle scars. *Abhandlungen der Geologischen Bundesanstalt* 57: 379–392.

Mutvei, H. (2002b). Connecting ring structure and its significance for classification of the orthoceratid cephalopods. *Acta Palaeontologica Polonica* 47: 157–168.

Mutvei, H., Zhang, Y.-B. & Dunca, E. (2007). Late Cambrian plectronocerid nautiloids and their role in cephalopod evolution. *Order: A Journal on the Theory of Ordered Sets and its Applications* 50: 1327–1333.

Mutvei, H. (2012). Siphuncular structure in Silurian discosorid and ascocerid nautiloids (Cephalopoda) from Gotland, Sweden: Implications for interpretation of mode of life and phylogeny. *GFF: Journal of the Geological Society of Sweden* 134: 27–37.

Mutvei, H. (2013). Characterization of nautiloid orders Ellesmerocerida, Oncocerida, Tarphycerida, Discosorida and Ascocerida: New superorder Multiceratoidea. *GFF: Journal of the Geological Society of Sweden* 135: 171–183.

Mutvei, H. (2015). Characterization of two new superorders Nautilosiphonata and Calciosiphonata and a new order Cyrtocerinida of the subclass Nautiloidea; siphuncular structure in the Ordovician nautiloid *Bathmoceras* (Cephalopoda). *GFF: Journal of the Geological Society of Sweden* 137: 164–174.

Mutvei, H. (2017). The new order Mixosiphonata (Cephalopoda: Nautiloidea) and related taxa; estimations of habitat depth based on shell structure. *GFF: Journal of the Geological Society of Sweden* 139: 219–232.

Mutvei, H. (2018). Cameral deposits in Paleozoic cephalopods. *GFF: Journal of the Geological Society of Sweden* 140: 254–263.

Mutvei, H. & Mapes, R. H. (2018). Carboniferous coleoids with mixed coleoid-orthocerid characteristics: A new light on cephalopod evolution. *GFF: Journal of the Geological Society of Sweden* 140: 11–24.

Naef, A. (1911). Studien zur generellen Morphologie der Mollusken. I. Teil: Über Torsion und Asymmetrie der Gastropoden. *Ergebnisse und Fortschritte der Zoologie* 3: 73–164.

Naef, A. (1913). Studien zur generellen Morphologie der Mollusken. II. Teil: Das Cölomsystem in seinen topographischen Beziehungen. *Ergebnisse und Fortschritte der Zoologie* 3: 329–462.

Naef, A. (1921). Die Cephalopoden. *Fauna und Flora des Golfes von Neapel,* Monographie 35. Systematik, I Teil, 1 Band, Fascicle I: 1–148. Translated in A. Mercado. 1972. Cephalopoda. Israel Program for Scientific Translations. Jerusalem.

Naef, A. (1922). *Die Fossilen Tintenfische; Eine Paläozoologische Monographie.* Jena, Fischer.

Naef, A (1923). Die Cephalopoden. *Fauna und Flora des Golfes von Neapel,* Monographie 35. Systematik, Part I, vol. 1, Fascicle II: 149–863. Translated in A. Mercado. 1972. Cephalopoda. Israel Program for Scientific Translations. Jerusalem.

Naef, A. (1926). Studien zur generellen Morphologie der Mollusken. III. Teil: Die typischen Beziehungen der Weichtiere untereinander und das verhältnis ihrer Urformen zu anderen Cälomaten. *Ergebnisse und Fortschritte der Zoologie* 6: 27–124.

Naef, A. (1928). *Die Cephalopoden. Embryologie.* Vol. 35. Berlin, R. Friedlander & Sohn.

Naegele, K.-L. & Hausdorf, B. (2015). Comparative phylogeography of land snail species in mountain refugia in the European Southern Alps. *Journal of Biogeography* 42: 821–832.

Naggs, F. (1994). The reproductive anatomy of *Paropeas achatinaceum* and a new concept of *Paropeas* (Pulmonata: Achatinoidea: Subulinidae). *Journal of Molluscan Studies* 60: 175–191.

Naglik, C., Rikhtegar, F. & Klug, C. (2016). Buoyancy of some Palaeozoic ammonoids and their hydrostatic properties based on empirical 3D-models. *Lethaia* 49: 3–12.

Nair, N. B. & Saraswathy, M. (1971). The biology of wood-boring teredinid molluscs. *Advances in Marine Biology* 9: 335–509.

Nakamura, H. K. (1985). A review of molluscan cytogenetic information based on the CISMOCH [computerized index system for molluscan chromosomes]: Bivalvia, Polyplacophora and Cephalopoda. *Venus* 44: 193–225.

Nakamura, K., Watanabe, H., Miyazaki, J., Takai, K., Kawagucci, S., Noguchi, T., Nemoto, S., Watsuji, T.-O., Matsuzaki, T. & Shibuya, T. (2012). Discovery of new hydrothermal activity and chemosynthetic fauna on the Central Indian Ridge at 18–20 S. *PLoS ONE* 7: e32965.

Nakano, T. & Ozawa, T. (2004). Phylogeny and historical biogeography of limpets of the order Patellogastropoda based on mitochondrial DNA sequences: Part 1. *Journal of Molluscan Studies* 70: 31–41.

Nakano, T. & Ozawa, T. (2007). Worldwide phylogeography of limpets of the order Patellogastropoda: Molecular, morphological and palaeontological evidence. *Journal of Molluscan Studies* 73: 79–99.

Nakano, T. & Sasaki, T. (2011). Recent advances in molecular phylogeny, systematics and evolution of patellogastropod limpets. *Journal of Molluscan Studies* 77: 203–217.

Nakashima, Y. (1995). Mucous trail following in 2 intertidal nudibranchs. *Journal of Ethology* 13: 125–128.

van Name, W. G. (1926). A new specimen of *Protobalanus*, supposed Paleozoic barnacle. *American Museum Novitates* 227: 1–6.

Naud, M. J. & Havenhand, J. N. (2006). Sperm motility and longevity in the giant cuttlefish, *Sepia apama* (Mollusca: Cephalopoda). *Marine Biology* 148: 559–566.

Navrot, J., Amiel, A. J. & Kronfeld, J. (1974). *Patella vulgata*: A biological monitor of coastal metal pollution – A preliminary study. *Environmental Pollution* 7: 303–308.

Neige, P., Rouget, I. & Moyne, S. (2007). Phylogenetic practices among scholars of fossil cephalopods, with special reference to cladistics, pp. 3–14 *in* N. H. Landman, Davis, R. A. & Mapes, R. H. (eds.), *Cephalopods Present and Past. New Insights and Fresh Perspectives. 6th International Symposium*, University of Arkansas, Fayetteville, 16–19 September 2004. Dordrecht, the Netherlands, Springer.

Neige, P., Lapierre, H. & Merle, D. (2016). New Eocene coleoid (Cephalopoda) diversity from Statolith remains: Taxonomic assignation, fossil record analysis, and new data for calibrating molecular phylogenies. *PloS ONE* 11: e0154062.

Neil, T. R. & Askew, G. N. (2018). Swimming mechanics and propulsive efficiency in the chambered nautilus. *Royal Society Open Science* 5: 170467 (170461–170469).

Nelson, G. (1974). Historical biogeography – Alternative formalization. *Systematic Zoology* 23: 555–558.

Nesis, K. N. (1975). Evolution of living forms of cephalopod mollusks. *Trudy Instituta Okeanologii Akademiya Nauk SSSR* 101: 124–142.

Nesis, K. N. (1995). Mating, spawning, and death in oceanic cephalopods: A review. *Ruthenica* 6: 23–64.

Neusser, T. P., Fukuda, H., Jörger, K. M., Kano, Y. & Schrödl, M. (2011). Sacoglossa or Acochlidia? 3D reconstruction, molecular phylogeny and evolution of Aitengidae (Gastropoda: Heterobranchia). *Journal of Molluscan Studies* 77: 332–350.

Nevesskaja, L. A., Scarlato, O. A., Starobogatov, I. & Eberzin, A. G. (1971). Novie predstavlenia o sisteme dvustvorchatikh molliuskov [New ideas on bivalve systematics]. *Paleontologicheskii Zhurnal* 5: 3–20.

Nevesskaja, L. A. (2009). Principles of systematics and the system of bivalves. *Paleontological Journal* 43: 1–11.

Newell, N. D. (1965). Classification of the Bivalvia. *American Museum Novitates* 2206: 1–25.

Newell, N. D. (1969). Classification of Bivalvia, pp. N205–N224 *in* R. C. Moore. *Treatise on Invertebrate Paleontology Part N*. Lawrence, KS, Geological Society of America and University of Kansas Press.

Newell, N. D. & Boyd, D. W. (1970). Oyster-like Permian Bivalvia. *Bulletin of the American Museum of Natural History* 143: 217–282.

Newell, N. D. & Boyd, D. W. (1990). Nacre in a Carboniferous pectinoid mollusc and a new subfamily Limipectininae. *American Museum Novitates* 2970: 1–7.

Newell, N. D. & Boyd, D. W. (1995). Pectinoid bivalves of the Permian-Triassic crisis. *Bulletin of the American Museum of Natural History* 227: 5–95.

Nezlin, L. P. (1997). The osphradium is involved in the control of egg-laying in the pond snail *Lymnaea stagnalis*. *Invertebrate Reproduction & Development* 32: 163–166.

Ng, T. P. T., Saltin, S. H., Davies, M. S., Johannesson, K., Stafford, R. & Williams, G. A. (2013). Snails and their trails: The multiple functions of trail-following in gastropods. *Biological Reviews* 88: 683–700.

Nicklas, N. L. & Hoffmann, R. J. (1981). Apomictic parthenogenesis in a hermaphroditic terrestrial slug, *Deroceras laeve* (Muller). *The Biological Bulletin* 160: 123–135.

Nicotri, M. E. (1974). *Resource partitioning, grazing activities, and influence on the microflora by intertidal limpets*. Ph.D. dissertation, University of Washington.

Nielsen, C. (1979). Larval ciliary bands and metazoan phylogeny. *Fortschritte der Zoologie Systematik und Evolutionsforschung* 1: 178–184.

Nielsen, C. (1995). Phylum Mollusca, pp. i, 110–123. *Animal Evolution: Interrelationships of the Living Phyla*. Oxford, UK, Oxford University Press.

Nielsen, C. (2001). *Animal Evolution: Interrelationships of the Living Phyla*. Oxford, UK, Oxford University Press.

Nielsen, C. (2004). Trochophora larvae: Cell-lineages, ciliary bands, and body regions. 1. Annelida and Mollusca. *Journal of Experimental Zoology Part B* 302: 35–68.

Nielsen, C., Haszprunar, G., Ruthensteiner, B. & Wanninger, A. (2007). Early development of the aplacophoran mollusc *Chaetoderma*. *Acta Zoologica* 88: 231–247.

Nigmatullin, C. M., Arkhipkin, A. I. & Sabirov, R. M. (1991). Structure of the reproductive system of the squid *Thysanoteuthis rhombus* (Cephalopoda: Oegopsida). *Journal of Zoology* 224: 271–283.

Nigmatullin, C. M. & Arkhipkin, A. I. (1998). A review of the biology of the diamondback squid, *Thysanoteuthis rhombus* (Oegopsida: Thysanoteuthidae), pp. 155–181 *in* T. Okutani (ed.), *Large Pelagic Squids*. Tokyo, Japan Marine Fishery Resource Research Center.

Nigmatullin, C. M. & Markaida, U. (2009). Oocyte development, fecundity and spawning strategy of large sized jumbo squid *Dosidicus gigas* (Oegopsida: Ommastrephinae). *Journal of the Marine Biological Association of the United Kingdom* 89: 789–801.

Nilsson, D. E., Warrant, E. J., Johnsen, S., Hanlon, R. T. & Shashar, N. (2012). A unique advantage for giant eyes in giant squid. *Current Biology* 22: 683–688.

Nishiguchi, M. K. & Mapes, R. H. (2008). Cephalopoda, pp. 163–199 *in* W. F. Ponder & Lindberg, D. R. (eds.), *Phylogeny and Evolution of the Mollusca*. Berkeley, CA, University of California Press.

Nishimura, S. (1976). *Dynoidella conchicola*, gen. et sp. nov. (Isopoda, Sphaeromatidae), from Japan, with a note on its association with intertidal snails. *Publications of the Seto Marine Biological Laboratory* 23: 275–282.

Nishiwaki, S. (1964). Phylogenetic study on the type of the dimorphic spermatozoa in Prosobranchia. *Science Reports. Tokyo Kyoiku Daigaku*: Section B 11: 237–275.

Nixon, M. (1979). Hole-boring in shells by *Octopus vulgaris* Cuvier in the Mediterranean. *Malacologia* 18: 431–443.

Nixon, M. (1980). The salivary papilla of *Octopus* as an accessory radula for drilling shells. *Journal of Zoology* 190: 53–57.

Nixon, M. (1988a). The buccal mass of fossil and Recent Cephalopoda, pp. 103–122 *in* M. R. Clarke & Trueman, E. R. (eds.), *Paleontology and Neontology of Cephalopods. The Mollusca*. Vol. 12. New York, Academic Press.

Nixon, M. (1988b). The feeding mechanisms and diets of cephalopods – Living and fossil, pp. 641–652 *in* J. Wiedmann & Kullmann, J. (eds.), *Cephalopods Present and Past. 2nd International Cephalopod Symposium*: O. H. Schindewolf Symposium, Tübingen 1985. Stuttgart, E. Schweizerbart'sche Verlagsbuchhandlung.

Nixon, M. (1995). A nomenclature for the radula of the Cephalopoda (Mollusca) – Living and fossil. *Journal of Zoology* 236: 73–81.

Nixon, M. (1996). Morphology of the jaws and radula in ammonoids, pp. 23–42 *in* N. H. Landman, Tanabe, K. & Davis, R. A.(eds.), *Ammonoid Paleobiology: From Anatomy to Ecology. Topics in Geobiology*. New York, Plenum Press.

Nobuhara, T., Onda, D., Sato, T., Aosawa, H., Ishimura, T., Ijiri, A., Tsunogai, U., Kikuchi, N., Kondo, Y. & Kiel, S. (2016). Mass occurrence of the enigmatic gastropod *Elmira* in the Late Cretaceous Sada Limestone seep deposit in southwestern Shikoku, Japan. *PalZ* 90: 701–722.

Noetling, F. (1884). Beiträge zur Kenntniss der Cephalopoden aus Silurgeschieben der Provinz Ost-Preussen. *Jahrbuch der Königlich Preussischen Geologischen Landesanstalt und Bergakademie zu Berlin* 1883: 101–135, plts 116–118.

Nolen, T. G., Johnson, P. M., Kicklighter, C. E. & Capo, T. (1995). Ink secretion by the marine snail *Aplysia californica* enhances its ability to escape from a natural predator. *Journal of Comparative Physiology A* 176: 239–254.

Nordsieck, H. (1985). The system of the Stylommatophora (Gastropoda) with special regard to the systematic position of the Clausiliidae. I. Importance of the excretory and genital systems. *Archiv für Molluskenkunde* 116: 1–24.

Nordsieck, H. (1986). The system of the Stylommatophora (Gastropoda), with special regard to the systematic position of the Clausiliidae. II. Importance of the shell and distribution. *Archiv für Molluskenkunde* 117: 93–116.

Nordsieck, H. (1992). Systematic revision of the Helicoidea (Gastropoda: Stylommatophora). *Harvard University Museum Of Comparative Zoology. Special Occasional Publication* 1992: 1–79.

Norell, M. A. (1992). Taxic origin and temporal diversity: The effect of phylogeny, pp. 89–118 in M. J. Novacek & Wheeler, Q. D. (eds.), *Phylogeny and Extinction*. New York, Columbia University Press.

Norman, E. (1976). The vertical distribution of the wood-boring molluscs *Teredo navalis* L., *Psiloteredo megotara* H. and *Xylophaga dorsalis* T. on the Swedish west coast. *Material und Organismen* 11: 303–316.

Norman, M. D. & Hochberg, F. G. (2005). The 'mimic octopus' (*Thaumoctopus mimicus* n. gen. and sp.), a new octopus from the tropical Indo-West Pacific (Cephalopoda: Octopodidae). *Molluscan Research* 25: 57–70.

Novack-Gottshall, P. M. & Burton, K. (2014). Morphometrics indicates giant Ordovician macluritid gastropods switched life habit during ontogeny. *Journal of Paleontology* 88: 1050–1055.

Nowak, H., Servais, T., Monnet, C., Molyneux, S. G. & Vandenbroucke, T. R. A. (2015). Phytoplankton dynamics from the Cambrian Explosion to the onset of the Great Ordovician Biodiversification Event: A review of Cambrian acritarch diversity. *Earth-Science Reviews* 151: 117–131.

Nowikoff, M. (1907). Uber die Ruckensinnesorgane der Placophoren nebst einigen Bemerkugen uber die Schale derselben. *Zeitschrift für wissenschaftliche Zoologie* 88: 154–186.

Nowikoff, M. (1912). Studien über das Knorpelgewebe von Wirbellosen. *Zeitschrift für wissenschaftliche Zoologie* 103: 661–717.

Nugranad, J., Bonnaud, L., Byrne, R. A., Chang, K.-Y., Hylleberg, J., Jivaluk, J., Kasugai, T., Lucero, M. T., Mather, J. A. & Miske, V. C. (2005). *Idiosepius*: Ecology, biology and biogeography of a mini-maximalist. *Phuket Marine Biology Centre Research Bulletin* 66: 11–22.

Nützel, A. (1997). Über die Stammesgeschichte der Ptenoglossa (Gastropoda). *Berliner Geowissenschaftliche Abhandlungen*: Reihe E 26: 1–229.

Nützel, A. & Bandel, K. (2000). Goniasmidae and Orthonemidae: Two new families of the Palaeozoic Caenogastropoda (Mollusca, Gastropoda). *Neues Jahrbuch für Geologie und Paläontologie-Monatshefte* 2000: 557–569.

Nützel, A., Erwin, D. H. & Mapes, R. H. (2000). Identity and phylogeny of the late Paleozoic Subulitoidea (Gastropoda). *Journal of Paleontology* 74: 575–598.

Nützel, A. & Mapes, R. H. (2001). Larval and juvenile gastropods from a Carboniferous black shale: Palaeoecology and implications for the evolution of the Gastropoda. *Lethaia* 34: 143–162.

Nützel, A. & Erwin, D. H. (2002). *Battenizyga*, a new Early Triassic gastropod genus with a discussion of the caenogastropod evolution at the Permian/Triassic boundary. *Palaeontologische Zeitschrift* 76: 21–27.

Nützel, A. (2005). A new Early Triassic gastropod genus and the recovery of gastropods from the Permian/Triassic extinction. *Acta Palaeontologica Polonica* 50: 19–24.

Nützel, A., Frýda, J., Yancey, T. E. & Anderson, J. R. (2007a). Larval shells of Late Palaeozoic naticopsid gastropods (Neritopsoidea: Neritimorpha) with a discussion of the early neritimorph evolution. *Palaeontologische Zeitschrift* 81: 213–228.

Nützel, A., Lehnert, O. & Frýda, J. (2007b). Origin of planktotrophy – Evidence from early molluscs: A response to Freeman and Lundelius. *Evolution & Development* 9: 313–318.

Nützel, A. (2010). A review of the Triassic gastropod genus *Kittliconcha* Bonarelli, 1927 – Implications for the phylogeny of Caenogastropoda. *Zitteliana* A50: 9–20.

Nützel, A. (2014). Larval ecology and morphology in fossil gastropods. *Palaeontology* 57: 479–503.

Nuwayhid, M. A., Davies, P. S. & Elder, H. Y. (1978). Gill structure in the common limpet *Patella vulgata*. *Journal of the Marine Biological Association of the United Kingdom* 58: 817–823.

Nybakken, J. & McDonald, G. (1981). Feeding mechanisms of west American nudibranchs feeding on Bryozoa Cnidaria and Ascidiacea, with special respect to the radula. *Malacologia* 20: 439–449.

O'Donoghue, J. (2008). The second coming. *New Scientist* 198: 34–37.

O'Dor, R. K. & Webber, D. M. (1986). The constraints on cephalopods: Why squid aren't fish. *Canadian Journal of Zoology* 64: 1591–1605.

O'Dor, R. K. (1988). Limitations on locomotor performance in squid. *Journal of Applied Physiology* 64: 128–134.

O'Dor, R. K. & Webber, D. M. (1991). Invertebrate athletes: Trade-offs between transport efficiency and power density in cephalopod evolution. *Journal of Experimental Biology* 160: 93–112.

O'Dor, R. K., Hoar, J. A., Webber, D. M., Carey, F. G., Tanaka, S., Martins, H. R. & Porteiro, F. M. (1994). Squid (*Loligo forbesi*) performance and metabolic rates in nature, pp. 163–177 in H. O. Pörtner, O'Dor, R. K. & Macmillan, D. L. (eds.), *Physiology of Cephalopod Molluscs: Lifestyle and Performance Adaptations. Marine and Freshwater Behaviour and Physiology*. Basel, Gordon and Breach Science Publishers SA.

O'Shea, S. (2004). The giant octopus *Haliphron atlanticus* (Mollusca: Octopoda) in New Zealand waters. *New Zealand Journal of Zoology* 31: 7–13.

O'Shea, S. & Bolstad, K. (2008). Giant and Colossal Squid Fact Sheet. *TONMO: The Octopus News Magazine Online*.

O'Sullivan, J. B., McConnaughey, P. R. & Huber, M. E. (1987). A blood-sucking snail: The Cooper's Nutmeg, *Cancellaria cooperi* Gabb, parasitizes the California Electric Ray, *Torpedo californica* Ayres. *Biological Bulletin* 172: 362–366.

ÓFoighil, D. & Thiriot-Quiévreux, C. (1991). Ploidy and pronuclear interaction in Northeastern Pacific *Lasaea* clones (Mollusca, Bivalvia). *Biological Bulletin* 181: 222–231.

ÓFoighil, D. & Graf, D. L. (2000). Prodissoconch morphology of the relict marine paleoheterodont *Neotrigonia margaritacea* (Mollusca: Bivalvia) indicates a non–planktotrophic prejuvenile ontogeny. *Journal of the Marine Biological Association of the United Kingdom* 80: 173–175.

Obermann, D., Bickmeyer, U. & Wägele, H. (2012). Incorporated nematocysts in *Aeolidiella stephanieae* (Gastropoda, Opisthobranchia, Aeolidoidea) mature by acidification shown by the pH sensitive fluorescing alkaloid Ageladine A. *Toxicon* 60: 1108–1016.

Ockelmann, K. W. (1958). Marine Lamellibranchiata: The zoology of East Greenland. *Meddelelser om Grønland* 122: 1–256.

Ockelmann, K. W. (1983). Description of mytilid species and definition of the Dacrydiinae n. subfam. (Mytilacea - Bivalvia). *Ophelia* 22: 81–123.

Odhner, N. H. (1919). Studies on the morphology, the taxonomy and the relations of recent Chamidae. *Kungliga Svenska Vetenskapsakademiens Handlingar* 59: 1–102.

Odiete, W. O. (1978). The cruciform muscle and its associated sense organ in *Scrobicularia plana* (da Costa). *Journal of Molluscan Studies* 44: 180–189.

Ogura, A., Yoshida, M.-a., Moritaki, T., Okuda, Y., Sese, J., Shimizu, K. K., Sousounis, K. & Tsonis, P. A. (2013). Loss of the six3/6 controlling pathways might have resulted in pinhole-eye evolution in *Nautilus*. *Nature. Scientific Reports* 3: 1432.

Ohkouchi, N., Tsuda, R., Chikaraishi, Y. & Tanabe, K. (2013). A preliminary estimate of the trophic position of the deep-water ram's horn squid *Spirula spirula* based on the nitrogen isotopic composition of amino acids. *Marine Biology* 160: 773–779.

Økland, S. (1980). The heart ultrastructure of *Lepidopleurus asellus* (Spengler) and *Tonicella marmorea* (Fabricius) (Mollusca: Polyplacophora). *Zoomorphology* 96: 1–19.

Økland, S. (1981). Ultrastructure of the pericardium in chitons (Mollusca: Polyplacophora), in relation to filtration and contraction mechanisms. *Zoomorphology* 97: 193–203.

Økland, S. (1982). The ultrastructure of the heart complex in *Patella vulgata* L. (Archaeogastropoda: Prosobranchia). *Journal of Molluscan Studies* 48: 331–341.

Okusu, A. (2002). Embryogenesis and development of *Epimenia babai* (Mollusca Neomeniomorpha). *Biological Bulletin* 203: 87–103.

Okusu, A. & Giribet, G. (2003). New 18S rRNA sequences from neomenioid aplacophorans and the possible origin of persistent exogenous contamination. *Journal of Molluscan Studies* 69: 385–387.

Okusu, A., Schwabe, E., Eernisse, D. J. & Giribet, G. (2003). Towards a phylogeny of chitons (Mollusca, Polyplacophora) based on combined analysis of five molecular loci. *Organisms Diversity & Evolution* 3: 281–302.

Okutani, T., Tagawa, M. & Horikawa, M. (1987). *Cephalopods from Continental Shelf and Slope around Japan*. Tokyo, Japan Fisheries Resource Conservation Association.

Okutani, T., Saito, H. & Hashimoto, J. (1989). A new neritacean limpet from a hydrothermal vent site near Ogasawara Islands, Japan. (*Shinkailepas kaikatensis* n. g., n. sp.). *Venus* 48: 223–230.

Oldfield, E. (1955). Observations on the anatomy and mode of life of *Lasaea rubra* (Montagu) and *Turtonia minuta* (Fabricius). *Proceedings of the Malacological Society of London* 31: 226–249.

Oldfield, E. (1964). The reproduction and development of some members of the Erycinidae and Monacutidae (Mollusca, Eulamellibranchia). *Proceedings of the Malacological Society of London* 36: 79–120.

Oliver, P. G. (1981). The functional morphology and evolution of recent Limopsidae (Bivalvia, Arcoidea). *Malacologia* 21: 61–93.

Oliver, P. G. (1982). A new species of cancellariid gastropod from Antarctica with a description of the radula. *British Antarctic Survey Bulletin* 57: 15–20.

Oliver, P. G. (2001). Functional morphology and description of a new species of *Amygdalum* (Mytiloidea) from the oxygen minimum zone of the Arabian Sea. *Journal of Molluscan Studies* 67: 225–241.

Oliver, P. G. & Holmes, A. M. (2004). Cryptic bivalves with descriptions of new species from the Rodrigues lagoon. *Journal of Natural History* 38: 3175–3227.

Oliver, P. G. & Holmes, A. M. (2006). The Arcoidea (Mollusca: Bivalvia): A review of the current phenetic-based systematics. *Zoological Journal of the Linnean Society* 148: 237–251.

Oliver, P. G. & Lützen, J. (2011). An anatomically bizarre, fluid-feeding galeommatoidean bivalve: *Draculamya porobranchiata* gen. et sp. nov. (Mollusca: Bivalvia). *Journal of Conchology* 40: 365–392.

Oliver, P. G., Southward, E. C. & Dando, P. R. (2012). Bacterial symbiosis in *Syssitomya pourtalesiana* Oliver, 2012 (Galeommatoidea: Montacutidae), a bivalve commensal with the deep-sea echinoid *Pourtalesia*. *Journal of Molluscan Studies* 79: 30–41.

Oliver, P. G. & Taylor, J. D. (2012). Bacterial symbiosis in the Nucinellidae (Bivalvia: Solemyida) with descriptions of two new species. *Journal of Molluscan Studies* 78: 81–91.

Oliver, P. G. (2013). Description of *Atopomya dolobrata* gen. et sp. nov.: First record of bacterial symbiosis in the Saxicavellinae (Bivalvia). *Journal of Conchology* 41: 359–368.

Olsson, A. A. (1970). The cancellariid radula and its interpretation. *Palaeontographica Americana* 7: 19–[32].

Ong, E., Hallas, J. M. & Gosliner, T. M. (2017). Like a bat out of heaven: The phylogeny and diversity of the bat-winged slugs (Heterobranchia: Gastropteridae). *Zoological Journal of the Linnean Society* 180: 755–789.

Orlov, Y. A. (Ed.) (1962). *Osnovy Paleontologii, Mollyuski - Golovonogie 1*. Moskva, Akademiya Nauk SSSR.

Ortiz, N. & Ré, M. E. (2006). First report of pseudohermaphroditism in cephalopods. *Journal of Molluscan Studies* 72: 321–323.

Osca, D., Templado, J. & Zardoya, R. (2015). Caenogastropod mitogenomics. *Molecular Phylogenetics and Evolution* 93: 118–128.

Oskars, T. R., Bouchet, P. & Malaquias, M. A. E. (2015). A new phylogeny of the Cephalaspidea (Gastropoda: Heterobranchia) based on expanded taxon sampling and gene markers. *Molecular Phylogenetics and Evolution* 89: 130–150.

Otaïza, R. D. & Santelices, B. (1985). Vertical distribution of chitons (Mollusca: Polyplacophora) in the rocky intertidal zone of central Chile. *Journal of Experimental Marine Biology and Ecology* 86: 229–240.

Ovchinnikov, Y. A., Abdulaev, N. G., Zolotarev, A. S., Artamonov, I. D., Bespalov, I. A., Dergachev, A. E. & Tsuda, M. (1988). *Octopus* rhodopsin – Amino-acid sequence deduced from C-DNA. *Federation of European Biochemical Societies (FEBS) Letters* 232: 69–72.

Owen, G., Trueman, E. R. & Yonge, C. M. (1953). The ligament in the Lamellibranchia. *Nature* 171: 73–75.

Owen, G. (1958). Shell form, pallial attachment and the ligament in the Bivalvia. *Proceedings of the Zoological Society of London* 131: 637–648.

Owen, G. (1973). The fine structure and histochemistry of the digestive diverticula of the protobranchiate bivalve *Nucula sulcata*. *Proceedings of the Royal Society B* 183: 249–264.

Owen, G. & McCrae, J. M. (1976). Further studies on the laterofrontal tracts of bivalves. *Proceedings of the Royal Society B* 194: 527–544.

Owen, G. (1978). Classification and the bivalve gill. *Philosophical Transactions of the Royal Society B* 284: 377–385.

Owen, G. & McCrae, J. M. (1979). Sensory cell/gland cell complexes associated with the pallial tentacles of the bivalve *Lima hians* (Gmelin) with a note on specialized cilia on the pallial curtains. *Philosophical Transactions of the Royal Society B* 287: 45–62.

Owen, R. (1878). On the relative positions to their construction of the chambered shells of cephalopods. *Proceedings of the Zoological Society of London* [1878]: 955–975.

Owre, H. B. (1964). Observations on the development of the heteropod molluscs *Pterotrachea hippocampus* and *Firoloida desmaresti*. *Bulletin of Marine Science* 14: 529–538.

Packard, A. (1972). Cephalopods and fish: The limits of convergence. *Biological Reviews* 47: 241–307.

Packard, A. & Trueman, E. R. (1974). Muscular activity of the mantle of *Sepia* and *Loligo* (Cephalopoda) during respiration and jetting and its physiological interpretation. *Journal of Experimental Biology* 61: 411–420.

Packard, A., Bone, Q. & Hignette, M. (1980). Breathing and swimming movements in a captive *Nautilus*. *Journal of the Marine Biological Association of the United Kingdom* 60: 313–328.

Packard, A. (1988). Visual tactics and evolutionary strategies, pp. 89–104 *in* J. Wiedmann & Kullmann, J. (eds.), *Cephalopods Present and Past. 2nd International Cephalopod Symposium: O. H. Schindewolf Symposium, Tübingen 1985*. Stuttgart, E. Schweizerbart'sche Verlagsbuchhandlung.

Packard, A. & Wurtz, M. (1994). An octopus, *Ocythoe*, with a swimbladder and triple jets. *Philosophical Transactions of the Royal Society B* 344: 261–275.

Pafort-van Iersel, T. & van der Spoel, S. (1979). The structure of the columellar muscle system in *Clio pyramidata* and *Cymbulia peroni* (Thecosomata, Gastropoda) with a note on the phylogeny of both species. *Bijdragen tot de Dierkunde* 48: 111–126.

Pafort-van Iersel, T. & van der Spoel, S. (1986). Schizogamy in the planktonic opisthobranch *Clio* – A previously undescribed mode of reproduction in the Mollusca. *International Journal of Invertebrate Reproduction & Development* 10: 43–50.

Page, K. N. (2008). The evolution and geography of Jurassic ammonoids. *Proceedings of the Geologists' Association* 119: 35–57.

Page, L. R. (1995). Similarities in form and developmental sequence for three larval shell muscles in nudibranch gastropods. *Acta Zoologica* 76: 177–191.

Page, L. R. (2011). Developmental modularity and phenotypic novelty within a biphasic life cycle: Morphogenesis of a cone snail venom gland. *Proceedings of the Royal Society B* 279: 77–83.

Paine, R. T. (1963). Food recognition and predation on opisthobranchs by *Navanax inermis* (Gastropoda, Opisthobranchia). *The Veliger* 6: 1–9.

Pakarinen, E. (1994a). Autotomy in arionid and limacid slugs. *Journal of Molluscan Studies* 60: 19–23.

Pakarinen, E. (1994b). The importance of mucus as a defence against carabid beetles by the slugs *Arion fasciatus* and *Deroceras reticulatum*. *Journal of Molluscan Studies* 60: 149–155.

Palmer, A. R. (1992). Calcification in marine molluscs: How costly is it? *Proceedings of the National Academy of Sciences of the United States of America* 89: 1379–1382.

Palmer, C. (1974). A supraspecific classification of the scaphopod Mollusca. *The Veliger* 17: 115–123.

Palmer, C. P. & Steiner, G. (1998). Class Scaphopoda – Introduction, pp. 431–443 *in* P. L. Beesley, Ross, G. J. B. & Wells, A. (eds.), *Mollusca: The Southern Synthesis. Part A. Fauna of Australia.* Melbourne, CSIRO Publishing.

Pan, H.-Z. & Erwin, D. H. (2002). Gastropods from the Permian of Guangxi and Yunnan provinces, South China. *Journal of Paleontology* 76: 1–49.

Paps, J., Baguñà, J. & Riutort, M. (2009). Lophotrochozoa internal phylogeny: New insights from an up-to-date analysis of nuclear ribosomal genes. *Proceedings of the Royal Society B* 276: 1245–1254.

Paraense, W. L. (1976). The sites of cross- and self-fertilization in planorbid snails. *Revista Brasileira de Biologia* 36: 535–539.

Park, J. K. & Ó Foighil, D. (2000). Sphaeriid and corbiculid clams represent separate heterodont bivalve radiations into freshwater environments. *Molecular Phylogenetics and Evolution* 14: 75–88.

Parkhaev, P. (1998). Siphonoconcha – A new class of Early Cambrian bivalved organisms. *Paleontological Journal* 32: 1–15.

Parkhaev, P. (2000). The functional morphology of the Cambrian univalved Mollusks – helcionellids. Part 1. *Paleontological Journal* 34: 392–399.

Parkhaev, P. (2001). The functional morphology of the Cambrian univalved mollusks – helcionellids. Part 2. *Paleontological Journal* 35: 470–475.

Parkhaev, P. (2002a). Phylogenesis and the system of the Cambrian univalved mollusks. *Paleontological Journal* 36: 25–36.

Parkhaev, P. (2002b). Muscle scars of the Cambrian univalved molluscs and their significance for systematics. *Paleontological Journal* 36: 453–459.

Parkhaev, P. (2004). New data on the morphology of shell muscles in Cambrian helcionelloid mollusks. *Paleontological Journal* 38: 254–256.

Parkhaev, P. (2006a). Adaptive radiation of the Cambrian helcionelloid mollusks (Gastropoda, Archaeobranchia), pp. 282–296 *in* S. V. Rozhnov (ed.), *Evolution of the Biosphere and Biodiversity. Towards the 70th Anniversary of A. Y. Rozanov.* Moscow, KMK Scientific Press.

Parkhaev, P. (2006b). On the genus *Auricullina* Vassiljeva, 1998 and shell pores of the Cambrian helcionelloid mollusks. *Paleontological Journal* 40: 20–33.

Parkhaev, P. (2007). The Cambrian 'basement' of gastropod evolution, pp. 415–421 *in* P. Vickers-Rich & Komarower, P. (eds.), *The Rise and Fall of the Ediacaran Biota. Geological Society, London, Special Publications.* Vol. 286.

Parkhaev, P. (2008). The Early Cambrian radiation of Mollusca, pp. 33–69 *in* W. F. Ponder & Lindberg, D. R. (eds.), *Phylogeny and Evolution of the Mollusca.* Berkeley, CA, University of California Press.

Parkhaev, P. & Demidenko, E. (2010). Zooproblematica and mollusca from the Lower Cambrian Meishucun section (Yunnan, China) and taxonomy and systematics of the Cambrian small shelly fossils of China. *Paleontological Journal* 44: 883–1161.

Parkhaev, P. & Karlova, G. A. (2011). Taxonomic revision and evolution of Cambrian mollusks of the genus *Aldanella* Vostokova, 1962 (Gastropoda: Archaeobranchia). *Paleontological Journal* 45: 1145–1205.

Parkhaev, P. (2014). Structure of shell muscles in the Cambrian gastropod genus *Bemella* (Gastropoda: Archaeobranchia: Helcionellidae). *Paleontological Journal* 48: 17–25.

Parkhaev, P. (2017). On the position of Cambrian archaeobranchians in the system of the class Gastropoda. *Paleontological Journal* 51: 453–463.

Pashchenko, S. V. & Drozdov, A. L. (1998). Morphology of gametes in five species of far-eastern chitons. *Invertebrate Reproduction & Development* 33: 47–56.

Passamaneck, Y. J., Schander, C. & Halanych, K. M. (2004). Investigation of molluscan phylogeny using large-subunit and small-subunit nuclear rRNA sequences. *Molecular Phylogenetics and Evolution* 32: 25–38.

Payne, J. L., Lehrmann, D. J., Wei, J., Orchard, M. J., Schrag, D. P. & Knoll, A. H. (2004). Large perturbations of the carbon cycle during recovery from the end-Permian extinction. *Science* 305: 506–509.

Payne, J. L. (2005). Evolutionary dynamics of gastropod size across the end-Permian extinction and through the Triassic recovery interval. *Paleobiology* 31: 269–290.

Pchelintsev, V. F. (1963). Briukhonogie Mezozoia Gornogo Kryma. [Mesozoic Gastropoda of the Crimean highlands] [in Russian]. *Geologicheskii Muzei Karpinskogo, Seriia Monograficheskaia* 4: 1–132.

Pearce, T. A. & Gaertner, A. (1996). Optimal foraging and mucus-trail following in the carnivorous land snail *Haplotrema concavum* (Gastropoda: Pulmonata). *Malacological Review* 29: 85–99.

Pearse, J. S. (1979). Polyplacophora, pp. 27–86 *in* A. C. Giese & Pearse, J. S. *Reproduction of Marine Invertebrates. Molluscs: Pelecypods and Lesser Classes*. Vol. 5. New York, Academic Press.

Pechenik, J. A. (1991). *Biology of the Invertebrates*, 2nd edition. Dubuque, IA, Wm. C. Brown.

Peckmann, J., Gischler, E., Oschmann, W. & Reitner, J. (2001). An Early Carboniferous seep community and hydrocarbon-derived carbonates from the Harz Mountains, Germany. *Geology* 29: 271–274.

Peebles, B., Gordon, K., Smith, A. & Smith, G. (2017). First record of carotenoid pigments and indications of unusual shell structure in chiton valves. *Journal of Molluscan Studies* 83: 476–480.

Peel, J. S. (1977). Relationship and internal structure of a new *Pilina* (Monoplacophora) from the Late Ordovician of Oklahoma. *Journal of Paleontology* 51: 116–122.

Peel, J. S. (1991a). The classes Tergomya and Helcionelloida, and early molluscan evolution. *Bulletin Grønlands Geologiske Undersøgelse* 161: 11–65.

Peel, J. S. (1991b). Functional morphology, evolution and systematics of Early Palaeozoic univalved molluscs (Geological Survey of Greenland). *Bulletin Grønlands Geologiske Undersøgelse* 161: 1–116.

Peel, J. S. (1991c). Functional morphology of the Class Helcionella nov., and the early evolution of the Mollusca, pp. 157–177 *in* S. Conway Morris & Simonetta, A. M. (eds.), *The Early Evolution of Metazoa and the Significance of Problematic Taxa: Proceedings of an International Symposium [University of Camerino, Italy]*. Cambridge, UK, Cambridge University Press.

Peel, J. S. (1993). Muscle scars and mode of life of *Carinaropsis* (Bellerophontoidea, Gastropoda) from the Ordovician of Tennessee. *Journal of Paleontology* 67: 528–534.

Peel, J. S. & Horný, R. J. (1999). Muscle scars and systematic position of the Lower Palaeozoic limpets *Archinacella* and *Barrandicella* gen. n. (Mollusca). *Journal of the Czech Geological Society* 44: 97–115.

Peel, J. S. (2004). *Pinnocaris* and the origin of scaphopods. *Acta Palaeontologica Polonica* 49: 543–550.

Peel, J. S. (2006). Scaphopodization in Palaeozoic molluscs. *Palaeontology* 49: 1357–1364.

Pekkarinen, M. & Englund, V. P. M. (1995). Description of unionid glochidia with a table aiding in their identification. *Archives of Hydrobiology* 134: 515–531.

Pelman, L. (1985). New stenothecoids from the Lower Cambrian of West Mongolia, pp. 103–114 *in* B. S. Sokolov & Zhuravleva, I. T. (eds.), *Problematics of the Late Precambrian and Paleozoic*. Vol. 632. Moscow, Trudy Instituta Geologii i Geofiziki, Akademiya Nauk SSSR, Sibirskoe Otdelenie.

Pelseneer, P. (1888). *Report on the Pteropoda collected by H.M.S. Challenger during the years 1873–76. Part III – Anatomy. Report of the Scientific Results of the Voyage of H.M.S. Challenger during the years 1873–76.* London, H.M Government. 23: 1–97.

Pelseneer, P. (1894). Recherches sur divers opisthobranches. *Mémoires couronnés et mémoires des savants étrangers publiés par l'Académie royale des sciences, des lettres, et des beaux-arts de Belgique. Extrait du tome* 53: 1–157.

Pelseneer, P. (1901). Études sur des gastéropodes pulmonés. *Mémoires de l'Académie royale des sciences, des lettres et des beaux-arts de Belgique* 54: 1–76.

Pelseneer, P. (1906). Mollusca, pp. 1–355 *in* E. R. Lankester (ed.), *A Treatise on Zoology*. London, V. Adam and Charles Black.

Pelseneer, P. (1920). Les Variations et Leur Hérédité Chez les Mollusques. *Académie Royal de Belgique, Classe des Sciences, Mémoires* 5: 1–826.

Pelseneer, P. (1929). La variabilité relative des sexes, particulierement dans les mollusques, pp. 186–187. *International Congress of Zoology (Xe Congrès International de Zoologie, Tenu à Budapest du 4 au 10 Septembre 1927)*. Vol. 10. Budapest, Imprimerie Stephaneus s.a.

Penney, B. K. (2002). Lowered nutritional quality supplements nudibranch chemical defense. *Oecologia* 132: 411–418.

Penney, B. K. (2006). Morphology and biological roles of spicule networks in *Cadlina luteomarginata* (Nudibranchia, Doridina). *Invertebrate Biology* 125: 222–232.

Penney, B. K. (2008). Phylogenetic comparison of spicule networks in cryptobranchiate dorid nudibranchs (Gastropoda, Euthyneura, Nudibranchia, Doridina). *Acta Zoologica* 89: 311–329.

Penney, B. K. (2009). A comment on F. Aguado & A. Marin: 'Warning coloration associated with nematocyst-based defences in aeolidioidean nudibranchs'. *Journal of Molluscan Studies* 75: 199–200.

Penney, B. K., Ehresmann, K. R., Jordan, K. J. & Rufo, G. (2018). Micro-computed tomography of spicule networks in three genera of dorid sea-slugs (Gastropoda: Nudipleura: Doridina) shows patterns of phylogenetic significance. *Acta Zoologica*: 1–19. https://doi.org/10.1111/azo.12266.

Pennington, J. T. & Chia, F.-S. (1985). Gastropod torsion: A test of Garstang's hypothesis. *Biological Bulletin* 169: 391–396.

Pérez-Huerta, A., Cusack, M., McDonald, S., Marone, F., Stampanoni, M. & Mackay, S. (2009). Brachiopod punctae: A complexity in shell biomineralisation. *Journal of Structural Biology* 167: 62–67.

Perkins, B. F. (1969). Rudist morphology, pp. N751–N764 *in* R. C. Moore (ed.), *Treatise on Invertebrate Paleontology Part N*. Vol. 2. Lawrence, KS, Geological Society of America and University of Kansas Press.

Pernet, B. & Kohn, A. J. (1998). Size-related obligate and facultative parasitism in the marine gastropod *Trichotropis cancellata*. *Biological Bulletin* 195: 349–356.

Pernice, M., Wetzel, S., Gros, O., Boucher-Rodoni, R. & Dubilier, N. (2007). Enigmatic dual symbiosis in the excretory organ of *Nautilus macromphalus* (Cephalopoda: Nautiloidea). *Proceedings of the Royal Society B* 274: 1143–1152.

Pernice, M. & Boucher-Rodoni, R. (2012). Occurrence of a specific dual symbiosis in the excretory organ of geographically distant nautiloid populations. *Environmental Microbiology Reports* 4: 504–511.

Perrier, R. (1889). Recherches sur l'anatomie et l'histologie du rein des gastéropodes. Prosobranchiata. *Annales des sciences naturelles. Zoologie* 8: 61–192.

Perrier, R. & Fischer, H. (1909). Sur la cavité palléale et ses dépendances chez les Bulléens. *Comptes rendus de l'Académie des sciences* 153: 1–3.

Perrier, R. & Fischer, H. (1911). Recherches anatomiques et histologiques sur la cavité palléale et ses dependances chez les Bulléens. (*Acteon, Hydatina, Scaphander, Akera*). *Annales des sciences naturelles. Zoologie* 14: 1–189.

Perron, F. E. (1981). The partitioning of reproductive energy between ova and protective capsules in marine gastropods of the genus *Conus. American Naturalist* 118: 110–118.

Person, P. & Philpott, D. E. (1969). The nature and significance of invertebrate cartilages. *Biological Reviews* 44: 1–16.

Peters, S. E. & Gaines, R. R. (2012). Formation of the 'Great Unconformity' as a trigger for the Cambrian explosion. *Nature* 484: 363–366.

Peters, S. E., Zhang, C., Livny, M. & Ré, C. (2014). A machine-compiled macroevolutionary history of Phanerozoic life. arXiv:1406.2963: 1–43.

Peters, W. (1972). Occurrence of chitin in Mollusca. *Comparative Biochemistry and Physiology Part B* 41: 541–550.

Petsios, E. & Bottjer, D. J. (2013). Turnover of dominant benthic taxa in the Early Triassic. *Geological Society of America Abstracts with Programs* 45: 882.

Pichon, D., Gaia, V., Norman, M. D. & Boucher-Rodoni, R. (2005). Phylogenetic diversity of epibiotic bacteria in the accessory nidamental glands of squids (Cephalopoda: Loliginidae and Idiosepiidae). *Marine Biology* 147: 1323–1332.

Pichon, Y., Mouëza, M. & Frenkiel, L. (1980). Mechanoreceptor properties of the sense organ of the cruciform muscle in a tellinacean lamellibranch, *Donax trunculus* L., an electrophysiological approach. *Marine Biology Letters* 1: 273–284.

Picken, G. B. & Allan, D. (1983). Unique spawning behavior by the Antarctic limpet *Nacella* (*Patinigera*) *concinna* (Strebel, 1908). *Journal of Experimental Marine Biology and Ecology* 71: 283–288.

Pickford, G. E. (1947). Untitled in 'Comments by readers'. *Science* 105: 522.

Pickford, G. E. (1949). *Vampyroteuthis infernalis* Chun an archaic dibranchiate cephalopod: II: External anatomy. *Dana-Report: The Carlsberg Foundation's Oceanographical Expedition Round the World 1928–30.* Copenhagen, Carlsberg Foundation. 32: 1–132.

Pignatti, J. S. & Mariotti, N. (1996). Systematics and phylogeny of the Coleoidea (Cephalopoda): A comment upon recent works and their bearing on the classification of the Aulacocerida. *Palaeopelagos* 5: 33–44.

Pigneur, L.-M., Hedtke, S. M., Etoundi, E. & van Doninck, K. (2012). Androgenesis: A review through the study of the selfish shellfish *Corbicula* spp. *Heredity* 108: 581–591.

Pilkington, J. B., Little, C. & Stirling, P. E. (1984). A respiratory current in the mantle cavity of *Amphibola crenata* (Mollusca, Pulmonata). *Journal of the Royal Society of New Zealand* 14: 327–334.

Pilsbry, H. A. (1891). *Manual of Conchology; Structural and Systematic. Acmeidae, Lepetidae, Patellidae, Titiscaniidae.* Vol. 13. Philadelphia, Conchological Section, Academy of Natural Sciences.

Pilsbry, H. A. (1892). *Manual of Conchology; Structural and Systematic. Polyplacophora, (Chitons,) Lepidopleuridae, Ischnochitonidae, Chitonidae, Mopaliidae.* Vol. 14. Philadelphia, Conchological Section, Academy of Natural Sciences.

Pilsbry, H. A. (1900a). On the zoological position of *Partula* and *Achatinella. Proceedings of the Academy of Natural Sciences of Philadelphia* 1900: 561–567.

Pilsbry, H. A. (1900b). New South American land snails. *Proceedings of the Academy of Natural Sciences of Philadelphia* 1900: 385–394.

Pilsbry, H. A. (1909). An internal septum in *Holospira bartschi. The Nautilus* 23: 32.

Pilsbry, H. A. (1948). *Land Mollusca of North America (north of Mexico).* Philadelphia, PA, Wickersham Printing Co.

Pinchuck, S. C. & Hodgson, A. N. (2009). Comparative structure of the lateral pedal defensive glands of three species of *Siphonaria* (Gastropoda: Basommatophora). *Journal of Molluscan Studies* 75: 371–380.

Plagányi, E. E. & Branch, G. M. (2000). Does the limpet *Patella cochlear* fertilize its own algal garden? *Marine Ecology Progress Series* 194: 113–122.

Plate, L. H. (1897). Die anatomie und phylogenie der Chitonen. *Zoologisches Jahrbuch Supplement* 4: 1–243.

Plate, L. H. (1899). Die anatomie und phylogenie der Chitonen. *Zoologisches Jahrbuch Supplement* 2: 15–216.

Plate, L. H. (1901). Die anatomie und phylogenie der Chitonen. *Zoologisches Jahrbuch Supplement* 2: 281–600.

Platnick, N. I. (1976). Concepts of dispersal in historical biogeography. *Systematic Zoology* 25: 294–295.

Plazzi, F. & Passamonti, M. (2010). Towards a molecular phylogeny of Mollusks: Bivalves' early evolution as revealed by mitochondrial genes. *Molecular Phylogenetics and Evolution* 57: 641–657.

Plazzi, F., Ceregato, A., Taviani, M. & Passamonti, M. (2011). A molecular phylogeny of bivalve mollusks: Ancient radiations and divergences as revealed by mitochondrial genes. *PLoS ONE* 6: e27147.

Počta, F. (1904). *Rukovět' palaeozoologie.* Vol. I. Invertebrata. Prague, Nákladem České akademie císaře Františka Josefa pro vědy, slovesnost a umění.

Podsiadlowski, L., Braband, A., Struck, T. H., Döhren, J., Von & Bartolomaeus, T. (2009). Phylogeny and mitochondrial gene order variation in Lophotrochozoa in the light of new mitogenomic data from Nemertea. *BMC Genomics* 10: 364.

Pohlo, R. (1969). Confusion concerning deposit feeding in the Tellinacea. *Proceedings of the Malacological Society of London* 38: 361–364.

Pohlo, R. (1982). Evolution of the Tellinacea (Bivalvia). *Journal of Molluscan Studies* 48: 245–256.

Pojeta, J. (1971). Review of Ordovician pelecypods. *US Geological Survey Professional Paper* 695: 1–46.

Pojeta, J., Runnegar, B. N., Morris, N. J. & Newell, N. D. (1972). Rostroconchia: A new class of bivalved mollusks. *Science* 177: 264–267.

Pojeta, J. & Palmer, T. J. (1976). The origin of rock boring in mytilacean pelecypods. *Alcheringa* 1: 167–179.

Pojeta, J. & Runnegar, B. N. (1976). The paleontology of rostroconch mollusks and the early history of the phylum Mollusca. *Geological Survey Professional Paper* 968: 1–88.

Pojeta, J. & Gilbert-Tomlinson, J. (1977). Australian Ordovician pelecypod molluscs. *Bureau of Mineral Resources, Geology and Geophysics Bulletin* 174: 1–64.

Pojeta, J. (1978). The origin and early taxonomic diversification of pelecypods. *Philosophical Transactions of the Royal Society B* 284: 225–246.

Pojeta, J. (1979). Geographic distribution of Cambrian and Ordovician rostroconch mollusks, pp. 27–36 *in* J. Gray & Boucot, A. J. (eds.), *Historical Biogeography, Plate Tectonics, and the Changing Environment.* Corvallis, OR, Oregon State University Press.

Pojeta, J. & Runnegar, B. N. (1979). *Rhytiodentalium kentuckyensis*, a new genus and new species of Ordovician scaphopod, and the early history of scaphopod mollusks. *Journal of Paleontology* 53: 530–541.

Pojeta, J. (1980). Molluscan phylogeny. *Tulane Studies in Geology and Paleontology* 16: 55–80.

Pojeta, J. (1985). Early evolutionary history of diasome mollusks, pp. 102–121 *in* T. W. Broadhead. *Mollusks: Notes for a Short Course organized by D. J. Bottjer, C. S. Hickman and P. D. Ward. University of Tennessee Studies in Geology.* Vol. 13. Knoxville TN, University of Tennessee Department of Geological Sciences.

Pojeta, J. & Runnegar, B. N. (1985). The early evolution of diasome molluscs, pp. 295–336 *in* E. R. Trueman & Clarke, M. R. (eds.), *Evolution. The Mollusca.* Vol. 10. New York, Academic Press.

Pojeta, J. (1987). Class Pelecypoda, pp. 386–435 *in* R. S. Boardman, Cheetham, A. H. & Rowell, A. J. (eds.), *Fossil Invertebrates.* Palo Alto, CA, Blackwell Scientific Publications.

Pojeta, J. (2000). Cambrian Pelecypoda (Mollusca). *American Malacological Bulletin* 15: 157–166.

Pojeta, J., Eernisse, D. J., Hoare, R. D. & Henderson, M. D. (2003). *Echinochiton dufoei*: A new spiny Ordovician chiton. *Journal of Paleontology* 77: 646–654.

Pojeta, J. & Dufoe, J. (2008). New information about *Echinochiton dufoei*, the Ordovician spiny chiton. *American Malacological Bulletin* 25: 25–34.

Pojeta, J., Vendrasco, M. J. & Darrough, G. (2010). Upper Cambrian Chitons (Mollusca, Polyplacophora) from Missouri, USA. *Bulletins of American Paleontology* 2010: 1–88.

Pokora, Z. (1989). Strobilation as a form of asexual reproduction in certain pteropods (Gastropoda, Opisthobranchia): Functional significance of this process in their life cycle. *Przeglad Zoologiczny* 33: 397–410.

Pokryszko, B. M. (1987). On the aphally in the Vertiginidae (Gastropoda: Pulmonata: Orthurethra). *Journal of Conchology* 32: 365–375.

Pola, M., Cervera, J. L. & Gosliner, T. M. (2003). The genus *Roboastra* Bergh, 1877 (Nudibranchia, Polyceridae, Nembrothinae) in the Atlantic Ocean. *Proceedings of the California Academy of Sciences* 54: 381–392.

Pola, M. & González Duarte, M. M. (2008). Is self-fertilization possible in nudibranchs? *Journal of Molluscan Studies* 74: 305–308.

Pola, M. & Gosliner, T. M. (2010). The first molecular phylogeny of cladobranchian opisthobranchs (Mollusca, Gastropoda, Nudibranchia). *Molecular Phylogenetics and Evolution* 56: 931–941.

Polechová, M. (2015). The youngest representatives of the genus *Ribeiria* Sharpe, 1853 from the late Katian of the Prague Basin (Bohemia). *Estonian Journal of Earth Sciences* 64: 84–90.

Ponder, W. F. & Warren, T. P. (1965). An illustrated guide to the genera of the land mollusca of New Zealand. *Tane* 11: 21–46.

Ponder, W. F. (1966). A new family of the Rissoacea from New Zealand. *Records of the Dominion Museum* 5: 177–184.

Ponder, W. F. (1968). The morphology of some small New Zealand prosobranchs. *Records of Dominion Museum Wellington, New Zealand* 6: 61–95.

Ponder, W. F. (1970a). The morphology of *Alcithoe arabica* (Gastropoda: Volutidae). *Malacological Review* 3: 127–165.

Ponder, W. F. (1970b). Some aspects of the morphology of four species of the neogastropod family Marginellidae with a discussion on the evolution of the toxoglossan poison gland. *Journal of the Malacological Society of Australia* 2: 55–81.

Ponder, W. F. (1971). Some New Zealand and subantarctic bivalves of the Cyamiacea and Leptonacea with description of new taxa. *Records of the Dominion Museum* 7: 119–141.

Ponder, W. F. (1972). The morphology of some mitriform gastropods with special reference to their alimentary and reproductive systems (Neogastropoda). *Malacologia* 11: 295–342.

Ponder, W. F. (1974). The origin and evolution of the Neogastropoda. *Malacologia* 12: 295–338.

Ponder, W. F. (1975). Lizard Island Malacological Workshop, Report. *Australian Shell News* 13: 1–3.

Ponder, W. F. & Gooding, R. U. (1978). Four new eulimid gastropods associated with shallow-water diadematid echinoids in the Western Pacific. *Pacific Science* 32: 157–181.

Ponder, W. F. & Yoo, E. K. (1978). A revision of the Eatoniellidae of Australia (Mollusca, Gastropoda, Littorinacea). *Records of the Australian Museum* 31: 606–658.

Ponder, W. F. & Yoo, E. K. (1980). A review of the genera of the Cingulopsidae with a revision of the Australian and tropical Indo-Pacific species (Mollusca: Gastropoda: Prosobranchia). *Records of the Australian Museum* 33: 1–88.

Ponder, W. F., Colman, P. H., Yonge, C. M. & Colman, M. H. (1981). The taxonomic position of *Hemidonax* with a review of the genus (Bivalvia Cardiacea). *Journal of the Malacological Society of Australia* 5: 41–64.

Ponder, W. F. (1983). Review of the genera of the Barleeidae (Mollusca: Gastropoda: Rissoacea). *Records of the Australian Museum* 35: 231–281.

Ponder, W. F. & Hall, S. J. (1983). Pelycidiidae, a new family of archaeogastropod mollusks. *The Nautilus* 97: 30–35.

Ponder, W. F. (1985). A review of the genera of the Rissoidae (Mollusca, Mesogastropoda: Rissoacea). *Records of the Australian Museum, Supplement* 4: 1–221.

Ponder, W. F. (1986). Glacidorbidae (Glacidorbacea: Basommatophora), a new family and superfamily of operculate freshwater gastropods. *Zoological Journal of the Linnean Society* 87: 53–83.

Ponder, W. F. (1988). The Truncatelloidean (= Rissoacean) radiation – A preliminary phylogeny, pp. 129–166 *in* W. F. Ponder, Eernisse, D. J. & Waterhouse, J. H. (eds.), *Prosobranch Phylogeny. Malacological Review Supplement.* Ann Arbor, MI, Malacological Review.

Ponder, W. F. & Clark, G. A. (1988). A morphological and electrophoretic examination of *Hydrobia buccinoides*, a variable brackish-water gastropod from temperate Australia (Mollusca: Hydrobiidae). *Australian Journal of Zoology* 36: 661–689.

Ponder, W. F. & Warén, A. (1988). Classification of the Caenogastropoda and Heterostropha – A list of the family-group names and higher taxa (Appendix), pp. 288–328 *in* W. F. Ponder, Eernisse, D. J. & Waterhouse, J. H. (eds.), *Prosobranch Phylogeny. Malacological Review Supplement.* Ann Arbor, MI, Malacological Review.

Ponder, W. F. (1990a). The anatomy and relationships of the Orbitestellidae (Gastropoda: Heterobranchia). *Journal of Molluscan Studies* 56: 515–532.

Ponder, W. F. (1990b). The anatomy and relationships of a marine valvatoidean (Gastropoda: Heterobranchia). *Journal of Molluscan Studies* 56: 533–555.

Ponder, W. F. (1991). Marine valvatoidean gastropods: Implications for early heterobranch phylogeny. *Journal of Molluscan Studies* 57: 21–32.

Ponder, W. F. & Taylor, J. D. (1992). Predatory shell drilling by two species of *Austroginella* (Gastropoda: Marginellidae). *Journal of Zoology* 228: 317–328.

Ponder, W. F., Clark, G. A., Miller, A. C. & Toluzzi, A. (1993). On a major radiation of freshwater snails in Tasmania and eastern Victoria: A preliminary overview of the *Beddomeia* group (Mollusca: Gastropoda: Hydrobiidae). *Invertebrate Taxonomy* 7: 501–750.

Ponder, W. F. (1994). The anatomy and relationships of three species of vitrinelliform gastropods (Caenogastropoda, Rissooidea) from Hong Kong, pp. 243–281 *in* B. Morton (ed.), *The Malacofauna of Hong Kong and Southern China, III: Proceedings of the Third International Workshop on the Malacofauna of Hong Kong and Southern China, Hong Kong,* 13 April–1 May 1992. Vol. 1. Hong Kong, Hong Kong University Press.

Ponder, W. F. & Lindberg, D. R. (1996). Gastropod phylogeny – Challenges for the 90s, pp. 135–154 in J. D. Taylor (ed.), *Origin and Evolutionary Radiation of the Mollusca.* Oxford, UK, Oxford University Press.

Ponder, W. F. & Lindberg, D. R. (1997). Towards a phylogeny of gastropod molluscs: An analysis using morphological characters. *Zoological Journal of the Linnean Society* 119: 83–265.

Ponder, W. F. & de Keyzer, R. G. (1998). Superfamily Cyamioidea, pp. 318–322 in P. L. Beesley, Ross, G. J. B. & Wells, A. (eds.), *Mollusca: The Southern Synthesis. Part A. Fauna of Australia.* Melbourne, VIC, CSIRO Publishing.

Ponder, W. F. (1999). Using museum collection data to assist in biodiversity assessment, pp. 253–256 in W. F. Ponder & Lunney, D. (eds.), *The Other 99%: The Conservation and Biodiversity of Invertebrates.* Sydney, NSW, Royal Zoological Society of New South Wales.

Ponder, W. F. & Lunney, D., Eds. (1999). *The Other 99%: The Conservation and Biodiversity of Invertebrates [Proceedings of a four-day meeting held at the Australian Museum, Sydney, 9–12 December 1997].* Transactions of the Royal Zoological Society of New South Wales. Sydney, NSW, Royal Zoological Society of New South Wales.

Ponder, W. F. & Avern, G. J. (2000). The Glacidorbidae (Mollusca: Gastropoda: Heterobranchia) of Australia. *Records of the Australian Museum* 52: 307–353.

Ponder, W. F. & Colgan, D. J. (2002). What makes a narrow range taxon? Insights from Australian freshwater snails. *Invertebrate Systematics* 16: 571–582.

Ponder, W. F., Colgan, D. J., Healy, J. M., Nützel, A., de Simone, L. R. L. & Strong, E. E. (2008). Caenogastropoda, pp. 331–383 in W. F. Ponder & Lindberg, D. R. (eds.), *Phylogeny and Evolution of the Mollusca.* Berkeley, CA, University of California Press.

Ponder, W. F. & Lindberg, D. R. (2008). Molluscan evolution and phylogeny: An introduction, pp. 1–17 in W. F. Ponder & Lindberg, D. R. (eds.), *Phylogeny and Evolution of the Mollusca.* Berkeley, CA, University of California Press.

Ponder, W. F., Carter, G. A., Flemons, P. & Chapman, R. R. (2010). Evaluation of museum collection data for use in biodiversity assessment. *Conservation Biology* 15: 648–657.

Poon, P. A. (1987). The diet and feeding behavior of *Cadulus tolmei* Dall, 1897 (Scaphopoda: Siphonodentaloida). *The Nautilus* 101: 88–92.

Popov, L. (1992). The Cambrian radiation of brachiopods, pp. 399–423 in J. H. Lipps & Signor, P. W. (eds.), *Origin and Early Evolution of the Metazoa. Topics in Geobiology.* New York, Springer Science + Business Media.

Porter, S. M. (2008). Skeletal microstructure indicates chancelloriids and halkieriids are closely related. *Palaeontology* 51: 865–879.

Poulicek, M. & Jeuniaux, C. (1981). La matrice organique de la coquille et position phylétique de *Neopilina galatheae* (Mollusques, Monoplacophores). *Annales de la Société royale zoologique de Belgique* 111: 143–150.

Poulicek, M., Voss-Foucart, M.-F. & Jeuniaux, C. (1991). Regressive shell evolution among opisthobranch gastropods. *Malacologia* 32: 223–232.

Powell, A. W. B. (1966). The molluscan families Speightiidae and Turridae. *Bulletin of the Auckland Institute and Museum* 5: 1–184.

Powell, A. W. B. (1973). The patellid limpets of the world (Patellidae). *Indo-Pacific Mollusca* 3: 75–206.

Powell, A. W. B. (1979). *The New Zealand Mollusca; Marine, Land and Freshwater Snails.* Auckland, New Zealand, Collins.

Powell, E. N. & Cummins, H. (1985). Are molluscan maximum life spans determined by long-term cycles in benthic communities? *Oecologia* 67: 177–182.

Powell, E. N. & Stanton, R. J. (1985). Estimating biomass and energy flow of molluscs in palaeo-communities. *Paleontology* 28: 1–34.

Prezant, R. S. (1981). The arenophilic radial glands of the Lyonsiidae (Bivalvia, Anomalodesmata), with notes on lyonsid evolution. *Malacologia* 20: 267–289.

Prezant, R. S. (1985). Derivations of arenophilic mantle glands in the Anomalodesmata. *Malacologia* 26: 273–276.

Prezant, R. S. (1998). Subclass Palaeoheterodonta – Introduction, pp. 289–294 in P. L. Beesley, Ross, G. J. B. & Wells, A. (eds.), *Mollusca: The Southern Synthesis. Part A. Fauna of Australia.* Melbourne, VIC, CSIRO Publishing.

Price, R. M. (2003). Columellar muscle of neogastropods: Muscle attachment and the function of columellar folds. *Biological Bulletin* 205: 351–366.

Proćków, M., Strzala, T., Kuźnik-Kowalska, E., Proćków, J. & Mackiewicz, P. (2017). Ongoing speciation and gene flow between taxonomically challenging *Trochulus* species complex (Gastropoda: Hygromiidae). *PLoS ONE* 12: e0170460.

Pruvot-Fol, A. (1937). Étude d'un prosobranche d'eau douce: *Helicostoa sinensis* Lamy. *Bulletin de la Société zoologique de France* 62: 250–257.

Pruvot, G. (1890). Sur le développement d'un Solénogastre. *Comptes rendus hebdomadaire des séances de l'Académie des sciences* 111: 689–692.

Przeslawski, R. (2004). A review of the effects of environmental stress on embryonic development within intertidal gastropod egg masses. *Journal of Molluscan Studies* 24: 43–63.

Przeslawski, R., Davis, A. R. & Benkendorff, K. (2004). Effects of ultraviolet radiation and visible light on the development of encapsulated molluscan embryos. *Marine Ecology Progress Series* 268: 151–160.

Przeslawski, R. (2011). Notes on the egg capsule and variable embryonic development of *Nerita melanotragus* (Gastropoda: Neritidae). *Molluscan Research* 31: 152–158.

Puillandre, N., Samadi, S., Boisselier, M.-C., Cruaud, C. & Bouchet, P. (2009). Molecular data provide new insights on the phylogeny of the Conoidea (Neogastropoda). *The Nautilus* 123: 202–210.

Puillandre, N., Kantor, Y. I., Sysoev, A. V., Couloux, A., Meyer, C., Rawlings, T., Todd, J. A. & Bouchet, P. (2011). The dragon tamed? A molecular phylogeny of the Conoidea (Gastropoda). *Journal of Molluscan Studies* 77: 259–272.

Purchon, R. D. (1956). The stomach in the Protobranchiata and Septibranchia (Lamellibranchia). *Proceedings of the Zoological Society of London* 127: 511–525.

Purchon, R. D. (1959). Phylogenetic classification of the Lamellibranchia, with special reference to the Protobranchia. *Proceedings of the Malacological Society of London* 33: 224–230.

Purchon, R. D. (1960). The stomach in the Eulamellibranchia: Stomach types IV and V. *Proceedings of the Zoological Society of London* 135: 431–489.

Purchon, R. D. (1963). Phylogenetic classification of the Bivalvia, with special reference to the Septibranchia. *Proceedings of the Malacological Society of London* 35: 71–80.

Purchon, R. D. (1978). An analytical approach to a classification of the Bivalvia. *Philosophical Transactions of the Royal Society B* 284: 425–436.

Purchon, R. D. (1987a). The stomach in the Bivalvia. *Philosophical Transactions of the Royal Society B* 316: 183–276.

Purchon, R. D. (1987b). Classification and evolution of the Bivalvia: An analytical study. *Philosophical Transactions of the Royal Society B* 316: 277–302.

Purchon, R. D. (1990). Stomach structure, classification and evolution of the Bivalvia, pp. 73–82 *in* B. Morton (ed.), *The Bivalvia. Proceedings of a Memorial Symposium in Honour of Sir Charles Maurice Yonge (1899–1986)*, Edinburgh, 1986. Hong Kong, Hong Kong University Press.

Quick, H. E. (1960). British slugs. *Bulletin of the British Museum (Natural History)* 6: 103–226.

Quinn, J. F. (1983). A revision of the Seguenziacea Verrill, 1884 (Gastropoda: Prosobranchia). I. Summary and evaluation of the superfamily. *Proceedings of the Biological Society of Washington* 96: 725–757.

Quinn, J. F. (1991). Systematic position of *Basilissopsis* and *Guttula*, and a discussion of the phylogeny of the Seguenzioidea (Gastropoda: Prosobranchia). *Bulletin of Marine Science* 49: 575–598.

Rabosky, D. L. (2013). Diversity-dependence, ecological speciation, and the role of competition in macroevolution. *Annual Review of Ecology, Evolution, and Systematics* 44: 481–502.

Ramírez, R., Borda, V., Romero, P., Ramírez, J., Congrains, C., Chirinos, J., Ramírez, P., Velásquez, L. E. & Mejía, K. (2012). Biodiversity and endemism of the western Amazonia land snails *Megalobulimus* and *Systrophia*. *Revista Peruana de Biología* 19: 59–74.

Rasetti, F. (1954). Internal shell structures in the Middle Cambrian gastropod *Scenella* and the problematic genus *Stenothecoides*. *Journal of Paleontology* 28: 59–66.

Rath, E. (1986). *Beiträge zur Anatomie und Ontogenie der Valvatidae (Mollusca: Gastropoda)*. Ph.D. dissertation, University of Vienna.

Raup, D. (1977). Removing sampling biases from taxonomic diversity data. *Journal of Paleontology* 51: 21.

Raup, D. M. & Sepkoski, J. J. (1982). Mass extinctions in the marine fossil record. *Science* 215: 1501–1503.

Raup, D. M. & Jablonski, D. (1993). Geography of end-Cretaceous marine bivalve extinctions. *Science* 260: 971–973.

Raut, S. K. (2004). Bacterial and non-microbial diseases in terrestrial gastropods, pp. 599–611 *in* G. M. Barker (ed.), *Natural Enemies of Terrestrial Molluscs*. Oxford, UK/Cambridge, MA, CABI Publishing.

Raven, C. P. (1964). Development, pp. 165–195 *in* K. M. Wilbur & Yonge, C. M. (eds.), *Physiology of Mollusca*. Vol. 1. New York, Academic Press.

Raven, C. P. (1966). *Morphogenesis: The Analysis of Molluscan Development*. Oxford, UK, Pergamon Press.

Raven, C. P. (1975). Development, pp. 367–400 *in* V. Fretter & Peake, J. (eds.), *Pulmonates: Functional Anatomy and Physiology*. Vol. 1. London, Academic Press.

Rawlings, T. A. (1994). Effect of elevated predation risk on the metabolic rate and spawning intensity of a rocky shore marine gastropod. *Journal of Experimental Marine Biology and Ecology* 181: 67–69.

Read, K. R. H. (1962). The hemoglobin of the bivalved mollusc, *Phacoides pectinatus* Gmelin. *Biological Bulletin* 123: 605–617.

Redfield, A. C. & Goodkind, R. (1929). The significance of the Bohr effect in the respiration and asphyxiation of the squid, *Loligo pealei*. *Journal of Experimental Biology* 6: 340–349.

Rees, J. F., Wergifosse, B., Noiset, O., Dubuisson, M., Janssens, B. & Thompson, E. (1998). The origins of marine bioluminescence: Turning oxygen defence mechanisms into deep-sea communication tools. *Journal of Experimental Biology* 201: 1211–1221.

Reeve, L. (1843–1878). *Conchologia iconica, or Illustrations of the Shells of Molluscous Animals*. London, Savill, Edwards and Co., Spottiswoode & Co.

Régnier, C., Fontaine, B. & Bouchet, P. (2009). Not knowing, not recording, not listing: Numerous unnoticed mollusk extinctions. *Conservation Biology* 23: 1214–1221.

Régnier, C., Achaz, G., Lambert, A., Cowie, R. H., Bouchet, P. & Fontaine, B. (2015). Mass extinction in poorly known taxa. *Proceedings of the National Academy of Sciences of the United States of America* 112: 7761–7766.

Régondaud, J., Brisson, P. & de Larambergue, M. (1974). Considerations sur la morphogénèse et l'évolution de la commissure viscérale chez les gastéropodes pulmonés. *Haliotis* 4: 49–55.

Reid, A. (2001). A new cuttlefish, *Sepia grahami*, sp nov (Cephalopoda: Sepiidae) from Eastern Australia. *Proceedings of the Linnean Society of New South Wales* 123: 159–172.

Reid, A., Jereb, P. & Roper, C. F. E. (2005). Family Sepiidae, pp. 57–152 *in* P. Jereb & Roper, C. F. E. (eds.), *Cephalopods of the World: An Annotated and Illustrated Catalogue of Cephalopod Species Known to Date. Volume 1: Chambered Nautiluses and Sepioids (Nautilidae, Sepiidae, Sepiolidae, Sepiadariidae, Idiosepiidae and Spirulidae)*. Rome, Food and Agriculture Organisation of the United Nations.

Reid, A. (2016). *Cephalopods of Australia and Sub-Antarctic Territories*. Melbourne, VIC, CSIRO Publishing.

Reid, D. G. (1986a). *Mainwaringia* Nevill, 1885, a littorinid genus from Asiatic mangrove forests, and a case of protandrous hermaphroditism. *Journal of Molluscan Studies* 52: 225–242.

Reid, D. G. (1986b). *The Littorinid Molluscs of Mangrove Forests in the Indo-Pacific Region: The Genus Littoraria (Mollusca, Gastropoda, Littorinidae)*. London, Butler & Tanner Ltd.

Reid, D. G. (1988). The genera *Bembicium* and *Risellopsis* (Gastropoda: Littorinidae) in Australia and New Zealand. *Records of the Australian Museum* 40: 91–150.

Reid, D. G. (1989). The comparative morphology, phylogeny and evolution of the gastropod family Littorinidae. *Philosophical Transactions of the Royal Society B* 324: 1–110.

Reid, D. G. (1998). Family Littorinidae, pp. 738–739 *in* P. L. Beesley, Ross, G. J. B. & Wells, A. (eds.), *Mollusca: The Southern Synthesis. Part B. Fauna of Australia*. Melbourne, VIC, CSIRO Publishing.

Reid, D. G., Aravind, N. A. & Madhyastha, N. A. (2013). A unique radiation of marine littorinid snails in the freshwater streams of the Western Ghats of India: The genus *Cremnoconchus* W. T. Blanford, 1869 (Gastropoda: Littorinidae). *Zoological Journal of the Linnean Society* 167: 93–135.

Reid, R. G. B. & Reid, A. M. (1974). The carnivorous habit of members of the septibranch genus *Cuspidaria* (Mollusca, Bivalvia). *Sarsia* 56: 47–56.

Reid, R. G. B. (1990). Evolutionary implications of sulphide-oxidizing symbioses in bivalves, pp. 127–140 *in* B. Morton (ed.), *The Bivalvia. Proceedings of a Memorial Symposium in Honour of Sir Charles Maurice Yonge (1899–1986)*, Edinburgh, 1986. Hong Kong, Hong Kong University Press.

Reid, R. G. B., McMahon, R. F., Ó Foighil, D. & Finnigan, R. (1992). Anterior inhalant currents and pedal feeding in bivalves. *The Veliger* 35: 93–104.

Reindl, S. & Haszprunar, G. (1994). Light and electron microscopical investigations on shell pores (caeca) of fissurellid limpets (Mollusca: Archaeogastropoda). *Journal of Zoology* 233: 385–404.

Reitner, J. (2009). Preserved gill remains in *Phragmoteuthis conocauda* (Quenstedt, 1846–49) (Toarcian, Southern Western Germany). *Berliner Paläobiologische Abhandlungen* 10: 289–295.

Rex, M. A. & Boss, K. J. (1976). Open coiling in Recent gastropods. *Malacologia* 15: 289–297.

Reynolds, K. C., Watanabe, H., Strong, E. E., Sasaki, T., Uematsu, K., Miyake, H., Kojima, S., Suzuki, Y., Fujikura, K., Kim, S. K. & Young, C. M. (2010). New molluscan larval form: Brooding and development in a hydrothermal vent gastropod, *Ifremeria nautilei* (Provannidae). *Biological Bulletin* 219: 7–11.

Reynolds, P. D. (1990a). Fine structure of the kidney and characterization of secretory products in *Dentalium rectius* (Mollusca, Scaphopoda). *Zoomorphology* 110: 53–62.

Reynolds, P. D. (1990b). Functional morphology of the perianal sinus and pericardium of *Dentalium rectius* (Mollusca: Scaphopoda) with a reinterpretation of the scaphopod heart. *American Malacological Bulletin* 7: 137–146.

Reynolds, P. D. (1992). Distribution and ultrastructure of ciliated sensory receptors in the posterior mantle epithelium of *Dentalium rectius* (Mollusca, Scaphopoda). *Acta Zoologica* 73: 263–270.

Reynolds, P. D. (2002). The Scaphopoda. *Advances in Marine Biology* 42: 137–236.

Reynolds, P. D. (2006). Scaphopoda: The tusk shells, pp. 229–237 *in* C. F. Sturm, Pearce, T. A. & Valdes, A. (eds.), *The Mollusks: A Guide to their Study, Collection, and Preservation*. Boca Raton, FL, Universal Publishers.

Reynolds, P. D. & Steiner, G. (2008). Scaphopoda, pp. 143–161 *in* W. F. Ponder & Lindberg, D. R. (eds.), *Phylogeny and Evolution of the Mollusca*. Berkeley, CA, University of California Press.

Richard, S. M., Pearthree, G., Aufdenkampe, A. K., Cutcher-Gershenfeld, J., Daniels, M., Gomez, B., Kinkade, D. & Percivall, G. (2014). Community-developed geoscience cyberinfrastructure. *Eos, Transactions American Geophysical Union* 95: 165–166.

Richling, I. (2004). Classification of the Helicinidae: Review of morphological characteristics based on a revision of the Costa Rican species and application to the arrangement of the Central American mainland taxa (Mollusca: Gastropoda: Neritopsina). *Malacologia* 45: 195–440.

Richling, I., Malkowsky, Y., Kuhn, J., Niederhofer, H.-J. & Boeters, H. D. (2017). A vanishing hotspot – The impact of molecular insights on the diversity of Central European *Bythiospeum* Bourguignat, 1882 (Mollusca: Gastropoda: Truncatelloidea). *Organisms Diversity & Evolution* 17: 67–85.

Richter, H.-P. (1986). Ultrastructure of follicular epithelia in the ovary of *Lepidochitona cinerea* (L.) (Mollusca: Polyplacophora). *Development, Growth & Differentiation* 28: 7–16.

Richter, U. (2002). Gewebeansatz-Strukturen auf Steinkernen von Ammonoideen. *Geologische Beiträge Hannover* 4: 1–113.

Rico, C., Antonio Cuesta, J., Drake, P., Macpherson, E., Bernatchez, L. & Marie, A. D. (2017). Null alleles are ubiquitous at microsatellite loci in the Wedge Clam (*Donax trunculus*). *PeerJ* 5: e3188.

Ridewood, W. G. (1903). On the structure of the gills of Lamellibranchia. *Philosophical Transactions of the Royal Society B* 195: 147–284.

Riding, R. (2006a). Microbial carbonate abundance compared with fluctuations in metazoan diversity over geological time. *Sedimentary Geology* 185: 229–238.

Riding, R. (2006b). Cyanobacterial calcification, carbon dioxide concentrating mechanisms, and Proterozoic–Cambrian changes in atmospheric composition. *Geobiology* 4: 299–316.

Riedel, F. (1995). Recognition of the superfamily Ficoidea Meek 1864 and definition of the Thalassocynidae fam. nov. (Gastropoda). *Zoologische Jahrbücher. Abteilung für Systematik Okologie und Geographie der Tiere* 121: 457–474.

Riedel, F. (2000). *Ursprung und Evolution der 'höheren' Caenogastropoda - eine paläobiologische Konzeption*. Vol. 32. Harxheim, Germany, ConchBooks.

Rigby, J. E. (1965). *Succinea putris*, a terrestrial opisthobranch mollusc. *Proceedings of the Zoological Society of London* 144: 445–487.

Rinkevich, B. (1993). Major primary stages of biomineralization in radular teeth of the limpet *Lottia gigantea*. *Marine Biology* 117: 269–277.

von Rintelen, T. & Glaubrecht, M. (2006). Rapid evolution of sessility in an endemic species flock of the freshwater bivalve *Corbicula* from ancient lakes on Sulawesi, Indonesia. *Biology Letters* 2: 73–77.

von Rintelen, T., Stelbrink, B., Marwoto, R. M. & Glaubrecht, M. (2014). A snail perspective on the biogeography of Sulawesi, Indonesia: Origin and intra-island dispersal of the viviparous freshwater Gastropod *Tylomelania*. *PLoS ONE* 9: e98917.

Risbec, J. (1955). Considérations sur l'anatomie comparée et la classification des Gastéropodes Prosobranches [Considerations on the comparative anatomy and the classification of prosobranch gastropods]. *Journal de Conchyliologie* 95: 45–82.

Rivest, B. R. (1984). Copulation by hypodermic injection in the nudibranchs *Palio zostera* and *P. dubia* (Gastropoda, Opisthobranchia). *Biological Bulletin* 167: 543–554.

Robertson, R. (1970). Systematics of Indo-Pacific *Philippia* (*Psilaxis*), architectonicid gastropods with eggs and young in the umbilicus. *Pacific Science* 24: 66–83.

Robertson, R., Scheltema, R. S. & Adams, F. W. (1970). The feeding, larval dispersal, and metamorphosis of *Philippia* (Gastropoda: Architectonicidae). *Pacific Science* 24: 55–65.

Robertson, R. (1978). Spermatophores of six Eastern North American pyramidellid gastropods and their systematic significance (with the new genus *Boonea*). *Biological Bulletin* 155: 360–382.

Robertson, R. (1983). Observations on the life history of the Wentletrap *Epitonium albidum* in the West Indies. *American Malacological Bulletin* 1: 1–12.

Robertson, R. (1985). Four characters and the higher category systematics of gastropods. *American Malacological Bulletin* 1: 1–22.

Robertson, R. (1989). Spermatophores of aquatic non-stylommatophoran gastropods: A review with new data on *Heliacus* (Architectonicidae). *Malacologia* 30: 341–364.

Robertson, R. (1993). Snail handedness – The coiling directions of gastropods. *National Geographic Research and Exploration* 9: 104–119.

Robertson, R. (2007). Taxonomic occurrences of gastropod spermatozeugmata and non-stylommatophoran spermatophores updated. *American Malacological Bulletin* 23: 11–16.

Robilliard, G. A. (1971). Predation by the nudibranch *Dirona albolineata* on three species of prosobranchs. *Pacific Science* 25: 429–435.

Robison, B., Seibel, B. A. & Drazen, J. (2014). Deep-sea octopus (*Graneledone boreopacifica*) conducts the longest-known egg-brooding period of any animal. *PLoS ONE* 9: e103437.

Robison, B. H., Reisenbichler, K. R., Hunt, J. C. & Haddock, S. H. D. (2003). Light production by the arm tips of the deep-sea cephalopod *Vampyroteuthis infernalis*. *Biological Bulletin* 205: 102–109.

Robson, G. C. (1930). Cephalopoda, I. Octopoda. *Discovery Reports* 2: 373–401.

Robson, G. C. (1931). *A Monograph of Recent Cephalopoda. Part II. Octopoda (excluding the Octopodinae)*. London, British Museum.

Rode, A. L. (2004). Phylogenetic revision of Leptodesma (Leiopteria) (Devonian: Bivalvia). *Postilla* 229: 1–26.

Rodhouse, P. G. K. & Nigmatullin, C. M. (1996). Role as consumers. *Philosophical Transactions of the Royal Society B* 351: 1003–1022.

Rodland, D. L. & Bottjer, D. J. (2001). Biotic recovery from the End-Permian Mass Extinction: Behavior of the inarticulate brachiopod *Lingula* as a disaster taxon. *Palaios* 16: 95–101.

Rodríguez-Tovar, F. J., Uchman, A. & Puga-Bernabéu, Á. (2015). Borings in gneiss boulders in the Miocene (Upper Tortonian) of the Sorbas Basin, SE Spain. *Geological Magazine* 152: 287–297.

Roeselers, G. & Newton, I. L. G. (2012). On the evolutionary ecology of symbioses between chemosynthetic bacteria and bivalves. *Applied Microbiology and Biotechnology* 94: 1–10.

Roger, L. M., Richardson, A. J., McKinnon, A. D., Knott, B., Matear, R. & Scadding, C. (2012). Comparison of the shell structure of two tropical Thecosomata (*Creseis acicula* and *Diacavolinia longirostris*) from 1963 to 2009: Potential implications of declining aragonite saturation. *ICES Journal of Marine Science* 69: 465–474.

Rogers-Lowery, C. L. & Dimock, R. V. (2006). Encapsulation of attached ectoparasitic glochidia larvae of freshwater mussels by epithelial tissue on fins of naive and resistant host fish. *Biological Bulletin* 210: 51–63.

Rogers, D. W. & Chase, R. (2001). Dart receipt promotes sperm storage in the garden snail *Helix aspersa*. *Behavioral Ecology and Sociobiology* 50: 122–127.

Rohr, D. M., Frýda, J. & Blodgett, R. B. (2003). *Alaskadiscus*, a new bellerophontoidean gastropod from the Upper Ordovician of the York and Farewell Terranes of Alaska. *Short Notes on Alaska Geology* 120: 95–99.

Rohr, D. M., Blodgett, R. B. & Baichtal, J. (2006). Scaphopoda from the Alexander Terrane, Southeast Alaska: The first occurrence of Scaphopoda in the Silurian. *Palaeoworld* 15: 211–215.

Romero, P. E., Pfenninger, M., Kano, Y. & Klussmann-Kolb, A. (2016). Molecular phylogeny of the Ellobiidae (Gastropoda: Panpulmonata) supports independent terrestrial invasions. *Molecular Phylogenetics and Evolution* 97: 43–54.

Roper, C. F. E. & Young, R. E. (1975). Vertical Distribution of Pelagic Cephalopods. *Smithsonian Contributions to Zoology* Vol. 209. Washington, DC, Smithsonian Institution Press.

Roper, C. F. E. & Hochberg, F. G. (1988). Behavior and systematics of cephalopods from Lizard Island, Australia, based on color and body patterns *Malacologia* 29: 153–194.

Roper, C. F. E. & Jereb, P. (2010). Family Cranchiidae, pp. 148–178 *in* P. Jereb & Roper, C. F. E. (eds.), *Cephalopods of the World: An Annotated and Illustrated Catalogue of Species Known to Date. No. 4, Vol. 2. Myopsid and Oegopsid Squids*. Vol. 2. Rome, FAO Species Catalogue for Fishery Purposes.

Rosen, M. D., Stasek, C. R. & Hermans, C. O. (1979). The ultrastructure and evolutionary significance of the ocelli in the larva of *Katharina tunicata* (Mollusca, Polyplacophora). *The Veliger* 22: 173–178.

Rosenberg, G. (1996). Independent evolution of terrestriality in Atlantic truncatellid gastropods. *Evolution* 50: 682–693.

Roth, B. (2001). Phylogeny of pneumostomal area morphology in terrestrial Pulmonata (Gastropoda). *The Nautilus* 115: 140–146.

Rothman, D. H., Fournier, G. P., French, K. L., Alm, E. J., Boyle, E. A., Cao, C. & Summons, R. E. (2014). Methanogenic burst in the end-Permian carbon cycle. *Proceedings of the National Academy of Sciences* 111: 5462–5467.

Rothschild, B. J., Ault, J., Goulletquer, P. & Heral, M. (1994). Decline of the Chesapeake Bay oyster population: A century of habitat destruction and overfishing. *Marine Ecology Progress Series* 111: 29–39.

Roule, M. L. (1891). Considérations sur l'embranchement des Trochozoaires. *Annales des sciences naturelles. Zoologie* 11: 121–178.

Rouse, G. W. (1999). Trochophore concepts: Ciliary bands and the evolution of larvae in spiralian Metazoa. *Biological Journal of the Linnean Society* 66: 411–464.

Roy, K., Jablonski, D., Valentine, J. W. & Rosenberg, G. (1998). Marine latitudinal diversity gradients: Tests of causal hypotheses. *Proceedings of the National Academy of Sciences of the United States of America* 95: 3699–3702.

Roy, K., Collins, A. G., Becker, B. J., Begovic, E. & Engle, J. M. (2003). Anthropogenic impacts and historical decline in body size of rocky intertidal gastropods in southern California. *Ecology Letters* 6: 205–211.

Roy, K., Hunt, G., Jablonski, D., Krug, A. Z. & Valentine, J. W. (2009a). A macroevolutionary perspective on species range limits. *Proceedings of the Royal Society B* 276: 1485–1493.

Roy, K., Jablonski, D. & Valentine, J. W. (2009b). Dissecting latitudinal diversity gradients: Functional groups and clades of marine bivalves. The Royal Society 267: 293–299.

Rozanov, A. & Zhuravlev, A. (1992). The lower Cambrian fossil record of the Soviet Union, pp. 205–282 *in* J. H. Lipps & Signor, P. W. (eds.), *Origin and Early Evolution of the Metazoa. Topics In Geobiology*. New York, Springer Science + Business Media.

Rozanov, A., Khomentovsky, V. V., Shabanov, Y., Karlova, G. A., Varlamov, A. I., Luchinina, V. A., Pegel, T. V., Demidenko, E., Parkhaev, P., Korovnikov, I. V. & Skorlotova, N. A. (2008). To the problem of stage subdivision of the Lower Cambrian. *Stratigraphy and Geological Correlation* 16: 1–19.

Rozov, S. N. (1984). Morphology, terminology and systematic affinity of stenothecoids. *Akademiya Nauk SSSR Sibirskoye Otdeleniye, Trudy* 597: 117–133.

Rudall, K. M. (1955). The distribution of collagen and chitin. *Symposia of the Society of Experimental Biology* 9: 49–71.

Rudman, W. B. (1971). Structure and functioning of the gut in the Bullomorpha (Opisthobranchia): Part 1. Herbivores. *Journal of Natural History* 5: 647–675.

Rudman, W. B. (1972a). The anatomy of the opisthobranch genus *Hydatina* and the functioning of the mantle cavity and alimentary canal. *Zoological Journal of the Linnean Society* 51: 121–139.

Rudman, W. B. (1972b). Structure and functioning of the gut in the Bullomorpha (Opisthobranchia): Part 3: Philinidae. *Journal of Natural History* 6: 459–474.

Rudman, W. B. (1972c). A comparative study of the genus *Philinopsis* (Pease, 1860) (Aglajidae, Opisthobranchia). *Pacific Science* 26: 381–399.

Rudman, W. B. (1972d). On *Melanochlamys* (Cheesman, 1881), a genus of the Aglajidae (Opisthobranchia, Gastropoda). *Pacific Science* 26: 50–62.

Rudman, W. B. (1972e). A study of the anatomy of *Pupa* and *Maxacteon* (Acteonidae, Opisthobranchia), with an account of the breeding cycle of *Pupa kirki*. *Journal of Natural History* 6: 603–619.

Rudman, W. B. (1972f). Structure and functioning of the gut in the Bullomorpha (Opisthobranchia). Part 2. Acteonidae. *Journal of Natural History* 6: 311–324.

Rudman, W. B. (1972g). Studies on the primitive opisthobranch genera *Bullina* (Férussac) and *Micromelo* (Pilsbry). *Zoological Journal of the Linnean Society* 51: 105–119.

Rudman, W. B. (1974). A comparison of *Chelidonura*, *Navanax* and *Aglaja* with other genera of the Aglajidae (Opisthobranchia: Gastropoda). *Zoological Journal of the Linnean Society* 54: 185–212.

Rudman, W. B. (1978). A new species and genus of the Aglajidae and the evolution of the philinacean opisthobranch molluscs. *Zoological Journal of the Linnean Society* 62: 89–107.

Rudman, W. B. (1981a). Further studies on the anatomy and ecology of opisthobranch molluscs feeding on the scleractinian coral *Porites*. *Zoological Journal of the Linnean Society* 71: 373–412.

Rudman, W. B. (1981b). The anatomy and biology of alcyonarian-feeding aeolid opisthobranch mollusks and their development of symbiosis with zooxanthellae. *Zoological Journal of the Linnean Society* 72: 219–262.

Rudman, W. B. (1982). The taxonomy and biology of further aeolidacean and arminacean nudibranch molluscs with symbiotic zooxanthellae. *Zoological Journal of the Linnean Society* 74: 147–196.

Rudman, W. B. (1984). The Chromodorididae (Opisthobranchia: Mollusca) of the Indo-West Pacific: A review of the genera. *Zoological Journal of the Linnean Society* 81: 115–273.

Rudman, W. B. (1986). The Chromodorididae (Opisthobranchia: Mollusca) of the Indo-West Pacific: The genus *Glossodoris* Ehrenbergh (=*Casella*, H. & A. Adams). *Zoological Journal of the Linnean Society* 86: 101–184.

Rudman, W. B. (1988). The Chromodorididae (Opisthobranchia: Mollusca) of the Indo-West Pacific: The genus *Ceratosoma* (J.E. Gray). *Zoological Journal of the Linnean Society* 93: 133–185.

Rudman, W. B. (1991a). Purpose in pattern: The evolution of colour in chromodorid nudibranchs. *Journal of Molluscan Studies* 57: 5–21.

Rudman, W. B. (1991b). Further studies on the taxonomy and biology of the octocoral-feeding genus *Phyllodesmium* (Ehrenberg, 1831) (Nudibranchia, Aeolidoidea). Part 2. *Journal of Molluscan Studies* 57: 167–203.

Rudman, W. B. (1998). *Autotomy*. http://www.seaslugforum.net/factsheet/defauto.

Rudman, W. B. & Willan, R. C. (1998). Opisthobranchia: Introduction, pp. 915–942 in P. L. Beesley, Ross, G. J. B. & Wells, A. (eds.), *Mollusca: The Southern Synthesis*. Part B. *Fauna of Australia*. Vol. 5. Melbourne, VIC, CSIRO Publishing.

Rudman, W. B. (2000). What eats sea slugs? *Sea Slug Forum*, 2013. Available from http://www.seaslugforum.net/factsheet/predrecord.

Rudman, W. B. & Bergquist, P. R. (2007). A review of feeding specificity in the sponge-feeding Chromodorididae (Nudibranchia: Mollusca). *Molluscan Research* 27: 60–88.

Rumi, A., Gutiérrez Gregoric, D. E., Landoni, N., Cárdenas Mancilla, J., Gordillo, S., Gonzalez, J. & Alvarez, D. (2015). Glacidorbidae (Gastropoda: Heterobranchia) in South America: Revision and description of a new genus and three new species from Patagonia. *Molluscan Research* 35: 143–152.

Runham, N. W., Thornton, P. R., Shaw, D. A. & Wayte, R. C. (1969). The mineralization and hardness of the radular teeth of the limpet *Patella vulgata* L. *Zeitschrift für Zellforschung und Mikroskopische Anatomie* 99: 608–626.

Runham, N. W. (1993). Mollusca, pp. 311–383 in K. G. Adiyodi & Adiyodi, R. G. (eds.), *Reproductive Biology of Invertebrates. Asexual Propagation and Reproductive Strategies. Reproductive Biology of Invertebrates*. Vol. 6. Chichester, John Wiley & Sons.

Runnegar, B. N. (1974). Evolutionary history of the bivalve subclass Anomalodesmata. *Journal of Paleontology* 48: 904–939.

Runnegar, B. N. & Pojeta, J. (1974). Molluscan phylogeny: The paleontological viewpoint. *Science* 186: 311–317.

Runnegar, B. N., Pojeta, J., Morris, N. J., Taylor, J. D., Taylor, M. E. & McClung, G. (1975). Biology of the *Hyolitha*. *Lethaia* 8: 181–191.

Runnegar, B. (1978). Origin and evolution of the class *Rostroconchia*. *Philosophical Transactions of the Royal Society B* 284: 319–333.

Runnegar, B. N., Goodhart, C. B. & Yochelson, E. L. (1978). Origin and evolution of the class Rostroconchia [and discussion]. *Philosophical Transactions of the Royal Society B* 284: 319–333.

Runnegar, B. N., Pojeta, J., Taylor, M. E. & Collins, D. (1979). New species of the Cambrian and Ordovician chitons *Matthevia* and *Chelodes* from Wisconsin and Queensland: Evidence for the early history of polyplacophoran mollusks. *Journal of Paleontology* 53: 1374–1394.

Runnegar, B. N. (1981). Muscle scars, shell form and torsion in Cambrian and Ordovician univalved molluscs. *Lethaia* 14: 311–322.

Runnegar, B. N. (1983). Molluscan phylogeny revisited. *Memoir of the Association of Australasian Paleontologists* 1: 121–144.

Runnegar, B. N. & Bentley, C. (1983). Anatomy, ecology and affinities of the Australian Early Cambrian bivalve *Pojetaia runnegari* (Jell). *Journal of Paleontology* 57: 73–92.

Runnegar, B. N. (1985). Shell microstructures of Cambrian molluscs replicated by phosphate. *Alcheringa* 9: 245–257.

Runnegar, B. N. & Pojeta, J. (1985). Origin and diversification of the Mollusca, pp. 1–57 in E. R. Trueman & Clarke, M. R. (eds.), *Evolution. The Mollusca*. Vol. 10. New York, Academic Press.

Runnegar, B. N. (1989). The evolution of mineral skeletons, pp. 75–94 in R. E. Crick (ed.), *Origin, Evolution, and Modern Aspects of Biomineralization in Plants and Animals*. New York, Plenum Press.

Runnegar, B. N. & Pojeta, J. (1992). The earliest bivalves and their Ordovician descendants. *American Malacological Union Bulletin* 9: 117–122.

Runnegar, B. N. (1996). Early evolution of the Mollusca: The fossil record, pp. 77–87 in J. D. Taylor (ed.), *Origin and Evolutionary Radiation of the Mollusca*. Oxford, UK, Oxford University Press.

Rupert, S. D. & Peters, W. S. (2011). Autotomy of the posterior foot in *Agaronia* (Caenogastropoda: Olividae) occurs in animals that are fully withdrawn into their shells. *Journal of Molluscan Studies* 77: 437–440.

Russell-Hunter, W. D. & McMahon, R. F. (1976). Evidence for functional protandry in a fresh-water basommatophoran limpet, *Laevapex fuscus*. *Transactions of the American Microscopical Society* 95: 174–182.

Russell-Hunter, W. D. (1988). The gills of chitons (Polyplacophora) and their significance in molluscan phylogeny. *American Malacological Bulletin* 6: 69–78.

Russell, M. P. (1991). Modern death assemblages and Pleistocene fossil assemblages in open coast high energy environments, San Nicolas Island, California. *Palaios* 6: 179–191.

Ruthensteiner, B. (1991). *Beiträge zur Entwicklung der Ellobiidae (Pulmonata, Gastropoda)*. Ph.D. dissertation, Universität Wien.

Ruthensteiner, B. & Schaefer, K. (1991). On the protonephridia and 'larval kidneys' of *Nassarius* (*Hinia*) *reticulatus* (Linnaeus) (Caenogastropoda). *Journal of Molluscan Studies* 57: 323–329.

Ruthensteiner, B. (1997). Homology of the pallial and pulmonary cavity of gastropods. *Journal of Molluscan Studies* 63: 353–367.

Ruthensteiner, B. (1999). Nervous system development of a primitive pulmonate (Mollusca: Gastropoda) and its bearing on comparative embryology of the gastropod nervous system. *Bollettino Malacologico* 34: 1–22.

Ruthensteiner, B., Wanninger, A. & Haszprunar, G. (2001). The protonephridial system of the tusk shell, *Antalis entalis* (Mollusca, Scaphopoda). *Zoomorphology* 121: 19–26.

Ruthensteiner, B. (2006). Redescription and 3D morphology of *Williamia gussonii* (Gastropoda: Siphonariidae). *Journal of Molluscan Studies* 72: 327–336.

Ruthensteiner, B., Schropel, V. & Haszprunar, G. (2010). Anatomy and affinities of *Micropilina minuta* (Warén, 1989) (Monoplacophora: Micropilinidae). *Journal of Molluscan Studies* 76: 323–332.

Saelen, G. (1989). Diagenesis and construction of the belemnite rostrum. *Palaeontology* 32: 765–798.

Safriel, U. N. (1969). Ecological segregation, polymorphism and natural selection in two intertidal gastropods of the genus *Nerita* at Elat (Red Sea, Israel). *Israel Journal of Zoology* 18: 205–231.

Sahney, S. & Benton, M. J. (2008). Recovery from the most profound mass extinction of all time. *Proceedings of the Royal Society B* 275: 759–765.

Sahney, S., Benton, M. J. & Falcon-Lang, H. J. (2010). Rainforest collapse triggered Carboniferous tetrapod diversification in Euramerica. *Geology* 38: 1079–1082.

Saito, H. & Okutani, T. (1990). Two new chitons (Mollusca: Polyplacophora) from a hydrothermal vent site of the Iheya Small Ridge, Okinawa Trough, East China Sea. *Venus* 49: 165–179.

Saito, H. & Okutani, T. (1992). Carnivorous habits of two species of the genus *Craspedochiton* (Polyplacophora: Acanthochitonidae). *Journal of the Malacological Society of Australia* 13: 55–63.

Saito, H. (2004). Phylogenetic significance of the radula in chitons, with special reference to the Cryptoplacoidea (Mollusca: Polyplacophora). *Bollettino Malacologico* 39: 83–104.

Saleuddin, A. S. M. (1965). The mode of life and functional anatomy of *Astarte* spp. (Eulamellibranchia). *Proceedings of the Malacological Society of London* 36: 229–257.

Saleuddin, A. S. M. (1971). Fine structure of normal and regenerated shell of *Helix*. *Canadian Journal of Zoology* 49: 37–41.

Salvazzi, E. (1984). Adaptive significance of shell torsion in mytilid bivalves. *Palaeontology* 27: 307–314.

Salvi, D., Macali, A. & Mariottini, P. (2014). Molecular phylogenetics and systematics of the bivalve family Ostreidae based on rRNA sequence-structure models and multilocus species tree. *PLoS ONE* 9: e108696.

Salvi, D. & Mariottini, P. (2017). Molecular taxonomy in 2D: A novel ITS 2 rRNA sequence structure approach guides the description of the oysters' subfamily Saccostreinae and the genus *Magallana* (Bivalvia: Ostreidae). *Zoological Journal of the Linnean Society* 179: 263–276.

Salvini-Plawen, L. von (1968). Über Lebendbeobachtungen an Caudofoveata (Mollusca, Aculifera), nebst Bemerkungen zum system der Klasse. *Sarsia* 31: 105–126.

Salvini-Plawen, L. von (1969). Solenogastres and Caudofoveata (Mollusca, Aculifera): Organisation and phylogenetic significance. *Malacologia* 9: 191–216.

Salvini-Plawen, L. von (1971). *Schild- und Furchenfüßer (Caudofoveata und Solenogastres)*. Vol. 441. Lutherstadt Wittenberg, Ziemsen Verlag.

Salvini-Plawen, L. von (1972). Zur morphologie und phylogenie der Mollusken: Die Beziehung der Caudofoveata und der Solenogastres als Aculifera, als Mollusca und als Spiralia. *Zeitschrift für Wissenschaftliche Zoologie* 184: 205–394.

Salvini-Plawen, L. von & Nopp, H. (1974). Presence of chitin in Caudofoveata (Mollusca), and derivation of their radula apparatus. *Zeitschrift für Morphologie der Tiere* 77: 77–86.

Salvini-Plawen, L. von (1975). Mollusca, Caudofoveata. *Marine Invertebrates of Scandinavia*. Oslo, Universitetsforlaget 4: 1–55.

Salvini-Plawen, L. von (1978). Antarktische und subantarktische Solenogastres (eine Monographie 1898–1974). *Zoologica (Stuttgart)* 128: 1–315.

Salvini-Plawen, L. von (1980). A reconsideration of systematics in the Mollusca (phylogeny and higher classification). *Malacologia* 19: 249–278.

Salvini-Plawen, L. von (1981). On the origin and evolution of the Mollusca. *Atti dei Convegni Lincei (Roma)* 49: 235–293.

Salvini-Plawen, L. von & Haszprunar, G. (1982). On the affinities of Septibranchia (Bivalvia). *The Veliger* 25: 83–85.

Salvini-Plawen, L. von (1985a). Early evolution and the primitive groups, pp. 59–150 *in* E. R. Trueman & Clarke, M. R. (eds.), *Evolution. The Mollusca*. Vol. 10. New York, Academic Press.

Salvini-Plawen, L. von & Haszprunar, G. (1987). The Vetigastropoda and the systematics of streptoneurous Gastropoda (Mollusca). *Journal of Zoology* 211: 747–770.

Salvini-Plawen, L. von (1988). The structure and function of molluscan digestive systems, pp. 301–379 *in* E. R. Trueman & Clarke, M. R. (eds.), *Form and Function. The Mollusca*. Vol. 11. New York, Academic Press.

Salvini-Plawen, L. von (1990). Origin, phylogeny and classification of the phylum Mollusca. *Iberus* 9: 1–33.

Salvini-Plawen, L. von (1991). The status of the Rhodopidae (Gastropoda: Euthyneura). *Malacologia* 32: 301–311.

Salvini-Plawen, L. von & Steiner, G. (1996). Synapomorphies and plesiomorphies in higher classification of Mollusca, pp. 29–51 *in* J. D. Taylor (ed.), *Origin and Evolutionary Radiation of the Mollusca*. Oxford, UK, Oxford University Press.

Salvini-Plawen, L. von (2003). On the phylogenetic significance of the aplacophoran Mollusca. *Iberus* 21: 67–97.

Sampson, L. V. (1985). The musculature of chiton. *Journal of Morphology* 11: 595–628; 631–633.

Sanchez, T. M. & Vaccari, N. E. (2003). Ucumariidae new family (Bivalvia, Anomalodesmata) and other bivalves from the Early Ordovician (Tremadocian) of northwestern Argentina. *Ameghiniana* 40: 415–424.

Sánchez, T. M. (2006). Taxonomic position and phylogenetic relationships of the bivalve *Goniophorina* Isberg, 1934, and related genera from the early Ordovician of northwestern Argentina. *Ameghiniana* 43: 113–122.

Sandberger, G. (1853). Einige Beobachtungen über Clymenien; mit besonderen Rücksicht auf die westfälischen Arten. *Verhandlungen Naturhist. Ver Preuss. Rheinlande Westphalia* 10: 171–216.

Santi, P. A. & Graziadei, P. P. C. (1975). A light and electron microscope study of intra-epithelial putative mechanoreceptors in squid suckers. *Tissue and Cell* 7: 689–702.

Sarasin, P. & Sarasin, F. (1899). *Die Land-Mollusken von Celebes*. Vol. 2. Wiesbaden, C. W. Kreidel's Verlag.

Sars, G. O. (1878). *Mollusca Regionis Arcticae Norwegiae*. Christiania, Norway, A. W. Brøgger.

Sartori, A. F., Passos, F. D. & Domaneschi, O. (2006). Arenophilic mantle glands in the Laternulidae (Bivalvia: Anomalodesmata) and their evolutionary significance. *Acta Zoologica* 87: 265–272.

Sasaki, T. (1998). Comparative anatomy and phylogeny of the Recent Archaeogastropoda (Mollusca: Gastropoda). *Bulletin of the University Museum*, University of Tokyo 38: i–vi, 1–223.

Sasaki, T. & Ishikawa, H. (2002). The first occurrence of a neritopsine gastropod from a phreatic community. *Journal of Molluscan Studies* 68: 286–288.

Sasaki, T., Okutani, T. & Fujikura, K. (2003). New taxa and new records of patelliform gastropods associated with chemoautosynthesis-based communities in Japanese waters. *The Veliger* 46: 189–210.

Sasaki, T. & Saito, H. (2005). Feeding of *Neomenia yamamotoi* Baba, 1975 (Mollusca: Solenogastres) on a sea anemone. *Venus* 64: 191–194.

Sasaki, T., Okutani, T. & Fujikura, K. (2006a). Anatomy of *Bathyacmaea secunda* Okutani, Fujikura & Sasaki, 1993 (Patellogastropoda: Acmaeidae). *Journal of Molluscan Studies* 72: 295–309.

Sasaki, T., Okutani, T. & Fujikura, K. (2006b). Anatomy of *Shinkailepas myojinensis* Sasaki, Okutani & Fujikura, 2003 (Gastropoda: Neritopsina). *Malacologia* 48: 1–26.

Sasaki, T., Warén, A., Kano, Y., Okutani, T. & Fujikura, K. (2010). Gastropods from Recent hot vents and cold seeps: Systematics, diversity and life strategies, pp. 169–254 *in* S. Kiel (ed.), *The Vent and Seep Biota: Aspects from Microbes to Ecosystems. Topics in Geobiology*. Dordrecht, the Netherlands, Springer Netherlands.

Saul, L. R. (1973). Evidence for the origin of the Mactridae (Bivalvia) in the Cretaceous. *University of California Publications in Geological Sciences* 97.

Saunders, H. L. & Allen, J. A. (1973). Studies on deep-sea Protobranchia (Bivalvia); prologue and the Pristiglomidae. *Bulletin of The Museum of Comparative Zoology* 145: 237–261.

Saunders, W. B. & Spinosa, C. (1978). Sexual dimorphism in *Nautilus* from Palau. *Paleobiology* 4: 349–358.

Saunders, W. B. & Richardson, E. S. (1979). Middle Pennsylvanian (Desmoinesean) Cephalopoda of the Mazon Creek Fauna, Northeastern Illinois, pp. 333–359 *in* M. H. Nitecki (ed.), *Mazon Creek Fossils*. New York, Academic Press.

Saunders, W. B. (1985). Studies of living *Nautilus* in Palau. *National Geographic Society Research Reports* 18: 669–682.

Saunders, W. B. & Shapiro, E. A. (1986). Calculation and simulation of ammonoid hydrostatics. *Paleobiology* 12: 64–79.

Saunders, W. B. & Ward, P. D. (1987). Ecology, distribution, and population characteristics of *Nautilus*, pp. 137–162 *in* W. B. Saunders & Landman, N. H. (eds.), *Nautilus: The Biology and Paleobiology of a Living Fossil. Topics in Geobiology*. New York, Springer.

Saunders, W. B., Greenfest-Allen, E., Work, D. M. & Nikolaeva, S. V. (2008). Morphologic and taxonomic history of Paleozoic ammonoids in time and morphospace. *Paleobiology* 34: 128–154.

Savazzi, E. (1982). Commensalism between a boring mytilid bivalve and a soft bottom coral in the Upper Eocene of northern Italy. *Paläontologische Zeitschrift* 56: 165–175.

Savazzi, E. (1996). Adaptations of vermetid and siliquariid gastropods. *Palaeontology* 39: 157–177.

Savazzi, E. (2000). Morphodynamics of *Bryopa* and the evolution of clavagellids, pp. 313–327 *in* E. M. Harper, Taylor, J. D. & Crame, J. A. (eds.), *Evolutionary Biology of the Bivalvia*. London, The Geological Society (Special Publication No. 177).

Savazzi, E. (2001). A review of symbiosis in the Bivalvia, with special attention to macrosymbiosis. *Paleontological Research* 5: 55–73.

Savazzi, E. (2005). The function and evolution of lateral asymmetry in boring endolithic bivalves. *Paleontological Research* 9: 169–187.

Scarabino, V. (1995). Scaphopoda of the tropical Pacific and Indian Oceans, with description of 3 new genera and 42 new species. *Mémoires du Muséum national d'Histoire naturelle* 167: 189–379.

Scarano, A. & Ituarte, C. (2009). First report of a case of occasional hermaphroditism in Polyplacophora. *Journal of Molluscan Studies* 75: 91–92.

Scarlato, O. A. & Starobogatov, I. (1978). Phylogenetic relations and the early evolution of the class Bivalvia. *Philosophical Transactions of the Royal Society B* 284: 217–224.

Schaefer, K. (1996). Development and homologies of the anal gland in *Haminaea navicula* (Da Costa, 1778) (Opisthobranchia, Bullomorpha), pp. 249–260 *in* J. D. Taylor (ed.), *Origin and Evolutionary Radiation of the Mollusca*. Oxford, UK, Oxford University Press.

Schaefer, K. & Haszprunar, G. (1996). Anatomy of *Laevipilina antarctica*, a monoplacophoran limpet (Mollusca) from Antarctic waters. *Acta Zoologica* 77: 295–314.

Schaefer, K. & Haszprunar, G. (1997). Organisation and fine structure of the mantle of *Laevipilina antarctica* (Mollusca, Monoplacophora). *Zoologischer Anzeiger* 236: 13–23.

Schaefer, K. (2000). The adoral sense organ in protobranch bivalves (Mollusca): Comparative fine structure with special reference to *Nucula nucleus*. *Invertebrate Biology* 119: 188–214.

Schalie, H., Van der (1965). Observations on the sex in *Campeloma* (Gastropoda: Viviparidae). *Occasional Papers. Museum of Zoology, University of Michigan* 641: 1–15.

Schander, C. (1997). *Taxonomy and Phylogeny of the Pyramidellidae (Mollusca, Gastropoda, Heterobranchia)*. Doctoral Thesis, Göteborg, Dept of Zoology, Göteborg University.

Schander, C. S. & Scheltema, A. H. (2003). Workshops on aplacophoran Mollusca, Report. *Bulletin of the Malacological Society of London* 40: 1.

Schander, C. & Willassen, E. (2005). What can biological barcoding do for marine biology? *Marine Biology Research* 1: 79–83.

Schander, C., Scheltema, A. H. & Ivanov, D. L. (2006). *Falcidens halanychi*, a new species of Chaetodermomorpha (=Caudofoveata) (Mollusca) from the northwest Atlantic Ocean. *Marine Biology Research* 2: 303–315.

Scheel, D., Chancellor, S., Hing, M., Lawrence, M., Linquist, S. & Godfrey-Smith, P. (2017). A second site occupied by *Octopus tetricus* at high densities, with notes on their ecology and behavior. *Marine and Freshwater Behaviour and Physiology* 50: 1–7.

Scheltema, A. H. (1978). Position of the class Aplacophora in the phylum Mollusca. *Malacologia* 17: 99–109.

Scheltema, A. H. (1981). Comparative morphology of the radulae and alimentary tracts in the Aplacophora. *Malacologia* 20: 361–383.

Scheltema, A. H. (1985). The aplacophoran family Prochaetodermatidae in the North American Basin, including *Chevroderma* n. g. and *Spathoderma* n. g. (Mollusca; Chaetodermomorpha). *Biological Bulletin* 169: 484–529.

Scheltema, A. H. (1987). Reproduction and rapid growth in a deep-sea aplacophoran mollusc, *Prochaetoderma yongei*. *Marine Ecology Progress Series* 37: 171–180.

Scheltema, A. H. (1988a). Ancestors and descendants: Relationships of the Aplacophora and Polyplacophora. *American Malacological Bulletin* 6: 57–68.

Scheltema, A. H. (1989). The primitive molluscan radula, p. 220 *in* K. Kerth & Kuzirian, A. M. (eds.), *Proceedings of the 10th International Malacological Congress, Tübingen*, 27 August–2 September 1989. Abstracts. Tübingen, Germany, University of Tübingen.

Scheltema, A. H. & Kuzirian, A. M. (1991). *Helicoradomenia juani* gen. et sp. nov., a Pacific hydrothermal vent Aplacophora (Mollusca, Neomeniomorpha). *The Veliger* 34: 195–203.

Scheltema, A. H. (1992). *The Aplacophora: History, taxonomy, phylogeny, biogeography, and ecology.* Ph.D. dissertation, University of Oslo.

Scheltema, A. H. (1993). Aplacophora as progenetic aculiferans and the coelomate origin of mollusks as the sister taxon of Sipuncula. *Biological Bulletin* 184: 57–78.

Scheltema, A. H. & Jebb, M. (1994). Natural history of a solenogaster mollusk from Papua New Guinea, *Epimenia australis* (Thiele) (Aplacophora, Neomeniomorpha). *Journal of Natural History* 28: 1297–1318.

Scheltema, A. H., Tscherkassy, M. & Kuzirian, A. M. (1994). Aplacophora, pp. 13–54 in F. W. Harrison & Kohn, A. J. (eds.), *Microscopic Anatomy of Invertebrates. Mollusca 1.* Vol. 5. New York, Wiley-Liss.

Scheltema, A. H. (1996). Phylogenetic position of Sipuncula, Mollusca and the progenetic Aplacophora, pp. 53–58 in J. D. Taylor. *Origin and Evolutionary Radiation of the Mollusca.* Oxford, UK, Oxford University Press.

Scheltema, A. H. (1998). Class Aplacophora, pp. 145–159 in P. L. Beesley, Ross, G. J. B. & Wells, A. (eds.), *Mollusca: The Southern Synthesis. Part A. Fauna of Australia.* Melbourne, VIC, CSIRO Publishing.

Scheltema, A. H. & Schander, C. (2000). Discrimination and phylogeny of solenogaster species through the morphology of hard parts (Mollusca, Aplacophora, Neomeniomorpha). *Biological Bulletin* 198: 121–151.

Scheltema, A. H. & Ivanov, D. L. (2002). An aplacophoran postlarva with iterated dorsal groups of spicules and skeletal similarities to Paleozoic fossils. *Invertebrate Biology* 121: 1–10.

Scheltema, A. H., Kerth, K. & Kuzirian, A. M. (2003). Original molluscan radula: Comparisons among Aplacophora, Polyplacophora, Gastropoda, and the Cambrian fossil *Wiwaxia corrugata. Journal of Morphology* 257: 219–245.

Scheltema, A. H. & Ivanov, D. L. (2009). A natural history of the deep-sea aplacophoran *Prochaetoderma yongei* and its relationship to confamilials (Mollusca, Prochaetodermatidae). *Deep Sea Research Part II: Topical Studies in Oceanography* 56: 1856–1864.

Scheltema, A. H. (2014). The original molluscan radula and progenesis in Aplacophora revisited. *Journal of Natural History* 48: 2855–2869.

Scheltema, R. S. (1971). The dispersal of the larvae of shoal-water benthic invertebrate species over long distances by ocean currents, pp. 7–28 in D. J. Crisp (ed.), Fourth European Marine Biology Symposium held at Bangor, North Wales in September 1969. Cambridge, UK, Cambridge University Press.

Scheltema, R. S. (1988b). Initial evidence for the transport of teleplanic larvae of benthic invertebrates across the East Pacific Barrier. *Biological Bulletin* 174: 145–152.

Scheltema, R. S. & Williams, I. P. (2009). Reproduction among protobranch bivalves of the family Nuculidae from sublittoral, bathyal, and abyssal depths off the New England coast of North America. *Deep Sea Research Part II: Topical Studies in Oceanography* 56: 1835–1846.

Schiaparelli, S. & Cattaneo-Vietti, R. (1999). Functional morphology of vermetid feeding-tubes. *Lethaia* 32: 41–46.

Schiaparelli, S., Albertelli, G. & Cattaneo-Vietti, R. (2006). Phenotypic plasticity of Vermetidae suspension feeding: A potential bias in their use as biological sea-level indicators. *Marine Ecology* 27: 44–53.

Schiemenz, P. (1889). Parasitische Schnecken. *Biologisches Zentralblatt* 9: 567–574 & 585–594.

Schileyko, A. A. (1979). The system of the order Geophila (=Helicida) (Gastropoda Pulmonata). *Transactions of the Zoological Institute of the Academy of Sciences, USSR* 80: 44–69.

Schileyko, A. A. (1998a). Treatise on Recent terrestrial pulmonate molluscs. Part 2. Gastrocoptidae, Hypselostomatidae, Vertiginidae, Truncatellinidae, Pachnodidae, Enidae, Sagdidae. *Ruthenica Supplement* 2: 127–261.

Schileyko, A. A. (1998b). Treatise on Recent terrestrial pulmonate molluscs. Part 1. Achatinellidae, Amastridae, Orculidae, Strobilopsidae, Spelaeodiscidae, Valloniidae, Cochlicopidae, Pupillidae, Chondrinidae, Pyramidulidae. *Ruthenica Supplement* 2: 1–127.

Schileyko, A. A. (1999a). Treatise on Recent terrestrial pulmonate molluscs. Part 3. Partulidae, Aillyidae, Bulimulidae, Orthalicidae, Megaspiridae, Urocoptidae. *Ruthenica Supplement* 2: 263–436.

Schileyko, A. A. (1999b). Treatise on Recent terrestrial pulmonate molluscs. Part 4. Draparnaudiidae, Caryodidae, Macrocyclidae, Acavidae, Clavatoridae, Dorcasiidae, Sculptariidae, Corillidae, Plectopylidae, Megalobulimidae, Strophocheilidae, Cerionidae, Achatinidae, Subulinidae, Glessulidae, Micractaeonidae, Ferussaciidae. *Ruthenica Supplement* 2: 437–564.

Schileyko, A. A. (2000a). Treatise on Recent terrestrial pulmonate molluscs. Part 6: Rhytididae, Chlamydephoridae, Systrophiidae, Haplotrematidae, Streptaxidae, Spiraxidae, Oleacinidae, Testacellidae. *Ruthenica Supplement* 2: 731–880.

Schileyko, A. A. (2000b). Treatise on Recent terrestrial pulmonate molluscs. Part 5. Clausilliidae [Clausiliidae]. *Ruthenica Supplement* 2: 565–729.

Schileyko, A. A. (2001). Treatise on Recent terrestrial pulmonate molluscs. Part 7: Endodontidae, Thyrophorellidae, Charopidae. *Ruthenica Supplement* 2: i–ii, 881–1034.

Schileyko, A. A. (2003a). Treatise on Recent terrestrial pulmonate molluscs. Part 10. Ariophantidae, Ostracolethidae, Ryssotidae, Milacidae, Dyakiidae, Staffordiidae, Gastrodontidae, Zonitidae, Daudebardiidae, Parmacellidae. *Ruthenica Supplement* 2: 1309–1466.

Schileyko, A. A. (2003b). Treatise on Recent terrestrial pulmonate molluscs. Part 11. Trigonochlamydidae, Papillodermidae, Vitrinidae, Limacidae, Bielziidae, Agriolimacidae, Boettgerillidae, Camaenidae. *Ruthenica Supplement* 2: 1467–1626.

Schilthuizen, M., Til, A., Van, Salverda, M., Liew, T.-S., James, S. S., bin Elahan, B. & Vermeulen, J. J. (2006). Microgeographic evolution of snail shell shape and predator behavior. *Evolution* 60: 1851–1858.

Schindewolf, O. H. (1937). Zur Stratigraphie und Palaontologie der Wocklumer Schichten (Oberdevon). *Abhandlungen der Preußischen Geologischen Landesanstalt* 178: 1–132.

Schipp, R., Mollenhauer, S. & Boletzky, S. Von (1979). Electron microscopical and histochemical studies of differentiation and function of the cephalopod gill (*Sepia officinalis* L.). *Zoomorphology* 93: 193–207.

Schlesinger, A., Kramarsky-Winter, E. & Loya, Y. (2009). Active nematocyst isolation via nudibranchs. *Marine Biotechnology* 11: 441–444.

Schmekel, L. (1971). Histologie und Feinstruktur der Genitalorgane von Nudibranchiern (Gastropoda, Euthyneura). *Zeitschrift für Morphologie der Tiere* 69: 115–183.

Schmekel, L. & Weischer, M. L. (1973). Die Blutdrüse der Doridacea (Gastropoda, Opisthobranchia) als Ort möglicher Hämocyanin-Synthese. *Zeitschrift für Morphologie der Tiere* 76: 261–284.

Schmekel, L. & Portmann, A. (1982). *Opisthobranchia des Mittelmeeres, Nudibranchia und Saccoglossa*, Berlin, Springer-Verlag.

Schmekel, L. (1985). Aspects of evolution within the opisthobranchs, pp. 221–267 *in* E. R. Trueman & Clarke, M. R. (eds.), *Evolution. The Mollusca*. Vol. 10. New York, Academic Press.

Schmidt-Nielsen, K., Taylor, C. R. & Shkolnik, A. (1971). Desert snails: Problems of heat, water and food. *Journal of Experimental Biology* 55: 385–398.

Schmitt, V., Anthes, N. & Michiels, N. K. (2007). Mating behaviour in the sea slug *Elysia timida* (Opisthobranchia, Sacoglossa): Hypodermic injection, sperm transfer and balanced reciprocity. *Frontiers in Zoology* 4: 1–9.

Schmitz, L., Motani, R., Oufiero, C. E., Martin, C. H., McGee, M. D., Gamarra, A. R., Lee, J. J. & Wainwright, P. C. (2013). Allometry indicates giant eyes of giant squid are not exceptional. *BMC Evolutionary Biology* 13: 45 (41–49).

Schneider, D. W. & Lyons, J. (1993). Dynamics of upstream migration in two species of tropical freshwater snails. *Journal of the North American Benthological Society* 12: 3–16.

Schneider, J. A. (1995). Phylogeny of the Cardiidae (Mollusca, Bivalvia): Protocardiinae, Laevicardiinae, Lahilliinae, Tulongocardiinae subfam. n. and Pleuriocardiinae subfam. n. *Zoologica Scripta* 24: 321–346.

Schneider, J. A. (2001). Bivalve systematics during the 20th century. *Journal of Paleontology* 75: 1119–1127.

Schneider, J. A. & Carter, J. G. (2001). Evolution and phylogenetic significance of cardioidean shell microstructure (Mollusca, Bivalvia). *Journal of Paleontology* 75: 607–643.

Schrödl, M., Linse, K. & Schwabe, E. (2006). Review on the distribution and biology of Antarctic Monoplacophora, with first abyssal record of *Laevipilina antarctica*. *Polar Biology* 29: 721–727.

Schrödl, M. & Neusser, T. P. (2010). Towards a phylogeny and evolution of Acochlidia (Mollusca: Gastropoda: Opisthobranchia). *Zoological Journal of the Linnean Society* 158: 124–154.

Schrödl, M., Jörger, K. M. & Wilson, N. G. (2011). Bye bye 'Opisthobranchia'! A review on the contribution of mesopsammic sea slugs to euthyneuran systematics. *Thalassas* 27: 101–112.

Schuett, C. & Doepke, H. (2013). Endobacterial morphotypes in nudibranch cerata tips: A SEM analysis. *Helgoland Marine Research* 67: 219–227.

Schuhmacher, H. (1973). Notes on occurrence, feeding, and swimming behavior of *Notarchus indicus* and *Melibe bucephala* at Elat, Red Sea (Mollusca: Opisthobranchia). *Israel Journal of Zoology* 22: 13–25.

Schwabe, E. (2005). A catalogue of Recent and fossil chitons (Mollusca: Polyplacophora). Addenda. *Novapex* 6: 89–105.

Schwabe, E. (2008). A summary of reports of abyssal and hadal Monoplacophora and Polyplacophora (Mollusca). *Zootaxa* 1866: 205–222.

Schwabe, E. (2010). Illustrated summary of chiton terminology. *Spixiana* 33: 171–194.

Schweigert, G. & Fuchs, D. (2012). First record of a true coleoid cephalopod from the Germanic Triassic (Ladinian). *Neues Jahrbuch für Geologie und Paläontologie-Abhandlungen* 266: 19–30.

Scott, P. J. B. (1988). Initial settlement behaviour and survivorship of *Lithophaga bisulcata* (d'Orbigny) (Mytilidae: Lithophaginae). *Journal of Molluscan Studies* 54: 97–108.

Seed, R. (1980). Shell growth and form in the Bivalvia, pp. 23–67 *in* D. C. Rhoads & Lutz, R. A. (eds.), *Skeletal Growth of Aquatic Organisms*. New York, Plenum.

Seibel, B. A., Thuesen, E. V., Childress, J. J. & Gorodezky, L. A. (1997). Decline in pelagic cephalopod metabolism with habitat depth reflects differences in locomotory efficiency. *Biological Bulletin* 192: 262–278.

Seibel, B. A., Thuesen, E. V. & Childress, J. J. (1998). Flight of the vampire: Ontogenetic gait-transition in *Vampyroteuthis infernalis* (Cephalopoda: Vampyromorpha). *Journal of Experimental Biology* 201: 2413–2424.

Seibel, B. A., Chausson, F., Lallier, F. H., Zal, F. & Childress, J. J. (1999). Vampire blood: Respiratory physiology of the vampire squid (Cephalopoda: Vampyromorpha) in relation to the oxygen minimum layer. *Experimental Biology Online* 4: 1–10.

Seibel, B. A. (2007). On the depth and scale of metabolic rate variation: Scaling of oxygen consumption rates and enzymatic activity in the Class Cephalopoda (Mollusca). *Journal of Experimental Biology* 210: 1–11.

Seilacher, A. (1972). Divaricate patterns in pelecypod shells. *Lethaia* 5: 325–343.

Seilacher, A. (1984a). Late Precambrian and Early Cambrian Metazoa: Preservational or real extinctions? pp. 159–168, in H. D. Holland & Trendall, A. F. (eds.), *Patterns of Change in Earth Evolution*. Berlin, Springer.

Seilacher, A. (1984b). Constructional morphology of bivalves: Evolutionary pathways in primary versus secondary soft-bottom dwellers. *Palaeontology* 27: 207–237.

Seilacher, A. (1990). Aberrations in bivalve evolution related to photo- and chemosymbiosis. *Historical Biology* 3: 289–311.

Seilacher, A. (1993). Ammonite aptychi: How to transform a jaw into an operculum? *American Journal of Science* 293-A: 20–32.

Seilacher, A. (1998). Rudists as bivalvian dinosaurs, pp. 423–436 *in* P. A. Johnston & Haggart, J. W. (eds.), *Bivalves: An Eon of Evolution*. Calgary, AB, University of Calgary Press.

Selman, B. J. & Jones, A. A. (2004). Microsporidia (*Microspora*) parasitic in terrestrial gastropods, pp. 579–598 *in* G. M. Barker (ed.), *Natural Enemies of Terrestrial Molluscs*. Oxford, UK/ Cambridge, MA, CABI Publishing.

Selwood, L. (1968). Interrelationships between developing oocytes and ovarian tissues in the chiton *Sypharochiton septentriones* (Ashby)(Mollusca, Polyplacophora). *Journal of Morphology* 125: 71–103.

Selwood, L. (1970). The role of the follicle cells during oogenesis in the chiton *Sypharochiton pelliserpentis septentriones* (Ashby) (Polyplacophora, Mollusca). *Zeitschrift für Zellforschung und mikroskopische Anatomie* 104: 178–192.

Seminoff, J. A., Resendiz, A. & Nichols, W. J. (2002). Diet of East Pacific green turtles (*Chelonia mydas*) in the Central Gulf of California, México. *Journal of Herpetology* 36: 447–453.

Sepkoski, J. J. (1981). A factor analytic description of the Phanerozoic marine fossil record. *Paleobiology* 7: 36–53.

Sepkoski, J. J. (1986). Phanerozoic overview of mass extinction, pp. 277–295 *in* D. M. Raup & Jablonski, D. (eds.), *Patterns and Processes in the History of Life*. Berlin/Heidelberg, Springer.

Sepkoski, J. J. (1998). Rates of speciation in the fossil record. *Philosophical Transactions of the Royal Society B* 353: 315–326.

Sepkoski, J. J., Jablonski, D. & Foote, M. (2002). A compendium of fossil marine animal genera. *Bulletins of American Paleontology* 363: 1–560.

Serova, M., de Gramont, A., Bieche, I., Riveiro, M. E., Galmarini, C. M., Aracil, M., Jimeno, J., Faivre, S. & Raymond, E. (2013). Predictive factors of sensitivity to elisidepsin, a novel Kahalalide F-derived marine compound. *Marine Drugs* 11: 944–959.

Servais, T., Lehnert, O., Li, J., Mullins, G. L., Munnecke, A., Nützel, A. & Vecoli, M. (2008). The Ordovician Biodiversification: Revolution in the oceanic trophic chain. *Lethaia* 41: 99–109.

Servais, T., Owen, A. W., Harper, D. A. T., Kröger, B. & Munnecke, A. (2010). The great Ordovician biodiversification event (GOBE): The palaeoecological dimension. *Palaeogeography, Palaeoclimatology, Palaeoecology* 294: 99–119.

Shadwick, R. E., O'Dor, R. K. & Gosline, J. M. (1990). Respiratory and cardiac function during exercise in squid. *Canadian Journal of Zoology* 68: 792–798.

Sharma, P. P., González, V. L., Kawauchi, G. Y., Andrade, S. C. S., Guzmán, A., Collins, T. M., Glover, E. A., Harper, E. M., Healy, J. M., Mikkelsen, P. M., Taylor, J. D., Bieler, R. & Giribet, G. (2012). Phylogenetic analysis of four nuclear protein-encoding genes largely corroborates the traditional classification of Bivalvia (Mollusca). *Molecular Phylogenetics and Evolution* 65: 64–74.

Sharma, P. P., Zardus, J. D., Boyle, E. E., González, V. L., Jennings, R. M., McIntyre, E., Wheeler, W. C., Etter, R. J. & Giribet, G. (2013). Into the deep: A phylogenetic approach to the bivalve subclass Protobranchia. *Molecular Phylogenetics and Evolution* 69: 188–204.

Sharp, B. (1888). Remarks on the Phylogeny of the Lamellibranchiata. *Proceedings of the Academy of Natural Sciences of Philadelphia* 40: 121–124.

Sheehan, P. M. (2001). The late Ordovician mass extinction. *Annual Review of Earth and Planetary Sciences* 29: 331–364.

Shevyrev, A. A. (2005). The cephalopod macrosystem: A historical review, the present state of knowledge, and unsolved problems: 1. Major features and overall classification of cephalopod mollusks. *Paleontological Journal* 39: 606–614.

Shevyrev, A. A. (2006). The cephalopod macrosystem: A historical review, the present state of knowledge, and unsolved problems: 3. Classification of Bactritoidea and Ammonoidea. *Paleontological Journal* 40: 150–161.

Shigemiya, Y. & Kato, M. (2001). Age distribution, growth, and lifetime copulation frequency of a freshwater snail, *Clithon retropictus* (Neritidae). *Population Ecology* 43: 133–140.

Shigeno, S., Sasaki, T. & Haszprunar, G. (2007). Central nervous system of *Chaetoderma japonicum* (Caudofoveata, Aplacophora): Implications for diversified ganglionic plans in early molluscan evolution. *Biological Bulletin* 213: 122–134.

Shigeno, S., Sasaki, T., Moritaki, T., Kasugai, T., Vecchione, M. & Agata, K. (2008). Evolution of the cephalopod head complex by assembly of multiple molluscan body parts: Evidence from *Nautilus* embryonic development. *Journal of Morphology* 269: 1–17.

Shimamoto, M. (1993). Shell microstructure and amino acid composition of organic matrix from venerid shell, pp. 89–97 *in* I. Kobayashi, Mutvei, H. & Sahni, A. (eds.), *Structure, Formation and Evolution of Fossil Hard Tissues. Symposium: 29th International Geological Congress Kyoto*, 24 August–3 September, 1992. Tokyo, Tokai University Press.

Shimek, R. L. (1988). The functional morphology of scaphopod captacula. *The Veliger* 30: 213–221.

Shimek, R. L. (1989). Shell morphometrics and systematics: A revision of the slender, shallow-water *Cadulus* of the Northeastern Pacific (Scaphopoda: Gadilida). *The Veliger* 32: 233–246.

Shimek, R. L. (1990). Diet and habitat utilization in a Northeastern Pacific Ocean scaphopod assemblage. *American Malacological Bulletin* 7: 147–169.

Shimek, R. L. (1997). A new species of Eastern Pacific *Fissidentalium* (Mollusca: Scaphopoda) with a symbiotic sea anemone. *The Veliger* 40: 178–191.

Shimek, R. L. & Steiner, G. (1997). Scaphopoda, pp. 719–781 *in* F. W. Harrison & Kohn, A. J. (eds.), *Microscopic Anatomy of Invertebrates. Mollusca 2*. Vol. 6B. New York, Wiley-Liss.

Shimek, R. L. (2008). Scaphopods. http://www.ronshimek.com/scaphopod.html.

Shipway, J. R., Altamia, M. A., Haga, T., Velásquez, M., Albano, J., Dechavez, R., Concepcion, G. P., Haygood, M. G. & Distel, D. L. (2018). Observations on the life history and geographic range of the giant chemosymbiotic shipworm *Kuphus polythalamius* (Bivalvia: Teredinidae). *The Biological Bulletin* 235: 167–177.

Shrock, R. R. & Twenhofel, W. H. (1953). *Principles of Invertebrate Paleontology*. New York, McGraw-Hill Book Company.

Shuto, T. (1974). Larval ecology of prosobranch gastropods and its bearing on biogeography and paleontology. *Lethaia* 7: 239–256.

Siddall, M. E., Fontanella, F. M., Watson, S. C., Kvist, S. & Erséus, C. (2009). Barcoding bamboozled by bacteria: Convergence to metazoan mitochondrial primer targets by marine microbes. *Systematic Biology* 58: 445–451.

Signor, P. W. & Lipps, J. H. (1982). Sampling bias, gradual extinction patterns and catastrophes in the fossil record. *Geological Society of America Special Papers* 190: 291–296.

Signor, P. W. & Brett, C. E. (1984). The mid-Paleozoic precursor to the Mesozoic marine revolution. *Paleobiology* 10: 229–245.

Signor, P. W. & Kat, P. W. (1984). Functional significance of columellar folds in turritelliform gastropods. *Journal of Paleontology* 58: 210–216.

Signor, P. W. & Vermeij, G. J. (1994). The plankton and the benthos: Origins and early history of an evolving relationship. *Paleobiology* 20: 297–319.

Sigwart, J. D. & Sutton, M. D. (2007). Deep molluscan phylogeny: Synthesis of palaeontological and neontological data. *Proceedings of the Royal Society B* 274: 2413–2419.

Sigwart, J. D. (2008). Gross anatomy and positional homology of gills, gonopores, and nephridiopores in "basal" living chitons (Polyplacophora: Lepidopleurina). *American Malacological Bulletin* 25: 43–49.

Sigwart, J. D., Stoeger, I., Knebelsberger, T. & Schwabe, E. (2013). Chiton phylogeny (Mollusca: Polyplacophora) and the placement of the enigmatic species *Choriplax grayi* (H. Adams & Angas). *Invertebrate Systematics* 27: 603–621.

Sigwart, J. D., Sumner-Rooney, L. H., Schwabe, E., Hess, M., Brennan, G. P. & Schrödl, M. (2014). A new sensory organ in 'primitive' molluscs (Polyplacophora: Lepidopleurida), and its context in the nervous system of chitons. *Frontiers in Zoology* 11: 7.

Sigwart, J. D. & Lindberg, D. R. (2015). Consensus and confusion in molluscan trees: Evaluating morphological and molecular phylogenies. *Systematic Biology* 64: 384–395.

Sigwart, J. D. & Sumner-Rooney, L. H. (2015). Mollusca: Caudofoveata, Monoplacophora, Polyplacophora, Scaphopoda, and Solenogastres, pp. 172–189 *in* A. Schmidt-Rhaesa, Harzsch, S. & Purschke, G. (eds.), *Structure and Evolution of Invertebrate Nervous Systems*. Oxford, UK, Oxford University Press.

Sigwart, J. D. (2016). Deep trees: Woodfall biodiversity dynamics in present and past oceans. *Deep Sea Research Part II: Topical Studies in Oceanography* 137: 282–287.

Sigwart, J. D. & Schwabe, E. (2017). Anatomy of the many feeding types in polyplacophoran molluscs. *Invertebrate Zoology* 14: 205–216.

Sigwart, J. D., Sumner-Rooney, L. H., Dickey, J. & Carey, N. (2017). The scaphopod foot is ventral: More evidence from the anatomy of *Rhabdus rectius* (Carpenter, 1864) (Dentaliida: Rhabdidae). *Molluscan Research* 37: 79–87.

Sigwart, J. D., Wicksten, M. K., Jackson, M. G. & Herrera, S. (2018). Deep-sea video technology tracks a monoplacophoran to the end of its trail (Mollusca, Tryblidia). *Marine Biodiversity* 48: 1–8.

Silantiev, V. V. (1998). New data on the Upper Permian non-marine bivalve *Palaeomutela* in European Russia, pp. 437–442 *in* P. A. Johnston & Haggart, J. W. (eds.), *Bivalves: An Eon of Evolution*. Calgary, AB, University of Calgary Press.

Silantiev, V. V., Chandra, S. & Urazaeva, M. N. (2015). Systematics of nonmarine bivalve mollusks from the Indian Gondwana Coal Measures (Damuda Group, Permian, India). *Paleontological Journal* 49: 1235–1274.

Silva-Souza, Â. T. & Eiras, J. C. (2002). The histopathology of the infection of *Tilapia rendalli* and *Hypostomus regani* (Osteichthyes) by lasidium larvae of *Anodontites trapesialis* (Mollusca, Bivalvia). *Memórias do Instituto Oswaldo Cruz* 97: 431–433.

Silva, E. P. & Russo, C. A. M. (2000). Techniques and statistical data analysis in molecular population genetics. *Hydrobiologia* 420: 119–135.

Simkiss, K. & Wilbur, K. M. (1989). Annelids – Glandular secretions, pp. 190–204 *in* K. M. Wilbur & Simkiss, K. (eds.), *Biomineralization: Cell Biology and Mineral Deposition*. San Diego, CA, Elsevier Science.

Simone, L. R. L. de (1995). *Rissoella ornata*, a new species of Rissoellidae (Mollusca: Gastropoda: Rissoelloidea) from the southeastern coast of Brazil. *Proceedings of the Biological Society of Washington* 108: 560–567.

Simone, L. R. L. de & Martins, C. M. (1995). *Annulobalcis aurisflamma*, a new species of Eulimidae (Gastropoda, Prosobranchia) parasitic on a crinoid from Brazil. *Journal of Conchology* 35: 223–235.

Simone, L. R. L. de (2001). Phylogenetic analyses of Cerithioidea (Mollusca, Caenogastropoda) based on comparative morphology. *Arquivos de Zoologia* 36: 147–263.

Simone, L. R. L. de (2004a). Comparative morphology and phylogeny of representatives of the superfamilies of architaenioglossans and the Annulariidae (Mollusca, Caenogastropoda). *Arquivos do Museu Nacional (Rio de Janeiro)* 62: 387–504.

Simone, L. R. L. de (2004b). *Morphology and Phylogeny of the Cypraeoidea (Mollusca, Caenogastropoda)*. Rio de Janeiro, Papel Virtual Editora.

Simone, L. R. L. de (2009). Comparative morphology among representatives of main taxa of Scaphopoda and basal protobranch Bivalvia (Mollusca). *Papéis Avulsos de Zoologia (São Paulo)* 49: 405–458.

Simone, L. R. L. de (2011). Phylogeny of the Caenogastropoda (Mollusca), based on comparative morphology. *Arquivos de Zoologia* 42: 83–323.

Simone, L. R. L. de, Bunioto, T. C., Avelar, W. E. P. & Hayashi, C. (2012). Morphology and biological aspects of *Gundlachia ticaga* from SE Brazil (Gastropoda: Basommatophora: Ancylidae). *Archiv für Molluskenkunde* 141: 21–30.

Simroth, H. (1892–94). Amphineura und Scaphopoda, pp. i–vii, 1–1056 *in* H. G. Bronn (ed.), *H. G. Bronn's Klassen und Ordnungen des Tier-Reichs wissenschaftlich dargestellt in Wort und Bild (1892–1894)*. Vol. 3, Mollusca, Part 1. Leipzig, C. F. Winter.

Sirenko, B. I. & Minichev, S. (1975). Développement ontogénétique de la radula chez les polyplacophores. *Cahiers de Biologie Marine* 16: 425–433.

Sirenko, B. I. (1993). Revision of the system of the order Chitonida (Mollusca: Polyplacophora) on the basis of correlation between the type of gills arrangement and the shape of the chorion processes. *Ruthenica* 3: 93–117.

Sirenko, B. I. (1997). The importance of the development of the articulamentum for taxonomy of chitons (Mollusca, Polyplacophora). *Ruthenica* 7: 1–24.

Sirenko, B. I. (1998). One more deep water chiton *Leptochiton vietnamensis* sp. nov. (Mollusca, Polyplacophora) living and feeding on sunken wood from the South China Sea. *Ruthenica* 8: 1–6.

Sirenko, B. I. (2006). New outlook on the system of chitons (Mollusca: Polyplacophora). *Venus* 65: 27–49.

Sirenko, B. I. (2015). The enigmatic viviparous chiton *Calloplax vivipara* (Plate, 1899) (Mollusca: Polyplacophora) and a survey of the types of reproduction in chitons. *Russian Journal of Marine Biology* 41: 24–31.

Sirenko, B. I. (2018). The larval development of the chiton *Deshayesiella curvata* (Carpenter in Dall, 1879)(Mollusca: Polyplacophora). *Russian Journal of Marine Biology* 44: 304–308.

Sirgel, W. F. (1973). Contributions to the morphology and histology of the genital system of the pulmonate *Agriolimax caruanae* Pollonera. *Annale van de Uniwersiteit van Stellenbosch Serie A II (Zoologie)* 48: 1–43.

Skawina, A. & Dzik, J. (2011). Umbonal musculature and relationships of the Late Triassic filibranch unionoid bivalves. *Zoological Journal of the Linnean Society* 163: 863–883.

Skelton, P. W. (1976). Functional morphology of the *Hippuritidae*. *Lethaia* 9: 83–100.

Skelton, P. W. (1985). Preadaptation and evolutionary innovation in rudist bivalves, pp. 159–173 *in* J. C. W. Cope & Skelton, P. W. (eds.), *Evolutionary Case Histories from the Fossil Record. Special Papers in Palaeontology*. London, Palaeontological Association.

Skelton, P. W., Crame, J. A., Morris, N. J. & Harper, E. M. (1990). Adaptive divergence and taxonomic radiation in post-Palaeozoic bivalves, pp. 91–117 *in* P. D. Taylor & Larwood, G. P. (eds.), *Major Evolutionary Radiations. The Systematics Association Special Volume*. Oxford, UK, Clarendon Press.

Skelton, P. W. & Smith, A. B. (2000). A preliminary phylogeny for rudist bivalves: Sifting clades from grades, pp. 97–127 *in* E. M. Harper, Taylor, J. D. & Crame, J. A. (eds.), *Evolutionary Biology of the Bivalvia*. London, The Geological Society (Special Publication No. 177).

Skovsted, C. B., Brock, G. A., Paterson, J. R., Holmer, L. E. & Budd, G. E. (2008). The scleritome of *Eccentrotheca* from the Lower Cambrian of South Australia: Lophophorate affinities and implications for tommotiid phylogeny. *Geology* 36: 171–174.

Skovsted, C. B., Balthasar, U., Brock, G. A. & Paterson, J. R. (2009). The tommotiid *Camenella reticulosa* from the Early Cambrian of South Australia: Morphology, scleritome reconstruction, and phylogeny. *Acta Palaeontologica Polonica* 54: 525–540.

Sleurs, W. J. M. (1985). *Ammonicera angulata* sp. nov. from Laing Island, Papua New Guinea, with comments on the genus *Ammonicera* Vayssiere, 1983 (Mollusca, Gastropoda). *Annales de la Société royale zoologique de Belgique* 115: 177–181.

Smale, M. J. (1996). Cephalopods as prey. IV. Fishes. *Philosophical Transactions of the Royal Society B* 351: 1067–1081.

Smith, A. & Kenny, R. (1987). Reproduction and development of *Onchidium damelii* Semper 1882. *Journal of the Malacological Society of Australia* 8: 37–39.

Smith, A. G. & Hoare, R. D. (1987). Paleozoic Polyplacophora: A checklist and bibliography. *Occasional Papers of the California Academy of Sciences* 146: 1–71.

Smith, A. M. & Spencer, H. G. (2016). Skeletal mineralogy of scaphopods: An unusual uniformity. *Journal of Molluscan Studies* 82: 344–348.

Smith, B. J. & Stanisic, J. (1998). Pulmonata: Introduction, pp. 1037–1061 *in* P. L. Beesley, Ross, G. J. B. & Wells, A. (eds.), *Mollusca: The Southern Synthesis. Part B. Fauna of Australia*. Melbourne, VIC, CSIRO Publishing.

Smith, E. H. (1967a). The neogastropod stomach, with notes on the digestive diverticula and intestine. *Transactions of the Royal Society of Edinburgh* 67: 23–42.

Smith, E. H. (1967b). The reproductive system of the British Turridae (Gastropoda: Toxoglossa). *The Veliger* 10: 176–187.

Smith, E. H. (1969). Functional morphology of *Penitella conradi* relative to shell penetration. *American Zoologist* 9: 869–888.

Smith, F. D. M., May, R. M., Pellew, R., Johnson, T. H. & Walter, K. R. (1993). How much do we know about the current extinction rate? *Trends in Ecology & Evolution* 8: 375–378.

Smith, F. G. W. (1935). The development of *Patella vulgata*. *Philosophical Transactions of the Royal Society of Edinburgh* 225: 95–125.

Smith, J. P. (1901). The larval coil of *Baculites*. *American Naturalist* 35: 39–49.

Smith, J. T. (1970). Taxonomy, distribution, and phylogeny of the cymatiid gastropods *Argobuccinum*, *Fusitriton*, *Mediargo*, and *Priene*. *Bulletins of American Paleontology* 56: 444–573.

Smith, M. R. & Caron, J.-B. (2010). Primitive soft-bodied cephalopods from the Cambrian. *Nature* 465: 469–472.

Smith, M. R. (2012). Mouthparts of the Burgess Shale fossils *Odontogriphus* and *Wiwaxia*: Implications for the ancestral molluscan radula. *Proceedings of the Royal Society B: Biological Sciences* 279: 4287–4295.

Smith, M. R. (2013). Ontogeny, morphology and taxonomy of the soft-bodied Cambrian 'mollusc' *Wiwaxia*. *Palaeontology* 57: 215–229.

Smith, S. (2008). Interpreting molluscan death assemblages on rocky shores: Are they representative of the regional fauna? *Journal of Experimental Marine Biology and Ecology* 366: 151–159.

Smith, S. A., Wilson, N. G., Goetz, F., Feehery, C., Andrade, S. C. S., Rouse, G. W., Giribet, G. & Dunn, C. W. (2011). Resolving the evolutionary relationships of molluscs with phylogenomic tools. *Nature* 480: 364–367.

Smith, S. A., Wilson, N. G., Goetz, F., Feehery, C., Andrade, S. C. S., Rouse, G. W., Giribet, G. & Dunn, C. W. (2012). Corrigendum: Resolving the evolutionary relationships of molluscs with phylogenomic tools. *Nature* 493: 708.

Smolensky, N., Romero, M. R. & Krug, P. J. (2009). Evidence for costs of mating and self-fertilization in a simultaneous hermaphrodite with hypodermic insemination, the opisthobranch *Alderia willowi*. *Biological Bulletin* 216: 188–199.

Sober, E. (1991). *Reconstructing the Past: Parsimony, Evolution, and Inference*. Cambridge, MA, MIT Press.

Solem, A. C. (1959). *Systematics and Zoogeography of the Land- and Fresh-Water Mollusca of the New Hebrides*. Chicago, IL, Chicago Natural History Museum.

Solem, A. C. (1973). Convergence in pulmonate radulae. *The Veliger* 15: 165–171.

Solem, A. C. (1974). *The Shell Makers. Introducing Mollusks*. New York, John Wiley & Sons.

Solem, A. C. & Richardson, E. S. (1975). *Paleocadmus*, a nautiloid cephalopod radula from the Pennsylvanian Francis Creek Shale of Illinois. *The Veliger* 17: 233–242.

Solem, A. C. (1978). Classification of the land Mollusca, pp. 49–98 *in* V. Fretter & Peake, J. (eds.), *Pulmonates: Systematics, Evolution and Ecology*. Vol. 2A. London, Academic Press.

Solem, A. C. & Yochelson, E. L. (1979). North American Paleozoic land snails, with a summary of other Paleozoic nonmarine snails. *United States Geological Survey. Professional* Paper 1072: 1–42.

Solem, A. C. (1983). Lost or kept internal whorls: Ordinal differences in land snails. *Journal of Molluscan Studies*: Supplement 12A: 172–178.

Soliman, G. N. (1987). A scheme for classifying gastropod egg masses with special reference to those from the northwestern Red Sea. *Journal of Molluscan Studies* 53: 1–12.

Song, H., Wignall, P. B., Tong, J. & Yin, H. (2013). Two pulses of extinction during the Permian-Triassic crisis. *Nature Geoscience* 6: 52–56.

Sorensen, F. E. & Lindberg, D. R. (1991). Preferential predation by American black oystercatchers on transitional ecophenotypes of the limpet *Lottia pelta* (Rathke). *Journal of Experimental Marine Biology and Ecology* 154: 123–136.

Sørensen, M. V. & Sterrer, W. (2002). New characters in the gnathostomulid mouth parts revealed by scanning electron microscopy. *Journal of Morphology* 253: 310–334.

Sørensen, M. V. (2003). Further structures in the jaw apparatus of *Limnognathia maerski* (Micrognathozoa), with notes on the phylogeny of the Gnathifera. *Journal of Morphology* 255: 131–145.

Sowerby (II), G. B. (1847–1877). *Thesaurus Conchyliorum: Or Monographs of Genera of Shells. Volumes 1–5*. London, Privately published.

Speiser, B. (2001). Food and feeding behaviour, pp. 259–288 *in* G. M. Barker (ed.), *The Biology of Terrestrial Molluscs*. Wallingford, UK, CABI Publishing.

Spengel, J. W. (1881). Die Geruchsorgane und das Nervensystem der Mollusken. *Zeitschrift für wissenschaftliche Zoologie* 35: 333–383.

van der Spoel, S. (1967). *Euthecosomata – A Group with Remarkable Developmental Stages (Gastropoda, Pteropoda)*. Gorinchem, the Netherlands, J. Noorduijn en Zoon N.V.

van der Spoel, S. (1973). Strobilation in a mollusk: The development of aberrant stages in *Clio pyramidata* Linnaeus, 1767 (Gastropoda, Pteropoda). *Bijdragen tot de Dierkunde* 43: 202–215.

Stanley, G. D. & Teichert, C. (1976). Lamellorthoceratids (Cephalopoda, Orthoceroidea) from the lower Devonian of New York. *University of Kansas Paleontological Contributions Paper* 86: 1–14.

Stanley, S. M. (1969). Bivalve mollusk burrowing aided by discordant shell ornamentation. *Science* 166: 634–635.

Stanley, S. M. (1968). Post-Paleozoic adaptive radiation of infaunal bivalve molluscs: A consequence of mantle fusion and siphon formation. *Journal of Paleontology* 42: 214–229.

Stanley, S. M. (1970). *Relation of Shell Form to Life Habits of the Bivalvia (Mollusca)*. Vol.125 Boulder, CO, Geological Society of America.

Stanley, S. M. (1972). Functional morphology and evolution of byssally attached bivalved molluscs. *Journal of Paleontology* 46: 165–212.

Stanley, S. M. (1973). An ecological theory for the sudden origin of multicellular life in the late Precambrian. *Proceedings of the National Academy of Sciences of the United States of America* 70: 1486–1489.

Stanley, S. M. (1975). Adaptive themes in the evolution of the Bivalvia (Mollusca). *Annual Review of Earth and Planetary Sciences* 3: 361–385.

Stanley, S. M. (1976). Fossil data and the Precambrian-Cambrian evolutionary transition. *American Journal of Science* 276: 56–76.

Stanley, S. M. & Waller, T. R. (1978). Aspects of the adaptive morphology and evolution of the Trigoniidae. *Philosophical Transactions of the Royal Society B* 284: 247–258.

Stanley, S. M., Addicott, W. O. & Chinzei, K. (1980). Lyellian curves in paleontology: Possibilities and limitations. *Geology* 8: 422–426.

Starmühlner, F. (1993). The mountain stream fauna of Sri Lanka with special reference to molluscs (A summarizing review), pp. 121–188 in W. Erdelen, Preu, C., Ishwaran, N. & Madduma Bandara, C. M. (eds.), *Ecology and Landscape Management in Sri Lanka: Proceedings of the International and Interdisciplinary Symposium. Colombo, Sri Lanka.* Weikersheim, Germany, Margraf Verlag.

Starobogatov, I. (1970). Systematics of Early Paleozoic Monoplacophora. *Paleontological Journal* 3: 293–302.

Starobogatov, I. (1974). Xenoconchias and their bearing on the phylogeny and systematics of some molluscan classes. *Paleontological Journal* 3: 3–18.

Starobogatov, I. (1983). System of Cephalopoda, pp. 4–7 in I. Starobogatov & Nesis, K. N. (eds.), *Systematics and Ecology of Cephalopods.* Leningrad, Zoological Institute of the Academy of Sciences of the USSR.

Starobogatov, I. & Sitnikova, T. (1983). Sistema otriada Littoriniformes (Gastropoda, Pectinibranchia) [The system of the order Littoriniformes] (in Russian) *Vsesoiuznoe soveshchanie po izucheniiu molliuskov [Leningrad]* 7: 18–22.

Starobogatov, I. & Moskalev, L. I. (1987). Systematics of the Monoplacophora, pp. 7–11 in I. Starobogatov, Golikov, A. N. & Likharev, I. M. (eds.), *Molluscs. Results and Perspectives of Investigation. Eighth Meeting on the Investigation of Molluscs. Abstracts of Communications* [in Russian, incl. English translation]. Leningrad, Zoological Institute of the Academy of Sciences of the USSR.

Starobogatov, I. (1992). Morphological basis for phylogeny and classification of Bivalvia. *Ruthenica* 2: 1–25.

Stasek, C. R. (1967). Autotomy in the Mollusca. *Occasional Papers of the California Academy of Science* 59: 1–44.

Stasek, C. R. (1972). The molluscan framework, pp. 1–44 in M. Florkin & Scheer, B. T. (eds.), *Chemical Zoology. Mollusca.* Vol. 7. New York, Academic Press.

Staubach, S. (2008). *The evolution of the cephalic sensory organs within the opisthobranchia.* Ph.D. dissertation, Johann Wolfang Goethe Universität.

Stechmann, A. & Schlegel, M. (1999). Analysis of the complete mitochondrial DNA sequence of the brachiopod *Terebratulina retusa* places Brachiopoda within the protostomes. *Proceedings of the Royal Society B* 266: 2043–2052.

Steiner, G. (1990). *Beiträge zur vergleichenden Anatomie und Systematik der Scaphopoda (Mollusca).* Ph.D. dissertation, Universität Wien Institut Zoologie.

Steiner, G. (1991). Observations on the anatomy of the scaphopod mantle and the description of a new family, the Fustiariidae. *American Malacological Bulletin* 9: 1–20.

Steiner, G. (1992a). The organisation of the pedal musculature and its connection to the dorsoventral musculature in Scaphopoda. *Journal of Molluscan Studies* 58: 181–197.

Steiner, G. (1992b). Phylogeny and classification of Scaphopoda. *Journal of Molluscan Studies* 58: 385–400.

Steiner, G. (1994). Variations in the number of intestinal loops in Scaphopoda (Mollusca). *Marine Ecology* 15: 165–174.

Steiner, G. (1998a). Order Dentaliida, pp. 439–443 in P. L. Beesley, Ross, G. J. B. & Wells, A. (eds.), *Mollusca: The Southern Synthesis. Part A. Fauna of Australia.* Melbourne, VIC, CSIRO Publishing.

Steiner, G. (1998b). Phylogeny of *Scaphopoda* (Mollusca) in the light of new anatomical data on the Gadilinidae and some Problematica, and a reply to Reynolds. *Zoologica Scripta* 27: 73–82.

Steiner, G. & Hammer, S. (2000). Molecular phylogeny of the Bivalvia inferred from 18S rDNA sequences with particular reference to the Pteriomorphia, pp. 11–29 in E. M. Harper,

Taylor, J. D. & Crame, J. A. (eds.), *Evolutionary Biology of the Bivalvia.* London, The Geological Society (Special Publication No. 177).

Steiner, G. & Kabat, A. R. (2001). Catalogue of supraspecific taxa of Scaphopoda (Mollusca). *Zoosystema* 23: 433–460.

Steiner, G. & von Salvini-Plawen, L. (2001). Invertebrate evolution - *Acaenoplax* - polychaete or mollusc? *Nature* 414: 601–602.

Steiner, G. & Dreyer, H. (2003). Molecular phylogeny of Scaphopoda (Mollusca) inferred from 18S rDNA sequences: Support for a Scaphopoda-Cephalopoda clade. *Zoologica Scripta* 32: 343–356.

Steiner, G. & Kabat, A. R. (2004). Catalog of species-group names of Recent and fossil Scaphopoda (Mollusca). *Zoosystema* 26: 549–726.

Steinova, M. (2012). Probable ancestral type of actinodont hinge in the Ordovician bivalve *Pseudocyrtodonta* Pfab, 1934. *Bulletin of Geosciences* 87: 333–346.

Stephen, D. A. & Stanton, R. J. (2002). Impact of reproductive strategy on cephalopod evolution. *Abhandlungen der Geologischen Bundesanstalt* 57: 151–155.

Steuber, T. (1999). Cretaceous rudists of Boeotia, Central Greece. *Special Papers in Palaeontology* 61: 1–229.

Steuber, T., Mitchell, S. F., Buhl, D., Gunter, G. & Kasper, H. U. (2002). Catastrophic extinction of Caribbean rudist bivalves at the Cretaceous-Tertiary boundary. *Geology* 30: 999–1002.

Stevens, G. R. (1988). Giant ammonites: A review, pp. 141–166 in J. Wiedmann & Kullmann, J. (eds.), *Cephalopods Present and Past. 2nd International Cephalopod Symposium: O. H. Schindewolf Symposium, Tübingen 1985.* Stuttgart, E. Schweizerbart'sche Verlagsbuchhandlung.

Stevens, K., Griesshaber, E., Schmahl, W., Casella, L. A., Iba, Y. & Mutterlose, J. (2017). Belemnite biomineralization, development, and geochemistry: The complex rostrum of *Neohibolites minimus. Palaeogeography, Palaeoclimatology, Palaeoecology* 468: 388–402.

Stewart, F. J. & Cavanaugh, C. M. (2006). Bacterial endosymbioses in *Solemya* (Mollusca: Bivalvia) – Model systems for studies of symbiont-host adaptation. *Antonie van Leeuwenhoek* 90: 343–360.

Stimson, J. (1970). Territorial behavior of the owl limpet, *Lottia gigantea. Ecology* 51: 113–118.

Stinchcomb, B. L. & Echols, D. J. (1966). Missouri Upper Cambrian Monoplacophora previously considered cephalopods. *Journal of Paleontology* 40: 647–650.

Stinchcomb, B. L. (1980). New information on Late Cambrian Monoplacophora *Hypseloconus* and *Shelbyoceras* (Mollusca). *Journal of Paleontology* 54: 45–49.

Stinchcomb, B. L. & Darrough, G. (1995). Some molluscan problematica from the Upper Cambrian-Lower Ordovician of the Ozark uplift. *Journal of Paleontology* 69: 52–65.

Stock, J. H., Iliffe, T. M. & Williams, D. (1986). The concept 'anchialine' reconsidered. *Stygologia* 2: 90–92.

Stöger, I., Sigwart, J. D., Kano, Y., Knebelsberger, T., Marshall, B. A. & Schwabe, E. (2013). The continuing debate on deep molluscan phylogeny: Evidence for Serialia (Mollusca, Monoplacophora + Polyplacophora). *BioMed Research International* 2013: 1–18.

Stöger, I., Kocot, K., Poustka, A., Wilson, N., Ivanov, D., Halanych, K. & Schrödl, M. (2016). Monoplacophoran mitochondrial genomes: Convergent gene arrangements and little phylogenetic signal. *BMC Evolutionary Biology* 16: 274.

Stoliczka, F. (1863–1866). The fossil Cephalopoda of the Cretaceous rocks of southern India. Ammonitidae with revision of the Nautilidae etc. Palaeontologia Indica. *Memoirs of the Geological Survey of India* 3 (Parts 1–13): 41–216.

Storey, K. B. & Storey, J. M. (1983). Carbohydrate metabolism in cephalopod molluscs, pp. 91–136 *in* P. W. Hochachka (ed.), *Metabolic Biochemistry and Molecular Biomechanics. The Mollusca.* Vol. 1. New York, Academic Press.

Stotz, W. B. (2000). *Informe Final Proyecto 97–36. Formulacion de una metodologia para el estudio de edad y crecimiento en el recurso loco*, Universidad Católica del Norte Facultad de Ciencias del Marina. Departo. de Biología Marina.

Strathmann, M. F. & Strathmann, R. R. (2007). An extraordinarily long larval duration of 4.5 years from hatching to metamorphosis for teleplanic veligers of *Fusitriton oregonensis. Biological Bulletin* 213: 152–159.

Streit, K., Geiger, D. L. & Lieb, B. (2006). Molecular phylogeny and the geographic origin of Haliotidae traced by haemocyanin sequences. *Journal of Molluscan Studies* 72: 105–110.

Stridsberg, S. (1984). Aptychopsid plates - jaw elements or protective operculum. *Lethaia* 17: 93–98.

Stridsberg, S. (1985). Silurian oncocerid cephalopods from Gotland. *Fossils and Strata* 18: 1–65.

Strong, E. E. & Harasewych, M. G. (1999). Anatomy of the hadal limpet *Macleaniella moskalevi* (Gastropoda, Cocculinoidea). *Invertebrate Biology* 118: 137–148.

Strong, E. E. (2003). Refining molluscan characters: Morphology, character coding and a phylogeny of the Caenogastropoda. *Zoological Journal of the Linnean Society* 137: 447–554.

Strong, E. E., Harasewych, M. G. & Haszprunar, G. (2003). Phylogeny of the Cocculinoidea (Mollusca, Gastropoda). *Invertebrate Biology* 122: 114–125.

Strong, E. E., Gargominy, O., Ponder, W. F. & Bouchet, P. (2008). Global diversity of gastropods (Gastropoda: Mollusca) in freshwater. *Hydrobiologia* 595: 149–166.

Strong, E. E., Colgan, D. J., Healy, J. M., Lydeard, C., Ponder, W. F. & Glaubrecht, M. (2011). Phylogeny of the gastropod superfamily Cerithioidea using morphology and molecules. *Zoological Journal of the Linnean Society* 162: 43–89.

Strong, E. E., Galindo, L. A. & Kantor, Y. I. (2017). Quid est *Clea helena*? Evidence for a previously unrecognized radiation of assassin snails (Gastropoda: Buccinoidea: Nassariidae). *PeerJ* 5: e3638.

Strong, E. E., Puillandre, N., Beu, A. G., Castelin, M. & Bouchet, P. (2019). Frogs and tuns and tritons – A molecular phylogeny and revised family classification of the predatory gastropod superfamily Tonnoidea (Caenogastropoda). *Molecular Phylogenetics and Evolution* 130: 18–34.

Strugnell, J. M., Norman, M. D., Drummond, A. J., Jackson, J. B. C. & Cooper, A. B. (2005). Molecular phylogeny of coleoid cephalopods (Mollusca: Cephalopoda) using a multigene approach: The effect of data partitioning on resolving phylogenies in a Bayesian framework. *Molecular Phylogenetics and Evolution* 37: 426–441.

Strugnell, J. M., Jackson, J., Drummond, A. J. & Cooper, A. (2006). Divergence time estimates for major cephalopod groups: Evidence from multiple genes. *Cladistics* 22: 89–96.

Strugnell, J. M. & Nishiguchi, M. K. (2007). Molecular phylogeny of coleoid cephalopods (Mollusca: Cephalopoda) inferred from three mitochondrial and six nuclear loci: A comparison of alignment, implied alignment and analysis methods. *Journal of Molluscan Studies* 73: 399–410.

Strugnell, J. M., Rogers, A. D., Prodöhl, P. A., Collins, M. A. & Allcock, A. L. (2008). The thermohaline expressway: The Southern Ocean as a centre of origin for deep-sea octopuses. *Cladistics* 24: 853–860.

Strugnell, J. M., Norman, M. D., Vecchione, M., Guzik, M. T. & Allcock, A. L. (2014). The ink sac clouds octopod evolutionary history. *Hydrobiologia* 725: 215–235.

Strugnell, J. M., Hall, N. E., Vecchione, M., Fuchs, D. & Allcock, A. L. (2017). Whole mitochondrial genome of the Ram's Horn Squid shines light on the phylogenetic position of the monotypic order Spirulida (Haeckel, 1896). *Molecular Phylogenetics and Evolution* 109: 296–301.

Stuber, R. A. & Lindberg, D. R. (1989). Is the radula of living monoplacophorans primitive? *Geological Society of America, 1989 Annual Meeting.* St. Louis, MO, Geological Society of America. 21: A289.

Sturm, C. F., Pearce, T. A. & Valdés, Á. (2006). The Mollusks: Introductory comments, pp. 1–7 *in* C. F. Sturm, Pearce, T. A. & Valdés, Á. (eds.), *The Mollusks: A Guide to Their Study, Collection, and Preservation.* Boca Raton, FL, Universal Publishers.

Stürmer, W. (1985). A small coleoid cephalopod with soft parts from the Lower Devonian discovered using radiography. *Nature* 318: 53–55.

Stützel, R. (1984). Anatomische und ultrastrukturelle Untersuchungen an der Napfschnecke *Patella* L. unter besonderer Berücksichtigung der Anpassung an den Lebensraum. *Zoologica (Stuttgart)* 46: 1–54.

Sumner-Rooney, L. H., Murray, J. A., Cain, S. D. & Sigwart, J. D. (2014). Do chitons have a compass? Evidence for magnetic sensitivity in Polyplacophora. *Journal of Natural History* 48: 3033–3045.

Sumner-Rooney, L. H., Schrödl, M., Lodde-Bensch, E., Lindberg, D. R., Hess, M., Brennan, G. P. & Sigwart, J. D. (2015). A neurophylogenetic approach provides new insight to the evolution of Scaphopoda. *Evolution & Development* 17: 337–346.

Sumner-Rooney, L. H. & Sigwart, J. D. (2015). Is the Schwabe organ a retained larval eye? Anatomical and behavioural studies of a novel sense organ in adult *Leptochiton asellus* (Mollusca, Polyplacophora) indicate links to larval photoreceptors. *PloS ONE* 10: e0137119.

Sumner-Rooney, L. & Sigwart, J. D. (2018). Do chitons have a brain? New evidence for diversity and complexity in the polyplacophoran central nervous system. *Journal of Morphology* 279: 936–949.

Sunnucks, P. (2000). Efficient genetic markers for population biology. *TREE* 15: 199–203.

Sutcharit, C. & Panha, S. (2006). Taxonomic review of the tree snail *Amphidromus* Albers, 1850 (Pulmonata: Camaenidae) in Thailand and adjacent areas: Subgenus *Amphidromus. Journal of Molluscan Studies* 72: 1–30.

Sutton, M., Perales-Raya, C. & Gilbert, I. (2016). A phylogeny of fossil and living neocoleoid cephalopods. *Cladistics* 32: 297–307.

Sutton, M. D., Briggs, D. E. G. & Siveter, D. J. (2001a). An exceptionally preserved vermiform mollusc from the Silurian of England. *Nature* 410: 461–463.

Sutton, M. D., Briggs, D. E. G., Siveter, D. J. & Siveter, D. J. (2001b). *Acaenoplax* - polychaete or mollusc? Reply. *Nature* 414: 602.

Sutton, M. D., Briggs, D. E. G. & Siveter, D. J. (2004). Computer reconstruction and analysis of the vermiform mollusc *Acaenoplax hayae* from the Herefordshire Lagerstätte (Silurian, England), and implications for molluscan phylogeny. Part 2. *Palaeontology* 47: 293–318.

Sutton, M. D., Briggs, D. E. G. & Siveter, D. J. (2006). Fossilized soft tissues in a Silurian platyceratid gastropod. *Proceedings of the Royal Society B* 273: 1039–1044.

Sutton, M. D. (2008). Tomographic techniques for the study of exceptionally preserved fossils. *Proceedings of the Royal Society B* 275: 1587–1593.

Sutton, M. D., Briggs, D. E. G., Siveter, D. J. & Sigwart, J. D. (2012). A Silurian armoured aplacophoran and implications for molluscan phylogeny. *Nature* 490: 94–97.

Sutton, M. D. & Sigwart, J. D. (2012). A chiton without a foot. *Palaeontology* 55: 401–411.

Suvorov, A. N. (1993). Problems of functional morphology of ostium in Pupillacea snails (Gastropoda: Pulmonata). *Ruthenica* 3: 141–152.

Suvorov, A. N. (1999). Functional relations between shell structures and soft body in lower Geophila. 1. Pupillina, Oleacinina. *Zoologicheskii Zhurnal* 78: 5–15.

Suvorov, A. N. (2000). Functional morphology of pneumostomal area in terrestrial Pulmonata (Gastropoda). *Ruthenica* 10: 89–104.

Symondson, W. O. C. (2004). Coleoptera (Carabidae, Staphylinidae, Lampyridae, Drilidae and Silphidae) as predators of terrestrial gastropods, pp. 37–84 in G. M. Barker (ed.), *Natural Enemies of Terrestrial Molluscs*. Oxford, UK/Cambridge, MA, CABI Publishing.

Szal, R. A. (1971). 'New' sense organ of primitive gastropods. *Nature* 229: 490–492.

Tajika, A., Naglik, C., Morimoto, N., Pascual-Cebrian, E., Hennhöfer, D. & Klug, C. (2015). Empirical 3D model of the conch of the Middle Jurassic ammonite microconch *Normannites*: Its buoyancy, the physical effects of its mature modifications and speculations on their function. *Historical Biology* 27: 181–191.

Tajika, A., Nützel, A. & Klug, C. (2018). The old and the new plankton: Ecological replacement of associations of mollusc plankton and giant filter feeders after the Cretaceous? *PeerJ* 6: e4219.

Takano, T. & Kano, Y. (2014). Molecular phylogenetic investigations of the relationships of the echinoderm-parasite family Eulimidae within Hypsogastropoda (Mollusca). *Molecular Phylogenetics and Evolution* 79: 258–269.

Takeda, N. & Tsuroka, H. (1979). A sex pheromone secreting gland in the terrestrial snail, *Euhadra peliomphala*. *Journal of Experimental Zoology Part A* 207: 17–26.

Takeda, N. (1980). Hormonal control of head-wart development in the snail, *Euhadra peliomphala*. *Journal of Embryology and Experimental Morphology* 60: 57–69.

Takeda, N. (1982). Notes on the fine structure of the head-wart in some terrestrial snails. *The Veliger* 24: 328–330.

Tan, K. S. & Lee, S. S. C. (2009). Neritid egg capsules: Are they all that different? *Steenstrupia* 30: 115–125.

Tanabe, K. & Fukuda, Y. (1983). Buccal mass structure of the Cretaceous ammonite *Gaudryceras*. *Lethaia* 16: 249–256.

Tanabe, K. & Fukuda, Y. (1987). Mouth part histology and morphology, pp. 313–322 in W. B. Saunders & Landman, N. H. (eds.), *Nautilus: The Biology and Paleobiology of a Living Fossil. Topics in Geobiology*. New York, Springer.

Tanabe, K., Tsukahara, J., Fukuda, Y. & Taya, Y. (1991). Histology of a living *Nautilus* embryo: Preliminary observations. *Journal of Cephalopod Biology* 2: 13–22.

Tanabe, K. & Mapes, R. H. (1995). Jaws and radula of the Carboniferous ammonoid *Cravenoceras*. *Journal of Paleontology* 69: 703–707.

Tanabe, K. (2000). Soft-part anatomy of the siphuncle in Permian prolecanitid ammonoids. *Lethaia* 33: 83–91.

Tanabe, K., Kruta, I. & Landman, N. H. (2015). Ammonoid buccal mass and jaw apparatus, pp. 439–494 in C. Klug, Korn, D., De Baets, K., Kruta, I. & Mapes, R. H. (eds.), *Ammonoid Paleobiology: From Anatomy to Ecology. Topics in Geobiology*. Dordrecht, the Netherlands/Heidelberg/New York/London, Springer.

Tang, C. M. (2002). Oxford Clay: England's Jurassic marine park Pp. 307–325 in D. J. Bottjer, Etter, W., Hagadorn, J. W. & Tang, C. M. (eds.), *Exceptional Fossil Preservation: A Unique View on the Evolution of Marine Life*. New York, Columbia University Press.

Tanner, A. R., Fuchs, D., Winkelmann, I. E., Gilbert, M. T. P., Pankey, M. S., Ribeiro, Â. M., Kocot, K. M., Halanych, K. M., Oakley, T. H., Da Fonseca, R. R., Pisani, D. & Vinther, J. (2017). Molecular clocks indicate turnover and diversification of modern coleoid cephalopods during the Mesozoic Marine Revolution. *Proceedings of the Royal Society B* 284. 10.1098/rspb.2016.2818.

Tardy, J. (1965). Spermatophores chez quelques espèces d' Aeolidiidae (Mollusques, Nudibranches). *Comptes rendus des séances de la Société de biologie et de ses filiales* 160: 369–371.

Tardy, J. (1969). Un nouveau genre de Nudibranche méconnu des côtes Atlantique et de la Manche: *Pruvotfolia* (nov. g.) *pselliotes*, (Labbé), 1923. *Vie et Milieu Série A: Biologie Marine* 20: 327–346.

Tardy, J. (1970). Organogenèse de l'appareil génital chez les mollusques. *Bulletin de la Société zoologique de France* 95: 407–428.

Taviani, M., Sabelli, B. & Candini, F. (1990). A fossil Cenozoic monoplacophoran. *Lethaia* 23: 213–216.

Taylor, J. D., Kennedy, W. J. & Hall, A. (1969). The shell structure and mineralogy of the Bivalvia. 1. Introduction, Nuculacea - Trigoniacea. *British Museum (Natural History) Bulletin. Zoology* 3: 1–125.

Taylor, J. D. & Layman, M. (1972). The mechanical properties of bivalve (Mollusca) shell structures. *Palaeontology* 15: 5.

Taylor, J. D. (1973). The structural evolution of the bivalve shell. *Palaeontology* 16: 519–534.

Taylor, J. D., Kennedy, W. J. & Hall, A. (1973). The shell structure and mineralogy of the Bivalvia. II. Lucinacea - Clavagellacea. Conclusions. *British Museum (Natural History) Bulletin. Zoology* 22: 255–294.

Taylor, J. D. (1978). Habitats and diet of predatory gastropods at Addu Atoll, Maldives. *Journal of Experimental Marine Biology and Ecology* 31: 83–103.

Taylor, J. D. & Morris, N. J. (1988). Relationships of neogastropods, pp. 167–179 in W. F. Ponder, Eernisse, D. J. & Waterhouse, J. H. (eds.), *Prosobranch Phylogeny. Malacological Review Supplement*. Ann Arbor, MI, Malacological Review.

Taylor, J. D. & Miller, J. A. (1989). The morphology of the osphradium in relation to feeding habits in meso- and neo-gastropods. *Journal of Molluscan Studies* 55: 227–237.

Taylor, J. D. (1990). The anatomy of the foregut and relationships in the Terebridae. *Malacologia* 32: 19–34.

Taylor, J. D. & Miller, J. A. (1990). A new type of gastropod proboscis: The foregut of *Hastula bacillus* (Gastropoda: Terebridae). *Journal of Zoology* 220: 603–617.

Taylor, J. D., Kantor, Y. I. & Sysoev, A. V. (1993). Foregut anatomy, feeding mechanisms, relationships and classification of the Conoidea (=Toxoglossa) (Gastropoda). *Bulletin of the British Museum (Natural History). Zoology* 59: 125–170.

Taylor, J. D. (1994). Foregut anatomy of the larger species of *Turrinae, Clavatulinae* and *Crassispirinae* (Gastropoda: Conoidea) from Hong Kong, pp. 185–213 in B. Morton (ed.), *The Malacofauna of Hong Kong and Southern China, III: Proceedings of the Third International Workshop on the Malacofauna of Hong Kong and Southern China, Hong Kong*. Vol. 3. Hong Kong, Hong Kong University Press.

Taylor, J. D. & Glover, E. A. (1997). New species and records of *Rastafaria* and *Megaxinus* (Bivalvia: Lucinidae) from the western Indian Ocean and Red Sea, with a reappraisal of *Megaxinus*. *Journal of Conchology* 36: 1–18.

Taylor, J. D. (1998). Understanding biodiversity: Adaptive radiations of predatory marine gastropods, pp. 187–206 in B. Morton (ed.), *The Marine Biology of the South China Sea 3. Proceedings of the Third International Conference on* the Marine Biology of the South China Sea, Hong Kong, 28 October–1 November 1996. Vol. 3. Hong Kong, Hong Kong University Press.

Taylor, J. D. & Glover, E. A. (2000). Functional anatomy, chemosymbiosis and evolution of the Lucinidae, pp. 207–225 in E. M. Harper, Taylor, J. D. & Crame, J. A. (eds.), *Evolutionary Biology of the Bivalvia*. London, The Geological Society (Special Publication No. 177).

Taylor, J. D., Glover, E. A., Peharda, M., Bigatti, G. & Ball, A. (2004). Extraordinary flexible shell sculpture; the structure and formation of calcified periostracal lamellae in *Lucina pensylvanica* (Bivalvia: Lucinidae). *Malacologia* 46: 277–294.

Taylor, J. D., Glover, E. A. & Williams, S. T. (2005). Another bloody bivalve: Anatomy and relationships of *Eucrassatella donacina* from south western Australia (Mollusca: Bivalvia: Crassatellidae), pp. 261–288 in F. E. Wells, Walker, D. I. & Kendrick, G. A. (eds.), *Proceedings of the Twelfth International Marine Biological Workshop: The Marine Flora and Fauna of Esperance, Western Australia*. Esperance, WA, Western Australian Museum.

Taylor, J. D. & Glover, E. A. (2006). Lucinidae (Bivalvia) – The most diverse group of chemosymbiotic molluscs. *Zoological Journal of the Linnean Society* 148: 421–438.

Taylor, J. D., Williams, S. T. & Glover, E. A. (2007a). Evolutionary relationships of the bivalve family Thyasiridae (Mollusca: Bivalvia), monophyly and superfamily status. *Journal of the Marine Biological Association of the United Kingdom* 87: 565.

Taylor, J. D., Williams, S. T., Glover, E. A. & Dyal, P. (2007b). A molecular phylogeny of heterodont bivalves (Mollusca: Bivalvia: Heterodonta): New analyses of 18S and 28S rRNA genes. *Zoologica Scripta* 36: 587–606.

Taylor, J. D., Glover, E. A. & Williams, S. T. (2008). Ancient chemosynthetic bivalves: Systematics of Solemyidae from eastern and southern Australia (Mollusca: Bivalvia). *Memoirs of the Queensland Museum - Nature* 54: 75–104.

Taylor, J. D., Glover, E. A. & Williams, S. T. (2009). Phylogenetic position of the bivalve family Cyrenoididae – Removal from (and further dismantling of) the superfamily Lucinoidea. *Journal of Molluscan Studies* 123: 9–13.

Taylor, J. D. & Glover, E. A. (2010). Chemosymbiotic bivalves, pp. 107–135 in S. Kiel (ed.), *The Vent and Seep Biota: Aspects from Microbes to Ecosystems. Topics in Geobiology*. Dordrecht, the Netherlands, Springer Netherlands.

Taylor, P. D. & Wilson, M. A. (2003). Palaeoecology and evolution of marine hard substrate communities. *Earth-Science Reviews* 62: 1–103.

Taylor, P. D., Vinn, O. & Wilson, M. A. (2010). Evolution of biomineralisation in 'lophophorates'. *Special Papers in Palaeontology* 84: 317–333.

Teasdale, L. C. (2017). *Phylogenomics of the pulmonate land snails*. Ph.D. dissertation, University of Melbourne.

Teichert, C. (1933). Der Bau der Actinoceroidean Cephalopoden. *Palaeontographica. Abteilung A* 78: 111–230.

Teichert, C. (1964a). Doubtful Taxa, pp. K484–K490 in R. C. Moore (ed.), *Treatise on Invertebrate Paleontology Part K*. Vol. Mollusca 3. Lawrence, KS, Geological Society of America and University of Kansas Press.

Teichert, C. (1964b). Morphology of Hard Parts, pp. K13–K53 in R. C. Moore (ed.), *Treatise on Invertebrate Paleontology Part K*. Vol. Mollusca 3. Lawrence, KS, Geological Society of America and University of Kansas Press.

Teichert, C. (1964c). Actinoceratoidea, pp. K190–K216 in R. C. Moore (ed.), *Treatise on Invertebrate Paleontology Part K*. Vol. Mollusca 3. Lawrence, KS, Geological Society of America and University of Kansas Press.

Teichert, C., Kummel, B., Sweet, W. C., Stenzel, H. B., Furnish, W. M., Glenister, B. F., Erben, H. K., Moore, R. C. & Nodine Zeller, D. E. (1964a). Cephalopoda – General features: Endoceratoidea, Actinoceratoidea, Nautiloidea, Bactritoidea, pp. K4–K505 in R. C. Moore (ed.), *Treatise on Invertebrate Paleontology Part K*. Vol. 3. Lawrence, KS, Geological Society of America and University of Kansas Press.

Teichert, C. & Moore, R. C. (1964). Classification and Stratigraphic Distribution, pp. K94–K106 in R. C. Moore (ed.), *Treatise on Invertebrate Paleontology Part K. Treatise on Invertebrate Paleontology*. Lawrence, KS, Geological Society of America and University of Kansas Press.

Teichert, C., Moore, R. C. & Nodine Zeller, D. E. (1964b). Rhyncholites, pp. K467–K484 in R. C. Moore (ed.), *Cephalopoda – General features, Endoceratoidea, Actinoceratoidea, Nautiloidea, Bactritoidea. Treatise on Invertebrate Paleontology Part K*. Vol. Mollusca 3. Lawrence, KS, Geological Society of America and University of Kansas Press.

Teichert, C. (1967). Major features of cephalopod evolution, pp. 162–210 in C. Teichert & Yochelson, E. L. (eds.), *Essays in Paleontology and Stratigraphy*. Lawrence, KS, University of Kansas Press.

Teichert, C. (1988). Main features of cephalopod evolution, pp. 11–79 in M. R. Clarke & Trueman, E. R. (eds.), *Paleontology and Neontology of Cephalopods. The Mollusca*. Vol. 12. New York, Academic Press.

Telford, M. J. & Budd, G. E. (2011). Invertebrate evolution: Bringing order to the molluscan chaos. *Current Biology* 21: R964–R966.

Tëmkin, I. (2006). Morphological perspective on the classification and evolution of Recent Pterioidea (Mollusca: Bivalvia). *Zoological Journal of the Linnean Society* 148: 253–312.

Templado, J. & Ortea, J. (2001). The occurrence of the shell-less neritacean gastropod *Titiscania limacina* in the Galápagos Islands. *The Veliger* 44: 404–406.

Tendal, O. S. (1985). Xenophyophores (Protozoa, Sarcodina) in the diet of *Neopilina galathea* (Mollusca, Monoplacophora). *Galathea Report* 16: 95–99.

Tendler, A., Mayo, A. & Alon, U. (2015). Evolutionary tradeoffs, Pareto optimality and the morphology of ammonite shells. *BMC Systems Biology* 9: 12.

Termier, H. & Termier, G. (1971). Les Prebelemnitida: Un nouvel ordre des cephalopods. *Annales de la Société geologique du Nord* 90: 109–112.

Terwilliger, R. C. & Read, K. R. H. (1969). The radular muscle myoglobins of the amphineuran mollusc, *Acanthopleura granulata* Gmelin. *Comparative Biochemistry and Physiology* 29: 551–560.

Terwilliger, R. C. & Terwilliger, N. B. (1985). Molluscan hemoglobins. *Comparative Biochemistry and Physiology Part B* 81: 255–261.

Tesch, J. J. (1904). *The Thecosomata and Gymnosomata of the Siboga Expedition*. Vol. 52. Leyden, E. J. Brill.

Tevesz, M. J. S. (1975). Structure and habits of the 'living fossil' pelecypod *Neotrigonia*. *Lethaia* 8: 321–327.

Tevesz, M. J. S. & Carter, J. G. (1979). Form and function in *Trisidos* (Bivalvia) and a comparison with other burrowing arcoids. *Malacologia* 19: 77–85.

Tevesz, M. J. S. & McCall, P. L. (1979). Evolution of substratum preference in bivalves (Mollusca). *Journal of Paleontology* 53: 112–120.

Theler, J. M., Castellucci, V. F. & Baertschi, A. J. (1987). Ultrastructure of the osphradium of *Aplysia californica*. *Cell and Tissue Research* 247: 639–649.

Thiele, J. (1891–1893). *Das Gebiß der Schnecken, zur Begründung einer Natürlichen Classification (Continuation of Work by F. H. Troschel).* Vol. 2 (7–8). Berlin, Nicolai.

Thiele, J. (1893). Über die Kiemensinnesorgane bei Patelliden. *Zoologischer Anzeiger* 16: 49–50.

Thiele, J. (1902). Die systematische Stellung der Solenogastres und die Phylogenie der Mollusken. *Zeitschrift für Wissenschaftliche Zoologie* 72: 249–466.

Thiele, J. (1903). Die Anatomie und systematische Stellung der Gattung *Cocculina. Wissenschaftliche Ergebnisse der Deutschen Tiefsee-Expedition auf dem Dampfer "Valdivia".* 7: 147–179.

Thiele, J. (1910). Über die Anatomie von *Hydrocena cattaronensis. Abhandlungen der Senckenbergischen Naturforschenden Gesellschaft* 32: 349–358.

Thiele, J. (1913). *Solenogastres.* Berlin, R. Friedländer und Sohn.

Thiele, J. (1925). Gastropoda, pp. 38–96 *in* T. Krumbach (ed.), *Handbuch der Zoologie.* Vol. 5. Leipzig, Walter de Gruyter & Co.

Thiele, J. (1929–1935). *Handbuch der Systematischen Weichtierkunde. Volume 1, Teil 1, Loricata and Gastropoda I: Prosobranchia (1929), Teil 2, Opisthobranchia & Pulmonata (1931); Volume 2, Teil 3, Scaphopoda, Bivalvia, Cephalopoda, Tiel 4, General and Corrections.* Jena, Gustav Fischer Verlag.

Thiem, H. (1917a). Beiträge zur Anatomie und Phylogenie der Docoglossen. I. Zur Anatomie von *Helcioniscus ardosiaeus* Hombron & Jaquinot unter Bezugnahme auf die Bearbeitung von Erich Schusterinden. *Jenaische Zeitschrift für Naturwissenschaft* 54: 333–404.

Thiem, H. (1917b). Beiträge zur Anatomie und Phylogenie der Docoglossen. II. Die Anatomie und Phylogenie der Monobranchen (Akmäiden und Scurriden nach der Sammlung Plates). *Jenaische Zeitschrift für Naturwissenschaft* 54: 405–630.

Thiriot-Quiévreux, C. (1983). Summer meroplanktonic prosobranch larvae occurring off Beaufort, North Carolina. *Estuaries* 6: 387–398.

Thomas, K. N., Robison, B. H. & Johnsen, S. (2017). Two eyes for two purposes: *In situ* evidence for asymmetric vision in the cockeyed squids *Histioteuthis heteropsis* and *Stigmatoteuthis dofleini. Philosophical Transactions of the Royal Society B* 372: 20160061–20160069.

Thomas, R. D. K. (1978). Shell form and the ecological range of living and extinct Arcoida. *Paleobiology* 4: 181–194.

Thomas, R. D. K., Vinther, J. & Matt, K. (2010). Paired chaetae associated with spiral shells of the late Early Cambrian mollusc *Pelagiella* from the Kinzers Formation: Taphonomy, functional morphology, and potential evolutionary relationships. *Geological Society of America Abstracts with Programs* 42: 633.

Thomas, R. D. K. & Vinther, J. (2012). Implications of the occurrence of paired anterior chaetae in the late Early Cambrian mollusc *Pelagiella* from the Kinzers Formation of Pennsylvania for relationships among taxa and early evolution of the Mollusca. *Geological Society of America Abstracts with Programs* 44: 326.

Thomas, R. F. (1973). Homing behaviour and movement rhythms in the pulmonate limpet, *Siphonaria pectinata* Linnaeus. *Proceedings of the Malacological Society of London* 40: 303–311.

Thompson, F. G. (1980). Proserpinoid land snails and their relationships within the Archaeogastropoda. *Malacologia* 20: 1–33.

Thompson, F. G. & Dance, S. P. (1983). Non-marine Mollusks of Borneo: II Pulmonata: Pupillidae, Clausiliidae: III Prosobranchia: Hydrocenidae, Helicinidae. *Bulletin of the Florida State Museum Biological Sciences* 29: 101–150.

Thompson, J. T. (2001). The evolution of mechanical function in the mantle of squids. *American Zoologist* 41: 1606–1607.

Thompson, T. E. (1958). The natural history, embryology, larval biology and post-larval development of *Adalaria proxima* (Alder and Hancock) (Gastropoda Opisthobranchia). *Philosophical Transactions of the Royal Society B* 242: 1–58.

Thompson, T. E. & Slinn, S. J. (1959). On the biology of the opisthobranch *Pleurobranchus membranaceus. Journal of the Marine Biological Association of the United Kingdom* 38: 507–524.

Thompson, T. E. (1960a). The development of *Neomenia carinata* Tullberg (Mollusca, Aplacophora). *Proceedings of the Royal Society B* 153: 263–278.

Thompson, T. E. (1960b). Defensive acid-secretion in marine gastropods. *Journal of the Marine Biological Association of the United Kingdom* 39: 115–122.

Thompson, T. E. (1961a). The importance of the larval shell in the classification of the *Sacoglossa* and the *Acoela* (Gastropoda, Opisthobranchia). *Proceedings of the Malacological Society of London* 34: 233–238.

Thompson, T. E. (1961b). The structure and mode of functioning of the reproductive organs of *Tritonia hombergi* (Gastropoda: Opisthobranchia). *Quarterly Journal of Microscopical Science* 102: 1–14.

Thompson, T. E. (1962). Studies on the ontogeny of *Tritonia hombergi* Cuvier (Gastropoda Opisthobranchia). *Philosophical Transactions of the Royal Society B* 245: 171–218.

Thompson, T. E. (1967). Adaptive significance of gastropod torsion. *Malacologia* 5: 423–430.

Thompson, T. E. & Bebbington, A. (1969). Structure and function of the reproductive organs of three species of *Aplysia* (Gastropoda: Opisthobranchia). *Malacologia* 7: 347–380.

Thompson, T. E. & Bennett, I. (1969). *Physalia* nematocysts: Utilized by mollusks for defense. *Science* 166: 1532–1533.

Thompson, T. E. & Bennett, I. (1970). Observations on Australian Glaucidae (Mollusca: Opisthobranchia). *Journal of the Linnean Society* 49: 187–197.

Thompson, T. E. (1972). Eastern Australian Dendronotoidea (Gastropoda: Opisthobranchia). *Zoological Journal of the Linnean Society* 51: 63–77.

Thompson, T. E. (1973). Euthyneuran and other molluscan spermatozoa: General characteristics of molluscan spermatozoa. *Malacologia* 14: 167–206.

Thompson, T. E. (1976). *Biology of Opisthobranch Molluscs.* Vol. 1. London, Ray Society.

Thompson, T. E. & Brown, G. H. (1976). *British Opisthobranch Molluscs: Keys and Notes for the Identification of the Species.* London, Academic Press.

Thompson, T. E. (1988). Acidic allomones in marine organisms. *Journal of the Marine Biological Association of the United Kingdom* 68: 499–518.

Thornhill, A. H., Mishler, B. D., Knerr, N. J., González-Orozco, C. E., Costion, C. M., Crayn, D. M., Laffan, S. W. & Miller, J. T. (2016). Continental-scale spatial phylogenetics of Australian angiosperms provides insights into ecology, evolution and conservation. *Journal of Biogeography* 43: 2085–2098.

Thorson, G. (1935). Studies on the egg-capsules and development of Arctic marine bottom invertebrates. *Meddelelser om Grønland* 100: 1–71.

Thorson, G. (1940). Studies on the egg masses and larval development of Gastropoda from the Iranian Gulf [Part 2]. *Danish Scientific Investigations in Iran* 2: 159–238.

Thorson, G. (1946). Reproduction and larval development of Danish marine bottom invertebrates, with special reference to the planktonic larvae in the Sound (Øresund). *Meddelelser fra Kommissionen for Danmarks Fiskeri-og Havunder Søgelser. Serie Plankton* 4: 1–523.

Tielecke, H. (1940). Anatomie, phylogenie und tiergeographie der Cyclophoriden. *Archiv für Naturgeschichte* 9: 317–371.

Tikasingh, E. S. & Pratt, I. (1961). The classification of endoparasitic gastropods. *Systematic Zoology* 10: 65–69.

Tikasingh, E. S. (1962). The microanatomy and histology of the parasitic gastropod, *Comenteroxenos parastichopoli* (Tikasingh). *Transactions of the American Microscopical Society* 81: 320–327.

Tillier, S. (1983). A new mountain *Platevindex* from Philippine Islands (Pulmonata: Onchidiidae). *Journal of Molluscan Studies*: Supplement 12A: 198–202.

Tillier, S. (1984a). Relationships of gymnomorph gastropods (Mollusca: Gastropoda). *Zoological Journal of the Linnean Society* 82: 345–362.

Tillier, S. (1984b). Patterns of digestive tract morphology in the limacisation of helicarionid, succineid and athoracophorid snails and slugs (Mollusca: Pulmonata). *Malacologia* 25: 173–192.

Tillier, S. (1989). Comparative morphology, phylogeny and classification of land snails and slugs (Gastropoda: Pulmonata: Stylommatophora). *Malacologia* 30: 1–304.

Tillier, S. & Ponder, W. F. (1992). New species of *Smeagol* from Australia and New Zealand, with a discussion of the affinities of the genus (Gastropoda: Pulmonata). *Journal of Molluscan Studies* 58: 135–155.

Tillier, S., Masselot, M., Guerdoux, J. & Tillier, A. (1994). Monophyly of major gastropod taxa tested from partial 28S rRNA sequences, with emphasis on Euthyneura and hot-vent limpets Peltospiroidea. *The Nautilus* 108: 122–140.

Tillier, S., Masselot, M. & Tillier, A. (1996). Phylogenetic relationships of the pulmonate gastropods from rRNA sequences, and tempo and age of the stylommatophoran radiation, pp. 267–284 *in* J. D. Taylor (ed.), *Origin and Evolutionary Radiation of the Mollusca*. Oxford, UK, Oxford University Press.

Tirado, C., Marina, P., Urra, J., Antit, M. & Salas, C. (2017). Reproduction and population structure of *Acanthocardia tuberculata* (Linnaeus, 1758) (Bivalvia: Cardiidae) in southern Spain: Implications for stock management. *Journal of Shellfish Research* 36: 61–68.

Tochimoto, T. (1967). Comparative histochemical study on the dimorphic spermatozoa of the Prosobranchia with special reference to polysaccharides. *Science Reports. Tokyo Kyoiku Daigaku*: Section B 13: 75–109.

Todd, C. D. (1981). The ecology of nudibranch molluscs. *Oceanography and Marine Biology Annual Review* 19: 142–234.

Todd, C. D., Hadfield, M. G. & Snedden, W. A. (1997). Juvenile mating and sperm storage in the tropical corallivorous nudibranch *Phestilla sibogae*. *Invertebrate Biology* 116: 322–330.

Todt, C. & von Salvini-Plawen, L. (2005). The digestive tract of *Helicoradomenia* (Solenogastres, Mollusca), aplacophoran molluscs from the hydrothermal vents of the East Pacific Rise. *Invertebrate Biology* 124: 230–253.

Todt, C., Büchinger, T. & Wanninger, A. (2008a). The nervous system of the basal mollusk *Wirenia argentea* (Solenogastres): A study employing immunocytochemical and 3D reconstruction techniques. *Marine Biology Research* 4: 290–303.

Todt, C., Okusu, A., Schander, C. & Schwabe, E. (2008b). Solenogastres, Caudofoveata and Polyplacophora, pp. 71–96 *in* W. F. Ponder & Lindberg, D. R. (eds.), *Phylogeny and Evolution of the Mollusca*. Berkeley, CA, University of California Press.

Todt, C. & Wanninger, A. (2010). Of tests, trochs, shells, and spicules: Development of the basal mollusk *Wirenia argentea* (Solenogastres) and its bearing on the evolution of trochozoan larval key features. *Frontiers in Zoology* 7: 1–17.

Todt, C. (2013). Aplacophoran mollusks – Still obscure and difficult? *American Malacological Bulletin* 31: 181–187.

Tomiyama, K. (1996). Mate-choice criteria in a protandrous simultaneously hermaphroditic land snail *Achatina fulica* (Férussac) (Stylommatophora: Achatinidae). *Journal of Molluscan Studies* 62: 101–111.

Tompa, A. S. (1976). A comparative study of the ultrastructure and mineralogy of calcified land snail eggs (Pulmonata: Stylommatophora). *Journal of Morphology* 150: 861–887.

Tompa, A. S. (1984). Land snails (Stylommatophora), pp. 47–140 *in* A. S. Tompa, Verdonk, N. H. & van den Biggelaar, J. A. M. (eds.), *Reproduction. The Mollusca*. Vol. 7. New York, Academic Press.

Tompsett, D. H. (1939). *Sepia*, pp. 1–191 *in* R. J. Daniel (ed.), *L. M. B. C. Memoirs on Typical British Marine Plants and Animals*. Liverpool/London, University Press of Liverpool.

Tong, L. K. Y. (1986). The population dynamics and growth of *Thais clavigera* and *Morula musiva* (Gastropoda: Muricidae) in Hong Kong. *Asian Marine Biology* 3: 145–162.

Torres, F. I., Ibáñez, C. M., Sanhueza, V. E. & Pardo-Gandarillas, M. C. (2018). Mollusk freaks: New teratological cases on marine mollusks from the South Pacific Ocean. *Latin American Journal of Aquatic Research* 46: 683–689.

Towe, K. M. (1978). *Tentaculites*: Evidence for a brachiopod affinity? *Science* 201: 626–628.

Tracey, S., Todd, J. A. & Erwin, D. H. (1993). Mollusca: Gastropoda, pp. 131–167 *in* M. J. Benton (ed.), *Fossil Record 2*. Vol. 2. London/New York, Chapman and Hall.

Treves, K., Traub, W., Weiner, S. & Addadi, L. (2003). Aragonite formation in the chiton (Mollusca) girdle. *Helvetica Chimica Acta* 86: 1101–1112.

Troncoso, J. S., Moreira, J. & Díaz-Agras, G. (2011). Proceedings of the 3rd International Workshop on Opisthobranchs, Vigo, Spain, 2010. *Thalassas* 27: 9–238.

Troschel, F. H. (1856–1879). *Das Gebiss der Schnecken, zur Begründung einer natürlichen Classification*. Vol. 1 (1–5), 2 (1–6). Berlin, Nicolai.

Trowbridge, C. D. (1995). Hypodermic insemination, oviposition, and embryonic development of a pool-dwelling ascoglossan (= sacoglossan) opisthobranch: *Ercolania felina* (Hutton, 1882) on New Zealand shores. *The Veliger* 38: 203–211.

Trueman, A. E. (1941). The ammonite body chamber, with special reference to the buoyancy and mode of life of the living ammonite. *Quarterly Journal of the Geological Society, London* 96: 339–383.

Trueman, E. R. (1968). The burrowing process of *Dentalium* (Scaphopoda). *Journal of Zoology* 154: 19–27.

Tsitrone, A., Jarne, P. & David, P. (2003). Delayed selfing and resource reallocations in relation to mate availability in the freshwater snail *Physa acuta*. *American Naturalist* 162: 474–488.

Turek, V. (1978). Biological and stratigraphical significance of the Silurian nautiloid *Aptychopsis*. *Lethaia* 11: 127–138.

Turner, R. D. (1954). The family Pholadidae in the Western Atlantic and Eastern Pacific. 1. Pholadidae. *Johnsonia* 3: 33–162.

Turner, R. D. (1966). *A Survey and Illustrated Catalogue of the Teredinidae (Mollusca: Bivalvia)*. Cambridge, UK, Museum of Comparative Zoology.

Turner, R. D. & Johnson, A. C. (1971). Biology of marine woodboring molluscs, pp. 259–301 *in* E. B. G. Jones & Eltringham, S. K. (eds.), *Marine Borers, Fungi and Fouling Organisms of Wood*. Paris, Organisation for Economic Co-operation and Development.

Turner, R. D. (1977). Genetic relations of deep-sea wood-borers. *Bulletin of the American Malacological Union Incorporated* 1977: 19–24.

Turner, R. D. & Yakovlev, M. (1983). Dwarf males in the Teredinidae (Bivalvia, Pholadacea). *Science* 219: 1077–1078.

Ubukata, T. (2000). Theoretical morphology of hinge and shell form in Bivalvia: Geometric constraints derived from space conflict between umbones. *Paleobiology* 26: 606–624.

Ubukata, T. (2005). Theoretical morphology of bivalve shell sculptures. *Paleobiology* 31: 643–655.

Ubukata, T., Tanabe, K., Shigeta, Y., Maeda, H. & Mapes, R. H. (2014). Wavelet analysis of ammonoid sutures. *Palaeontologia Electronica* 17: 1–17.

Unabia, C. R. C. (1996). *Radular structure in the gastropod order Neritopsina; a source of phylogenetic information.* Ph.D. dissertation, University of Hawaii.

Unabia, C. R. C. (2011). The snail *Smaragdia bryanae* (Neritopsina, Neritidae) is a specialist herbivore of the seagrass *Halophila hawaiiana* (Alismatidae, Hydrocharitaceae). *Invertebrate Biology* 130: 100–114.

Underwood, A. J. (1972). Spawning, larval development and settlement behaviour of *Gibbula cineraria* (Gastropoda: Prosobranchia) with a reappraisal of torsion in gastropods. *Marine Biology* 17: 341–349.

Underwood, A. J. (1977). Movements of intertidal gastropods. *Journal of Experimental Marine Biology and Ecology* 26: 191–201.

Underwood, A. J. (1979). The ecology of intertidal gastropods. *Advances in Marine Biology* 16: 111–210.

Urazaeva, M. N., Silantiev, V. V. & Usmanova, R. R. (2015). Revision of Late Permian nonmarine bivalves of the genus *Verneuilunio* Starobogatov, 1987 and its type species *Naiadites verneuili* Amalitzky, 1892. *Paleontological Journal* 49: 1174–1183.

Urgorri, V., García-Álvarez, O. & Luque, Á. (2005). *Laevipilina cachuchensis*, a new neopilinid (Mollusca: Tryblidia) from off north Spain. *Journal of Molluscan Studies* 71: 59–66.

Uribe, J. E., Colgan, D., Castro, L. R., Kano, Y. & Zardoya, R. (2016a). Phylogenetic relationships among superfamilies of Neritimorpha (Mollusca: Gastropoda). *Molecular Phylogenetics and Evolution* 104: 21–31.

Uribe, J. E., Kano, Y., Templado, J. & Zardoya, R. (2016b). Mitogenomics of Vetigastropoda: Insights into the evolution of pallial symmetry. *Zoologica Scripta* 45: 145–159.

Uribe, J. E., Williams, S. T., Templado, J., Abalde, S. & Zardoya, R. (2017). Denser mitogenomic sampling improves resolution of the phylogeny of the superfamily Trochoidea (Gastropoda: Vetigastropoda). *Journal of Molluscan Studies* 83: 111–118.

Uribc, J. E., Irisarri, I., Templado, J. & Zardoya, R. (2019). New patellogastropod mitogenomes help counteracting long-branch attraction in the deep phylogeny of gastropod mollusks. *Molecular Phylogenetics and Evolution* 133: 12–23.

Vaccaro, E. & Waite, J. H. (2001). Yield and post-yield behavior of mussel byssal thread: A self-healing biomolecular material. *Biomacromolecules* 2: 906–911.

Vahldiek, B. W. & Schweigert, G. (2007). Oldest record of wood-boring bivalves. *Neues Jahrbuch für Geologie und Paläontologie-Abhandlungen* 244: 261–271.

Vail, V. A. (1977). Comparative reproductive anatomy of three viviparid gastropods. *Malacologia* 16: 519–540.

Valdés, Á. & Gosliner, T. M. (1999). Phylogeny of the radula-less dorids (Mollusca, Nudibranchia), with the description of a new genus and a new family. *Zoologica Scripta* 28: 315–360.

Valdés, Á. & Gosliner, T. M. (2001). Systematics and phylogeny of the caryophyllidia-bearing dorids (Mollusca, Nudibranchia), with descriptions of a new genus and four new species from Indo-Pacific deep waters. *Zoological Journal of the Linnean Society* 133: 103–198.

Valdés, Á. (2002). Phylogenetic systematics of "*Bathydoris*" s.l. Bergh, 1884 (Mollusca, Nudibranchia), with the description of a new species from New Caledonian deep waters. *Canadian Journal of Zoology* 80: 1084–1099.

Valdés, Á. (2004). Phylogeography and phyloecology of dorid nudibranchs (Mollusca, Gastropoda). *Biological Journal of the Linnean Society* 83: 551–559.

Valdés, Á., Blanchard, L. & Marti, W. (2013). Caught naked: First report [of] a nudibranch sea slug attacked by a cone snail. *American Malacological Bulletin* 31: 337–338.

Valen, L. M. v. (1984). A resetting of Phanerozoic community evolution. *Nature* 307: 50–52.

Valentine, J. W. (1973). *Evolutionary Paleoecology of the Marine Biosphere.* Englewood Cliffs, NJ, Prentice-Hall.

Valiente-Banuet, A., Aizen, M. A., Alcántara, J. M., Arroyo, J., Cocucci, A., Galetti, M., García, M. B., García, D., Gómez, J. M. & Jordano, P. (2015). Beyond species loss: The extinction of ecological interactions in a changing world. *Functional Ecology* 29: 299–307.

Vallès, Y. & Gosliner, T. M. (2006). Shedding light onto the genera (Mollusca: Nudibranchia) *Kaloplocamus* and *Plocamopherus* with description of new species belonging to these unique bioluminescent dorids. *The Veliger* 48: 178–205.

Van Bocxlaer, B. & Strong, E. E. (2016). Anatomy, functional morphology, evolutionary ecology and systematics of the invasive gastropod *Cipangopaludina japonica* (Viviparidae: Bellamyinae). *Contributions to Zoology* 85: 235–236.

Vaney, C. (1913). L'adaptation des gastropodes [sic] au parasitisme. *Bulletin scientifique de la France et de la Belgique* 47: 1–87.

Vannier, J., Wang, S. Q. & Coen, M. (2001). Leperditicopid arthropods (Ordovician-Late Devonian): Functional morphology and ecological range. *Journal of Paleontology* 75: 75–95.

Vaughn, C. C. (2010). Biodiversity losses and ecosystem function in freshwaters: Emerging conclusions and research directions. *BioScience* 60: 25–35.

Vaziri, S. H., Majidifard, M. R. & Laflamme, M. (2018). Diverse assemblage of Ediacaran fossils from Central Iran. *Scientific Reports* 8: 5060.

Vecchione, M., Young, R. E., Donovan, D. T. & Rodhouse, P. G. K. (1999). Reevaluation of coleoid cephalopod relationships based on modified arms in the Jurassic coleoid *Mastigophora*. *Lethaia* 32: 113–118.

Vendrasco, M. J. & Runnegar, B. N. (2004). Late Cambrian and early stem group chitons (Mollusca: Polyplacophora) from Utah and Missouri. *Journal of Paleontology* 78: 675–689.

Vendrasco, M. J., Wood, T. E. & Runnegar, B. N. (2004). Articulated Palaeozoic fossil with 17 plates greatly expands disparity of early chitons. *Nature* 429: 288–291.

Vendrasco, M. J., Li, G., Porter, S. M. & Fernandez, C. Z. (2009). New data on the enigmatic *Ocruranus-Eohalobia* group of Early Cambrian small skeletal fossils. *Palaeontology* 52: 1373–1396.

Vendrasco, M. J., Porter, S. M., Kouchinsky, A., Li, G. & Fernandez, C. Z. (2010). New data on molluscs and their shell microstructures from the Middle Cambrian Gowers Formation, Australia. *Palaeontology* 53: 97–135.

Vendrasco, M. J., Checa, A. G. & Kouchinsky, A. V. (2011a). Shell microstructure of the early bivalve *Pojetaia* and the independent origin of nacre within the mollusca. *Palaeontology* 54: 825–850.

Vendrasco, M. J., Kouchinsky, A. V., Porter, S. M. & Fernandez, C. Z. (2011b). Phylogeny and escalation in *Mellopegma* and other Cambrian molluscs. *Palaeontologia Electronica* 14: 1–44.

Vendrasco, M. J. (2012). Early evolution of molluscs, pp. 1–43 *in* A. Fyodorov & Yakovlev, H. (eds.), *Mollusks: Morphology, Behaviour and Ecology.* New York, Nova Science Publishers, Inc.

Vendrasco, M. J. & Checa, A. G. (2015). Shell microstructure and its inheritance in the calcitic helcionellid *Mackinnonia*. *Estonian Journal of Earth Sciences* 64: 99–99.

Vermeij, G. J. (1975). Evolution and distribution of left-handed and planispiral coiling in snails. *Nature* 254: 419–420.

Vermeij, G. J. (1977). The Mesozoic marine revolution: Evidence from snails, predators and grazers. *Paleobiology* 3: 245–258.

Vermeij, G. J. (1978). *Biogeography and Adaptation*. Cambridge, MA, Belknap Press.

Vermeij, G. J. & Petuch, E. J. (1986). Differential extinction in tropical American mollusks - endemism, architecture, and the Panama land-bridge. *Malacologia* 27: 29–41.

Vermeij, G. J. (1987). *Evolution and Escalation*. Princeton, NJ, Princeton University Press.

Vermeij, G. J. (1989). The origin of skeletons. *Palaios* 4: 585–589.

Vermeij, G. J. (1992). Time of origin and biogeographical history of specialized relationships between northern marine plants and herbivorous molluscs. *Evolution* 46: 657–664.

Vermeij, G. J. & Signor, P. W. (1992). The geographic, taxonomic and temporal distribution of determinate growth in marine gastropods. *Biological Journal of the Linnean Society* 47: 233–247.

Vermeij, G. J. & Kool, S. P. (1994). Evolution of labral spines in *Acanthais*, new genus, and other rapanine muricid gastropods. *The Veliger* 37: 414–424.

Vermeij, G. J. (1995). Economics, volcanoes, and phanerozoic revolutions. *Paleobiology* 21: 125–152.

Vermeij, G. J. & Dudley, R. (2000). Why are there so few evolutionary transitions between aquatic and terrestrial ecosystems? *Biological Journal of the Linnean Society* 70: 541–554.

Vermeij, G. J. & Lindberg, D. R. (2000). Delayed herbivory and the assembly of marine benthic ecosystems. *Paleobiology* 26: 419–430.

Vermeij, G. J. (2001). Innovation and evolution at the edge: Origins and fates of gastropods with a labral tooth. *Biological Journal of the Linnean Society* 72: 461–508.

Vermeij, G. J. (2007). The ecology of invasion: Acquisition and loss of the siphonal canal in gastropods. *Paleobiology* 33: 469–493.

Vermeij, G. J. & Frey, M. A. (2008). Almost terrestrial: Small supratidal species of *Nerita* (Gastropoda, Neritidae) in the western Pacific. *Basteria* 72: 253–261.

Vermeij, G. J. & Raven, H. (2009). Southeast Asia as the birthplace of unusual traits: The Melongenidae (Gastropoda) of northwest Borneo. *Contributions to Zoology* 78: 113–127.

Vermeij, G. J. (2013). Molluscan marginalia: Hidden morphological diversity at the bivalve shell edge. *Journal of Molluscan Studies* 79: 283–295.

Vermeij, G. J. (2016). The limpet form in gastropods: Evolution, distribution, and implications for the comparative study of history. *Biological Journal of the Linnean Society* 120: 22–37.

Vestergaard, K. & Thorson, G. (1938). Über den Laich und die Larven von *Duvaucelia plebeia*, *Polycera quadrilineata*, *Eubranchus pallidus* und *Limapontia capitata* (Gastropoda, Opisthobranchia). *Zoologischer Anzeiger* 124: 129–138.

Vicente, N. & Gasquet, M. (1970). Étude du système nerveux et de la neurosécrétion chez quelques mollusques polyplacophores. *Téthys* 2: 515–546.

Villanueva, R. (1992). Continuous spawning in the cirrate octopods *Opisthoteuthis agassizii* and *O. vossi*: Features of sexual maturation defining a reproductive strategy in cephalopods. *Marine Biology* 275: 265–275.

Villanueva, R., Segonzac, M. & Guerra, A. (1997). Locomotion modes of deep-sea cirrate octopods (Cephalopoda) based on observations from video recordings on the Mid-Atlantic Ridge. *Marine Biology* 129: 113–122.

Villanueva, R., Perricone, V. & Fiorito, G. (2017). Cephalopods as predators: A short journey among behavioral flexibilities, adaptions, and feeding habits. *Frontiers in Physiology* 8: 598.

Villarroel, M. & Stuardo, J. (1998). Protobranchia (Mollusca: Bivalvia) of recent distribution in Chile and some fossils. *Malacologia* 40: 113–229.

de Villiers, C. J. & Hodgson, A. N. (1987). The structure of the secondary gills of *Siphonaria capensis* (Gastropoda: Pulmonata). *Journal of Molluscan Studies* 53: 129–138.

Vinn, O. & Mutvei, H. (2005). Observations on the morphology and affinities of cornulitids from the Ordovician of Anticosti Island and the Silurian of Gotland. *Journal of Paleontology* 79: 726–737.

Vinn, O., Ten Hove, H. A., & Mutvei, H. (2008). On the tube ultrastructure and origin of calcification in sabellids (Annelida, Polychaeta). *Palaeontology* 51: 295–301.

Vinn, O. & Mutvei, H. (2009). Calcareous tubeworms of the Phanerozoic. *Estonian Journal of Earth Sciences* 58: 286–296.

Vinn, O. & Zatoń, M. (2012). Phenetic phylogenetics of tentaculitoids - extinct, problematic calcareous tube-forming organisms. *GFF: Journal of the Geological Society of Sweden* 134: 145–156.

Vinther, J. & Nielsen, C. (2005). The early Cambrian *Halkieria* is a mollusc. *Zoologica Scripta* 34: 81–89.

Vinther, J. (2009). The canal system in sclerites of Lower Cambrian *Sinosachites* (Halkieriidae: Sachitida): Significance for the molluscan affinities of the sachitids. *Palaeontology* 52: 689–712.

Vinther, J., Sperling, E. A., Briggs, D. E. G. & Peterson, K. J. (2011). A molecular palaeobiological hypothesis for the origin of aplacophoran molluscs and their derivation from chiton-like ancestors. *Proceedings of the Royal Society B* 279: 1259–1268.

Vinther, J. (2014). A molecular palaeobiological perspective on aculiferan evolution. *Journal of Natural History* 48: 2805–2823.

Vinther, J. (2015). The origins of molluscs. *Palaeontology* 58: 19–34.

Vinther, J., Parry, L., Briggs, D. E. & Van Roy, P. (2017). Ancestral morphology of crown-group molluscs revealed by a new Ordovician stem aculiferan. *Nature* 542: 471.

Visser, M. H. C. (1977). The morphology and significance of the sperm oviduct and prostate in the evolution of the reproductive system of the Pulmonata. *Zoologica Scripta* 6: 43–54.

Visser, M. H. C. (1981). Monauly versus diauly as the original condition of the reproductive system of Pulmonata and its bearing on the interpretation of the terminal ducts. *Zeitschrift für Zoologische Systematik und Evolutionsforschung* 19: 59–68.

Visser, M. H. C. (1988). The significance of terminal duct structures and the role of neoteny in the evolution of the reproductive system of Pulmonata. *Zoologica Scripta* 17: 239–252.

Vöcking, O., Kourtesis, I. & Hausen, H. (2015). Posterior eyespots in larval chitons have a molecular identity similar to anterior cerebral eyes in other bilaterians. *EvoDevo* 6: 40.

Vogel, K. (1975). Endosymbiotic algae in rudists? *Palaeogeography, Palaeoclimatology, Palaeoecology* 17: 327–332.

Vogt, C. (1866). *Lehrbuch der Geologie und Petrefactenkunde: Zum Gebrauche bei Vorlesungen und zum Selbstunterrichte; in zwei Bänden*. Vol. 3. Braunschweig, Friedrich Vieweg und Sohn.

Voight, J. R., Pörtner, H. O. & O'Dor, R. K. (1994). A review of ammonia-mediated buoyancy in squids (Cephalopoda: Teuthoidea), pp. 193–203 *in* H. O. Pörtner, O'Dor, R. K. & Macmillan, D. L. (eds.), *Physiology of Cephalopod Molluscs: Lifestyle and Performance Adaptations. Marine and Freshwater Behaviour and Physiology*. Basel, Gordon and Breach Science Publishers SA.

Voight, J. R. (1997). Cladistic analysis of the octopods based on anatomical characters *Journal of Molluscan Studies* 63: 311–325.

Vokes, H. E. (1946). Contributions to the paleontology of the Lebanon Mountains, Republic of Lebanon. Part 3. The pelecypod fauna of the 'Olive locality' (Aptian) at Abeih. *Bulletin of the American Museum of Natural History* 87: 141–215.

Voltzow, J. (1988). The organization of limpet pedal musculature and its evolutionary implications for the Gastropoda, pp. 273–283 *in* W. F. Ponder, Eernisse, D. J. & Waterhouse, J. H. (eds.), *Prosobranch Phylogeny. Malacological Review Supplement.* Ann Arbor, MI, Malacological Review.

Voltzow, J. (1994). Gastropoda: Prosobranchia, pp. 111–252 *in* F. W. Harrison & Kohn, A. J. (eds.), *Microscopic Anatomy of Invertebrates. Mollusca 1.* Vol. 5. New York, Wiley-Liss.

Voronezhskaya, E. E., Tyurin, S. A. & Nezlin, L. P. (2002). Neuronal development in larval chiton *Ischnochiton hakodadensis* (Mollusca: Polyplacophora). *The Journal of Comparative Neurology* 444: 25–38.

Vortsepneva, E. V. & Tzetlin, A. B. (2014). New data on the fine structure of hooks in *Clione limacina* (Gastropoda, Opisthobranchia) and diversity of the jaw apparatus in gastropods. *Zoologicheskii Zhurnal* 93: 466–478.

Voss, G. L. (1988). Evolution and phylogenetic relationships of deep-sea octopods (Cirrata and Incirrata), pp. 253–276 *in* M. R. Clarke & Trueman, E. R. (eds.), *Paleontology and Neontology of Cephalopods. The Mollusca.* Vol. 12. New York, Academic Press.

Voss, G. L. & Pearcy, W. G. (1990). Deep-water octopods (Mollusca: Cephalopoda) of the northeastern Pacific. *Proceedings of the California Academy of Sciences* 47: 47–94.

Wächtler, K., Dreher-Mansur, M. C. & Richter, T. (2001). Larval types and early postlarval biology in naiads (Unionoida), pp. 93–125 *in* G. Bauer & Wächtler, K. (eds.), *Ecology and Evolution of the Freshwater Mussels Unionoida.* Berlin, Springer-Verlag.

Wade, M. (1972). Hydrozoa and Scyphozoa and other medusoids from the Precambrian Ediacara fauna, South Australia. *Palaeontology* 15: 197–225.

Wade, M. (1988). Nautiloids and their descendants: Cephalopod classification in 1986. *New Mexico Bureau of Mines and Mineral Resources Memoir* 44.

Wade, M. & Stait, B. (1998). Subclass Nautiloidea – Introduction and fossil record, pp. 485–493 *in* P. L. Beesley, Ross, G. J. B. & Wells, A. (eds.), *Mollusca: The Southern Synthesis. Part A. Fauna of Australia.* Melbourne, VIC, CSIRO Publishing.

Wägele, H. (1989a). Die gattung *Bathydoris* (Bergh, 1884) (Gnathodoridacea) im phylogenetischen System der Nudibranchia (Opisthobranchia, Gastropoda). *Zeitschrift für Zoologische Systematik und Evolutionsforschung* 27: 273–281.

Wägele, H. (1989b). A revision of the Antarctic species of *Bathydoris* (Bergh,1884) and comparison with other known Bathydorids (Opisthobranchia, Nudibranchia). *Journal of Molluscan Studies* 55: 343–364.

Wägele, H. (1989c). On the anatomy and zoogeography of *Tritoniella belli* (Eliot, 1907) (Opisthobranchia, Nudibranchia) and the synonymy of *T. sinuata* (Eliot, 1907). *Polar Biology* 9: 235–243.

Wägele, H. (1990a). Revision of the Antarctic genus *Notaeolidia* (Gastropoda, Nudibranchia), with a description of a new species. *Zoologica Scripta* 19: 309–330.

Wägele, H. (1990b). Revision of the genus *Austrodoris* Odhner, 1926 (Gastropoda, Opisthobranchia). *Journal of Molluscan Studies* 56: 163–180.

Wägele, H. (1998). Histological investigation of some organs and specialised cellular structures in Opisthobranchia (Gastropoda) with the potential to yield phylogenetically significant characters. *Zoologischer Anzeiger* 236: 119–131.

Wägele, H., Brodie, G. D. & Klussmann-Kolb, A. (1999). Histological investigations on *Dendrodoris nigra* (Stimpson, 1855) (Gastropoda, Nudibranchia, Dendrodorididae). *Molluscan Research* 20: 79–94.

Wägele, H. & Willan, R. C. (2000). Phylogeny of the Nudibranchia. *Zoological Journal of the Linnean Society* 130: 83–181.

Wägele, H. (2004). Potential key characters in Opisthobranchia (Gastropoda, Mollusca) enhancing adaptive radiation. *Organisms Diversity & Evolution* 4: 175–188.

Wägele, H. & Klussmann-Kolb, A. (2005). Opisthobranchia (Mollusca, Gastropoda) – More than just slimy slugs: Shell reduction and its implications on defence and foraging. *Frontiers in Zoology* 2: 1–18.

Wägele, H., Ballesteros, M. & Avila, C. (2006). Defensive glandular structures in opisthobranch molluscs – From histology to ecology. *Oceanography and Marine Biology* 44: 197–276.

Wägele, H., Klussmann-Kolb, A., Vonnemann, V. & Medina, M. (2008). Heterobranchia I: The Opisthobranchia, pp. 385–408 *in* W. F. Ponder & Lindberg, D. R. (eds.), *Phylogeny and Evolution of the Mollusca.* Berkeley, CA, University of California Press.

Wägele, H., Raupach, M. J., Burghardt, I., Grzymbowski, Y. & Händeler, K. (2010). Solar powered seaslugs (Opisthobranchia, Gastropoda, Mollusca): Incorporation of photosynthetic units: A key character enhancing radiation? pp. 263–282 *in* M. Glaubrecht (ed.), *Evolution in Action: Case Studies in Adaptive Radiation, Speciation and the Origin of Biodiversity.* Special volume from contributions to SPP 1127 of the Deutsche Forschungsgemeinschaft. Heidelberg, Springer Verlag.

Wägele, H., Klussmann-Kolb, A., Verbeek, E. & Schrödl, M. (2014). Flashback and foreshadowing – A review of the taxon Opisthobranchia. *Organisms Diversity & Evolution* 14: 133–149.

Wägele, J. W., Letsch, H., Klussmann-Kolb, A., Mayer, C., Misof, B. & Wägele, H. (2009). Phylogenetic support values are not necessarily informative: The case of the Serialia hypothesis (a mollusk phylogeny). *Frontiers in Zoology* 6: 1–15.

Wagner, P. J. (1995). Diversity patterns among early gastropods: Contrasting taxonomic and phylogenetic descriptions. *Paleobiology* 21: 410–439.

Wagner, P. J. (1997). Patterns of morphologic diversification among the Rostroconchia. *Paleobiology* 23: 115–150.

Wagner, P. J. (1999). The utility of fossil data in phylogenetic analyses: A likelihood example using Ordovician-Silurian species of the Lophospiridae (Gastropoda: Murchisoniina). *American Malacological Bulletin* 15: 1–31.

Wagner, P. J. (2001). Gastropod phylogenetics: Progress, problems, and implications. *Journal of Paleontology* 75: 1128–1140.

Wagner, P. J. (2002). Phylogenetic relationships of the earliest anisostrophically coiled gastropods. *Smithsonian Contributions to Paleobiology* 88: 1–152.

Wagner, P. J., Kosnik, M. A. & Lidgard, S. (2006). Abundance distributions imply elevated complexity of post-Paleozoic marine ecosystems. *Science* 314: 1289–1292.

Wagner, P. J. (2000). Exhaustion of morphologic character states among fossil taxa. *Evolution* 54: 365–386.

Wahlman, G. P. (1985). *Middle and Upper Ordovician Monoplacophora and bellerophontacean Gastropoda of the Cincinnati Arch region.* Doctoral thesis, University of Cincinnati.

Wahlman, G. P. (1992). Middle and Upper Ordovician symmetrical univalved mollusks (Monoplacophora and Bellerophontina) of the Cincinnati Arch region. United States Geological Survey. *Professional* Paper 1066-O: O1–O213.

Waite, J. H. (1983). Adhesion in byssally attached bivalves. *Biological Reviews* 58: 209–231.

Waite, J. H. & Qin, X. X. (2001). Polyphosphoprotein from the adhesive pads of *Mytilus edulis. Biochemistry* 40: 2887–2893.

Waite, J. H., Lichtenegger, H. C., Stucky, G. D. & Hansma, P. (2004). Exploring molecular and mechanical gradients in structural bioscaffolds. *Biochemistry* 43: 7653–7662.

Waite, R. & Allmon, W. D. (2013). Observations on the biology and sclerochronology of *Turritella leucostoma* (Valenciennes, 1832; Cerithioidea: Turritellidae) from the Gulf of California. *American Malacological Bulletin* 31: 297–310.

Walcott, C. D. (1911). Middle Cambrian annelids. *Smithsonian Miscellaneous Collections* 57: 110–145.

Walker, C. G. (1968). Studies on the jaw, digestive system, and coelomic derivatives in representatives of the genus *Acmaea*. *The Veliger* 11: 88–97.

Walker, K. F. (1981). The ecology of freshwater mussels in the River Murray. *Australian Water Research Council Technical Papers* 63: 1–119.

Wallace, C. (1992). Parthenogenesis, sex and chromosomes in *Potamopyrgus*. *Journal of Molluscan Studies* 58: 93–107.

Waller, T. R. (1978). Morphology, morphoclines and a new classification of the Pteriomorphia (Mollusca: Bivalvia). *Philosophical Transactions of the Royal Society of London B* 284: 345–365.

Waller, T. R. (1980). Scanning electron microscopy of shell and mantle in the order Arcoida (Mollusca: Bivalvia). *Smithsonian Contributions to Zoology* 313: 1–58.

Waller, T. R. (1990). The evolution of ligament systems in the Bivalvia, pp. 49–71 *in* B. Morton (ed.), *The Bivalvia. Proceedings of a memorial symposium in honour of Sir Charles Maurice Yonge (1899–1986), Edinburgh, 1986*. Edinburgh, Scotland, UK, Hong Kong University Press.

Waller, T. R. (1998). Origin of the molluscan class Bivalvia and a phylogeny of major groups, pp. 1–45 *in* P. A. Johnston & Haggart, J. W. (eds.), *Bivalves: An Eon of Evolution. Paleobiological Studies honoring Norman D. Newell*. Calgary, AB, University of Calgary Press.

Waller, T. R. & Stanley, G. D. (2005). Middle Triassic pteriomorphian Bivalvia (Mollusca) from the New Pass Range, west-central Nevada: Systematics, biostratigraphy, paleoecology, and paleobiogeography. *Journal of Paleontology* 79: 1–58.

Waller, T. R. (2006). Phylogeny of families in the Pectinoidea (Mollusca: Bivalvia): Importance of the fossil record. *Zoological Journal of the Linnean Society* 148: 313–342.

Walsby, J. R., Morton, J. E. & Croxall, J. P. (1973). The feeding mechanism and ecology of the New Zealand pulmonate limpet *Gadinalea nivea*. *Journal of Zoology* 171: 257–283.

Wang, J.-R., He, W.-F. & Guo, Y.-W. (2013). Chemistry, chemoecology, and bioactivity of the South China Sea opisthobranch molluscs and their dietary organisms. *Journal of Asian Natural Products Researches* 15: 185–197.

Wanninger, A. & Haszprunar, G. (2001). The expression of an engrailed protein during embryonic shell formation of the tusk-shell, *Antalis entalis* (Mollusca, Scaphopoda). *Evolution & Development* 3: 312–321.

Wanninger, A. & Haszprunar, G. (2002a). Muscle development in *Antalis entalis* (Mollusca: Scaphopoda) and its significance for scaphopod relationships. *Journal of Morphology* 254: 53–64.

Wanninger, A. & Haszprunar, G. (2002b). Chiton myogenesis: Perspectives for the development and evolution of larval and adult muscle systems in molluscs. *Journal of Morphology* 251: 103–113.

Wanninger, A., Fuchs, J. & Haszprunar, G. (2007). The anatomy of the serotonergic nervous system of an entoproct creeping-type larva and its phylogenetic implications. *Invertebrate Biology* 126: 268–278.

Ward, P., Greenwald, L. & Greenwald, O. E. (1980). The buoyancy of the chambered Nautilus. *Scientific American* 243: 190–203.

Ward, P. (1981). Shell sculpture as a defensive adaptation in ammonoids. *Paleobiology* 7: 96–100.

Ward, P., Greenwald, L. & Magnier, Y. (1981). The chamber formation cycle in *Nautilus macromphalus*. *Paleobiology* 7: 481–493.

Ward, P. (1982). The relationship of siphuncle size to emptying rates in chambered cephalopods: Implications for cephalopod paleobiology. *Paleobiology* 8: 426–433.

Ward, P. D. (1985). Periodicity of chamber formation in chambered cephalopods: Evidence from *Nautilus macromphalus* and *Nautilus pompilius*. *Paleobiology* 11: 438–450.

Ward, P. D. (1986). Cretaceous ammonoid shell shapes. *Malacologia* 27: 3–28.

Ward, P. D. & Saunders, W. B. (1997). *Allonautilus*: A new genus of living nautiloid cephalopod and its bearing on phylogeny of the Nautilida. *Journal of Paleontology* 71: 1054–1064.

Ward, P. D., Botha, J., Buick, R., De Kock, M. O., Erwin, D. H., Garrison, G. H., Kirschvink, J. L. & Smith, R. (2005). Abrupt and gradual extinction among Late Permian land vertebrates in the Karoo Basin, South Africa. *Science* 307: 709–714.

Ward, P. D. & Greenwald, L. (2009). Chamber refilling in *Nautilus*. *Journal of the Marine Biological Association of the United Kingdom* 62: 469.

Warén, A. (1983a). An anatomical description of *Eulima bilineata* Alder with remarks on and a revision of *Pyramidelloides* Nevill (Mollusca, Prosobranchia, Eulimidae). *Zoologica Scripta* 12: 273–294.

Warén, A. (1983b). A generic revision of the family Eulimidae (Gastropoda, Prosobranchia). *Journal of Molluscan Studies*: Supplement 13: 1–95.

Warén, A. & Bouchet, P. (1989). New gastropods from east Pacific hydrothermal vents. *Zoologica Scripta* 18: 67–102.

Warén, A. (1991). New and little known mollusca from Iceland and Scandinavia. [Part 1]. *Sarsia* 76: 53–124.

Warén, A. & Ponder, W. F. (1991). New species, anatomy, and systematic position of the hydrothermal vent and hydrocarbon seep gastropod family Provannidae fam. n. (Caenogastropoda). *Zoologica Scripta* 20: 27–56.

Warén, A. (1992). New and little known 'skeneimorph' gastropods from the Mediterranean Sea and the adjacent Atlantic Ocean. *Bollettino Malacologico* 27: 149–248.

Warén, A. & Hain, S. (1992). *Laevipilina antarctica* and *Micropilina arntzi*, new species, two new monoplacophorans from the Antarctic. *The Veliger* 35: 165–176.

Warén, A. (1993). New and little known Mollusca from Iceland and Scandinavia. Part 2. *Sarsia* 78: 159–201.

Warén, A. & Bouchet, P. (1993). New records, species, genera, and a new family of gastropods from the hydrothermal vents and hydrocarbon seeps. *Zoologica Scripta* 22: 1–90.

Warén, A. (1996). Description of *Bathysciadium xylophagum* Warén & Carrozza, sp. n. and comments on *Addisonia excentrica* (Tiberi) two Mediterranean cocculiniform gastropods. *Bollettino Malacologico* 31: 231–266.

Warén, A. & Gofas, S. (1996). A new species of Monoplacophora, redescription of the genera *Veleropilina* and *Rokopella*, and new information on three species of the class. *Zoologica Scripta* 25: 215–232.

Warén, A. & Bouchet, P. (2001). Gastropoda and Monoplacophora from hydrothermal vents and seeps: New taxa and records. *The Veliger* 44: 116–231.

Warén, A., Bengtson, S., Goffredi, S. K. & Van Dover, C. L. (2003). A hot-vent gastropod with iron sulfide dermal sclerites. *Science* 302: 1007.

Warén, A. & Bouchet, P. (2009). New gastropods from deep-sea hydrocarbon seeps off West Africa. *Deep Sea Research Part II: Topical Studies in Oceanography* 57: 2326–2349.

Warén, A. (2013). Murchisonellidae: Who are they, where are they and what are they doing? (Gastropoda, lowermost Heterobranchia). *Vita Malacologica* 11: 1–14.

Warmke, G. L. & Almodóvar, L. S. (1972). Observation on the life cycle and regeneration in *Oxynoe antillarium* Mörch, an ascoglossan opisthobranch from the Caribbean. *Bulletin of Marine Science* 22: 67–74.

Warnke, K. M. & Keupp, H. (2005). *Spirula* – A window to the embryonic development of ammonoids? Morphological and molecular indications for a palaeontological hypothesis. *Facies* 51: 60–65.

Warnke, K. M. (2007). On the species status of *Spirula spirula* (Linné, 1758) (Cephalopoda): A new approach based on divergence of amino acid sequences between the Canaries and New Caledonia, pp. 144–155 *in* N. H. Landman, Davis, R. A. & Mapes, R. H. (eds.), *Cephalopods Present and Past. New Insights and Fresh Perspectives. 6th International Symposium*, University of Arkansas, Fayetteville, 16–19 September 2004. Dordrecht, the Netherlands, Springer.

Warnke, K. M., Meyer, A., Ebner, B. & Lieb, B. (2011). Assessing divergence time of Spirulida and Sepiida (Cephalopoda) based on hemocyanin sequences. *Molecular Phylogenetics and Evolution* 58: 390–394.

Watanabe, J. M. & Cox, L. R. (1975). Spawning behavior and larval development in *Mopalia lignosa* and *Mopalia mucosa* (Mollusca, Polyplacophora) in central California. *The Veliger* 18: 18–27.

Waterhouse, J. B. (2008). Aspects of the evolutionary record for fossils of the bivalve subclass Pteriomorphia Beurlen. *Earthwise* 8: 1–219.

Watson, M. E. & Signor, P. W. (1986). How a clam builds windows: Shell microstructure in *Corculum* (Bivalvia: Cardiidae). *The Veliger* 28: 348–355.

Watters, G. T. & O'Dee, S. H. (1998). Metamorphosis of freshwater mussel *Glochidia* (Bivalvia: Unionidae) on amphibians and exotic fishes. *American Midland Naturalist* 139: 49–57.

Waugh, J. (2007). DNA barcoding in animal species: Progress, potential and pitfalls. *BioEssays* 29: 188–197.

Wealthall, R. J., Brooker, L. R., Macey, D. J. & Griffin, B. J. (2005). Fine structure of the mineralized teeth of the chiton *Acanthopleura echinata* (Mollusca: Polyplacophora). *Journal of Morphology* 265: 165–175.

Webb, W. M. (1893). On the manner of feeding in *Testacella scutulum*. *The Zoologist* 17: 281–289.

Webby, B. D., Paris, F., Droser, M. L. & Percival, I. G. (2004). *The Great Ordovician Biodiversification Event*. New York, Columbia University Press.

Webers, G. F., Yochelson, E. L. & Kase, T. (1991). Observations on a Late Cambrian cephalopod. *Lethaia* 24: 347–348.

Webster, N. B. & Vermeij, G. J. (2017). The varix: Evolution, distribution, and phylogenetic clumping of a repeated gastropod innovation. *Zoological Journal of the Linnean Society* 180: 732–754.

Wefer, G. & Killingley, J. S. (1980). Growth histories of strombid snails from Bermuda recorded in their O-18 and C-13 profiles. *Marine Biology* 60: 129–135.

Wegener, B. J., Stuart-Fox, D., Norman, M. D. & Wong, B. B. M. (2013). Spermatophore consumption in a cephalopod. *Biology Letters* 9: 20130192.

Wei, F., Gong, Y. & Yang, H. (2012). Biogeography, ecology and extinction of Silurian and Devonian tentaculitoids. *Palaeogeography, Palaeoclimatology, Palaeoecology* 358: 40–50.

Weigand, A. M., Jochum, A. & Klussmann-Kolb, A. (2014). DNA barcoding cleans house through the Carychiidae (Eupulmonata, Ellobioidea). *American Malacological Bulletin* 32: 236–245.

Weiss, K. M. & Wägele, H. (1998). On the morphology, anatomy and histology of three species of *Onchidella* (Gastropoda: Gymnomorpha: Onchidiida). *Archiv für Molluskenkunde* 127: 69–91.

Weitschat, W. & Bandel, K. (1991). Organic components in phragmocones of boreal Triassic ammonoids: Implications for ammonoid biology. *Palaeontologische Zeitschrift* 65: 269–303.

Wells, F. E. (2002). Malacological results of the western Australian marine biological workshop series. *Abstracts and Souvenir Program of the Seventh International Congress on Medical and Applied Malacology*, 21–24 October 2002, Los Banos, Laguna, Philippines. Philippines, Malacological Society of the Philippines.

Wells, M. J. & Wells, J. (1982). Ventilatory currents in the mantle of cephalopods. *Journal of Experimental Biology* 99: 315–330.

Wells, M. J. & Wells, J. (1985a). Ventilation and oxygen uptake by *Nautilus*. *Journal of Experimental Biology* 118: 297–312.

Wells, M. J. & Wells, J. (1985b). Ventilation frequencies and stroke volumes in acute hypoxia in *Octopus*. *Journal of Experimental Biology* 118: 445–448.

Wells, M. J. (1987). Ventilation and oxygen extraction by *Nautilus*, pp. 339–350 *in* W. B. Saunders & Landman, N. H. (eds.), *Nautilus: The Biology and Paleobiology of a Living Fossil*. Topics in Geobiology. New York, Springer.

Wells, M. J. & Smith, P. J. S. (1987). The performance of the octopus circulatory system: A triumph of engineering over design. *Experientia* 43: 487–499.

Wells, M. J. (1988). The mantle muscle and mantle cavity of cephalopods, pp. 287–300 *in* E. R. Trueman & Clarke, M. R. (eds.), *Form and Function. The Mollusca*. Vol. 11. New York, Academic Press.

Wells, M. J., Hanlon, R. T., Lee, P. G. & Dimarco, F. P. (1988). Respiratory and cardiac performance in *Lolliguncula brevis* (Cephalopodia, Myopsida) – The effects of activity, temperature and hypoxia. *Journal of Experimental Biology* 138: 17–36.

Wells, M. J. (1990). Oxygen extraction and jet propulsion in cephalopods. *Canadian Journal of Zoology* 68: 815–824.

Wells, M. J. & Wells, J. (1995). Ventilation and oxygen uptake by *Nautilus*. *Journal of Experimental Biology* 118: 297–312.

Welsch, U. & Storch, V. (1969). Über das Osphradium der prosobranchen Schnecken *Buccinum undatum* (L.) und *Neptunea antiqua* (L.). *Zeitschrift für Zellforschung und Mikroskopische Anatomie* 95: 317–330.

Wen, Y. (1979). Earliest Cambrian monoplacophorans and gastropods from western Hubei with their biostratigraphical significance (in Chinese with English abstract). *Acta Palaeontologica Sinica* 18: 233–270.

Wen, Y. (1981). New Earliest Cambrian monoplacophorans and gastropods from W[est] Hubei and E[ast] Yunnan. *Acta Palaeontologica Sinica* 20: 552–556.

Wen, Y. (1984). Early Cambrian molluscan faunas of Meishucun Stage with special reference to Precambrian-Cambrian boundary. *Developments in Geoscience. Contribution to 27th International Geological Congress, July, 1984, Moscow*: 21–35.

Wen, Y. (1990). The first radiation of shelled molluscs. *Palaeontologia Cathayana* 5: 139–170.

Wen, Y. & Yochelson, E. L. (1999). Some Late Cambrian molluscs from Liaoning Province. *Records of the Western Australian Museum* 19: 379–389.

Wen, Y. (2001). The earliest Cambrian polyplacophorans from China. *Records of the Western Australian Museum* 20: 167–185.

Wen, Y. (2008). On the genus *Yangtzemerisma* and related genera (Mollusca: Merismoconchia). *Records of the Western Australian Museum* 24: 181–194.

Wenz, W. (1938–1944). Gastropoda: Teil I: Allgemeiner Teil und Prosobranchia, pp. i–viii, 1–1639 in O. H. Schindewolf (ed.), *Handbuch der Paläontologie*. Vol. 6, Teil 1. Berlin, Gebrüder Borntraeger.

Wenz, W. (1940). Ursprung und frühe Stammesgeschichte der Gastropoden. *Archiv für Molluskenkunde* 72: 1–10.

West, T. L. (1990). Feeding behavior and functional morphology of the epiproboscis of *Mitra idae* (Mollusca, Gastropoda, Mitridae). *Bulletin of Marine Science* 46: 761–779.

Westermann, G. E. G. (1971). Form, structure and function of shell and siphuncle in coiled Mesozoic ammonoids. *Royal Ontario Museum. Life Sciences. Contributions* 78: 1–39.

Westermann, G. E. G. (1973). Strength of concave septa and depth limits of fossil cephalopods. *Lethaia* 6: 383–403.

Westermann, G. E. G. (1996). Ammonoid life and habitat, pp. 607–707 in N. H. Landman, Tanabe, K. & Davis, R. A. (eds.), *Ammonoid Paleobiology: From Anatomy to Ecology. Topics in Geobiology*. New York, Plenum Press.

Westermann, G. E. G. (1999). Life habits of nautiloids, pp. 263–298 in E. Savazzi (ed.), *Functional Morphology of the Invertebrate Skeleton*. Chichester, UK, John Wiley & Sons.

Westheide, W. & Rieger, R., Eds. (1996). *Spezielle Zoologie – Erster Teil: Einzeller und Wirbellose*. Stuttgart, Jena und New York, Gustav Fischer.

White, T. R., Pagels, A. K. W. & Fautin, D. G. (1999). Abyssal sea anemones (Cnidaria: Actiniaria) of the northeast Pacific symbiotic with molluscs: *Anthosactis nomados*, a new species, and *Monactis vestita* (Gravier, 1918). *Proceedings of the Biological Society of Washington* 112: 637–651.

White, T. R., Conrad, M. M., Tseng, R., Balayan, S., Golding, R., de Frias Martins, A. M. & Dayrat, B. A. (2011). Ten new complete mitochondrial genomes of pulmonates (Mollusca: Gastropoda) and their impact on phylogenetic relationships. *BMC Evolutionary Biology* 11: 295.

Whyte, M. A. (1992). Phosphate gill supports in living and fossil bivalves, pp. 427–431 in S. Suga & Nakahara, H. (eds.), *Mechanisms and Phylogeny of Mineralisation in Biology Systems*. Tokyo, Springer.

Wiedmann, J. (1988). Ammonoid extinction and the 'Cretaceous-Tertiary Boundary Event', pp. 117–140 in J. Wiedmann & Kullmann, J. (eds.), *Cephalopods Present and Past. 2nd International Cephalopod Symposium: O. H. Schindewolf Symposium, Tübingen 1985*. Stuttgart, E. Schweizerbart'sche Verlagsbuchhandlung.

Wiens, J. J. (2007). Species delimitation: New approaches for discovering diversity. *Systematic Biology* 56: 875–878.

Wiesenauer, E. (1976). Vollständige Belemnitentiere aus dem Holzmadener Posidonienschiefer. *Neues Jahrbuch für Geologie und Paläontologie-Monatshefte* 1976: 603–608.

Wignall, P. B. & Twitchett, R. J. (2002). Extent, duration, and nature of the Permian-Triassic superanoxic event. *Geological Society of America* Special Papers 356: 395–413.

Wiktor, A. (1987). Spermatophores in Milacidae and their significance for classification (Gastropoda, Pulmonata). *Malakologische Abhandlungen* 12: 85–100.

Will, K. W., Mishler, B. D. & Wheeler, Q. D. (2005). The perils of DNA barcoding and the need for integrative taxonomy. *Systematic Biology* 54: 844–851.

Willan, R. C. (1984). A review of diets in the Notaspidea (Mollusca: Opisthobranchia). *Journal of the Malacological Society of Australia* 6: 125–142.

Willan, R. C. (1987). Phylogenetic systematics of the Notaspidea (Opisthobranchia) with reappraisal of families and genera. *American Malacological Bulletin* 5: 215–241.

Willcox, M. A. (1898). Zur Anatomie von *Acmaea fragilis* Chemnitz. *Jenaische Zeitschrift für Naturwissenschaft* 32: 411–456.

Willem, V. (1892). Contributions à l'étude physiologique des organes des sens chez les mollusques. III. Observations sur la vision et les organes visuels de quelques mollusques prosobranches et opisthobranches. *Archives de Biologie* 12: 123–149.

Willey, A. (1902). Contribution to the natural history of the pearly nautilus, pp. 691–830. *Zoological Results based on Material from New Britain, New Guinea, Loyalty Islands and elsewhere, collected during the years 1895, 1896 and 1897 by Arthur Willey*. Vol.[part] 6. Cambridge, UK, Cambridge University Press.

Williams, A. (1997). Brachiopoda: Introduction and integumentary system, pp. 237–296 in F. W. Harrison & Woollacott, R. M. (eds.), *Microscopic Anatomy of Invertebrates. Lophophorates, Entoprocta, and Cycliophora*. Vol. 13. New York, Wiley-Liss.

Williams, E. H. & Williams, L. B. (1986). The first association of an adult mollusk (Nudibranchia: Doridae) and a fish (Perciformes: Gobiidae). *Venus* 45: 210–211.

Williams, S. T., Taylor, J. D. & Glover, E. A. (2004). Molecular phylogeny of the Lucinoidea (Bivalvia): Non-monophyly and separate acquisition of bacterial chemosymbiosis. *Journal of Molluscan Studies* 70: 187–202.

Williams, S. T. & Ozawa, T. (2006). Molecular phylogeny suggests polyphyly of both the turban shells (family Turbinidae) and the superfamily Trochoidea (Mollusca: Vetigastropoda). *Molecular Phylogenetics and Evolution* 39: 33–51.

Williams, S. T., Karube, S. & Ozawa, T. (2008). Molecular systematics of Vetigastropoda: Trochidae, Turbinidae and Trochoidea redefined. *Zoologica Scripta* 37: 483–506.

Williams, S. T., Foster, P. G. & Littlewood, D. T. J. (2014). The complete mitochondrial genome of a turbinid vetigastropod from MiSeq Illumina sequencing of genomic DNA and steps towards a resolved gastropod phylogeny. *Gene* 533: 38–47.

Williams, S. T., Foster, P. G., Hughes, C., Harper, E. M., Taylor, J. D., Littlewood, D. T. J., Dyal, P., Hopkins, K. P. & Briscoe, A. G. (2017). Curious bivalves: Systematic utility and unusual properties of anomalodesmatan mitochondrial genomes. *Molecular Phylogenetics and Evolution* 110: 60–72.

Wilson, B. R. (2006). A new generic name for a burrowing mytilid (Mollusca: Bivalvia: Mytilidae). *Molluscan Research* 26: 89–97.

Wilson, M. A., Palmer, T. J., Guensburg, T. E., Finton, C. D. & Kaufman, L. E. (1992). The development of an Early Ordovician hard ground community in response to rapid seafloor calcite precipitation. *Lethaia* 25: 19–34.

Wilson, N. G., Huang, D., Goldstein, M. C., Cha, H., Giribet, G. & Rouse, G. W. (2009). Field collection of *Laevipilina hyalina* McLean, 1979 from southern California, the most accessible living monoplacophoran. *Journal of Molluscan Studies* 75: 195–197.

Wilson, N. G., Rouse, G. W. & Giribet, G. (2010). Assessing the molluscan hypothesis Serialia (Monoplacophora + Polyplacophora) using novel molecular data. *Molecular Phylogenetics and Evolution* 54: 187–193.

Wilson, N. G., Jörger, K. M., Brenzinger, B. & Schrödl, M. (2017). Phylogenetic placement of the enigmatic worm-like Rhodopemorpha slugs as basal Heterobranchia. *Journal of Molluscan Studies* 83: 399–408.

Wingstrand, K. G. (1985). On the anatomy and relationships of Recent Monoplacophora. *Galathea Report* 16: 1–94.

Winik, B. C., Catalan, N. M. Y. & Schlick, O. C. (2001). Genesis of the apyrene parasperm in the apple snail *Pomacea canaliculata* (Gastropoda: Ampullariidae): An ultrastructural study. *Journal of Molluscan Studies* 67: 81–94.

Winter, A. J. de (2008). Redefinition of *Thapsia* Albers, 1860, and description of three more helicarionoid genera from western Africa (Gastropoda, Stylommatophora). *Zoologische Mededelingen (Leiden)* 82: 441–477.

Wirén, A. (1892a). Studien über Solenogastres I. Monographie der *Chaetoderma nitidulum* Lovén. *Kungliga Svenska Vetenskapsakademiens Handlingar* 24: 1–66.

Wirén, A. (1892b). Studien über Solenogastres II (*Chaetoderma productum, Neomenia, Proneomenia acuminata*). *Kungliga Svenska Vetenskapsakademiens Handlingar* 25: 1–9.

Wise, J. B. (1999). Reassignment of *Henrya morrisoni* Bartsch, 1947 from the family Aclididae to the Ebalidae (Gastropoda: Heterobranchia). *The Nautilus* 113: 64–70.

Wolter, K. (1992). Ultrastructure of the radula apparatus in some species of aplacophoran molluscs. *Journal of Molluscan Studies* 58: 245–256.

Wondrak, G. (1981). Ultrastructure of the supporting cells in the chemoreceptor areas of the tentacles of *Pomatias elegans* (Mollusca, Prosobranchia) and the ommatophore of *Helix pomatia* (Mollusca, Pulmonata). *Journal of Morphology* 167: 211–230.

Wood, E. M. (1974). Some mechanisms involved in host recognition and attachment of the glochidium larva of *Anodonta cygnea* (Mollusca: Bivalvia). *Journal of Zoology* 173: 15–30.

Wood, J. B., Kenchington, E. L. R. & O'Dor, R. K. (1998). Reproduction and embryonic development time of *Bathypolypus arcticus*, a deep-sea octopod (Cephalopoda: Octopoda). *Malacologia* 39: 11–20.

Wood, J. B. & O'Dor, R. K. (2000). Do larger cephalopods live longer? Effects of temperature and phylogeny on interspecific comparisons of age and size at maturity. *Marine Biology* 136: 91–99.

Wood, J. B. (2003). *CephBase: A database-driven web site on all living cephalopods (octopus, squid, cuttlefish and nautilus).* http://cephbase.eol.org/

Woods, A. D., Bottjer, D. J., Mutti, M. & Morrison, J. (1999). Lower Triassic large sea-floor carbonate cements: Their origin and a mechanism for the prolonged biotic recovery from the end-Permian mass extinction. *Geology* 27: 645–648.

Woodward, M. F. (1901). The anatomy of *Pleurotomaria beyrichii*, Hilg. *Quarterly Journal of Microscopical Science* 44: 215–268.

Woodward, S. P. (1851–1856). *A Manual of the Mollusca: Or, A Rudimentary Treatise of Recent and Fossil Shells.* London, J. Weale.

Worm, B., Barbier, E. B., Beaumont, N., Duffy, J. E., Folke, C., Halpern, B. S., Jackson, J. B., Lotze, H. K., Micheli, F. & Palumbi, S. R. (2006). Impacts of biodiversity loss on ocean ecosystem services. *Science* 314: 787–790.

Wouters-Tyrou, D., Martin-Ponthieu, A., Briand, G., Sautiére, P. & Biserte, G. (1982). The amino-acid-sequence of histone H2a from cuttlefish *Sepia officinalis*. *European Journal of Biochemistry* 124: 489–498.

Wyeth, R. C. & Willows, A. O. D. (2006a). Field behavior of the nudibranch mollusc *Tritonia diomedea*. *Biological Bulletin* 210: 81–96.

Wyeth, R. C. & Willows, A. O. D. (2006b). Odours detected by rhinophores mediate orientation to flow in the nudibranch mollusc *Tritonia diomedea*. *Journal of Experimental Biology* 209: 1441–1453.

Wyeth, R. C., Woodward, O. M. & Willows, A. O. D. (2006). Orientation and navigation relative to water flow, prey, conspecifics, and predators by the nudibranch mollusc *Tritonia diomedea*. *Biological Bulletin* 210: 97–108.

Yabe, H. (1904). Cretaceous "Cephalopoda" from the Hokkaidō. Part II: *Turrilites, Helicoceras, Heteroceras, Nipponites, Olcostephanus, Desmoceras, Hauericeras*, and an undetermined genus. *Journal of the College of Science, Imperial University, Tōkyō* 20: 1–45, 46 plts.

Yahagi, T., Kayama Watanabe, H., Kojima, S. & Kano, Y. (2017). Do larvae from deep-sea hydrothermal vents disperse in surface waters? *Ecology* 98: 1524–1534.

Yates, A. M., Gowlett-Holmes, K. L. & McHenry, B. J. (1992). *Triplicatella disdoma* Conway Morris, 1990, reinterpreted as the earliest known polyplacophoran (abstract). *Journal of the Malacological Society of Australia* 13: 71.

Yochelson, E. L. (1967). Quo vadis, *Bellerophon*? Pp. 141–161 *in* C. Teichert & Yochelson, E. L. *Essays in Paleontology and Stratigraphy.* Lawrence, KS, University of Kansas Press.

Yochelson, E. L. (1969). Stenothecoida, a proposed new class of Cambrian Mollusca. *Lethaia* 2: 49–62.

Yochelson, E. L. (1971). A new Late Devonian gastropod and its bearing on problems of open coiling and septation. *Smithsonian Contributions to Paleobiology* 3: 231–241.

Yochelson, E. L. & Linsley, R. M. (1972). Opercula of two gastropods from the Lilydale Limestone (Early Devonian) of Victoria, Australia. *Memoirs of the National Museum of Victoria* 33: 1–14.

Yochelson, E. L., Flower, R. H. & Webers, G. F. (1973). The bearing of the new Late Cambrian monoplacophoran genus *Knightoconus* upon the origin of the Cephalopoda. *Lethaia* 6: 275–310.

Yochelson, E. L. (1975). Discussion of Early Cambrian 'mollusks'. *Journal of the Geological Society of London* 131: 661–662.

Yochelson, E. L. (1978). An alternative approach to the interpretation of the phylogeny of ancient mollusks. *Malacologia* 17: 165–191.

Yochelson, E. L. & Richardson, E. S. (1979). Polyplacophoran molluscs of the Essex fauna (middle Pennsylvanian, Illinois), pp. 321–332 *in* M. H. Nitecki (ed.), *Mazon Creek Fossils.* New York, Academic Press.

Yochelson, E. L. (1981). "*Fordilla troyensis* Barrande: The oldest known pelecypod" may not be a pelecypod. *Journal of Paleontology* 1: 113–125.

Yochelson, E. L. (1988). A new genus of Patellacea (Gastropoda) from the Middle Ordovician of Utah: The oldest known example of the superfamily. *New Mexico Bureau of Mines and Mineral Resources Memoir* 44: 195–200.

Yochelson, E. L. (1999). Scaphopoda, pp. 363–367 *in* E. Savazzi. *Functional Morphology of the Invertebrate Skeleton.* Chichester, UK, John Wiley & Sons Ltd.

Yochelson, E. L. (2000). Concerning the concept of extinct classes of Mollusca: Or what may/may not be a class of mollusks. *American Malacological Bulletin* 15: 195–202.

Yochelson, E. L. (2004). The record of the early 'Scaphopoda' (?Mollusca) reevaluated. *Annalen des Naturhistorischen Museums in Wien Serie* A 106: 13–31.

Yochelson, E. L. & Holland, C. H. (2004). *Dentalium saturni* Goldfuss, 1841 (Eifelian: Mollusca): Complex issues from a simple fossil. *Palaeontologische Zeitschrift* 78: 97–102.

Yochelson, E. L., Mapes, R. H. & Heidelberger, D. (2007). An enigmatic molluscan fossil from the Devonian of Germany: Scaphopod or cephalopod? *Palaeontologische Zeitschrift* 81/82: 118–122.

Yom-Tov, Y. (1971). The biology of two desert snails *Trochoidea* (*Xerocrassa*) *seetzeni* and *Sphincterochila boissieri*. *Israel Journal of Zoology* 20: 231–248.

Yonge, C. M. (1926). Structure and physiology of the organs of feeding and digestion in *Ostrea edulis*. *Journal of the Marine Biological Association of the United Kingdom* 14: 295–386.

Yonge, C. M. (1930). The crystalline styles of the Mollusca and a carnivorous habit cannot normally co-exist. *Nature* 125: 444–445.

Yonge, C. M. (1932). Notes on feeding and digestion in *Pterocera* and *Vermetus,* with a discussion on the occurrence of the crystalline style in the Gastropoda. *Scientific Reports of the Great Barrier Reef Expedition.* London, British Museum (Natural History) 1: 259–281.

Yonge, C. M. (1936). Mode of life, feeding, digestion and symbiosis with zooxanthellae in the Tridacnidae. *Scientific Reports of the Great Barrier Reef Expedition.* London, British Museum (Natural History) 1: 283–321.

Yonge, C. M. (1938). Evolution of ciliary feeding in the Prosobranchia, with an account of feeding in *Capulus ungaricus*. *Journal of the Marine Biological Association of the United Kingdom* 22: 453.

Yonge, C. M. (1939a). The protobranchiate Mollusca: A functional interpretation of their structure and evolution. *Philosophical Transactions of the Royal Society B* 230: 79–147.

Yonge, C. M. (1939b). On the mantle cavity and its contained organs in the Loricata (Placophora). *Quarterly Journal of Microscopical Science* 81: 367–390.

Yonge, C. M. & Iles, E. J. (1939). On the mantle cavity, pedal gland, and evolution of mucous feeding in the Vermetidae. *Annals and Magazine of Natural History* 11: 536–556.

Yonge, C. M. (1947). The pallial organs in the aspidobranch Gastropoda and their evolution throughout the Mollusca. *Philosophical Transactions of the Royal Society B* 232: 443–518.

Yonge, C. M. (1948a). Formation of siphons in Lamellibranchia. *Nature* 161: 198–199.

Yonge, C. M. (1948b). Cleansing mechanisms and the function of the fourth pallial aperture in *Spisula subtruncata* (Da Costa) and *Lutraria lutraria* (L.). *Journal of the Marine Biological Association of the United Kingdom* 27: 585–596.

Yonge, C. M. (1949). The structure and adaptations of the Tellinacea, deposit feeding Eulamellibranchia. *Philosophical Transactions of the Royal Society B* 234: 29–76.

Yonge, C. M. (1952a). The mantle cavity in *Siphonaria alternata* Say. *Proceedings of the Malacological Society of London* 29: 190–199.

Yonge, C. M. (1952b). Studies on Pacific coast mollusks IV. Observations on *Siliqua patula* Dixon and on evolution within the Solenidae. *University of California Publications in Zoology* 55: 421–438.

Yonge, C. M. (1953a). The monomyarian condition in the Lamellibranchia. *Transactions of the Royal Society of Edinburgh* 62: 443–478.

Yonge, C. M. (1953b). Mantle chambers and water circulation in the Tridacnidae (Mollusca). *Proceedings of the Zoological Society of London* 123: 551–561.

Yonge, C. M. (1953c). Observations on *Hipponix antiquatus* (Linnaeus). *Proceedings of the California Academy of Sciences* 28: 1–24.

Yonge, C. M. (1957). Mantle fusion in the Lamellibranchia. *Pubblicazioni della Stazione Zoologica di Napoli* 29: 151–171.

Yonge, C. M. (1958). Observations in life on the pulmonate limpet *Trimusculus* (*Gadinia*) *reticulatus* (Sowerby). *Proceedings of the Malacological Society of London* 33: 31–37.

Yonge, C. M. (1959a). The status of the Protobranchia in the bivalve Mollusca. *Journal of Molluscan Studies* 33: 210–214.

Yonge, C. M. (1959b). On the structure, biology and systematic position of *Pharus legumen*. *Journal of the Marine Biological Association of the United Kingdom* 38: 277–290.

Yonge, C. M. (1962a). On the biology of the mesogastropod *Trichotropis cancellata* (Hinds), a benthic indicator species. *Biological Bulletin* 122: 160–181.

Yonge, C. M. (1962b). On the primitive significance of the byssus in the Bivalvia and its effects in evolution. *Journal of the Marine Biological Association of the United Kingdom* 42: 113–125.

Yonge, C. M. (1963). Rock-boring organisms, pp. 1–24 *in* R. F. Sognnæs (ed.), *Mechanisms of Hard Tissue Destruction: A Symposium presented at the Philadelphia Meeting of the American Association for the Advancement of Science, December 29 and 30, 1962.* Vol. 75. Washington, DC, American Association for the Advancement of Science.

Yonge, C. M. (1967a). Form, habit and evolution in the Chamacea (Bivalvia) with special reference to conditions in the rudists (Hippuritacea). *Philosophical Transactions of the Royal Society B* 252: 49–105.

Yonge, C. M. (1967b). Observations on *Pedum spondyloideum* (Chemnitz) Gmelin, a scallop associated with reef building corals. *Proceedings of the Malacological Society of London* 37: 311–323.

Yonge, C. M. (1969). Functional morphology and evolution within the Carditacea (Bivalvia). *Journal of Molluscan Studies* 38: 493–527.

Yonge, C. M. (1976). Primary and secondary ligaments with the lithodesma in the Lyonsiidae (Bivalvia, Pandoracea). *Journal of Molluscan Studies* 42: 395–408.

Yonge, C. M. & Thompson, T. E. (1976). *Living Marine Molluscs.* London, Collins.

Yonge, C. M. (1977). Form and evolution in Anomiacea (Mollusca: Bivalvia) – *Pododesmus, Anomia, Patro, Enigmonia* (Anomiidae) – *Placunanomia, Placuna* (Placunidae fam. nov.). *Philosophical Transactions of the Royal Society B* 276: 453–527.

Yonge, C. M. (1979). Cementation in bivalves, pp. 83–106 *in* S. van der Spoel, van Bruggen, A. C. & Lever, J. (eds.), *Pathways in Malacology. 6th International Congress of Unitas Malacologica Europaea,* 15–20 August, 1977. Utrecht & The Hague, the Netherlands, Bohn, Scheltema & Holkema & W. Junk.

Yonge, C. M. & Morton, B. (1980). Ligament and lithodesma in the Pandoracea and the Poromyacea, with a discussion on evolutionary history in the Anomalodesmata (Mollusca: Bivalvia). *Journal of Zoology* 191: 263–292.

Yonge, C. M. (1982). Mantle margins with a revision of siphonal types in the Bivalvia. *Journal of Molluscan Studies* 48: 102–103.

Yonge, C. M. & Allen, J. A. (1985). On significant criteria in establishment of superfamilies in the Bivalvia: The creation of the superfamily Mesodesmatidae. *Journal of Molluscan Studies* 51: 345–349.

Yoo, E. K. (1994). Early Carboniferous Gastropoda from the Tamworth Belt, New South Wales, Australia. *Records of the Australian Museum* 46: 63–120.

Yoshioka, E. (1989). Experimental analysis of the diurnal and tidal spawning rhythm in the chiton *Acanthopleura japonica* (Lischke) by manipulating conditions of light and tide. *Journal of Experimental Marine Biology and Ecology* 133: 81–91.

Young, D. K. (1969a). The functional morphology of the feeding apparatus of some Indo-West-Pacific dorid nudibranchs. *Malacologia* 9: 421–445.

Young, D. K. (1969b). *Okadaia elegans*, a tube-boring nudibranch mollusc from the central and west Pacific. *American Zoologist* 9: 903.

Young, J. Z. (1965). The central nervous system of *Nautilus*. *Philosophical Transactions of the Royal Society B* 249: 1–25.

Young, J. Z. (1988). Evolution of the cephalopod brain, pp. 215–228 *in* M. R. Clarke & Trueman, E. R. (eds.), *Paleontology and Neontology of Cephalopods. The Mollusca*. Vol. 12. New York, Academic Press.

Young, R. E. (1967). Homology of retractile filaments of vampire squid. *Science* 156: 1633–1634.

Young, R. E. (1972). Brooding in a bathypelagic octopus. *Pacific Science* 26: 400–403.

Young, R. E. (1977). Ventral bioluminescent countershading in midwater cephalopods, pp. 161–190 *in* M. Nixon & Messenger, J. B. (eds.), *The Biology of Cephalopods. Symposia of The Zoological Society of London*. London, Academic Press.

Young, R. E. & Mencher, F. M. (1980). Bioluminescence in mesopelagic squid: Diel color change during counterillumination. *Science* 208: 1286–1288.

Young, R. E. & Arnold, J. M. (1982). The functional morphology of a ventral photophore from the mesopelagic squid, *Abralia trigonura*. *Malacologia* 23: 135–163.

Young, R. E. & Bennett, T. M. (1988). Photophore structure and evolution within the Enoploteuthinae (Cephalopoda), pp. 241–251 *in* M. R. Clarke & Trueman, E. R. (eds.), *Paleontology and Neontology of Cephalopods. The Mollusca*. Vol. 12. New York, Academic Press.

Young, R. E. & Vecchione, M. (1996). Analysis of morphology to determine primary sister-taxon relationships within coleoid cephalopods. *American Malacological Bulletin* 12: 91–112.

Young, R. E., Vecchione, M. & Donovan, D. T. (1998). The evolution of coleoid cephalopods and their present biodiversity and ecology. *South African Journal of Marine Science* 20: 393–420.

Young, R. E. & Vecchione, M. (1999). Morphological observations on a hatchling and a paralarva of the vampire squid, *Vampyroteuthis infernalis* Chun (Mollusca: Cephalopoda). *Proceedings of the Biological Society of Washington* 112: 661–666.

Young, R. E. & Vecchione, M. (2002). Evolution of the gills in the octopodiformes. *Bulletin of Marine Science* 71: 1003–1017.

Young, R. E. & Vecchione, M. (2009). Lepidoteuthidae Pfeffer 1912. *The Tree of Life Web Project*. http://tolweb.org/Lepidoteuthis_grimaldii/19833.

Young, R. E., Vecchione, M. & Mangold, K. M. (2018). Cephalopoda Cuvier 1797. Octopods, squids, nautiluses, etc. Version 20 February 2018 (under construction). *ToL project*. http://tolweb.org/Cephalopoda/19386/2018.02.20.

Yusa, Y. (2008). Size-dependent sex allocation and sexual selection in *Aplysia kurodai*, a hermaphrodite with nonreciprocal mating. *Invertebrate Biology* 127: 291–298.

Zapata, F., Wilson, N. G., Howison, M., Andrade, S. C. S., Jörger, K. M., Schrödl, M., Goetz, F. E., Giribet, G. & Dunn, C. W. (2014). Phylogenomic analyses of deep gastropod relationships reject Orthogastropoda. *Proceedings of the Royal Society B* 281: 20141739.

Zardus, J. D. (2002). Protobranch bivalves. *Advances in Marine Biology* 42: 1–65.

Zatylny, C., Marvin, L., Gagnon, J. & Henry, J. (2002). Fertilization in *Sepia officinalis*: The first mollusk sperm-attracting peptide. *Biochemical and Biophysical Research Communications* 296: 1186–1193.

Zbinden, M., Marqué, L., Gaudron, S. M., Ravaux, J., Léger, N. & Duperron, S. (2015). Epsilonproteobacteria as gill epibionts of the hydrothermal vent gastropod *Cyathermia naticoides* (North East-Pacific Rise). *Marine Biology* 162: 435–448.

Zeidberg, L. D. (2009). First observations of 'sneaker mating' in the California market squid, *Doryteuthis opalescens* (Cephalopoda: Myopsida). *Marine Biodiversity Records* 2: e6.

Zelditch, M. L., Swiderski, D. L. & Sheets, H. D. (2012). *Geometric Morphometrics for Biologists: A Primer*. Amsterdam, the Netherlands/Boston, MA, Elsevier Academic Press.

Zhang, K., Wang, D., Shen, H., Qian, J., Guan, J., Wu, H. & Gao, Y. (2017). Redescription of *Platevindex mortoni* (Gastropoda: Eupulmonata: Onchidiidae) from China. *Molluscan Research* 37: 72–78.

Zhukova, N. V. & Eliseikina, M. G. (2012). Symbiotic bacteria in the nudibranch mollusk *Dendrodoris nigra*: Fatty acid composition and ultrastructure analysis. *Marine Biology* 159: 1783–1794.

Zhuravleva, F. A. (1972). Devonskie nautiloidei. Otryad Discosorida. *Trudy Paleontologicheskogo Instituta Akademiia Nauk SSSR* 134: 1–311.

Zhuravleva, F. A. (1994). The order Dissidocerida (Cephalopoda). *Paleontological Journal* 28: 115–133.

Zhuravleva, F. A. & Doguzhaeva, L. A. (2002). Astrovioidea: A new superorder of Paleozoic cephalopods. *Paleontological Journal*, Supplement 38: S1–S73.

Zielske, S., Glaubrecht, M. & Haase, M. (2011). Origin and radiation of rissooidean gastropods (Caenogastropoda) in ancient lakes of Sulawesi. *Zoologica Scripta* 40: 221–237.

Ziclske, S. & Haase, M. (2014). New insights into tateid gastropods and their radiation on Fiji based on anatomical and molecular methods (Caenogastropoda: Truncatelloidea). *Zoological Journal of the Linnean Society* 172: 71–102.

Zittel, K. A. (1868). *Diploconus*, ein neuer Genus aus der Familie der Belemnitiden. *Neues Jahrbuch für Mineralogie, Geologie und Paläontologie* 1868: 548–552.

Zittel, K. A. V. & Eastman, C. R. (1913). *Text-Book of Paleontology*. Vol. 1. London, Macmillan and Co. Limited.

Zou, S., Li, Q. & Kong, L. (2011). Additional gene data and increased sampling give new insights into the phylogenetic relationships of Neogastropoda, within the caenogastropod phylogenetic framework. *Molecular Phylogenetics and Evolution* 61: 425–435.

Index

This index includes terms, taxonomic names and various features, chemicals, structure etc. mentioned in the text. It does not include authors names, geographic locations or names for geological periods or stages.

Families referred to in the text as, for example, trochids, are combined under one entry, Trochidae. Similarly, for superfamilies (e.g., trochoideans, as Trochoidea) and other higher taxa are treated in the same way.

The index was constructed using *TExtract®* and we thank the developer of that program, Harry Bego, for his kind assistance.

A

abalone (*see also* Haliotidae), 46, 289, 324, 334, 336, 377–378, 439, 524, 534
abalone shell, 46
abanal gill condition (chitons), 69, **71**, 72, 549
abdominal sense organs, 145
Abdopus abaculas, **197**
Abralia, 233
Abralia trigonura, **232**
Abronicidae, 700
absorption, 152, 189, 233, 252, 308, 344, 350
Abyssochitonidae, **68**, **74**, 548
Abyssochrysidae, 674
Abyssochrysoidea, 366, 379, 417, 674
Acaenoplax, **25**, 58–59, 69, 107, 549
Acaenoplax hayae, 15, **58**
Acanthaster planci, 159
Acanthoceras, **620**
Acanthoceratidae, 620
Acanthoceratoidea, 620
Acanthochitona, **74**, **549**
Acanthochitona fascicularis, **68**
Acanthochitonidae, **68**, **74**, 82, 549
Acanthochitonina, 69, **77**, 549
Acanthoclymeniidae, 605
Acanthodii, 55
Acanthodoris, 510, **512**
Acanthodoris planca, **426**
Acanthomeniidae, 550
Acanthonematidae, 666
Acanthopleura echinata, 71, **537**
Acanthopleura granulata, 70
Acanthoteuthis, **206**
Acar, 153
Acaroceratidae, 590
Acavidae, 507, 721
Acavoidea, 721
Acavus, 543, **721**
accessory boring organs, 405, 679, 687
accessory bursa, 480
accessory salivary glands. *See* salivary glands, accessory
accretionary growth. *See* growth, accretionary
Acephala, 109, 552
Acéphales, 1
Acerentomon, **14**
acetabulum, 205, 207
Achatina, 470, 475–476, 478, **501**, 502, **714**
Achatina achatina, 289
Achatina fulica, **432**, **539**
Achatinella, 491
Achatinellidae, 499, 514, 718–719
Achatinidae, **432**, 525, 714
Achatinina, 421, 714

Achatinoidea, 421, **501**, 714
Acicula, **397**, 404
Aciculidae, **397**, 415, 669
Aciculoidea, 668
acid, hydrochloric, 447
acid secretions, 404, 412, 447, 453, 516
acidification, ocean, 65, 115, 439, 536, 540
Acila, **125**
Acila castrensis, **539**
Acleistoceratidae, 592
Aclididae, 378, 385, 389, 393–394, 400, 403–404, **410**, **411**, 413, 417, 676–678, 681
Aclis, **678**
Acmaea, 313, **318**, 321–322, **653**
Acmaea mitra, **314**
Acmaeidae, 313, **314**, 315, 317, **318**, 319, 321–322, 653
Acochlida, 421, **422**, 479, 709
Acochlidia/acochlidians, 95, 420–421, **428**, 429, **433**, **434**, 435, 444, 447–448, 454, **464**, 472–474, 479, 482, 494, 496, 503–504, 508–509, 516, 522–523, 695, 706, 709
Acochlidiidae, **428**, **464**, 508, 709
Acochlidimorpha, 709
Acochlidioidea, 421, 458, 461, **464**, 482, 516, 709
Acochlidium, **428**, **709**
Acoela, 6, **7**
Aconeceras, **209**, **236**
Acostaea, 170, **570**
Acrioceras, 265
Acrochordiceratidae, 614
Acroloxidae, 439, 451, **465**, **469**, 524, 711
Acroloxoidea, **465**, 711
Acroloxus, 461, **465**, **469**, 470, 544, **711**
Acroreiidae, 709
acrosome, 78, 104, 320, 481–482
Acrostaea, 170
Acroteuthis, **632**
Acteocina, **424**, 458, **460**, 510, **512**
Acteocinidae, **424**, **460**, 702
Acteon, 298, **436**, 446, 450, 458, **459**, **468**, **477**, 481, **487**, 503, **539**, 696
Acteonella, **693**
Acteonellidae, 693
Acteonelloidea, 692–693
Acteonida, 420, **422**, 522, 696
Acteonidae, 421, **424**, 440, 447, 451, **459**, 468, 472, 475, 486, 491–492, 522, 696
Acteonimorpha, 420, **422**, **468**, 522, 695–696
Acteonina, **671**
Acteoninidae, 467, 479, 490, 671
Acteoninoidea, 671

Acteonoidea, **298**, 420–421, **424**, **433**, **434**, 435, **436**, 442, 451, 453–454, 458, **459**, **468**, 474, **477**, 479, 485–486, **487**, 490, 494, 505, 521–522, 696–697, 701
Actiniaria, 105, 505–506
Actinoceras, **600**
Actinoceratia, 62, 200, 204, 213, 219, **223**, 225, 256–259, 261–262, 599
Actinoceratida, 200, 593, 596, 600
Actinoceratidae, **600**
Actinoceratoidea, 200, 599–600
Actinocerida, 600
Actinocyclidae, 454, 698
Actinodonta, 172, **571**
Actinodontida, 109, 120, 161, 559, 571
Actinodontidae, **121**, 571
Actinodontoida, 559
Actinodontoidea, 559, 571
Actinopterygii/actinopterygians, 59, 256
Actinosepia, **280**
Actinosepiidae, 644
Actophila, 712
Aculifera/aculiferan, 1, 3, **4**, 5, 9, 38, 59, 69, 183
Acutichitonidae, 548
Adacnarca, **144**
Adalaria, 463, 475, 505, **512**
Adamnestia, **703**
adanal gill condition (chitons), 69, **71**, 72, 548
Adapedonta, 578
adaptation, 95, 117, 160, 167, 217, 220, 253, 339, 407, 470, 516, 524
Addisonia, 299, **340**, 343, 346, **660**
Addisoniidae, **299**, **340**, 345, 660
adductors. *See* muscles, adductors
Adelacerithiinae, 384
Adelacerithium, **383**, 385
Adelomelon beckii, 408
adelphophagy, 408
Adenogastropoda, 680–681
Adenopoda, 3, **4**, 99
adhesion (by foot), 70, 77, 117, 445
Adrianites, **608**
Adrianitidae, 608
Adrianitoidea, 608
Adygeya, **634**
Adygeyidae, 284, 634
Aegires, 458, **463**
Aegiridae, 444, **463**, 698
Aegoceras, **622**
Aenigmoconchidae, 572
Aeolidia, **463**, **468**
Aeolidia loui, **427**
Aeolidida, 470–471, 505
Aeolidiella, **463**, 496, **701**

9 781032 173542